Lecture Notes in Computer Science 9022

Allan Hanbury   Gabriella Kazai
Andreas Rauber   Norbert Fuhr (Eds.)

Advances in Information Retrieval

37th European Conference on IR Research, ECIR 2015
Vienna, Austria, March 29 - April 2, 2015
Proceedings

Springer

Volume Editors

Allan Hanbury
Andreas Rauber
Vienna University of Technology
Institute of Software Technology
and Interactive Systems
Vienna, Austria
E-mail: {hanbury, rauber}@ ifs.tuwien.ac.at

Gabriella Kazai
Lumi, Semion Ltd.
London, UK
E-mail: gabriella.kazai@gmail.com

Norbert Fuhr
Universität Duisburg-Essen
Duisburg, Germany
E-mail: norbert.fuhr@uni-due.de

ISSN 0302-9743 e-ISSN 1611-3349
ISBN 978-3-319-16353-6 e-ISBN 978-3-319-16354-3
DOI 10.1007/978-3-319-16354-3
Springer Cham Heidelberg New York Dordrecht London

Library of Congress Control Number: 2015933019

LNCS Sublibrary: SL 3 – Information Systems and Application, incl. Internet/Web and HCI

Typesetting: Camera-ready by author, data conversion by Scientific Publishing Services, Chennai, India

Printed on acid-free paper

Springer is part of Springer Science+Business Media (www.springer.com)

Preface

These proceedings contain the full papers, short papers, and demonstrations selected for presentation at the 37th European Conference on Information Retrieval (ECIR 2015). The event was organized by the Institute of Software Technology and Interactive Systems of the Vienna University of Technology in cooperation with the Austrian Computer Society (OCG). The conference was held from March 29 to April 2, 2015, in Vienna, Austria.

ECIR 2015 received a total of 305 submissions in three categories: 190 full papers, 103 short papers, and 12 demonstrations. The geographical distribution of the submissions is as follows: 54% were from Europe, 18% from Asia, 17% from North and South America, 5% from Australasia, and 6% from North Africa and the Middle East. All submissions were reviewed by at least three members of an international two-tier Program Committee. Of the full papers submitted to the conference, 44 were accepted for oral presentation (23%). Of the short papers submitted to the conference, 39 were accepted for poster presentation (38%). In addition, seven demonstrations (58%) were accepted. The accepted contributions represent the state of the art in information retrieval, cover a diverse range of topics, propose novel applications, and indicate promising directions for future research. We thank all Program Committee members for their time and effort in ensuring a high-quality level of the ECIR 2015 program.

An innovation of the ECIR 2015 was the creation of a pilot Reproducible IR track as a sub-track of the full-paper track. Reproducibility is key for establishing research to be reliable, referenceable, and extensible for the future. Experimental papers are therefore most useful when their results can be tested and generalized by peers. This track specifically invited the submission of papers reproducing a single paper or a group of papers from a third party, where the authors were not directly involved in the original paper. Authors were requested to emphasize the motivation for selecting the papers to be reproduced, the process of how results were attempted to be reproduced (successful or not), the communication that was necessary to gather all information, the potential difficulties encountered, and the result of the process. Of the seven papers submitted to this track, three were accepted. A panel at the ECIR, including members of the Program Committee of this track, discussed experiences and recommendations for continuing this track, which was generally felt to be a valuable contribution to the ECIR.

Additionally, ECIR 2015 hosted five tutorials and five workshops covering a range of information retrieval topics. These were selected by workshop and tutorial committees. The workshops were:

- Second International Workshop on Bibliometric-Enhanced Information Retrieval (BIR)
- Fifth Workshop on Context-Awareness in Retrieval and Recommendation (CaRR)

- Second International Workshop on Gamification for Information Retrieval (GamifIR)
- Multimodal Similar Case Retrieval in the Medical Domain (MRDM)
- Supporting Complex Search Tasks

The following ECIR 2015 tutorials were selected:

- Visual Analytics for Information Retrieval Evaluation (VAIRE 2015)
- Measuring Document Retrievability
- A Formal Approach to Effectiveness Metrics for Information Access: Retrieval, Filtering, and Clustering
- Statistical Power Analysis for Sample Size Estimation in Information Retrieval Experiments with Users
- Join the Living Lab: Evaluating News Recommendations in Real Time

Short descriptions of these workshops and tutorials are included in the proceedings.

We would like to thank our invited speakers for their contributions to the program: Marti Hearst (University of California at Berkeley), Ryen White (Microsoft Research), and Stefan Thurner (Medical University of Vienna). We are very grateful to a committee led by Stefan Rüger for selecting the winner of the 2014 Karen Spärck-Jones Award, and we congratulate Ryen White for receiving this award.

The final day of the conference was an Industry Day. The focus was on startups and small companies in the information retrieval domain, with presentations by company founders on the successes, challenges, and stumbling blocks in setting up and running a company.

Finally, ECIR 2015 would not have been possible without the generous financial support from our sponsors: Google (gold level); Yahoo! Labs and Yandex (silver level); Precognox and max.recall (bronze level). The conference was supported by the Information Retrieval Specialist Group at the British Computer Society (BCS-IRSG), the ELIAS Research Network Program of the European Science Foundation, and the City of Vienna.

January 2015

Allan Hanbury
Gabriella Kazai
Andreas Rauber
Norbert Fuhr

Organization

General Chairs

Norbert Fuhr	University of Duisburg-Essen, Germany
Andreas Rauber	Vienna University of Technology, Austria

Program Chairs

Gabriella Kazai	Lumi.do, Semion Ltd., UK
Allan Hanbury	Vienna University of Technology, Austria

Industry Day Chairs

Jussi Karlgren	Gavagai and KTH, Sweden
Paul Ogilvie	LinkedIn, USA

Tutorial Chairs

Birger Larsen	Aalborg University, Denmark
Adrian Iftene	"Al.I. Cuza" University Iasi, Romania

Workshop Chairs

Guido Zuccon	Queensland University of Technology, Australia
András Benczur	Hungarian Academy of Sciences, Hungary

Demonstration Chairs

João Magalhães	Universidade Nova de Lisboa, Portugal
Jan Šnajder	University of Zagreb, Croatia

Student Mentor Chairs

Pia Borlund	Royal School of Library and Information Science, Denmark
Michal Laclavík	Slovak Academy of Sciences, Slovakia

Publicity Chair

Ralf Bierig	Vienna University of Technology, Austria

Best Paper Award Chair

John Tait	johntait.net Ltd., UK

Local Organizers

Linda Andersson	Vienna University of Technology, Austria
Aldo Lipani	Vienna University of Technology, Austria
Mihai Lupu	Vienna University of Technology, Austria
João Palotti	Vienna University of Technology, Austria
Florina Piroi	Vienna University of Technology, Austria
Navid Rekabsaz	Vienna University of Technology, Austria
Serwah Sabetghadam	Vienna University of Technology, Austria
Veronika Stefanov	Vienna University of Technology, Austria

Program Committee

Full Paper Meta-Reviewers

Eugene Agichtein	Emory University, USA
Giambattista Amati	Fondazione Ugo Bordoni, Italy
Jaime Arguello	University of North Carolina at Chapel Hill, USA
Leif Azzopardi	University of Glasgow, UK
Krisztian Balog	University of Stavanger, Norway
Nicholas Belkin	Rutgers University, USA
Paul Clough	University of Sheffield, UK
Bruce Croft	University of Massachusetts Amherst, UK
David Elsweiler	University of Regensburg, Germany
Eric Gaussier	Laboratory of Informatics of Grenoble (LIG), Université Joseph Fourier, France
Ayse Goker	Robert Gordon University, UK
Cathal Gurrin	Dublin City University, Ireland
Hideo Joho	University of Tsukuba, Japan
Gareth Jones	Dublin City University, Ireland
Franciska De Jong	University of Twente, The Netherlands
Joemon Jose	University of Glasgow, UK
Jaap Kamps	University of Amsterdam, The Netherlands
Evangelos Kanoulas	Google Inc., Switzerland
Liadh Kelly	Trinity College Dublin, Ireland
Udo Kruschwitz	University of Essex, UK
David Losada	University of Santiago de Compostela, Spain
Stefano Mizzaro	University of Udine, Italy

Josiane Mothe	Institut de Recherche en Informatique de Toulouse, France
Maarten de Rijke	University of Amsterdam, The Netherlands
Paolo Rosso	Polytechnic University Valencia, Spain
Stefan Rueger	Knowledge Media Institute, UK
Fabrizio Sebastiani	Istituto di Scienza e Tecnologie dell'Informazione, Consiglio Nazionale delle Ricerche, Italy
Pavel Serdyukov	Yandex, Russian Federation
Fabrizio Silvestri	Yahoo Labs, Spain
Arjen de Vries	Centrum Wiskunde & Informatica (CWI), The Netherlands
Jun Wang	University College London, UK

Full Paper, Short Paper, and Demonstration Reviewers

Mikhail Ageev	Moscow State University, Russia
Dirk Ahlers	Norwegian University of Science and Technology, Norway
Ahmet Aker	University of Sheffield, UK
Elif Aktolga	University of Massachusetts Amherst, USA
M-Dyaa Albakour	University of Glasgow, UK
Omar Alonso	Microsoft, USA
Ismail Sengor Altingovde	Middle East Technical University, Turkey
Linda Andersson	Vienna University of Technology, Austria
Avi Arampatzis	Democritus University of Thrace, Greece
Javed Aslam	Northeastern University, USA
Alvaro Barreiro	University of A Coruña, Spain
Roberto Basili	University of Roma Tor Vergata, Italy
Srikanta Bedathur Jagannath	IBM Research, India
Michel Beigbeder	Ecole Nationale Superieure des Mines de Saint-Etienne, France
Alejandro Bellogin	Universidad Autonoma de Madrid, Spain
Patrice Bellot	LSIS - University of Marseille, France
András Benczur	Hungarian Academy of Sciences, Hungary
Klaus Berberich	Max Planck Institute for Informatics, Germany
Bettina Berendt	K.U. Leuven, Belgium
Ralf Bierig	Vienna University of Technology, Austria
Toine Bogers	Aalborg University Copenhagen, Denmark
Gloria Bordogna	Consiglio Nazionale delle Ricerche, Italy
Paul Buitelaar	DERI - National University of Ireland, Galway, Ireland
Fidel Cacheda	Universidad de A Coruña, Spain
Pável Calado	IST/INESC-ID, Portugal
Fazli Can	Bilkent University, Turkey

Mark Carman	Monash University, Australia
Claudio Carpineto	Fondazione Ugo Bordoni, Italy
Marc Cartright	University of Massachusetts Amherst, USA
Paul-Alexandru Chirita	Adobe Systems Inc., Romania
Fabio Crestani	University of Lugano, Switzerland
Bin Cui	Peking University, China
Alfredo Cuzzocrea	ICAR-CNR and University of Calabria, Spain
Pablo de La Fuente	Universidad de Valladolid, Spain
Adriel Dean-Hall	University of Waterloo, Canada
Thomas Demeester	Ghent University, Belgium
Romain Deveaud	University of Glasgow, UK
Giorgio Maria Di Nunzio	University of Padua, Italy
Vladimir Dobrynin	St. Petersburg State University, Russia
Huizhong Duan	@WalmartLabs, USA
Eva D'hondt	Laboratoire d'Informatique pour la Mécanique et les Sciences de l'Ingénieur (LIMSI), France
Carsten Eickhoff	ETH Zurich, Switzerland
Liana Ermakova	Institut de Recherche en Informatique de Toulouse (IRIT), France, Perm State National Research University, Russian Federation
Hui Fang	University of Delaware, USA
Yi Fang	Santa Clara University, USA
Juan M. Fernández-Luna	University of Granada, Spain
Nicola Ferro	University of Padua, Italy
Luanne Freund	University of British Columbia, Canada
Karin Friberg Heppin	University of Gothenburg, Sweden
Ingo Frommholz	University of Bedfordshire, UK
Patrick Gallinari	LIP6 - University of Paris 6, France
Kavita Ganesan	University of Illinois at Urbana-Champaign, USA
Giorgos Giannopoulos	Imis Institute, "Athena" R.C., Greece
Richard Glassey	Robert Gordon University, UK
Lorraine Goeuriot	Université Joseph Fourier, Grenoble, France
David Grossman	Illinois Institute of Technology, USA
Antonio Gulli	Elsevier, The Netherlands
Matthias Hagen	Bauhaus University Weimar, Germany
Preben Hansen	Stockholm University, Sweden
Donna Harman	NIST, USA
Morgan Harvey	University of Lugano (USI), Switzerland
Claudia Hauff	Delft University of Technology, The Netherlands
Jer Hayes	IBM, Ireland

Ben He	University of Chinese Academy of Sciences, China
Daqing He	University of Pittsburgh, USA
Jiyin He	University of Amsterdam, The Netherlands
Yulan He	Aston University, UK
Nathalie Hernandez	Institut de Recherche en Informatique de Toulouse (IRIT), France
Katja Hofmann	Microsoft, UK
Andreas Hotho	University of Würzburg, Germany
Gilles Hubert	Institut de Recherche en Informatique de Toulouse (IRIT), France
Adrian Iftene	"Al.I.Cuza" University Iasi, Romania
Dmitry Ignatov	National Research University Higher School of Economics, Russian Federation
Shen Jialie	Singapore Management University, Singapore
Jiepu Jiang	University of Massachusetts Amherst, USA
Richard Johansson	University of Gothenburg, Sweden
Frances Johnson	Manchester Metropolitan University, UK
Kristiina Jokinen	University of Helsinki, Finland
Nattiya Kanhabua	L3S Research Center, Germany
Mostafa Keikha	University of Massachusetts Amherst, USA
Diane Kelly	University of North Carolina, USA
Yiannis Kompatsiaris	CERTH - ITI, Greece
Marijn Koolen	University of Amsterdam, The Netherlands
Alexander Kotov	Wayne State University, USA
Oren Kurland	Technion, Israel Institute of Technology, Israel
Kyumin Lee	Utah State University, USA
Wang-Chien Lee	The Pennsylvania State University, USA
Monica Lestari Paramita	The University of Sheffield, UK
Johannes Leveling	Dublin City University (DCU), Ireland
Liz Liddy	Syracuse University, USA
Christina Lioma	University of Copenhagen, Denmark
Xiaozhong Liu	Indiana University Bloomington, USA
Elena Lloret	University of Alicante, Spain
Fernando Loizides	Cyprus University of Technology, Cyprus
Bernd Ludwig	University of Regensburg, Germany
Mihai Lupu	Vienna University of Technology, Austria
Yuanhua Lv	Microsoft Research, USA
Craig Macdonald	University of Glasgow, UK
Andrew Macfarlane	City University London, UK
Walid Magdy	Qatar Computing Research Institute, Qatar
Marco Maggini	University of Siena, Italy
Thomas Mandl	University of Hildesheim, Germany
Stephane Marchand-Maillet	University of Geneva, Switzerland

Ralf Schenkel	Universität Passau, Germany
Falk Scholer	RMIT University, Australia
Florence Sedes	Institut de Recherche en Informatique de Toulouse (IRIT), University Paul Sabatier, France
Giovanni Semeraro	University of Bari Aldo Moro, Italy
Jangwon Seo	University of Massachusetts Amherst, USA
Azadeh Shakery	University of Tehran, Iran
Milad Shokouhi	Microsoft Research, UK
Alan Smeaton	Dublin City University, Ireland
Jan Šnajder	University of Zagreb, Croatia
Parikshit Sondhi	University of Illinois at Urbana Champaign, USA
Yang Song	Microsoft Research, USA
Simone Stumpf	City University London, UK
L. Venkata Subramaniam	IBM Research, India
Lynda Tamine	Paul Sabatier University, France
Martin Theobald	University of Antwerp, Belgium
Bart Thomee	Yahoo! Research, Spain
Ilya Tikhomirov	Institute for systems analysis RAS, Russian Federation
Marko Tkalcic	Johannes Kepler University, Austria
Anastasios Tombros	Queen Mary University of London, UK
Dolf Trieschnigg	University of Twente, The Netherlands
Christos Tryfonopoulos	University of the Peloponnese, Greece
Ming-Feng Tsai	National Chengchi University, Taiwan
Theodora Tsikrika	Information Technologies Institute, CERTH, Greece
Denis Turdakov	Institute for System Programming RAS, Russian Federation
Ata Turk	Yahoo Labs, Spain
Yannis Tzitzikas	University of Crete and FORTH-ICS, Greece
David Vallet	Universidad Autónoma de Madrid, Spain
Marieke Van Erp	VU University Amsterdam, The Netherlands
Jacco van Ossenbruggen	CWI & VU University Amsterdam, The Netherlands
Natalia Vassilieva	HP Labs, Russian Federation
Sumithra Velupillai	Stockholm University, Sweden
Suzan Verberne	Radboud University Nijmegen, The Netherlands
Robert Villa	University of Sheffield, UK
Stefanos Vrochidis	Information Technologies Institute, Greece
Jeroen Vuurens	Delft University of Technology, The Netherlands
V.G.Vinod Vydiswaran	University of Michigan, USA

Xiaojun Wan	Peking University, China
Hongning Wang	University of Virginia, USA
Lidan Wang	University of Illinois, Urbana-Champaign, USA
Wouter Weerkamp	904Labs, The Netherlands
Thijs Westerveld	WizeNoze, The Netherlands
Christa Womser-Hacker	University of Hildesheim, Germany
Tao Yang	Ask.com and UCSB, USA
David Zellhoefer	BTU Cottbus, Germany
Dan Zhang	Facebook, USA
Duo Zhang	University of Illinois at Urbana-Champaign, USA
Lanbo Zhang	University of California, Santa Cruz, USA
Ke Zhou	University of Glasgow, UK
Guido Zuccon	Queensland University of Technology, Australia

Reproducible IR Track Reviewers

Norbert Fuhr	University of Duisburg-Essen, Germany
Lorraine Goeuriot	Université Joseph Fourier, Grenoble, France
Jaap Kamps	University of Amsterdam, The Netherlands
Monica Landoni	Universitá della Svizzera italiana (USI), Switzerland
Mihai Lupu	Vienna University of Technology, Austria
Alistair Moffat	The University of Melbourne, Australia
Martin Potthast	Bauhaus University Weimar, Germany
Andreas Rauber	Vienna University of Technology, Austria
Jan Šnajder	University of Zagreb, Croatia
Justin Zobel	University of Melbourne, Australia

Tutorial Selection Committee

Lenuta Alboaie	"Al.I.Cuza" University Iasi, Romania
Mihaela Breaban	"Al.I.Cuza" University Iasi, Romania
Corina Forascu	"Al.I.Cuza" University Iasi, Romania
Claudia Hauff	TU Delft, The Netherlands
Mihai Alex Moruz	"Al.I.Cuza" University Iasi, Romania
Hideo Joho	University of Tsukuba, Japan
Marijn Koolen	University of Amsterdam, The Netherlands
Monica Landoni	Universitá della Svizzera italiana (USI), Switzerland
Mihai Lupu	Vienna University of Technology, Austria
Henning Müller	University of Applied Sciences Western Switzerland (HES-SO), Switzerland
Alan Said	Recorded Future, Netherlands
Theodora Tsikrika	Centre for Research and Technology Hellas, Greece
Ke (Adam) Zhou	Yahoo Labs London, UK
Guido Zuccon	Queensland University of Technology, Australia

Workshop Selection Committee

Leif Azzopardi	University of Glasgow, UK
Krisztian Balog	University of Stavanger, Norway
Roi Blanco	Yahoo! Research, Spain
Peter Bruza	Queensland University of Technology, Australia
Bevan Koopman	CSIRO, Australia
Oren Kurland	Technion, Israel Institute of Technology, Israel
Christina Lioma	University of Copenhagen, Denmark
Emine Yilmaz	Microsoft Research Cambridge, UK
Justin Zobel	University of Melbourne, Australia

Additional Reviewers

Muhammad Kamran Abbasi
Rafik Abbes
Nitish Aggarwal
Mohammad Allaho
Kartik Asooja
Pierpaolo Basile
Martin Becker
Claudio Biancalana
Janez Brank
Annalina Caputo
Elisavet Chatzilari
Zhiyong Cheng
Shruti Chhabra
Pantelis Chronis
David Corney
Humberto Corona
Alexander Dallmann
Elena Demidova
Charalampos Doulaverakis
George Drosatos
Pavlos Fafalios
Soude Fazeli
Diego Fernández Iglesias
Blaz Fortuna
Tao-Yang Fu
Tianshi Gao
Anastasia Giachanou
Tatiana Gossen
Jheser Guzman
Morgan Harvey
Hui-Ju Hung
Houssem Jerbi
Xin Jin
Mario Karlovcec
Arlind Kopliku
Michael Kotzyba
Monica Landoni
Philip Leroux
Dimitris Liparas
Babak Loni
Thomas Low
Kuang Lu
Maria Maistro
Graham Mcdonald
Alexandra Moraru
Cataldo Musto
Fedelucio Narducci
Sapna Negi
Luis Nieto Piña
Rifat Ozcan
Panagiotis Papadakos
Javier Parapar
Bianca Pereira
Jing Ren
Georgios Rizos
Giuseppe Sansonetti
Gianmaria Silvello
Marcin Skowron
Laure Soulier
Abdel Aziz Taha
Chiraz Trabelsi
Daniel Valcarce

Matteo Venanzi
Thanasis Vergoulis
Liang Xiong
Tianbing Xu
Sergej Zerr
Xiaofei Zhu
Daniel Zoller

Sponsoring Institutions

Platinum Sponsor	ELIAS Research Network Programme
Gold Sponsor	Google
Silver Sponsors	Yandex
	Yahoo! Labs
Bronze Sponsors	Precognox
	max.recall

Invited Papers

Still Haven't Found What I'm Looking for: Suggestions for Search Research

Marti A. Hearst

UC Berkeley

1 Abstract

What's even more fun than doing search research? Suggesting what other people should do search research on!

So in this talk I will pose suggestions about what I dream of seeing in next year's ECIR list of accepted papers.

Topics will include Orphan Search Problems, Should Have Been Solved Years Ago, Solved When We Weren't Looking, Hard But of Increasing Importance, and the Upcoming Text Divide.

2 Biography

Dr. Marti Hearst is a professor in the School of Information at UC Berkeley with an affiliate appointment in the CS department. Her primary research interests are user interfaces for search engines, information visualization, natural language processing, and improving MOOCs. She wrote the first academic book on Search User Interfaces. Prof. Hearst was named a Fellow of the ACM in 2013 and has received an NSF CAREER award, an IBM Faculty Award, two Google Research Awards, three Excellence in Teaching Awards, and has been principal investigator for more than \$3.5M in research grants. Prof. Hearst has served on the Advisory Council of NSF's CISE Directorate and is currently on the Web Board for CACM, member of the Usage Panel for the American Heritage Dictionary, and on the Edge.org panel of experts. She is on the editorial board of ACM Transactions on Computer-Human Interaction and was formerly on the boards of ACM Transactions on the Web, Computational Linguistics, ACM Transactions on Information Systems, and IEEE Intelligent Systems. Prof. Hearst received BA, MS, and PhD degrees in computer science from the University of California at Berkeley, and she was a Member of the Research Staff at Xerox PARC from 1994 to 1997.

Mining and Modeling Online Health Search

Ryen W. White

Microsoft Research
One Microsoft Way
Redmond, WA 98052 USA
ryenw@microsoft.com

1 Abstract

People frequently search the Web for health information and this can have significant consequences for their health and wellbeing. Over the past few years, in collaboration with colleagues, I have been exploring various aspects of online health search. Our research has focused on a number of areas, ranging from characterizing aspects of general health seeking via search engine log data (and user studies / surveys), examining potential anxieties and biases which may arise during health search, and the application of population-scale analyses of health search activity for monitoring and improving public health. In this talk, I discuss some highlights of our research in this area, with a focus on four aspects: (1) patterns of health search within sessions and over time, including self-diagnosis and "web-to-world" transitions from health search to the pursuit of professional medical attention; (2) anxieties in health search, including evidence of escalations in health concerns during searching (so-called "cyberchondria"), and searcher preferences for potentially alarming content; (3) biases in both searcher cognition and in online health content; and (4) applications of aggregated health search query log data in scenarios such as monitoring nutritional intake in populations and detecting adverse drug reactions and interactions. The talk underscores the criticality of research in health search and presents opportunities for further work in this area. More broadly, I also discuss related challenges and opportunities in behavioral analysis and search result provision that have implications far beyond the health search domain.

2 Biography

Ryen White is a Senior Researcher at Microsoft Research. His research interests lie in understanding search interaction and in developing tools to help people search more effectively. He received his Ph.D. in Interactive Information Retrieval from the Department of Computing Science, University of Glasgow, United Kingdom, in 2004. Ryen has published many conference papers and journal articles in Web search, log analysis, and user studies of search systems. He has received eight best-paper awards in conferences and journals, including at ACM SIGIR (2007, 2010, and 2013), ACM CIKM (2014), ACM SIGCHI (2011), and in JASIST (2010). His doctoral research received the British Computer Society's (BCS) Distinguished Dissertation Award for the best Computer Science Ph.D. dissertation in the United Kingdom in 2004/2005. In 2014, Ryen received the Microsoft BCS/BCS IRSG Karen Spärck Jones Award for contributions to Information Retrieval. He has co-organized many workshops on information seeking, especially exploratory search, including an NSF-sponsored invitational workshop, and has guest co-edited special issues in these areas for a variety of outlets, including Communications of the ACM and IEEE Computer. From 2008–2013, Ryen co-organized the HCIR Symposium. He has served as area chair for top conferences such as SIGIR, WSDM, WWW, and CIKM, and currently serves on the editorial board of ACM TOIS, ACM TWEB, the Information Retrieval Journal, and other journals. Ryen chairs the steering committee for the ACM SIGIR Conference on Human Information Interaction and Retrieval (CHIIR, chiir.org). He is short papers co-chair of SIGIR 2015 and PC co-chair of SIGIR 2017. In addition to academic impact, Ryen's research has shipped in many Microsoft products, including Bing, Xbox, Internet Explorer, and Lync.

What to Do If You Know Everything? Studying Human Behavior in a Virtual World

Stefan Thurner

Section for Science of Complex Systems, Medical University of Vienna, and Sante Fe Institute

1 Abstract

We use a massive multiplayer online game to study human interactions and social behaviour. We have complete information on every action carried out by each of the 480.000 players in the game. This complete information on a human society, in particular its time varying social networks of several types allows us to quantify how humans form social bounds, how humans organise, how behaviour is gender specific, and how wealth of players is related to positions in their social multiplex networks.

2 Biography

Stefan Thurner is full professor for Science of Complex Systems at the Medical University of Vienna. Since 2007 he is external professor at the Santa Fe Institute and since 2010 a part time researcher at IIASA (International Institute for Applied Systems Analysis). He obtained a PhD in theoretical physics from the Technical University of Vienna, a second PhD in economics from the University of Vienna and his habilitation in theoretical physics. Thurner has published more than 170 scientific articles in fundamental physics, applied mathematics, complex systems, network theory, evolutionary systems, life sciences, economics and lately in social sciences. He holds 2 patents. Thurner has (co-)organized many international workshops, conferences and summer schools, and has himself presented more than 200 talks. His work has received broad interest from the media such as the New York Times, BBC world, Nature, New Scientist, Physics World and is featured in more than 400 newspaper, radio and television reports. He has coordinated many national and international research projects, and is part of many European science initiatives. Thurner serves as a member of many scientific and editorial boards.

Table of Contents

Aggregated Search and Diversity

Classification

Cross-Lingual and Discourse

Efficiency

Evaluation

Event Mining and Summarisation

Information Extraction

Recommender Systems

Semantic and Graph-Based Models

Sentiment and Opinion

Social Media

Specific Search Tasks

Temporal Models and Features

Topic and Document Models

User Behavior

Reproducible IR

Demonstrations

Tutorials

Workshops

Towards Query Level Resource Weighting for Diversified Query Expansion

Arbi Bouchoucha, Xiaohua Liu, and Jian-Yun Nie

Dept. of Computer Science and Operations Research
University of Montreal
Montreal, Quebec, Canada
{bouchoar,liuxiao,nie}@iro.umontreal.ca

Abstract. Diversifying query expansion that leverages multiple resources has demonstrated promising results in the task of search result diversification (SRD) on several benchmark datasets. In existing studies, however, the weight of a resource, or the degree of the contribution of that resource to SRD, is largely ignored. In this work, we present a query level resource weighting method based on a set of features which are integrated into a regression model. Accordingly, we develop an SRD system which generates for a resource a number of expansion candidates that is proportional to the weight of that resource. We thoroughly evaluate our approach on TREC 2009, 2010 and 2011 Web tracks, and show that: 1) our system outperforms the existing methods without resource weighting; and 2) query level resource weighting is superior to the non-query level resource weighting.

1 Introduction

Diversified query expansion (DQE) has been proposed as a way to generate diversified search results, motivated by the fact that the initial search results of the original query may not be diverse enough and some of the subtopics of the original query may be missing [3,4]. One critical step of DQE is to expand the original query in different directions so as to identify better diversified results. This expansion step often relies on one or multiple external resources, *e.g.*, ConceptNet, Wikipedia, and query logs. Since multiple resources tend to complement each other for DQE, integrating multiple resources can often yield substantial improvements (better diversified search results) compared to using one single resource, as demonstrated by [2,4]. Consider the query "avp" for example, the #52 query from the TREC 2010 Web track. This query has seven subtopics[1]. Using different resources - Wikipedia, query logs and feedback documents, we can respectively cover the following subsets of subtopics: {1, 5, 6, 7}, {2, 3, 6, 7} and {4, 6}[2]. It can be seen that each single resource can cover only part of the subtopics and by combining all these resources, one may expect to get better coverage of all the query subtopics.

[1] http://trec.nist.gov/data/web/10/wt2010-topics.xml

[2] For this query, ConceptNet does not cover any of the subtopics.

A. Hanbury et al. (Eds.): ECIR 2015, LNCS 9022, pp. 1–12, 2015.

When multiple resources are considered, DQE faces the challenge of properly weighting a resource, or computing a non-negative real number for a resource which indicates the degree of the contribution of that resource to the SRD performance. This challenge arises for two reasons. On one hand, the usefulness of a resource can greatly change depending on the queries. On the other hand, the weight of a resource is a key factor in selecting expansion term candidates: the expansion terms recommended by a resource with a great weight should be preferred since they are more likely to be related to one or several subtopics of the query, and their combination tends to cover a good part of all the subtopics.

Existing studies that combine multiple resources to perform DQE based SRD largely overlooked at this problem. Different resources were either simply merged together [12] or be assigned the same weight [4], regardless to the resource and to the query. Even though using several resources can potentially increase the coverage of subtopics, the lack of a proper resource weighting can jeopardize the real impact of the resources. Intuitively, a proper utilization of different resources depending on the query will yield better SRD performance. To be convinced, let us examine again the example of the query "avp" that we described before. Table 1 shows 2 sets of expansion terms corresponding to this query. These terms are selected from 2 resources (Wikipedia and feedback documents) following [3,4].

Table 1. Two sets of expansion terms selected for the query "avp", from Wikipedia and feedback documents, respectively

Wikipedia	*volleyball, enterprise, alien, violence, avon, film, beach, pennsylvania, wilkes-barre, casting.*
Feeback documents	*news, price, product, planet, movie, game, world, version, alien, download.*

From Table 1, we clearly observe that the expansion terms from Wikipedia are more related to the query[3] than the ones selected from feedback documents. This means that Wikipedia is a good resource for the query "avp", while the feedback documents seem less appropriate for the same query. With the absence of a proper weighting of these two resources, one can only select terms uniformly from both resources, thus introducing noise terms (those that are irrelevant to the query). To benefit from the high-quality of expansion terms obtained from Wikipedia, one should assign a higher importance to it.

In this work, we advocate that we should assign proper resource weights while building a DQE based SRD system with multiple resources. This paper focus on the problem of proper resource weighting. Once the resources are weighted, we use the approach proposed in [4] to incorporate the resources into the SRD system, *i.e.*, selecting a number of expansion candidates from a resource that is proportional to the weight of that resource, and using resource weights to adjust

[3] These terms are more closely related to the subtopics manually identified by TREC assessors for the query "avp". This query has several subtopics: 'association of volleyball professionals', 'alien vs predator', 'wilkes-barre airport in Pennsylvania', etc.

the weights of the finally selected expansion terms (see Section 3.1 for details). One straightforward approach to modeling resource weight is to compute the average contribution of a resource to SRD on all the queries for training. Experimentally, we find this overall resource weighting approach, though simple, significantly improves the α-nDCG [7] and S-recall [18] scores on three topic sets. However this approach suffers from one issue: it ignores the fact the contribution of a resource to SRD varies depending on the query. To address this limitation, we develop a linear regression model to compute query level resource weighting, which considers 39 features. Although more sophisticated regression methods could be used in the future, our approach already shows that the SRD performance can be further improved using query-dependent resource weighting.

In our experiments, the ideal weight of a resource for a query is not annotated. To estimate this weight, for each training query, we use an exhaustive grid search to obtain the ideal resource weights that yield the best α-nDCG@20 for that query and consider them as the ground-truth resource weights.

We evaluate the query level resource weighting method using TREC 2009, 2010 and 2011 Web tracks, with the following four resources: ConceptNet, query logs, Wikipedia and feedback documents. The experimental results show that this method produces significantly better diversified search results than the overall resource weighting method.

In summary, we make the following contributions in this work. First, we introduce the resource weighting task to a DQE based SRD system with multiple resources, which has been largely ignored by existing studies. Second, we propose a linear regression model to learn the weight of a resource for each query. Third, we extensively evaluate our resource weighting model on three query sets using a publicly available datasets, and experimentally show the advantage of our method over uniform weighting and non-query level resource weighting.

The remainder of this paper is organized as follows. In Section 2, we briefly survey some related studies. Section 3 will be dedicated to present details of our proposed framework, and Section 4 describes our experimental settings and our results. In Section 5, we conclude the work and show some possible future work.

2 Related Work

The work described in this paper aims to improve the diversity of search results. SRD has been extensively studied in the recent years (*e.g.*, [6,9,12,13,16] to name just a few). Most of the proposed approaches consider SRD as an optimization problem, which aims at re-ranking search results, in order to maximize novelty and/or coverage. While some of these approaches use a single resource (*e.g.*, [13,16]), a number of other approaches combine several resources (*e.g.*, [4,12]) to help select better documents from the initial retrieval results covering more subtopics. However, most of the approaches start with an initial retrieval result using the initial query. They are thus naturally limited by the coverage of subtopics by the initial retrieval result. In particular, we observe that for a number of queries, some subtopics are not correctly covered by these documents; thus a re-ranking of these documents has a limited impact on result diversity. In this paper, our goal is to expand the initial query by adding diversified terms so that the initial retrieval result is better diversified. Using the same document re-ranking method on the retrieval result, we can better cover different subtopics.

Some existing studies on SRD show the usefulness of weighting the query aspects in explicit SRD. For instance, Santos et al. [16] estimate the sub-query importance in order to promote aspects of interest to the user, and show that weighting query aspects improves both relevance and diversity of search results. In the same context, Ozdemiray and Altingovde [14] use post-retrieval query performance predictors to estimate aspects' weights based on the retrieval effectiveness on the document set. They experimentally show that weighting query aspects improves the state-of-the-art SRD approaches.

External resources have been widely used in different fields in information retrieval (*e.g.*, to collect good expansion terms for query expansion), such as ConceptNet [3], query logs [13], or a combination of multiple resources (*e.g.*, [2,4,10,12]). For instance, Diaz and Metzler [10] present a mixture of relevance models, by which they found that combining multiple external resources improves the relevance of the results in terms of precision. Bendersky et al. [2] collect expansion terms (concepts) from newswire and Web corpora. He et al. [12] propose the combination of click-logs, anchor text and web n-grams to generate related terms for query expansion, and show that combining several resources allows to select high-quality expansion terms.

More recently, multiple resources have also been used for the purpose of SRD, more specifically for DQE. Bouchoucha et al. [4] show that integrating multiple resources can improve the diversity of search results and the coverage of the query aspects. During their participation to the NTCIR IMine task [5], the authors combine five different resources and observe that the more resources we consider, the more aspects of the query we can cover. However, the resources are weighted uniformly in that work. Indeed, no previous study has proposed to properly weight different resources for the purpose of SRD, as we propose in this paper. Our work is thus an extension of [4] by proposing a query-dependent resource weighting method. We show that such a proper weighting can lead to significant gains in retrieval effectiveness.

3 Proposed Framework

In this section, we first give a formal definition of our task. Then we present the details of our query level resource weighting framework based on linear regression. Finally, we describe the set of features used to learn the regression model for resource weighting.

3.1 Task of Resource Weighting

In the context of DQE based SRD with multiple resources, given a query and a set of resources as input, the task of resource weighting outputs a non-negative and normalized real number for each resource that is proportional to the degree to which that resource can help to diversify the search results for that query. Hereafter, we will use q to denote the query, r a resource, R the set of resources under consideration, and $w(q, r)$ the weight of resource r for query q.

In this study, resource weights are used in the same way as in MMRE (Maximal Marginal Relevance based Expansion) [3,4], which is one of the sate-of-the-art DQE approaches. In particular, we generate a set of candidate expansion terms from each resource $r \in R$, which has a strong relation with the query (query terms). The similarity of a candidate expansion term e to an original query q (denoted by $s_r(q, e)$ hereafter) is measured according to the resource r as in [3,4]. For example, ConceptNet can suggest terms that are connected to query terms in the ConceptNet graph; feedback documents can suggest terms that co-occur often in text windows with query terms; and query logs suggest terms that appear in the same query sessions as the query.

Afterwards, we decide the number of expansion terms ($n(q, r)$) that we should keep from each resource, which is proportional to the weight of that resource $w(q, r)$ (which is to be determined by a regression method), as follows:

$$n(q, r) = \lceil \frac{w(q, r)}{\sum_{r' \in R} w(q, r')} \cdot N \rceil \tag{1}$$

where N is the total number of expansion terms to select (we select 10 in our experiments). Eq. 1 encodes our intuition that the more a resource is important for a query, the more we should select terms from it.

With the above proportion determined, we apply the MMRE method to select expansion terms iteratively as follows: the number $n(q, r)$ expansion terms are to be selected from each resource, starting from the most important resource. To select the next expansion term, we rely on the following MMRE formula, which combines relevance and diversity:

$$\arg\max_{e \in E-ES} \{\beta \cdot sim(e, q) - (1 - \beta) \cdot \max_{e' \in ES} sim(e, e')\} \tag{2}$$

Here, E represents the set of candidate expansion terms suggested from different resources, ES is the set of expansion terms already selected, and $\beta \in [0,1]$ controls the trade-off between relevance and redundancy of the expansion terms. $sim(e, q)$ (resp. $sim(e, e')$) is a function that computes the similarity score between a term e and the original query q (resp. a term e') as already proposed in [4].

Finally, a selected expansion term e is assigned a weight which is computed according to Eq. 3, with the intention to promote expansion terms from highly weighted resources.

$$w(q, e) = \sum_{r \in R \wedge e \in E_r(q)} w(q, r) \cdot s_r(q, e) \tag{3}$$

where $s_r(q, e)$ denotes the similarity score between query q and expansion term e based on resource r, as defined in [4].

The weighted expansion terms are then used to construct a new search query, which is sent to an information retrieval system to obtain a diversified set of search results. Note that the retrieved results are not processed by any additional document selection process (such as MMR [6] or xQuAD [16]) for further diversification, although this is possible.

3.2 Regression Model for Resource Weighting

A simple model of resource weighting is to assign the same weight to all the resources, *e.g.*, $w(q,r) = \frac{1}{|R|}$. This model totally ignores the contribution differences among resources. Another model is to give a query independent constant weight to each resource, for example, weighting a resource according to the average performance of a SRD system using that resource on all the training queries. This model considers the overall contribution difference among resources, but ignores the differences between individual queries. Here we present a query level resource weighting model based regression which removes the above limitation.

First, we characterize the resource weighting task by a set of features. One example feature can be the number of different expansion candidates generated by a resource (*i.e.*, the number of terms that are judged similar to query terms using the resource). Let x_i denote the ith feature derived from resource query pair (q,r), and ω_i the weight of the ith feature, then $w(q,r)$ can be expressed as the weighted combination of all the features plus an offset (denoted by b), as defined in Eq. 4.

$$w(q,r) = \sum_i \omega_i \cdot x_i + b \tag{4}$$

Then, we learn the feature weights by using Support Vector Regression (SVR) [17], *i.e.*, resolving the following optimization problem as defined in Eq. 5.

$$\arg\min_{\omega_i} \frac{1}{2} \cdot \sum_i \omega_i^2 + C \cdot \sum_{r \in R, q \in Q} (\xi_{q,r} + \xi_{q,r}^*) \tag{5a}$$

$$s.t. \begin{cases} w_{q,r} - w(q,r) \leq \varepsilon + \xi_{q,r}, \\ w(q,r) - w_{q,r} \leq \varepsilon + \xi_{q,r}^*, \\ \xi_{q,r}, \xi_{q,r}^* \geq 0. \end{cases} \tag{5b}$$

where Q denotes the queries for training, $w_{q,r}$ denotes the ideal weight of resource r for query q, the constant C determines the trade-off between the L_2 regularization on the resource weights and the ε-insensitive loss on the observations. This optimization problem is convex, and can be efficiently resolved.

For the above linear regression, we need training queries, *i.e.*, the features and the corresponding ideal weight $w_{q,r}$ of each resource. To obtain the ideal weights, for each $q \in Q$, we run the SRD procedure introduced in Section 3.1, with all possible resource weights, *i.e.*, $(w_{q,r_1}, w_{q,r_2}, \cdots, w_{q,r_{|R|}})$, where $w_{q,r_i} \geq 0$ and $\sum_{i=1,\cdots,|R|} w_{q,r_i} = 1$, and select the resource weight sequence that yields the best α-nDCG@20. In our experiments, we consider a grid search of step 0.05.

3.3 Resource Weighting Features

We derive a set of features related to the contribution of resource r to diversifying the search results of query q. Table 2 describes all the features, which are organized into two groups: features common to all resources and resource specific features. These latter are further organized into four categories, depending on which resource they are derived from (Wikipedia, ConceptNet, query logs or feedback documents). It is worth noting that, in case a resource cannot generate a resource specific feature, the value of that feature is set to 0. For example, for the resource *feedback documents*, we will have 3 resource nonspecific features and 5 resource specific features. All the other features will have 0 values.

Note that resource weights are *independently* learnt by our proposed regression model. However, in practice, the weights are not independent[4]. To tackle this problem, we perform a normalization of the learnt weights (similar to Eq. 1) to ensure that the sum of weights of all resources w.r.t. one query equals to 1.

Resource Nonspecific Features. For the features that are common to all resources, we use the number of different candidate expansion terms suggested by each resource (`DiffExpanTerms`), since we believe that the more a resource suggests different expansion terms, the more it is likely to cover the different aspects of the query. The average Inverse Document Frequency (`AvgIDF`) of these terms could also be a good indicator of the quality of expansion terms obtained from each resource.

A new feature that we define in this work is `ContribExpan` ($c(q,r)$) which denotes the aggregated contributions of all the suggested expansion terms by resource r to the diversity of the search results of a given query q. A greater $c(q,r)$ indicates that resource r is more effective to SRD for query q. $c(q,r)$ is normalized into $[0,1]$, and meets the constraint that the contribution scores of all considered resources sum up to 1. $c(q,r)$ is computed using Eq. 6:

$$c(q,r) \propto \frac{1}{|gen_r(q)|} \sum_{k=1}^{|gen_r(q)|} c(e_k,r) \tag{6}$$

where e_k denotes the kth expansion term for query q when using resource r, and $gen_r(q)$ is the set of candidate expansion terms generated using resource r. Following [9], we use Eq. 7 to compute the contribution of an expansion term:

$$c(e_k,r) = \max\{0, p(e_k|q) - \sum_{j=1}^{k-1} p(e_k|e_j)\} \tag{7}$$

where $p(e_k|q)$ represents the individual contribution of e_k to q, and $p(e_k|e_j)$ denotes the probability of e_k being predicted given e_j, which is estimated based on the co-occurrences between the two terms calculated on the whole

[4] If we give a high weight to a weak resource, then the stronger resources should have higher weights.

Table 2. All features computed in this work for automatically weighting resources. (Here, q denotes an original query, D denotes the set of top 50 retrieval results of q, and r denotes a resource that could be Wikipedia, ConceptNet, query logs, or feedback documents).

Category	Description	Total
***** Resource nonspecific***		
DiffExpanTerms	Number of different candidate expansion terms suggested by resource r	4
AvgIDF	Average IDF score of the top 10 expansion terms obtained from resource r	4
ContribExpan	Contribution score to q after being expanded using top 10 expansion terms from resource r	4
***** Resource specific***		
** Feedback documents:*		
PropFD	Proportion of the feedback documents that contain the terms of q, computed on D	1
AvgPMI	Average mutual information score between the terms of q and the top 10 terms that co-occur a lot with the terms of q in D	1
ClarityScore	Clarity score of q computed on D and the whole document collection [8]	1
CoocFreq	Co-occurrence frequency of the query terms computed at window of size 15 on D	1
TFIDF	TFxIDF score of the terms of q computed on D	1
** Wikipedia:*		
PropWiki	Proportion of the terms of q having an exact Wikipedia matching page	1
PageRank	PageRank score [15] of the Wikipedia page that matches q	1
NumInterp	Number of (possible) interpretations of q in the Wikipedia disambiguation page of q	1
WikiLength	Wikipedia page length (number of words) that matches with q	1
** ConceptNet:*		
PropConcep	Proportion of the terms of q that correspond to a node in the graph of ConceptNet	1
NumDiffNodes	Number of different adjacent nodes that are related to the nodes of the graph of q	1
AvgCommonNodes	Average number of common nodes shared between the nodes of the graph of q (*i.e.*, nodes that are connected to at least two edges)	1
NumDiffRelations	Number of different relation types defined between the adjacent nodes in the graph of q	1
** Query logs:*		
PropQL	Proportion of the terms of q that appear in the query logs	1
NumClicks	Max, Min and average number of clicked URLs for q in all the sessions	3
PercentageClicks	Percentage of shared clicked URLs between different users who issued q	1
ClickEntropy	Click entropy of the query q [11]	1
NumSessions	Total number of sessions with q	1
SessionLength	Max, Min and average session duration (in seconds) with q	3
NumTermsReform	Total number of different terms added by users to reformulate q in all the sessions	1
ReformLength	Max, Min and average number of terms added by users to reformulate q in all the sessions	3
Grand Total		**39**

document collection. Now, to estimate $p(e_k|q)$, we divide the computation into two parts[5]:

$$p(e_k|q) = p(e_k|q, r) \cdot p(r) \tag{8}$$

where $p(r)$ corresponds to the a priori contribution of the resource, which is approximated by the average contribution of resource r on the set of training queries. $p(e_k|q, r)$ is the importance of expansion term e_k in the query q, with respect to the resource r, which is estimated as follows:

$$p(e_k|q, r) = \max_{s \in q} s_r(s, e_k) \cdot \frac{|s|}{|q|} \tag{9}$$

where s is a sub-string of q, $|s|$ denotes the number of words in s, and $s_r(s, e_k)$ is the similarity between s and e_k according to r, as described in Section 3.1. Eq. 9 is based on our intuition that an expansion term that corresponds to a large part of the query should be attributed a high importance.

[5] We marginalise $p(e_k|q)$ over all resources.

Resource Specific Features. Most of the features in this category are straightforward and have been used in previous studies. So we only provide a brief explanation here. All the feedback documents-based features are computed on the top 50 results returned for the original query. These features are useful to assess the quality of search results in terms of adhoc retrieval and diversity, which help to decide whether we should rely on these results. *E.g.*, the clarity score introduced in [8] is a good indicator of the ambiguity level of a query. It was shown that the returned search results of an ambiguous query are in general ineffective [8].

For Wikipedia, we use the pages that match with the original query (or a part of the query terms) to derive our features. For example, PageRank score [15] is adopted to measure the importance of the Wikipedia pages corresponding to the query: the more important a Wikipedia page is, the more we expect selecting candidate expansion terms from it that are relevant to the query.

On query logs, we develop a number of additional features that are derived from the query reformulations, the click-through data and the query sessions. By investigating the past usage of the original query in the log, one can expect to get candidate expansion terms corresponding to the user intents.

Finally, for ConceptNet, we construct a graph for each query, such that the nodes of the graph are those connected to the query terms or a part of the query terms, from the graph of ConceptNet. The four considered features based on ConceptNet are then computed based on the graph of the query.

4 Experiments

In this section, we evaluate our proposed query level resource weighting method based on regression (denoted by QL-RW hereafter) for SRD. We compare our method to uniform resource weighting and non-query level resource weighting, which have been used in previous studies [4] and have shown competitive effectiveness against other state-of-the-art approaches.

4.1 Experimental Setup

Data and System. We conduct experiments on the ClueWeb09 (category B) dataset, which has 50,220,423 documents (about 1.5 TB), and use the test queries from TREC 2009, 2010 and 2011 Web tracks (hereafter denoted by WT09, WT10 and WT11, respectively). Indri is used as a basic IR system for indexing and retrieval. We use the query likelihood language model with Dirichlet smoothing. Weighted expansion terms are added into the query using *#weight* operator. The resources we use are: the log data of Microsoft Live Search 2006 which spans over one month (starting from May 1st) consisting of almost 14.9M queries shared between around 5.4M user sessions; the last version of ConceptNet[6]; the English Wikipedia dumps of July 8th, 2013; and the top 50 results returned for the original query. Since spam filtering is known to be an important component of

[6] `http://conceptnet5.media.mit.edu`

Web retrieval, we have applied the publicly available Waterloo spam ranking[7] to the ClueWeb09 collection. We consider a percentile of 60% which is shown to be optimal for the ClueWeb dataset [1].

Reference Systems and Parameter Setting. For comparison purpose, we consider the following two reference systems: nQL-RW, non query level-resource weighting, which assigns to each resource a query independent constant proportional to the average contribution of resource r for an SRD system on the whole training queries; U-RW, uniform resource weighting, which assigns uniform weights to the resources for all queries. Note that nQL-RW, U-RW, and QL-RW use the same SRD framework, the same resources, and the same parameter settings as in [4]. Besides, for a fair comparison between the three methods, each query is expanded with exactly the same words, but with different weights according to the method[8]. Both parameter β in Eq. 2 and parameter C in Eq. 5 are set using 3-fold cross validation: we use in turn each of the query sets from WT09, WT10 and WT11 for test while the other two sets for training. During this procedure, we optimize for α-nDCG@20. To resolve the optimization problem described in Section 3.2, we directly use SVM-Light tool[9] with option "-z r" [10]. Finally, the co-occurrences between pairs of terms in the feedback documents are computed using text windows of size 15.

Evaluation Metrics. We consider the following official measures as performance metrics: nDCG and ERR for adhoc relevance performance, α-nDCG [7] (in our experiments, α=0.5), ERR-IA, NRBP and Prec-IA for diversity measure, and S-recall [18] to measure the ratio of covered subtopics for a given query. All the results presented in this paper are computed at cutoff 20.

4.2 Results

We report the performance numbers in Table 3 on queries of WT09, WT10, and WT11, respectively. From Table 3, we observe that nQL-RW performs better than U-RW. This shows that a global average weighting is more appropriate than a uniform weighting. We also observe that our method (QL-RW) consistently significantly outperforms the other two reference systems, in both relevance and diversity measures, on all datasets. This observation confirms that resource weighting plays an important role in adhoc and diversity tasks, and suggests that resources should be incorporated according to their possible impact on the given query, rather than using query-independent or uniform weights.

---

[7] `https://plg.uwaterloo.ca/~gvcormac/clueweb09spam`

[8] We fix the expansion terms and change their weights in different methods according to the weights of resources. The different weights are assigned to the terms directly.

[9] `http://svmlight.joachims.org`

[10] Parameter C in Eq. 5.a is set to 1.5 using 3-fold cross validation. For the other parameters in SVM-Light, their default values are used in our experiments.

Table 3. Results of different methods on TREC Web tracks query sets. U and N indicate significant improvement (p <0.05 in two-tailed T-test) over U-RW and nQL-RW, respectively.

Queries	Method	nDCG	ERR	α-nDCG	ERR-IA	NRBP	Prec-IA	S-recall
WT09	U-RW	0.380	0.156	0.367	0.237	0.205	0.155	0.544
	nQL-RW	0.393^{U}	0.159	0.386^{U}	0.251^{U}	0.219^{U}	0.163	0.587^{U}
	QL-RW	$\mathbf{0.413}^{UN}$	$\mathbf{0.169}^{UN}$	$\mathbf{0.428}^{UN}$	$\mathbf{0.274}^{UN}$	$\mathbf{0.243}^{UN}$	$\mathbf{0.172}^{UN}$	$\mathbf{0.628}^{UN}$
WT10	U-RW	0.239	0.175	0.391	0.246	0.236	0.219	0.592
	nQL-RW	0.258^{U}	0.179	0.405^{U}	0.259^{U}	0.241^{U}	0.236^{U}	0.627^{U}
	QL-RW	$\mathbf{0.283}^{UN}$	$\mathbf{0.192}^{UN}$	$\mathbf{0.429}^{UN}$	$\mathbf{0.285}^{UN}$	$\mathbf{0.253}^{UN}$	$\mathbf{0.258}^{UN}$	$\mathbf{0.664}^{UN}$
WT11	U-RW	0.371	0.169	0.611	0.522	0.459	0.287	0.802
	nQL-RW	0.387^{U}	0.176	0.629^{U}	0.540^{U}	0.463	0.298^{U}	0.821^{U}
	QL-RW	$\mathbf{0.402}^{UN}$	$\mathbf{0.187}^{UN}$	$\mathbf{0.657}^{UN}$	$\mathbf{0.575}^{UN}$	$\mathbf{0.476}^{UN}$	$\mathbf{0.323}^{UN}$	$\mathbf{0.851}^{UN}$

4.3 Feature Effects

In this section, we investigate the usefulness of each group of features that we derived in this paper. Table 4 shows the performance of each group of features, in terms of nDCG@20 and α-nDCG@20, computed on the set of 144 queries [9]. In each row, only features of the corresponding category are selected (*e.g.*, QL-RW (Wikipedia) uses only features based on Wikipedia). Recall that U-RW uses a uniform weighting and corresponds to the approach with no feature selection.

First, we observe that every category of features produces some positive impact on the results, compared to U-RW. This highlights the role that our features play. Also, it is clear that considering all features yields larger improvements than using only a single group of features. Second, resource nonspecific features constitute the most robust group of features, yielding the best performance among the groups. In particular, our feature `ContribExpan` has been assigned a high importance. Finally, when comparing the groups of resource specific features, we observe that the features derived from query logs contribute more than the others. A possible reason is that the 144 queries used in this experiment are all well covered by the query logs, which may not be the case for the other resources.

Table 4. Performance with different feature sets in terms of nDCG and α-nDCG

Feature set	nDCG	α-nDCG
U-RW	0.326	0.451
QL-RW (resource nonspecific)	0.350	0.493
QL-RW (feedback documents)	0.331	0.471
QL-RW (Wikipedia)	0.338	0.479
QL-RW (ConceptNet)	0.335	0.478
QL-RW (query logs)	0.346	0.489
QL-RW (all features)	**0.359**	**0.504**

5 Conclusion and Future Work

In this paper, we propose a new query-level resource weighting method in the context of diversified query expansion. For that, we develop a regression model enabling us to learn, for each query, the weights of resources based on a set of features. We evaluated our approach on three topic sets, and using four representative resources. Our results demonstrate the advantage of our method over uniform weighting and non-query level resource weighting.

In this work, we considered four external resources. We believe that other resources could also be effective in our task, such as WordNet, anchor text collections and other resources, from which we can derive additional features for resource weighting. Another aspect where further improvement can be gained is the learning method: instead of using linear regression, other algorithms could be tested, such as those implemented in Weka.

References

1. Bendersky, M., Fisher, D., Croft, W.B.: Umass at trec 2010 web track: Term dependence, spam filtering and quality bias. In: Proc. of TREC (2010)
2. Bendersky, M., Metzler, D., Croft, W.B.: Effective query formulation with multiple information sources. In: Proc. of WSDM, Washington, USA, pp. 443–452 (2012)
3. Bouchoucha, A., He, J., Nie, J.Y.: Diversified query expansion using conceptnet. In: Proc. of CIKM, Burlingame, USA, pp. 1861–1864 (2013)
4. Bouchoucha, A., Liu, X., Nie, J.-Y.: Integrating multiple resources for diversified query expansion. In: de Rijke, M., Kenter, T., de Vries, A.P., Zhai, C., de Jong, F., Radinsky, K., Hofmann, K. (eds.) ECIR 2014. LNCS, vol. 8416, pp. 437–442. Springer, Heidelberg (2014)
5. Bouchoucha, A., Nie, J.Y., Liu, X.: Universite de montreal at the ntcir-11 imine task. In: Proc. of NTCIR IMine Task, pp. 28–35 (2014)
6. Carbonell, J., Goldstein, J.: The use of mmr, diversity-based reranking for reordering documents and producing summaries. In: Proc. of SIGIR, pp. 335–336 (1998)
7. Clarke, C.L., Kolla, M., Cormack, G.V., Vechtomova, O., Ashkan, A., Büttcher, S., MacKinnon, I.: Novelty and diversity in information retrieval evaluation. In: Proc. of SIGIR, Singapore, pp. 659–666 (2008)
8. Cronen-Townsend, S., Zhou, Y., Croft, W.B.: Predicting query performance. In: Proc. of SIGIR, NY, USA, pp. 299–306 (2002)
9. Dang, V., Croft, B.W.: Term level search result diversification. In: Proc. of SIGIR, NY, USA, pp. 603–612 (2013)
10. Diaz, F., Metzler, D.: Improving the estimation of relevance models using large external corpora. In: Proc. of SIGIR, NY, USA, pp. 154–161 (2006)
11. Dou, Z., Song, R., Wen, J.R.: A large-scale evaluation and analysis of personalized search strategies. In: Proc. of WWW, NY, USA, pp. 581–590 (2007)
12. He, J., Hollink, V., de Vries, A.: Combining implicit and explicit topic representations for result diversification. In: Proc. of SIGIR, NY, USA, pp. 851–860 (2012)
13. Liu, X., Bouchoucha, A., Sordoni, A., Nie, J.-Y.: Compact aspect embedding for diversified query expansions. In: Proc. of AAAI, pp. 115–121 (2014)
14. Ozdemiray, A., Altingovde, I.: Query performance prediction for aspect weighting in search result diversification. In: Proc. of CIKM, NY, USA, pp. 871–874 (2014)
15. Page, L., Brin, S., Motwani, R., Winograd, T.: The pagerank citation ranking: Bringing order to the web. Technical report (1999)
16. Santos, R.L., Macdonald, C., Ounis, I.: Exploiting query reformulations for web search result diversification. In: Proc. of WWW, Raleigh, USA, pp. 881–890 (2010)
17. Smola, A.J., Schölkopf, B.: A tutorial on support vector regression. Statistics and Computing 14(3), 199–222 (2004)
18. Zhai, C.X., Cohen, W.W., Lafferty, J.: Beyond independent relevance: Methods and metrics for subtopic retrieval. In: Proc. of SIGIR, NY, USA, pp. 10–17 (2003)

Exploring Composite Retrieval from the Users' Perspective

Horaţiu Bota[1], Ke Zhou[2], and Joemon M. Jose[1]

[1] University of Glasgow, Lilybank Gardens, Glasgow, UK
h.bota.1@research.gla.ac.uk, joemon.jose@glasgow.ac.uk
[2] Yahoo Labs London, 125 Shaftesbury Avenue, London, UK
kezhou@yahoo-inc.com

Abstract Aggregating results from heterogeneous sources and presenting them in a blended interface – *aggregated search* – has become standard practice for most commercial Web search engines. *Composite retrieval* is emerging as a new search paradigm, where users are presented with semantically aggregated information objects, called *bundles*, containing results originating from different verticals. In this paper we study *composite retrieval* from the user perspective. We conducted an exploratory user study where 40 participants were required to manually generate *bundles* that satisfy various information needs, using heterogeneous results retrieved by modern search engines. Our main objective was to analyse the contents and characteristics of user-generated bundles. Our results show that users generate bundles on common *subtopics*, centred around *pivot* documents, and that they favour bundles that are *relevant*, *diverse* and *cohesive*.

Keywords: Composite retrieval, bundle, vertical, diversity, relevance, cohesion.

1 Introduction

The past three decades have seen an explosion of information on the Web, in terms of both quantity and diversity of content. Modern Web search engines aggregate results from heterogeneous information sources – so called *verticals* – in order to satisfy complex user information needs [22]. Different approaches to aggregating information on the Web have been proposed and studied, such as *federated search* or *aggregated search*[9]. In general, these approaches focused on merging results from multiple homogeneous text collections into one ranked list or inserting blocks of results from different heterogeneous information sources within a standard search engine results page (SERP). As the Web is becoming more diverse, it is important to return to users semantically assembled information objects, containing information extracted from different sources. *Composite retrieval* has recently been proposed[4,15] as a solution for organising search results into *bundles* that support complex information needs. Each bundle on the SERP reflects one aspect (or *subtopic*) of the information need and potentially consists of documents originating from multiple verticals.[1].

Consider the following user information need: *travelling to Austria*. Finding all the information that satisfies this need typically involves submitting several queries, each

[1] We use the terms "facet", "aspect" and "subtopic" interchangeably throughout the paper.

A. Hanbury et al. (Eds.): ECIR 2015, LNCS 9022, pp. 13–24, 2015.

focusing on different aspects of travelling, such as directions, accommodation or points of interest. Composite retrieval on the Web aims to address the limitation of current search paradigms, such as aggregated search, and return to users semantically organised bundles of results, each satisfying different aspects of such a complex need.

Prior research on composite retrieval has focused on either analysing the algorithmic formulations of generating bundles or formalising the desirable properties of bundles[15]. Bota el al.[4] argued that bundles should be *relevant*, *cohesive* and *diverse*, and evaluated their proposed composite retrieval systems on each of these criteria independently. Although interesting, little has been understood in composite retrieval from the user perspective. For example, what are the most important criteria for users when assessing bundle quality? How do users formulate bundles? Answering these questions can align future developments of composite retrieval systems to user expectations. Therefore, we pursue this line of research and conduct an exploratory user study which allows us to investigate the contents, characteristics and topical focus of user-generated bundles. Essentially, our study participants were shown search results originating from various heterogeneous sources and were required to manually generate bundles that satisfy their information needs. After building bundles, different assessments of bundle characteristics are collected from users, in order to understand their preference when evaluating bundle quality. We specifically aim to answer: (**RQ1**) Do users agree with each other on the *subtopics* they form bundles on? (**RQ2**) How do users aggregate information to build bundles? How vertically diverse are the bundles generated by users? (**RQ3**) Which bundle characteristics are most important to users? What are the interactions between these characteristics?

Although our study tasks did not consist of explicit search interactions, but rather *composition* interactions, the analysis of user-built bundles and accompanying assessments offers significant insights into user expectations from composite objects and describes new directions in understanding composite retrieval. The main contributions of our study are: (i) we conducted the first investigation aimed at understanding user behaviour in a composite retrieval context; and (ii) our results provide valuable insights on user preference and the complexity of evaluation for composite retrieval, an essential first step towards a unified evaluation metric for composite retrieval.

2 Related Work

Composite Retrieval. Responding to information retrieval queries by presenting composite items has been proposed and investigated in a number of recent papers [1,8,11,15]. Many of the above papers have provided contributions on the theoretical side, studying the complexity of evaluating queries with constraints, and proposing different algorithmic formulations. Mendez-Diaz et al. [15] studied the complexity of composing bundles with constraints (such as budget), and proposed different algorithmic formulations to solve the problem of bundle generation. In many ways, composite retrieval on the Web is similar to category based search result clustering, which has been extensively studied in previous work[14,20]. It has been shown that hierarchical presentations of results improve navigation of results and is more effective, in terms of search time, exploration of results and discovery of content, than traditional ranked lists[14]. Recently, Bota et al.[4] employed composite retrieval in a Web-search context and showed that it can improve retrieval performance, in terms of traditional topical relevance, in a heterogeneous Web setting, while promoting both topical and vertical diversity.

User Studies. Prior work has looked at user search behaviour and motivation at length [12,17,18], mainly focusing on behaviour in traditional search environments. Our work aims to go beyond traditional search scenarios and investigates user behaviour in a result composition setting. User behaviour in exploratory collaborative Web search has been studied in work related to ours, specifically focused on modelling user search processes[21]. Furthermore, [20] provides a systematic guide for designing taxonomy-based search systems. Significant effort has been made to understand user behaviour in an aggregated search setting [2,5,23]. In particular, [2] investigated different aspects related to results page coherence that influence search behaviour in an aggregated search scenario. User preference of result aggregation methods is investigated by [5], where it is shown that users prefer heterogeneous blocks blended into traditional lists over tabbed displays when trying to obtain an overview of the available information space. This indicates that composition of results is beneficial in exploratory tasks, and motivates our efforts to investigate more elaborate aggregation techniques, such as composite retrieval. Although composite retrieval is similar to aggregated search, user behaviour in a composite setting has not been studied. Our aim is to investigate user behaviour in a composition scenario and analyse the contents and assessments of manually generated bundles to get an insight into user expectations from a composite search system.

3 User Study

Our objective was to determine the contents and characteristics of user-generated bundles of search results. In light of this objective, we ran a laboratory-based user study where participants constructed composite items using search results originating from different verticals and assessed their own bundles in terms of several criteria. For the study, we employed 40 participants (17 female, 23 male) with an average age of 24 ($mean = 24.75$, $stdev = 5.42$). Each participant was compensated with £10 for their help. Half of the participants were undergraduate students at the time of the study, 17 were postgraduates, and 3 were in active employment. In terms of background, 60% of them had obtained, or were interested in obtaining, a technical degree. Participants were given 4 different *composition* tasks and were asked to construct bundles, as described below, using results cached from several existing search engines. Each task was completed in approximately 15 minutes ($mean = 15.55$, $stdev = 8.80$). We used 40 different topics, collected from various aggregated search collections [7,22]. Topics were assigned randomly to participants, the only constraint being that each topic needed to be assigned to exactly 4 different users. Overall, each of the 40 participants performed 4 separate tasks, for a total of 160 bundle building tasks[2].

Task Design. To reflect complex exploratory information needs suited to composite retrieval, participants were asked to imagine that they are bloggers, preparing a series of blog posts on different aspects related to a given topic (e.g. living in India). Their

[2] More information on the user study (e.g. search tasks used) is available at dcs.gla.ac.uk/~horatiu

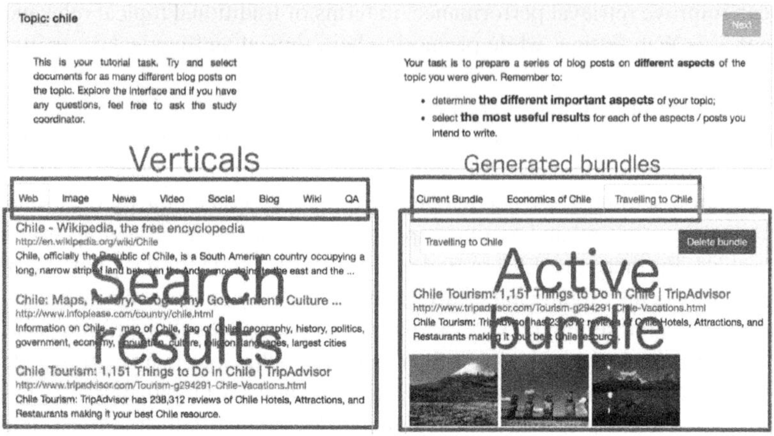

Fig. 1. Web based interface used by our participants to build bundles

choice of aspects (or *subtopics*) to focus on was unrestricted, however the subtopics were required to be distinct. For each subtopic, they were instructed to select the *most useful search results* – that they considered to be the most helpful for writing the blog post – and place them in a *bundle* of search results. Although they were required to title their bundle, they were not required to write an actual blog post, only to pre-select search results that might be useful for writing it.

During the study, participants were first shown a description of their general task, that of building bundles, and were guided through the system interface. The interface allowed participants to explore eight different verticals (shown in Figure 1), each containing 50 pre-retrieved results, for the topic they were assigned. All text-based results were presented using a standard Web search engine style, namely a highlighted title above a short snippet. Hovering over any search result displayed a tooltip window that contained additional information about the result. For example, hovering over *Video* results played a 10 second extract from the actual video result.

Figure 1 shows the system interface for building bundles. The verticals were presented as part of a tabbed section which occupied the left half of the interface. Search results were displayed in a traditional search engine layout – text-based documents were displayed in a ranked list of results, whereas *Images* and *Video* were displayed in a grid of thumbnails. The right section of the interface was occupied by the "bundling" area, where participants could create bundles by adding documents from any of the verticals, and assign titles to their bundles. There were no restrictions imposed on the number or size of bundles. After the bundle building phase, participants were required to assess each of their bundles in terms of the five criteria described below. They were also required to assign relevance labels (non-relevant, relevant, highly relevant, key and navigational) to each of the documents contained by the bundles.

Finally, participants were presented with pairs of their own bundles in a side-by-side view and asked to make a preference judgement between the two bundles. When indicating preference, they were also required to indicate the motive behind their preference in both free-form text and by indicating one of the five bundle criteria as being most influential on their choice (options *None* and *Overall* were also available). Bundle preference assessments allow us to determine which bundle characteristics are the most frequent indicators of preference, and also determine the degree to which our participants effectively assess bundle characteristics independently.

Bundle Characteristics. After generating the bundles, participants were required to rate them on five different criteria, using a five point scale (very, fairly, somewhat, slightly, not at all). Our choice of evaluation criteria was inspired by previous work on evaluating search results in context[10,3], where it has been shown that certain aspects of search results relevance are difficult or impossible to judge in isolation. In line with previous work, and given that prior work on composite retrieval [4] has already proposed basic evaluation metrics for bundle relevance, cohesion and diversity, we focus our evaluation of bundle characteristics on the five criteria described below:

Relevance – Are the documents in your bundle relevant to the topic?
Diversity – Does the bundle contain a diverse set of documents?
Cohesion – Are the documents in your bundle related and about one specific aspect of the topic?
Freshness – Is the bundle interesting and current?
Overall – How satisfied are you with your bundle?

Limitations. We chose to present results in a traditional layout to maintain user interface familiarity. Bundled results were presented using the same type of layout because there is limited understanding of how composite objects can be presented effectively on a search results page, without confusing searchers. Because we are interested in the contents and characteristics of bundles, not in the actual search interaction with bundles of results, we believe the presentation of bundles only minimally influenced their contents. We leave the investigation of bundle presentation issues for future work.

Verticals were presented in a fixed tabbed interface, in a predefined order – e.g. *General Web* occupied the first tab, followed by *Image*, *Video* and other types of documents. Even though the ordering of verticals may have had a biasing effect on document selection, we believe it was minimal: on one hand because participants were explicitly encouraged to explore all verticals before generating bundles; on the other hand, our interaction logs show that study participants explored an average of 7 ($mean = 7.275$, $stdev = 1.584$) verticals per tasks, suggesting they were at least acquainted with the top results in the majority of verticals.

One of the limitations of our study is the fact that participants were unable to explore actual documents, but were constrained to generating bundles using search results. Because we wanted to minimise the cognitive load on our participants, as well as keep task duration manageable, we chose not to allow participants access to actual documents. Even so, we consider result snippets to be highly representative of actual documents, and partially mitigated this limitation by allowing users to view highlighted document snippets, and for *Video*, short excerpts from the actual material.

System. Search results for all topics were cached on our server. The *General Web*, *Image* and *News* results were retrieved using the Bing Web Search API; the *Video* vertical was populated using the YouTube API; the *Social* vertical was populated using the

Twitter API; all other verticals were populated using the Google Custom Search API over specific websites[3] that matched the vertical profile, sourced from[7].

4 Results

As mentioned, our aim was to examine composite retrieval from two different perspectives: on one hand, we intended to analyse the contents and structure of composite objects by analysing the types of documents they contain and their topical focus; on the other hand, we were interested in determining how users assess bundles in terms of the five criteria we outlined in Section 3. In particular, we were interested in determining which criteria are most important to users. The main questions we aim to answer are:

- What documents do user-generated bundles contain?
- What bundle properties do users consider most important?

The following sections describe our findings. Section 4.1 provides an analysis of bundle contents, as well as an analysis of bundle subtopic agreement among users. It also includes a survey of potential *functions* that documents perform within bundles. Section 4.2 presents our results regarding user-assessed bundle characteristics.

4.1 Bundle Contents

Subtopics. Participants were asked to construct bundles using available documents, focusing each of their bundles on a specific aspect, or subtopic, of the topic they were given. The choice of subtopic was unrestricted as long as it was pertinent to the general topic. They were also required to title each of their bundles. Therefore, we define the *subtopic* of a bundle as *the facet of a specific topic around which a bundle is focused, as reflected by its title*. We employ this specific definition of bundle subtopic because our intention is to determine bundles that are similar, in terms of their topical focus, and analyse their contents and properties. We use bundle titles, as assigned by users, as proxies for evaluating the topical similarity of bundles.

To determine the semantic similarity of titles (and ultimately, bundles), we used a directional similarity metric – similarity of title t_i with respect to title t_j – inspired from [6]. Bundle titles are tokenised and part-of-speech tagged, and because they are relatively short ($mean$ length of 2.55 words, $stdev = 1.61$) we annotate each non-stopword in a title with a subset of its most likely synonyms, as determined by WordNet[16]. Starting with one title, for each word in its word class set, we determine the most similar word (using the Jiang-Conrath[13] similarity) from the corresponding set in the other title. We use the word similarity scores, weighed by the idf^4 scores of corresponding words and normalised by the idf scores of starting words, to compute the directional similarity of two bundle titles, as elaborated in [6]. We use the directionality of the metric to determine whether two bundles are mutually about the same subtopic.

[3] For example, the *Blog* vertical was populated using a Google custom search engine over the following domains: *wordpress.com*, *medium.com*, *tumblr.com* and *blogspot.com*.

[4] The British National Corpus was used to derive document frequency counts. www.natcorp.ox.ac.uk

Given two bundles $b_i, b_j \in B$, with their respective titles t_i, t_j, where B is the set of all user-generated bundles on a given topic, we assume that two bundles focus on the same subtopic if their titles mutually have the highest semantic similarity score:

$$max(\{\ \forall b_k \in B,\ i \neq k \mid sim(\mathbf{t_i}, \mathbf{t_k})\ \}) = sim(\mathbf{t_i}, \mathbf{t_j})$$
$$max(\{\ \forall b_k \in B,\ j \neq k \mid sim(\mathbf{t_j}, \mathbf{t_k})\ \}) = sim(\mathbf{t_j}, \mathbf{t_i})$$

This measure of title similarity is used to determine participant agreement on bundle subtopic: we want to determine whether participants building bundles on a given topic choose to focus their bundles on common subtopics. Because we want to determine different levels of subtopic agreement among users – e.g. 2 out of 4 (50%) participants generating bundles on the same topic agree on at least one common subtopic – we can restrict set B to include only bundles generated by a subset of users.

Table 1. Bundle subtopic agreement based on semantic similarity of bundle titles

		Proportion of participants / topic involved in determining subtopic agreement		
		100%	[75%, 100%)	[50%, 75%)
Percentage of bundles "about" same subtopic		~12%	~14%	~16%
Percentage of topics with	at least 1 common subtopic	32%	75%	90%
	at least 2 common subtopics	0%	32%	85%
	at least 3 common subtopics	0%	5%	60%

Our results (Table 1) show that, on average, users build 3 bundles ($mean = 3.2$, $stdev = 0.9$) of search results, focused on 3 distinct subtopics. Furthermore, there is a general tendency for user agreement on at least one common subtopic for a given topic — half the users build bundles on at least two common subtopics, for 85% of the topics used in our study. As an example, for the topic *"living in India"*, the following bundles from different users were determined to be similar based on their titles: *"Cost of living in India"*, *"average prices in india"* and *"Employment, Cost and Standard of Living"*.

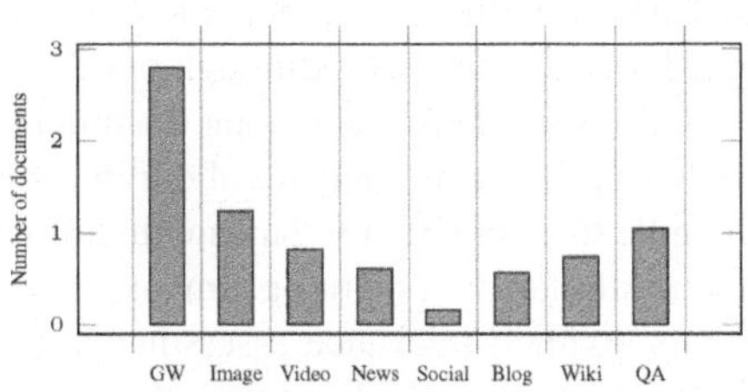

Fig. 2. Average vertical diversity

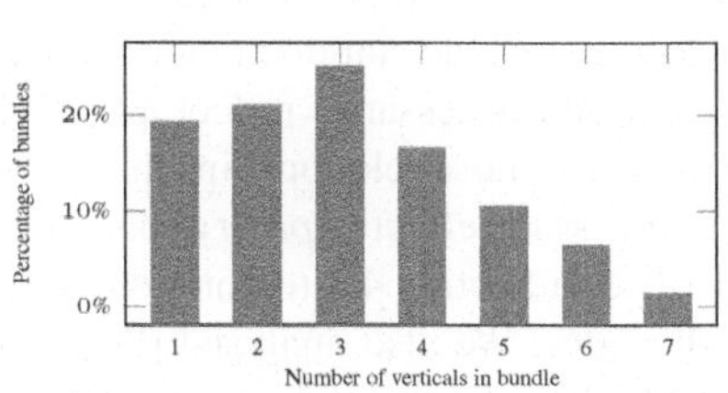

Fig. 3. Number of verticals in bundles

Vertical Composition. One of our main research objectives was to analyse the type of documents user-generated bundles contain. Figure 2 shows the average vertical composition of a bundle. The average number of documents contained by a bundle is 7 ($mean = 7.827, stdev = 5.577$), originating from 3 different verticals ($mean =$

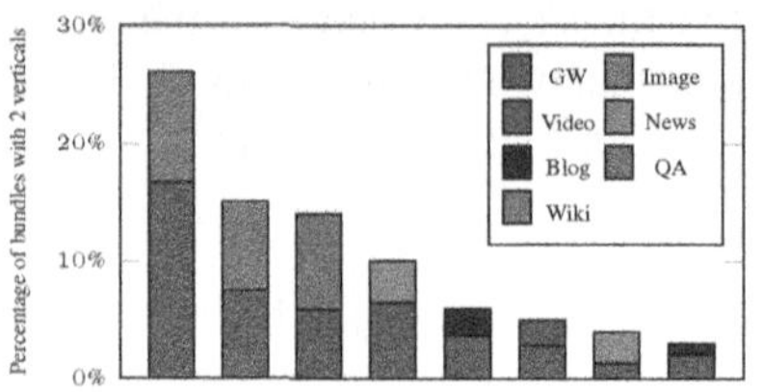

Fig. 4. Distribution of verticals in two-vertical bundles

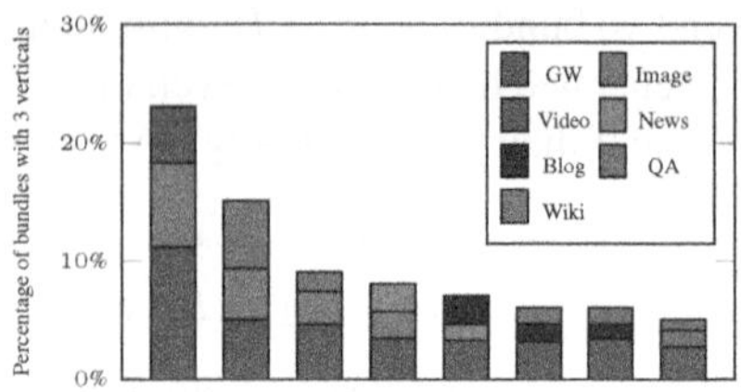

Fig. 5. Distribution of verticals in three-vertical bundles

$3.027, stdev = 1.535$). *General Web (GW)* is the most represented vertical in user-generated bundles, which is not unexpected given its intended, indeed highly optimised, purpose of satisfying wide ranges of information needs. The multimedia verticals – *Image* and *Video* – are also well represented, with roughly one document in each bundle. This is a reflection of the vertical orientation of topics used in the study and a potential click bias towards this type of media[19], but also suggests the importance of vertical diversity for users when building bundles. Even more, user inclination towards vertical diversity is suggested by the vertical composition of bundles, given that the majority of bundles contain more than two verticals (Figure 3). Note that we only instructed participants to bundle the most useful results and did not explicitly encourage vertical diversity in our instructions. Vertical distributions is detailed in Figures 4 and 5.

Document Roles. Prior work on composite retrieval on the Web constructed bundles by attaching search results to a central document, or set of documents, also called a *pivot*[4]. Inspired by this approach, we analyse different *functions* that documents perform within bundles. We distinguish between two separate functions:

- **Pivot documents** – Set of documents that appear in multiple bundles on the same subtopic, where bundle subtopic agreement is established as previously described.
- **Ornament documents** – Set of documents which originate from different verticals than pivot documents, which are not assessed as completely irrelevant by users.

Given our definitions of document functions, it is clear that not all documents within bundles (e.g. irrelevant documents) are assigned a function label. Although our definitions do not necessarily reflect all possible relationships between among documents, we highlight those relationships that are immediately distinctive and are of our interest.

To assess the effect of *pivot* documents on bundle structure, we used the methodology described in Section 4.1 to compute the semantic similarity of pivot document titles to bundle titles. We also analysed pivot documents' explicit relevance assessments. Related bundles, determined by similarity of their titles and for which subtopic agreement existed between at least 50% of users, were used to identify and analyse pivot documents. In total, 47% of the bundles we determined as being about the same subtopic contained at least one pivot document. On average, the bundles contained one pivot document ($mean = 1.27, stdev = 0.56$), with the largest pivot document set containing 4 documents. Our results show that pivot document titles are significantly (determined using one-way ANOVA testing: $F(1, 871) = 31.764$, $p < 0.01$) more similar to bundle titles, and are also significantly ($F(1, 871) = 70.831$, $p < 0.01$) more relevant than other documents within bundles This suggests that ~21% of user-generated bundles are constructed around at least one pivot document which is central to the composition process and determines the bundle subtopic (represented by the title).

In addition, we analysed the vertical origin of pivot document sets. It is perhaps not surprising that the majority of pivot documents originated from *General Web* (61% of pivots) and *Wiki* (23% of pivots) verticals, considering their broader scope and perhaps higher semantic load than multimedia, *QA* or *Blog* documents.

Table 2. Ornament diversity under different pivot types

	Bundle pivot type	
	GW	*Wiki*
GW	–	24.6%
Image	23.5%	31.1%
Video	21.3%	18%
News	7.1%	1.6%
Social	<1%	6.6%
Blog	9%	11.5%
QA	17.4%	4.9%
Wiki	19.7%	–

Table 3. Average document relevance in multi-vertical bundles

	Verticals in bundle	
	2 verticals	3 verticals
GW	3.872	3.667
Image	3.208	3.352
Video	3.228	3.649
News	2.954	3.156
Social	2.667	2.200
Blog	3.593	3.402
QA	2.560	2.652
Wiki	3.553	3.584

To determine the *ornament* composition of bundles, we analysed similar bundles that contained at least one pivot document and extracted documents that originated from other verticals than the pivots, and which were assessed by users as not completely irrelevant. Our intention was to determine which documents provide value through "composition" rather than explicit relevance, by complementing bundle contents. Our results show that similar bundles, which contain at least one pivot, have an average of 4 documents ($mean = 3.60$, $stdev = 4.26$) that match our ornament definition, originating from the *Image* (23%), *Video* (19%), *Wiki* (17%) and *QA* (16%) verticals.

We investigated the relationship between pivots and ornaments by analysing the vertical distributions of ornaments in bundles with different types of pivots. In particular, we focused on the two main pivot types (*GW* and *Wiki*) and analysed the types of ornaments associated with these types of pivots. Table 2 shows that pivot type affects ornament diversity to an extent. Even though only *Social* ornaments were significantly (determined using one-way ANOVA testing: $F(1, 82) = 4.442, p < 0.05$) more frequent in bundles with *Wiki* pivots, there is a trend that suggest a complementarity relationships between different types of pivots and ornaments. *Images* appear more frequently in bundles centred around *Wiki* pivots, whereas *QA* documents complement *General Web* pivots more often. Table 3[5] also shows that, as bundle vertical diversity increases, ornament documents (such as *Video*) tend to be assessed as more relevant, whereas *GW* are assessed as being less relevant. This suggests that relevance becomes distributed across different verticals in more diverse bundles.

[5] Significant trends are highlighted.

4.2 Bundle Characteristics

In addition to examining bundle contents, our research objectives include analysing user assessments of bundle characteristics – *Relevance, Diversity, Cohesion* and *Freshness*. Our intention was to assess how users rate bundles in relation to these four criteria and determine a potential hierarchy of bundles characteristics, as well as uncover correlations among these characteristics.

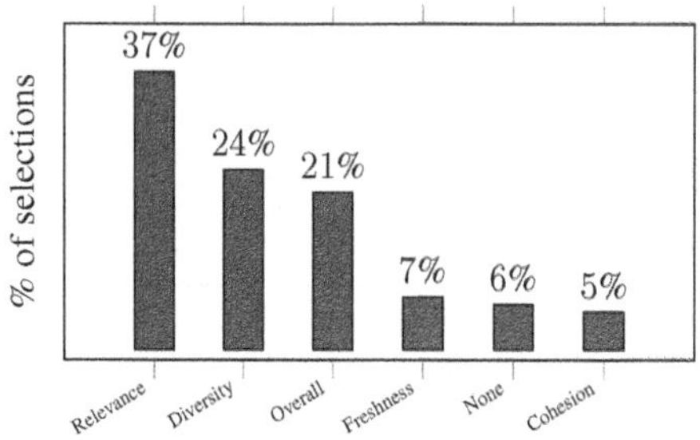

Fig. 6. User indicated most influential criterion for bundle preference

Table 4. Correlation of criteria assessments with bundle preference

	Pearson's R	
Criterion	**All**	**Chosen**
Relevance	0.332	0.496
Cohesion	0.228	0.432
Diversity	0.334	0.487
Freshness	0.208	0.213
Overall	0.453	0.454

Participants were also required to make explicit preference judgments between bundle pairs and motivate their preference by indicating one of the criteria above as influential on their choice (options *None* and *Overall* were also available). As shown in Figure 6, *Relevance* and *Diversity* were most frequently indicated as the criterion that influences preference. In addition, 21% of the participants indicate *Overall* (all of the criteria) as the motivation for their preference.

Table 5. Correlation of bundle properties (based on bundle assessments)

	Relevance	*Diversity*	*Cohesion*	*Freshness*	*Overall*
Relevance	–	0.272	**0.538**	0.334	**0.630**
Diversity	0.272	–	0.144	0.485	0.478
Cohesion	**0.538**	0.144	–	0.250	**0.548**
Freshness	0.334	0.485	0.250	–	**0.537**
Overall	**0.630**	0.478	**0.548**	**0.537**	–

We also analysed the correlation between bundle preference and bundle characteristics, as shown in Table 4. In particular, we examined whether preference of bundle A over bundle B correlates with higher criteria assessments for bundle A than for bundle B in **All** cases. Additionally, we examined this correlation taking into account the criterion **Chosen** by users as most influential — i.e. if *Relevance* is indicated by the user as the influential criterion for preference of bundle A over bundle B, is the user assessed *Relevance* of A higher than that of B? Table 4 shows that there is modest correlation between preference and user assessment of bundle characteristics, even in cases where the specific characteristics are indicated as the reason behind bundle preference. In roughly 50% of the cases, even though users explicitly mention *Relevance* as the motive behind their preference, they prefer the bundle assessed as less relevant. Although part of this can be due to noise in our data, this highlights the difficulty of determining the bundle characteristics that are most important to users.

Finally, we investigated the correlation between different pairs of bundle characteristics, shown in Table 5. It is worth noting that there is strong correlation between several bundle characteristics, the strongest correlation being that between *Relevance* and *Overall*. This suggests that bundle characteristics are difficult to assess independently and, combined with results mentioned above, collectively contribute to user preference. Even so, *Relevance*, *Cohesion* and *Diversity* are correlated with both user preference and among themselves, which demonstrates their combined significance to user experience in a composite Web environment.

5 Conclusion and Discussions

In this paper, we analysed the contents and characteristics of user-generated bundles of documents. Our primary interest was to determine how bundles are generated by users with regard to their topical focus, document composition and user-assessed characteristics. Our results suggest the following trends:

- To answer **RQ1**, we found there is an agreement between users on the topical focus of bundles. This suggests that composition of search results can be focused on distinctive facets of given topics and the composite retrieval systems should utilize the interests of a population of users to determine the most interesting subtopics to present.
- To answer **RQ2**, we observe there is a trend for bundles to contain central documents, or pivots, that are more relevant and reflect the bundle contents (subtopic). These documents tend to originate from verticals with higher semantic load (such as *General Web* or *Wiki*). Furthermore, ornament documents, which tend to be less relevant than pivots and more vertically diverse, are also included in bundles. The *Image*, *Video* & *QA* verticals are the most popular origins of ornament documents. The above results suggest that one effective strategy for composite retrieval system for bundling is to first select a small subset of key pivot documents, and then explicitly attach to bundles other documents complementing pivots, in order to boost complementarity and diversity.
- To answer **RQ3**, although our results do not establish a clear hierarchy of bundle characteristics, we confirm assumptions made by previous work [4] and determine that *Relevance*, *Cohesion* and *Diversity* are important to participants, but are difficult to assess independently. Corroborated with the above-mentioned insights on vertical diversity, this implies that, although explicit relevance is crucial to users, composition of diverse results can generate additional value.

In terms of future work, many open questions remain. Our work so far has investigated user generation of bundles, but has not explored composite retrieval in an actual search scenario. To explore the search potential of bundles, further work is needed to understand the complex aspects related to the presentation of a composite results page. It is likely that presentation factors can influence both the perceived relevance and user interaction with bundled documents, and we aim to analyse these factors in future work.

Acknowledgments. This work was partially funded by the Linguistically Motivated Semantic Aggregation Engines (www.limosine-project.eu) EU project.

References

1. Angel, A., Chaudhuri, S., Das, G., Koudas, N.: Ranking objects based on relationships and fixed associations. In: EDBT 2009, pp. 910–921 (2009)
2. Arguello, J., Capra, R.: The effect of aggregated search coherence on search behavior. In: CIKM 2012, pp. 1293–1302 (2012)
3. Bailey, P., Craswell, N., White, R.W., Chen, L., Satyanarayana, A., Tahaghoghi, S.M.: Evaluating search systems using result page context. In: IIiX 2010, pp. 105-114 (2010)
4. Bota, H., Zhou, K., Jose, J.M., Lalmas, M.: Composite retrieval of heterogeneous web search. In: WWW 2014, pp. 119–130 (2014)
5. Bron, M., van Gorp, J., Nack, F., Baltussen, L.B., de Rijke, M.: Aggregated search interface preferences in multi-session search tasks. In: SIGIR 2013, pp. 123–132 (2013)
6. Corley, C., Mihalcea, R.: Measuring the semantic similarity of texts. In: EMSEE 2005, pp. 13–18 (2005)
7. Demeester, T., Trieschnigg, D., Nguyen, D., Hiemstra, D.: Overview of the trec 2013 federated web search track. In: Proceedings of the Text Retrieval Conference, pp. 1–11 (2013)
8. Deng, T., Fan, W., Geerts, F.: On the complexity of package recommendation problems. In: PODS 2012, pp. 261-272 (2012)
9. Diaz, F., Lalmas, M., Shokouhi, M.: From federated to aggregated search. In: SIGIR 2010, p. 910 (2010)
10. Golbus, P.B., Zitouni, I., Kim, J.Y., Hassan, A., Diaz, F.: Contextual and dimensional relevance judgments for reusable serp-level evaluation. In: WWW 2014, pp. 131–142 (2014)
11. Guo, X., Ishikawa, Y.: Multi-objective optimal combination queries. In: Hameurlain, A., Liddle, S.W., Schewe, K.-D., Zhou, X. (eds.) DEXA 2011, Part I. LNCS, vol. 6860, pp. 47–61. Springer, Heidelberg (2011)
12. Jansen, B.J., Pooch, U.: A review of web searching studies and a framework for future research. J. Am. Soc. Inf. Sci. Technol. 52(3), 235–246 (2001)
13. Jiang, J.J., Conrath, D.W.: Semantic similarity based on corpus statistics and lexical taxonomy. CoRR, cmp-lg/9709008 (1997)
14. Käki, M.: Findex: Search result categories help users when document ranking fails. In: CHI 2005, pp. 131–140 (2005)
15. Mendez-Diaz, I., Zabala, P., Bonchi, F., Castillo, C., Feuerstein, E., Amer-Yahia, S.: Composite retrieval of diverse and complementary bundles. IEEE Transactions on Knowledge and Data Engineering 99(preprints), 1 (2014)
16. Miller, G.A.: Wordnet: A lexical database for english. Commun. ACM 38(11), 39–41 (1995)
17. Rose, D.E., Levinson, D.: Understanding user goals in web search. In: WWW 2004, pp. 13-19 (2004)
18. Spink, A., Jansen, B.J., Wolfram, D., Saracevic, T.: From e-sex to e-commerce: Web search changes. Computer 35(3), 107–109 (2002)
19. Sushmita, S., Joho, H., Lalmas, M., Villa, R.: Factors affecting click-through behavior in aggregated search interfaces. In: CIKM 2010, pp. 519–528 (2010)
20. Wilson, M.L., Kules, B., Schraefel, M.C., Shneiderman, B.: Designing future search interfaces for the web. Found. Trends Web Sci. 2(1), 1–97 (2010)
21. Yue, Z., Han, S., He, D.: Modeling search processes using hidden states in collaborative exploratory web search. In: CSCW 2014, pp. 820–830 (2014)
22. Zhou, K., Cummins, R., Lalmas, M., Jose, J.M.: Evaluating aggregated search pages. In: SIGIR 2012, pp. 115–124. ACM (2012)
23. Zhou, K., Cummins, R., Lalmas, M., Jose, J.M.: Which vertical search engines are relevant? In: WWW 2013, pp. 1557-1568 (2013)

Improving Aggregated Search Coherence

Jaime Arguello

University of North Carolina at Chapel Hill
jarguello@unc.edu

Abstract. Aggregated search is that task of blending results from different search services, or *verticals*, into the core web results. Aggregated search coherence is the extent to which results from different sources focus on similar senses of an ambiguous or underspecified query. Prior research studied the effect of aggregated search coherence on search behavior and found that the query-senses in the vertical results can affect user interaction with the web results. In this work, we develop and evaluate algorithms for *vertical results selection*—deciding which results from a particular vertical to display. Results from a large-scale user study suggest that algorithms that improve the level of coherence between the vertical and web results influence users to make more productive decisions with respect to the web results—to engage with the web results when at least one of them is relevant and, to a lesser extent, to *avoid* engaging with the web results otherwise.

1 Introduction

Commercial search portals such as Google, Bing, and Yahoo! provide access to a wide range of specialized search services or *verticals*. Example verticals include search engines for a specific type of media (images, videos, books) or a specific type of search task (search for news, local businesses, on-line products). The goal of *aggregated search* is to integrate results from different verticals into the core web results. From a system perspective, aggregated search is a two-part task: (1) predicting *which* verticals to present for a given query (*vertical selection*) and (2) predicting *where* to present those verticals selected (*vertical presentation*). Typically, a vertical is presented by blending a few of its top results somewhere in the first page of web results.

In this work, we study a phenomenon called *aggregated search coherence*. Given an ambiguous or underspecified query (e.g., "saturn"), a common strategy for a search engine is to diversify its results (e.g., to return results about "saturn" the planet, the car, and the Roman god). Aggregated search coherence is the extent to which results from different sources focus on similar senses of the query. Suppose that a user enters the query "saturn" and the system decides to integrate image vertical results into the web results. If the web results focus on the car, but the blended images focus on the planet, then the aggregated results have a *low* level of coherence. Conversely, if both sets of results focus on the same query-sense(s), then the aggregated results have a *high* level of coherence.

A. Hanbury et al. (Eds.): ECIR 2015, LNCS 9022, pp. 25–36, 2015.

Prior work investigated the effects of aggregated search coherence on search behavior. Specifically, Arguello and Capra [2,4,3] found that users are more likely to interact with the web results when the vertical results are more consistent with the user's intended query-sense. That is, a user looking for "saturn" the car is more likely to interact with the web results if the vertical results blended on the SERP include results about the car versus the planet. This is referred to as a "spill-over" effect. The spill-over effect suggests that while the vertical results come from a completely independent system, they can still influence user engagement with other components on the SERP (e.g., the web results).

Modeling cross-component effects is an important, yet understudied problem in aggregated search. If a user wants results from multiple sources (e.g., vertical and web results) or wants web results instead of vertical results, it is important for the system to display vertical results that show how the vertical is relevant to the query, but do not negatively affect user engagement with other components on the SERP. In this paper, we evaluate algorithms for *vertical results selection*—deciding which results from a particular vertical to display. We focus on algorithms that improve the level of coherence between the vertical and web results and show that these methods avoid negatively affecting user engagement with the web results.

There are two ways in which incoherent vertical results can negatively affect user engagement with the web results. First, if the vertical results contain the user's intended query-sense, but the web results do not, then the vertical results may influence the user to engage with the web results in vain. A more productive decision would be to quickly reformulate the query. Second, if the vertical results *do not* contain the intended query-sense, but the web results do, then the vertical results may influence the user to unnecessarily reformulate the query. A more productive decision would be to engage with the web results. If we treat user engagement with the web results as a binary decision, then these two situations represent false-positive and false-negative decisions by the user, respectively.

We evaluate several different vertical results selection algorithms across four verticals: images, news, shopping, and video. Results from a large-scale user study suggest that algorithms that improve the level of coherence between the vertical and web results influence users to make more productive decisions with respect to the web results—to engage with the web results when there is a relevant web result on the SERP and, to a lesser extent, to *avoid* engaging with the web results otherwise.

2 Related Work

Current methods for aggregated search prediction and evaluation do not *explicitly* favor coherent results. Algorithms for vertical selection and presentation use machine learning to combine a wide range of features. Prior work investigated features derived from the query string [6,11,14], from the vertical results [5,6,10,11], from the vertical query-log [5,6,10,11], and from historic click-through rates on the vertical results [14]. None of these features consider the relationship between the vertical results and those from other components on the SERP. Evaluation methods for aggregated search fall under three categories: on-line, test-collection, and whole-page evaluation methods. On-line methods are used to evaluate systems in a live environment using implicit feedback (i.e., vertical *clicks* and *skips*). One limitation of these methods is all false positive vertical predictions (signaled by a *skip*) are treated equally. Prior work found that, depending on the vertical results, displaying a non-relevant vertical can also affect engagement with other components on the SERP [2,4,3]. An aggregated search test-collection includes a set of queries, cached results from different sources, and relevance judgements. Zhou *et al.* [20] proposed an evaluation metric that considers three distinguishing properties between verticals: (1) its relevance to the task, (2) the visual salience of the vertical results, and (3) the effort required to assess their relevance. Our research suggests a fourth aspect to consider: the expected spill-over from the vertical results to other components. Bailey *et al.* [7] proposed an evaluation method that elicits human judgements on the whole SERP. While cross-component coherence is mentioned an important aspect of whole-page quality, its effect on search behavior was not investigated.

Incoherent results occur when the different aggregated components focus on different senses of an ambiguous query. A natural question is: How often does this happen? Sanderson [16] analyzed a large commercial query-log and found that about 4% of all unique queries and 16% of all unique head queries corresponded to ambiguous entities in Wikipedia and WordNet. This result suggests that ambiguous queries are common. Given an ambiguous query, incoherent results are more likely when results from different sources favor different senses. The analysis by Santos *et al.* [18] suggests that this is often the case. Santos *et al.* considered the different senses for a set of ambiguous entities and compared their frequencies in query-logs from a commercial web search engine and three verticals. Results found that different sources are often skewed towards different senses (e.g., the shopping vertical had more queries about "amazon" the company, while the images vertical had more queries about the rainforest).

One strategy for improving aggregated search coherence is to diversify results from different components across similar query-senses. Approaches for search result diversification fall under two categories: *implicit* and *explicit*. Implicit approaches diversify results by minimizing redundancy in the top ranks [8]. Explicit approaches diversify results by directly targeting results about different aspects of the query. Prior work investigated predicting the different query-aspects using topic categorization [1], a clustering of the collection [9], query reformulations in a query-log [15], and query suggestions from an on-line "related queries" API [17]. In this work, we focus on methods for selecting vertical results on the same query-senses as the web results and include Maximal Marginal Relevance [8] (an implicit diversification method) as a baseline for comparison.

3 Algorithms for Vertical Results Selection

Preliminaries. We describe our algorithms using the following notation. First, we assume that each vertical v is associated with some number τ_v of results that are blended into the web results if the vertical is presented. We considered four verticals. For the images, shopping, and video verticals, $\tau_v = 5$. For the news vertical, $\tau_v = 3$. Let $\mathcal{R}_q^v$ denote the original retrieval from vertical v in response to query q. All the algorithms described below take $\mathcal{R}_q^v$ as the input and produce a new ranking denoted as $\tilde{\mathcal{R}}_q^v$. The goal for the system is to decide which t_v results from $\mathcal{R}_q^v$ to include in $\tilde{\mathcal{R}}_q^v$ and in what order. Next, let $\mathcal{R}_q^w$ denote the top 10 web results for query q and $\tilde{\mathcal{R}}_q^w$ denote a diversified re-ranking of $\mathcal{R}_q^w$. One of our algorithms uses $\tilde{\mathcal{R}}_q^w$ internally to diversify the vertical results. Also, let $\mathcal{R}_q^*(k)$ denote the result at rank k in $\mathcal{R}_q^*$. Finally, all the algorithms described below require measuring the similarity between pairs of web and/or vertical documents. This similarity function is denoted as $\phi(d_i, d_j)$ and is explained later.

Maximal Marginal Relevance. MMR diversifies results by minimizing redundancy in the top ranks [8]. Given an initial ranking $\mathcal{R}_q$, it constructs a new ranking $\tilde{\mathcal{R}}_q$ by iteratively appending documents that are similar to the query (relevant) and dissimilar to those already in $\tilde{\mathcal{R}}_q$ (novel).

Our implementation of MMR assumes that the relevance of every vertical result in $\mathcal{R}_q^v$ is constant. Thus, vertical results are appended to $\tilde{\mathcal{R}}_q^v$ solely based on their dissimilarity to those already in $\tilde{\mathcal{R}}_q^v$. We first initialize $\tilde{\mathcal{R}}_q^v$ by appending the top vertical result in $\mathcal{R}_q^v$ and then iteratively append vertical results from $\mathcal{R}_q^v$ with the lowest similarity with the most similar ones already in $\tilde{\mathcal{R}}_q^v$.

MMR may improve coherence if the web results are diversified and the top vertical results are initially skewed towards a particular query sense. However, MMR selects vertical results *independently* from the web results. The next three approaches explicitly select vertical results that are similar to the web results.

Web Similarity. WebSim (Algorithm 1) aims to diversify the vertical results in $\tilde{\mathcal{R}}_q^v$ across the same query-senses in the top τ_v web results. Specifically, it iteratively appends vertical results to $\tilde{\mathcal{R}}_q^v$ such that $\tilde{\mathcal{R}}_q^v(k)$ corresponds to the vertical result in $\mathcal{R}_q^v$ most similar to $\mathcal{R}_q^w(k)$ (lines 3-6).

A possible disadvantage of WebSim is that the top τ_v web results may not cover all the query-senses in the top 10 web results. For example, the top 10 web results may include results about "saturn" the planet and the car, but the top

Algorithm 1. Web Similarity

WebSim($\mathcal{R}_q^v, \mathcal{R}_q^w, \tau_v$)
1: $\tilde{\mathcal{R}}_q^v \leftarrow \emptyset$; $k \leftarrow 1$
2: **while** $|\tilde{\mathcal{R}}_q^v| < \tau_v$ **do**
3:     **for all** $d_i \in \mathcal{R}_q^v$ **do**
4:         $sim(d_i) \leftarrow \phi\left(d_i, \mathcal{R}_q^w(k)\right)$
5:     **end for**
6:     $d^* \leftarrow \arg\max_{d_i} sim(d_i)$
7:     $\tilde{\mathcal{R}}_q^v \leftarrow \tilde{\mathcal{R}}_q^v \cup \{d^*\}$; $\mathcal{R}_q^v \leftarrow \mathcal{R}_q^v \backslash \{d^*\}$; $k \leftarrow k+1$
8: **end while**
9: **return** $\tilde{\mathcal{R}}_q^v$

τ_v web results may all be about the planet. The next two approaches attempt to address this issue.

Web Similarity MMR. WebSimMMR (Algorithm 2) is almost identical to WebSim. However, instead of selecting the vertical results most similar to the top τ_v results in $\mathcal{R}_q^w$, it first uses MMR to re-rank $\mathcal{R}_q^w$ into $\tilde{\mathcal{R}}_q^w$ (line 1). Then, it iteratively appends vertical results to $\tilde{\mathcal{R}}_q^v$ such that $\tilde{\mathcal{R}}_q^v(k)$ corresponds to the vertical result in $\mathcal{R}_q^v$ most similar to $\tilde{\mathcal{R}}_q^w(k)$ (lines 3-6). The goal of internally re-ranking the web results using MMR is to have the top τ_v results in $\tilde{\mathcal{R}}_q^w$ represent different query-senses present in the top 10 web results.

Algorithm 2. Web Similarity (MMR)

WebSimMMR($\mathcal{R}_q^v, \mathcal{R}_q^w, \tau_v$)

1: $\tilde{\mathcal{R}}_q^v \leftarrow \emptyset$; $k \leftarrow 1$; $\tilde{\mathcal{R}}_q^w \leftarrow$ MMR($\mathcal{R}_q^w$)$\triangleright$ Re-rank top 10 web results ($\mathcal{R}_q^w$) with MMR.
2: **while** $|\tilde{\mathcal{R}}_q^v| < \tau_v$ **do**
3:     **for all** $d_i \in \mathcal{R}_q^v$ **do**
4:         $sim(d_i) \leftarrow \phi\left(d_i, \tilde{\mathcal{R}}_q^w(k)\right)$
5:     **end for**
6:     $d^* \leftarrow \arg\max_{d_i} sim(d_i)$
7:     $\tilde{\mathcal{R}}_q^v \leftarrow \tilde{\mathcal{R}}_q^v \cup \{d^*\}$; $\mathcal{R}_q^v \leftarrow \mathcal{R}_q^v \backslash \{d^*\}$; $k \leftarrow k + 1$
8: **end while**
9: **return** $\tilde{\mathcal{R}}_q^v$

A potential disadvantage of WebSimMMR is that the ordering of vertical results in $\tilde{\mathcal{R}}_q^v$ is somewhat arbitrary. Our final approach attempts to order the vertical results based on the proportion of top 10 web results on that query-sense.

Web Cluster Similarity. WebClusterSim (Algorithm 3) first clusters the top 10 web results into τ_v clusters (line 1). We used complete-link agglomerative clustering. The resulting clusters ($\mathcal{C}_q^w$) are ordered by size such that $\mathcal{C}_q^w(k)$ corresponds to the kth largest cluster. Then, WebClusterSim iteratively appends vertical results to $\tilde{\mathcal{R}}_q^v$ such that $\tilde{\mathcal{R}}_q^v(k)$ corresponds to the vertical result in $\mathcal{R}_q^v$ with the greatest average similarity with the web results assigned to cluster $\mathcal{C}_q^w(k)$ (lines 3-6). The goal of WebClusterSim is to have vertical result $\tilde{\mathcal{R}}_q^v(k)$ be about the kth most frequent query-sense in the top 10 web results.

Implementation Details. All of the above algorithms required measuring the similarity between pairs of web and/or vertical documents (denoted as function ϕ in Algorithms 1-3). To this end, we represented documents using their topical distribution.[1] First, we identified 128 second-level categories from the Open Directory Project (ODP) hierarchy and crawled 2,000 random webpages from each category.[2] Then, we trained 128 logistic regression classifiers using the Liblinear Toolkit.[3] We adopted a simple TF.IDF representation with stemming and

[1] All results had a textual representation. The web and news results had a title and summary snippet, while the image, shipping, and video results had a title.
[2] `http://www.dmoz.org/`
[3] `http://www.csie.ntu.edu.tw/~cjlin/liblinear/`

Algorithm 3. Web Cluster Similarity

$\textsc{WebClusterSim}(\mathcal{R}_q^v, \mathcal{R}_q^w, \tau_v)$

$\tilde{\mathcal{R}}_q^v \leftarrow \emptyset$; $k \leftarrow 1$; $\mathcal{C}_q^w \leftarrow Cluster(\mathcal{R}_q^w)$ ▷ Cluster top 10 web results into t_v clusters.

while $|\tilde{\mathcal{R}}_q^v| < \tau_v$ **do**

  for all $d_i \in \mathcal{R}_q^v$ **do**

    $sim(d_i) \leftarrow \phi_{avg}\left(d_i, \mathcal{C}_q^w(k)\right)$ ▷ Compute average similarity.

  end for

  $d^* \leftarrow \arg\max_{d_i} sim(d_i)$

  $\tilde{\mathcal{R}}_q^v \leftarrow \tilde{\mathcal{R}}_q^v \cup \{d^*\}$; $\mathcal{R}_q^v \leftarrow \mathcal{R}_q^v \backslash \{d^*\}$; $k \leftarrow k+1$

end while

return $\tilde{\mathcal{R}}_q^v$

stopwords removed, and normalized documents to unit length. Finally, we used the mass-normalized prediction confidence values from each classifier to generate a topical distribution for a each web and vertical document. Document similarity was measured using the symmetrized Kullback-Leibler divergence (KLD) [12].[4]

4 User Study

Experimental Protocol. Our goal was to study search behavior under the following scenario: First, a user has a particular search task in mind (e.g., "Find scientific information about Saturn the planet.") and enters an ambiguous query (e.g., "saturn"). Then, in response to this query, the system decides to integrate results from a particular vertical (e.g., images) into the web results. While the vertical results may be relevant to a different user, this particular user's information need is better satisfied by web results. Finally, based on the vertical and web results presented, the user must decide whether to engage with the web results or reformulate the query. We evaluate algorithms for deciding which vertical results to display. The goal is to influence the user to make a productive decision with respect to the web results—to engage with the web results if at least one of them is relevant and to *avoid* engaging otherwise.

The experimental protocol proceeded as follows. Participants were given a search task and were asked to use a live search engine to find a webpage containing the requested information. Search tasks had the form "Find information about `<entity>`", for example, "Find tourism information about Washington State." In order to do a controlled study of the scenario described above, participants were told that "to help get you started with the search task, you will be provided with an initial query and a set of results." This starting point SERP, called the *initial SERP*, is where the experimental manipulation took place.

The initial SERP included a search task description, an initial query, and a set of results, supposedly returned in response to the initial query. As described in detail below, the initial query was purposely ambiguous (e.g., "washington",

[4] KLD measures distance (i.e., smaller values indicate greater similarity). Thus, all of the above algoritms used the negative KLD to measure similarity.

which could mean the city, state, or historical figure) and the search results included web results and blended results from one of four verticals (images, news, shopping, or video). The web results corresponded to the top 10 results returned by the Bing Web Search API (in their original order) and the vertical results were determined by one of the algorithms described in Section 3. The vertical results were always blended between the third and fourth web result.

From the initial SERP, participants were asked to search naturally by examining the results provided or entering their own queries. Participant queries returned results using the Bing Web Search API without vertical results. Clicking on a result opened the landing page inside an HTML frame, with a button above the frame labeled: "Click here if the page contains the requested information." Clicking this button ended the search task. The goal of the study was disguised by telling participants that we were testing a new search engine.

Verticals. We experimented with four verticals: images, news, shopping, and video. Results for the images, news, and video verticals were obtained using Bing APIs and results for the shopping vertical were obtained using the eBay API. Vertical results were presented similarly to how they are presented in commercial systems. For the image, shopping, and video verticals, we blended five results horizontally on the SERP ($\tau_v = 5$), and for the news vertical, we blended three results vertically ($\tau_v = 3$). Image results were presented using thumbnails; news results were presented using the article title, summary, news source, and publication age; shopping results were presented using the product title, price, condition, and a thumbnail of the product; and video results were presented using the title, duration, and a keyframe of the video.

Search Tasks. Each vertical was associated with its own set of search tasks. For the purpose of our study, we extended the set of search tasks used in Arguello and Capra [4]. Next, we describe how the original search tasks were created and how we added new tasks.

Each search task was associated with two components: the search task description and the initial query. The search task description was a simple request for information and the initial query was purposely ambiguous. Arguello and Capra [4] created 300 search tasks (75 per vertical) using the following process. The first step was to gather a large set of ambiguous queries. To this end, the authors identified all entities associated with a Wikipedia disambiguation page that also appear as a query in the AOL query-log. The next step was to identify queries with a strong orientation towards one of the four verticals considered. To accomplish this, each candidate initial query was issued to Bing and four (possibly overlapping) sets of queries were gathered based on whether the query triggered the image, news, shopping, and/or video vertical in the Bing results. Finally, the authors identified 75 queries per vertical that returned multiple senses from its corresponding vertical search API. For each query, the search task was constructed about one of the senses in the vertical results.

To conduct a more robust evaluation, we aimed to double the number of search tasks. For each initial query, we tried to create a new search task based on a different query-sense in the vertical results. We were unable to construct a new search task for 29 initial queries because the other query-senses in the vertical results were too obscure. We ended up with a total of 571 search tasks. In order to study the spill-over effect from the vertical to the web results, search tasks were designed to require web results instead of vertical results. See Arguello and Capra [4] (Table 1) for a few example tasks from the original set.

User Study Implementation. The study was run as a remote study using Amazon's Mechanical Turk (MTurk). Each MTurk Human Intelligence Task (HIT) was associated with a single search task. We evaluated a total of five algorithms: the four algorithms described in Section 3 and, as a baseline for comparison, an approach that simply presented the top τ_v results returned by the corresponding vertical API for the initial query. Additionally, we collected data by showing participants only the web results (without any vertical results). In total, this resulted in 3,426 experimental conditions (571 search tasks $\times$ (5 algorithms + 1 no vertical) = 3,426). Finally, we collected data from 6 redundant participants for each experimental condition, for a total of 3,426 $\times$ 6 = 20,556 trials or HITs. Each HIT was priced at $0.10 USD.

Our HITs were implemented as *external* HITs, meaning that everything besides recruitment and compensation was managed by our own server. Hosting our HITs externally allowed us to control the assignment of MTurk workers to experimental conditions. Workers were assigned to experimental conditions randomly, except for two constraints. First, participants were not allowed to complete search tasks for the same initial query. Second, in order to obtain interaction data from a large number of participants, workers were not allowed to complete more than 60 HITs. We collected data from 1,135 participants.

MTurk studies require quality control and we addressed this in three ways. First, we restricted our HITs to workers with a 95% acceptance rate or greater. Second, to help ensure English language proficiency, we limited our HITs to workers in the US. Finally, using an external HIT design allowed us to do quality control dynamically. Prior to the experiment, we conducted a preliminary study to judge the relevance of each web result on an initial SERP. During the experiment, participants who selected three non-relevant web results from an initial SERP as being relevant were not allowed to do more HITs.

Evaluation Methodology. We evaluate algorithms for deciding which results from a particular vertical to display. Algorithms were evaluated based on their ability to influence our study participants to make productive decisions with respect to the web results on the initial SERP. If we view user engagement with the web results as a binary decision, there are two ways users can make a productive decision: (1) they can engage with the web results if at least one of them is relevant or (2) they can *avoid* engaging with the web results otherwise. These correspond to true-positive and true-negative decisions, respectively.

To facilitate our analysis, it was first necessary to determine the relevance of each web result on an initial SERP. We collected relevant judgements using MTurk. We collected 10 redundant judgements per web-result/search-task pair for a total of 57,100 judgements (571 search tasks $\times$ 10 web results per task $\times$ 10 redundant judgements). The Fleiss' Kappa agreement was $\kappa_f = 0.595$, which is approaching *substantial* agreement (i.e., $\kappa_f = 0.600$) [13]. We aggregated relevance judgements using a majority vote—a web result was considered relevant if more than five MTurk workers marked it as relevant.

Engagement with the web results on an initial SERP was operationalized using clicks. We say that a participant engaged with the web results if he/she clicked on *at least one* and did not engage with the web results otherwise. Algorithms were evaluated using three metrics: (1) *accuracy* measures the percentage of true-positive and true-negative decisions (i.e., the participant clicked on a web result on the initial SERP and at least one of them was relevant *or* did not click on any and none of them were relevant), (2) the *true positive rate* measures the percentage of times there was a relevant web result on the initial SERP and the participant clicked on at least one, and (3) the *true negative rate* measures the percentage of times there were no relevant web results on the initial SERP and the participant did not click on any. Each experimental condition (i.e., search-task/algorithm pair) was completed by 6 redundant participants. We report performance by macro-averaging across search tasks and computed statistical significance using an approximation of Fisher's randomization test [19].

5 Results and Discussion

Results are presented in Tables 1-3 in terms of accuracy, true positive rate (TPR), and true negative rate (TNR). We were interested in measuring performance overall and for each vertical independently. Thus, we present macro-averaged performance across all search tasks (i.e., combining those from every vertical) and separately for those search tasks specific to each vertical. NoVertical gives the performance obtained from showing participants only the web results (without any vertical results) and Algo gives to the performance obtained from showing participants the top t_v results from the corresponding vertical search API. The Algo approach represents an aggregated search system that does not perform *vertical results selection*. The percentages indicate the percent change compared to NoVertical. The symbols $^{\vartriangle}(^{\triangledown})$ denote a statistically significant increase(decrease) in performance compared to NoVertical and the symbols $^{\blacktriangle}(^{\blacktriangledown})$ denote a statistically significant increase(decrease) in performance compared to Algo. The gray cells indicate the best performing algorithm within each column. Next, we discuss the differences in performance between algorithms, verticals, and evaluation metrics.

Algorithms. In terms of accuracy and TPR, ClusterWebSim was the best-performing algorithm. ClusterWebSim outperformed NoVertical for images, shopping and video, and performed only slightly worse for news (not significant). Moreover, ClusterWebSim outperformed Algo for shopping and video, performed at the same level for images, and only slightly worse for news (not significant).[5]

[5] For the video vertical, the improvement of ClusterWebSim over Algo was marginally significant in terms of accuracy ($p = 0.059$) and TPR ($p = 0.060$).

In terms of TNR, there was no clear winner—different algorithms performed better for different verticals. That said, ClusterWebSim was statistically indistinguishable from NoVertical and Algo for all verticals. It should also be noted that the differences between algorithms were less pronounced for TNR than for the other two metrics. We return to this point below.

It is also worth noting that ClusterWebSim outperformed MMR across all verticals and metrics. These two algorithms represent two different types of approaches to vertical results selection. ClusterWebSim selects results that are similar to the web results on the SERP and MMR selects results independently from the web results. Our results suggest that selecting vertical results that are similar to the web results can influence users to make more productive decisions with respect to the web results.

Verticals. In terms of accuracy and TPR (the metrics with the greatest variance), performance varied widely across verticals. The vertical results had a stronger effect for images and shopping than for news and video. For example, in terms of accuracy, the greatest improvement over NoVertical was greater for images (11.29%) and shopping (6.57%) than for news (2.74%) and video (3.78%). A similar trend was observed in terms of TPR. This trend is consistent with the results from Arguello and Capra [4]. Arguello and Capra found that users are more likely to interact with the web results when the vertical results are more consistent with the intended query-sense. However, the spill-over effect was only significant for images and shopping and not for news and video. Results from one of their studies suggests that images and shopping had more spill-over because their results are more salient and require less cognitive effort to process.

Metrics. Performance across algorithms varied widely in terms of TPR, but was fairly stable in terms of TNR. There are two possible explanations for this. First, it may be that the vertical results had a stronger effect in causing participants to engage with the web results than in causing participants to *avoid* engaging with the web results. In other words, seeing the relevant query-sense in the vertical results may have a strong positive effect on users, but *not* seeing

Table 1. Accuracy

	All Verticals	Images	News	Shopping	Video
NoVertical	0.573	0.549	0.583	0.578	0.582
Algo	0.587 (2.44%)	0.610 (11.11%)$^{\vartriangle}$	0.592 (1.54%)	0.569 (-1.56%)	0.577 (-0.86%)
MMR	0.580 (1.22%)	0.576 (4.92%)	0.575 (-1.37%)	0.578 (0.00%)	0.592 (1.72%)
WebSim	0.592 (3.32%)$^{\vartriangle}$	0.601 (9.47%)$^{\vartriangle}$	0.589 (1.03%)	0.588 (1.73%)	0.590 (1.37%)
WebSimMMR	0.581 (1.40%)	0.566 (3.10%)$^{\blacktriangledown}$	0.599 (2.74%)	0.574 (-0.69%)	0.582 (0.00%)
ClusterWebSim	0.602 (5.06%)$^{\vartriangle}$	0.611 (11.29%)$^{\vartriangle}$	0.580 (-0.51%)	0.616 (6.57%)$^{\vartriangle\blacktriangle}$	0.604 (3.78%)

Table 2. True Positive Rate (TPR)

	All Verticals	Images	News	Shopping	Video
NoVertical	0.395	0.422	0.393	0.377	0.386
Algo	0.415 (5.06%)	0.500 (18.48%)$^{\vartriangle}$	0.398 (1.27%)	0.374 (-0.80%)	0.375 (-2.85%)
MMR	0.408 (3.29%)	0.461 (9.24%)	0.381 (-3.05%)	0.383 (1.59%)	0.403 (4.40%)
WebSim	0.420 (6.33%)$^{\vartriangle}$	0.481 (13.98%)$^{\vartriangle}$	0.401 (2.04%)	0.405 (7.43%)	0.383 (-0.78%)
WebSimMMR	0.410 (3.80%)	0.458 (8.53%)$^{\blacktriangledown}$	0.415 (5.60%)	0.380 (0.80%)	0.379 (-1.81%)
ClusterWebSim	0.435 (10.13%)$^{\vartriangle}$	0.502 (18.96%)$^{\vartriangle}$	0.387 (-1.53%)	0.429 (13.79%)$^{\vartriangle\blacktriangle}$	0.415 (7.51%)

Table 3. True Negative Rate (TNR)

	All Verticals	Images	News	Shopping	Video
NoVertical	0.950	0.960	0.959	0.951	0.935
ALGO	0.952 (0.21%)	0.965 (0.52%)	0.977 (1.88%)	0.932 (-2.00%)	0.941 (0.64%)
MMR	0.945 (-0.53%)	0.944 (-1.67%)	0.960 (0.10%)	0.941 (-1.05%)	0.935 (0.00%)
WEBSIM	0.957 (0.74%)	0.985 (2.60%)$^{\triangle}$	0.960 (0.10%)	0.929 (-2.31%)	0.964 (3.10%)$^{\triangle}$
WEBSIMMMR	0.942 (-0.84%)	0.914 (-4.79%)$^{\triangledown\blacktriangledown}$	0.963 (0.42%)$^{\blacktriangle}$	0.934 (-1.79%)	0.947 (1.28%)
CLUSTERWEBSIM	0.957 (0.74%)	0.960 (0.00%)	0.963 (0.42%)	0.963 (1.26%)	0.944 (0.96%)

the relevant query-sense may have only a weak *negative* effect. Alternatively, the stability in TNR performance might be explained by our use of *clicks* as a proxy for user engagement with the web results. It may be that participants were often misled by incoherent vertical results, but were still effective at *not* clicking on a non-relevant web result based on its surrogate. Future work might consider a less conservative proxy for user engagement, for example, derived from browsing behavior (e.g., Did the participant scroll down the initial SERP?). A less conservative proxy might reveal greater differences in terms of TNR.

6 Conclusion

We developed and evaluated algorithms for vertical results selection—deciding which results from a particular vertical to display. Algorithms were evaluated based on their ability to influence users to make productive decisions with respect to the web results on the SERP. Results from our user study suggest the following trends. First, our best-performing algorithm (CLUSTERWEBSIM) selects vertical results that are similar to the web results. This algorithm performed better than simply presenting the top vertical results (ALGO) and diversifying the vertical results independently from the web results (MMR). We treat this as evidence that improving the level of coherence between the vertical and web results can influence users to make more productive decisions with respect to the web results. Second, the vertical results had as stronger effect for some verticals (images, shopping) than others (news, video). This is consistent with prior work and may be due to the vertical surrogate representation. Finally, we observed that the vertical results had a greater effect on users discovering relevant web results on the SERP than on users avoiding non-relevant ones. We used clicks as a proxy for user engagement with the web results. It remains to be seen whether this trend holds true for a less conservative measurement of engagement.

Our findings have important implications for aggregated search. Current methods for vertical selection and presentation do not explicitly ensure coherence with other components on the SERP. We show that relatively simple algorithms for vertical results selection can help avoid negative cross-component effects. In this work, we focused on search tasks that favored web results and performed vertical results selection to ensure coherence with the web results. Future work will develop a unified framework that performs *results selection* to ensure coherence with the most confidently relevant component(s).

References

1. Agrawal, R., Gollapudi, S., Halverson, A., Ieong, S.: Diversifying search results. In: WSDM, pp. 5–14 (2009)
2. Arguello, J., Capra, R.: The effect of aggregated search coherence on search behavior. In: CIKM, pp. 1293–1302 (2012)
3. Arguello, J., Capra, R.: The effects of vertical rank and border on aggregated search coherence and search behavior. In: CIKM, pp. 539–548 (2014)
4. Arguello, J., Capra, R., Wu, W.-C.: Factors affecting aggregated search coherence and search behavior. In: CIKM, pp. 1989–1998 (2013)
5. Arguello, J., Diaz, F., Callan, J.: Learning to aggregate vertical results into web search results. In: CIKM, pp. 201–210 (2011)
6. Arguello, J., Diaz, F., Callan, J., Crespo, J.-F.: Sources of evidence for vertical selection. In: SIGIR, pp. 315–322 (2009)
7. Bailey, P., Craswell, N., White, R.W., Chen, L., Satyanarayana, A., Tahaghoghi, S.M.: Evaluating search systems using result page context. In: IIiX, pp. 105–114 (2010)
8. Carbonell, J., Goldstein, J.: The use of mmr, diversity-based reranking for reordering documents and producing summaries. In: SIGIR, pp. 335–336 (1998)
9. Carterette, B., Chandar, P.: Probabilistic models of ranking novel documents for faceted topic retrieval. In: CIKM, pp. 1287–1296 (2009)
10. Diaz, F.: Integration of news content into web results. In: WSDM, pp. 182–191 (2009)
11. Diaz, F., Arguello, J.: Adaptation of offline vertical selection predictions in the presence of user feedback. In: SIGIR, pp. 323–330 (2009)
12. Jeffreys, H.: An invariant form for the prior probability in estimation problems. Proceedings of the Royal Society of London. Series A, Mathematical and Physical Sciences 186(1007), 453–461 (1946)
13. Landis, J.R., Koch, G.G.: The measurement of observer agreement for categorical data. Biometrics 33(1), 159–174 (1977)
14. Ponnuswami, A.K., Pattabiraman, K., Wu, Q., Gilad-Bachrach, R., Kanungo, T.: On composition of a federated web search result page: Using online users to provide pairwise preference for heterogeneous verticals. In: WSDM, pp. 715–724 (2011)
15. Radlinski, F., Dumais, S.: Improving personalized web search using result diversification. In: SIGIR, pp. 691–692 (2006)
16. Sanderson, M.: Ambiguous queries: test collections need more sense. In: SIGIR, pp. 499–506 (2008)
17. Santos, R.L.T., Macdonald, C., Ounis, I.: Exploiting query reformulations for Web search result diversification. In: WWW, pp. 881–890 (2010)
18. Santos, R.L.T., Macdonald, C., Ounis, I.: Aggregated search result diversification. In: ITCIR, pp. 250–261 (2011)
19. Smucker, M.D., Allan, J., Carterette, B.: A comparison of statistical significance tests for information retrieval evaluation. In: CIKM, pp. 623–632 (2007)
20. Zhou, K., Cummins, R., Lalmas, M., Jose, J.M.: Evaluating aggregated search pages. In: SIGIR, pp. 115–124 (2012)

On-topic Cover Stories from News Archives

Christian Schulte[1], Bilyana Taneva[2], and Gerhard Weikum[1]

[1] Max-Planck Institute for Informatics, Saarbrücken, Germany
{cschulte,weikum}@mpi-inf.mpg.de
[2] CNRS-LIG, Grenoble, France
bilyana.taneva@imag.fr

Abstract. While Web or newspaper archives store large amounts of articles, they also contain a lot of near-duplicate information. Examples include articles about the same event published by multiple news agencies or articles about evolving events that lead to copies of paragraphs to provide background information. To support journalists, who attempt to read all information on a given topic at once, we propose an approach that, given a topic and a text collection, extracts a set of articles with broad coverage of the topic and minimum amount of duplicates.

We start by extracting articles related to the input topic and detecting duplicate paragraphs. We keep only one instance from each group of duplicates by using a weighted quadratic optimization problem. It finds the best position for all paragraphs, such that some articles consist mainly of distinct paragraphs and others consist mainly of duplicates. Finally, we present to the reader the articles with more distinct paragraphs. Our experiments show the high precision and recall of our approach.

1 Introduction

Web archives such as the Internet Archive (archive.org) or the Gigaword archive [6] store large amounts of articles many of which however are near-duplicates. For example, Gigaword contains articles published by multiple news agencies that often describe the news events in similar words. Moreover, articles about events spread over time (e.g., a political scandal discussed for several months) often repeat previously published paragraphs that introduce the topic to new readers, but do not deliver novel information to readers familiar with the event.

This is why journalists or historians, who attempt to read all information on a given topic at once, waste time reading duplicate information rather than focusing on a set of selected articles with broad coverage and minimum repetitions. For example, for a given event such articles would have wide time span, discuss related events and people, or explain its cause and potential effect.

Problem Statement. Given a topic of interest and a text collection, our goal is to extract a set of articles with broad coverage of the topic and minimum amount of repetitions. We assume that the topic is given by the user as a *seed text* that could be an entire article or a few paragraphs describing the topic. The input collection could be a newspaper archive with articles by different news agencies, or any collection of articles.

A. Hanbury et al. (Eds.): ECIR 2015, LNCS 9022, pp. 37–42, 2015.

Our problem resembles the Max-Min or Max-Avg facility dispersion problems [7]. They select a set of k articles such that the minimum (or the average) of their pairwise distances is maximized, where k is a user-specified parameter. The distance between two articles is small if they have similar contents, and large otherwise. However, these approaches have certain limitations: (1) although diverse, the selected articles do not necessarily cover all details on the topic, and (2) it is difficult to adjust the parameter k without prior knowledge on the topic.

In contrast, our approach selects diverse articles that cover all details on the topic, and does not require a specified value for the number of output articles. We output more articles for widely discussed topics, and less articles otherwise.

Approach and Contribution. In this paper, we develop a method for extracting a set of articles on a given topic with broad coverage and minimum amount of repetitions. We call this set a *cover story.*

Our approach works as follows. First, we extract articles related to the input topic. To this end, we compare the topic's seed text to all articles in the input collection using *tf-idf* weighting and cosine similarity. In a similar fashion, across the seed-related articles we detect duplicate paragraphs and group them (Section 2). Ideally, we would like to keep in the cover story only one instance from each group of duplicate paragraphs. To decide which is the best paragraph to keep, we use a weighted quadratic optimization problem that finds the best article that should contain it. As a result some articles consist mainly of distinct paragraphs and others consist mainly of duplicates (Section 3). For the cover story, we choose the articles with more distinct paragraphs and present them to the reader.

In summary, this paper makes the following novel contributions: (1) formulating and modeling the problem of extracting a set of articles on a given topic with broad coverage and minimum repetitions, (2) devising an algorithm for building cover stories by solving a weighted quadratic optimization problem, and (3) conducting experiments that show the benefits of our approach for readers.

2 Building a Cover Story

Given a seed text about a topic and a collection of articles on various topics, first we retrieve a set of seed-related articles. To this end, we compute the cosine similarity between the seed and all articles in the collection using their *tf-idf* (term frequency - inverse document frequency) vectors. As seed-related, we retrieve all articles that have cosine similarity with the seed larger than a threshold θ_{seed}. To collect articles that are not only similar to the seed, but also contain new information about the topic, we use low value of θ_{seed}, namely $\theta_{seed} = 0.2$. We determined this value by experimenting with 10 topics and their related articles.

The seed-related articles could contain duplicate paragraphs, sections, or even complete copies of articles. To help the user focus only on the important content without the need to read the same information multiple times, we detect and remove the duplicate paragraphs. We mark as duplicates all paragraphs that have cosine similarity between their *tf-idf* vectors larger than a specified threshold θ_{par}. We show experiments with different values of θ_{par} in Section 4.

At this stage, we would like to minimize the duplicate paragraphs. One option would be to remove all instances but one from each group of duplicate paragraphs by keeping only the paragraph in the article published first. However, as a result the non-duplicate paragraphs are scattered across all articles (left example on Figure 1). Thus, we would either have to present to the user separate paragraphs without clear context, or complete articles with many duplicate paragraphs.

3 Optimization Problem

We propose an alternative algorithm for removing duplicate paragraphs using weighted quadratic optimization. We keep only one paragraph from each group of duplicates, but in contrast to Section 2, we try to place as many distinct paragraphs as possible into complete articles. In addition, we consider the length of articles in order to "attract" more paragraphs into longer articles. For the final cover story, we choose the articles that consist mainly of distinct paragraphs.

Our problem formulation needs to capture the following requirements: (1) keep one instance for each group of duplicate paragraphs, (2) group unique paragraphs into complete articles, and (3) prefer longer articles.

We introduce the following binary variables and weights. We denote all seed-related articles by $A = \{a_1, \ldots, a_n\}$ and all unique paragraphs across the articles in A by $P = \{p_1, \ldots, p_m\}$. We use the binary variables X_{ij} to model the paragraph p_j with respect to the article a_i. Since, not all articles in A contain all paragraphs from P, some X_{ij} variables are set to 0 before running the optimization program. For each article a_i that contains a duplicate of paragraph p_j, X_{ij} is a binary variable. X_{ij} is set to 1, if paragraph p_j is kept in a_i, and X_{ij} is set to 0, if p_j is removed or disabled from a_i. In addition, we consider a weight w_i for each article $a_i \in A$, which corresponds to the length of a_i in terms of words.

We define our weighted quadratic optimization problem as follows:

$$\begin{aligned} \text{maximize} \quad & \sum_{i=1}^{n} \Big(\sum_{j=1}^{m} w_i X_{ij} \Big)^2 \\ \text{subject to} \quad & \sum_{i=1}^{n} X_{ij} = 1, \text{ for each paragraph } p_j \in P \end{aligned}$$

Being quadratic, the objective function is maximized when we place all distinct paragraphs in as few articles as possible. A linear objective $\sum_{i=1}^{n} \sum_{j=1}^{m} w_i X_{ij}$, on the other hand, is less sensitive to the paragraph placements: if all articles have equal weights, the objective remains constant regardless of the values X_{ij}.

The constraints refer to the requirement that only one instance of a group of duplicate paragraphs is kept in the final result. By considering the article weights w_i, we place this instance in a textually longer article. In Figure 1 we compare this approach to greedily removing duplicates by publication date (Section 2). We implemented our quadratic optimization problem with Gurobi (gurobi.com).

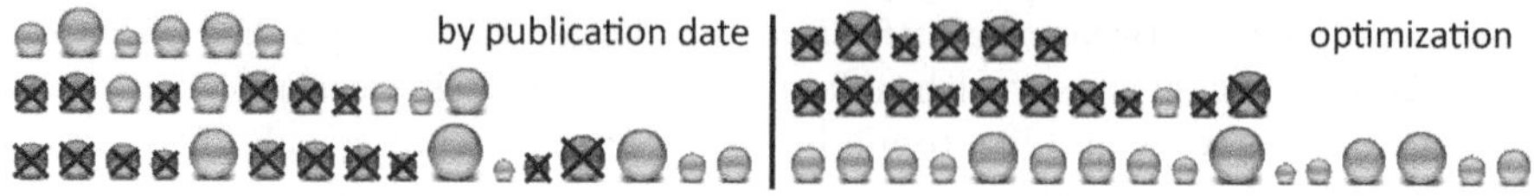

Fig. 1. Removal of duplicate paragraphs. Each line of globes shows the paragraphs of one article. The size of a globe reflects the paragraph's length. Gray crossed globes indicate duplicates.

4 Experiments

Experimental Data. We compiled a set of 43 topics from Gigaword [6]. Examples are "Missing bomber in Israel (happened in 05/2003)" and "US firm makes bid for Waterford Crystal (02/2009)". The seed texts were complete news articles. We extracted the seed-related articles from "Associated Press Worldstream, English Service (apw-eng)" in Gigaword, following Section 2 with $\theta_{seed} = 0.2$.

Compared Methods. We compare two methods:
- **OPT**: our approach using the optimization problem from Section 3, and
- **DISP**: the Max-Min dispersion algorithm from [7].

We aim to present to the user a cover story only with informative articles. To this end, for OPT we choose all articles with > 80% of distinct (enabled) paragraphs. We set the number of selected articles k by DISP to be equal to the number of articles found with our approach OPT. Thus k is different for each topic: it is larger for topics with less duplicate content or such with many related articles.

To detect duplicate paragraphs, we experimented with different values for the threshold θ_{par}: 0.4, 0.5, and 0.6. The respective average percentage of distinct paragraphs across the seed-related articles of the topics is 0.540, 0.597, and 0.635. Small θ_{par} means that more paragraphs are detected as duplicates, and large θ_{par} means that the detection is more restrictive. Thus the percentage of distinct paragraphs increases when θ_{par} increases.

The approximation algorithm for Max-Min dispersion from [7] initializes the set S of articles in the cover story with the two most distant articles. We define distance between two articles a and b to be $d(a, b) = 1 - \textit{cos-sim}(a, b)$, where $\textit{cos-sim}(a, b)$ is the cosine similarity between the *tf-idf* vectors of a and b. The distance between an article $a \notin S$ and S is defined as $d(a, S) = \min_{b \in S} d(a, b)$. The algorithm greedily adds to S the next article x, where $x = \arg\max_{y \notin S} d(y, S)$, until $|S| = k$. Note that the output does not depend on a threshold for similarity, as the algorithm greedily chooses the next most distant article to the set S.

Our data, including seed texts, seed-related articles, extractions for OPT and DISP with $\theta_{par} = 0.5$, detailed results, and source code is publicly available[1].

Quality Metrics. We automatically compiled a ground-truth data with groups of duplicate paragraphs $G = \{g_1, \ldots, g_m\}$. To find the groups g_i, we used cosine

[1] `http://resources.mpi-inf.mpg.de/d5/cover-stories-ECIR2015/public_data.zip`

similarity with $\theta_{par} = 0.5$, and compared all pairs of paragraphs from the seed-related articles. We determined $\theta_{par} = 0.5$ after experimenting with values in the range of [0.3, 0.7] and manually inspecting sets of duplicate paragraphs from several topics. We use the following metrics:

– **Recall** measures the fraction of covered groups in G: $recall = \sum_{i=1}^{m} isCovered(g_i, S)/m$, where S is a cover story. $isCovered(g_i, S) = 1$, if there is a paragraph in S that belongs to g_i, and $isCovered(g_i, S) = 0$ otherwise.

– **Precision** measures the fraction of paragraphs in S that belong to distinct groups in G: $precision = \sum_{i=1}^{m} isCovered(g_i, S)/|S|$, where $|S|$ is the number of all paragraphs in S.

Results. Table 1 shows the recall and precision results for OPT and DISP for different values of θ_{par}. OPT has higher recall than the baseline DISP with almost 10% improvement, while the two methods achieve comparable precision. We performed two-sided paired t-tests to compare OPT and DISP in terms of recall for all values of θ_{par}: the p-values were < 0.001. OPT aims to select a maximum number of distinct paragraphs for each topic. In contrast, DISP selects a set of articles which, although diverse, does not always cover all distinct paragraphs. This is why, OPT achieves significantly higher recall than DISP. The precision for OPT and DISP is comparable, since both approaches select articles with maximally different contents, and thus minimum duplicates.

Table 1. Evaluation for cover stories.

Method	Metric	$\theta_{par} = 0.4$	$\theta_{par} = 0.5$	$\theta_{par} = 0.6$
OPT	recall	0.677	0.796	0.849
	precision	0.981	0.966	0.945
DISP	recall	0.580	0.697	0.770
	precision	0.975	0.956	0.944

The average number of articles in the cover stories for $\theta_{par} = 0.4, 0.5$, and 0.6 is 41, 54, and 61, respectively. It increases when θ_{par} increases as there is less duplicate content and there are more selected articles in the stories. For example, the cover story for an event about music and video piracy in Malaysia from 07/2004 contains 19, 29, and 34 articles for $\theta_{par} = 0.4, 0.5$, and 0.6, respectively.

5 Related Work

Our problem is related to *diversification* [1,3,4] and *facility dispersion* [7], where the goal is to select a set of k diverse articles for a user-specified parameter k. In contrast to our approach however, the selected articles do not cover all details of the topic and the parameter k is difficult to adjust for the various topics.

Extractive text summarization [2,5,9] is the process of selecting sentences and paragraphs that represent a given text. The Maximal Marginal Relevance approach [2] sequentially selects sentences by penalizing the ones that are similar to already selected sentences. The approach in [9] views the problem as a maximum coverage problem with knapsack constraint. However, such summaries are often hard to read as they consist of separate linguistic units without clear context and flow. In contrast, we select entire articles that are easy to read.

Our work is also related to *content enrichment* [8,10], where the task is to extend a given seed text with related content. [8] extracts segments from external sources and compiles a pseudo-document with the most related pieces. In contrast to our work, the result is not intended for humans. Despite being related, the problem in [10] is very different than ours, namely, to extract text pieces related to the seed, using minimal assumptions on the sources while meeting the constraint that the user is willing to read only a certain amount of information.

6 Conclusion

In this paper we presented an approach that aids journalists, historians, and encyclopedia editors to easily collect and comprehend information on specific topics. Our method extracts, from large data repositories, a set of well chosen articles that do not contain repetitions and also cover all details on the topic. Our experimental results show the good quality of the extracted cover stories and the improvement over a state-of-the-art baseline.

References

1. Agrawal, R., Gollapudi, S., Halverson, A., Ieong, S.: Diversifying search results. In: WSDM, pp. 5–14 (2009)
2. Carbonell, J., et al.: The use of MMR, diversity-based reranking for reordering documents and producing summaries. In: SIGIR, pp. 335–336 (1998)
3. Drosou, M., Pitoura, E.: Search result diversification. SIGMOD Rec. 39(1), 41–47 (2010)
4. Gollapudi, S., Sharma, A.: An axiomatic approach for result diversification. In: WWW, pp. 381–390 (2009)
5. Nenkova, A., McKeown, K.: Automatic summarization. Foundations and Trends in Information Retrieval 5(2-3), 103–233 (2011)
6. Parker, R., et al.: English Gigaword, 5th edn., Linguistic Data Consortium (2011)
7. Ravi, S.S., Rosenkrantz, D.J., Tayi, G.K.: Heuristic and special case algorithms for dispersion problems. Operations Research 42(2), 299–310 (1994)
8. Schlaefer, N., Chu-Carroll, J., Nyberg, E., Fan, J., Zadrozny, W., Ferrucci, D.: Statistical source expansion for question answering. In: CIKM, pp. 345–354 (2011)
9. Takamura, H., Okumura, M.: Text summarization model based on maximum coverage problem and its variant. In: EACL, pp. 781–789 (2009)
10. Taneva, B., Weikum, G.: Gem-based entity-knowledge maintenance. In: CIKM, pp. 149–158 (2013)

Multi-emotion Detection in User-Generated Reviews

Lars Buitinck[1,2], Jesse van Amerongen[2], Ed Tan[2], and Maarten de Rijke[2]

[1] Netherlands eScience Center, Amsterdam, The Netherlands
l.buitinck@esciencecenter.nl
[2] University of Amsterdam, Amsterdam, The Netherlands
{j.vanamerongen,e.tan,derijke}@uva.nl

Abstract. Expressions of emotion abound in user-generated content, whether it be in blogs, reviews, or on social media. Much work has been devoted to detecting and classifying these emotions, but little of it has acknowledged the fact that emotionally charged text may express multiple emotions at the same time. We describe a new dataset of user-generated movie reviews annotated for emotional expressions, and experimentally validate two algorithms that can detect multiple emotions in each sentence of these reviews.

1 Introduction

The problem of emotion detection in written language has received much attention in recent years, as part of a larger trend toward "affective computing." Few researchers, though, seem to have tried to tackle the full problem of simultaneously detecting emotionally charged phrases and classifying them according to which emotion they mention or express. Instead, attention is usually focused on simple *valence* classification and opinion mining (positive vs. negative, sometimes with the addition of a neutral class) or the classification of utterances that are known to be emotionally charged a priori.

In the present work, we consider the combined problem of detecting and classifying expressions of emotion in the context of movie reviews. This work was borne of basic research into film, using reviews as reflections of the complexity of viewer emotions, but its results may find applications in product search and recommendation for films and other artistic products; e.g., clustering products by emotional charge.

We phrase the problem as *multi-label classification*: we label individual sentences from reviews with a subset (possibly empty) of a predetermined set of emotion labels. Our research question is how to tackle this problem in a supervised way. We contrast two methods that reduce multi-label learning to familiar binary and (disjoint) multi-class classification: one-vs.-rest and an ensemble method that learns from correlations between labels. Both methods use textual features only, where much other research into emotion detection has focused on facial expressions and pitch features in spoken language [4]. Our label set consists of the seven basic emotions identified in the hierarchical cluster analysis of Shaver et al. [12], with the emotion "interest" added [16].

We first survey the state of the art in emotion recognition in Section 2, then discuss a new purpose-built dataset in Section 3. Section 4 contains a description of our feature extraction and learning algorithms, with particular attention to parameter tuning in the multi-label setting. Experimental results are given in Section 5. Section 6 wraps up with conclusions and plans for future research.

A. Hanbury et al. (Eds.): ECIR 2015, LNCS 9022, pp. 43–48, 2015.

2 Related Work

Affective computing has been the focus of much research in the past two decades; a survey of emotion/affect detection in writing, spoken language, and other modalities is given by Calvo and D'Mello [4]. Much of the initial work on written text (e.g., [9]) has focused on valence classification, also known as sentiment analysis or opinion mining, where the two allowed emotions are "positive" and "negative."

Aman and Szpakowicz [2], for example, perform binary classification of sentences in blog posts as emotional/non-emotional. Alm et al. [1] extend the scheme to a three-way classification of sentences as expressing positive, negative, or no emotions. Yang et al. [20] perform classification of blog posts into four categories, "happy," "joy," "sad" and "angry," apparently using the occurrence of certain emoticons as ground truth labels. Their work can be considered to be a finer-grained version of valence detection.

At SemEval 2007 [14], various systems where benchmarked on the task of classifying news headlines according to a six-label annotation scheme, viz. anger, disgust, fear, joy, sadness and surprise. The focus was on unsupervised methods; Strapparava and Mihalcea [15] additionally tested a weakly supervised transfer learning approach.

Closer to our work is that of Danisman and Alpkocak [5], who perform supervised learning of emotion labels at the sentence/snippet level. They show that a simple nearest centroid classifier using bag-of-words features and tf–idf weighting can achieve an F_1 score of 32.22% in a five-way multiclass prediction problem using a set of 7,666 text snippets. D'Mello et al. [6] achieve higher scores, but in a problem that only involves three emotional states. Neither of these works takes into account a neutral state.

Our work differs from the work listed above in the following important ways. First, we do not make the simplifying assumption that emotional states are mutually exclusive. Second, while we use supervised learning and manual annotation, we use only a small labeled training set of a few hundred sentences, where earlier attempts have typically used thousands of training samples.

3 Dataset

We hand-labeled 44 movie reviews using the BRAT annotation interface [13], identifying emotionally charged phrases. The reviews were taken from IMDB and concern the films *American History X*, *The Bourne Identity*, *Earth* (2007), *The Godfather*, *Little Miss Sunshine*, *The Notebook*, *SAW*, and *Se7en*; all Hollywood productions, but of varying genres. Each film is covered by six reviews, except for *The Godfather* (two reviews, due to time constraints).[1]

We perform sentence splitting on each of the reviews, and turn the problem into a multi-label classification problem by assigning to each sentence the set of labels used to label any string of words within the sentence. Doing so yielded 629 sentences containing 13,409 tokens, distributed over the various films as shown in Table 1, with the label distribution given in Table 2.

[1] `https://github.com/NLeSC/spudisc-emotion-classification`

Table 1. Samples per film

Title	Sent.	Emot.
American History X	77	63
The Bourne Identity	90	41
Earth	63	45
The Godfather	18	18
Little Miss Sunshine	95	51
The Notebook	107	73
SAW	65	54
Se7en	114	75

Table 2. Absolute label frequencies

Label	All	Test
Anger	18 (24)	6
Disgust/contempt	6 (0)	–
Fear	37	11
Interest	69	20
Joy	47	9
Love	272	48
Sadness	35	10
Surprise	80	16

Of the 629 sentences, 420 have at least one label, showing how prevalent the expression of emotions in film reviews is. The average number of labels per sentence is 0.887, while the maximum is five (the combination "Joy–Sadness–Love–Interest–Surprise," which occurs once). We reserve roughly 20% of our sentences as a test set, using the remainder for classifier training and tuning. Because the "Disgust/contempt" label has only six samples, we replace it with "Anger."

4 Classification Algorithms

We tested two algorithms for performing multi-label classification. Both use standard bag-of-words features with stop word removal and optional tf–idf weighting, and reduce the multi-label problem to either binary or multiclass learning, for which we use linear support vector machines. We implement these using scikit-learn [3, 10], which includes the linear SVM learner of Fan et al. [7].

4.1 Reduction to Binary Classifiers

The first algorithm we consider reduces the K-way multi-label classification problem to K independent binary SVMs that learn to distinguish one emotion from all others. This is variously called the *one-vs.-rest*, or *binary relevance* reduction [18]. While this problem reduction cannot take advantage of correlations between labels, it has the advantage that we can separately tune the settings of each SVM, so that we end up with an optimal model for each binary sub-problem.

I.e., for each label separately, we do a parameter sweep and select the parameter settings that result in the maximum F_1 score for that label according to five-fold stratified cross-validation on the training set. We try all parameter settings in the grid defined by $C \in \{.1, 1, 10, 100, 1000\}$, L_1 or L_2 regularization, linear or logarithmic tf, whether to use tf or tf–idf, and whether to oversample the minority class in each sub-problem. These settings were chosen based on experience with other text classification problems.

4.2 Learning from Label Dependencies

As an alternative to the one-vs.-rest reduction just sketched, we also benchmark the random k-labelsets (RAKEL) algorithm [17, 19]. To understand this method, we must first introduce an alternative problem reduction strategy for multi-label classification, the *label powerset* method. A label powerset model is a regular classifier trained using all subsets of a multi-label problem's set of labels as its classes, so in our problem, the triple (Fear, Love, Surprise) would be one class. This method is very powerful in that it can learn dependencies between labels, but it requires solving an exponential-sized multiclass problem.

To prevent this combinatorial blowup, RAKEL builds an ensemble of label powerset classifiers, each trained on a subset of labels of fixed size k, chosen at random without replacement. Prediction proceeds by a voting scheme: for each randomly generated subset J of all labels, its associated classifier predicts a subset $f(x) \subseteq J$ to which sample x should belong. Each $j \in f(x)$ gets a positive vote; each $j \notin f(x)$ a negative vote. A positive tally for a label means a positive prediction in the full multi-label problem. We use the RAKEL algorithm with linear SVMs as its base learners.

A problem with RAKEL is that it is not clear how to tune its parameters with a small amount of training data. We might like to apply the same tuning as for the one-vs.-rest strategy, i.e., optimize each base learner separately before combining them; but this is infeasible, because the label powerset classifiers must solve overly sparse sub-problems. Some label subsets, such as (Interest, Love, Sadness), occur only once in the training set, making proper stratification impossible. Fitting multiple RAKEL ensembles in a stratified CV setting may be possible with the multi-label stratification strategy of Sechidis et al. [11], but time constraints prevented us from implementing it. We therefore use tf–idf weighting with logarithmic tf, automatic oversampling, and a fixed regularization parameter $C = 1$ for all SVMs.

We let $k = 3$ be the size of the label subsets in RAKEL, which has the effect of undoing the randomization: only 35 size-k subsets of our label set occur in the training set, so we can simply fit a classifier to each of them.

5 Results

Our main research question is to find out how a relatively simple but carefully tuned one-vs.-rest baseline compares against a more advanced multi-label classification method on the task of emotion classification. To answer this question, we empirically evaluate the algorithms from the previous section on the dataset described in Sect. 3.

We report accuracy and F_1 scores per class and averaged over all classes. We compute the overall accuracy score as defined by Godbole and Sarawagi [8], i.e., one minus the Hamming loss. Since accuracy has the problem of overestimating performance in highly-unbalanced classification problems, we consider F_1 score to be our main evaluation metric. All scores are averaged over ten runs of each training algorithm to account for the randomization in both; in the case of RAKEL, the results of all runs achieved the exact same scores despite randomization in the SVM learner [7].

Our main results are shown in Table 3. We see that RAKEL achieves slightly, but significantly, better overall F_1 score. Because its parameters are fixed, it also achieves this result noticeably faster than OvR: the expensive tuning of OvR takes many minutes of computing time, whereas RAKEL finishes in mere seconds.

However, RAKEL is not superior on all labels, and in particular does not learn to predict the "Anger" label at all. The OvR learner similarly shows difficulty with this

Table 3. Sentence-level accuracy and F_1 score for one-vs.-rest (OvR) and RAKEL. Differences in F_1 score between the two algorithms were tested using Welch's one-sided t-test. $\triangle$: significantly better at $\alpha = .05$, $\blacktriangle$: significantly better at $\alpha = .001$, or consistently better with zero variance.

	Algorithm/performance metric			
	OvR accuracy	OvR F_1 score	RAKEL acc.	RAKEL F_1
Anger	$.940 \pm .018$	$.105 \pm .129^{\triangle}$	.937	.000
Fear	$.910 \pm .015$	$.267 \pm .039$	.921	$.546^{\blacktriangle}$
Interest	$.802 \pm .000$	$.359 \pm .000^{\blacktriangle}$	.818	.343
Joy	$.939 \pm .005$	$.494 \pm .056$	.929	.471
Love	$.706 \pm .000$	$.626 \pm .000^{\blacktriangle}$	.675	.586
Sadness	$.849 \pm .000$	$.296 \pm .000$	.905	.400
Surprise	$.740 \pm .051$	$.231 \pm .029$	.794	$.278^{\blacktriangle}$
Overall	$.841 \pm .007$	$.432 \pm .008$	.854	$.456^{\blacktriangle}$

label, achieving $F_1 \geq .25$ in four runs, but zero in the remaining six. Inspection of the dataset indicates that the problem with the "Anger" label is that it is often used to mark disappointment or criticism, and reviews tend to express this disappointment in a subtle and indirect way. Words like "frustrated" or "contrived" are rare, and reviewers may express their disappointment by praising a movie that they preferred over the one being reviewed, using a positive register of expression.

6 Conclusion

We have shown how the problem of emotion detection and classification at the sentence level can viably be tackled as one of supervised classification, even with relatively small labeled datasets, using standard bag-of-words features and while allowing for multiple emotion labels per sentence. We have shown that careful tuning of a baseline method can make it almost as strong as the more advanced RAKEL algorithm; tuning of RAKEL is an interesting problem that requires further attention.

In future work, we intend to further classify emotionally charged utterances according to the trigger of the emotion: either a film regarded as an artifact, or the content (storyline) of the film. E.g., we intend to automatically determine whether anger is caused by a bad performance on the part of actors or directors, or by a good performance that evokes genuine anger at the "bad guy" in the plot. This should decouple emotion from opinion, and provide further insight into the emotional response that films evoke.

Acknowledgements. This research was partially supported by the European Community's Seventh Framework Programme (FP7/2007-2013) under grant agreement nr 312827 (VOX-Pol), the Netherlands Organisation for Scientific Research (NWO) under project nrs 727.011.005, 612.001.116, HOR-11-10, 640.006.013, 612.066.930, the Center for Creation, Content and Technology (CCCT), the Dutch national program COMMIT, the ESF Research Network Program ELIAS, the Elite Network Shifts project funded by the Royal Dutch Academy of Sciences (KNAW), the Netherlands eScience Center under project number 027.012.105, the Yahoo! Faculty Research and Engagement Program, the Microsoft Research PhD program, and the HPC Fund.

References

[1] Alm, C.O., Roth, D., Sproat, R.: Emotions from text: machine learning for text-based emotion prediction. In: Proc. HLT–EMNLP, pp. 579–586 (2005)
[2] Aman, S., Szpakowicz, S.: Identifying expressions of emotion in text. In: Matoušek, V., Mautner, P. (eds.) TSD 2007. LNCS (LNAI), vol. 4629, pp. 196–205. Springer, Heidelberg (2007)
[3] Buitinck, L., Louppe, G., Blondel, M., Pedregosa, F., Müller, A., Grisel, O., Niculae, V., Prettenhofer, P., Gramfort, A., Grobler, J., Layton, R., Vanderplas, J., Joly, A., Holt, B., Varoquaux, G.: API design for machine learning software: experiences from the scikit-learn project. In: ECML PKDD Workshop on Languages for Machine Learning (2013)
[4] Calvo, R.A., D'Mello, S.K.: Affect detection: an interdisciplinary review of models, methods, and their applications. IEEE Trans. on Affective Computing 1(1), 18–37 (2010)
[5] Danisman, T., Alpkocak, A.: Feeler: emotion classification of text using vector space model. In: Proc. AISB Convention (2008)
[6] D'Mello, S.K., Craig, S.D., Sullins, J., Graesser, A.C.: Predicting affective states expressed through an emote-aloud procedure from AutoTutor's mixed-initiative dialogue. Int'l J. AI in Education 16, 3–28 (2006)
[7] Fan, R.E., Chang, K.W., Hsieh, C.J., Wang, X.R., Lin, C.J.: Liblinear: A library for large linear classification. JMLR 9, 1871–1874 (2008)
[8] Godbole, S., Sarawagi, S.: Discriminative methods for multi-labeled classification. In: Dai, H., Srikant, R., Zhang, C. (eds.) PAKDD 2004. LNCS (LNAI), vol. 3056, pp. 22–30. Springer, Heidelberg (2004)
[9] Pang, B., Lee, L.: A sentimental education: Sentiment analysis using subjectivity. In: Proc. ACL (2004)
[10] Pedregosa, F., Varoquaux, G., Gramfort, A., Michel, V., Thirion, B., Grisel, O., Blondel, M., Prettenhofer, P., Weiss, R., Dubourg, V., Vanderplas, J., Passos, A., Cournapeau, D., Brucher, M., Perrot, M., Duchesnay, E.: Scikit-learn: Machine learning in Python. JMLR 12 (2011)
[11] Sechidis, K., Tsoumakas, G., Vlahavas, I.: On the stratification of multi-label data. In: Gunopulos, D., Hofmann, T., Malerba, D., Vazirgiannis, M. (eds.) ECML PKDD 2011, Part III. LNCS, vol. 6913, pp. 145–158. Springer, Heidelberg (2011)
[12] Shaver, P., Schwartz, J., Kirson, D., O'Connor, C.: Emotion knowledge: further exploration of a prototype approach. J. Personality and Social Psychology 52(6) (1987)
[13] Stenetorp, P., Pyysalo, S., Topić, G., Ohta, T., Ananiadou, S., Tsujii, J.: BRAT: a web-based tool for NLP-assisted text annotation. In: Demos at 13th Conf. EACL, pp. 102–107 (2012)
[14] Strapparava, C., Mihalcea, R.: SemEval-2007 task 14: Affective text. In: Proc. 4th Int'l Workshop on Semantic Evaluations, pp. 70–74 (2007)
[15] Strapparava, C., Mihalcea, R.: Learning to identify emotions in text. In: Proc. ACM Symp. Applied Computing, pp. 1556–1560 (2008)
[16] Tan, E.: Emotion and the structure of narrative film. Erlbaum, Mahwah (1996)
[17] Trohidis, K., Tsoumakas, G., Kalliris, G., Vlahavas, I.: Multi-label classification of music into emotions. In: Proc. Int'l Conf. on Music IR, pp. 325–330 (2008)
[18] Tsoumakas, G., Katakis, I.: Multi-label classification: An overview. Int'l J. Data Warehousing and Mining 3(3), 1–13 (2007)
[19] Tsoumakas, G., Vlahavas, I.: Random *k*-labelsets: An ensemble method for multilabel classification. In: Kok, J.N., Koronacki, J., Lopez de Mantaras, R., Matwin, S., Mladenič, D., Skowron, A. (eds.) ECML 2007. LNCS (LNAI), vol. 4701, pp. 406–417. Springer, Heidelberg (2007)
[20] Yang, C., Lin, K.H.Y., Chen, H.H.: Emotion classification using web blog corpora. In: IEEE/WIC/ACM Int'l Conf. on Web Intelligence, pp. 275–278 (2007)

Classification of Historical Notary Acts with Noisy Labels

Julia Efremova[1], Alejandro Montes García[1], and Toon Calders[1,2]

[1] Eindhoven University of Technology, The Netherlands
[2] Université Libre de Bruxelles, Belgium

Abstract. This paper approaches the problem of automatic classification of real-world historical notary acts from the 14th to the 20th century. We deal with category ambiguity, noisy labels and imbalanced data. Our goal is to assign an appropriate category for each notary act from the archive collection. We investigate a variety of existing techniques and describe a framework for dealing with noisy labels which includes category resolution, evaluation of inter-annotator agreement and the application of a two level classification. The maximum accuracy we achieve is 88%, which is comparable to the agreement between human annotators.

1 Introduction

Text Classification (TC) is the problem of assigning one or several predefined categories to text documents [5]. TC is a relevant research question, given the large amount of uncategorized digital text documents. It is widely used to solve text mining problems (e. g. topic detection, spam filtering, folktale classification, news analysis, SMS mining, etc. [7,4,6]).

The TC has been studied by many researchers. Sebastiani [8] presented a detailed survey about supervised TC techniques. Later Ikonomakis [5] extended his work and summarised available machine learning approaches for the overall TC process. Recently Aggarwal [1] provided a survey of a wide variety of TC algorithms. Constantopoulos et al. [3] designed a digital library for historical documents that includes indexing techniques for the document annotation and retrieval.

In our case we have to deal with historical data and we use a number of machine learning algorithms together with the extraction of names, places and lexical information. Archived documents, presented in the form of unstructured text, contain a large amount of information about legal events. In many cases they are the only source of historical facts.

In this paper we develop of a classification framework for a large collection of Dutch notary acts from the 14th to the 20th century, examine the influence of lexical features, namely Parts-Of-Speech, as well as personal information elimination on the classification process, and provide an annotated corpus to the research community[1].

[1] `http://wwwis.win.tue.nl/~amontes/ecir2015/dataset.zip`

A. Hanbury et al. (Eds.): ECIR 2015, LNCS 9022, pp. 49–54, 2015.

2 Data Description and General Approach

Our dataset is comprised of notary acts provided by the Brabants Historical Information Center. The documents contain information about people involved in property transfers, loans, wills, etc. They were written between the 14th and the 20th century. An example can be found on `http://goo.gl/NhdFeq`. 115 967 documents out of 234 325 documents were labelled by volunteers with a single category for each document that describes their content. The assigned categories often contain spelling errors and duplicates and the collection is unbalanced.

The original dataset contains 455 categories identified by volunteers and around 20% of the classified documents belong to only one category.

To preprocess the documents we remove from the raw data punctuation marks or non-alphabetical symbols and transform the text to lower case. Then we split the original documents into sets of words called *tokens* and remove Dutch stop-words. We explore *personal information elimination* (PIE) by removing person names and locations. To do so we use person name and location dictionaries obtained from the database of the Meertens Institute[2] and the Historical Sample of the Netherlands[3]. Moreover, we apply stemming [1].

Then we create a feature for each remaining token, and set their values using the *term frequency inverse document frequency* (TF-IDF) [5]. The output of the feature extraction step is a set of numerical features. The entire vocabulary of the overall collection of notary acts is very large, the resulting feature set is sparse. Table 1 demonstrates the number of unique features for each experimental setup.

To overcome the sparsity problem we use different feature selection techniques, namely *Pearson's chi-squared test* [5] and *Latent semantic analysis* [1] and choose the 2000 most representative features for the whole corpus. In addition, we investigate the role of *part of speech (POS) lexical features*: nouns, verbs and adjectives. To obtain POS fragments we use the Frog tool[4] which is a morpho-syntactic tagger and parser for Dutch text [2].

The last step of the overall TC process is learning the model and classification. We apply and evaluate the *Support Vector Machines* (SVM) [1] classifier use from the scikit-learn python tool[5] with a linear basis kernel function. Then the algorithm is ready to classify the documents [1,5].

Table 1. The number of unique words-features in each experiment

Stemming	Personal Information Elimination	Number of features
✗	✗	49967
✗	✓	38670
✓	✗	42383
✓	✓	31106

2 `http://www.meertens.knaw.nl/nvb/`
3 `http://www.iisg.nl/hsn/data/`
4 `http://ilk.uvt.nl/frog/`
5 `http://scikit-learn.org/`

3 Dealing with Noisy Labels

To identify duplicated categories we generate pairs of categories which can be candidates for merging using a confusion matrix $\mathcal{M}$. Fig. 1 shows a part of the confusion matrix for eleven randomly selected categories. The complete $\mathcal{M}$ has 455 rows and columns. The confusion means that one category was incorrectly predicted as another category. The matrix is obtained by the SVM classifier applied to notary acts without stemming, PIE or feature selection (see Experiment 1, Section 5). We analyse the confusion matrix to identify categories that were duplicated and perform a category resolution with the help of an expert.

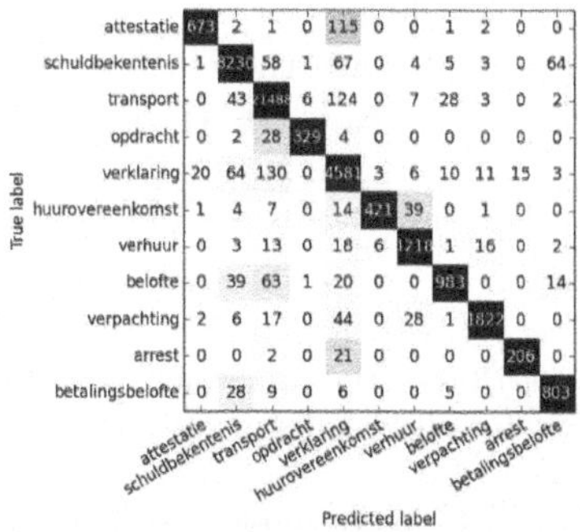

Fig. 1. Confusion matrix for randomly selected categories

We have developed a web interface for a historian-expert which for each category recommends the list of typically confused categories. The expert had to review each category and decide: keep a category as it is, merge it with another category or drop the category and relabel the related documents. After reviewing manually the list of categories we obtained 88 final categories.

In addition, we evaluate the agreement between human annotators. We consider the inter-annotator agreement in category assignment as a level of performance that may be achieved by automatic documents classifiers. We randomly selected 2000 labelled notary acts and asked another human annotator to assign a category after removing the label. Then we evaluated the pairwise agreement between annotators using *Cohen's kappa coefficient*. According to the weighted average kappa coefficient the annotators agree on 88.49%. The disagreement occurs because there are no clear borders between some categories.

4 Two Level Classification

We are interested in obtaining accurate results as well as predicting rare categories in the collection of documents $\mathcal{D}$. The prediction of frequent categories will allow us to get high performance results, but in many cases the smaller categories will be confused with the larger ones. Therefore we design an approach that takes into account the category frequency information. We introduce the following definitions:

Definition 1. The support of a category in a set of documents is its proportional size in the set.

Definition 2. The category $c \in \mathcal{C}$ is *frequent* if $sup(c)$ is above a minimum defined threshold min_sup, otherwise c is *non-frequent*:

$$sup(c) > min\_sup \tag{1}$$

At the first level, all infrequent categories are joined to form one cluster with the smallest categories, whereas the frequent ones make up their own cluster. The minimum support can be learnt during a training phase. We used 2%. The output of this level is a set of cluster-labels $\{f_1, ..., f_n\}$ associated with each document $d \in D$ and the set of clusters $\mathcal{F}$.

At the second level we incorporate the clustering results into a prediction model and the TC process. This idea is described in Algorithm 1.

Algorithm 1. Building prediction model and TC classification

Input: Training set $\mathcal{D} = \{d_1, ..., d_n\}$ with category-labels $\{c_1, ..., c_k\}$ and cluster-labels $\{f_1, ..., f_n\}$. Test set $\mathcal{T} = \{t_1, ..., t_h\}$. Set of categories $\mathcal{C} = \{c_1, ..., c_k\}$ and set of clusters $\mathcal{F}$. Learning algorithm of the prediction model $\mathcal{L}$

Output: Predicted labels $\mathcal{N}$ for all test instances $\mathcal{T}$

$\mathcal{N} \leftarrow \emptyset$
$\mathcal{M} \leftarrow TrainModel(\mathcal{D}, \mathcal{F}, \mathcal{L})$ # Learn model on cluster labels
$\mathcal{N}^* \leftarrow Classify(\mathcal{T}, \mathcal{M})$ # Classify test data with cluster-labels
for each cluster f_i in $\mathcal{F}$ **do**
    $\mathcal{D}_i \in \mathcal{D}, \mathcal{T}_i \in \mathcal{T}, \mathcal{C}_i \in \mathcal{C}$ # Associate data with the cluster
    $\mathcal{M}_i \leftarrow TrainModel(\mathcal{D}_i, \mathcal{C}_i, \mathcal{L})$ # Learn model on category labels
    $\mathcal{N}_i \leftarrow Classify(\mathcal{T}_i, \mathcal{M}_i)$ # Classify data with final categories
    $\mathcal{N} \leftarrow \mathcal{N} \cup \mathcal{N}_i$
end for
return $\mathcal{N}$

5 Experiments and Results

We conducted experiments on the annotated datasets described in Section 2. We have three sets of experiments. In order to assess the performance of our results, we apply 10-fold cross-validation. Due to lack of space, a more detailed and graphical view of the experiments is available on the web[6]

Experiment 1: TC Results before Category Resolution. Table 2 presents the overall accuracy for each experimental setup before category resolution (i.e. with 455 categories). The best results were achieved with an SVM classifier using the complete lexical vocabulary as a features without stemming procedure and named entity elimination. We expected that the elimination of person names and locations would affect the accuracy of the classifier positively, but from the results we see the opposite: there is a small correlation between locations and person names and type of notarial acts. However, despite the

[6] `http://wwwis.win.tue.nl/~amontes/ecir2015/results.html`

Table 2. Performance accuracy in the experiment 1 before category resolution

Model	Feature	Stemming	PIE	all features	chi-sq.	lsa	POS
SVM, lin. kernel	tf-idf	✗	✗	**86.84**	84.70	84.03	86.32
SVM, lin. kernel	tf-idf	✗	✓	85.67	84.05	83.89	84.52
SVM, lin. kernel	tf-idf	✓	✗	86.55	85.22	84.02	85.11
SVM, lin. kernel	tf-idf	✓	✓	86.38	84.45	83.85	84.52

promising overall results 307 categories are completely ignored by the classifier. Therefore, we performed the category resolution described in Section 3.

Experiment 2: TC Techniques after Category Resolution. Table 3 presents the accuracy results for each experimental setup after category resolution. The best results again are achieved by applying a SVM classifier and using a complete sparse lexical vocabulary as feature vector without named entity elimination. The classifier in this case is not sensitive to the stemming procedure. In this experiment we achieved a maximum accuracy of 87.79% which is 0.95% higher than before. The number of categories with an absolute zero f-score is reduced to 17. That can be explained by the very few examples in each category.

Table 3. Performance accuracy in the experiment 2 after category resolution

Model	Feature	Stemming	PIE	all features	chi-sq.	lsa	POS
SVM, lin. kernel	tf-idf	✗	✗	**87.79**	86.56	84.86	87.40
SVM, lin. kernel	tf-idf	✗	✓	86.38	85.50	84.95	85.65
SVM, lin. kernel	tf-idf	✓	✗	**87.79**	86.80	85.02	87.42
SVM, lin. kernel	tf-idf	✓	✓	86.25	85.53	84.93	85.65

Experiment 3: TC Using Two Level Classification. Table 4 presents the accuracy results of the proposed framework for each experimental setup. The maximum accuracy is increased up to 88.08% which is 0.3% higher than before. There is also a slight improvement in the number of unidentified categories, it is reduced to 14 compared to 17 unidentified categories in the previous experiment. Still the very few examples in rare categories does not allow to the classifier to recognise them all. Nevertheless the proposed simple clustering technique as a framework to the overall classification process shows promising results.

Table 4. Performance accuracy in the experiment 3 using two level classification

Model	Feature	Stemming	PIE	all features	chi-sq.	lsa	POS
SVM, lin. kernel	tf-idf	✗	✗	**88.08**	87.50	85.60	87.51
SVM, lin. kernel	tf-idf	✗	✓	86.51	85.93	85.42	85.77
SVM, lin. kernel	tf-idf	✓	✗	88.07	87.60	85.73	87.52
SVM, lin. kernel	tf-idf	✓	✓	86.39	85.91	85.52	85.77

Comparative Evaluation. We performed a comparative analysis of our two-level classification algorithm versus human agreement (see Fig. 2). Since we have two annotators, we consider the labelling results from the 1st annotator as the ground truth and evaluate the results of the 2nd annotator. In Fig. 2a, we compare the results for each category for the manual and the automatic evaluation method. The small categories are recognised much better by people, while for larger categories the results are comparable. Fig. 2b shows that the performance of humans for most of the categories correlates with the automatic classification. However there is a number of categories where humans significantly outperforms our algorithm. Those categories have a very small support value.

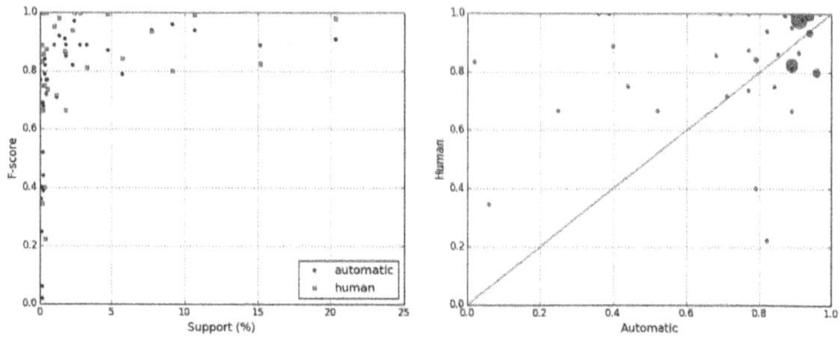

(a) F-score vs support per category (b) F-score for human vs automatic TC

Fig. 2. Evaluation of f-score for individual categories for automatic TC and human

6 Conclusions

In this paper we described a framework for dealing with noisy labels. We examined existing text classification algorithms, studied the influence of lexical information and analyzed a number of feature selection methods. Then we created a two level classification approach that slightly improved the results, achieving a performance close to the inter-annotator agreement. The developed methods can be applied for classification of narrative data in different domains.

References

1. Aggarwal, C.C., Zhai, C.: A survey of text classification algorithms. In: Mining Text Data, pp. 163–222. Springer (2012)
2. Van den Bosch, A., Busser, B., Canisius, S., Daelemans, W.: An efficient memory-based morphosyntactic tagger and parser for dutch. In: CLIN, pp. 99–114 (2007)
3. Constantopoulos, P., Doerr, M., Theodoridou, M., Tzobanakis, M.: Historical documents as monuments and as sources. In: Proceedings of Computer Applications and Quantitative Methods in Archaeology Conference (2002)
4. Iglesias, J.A., Tiemblo, A., Ledezma, A.I., Sanchis, A.: News mining using evolving fuzzy systems. In: Corchado, E., Lozano, J.A., Quintián, H., Yin, H. (eds.) IDEAL 2014. LNCS, vol. 8669, pp. 327–335. Springer, Heidelberg (2014)
5. Ikonomakis, M., Kotsiantis, S., Tampakas, V.: Text classification using machine learning techniques (2005)
6. Leong, C.K., Lee, Y.H., Mak, W.K.: Mining sentiments in sms texts for teaching evaluation. Expert Systems with Applications 39(3), 2584–2589 (2012)
7. Nguyen, D., Trieschnigg, D., Theune, M.: Folktale classification using learning to rank. In: Serdyukov, P., Braslavski, P., Kuznetsov, S.O., Kamps, J., Rüger, S., Agichtein, E., Segalovich, I., Yilmaz, E. (eds.) ECIR 2013. LNCS, vol. 7814, pp. 195–206. Springer, Heidelberg (2013)
8. Sebastiani, F.: Machine learning in automated text categorization. ACM Comput. Surv. 34(1), 1–47 (2002)

ConceptFusion: A Flexible Scene Classification Framework

Mustafa Ilker Sarac[1], Ahmet Iscen[2], Eren Golge[1], and Pinar Duygulu[1,3]

[1] Department of Computer Engineering, Bilkent University, Ankara, Turkey
[2] Inria, Rennes, France
[3] Carnegie Mellon University, PA, USA

Abstract. We introduce ConceptFusion, a method that aims high accuracy in categorizing large number of scenes, while keeping the model relatively simpler and efficient for scalability. The proposed method combines the advantages of both low-level representations and high-level semantic categories, and eliminates the distinctions between different levels through the definition of concepts. The proposed framework encodes the perspectives brought through different concepts by considering them in concept groups that are ensembled for the final decision. Experiments carried out on benchmark datasets show the effectiveness of incorporating concepts in different levels with different perspectives.

Keywords: Scene recognition, Concepts, Ensemble of Classifiers.

1 Introduction

With the recent advancements in capturing devices, billions of images have been stored in personal collections and shared in social networks. Due to limitation and subjectivity of the tags, visual categorisation of images is desired to manage huge volume of data.

As an important visual content, scenes have been considered in many studies to retrieve images. Low-level features are commonly used to classify scenes, such as for indoor versus outdoor, or city versus landscape [9, 11, 13–15]. Alternatively, object detector responses have been used as high-level features to represent semantics [8]. While the number of objects could reach to hundreds and thousands with the recent detectors that can be generalised to variety of catagories, the main drawback of object-based approaches is the requirement for manual labeling to train the object models. Moreover, it may be difficult to describe some images through specific objects. Recently, a set of mid-level attributes that are shared between object categories, such as object parts (wheels, legs) or adjectives (round, striped) [3, 6], have been used. However, these methods also heavily depend on training to model human-defined attributes. The main question is how can we melt representations with different characteristics in the same pool? Moreover, how can we scale it to large number of concepts?

In this study, we introduce ConceptFusion, in which we use the term concept for any type of intermediate representation, ranging from visual words to attributes and objects. We handle the variations between different levels of concepts, by putting them into concept groups. Separate classifiers are trained for each concept group. The contributions

A. Hanbury et al. (Eds.): ECIR 2015, LNCS 9022, pp. 55–60, 2015.

of each concept group to the final categorization are provided in the form of confidence values that are ensembled for the final decision. The framework is designed to be generalised to large number of different concepts. While early and late fusuion techniques have been studied for a long time, the spirit of our work differs from the others in the following aspects.

- We do not restrict ourselves to only semantic categories that can be described by humans, but also map low-/mid-level representations into concepts.
- Motivated by the recent studies in learning large number of concepts from weakly labeled and noisy web images, the framework is designed to be scaled through the introduction of concept groups.

2 Our Method

ConceptFusion brings the ability of using different levels of descriptors through the definition of concepts and concept groups (see Figure 1). Low-level local or global descriptors could be quantized to obtain concepts in the form of visual words, and then concept group can be represented as Bag-of-Words. On the other hand, each object category could correspond to a concept, and as a whole the concept group could be represented through a vector of confidence values of object detectors. ConceptFusion is designed to allow the integration of different *concept groups* for classification. Concept groups are not required to have any semantic meaning; we suppose that, each concept group can add a different perspective for classification. The classification has two main parts; *individual classification* and *ensemble of classifiers*. Individual classification is applied to each concept group separately, and classification results are combined in ensemble of classifiers stage before making a final prediction.

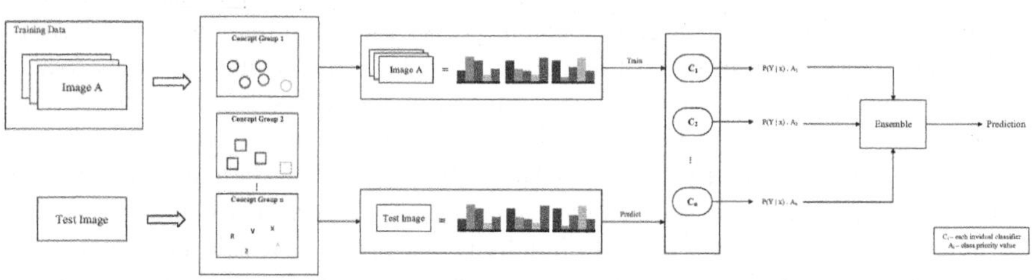

Fig. 1. Overview of ConceptFusion. Individual classifiers are trained for each concept group, and for each individual classifier a concept-priority value is computed. A test image, represented by concepts, is fed to the individual classifiers. Class-confidence values incorporated with the concept-priority values, are combined in the ensemble stage for final prediction.

Individual Classification: To support our hypothesis of trying to examine the different perspectives of each concept group, we consider each group independently. That is, we assume that the individual classification performance of a concept group has no effect on another, and should therefore be treated completely separately. This also allows us to have an agnostic classification method that can be used with any type of *concepts*. To implement this idea, we train a separate, individual probabilistic classifier for each concept group. For a given image query, the role of each individual classifier is to give the probability of the image belonging to each class. We use probabilistic Support Vector Machine (SVM) as classifier.

Ensemble of Classifiers: After training a separate classifier for each concept group, we must be able to combine them properly before making a final decision. Since we cannot guarantee that each individual classifier will perform well, especially in the case of classifiers trained from weakly labeled web images, we decide to explore giving priorities to each individual classifier. To decide which individual classifier gets which priority, we should estimate how a classifier would work on unseen data, so we can assign more weight to decisions of those that are expected perform well, and less weight to those that are predicted to perform poorly.

We introduce the notation of *concept-priority value* as an estimate of how each classifier would perform generally. We find this value by performing cross-validation on the training set using each classifier and assigning the average accuracy value as the *concept-priority value* of the corresponding individual classifier. Now that we have a generalized estimation for the performance of each individual classifier, we can weight their outputs accordingly. Probability outputs of each single classifier is multiplied by its *concept-priority value*. After obtaining the weighted class-confidence probabilities from each classifier, we ensemble them together in the final step. At the end, the class that obtains the highest value is selected as the final prediction.

To demonstrate the ConceptFusion idea, it is desired to include concepts at different levels. To eliminate effort for the manual labeling of objects or attributes, we take the advantage of two benchmark datasets where the semantic categories are already available in some form: MIT Indoor [12] and SUN Attribute Dataset [10].

3 Evaluation of ConceptFusion Framework

In this section, we evaluate ConceptFusion framework to understand the effect of different ensemble techniques, number of concepts and different classifiers.

First, we evaluate the possibilities of using different ensemble methods to combine vectors from different concept groups: *(i) Confidence summation without weighted classifier ensemble* which simply sums the confidence values obtained from classifiers of different concept groups, that is we treat each classifier with equal importance and do not consider any weighting to their results. *(ii) Confidence summation with weighted classifier ensemble* in which before combining the confidence values of each classifier in the summation step, we multiply each of them by the corresponding class priority value. *(iii) Ranking without weighted classifier ensemble*, in which we integrate a classic ranking system [5] to combine different features. Instead of using exact confidence values, we sort the confidence values of each class and rank each class in the order of preference. Then we sum their ranks to come up with a final decision. *(iv) Ranking with weighted classifier ensemble* which weigths the class ranks from classifier by its *concept-priority* value, in order to avoid the possible issues that can rise from treating each classifier equally. *(v) Two-layer classifier as ensemble* where the input of the classifier would be the output of the previous classifiers concatenated together.

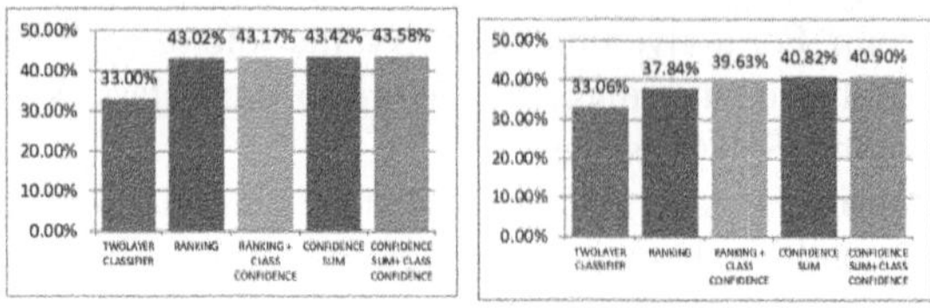

Fig. 2. The effect of ensemble techniques in Sun Attribute (left) and MIT Indoor (right) datasets

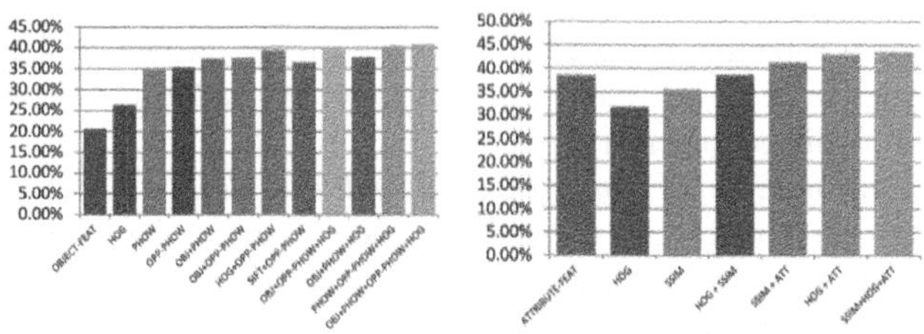

Fig. 3. Comparing different number of concept groups on MIT (left) and SUN (right) datasets

As seen in Figure 2, although changing the ensemble method did not have much effect in the Sun Attribute Dataset, the results of the MIT Indoor Dataset are more distinct. In MIT Indoor, ensembling concept groups using confidence summation and weighted methods is clearly more advantageous than using a ranking system or a non-weighted system. Using confidence- based methods reduces the probability of losing information classifier information, and class-priorities give each classifier their assumed generalized performance rate. We can argue for the same trend in SUN Attribute Dataset, but the difference of accuracies is much less. Two-layer classifiers gives us the worst results for both datasets, because the second level classifier is extremely prone to over-fitting the output of the first layer classifier during the training stage, hence not working well in the testing stage.

Secondly, we evaluated ConceptFusion by changing the number of different concepts used in each dataset. For ensemble of classifiers phase, we use the weighted versions. SVM parameters are set using cross-validation on training data. Results for both datasets are reported in Figure 3. We observe that the accuracy of the classifier also generally increases as we add more concept groups to our system. We obtain the best results by using the highest amount of concept groups. This shows that the combination features from completely different concept groups can be beneficial to the overall classifier, and that our method makes use of this relation in a meaningful way.

Finally, to evaluate the effect of using different classifiers, we used a fixed ensemble configuration and changed the type of our classifier in order to observe any different behaviors. We originally designed ConceptFusion with SVM classifier, however we believe it would also be necessary to see the performance of our framework using two other classifiers: Ada- Boost [4] and Random Forests [1]. As seen on Table 1, LIBSVM's [2] implementation of SVM outperforms the other two classifiers with its capability of constructing non-linear decision boundaries.

Table 1. Comparison of different classifiers on MIT and SUN dataset

	MIT Indoor		SUN Attribute	
	Confidence	Ranking	Confidence	Ranking
Random Forests	37.3%	32.3%	32.7%	33.3%
Ada-Boost	35.8%	33.9%	33.2%	34.7%
SVM	43.6%	43.2%	40.9%	39.6%

4 Comparison with Other Methods

We compare the results of ConceptFusion with a baseline method, and with the state-of-the-art Object Bank method [7] (see Table 2). As the baseline we combine different concepts or features just by concatenating them. This method is extremely simple and widely used, but it can have many disadvantages, such as resulting features being in very high dimensions. Also, combining features from very different concepts, such as low- level and high-level features, does not necessarily add any meaning for classification purposes, and can provide low results. Object Bank [7] is a well known method with the idea of having a higher semantic level description of images, exposing scene's semantic structure similar to human understanding of views. Although ObjectBank provides a good interpretation of the image, it produces a very high dimensional vectors, and concatenation of large number of features does not to perform well.

Table 2. Comparisons with feature concatenation and Object Bank [7] on MIT dataset

Method	Accuracy
Feature Concatenation	9.48%
OB-LR [7]	37.6%
ConceptFusion	40.9%

5 Discussion and Future Work

We proposed ConceptFusion as a framework for combining concept groups from many different levels and perspectives for the purpose of scene categorization. The proposed framework provides flexibility for supporting any type of concept groups, such as those that have semantic meanings like objects and attributes, or low-level features that have no meanings semantically but can provide important information about the structure of an image. There is no limit in the definition of concepts, and it is easy to be expanded through inclusion of any other intermediate representation describing the whole or part of the image in content or semantics.

Current framework examines each concept group on the same level, by assuming that their classification models are completely independent from each other. We plan to extend our framework by modifying this idea, and establishing dependence between each concept group by their semantic meanings.

Acknowledgement. This study was partially supported by TUBITAK project with grant no 112E174 and CHIST-ERA MUCKE project. This paper was also partially supported by the US Department of Defense, U. S. Army Research Office (W911NF-13-1-0277) and by the National Science Foundation under Grant No. IIS-1251187. The U.S. Government is authorized to reproduce and distribute reprints for Governmental purposes notwithstanding any copyright annotation thereon. Disclaimer: The views and conclusions contained herein are those of the authors and should not be interpreted as necessarily representing the official policies or endorsements, either expressed or implied, of ARO, the National Science Foundation or the U.S. Government.

References

[1] Breiman, L.: Random forests. Mach. Learn. 45(1), 5–32 (2001)
[2] Chang, C.-C., Lin, C.-J.: LIBSVM: A library for support vector machines. ACM Transactions on Intelligent Systems and Technology, 2, 27:1–27:27 (2011)
[3] Farhadi, A., Endres, I., Hoiem, D., Forsyth, D.: Describing objects by their attributes. In: CVPR (2009)
[4] Freund, Y., Schapire, R.E.: A decision-theoretic generalization of on-line learning and an application to boosting. In: Vitányi, P.M.B. (ed.) EuroCOLT 1995. LNCS, vol. 904, pp. 23–37. Springer, Heidelberg (1995)
[5] Ho, T.K., Hull, J.J., Srihari, S.N.: Decision combination in multiple classifier systems. IEEE PAMI 16(1), 66–75 (1994)
[6] Lampert, C.H., Nickisch, H., Harmeling, S.: Learning to detect unseen object classes by betweenclass attribute transfer. In: CVPR (2009)
[7] Li, L.-J., Su, H., Fei-Fei, L., Xing, E.P.: Object bank: A high-level image representation for scene classification & semantic feature sparsification. In: NIPS (2010)
[8] Li, L.-J., Su, H., Lim, Y., Fei-Fei, L.: Objects as attributes for scene classification. In: Kutulakos, K.N. (ed.) ECCV 2010 Workshops, Part I. LNCS, vol. 6553, pp. 57–69. Springer, Heidelberg (2012)
[9] Oliva, A., Torralba, A.: Modeling the shape of the scene: A holistic representation of the spatial envelope. Int. J. Comput. Vision 42(3), 145–175 (2001)
[10] Patterson, G.: Sun attribute database: Discovering, annotating, and recognizing scene attributes. In: CVPR (2012)
[11] Payne, A., Singh, S.: Indoor vs. outdoor scene classification in digital photographs. Pattern Recogn. 38(10), 1533–1545 (2005)
[12] Quattoni, A., Torralba, A.: Recognizing indoor scenes (2007)
[13] Quelhas, P., Monay, F., Odobez, J.M., Gatica-Perez, D., Tuytelaars, T.: A thousand words in a scene. IEEE PAMI 29(9), 1575–1589 (2007)
[14] Serrano, N., Savakis, A.E., Luo, J.: A computationally efficient approach to indoor/outdoor scene classification. In: ICPR (4) (2002)
[15] Vailaya, A., Member, A., Figueiredo, M.A.T., Jain, A.K., Zhang, H.-J., Member, S.: Image classification for content-based indexing. IEEE Transactions on Image Processing 10, 117–130 (2001)

An Audio-Visual Approach to Music Genre Classification through Affective Color Features

Alexander Schindler[1,2] and Andreas Rauber[1]

[1] Department of Software Technology and Interactive Systems
Vienna University of Technology
rauber@ifs.tuwien.ac.at

[2] Information management, Digital Safety and Security Department
AIT Austrian Institute of Technology
alexander.schindler@ait.ac.at

Abstract. This paper presents a study on classifying music by affective visual information extracted from music videos. The proposed audio-visual approach analyzes genre specific utilization of color. A comprehensive set of color specific image processing features used for affect and emotion recognition derived from psychological experiments or art-theory is evaluated in the visual and multi-modal domain against contemporary audio content descriptors. The evaluation of the presented color features is based on comparative classification experiments on the newly introduced 'Music Video Dataset'. Results show that a combination of the modalities can improve non-timbral and rhythmic features but show insignificant effects on high performing audio features.

1 Introduction

Over the past decades music videos distinctively influenced our pop-culture and became a significant part of it. Since their inception in the early 1980-ies music videos emerged from a promotional support medium into an art form of their own. The effort invested to produce a video creates enough information such that many music genres can be predicted by the moving pictures only. This potential of information provided was demonstrated in previous work on music video based artist identification [13], where a precision improvement of 27% could be observed over conventional audio features. Harnessing this potential presents a new way to approach existing Music Information Retrieval (MIR) problems such as an audio-visual approach to music video segmentation [4]. Approaches to affective content analysis of music videos are provided by [19] and [20]. In order to use the visual domain for music retrieval tasks, it has to be linked to the acoustic domain. Since substantial research on audio-visual correlations in music videos is yet scarce or not available, we base our approach on the simplified assumption that both layers intend to express the same emotions. In this paper we evaluate if this information - and more specifically the color information - is sufficient to discriminate music genres. Using color in content-based image retrieval has been extensively studied [9,10,12] and is yet described as problematic since it is

A. Hanbury et al. (Eds.): ECIR 2015, LNCS 9022, pp. 61–67, 2015.

Table 1. Overview of all features. The column '#' indicates the dimensionality of the corresponding feature set.

	Short Name	#	Descriptiom
Audio	Statistical Spectrum Descriptors (SSD)	168	Statistical description of a psycho-accoustic transformed audio spectrum
	Rhythm Patterns (RP)	1024	Description of spectral fluctuations
	Rhythm Histograms (RH)	60	Aggregated Rhythm Patterns
	Temporal SSD and RH		Temporal variants of RH (TRH #420), SSD (TSSD #1176)
	MFCC	12	Mel Frequency Cepstral Coefficients
	Chroma	12	12 distinct semitones of the musical octave
Visual	Global Color Statistics	6	mean saturation and brightness, mean angular hue, angular deviation, with/without saturation weighting
	Colorfulness	1	colorfulness measure based on Earth Movers Distance
	Color Names	8	Magenta, Red,Yellow,Green,Cyan Blue, Black, White
	Pleasure, Arousal, Dominance	3	approx. emotional values based on brightness and saturation
	Itten Contrasts	4	Contrast of Light and Dark, Contrast of Saturation, Contrast of Hue and Contrast of Warm and Cold
	Wang Emotional Factors	18	Features for the 3 affective factors by Wang et al. [17]
	Lightness Fluctuation Patterns	80	Rhythmic fluctuations in video lightness

highly influenced by lighting conditions during image acquisition. In music videos different illumination settings and colors are usually desired artistic effects. In the following section we introduce seven feature sets that derive from psychological experiments, art-theory or try to model human perception. Section 3 lays out the evaluation and introduces the Music Video Dataset to foster further research. After discussing the results in Section 4 conclusions and outlooks to future work are provided in Section 5.

2 Method

Audio features are extracted from the separated audio channel of the music videos. Visual features are extracted from each frame of a video and aggregated during post-processing by calculating the statistical measures mean, median, standard deviation, min, max skewness, kurtosis. As a pre-processing step black bars at the borders of video frames, also called *Letterboxing* or *Pillarboxing*, are removed.

2.1 Audio Features

Psycho-accoustic Music Descriptors as proposed by [7] are based on a psycho-acoustically modified Sonogram representation that reflects human loudness sensation. *Statistical Spectrum Descriptors (SSD)* subsequently compute statistical moments for the 24 critical bands of hearing. *Rhythm Patterns (RP)* describe fluctuations in modulation frequency which provide a rough interpretation of the rhythmic energy of a song. *Rhythm Histograms (RH)* aggregate the modulation amplitude values of the individual critical bands computed in a RP. *Temporal Variants (TSSD, TRH)* describe variations over time through statistical moments calculated from consecutive segments of a track. For the extraction, we employed the Matlab-based implementation, version 0.6411.

Mel Frequency Cepstral Coefficients (MFCC) are well known audio features derived from speech recognition. **Chroma** features project the spectrum onto 12 bins representing the semitones of the musical octave. We utilized MARSYAS [14] version 0.4.5.

2.2 Visual Features

Global Color Statistics calculate *Mean Saturation* and *Mean Bightness* based on the Improved Hue, Luminance and Saturation (IHLS) color space [18] which has the advantages of low saturation values of achromatic pixels and independence of saturation from the brightness function. Hue in IHLS is an angular value. Circular statistics has to be applied [5] to assess *angular mean Hue* and *angular deviation of Hue. Saturation weighted mean Hue and deviation of Hue* are more robust towards weakly saturated colors.

Global Emotion values refer to a Pleasure-Arousal-Dominance model based on investigated emotional reactions presented in [15]. The introduced relationship between saturation (S) and brightness (B) is calculated from the corresponding IHLS channels:

$$Pleasure = 0.69 * B + 0.22 * S \tag{1}$$

$$Arousal = -0.31 * B + 0.60 * S \tag{2}$$

$$Dominance = 0.76 * B + 0.32 * S \tag{3}$$

Colorfulness is one of the features used in [2] to computationally describe aesthetics in photographies. The proposed method is based on a partitioned RGB palette using Earth Mover's Distance (EMD) [11] to calculate the dissimilarity of a supplied image to an *ideal* color distribution of a *colorful* image.

Wang Emotional Factors Wang et al. [17] identified three factors based on emotional word correlations that are relevant for image retrieval based on emotion semantics. Three feature sets are calculated using fuzzy membership functions to assign values of the perceptual psychology motivated L*C*H* color space to discrete semantic words. *Feature One* includes lightness description of a segmented image ranging from *very dark* to *very bright*. These are combined with the calculated hue labels *cold* and *warm*. *Feature Two* provides a description of warm or cool regions with respect to different saturations as well as a description of contrast. *Feature Three* combines lightness contrast with an sharpness estimation. A no-reference perceptual blur measure [1] was used. The sharpness is further calculated by $1 - blurIndex$. The contrast description overlaps with the *Itten contrasts* and is omitted.

Itten's Contrasts are a set of art-theory concepts defined by Johannes Itten [6] for combining colors to induce emotions based on an proportional opponent color model. The contrast calculation is aligned to the method presented in [8] which uses Wang's feature extraction [17] as a predecessor. Instead of a waterfall segmentation we used a Quick Shift [16] approach due to better performance at reasonable processing time. We calculated the following contrasts: *Contrast of Light and Dark*, *Contrast of Saturation*, *Contrast of Hue* and *Contrast of Warm and Cold.*

Color Names describe color distributions of the reduced Web-safe Elementary-color palette consisting of the 8 elementary colors Magenta, Red, Yellow, Green, Cyan, Blue, Black and White. To map a frame of a video to this palette it is converted to Hue Value Saturation (HSV) color-space. *Contrast, brightness and color enhancement* is applied through application of Contrast Limited Adaptive Histogram Equalization (CLAHE) [21]. *Color Quantization* to reduce the number of distinct colors of the frame to the desired palette is obtained by applying *error diffusion* which computes the mean square error between the original pixel value and its closest match which is then propagated locally to its surrounding pixels. *Ordered Dithering* was used since it reduces the effect of contouring but stays more consistent with the original colors. A 32x32 Bayer pattern matrix was used as threshold map. *Feature Calculation* is concluded by calculating the statistical moments mean, median, variance, min, max, skew and kurtosis of the reduced palette.

Lightness Fluctuation Patterns are calculated analogous to the music feature Rhythm Patterns (RP) [7] from the perceptually uniform LAB color space. For each frame a 24 bin histogram of the lightness channel is calculated. Fast Fourier Transform (FFT) is applied to the histogram space of all video frames. This results in a time-invariant representation of the 24 lightness levels capturing reoccurring patterns in the video. Only amplitude modulations in the range from 0 to 10 Hz are used for the final feature set, since rhythm cannot be perceived from higher modulation frequencies. Based on the observation that light effects, motions and shots are usually beat synchronized in music videos, LFPs can be assumed to express rhythmic structures of music videos.

3 Evaluation - The Music Video Dataset

The empirical evaluation is based on the Music Video Dataset (MVD). We use empirical classification experiments and Chi-square feature selection to analyze the performance of the visual and audio-visual feature-spaces. The MVD is a collection of carefully selected music videos. It consists of different subsets that can be combined to bigger data-sets. The following sub-sets of the MVD are used to evaluate the features presented in Section 2:

MVD-VIS: The *M*usic *V*ideo *D*ataset for *VIS*ual content analysis and classification is intended for classifying music videos by their visual properties only. Special emphasis has been set on minimizing the intra- and maximising the inter-class variance in the acoustic domain of the dataset. Non overlapping sub-genres were chosen and tracks within a certain class share very similar musical characteristics. Music genre classification based on conventional audio features provides accuracy above-average (see Table 2) compared to current benchmarks of the Music Information Retrieval domain [3].

MVD-MM: The *M*usic *V*ideo *D*ataset for *M*ulti*M*odal content analysis and classification is intended for multi-modal classification and retrieval tasks. The overlapping classes have high inter and intra class variance. Genre classification based on audio features provides average results and serves as starting point for multi-modal approaches.

MVD-MIX: The MVD-MIX data-set is a combination of the data-sets MVD-VIS and MVD-MM. The distinct genres of the sub-sets have been selected in a way, that a union of the two sets provides a non-overlapping bigger set. Consequently the inter-class variance increases while the intra-class variance remains the same as for the individual sets. While the sub-sets are intended for developing content descriptors, the MVD-MIX should be used for audio-visual evaluations.

The dataset creation was preceded by the selection of the non-overlapping genres respectively to enable the combination of the two subsets into the bigger *MVD-MIX* dataset. Each genre consists of 100 selected videos. Resulting in dataset sizes of 800 music videos for *MVD-VIS* and *MVD-MM* each as well as 1600 for the MV-MIX dataset. Music videos were selected primarily by their audible properties. A set of selection criteria has been applied such as quality criteria of at least 90 kBits/s audio encoding and video resolution ranging from QVGA to VGA. Only official music videos were selected, no live performance, abstract or animated videos. Artist stratification is provided by selecting only two tracks per artist.

Data Provision: Due to copyright restrictions it is not possible to redistribute music videos or audio files. Yet, all videos have been retrieved from Google's Youtube platform and a list of corresponding Youtube video-ids is provided. It should be stated that the availability of these videos cannot be guaranteed and that some may vanish over time. To ensure comparability of results and reproducibility of the experiments, all features of this publication including a range of standard visual and acoustic features are being provided and customized features will be extracted and provided on request. All extracted features are made available for download at: `http://www.ifs.tuwien.ac.at/mir/mvd/`.

4 Results

Table 2 summarizes the results of the comparative classification experiments. The top segment of the table provides audio only results which serve as baseline for evaluating the visual and audio-visual approaches. Using visual features only an accuracy of 50.13% could be reached for Support Vector Machines (SVM) for the MVD-VIS set. Accuracies for other sets or classifiers range from 17.89% to 39.38%. Because all classes equal in size these results are above a baseline of 12.5% or 6.25% respectively. Yet, the performance of the visual features alone is not representative. The audio-visual results show interesting effects. Generally, there is insignificant or no improvement of the performance over the top performing audio features. The results show that combining the visual features with chroma and rhythm descriptors has a positive effect on the accuracy while it is negative with spectral and timbral features. Applying ranked Chi-square attribute selection on the visual features shows, that affective features as well as the frequencies of black and white pixels have highest values. Further, more information is provided by variance and min/max aggregated values than by mean values.

Table 2. Classification results for audio, visual and audio-visual features showing accuracies for Support Vector Machines (SVM), K-Nearest Neighbors (KNN), Random Forest (RF) and Naive Bayes (NB) classifiers. Bold-faced values highlight improvements of audio-visual approaches over audio features.

		MVD-VIS				MVD-MM				MVD-MIX			
		SVM	KNN	RF	NB	SVM	KNN	RF	NB	SVM	KNN	RF	NB
Audio	TSSD-RP-TRH	93.79	80.85	77.13	71.46	74.76	55.00	55.84	52.20	75.91	54.16	49.80	48.32
	TSSD	86.81	72.58	70.72	62.61	69.97	53.33	56.16	53.65	66.19	47.40	45.33	44.22
	RP	87.26	69.81	71.29	64.04	60.35	42.38	43.85	41.63	63.19	43.06	42.53	41.39
	SSD	85.78	73.18	72.80	58.81	68.74	50.28	54.43	48.41	65.11	44.64	46.18	38.92
	TRH	71.04	55.83	55.16	53.86	49.50	38.28	37.66	39.66	46.61	33.02	30.54	35.70
	MFCC	62.28	48.58	49.04	46.95	42.14	29.16	32.50	34.17	37.02	26.60	25.57	27.11
	Chroma	36.34	28.09	34.41	23.03	25.26	20.11	23.16	19.41	19.64	14.68	16.52	12.08
	Visual Features	50.13	34.04	38.60	39.38	31.69	21.16	22.86	23.38	32.22	17.89	19.36	21.16
Audio-Visual	TSSD-RP-TRH	**94.86**	**81.38**	76.51	**71.65**	**75.69**	**55.78**	54.36	51.36	**76.53**	**55.76**	49.15	**49.08**
	TSSD	**88.45**	71.65	68.80	**64.75**	**70.55**	52.60	54.34	52.25	**69.46**	46.15	43.12	**45.16**
	RP	**89.80**	**71.99**	69.90	**65.78**	**62.79**	**43.93**	43.74	41.61	**66.59**	**44.47**	40.61	**41.68**
	SSD	85.25	62.05	66.22	57.80	65.34	42.28	48.53	44.24	**65.21**	36.13	39.64	**38.76**
	TRH	**77.84**	**55.98**	**57.21**	**59.71**	**58.50**	32.79	35.60	**41.40**	**56.31**	31.39	**31.18**	40.09
	MFCC	**63.71**	41.53	45.78	46.28	**42.88**	24.38	27.40	27.35	**43.11**	22.33	22.89	**25.62**
	Chroma	**55.70**	**39.28**	**42.78**	**43.13**	**35.29**	**24.16**	**26.04**	**25.51**	**35.43**	**20.10**	**21.91**	**24.14**

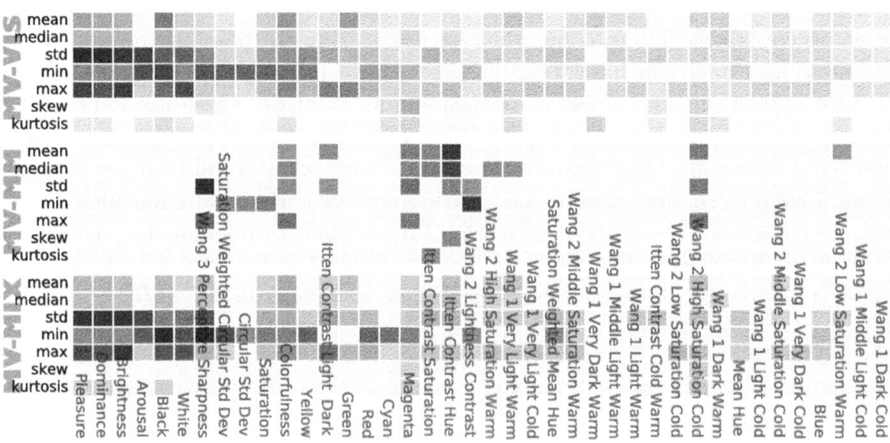

Fig. 1. Chi Square Feature Evaluation in descending order from left to right. Dark blue areas correspond with high χ^2 values.

5 Conclusions and Future Work

We presented a comparative evaluation of audio-visual music classification that focused on the color information of music videos. A set of diverging approaches based on psychological or perceptive models has been applied to extract different kinds of semantic information. We further introduced a descriptor that captures rhythmical changes in illumination. The performance of the color features is generally noticed as weak, while some interesting effects on chromatic and rhythmic features in the audio-visual domain are observed.

Future work on music videos will extend the semantic space to include texture, local features and object detection. Results are expected to provide information about how appropriate these methods are to solve MIR problems and how they can be used to connect the audio with the visual domain to facilitate new scenarios such as query-by-image.

References

1. Crete, F., et al.: The blur effect: Perception and estimation with a new no-reference perceptual blur metric. In: Electronic Imaging 2007, p. 64920 (2007)
2. Datta, R., Joshi, D., Li, J., Wang, J.Z.: Studying aesthetics in photographic images using a computational approach. In: Leonardis, A., Bischof, H., Pinz, A. (eds.) ECCV 2006, Part III. LNCS, vol. 3953, pp. 288–301. Springer, Heidelberg (2006)
3. Fu, Z., Lu, G., Ting, K.M., Zhang, D.: A survey of audio-based music classification and annotation. IEEE Transactions on Multimedia 13(2), 303–319 (2011)
4. Gillet, O., Essid, S., Richard, G.: On the correlation of automatic audio and visual segmentations of music videos. IEEE Trans. on Circuits and Sys. for Video Tech. (2007)
5. Hanbury, A.: Circular statistics applied to colour images. In: 8th Computer Vision Winter Workshop, vol. 91, pp. 53–71. Citeseer (2003)
6. Itten, J., Van Haagen, E.: The art of color: The subjective experience and objective rationale of color. Van Nostrand Reinhold, New York (1973)
7. Lidy, T., Rauber, A.: Evaluation of feature extractors and psycho-acoustic transformations for music genre classification. In: ISMIR (2005)
8. Machajdik, J., Hanbury, A.: Affective image classification using features inspired by psychology and art theory. In: Proc. Int. Conf. on Multimedia, pp. 83–92 (2010)
9. Manjunath, B.S., Ohm, J.-R., Vasudevan, V.V., Yamada, A.: Color and texture descriptors. IEEE Trans. on Circuits and Sys. for Video Tech. 11(6), 703–715 (2001)
10. Plataniotis, K.N., Venetsanopoulos, A.N.: Color image proc. and applications (2000)
11. Rubner, Y., Tomasi, C., Guibas, L.J.: The earth mover's distance as a metric for image retrieval. International Journal of Computer Vision 40(2), 99–121 (2000)
12. Schettini, R., Ciocca, G., Zuffi, S., et al.: A survey of methods for colour image indexing and retrieval in image databases. In: Color Imaging Science: Exploiting Digital Media (2001)
13. Schindler, A., Rauber, A.: A music video information retrieval approach to artist identification. In: 10th Symp. on Computer Music Multidisciplinary Research (2013)
14. Tzanetakis, G., Cook, P.: Marsyas: A framework for audio analysis. Organised Sound (2000)
15. Valdez, P., Mehrabian, A.: Effects of color on emotions. Journal of Experimental Psychology: General 123(4), 394 (1994)
16. Vedaldi, A., Soatto, S.: Quick shift and kernel methods for mode seeking. In: Forsyth, D., Torr, P., Zisserman, A. (eds.) ECCV 2008, Part IV. LNCS, vol. 5305, pp. 705–718. Springer, Heidelberg (2008)
17. Wei-ning, W., Ying-lin, Y., Sheng-ming, J.: Image retrieval by emotional semantics: A study of emotional space and feature extraction. In: IEEE International Conference on Systems, Man and Cybernetics (2006)
18. Wildenauer, H., Blauensteiner, P., Hanbury, A., Kampel, M.: Motion detection using an improved colour model. In: Bebis, G., et al. (eds.) ISVC 2006. LNCS, vol. 4292, pp. 607–616. Springer, Heidelberg (2006)
19. Yazdani, A., Kappeler, K., Ebrahimi, T.: Affective content analysis of music video clips. In: Music Information Retrieval with User-Centered and Multimodal Strategies (2011)
20. Zhang, S., Huang, Q., Jiang, S., Gao, W., Tian, Q.: Affective visualization and retrieval for music video. IEEE Transactions on Multimedia 12(6), 510–522 (2010)
21. Zuiderveld, K.: Contrast limited adaptive histogram equalization. In: Graphics Gems IV, pp. 474–485. Academic Press Professional, Inc. (1994)

Multi-modal Correlated Centroid Space for Multi-lingual Cross-Modal Retrieval

Aditya Mogadala and Achim Rettinger

Institute AIFB, Karlsruhe Institute of Technology, Germany
{aditya.mogadala,rettinger}@kit.edu

Abstract. We present a novel cross-modal retrieval approach where the textual modality is present in different languages. We retrieve semantically similar documents across modalities in different languages using a correlated centroid space unsupervised retrieval (C^2SUR) approach. C^2SUR consists of two phases. In the first phase, we extract heterogeneous features from a multi-modal document and project it to a correlated space using kernel canonical correlation analysis (KCCA). In the second phase, correlated space centroids are obtained using clustering to retrieve cross-modal documents with different similarity measures. Experimental results show that C^2SUR outperforms the existing state-of-the-art English cross-modal retrieval approaches and achieve similar results for other languages.

1 Introduction

Digital items often comprise different modalities represented by text, image, video or an audio. Sometimes one or more modalities represent a multi-modal document as found in on-line news articles. They are either embedded with a video or an image along with the text in different languages. Figure 1 show images[1] taken from news articles describing the same incident written in English[2], German[3] and Spanish[4] respectively. Similar multi-modal articles are also found in blogs, social networks, Wikipedia and personal websites.

Fig. 1. Images from three semantically related news articles written in different languages

[1] Images are of different resolution.
[2] `http://bit.ly/1AUcpqG`
[3] `http://bit.ly/1rA3kCq`
[4] `http://bit.ly/VQDo6K`

A. Hanbury et al. (Eds.): ECIR 2015, LNCS 9022, pp. 68–79, 2015.

Mining multi-modal documents poses numerous challenges. In the recent years, multimedia and computer vision communities published considerable research in bridging the gap between modalities to facilitate cross-modal applications [1]. Their research aims to address the problems of automatic image tagging with class labels [3], usage of image queries for text retrieval [4] or vice-versa. From the old Chinese proverb and its interpretations [8], we understand that "A picture is worth 10,000 words". This principle has been well adopted by existing multi-modal learning [9,11] approaches for cross-modal retrieval. They combine visual information with text for both image and text based retrievals.Other cross-modal approaches [5] leverage other modalities like video and audio. But, most of the work pertaining to text is limited to English.

Similarly, natural language processing(NLP) and information retrieval(IR) communities which work on different cross-lingual applications [2] concentrate only on text and diminish the importance of other modalities present in a multi-modal document. Also, some of the cross-language retrieval systems are highly dependent on transliteration or translation tools [6] and support only keyword based queries.

In this paper, we want to tackle the problem of cross-modal retrieval in a multilingual setting. We aim to design a cross-modal retrieval approach which is invariant to the languages present in a multi-modal document. Something similar to our work was done by Wu [10] using cross-lingual news stories to identify novelty and redundancy with visual duplicates in videos. Our contributions can be broadly summarized as:

- Designed a novel approach to link text in multiple languages with visual content and vice-versa to facilitate multi-lingual cross-modal retrieval.
- Extended an existing dataset [5] to multiple languages to facilitate multi-lingual cross-modal research.
- Empirical evidence showed that C^2SUR outperforms existing state-of-the-art mono-lingual (English) cross-modal retrieval approaches.

The remainder of this paper is organized into the following sections. Related work is mentioned in section 2. The section 3 presents the research question and describes our approach to perform unsupervised cross-modal retrieval. The experimental setup, dataset and evaluation metrics used for the approach are described in section 4. The section 5 details the experiments performed on different languages, while results are analyzed in section 6. Conclusion and future work is discussed in section 7.

2 Related Work

Several approaches have been proposed in bridging modalities with joint dimensionality reduction approaches [9,11] using extended CCA with semantic class

[5] `http://www.svcl.ucsd.edu/projects/crossmodal/`

labels. Some approaches formulate an optimization problem [12] where correlation between modalities is found by separating the classes in their respective feature spaces. As cross-modal data involves heterogeneous features, most of the approaches [14] aim in learning these features implicitly without any external representation. Zhai [13] focus on joint representation of multiple media types using joint representation learning which incorporates sparse and graph regularization. We use KCCA for maximizing pair-wise correlation between different media as Blaschko [15] used for correlational spectral clustering.

3 Approach

In this section, we formulate our research question formally and present our approach.

3.1 Problem Formulation

As discussed in the previous section, multi-modal documents on the web are found in the form of pair-wise modalities. Sometimes, there can be multiple instances of modalities present in the documents. To reduce the complexity, we assume a multi-modal document $D_i = (Text, Media)$ to contain a single media item either an image, video or audio embedded with a text description. A collection $C_j = \{D_1, D_2...D_i...D_n\}$ of these documents in different languages $L = \{L_{C_1}, L_{C_2}...L_{C_j}...L_{C_m}\}$ are spread across web. Formally, our research question is to find a cross-modal semantically similar document across language collections L_{C_o} using unsupervised similarity measures on low-dimension correlation space representation. Figure 2 shows broad visualization of the approach.

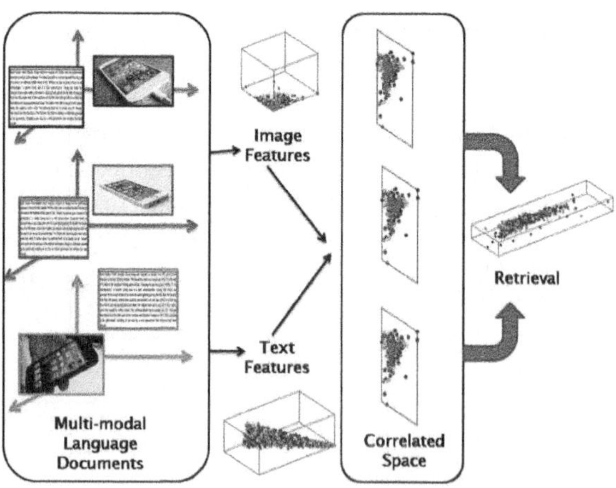

Fig. 2. Correlated Space Retrieval

3.2 Background: CCA

We build a low-dimension correlated space vectors of two different modalities using canonical correlation analysis(CCA) [16]. Given any two sets of multivariate random variables $T \in R^{d_t}$ and $I \in R^{d_m}$ representing text and image modality respectively, CCA aims to find the projection vectors $U \in R^{d_1}$ and $V \in R^{d_1}$ such that T and I are highly correlated in the projected space. The transformation can be visualized in the Equation 1.

$$(T, I) \rightarrow (UT, VI) \tag{1}$$

where UT represents the text projection, while VI represents an image projection. In order to maximize this correlation, we build an optimization function using Equation 2 with certain constraints as shown in Equation 3. We can observe that optimization function is invariant to scaling. Also projections are constrained to unit variance [17].

$$(correlation) = \underset{U,V}{\arg\max} \frac{U' \Sigma_{tm} V}{\sqrt{U' \Sigma_{tt} U} \sqrt{V' \Sigma_{mm} V}} \tag{2}$$

$$(correlation) = \underset{U' \Sigma_{tt} U = V' \Sigma_{mm} V = 1}{\arg\max} U' \Sigma_{tm} V \tag{3}$$

where Σ_{tt} represent the covariance matrix of a text modality and Σ_{mm} represent the covariance matrix of a image modality; while Σ_{tm} is cross-covariance matrix between text and image modalities. Equation 2 is solved using a generalized eigenvalue problem to maximize the correlation by learning projection vectors U and V given by Equation 4 and Equation 5 respectively. Here, λ represent an eigenvalue.

$$\Sigma_{tt}^{-1} \Sigma_{tm} \Sigma_{mm}^{-1} \Sigma_{mt} U = \lambda^2 U \tag{4}$$

$$\Sigma_{mm}^{-1} \Sigma_{tm} \Sigma_{tt}^{-1} \Sigma_{tm} V = \lambda^2 V \tag{5}$$

3.3 Background: Kernel CCA (KCCA)

Kernelization of CCA is helpful in finding the correlation between non-linear relationships [18]. Given any two sets of multivariate random variables $T \in R^{d_t}$ and $I \in R^{d_m}$ representing text and image modalities respectively. We find the kernel functions $K_T = k_T(t_i, t_j)$ and $K_I = k_I(m_i, m_j)$, such that K_T,$K_I \in R^{n \times n}$ are both positive semi-definite kernel matrices. To find the correlation between the transformed kernel matrices, we follow the similar optimization approach as of CCA given by Equation 6 and Equation 7.

$$(correlation) = \underset{X,Y}{\arg\max} \frac{X' K_T K_I Y}{\sqrt{X' K_T^2 X} \sqrt{Y' K_I^2 Y}} \tag{6}$$

$$(correlation) = \underset{X' K_T^2 X = Y' K_I^2 Y = 1}{\arg\max} X' K_T K_I Y \tag{7}$$

where $X \in R^{d_2}$ and $Y \in R^{d_2}$ are the projected vectors of T and I respectively in the projected correlated space.

3.4 Correlated Space Unsupervised Retrieval (CSUR)

Correlated low-dimension space of heterogeneous features obtained using KCCA is now used to find semantically similar cross-modal documents using different unsupervised similarity measures. Lots of similarity measures like cosine similarity, normalized correlation, minkowski distance, etc. have been well adopted for clustering and other semantic similarity tasks. We use 5 of these similarity measures mainly cosine, correlation, minkowski, mahalanobis and chebyshev for correlated space unsupervised retrieval (CSUR).

3.5 Correlated Centroid Space Unsupervised Retrieval (C²SUR)

In this approach, we modify the correlated space of text and image training documents. Correlated low-dimension space of each text and image sample is replaced with its closest centroids obtained using k-means clustering.

Let $m_T = \{m_{T_1}...m_{T_k}\}$ and $m_I = \{m_{I_1}...m_{I_k}\}$ denote the initial k centroids for the correlated text and image space respectively. Iterating over the samples of the training data, we perform assignment and update steps to obtain final k centroids. The assignment step assigns the each observed sample to its closest mean, while the update step calculates the new means that will be a centroid.

Correlated low-dimension space of text and image samples of the training data is given by CS_{Tr_T} and CS_{Tr_I} respectively. Choice of k is dependent on number of classes in the training data, while p represents the total training samples. $S_{T_i}^{(t)}$ and $S_{I_i}^{(t)}$ denote new samples of text and image modalities assigned to its closest mean. Algorithm 1 lists the procedure. Now the modified feature space is used for cross-modal retrieval similar to CSUR.

Algorithm 1. Correlated Centroid Space

Require: $CS_{Tr_T} = x_{T_1}...x_{T_p}$, $CS_{Tr_I} = x_{I_1}...x_{I_p}$
Ensure: $p > 0$ {**Output**: Final K-Centroids}
Assignment Step:
$S_{T_i}^{(t)} = x_{T_j} : ||x_{T_j} - m_{T_i}|| \leq ||x_{T_j} - m_{T_{i*}}|| \forall i^* = 1...k$
$S_{I_i}^{(t)} = x_{I_j} : ||x_{I_j} - m_{I_i}|| \leq ||x_{I_j} - m_{I_{i*}}|| \forall i^* = 1...k$
Update Step:
$$m_{T_i}^{(t+1)} = \frac{\sum_{x_{T_j} \in S_{T_i}^{(t)}} x_{T_j}}{|S_{T_i}^{(t)}|}, \quad m_{I_i}^{(t+1)} = \frac{\sum_{x_{I_j} \in S_{I_i}^{(t)}} x_{I_j}}{|S_{I_i}^{(t)}|}$$

4 Experimental Setup

In this section, we provide details about the dataset that is used and created to perform the experiments. Also, we describe features that are extracted from text and image modalities to learn a correlated space. It is then followed by methods used to evaluate the approach.

4.1 Dataset Creation

We used Wiki dataset[6] created for English texts and images using Wikipedia's featured articles. It has 2866 documents containing selected text paragraph and image pairs belonging to 10 semantic categories taken from art, biology, sport etc. We expanded the dataset into two more languages, mainly German and Spanish, using the Yandex machine translation API[7], while keeping the original images for every language. Thus, the expanded dataset consists of text and image pairs in three different languages.

We relied on machine translation, as it is the most efficient way to create such a corpus.[8]

4.2 Feature Extraction

Features extracted from the dataset provide a representation of information distribution in text or image. For the text, we used polylingual topic models (PTM) [19] to extract features as a distribution of topics in multiple languages. We leveraged the large collections that have interlingual connections like Wikipedia to train the PTM across languages. A trained PTM model on Wikipedia provides the same topic distribution on English, German and Spanish. We have trained PTM model for 10, 100, and 200 topics using the text of around 250k wikipedia articles in each language. The concentration parameter α is initialized to 1T. Using the training and testing parts of our dataset, each text document is represented as 10, 100 and 200 dimension topic distribution vectors.

Similarly, each image is represented as 128-dimension SIFT descriptor histograms as used in earlier works [9,11].

4.3 Evaluation

We evaluated cross-modal retrieval using mean average precision (MAP) [9,11] and mean reciprocal rank (MRR) scores. Experiments were repeated 10 times with different combinations of training and testing data to reduce selection bias. We used the same split as in Rasiwasia [9] for all languages to create 2173 training documents and 693 testing documents.

5 Experiments

Using the datasets created for different languages, we segregate the tasks and evaluate them separately. First, we see the MAP and MRR scores obtained for text and image queries using 10 text topics and 128-dimension SIFT descriptor histograms. Then, we show the variation in MAP scores when changing the number of topics.

[6] `http://www.svcl.ucsd.edu/projects/crossmodal/`

[7] `http://api.yandex.com/translate/`

[8] Please note, that the approach is invariant to machine translation and capable of *cross-lingual* cross-modal retrieval.

5.1 Text Query - Image Retrieval

We used the text queries from testing data to find semantically similar images present in training data. Text from testing data is projected into correlated space of images and text present in training data to retrieve images belonging to the same semantic category. Table 1 and Table 2 shows the results[9] obtained for English, German and Spanish using CCA, Polynomial kernel with degree 2(poly-2) CCA and RBF kernel CCA with CSUR and C^2SUR approach respectively.

Table 1. Text Query - Image Retrieval **(CSUR)**

Text Query-Image Retrieval(Method)		MAP	MRR
English	CCA-Mahalanobis	0.224 ± 0.002	0.241 ± 0.001
	(Poly-2)CCA-Correlation	0.233 ± 0.001	0.247 ± 0.002
	(RBF)CCA-Correlation	**0.235 ± 0.005**	**0.250 ± 0.003**
German	CCA-Cosine	0.219±0.003	0.242 ± 0.002
	(Poly-2)CCA-Chybyshev	**0.256 ± 0.001**	**0.308 ± 0.002**
	(RBF)CCA-Correlation	0.246 ± 0.003	0.272 ± 0.001
Spanish	CCA-Cosine	0.208 ± 0.002	0.223 ± 0.001
	(Poly-2)CCA-Cosine	**0.249 ± 0.002**	**0.283 ± 0.003**
	(RBF)CCA-Correlation	0.229 ± 0.002	0.249 ± 0.003

Table 2. Text Query - Image Retrieval **(C^2SUR)**

Text Query-Image Retrieval(Method)		MAP	MRR
English	CCA-Correlation	0.245 ± 0.003	0.273 ± 0.002
	(Poly-2)CCA-Chebyshev	0.245 ± 0.002	0.259 ± 0.001
	(RBF)CCA-Correlation	**0.262 ± 0.003**	**0.277 ± 0.001**
German	CCA-Correlation	0.215 ± 0.001	0.246 ± 0.002
	(Poly-2)CCA-Correlation	**0.263 ± 0.003**	**0.265 ± 0.002**
	(RBF)CCA-Chebyshev	0.226 ± 0.002	0.255 ± 0.003
Spanish	CCA-Chebyshev	0.230 ± 0.003	0.255 ± 0.002
	(Poly-2)CCA-Chebyshev	0.259 ± 0.002	0.267 ± 0.001
	(RBF)CCA-Correlation	**0.268 ± 0.002**	**0.268 ± 0.002**

For the text query, we performed "unpaired t-test" between best performing methods of CSUR and C^2SUR for testing statistical significance. The two-tailed P value is less than 0.0001 for all languages, which is considered to be extremely statistically significant.

5.2 Image Query - Text Retrieval

Images from testing data is projected into common space of images and text present in training data to retrieve text belonging to same semantic category.

[9] Tables show only those similarity measures which obtained best results for each of the given kernels.

Table 3. Image Query - Text Retrieval **(CSUR)**

Image Query-Text Retrieval(Method)		MAP	MRR
English	CCA-Minkowski	0.241 ± 0.002	0.263 ± 0.001
	(Poly-2)CCA-Correlation	0.239 ± 0.002	0.256 ± 0.002
	(RBF)CCA-Mahalanobis	**0.273 ± 0.003**	**0.311 ± 0.002**
German	CCA-Mahalanobis	0.219 ± 0.001	0.233 ± 0.002
	(Poly-2)CCA-Minkowski	**0.282 ± 0.001**	**0.275 ± 0.001**
	(RBF)CCA-Mahalanobis	0.248 ± 0.002	0.271 ± 0.001
Spanish	CCA-Chebyshev	0.220 ± 0.002	0.234 ± 0.001
	(Poly-2)CCA-Cosine	**0.238 ± 0.001**	**0.257 ± 0.003**
	(RBF)CCA-Cosine	0.225 ± 0.004	0.238 ± 0.002

Table 4. Image Query - Text Retrieval **(C^2SUR)**

Image Query-Text Retrieval(Method)		MAP	MRR
English	CCA-Chebyshev	0.253 ± 0.002	0.257 ± 0.003
	(Poly-2)CCA-Chebyshev	**0.273 ± 0.002**	**0.293 ± 0.002**
	(RBF)CCA-Chebyshev	0.263 ± 0.003	0.287 ± 0.002
German	CCA-Chebyshev	0.226 ± 0.003	0.252 ± 0.002
	(Poly-2)CCA-Minkowski	0.231 ± 0.001	0.241 ± 0.002
	(RBF)CCA-Correlation	**0.284 ± 0.002**	**0.274 ± 0.001**
Spanish	CCA-Minkowski	**0.250 ± 0.001**	**0.284 ± 0.002**
	(Poly-2)CCA-Correlation	0.231 ± 0.003	0.258 ± 0.002
	(RBF)CCA-Chebyshev	0.219 ± 0.002	0.244 ± 0.003

Table 3 and Table 4 shows the results[10] obtained for English, German and Spanish using CCA, Polynomial kernel with degree 2(poly-2) CCA and RBF kernel CCA with CSUR and C^2SUR approach respectively. For the image query, "unpaired t-test" between best performing methods of CSUR and C^2SUR showed that two-tailed P value equals 0.0111 for German and less than 0.0001 for Spanish. Although, there was no significant improvement for English. Topic distribution of text can show influence on the cross-modal retrieval. To apprehend it, we evaluated C^2SUR approach on various kernels with different topic distributions. Figure 3, Figure 4 and Figure 5 shows average of MAP scores obtained for text and image queries using different similarity measures.

5.3 Cross-Modal Retrieval Comparison

Most of the earlier works [9,12,11] performed cross-modal experiments only on English text with 10-topics and 128-dimension SIFT image features. We compared the best methods of CSUR and C^2SUR with the existing approaches[11].

[10] Tables only show those similarity measures which obtained best results for each of the given kernels.

[11] Cluster-CCA [11] and Cluster-KCCA [11] approaches are not directly comparable with ours. They compare the cluster labels of instances, while we compare the original semantic category labels.

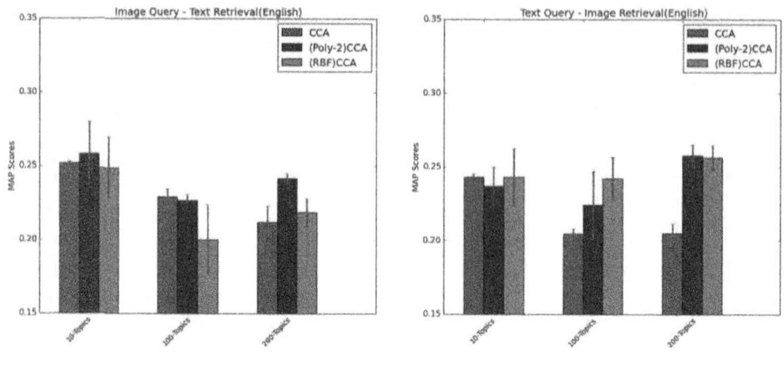

Fig. 3. English-C^2SUR

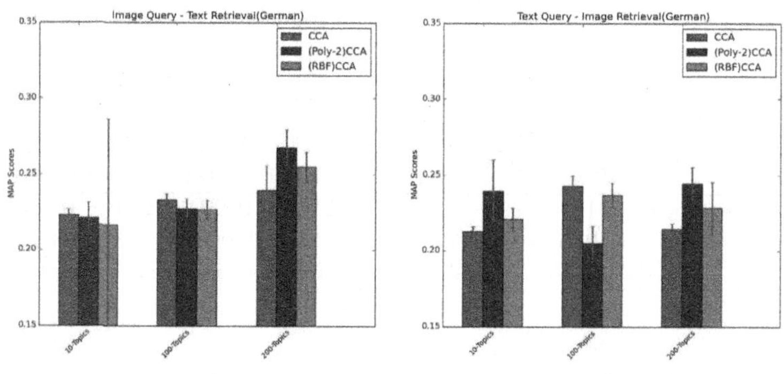

Fig. 4. German-C^2SUR

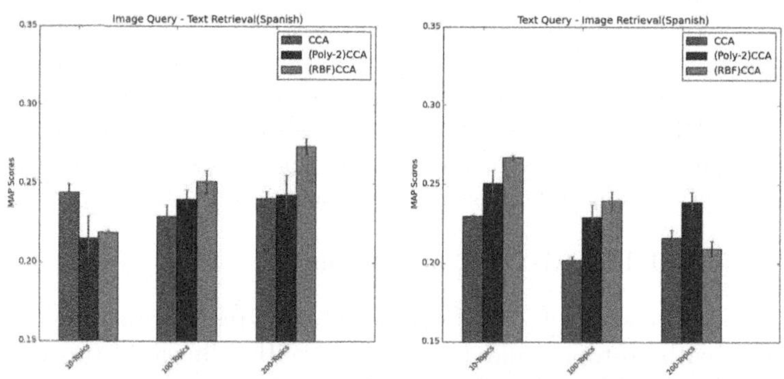

Fig. 5. Spanish-C^2SUR

Table 5 shows the comparison on text and image queries for English, German and Spanish on the Wiki dataset. We show the best MAP scores for CSUR and C^2SUR for German and Spanish with different topic variations. For Example,

Table 5. Text and Image Query Comparison **(Wiki)**

(Language)System		**Image Query**	**Text Query**	**Average (MAP)**
English	SM [9]	0.225	0.223	0.224
	Mean-CCA [11]	0.246 ± 0.005	0.194 ± 0.005	0.220 ± 0.005
	SCDL [20]	0.252	0.198	0.225
	SliM^2 [21]	0.255	0.202	0.229
	GMLDA [12]	0.272	0.232	0.252
	CSUR-10	**0.273 ± 0.003**	0.235 ± 0.005	0.254 ± 0.004
	C^{2}SUR-10	**0.273 ± 0.002**	**0.262 ± 0.003**	**0.268 ± 0.003**
German	CSUR-10	0.282 ± 0.001	0.256 ± 0.001	0.269 ± 0.001
	CSUR-100	0.230 ± 0.002	0.242 ± 0.004	0.236 ± 0.003
	CSUR-200	0.240 ± 0.002	0.243 ± 0.004	0.241 ± 0.003
	C^{2}SUR-10	**0.284 ± 0.002**	**0.263 ± 0.003**	**0.276 ± 0.003**
	C^{2}SUR-100	0.236 ± 0.004	0.250 ± 0.008	0.243 ± 0.006
	C^{2}SUR-200	0.278 ± 0.002	0.253 ± 0.002	0.266 ± 0.002
Spanish	CSUR-10	0.238 ± 0.001	0.249 ± 0.002	0.244 ± 0.002
	CSUR-100	0.254 ± 0.003	0.236 ± 0.003	0.245 ± 0.003
	CSUR-200	0.259 ± 0.002	0.231 ± 0.002	0.245 ± 0.002
	C^{2}SUR-10	0.250 ± 0.001	**0.268 ± 0.002**	**0.259 ± 0.002**
	C^{2}SUR-100	0.258 ± 0.008	0.243 ± 0.004	0.251 ± 0.006
	C^{2}SUR-200	**0.267 ± 0.003**	0.244 ± 0.002	0.256 ± 0.003

CSUR-10 represent 10-topics. Please note, that the related work can only be applied to English text.

6 Result Analysis

In this section, we analyzed the results obtained using our proposed approaches to perform cross-modal retrieval.

Table 1 and Table 2 shows the results obtained using text queries for image retrieval with CSUR and C^{2}SUR approaches respectively. It can be inferred that kernel versions of CCA (KCCA) in both the approaches outperformed baseline CCA on MAP scores. Best performing kCCA used in CSUR and C^{2}SUR approaches had an average improvement of 0.029 and 0.034 respectively over baseline CCA in all languages. It shows the presence of non-linearity in the data. Also, the best approach in C^{2}SUR achieved an average improvement of 0.017 over the best approach of CSUR in all languages. It exhibits the efficiency of C^{2}SUR in eliminating the noisy information from the correlated space of text and image. Similar analysis can be performed on the image queries.

Table 3 and Table 4 shows the results obtained using image queries for text retrieval with CSUR and C^{2}SUR respectively. Alike to text query, best performing kCCA used in CSUR and C^{2}SUR approaches had an average improvement of 0.037 and 0.019 respectively over baseline CCA in all languages. Also, the best performing approach of C^{2}SUR attained an average improvement of 0.007 over best approach of CSUR in all languages.

Effect of text topic distribution on C^2SUR approach is evaluated with different text topic distributions and fixed 128-dimension SIFT image features. It can be observed from the Figure 3 that increase in number of topics can have a negative effect. Possible explanation is due to padding of zeros in the correlated space of training data to carry out similarity measures with the testing data. To negate this behavior, dimensions of the images also have to be increased with SIFT features.

We also compared our best performing approach with the existing approaches based on MAP scores for English cross-modal retrieval. Table 5 shows that C^2SUR outperforms existing approaches on the average MAP scores. We assume this is due to the ability of C^2SUR to efficiently reduce the error in correlation space by improving the classification of borderline samples. In addition, performance on German and Spanish was comparable to English in finding semantically similar documents across modalities.

7 Conclusion and Future Work

In this paper, we presented a novel approach C^2SUR to perform the cross-modal retrieval in multiple languages. We built a common space for the heterogeneous features of a multi-modal document using kernel correlation analysis(KCCA), which is further modified with K-Means centroids to retrieve similar documents. We found that C^2SUR is effective in finding semantically similar multi-modal documents across languages.

In future, we aim to extend the approach to more than two modalities or use other modalities like video and audio to perform cross-modal analysis. We also aim to perform experiments on large scale real world datasets to evaluate the scalability of approach.

Acknowledgments. The research leading to these results has received funding from the European Union Seventh Framework Programme (FP7/2007-2013) under grant agreement no. 611346.

References

1. Rafailidis, D., Manolopoulou, S., Daras, P.: A unified framework for multimodal retrieval. Pattern Recognition 46(12), 3358–3370 (2013)
2. Peters, C., Braschler, M., Clough, P.: Cross-Language Information Retrieval. Multilingual Information Retrieval, 57–84 (2012)
3. Moran, S., Lavrenko, V.: Sparse Kernel Learning for Image Annotation. In: Proceedings of International Conference on Multimedia Retrieval (2014)
4. Mishra, A., Alahari, K., Jawahar, C.V.: Image Retrieval using Textual Cues. In: IEEE International Conference on Computer Vision (ICCV) (2013)
5. Metze, F., Ding, D., Younessian, E., Hauptmann, A.: Beyond audio and video retrieval: Topic-oriented multimedia summarization. International Journal of Multimedia Information Retrieval 2(2), 131–144 (2013)

6. Shakery, A., Zhai, C.X.: Leveraging comparable corpora for cross-lingual information retrieval in resource-lean language pairs. Information Retrieval 16(1), 1–29 (2013)
7. Hassan, S., Mihalcea, R.: Cross-lingual semantic relatedness using encyclopedic knowledge. In: Proceedings of Empirical Methods in Natural Language Processing (EMNLP), pp. 1192–1201 (2009)
8. Larkin, J.H., Simon, H.A.: Why a diagram is (sometimes) worth ten thousand words. Cognitive Science 11(1), 65–100 (1987)
9. Rasiwasia, N., Pereira, J.C., Coviello, E., Doyle, G., Lanckriet, G.R., Levy, R., Vasconcelos, N.: A new approach to cross-modal multimedia retrieval. In: Proceedings of the International Conference on Multimedia, pp. 251–260 (2010)
10. Wu, X., Hauptmann, A.G., Ngo, C.-W.: Novelty detection for cross-lingual news stories with visual duplicates and speech transcripts. In: Proceedings of the 15th International Conference on Multimedia (2007)
11. Rasiwasia, N., Mahajan, D., Mahadevan, V., Aggarwal, G.: Cluster Canonical Correlation Analysis. In: Proceedings of the Seventeenth AISTATS, pp. 823–831 (2014)
12. Sharma, A., Kumar, A., Daume, H., Jacobs, D.: Generalized multiview analysis: A discriminative latent space. In: Computer Vision and Pattern Recognition (CVPR) (2012)
13. Zhai, X., Peng, Y., Xiao, J.: Learning Cross-Media Joint Representation with Sparse and Semi-Supervised Regularization. IEEE Journal (2013)
14. Zhai, X., Peng, Y., Xiao, J.: Effective Heterogeneous Similarity Measure with Nearest Neighbors for Cross-Media Retrieval. In: Schoeffmann, K., Merialdo, B., Hauptmann, A.G., Ngo, C.-W., Andreopoulos, Y., Breiteneder, C. (eds.) MMM 2012. LNCS, vol. 7131, pp. 312–322. Springer, Heidelberg (2012)
15. Blaschko, M.B., Lampert, C.H.: Correlational spectral clustering. In: Computer Vision and Pattern Recognition (CVPR) (2008)
16. Hotelling, H.: Relations between two sets of variates. Biometrika 28(3/4), 321–377 (1936)
17. Andrew, G., Arora, R., Bilmes, J., Livescu, K.: Deep Canonical Correlation Analysis. In: Proceedings of The 30th International Conference on Machine Learning, pp. 1247–1255 (2013)
18. Hardoon, D., Szedmak, S., Shawe-Taylor, J.: Canonical correlation analysis: An overview with application to learning methods. Neural Computation 16(12), 2639–2664 (2004)
19. Mimno, D., Wallach, H.M., Naradowsky, J., Smith, D.A., McCallum, A.: Polylingual topic models. In: Proceedings of Empirical Methods in Natural Language Processing (EMNLP), vol. 2, pp. 880–889 (2009)
20. Wang, S., Zhang, L., Liang, Y., Pan, Q.: Semi-coupled dictionary learning with applications to image super-resolution and photo-sketch synthesis. In: Computer Vision and Pattern Recognition (CVPR), pp. 2216–2223 (2012)
21. Zhuang, Y., Wang, Y., Wu, F., Zhang, Y., Lu, W.: Supervised coupled dictionary learning with group structures for multi-modal retrieval. In: Proceedings of 25th AAAI (2013)

A Discourse Search Engine Based on Rhetorical Structure Theory

Pascal Kuyten, Danushka Bollegala, Bernd Hollerit, Helmut Prendinger, and Kiyoharu Aizawa

National Institute of Informatics, 2-1-2 Hitotsubashi, Chiyoda-ku, Tokyo, 101-8430, Japan
The University of Tokyo, 7-3-1 Hongo, Bunkyo-ku, Tokyo, 113-8656, Japan
The University of Liverpool, Liverpool, L693BX, United Kingdom
pascal@kuyten.com, hollerit@gmail.com, helmut@nii.ac.jp

Abstract. Representing a document as a bag-of-words and using keywords to retrieve relevant documents have seen a great success in large scale information retrieval systems such as Web search engines. Bag-of-words representation is computationally efficient and with proper term weighting and document ranking methods can perform surprisingly well for a simple document representation method. However, such a representation ignores the rich discourse structure in a document, which could provide useful clues when determining the relevancy of a document to a given user query. We develop the first-ever *Discourse Search Engine* (DSE) that exploits the discourse structure in documents to overcome the limitations associated with the bag-of-words document representations in information retrieval. We use Rhetorical Structure Theory (RST) to represent a document as a discourse tree connecting numerous elementary discourse units (EDUs) via discourse relations. Given a query, our discourse search engine can retrieve not only relevant documents to the query, but also individual statements from those relevant documents that describe some discourse relations to the query. We propose several ranking scores that consider the discourse structure in the documents to measure the relevance of a pair of EDUs to a query. Moreover, we combine those individual relevance scores using a random decision forest (RDF) model to create a single relevance score. Despite the numerous challenges of constructing a rich document representation using the discourse relations in a document, our experimental results show that it improves the F-score in an information retrieval task. We publicly release our manually annotated test collection to expedite future research in discourse-based information retrieval.

1 Introduction

In a typical bag-of-words (BOW) approach to document representation, first a document is tokenized into a set of tokens (often unigrams or bigrams), next a pre-defined set of stop words is removed from the tokens, and finally the remainder of the tokens are used as index entries to build an inverted index. When a user of a search engine enters keywords (often one or two words) describing her information need, those keywords are matched against the inverted index, and matching documents are returned to the user. If the number of search results is large as in a typical web search engine, accurate ranking of search results, considering the relevance of a document to the user query,

A. Hanbury et al. (Eds.): ECIR 2015, LNCS 9022, pp. 80–91, 2015.

(a) A scientific paper published at a medical journal revealed that there is a high correlation between **coffee** drinking and **cancer**. The correlation between amount of coffee drank by a group of subjects and the cancers detected was statistically significantly.

(b) **Coffee** is a great drink if you want to stay up late. However, deprivation of sleep can lead to memory loss and even brain **cancer** in the long run.

Fig. 1. Two documents mentioning the two terms *coffee* and *cancer*. Document (a) describes an EVIDENCE discourse relation between the two sentences, whereas in document (b), there is a CONTRAST discourse relation between the two sentences. For a user who searches for evidence that supports the claim *coffee causes cancer*, document (a) is more relevant than (b).

becomes important. Although the BOW representation is attractive for its robustness and efficiency, which are indeed vital factors when considering the scale and the quality of the documents found on the Web, a natural question is *whether IR can benefit from linguistically rich document representations beyond the BOW approach?*. We address this question by proposing and evaluating a document representation method based on the discourse relational structure in a document.

Despite its simplicity and popularity, the BOW representation of documents in IR systems ignores the rich discourse structure embedded in the documents, which can provide useful clues when determining the relevance of a document to a user query. For example, consider the two documents shown in Fig. 1. A user who is interested in evidence that supports the claim *coffee causes cancer* would benefit from the document (a) than the document (b). However, the BOW representations for each document contain both words *coffee* and *cancer*. Consequently, a search engine that indexes documents represented as bags-of-words will be unable to differentiate the subtle differences of the relevancies of the documents to user queries.

Discourse theories such as the Rhetorical Structure Theory (RST) [14] represent a document using a set of discourse units, connected via a pre-defined set of discourse relations (e.g. ELABORATION, CONTRAST, and JUSTIFICATION). For example in Fig. 1, the two sentences in document (a) and (b) are connected respectively through ELABORATION and CONTRAST relations. An IR system that utilizes the discourse structure of a document will be able to rank document (a) higher than (b) for the query *coffee causes cancer*, thereby improving the user satisfaction. Discourse information has shown to improve performance in numerous related tasks in natural language processing (NLP) such as text summarization [13].

Despite the benefits to an IR system from a discourse-based representation of documents, building discourse-aware IR systems is a challenging task due to several reasons. First, accurately identifying the discourse relations in natural language texts is difficult. Discourse markers such as *however, but, contrastingly, therefore,* etc. can be ambiguous with respect to the discourse relations they express [7]. It is inadequate to classify discourse relations purely based on discourse markers, and discourse parsers that use more advanced NLP methods are required [5,6,9,10,12,19,22]. Second, not all types of natural language texts are amenable to discourse parsers. For example, unlike newspaper articles, scientific publications, or Wikipedia articles that are logically structured and proofread, most texts found on the Web do not possess a well-organized discourse structure. Third, relevance measures that capture the underlying discourse structure of documents are lacking. It is non-obvious as to which discourse relations are useful for IR. Fourth, there does not exist any benchmark test collections that are annotated with discourse information for IR. It is difficult to empirically evaluate the pros and cons of discourse-motivated IR systems at larger scales without having access to discourse-annotated test collections.

We propose *Discourse Search*, a novel search paradigm that goes beyond the simple BOW representations of documents and captures the rich discourse structure present in the documents. First, we segment each document into Elementary Discourse Units (EDUs). An EDU is defined as a single unit of discourse and can be either a clause, a single sentence, or a set of sentences. Next, we identify EDU pairs that have some discourse relations according to RST. Discourse relations proposed in RST are directional relations and distinguish the main and the subordinate EDUs involved, referred to as respectively the *nucleus* and the *satellite*. Finally, all EDUs are arranged into a single binary tree structure covering the entire document. We index each sub-tree consisting of a pair of EDUs and a discourse relation. During retrieval time, we match a user query against this index and return pairs of EDUs as search results to the user. In particular, our discourse search engine goes beyond document-level IR and can retrieve the exact statements from the relevant documents. This is particularly useful when a single document expresses various opinions about a particular topic.

Our contributions in this paper can be summarized as follows.

- We develop a *Discourse Search Engine* (DSE) that considers the discourse structure present in documents to measure the relevance to a given user query. To our knowledge, ours is the first-ever IR system that uses RST to build a DSE.
- We propose three discourse proximity scores to measure the relevance of a pair of EDUs to a user query, considering the discourse structure in a document. Moreover, we learn the optimal combination of those three scores using random decision forests.
- We create a test collection annotated with discourse information to evaluate discourse-based IR systems. Specifically, for each test query, the created test collection contains a ranked list of EDU pairs indicating their relevance to the query. Considering the immense impact that test collections such as TREC benchmarks has had upon the progress of the research in IR, we publicly release the created test collection to expedite the future research in discourse-based IR.

2 Related Work

The use of discourse analysis as a tool for studying the interaction between a user and an IR system dates back to early 80's work of Brooks and Belkin. [3]. The task of retrieving information related to a particular information need is seldom a one-step process, and requires multiple interactions with the IR system. By analyzing this dialogue between a user and an IR system, we can improve the relevance of the retrieved search results. For example, by using search session data, it is possible to accurately predict the user intent [16]. Our work in this paper is fundamentally different from this line of prior work, because we are analyzing the discourse structure in the *documents* instead in the *dialgoues* between a user and a search engine.

Wang et al. [21] classified queries based on their discourse types and proposed a graph-based re-ranking method. In particular, they considered queries that describe an information need related to the advantages and disadvantages of a particular decision (e.g. *What are the advantages and disadvantages of same-sex schools?*). The relevance between a query and a document is measured using a series of proximity-based measures. However, unlike our work, they do not consider the discourse structure present in the documents. Moreover, our DSE is not limited to a particular type of discourse queries, and supports a wide-range of queries.

Using semantic relations that exist between entities in a document to improve IR has received wide-attention in the NLP community. For example, in Latent Relational Search [4], given the two entities *YouTube* and *Google* as the query, the objective is to retrieve other pairs of entities between which the same semantic relations exist. Here, the semantic relation ACQUISITION holds between *YouTube* and *Google*. Therefore, other such pairs of entities where one of the entities is acquired by the other such as, *Powerset* and *Microsoft* are considered as relevant search results. Latent relational search can be classified as an instance of analogical search, where the focus is on the semantic relations between the entities and not the entities themselves. Latent relational search engines represent the semantic relations between two entities using a vector of lexical pattern frequencies, and measure the relational similarity between two pairs of entities by the cosine similarity between the corresponding lexical pattern frequency vectors. Interestingly, this approach can be extended to cross-language relational search as well.

Miyao et al. [15] developed a search engine for Bio-medical IR by extracting the semantic relations that are common in the Bio-medical domain such as, the interaction between proteins, or side-effects of a drug. First, they apply a term extraction method to detect Bio-medical terms in the documents, and extract numerous features from an Head-driven Phrase Structure Grammar (HPSG) parse tree of a sentence. A bio-medical relation classifier is trained using the extracted features. Although semantic relations are useful as an alternative to the BOW representation, it is complementary to the discourse structure that we exploit in our DSE. Indeed, an interesting future research direction would be to combine both semantic relations and discourse relations to further improve the performance of IR systems.

3 Rhetorical Structure Theory

We briefly review Rhetorical Structure Theory (RST) [14] that defines the discourse structure that we use in our document representation. In RST, documents are segmented into non-overlapping elementary discourse units (EDUs). EDUs are related by a discourse relation, where the head EDU (*nucleus*), has a relation with the subordinate EDU (*satellite*). EDUs are arranged into a binary tree to create a *discourse tree* for a document. Directed edges of a discourse tree point from a satellite to a nucleus and are labeled with a discourse relation. In RST, nuclei and satellites may consist of single or multiple EDUs in the latter case, the individual EDUs are related by a path of

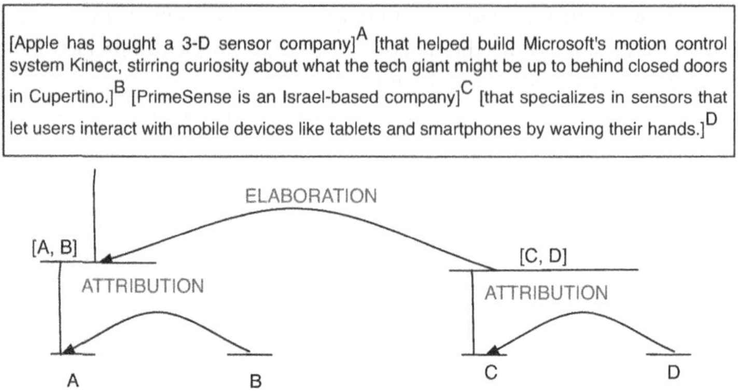

Fig. 2. A discourse tree covering four EDUs

discourse relations. An example of a discourse tree is shown in Fig. 2, covering four EDUs, where ATTRIBUTION relations exist between the two EDUs in each sentence, and an ELABORATION relation holds between the two sentences.

Discourse trees can be automatically generated using discourse parsers such as SPADE [18], or HILDA [8]. SPADE produces sentence level discourse structures, whereas a complete discourse tree covering all the sentences in a document can be generated using HILDA. Because our goal is to represent entire documents considering their discourse structures, we use HILDA as our preferred discourse parser. HILDA builds a single discourse tree by segmenting the document into EDUs using a Hidden Markov Model (HMM). Next, from these EDUs, a single discourse tree is built by discourse relation classification using Support Vector Machines (SVMs) [5]. HILDA's classifies 18 discourse relation types such as, ELABORATION, ATTRIBUTION, and CONTRAST.

4 Discourse Search Engine

Let us denote the discourse tree of a document d by $\mathcal{T}(d) = \{(\delta_n, \delta_s, r(\delta_n, \delta_s))\}$, where a discourse relation $r(\delta_n, \delta_s)$ holds between a nucleus δ_n and a satellite δ_s in the discourse tree. To simplify the notation, we will write r in place of $r(\delta_n, \delta_s)$, when it is clear from the context as to r holds between which two EDUs. For example, the document shown in Fig. 2 is represented by the set consisting of the three elements: $(A, B, \text{ATTRIBUTION})$, $(C, D, \text{ATTRIBUTION})$, and $([A, B], [C, D], \text{ELABORATION})$. Here, $[A, B]$ denotes the parent vertex of EDUs A and B. Given d, a discourse parser can be used to generate $\mathcal{T}(d)$.

Likewise, we define a discourse query Q as a three-valued tuple $(q_n, q_s, r(q_n, q_s))$, where a discourse relation $r(q_n, q_s)$ holds between the nucleus q_n and the satellite q_s of Q. For example, the query *coffee causes cancer* is mapped to the tuple (*coffee*, *cancer*, EVIDENCE). Information needs of a user can be mapped into a discourse query by several methods. Given a natural language input such as *coffee causes cancer*, a discourse parser can be used to generate a discourse query. Alternatively, we could train a sequence labeller such as a conditional random field [11], to extract the two EDUs and the discourse relation between them. A more manual approach would be to provide a search front end where a user can enter the nucleus, satellite and select a discourse relation from a drop-down list. The DSE we propose can be easily incorporated with all of those approaches.

We model the relevancy of a discourse query Q to a document d as a function $f(Q, d)$, which is the summation of the product between a **discourse relation selector**, $\phi(Q, \delta_n, \delta_s, r) \in [0, 1]$, and a **discourse proximity score**, $\psi(Q, \delta_n, \delta_s, \mathcal{T}(d)) \in [0, 1]$, over all EDU pairs in the discourse tree $\mathcal{T}(d)$ as follows

$$f(Q, d) = \sum_{(\delta_n, \delta_s, r) \in \mathcal{T}(d)} \phi(Q, \delta_n, \delta_s, r)\psi(Q, \delta_n, \delta_d, \mathcal{T}(d)). \quad (1)$$

For the document shown in Fig. 2, all possible combinations between δ_n and δ_s are listed in Table 2. Next, we will discuss each of those factors in detail.

A DSE must consider the agreement between the discourse relation $r(q_n, q_s)$ in the query, and the relation $r(\delta_n, \delta_s)$ between two EDUs δ_n, δ_s in the document. Moreover, not all words are equally significant when considering the relevance between a query and a document. For example, frequent non-content words are removed from the queries using a pre-defined stop-words list by most search engines, and term-weighting scores such as tfidf, or BM25 are used to detect salient matches. We propose discourse relation selector, $\phi(Q, \delta_n, \delta_s, r)$, as a function that captures those two requirements. It is defined as follows:

$$\phi(Q, \delta_n, \delta_s, r) = s(q_n, \delta_n)s(q_s, \delta_s)\mathrm{I}[r(q_n, q_s) \in l(\delta_n, \delta_s, \mathcal{T}(d))]. \quad (2)$$

Here, $s(w, \delta)$ is a salience score such as, tfidf or BM25 indicating the salience of a word w in an EDU δ in the discourse tree $\mathcal{T}(d)$, and $\mathrm{I}[r(q_n, q_s) \in l(\delta_n, \delta_s, \mathcal{T}(d))]$ is the indicator function which returns 1 if the discourse relation $r(q_n, q_s)$ between q_n and q_s appears in the discourse path $l(\delta_n, \delta_s, \mathcal{T}(d))$ between EDUs δ_n, δ_s in the document, and 0 otherwise. For example, the discourse path between EDUs A and C shown in Fig. 2 is $A \rightarrow [A, B] \xrightarrow{\text{ELABORATION}} [C, D] \leftarrow C$. It contains the ELABORATION discourse relation between A and C. In our experiments, we used tfidf as the salience score $s(w, \delta)$, and consider the occurrences of query words in the EDUs extracted from the documents. Because a single document can contain multiple topics, we found it is more accurate to compute tfidfs over EDUs than entire documents. Using the Porter's stemming algorithm[1] we perform stemming on the words in the documents before computing tfidf scores.

The location of the words used in the query in their appearance in the document is an important feature that influences the relevance of the document to the query. For example, if all the words used in the query appear within close proximity in the document, the higher is the relevance [1]. We adopt this intuition to discourse trees by proposing three types of discourse proximity scores for $\psi(Q, \delta_n, \delta_s)$ as we describe next.

The first of the three discourse proximity scores we propose, **segment proximity**, $\psi_{seg}(Q, \delta_n, \delta_s, \mathcal{T}(d))$, measures the distance between two EDUs δ_n, δ_s as the number of discourse segments (EDUs) that appear in between δ_n and δ_s in the document. The segment proximity is given by,

[1] http://tartarus.org/martin/PorterStemmer/

$$\psi_{seg}(Q, \delta_n, \delta_s, \mathcal{T}(d)) = 1 - \frac{|t(\delta_n, \mathcal{T}(d)) - t(\delta_s, \mathcal{T}(d))| - 1}{\mathbf{E}(d) - 2}. \quad (3)$$

Here, $t(\delta, \mathcal{T}(d))$ indicates the segment number (starting with 1 and counted from the beginning of the document) of the EDU δ in the discourse tree $\mathcal{T}(d)$, and $\mathbf{E}(d)$ denotes the total number of EDUs in the document. $\psi_{seg}(Q, \delta_n, \delta_s)$ is normalized by dividing from $\mathbf{E}(d)$ to remove any biases due to differences in document lengths. If two EDUs δ_n, δ_s are closer to each other in the document, the higher their segment proximity will be. For the example shown in Fig. 2, the four EDUs appear in the order $t(A, \mathcal{T}(d)) = 1$, $t(B, \mathcal{T}(d)) = 2$, $t(C, \mathcal{T}(d)) = 3$, and $t(D, \mathcal{T}(d)) = 4$ in the document text. Therefore, for example, $\psi_{seg}(Q, A, C) = 1 - ((|1 - 3| - 1)/(4 - 2)) = 1/2$.

Two EDUs that appear in distant locations on the surface text of a document, might have a direct discourse relation between them. Such EDUs appear close together on the discourse tree, despite being located far apart on the surface text of the document. The segment proximity would assign a low score for such related EDUs because it only considers the surface text and ignores the discourse tree structure. We propose **path proximity**, $\psi_{path}(Q, \delta_n, \delta_s, \mathcal{T}(d))$, as a measure that computes the closeness between two EDUs δ_n, δ_s over the discourse tree $\mathcal{T}(d)$ by the length of the discourse path $l(\delta_n, \delta_s, \mathcal{T}(d))$ that connects δ_n to δ_s. Specifically, path proximity is given by,

$$\psi_{path}(Q, \delta_n, \delta_s, \mathcal{T}(d)) = 1 - \frac{|l(\delta_n, \delta_s, \mathcal{T}(d)| - 1}{\log_2 \mathbf{E}(d)}. \quad (4)$$

Here, $|l(\delta_n, \delta_s, \mathcal{T}(d)|$ denotes the length of the discourse path connecting δ_n to δ_s, and is measured by the number of discourse relations (ignoring the directions) along the discourse path. For example, the discourse path between EDUs A and C shown in Fig. 2, $A \rightarrow [A, B] \xrightarrow{\text{ELABORATION}} [C, D] \leftarrow C$, contains one discourse relation, ELABORATION, resulting in a length of 1. The $\log_2 \mathbf{E}(d)$ term in the denominator is the diameter of the discourse tree (i.e. maximum distance between any two vertices), and is derived using the property that discourse trees are binary trees. For example, the path proximity $\psi_{path}(Q, A, C, \mathcal{T}(d))$ between A and C is computed as,

$$\psi_{path}(Q, A, C, \mathcal{T}(d)) = 1 - \frac{1 - 1}{\log_2 4} = 1.$$

The first occurrence of an entity in a document often contains its definition. For example, in news text summarization, the first sentence baseline where the first sentence (also known as the lead sentence) is used as the summary of the document [13]. We translate this heuristic to measure the relevance of a query to a discourse tree by considering the shortest segment distance from the first EDU to the two discourse units δ_n and δ_s that contain respectively q_n and q_s. We refer to this relevance score as the **Lead EDU Proximity** score, which is given by,

$$\psi_{lead}(Q, \delta_n, \delta_s, \mathcal{T}(d)) = 1 - \frac{\min\left(t(\delta_n, \mathcal{T}(d)), t(\delta_s, \mathcal{T}(d))\right) - 1}{\mathbf{E}(d) - 2}. \quad (5)$$

Similar to the segment proximity, we normalize the lead EDU proximity by dividing from the number of EDUs in the discourse tree to remove any bias due to the differences in document lengths. As an example, we compute the lead EDU proximity, $\psi_{lead}(Q, A, C, \mathcal{T}(d))$, between the two EDUs A and C in Fig. 2 as,

$$\psi_{lead}(Q,A,C,\mathcal{T}(d)) = 1 - \frac{\min(t(A,\mathcal{T}(d)),t(C,\mathcal{T}(d)) - 1}{4-2} = 1 - \frac{\min(1,3)-1}{2} = 0.$$

Recall that the overall relevance of a query Q to a document d is given by Equation 1 as the sum over the product of discourse relation selector, $\phi(Q,\delta_n,\delta_s,r)$, and each one of the three discourse proximity scores, $\psi(Q,\delta_n,\delta_s,\mathcal{T}(d))$. If there are no matching discourse relations between the query and a pair of discourse units selected from the document, then $f(Q,d)$ will be zero. We can use this fact to speed up the computation of $f(Q,d)$ in Equation 1 by not computing the discourse proximity scores for EDU pairs δ_n,δ_s for which $\phi(Q,\delta_n,\delta_s,r)$ is zero.

4.1 Combining Different Discourse Proximity Scores

Although we proposed three different discourse proximity scores for computing the relevance between a discourse query and a document it is not obvious as to the optimal combination of those discourse proximity scores that gives the best relevancy model. We model the problem of learning the optimal combination of discourse proximity scores as a learning-to-rank problem. Specifically, using a manually labeled dataset that lists a set of relevant documents for a discourse query, we follow a pairwise rank learning approach and train a binary classifier to detect relevant query-document pairs (positive class) from the irrelevant ones (negative class). Each query-document pair (Q,d) is represented by a three-valued feature vector using the relevance scores $f(Q,d)$ computed using each discourse proximity score in turn. Next, a Random Decision Forest (RDF) [2] is trained using the ALGIB[2] tool. The posterior probability, $p(+1|(Q,d))$, indicating the degree of relevance of Q to d is used as the combined relevancy score for the purpose of ranking documents retrieved for a discourse query[3]. All parameters of the RDF classifier are set to their default values as specified in ALGLIB.

4.2 Indexing and Query Processing

To efficiently process discourse queries, we create two inverted indexes: (1) an inverted index between all distinct n-grams in EDUs and the EDU IDs (similar to document IDs (urls) in traditional IR systems, we assign each EDU a unique ID) of the EDUs in which those n-grams occur, (2) an inverted index between nuclei EDU IDs and their corresponding satellite EDU IDs paired with the corresponding discourse relations. For the document shown in Fig. 2, an excerpt of the first inverted index is shown in Table 1, whereas Table 2 shows the corresponding second inverted index. Given a user query Q, we match the terms in q_n and q_s against the first index to find the matching EDUs. Next, we use the second index to compute the discourse proximity scores. Finally, the set of EDUs that matches with the user query is ranked according to the relevance score $f(Q,d)$, computed using Equation 1 and returned to the user.

[2] `http://www.alglib.net/dataanalysis/decisionforest.php#header3`

[3] Similarly, in a multi-class classifier, the posterior probability for the most probable class can be used as the ranking score.

Table 1. Excerpt of the inverted index between n-grams and EDU IDs for the document in Fig. 2

Term	EDU ID
Apple	A
company	A, C
PrimeSense	C

Table 2. Inverted index between nucleus EDU IDs and their corresponding satellite EDU IDs with discourse relations

Nucleus EDU ID	(Satellite EDU ID, discourse relations)
A	(B, ATTRIBUTION), (C, ATTRIBUTION, ELABORATION), (D, ATTRIBUTION, ELABORATION)
B	(C, ELABORATION, ATTRIBUTION), (D, ATTRIBUTION, ELABORATION)
C	(D, ATTRIBUTION)

5 Evaluation

Evaluating an information retrieval system is a complex task involving numerous aspects such as, efficiency, accuracy, latency (indexing vs. query processing), scalability, and user satisfaction. Compared to keyword-based IR systems that have established evaluation measures and large test collections, discourse search is still in its early stages. To our knowledge, there does not exist an IR system that uses a document representation based on discourse relations, nor there exist benchmark test collections for discourse search. Therefore, an important contribution of our work is to create a test collection for evaluating discourse search engines for their accuracy. Section 5.1 describes the test collection we created for this purpose.

5.1 Dataset

We selected 10 online news articles covering news events related to major players in the IT industry such as (*Apple*, *Google*, *Microsoft*, *Facebook* and *Twitter*). We select major players in the IT industry to ensure our annotators, all graduate Computer Science students, would be familiar with the topic. Next, we generate a discourse tree, $\mathcal{T}(d)$, from each document d using the HILDA [8] discourse parser. Then, for each document we formulated a relevant query $Q(q_n, q_s, r(q_n, q_s))$ as (*main entity*, *related entity*, DISCOURSE RELATION). For example, a news article about *Microsoft* that introduces *Apple* as a competitor would result in the discourse query (*Microsoft*, *Apple*, ELABORATION). Finally, we extract multiple candidate EDU pairs (δ_n, δ_s) from each document that are connected by some discourse relation. HILDA segmented each document d into ca. 56 EDUs (min = 42, median = 57, max = 69), and ca. 6 candidate EDU pairs are selected from each $\mathcal{T}(d)$ (min = 4, median = 7, max = 10).

Six annotators individually read and rank 3 to 5 documents using a web interface during a 45 minute session. Documents were randomly distributed among the annotators, and we ensured each document was annotated by 3 to 5 annotators. The web

Table 3. Median values for discourse proximity scores

Grading	$\rho_{\mu-n}$	total no. of instances	ψ_{seg}	ψ_{path}	ψ_{lead}
$n = 2$ (irrelevant)	0	17	0.54	0.67	0.27
$n = 2$ (relevant)	1	40	0.09	0	0.35
$n = 4$ (irrelevant)	0	17	0.54	0.67	0.27
$n = 4$ (less relevant)	$\frac{1}{3}$	18	0.06	0.42	0.87
$n = 4$ (moderately relevant)	$\frac{2}{3}$	13	0.15	0	0.14
$n = 4$ (highly relevant)	1	9	0.25	0	0

Table 4. RDF performance for classifying EDU pairs for a query

Features	$F(n = 2)$	$F(n = 4)$
$\psi_{seg}, \psi_{path}, \psi_{lead}$	0.75	0 .60
ψ_{seg}, ψ_{path}	**0.77**	0.54
ψ_{seg}, ψ_{lead}	0.75	0.58
ψ_{path}, ψ_{lead}	0.65	**0.61**
ψ_{seg}	0.74	0.32
ψ_{path}	0.68	0.26
ψ_{lead}	0.70	0.49

interface first showed the instructions, then the annotators were asked to read a document. When an annotator clicked a button stating the document has been read, new instructions were presented. Next, a query and a set of candidate EDU pairs extracted from the document were presented. Annotators will mark an EDU pair as either relevant or irrelevant to a given query. Moreover, EDU pairs that are considered as relevant are further ordered according to their relevance to the query. Candidate EDU pairs were presented as complete sentences instead of segments by expanding the nucleus and the satellite to cover the entire sentences. For example, the EDU pair (A, C) in Fig. 2 is presented as (AB, CD) to the annotators. If both EDUs are in the same sentence only one sentence is presented. Our dataset is publicly available[4].

5.2 Results

Using the manually annotated dataset we created in Section 5.1, we evaluate the performance of the discourse proximity scores $\psi(Q, \delta_n, \delta_s, \mathcal{T}(d))$ we proposed in Section 4, by measuring the agreement between human annotations in the dataset and the relevance scores predicted by $f(Q, d)$. We denote the reciprocal of the rank given by the annotator a_i for the pair of EDUs (δ_n, δ_s), indicating its relevance to a query Q by $\pi(a_i, Q, \delta_n, \delta_s)$. The set of reciprocal ranks assigned by all annotators for a pair of EDUs (δ_n, δ_s) indicating its relevance to a query Q is denoted by $\rho(Q, \delta_n, \delta_s) = \{\forall_i | \pi(a_i, Q, \delta_n, \delta_s)\}$. We consider the majority vote, $\rho_\mu(Q, \delta_n, \delta_s)$, over the set of reciprocal ranks as the final relevance score of an EDU pair (δ_n, δ_s) to a query Q. Ties are resolved by selecting randomly between the majority reciprocal ranks. For example, if $\rho(Q, A, B) = \{\frac{1}{2}, \frac{1}{2}, 0\}$ then $\rho_\mu(Q, A, B) = \frac{1}{2}$. Considering the majority vote instead

[4] `http://t2d.globallabproject.net/files/ECIR15.zip`

of the arithmetic mean has shown to improve the reliability when aggregating human ratings in annotation tasks [17]. For each query Q, we normalize the $\rho_\mu(Q, \delta_n, \delta_s)$ values for all candidate EDUs (δ_n, δ_s) retrieved for Q to the range $[0, 1]$ by fitting a uniform distribution. For example, given four EDU pairs, (A, B), (C, D), (E, F), and (G, H) retrieved for a query Q, if an annotator a labeled (A, B) as irrelevant and ranked $(C, D) \prec (E, F) \prec (G, H)$ in the ascending order of their relevancy, then the normalized values of the four EDU pairs (A, B), (C, D), (E, F), and (G, H) will be respectively 0, $\frac{1}{3}$, $\frac{2}{3}$, and 1.

To measure median values for ψ_{seg}, ψ_{path} and ψ_{lead} over all candidate EDU pairs of all documents, we group instances (Q, δ_n, δ_s) into n categories of ρ_μ denoted by $\rho_{\mu-n}$. We consider two groups in particular: $n = 2$ (two-valued grading system indicating relevant vs. irrelevant instances), and $n = 4$ (four-valued grading system indicating irrelevant, less relevant, moderately relevant, and highly relevant instances). By considering a coarse two-valued grading and a more finer four-valued grading, we can evaluate the ability of the proposed discourse proximity scores to detect different granularities of relevancies. Table 3 shows the median values of the three discourse proximity scores. We see that for $n = 2$, the EDU pairs ranked as relevant have a smaller median ψ_{seg}, have a smaller median ψ_{path}, and have a smaller median ψ_{lead} in the document. This outcome mirrors the results from [20], where correlations have been found on proximity of query terms in text and document relevance. $n = 4$ case shows similar trends for ψ_{path} and ψ_{lead}. However, for ψ_{seg} we see that EDU pairs ranked as highly relevant have a larger median ψ_{seg} than EDU pairs ranked as less relevant.

We train an RDF classifier as described in Section 4.1, with the test collection as described in Section 5.1, using different combinations of discourse proximity scores as shown in Table 4. In $n = 2$ setting, we train a binary classifier, whereas a multi-class classifier is trained for the $n = 4$ setting. From the leave-one-out F scores shown in Table 4 we see that the combination of ψ_{seg} and ψ_{path} gives the best performance for the $n = 2$ setting, whereas the combination of ψ_{path} and ψ_{lead} gives the best performance for the $n = 4$ setting. In particular, path discourse proximity is found to be a useful feature for detecting relevancy in both settings, which supports our proposal to use discourse trees to represent documents in information retrieval systems.

6 Conclusion

We proposed a discourse search engine that considers the discourse structure in documents to measure the relevance between a query and a document. Three discourse proximity measures that capture different aspects of relevance within the context of a discourse tree were proposed. A random decision forest (RDF) was trained to combine the different discourse proximity scores. We create a test collection for evaluating discourse-based IR systems. Our experiments show the usefulness of the proposed discourse proximity measures. In future, we plan to incorporate the semantic relations between entities in documents within our discourse relevance model; and add TREC collections to the test collection to further improve its performance.

References

1. Baeza-Yates, R.A., Ribeiro-Neto, B.: Modern Information Retrieval. Addison-Wesley Longman Publishing Co., Inc., Boston (1999)
2. Breiman, L.: Random forests. Machine Learning 45(1), 5–32 (2001)
3. Brooks, H.M., Belkin, N.J.: Using discourse analysis for the design of information retrieval interaction mechanisms. In: SIGIR, pp. 31–47 (1983)
4. Duc, N.T., Bollegala, D., Ishizuka, M.: Cross-language latent relational search: Mapping knowledge across languages. In: AAAI, pp. 1237 – 1242 (2011)
5. duVerle, D.A., Prendinger, H.: A novel discourse parser based on support vector machine classification. In: ACL, pp. 665–673 (2009)
6. Feng, V.W., Hirst, G.: Text-level discourse parsing with rich linguistic features. In: ACL, pp. 60–68 (2012)
7. Hernault, H., Bollegala, D., Ishizuka, M.: A semi-supervised approach to improve classification of infrequent discourse relations using feature vector extension. In: EMNLP, pp. 399–409 (2010)
8. Hernault, H., Prendinger, H., duVerle, D., Ishizuka, M.: Hilda: A discourse parser using support vector machine classification. Dialogue and Discourse. An International Journal 1(3), 1–33 (2010)
9. Joty, S., Carenini, G., Ng, R.: A novel discriminative framework for sentence-level discourse analysis. In: EMNLP, pp. 904–915 (2012)
10. Joty, S., Carenini, G., Ng, R., Mehdad, Y.: Combining intra- and multi-sentential rhetorical parsing for document-level discourse analysis. In: ACL, pp. 486–496 (2013)
11. Lafferty, J., McCallum, A., Pereira, F.C.: Conditional random fields: Probabilistic models for segmenting and labeling sequence data (2001)
12. Lan, M., Xu, Y., Niu, Z.: Leveraging synthetic discourse data via multi-task learning for implicit discourse relation recognition. In: ACL, pp. 476–485 (2013)
13. Louis, A., Joshi, A., Nenkova, A.: Discourse indicators for content selection in summarization. In: SIGDIAL, pp. 147–156 (2010)
14. Mann, W.C., Thompson, S.A.: Rhetorical structure theory: Toward a functional theory of text organization. Text 8(3), 243–281 (1988)
15. Miyao, Y., Ohta, T., Masuda, K., Tsuruoka, Y., Yoshida, K., Ninomiya, T., Tsujii, J.: Semantic retrieval for the accurate identification of relational concepts in massive textbases. In: ACL, pp. 1017–1024 (2006)
16. Sadikov, E., Madhavan, J., Wang, L., Halevy, A.: Clustering query refinements by user intent. In: WWW, pp. 841–850 (2010)
17. Snow, R., O'Connor, B., Jurafsky, D., Ng, A.Y.: Cheap and fast - but is it good? evaluating non-expert annotations for natural language tasks. In: EMNLP, pp. 254–263 (2008)
18. Soricut, R., Marcu, D.: Sentence level discourse parsing using syntactic and lexical information. In: NAACL, pp. 149–156 (2003)
19. Subba, R., Eugenio, B.D.: An effective discourse parser that uses rich linguistic information. In: HLT-NAACL, pp. 566–574 (2009)
20. Tao, T., Zhai, C.: An exploration of proximity measures in information retrieval. In: SIGIR, pp. 295–302 (2007)
21. Wang, D.Y., Luk, R.W.P., Wong, K.F., Kwok, K.L.: An information retrieval approach based on discourse type. In: Kop, C., Fliedl, G., Mayr, H.C., Métais, E. (eds.) NLDB 2006. LNCS, vol. 3999, pp. 197–202. Springer, Heidelberg (2006)
22. Zhou, L., Li, B., Gao, W., Wei, Z., Wong, K.F.: Unsupervised discovery of discourse relations for eliminating intra-sentence polarity ambiguities. In: EMNLP, pp. 162–171 (2011)

Knowledge-Based Representation for Transductive Multilingual Document Classification

Salvatore Romeo[1], Dino Ienco[2,3], and Andrea Tagarelli[1]

[1] DIMES, University of Calabria, Italy
{sromeo,tagarelli}@dimes.unical.it
[2] IRSTEA, UMR TETIS, Montpellier, France
dino.ienco@irstea.fr
[3] LIRMM, Montpellier, France

Abstract. Multilingual document classification is often addressed by approaches that rely on language-specific resources (e.g., bilingual dictionaries and machine translation tools) to evaluate cross-lingual document similarities. However, the required transformations may alter the original document semantics, raising additional issues to the known difficulty of obtaining high-quality labeled datasets. To overcome such issues we propose a new framework for multilingual document classification under a transductive learning setting. We exploit a large-scale multilingual knowledge base, BabelNet, to support the modeling of different language-written documents into a common conceptual space, without requiring any language translation process. We resort to a state-of-the-art transductive learner to produce the document classification. Results on two real-world multilingual corpora have highlighted the effectiveness of the proposed document model w.r.t. document representations usually involved in multilingual and cross-lingual analysis, and the robustness of the transductive setting for multilingual document classification.

1 Introduction

Textual data constitutes a huge, continuously growing source of information, as everyday millions of documents are generated. This is partly explained by the increased popularity of tools for collaboratively editing through contributors across the world, which eases the production of different language-written documents, leading to a new phenomenon of *multilingual information overload*. Analyzing multilingual document collections is getting increased attention as it can support a variety of tasks, such as building translation resources [20,14], detection of plagiarism in patent collections [1], cross-lingual document similarity and multilingual document classification [18,16,6,2,5].

In this paper, we focus on the latter problem. Existing methods in the literature can mainly be characterized based on the language-specific resources they use to perform cross-lingual tasks. A common approach is resorting to machine translation techniques or bilingual dictionaries to map a document to the target

A. Hanbury et al. (Eds.): ECIR 2015, LNCS 9022, pp. 92–103, 2015.

language, and then perform cross-lingual document similarity and categorization (e.g., [6,9]). Some works (e.g., [16,2]) have also used Wikipedia as benchmark or knowledge base. However, in a cross-lingual supervised setting, the classification performance can significantly vary by exchanging documents from source to target languages. The language-specific machine translation systems typically introduce noise in understanding the document semantics, thus negatively affecting the final results. Furthermore, the classification performance will depend on the number and quality of the multilingual documents obtained by a single yet non-ontological knowledge base like Wikipedia.

We address the multilingual document classification problem differently from the above mentioned approaches. First, we are not restricted to deal with bilingual corpora dependent on machine translation. In this regard, we exploit a large, publicly available knowledge base specifically designed for multilingual retrieval tasks: *BabelNet* [14]. BabelNet embeds both the lexical ontology capabilities of WordNet and the encyclopedic power of Wikipedia. Second, our view is different from the standard inductive learning setting: in multilingual corpora often documents are all available at the same time and the classifications for the unlabeled instances need to be provided contextually to the learning of the current document collection. Examples of such tasks are relevance feedback, online news filtering, and reorganization of a document collection, where the system needs to automatically label documents in a collection starting from few labeled ones supplied by the user. Finally, high-quality labeled datasets are difficult to obtain due to costly and time-consuming annotation processes. This particularly holds for the multilingual scenario where language-specific experts need to be involved in the annotation process. To deal with these issues, *transductive learning* [7] offers an effective approach to supplying contextual classification of unlabeled documents by using a relatively small set of labeled ones. This learning setting fits well real-world applications and it can be very helpful in multilingual text analysis, where document labels are more difficult to obtain than in the monolingual counterpart and the classification decisions should not be made separately from learning the current data.

Motivated by the above considerations, in this work we propose a new framework for multilingual document classification under a transductive learning setting. By exploiting BabelNet, we model the multilingual documents using a common conceptual feature space. This representation model does not impose any methodological limitation on the number of languages of the documents. We then employ a state-of-the-art transductive learner [10] to produce the document classification. Using RCV2 and Wikipedia document collections, we compare our proposal w.r.t. document representations usually involved in multilingual and cross-lingual analysis. To the best of our knowledge, this is the first work that analyzes multilingual documents using a transductive learner through the lens of BabelNet. Note that transductive learning is also considered in [6], however only for bilingual analysis. Moreover, [5] also exploits BabelNet, although to propose a bilingual similarity measure, while our approach can effectively deal with comparable corpora in more than two languages.

2 Background on BabelNet

BabelNet [14] is a multilingual semantic network obtained by linking Wikipedia with WordNet, that is, the largest multilingual Web encyclopedia and the most popular computational lexicon. The linking of the two knowledge bases was performed through an automatic mapping of WordNet synsets and Wikipages, harvesting multilingual lexicalization of the available concepts through human-generated translations provided by the Wikipedia inter-language links or through machine translation techniques.

It should be noted that the large-scale coverage of both lexicographic and encyclopedic knowledge represents a major advantage of BabelNet versus other knowledge bases that could in principle be used for cross-lingual or multilingual retrieval tasks. For instance, the multilingual thesaurus EUROVOC (created by the European Commission's Publications Office) was used in [18] for document similarity purposes; however, EUROVOC utilizes less than 6 000 descriptors, which leads to evident limits in semantic coverage. Furthermore, other knowledge bases such as EuroWordNet [20] only utilize lexicographic information, while conversely studies that focus on Wikipedia (e.g., [16,2]) cannot profitably leverage on lexical ontology knowledge.

Multilingual knowledge in BabelNet is represented as a labeled directed graph in which nodes are concepts or named entities and edges connect pairs of nodes through a semantic relation. Each edge is labeled with a relation type (is-a, part-of, etc.), while each node corresponds to a *BabelNet synset*, i.e., a set of lexicalizations of a concept in different languages. BabelNet also provides functionalities for graph-based word sense disambiguation in a multilingual context. Given an input set of words, a semantic graph is built by looking for related synset paths and by merging all them in a unique graph. Once the semantic graph is built, the graph nodes can be scored with a variety of algorithms. Finally, this graph with scored nodes is used to rank the input word senses by a graph-based approach.

3 Transductive Multilingual Document Classification

3.1 Text Representation Models

Bag-of-Synset Representation. We model the multilingual documents into a common *conceptual feature space*, which is built using the multilingual lexical knowledge of BabelNet [17]. We will refer to this representation as *BoS* (i.e., *bag-of-synsets*), since conceptual features of the documents correspond to BabelNet synsets.

The input document collection is subject to a two-step processing phase. In the first step, each document is broken down into a set of lemmatized and POS-tagged sentences, in which each word is replaced with related lemma and associated POS-tag. Let us denote with $\langle w, POS(w)\rangle$ a lemma and associated POS-tag occurring in any sentence s of the document. In the second step, a word sense disambiguation (WSD) method is applied to each pair $\langle w, POS(w)\rangle$ to detect the most appropriate BabelNet synset σ_w for $\langle w, POS(w)\rangle$ contextually to s. The WSD algorithm is carried out in such a way that all words from all languages are disambiguated over the same concept space, producing a language-independent feature space for the whole multilingual corpus. Each document is finally modeled as a $|\mathcal{BS}|$-dimensional vector of BabelNet synset frequencies, being $\mathcal{BS}$ the set of retrieved BabelNet synsets.

As previously discussed in Section 2, BabelNet provides WSD algorithms for multilingual corpora. The authors in [15] suggest to use the degree ranking algorithm (i.e., given a semantic graph for the input context, it simply selects the sense of the target word with the highest vertex degree), as it has shown to yield highly competitive performance in the multilingual context. Clearly, other methods for (unsupervised) WSD, particularly PageRank-style methods (e.g., [12,21]), can be plugged in to perform multilingual WSD based on BabelNet; however, this subject is out of the scope of this paper.

Bag-of-words and Machine-translation Based Models. The *bag-of-words* model has been employed also in the context of multilingual documents [11]. Hereinafter we use notation *BoW* to refer to the term-frequency vector representation of documents over the union of language-specific term vocabularies.

However, in the multilingual setting, the use of *BoW* poses additional issues as it tends to exacerbate the sparsity in the document modeling, i.e., the language-specific vocabularies are generally very different, making the cross-lingual document similarity hard to compute. To overcome this issue, a common solution adopted in the literature is to translate all documents to a unique anchor language and represent the translated documents with the *BoW* model [11,6]. In this work, we have considered three settings corresponding to the use of *English*, *French* or *Italian* as anchor language; the resulting representation models will be referred to as *BoW-MT-en*, *BoW-MT-fr* and *BoW-MT-it*, respectively. As an alternative model, we resort to a dimensionality reduction approach via Latent Semantic Indexing (LSI) [4] over the *BoW* representation. Recall that, given the document-term matrix obtained using *BoW*, LSI consists in computing the SVD decomposition of that matrix and representing the documents with low-dimensional vectors. We will refer to this model as *BoW-LSI*.

3.2 Transductive Setting and Label Propagation Algorithm

Given a document collection $\mathcal{D} = \{d_i\}_{i=1}^{N}$, let us denote with $\mathcal{L}$ the subset of $\mathcal{D}$ comprised of labeled documents, and with $\mathcal{U} = \mathcal{D} \setminus \mathcal{L}$ the subset of unlabeled documents. Note that $\mathcal{U}$ can in principle have any proportion w.r.t. $\mathcal{L}$, but in many real cases $\mathcal{U}$ is much larger than $\mathcal{L}$. Every document in $\mathcal{L}$ is assigned a label that refers to one of the known M classes $\mathcal{C} = \{C_j\}_{j=1}^{M}$. We also denote with $\mathbf{Y}$ a $N \times M$ matrix such that $\mathbf{Y}_{ij} = 1$ if C_j is the label assigned to document d_i, 0 otherwise.

The goal of a *transductive learner* is to make an inference "from particular to particular", i.e., given the classifications of the instances in the training set $\mathcal{L}$, it aims to guess the classifications of the instances in the test set $\mathcal{U}$, rather than inducing a general rule that works out for classifying new unseen instances [19]. Transduction is naturally related to the class of case-based learning algorithms, whose most well-known algorithm is the k-nearest neighbor (kNN) [8].

To the best of our knowledge, we bring for the first time a transductive learning approach to a multilingual document classification. We use a particularly effective transductive learner, named Robust Multi-class Graph Transduction (*RMGT*) approach [10]. RMGT has shown to outperform all the other state-of-the-art transductive classifiers in the recent evaluation study by Sousa et al. [3]. Essentially, RMGT implements a graph-based label propagation approach, which exploits a kNN graph built over the entire document collection to propagate the class information from the labeled to the unlabeled documents. In the following we describe in detail the mathematics behind RMGT.

Let $\mathcal{G} = \langle \mathcal{V}, \mathcal{E}, w \rangle$ be an undirected graph whose vertex set is $\mathcal{V} = \mathcal{D}$, edge set is $\mathcal{E} = \{(d_i, d_j) | d_i, d_j \in \mathcal{D} \wedge sim(d_i, d_j) > 0\}$, and edge weighting function is $w = sim(d_i, d_j)$. Given a positive integer k, consider the kNN graph $\mathcal{G}_k = \langle \mathcal{V}, \mathcal{E}_k, w \rangle$ derived from $\mathcal{G}$ and such that $\mathcal{E} = \{(d_i, d_j) | d_j \in N_i\}$, where N_i denotes the set of d_i's k-nearest neighbors. A weighted sparse matrix is obtained as $\mathbf{W} = \mathbf{A} + \mathbf{A}^{\mathrm{T}}$, where $\mathbf{A}$ is the weighted adjacency matrix of $\mathcal{G}_k$ and $\mathbf{A}^{\mathrm{T}}$ is the transpose of $\mathbf{A}$; the matrix $\mathbf{W}$ represents a *symmetry-favored* kNN graph [10]. Moreover, let $\mathbf{L} = \mathbf{I}_N - \mathbf{D}^{-1/2}\mathbf{W}\mathbf{D}^{-1/2}$ the normalized Laplacian of $\mathbf{W}$, where $\mathbf{I}_N$ is the $N \times N$ identity matrix and $\mathbf{D} = diag(\mathbf{W}\mathbf{1}_N)$. Without loss of generality, we can rewrite $\mathbf{L}$ and $\mathbf{W}$ as subdivided into four and two submatrices, respectively:

$$\mathbf{L} = \begin{bmatrix} \mathbf{\Delta}_{\mathcal{L}\mathcal{L}} \; \mathbf{\Delta}_{\mathcal{L}\mathcal{U}} \\ \mathbf{\Delta}_{\mathcal{U}\mathcal{L}} \; \mathbf{\Delta}_{\mathcal{U}\mathcal{U}} \end{bmatrix}, \quad \mathbf{Y} = \begin{bmatrix} \mathbf{Y}_{\mathcal{L}} \\ \mathbf{Y}_{\mathcal{U}} \end{bmatrix} \tag{1}$$

where $\mathbf{\Delta}_{\mathcal{L}\mathcal{L}}$ and $\mathbf{Y}_{\mathcal{L}}$ are the submatrices of $\mathbf{L}$ and $\mathbf{Y}$, respectively, corresponding to the labeled documents, and analogously for the other submatrices. The RMGT learning algorithm finally yields a matrix $\mathbf{F} \in \mathbb{R}^{N \times M}$ defined as:

$$\mathbf{F} = -\mathbf{\Delta}_{\mathcal{U}\mathcal{U}}^{-1}\mathbf{\Delta}_{\mathcal{U}\mathcal{L}}\mathbf{Y}_{\mathcal{L}} + \frac{\mathbf{\Delta}_{\mathcal{U}\mathcal{U}}^{-1}\mathbf{1}_u}{\mathbf{1}_u^{\mathrm{T}}\mathbf{\Delta}_{\mathcal{U}\mathcal{U}}^{-1}\mathbf{1}_u}(N\omega - \mathbf{1}_l^{\mathrm{T}}\mathbf{Y}_{\mathcal{L}} + \mathbf{1}_u^{\mathrm{T}}\mathbf{\Delta}_{\mathcal{U}\mathcal{U}}^{-1}\mathbf{\Delta}_{\mathcal{U}\mathcal{L}}\mathbf{Y}_{\mathcal{L}}) \tag{2}$$

where $\omega \in \mathbb{R}^M$ is the class prior probabilities.

The transductive learning scheme used by RMGT employs spectral properties of the kNN graph to spread the labeled information over the set of test documents. Specifically, the label propagation process is modeled as a constrained convex optimization problem where the labeled documents are employed to constrain and guide the final classification. The mathematical formulation given in Eq. (2) enables a closed form solution of this optimization problem. After the propagation step, every unlabeled document d_i is associated to a vector (i.e., the i-th row of $\mathbf{F}$) representing the likelihood of the document d_i for each of the classes; therefore, d_i is assigned to the class that maximizes the likelihood.

Algorithm 1 sketches the main steps of our multilingual document classification framework based on the RMGT learning approach. Initially, a pre-processing step is required to model every document in the collection using our proposed

Algorithm 1. Transductive classification of multilingual documents

Input: A collection of multilingual documents $\mathcal{D}$, with labeled documents $\mathcal{L}$ and unlabeled documents $\mathcal{U}$ (with $\mathcal{D} = \mathcal{L} \cup \mathcal{U}$ and $\mathcal{L} \cap \mathcal{U} = \emptyset$); a set of labels $\mathcal{C} = \{C_j\}_{j=1}^{M}$ assigned to the documents in $\mathcal{L}$; a positive integer k for the neighborhood selection.
Output: A classification over $\mathcal{C}$ for the documents in $\mathcal{U}$.
1. Model each document in $\mathcal{D}$ using *BoS* or alternative representations. /* *Section 3.1* */
2. Build the similarity graph $\mathcal{G}$ for the document collection $\mathcal{D}$.
3. Extract the k-nearest neighbor graph $\mathcal{G}_k$ from $\mathcal{G}$. /* *Section 3.2* */
4. Build the matrix $\mathbf{W}$ from $\mathcal{G}_k$, which represents the symmetry-favored k-nearest neighbor graph. /* *Section 3.2* */
5. Compute the normalized Laplacian of $\mathbf{W}$. /* *Section 3.2* */
6. Compute the *RMGT* solution $\mathbf{F}$. /* *Eq. (2)* */
7. Assign document $d_i \in \mathcal{U}$ to the class C_{j*} that maximizes the class likelihood, $j^* = \arg\max_j \mathbf{F}_{ij}$.

BoS representation or alternative representations (Line 1). Upon the computation of the similarity matrix over all documents in the collection (Line 2), the graph-based label propagation process requires the construction of the kNN graph (Line 3) and its symmetry-favored transformation (Line 4). Concerning the $sim(\cdot,\cdot)$ function, we employ the cosine similarity as standard measure in document classification, but other measures can alternatively be utilized. Moreover, the class priors (ω) used in Eq. (2) are defined as uniformly distributed.

4 Experimental Evaluation

4.1 Data

We used two document collections, built from the *RCV2* corpus[1] and from the *Wikipedia* online encyclopedia. Both datasets were constructed to contain documents in three different languages, namely *English*, *French*, and *Italian*. Six topic-classes were identified, which correspond to selected values of TOPICS Reuters field in RCV2 and to selected Wikipage titles in Wikipedia. Our choice of languages and topics allowed us to obtain a significant topical coverage in all languages. Moreover, according to [17], we considered a *balanced* way for the document assignment to each topic-language pair; specifically, 850 and 1000 documents per pair, in RCV2 and Wikipedia, respectively. RCV2 contains 15 300 documents represented over a space of 12 698 terms, for the *BoW* model, and 10 033 synsets, for the *BoS* model, with density (i.e., the fraction of non-zero entries in the document-term matrix, resp. document-synset matrix) of 4.56E-3 for *BoW* and 3.87E-3 for *BoS*. Wikipedia is comprised of 18 000 documents, with 15 634 terms and 10 247 synsets, and density of 1.61E-2 for *BoW* and 1.81E-2 for *BoS*.[2]

Note that although the two datasets were built using the same number of languages and topics, they can be distinguished by an important aspect: in RCV2, the different language-written documents belonging to the same topic-class do

[1] `http://trec.nist.gov/data/reuters/reuters.html`.

[2] Datasets are made publicly available at `http://uweb.dimes.unical.it/tagarelli/data/`.

not necessarily share the content subjects; by contrast, the encyclopedic nature of Wikipedia favors a closer correspondence in content among the different language-specific versions of articles discussing the same Wikipedia concept (although, these versions are not translation of each other). We underline that both corpora have not been previously used in a multilingual transductive scenario.

Every document was subject to lemmatization and, in the *BoS* case, to POS-tagging as well. All text processing steps were performed using the Freeling library tool.[3] To setup the transductive learner, we used $k = 10$ for the kNN graph construction, and we evaluated the classification performance by varying the percentage of labeled documents from 1% to 20% with a step of 1% for both datasets. Results were averaged over 30 runs (to avoid sampling bias) and assessed by using standard F-measure, Precision and Recall criteria [11].

4.2 Evaluation of BabelNet Coverage

The extent to which our approach will actually lead to good solutions depends on the semantic coverage capabilities of the multilingual knowledge base as well as on the corpus characteristics. Therefore, we initially investigated how well BabelNet allows us to represent the concepts discussed in each of the datasets.

For every document, we calculated the BabelNet coverage as the fraction of words belonging to the document whose concepts are present as entries in BabelNet. We then analyzed the distribution of documents over different values of BabelNet coverage. Figures 1(a)–1(b) show the probability density function (pdf) of BabelNet coverage for each of the topic-classes, on RCV2 and Wikipedia, respectively; analogously, Figs. 1(c)–1(d) visualize the distributions per language.

Generally, we observe roughly bi-modal distributions in both evaluation cases and for both datasets. Considering the per-topic distributions, all of them tend to have a peak around coverage of 0.5 and a lower peak around 0.84, following the overall trend with no evident distinctions. By contrast, the per-language distributions (Fig. 1(c)–1(d)) supply more helpful clues to understand the BabelNet coverage capabilities. In fact we observe that both French and Italian documents determine the left peak of the overall distributions, actually corresponding to roughly normal distributions; on the contrary, the English documents correspond to negatively skewed (i.e., left-tailed) distributions, thus characterizing the right peak of the overall distributions. Interestingly, these remarks hold for both RCV2 and Wikipedia datasets, which indicates that BabelNet provides a more complete coverage for English documents than for French/Italian documents.

4.3 Classification Performance

In this section we assess the impact of *BoS* and the other document models on the performance of our transductive multilingual classification approach. In order to inspect the models' behavior under different corpus characteristics, in this stage of evaluation we also produced unbalanced versions of the datasets,

[3] `http://nlp.lsi.upc.edu/freeling/`.

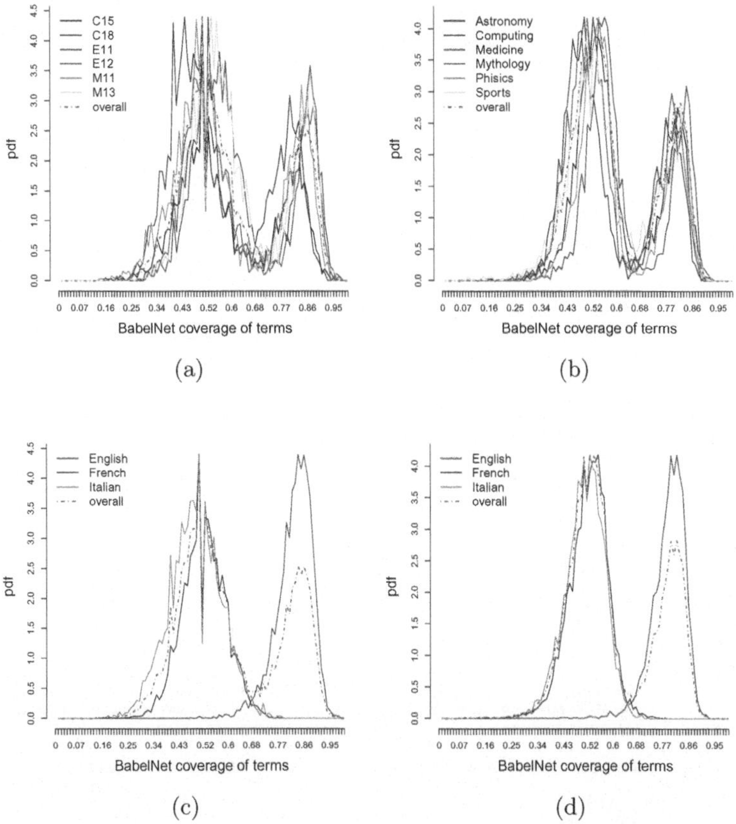

Fig. 1. BabelNet coverage: (a) RCV2, (b) Wikipedia per topic-class, and (c) RCV2, (d) Wikipedia per language. (Better viewed in the electronic version.)

hereinafter referred to *unbalanced RCV2* and *unbalanced Wikipedia*. Specifically, in each of the two original datasets we kept the subset of English documents while sampling half of the French and half of the Italian subsets. In the light of the remarks that stand out from the previous analysis on the BabelNet coverage, we aim here to understand how much the classification performance varies when using an English-biased multilingual corpus.

In the following, we present results obtained in the two distinguished cases of balanced and unbalanced datasets. Figure 2 shows the methods' performance (F-measure) by varying the training percentage of the transductive learning algorithm, while Table 1 summarizes the best performances in terms of F-measure, Precision and Recall. We begin with evaluation on the balanced case, which we then couple with an inspection of the intra-class and inter-class similarity of the datasets. This will allow us to unveil important aspects of the behaviors of the *BoS* model and competing ones that eventually advocate the significance of our further evaluation on unbalanced datasets.

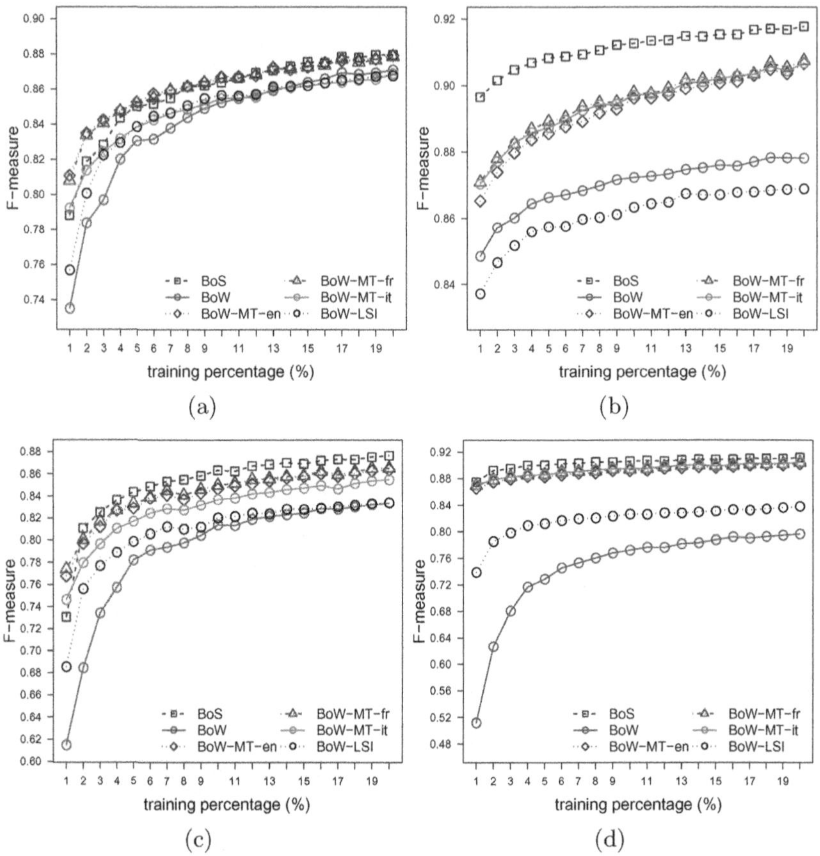

Fig. 2. F-measure for (a) RCV2, (b) Wikipedia, and (c) unbalanced RCV2, (d) unbalanced Wikipedia. (Better viewed in the electronic version.)

Evaluation on Language-balanced Corpora. On RCV2 (Fig. 2(a)), we observe that *BoS* follows an increasing trend, similarly to those shown by the other models and performing (for training percentage values above 4%) comparably to the best of the competing models, which are *BoW-MT-en* and *BoW-MT-fr*. The *BoW-MT-it* and *BoW-LSI* achieve lower F-measures, which become very close to the basic *BoW* for higher values of training percentage.

A different scenario is instead depicted in Fig. 2(b) for Wikipedia. *BoS* clearly outperforms the other document representation models, including *BoW-MT-en* which in this case achieves results that are similar to (or slightly lower than) those obtained by *BoW-MT-fr* and *BoW-MT-it*. *BoW-LSI* and *BoW* also show a performance gap from the other models, which is much more significant than in the RCV2 case.

As a general remark it should be noted that *BoS* not only performs comparably or significantly better than the other models—this is confirmed by the best-performance evaluation reported in Table 1—but also it exhibits a performance

Table 1. Summary of best performance results of the various representation methods. Bold values correspond to the best performance per dataset and assessment criterion, whereas italic values refer to *BoW* related methods.

	Balanced RCV2			*Balanced Wikipedia*			*Unbalanced RCV2*			*Unbalanced Wikipedia*		
	FM	*P*	*R*	*FM*	*P*	*R*	*FM*	*P*	*R*	*FM*	*P*	*R*
BoS	**0.880**	**0.883**	**0.881**	**0.912**	**0.915**	**0.912**	**0.877**	**0.880**	**0.878**	**0.912**	**0.915**	**0.912**
BoW	0.871	0.876	0.872	0.872	0.876	0.872	0.834	0.839	0.836	0.797	0.817	0.794
BoW-MT-en	*0.879*	*0.881*	*0.880*	0.895	0.896	0.895	0.864	*0.867*	0.865	0.902	0.903	0.902
BoW-MT-fr	*0.879*	0.879	0.879	*0.898*	*0.899*	*0.898*	*0.865*	0.866	*0.866*	0.904	0.906	0.904
BoW-MT-it	0.869	0.870	0.870	0.897	*0.899*	0.897	0.855	0.856	0.856	*0.905*	*0.907*	*0.905*
BoW-LSI	0.868	0.872	0.869	0.863	0.867	0.863	0.834	0.840	0.837	0.838	0.845	0.838

trend that is not affected by issues related to the language specificity. In fact, the machine-translation based models have relative performance that may vary on different datasets, since a language that leads to better results on a dataset can perform worse than other languages on another dataset.

Intra-class and Inter-class Document Similarity. The differences observed in the relative trends exhibited by *BoS* and the other models on RCV2 compared with Wikipedia, prompted us to investigate the topic homogeneity and topic separation on the datasets, over the various topic-classes and languages.

Figure 3 compares the similarity matrices for the balanced datasets obtained using *BoS*. Note that the main diagonal on each matrix corresponds to the intra-class document similarity, while the remaining cells refer to similarity between two different topic-classes (i.e., inter-class similarity). On every cell, the hue toward red (resp. blue) indicates higher (resp. lower) cosine similarity.

A first remark common to RCV2 and Wikipedia (Fig. 3(a)–(b)) is that both the intra-class and inter-class is low when only French and Italian documents are considered. This might be explained by a different support of *BabelNet* to the conceptual representation of documents in non-English languages; in particular, as discussed in [17], French and Italian documents have a significantly lower dimensional representation according to the *BoS* model, which would hence affect both intra- and inter-class document similarities.

Looking at the upper left blocks of the matrices, which correspond to English document classes, we observe that on RCV2 the intra-class similarity is high for three topics (i.e., "E12", "M11", "M13"), and, in general, higher than on Wikipedia; however, also the inter-class similarity is higher (i.e., worse) than on Wikipedia. The topic separation between English and French/Italian documents is lower on RCV2. The above findings would indicate that RCV2 appears to be a harder testbed than Wikipedia for our proposed *BoS* model.

Evaluation on Language-unbalanced Corpora. Here we quantify how the methods' performance change when the English written portion in the corpus varies (i.e., is double) relatively to the other languages' portions. Figure 2(c) and Table 1 show that the *BoS* results are always higher (though slightly) than

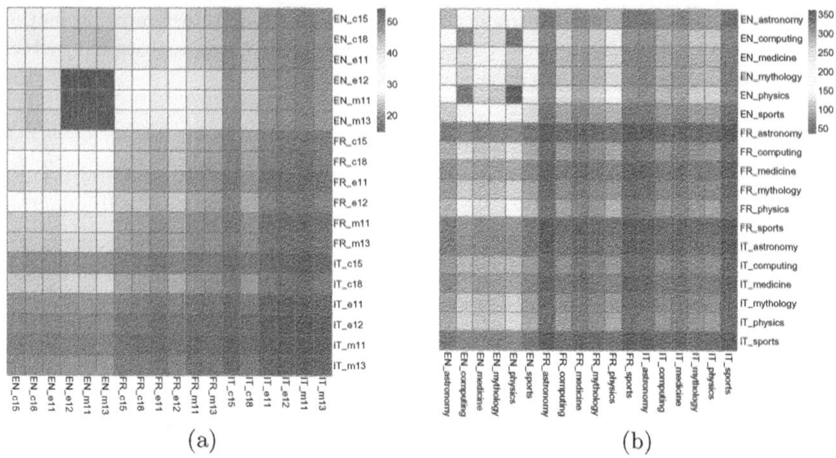

Fig. 3. Similarity matrices of *BoS*-modeled documents grouped by language and class for balanced datasets: (a) RCV2 and (b) Wikipedia. (Better viewed in the electronic version.)

the best competing methods. More interestingly, the advantage taken by *BoS* is actually explained by a decreased performance of the other models, which would indicate a higher robustness of the *BoS* model w.r.t. the corpus characteristics.

Note also that on Wikipedia (Fig. 2(d)), the relative performance between *BoS* and the machine-translation based models is not changed w.r.t. the balanced case, and the finer scale-grain of the y-axis gives evidence of the decreased performance of *BoW* and *BoW-LSI*.

5 Conclusion

We have proposed a new framework for multilingual document classification under a transductive setting and with the support of the BabelNet knowledge base. Our proposed conceptual representation model for multilingual documents, *BoS*, has shown to be effective for multilingual comparable corpora: *BoS* not only leads to generally better results than various language-dependent representations, but it has also shown to preserve its performance on both balanced and unbalanced datasets. This aspect highlights the robustness of our knowledge-based representation, paving the way for future analysis of multilingual documents. Furthermore, the transductive learning approach has shown to be useful in the multilingual scenario, obtaining good classification performance with a quite small (5%) portion of labeled documents.

As future work we plan to exploit more types of information provided in BabelNet (i.e., relations among the synsets) to enrich our multilingual document model. We are also interested in combining transductive with active learning, which can aid solicit user interaction in order to guide the labeling process.

References

1. Barrón-Cedeño, A., Gupta, P., Rosso, P.: Methods for cross-language plagiarism detection. Knowl.-Based Syst. 50, 211–217 (2013)
2. Barrón-Cedeño, A., Paramita, M.L., Clough, P., Rosso, P.: A comparison of approaches for measuring cross-lingual similarity of wikipedia articles. In: de Rijke, M., Kenter, T., de Vries, A.P., Zhai, C., de Jong, F., Radinsky, K., Hofmann, K. (eds.) ECIR 2014. LNCS, vol. 8416, pp. 424–429. Springer, Heidelberg (2014)
3. de Sousa, C.A.R., Rezende, S.O., Batista, G.E.A.P.A.: Influence of graph construction on semi-supervised learning. In: Blockeel, H., Kersting, K., Nijssen, S., Železný, F. (eds.) ECML PKDD 2013, Part III. LNCS, vol. 8190, pp. 160–175. Springer, Heidelberg (2013)
4. Deerwester, S., Dumais, S.T., Furnas, G.W., Landauer, T.K., Harshman, R.: Indexing by latent semantic analysis. Journal of the American Society for Information Science 41(6), 391–407 (1990)
5. Franco-Salvador, M., Rosso, P., Navigli, R.: A knowledge-based representation for cross-language document retrieval and categorization. In: Proc. EACL, pp. 414–423 (2014)
6. Guo, Y., Xiao, M.: Transductive representation learning for cross-lingual text classification. In: Proc. ICDM, pp. 888–893 (2012)
7. Joachims, T.: Transductive inference for text classification using support vector machines. In: Proc. ICML, pp. 200–209 (1999)
8. Joachims, T.: Transductive Learning via Spectral Graph Partitioning. In: Proc. ICML (2003)
9. Klementiev, A., Titov, I., Bhattarai, B.: Inducing Crosslingual Distributed Representations of Words. In: Proc. COLING, pp. 1459–1474 (2012)
10. Liu, W., Chang, S.: Robust multi-class transductive learning with graphs. In: Proc. CVPR, pp. 381–388 (2009)
11. Manning, C.D., Raghavan, P., Schütze, H.: Introduction to Information Retrieval. Cambridge University Press, New York (2008)
12. Mihalcea, R., Tarau, P., Figa, E.: PageRank on semantic networks, with application to word sense disambiguation. In: Proc. COLING (2004)
13. Navigli, R., Lapata, M.: An experimental study of graph connectivity for unsupervised word sense disambiguation. IEEE TPAMI 32(4), 678–692 (2010)
14. Navigli, R., Ponzetto, S.P.: Babelnet: The automatic construction, evaluation and application of a wide-coverage multilingual semantic network. Artif. Intell. 193, 217–250 (2012)
15. Navigli, R., Ponzetto, S.P.: Multilingual WSD with just a few lines of code: the babelnet API. In: Proc. ACL, pp. 67–72 (2012)
16. Ni, X., Sun, J., Hu, J., Chen, Z.: Cross lingual text classification by mining multilingual topics from wikipedia. In: Proc. WSDM, pp. 375–384 (2011)
17. Romeo, S., Tagarelli, A., Ienco, D.: Semantic-Based Multilingual Document Clustering via Tensor Modeling. In: Proc. EMNLP, pp. 600–609 (2014)
18. Steinberger, R., Pouliquen, B., Hagman, J.: Cross-lingual document similarity calculation using the multilingual thesaurus EUROVOC. In: Gelbukh, A. (ed.) CICLing 2002. LNCS, vol. 2276, pp. 415–424. Springer, Heidelberg (2002)
19. Vapnik, V.: Statistical learning theory. Wiley (1998)
20. Vossen, P.: EuroWordNet: A multilingual database of autonomous and language-specific WordNets connected via an inter-lingual index. International Journal of Lexicography 17(2), 161–173 (2004)
21. Yeh, E., Ramage, D., Manning, C.D., Agirre, E., Soroa, A.: Wikiwalk: Random walks on wikipedia for semantic relatedness. In: Workshop on Graph-based Methods for Natural Language Processing, pp. 41–49 (2009)

Distributional Correspondence Indexing for Cross-Language Text Categorization

Andrea Esuli and Alejandro Moreo Fernández

Istituto di Scienza e Tecnologie dell'Informazione "A. Faedo"
Consiglio Nazionale delle Ricerche - Pisa, Italy
{andrea.esuli,alejandro.moreo}@isti.cnr.it

Abstract. Cross-Language Text Categorization (CLTC) aims at producing a classifier for a target language when the only available training examples belong to a different source language. Existing CLTC methods are usually affected by high computational costs, require external linguistic resources, or demand a considerable human annotation effort. This paper presents a simple, yet effective, CLTC method based on projecting features from both source and target languages into a common vector space, by using a computationally lightweight distributional correspondence profile with respect to a small set of pivot terms. Experiments on a popular sentiment classification dataset show that our method performs favorably to state-of-the-art methods, requiring a significantly reduced computational cost and minimal human intervention.

Keywords: Cross-Language Text Categorization, Distributional Semantics, Sentiment Analysis.

1 Introduction

Automated Text Categorization methods usually rely on a *training set* of labeled examples to learn a classifier that will then predict the categories of unlabeled documents. The creation of a training set requires substantial human effort, and it is inherently language-dependent. Cross-Language Text Categorization (CLTC [1]) aims at using the labeled examples available for a *source* language to learn a classifier for a different *target* language, thus reducing, or completely avoiding, the need for human labeling of examples in the target language. A practical scenario for CLTC is to exploit the labeled examples freely available on the Web for the prevailing languages (e.g., English star-rated reviews) to build classifiers for languages for which the amount of labeled examples is much smaller.

A number of different approaches to CLTC have been presented in literature. The use of *Machine Translation* (MT) [8,10] to reduce all the documents to a single language is a straightforward solution, but it is bound to the availability of MT systems/services for the relevant languages, and it suffers from the cost, economical and of time, of translating a large number of documents.

Methods exploiting *parallel corpora* [3,5,11] are usually affected by the high computational costs derived from the use of a sophisticated statistical analysis, e.g.,

A. Hanbury et al. (Eds.): ECIR 2015, LNCS 9022, pp. 104–109, 2015.

Principal Component Analysis (PCA), and are bound to the availability of a parallel corpus between the relevant languages.

Structural Correspondence Learning (SCL [2]) was applied to the cross-language setting (CL-SCL [6,7]) by using a word-translator oracle in order to create a set of word pairs (dubbed *pivots*). The pivots are later used to discover structural analogies between the source and target languages through unlabeled corpora. Even though CL-SCL succeeded in alleviating the problems posed by the use of MT tools, it still has a considerable computational cost, deriving from the intermediate optimizations of the *structural problems* (i.e., pivot predictors), and from the use of Latent Semantic Analysis (LSA).

Our method takes the CL-SCL idea as an inspiration, but it follows a different, simpler approach, with a more direct application of the *distributional hypothesis*, which states that words with similar distributions of use in text are likely to have similar meanings. Given a small sets of pivots, textual features extracted from both languages are projected into a common vector space (feature representation transfer [4]) in which each dimension reflects the *distributional correspondence* between the feature being projected and a pivot. The distributional correspondence is efficiently estimated on sets of unlabeled documents for each language. There is no need for a parallel corpus, and computationally-expensive statistical techniques are avoided.

Despite being simple, this method compares favorably to the state of the art in experiments on a popular sentiment classification dataset, sporting a significantly reduced computational cost, and also requiring less human intervention.

2 Distributional Correspondence Indexing

In the traditional bag-of-words model each word is mapped into a dedicated dimension of the vector space. Without resorting to translation or other source of external knowledge, words like the English "beautiful" and its German equivalent "*schöne*" point to orthogonal directions in the vector space, while their vectorial representation should be aligned in order to model their correspondence.

Our *Distributional Correspondence Indexing* (DCI) method profiles each feature with respect to its distributional correspondence to the pivots. As word pairs defining the pivots are expected to behave similarly in their respective language, semantically related words from the source and target languages should present similar distributions to them, thus obtaining similar representations.

Pivot selection. Words from the source training set are ranked by their relevance with respect to the classification task by means of a supervised feature selection function; similarly to [7], we use mutual information. The oracle is then requested to translate each source word t_S into its translation-equivalent word t_T in the target language, to form the pivot pairs $p = \langle t_S, t_T \rangle$. Following [7] the set of pivots consists of the top-m pivots with a *support* (occurrences in the unlabeled corpora) greater than a given threshold ϕ.

Feature profiles. Differently from [7], we propose to represent each source and target feature f (including pivots) as an m-dimensional profile vector:

$$\overrightarrow{f} = (\eta(f,p_1), \eta(f,p_2), ..., \eta(f,p_m)) \tag{1}$$

where p_i is the source or target word in the ith pivot, and η denotes the *distributional correspondence function* between the feature f and p_i, that we model with a probability-based linear function[1] that requires minimal computation:

$$\eta(f,p) = P(f|p) - P(f|\overline{p}) \tag{2}$$

where $P(f|p)$ denotes the conditional probability of finding f in documents containing p, and $P(f|\overline{p})$ is conditioned on documents not containing p. Both probabilities are estimated on the set of unlabeled documents for the pertinent language. All feature profile vectors $\overrightarrow{f_i}$ are then normalized to unit length.

Unification. As we assume pivot terms behave similarly in both languages, we *unify* their feature profiles by averaging them. Unification is also applied to profiles of words that the source and target languages have in common (e.g., proper nouns or non-lexicalized terms) having a support greater than ϕ.

Document indexing. Finally, train and test documents are represented into the cross-lingual space as the weighted sum of all profile vectors associated to their features. That is, document d_j is represented as the m-dimensional vector

$$\overrightarrow{d_j} = \sum_{f_i \in d_j} w_{ij} \cdot \overrightarrow{f_i} \tag{3}$$

where w_{ij} is the weight of feature f_i in document d_j. We used the normalized $tf \cdot idf$ weighting criterion in our implementation.

3 Experiments

We test our method[2] on the publicly available Webis-CLS-10 Cross-Lingual Sentiment collection[3] proposed in [6]. The dataset consists of Amazon product reviews written in four languages (**E**nglish, **G**erman, **F**rench, and **J**apanese), covering three product categories (**B**ooks, **D**VDs, and **M**usic). For each language-category pair there are 2,000 training documents, 2,000 test documents, and from 9,000 to 50,000 unlabeled documents depending on the language-category combination. Following [6], we consider English as the source language, and German, French, and Japanese as the target ones. Documents are either labeled as *Positive* or *Negative* (binary classification), and any train or test set contains an equal amount of positive and negative examples. The evaluation measure is *accuracy*, which is adequate since labels are always balanced in the dataset.

In our implementation we set $\phi = 30$, following the results of [6]. We test our method on three sizes for the pivot set: $m = 450$, which is the best-performing

---

1 We also investigated other alternatives coming from information theory including Information Gain, χ^2, and Odds ratio, with negative or unstable results.

2 The code to replicate our experiments is available at `http://hlt.isti.cnr.it/dci/`

3 `http://www.uni-weimar.de/en/media/chairs/webis/research/corpora/corpus-webis-cls-10/`

Table 1. Accuracy for cross-lingual sentiment analysis in the Webis-CLS-10 collection. Acronyms indicate source/target/product: "EGB" stands for English/German/Books.

	Upper	MT	SCL	LSI	KCCA	OPCA	SSMC	DCI_{450}	DCI_{100}	DCI_{20}
EGB	86.75	79.68	**83.34**	77.59	79.14	74.72	81.88	76.25	81.40	79.50
EGD	83.50	77.92	80.89	79.22	76.73	74.59	**82.25**	80.40	79.95	77.75
EGM	85.90	77.22	82.90	73.81	79.18	74.45	81.30	75.20	**83.30**	73.70
EFB	86.15	80.76	81.27	79.56	77.56	76.55	**83.05**	82.95	82.30	75.15
EFD	87.15	78.83	80.43	77.82	78.19	70.54	82.70	**84.10**	82.40	64.35
EFM	88.95	75.78	78.05	75.39	78.24	73.69	80.46	**81.90**	81.05	75.80
EJB	81.15	70.22	77.00	72.68	69.46	71.41	73.76	73.90	**79.10**	74.50
EJD	83.40	71.30	76.37	72.55	74.79	71.84	77.58	81.55	**82.25**	80.25
EJM	84.20	72.02	77.34	73.44	73.54	74.96	77.53	78.45	**82.00**	79.30

setup for SCL [6], $m = 100$, which is the minimal number of pivots tested in [6], and a minimal setup using just $m = 20$ pivots. To emulate the word-oracle – and for the sake of a fair comparison – we used the bilingual dictionary provided by [6]. We used the popular SVM^{light} implementation[4] of Support Vector Machines as the learning device, with default parameters.

In order to have an upper reference to accuracy, we implemented a method that trains the SVM classifier on the training set of the target language (Upper). We also report the MT baseline (MT) of [7], which first translates the target examples set into the source language. In Table 1 we compare DCI to the results published on the same dataset, same configuration, for five CLTC methods: structural correspondence learning (SCL [7]), latent semantic indexing (LSI [3]), kernel canonical correlation analysis (KCCA [9]), oriented principal component analysis (OPCA [5]), and semi-supervised matrix completion (SSMC [11]).

DCI_{450} obtains good results, performing better than the compared methods in four cases out of nine. DCI_{100} performs even better (five out of nine, and four highest results). DCI_{100} performs better than SCL in seven cases out of nine, with SCL requiring 450 calls to a word-oracle, 450 structural optimization problems, and LSA. DCI_{100} instead only needs 100 word-translations plus feature profile calculation and document indexing, which is extremely efficient[5]. SSMC performs better than DCI_{100} on German and French. SSMC algorithm requires however a parallel corpus, a double-sized source training set, and some labeled examples from the target language. Figure 1 shows how accuracy varies when varying m in the range between 15 and 500.

We noted that DCI performs much better than the other methods when Japanese is the target language. Given that DCI is applied to the same textual features used by all the other methods, and adopts the same SVM learner of Upper, with exactly the same parameters, we deem this difference to a better

[4] Available at `http://svmlight.joachims.org/`

[5] It took 22.2s, 15.3s, and 11.2s on average in the Books, DVDs, and Music tasks, respectively, to create the feature profiles and build the training index on a single threaded process on a 1.6GHz processor.

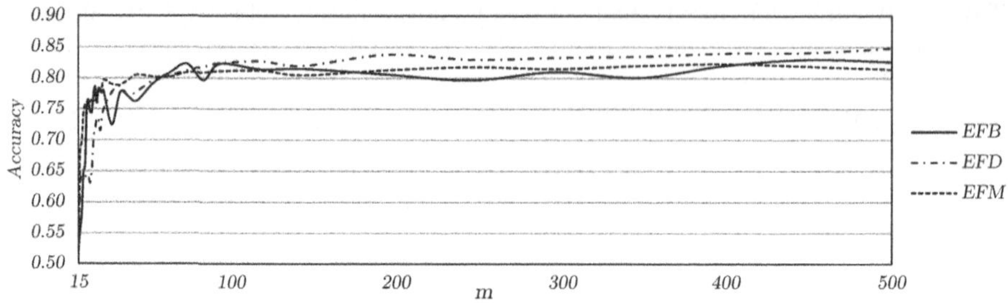

Fig. 1. Variation of accuracy at the variation of the number of pivots for EF* setups

Table 2. Five most similar words in a target language given a word in English

beautifully	classical	delightful
schöne (beautiful) 0.635	*adagio* 0.767	魅力(attractive) 0.610
liebevoll (loving) 0.596	*Martenot* 0.746	描き出さ(portrayed) 0.546
sehnsucht (longing) 0.533	*Charles-Marie* 0.736	風景(scenes) 0.545
ungewöhnlich (unusual) 0.510	*violoncelle* (cello) 0.727	繊細(delicate) 0.542
phantastisch (fantastic) 0.507	*soliste* (soloist) 0.720	味わえる(taste) 0.538

ability of DCI to embed the dispersed knowledge contained in less informative features, though this is a point left open to future investigation.

Statistical significance tests (paired t-test on the accuracy values) report that both DCI_{100} and DCI_{450} are significantly better, respectively with $p < 0.001$ and $p < 0.05$, than LSI, KCCA, and OPCA. There are no statistically significant differences between DCI, SCL and SSMC, so the comparison substantially ends with a tie, which is already a good result for a method so lightweight as DCI.

DCI obtains good results with just $m = 20$ pivots. For this value the list of source words to be translated is so small and composed by common-use words that even a user with an average proficiency in the foreign language could translate them without requiring external knowledge sources[6].

As a final note, we explored the ability of our feature profiles to capture the semantic relatedness of words, considering them as "cheap" word embeddings [12]. Table 2 illustrates the semantic properties captured by our feature profiles; it lists the most similar (cosine similarity) target words to a given source word.

4 Conclusions and Future Work

We have proposed Distributional Correspondence Indexing, an efficient feature-representation-transfer method for CLTC that creates feature profiles based on their distributional correspondence to a small set of pivots. The method indexes

[6] For example, for the EJD task the words to be translated were: great, worst, bad, awful, horrible, disappointed, terrible, love, wonderful, worse, disappointing, why, favorite, fun, performance, poor, collection, money, please, and enjoy.

documents in different languages into a common vector space where they become comparable. Empirical evaluation demonstrated our method performs comparably, and even better in some cases, to state-of-the-art methods. However, DCI has a much lower computational cost, and requires less human intervention.

DCI is a promising method, with many aspects worth being investigated: e.g., more sophisticated distributional correspondence functions; how to determine the optimal pivot set; testing DCI on imbalanced classes.

References

1. Bel, N., Koster, C.H.A., Villegas, M.: Cross-lingual text categorization. In: Koch, T., Sølvberg, I.T. (eds.) ECDL 2003. LNCS, vol. 2769, pp. 126–139. Springer, Heidelberg (2003)
2. Blitzer, J., McDonald, R., Pereira, F.: Domain adaptation with structural correspondence learning. In: Proceedings of the 2006 Conference on Empirical Methods in Natural Language Processing (EMNLP), pp. 120–128 (2006)
3. Dumais, S.T., Letsche, T.A., Littman, M.L., Landauer, T.K.: Automatic cross-language retrieval using latent semantic indexing. In: AAAI Spring Symposium on Cross-language Text and Speech Retrieval, p. 21 (1997)
4. Pan, S.J., Yang, Q.: A survey on transfer learning. IEEE Transactions on Knowledge and Data Engineering 22(10), 1345–1359 (2010)
5. Platt, J.C., Toutanova, K., Yih, W.T.: Translingual document representations from discriminative projections. In: Proceedings of the 2010 Conference on Empirical Methods in Natural Language Processing (EMNLP), pp. 251–261 (2010)
6. Prettenhofer, P., Stein, B.: Cross-language text classification using structural correspondence learning. In: Proceedings of the 48th Annual Meeting of the Association for Computational Linguistics (ACL), pp. 1118–1127 (2010)
7. Prettenhofer, P., Stein, B.: Cross-lingual adaptation using structural correspondence learning. ACM Transactions on Intelligent Systems and Technology (TIST) 3(1), 13 (2011)
8. Rigutini, L., Maggini, M., Liu, B.: An EM-based training algorithm for cross-language text categorization. In: Proceedings of the 3rd IEEE/WIC/ACM International Conference on Web Intelligence (WI), pp. 529–535 (2005)
9. Vinokourov, A., Shawe-Taylor, J., Cristianini, N.: Inferring a semantic representation of text via cross-language correlation analysis. In: Proceedings of the 16th Annual Conference on Neural Information Processing Systems (NIPS), pp. 1473–1480 (2002)
10. Wan, X.: Co-training for cross-lingual sentiment classification. In: Proceedings of the Joint Conference of the 47th Annual Meeting of the ACL and the 4th International Joint Conference on Natural Language Processing of the AFNLP: Volume 1-Volume 1, pp. 235–243 (2009)
11. Xiao, M., Guo, Y.: Semi-supervised matrix completion for cross-lingual text classification. In: Twenty-Eighth AAAI Conference on Artificial Intelligence (2014)
12. Zou, W.Y., Socher, R., Cer, D.M., Manning, C.D.: Bilingual word embeddings for phrase-based machine translation. In: Proceedings of the 2013 Conference on Empirical Methods in Natural Language Processing (EMNLP), pp. 1393–1398 (2013)

Adaptive Caching of Fresh Web Search Results

Liudmila Ostroumova Prokhorenkova, Yury Ustinovskiy, Egor Samosvat, Damien Lefortier, and Pavel Serdyukov

Yandex, Moscow, Russia
{ostroumova-la,yuraust,sameg,damien,pavser}@yandex-team.ru

Abstract. In this paper, we study the problem of caching search results with a rapid rate of their degradation. We suggest a new caching algorithm, which is based on queries' frequencies and the predicted staleness of cached results. We also introduce a new performance metric of caching algorithms called *staleness degree*, which measures the level of degradation of a cached result. In the case of frequently changing search results, this metric is more sensitive to those changes than the previously used *stale traffic ratio.*

Keywords: SERP caching, staleness degree, fresh vertical.

1 Introduction

Modern search engines have to continuously process a large number of queries by evaluating them against huge document indexes. This may cause an overload of the backend servers, an increase in the latency of query processing, and hence, eventually, user dissatisfaction. Search results caching helps to reduce query traffic to backend servers and thus to avoid overloading them. When queries with cached results are issued again, these results may be served from the cache, as it is much faster than to process those queries again. Unfortunately, it may as well lead to user dissatisfaction due to a chance to serve users with stale results. Therefore, cached results should be updated, when they are supposed to be stale. However, reissuing queries to update all cached results too often may lead to the immediate overload of the backend servers.

In previous works on search result caching, a cached result is thought to be stale, if the ordered set of top-k documents has changed even slightly: either their order has changed, or some documents have been added or deleted [4,6,9]. In this paper we show that this notion of staleness should be more flexible. First of all, the "time-to-die" for a cache entry should not necessarily come immediately after the corresponding SERP changed even a bit. SERPs for recency sensitive queries [7] may change very fast, what makes it impossible or, at least, impractical to update their caches every time when these small result changes take place. It is also often counterintuitive to update caches after even slightest search result changes, since some of them are often insignificant in terms of their influence on user satisfaction. Followed by these requirements and intuitions, our method, in contrast to its predecessors, tries to predict not only the fact that the result set for a query is changed at some point, but also the magnitude of its change. The main contributions of our paper are the following:

A. Hanbury et al. (Eds.): ECIR 2015, LNCS 9022, pp. 110–122, 2015.

- We argue that the "time-to-die" for a SERP does not need to come immediately with every change of that SERP. Being interested in the magnitude of SERP's changes, we define a new measure of caching algorithms' performance taking this intuition into account.
- We propose a new caching algorithm specially designed for search engines whose results change extremely fast. Our experiments demonstrate its advantage over the existing caching methods.
- The proposed algorithm is built upon the optimization framework that dynamically derives an optimal cache update policy based on frequencies of queries and the predicted degree of staleness of cache entries.

The rest of the paper is organized as follows. In Section 2, we discuss previous research on search results caching. In Section 3, we describe the caching framework we rely on, describe the data we use for the experiments, and discuss the new measure of caching algorithms' performance. We present the baseline algorithms and new caching strategies in Sections 4 and 5. The experimental results are presented in Section 6. Finally, we conclude the paper and outline directions for future research.

2 Related Work

Usually, a cached SERP is thought to be stale, if top-k documents have been changed. A cache invalidation policy is needed to detect if the cached query result is stale or not before processing the query. There are two groups of approaches to the invalidation of cached results. The first group uses knowledge about index changes [1,4]. These approaches are effective, but hard to implement in practice. First of all, they are computationally expensive due to the necessity of accurate and timely determination of changes in the index. Second, it is usually hard to observe how index changes affected cached SERPs and avoid costly processing of the corresponding queries. The algorithm proposed in this paper belongs to the second group of approaches, which do not monitor any index changes. Such algorithms rely just on the query log and the history of SERPs changes and can be of two types: active and passive. Passive methods update stale cache entries only in response to a user request [2]. Active methods update stale cache entries whenever they have available resources to do that [6,9]. Previous studies show that active methods outperform passive [10]. Hence, this paper aims at advancing the state-of-the-art active cache updating policies.

Passive methods use only TTL (time-to-live) values to invalidate cached results. If the age of a cache entry is greater than its TTL, then this entry is marked as expired. Often, each entry is associated with a fixed TTL [1,4,5,6], although, the methods with adaptive TTLs were also suggested [2]. Active policies are often also supplemented with TTLs in order to set limits on the age of shown results. Note that in our case we cannot use the adaptive TTLs from [2], which considered an idealized setting with no constraints on computing resources. The reason is that the sets of top documents from a fresh vertical for almost all queries change extremely frequently and this leads to very small adaptive TTLs according to [2]. Small TTLs inevitably lead to prohibitively frequent updates of cache entries, and that is what a search engine always tries to avoid.

This motivated us to predict the magnitude of SERP changes instead of just predicting whether the SERP has changed or not.

An active caching policy was suggested in [9]. Here a machine learning method is used to estimate the arrival times of future queries. Based on this information, some cached results are chosen for updating. Since in our case many queries are issued several times per second (see Section 3.3), we predict not the arrival time of queries, but their frequencies. Moreover, as we discuss in Section 5, even the precise knowledge of when and how many queries will be issued in the near future gives a rather small improvement. Another active caching policy with proactive prefetching of cached results was suggested in [6]. It also uses TTLs to invalidate cached results, but, besides, leverages idle cycles of the backend servers to re-process queries and refresh cache entries proactively, even before they expire according to their TTLs. A cached result is chosen to get updated based on the product of its age and the frequency of the respective query. Given all the above-mentioned constraints, we regard this method as the only appropriate baseline for our study.

3 General Framework

3.1 System Architecture

In this section, we describe the system architecture we consider in this paper. All queries are issued to the front-end of a search engine. A query result can be either taken from the cache or retrieved by forwarding the query further to the back-end search cluster. All previously unseen queries are always forwarded to the back-end cluster, since there are no cached results for them. Also, if TTL values are used by the search engine, all expired entries are deleted from the cache and the corresponding queries are considered as unseen. Results for other queries are served to users from the cache. The cache refreshing scheduler decides which SERP's cache to update during idle cycles of the search cluster (by using spare computing resources of the search engine).

The final SERP shown to the user usually contains results from different verticals with different indexes [3]. In this paper, we are interested in the vertical serving fresh content, removing any document older than 10 days from its index. This threshold is chosen in accordance with [7], where it was observed that most relevant documents for recency sensitive queries have ages within 10 days. However, such a threshold may depend on the current search engine's settings and its understanding of how old the content should be to be considered "fresh". For all queries seeking documents from this vertical, the list of top relevant documents changes extremely fast, therefore it is very important to have an efficient policy for caching the results served by this vertical.

3.2 Metrics

The primary requirements of caching algorithms with infinite cache capacity are high freshness of served results and reduced load of the search engine's back-end. The common evaluation measures of these algorithms are *stale traffic ratio*, i.e., the fraction of stale query results shown to users and *false positive ratio*, i.e., the fraction of redundant cache updates [2,4,6].

Most previous studies consider that a cached result page $S_c(q)$ (i.e., the list of k most relevant documents) served for the query q can be in just two states, either fresh or stale: $I_S(q) = 0$ if $S_c(q) = S_a(q)$ (fresh state) and $I_S(q) = 1$ if $S_c(q) \neq S_a(q)$ (stale state), where $S_a(q)$ is the actual up-to-date list of top-k documents obtained by processing the query [4,6,9]. Then stale traffic ratio ST is the average of binary staleness $I_S(q)$ over all queries $q \in Q$ issued within a given period of time $[t_0, t_1]$: $ST([t_0, t_1]) = \frac{1}{|Q|} \sum_{q \in Q} I_S(q)$.

We argue that *staleness* of a cached result page is *not necessarily a binary property*. Clearly, all stale result pages $S_c(q)$ are stale to a varying degree. Therefore, we propose to study more discriminative measures of staleness than ST metric. We decided to focus on the NDCG-like measure, which we suppose to be the most suitable for the task of fresh vertical results caching. First, we introduce *staleness degree* $d(S_c(q), S_a(q))$ of a single served result $S_c(q)$ — a non-binary alternative to $I_S(q)$. Second, we define *staleness degree ratio*, similarly to ST, as the average of $d(S_c(q), S_a(q))$ over queries Q issued within a period $[t_0, t_1]$:

$$StDeg([t_0, t_1]) = \frac{1}{|Q|} \sum_{q \in Q} d(S_c(q), S_a(q)). \tag{1}$$

To measure staleness degree $d(S_c(q), S_a(q))$ we compare top-k documents shown to users in $S_c(q)$ with the actual up-to-date top-k most relevant documents $S_a(q)$. The parameter k will be referred to as *cut-off parameter*. Motivated by the principles embodied in the classical NDCG [8] measure, we introduce the notions of "gain" and "discount" to compute the quality of a cached search result. Let $pos_c(u)$ and $pos_a(u)$ denote the positions of a document u in $S_c(q)$ and $S_a(q)$ respectively. As in NDCG measure, the gain of a document u is some increasing function of its *relevance* and its discount is a decreasing function of its position. Since the only information about the current relative relevance of the documents in $S_c(q)$ we normally have is their positions in $S_a(q)$, we assume that the relevance of document u depends solely on $pos_a(u)$. The resulting quality measure of $S_c(q)$ is $\mathcal{M}(S_c(q), S_a(q)) = \sum_{u \in S_c(q)} \text{gain}(pos_a(u))\text{discount}(pos_c(u))$. If $u \in S_c(q)$, but $u \notin S_a(q)$, u could have been ranked at any position $\geq k+1$. In that case, u is assigned the $k+1$ position: $pos_a(u) := k+1$. As in NDCG, we define staleness degree $d(S_c(q), S_a(q))$ by normalizing $\mathcal{M}(S_c(q), S_a(q))$ to the unit segment ($\mathcal{M}_{max}$ and $\mathcal{M}_{min}$ are maximal and minimal possible values of $\mathcal{M}$, they depend on the choice of gain and discount):

$$d(S_c(q), S_a(q)) = 1 - \frac{\mathcal{M}(S_c(q), S_a(q)) - \mathcal{M}_{min}}{\mathcal{M}_{max} - \mathcal{M}_{min}}. \tag{2}$$

In this paper, we set $gain(u) = 1/(pos_a(u)+1)$ and $discount(u) = 1/(pos_c(u)+1)$. In Section 6, we analyze how the choice of different values of k (the number of documents in $S_c(q)$ and $S_a(q)$) affects the performance of a caching algorithm optimized for $StDeg$. Further, if not specified otherwise, we use the fixed cut-off parameter $k = 10$. Here are some examples of staleness degree values (for $k = 10$): $d(S_c(q), S_a(q)) = 0.52$ if the first document of $S_a(q)$ is replaced by another (irrelevant) document in $S_c(q)$, $d(S_c(q), S_a(q)) = 0.36$ if the first document of $S_a(q)$ (the most relevant) is absent in $S_c(q)$ (all other documents are moved up by one position), $d(S_c(q), S_a(q)) = 0.15$ if the second document is absent.

3.3 Data

Our experiments are based on the query log of the most popular among the search engines operating in Russia — Yandex (*yandex.com*, *yandex.ru*). First, we collected the dataset D_1 required to conduct our motivating experiments (see Section 4) and tune several parameters of our algorithm. We sampled 6K random unique queries from the stream of queries issued on December 29, 2012. Then we monitored all issues of these queries by real users from December 29, 2012 to February 18, 2013 (~700M issues). We also collected the dataset D_2 required to evaluate the performance of our algorithms. For that purpose, we sampled another set of 6K random queries issued on February 25, 2013 and monitored all issues of these queries from February 25, 2013 to March 4, 2013 (~85M issues). The most frequent query was issued 20 times per second on average. Also, 23% of queries were, at some point, issued several times per second. On the other hand, 75% of queries were issued less then 5 times per hour each on average.

Note that, like the standard *ST* metric, *StDeg* metric can be used only for offline tuning since it requires the computation of actual query results in the search backend. In order to perform an offline evaluation and to train our predictor of a search result change, we needed to understand the dynamics of SERP changes. Therefore, we issued all 12K selected queries every 10 minutes during the corresponding above-mentioned time periods and saved the result pages of the vertical serving fresh content. Finally, for the first period we saved 45M SERPs and for the second period we saved 7M SERPs. The size of our dataset is comparable or exceeds the sizes of the datasets used in the previous studies of SERP caching with offline evaluation. In [2], e.g., 4,500 queries were issued once a day for a period of 120 days and the top 10 results were saved, so only 540K SERPs were saved.

4 Algorithms

In this section, we describe a general framework of all the algorithms we consider. All the algorithms have a limited quota for cache updates N: the number of query results which can be computed in the search backend per second. Note that N can be any positive real number. For example, if $N < 1$ then it is allowed to process only one query per $1/N$ seconds. As in the previous studies on caching [1,4,5,6,9], we also use TTL in order to always avoid showing too old results to users. As we already mentioned, all cache entries which have been in the cache for longer than TTL are marked as expired, and, if the corresponding queries are issued by users again, then they are passed directly to the back-end cluster. The cache entries for expired-but-reissued queries always have the highest priority and updated before any other cache entries, as proposed in [6]. Apparently, the more queries with expired TTL are issued, the less spare resources we have and the less cached results with non-expired TTL can be updated. In that way, the number of allowed cache updates per second in our system is dynamic.

The core part of the caching algorithms in our architecture is the refreshing scheduler, which updates cached results using spare computing resources. At every moment of time, we have a set of triples in the cache, each characterized by a query, its cached result and the time of its last update: $T_q = \{q, S_c(q), t_u\}$. Every τ seconds (re-ranking period) we rank all the triples $\{T_q\}$ according to their priorities in our refreshing scheduler — forming a queue of results $S_c(q)$ to update. As soon as we have spare resources, at time t_{now}, we take top queries from this queue, update their cache entries (i.e., for a query q, we replace its cached SERP $S_c(q)$ with the actual one $S_a(q)$), eliminate them from the queue.

The quality of our algorithm essentially depends on the way we prioritize cached results to be updated. In general, this prioritization should continuously estimate the cost of keeping the outdated cache entry in the cache in the next period of time, what can be also roughly described as the need to rank triples corresponding to frequent queries with highly outdated results S_c higher. So, the refreshing scheduler should periodically perform the following steps: 1) estimate the frequency of query q; 2) estimate integral staleness of the cached result S_c in the next τ seconds following the current batch cache update; 3) combine these quantities into a ranking function. Further in this section, we describe three approaches to implement a refreshing scheduler.

4.1 Baseline

We implement the *age-frequency* (AF) strategy from [6] as our baseline. In [6] the scheduler ranks all cached results according to the value $f(q)\Delta t$, where $f(q)$ is the frequency of q and $\Delta t = t_{now} - t_u$ is the age of the cached SERP $S_c(q)$. In some sense, AF estimates the staleness of S_c simply as Δt and combines staleness Δt and frequency $f(q)$ simply by taking their product. In the next section we show, that on real data the staleness of $S_c(q)$ as a function of Δt is neither linear, nor query-independent. This observation motivated us to propose a refreshing scheduler that takes these observations into account.

4.2 Staleness-Frequency Strategy

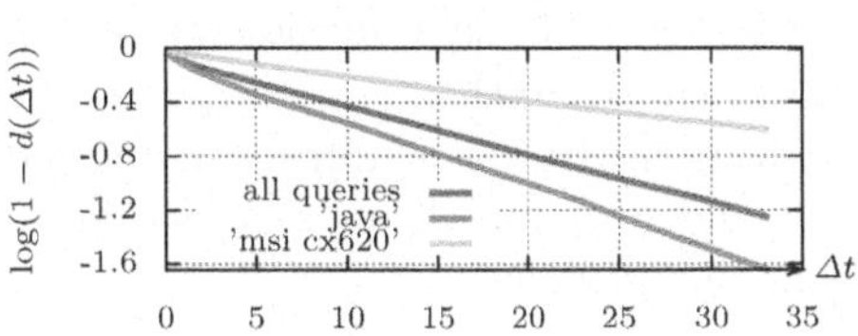

Fig. 1. Freshness of Δt old cached results

We start with some motivating experiments. For all queries we computed the average freshness (i.e., $1 - d(S_a(q), S_c(q))$) of cached results with $10, 20, \ldots, 2000$ minutes age. Figure 1 shows the obtained results averaged over all queries ('All queries') and two exemplary queries. There is freshness in logarithmic scale on y-axis and the age of cached result in hours on x-axis. It follows from this figure that the freshness of the cached result for the query q can be approximated as

$$1 - d(S_c(q), S_a(q)) = e^{-\theta(q)\Delta t}, \tag{3}$$

where $\theta(q)$ is the degradation rate of a cached result $S_c(q)$ created at time t_u, Δt is its age, and $S_a(q)$ is the actual result for the query q at time $t_u + \Delta t$.

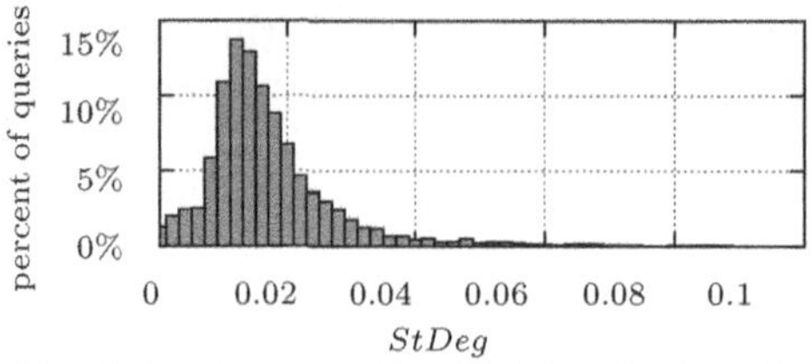

Fig. 2. Staleness of 10 min old cached results

Apart from the evolution of freshness for the 'average query', Figure 1 demonstrates that the de-gradation rate of the cached result essentially depends on the query: $\theta = \theta(q)$. Figure 2 gives additional evidence supporting this assertion. It displays the distribution of staleness degree $d(S_c(q), S_a(q))$ over all 10-min old cached results. Altogether, these observations allow to reduce the problem of estimation of staleness of $S_c(q)$ at a certain moment of time to the problem of estimation of $\theta(q)$.

The *staleness-frequency* (SF) strategy aims at minimizing the staleness degree of all results shown to users within some period of time $[t_0, t_1]$. Assume that: (1) we aim at minimizing $StDeg([t_0, t_1])$, (2) query issues are uniformly distributed within the time interval $[t_0, t_1]$ ($f(q)$ issues per second), (3) interval $[t_0, t_1]$ is 'small' (see details below). If we do not update triple $T_q = \{q, S_c(q), t_u\}$ during $[t_0, t_1]$, its contribution into $StDeg([t_0, t_1])$ is ($1/|Q|$ is the factor from the definition of $StDeg$):

$$L(q) = \frac{1}{|Q|}\int_{t_0}^{t_1} d(S_c(q), S_{a(t)}(q))f(q)\,dt = \frac{1}{|Q|}\int\limits_{t_0-t_u}^{t_1-t_u}\left(1 - e^{-\theta(q)x}\right)f(q)\,dx \simeq$$

$$\simeq \frac{t_1 - t_0}{|Q|}f(q)\left(1 - e^{-\theta(q)(t_0 - t_u)}\right). \quad (4)$$

In the latter equality, we assume that $\theta(q)(t_1 - t_0) \ll 1$, hence $1 - e^{-\theta(q)x}$ is almost constant on $[t_0 - t_u, t_1 - t_u]$. The greedy solution to the $StDeg([t_0, t_1])$-minimization problem always updates the cache entry with the maximal loss value $L(q)$. In our framework we reorder the queue of non-expired cache entries every τ seconds, therefore $StDeg$ and $L(q)$ in Equation (4) are computed over interval $[t_{now}, t_{now} + \tau]$. Note that the assumption $\theta(q)(t_1 - t_0) \ll 1$ is automatically satisfied, if τ is small enough. The refreshing scheduler sorts triples T_q according to the value $L(q)$ forming a queue. Basically, it sorts queries according to the value $f(q)(1 - e^{-\theta(q)\Delta t})$, since the factor $\frac{t_1 - t_0}{|Q|}$ in Equation (4) is the same for all triples. Afterwards, we greedily optimize $StDeg([t_{now}, t_{now} + \tau])$ by continuously updating the cache entries waiting in this queue during the period τ. The computation of $L(q)$ requires $f(q)$ and $\theta(q)$ for all queries. Section 5 describes how to estimate these values from the query log.

4.3 Log-Staleness-Frequency Strategy

Although the SF strategy optimizes our performance measure $StDeg$ greedily and directly, its performance may be heavily affected by our TTL policy. The reason is that SF policy tends to update frequent queries rather often, while rare queries with completely outdated results always have lower priority. For instance, assume that for queries q_1 and q_2 we have $f(q_1) \gg f(q_2)$. The staleness degree after 10 minutes on our data is greater or equal to 0.001 for 99% of queries, thus if $f(q_1)/f(q_2) > 1000$, the query q_1 will be ranked higher irrespective of degradation of q_2 for the majority of queries q_1 and q_2. Given that moderately frequent and rare queries constitute a large share of total query volume, continuous serving of stale results for them will soon result into a dramatic increase in user dissatisfaction. The TTL mechanism, described earlier, allows to upper-bound the age of cached results, but, eventually, due to the above-mentioned imbalance, leads to overabundance of expired cache entries.

This observation motivated us to develop a strategy, which gives more weight to the highly stale results in the queue and hence updates them before their expiration more often. This strategy is based on the greedy optimization of a new intermediate objective function $\mathcal{P}$. As for SF strategy, we assume, that staleness $d(S_c(q), S_a(q))$ does not change within a short time interval $[t_0, t_1]$ under consideration, i.e., $\theta(q)(t_1 - t_0) \ll 1$. We define $\mathcal{P}$ as the product of freshnesses (which is one minus staleness) of all results Q shown within $[t_0, t_1]$ and maximize it: $\mathcal{P}([t_0, t_1]) = \prod_{q \in Q}(1 - d(S_c(q), S_a(q))^{f(q)(t_1 - t_0)} \to \max$.

As we have discussed in the previous section, the function $e^{-\theta(q)\Delta t}$ gives a reasonable estimation for freshness of $S_c(q)$ at time t_{now}, where the age of the cache entry $\Delta t = t_{now} - t_u$. Thus, the maximization reduces to the following minimization problem:

$$- \log \mathcal{P}([t_0, t_1]) = -(t_1 - t_0) \sum_q f(q) \log(1 - d(S_c(q), S_a(q)) =$$

$$= (t_1 - t_0) \sum_q f(q)\theta(q)\Delta t \to \min . \quad (5)$$

As in the SF strategy (4), a greedy solution to this minimization problem on the time interval $[t_{now}, t_{now} + \tau]$ updates results with the maximal value of $\theta(q)f(q)\Delta t$. Note that for small values of Δt, $\theta(q)\Delta t$ is close to $1 - e^{-\theta(q)\Delta t} = d(S_a, S_c)$. So, this strategy is similar to SF for those cache entries, which are often updated. What is more important, it tends to update old cached results more often, since $1 - e^{-\theta(q)\Delta t} \ll \theta(q)\Delta t$ for large Δt. In Section 6 we show that this algorithm, referred to as LSF (log-staleness-frequency), indeed outperforms SF.

5 Prediction of Staleness and Query Frequency

The proposed algorithms require the estimation of query's frequency $f(q)$ and degradation rate $\theta(q)$. In this section, we describe our estimation methods.

5.1 Frequency

The frequency of the query is calculated over the whole period of past observations. The same historical frequency was used in [6]. We also noticed that knowing the true number of times the query will be issued in the next τ seconds (oracle strategies) improves the quality of all algorithms only by $\sim 1\%$, thus the estimation of frequency $f(q)$ represents a much less challenging problem than the estimation of degradation rate $\theta(q)$.

5.2 Staleness

Query-Independent Estimation. Firstly, we estimate the query-independent rate of staleness on the training data. That is, for every temporally consecutive pair of search engine result pages $S_1(q)$ at time t_1 and $S_2(q)$ at time t_2 we estimate the value of $\theta(q)$ as $\theta(q; T_q^1, T_q^2) = \frac{\log(1-d(S_c^1(q), S_c^2(q)))}{t_2 - t_1}$. In order to define $\widehat{\theta}$ we first average these estimates over all temporally consecutive pairs $S_1(q)$ and $S_2(q)$ for a given query q and then average obtained values over all queries. Let us remind that the interval $t_2 - t_1$ between two temporally consecutive search results is 10 minutes in our study (see Section 3.3). Basically, $\widehat{\theta}$ is just the slope of the 'All queries' blue line on Figure 1.

Historical Estimation. Now we proceed with the description of the estimation of query-dependent parameter $\theta(q)$. At any moment of time for each query q we have a sequence of cached results at the preceding moments of time $\mathcal{H}(q) = \{(S_c^i(q), t_u^i)\}_{i=0}^H$, where t_u^i are the moments at which we updated the cache entry for query q. Note that both the moments t_u^i and the number of cached results H depend on our caching strategy. We use these historical data in order to make the historical estimation of $\theta(q)$. Namely, for each adjacent pair $(S_c^{i-1}(q), t_u^{i-1}), (S_c^i(q), t_u^i)$ in $\mathcal{H}(q)$, we calculate the i-th estimate of $\theta(q)$:

$$\theta^i(q) := \frac{\log(1 - d(S_c^{i-1}(q), S_c^i(q)))}{t_i - t_{i-1}}. \tag{6}$$

We derive H-th historical estimation $\widehat{\theta}_H(q)$ of $\theta(q)$ from the sequence $\theta^i(q)$ by considering the historical average $\widehat{\theta}_H(q) = \frac{1}{H}\sum_{i=1}^H \theta^i(q)$ over $\leq T$ hours old cached results $S_c^i(q)$. For the experiments we took $T = 24$. We tried other values of T (12h, 48h) on the datased D_1 and noticed almost no influence of this parameter ($\pm 1\%$) on *StDeg*. We also tried to use exponential moving average instead of historical average and obtained better performance for historical average.

Since our approach requires sufficient historical data $\mathcal{H}(q)$, in order to construct a reasonable estimation of $\theta(q)$, we need some prior estimate for new queries and for queries with little history $\mathcal{H}(q)$. For that purpose, we combine historical estimation of $\widehat{\theta}(q)$ with query-independent estimation $\widehat{\theta}$:

$$\widehat{\theta}_C(q) = \frac{H}{w + H}\widehat{\theta}(q) + \frac{w}{w + H}\widehat{\theta}, \tag{7}$$

where w is the parameter of our algorithm accounting for the weight of query-independent estimation $\widehat{\theta}$. Moreover, the more cached results H for the query q we have, the more reliable historical estimation $\widehat{\theta}(q)$ we get and the higher weight to it is given in Equation (7), given the parameter w.

6 Experiments

In this section, we compare our algorithms SF and LSF with the baseline strategy AF and analyze the influence of parameters on their performance. Our algorithm has several parameters. Number of allowed cache updates per second N and TTL are usually defined by the requirements of a search engine: N directly corresponds to the maximum possible load of the back-end cluster and TTL limits the age of shown results. Further in this section, we analyze the influence of both parameters on the performance of the algorithms.

Re-Ranking Period τ. Re-ranking period is a parameter of all the algorithms. On the one hand, small values of τ lead to higher flexibility and better estimation of parameters. On the other hand, according to our framework, τ is the lower bound for the update frequency of each cache entry: we cannot update an entry more than once during one cycle. Therefore, the influence of τ is not necessarily monotone. In further experiments, for each algorithm and for every combination of other parameters we tune τ on the dataset D_1.

The Estimation of Frequency and Staleness. For the algorithms SF and LSF we tuned the weight w (see Equation 7) on the dataset D_1. We noticed that the performance of these algorithms is the best and almost constant if w is in the interval $[5, 100]$. For the rest of the experiments we fix $w = 10$.

Comparison with the Baseline. We compare all three strategies on the one-week test query log (dataset D_2, see Section 3.3). For every moment of time $t \in [t_s, t_e]$, where t_s and t_e correspond to the start and the end of our testing period, we define staleness degree metric $StDeg(t)$ as follows (see Equation (1)): $StDeg(t) := StDeg([t, t+24h])$. Then for each algorithm $\mathcal{A} \in \{\text{AF, SF, LSF}\}$ let $StDeg_{\mathcal{A}}(t)$ be the corresponding metric. Figure 3 demonstrates relative improvement of $StDeg_{\mathcal{A}}(t)$ over AF strategy. As one can see from the figure, both SF and LSF outperform AF strategy according to $StDeg(t)$ during the entire test period, except the short warm-up interval. Throughout this interval, SF and LSF accumulate information on query-specific historical degradation to better predict the staleness of the corresponding results. Apparently, the more historical data for a given query we collect, the more accurate prediction $\widehat{\theta(q)}$ of the degradation rate we make with Equation (7), improving the overall quality of both methods.

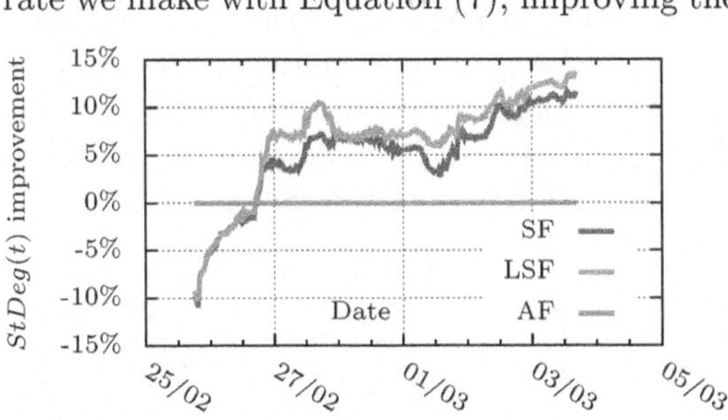

Fig. 3. Relative improvement of $StDeg(t)$

Furthermore, as we mentioned in Section 4.3, LSF often outperforms SF strategy. However, most of the time both methods improve *StDeg* with respect to AF by 5-15%. The effect of our caching algorithms on *false positive, stale traffic* and *staleness degree* measures over the test period $[t_s, t_e]$, except 1-day warm-up period, is given in Table 1 (see $N = 1$).

Let us present an interpretation of the value $StDeg = 0.02$ (see Section 3.2). $StDeg = 0.02$ if for 2% of queries completely irrelevant list of results is shown, or for 6% of queries the most relevant document is missed, or for 13% of queries the second most relevant document is missed.

Number of Allowed Updates Per Second. We evaluated all three algorithms with various values of N on the one-week test log (dataset D_2). For realistic experiments, this parameter should be proportional to the size of the analyzed query sample and should be adequate for the needs of a certain vertical. Since our dataset contains 6000 unique queries, we considered rather small values of N ($N \leq 1$). Note that if it is allowed to update only one cache entry per second ($N = 1$), then the cache for all 6000 queries can be updated in 100 minutes, i.e., it is possible to keep all the cached results to be not older than 2 hours, which is acceptable for a vertical serving fresh content. For example, parameters used in [6] allow to update cache for all unique queries in 3.5-7.5 days.

For every method we computed the average values of a certain metric during the one-week test period, except for the one day warm-up period. Exclusion of the warm-up period is very natural, since our algorithms are highly unstable in the beginning. Table 1 demonstrates the growth of *StDeg* with the decrease of N. As we expected, the influence of parameter N on the quality of caching algorithm is much stronger than, say, the choice of the refreshing scheduler. Indeed, with a small value of N we spend almost all available resources on queries with the expired cached entries, instead of updating non-expired cache entries queued by our caching algorithm, as long there are no idle cycles left to pro-actively update non-expired cache entries.

Table 1. Influence of N

Alg.\Metric	FP	ST	StDeg
AF, $N = 1$	0.23	0.067	0.021
SF, $N = 1$	**0.21**	0.064	**0.019**
LSF, $N = 1$	**0.21**	**0.063**	**0.019**
AF, $N = 1/2$	0.12	0.092	0.030
SF, $N = 1/2$	0.12	0.091	0.029
LSF, $N = 1/2$	**0.11**	**0.090**	**0.028**
AF, $N = 1/3$	0.079	0.12	0.039
SF, $N = 1/3$	0.075	0.12	0.038
LSF, $N = 1/3$	**0.073**	**0.11**	**0.037**
AF, $N = 1/5$	0.025	0.18	0.057
SF, $N = 1/5$	**0.024**	0.18	0.058
LSF, $N = 1/5$	**0.024**	**0.17**	**0.055**

It is interesting to note, that for larger values of allowed updates per second (e.g., $N = 1$), both SF and LSF policies perform relatively well improving over the AF's quality (see Table 1). On the contrary, small values of N result into the evident degradation of greedy SF algorithm in terms of *StDeg* in comparison with both AF and LSF methods. This observation was quite surprising to us, since SF directly relies and improves the objective measure *StDeg*. In fact, the results of this comparison of AF with SF motivated us to develop the modification of SF — LSF. We discuss this observation in details in Section 4.3.

TTL. Table 2 shows the influence of TTL on the performance of the algorithms. As expected, too small values of TTL make the performance of all algorithms worse. The reason is that all algorithms spend too many resources updating expired cached entries. When TTL becomes larger, performance stabilizes, since there are not too many expired entries and the algorithms are able to follow their main strategies. Note that all values of TTL are comparable, but are also slightly smaller than previously used TTL values [6]. We consider smaller values due to the specificity of fresh vertical, whose users have lower tolerance to stale results than users of an average vertical.

Table 2. Influence of TTL

Alg.\Metric	FP	ST	StDeg
AF, TTL = 2 h	0.16	0.086	0.026
SF, TTL = 2 h	**0.15**	**0.084**	**0.025**
LSF, TTL = 2 h	**0.15**	**0.084**	**0.025**
AF, TTL = 5 h	0.22	0.069	0.022
SF, TTL = 5 h	**0.21**	**0.066**	**0.020**
LSF, TTL = 5 h	**0.21**	**0.066**	**0.020**
AF, TTL = 10 h	0.23	0.067	0.021
SF, TTL = 10 h	**0.21**	0.064	**0.019**
LSF, TTL = 10 h	**0.21**	**0.063**	**0.019**

Table 3. Relative improvements of $StDeg_k$ of SF/LSF over AF for various k

Alg.\ k	1	2	3	4	5	6	7	8	9	10
LSF	16.0%	7.6%	7.1%	6.2%	5.6%	4.6%	5.6%	6.0%	10.9%	9.4%
SF	12.9%	6.8%	4.7%	4.7%	4.7%	3.3%	4.4%	3.4%	10.6%	6.8%

Cut-Off Parameter. The choice of the cut-off parameter k affects both the metric and our algorithms. Previously, we fixed $k = 10$, now for each $k = 1, \ldots, 10$ we run corresponding SF and LSF algorithms and measure their improvement over the baseline algorithm. For each k we denote the *StDeg*-metric as $StDeg_k$. Since for different k the algorithms SF and LSF aim at optimizing $StDeg_k$, we report improvements according to these metrics, see Table 3. One can see that despite of the choice of cut-off k, all our algorithms still significantly outperform the baseline according to the corresponding metric.

Estimation of $\theta(q)$. It was also interesting to know how much it is possible to improve the caching algorithms by improving our estimation of $\theta(q)$. To answer this question, we define and evaluate the Oracle based prediction of staleness. It takes the real staleness at the moment $d(S_c(q), S_a(q))$ and uses it in Equation (4) instead of $1 - e^{-\theta(q)(t_1 - t_0)}$ and in Equation (5). On Figure 4 we demonstrate $StDeg(t)$ for the AF baseline, for ordinary SF and LSF methods, and for SF and LSF algorithms employing Oracle estimation of staleness (again, $N = 1$, TTL $= 10h$). As one can see, the knowledge of real staleness extremely improves the performance of the algorithm, indicating that staleness prediction algorithms are a promising subject of research in the future.

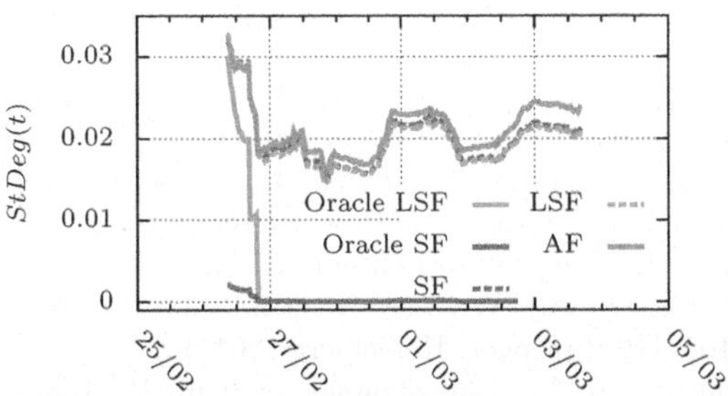

Fig. 4. Oracle prediction of staleness

7 Conclusion and Future Work

In this paper, we focus on the algorithms for caching results for the vertical serving fresh content, where top documents for a query change extremely fast. This motivated us to introduce and measure a new highly discriminative metric of cache entry quality: staleness degree. The algorithms we suggest are based on the minimization of the new metric — average staleness degree of results presented to users. The observed properties of this metric allow to solve the minimization problem greedily and directly. Our experimental results show that, independent of specific settings of various common parameters of algorithms, our methods outperform the baseline. The core part of both of our methods is the query-specific estimation of the degradation rate of a cache entry. In additional experiments we demonstrate that our approach has the potential to be improved by more accurate estimation of the degradation rate, which reveals a novel and promising direction in this research area.

We have also noticed that the staleness $d(S_c(q), S_a(q))$ is not only well approximated by an exponential function (see Equation (3)) with the parameter taking different values for different queries, but also this parameter changes in time. We noticed that the degradation rate is smaller during the weekends and at the night, as long as, indeed, new content usually appears and web pages are updated more often during business hours. Our way of degradation rate estimation utilizes rather small windows of historical averages and hence is able to adapt to daily trends dynamically. However, in our future work, we are going to experiment with time-dependent estimation of $\theta(q)$ which takes into account daily and weekly fluctuations.

References

1. Alici, S., Altingovde, I., Ozcan, R., Cambazoglu, B., Ulusoy, O.: Timestamp-based result cache invalidation for web search engines. In: Proc. SIGIR 2011 (2011)
2. Alici, S., Altingovde, I.S., Ozcan, R., Cambazoglu, B.B., Ulusoy, O.: Adaptive time-to-live strategies for query result caching in web search engines. In: Proc. 34th ECIR Conf., pp. 401–412 (2012)
3. Arguello, J., Diaz, F., Callan, J.: Learning to aggregate vertical results into web search results. In: Proc. 20th ACM CIKM Conf., pp. 201–210 (2011)
4. Blanco, R., Bortnikov, E., Junqueira, F., Lempel, R., Telloli, L., Zaragoza, H.: Caching search engine results over incremental indices. In: Proc. SIGIR 2010 (2010)
5. Bortnikov, E., Lempel, R., Vornovitsky, K.: Caching for realtime search. In: Clough, P., Foley, C., Gurrin, C., Jones, G.J.F., Kraaij, W., Lee, H., Mudoch, V. (eds.) ECIR 2011. LNCS, vol. 6611, pp. 104–116. Springer, Heidelberg (2011)
6. Cambazoglu, B., Junqueira, F., Plachouras, V., Banachowski, S., Cui, B., Lim, S., Bridge, B.: A refreshing perspective of search engine caching. In: WWW 2010 (2010)
7. Dong, A., Chang, Y., Zheng, Z., Mishne, G., Bai, J., Zhang, R., Buchner, K., Liao, C., Diaz, F.: Towards recency ranking in web search. In: Proc. WSDM 2010 (2010)
8. Järvelin, K., Kekäläinen, J.: Cumulated gain-based evaluation of IR techniques. ACM Transactions on Information Systems (TOIS) 20(4), 422–446 (2002)
9. Jonassen, S., Cambazoglu, B., Silvestri, F.: Prefetching query results and its impact on search engines. In: Proc. SIGIR 2012, pp. 631–640 (2012)
10. Kroeger, T.M., Long, D.D.E., Mogul, J.C.: Exploring the bounds of web latency reduction from caching and prefetching. In: 1st USITS (1997)

Approximating Weighted Hamming Distance by Probabilistic Selection for Multiple Hash Tables

Chiang-Yu Tsai, Yin-Hsi Kuo, and Winston H. Hsu

National Taiwan University, Taipei, Taiwan
{fishy,kuonini}@cmlab.csie.ntu.edu.tw, whsu@ntu.edu.tw

Abstract. With the large growth of photos on the Internet, the need for large-scale, real-time image retrieval systems is emerging. Current state-of-the-art approaches in these systems leverage binary features (e.g., hashed codes) for indexing and matching. They usually (1) index data with multiple hash tables to maximize recall, and (2) utilize weighted hamming distance (WHD) to accurately measure the hamming distance between data points. However, these methods pose several challenges. The first is in determining suitable index keys for multiple hash tables. The second is that the advantage of bitwise operations for binary features is offset by the use of floating point operations in calculating WHD. To address these challenges, we propose a probabilistic selection model that considers the weights of hash bits in constructing hash tables, and that can be used to approximate WHD (AWHD). Moreover, it is a general method that can be applied to any binary features with predefined (learned) weights. Experiments show a time savings of up to 95% when calculating AWHD compared to WHD while still achieving high retrieval accuracy.

1 Introduction

The last decade has seen explosive growth in the number of images on the Internet, driven by the widespread use of smartphones. The image sharing website Instagram has stated that their users post 60 million photos daily [1]. Because of these trends, mobile visual retrieval [2] has become an increasingly popular application.

When an image query is issued by a user, a retrieval system looks for similar images to return. Finding similar images uses a nearest neighbors search between the image being queried on and the pool of available images that can be returned. Most image searches involve binary signatures (e.g., [3]) to represent an image because of the fast computation speed, storage efficiency, and reduced bandwidth requirements associated with signatures.

Traditional nearest neighbor search methods usually use a single hash table to index data. However, this can lead to low recall in search results if inappropriate index keys are used. To deal with this, current state-of-the-art methods typically use a union of results from multiple hash tables [3]. Accurate results will be returned if at least one index key from the hash tables matches the index key of

A. Hanbury et al. (Eds.): ECIR 2015, LNCS 9022, pp. 123–134, 2015.

the queried data. This can lead to improved recall in search results, as will be shown in Section 5.2.

However, there is a challenge when using binary signatures with multiple hash tables: Determining suitable index keys for the hash tables. Index keys should ideally be generated by hashing functions with a high discriminative power in order to group similar data into a bucket.

State-of-the-art methods for nearest neighbors search typically use weighted hamming distance (WHD) to help determine similarity. Hamming distance (HD) is a simple and commonly used method to measure the similarity between two binary signatures. However, it cannot distinguish the relative importance of different bits. WHD addresses this by assigning weights to each bit [4] in calculating distance. But the calculation speed of WHD is much slower than HD due to the use of floating point operations.

To address these challenges, we utilize several selection processes to represent the discriminative power of each bit. We refer to the selection process as a probabilistic selection model. This model can produce different index keys based on the given weights. Using these index keys for multiple hash tables can lead to higher recall in search results. This model can also generate an approximate weighted hamming distance (AWHD) that can track closely to WHD without needing to use floating point operations. This can save up to 95% of the time normally needed to calculate WHD and still return accurate results. Such characteristics make a retrieval system more scalable with respect to big data, particularly the increasing number of images on the Internet.

2 Related Work

Content-based image retrieval (CBIR) has been researched since the 1990s and it is still a popular research area today because of the large growth in digital photos. Nowadays, a state-of-the-art image retrieval system utilizes bag-of-visual-words (BoVW) to efficiently return accurate results. Due to the widespread use of smartphones and tablets, performing CBIR on mobile devices and providing better capabilities for mobile visual search have been areas of huge interest. BoVW has shown promise [5] because it is generated from local features (e.g., SIFT [6], SURF [7]) that can describe the detailed information in each image. Moreover, it can integrate with inverted indexing structures for efficient large-scale image retrieval and adopt spatial verification to verify matching results [8].

However, because of the limited amount of memory in mobile devices, it may not be feasible to store a large vocabulary tree (e.g., 1M tree [5]). Hence, the authors in [3] adopt hash-based methods (e.g., [9]) to generate binary signatures on each local feature. These hashing methods consume a small amount of memory to store the projection matrix (i.e., hashing functions) [10][11]. Mobile visual search systems utilize multiple hash tables on indexing servers (e.g., [3] and [12]) to increase the number of result candidates from hash tables and achieve higher search accuracy.

Since binary representations are used in mobile visual search, measuring the similarity between two binary signatures is of great importance. Calculating the

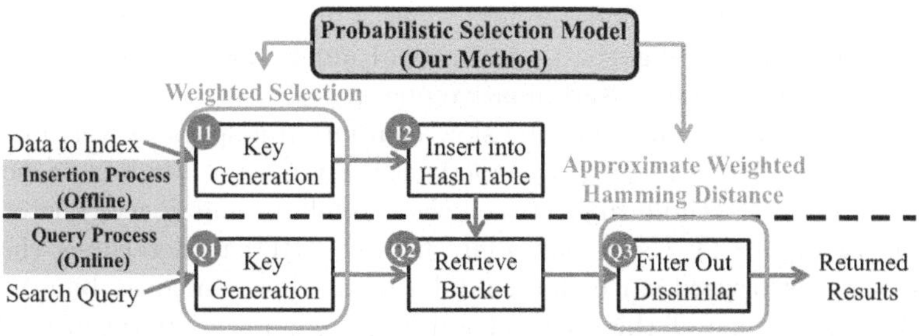

Fig. 1. A flowchart showing the insertion (I1 and I2) and query steps (Q1, Q2 and Q3) of a nearest neighbors search with hash tables. We propose to use a probabilistic selection model (Section 4) with weighted selection when generating index keys during the insertion and query processes (I1 and Q1). The model is also used in calculating an approximate weighted hamming distance (AWHD) between signatures in Q3 to improve the speed of determining similarity.

hamming distance between signatures is fast but may not be accurate in determining similarity because different bits can have different discriminative powers. Weighted hamming distance (WHD) addresses this concern. In [4], the authors attempt to weight the hamming distance of local features for image matching. In [13], the authors assign two weights to each hash bit and define a score function to measure the similarity between binary signatures in developing a ranking algorithm. In [14], the authors use a WHD ranking algorithm to learn the data-adaptive and query-sensitive weight for each bit. The common point in these papers is that the authors use weights (in the form of floating point numbers) on feature dimensions to provide more accuracy when calculating distance.

3 Problems in Hash Indexing and Distance Computation

A hash table has multiple buckets, with each bucket represented by a unique index key. A flowchart illustrating how a hash table is used to index data in a multimedia retrieval system is shown in Figure 1. In the insertion process (offline), an index key for each data entry is generated by a predefined hashing function, and the data is inserted into a corresponding bucket of the hash table. In the query process (online), an index key is generated for the queried data by the same hashing function and is used to retrieve a bucket from the hash table. The data in this bucket contains the nearest neighbor candidates. The similarity between the candidates and the queried data is determined, and some of the candidates deemed dissimilar are filtered out. The remaining candidates are defined to be the nearest neighbors (matched features) of the query.

Finding the nearest neighbors of binary signatures in a hash table poses several challenges. The first involves the hashing function used for index key generation (Step I1 and Q1 in Figure 1). Because of the low recall rate of a single hash table, multiple hash tables are often used to improve recall. However, this means that multiple index keys need to be generated for the hashed features across the hash tables. In [3], the authors randomly select bits to form an index key. This method though may not generate an optimal index key when bits (or feature dimensions) have different discriminative powers.

The second challenge involves the slow computation speed of weighted hamming distance (WHD). To provide some background, after obtaining candidates from hash tables, the similarity between the candidates and query is determined (Step Q3 in Figure 1). Using hamming distance (HD) to determine the similarity is straightforward since the candidates are represented as binary signatures. The advantage of this method is its fast computation speed as it only requires two binary operations, XOR ($\oplus$) and POPCNT (counting the number of 1 bits in the given argument). A d-bit binary signature can be represented as $f = [b_0, b_1, \ldots, b_{d-1}]$ where $b_i = \{0, 1\}$. The HD between f_p and f_q is thus defined as:

$$\mathrm{HD}(f_p, f_q) = \mathrm{POPCNT}(f_p \oplus f_q). \tag{1}$$

As mentioned in Section 2, since HD may not provide an accurate distance measurement, WHD is often used as an alternative. A weight vector $w = [w_0, w_1, \ldots, w_{d-1}]$ is given, where w_i is the weight for dimension i. Calculating WHD between f_p and f_q is thus an inner product of $(f_q \oplus f_t)$ and w. It can be formulated as:

$$\mathrm{WHD}(f_p, f_q) = (f_p \oplus f_q) \cdot w = \sum_{i}^{\#bits} [(b_i \text{ on } f_p) \oplus (b_i \text{ on } f_q)]\, w_i \tag{2}$$

When calculating distance, each bit is given a weight value w_i based on its discriminative power, and these weight values for the corresponding dimensions are used to calculate WHD. This can provide a more accurate distance measurement versus HD by avoiding the ambiguity caused when different bits in signatures result in the same HD distance values. But as weight values are usually provided as floating point numbers, calculating WHD can be much slower than that for HD.

In summary, there are two challenges: (1) Defining a suitable hashing function for index key generation; (2) Improving on the calculation time of WHD while retaining accurate distance measurements that will help determine similarity. In the following sections, we will describe our method to address these.

4 Proposed Method – Probabilistic Selection Model

To deal with the two major challenges, we propose a probabilistic selection model that is based on *bit expansion*. If the weight of a bit is higher than others, this bit can be expanded with more bits of the same value to reflect its relative importance. For example, given a binary signature of 10, if the left bit is weighted twice that of the right bit ($w = [2, 1]$), the signature can be expanded to 110. Or for the same signature of 10, if the right bit is weighted three times that of the left bit ($w = [1, 3]$), the signature can be expanded to 1000.

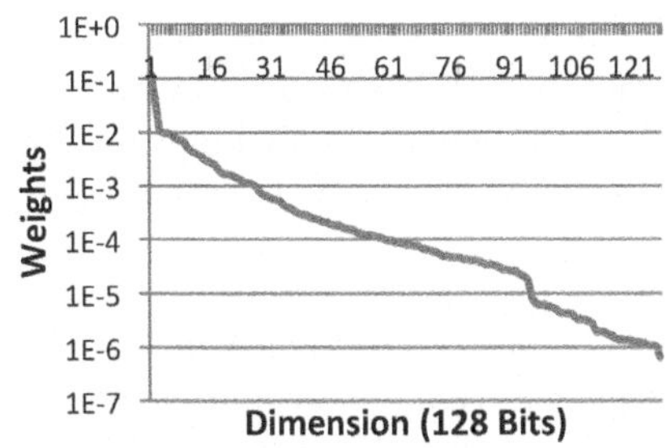

(a) Weight vector. We use a 128-bit PCA-based method and its sorted eigenvalues as weights in this example.

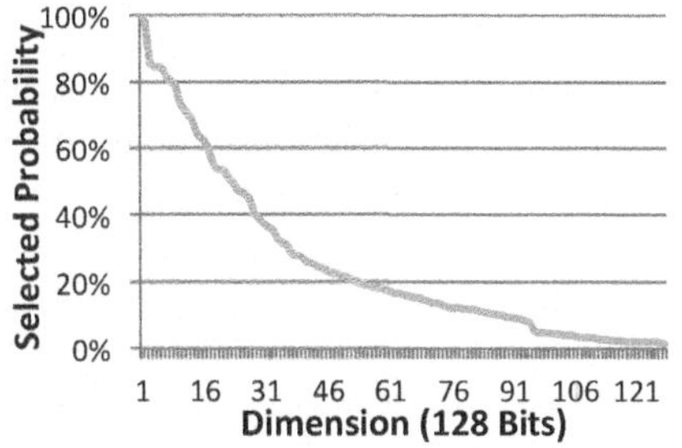

(b) The selected probabilities for each dimension when 32 bits are selected in a selection vector.

(c) The 20 generated selection vectors

Fig. 2. An example for generating selection vectors. (a) represents the weight vector w. The selection vectors (with 32 bits to be selected) are generated based on the probability vector p, which is normalized from w. The value of each dimension in (b) represents the probability of the corresponding bit to be selected in a selection vector when using the above settings. Twenty example selection vectors generated by p are shown in (c). Each selection vector is represented as a row and the selected bits are marked in black.

We use several selection processes that incorporate this bit expansion behavior. This does not involve floating point operations so the process is more efficient. In Section 4.1, we first generate selection vectors based on the weight of each bit in binary signatures to identify which bits should be selected. In Section 4.2 and 4.3, we will use selection vectors to address the stated challenges.

4.1 Selection Vectors for the Probabilistic Selection Model

Current state-of-the-art methods randomly select dimensions to generate an index key [3]. However, they do not consider the discriminative power of each bit. We propose using selection vectors to select dimensions as an alternative to random selection. Each selection vector is represented as $s = [s_0, s_1, \ldots, s_{d-1}]$ with $s_i = \{0, 1\}$, where 1 represents a dimension to be selected and 0 represents an unselected one. The idea is to use the vectors to select the essential dimensions for indexing. These selector vectors will be used later to generate index keys for hash tables (Section 4.2) and to calculate an approximate weighted hamming distance (Section 4.3).

To form selection vectors, we use a probability vector $p = [p_0, p_1, \ldots, p_{d-1}]$ where $p_i \in [0, 1]$ is a probability value, to describe the discriminative power (weight) of each dimension. Here p is normalized from a weight vector w and determines the selected dimensions. A selection vector is generated by repeating the process until the desired number of bits (the length of an index key) are obtained. By repeating the generation process N times, N selection vectors $s^{(1)}, s^{(2)}, \ldots, s^{(N)}$ can be obtained.

Figure 2 shows an example of the generation process for selection vectors. In this example, the weights are provided by a PCA-based hashing feature (128-bit) and its eigenvalues. We select 32 bits for each selection vector and generate 20 of them. Since the bits are sorted by their eigenvalues, the selection behavior of the 20 results is biased in favor of the left bits (larger weights). This is an example using the PCA-based hashing feature. Note that we propose a general method for binary features with weights that can adopt other types of learning methods (e.g., [15]).

4.2 Weighted Selection for Index Keys on Hash Tables

A critical aspect of hash tables is in generating a k-bit key for data to be indexed. Because the feature f being indexed is a binary signature, the index key generation process is actually a bit selection process of d bits like in the example given in Section 4.1. All d bits are not used as an index key as d is usually too large (e.g. larger than 32 or 64 bits). In the scheme of a single hash table, a common method to generate an index key is to select the first k-th discriminative dimensions [16] (e.g., the top-k eigenvalues for a PCA-based method).

In the scheme of multiple hash tables, one method of generating an index key is to use random selection [3]. For each hash table, the authors in [3] randomly select k dimensions to generate an index key. In [12], the authors use a sequential grouping method where the first hash table uses the first k dimensions ($b_0 \sim b_{k-1}$), the second hash table uses the second k dimensions ($b_k \sim b_{2k-1}$), and so on and so forth.

The method of random selection does not consider the discriminative power of dimensions while the sequential grouping method may generate undiscriminated index keys as the number of hash tables increase. Our proposed method involves a weighted selection in which each index key is generated according to the discriminative power (weights) of dimensions. Taking into account the weights of different bits in a binary signature, a selection vector s is obtained as described in Section 4.1. The generation process produces N selection vectors $s^{(1)}, s^{(2)}, \ldots, s^{(N)}$ for constructing N hash tables. In Figure 3, an example is shown of four selection vectors and a binary feature $f = 110011$. The index key for f in the i-th hash table is the concatenation of the bits which correspond to the dimensions marked as 1 in $s^{(i)}$. Therefore, the index keys for f in the 4 hash tables are 110, 111, 101 and 101 respectively.

4.3 Approximate Weighted Hamming Distance (AWHD)

Using WHD in a search system provides more accurate results than using HD (as illustrated in Section 5.5). But it is also more time consuming to calculate

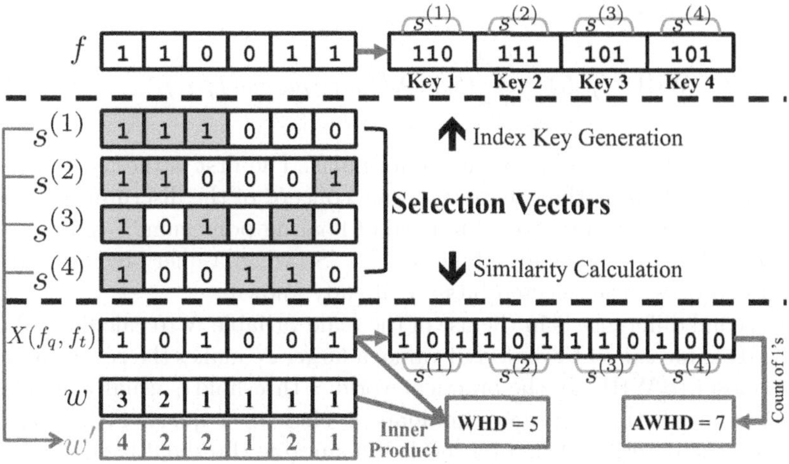

Fig. 3. An example of index key generation and similarity calculation using selection vectors. The feature dimension in this example is 6 and there are 4 selection vectors $s^{(1)}$ to $s^{(4)}$. At the top, the selection vectors are used to generate index keys to use in 4 hash tables for f. At the bottom, a comparison between calculating WHD and AWHD is done for a given w and $X(f_q, f_t)$.

due to the involved floating point operations. The notion of weighting is used in calculating an approximate weighted hamming distance (AWHD) as illustrated in Figure 3 and described in detail as follows:

1. Obtain $X(f_q, f_t)$, the XOR result of query f_q and target f_t.
2. Select k bits from the d bits in $X(f_q, f_t)$ by using a selection vector s (obtained as described in Section 4.1).
3. Repeat step 2 N times, using $s^{(i)}$ as the selection vector in the i-th iteration.
4. Sum up the selected k bits in every N iterations to form AWHD.

If a bit b_j in $X(f_q, f_t)$ and the corresponding s_j are both 1, it will contribute 1 to AWHD, which can be formulated as:

$$\text{AWHD}(f_q, f_t) = \sum_{i=0}^{N-1} \sum_{j=0}^{d-1} (b_j \text{ in } X(f_q, f_t)) s_j^{(i)} \tag{3}$$

The generation of selection vector s is based on probability vector p; thus, expected value can be used to estimate the proposed distance. Each bit of $X(f_q, f_t)$ has N chances and each chance has a probability p_i of being selected. A bit with a higher probability of being selected will contribute more to the distance. After replacing $s_j^{(i)}$ with p_i, the expected value of AWHD becomes:

$$\text{E}[\text{AWHD}(f_q, f_t)] = N(X(f_q, f_t) \cdot p) = N[(f_q \oplus f_t) \cdot p] \tag{4}$$

The formula for WHD is:

$$\mathrm{WHD}(f_q, f_t) = \sum_{i=0}^{d-1} (b_i \text{ in } X(f_q, f_t)) w_i = (f_q \oplus f_t) \cdot w \tag{5}$$

There are two differences between the formulas. The first is the use of p versus w. As p is based on w, they share the same concept of the discriminative power of bits. Although p is a vector comprised of floating point numbers, the process of selecting bits is done as an offline process. So when a query comes in during the online process, the selected bits are already known.

The second difference is in the use of the scale variable N to calculate AWHD. As AWHD is measured by expected value, a larger N makes the result more stable and closer to WHD. In the practical world, a threshold is usually maintained to determine whether a target candidate is close to a query. So N has little impact on the determination of a result if a suitable threshold is chosen.

An example of calculating AWHD versus WHD is illustrated at the bottom of Figure 3 for a given weight vector w and XOR result $X(f_q, f_t)$. The WHD of $X(f_q, f_t)$ for the given w is 5, which is the result of their inner product. When calculating AWHD, a 12-bit signature is generated from 4 selection vectors (4 selection vectors $\times$ 3 bits selected in each vector). The POPCNT instruction is used to count the number of 1's in this signature and obtain an AWHD of 7.

A criterion to observe the approximation behavior of AWHD is thru w', which is the sum of the selection vectors. From another viewpoint, AWHD can also be calculated from the inner product of $X(f_q, f_t)$ and w'. This means that if w' is similar to w as measured by cosine similarity, AWHD can approximate WHD with a scale factor N (the previously mentioned scale variable N). The approximation can be increased by increasing the number of selection vectors used.

5 Experiments

In this section, a series of experiments are conducted to show our proposed method is a preferred alternative to current state-of-the-art methods. In Section 5.1, the datasets used for the experiments are introduced. In Sections 5.2 to 5.5, the experiments are described and the outcomes are analyzed.

5.1 Datasets

We use image datasets including Stanford Mobile Visual Search (MVS) [17], 1 million images from Flickr (referred to as Flickr1M) and European Cities 1M [18]. The MVS dataset is an image retrieval dataset created specifically for mobile visual search research. It provides 3,300 queries for 1,200 target images across 8 categories, including CD covers, DVD covers, books, video clips, landmarks, business cards, text documents and printings. We will use recall of positive images in the top M results to evaluate our method's effectiveness. Re-ranking algorithms (e.g., spatial verification [19]) can be applied post-process to obtain a better sorted set of results within the top M. Therefore, the important thing is to bring the positive images to top M. In order to scale up the dataset, we sample 500,000 images each from Flickr1M and European Cities 1M. With these images used as backgrounds to the MVS images, we produce a working dataset that contains approximately one million images.

Since our method operates on binary signatures and we need to verify the accuracy of it in the binary space, a feature dataset is required for feature-wise experiments. We collect 1 million 128-d SURF features from the 1-million-image working dataset and reduce the feature dimensions to binary space. Ten thousand of these features are queries and the rest are targets. We have two feature datasets:

1. **256-bit version.** Random projection [20] with binarization is used to reduce features to 256 bits. These features are used to evaluate recall for different numbers of hash tables. Ground truth is defined as the closest target features (hamming distance ≤ 24) in binary space in order to prevent the effect of quantization errors in the process of dimension reduction.
2. **128-bit version.** PCA-hashing [4] is used to reduce the feature to 128 bits. These features are used to evaluate the ability of retrieving similar data in the original feature space. Ground truth is defined as the closest target feature in the SURF domain as measured by cosine similarity.

Notice that we use SURF features instead of SIFT. Although SIFT features are popular and widely used in computer vision because of its strong representational capability for interesting points in images, we chose to use SURF because of its speed performance advantage. This is especially important on mobile devices, which have limited processing power compared to a server. Although SURF features are about 10% less accurate than SIFT, SURF performs almost three times faster. This allows for the highly desirable user experience of getting a fast response. However, our proposed method can be applied to any feature types that can be compressed to a binary signature.

5.2 Performance Comparison of Single and Multiple Hash Tables

As stated in Section 3, using multiple hash tables to index data can achieve higher recall. We use the 256-bit feature dataset in comparing the recall and build/query time for different numbers of hash tables. In Figure 4, the recall when using a single hash table is shown to be low (less than 0.5) and increases as the number of hash tables used increases. However, the increase in recall follows a logarithmic-like growth rate as opposed to the more linear rate of increase in the time needed to build and query the hash tables. The number of hash tables to use will depend on deciding an acceptable recall for the increased latency trade-off and having a proper balance between these two performance attributes.

5.3 The Effect on Bit (Index Key) Selection

Suitable index keys need to be found for the multiple hash tables. In Section 4.2, we proposed a weighted selection in using selection vectors s to obtain index

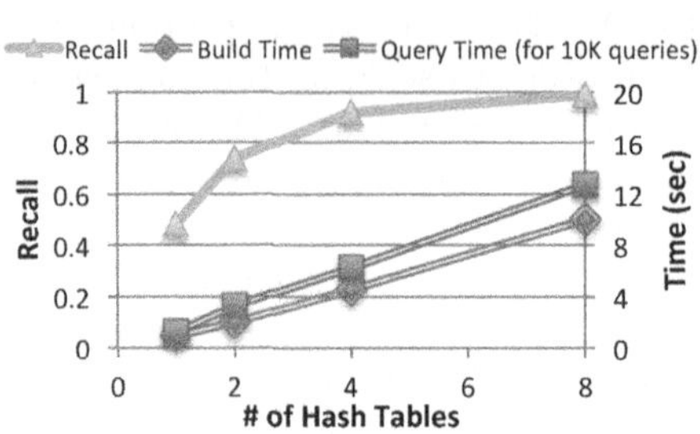

Fig. 4. Recall and build/query time for different numbers of hash tables using the 256-bit feature dataset. Multiple hash tables improve recall at a cost of more build and query time.

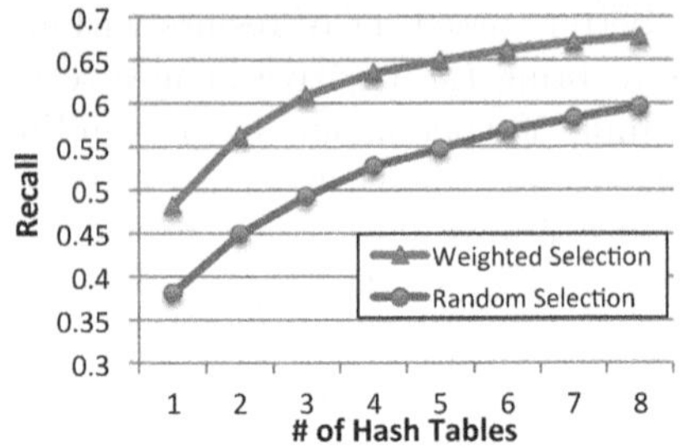

Fig. 5. Comparison of weighted and random selection vectors using the 128-bit feature dataset. The weighted selection model has higher recall for any number of hash tables.

keys for the hash tables. To evaluate this, we use the 128-bit feature dataset and the first 8 selection vectors from Figure 2(c) to measure recall when using up to 8 hash tables. In Figure 5, we compare our method with random selection [3]. The results show our method provides higher recall for any number of hash tables. This is due to the use of more optimal index keys that take into account the weights of bits in features.

5.4 Speed Comparison of AWHD and WHD

One of the disadvantages of WHD is its slower computation time compared with hamming distance. Our proposed method of calculating AWHD as an alternative avoids floating point operations to save time. Using a Xeon E5-2650 v2 machine and the 128-bit feature dataset, we randomly sample two features from the dataset, and calculate the WHD and AWHD between them. This process is repeated one million times. The results show that calculating WHD takes on average 1,219 nanoseconds while it only takes on average 55 nanoseconds to calculate AWHD when using a single selection vector (bit length = 32 bits). This is a significant time savings of 95% of the WHD calculation time.

5.5 Accuracy of Approximate Weighted Hamming Distance

Although our method of calculating AWHD performs much faster, it is still an approximation of WHD and the recall when using AWHD is bounded by the recall achieved when using WHD. But if the number of selection vectors used when calculating AWHD large enough, the recall can theoretically approximate WHD.

Figure 6 illustrates our findings comparing recall and computation time when using WHD, AWHD (with different numbers of selection vectors) and 128-bit HD on the 128-bit feature dataset. Figure 6(a) shows that the recall achieved when using our method approaches that when using WHD as the number of selection vectors used to calculate AWHD increases from 4 to 10. When 10

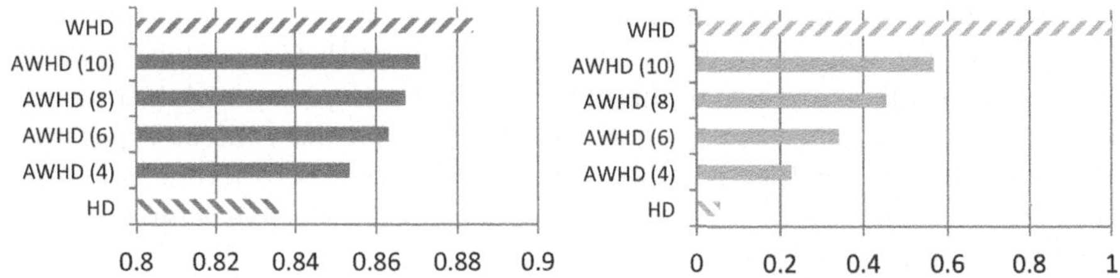

(a) 1 NN Recall@50 (b) Time (WHD Time = 1)

Fig. 6. Comparison of recall and computation time between HD, WHD and AWHD using the 128-bit feature dataset. A different number of selection vectors (4, 6, 8, 10) are used to evaluate AWHD. The recall using AWHD can approximate that when using WHD with much lower query latency.

selection vectors are used to obtain the recall for our method in the experiment, there is a time savings of almost 50% compared to WHD as seen in Figure 6(b).

Figure 7 shows the recall comparison in the image domain using the 1-million-image working dataset described in Section 5.1. The proposed method in this experiment uses 4 selection vectors, i.e. AWHD (4). The recall achieved using AWHD (4) closely approximates that using WHD for different numbers of images from the working dataset. This shows that in the image domain, fewer selection vectors are needed (as compared to in the feature domain) in order to approximate recall when using WHD. Using fewer selection vectors means that less time is needed to calculate an AWHD. In this case of using 4 selection vectors, about 78% less time is needed to calculate AWHD vs WHD.

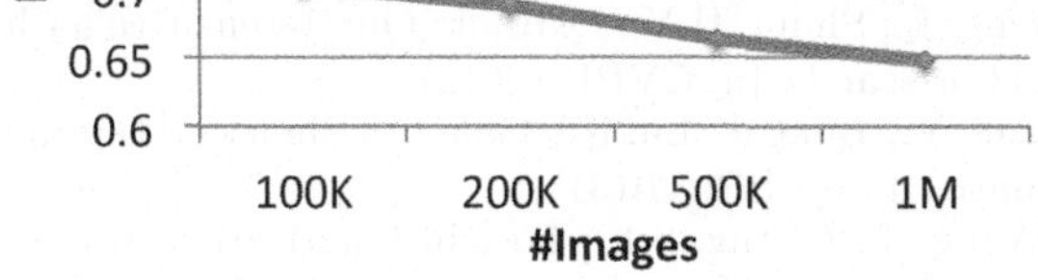

Fig. 7. Recall comparison of AWHD and WHD on the 1-million-image working dataset

6 Conclusions

Binary signatures and weighted hamming distance with multiple hash tables are widely used in nearest neighbors search algorithms for mobile visual search. We propose a probabilistic selection model and use selection vectors to address the challenge of choosing suitable index keys for multiple hash tables. Our results show this outperforms random selection. Based on selection vectors, we propose an approximate weighted hamming distance, a simple and efficient method that can substitute for weighted hamming distance. This has significant computation time savings yet tracks closely to weighted hamming distance while retaining high recall. In this paper, we evaluate the method using images. A future evaluation could also apply the method to other multimedia content types such as video or audio. Another possible area of exploration is to leverage learning methods to obtain more discriminative bits (weights, w) which may further boost retrieval accuracy.

References

1. Instagram, `http://instagram.com/press/`
2. Girod, B., Chandrasekhar, V., Chen, D.M., Cheung, N.M., Grzeszczuk, R., Reznik, Y., et al.: Mobile visual search. In: IEEE SPM (2011)
3. He, J., Feng, J., Liu, X., Cheng, T., Lin, T.H., Chung, H., Chang, S.F.: Mobile product search with bag of hash bits and boundary reranking. In: CVPR (2012)
4. Wang, X.J., Zhang, L., Jing, F., Ma, W.Y.: Annosearch: Image auto-annotation by search. In: CVPR (2006)
5. Philbin, J., Chum, O., Isard, M., Sivic, J., Zisserman, A.: Object retrieval with large vocabularies and fast spatial matching. In: CVPR (2007)
6. Lowe, D.G.: Distinctive image features from scale-invariant keypoints. In: IJCV (2004)
7. Bay, H., Tuytelaars, T., Van Gool, L.: SURF: Speeded up robust features. In: Leonardis, A., Bischof, H., Pinz, A. (eds.) ECCV 2006, Part I. LNCS, vol. 3951, pp. 404–417. Springer, Heidelberg (2006)
8. Jegou, H., Douze, M., Schmid, C.: Hamming embedding and weak geometric consistency for large scale image search. In: Forsyth, D., Torr, P., Zisserman, A. (eds.) ECCV 2008, Part I. LNCS, vol. 5302, pp. 304–317. Springer, Heidelberg (2008)
9. Wang, J., Kumar, S., Chang, S.F.: Semi-supervised hashing for large-scale search. In: TPAMI (2012)
10. Achlioptas, D.: Database-friendly random projections: Johnson-Lindenstrauss with binary coins. J. Comput. Sys. Sci. 66(4), 671–687 (2003)
11. Charikar, M.S.: Similarity estimation techniques from rounding algorithms. In: ACM STOC (2002)
12. Cai, J., Liu, Q., Chen, F., Joshi, D., Tian, Q.: Scalable Image Search with Multiple Index Tables. In: ICMR (2014)
13. Zhang, X., Zhang, L., Shum, H.Y.: QsRank: Query-sensitive hash code ranking for efficient ϵ-neighbor search. In: CVPR (2012)
14. Zhang, L., Zhang, Y., Tang, J., Lu, K., Tian, Q.: Binary code ranking with weighted hamming distance. In: CVPR (2013)
15. Jiang, Y.G., Wang, J., Chang, S.F.: Lost in binarization: query-adaptive ranking for similar image search with compact codes. In: ICMR (2011)
16. Zhou, W., Lu, Y., Li, H., Tian, Q.: Scalar quantization for large scale image search. ACM Multimedia (2012)
17. Chandrasekhar, V.R., Chen, D.M., Tsai, S.S., Cheung, N.M., Chen, H., Takacs, G., et al.: The stanford mobile visual search data set. ACM MMSys (2011)
18. Avrithis, Y., Kalantidis, Y., Tolias, G., Spyrou, E.: Retrieving landmark and non-landmark images from community photo collections. ACM Multimedia (2010)
19. Tsai, S.S., Chen, D., Takacs, G., Chandrasekhar, V., Vedantham, R., Grzeszczuk, R., Girod, B.: Fast geometric re-ranking for image-based retrieval. In: ICIP (2010)
20. Li, P., Hastie, T.J., Church, K.W.: Very sparse random projections. In: ACM SIGKDD (2006)

Graph Regularised Hashing

Sean Moran and Victor Lavrenko

School of Informatics, University of Edinburgh, UK
sean.moran@ed.ac.uk, vlavrenk@inf.ed.ac.uk

Abstract. Hashing has witnessed an increase in popularity over the past few years due to the promise of compact encoding and fast query time. In order to be effective hashing methods must maximally preserve the similarity between the data points in the underlying binary representation. The current best performing hashing techniques have utilised supervision. In this paper we propose a two-step iterative scheme, Graph Regularised Hashing (GRH), for incrementally adjusting the positioning of the hashing hypersurfaces to better conform to the supervisory signal: in the first step the binary bits are regularised using a data similarity graph so that similar data points receive similar bits. In the second step the regularised hashcodes form targets for a set of binary classifiers which shift the position of each hypersurface so as to separate opposite bits with maximum margin. GRH exhibits superior retrieval accuracy to competing hashing methods.

1 Introduction

Nearest neighbour search (NNS) is the problem of retrieving the most similar item(s) to a query point $\mathbf{q}$ in a database of N items $\mathbf{X} = \{\mathbf{x}_1, \mathbf{x}_2 \ldots, \mathbf{x}_N\}$. NNS is a fundamental operation in many applications - for example, the annotation of images with semantically relevant keywords [12]. The naïve approach to solving this problem would be to compare the query exhaustively to every single item in our dataset yielding a linear scaling in the query time. Unfortunately this brute-force approach is impractical for all but the smallest of datasets - in the modern age of *big data* considerably more efficient methods for NNS are required. Hashing-based approximate nearest neighbour (ANN) search is a proven and effective approach for solving the NNS problem in a constant time per query.

Hashing-based ANN search has witnessed a sharp rise in popularity due to explosion in the amount of multimedia data being produced, distributed and stored worldwide. It has been estimated, for example, that Facebook has on the order of 300 million images uploaded per day[1] - clearly efficient search methods are required to manage such vast collections of data. Hashing-based ANN search meets this requirement by compressing our data points into similarity preserving binary codes which can be used as the indices into the buckets of a hash table for constant time search. Many hashing methods employ hypersurfaces to partition the feature space into disjoint regions which constitute the buckets of a hash

[1] Velocity 2012: Jay Parikh, "Building for a Billion Users".

A. Hanbury et al. (Eds.): ECIR 2015, LNCS 9022, pp. 135–146, 2015.

table. The generation of similarity preserving binary codes can be viewed as involving two distinct steps: *projection* and *quantisation* - both steps when taken together effectively determine which sides of the hypersurfaces our query point inhabits.

Typically the projection stage involves a dot product onto the normal vectors of a set of hyperplanes (linear hypersurfaces) positioned either randomly or in data-aware positions in the feature space. The hyperplanes tessellate the space in a manner that gives a higher likelihood that similar data points will fall within the same region, and therefore are assigned the same binary encoding. In the second step the real-valued projections are quantised into binary by thresholding the corresponding projected dimensions [13]. Most research into hashing-based ANN involves maximising the *neighbourhood preservation* - that is the preservation of the distances in the original feature space - of one or both of these steps, as this directly translates into compact binary codes that are more similar for similar data points. Ideally this criterion should be met with the shortest possible length of hashcode.

Hashing-based ANN has shown great promise in terms of efficient query processing and data storage reduction across a wide range of research domains involving both textual and image-based data. For example, in [15], the authors present an efficient method for event detection in Twitter that scales to unbounded streams through a novel application of Locality Sensitive Hashing (LSH), a seminal randomised approach for ANN search [8]. In the streaming scenario the $\mathcal{O}(N)$ worst case complexity of inverted indexing is undesirable, motivating the use of LSH to maintain a hard constant $\mathcal{O}(1)$ query time upper bound. Hashing-based ANN has also proved particularly useful for search over dense and lower dimensional feature vectors, such as GIST [14], that are commonly employed in the field Computer Vision. For example, hashcodes have been successfully applied to image retrieval [17].

We propose a novel supervised hashing model, dubbed Graph Regularised Hashing (GRH), that achieves state-of-the-art performance with a straightforward optimisation framework. Our model employs graph regularisation [5], related to the *Cluster Hypothesis* of Information Retrieval (IR) which states that *"closely associated documents tend to be relevant to the same requests"* [18]. In our work graph regularisation smooths the distribution of binary bits so that neighbouring points are more likely to be assigned identical bits. The regularised bits are then used as targets for a set of binary classifiers that separate opposing bits with maximum margin. Iterating these two steps permits the hashing hypersurfaces to evolve into positions that better separate opposing bits, leading to superior retrieval accuracy over state-of-the-art hashing schemes.

2 Related Work

The field of hashing-based ANN search can be usefully divided into *data-independent* and *data-dependent* hashing models. Both fields are united in their use of hypersurfaces to partition the data-space into disjoint regions (or buckets). Data-independent hashing techniques position the hashing hypersurfaces randomly in the data space, making no assumptions on the data distribution. They also typically come with an asymptotic guarantee that as the number of hypersurfaces increase the distance in the Hamming space will converge to some specific measure of distance in the original data-space (e.g. Euclidean distance). Locality Sensitive Hashing (LSH) represents the seminal work in the data-independent hashing field [8] employing random projections for hash function generation. LSH has since been extended to kernel similarity [16].

Data-independent hashing methods such as LSH have the advantage that the hash function training stage is fast, effectively negligible - random hypersurface generation is a computationally inexpensive operation. This has made LSH, for example, the method of choice for real-time streaming-based applications where there is a strict bound on the indexing time [15]. On the downside, data-independent schemes usually require long hashcodes for precision and many hash tables in order to attain an acceptable level of recall. Random hypersurfaces can erroneously partition dense areas of the data space which may separate many true NNs and lead to lower retrieval accuracy.

Recently researchers have developed methods that introduce a degree of data dependency into the hypersurface generation, for example by using machine learning methods [19,20,7,10,9,21]. These models attempt to avoid placing hypersurfaces that partition related data points. Data-dependent hashing models can usefully be categorised into *supervised* or *unsupervised* methods. The unsupervised techniques commonly employ a dimensionality reduction step prior to quantisation: for example, principal component analysis (PCA) has been used extensively in seminal work including PCA hashing (PCAH) [19], Spectral Hashing (SH) [20] and Iterative Quantisation (ITQ) [7]. These techniques preserve the distances in the original feature space through an eigenvector formulation, effectively using the principal directions of the data as the hashing hypersurfaces.

The unsupervised data-dependent hashing models may generate hypersurfaces that do not respect the semantic similarity of the data-points. Supervised data-dependent hashing methods exhibit the highest retrieval accuracy by exploiting a supervisory signal, either in the form of a pairwise affinity matrix derived from metric nearest neighbours or through class labels. Representative approaches in this field include Supervised Hashing with Kernels (KSH) [10], Binary Reconstructive Embedding (BRE) [9] and Self-Taught Hashing (STH) [21]. Most of the supervised hashing models frame the generation of hashcodes as an optimisation problem where a set of hypersurfaces form the adjustable model parameters. The optimisation adjusts the hypersurfaces so that the resulting smoothed approximation to the Hamming distances are close to metric distances or class-based supervision.

To the best of our knowledge, the closest supervised method to our approach is the STH model of [21]. In STH, the authors also employ a two-step approach to generating binary codes: in the first step they construct a supervised low-dimensional embedding through the Laplacian Eigenmap [1], which is then followed by a step that learns a set of SVM classifiers using the resulting binarised dimensions as labels. Our method, GRH, is distinct from STH and previous work: firstly we are the first to integrate and explore *graph regularisation* in a hashing method. Secondly, in contrast to STH, GRH is *iterative* in nature incrementally evolving the positioning of the hypersurfaces as the distribution of hashcode bits are gradually smoothed over multiple iterations. By comparing directly to STH we show that our formulation of graph regularisation is critical for the superior retrieval accuracy of GRH.

3 Graph Regularised Hashing (GRH)

3.1 Problem Definition

We are given a dataset of N points $\mathbf{X} = \{\mathbf{x}_1 \ldots \mathbf{x}_N\}$, where each point $\mathbf{x}_i$ is a D-dimensional vector of real-valued features. Our goal is to represent each item with a binary hashcode $\mathbf{b}_i$ consisting of K bits. The aim is to select the bits in such a way that neighbouring points $\mathbf{x}_i, \mathbf{x}_j$ will have similar hashcodes $\mathbf{b}_i, \mathbf{b}_j$, as measured by the Hamming distance. The neighbourhood structure is encoded in a pairwise affinity matrix $\mathbf{S}$, where $S_{ij} = 1$ if points $\mathbf{x}_i$ and $\mathbf{x}_j$ are considered neighbours, and $S_{ij} = 0$ otherwise.

3.2 Overview of the Approach

Our approach is based on iteratively performing two steps: **(A) regularisation**, where we make the hashcodes $\mathbf{b}_1 \ldots \mathbf{b}_N$ more consistent with the affinity matrix $\mathbf{S}$; and **(B) partitioning**, where we learn a set of hypersurfaces $\mathbf{h}_1 \ldots \mathbf{h}_K$ that subdivide the space $\mathbb{R}^D$ into regions that are consistent with the hashcodes. These hypersurfaces are needed to efficiently compute the hashcodes for testing points $\mathbf{x}$, where we have no affinity information.

We initialise the hashcodes $\mathbf{b}_1 \ldots \mathbf{b}_N$ by running our points $\mathbf{x}_1 \ldots \mathbf{x}_N$ through any existing fingerprinting algorithm, such as LSH [8] or ITQ+CCA [7]. We then iterate the regularisation and partitioning steps in a way reminiscent of the *EM algorithm* [4]: the regularised hashcodes from step A adjust the hypersurfaces in step B, and these surfaces in turn generate new hashcodes for step A. We run the algorithm for a fixed number of iterations (M), and leave the analysis of convergence to future work. We now provide the details of steps A and B.

3.3 Step A: Regularisation

We take a graph-based approach to regularising the hashcodes. The nodes of the graph correspond to the points $\mathbf{x}_1 \ldots \mathbf{x}_N$. The affinity matrix $\mathbf{S}$ plays the role of an adjacency matrix: we insert an undirected edge between nodes i and j if and only if $S_{ij} = 1$. Each node i is annotated with K binary labels, corresponding to the K bits of the hashcode $\mathbf{b}_i$. Our aim is to increase the similarity of the label sets at the opposite ends of each edge in the graph. We achieve this by averaging the label set of each node with the label sets of its immediate neighbours. This is similar to the *score regularisation* method of [5], although our update equation is slightly different.

Figure 1 illustrates our approach. In the left side, we show a graph with 8 nodes $a \ldots h$ and edges showing the nearest-neighbour constraints. Each node is annotated with 3 labels which reflect the initial hashcode of the node (zero bits are converted to labels of -1). On the right side of Figure 1 we show the effect of label propagation for nodes c and e (which are immediate neighbours). Node e has initial labels $[+1, -1, -1]$ and 3 neighbours with the following label sets: c:$[+1, +1, +1]$, f:$[+1, +1, +1]$ and g:$[+1, +1, -1]$. We aggregate these four sets and look at the sign of the result to obtain a new set of labels for node e: $\text{sgn}[\frac{+1+1+1+1}{4}, \frac{-1+1+1+1}{4}, \frac{-1+1+1-1}{4}] = [+1, +1, -1]$. Note that the second label of e has become more similar to the labels of its immediate neighbours.

Formally, we regularise the labels via the following equation:

$$\mathbf{L} \leftarrow \text{sgn}\left(\alpha\ \mathbf{S}\mathbf{D}^{-1}\mathbf{L} + (1-\alpha)\mathbf{L}\right) \tag{1}$$

Here $\mathbf{S}$ is the adjacency matrix and $\mathbf{D}$ is a diagonal matrix containing the degree of each node in the graph. $\mathbf{L} \in \{-1, +1\}^{N \times K}$ represents the labels assigned to every node at the previous step of the algorithm, and α is a scalar smoothing parameter. sgn represents the sign function, modified so that $\text{sgn}(0) = -1$.

3.4 Step B: Partitioning

At the end of step A, each point $\mathbf{x}_i$ has K binary labels $\{-1, +1\}$. We will use these labels to learn a set of hypersurfaces $\mathbf{h}_1 \ldots \mathbf{h}_K$. Each surface $\mathbf{h}_k$ will partition the space $\mathbb{R}^D$ into two disjoint regions: *positive* and *negative*. The positive region of $\mathbf{h}_k$ should envelop all points $\mathbf{x}_i$ for which the k'th label was $+1$; while the negative region should contain all the $\mathbf{x}_i$ for which $L_{ik} = -1$. For simplicity, we restrict our discussion to linear hypersurfaces (hyperplanes), but a non-linear generalisation is straightforward via the kernel trick. We compare the performance of linear and non-linear boundaries in Section 4.

A hyperplane is defined by the normal vector $\mathbf{h}_k \in \mathbb{R}^D$ and a scalar bias b_k. Its positive region consists of all points $\mathbf{x}$ for which $\mathbf{h}_k^{\intercal}\mathbf{x} + b_k > 0$. We position each hyperplane $\mathbf{h}_k$ to maximise the margin, i.e. the separation between the points $\mathbf{x}_i$ that have $L_{ik} = -1$ and those that have $L_{ik} = +1$. We find the maximum-margin hyperplanes by independently solving K constrained optimisation problems:

$$\begin{aligned} \text{for } k = 1 \ldots K: \ \min \ & ||\mathbf{h}_k||^2 + C \textstyle\sum_{i=1}^{N} \xi_{ik} \\ \text{s.t. } \ & L_{ik}(\mathbf{h}_k^{\intercal}\mathbf{x}_i + b_k) \geq 1 - \xi_{ik} \quad \text{for } i = 1 \ldots N \end{aligned} \tag{2}$$

Here ξ_{ik} are slack variables that allow some points $\mathbf{x}_i$ to fall on the wrong side of the hyperplane $\mathbf{h}_k$; and C is a parameter that allows us to trade off the size of the margin $\frac{1}{||\mathbf{h}_k||}$ against the number of points misclassified by $\mathbf{h}_k$. We solve the optimisation problem in equation (2) using `liblinear` [6] and `libSVM` [2] for linear and non-linear hypersurfaces respectively.

Figure 2 illustrates step B for linear hypersurfaces. On the left side, we show the hyperplane $\mathbf{h}_1$ that partitions the points $a \ldots h$ using their first label as the target. Nodes a, b, c, d have the first label set to -1, while e, f, g, h are labelled as $+1$. The hyperplane $\mathbf{h}_1$ is a horizontal line, equidistant from points c and e: this provides maximum possible separation between the positives and the negatives. No points are misclassified, so all the slack variables $\xi_{i,1}$ are zero. The right side of Figure 2 shows the maximum-margin hyperplane $\mathbf{h}_2$ that partitions the points based on their second label. In this case, perfect separation is not possible, and $\xi_{i,2}$ is non-zero (nodes g and d are on the wrong side of $\mathbf{h}_2$).

Algorithm 1. Graph Regularised Hashing (GRH)

Input: Training dataset $\mathbf{X}$, training affinity matrix $\mathbf{S}$, degree matrix $\mathbf{D}$, interpolation parameter α, number of iterations M
Output: Hyperplanes $\mathbf{h}_1 \ldots \mathbf{h}_K$, biases $b_1 \ldots b_K$
Initialise $L \in \{0, 1\}$ via LSH/ITQ+CCA from $\mathbf{X}$
$L = \text{sgn}(L - \frac{1}{2})$
for $m = 1 : M$ **do**
  $\mathbf{L} = \text{sgn}\left(\alpha \mathbf{S} \mathbf{D}^{-1} \mathbf{L} + (1 - \alpha)\mathbf{L}\right)$
  for $k = 1 : K$ **do**
    $l_k = \mathbf{L}(:, k)$
    Train SVM$_k$ with l_k as labels, training dataset $\mathbf{X}$
    obtain hyperplane $\mathbf{h}_k$ and bias b_k
  end for
  $L_{ik} = \text{sgn}(\mathbf{h}_k{}^\intercal \mathbf{x}_i + b_k)$ for i=1...N and k=1...K
end for

The estimated hyperplanes $\mathbf{h}_1 \ldots \mathbf{h}_K$ are used to re-label the data-points:

$$L_{ik} = \text{sgn}(\mathbf{h}_k{}^\intercal \mathbf{x}_i + b_k) \quad \text{for } i=1\ldots N \text{ and } k=1\ldots K \tag{3}$$

The effect of this step is that points which could not be classified correctly will now be re-labelled to make them consistent with all hyperplanes. For example, the second label of node g in Figure 2 will change from -1 to $+1$ to be consistent with $\mathbf{h}_2$. These new labels are passed back into step A for the next iteration of the algorithm. After the last iteration, we use the hyperplanes $\mathbf{h}_1 \ldots \mathbf{h}_K$ to predict hashcodes for new instances $\mathbf{x}$: the k'th bit in the code is set to 1 if $\mathbf{h}_k{}^\intercal \mathbf{x} + b_k > 0$, otherwise it is zero. Algorithm 1 presents the pseudo-code for our approach.

3.5 Algorithm Analysis

Let T denote the number of *training* data-points. Graph regularisation is of $\mathcal{O}(T^2K)$. Training a linear SVM takes $\mathcal{O}(TDK)$ time while prediction (test time) is $\mathcal{O}(TDK)$. Therefore linear GRH is $\mathcal{O}(MT^2K)$ for M iterations. Typically $\mathbf{S}$ is sparse, $T \ll N$ and K is small ($\leq$ 64 bits) thereby ensuring GRH is scalable.

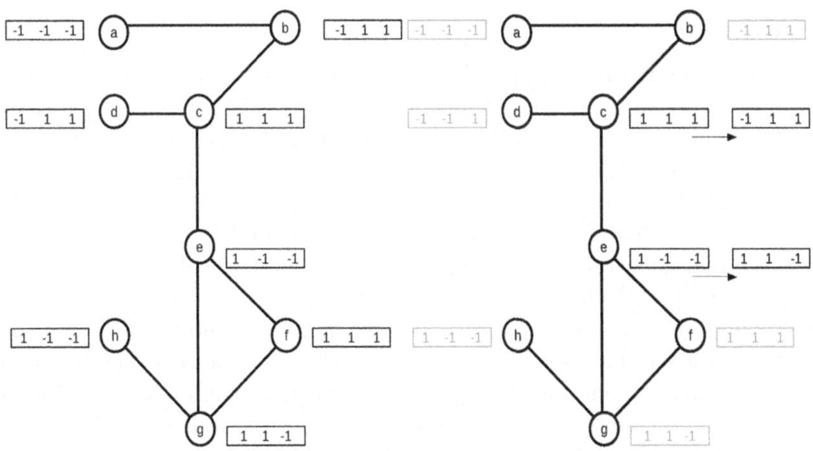

Fig. 1. The regularisation step. Nodes represent data-points and arcs represent neighbour relationships. The 3-bit hashcode assigned to a given node is shown in the boxes. We show the hashcode update for nodes c and e.

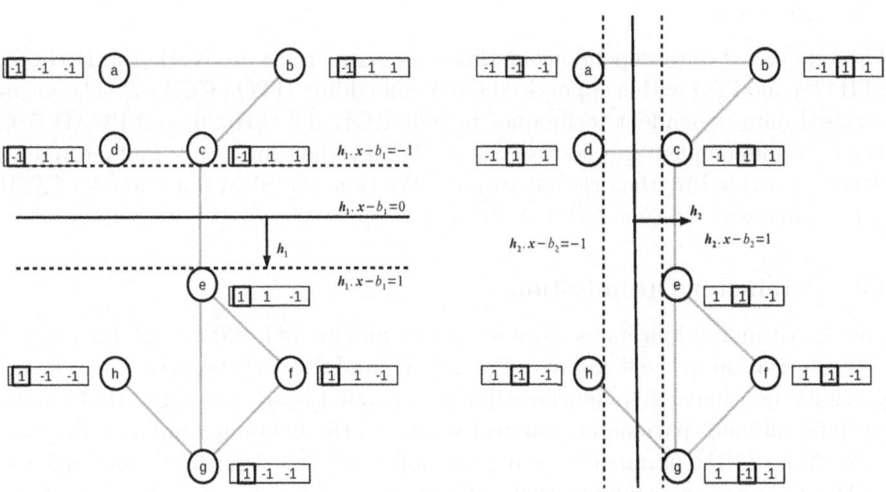

Fig. 2. The partitioning step. In this stage, the regularised hashcodes are used to re-position the hashing hyperplanes. **Left:** First bit of hashcode. **Right:** Second bit.

4 Experiments

4.1 Datasets

We evaluate on CIFAR-10[2], MNIST digits[3] and NUS-WIDE[4]. The datasets have been extensively used in related hashing research [10,11,9]. CIFAR-10 consists of 60,000 images sourced from the 80 million Tiny Images dataset. The images are encoded using 512-D GIST descriptors. The MNIST digits dataset contains 70,000, 28x28 greyscale images of written digits from '0' to '9'. NUS-WIDE consists of 269,658 Flickr images annotated with multiple classes from an 81 class vocabulary. We only use those images associated with the 21 most frequent classes as per [11]. Each image is represented as a 500-D bag of words.

Following previous related work [10,7], we define ground truth nearest neighbours based on the semantic labels supplied with the datasets - that is, if two images share a class in common they are regarded as true neighbours. We also follow previous work in constructing our set of queries and training/database subsets. We randomly sample 100 images (CIFAR/MNIST) or 500 images (NUSWIDE) from each class to construct our test queries. The remaining images form the database of images to be ranked. We randomly sample 100/200/500 images per class from the database to form the training dataset (T). Our validation dataset is created by sampling 100/500 images per class from the database.

4.2 Baselines

The supervised data-dependent methods we compare to are KSH [10], BRE [9], STH [21] and ITQ with a supervised CCA embedding (ITQ+CCA)[7]. The unsupervised data-dependent techniques include AGH [11], SH [20] and PCAH [19]. The data-independent method is LSH [8]. We use the source code and parameter settings provided by the original authors. We tune the SVM parameters of STH in the same way we tune GRH (Section 4.3).

4.3 Parameter Optimisation

The algorithm has four meta-parameters: the number of iterations M, the amount of regularisation α, the flexibility of margin C, and the surface curvature γ, which arises for non-linear hypersurfaces based on radial-basis functions (RBFs). We optimise all meta-parameters via grid search on the held-out validation dataset.

We tune GRH parameters using the following strategy: firstly holding the SVM parameters constant at their default values ($C = 1$, $\gamma = 1.0$), we perform a grid search over $M \in \{1 \dots 5\}$ and $\alpha \in \{0.1, \dots, 0.9, 1.0\}$, selecting the overall configuration that leads to the highest *validation* dataset mAP. We then hold M and α constant at their optimised values, and perform a coarse logarithmic grid search over $\gamma \in \{0.001, 0.01, 0.1, 1.0, 10.0\}$ and $C \in \{0.01, 0.1, 1.0, 10, 100\}$. We equally weigh both classes (-1 and 1) in the SVM.

[2] `http://www.cs.toronto.edu/~kriz/cifar.html`
[3] `http://yann.lecun.com/exdb/mnist/`
[4] `http://lms.comp.nus.edu.sg/research/NUS-WIDE.htm`

Table 1. Hamming ranking mAP on CIFAR-10. *lin*: linear kernel, *rbf*: RBF kernel, *lsh*: LSH initialisation, *cca*: ITQ+CCA initialisation.

	CIFAR-10 (60K)							
Method	$T = 1,000$				$T = 2,000$			
	16 bits	32 bits	48 bits	64 bits	16 bits	32 bits	48 bits	64 bits
LSH	0.1290	0.1394	0.1463	0.1525	–	–	–	–
PCAH	0.1322	0.1291	0.1256	0.1234	–	–	–	–
SH	0.1306	0.1296	0.1346	0.1314	–	–	–	–
AGH	0.1616	0.1577	0.1599	0.1588	–	–	–	–
ITQ+CCA	0.2015	0.2130	0.2208	0.2237	0.2469	0.2610	0.2672	0.2664
STH$_{lin}$	0.1843	0.1872	0.1889	0.1835	0.1933	0.2041	0.2006	0.2144
STH$_{rbf}$	0.2352	0.2072	0.2118	0.2000	0.2468	0.2468	0.2481	0.2438
BRE	0.1659	0.1784	0.1904	0.1923	0.1668	0.1873	0.1941	0.2018
KSH	0.2440	0.2730	0.2827	0.2905	0.2721	0.3006	0.3119	0.3236
GRH$_{lin,lsh}$	0.2195	0.2264	0.2475	0.2490	0.2342	0.2569	0.2554	0.2639
GRH$_{rbf,lsh}$	0.2848	0.3013	0.3129	0.3015	0.3191	0.3475	0.3542	0.3646
GRH$_{lin,cca}$	0.2292	0.2563	0.2566	0.2593	0.2646	0.2772	0.2861	0.2900
GRH$_{rbf,cca}$	**0.2976**	**0.3161**	**0.3171**	**0.3209**	**0.3435**	**0.3675**	**0.3722**	**0.3688**

4.4 Evaluation Protocol

Following previous work [10,11,7,21,9], we evaluate the performance of our model using the widely accepted *Hamming ranking* evaluation paradigm. In this scenario, binary codes are generated for both the query and the database images. The Hamming distance is then computed from the query images to all of the database images, with the database dataset images ranked in ascending order of the Hamming distance. We evaluate the accuracy of retrieval using mean average precision (mAP) and the precision within Hamming radius 2. Our reported figures are the average over five random query/database partitions.

4.5 Discussion

In this paper we examine a single hypothesis that targets the core novelty of our work: namely, graph regularisation embedded in our iterative two-step algorithm is crucial for achieving high retrieval accuracy with hashcodes. Our results are presented in Tables 1-3 and Figures 3-4.

We explore four variants of our GRH model - GRH$_{lin,lsh}$, GRH$_{lin,cca}$ which construct linear hypersurfaces $\mathbf{h}_k$ and initialise the bits from either LSH or supervised initialisation with ITQ+CCA; and GRH$_{rbf,lsh}$, GRH$_{rbf,cca}$ which use non-linear hypersurfaces based on the RBF kernel. If we compare GRH directly to STH across both datasets we observe that GRH substantially outperforms STH with a linear SVM kernel (STH$_{lin}$) and an RBF kernel (STH$_{rbf}$). As STH also uses SVMs trained with hashcodes as targets, this result suggests that the gain realised by GRH must be due to our two-step iterative algorithm involving graph regularisation and not simply due to the use of SVMs.

Table 2. Hamming ranking mAP. **Left:** MNIST. **Right:** NUS-WIDE.

Method	MNIST (70K)				NUS-WIDE (270K)			
	$T = 1,000$				$T = 10,500$			
	16 bits	32 bits	48 bits	64 bits	16 bits	32 bits	48 bits	64 bits
LSH	0.2151	0.2704	0.3003	0.3147	0.3784	0.3860	0.3863	0.3879
PCAH	0.2683	0.2459	0.2257	0.2128	0.3890	0.3863	0.3829	0.3804
SH	0.2709	0.2626	0.2468	0.2510	0.3734	0.3751	0.3760	0.3751
AGH	0.5254	0.5583	0.5415	0.5310	0.3820	0.3809	0.3782	0.3767
ITQ+CCA	0.4532	0.4894	0.5325	0.5091	0.4268	0.4186	0.4161	0.4101
STH$_{lin}$	0.5051	0.5017	0.4938	0.4840	0.4458	0.4602	0.4626	0.4629
STH$_{rbf}$	0.5405	0.5400	0.5273	0.5224	0.4320	0.4499	0.4322	0.4305
BRE	0.4808	0.5442	0.5744	0.5904	0.4476	0.4650	0.4736	0.4776
KSH	0.7577	0.8011	0.8202	0.8268	0.4981	0.5107	0.5189	0.5144
GRH$_{lin,lsh}$	0.6473	0.7019	0.7187	0.7203	0.4799	0.4880	0.4937	0.5018
GRH$_{rbf,lsh}$	0.8386	0.8664	0.8756	0.8804	0.4974	0.4969	0.5090	0.5096
GRH$_{lin,cca}$	0.6705	0.7144	0.7290	0.7309	0.4886	0.4916	0.4999	0.4935
GRH$_{rbf,cca}$	**0.8632**	**0.8893**	**0.9066**	**0.9000**	**0.4996**	**0.5144**	**0.5217**	**0.5269**

On all datasets we find that the GRH model with a supervised embedding and non-linear hypersurfaces (GRH$_{rbf,cca}$) outperforms all baseline hashing methods. For example, GRH$_{rbf,cca}$ at 32 bits on CIFAR-10 achieves a relative gain in mAP of 16% versus KSH. GRH dominates the baselines when examining the precision-recall and precision at Hamming distance 2 curves (Figures 3-4).

We note the higher performance possible through running GRH on top of a supervised embedding (GRH$_{lin,cca}$, GRH$_{rbf,cca}$) versus a random initialisation (GRH$_{lin,lsh}$, GRH$_{rbf,lsh}$). This is particularly noticeable when more supervision is used ($T = 2000$) in Table 1. Here, for example, the mAP of linear GRH is increased by 8-13% when comparing GRH$_{lin,lsh}$ to GRH$_{lin,cca}$ from 16-64 bits.

Table 3. Timings and validation mAP vs. Iterations (CIFAR-10 @ 32 bits, GRH$_{lin,lsh}$)

Timings (s)			
Method	**Train**	**Test**	**Total**
GRH$_{lin,lsh}$	42.68	0.613	43.29
KSH	81.17	0.103	82.27
BRE	231.1	0.370	231.4

α	Iteration (M)					
	0	**1**	**2**	**3**	**4**	**5**
0.8	0.1394	0.1978	0.2051	0.2080	0.2089	0.2096
0.9	0.1394	0.2215	0.2319	0.2343	0.2353	0.2353
1.0	0.1394	0.2323	0.2318	0.2318	0.2318	0.2318

The linear variant of GRH is competitive in training and test time to the baseline hashing schemes (Table 3). For example on CIFAR-10 at 32 bits, GRH$_{lin,lsh}$ with $M = 4$ requires only 50% of the training time of KSH and only 20% of BRE while having a similar sub-second prediction (test) time to both baselines[5].

Table 3 details the behaviour of GRH$_{lin,lsh}$ on CIFAR-10 at 32 bits versus M and α. The mAP depends heavily on the value of α, and less so on M. The optimal M depends on the manner of initialisation - with random hyperplanes

[5] Benchmark system: Matlab 16Gb, single core CPU (Intel 2.7GHz).

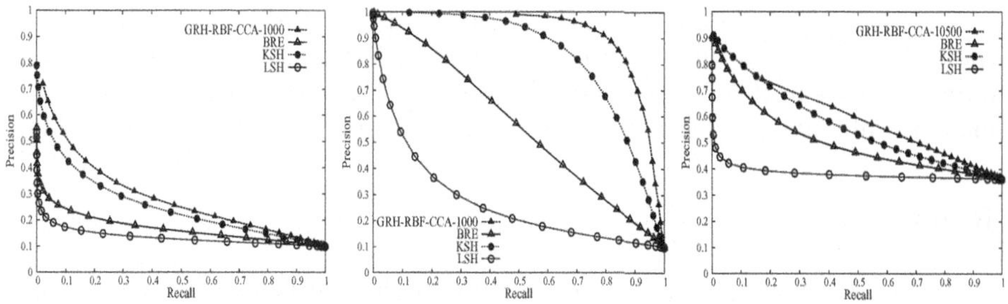

Fig. 3. PR curve @ 32 bits. **Left:** CIFAR. **Middle:** MNIST. **Right:** NUS-WIDE.

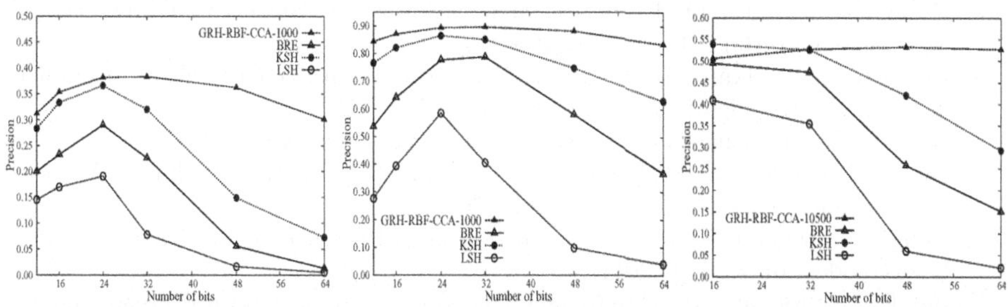

Fig. 4. Precision @ Radius 2. **Left:** CIFAR. **Middle:** MNIST. **Right:** NUS-WIDE.

(LSH), we find our method reaches the highest validation dataset retrieval accuracy within 3-4 iterations. With a supervised embedding (ITQ+CCA) only 1 iteration is typically needed due to the better initialisation of the hypersurfaces.

5 Conclusions and Future Work

In this paper we have introduced a novel two-step iterative hashing method, *Graph Regularised Hashing (GRH)* - in the first step we apply graph regularisation to enforce the constraint that similar data points have similar hashcodes. In the second step the regularised hashcodes form the labels for a set of binary classifiers, which has the effect of evolving the positioning of the hypersurfaces so as to separate opposing bits with maximum margin. GRH combines simplicity of implementation, competitive training time and state-of-the-art retrieval accuracy. These factors make GRH an ideal candidate for big data applications.

In our experimental validation we found GRH with *linear* hypersurfaces outperformed a broad selection of existing supervised hashing methods, and approaches closely the performance of the state-of-the-art *non-linear* Supervised Hashing with Kernels (KSH) method. This is encouraging as it means we can benefit from the lower computational cost of linear kernel learning, while sacrificing a modicum of retrieval accuracy. If spare CPU cycles are available and the highest retrieval accuracy is important, GRH can be used with non-linear hypersurfaces - this configuration outperformed all baseline hashing methods.

GRH is agnostic to the type of classifier used to learn the hypersurfaces. In the future we would be interested in porting GRH to a large-scale streaming data scenario - in this case a *passive aggressive* classifier [3] would be capable of incrementally updating the hypersurfaces in a computationally scalable fashion.

References

1. Belkin, M., Niyogi, P.: Laplacian eigenmaps for dimensionality reduction and data representation. In: NC (2003)
2. Chang, C.-C., Lin, C.-J.: Libsvm: A library for support vector machines. In: TIST (2011)
3. Crammer, K., Dekel, O., Keshet, J., Shalev-Shwartz, S., Singer, Y.: Online passive-aggressive algorithms. In: JMLR (2006)
4. Dempster, A.P., Laird, N.M., Rubin, D.B.: Maximum likelihood from incomplete data via the EM algorithm. In: JRSS, Series B (1977)
5. Diaz, F.: Regularizing query-based retrieval scores. In: IR (2007)
6. Fan, R.-E., Chang, K.-W., Hsieh, C.-J., Wang, X.-R., Lin, C.-J.: Liblinear: A library for large linear classification. In: JLMR (2008)
7. Gong, Y., Lazebnik, S.: Iterative quantization: A Procrustean approach to learning binary codes. In: CVPR (2011)
8. Indyk, P., Motwani, R.: Approximate nearest neighbors: Towards removing the curse of dimensionality. In: STOC (1998)
9. Kulis, B., Darrell, T.: Learning to hash with binary reconstructive embeddings. In: NIPS (2009)
10. Liu, W., Wang, J., Ji, R., Jiang, Y., Chang, S.: Supervised hashing with kernels. In: CVPR (2012)
11. Liu, W., Wang, J., Kumar, S., Chang, S.: Hashing with graphs. In: ICML (2011)
12. Moran, S., Lavrenko, V.: Sparse kernel learning for image annotation. In: ICMR (2014)
13. Moran, S., Lavrenko, V., Osborne, M.: Neighbourhood preserving quantisation for LSH. In: SIGIR (2013)
14. Oliva, A., Torralba, A.: Modeling the shape of the scene: A holistic representation of the spatial envelope. In: IJCV (2001)
15. Petrović, S., Osborne, M., Lavrenko, V.: Streaming first story detection with application to twitter. In: HLT (2010)
16. Raginsky, M., Lazebnik, S.: Locality-sensitive binary codes from shift-invariant kernels. In: NIPS (2009)
17. Torralba, A., Fergus, R., Freeman, W.T.: 80 million tiny images: A large data set for nonparametric object and scene recognition. In: PAMI (2008)
18. van Rijsbergen, C.J.: Information Retrieval. Butterworth (1979)
19. Wang, J., Kumar, S., Chang, S.: Semi-supervised hashing for large-scale search. In: PAMI (2012)
20. Weiss, Y., Torralba, A., Fergus, R.: Spectral hashing. In: NIPS (2008)
21. Zhang, D., Wang, J., Cai, D., Lu, J.: Self-taught hashing for fast similarity search. In: SIGIR (2010)

Approximate Nearest-Neighbour Search with Inverted Signature Slice Lists

Timothy Chappell, Shlomo Geva, and Guido Zuccon

Queensland University of Technology, Brisbane, Australia
{timothy.chappell,s.geva,g.zuccon}@qut.edu.au

Abstract. In this paper we present an original approach for finding approximate nearest neighbours in collections of locality-sensitive hashes. The paper demonstrates that this approach makes high-performance nearest-neighbour searching feasible on Web-scale collections and commodity hardware with minimal degradation in search quality.

Keywords: Locality-sensitive hashing, Hamming distance, Clustering.

1 Introduction

To determine the similarity between two documents in term vector space for nearest neighbour search, a cosine similarity calculation or similar measure must be performed for every term they share. To rank all documents in a collection by their distance from a given document, this must be repeated for all documents, rendering this operation infeasible for large collections with large vocabularies.

Locality-sensitive hashing ameliorates this issue by reducing the dimensionality of the term vector space in which these documents are stored and by representing these document vectors as binary strings. This allows expensive vector space similarity calculations to be replaced with cheaper Hamming distance calculations [16] that preserve pairwise relationships between document vectors.

Hashing is capable of reducing the costs of the individual document similarity computations; however, in Web-scale collections of hundreds of millions of documents, reducing the per-document processing time is not sufficient to make nearest-neighbour searching feasible. Efficient methods of computing document similarity are necessary for tasks such as near-duplicate detection (the discovery of pairs of documents that differ only marginally), e.g. for the purposes of removing redundant results while web crawling and plagiarism detection [12,13].

In this paper we consider the problem of performing efficient near-duplicate detection using document signatures, i.e. locality-sensitive hashes used to represent documents for the purpose of searching. Faloutsos and Christodoulakis [4] pioneered the use of superimposed coding signatures with an approach similar to Bloom filters, where signatures would be created directly from the filters and compared for similarity by masking them against other filters and counting the

A. Hanbury et al. (Eds.): ECIR 2015, LNCS 9022, pp. 147–158, 2015.

bits that remained. While this approach was shown to be inferior to the inverted file approach for ad hoc retrieval [18], recent work has since shown improvements on that original approach [7], leading to effectiveness comparable to inverted file approaches.

We introduce a novel approach for efficient near-duplicate detection that involves the generation of posting lists associated with a particular signature collection, making it possible to rapidly identify signatures that are close to a given search signature and discard those that are farther away. Our approach is empirically validated on ClueWeb09[1], a standard, publicly available, information retrieval collection. These experiments show that our approach is capable of performing near-duplicate detection on web-scale collections such as ClueWeb09 (500 million English-language documents) in 50 milliseconds on a commodity desktop PC costing under $10,000.

2 Locality-Sensitive Hashing

A hash function takes an arbitrary input object and produces a binary string (hash) of a fixed length. A standard property of conventional hash functions is that the same input will always produce the same hash, while a different input is almost certain to produce a vastly different hash. These binary strings can be much smaller than the original inputs, so comparing them for equality can be much faster. This makes them useful for applications such as verifying that a large file was transmitted correctly without needing to retransmit the entire file.

A frequently valued property of hash functions is the *avalanche effect*, where similar (but not identical) inputs produce entirely different hashes [6]. This is valued as it makes malicious attacks that rely on producing a certain hash more difficult. It also means that visual inspection of the hashes of two similar inputs will make it clear that there is a difference. By contrast, the locality-sensitive hash exhibits the reverse of this property: when a locality-sensitive hash function receives two slightly different inputs, the resultant hashes will be either identical or highly similar. This makes locality-sensitive hashing appropriate when it is desirable to match inputs that are similar.

For instance, when creating a collection of documents by crawling the Web it may be desirable to eliminate duplicate pages, as they contain no additional information and will consume extra space [2,12]. Because comparing a newly-downloaded web page to every web page downloaded so far could be very expensive, it may be desirable to hash them to make these comparisons faster. However, in the context of building a web collection, two pages that only differ in title or metadata are still essentially duplicates. With a locality-sensitive hash function, these two almost-identical web pages will have identical or almost-identical hashes, making it possible to detect these when comparing hashes.

This approach can be extended to the more general problem of determining object similarity. The similarity between two locality-sensitive hashes determines how

[1] `http://www.lemurproject.org/clueweb09.php/`, last visited February 13, 2015.

similar two objects are: hashes are used as a proxy for computing similarity using the Hamming distance [8] (the number of bits that differ between the two strings).

3 Related Work

Creating document signatures that can be compared for similarity with a Hamming distance calculation is a well-established use of locality-sensitive hashing.

Broder's Minhash [1] is one example of a locality-sensitive hashing algorithm that has been used successfully in the AltaVista search engine [2] for the purpose of discarding duplicate documents. Simhash [15], a more recent locality-sensitive hashing approach, has also been successfully used in this area. The main limiting factor in the scalability of these approaches is that, although Hamming distance computations can be performed extremely quickly, the execution time required to perform these computations over millions of signatures can quickly add up when dealing with web-scale collections.

Lin and Faloutsos [11] introduced frame-sliced signature files to improve on the performance of signature files without compromising on insertion speed the way Faloutsos' earlier bit-sliced signature files[5] did. In frame-sliced signature files, rather than each term setting bits throughout the signature, the bits set by each term are all set entirely within a randomly chosen frame in the signature. The signatures are then stored vertically frame-wise, requiring only the lists of frames corresponding to the frame positions used by terms in the search query to be processed.

Other attempts have been made to work around the scalability problems inherent to these approaches. Broder [2] found that storing the min hashes of each item in sorted order made searching for near duplicates an $O(n \log n)$ task as opposed to an $O(n^2)$ task. Recent work by Sood and Loguinov [17] makes use of the probabilistic nature of Simhash [15] to perform fuzzier searches without needing to scan the entire collection. In the field of image searching, Chum and Matas [3] use an inverted file approach to optimise the generation of Minhash document signatures for large image collections. We distinguish our approach from that used by Chun and Matas by using the inverted files directly to make searching the already-generated signatures more efficient.

4 Corpus Filtering Approaches

One way to avoid calculating Hamming distances for the entire collection is to remove from consideration signatures that are unlikely to be close to the search signature early on. One example is to use signatures small enough such that two documents that are similar enough to count as duplicates produce the same hash. The documents that correspond to each hash can then be stored in a list associated with that hash, immediately filtering out all the documents that do not have a matching signature.

This approach could be highly efficient, but is limited by the hashing function only supporting one level of discrimination, namely the exact match, which needs to be tuned to balance the frequency of type I and II errors. This tuning can only be applied per-collection, not per-document, as the search signature must be tuned with the same parameters. The inability to discriminate also prevents it from being used for k-nearest-neighbour searching as the threshold cannot be dynamically tuned for k.

5 Inverted Signature Slice Lists

The approach we propose in this paper, the **inverted signature slice list**, is similar to inverted files [14], but applied to the binary signature, not the original document. The document signature is subdivided into *bit slices*, each of a fixed length. The value of each bit slice and its position are then used to index into an array of lists. The list associated with this slice's value and position provide constant-time lookup of this signature and any others that share the same bit slice. Building these lists from a collection of signatures is very time-efficient because record lengths are fixed and text parsing is unnecessary.

Once the lists have been generated, searching is simply a matter of slicing the search signature and looking up the documents that share slices (both exact matches and close matches). The number of times a given signature appears in these lists and the quality of those occurrences (exact matches being more valuable than near matches) give an indication of how close the document signature is to the search signature. The top-k results can then be extracted from close candidates.

5.1 List Generation

The document signatures that comprise a signature collection are fixed-length signatures created as the output of a locality-sensitive hash function applied over all the documents in the original document collection. Document signatures are binary strings of a length that is fixed per collection. Shorter signatures require less storage space and are faster to process. Longer signatures can produce results of a higher quality due to minimising feature crosstalk, as one effect of the dimensionality reduction is that document features are all compressed and intermingled in the signature representation.

Typical signatures used for near-duplicate detection are short (32 or 64 bits long) while those used for image and document similarity comparisons are longer (e.g. Kulis and Grauman use 300-bit signatures [10]).

Signature Slicing. Generating the posting lists involves reading each signature in the collection, dividing that signature up into slices and adding its id to the lists associated with each slice. This process is very similar to the construction of a typical inverted file, but with two key differences:

- The position of the slice is stored along with the content of the slice to make up the corresponding term. For example, if a slice `00110011` makes up the

first 8 bits of a signature, and the (identical) slice `00110011` makes up the last 8 bits of a signature, the two slices have no relation to one another and hence correspond to entirely different inverted lists.
– While inverted files make use of an associative container for looking up terms, it is simpler and more efficient to use the slice's value directly as an array index. For instance, the slice `00110011` has its value (51 in decimal) used as an index into an array large enough to store all 256 possible slices.

In the proposed approach one of these arrays is created for every possible slice position. If signatures are 64 bits wide, there are 8 possible positions this slice could appear in, hence a total of 8 arrays capable of storing up to 256 slices. This can potentially represent a significant waste of memory if the collection is too small to cover most of the indices; as such, it is important to tune the slice width to suitably match the collection size.

Increasing the slice width reduces the load on any particular [value, position] pair by half; as there are more possible values of each slice, each slice value would cover less of the collection. and hence represents the most effective way of improving search performance.

The most efficient slice width for a particular collection may not necessarily be a power of 2. Furthermore, it may not divide evenly into the signature size. In those cases, when w-bit slices divide unevenly into the n-bit signature, $(w-1)$-bit slices may be included alongside the w-bit slices for some positions to ensure that the slices remain largely uniform in width and that they cover the entire signature. For instance, a 63-bit signature with 32-bit slices may have slice position 0 covered by a 32-bit slice and slice position 1 covered by a 31-bit slice. This means the corresponding table for that slice width may be jagged, with certain columns shorter than others. This has negligible implications for performance; uneven slice widths prove to work just as well in practice as even ones.

Storage Considerations. The slice lists only need to be generated once for each collection. After generation, the lists can be stored on disk and loaded into memory by the search tool. To minimise loading times, we store the slice lists in a block that can be loaded into memory and used as-is.

The amount of disk space (and, when searching, memory) consumed by the posting lists file is influenced by slice width, the number of slices per signature and the collection size. Low slice widths result in a smaller table structure, but more signatures being referenced in each list. Higher slice widths increase the size of the table, spreading the signature references across more lists.

A reference to every signature in the collection must appear in each column of the table (as every signature will match at least one pattern for every slice position). When the slice width is too small, increasing the slice width can actually reduce the disk space required to store the posting lists. As the slice width continues to increase, however, the amount of space taken up by the supporting structure will also increase, overwhelming the benefits from reducing the number of entries in the posting lists. As a result, for a given collection size and signature size there is a slice width for optimal memory consumption; increasing or decreasing that slice width will increase the amount of memory needed to store the file.

Slice list generation has little impact on the overall computational time efficiency of our approach. For example, creating the 26-bit slice lists for the English-language subset of ClueWeb09 (approximately 500 million signatures) on a 2.40GHz Intel Xeon computer took under 3 hours single-threaded (using 1024-bit signatures). Generation can be trivially parallelised by having each thread build slice lists for different subsets of the collection and merging them at the end.

5.2 List Searching

Searching the slice lists is a more complicated process than indexing them because the search component is responsible for handling slices that do not match the query slices exactly. Initially, the query signature is divided into slices in an identical fashion to the indexed signatures. This may mean uneven slice widths if the desired slice width does not divide evenly into the signature size, in which case it is important that the query signature is sliced in the exact same way.

Neighbourhood Expansion. The [value, position] pair associated with each slice is looked up in the array of posting lists, as done when indexing. Unlike indexing though, we expand the Hamming neighbourhood of each search and bring in similar signatures, under the assumption that even very similar signatures may not match any of the slices exactly. As an example, the 16-bit signature with two 8-bit slices `10110011 01010001` does not have any slice that exactly matches the search signature `00110011 01010101`, even though there are only 2 different bits and this may well be considered similar enough to match.

To expand the Hamming neighbourhood, after consulting the `[00110011, 0]` list looking for candidate documents to consider, we also consult every other possible slice value within a certain Hamming distance from the original query. For example, to perform a 1-bit Hamming expansion, we would include not only `00110011` but also the 8 other possible slice values that exist one bit away. This includes `10110011` from the example earlier, so this signature would be picked up, as would any other signature that contains a slice within a Hamming distance of 1 from the respective slice in the search signature.

We can continue expanding the Hamming neighbourhood of our search signature by bringing in slices that are farther away. This allows less precise matches to be made at the cost of additional search time. The number of posting lists that must be considered at each expansion is the binomial coefficient of the Hamming distance and the slice width, making the total number of posting lists considered the sum of all Hamming distances up to that point, or $\sum_{i=0}^{h} \binom{i}{w}$ where w is the slice width and h is the Hamming distance to expand the neighbourhood.

To illustrate the interaction between Hamming neighbourhood expansion and slice width, consider two documents with 24-bit signatures, one just different enough from the other to have 2 bits that differ (their Hamming distance is 2). This signature could be sliced up in a number of ways; e.g., into 8 or 12-bit

```
for all slice position ∈ query signature do
   query value ← query signature[slice position]
   for all v ∈ values with 0-n bits set do
      distance ← popcount(query value ⊕ v)
      similarity ← slice width − distance
      signature ← list[query value ⊕ v, slice position]
      score[signature] ← score[signature] + similarity
   end for
end for
```

Fig. 1. Pseudo-code algorithm for list searching

slices. If 12-bit slices are used, there is a $\frac{12}{23}$ probability that both differing bits will end up in different slices and an $\frac{11}{23}$ probability that they end up in the same slice. In the latter case, there is no need to expand the neighbourhood as one of the slices will match exactly. In the former case, a 1-bit expansion is necessary.

With 8-bit slices, there will always be at least one slice that is identical between the two signatures. As such, while neighbourhood expansion is unnecessary for the identification of all signatures a Hamming distance of 2 away when using 8-bit slices, 12-bit slices can only be expected to identify $\frac{11}{23}$ of them without expanding the neighbourhood.

It should be noted that 12-bit slices will have posting lists $\frac{1}{16}$ of the length of 8-bit slices, meaning that moving to a 1-bit neighbourhood expansion (and hence needing to process 13× the number of posting lists) would still improve performance over using 8-bit slices and no neighbourhood expansion.

In summary, while increasing the slice width does trade search accuracy for an increase in retrieval speed, the trade-off is sufficiently worthwhile that even expanding the Hamming neighbourhood to fully counteract the reduced search accuracy is often a more attractive option than leaving the slice width the same. However, given that the improvement in retrieval speed plateaus after the search table reaches a collection-dependent level of sparsity, retrieval time efficiency can only be increased up to a point while maintaining a given level of search accuracy.

Hamming Distance Estimation. Processing these lists up to the desired neighbourhood expansion allows the search tool to not only obtain a subset of the collection containing most of the close signatures, but also to use this same information for calculating optimistic and pessimistic Hamming distances. This can make it possible to cull the subset further before calculating true Hamming distances. Algorithm 1 shows the approach we use, with approximate Hamming distance similarity referred to as *score*. After processing the posting lists, the highest-scoring signatures are likely to be the signatures with the lowest Hamming distances from the query signature.

To illustrate this, consider the case of 32-bit signatures and four 8-bit slices before neighbourhood expansion. After consulting the posting lists for all slices, the potential range of each signature's Hamming distance can be calculated. A signature that appears in all 4 slices is one that has exactly matched the search signature and as a result has a Hamming distance of 0. A signature that appears in none of the slices cannot have a Hamming distance of less than 4 as it would appear in at least one slice otherwise. Therefore, its optimistic Hamming distance can be calculated as 4 (a case in which every slice had 1 bit differing from the search signature) and its pessimistic distance calculated at 32 (a case in which no slice had any bits in common with the search signature.) In the same way, a signature that appears in 3 of the slices has an optimistic Hamming distance of 1 (if the slice the signature did not appear in had 1 bit that differed from the respective slice in the search signature) and a pessimistic Hamming distance of 8 (that same slice containing all differing bits.)

The range between optimistic and pessimistic Hamming distances can be narrowed through neighbourhood expansion. In the previous example, one signature did not appear in any of the slices and hence could have had a Hamming distance of anything from 4 to 32. On expanding the neighbourhood by 1 bit, if the signature still never appears in any of the slices, the possible range of values its Hamming distance could occupy is reduced to 8-32.

Expanding the Hamming neighbourhood increases the quality of these estimations at the expense of more search time, but also reducing the subset of signatures that fall within the desired range, allowing these signatures to be skipped when calculating true distances later. Based on user requirements, the signature size, slice width, neighbourhood expansion and heuristics for discarding signatures based on their optimistic and/or pessimistic Hamming distances can be tuned to produce the desired trade-offs between performance, memory usage and quality of results.

6 Evaluation

Search accuracy and retrieval time are the most important factors when judging the efficacy of any search approach. Tuning parameters for the inverted signature slice list approach involves making speed-accuracy trade-offs. To judge whether certain trade-offs are worthwhile or not, it is necessary to be able to judge the correctness of the results returned.

Experiments are conducted on a subset of 500 million English-language documents from the ClueWeb09 Category A. We have used 1024-bit TOPSIG [7] signatures; while signature width has an impact on search quality this impact has been explored elsewhere [7] and is not the topic of our research, which is more concerned between the comparative quality between ISSL searches and searches of the raw signatures. As the inverted signature slice list approach is designed to retrieve the signatures with the closest Hamming distances to the query, we are using an exhaustive Hamming distance search that retrieves the closest results without fail as an approach to compare against. By definition, the closer the results retrieved by this approach are to the exhaustive results, the more correct they are.

Making a search quality judgement therefore requires a quantitative way of analysing one set of search results in terms of how closely it matches a second

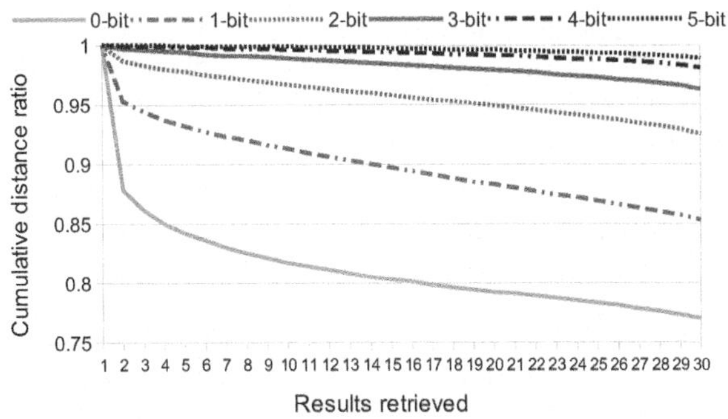

Fig. 2. The impact of neighbourhood expansion (0-bit meaning no expansion) on cumulative distance ratio

set of search results. We introduce the *cumulative distance ratio* metric, which is akin to a graded relevance metric designed for evaluating lists of Hamming distances. This metric considers two lists of equal length; one a list of the signatures returned by some retrieval method, the other being the definitive list of closest signatures (obtained using an exhaustive Hamming distance search with every signature in the collection). It ought to be remembered that, as the inverted signature slice list approach is only concerned with returning the top-k nearest neighbours, here we are measuring its accuracy compared to the definitive list of top-k nearest neighbours.

The distance ratio at position p is calculated as the ratio between the cumulative sums of the Hamming distances of the retrieved documents up to position p: $DR(p) = \frac{\sum_{i=1}^{p} T(i)}{\sum_{i=1}^{p} D(i)}$, where $D(i)$ is the Hamming distance of the ith result from the algorithm being evaluated and $T(i)$ is the Hamming distance of the ith closest signature. For the purposes of calculating the distance ratio, we let $0 \div 0 = 1$: this can be a common occurrence as it happens every time there is an exact duplicate in the collection (Hamming distance of 0) and the search algorithm finds it. From this, we can calculate the cumulative distance ratio $CDR(p) = \sum_{i=1}^{p} DR(i)/p$.

6.1 Hamming Neighbourhood Expansion

Expanding the Hamming neighbourhood, as described earlier, causes more posting lists to be consulted for each search. This increases the pool of candidates and hence search quality at the cost of increased retrieval time. As Figure 2 shows, only a few bits of neighbourhood expansion are needed to greatly improve search quality and expanding beyond that not only provides increasingly diminishing returns but also comes with a substantial impact to performance (search time: 3-bit = 5.084ms, 4-bit = 12.534, 5-bit = 27.865, 8-bit = 130.887ms). This is due to the number of posting lists increasing binomially while the number of close

Table 1. Searching a 1 million document subset of Wikipedia (1024-bit signatures, 16-bit slices, 20 threads, $k = 30$) with the smaller candidate threshold. (i = distance beyond which to stop considering posting lists. j = distance beyond which to stop extending the list of candidate signatures)

i	j	Search time	CDR@10
0	0	0.040ms	0.817
1	0	0.112ms	0.869
	1	0.193ms	0.913
2	0	0.568ms	0.896
	1	0.703ms	0.951
	2	1.399ms	0.967

i	j	Search time	CDR@10
3	0	2.242ms	0.911
	1	2.452ms	0.967
	2	3.251ms	0.985
	3	5.080ms	0.989
4	0	7.011ms	0.913
	1	7.258ms	0.971
	2	8.517ms	0.99
	3	11.483ms	0.995
	4	12.744ms	0.996

signatures remaining in the collection is soon depleted, causing the cumulative distance ratio to quickly plateau.

6.2 Slicing Optimisations

One optimisation we have implemented to gain some of the benefits from an expanded Hamming neighbourhood (specifically, the more precise Hamming ranges of the signatures found early on) is to define an earlier Hamming range, beyond which any signatures only seen for the first time will not be considered.

In other words, when processing posting lists beyond this Hamming distance, any documents that are seen and have already accrued score from earlier posting lists will have their score increased as normal. However, signatures that have not yet been seen and do not yet have a score will be ignored. This allows expensive write operations for signatures with a low likelihood of being close enough to the search query to be elided, saving that processing time as well as the processing time required to analyse the score table at the end and extract the top results.

Table 1 demonstrates that this can provide strong improvements in efficiency, but with a corresponding drop in search accuracy that may not be worthwhile under other circumstances.

While the most effective slice width for a given collection size will depend on a number of factors, including available memory, a good rule of thumb is to increase the slice width by one bit each time the collection doubles in size. Doubling the size of the collection will result in the average posting list length doubling in size too, which will make lookups far slower. Increasing the slice width, on the other hand, will cause the average posting list length to halve, the two effectively cancelling each other out.

Figure 3 captures the most significant aspect of the inverted slice signature lists approach. Note that each point on the curve corresponds to a different slice width, and a successive doubling of the collection size. As the collection size is increased 1024-fold along the x-axis, the search time is only increased by less

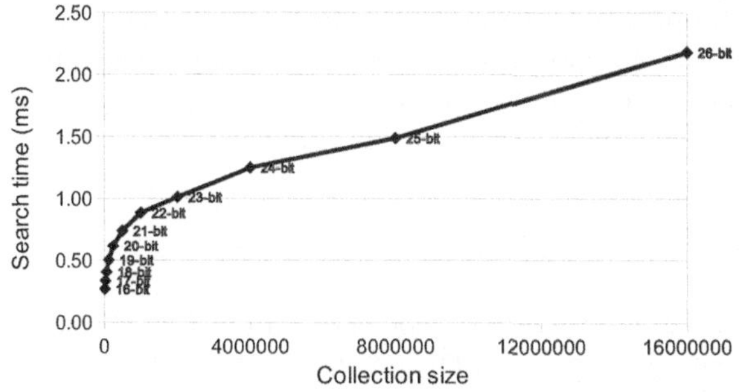

Fig. 3. Keeping the slice width in line with collection growth to reduce the corresponding growth in search times

Table 2. Searching ClueWeb09 (500 million documents). 3-bit Hamming expansion, 20 threads

Slice width	Search time	Memory (MB)	CDR@10
23-bit	199.738ms	180029.43	0.925
24-bit	112.783ms	177417.01	0.915
25-bit	66.753ms	176260.59	0.902
26-bit	56.955ms	177346.38	0.894
Exhaustive	2843.619ms	92266.03	1.000

than 10-fold. This is what makes it possible to search the English ClueWeb09 for top-k nearest in about 57ms. By comparison, an exhaustive signature search takes about 2.8s (see Table 2); we achieve approximately a 50-fold speedup with the inverted signature slice list approach.

7 Conclusion

We have presented an approach to improving the speed of nearest-neighbour signature searching without a considerable loss to search fidelity. While it is difficult to make direct comparisons to other systems, most of which have been designed for different purposes and for which publicly available code and/or data are not provided, none of the systems we have surveyed [3,9,12] work on web-scale collections with (high-end) consumer-level hardware. The field of prior research in this area seems largely divided into two camps: groups using consumer-level hardware searching non-web-scale collections (hundreds or thousands of documents or low millions) [3,9]; and groups searching web-scale collections with highly efficient networks of Hadoop clusters [12]. The former are working in an entirely different problem space while the latter are difficult to benchmark against, particularly if the code and computational platforms are not available.

We consider here that 50-millisecond search of a 500 million document collection on consumer-level hardware is a compelling justification for the modest loss of precision. The effective use of inverted signature slice lists may be limited to certain applications (near-duplicate detection, clustering etc.), in those situations they can provide great performance improvements over exhaustive approaches. The implementation described in this paper is available under an open-source license and distributed at `http://www.topsig.org`.

References

1. Broder, A.: On the resemblance and containment of documents. In: Compression and Complexity of Sequences 1997, pp. 21–29. IEEE (1997)
2. Broder, A.: Identifying and filtering near-duplicate documents. In: Giancarlo, R., Sankoff, D. (eds.) CPM 2000. LNCS, vol. 1848, pp. 1–10. Springer, Heidelberg (2000)
3. Chum, O., Matas, J.: Fast computation of min-hash signatures for image collections. In: CVPR 2012, pp. 3077–3084 (2012)
4. Faloutsos, C., Christodoulakis, S.: Signature files: An access method for documents and its analytical performance evaluation. TOIS 2(4), 267–288 (1984)
5. Faloutsos, C., Chan, R.: Fast text access methods for optical and large magnetic disks: Designs and performance comparison. VLDB 88, 280–293 (1988)
6. Feistel, H.: Cryptography and computer privacy. Sci. Am. 228, 15–23 (1973)
7. Geva, S., De Vries, C.: Topsig: topology preserving document signatures. In: CIKM 2011, pp. 333–338 (2011)
8. Hamming, R.: Error detecting and error correcting codes. Bell System Tech. J. 29(2), 147–160 (1950)
9. Jiang, Q., Sun, M.: Semi-supervised simhash for efficient document similarity search. In: ACL 2011, pp. 93–101 (2011)
10. Kulis, B., Grauman, K.: Kernelized locality-sensitive hashing for scalable image search. In: ICCV 2009, pp. 2130–2137 (2009)
11. Lin, Z., Faloutsos, C.: Frame-sliced signature files. IEEE Transactions on Knowledge and Data Engineering 4(3), 281–289 (1992)
12. Manku, G., Jain, A., Das Sarma, A.: Detecting near-duplicates for web crawling. In: WWW 2007, pp. 141–150 (2007)
13. Potthast, M., Stein, B.: New issues in near-duplicate detection. In: Data Analysis, Machine Learning and Applications, pp. 601–609. Springer (2008)
14. Van Rijsbergen, C.J.: Information Retrieval. In: Butterworth (1979)
15. Sadowski, C., Levin, G.: Simhash: Hash-based similarity detection. Technical report, Google Tech. Rep. (2007)
16. Slaney, M., Casey, M.: Locality-sensitive hashing for finding nearest neighbors. Signal Processing Magazine 25(2), 128–131 (2008)
17. Sood, S., Loguinov, D.: Probabilistic near-duplicate detection using simhash. In: CIKM 2011, pp. 1117–1126 (2011)
18. Zobel, J., Moffat, A., Ramamohanarao, K.: Inverted files versus signature files for text indexing. TODS 23(4), 453–490 (1998)

A Discriminative Approach to Predicting Assessor Accuracy

Hyun Joon Jung and Matthew Lease

School of Information
University of Texas at Austin, USA
{hyunJoon,ml}@utexas.edu

Abstract. Modeling changes in individual relevance assessor performance over time offers new ways to improve the quality of relevance judgments, such as by dynamically routing judging tasks to assessors more likely to produce reliable judgments. Whereas prior assessor models have typically adopted a single generative approach, we formulate a discriminative, flexible feature-based model. This allows us to combine multiple generative models and integrate additional behavioral evidence, enabling better adaptation to temporal variance in assessor accuracy. Experiments using crowd assessor data from the NIST TREC 2011 Crowdsourcing Track show our model improves prediction accuracy by 26-36% across assessors, enabling 29-47% improved quality of relevance judgments to be collected at 17-45% lower cost.

Keywords: search evaluation, crowdsourcing, machine learning and modeling.

1 Introduction

Recent efforts in efficiently collecting relevance judgments at scale have focused on how to collect high-quality relevance judgments with crowdsourcing [1] [2] [3]. Since quality of relevance judgments critically influences the results of IR system evaluation [4], a great deal of research has focused quality improvement of relevance judgments via various approaches: multiple labeling and aggregation [5], behavioral effects investigation [6], letting assessors select which tasks to work on [7], and efficient HIT (Human Intelligence Tasks) design [8].

Predicting the quality of judgments represents another opportunity to improve quality of crowdsourced relevance judgments. For instance, task routing in crowdsourcing [7] requires a method to match a worker to a task. One can route a specific judgment task to a specific assessor based on the prediction of a probability of an assessor's next judgment correctness, and expect improved quality of relevance judgments.

Prior work in predicting assessors' annotation performance has typically assumed that an assessor's judgments are independent and identically distributed (i.i.d) over time [9]. In other words, prior work has not considered temporal effects among judgments. To solve this problem, Donmez et al. [10] and Jung et al [11] proposed time-series models. However, while one could imagine many features characterizing an assessor's behavior, their models still rely upon a single generative model at time t.

A. Hanbury et al. (Eds.): ECIR 2015, LNCS 9022, pp. 159–171, 2015.

To address this problem, we build a Generalizable feature-based Assessor Model (GAM) that allows us to flexibly capture a wider range of assessor behaviors by incorporating features which model different aspects of this behavior. We integrate various features from prior studies which were used mainly or only for the estimation of crowd assessor's annotation performance [11] or judgment simulation [4]. In addition, we devise several new behavioral features indicating an assessor's annotation performance over time and integrate them with the existing features selected from prior studies.

We investigate this predictive model with the public NIST TREC 2011 Crowdsourcing Track dataset[1]. Firstly, we evaluate prediction quality, both in terms of hard prediction (binary correct or not) and soft prediction (probability of making a correct label). In particular, we study the effect of a *decision reject option*, which improves prediction accuracy by sacrificing prediction coverage, providing a tuning parameter for aggressive vs. conservative prediction given model confidence. In the second experiment, we conduct an in-depth feature analysis in order to compare the relative importance of each feature. Finally, we evaluate the effectiveness of our predictive model for crowdsourced judgment quality improvement under a realistic scenario assuming task routing and label aggregation. Our empirical evaluation demonstrates that our model improves prediction accuracy by 26-36% across 54 assessors. In addition, our experiments show that the quality of relevance judgments by our prediction model-based task routing improves its accuracy by 29-47% with lower cost (17-45%). Our research questions are:

RQ1: Feature Design for Prediction Model. *When we build a discriminative, feature-based learning framework for predicting work quality, what features are useful to include, and what is their relative importance?*

RQ2: Prediction Performance Improvement. *Does our prediction model improve prediction performance? How does decision rejection trade-off coverage vs. accuracy of prediction model in comparison to other baselines?*

RQ3: Impact on Judgment Quality and Cost. *Can our prediction model improve the quality of relevance judgments and/or decrease cost of collecting judgments?*

2 Problem

Estimating and predicting crowd assessors' performance has gained relatively little attention in IR system evaluation. Most prior work in crowd assessor modeling has focused on simple estimation of assessors' performance via metrics such as accuracy and F1 [12] [13]. Unlike other studies, Caterette and Soboroff presented several assessor models based on Bayesian-style accuracy with various types of Beta priors [4]. Recently, Ipeirotis and Gabrilovich presented a similar type of Bayesian style accuracy with a different Beta prior in order to measure assessors' performance [8]. However, neither investigated prediction of an assessor's judgment quality.

Figure 1 shows two real examples of failures of existing assessor models in predicting assessor's judgment correctness. The more accurate left assessor (a) begins with very strong accuracy (0.8) which continually degrades over time, whereas accuracy of the right assessor (b) hovers steadily around 0.5. Suppose that a crowd worker's next

[1] `https://sites.google.com/site/treccrowd/`

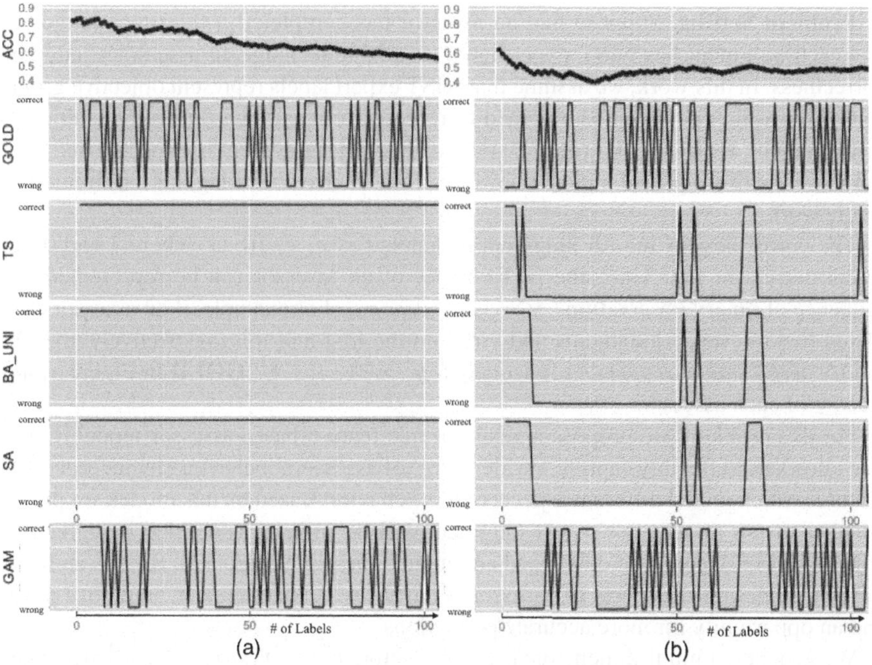

Fig. 1. Two examples of failures of existing assessor models and success of our proposed model, GAM in predicting the correctness of assessors' next label ((a) high accuracy assessor and (b) low accuracy assessor). While the agreement of a crowd assessor's judgments with that of the original NIST topic authority (GOLD) oscillates over time, the existing assessor models (Time-series (TS) [11]), Sample Running Accuracy (SA), Bayesian uniform beta prior (BA-UNI [8]) do not follow the temporal variation of the assessors' agreement with the gold labels. On the contrary, GAM is sensitive to such dynamics of labels over time for higher quality prediction.

label quality (y_t) is binary (correct/wrong) with respect to ground truth. While y_t oscillates over time, the existing models are not able to capture such temporal dynamics and thus prediction based on these models is almost always wrong. In particular, when an assessor's labeling accuracy is greater than 0.5 (eg., average accuracy = 0.67 in Figure 1 (a)), the prediction based on the existing models are always 1 (correct) even though the actual assessor's next label quality oscillates over time. A similar problem happens in Figure 1 (b) with another worker whose average accuracy is below 0.5.

In crowdsourcing and human computation, significant research has focused on the estimation or prediction of crowd workers' behavior or performance [14] [15]. However, most studies assumed that each annotation is independent and identically distributed (i.i.d) over time even though crowd worker behavior can have temporal dynamics as shown in Figure 1. Donmez et al. [10] was the first to propose a time-series model. Jung et al. [11] presented a temporal model to estimate asymptotic worker accuracy. However, while there exist many features characterizing a crowd assessor's behavior, these models only rely on the observation of labels [10] or labels' correctness [11]. For this reason, existing time-series models remain limited in terms of predicting an assessor's next judgment correctness as shown in Figure 1.

Problem Setting. Suppose that an assessor has completed n relevance judgments and each judgment has NIST expert labels available to judge an assessor's judgment correctness. In this work, we assume that NIST expert labels represent objective ground truth from which deviation is assumed to represent error, rather than valid, subjective disagreement. However, in practice, some level of disagreement is expected and common, even with simplified topical relevance [16]. We leave relaxing this assumption for future work.

The correctness of the ith judgment is denoted as $y_i \in \{0, 1\}$, where 1 and 0 represent correct or not. Thus, the performance of an assessor can be represented as a sequence of binary observations, $\mathbf{y} = \begin{bmatrix} y_1 \; y_2 \; \ldots \; y_n \end{bmatrix}$. For example, if an assessor completed five relevance judgments and erred on the first and third respectively, then his *binary performance sequence* is encoded as $\mathbf{y} = \begin{bmatrix} 0 \; 1 \; 0 \; 1 \; 1 \end{bmatrix}$. **GOLD** in Figure 1 indicates $\mathbf{y}$ of each assessor.

For this problem, we propose a generalizable feature-based assessor model (GAM) that allows us to flexibly capture a wider range of assessors' behaviors by incorporating features which model different aspects of this behavior. Based on this model, we predict whether or not an assessor's next judgment will be correct, as defined by agreement with the NIST expert who developed and judged the topic originally. By this ability to flexibly model more aspects of assessor behavior, we expect greater predictive power and an opportunity for more accurate predictions.

We generate a multi-dimensional feature vector, $x_i = \begin{bmatrix} x_{1i} \; x_{2i} \; \ldots \; x_{mi} \end{bmatrix}$ per time i and use x_i as an input of a prediction function f. Prior assessor models only consider a simple feature measure x_i by a single metric, accuracy, and then use this feature as an input of simple link function $y_{i+1} = \mathit{roundOff}(x_i)$. Instead, our proposed model incorporates a multi-dimensional feature vector x_i and uses this feature vector with a learning framework $f(x_i, y_i) = y_{i+1}$. The bottom plot of Figure 1 shows how GAM is able to track the assessor's varying correctness with greater fidelity.

3 Method: Generalized Time-Varying Assessor Model (GAM)

In this section, we present a generalizable feature-based assessor model that incorporates various observable and latent features modeling different aspects of assessors' behavior. We first examine feature generation and integration, and then discuss learning a predictive model with the generated features.

3.1 Feature Generation and Integration

An assessor's behavior and annotation performance may be captured by various types of features. In this study, we generate and integrate two types of features shown in Table 1: observable and latent features. Bayesian-style features have various forms in prior work according to different Beta prior settings. Among them, we adopt *optimistic* (a Beta prior $\alpha = 16, \beta = 1$) and *pessimistic* (a Beta prior $\alpha = 1, \beta = 16$) assessor models from Carterette and Soboroff's study [4]. In addition, we adopt a Bayesian style accuracy from Ipeirotis and Gabrilovich's study which assumes a Beta prior ($\alpha = 0.5, \beta = 0.5$), referred to here as the *uniform* assessor model. In these assessor models, each Beta prior characterizes each assessor's annotation performance.

Table 1. Features of generalized assessor model (GAM). n is the number of total judgments and x is the number of relevance judgments at time t.

	Feature Name	Description
Observable	Bayesian Optimistic Accuracy (BA_{opt}) [4]	a Bayesian style accuracy with a prior $Beta$ (16,1) $BA_{opt} = (x_t + 16)/(n_t + 17)$
	Bayesian Pessimistic Accuracy (BA_{pes}) [4]	a Bayesian style accuracy with a prior $Beta$ (1,16) $BA_{pes} = (x_t + 1)/(n_t + 17)$
	Bayesian Uniform Accuracy (BA_{uni}) [8]	a Bayesian style accuracy with a prior $Beta$ (0.5,0.5) $BA_{uni} = (x_t + 0.5/(n_t + 1)$
	Sample Running Accuracy (SA)	$SA_t = x_t/n_t$
	CurrentLabelQuality	a binary value indicating whether a current label is correct or wrong.
	TaskTime	time to spend in completing this judgment task. (ms)
	AccuracyChangeDirection (ACD)	a binary value indicating the absolute difference between $SA_{t-1} - SA_t$.
	TopicChange	a binary value indicating a topic change between time $t-1$ and time t.
	NumLabels	a cumulative number of completed relevance judgments at time t.
	TopicEverSeen	a real value [0∼1] indicating the familiarity of a topic. $\frac{1}{a\ number\ of\ judgments\ on\ topic\ k\ at\ time\ t}$
Latent	Asymptotic Accuracy (AA) [11]	a time-series accuracy estimated by latent time-series model proposed by Jung et al. $\frac{c}{1-\phi}$.
	ϕ [11]	a temporal correlation indicating how frequently a sequence of correct/wrong observations has changed over time.
	c [11]	a variable indicating the direction of judgments between correct and wrong.

For instance, the *optimistic* assessor model indicates that an assessor is likely to make a relevance judgment in a permissive fashion, while the *pessimistic* model tends to make more non-relevant judgments than relevant judgments. The *uniform* model has an equal chance of making a relevant or non-relevant judgment. Note that Bayesian style accuracies (BA_{opt}, BA_{pes}, BA_{uni}) were only used as a way of simulating judgments or estimating an assessor's performance in the original studies. In this study, we instead used these accuracies as a feature of estimating an assessor's annotation performance as well as predicting an assessor's next judgment's correctness. Other observable features include measurable features from a sequence of relevance judgments from an assessor. Among them, $TaskTime$ and $NumLabels$ are designed to capture an assessor's behavioral transition over time. $TopicChange$ checks the sensitivity of an assessor to topic variation over time. The $TopicEverSeen$ feature is designed to consider the effect of growing topic familiarity over time. The value is discounted by increased exposure to topic k.

Latent features are adopted from Jung et al's [11] model of temporal dynamics of assessor behavior (ϕ and c). While they only used *asymptotic accuracy* (AA) as an indicator of an assessor's annotation performance, we integrate all three features (AA, ϕ, and c) into our generalized assessor model. Our intuition is that each feature may capture a different aspect of an assessor's annotation performance and thus the integration of various features enabling greater predictive power for more accurate predictions.

3.2 Predicting Judgments Quality

To select a learning model, we adopt **L1-regularized logistic regression** due to several reasons. Firstly, it supports probabilistic classification as well as binary prediction by logistic function. In our problem setting, we conflate graded relevance judgments into binary values (0 or 1), and thus logistic regression is the best fit in order to handle such a binary classification problem. In addition, a logistic regression model allows us obtain the odds ratio, defined as the ratio of the probability of correct over incorrect relevance judgments. Secondly, L1-regularized logistic regression prevents over-fitting in learning models due to either co-linearity of the covariates or high-dimensionality. The regularized regression shrinks the estimates of the regression coefficients towards zero relative to the maximum likelihood estimate. Finally, logistic regression is relatively simple and fast. In practice, one of the challenging issues to run learning algorithms is that it takes too much time to update parameters and predict output values once a new label comes. However, this model is quite efficient.

In prediction, we consider a supervised learning task where we are given N training instances $\{(x_i, y_i), i = 1, ..., N\}$. Here, each $x_i \in \mathbb{R}^M$ is an M-dimensional feature vector, and $y_i \in 0, 1$ is a class label indicating whether an assessor's next judgment is correct (1) or wrong (0). Before fitting a model to our feature and target labels, we first normalize our features in order to ensure that normalized feature values implicitly weight all features equally in a model learning process. Logistic regression models the probability distribution of the class label y given a feature vector X as follows:

$$p(y = 1|x; \theta) = \sigma(\theta^T x) = \frac{1}{1 + exp(-\theta^T x)} \quad (1)$$

Here $\theta = \{\beta_0, \beta_1^T, ..., \beta_M^T\}$ are the parameters of the logistic regression model; $\sigma(\cdot)$ is the sigmoid function, defined by the second equality. The following function attempts to maximize the log-likelihood in order to fit a model to a given training data.

$$\max_{\theta}\{\sum_{i=1}^{N} [y_i(\beta_0 + \beta^T x_i) - log(1 + e^{\beta_0 + \beta^T x_i})] - \lambda \sum_{j=1}^{M} |\beta_j|\}. \quad (2)$$

3.3 Prediction with Decision Reject Option

Our predictive model can generate two types of outputs: a binary value predicting the correctness of an assessor's judgment (0 or 1) and a continuous value ($y_{i+1} \in [0, 1]$) indicating the probability of making a correct judgment. While a binary predictive value (*hard prediction*) can be used as it is, a probabilistic predicted value (*soft prediction*) can be used after a transformation, such as rounding-off. For instance, if an original predicted value is 0.76, we could round this to a binary predictive value of 1.

In term of soft prediction, there exists room for improving its quality by taking account of prediction confidence. For instance, if a value of soft prediction is close to 0.5, it fundamentally indicates very low confidence. Therefore, we may avoid the risk of getting noisy predictions by adopting a *decision rejection option* [17]. In this study, we round off a probabilistic predictive value with a decision reject option as follows. If $y_{i+1} < 0.5 - \delta$ or $y_{i+1} \geq 0.5 + \delta$ then y_{i+1} does not need any transformation and use its original value.

If $y_{i+1} \geq 0.5 - \delta$ or $y_{i+1} < 0.5 + \delta$ then y_{i+1} is $null$, indicating the reject of decision. δ is a parameter to control the limits of decision reject option $\in [0, 0.5]$. High δ indicates a conservative prediction which increases the range of decision rejection while sacrificing coverage. On the other hand, low δ allows prediction in a permissive manner, decreasing the threshold of decision rejection and increasing coverage.

4 Evaluation

Experimental Settings

Dataset. Data from the NIST TREC 2011 Crowdsourcing Track Task 2 is used. The dataset contains 89,624 *graded relevance judgments (2: strongly relevant, 1: relevant, 0: non-relevant)* collected from 762 workers rating the relevance of different Webpages to different search queries [18]. We conflate judgments into a binary scale (relevant / non-relevant), leaving prediction of graded judgment accuracy for future work. We processed this dataset to extract the original temporal order of the assessor's relevance judgments. We include 3,275 query-document pairs which have expert judgments labeled by NIST assessors, and we exclude workers making < 20 judgments to ensure stable estimation. Moreover, since the goal of our work is to predict assessors' next judgment quality, we intentionally focus on prolific workers who will continue to do this work in the future, for whom such predictions will be useful. 54 sequential relevance judgment sets are obtained, one per crowd worker. The average number of labels (i.e., sequence length) per worker is 154.

Metrics. Prior to measurement, we collect **gold** labels for each assessor by computing the agreement of a crowd assessor's judgments with that of the original NIST topic authority. We evaluate the performance of our prediction model with two metrics. Firstly, we measure the prediction performance with accuracy and *Mean Absolute Error* (MAE). Predicted probabilistic values (soft prediction) produced by our model are measured with MAE, indicating the absolute difference between a predicted value vs. original binary value indicating the correctness of an assessor's judgment: $MAE = \frac{1}{n}\sum_{i=1}^{n} |pred_i - gold_i|$, where n is the number of judgments. Rounded binary labels (hard labels) are evaluated by accuracy. Secondly, accuracy is used for measuring the prediction performance of the binary probabilistic values from our prediction method. Since our extracted dataset is well-balanced in terms of a ratio between relevant vs. non-relevant judgments, use of accuracy is appropriate.

Models. We evaluate our proposed Generalized Assessor Model (GAM) under various conditions of *decision reject options* with two metrics. Our initial model uses no decision reject option, setting $\delta = 0$. In order to examine the effect of *decision reject options*, we vary $\delta \in [0, 0.25]$ by 0.05 step-size. Since we have 54 workers, we build 54 different predictive models and evaluate their prediction performance and final judgment quality improvement.

Our model works in a sequential manner that updates the model parameter θ once a new binary observation value (correct/wrong) comes. We use each worker's first 20 binary observation values as an initial training set. For instance, suppose a worker has 50 sequential labels. We first collect a sequence of binary observation values (correct/wrong) by comparing a worker's label with a corresponding ground truth judged by

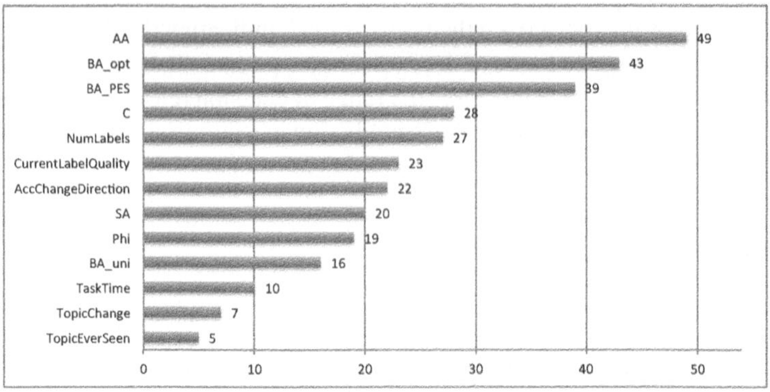

Fig. 2. Summary of relative feature importance across 54 regression models

NIST experts. Next, our prediction model takes the first 20 binary observation values and then predicts the 21st label's quality (correct/wrong) of this worker. Once actual 21st label comes from this worker, we measure the accuracy and MAE by comparing the label with a corresponding ground truth from NIST experts. For the following 29 judgments we repeat the same process in a sequential manner, predicting the quality of each label one-by-one.

To learn our logistic regression model, we choose the regularization parameter λ as 0.01 after the investigation of prediction performance with varying parameter values $\{0.1, 0.01, 0.001\}$ over the initial training set of each worker. For feature normalization, we apply standard min-max normalization to the 13 features defined in Section 3.1. Note that λ is the only model parameter we tune, and all settings of decision-reject parameter are reported in results.

As a baseline, we consider several assessor models proposed by prior studies [4] [8] [11] (Section 3.1). We adopt two assessor models from Carterette and Soboroff's study, *optimistic* assessor (BA_{opt}) and *pessimistic* assessor (BA_{pes}), and one assessor model of Bayesian accuracy (BA_{uni}) used in Ipeirotis and Gabrilovich's study (see Table 1). In addition, we test the performance of a time-series model (TS) proposed by Jung et al [11] and sample running accuracy (SA) as defined by Table 1. All of the baseline methods predict the binary correctness of the next judgment y_{i+1} by rounding off the worker's estimated accuracy at time i. *Decision reject options* are equally applied to all of the baseline methods.

4.1 Experiment 1 (RQ1): Feature Selection and Importance

Our first experiment is to figure out which features are relatively more important than others. Intuitively, having more features leads to more predictive power. However, in practice, excessive features may lead to over-fitting. Thus, we investigate relative feature importance by evaluating feature subsets.

We adopt the *bestglm* r package[2] and run the BICg model in order to find the best subset regression models. Since we have 54 assessors, we run this method for all of the

[2] `http://cran.r-project.org/web/packages/bestglm/vignettes/bestglm.pdf`

Table 2. Prediction performance (Accuracy and Mean Average Error) of different predictive models. *% Improvement* indicates an improvement in prediction performance between GAM vs. each baseline ($\frac{(GAM-baseline)}{baseline}$). *# of Wins* indicates the number of assessors that GAM outperforms a baseline method while *# of Losses* indicates the opposite of *# of Wins*. *# of Ties* indicates the number of assessors that both a method and GAM show the same prediction performance for an assessor. (*) indicates that GAM prediction outperforms the other six methods with a high statistical significance (p<0.01).

Metric	GAM	TS	BA_{uni}	BA_{opt}	BA_{pes}	SA
Accuracy	0.802*	0.621	0.599	0.601	0.522	0.599
% Improvement	NA	29.1	33.9	33.4	53.6	33.9
# of Wins	NA	50	52	50	54	52
# of Ties	NA	3	1	3	0	1
# of Losses	NA	1	1	1	0	1
MAE	0.340*	0.444	0.459	0.448	0.488	0.458
% Improvement	NA	23.4	25.9	24.1	33.0	25.8
# of Wins	NA	53	53	53	54	53
# of Losses	NA	1	1	1	0	1

54 original regression models. Next, we observe the selected features of each subset model, and count the cumulative selection of each feature across 54 regression models. Figure 2 shows the relative feature importance across 54 regression models for all of the assessors. Asymptotic accuracy (AA) is selected in 49 of 54 models, followed by BA_{opt} and BA_{pes} at 43 and 39, respectively. $Numlabels$ is selected in the half of the cases (27), which implicitly indicates that the increase in the quantity of the given tasks affects an assessor's next judgment correctness. On the contrary, the quality of next judgments of the 54 assessors in our dataset does not appear to be sensitive to topic change and topic familiarity. In addition, sample accuracy (SA) appears relatively less important than the other accuracy-based metrics such as AA, BA_{opt} and BA_{pes}. Interestingly, GAM model with only the top five features still shows little degraded performance (7-10% less) vs. the original regression models and outperforms all baselines.

4.2 Experiment 2 (RQ2): Prediction Performance Improvement

To answer our second research question, we first compare the overall prediction performance (Accuracy, MAE) of GAM with the baseline models across 54 crowd assessors. Table 2 shows that GAM prediction performance outperforms all of the baseline methods across 50-54 assessors in accuracy and 53-54 assessors in MAE. GAM improves the prediction accuracy (hard label) and MAE (soft label) by 26-36% on average. GAM prediction errs for only one assessor vs. the baselines. However, even for this assessor, GAM only made one or two more prediction errors in comparison to the other baselines.

Figure 3 shows the relationship between assessors' labeling accuracy (sample running accuracy) vs. prediction accuracy of GAM and the baseline models. While the baseline models show low accuracy against assessors whose labeling accuracy is near 0.5, GAM significantly improves prediction error for those assessors in particular.

Lastly, we examine the effects of *decision reject options* on GAM prediction. Figure 4 demonstrates that the baseline models show sharp decline of coverage in prediction in order to significantly improve their prediction accuracies. However, the

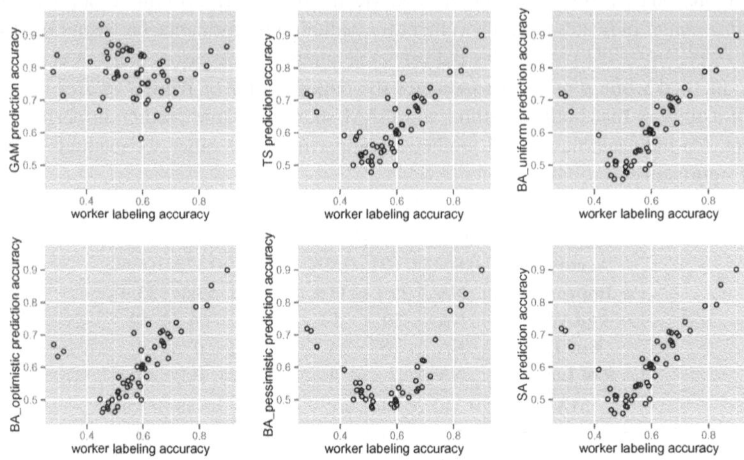

Fig. 3. Prediction accuracy of workers' next label by different methods ($\delta = 0$). While other methods show low accuracy against assessors with labeling accuracy near 0.5, the proposed model (GAM) shows significant improvement in predicting the correctness of workers' next judgments.

coverage of GAM prediction only gently decreases; even with the second strongest reject option ($\delta = 0.2$), it still covers almost the half of prediction. In sum, GAM prediction not only outperforms the baseline models in terms of prediction accuracy, but it also shows less sensitivity to the increase of the decision reject option.

4.3 Experiment 3 (RQ3): Impact on Judgment Quality and Cost

Our last experiment is to examine quality effects on relevance judgments via the proposed prediction model. We conduct an experiment based on task routing. For instance, if the prediction of an assessor's next judgment indicates that the assessor is expected to be correct, we route the given topic-document pair to this assessor and measure actual judgment quality against ground truth labeled by NIST. From our dataset, we only use 826 topic-document pairs that have more than three judgments per topic-document pair. Since the average number of judges per query is about 3.7, we test the cost saving effect with varying three task routing scenarios (*Number of Judges* = $\{1, 2, 3\}$). Judgment quality is measured with accuracy, and a paired t-test is conducted to check whether quality improvement is statistically significant.

Table 3 shows the results of judgment quality via predictive model-based task routing. GAM substantially outperforms the other baselines across three task routing cases. The improvement of final judgment quality grows with the increase of the number of judges per query-document pair (*Number of Judges*) from 29-32% to 36-47%. Notice that GAM with only two routed judges achieves 29% quality improvement. Moreover, GAM provides high-quality relevance judgments (accuracy > 0.8) with only 54% = ($\frac{2}{3.7}$) of the original assessment cost. In contrast, we see that task routing with baselines alone (BA_{uni},BA_{pes},SA) may not be any better than random assignment.

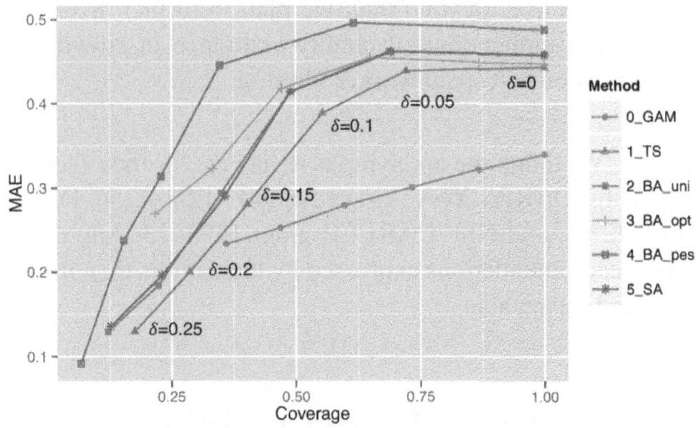

Fig. 4. Prediction performance (MAE) of assessors' next judgments and corresponding coverage across varying decision rejection options (δ=[0-0.25] by 0.05). While the other methods show a significant decrease in coverage, under all of the given reject options, GAM shows better coverage as well as prediction performance.

Table 3. Accuracy of relevance judgments via predictive models. *Number of Judges* indicates the number of judges per query-document pair. When the *Number of Judges* > 1, majority voting is used for label aggregation. Accuracy is measured against NIST expert gold labels. *% Improvement* indicates an improvement in label accuracy between GAM vs. each baseline ($\frac{(GAM-baseline)}{baseline}$). The average number of judges per query-document pair is 3.7. (*) indicates that GAM prediction outperforms the other six methods with high statistical significance (p<0.01).

	Prediction Models for Task routing							No Routing
Number of Judges	GAM	TS	BA_{uni}	BA_{opt}	BA_{pes}	SA	**Random**	**All labels**
1	0.786*	0.604	0.578	0.582	0.558	0.569	0.556	0.595
% Improvement	NA	30.1	36.0	35.1	40.9	38.1	41.4	
2	0.816**	0.617	0.592	0.595	0.574	0.582	0.572	
% Improvement	NA	32.3	37.8	37.1	42.2	40.2	42.7	
3	0.880*	0.647	0.608	0.623	0.598	0.608	0.581	
% Improvement	NA	36.0	44.7	41.3	47.2	44.7	51.5	

5 Conclusion and Future Work

Despite recent efforts of quality improvement in crowdsourced relevance judgment, prior work in crowd assessor modeling cannot adequately predict an assessor's next judgment quality since it simply measures assessor performance via a single generative model without considering temporal effects among relevance judgments. We present a general discriminative learning framework for integrating arbitrary and diverse evidence for temporal modeling and prediction of crowd work accuracy. Our experiments demonstrate that the proposed model improves prediction performance by 26-36% as well as crowdsourced relevance judgment quality by 29-47% at 17-45% lower cost.

As a next step, we plan to relax our restrictive assumption of the existence of NIST expert labels to judge the correctness of an assessor's judgments. In addition, we want to examine how to evaluate the correctness of judgments in recognition that even topical judgments are still subjective. Beyond that, we plan to further investigate how to use this model for different applications of quality assurance in crowdsourcing, such as weighted label aggregation and spam worker filtering.

Acknowledgments. We thank the anonymous reviewers for their feedback. This work is supported in part by DARPA YFA Award N66001-12-1-4256, IMLS Early Career grant RE-04-13-0042-13, and NSF CAREER grant 1253413. Any opinions, findings, and conclusions or recommendations expressed by the authors do not express the views of the supporting funding agencies.

References

1. Alonso, O., Rose, D.E., Stewart, B.: Crowdsourcing for relevance evaluation. ACM SIGIR Forum 42(2), 9–15 (2008)
2. Vuurens, J.B., de Vries, A.P.: Obtaining High-Quality Relevance Judgments Using Crowdsourcing. IEEE Internet Computing 16, 20–27 (2012)
3. Lease, M., Kazai, G.: Overview of the TREC 2011 Crowdsourcing Track (Conference Notebook). In: 20th Text Retrieval Conference (TREC) (2011)
4. Carterette, B., Soboroff, I.: The effect of assessor error on IR system evaluation. In: Proceedings of the 33rd International ACM SIGIR Conference on Research and Development in Information Retrieval, SIGIR 2010, pp. 539–546 (2010)
5. Hosseini, M., Cox, I.J., Milić-frayling, N.: On aggregating labels from multiple crowd. In: Proceedings of the 34th European Conference on Advances in Information Retrieval, ECIR 2012, pp. 182–194 (2012)
6. Kazai, G., Kamps, J., Milic-Frayling, N.: The Face of Quality in Crowdsourcing Relevance Labels: Demographics, Personality and Labeling Accuracy. In: Proceedings of the 21st ACM International Conference on Information and Knowledge Management, CIKM 2012, pp. 2583–2586 (2012)
7. Law, E., Bennett, P., Horvitz, E.: The effects of choice in routing relevance judgments. In: Proceedings of the 34th ACM SIGIR Conference on Research and Development in Information, SIGIR 2011, pp. 1127–1128 (2011)
8. Ipeirotis, P.G., Gabrilovich, E.: Quizz: targeted crowdsourcing with a billion (potential) users. In: Proceedings of the 23rd International Conference on World Wide Web, WWW 2014, pp. 143–154 (2014)
9. Yuen, M., King, I., Leung, K.S.: Task recommendation in crowdsourcing systems. In: Proceedings of the First International Workshop on Crowdsourcing and Data Mining, pp. 22–26 (2012)
10. Donmez, P., Carbonell, J., Schneider, J.: A probabilistic framework to learn from multiple annotators with time-varying accuracy. In: Proceedings of the SIAM International Conference on Data Mining, pp. 826–837 (2010)
11. Jung, H.J., Park, Y., Lease, M.: Predicting Next Label Quality: A Time-Series Model of Crowdwork. In: Proceedings of the 2nd AAAI Conference on Human Computation, HCOMP 2014, pp. 87–95 (2014)
12. Kazai, G.: In search of quality in crowdsourcing for search engine evaluation. In: Proceedings of the 30th European Conference on Advances in Information Retrieval. ECIR 2011, pp. 165–176 (2011)
13. Smucker, M.D., Jethani, C.P.: Measuring assessor accuracy: a comparison of NIST assessors and user study participants. In: Proceedings of the 34th International ACM SIGIR Conference on Research and Development in Information Retrieval, SIGIR 2011, pp. 1231–1232 (2011)

14. Raykar, V., Yu, S.: Eliminating spammers and ranking annotators for crowdsourced labeling tasks. Journal of Machine Learning Research 13, 491–518 (2012)
15. Rzeszotarski, J.M., Kittur, A.: Instrumenting the crowd: Using implicit behavioral measures to predict task performance. In: Proceedings of the 24th Annual ACM Symposium on User Interface Software and Technology, UIST 2011, pp. 13–22 (2011)
16. Voorhees, E.M.: Variations in relevance judgments and the measurement of retrieval effectiveness. Information Processing and Management 36, 697–716 (2000)
17. Pillai, I., Fumera, G., Roli, F.: Multi-label classification with a reject option. Pattern Recognition 46, 2256–2266 (2013)
18. Buckley, C., Lease, M., Smucker, M.D.: Overview of the TREC 2010 Relevance Feedback Track (Notebook). In: 19th Text Retrieval Conference, TREC (2010)

WHOSE – A Tool for Whole-Session Analysis in IIR

Daniel Hienert, Wilko van Hoek, Alina Weber, and Dagmar Kern

GESIS – Leibniz Institute for the Social Sciences, Cologne, Germany
firstname.lastname@gesis.org

Abstract. One of the main challenges in Interactive Information Retrieval (IIR) evaluation is the development and application of re-usable tools that allow researchers to analyze search behavior of real users in different environments and different domains, but with comparable results. Furthermore, IIR recently focuses more on the analysis of whole sessions, which includes all user interactions that are carried out within a session but also across several sessions by the same user. Some frameworks have already been proposed for the evaluation of controlled experiments in IIR, but yet no framework is available for interactive evaluation of search behavior from real-world information retrieval (IR) systems with real users. In this paper we present a framework for whole-session evaluation that can also utilize these uncontrolled data sets. The logging component can easily be integrated into real-world IR systems for generating and analyzing new log data. Furthermore, due to a supplementary mapping it is also possible to analyze existing log data. For every IR system different actions and filters can be defined. This allows system operators and researchers to use the framework for the analysis of user search behavior in their IR systems and to compare it with others. Using a graphical user interface they have the possibility to interactively explore the data set from a broad overview down to individual sessions.

Keywords: Interactive Information Retrieval, Sessions, Analysis, Evaluation, Logging.

1 Introduction

Kelly et al. [12] summarize the challenges and problems that arise in the evaluation of Interactive Information Retrieval (IIR) systems. One main goal should be the development of re-usable tools that enable researchers from different domains to investigate search behavior of real users in different environments and produce comparable results. Initial work on this task has been done and frameworks and toolkits have been proposed that allow controlled experiments in different settings [3, 5, 9]. This means that with the help of these frameworks researchers can design, create and conduct laboratory experiments for different domains, different data sets and carefully chosen user groups. Our aim in this work is to extend these set of tools with a tool that (1) supports the analysis of controlled and uncontrolled data sets from real-world IR systems and therefore from real users, (2) can either use existing log files or newly recorded data, (3) is based on whole-sessions and multiple sessions and (4) supports the overall process from logging over processing to interactive analysis.

A. Hanbury et al. (Eds.): ECIR 2015, LNCS 9022, pp. 172–183, 2015.

The topic of whole-session evaluation has been recently discussed in a seminar on "Whole Session Evaluation of Interactive Information Retrieval Systems"[1] which has been conducted by members of the IIR community. The main claim of the workshop output is that IR research has concentrated so far on how well an IR system responds to a single query, for example, by presenting a well-ranked result list. However, user interaction in an IR system takes place in the context of a search session. A session is not limited to a single query and some matching documents, but comprises all interactions, queries, resulting documents as well as the user's learning process about the topic and the system.

In this paper we present an analysis tool for whole-session analysis (WHOSE[2]) that concentrates on the inclusion and application in arbitrary IR systems with different functionality and technology stacks. It allows session-based analysis of user behavior in different systems, in different domains and with different domain knowledge. In WHOSE a whole-session is considered technically as a collection of actions a user performed from starting the system until closing the web browser session. System operators can define actions and filters to meet their individual requirements. All pre-processing, management and presentation of data is then handled by WHOSE. How this can be done is shown in section 4 where we report on experiences we made while applying WHOSE for analyzing log data from Sowiport[3]. WHOSE's graphical user interface consists of an interactive visualization, several filters and detailed session lists. It allows researchers to interactively explore user search behavior based on session data.

2 Related Work

The classical IR approach handles the search process as a single-query and multiple documents problem and is for example measured by the TREC evaluation campaign [23]. A more complex scenario arises by the investigation of user sessions. After posing an initial query, users often reformulate their search query until they are satisfied with the results. These multi-query sessions need other evaluation metrics [11]. Furthermore, each search session contains subtasks with explicit cognitive costs (e.g. scanning result lists), which can be addressed using a cost model based on time [1]. Belkin [4] proposes the measure of usefulness for the evaluation of entire information seeking episodes. He distinguishes usefulness in respect to (1) the entire task, (2) each step of interaction and (3) the system's support for each of these steps.

Longitudinal tasks over several sessions can be identified either by unique user ids or by machine algorithms. Jones et al. [10], for example, identified fine-grained task boundaries in a web search log by using different classifiers and machine learning. Kotov et al. [13] also tried to identify longitudinal tasks which are distributed over several search sessions. They used supervised machine learning with different classifiers to handle

[1] http://www.nii.ac.jp/shonan/blog/2012/03/05/whole-session-evaluation-of-interactive-information-retrieval-systems/
[2] Open Source code is available at https://git.gesis.org/public
[3] http://sowiport.gesis.org

identification of cross-session search behavior in web logs. Liao et al. [15] extract task trails from web search logs in contrast to search sessions. They found that user tasks can be mixed up in search logs because of the chronological order and the behavior of users to conduct concurrent tasks in multiple tabs or browsers. Identified tasks seemed to be more precise in determining user satisfaction in web search.

There are different measures and indicators that have been found to be important for session behavior. Fox et al. [7] conducted a user study in web search to find implicit measures that correlated best with sufficiently determing the user satisfaction. It was found that a combination of clickthrough, time spending on the search result page and how a user exited a result or search session correlated best with user satisfaction. Liu et al. [16] conducted a laboratory experiment in which they checked different measures influencing session behavior for different tasks. Three main behavioral measures were identified as important for document usefulness: dwell time on documents, the number of times a page has been visited during a session and the timespan before the first click after an query is issued. Dwell time showed to be most important, however, differs much in cut-off time and needs to be adaptive to different task types. Predictive models has then been applied to the TREC 2011 Session Track and showed improvement over the baseline by using pseudo relevance feedback on the last queries in each session.

The Interactive Probability Ranking Principle (IPRP) [8] is a theoretical framework for interactive information retrieval. It models the search process as transitions between situations. A list of choices is presented to the user in each situation, which can be e.g. a list of query reformulations, related terms or a document ranking. The user decides for one choice and is moved to the next situation. Each choice is connected to the parameters (i) effort, (ii) acceptance probability and (iii) resulting benefit. The overall goal of IPRP is to maximize the expected benefit by optimizing the ranking of choices. IPRP parameters can be derived from observation data like search logs, eye tracking [22] or mouse tracking. Resulting transition models for domains, tasks or subtasks can be visualized with Markov chains. Another popular visualization type that has been used in the field of website analysis for the visualization of user paths is node-link diagrams [6, 17, 24]. Very related to the area of whole-session analysis in IIR is also the field of visual web session log analysis, e.g. for the analysis of website behavior [14] or search usage behavior [19]. One goal of this kind of tools is to identify usage patterns that lead to successful completion of sessions, e.g. to finish a certain task in e-commerce.

There are already a number of frameworks to conduct controlled IIR evaluations. The Lemur Query Log Toolbar[4] is a web browser plug-ins that can capture user search and browse behavior as well as mouse clicks and scrolling events for web search sessions. ezDL [3] is an interactive search and evaluation platform. It supports searching heterogeneous collections of digital libraries or other sources, can be customized and extended, and provides extensive support for search session evaluation including mouse, gaze and eye tracking. Bierig et al. [5] present a framework to design and conduct task-based evaluations in Interactive Information Retrieval.

---

[4] `http://www.lemurproject.org/toolbar.php`

The system focuses on handling multiple inputs from mouse, keyboard and eye tracking. Hall and Toms [9] suggest a common framework for IIR evaluation which also includes components for logging user actions. The task workbench can handle pluggable components like a search box, search results etc. that can communicate with each other. The result is a rich log file where each component contributes detailed information. WiIRE [21] is a web-based system for configuration and conducting IIR experiments which incorporates essential components such as user access, task and questionnaire provision, and data collection. The same idea has been taken by SCAMP [18], a freely available web-based tool for designing and conducting lab-based IIR experiments, which includes all major processes from participant registration to logging and tracking of tasks. The intended benefit of all these frameworks is mainly for controlled IIR experiments in which users conduct several tasks in a laboratory setting. Evaluation data is recorded with logs, mouse, gaze, eye tracking and questionnaires. These controlled data sets can then be used for analysis of search behavior in a single system. However, these toolkits are not intended for the integration into existing IR systems, for the use of uncontrolled log data, their processing and the interactive analysis of user search behavior.

3 The Whole-Session Analysis Tool

In this section we present the general functionality of the analysis tool WHOSE: how user interaction data can be logged easily in different environments, how it can be mapped to actions and how data is preprocessed. Finally, we give a general overview of the user interface.

3.1 Logging Interaction Data

In IR systems user interactions can be logged in different ways. User interaction data can be recorded anew in various formats with different information depth or may already exist e.g. in form of web server log files.

A common approach is to record user actions in a well-defined schema (e.g. as in [9]). Here, the use of a certain schema has to be fixed and a list of possible interactions with its parameters has to be determined in advance by system experts. Then, the IR system has to trigger a new record to the log if the user applies a certain action. This can be quite a challenge in a real-world IR system if it is proprietary software, closed source or older code, because interceptors need to be implemented at various points in the source code which catch dozens of different user actions.

To overcome this issue, we implemented a logging approach that can handle uncontrolled data either from (1) function calls or (2) from existing log files and later maps them to a structured schema. The benefits here are that the logging component can be very easily implemented into existing software at only one central point in the source code and that existing uncontrolled log files like from the web server or application server can be used for analysis.

In web-based IR systems function calls are often realized by reloading the web page with additional GET/POST-parameters, calling JavaScript or internal AJAX/Servlet or other calls. Function calls contain a string which identifies the action (via function name or parameter) and several additional parameters. For example, in the discovery framework VuFind[5] (used in Sowiport) a simple search can be identified by the URL parameter "lookfor=" followed by a keyword. Similarly other user actions like exporting or adding an item to favorites can be identified. We found that, for example, in VuFind up to 90% of all user interactions can be identified by URL parameters, few interactions are conducted by AJAX or JavaScript calls.

Technically, all function calls can easily be intercepted by some lines of code and can be logged in a database or to a file. Function calls are handled as simple strings and no parsing or extraction is carried out. This makes the logging process very simple and adaptive for the application into many different contexts, be it a different domain, a different technical system or a different functionality. We used this approach in the new version of our IR system Sowiport that has been launched in April 2014. Here, we added an interceptor function at the main class that adds a new entry to the logging table in the database with every reload of the web page. The logging schema for WHOSE only contains very basic fields: "session-id", "user-id", "timestamp", "resultlist_ids", "url" and "referrer-url". "Session-id" is a unique session identifier which is generated in most IR system software. "User-id" is a unique user identifier provided by the IR system. "Resultlist_ids" contains a list of document identifiers from the result list if a search has been conducted. The field "url" contains the string with the requested URL, AJAX or other function calls. The field "referrer-url" contains the URL the systems user requested before the current action.

To test the other approach of handling data from log files, we used an existing database table from an older version of Sowiport with seven years of user data consisting of eleven million rows (with a size of 2GB) and transformed it easily into the necessary table structure.

3.2 Mapping Actions

In a next step the logged action data have to be mapped to concrete user actions. Every IR system provides different functionality and the representation in function calls or other uncontrolled data may be implemented differently. Therefore, WHOSE requested a mapping table in which a system expert can specify the mapping between defined user actions and corresponding parameters in the log data. For example, the user action "request search results for search term 'religion' is mapped to the log data entry www.xy.com/results?searchterm=religion". The goal is that the whole logic which is specific for an IR system is collected and defined in this table.

The mapping table is a simple table in CSV format that can be edited in any spreadsheet software. For every action in the IR system (such as searching, filtering or opening the detailed view) the expert needs to define a mapping. Actions are described with an internal and language specific labels and are identified by the system

---

[5] `http://vufind.org/`

with regular expression patterns. Table 1 shows a row from the mapping table that identifies a simple search action from the homepage. To identify the action the "url" and the "referrer" field from the logging table needs to match to the regular expressions defined in the "url_param" and "referer_param" fields.

In addition to the action mappings a group of mappings exist to extract entities like search terms, document ids or result list ids from function calls or strings. So far, we have implemented two operations for entity extraction: (1) *text* means that strings are extracted by the regular expression group functionality, e.g. for extracting query terms; (2) *field* means that the field is directly taken from the logging table into the analysis table, e.g. for logged document ids from the result list.

Table 1. Mapping rule for a simple search

Referer URL (referer_param)	URL (url_param)	Action
http:\/\/xy.com\/$	\/search\/results\?	Simple search from the homepage

3.3 Data Preprocessing

In a preprocessing step WHOSE used the mapping table to transform every row from the logging table into one or several user actions. In addition, further data such as session or action duration are computed and entities like search terms or document ids are extracted. The preprocessing step allows the reduction of data complexity, the mapping to simple actions and the creation of an analysis table. WHOSE can then utilize database functionalities like querying, grouping, indexing and calculation to query subsets and compute additional parameters much faster.

The computational effort for preprocessing can be quite high. Every row from the logging table needs to be matched against all mapping and extraction rules. Here the flexibility of regular expressions results in high computational costs. To improve performance the WHOSE tool uses the Java 6 Concurrency Library to split the work to multiple cores and threads.

3.4 User Interface

WHOSE's user interface consists of three parts: (a) filters for time, session and action parameters, (b) an overview visualization and (c) the detailed session list (see Figure 1). In general the design mantra of Shneiderman from the field of Information Visualization is applied: "Overview first, then filter and zoom, details on demand." [20]. This means, users can first get an impression of the overall session dataset with the overview visualization, and can then use filters and time restrictions to filter the data set to specific situations. Filtered user sessions can then be overviewed again in the visualization and in detail in the session list.

The upper part of the user interface contains components to filter the data set by time (Figure 1a). Users can choose from a list of time units (all, last 7 days or 30 days, etc.) or they can set the start and end date explicitly. Directly below, a series of filters are shown which allows the user to filter the whole data set. So far, we have implemented the following set of filters: (1) session contains text (e.g. search terms, facets etc.), (2) session duration, (3) show only sessions of users that are logged in, (4) sessions with a specific user-id, (5) sessions with more than x actions, (6) sessions that contain a certain action, and (7) action duration. Filters can be combined, which means for example that the data set can be filtered for all sessions which contain a certain keyword, and with a document view dwell time over 30 seconds. Additional filters could be implemented easily since the filter functionality relies on SQL-Filtering.

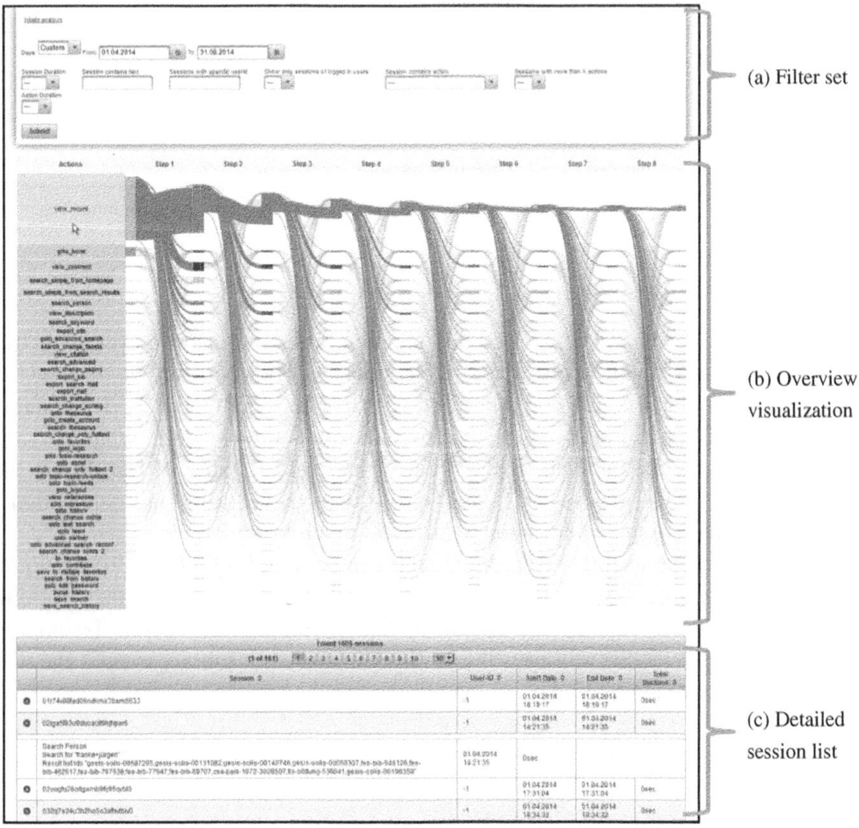

Fig. 1. Screenshot of the user interface with session data from Sowiport

We chose Sankey diagrams as an overview visualization (Figure 1b) for the actual set of sessions. Each row represents a specific action (e.g. a simple search), each column represents an ongoing search step in the session (first, second, third and so on). Actions are ordered from top to bottom by their highest occurrence within the first eight search steps. The height of the boxes at each search step represents the share of how often this action has been performed in this step. Bézier curves between the boxes show which portion goes to which action in the next step. Hovering with the mouse over an action label highlights the flows for this action and shows which actions have been performed in subsequent steps. The overview visualization in combination with filters can be used to identify user behavior patterns for specific situations.

The session list (Figure 1c) contains all user sessions that fit to the actual time span and filters. Here, the tool user can analyze in detail which actions including their parameters within a session have been performed. Sessions are ordered by descending date and can be unfolded to show all actions within a session.

4 Case Study: A First Look into User Behavior in Sowiport

In the following section we present how the tool can be used to analyze a large data set from a real-world IR system. Sowiport is a Digital Library for Social Science information. It contains more than 8 million literature references, 50,000 research projects, 9,000 institutions and 27,000 open access full texts from 18 different databases. Sowiport is available in English and German and reaches about 20,000 unique users per week. The majority of Sowiport's users are German-speaking. The portal has started in 2007 with a major redevelopment in April 2014 based on the VuFind framework and several extensions.

4.1 Data Preparation

Every search action in Sowiport is recorded in a logging table with fields like "timestamp", "url", "referrer-url", "result-list-ids" and "user-id". We used data logged between April 2014 and August 2014 consisting of around 2.5 million rows (about 800MB data). A mapping table has been created by system experts which defines about fifty actions and mapping rules specifically for Sowiport. The mapping rules have been tested with regard to completeness and correctness by comparing the system's logging data with screen recording data of six participants who were asked to use Sowiport over a time period of 10 minutes.

4.2 Data Analysis

The data analysis starts with a click on the "Submit"-Button that prepares all data for creating the visualization and the session list shown in Figure 1. At the beginning a broad overview of the dataset is provided by showing all log data. The diagram in Figure 1 shows that a large portion of users start their session with the action "view record". These are users that enter Sowiport directly from web search engines, where all detailed views of metadata records are indexed as individual web pages. The four main actions following step 1 can be identified as looking at another record, looking at the comments, looking at the abstract or initiating a new search. This pattern then reoccurs in the following steps.

The data can be filtered by different situations with specific attributes. For example, it can be checked if the search behavior of users that are logged in differs from those who are not. Figure 2 illustrates the results after applying the filter "show only sessions from logged in users". The main entry point for the filtered dataset is the homepage. Then, a large part of users continue with a simple search. In the third

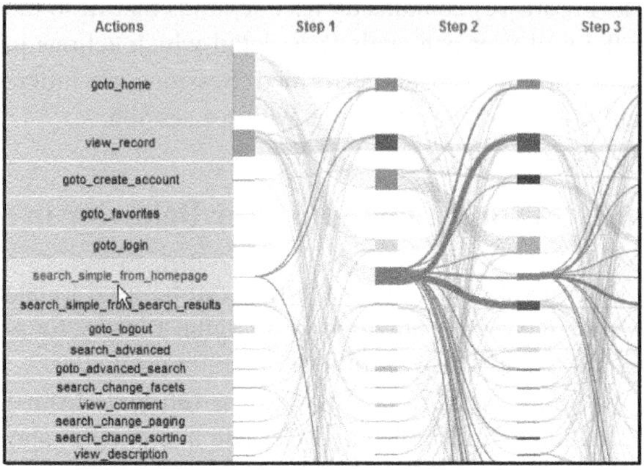

Fig. 2. The overview diagram shows action patterns for sessions filtered to logged in users with focus on the action "simple search"

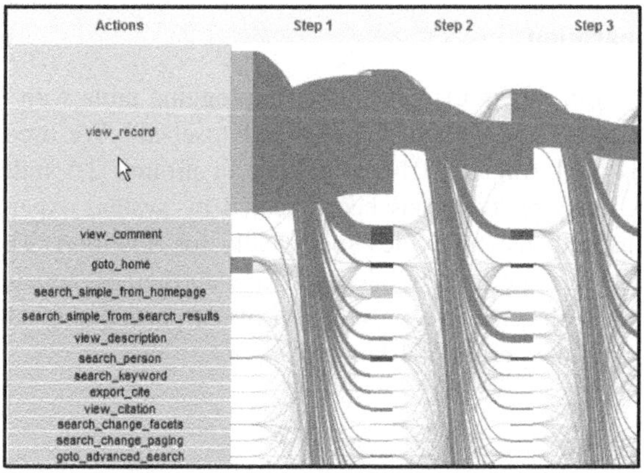

Fig. 3. Sessions that include a full view of more than 30 seconds

step, logged-in users go to a detailed view of a record, conduct another search or restart from the homepage.

In Figure 3 another example is illustrated. A researcher wants to find out when a session can be considered to be successful. This can be for example, a detailed view of a record, exporting the record or adding the record to a favorite folder. Any of these cases can be described with filters. For example, in our case all sessions are displayed in which records were viewed longer than 30 seconds. The researcher can check which action patterns lead to these situations. Finally, the resulting sessions can be further inspected in the detailed session list. Individual sessions can be opened and all its actions with parameters and durations are shown (see an unfolded session in Figure 1c).

4.3 Expert Evaluation

To gain insight on the way the tool can be used, we performed a first user study of WHOSE with two information science lecturers (one female and one male participant) from the Cologne University of Applied Sciences. We decided to use a real-world scenario so that our participants did not have to speculate about the intention of the users and could better concentrate on providing feedback to WHOSE. During a lecture in 2014 their students were assigned to perform a research task with Sowiport to a self-selected topic over a period of four weeks. Our participants used the tool to find out how the students used Sowiport to fulfill their assignments. The objective was not to analyze every single student's behavior but to identify typical search strategies or particularities. The test took about 45 minutes and the participants were asked to use WHOSE and to tell the experimenter everything they noticed, what they considered to be good or problematic, and what would further be needed for improvement.

Their comments give us valuable starting points for the further development of WHOSE. For example, they stated that the Sankey diagram is very complex at first sight and that more interaction opportunities on the diagram would be needed to show and hide selected paths or actions. Furthermore, it is essential for them that the diagram and the detail session list are well connected. Selecting an action in the list for example should trigger highlighting the path within the diagram. Vice versa selecting an action in the diagram should result in updating the table. Also it was observed, that the Sankey diagram helped the participants to identify main paths and to identify actions that are not often used but it did not provide information about absolute action frequencies. The participants suggested that providing a selection of several diagram types would help to be able to assess different aspects in more detail. As the Sankey diagram currently only shows a chronology of search steps, one participant asked for an opportunity to analyze the context of an individual action. She wanted to know which actions have led to a specific action and what the next actions were, independently of the point at which the action has been performed during the search session. On the whole, both participants saw high potential in using a further developed version of WHOSE for tracking typical or individual search steps, for identifying search strategies and furthermore for providing hints for usability problems.

5 Conclusion and Future Work

In this paper we introduced WHOSE, a tool for whole session evaluation. The goal is to analyze user search behavior in arbitrary IR systems. The presented mapping concept, based on function calls and user actions, allows not only looking at new recorded log data but also to analyze older log files possibly in different formats and from different systems. This makes it possible to compare or even to aggregate user search behavior of several IR systems in a uniform manner. The graphical user interface allows analyzing sessions of several users at the same time as well as on single user basis. Different filters are provided to reduce the amount of data to specific search situations. Thus, WHOSE can help domain experts and researchers for example to identify situations in which a session is successfully terminated and furthermore which behavioral patterns may lead to these situations. This can be a profound basis to understand at which points in the search process certain difficulties exist and the user can be further supported. Difficulties in the search process can arise from simple usability problems to more complex problems like missing search or domain knowledge. In the sense of the IPRP model [8] the latter problems can be addressed with a list of choices that suggests certain moves in the sense of Bates [2] up to different search strategies and value-added services that supports the user in successfully continuing the search process. In future work, we want to address this problem by automatically identifying critical situations and suggesting supporting services to the user. In addition, we plan to conduct a more comprehensive user study with the next version of WHOSE as well as to perform an expert workshop to identify a first set of typical user search behavior patterns.

Acknowledgements. The authors thank our colleagues from the department CSS for a working version of the mapping table for Sowiport.

References

1. Baskaya, F., et al.: Time Drives Interaction: Simulating Sessions in Diverse Searching Environments. In: Proceedings of the 35th International ACM SIGIR Conference on Research and Development in Information Retrieval, pp. 105–114. ACM, New York (2012)
2. Bates, M.J.: Where Should the Person Stop and the Information Search Interface Start? Inf. Process Manage. 26(5), 575–591 (1990)
3. Beckers, T., et al.: ezDL: An Interactive Search and Evaluation System. In: Proceedings of the SIGIR 2012 Workshop on Open Source Information Retrieval, Department of Computer Science, University of Otago, Dunedin, New Zealand, pp. 9–16 (2012)
4. Belkin, N.J.: On the evaluation of Interactive Information Retrieval Systems (2010)
5. Bierig, R., et al.: A User-Centered Experiment and Logging Framework for Interactive Information Retrieval. In: Underst. User - Workshop Conjuction SIGIR 2009 (2009)
6. Cugini, J., Scholtz, J.: VISVIP: 3D Visualization of Paths Through Web Sites. In: Proceedings of the 10th International Workshop on Database & Expert Systems Applications, p. 259. IEEE Computer Society, Washington, DC (1999)
7. Fox, S., et al.: Evaluating Implicit Measures to Improve Web Search. ACM Trans. Inf. Syst. 23(2), 147–168 (2005)
8. Fuhr, N.: A Probability Ranking Principle for Interactive Information Retrieval. Inf. Retr. 11(3), 251–265 (2008)
9. Hall, M.M., Toms, E.: Building a Common Framework for IIR Evaluation. In: Forner, P., Müller, H., Paredes, R., Rosso, P., Stein, B., et al. (eds.) CLEF 2013. LNCS, vol. 8138, pp. 17–28. Springer, Heidelberg (2013)
10. Jones, R., Klinkner, K.L.: Beyond the Session Timeout: Automatic Hierarchical Segmentation of Search Topics in Query Logs. In: Proceedings of the 17th ACM Conference on Information and Knowledge Management, pp. 699–708. ACM, New York (2008)
11. Kanoulas, E., et al.: Evaluating Multi-query Sessions. In: Proceedings of the 34th International ACM SIGIR Conference on Research and Development in Information Retrieval, pp. 1053–1062. ACM, New York (2011)

12. Kelly, D., et al.: Evaluation challenges and directions for information-seeking support systems. Computer 42(3), 60–66 (2009)
13. Kotov, A., et al.: Modeling and Analysis of Cross-session Search Tasks. In: Proceedings of the 34th International ACM SIGIR Conference on Research and Development in Information Retrieval, pp. 5–14. ACM, New York (2011)
14. Lam, H., et al.: Session Viewer: Visual Exploratory Analysis of Web Session Logs. In: IEEE VAST, pp. 147–154. IEEE (2007)
15. Liao, Z., et al.: Evaluating the Effectiveness of Search Task Trails. In: Proceedings of the 21st International Conference on World Wide Web, pp. 489–498. ACM, New York (2012)
16. Liu, C., et al.: Personalization of Search Results Using Interaction Behaviors in Search Sessions. In: Proceedings of the 35th International ACM SIGIR Conference on Research and Development in Information Retrieval, pp. 205–214. ACM, New York (2012)
17. Pitkow, J., Bharat, K.A.: Webviz: A Tool For World-Wide Web Access Log Analysis. In: Proceedings of the First International World-Wide Web Conference, pp. 271–277 (1994)
18. Renaud, G., Azzopardi, L.: SCAMP: A Tool for Conducting Interactive Information Retrieval Experiments. In: Proceedings of the 4th Information Interaction in Context Symposium, pp. 286–289. ACM, New York (2012)
19. Shen, Z., et al.: Visual analysis of massive web session data. In: Barga, R.S., et al. (eds.) LDAV, pp. 65–72. IEEE (2012)
20. Shneiderman, B.: The Eyes Have It: A Task by Data Type Taxonomy for Information Visualizations. In: Proceedings of the 1996 IEEE Symposium on Visual Languages, pp. 336–343. IEEE Computer Society, Washington, DC (1996)
21. Toms, E.G., et al.: WiIRE: the Web interactive information retrieval experimentation system prototype. Inf. Process. Manag. 40(4), 655–675 (2004)
22. Tran, T.V., Fuhr, N.: Quantitative Analysis of Search Sessions Enhanced by Gaze Tracking with Dynamic Areas of Interest. In: Zaphiris, P., Buchanan, G., Rasmussen, E., Loizides, F. (eds.) TPDL 2012. LNCS, vol. 7489, pp. 468–473. Springer, Heidelberg (2012)
23. Voorhees, E.M., Harman, D.K.: TREC: Experiment and Evaluation in Information Retrieval (Digital Libraries and Electronic Publishing). The MIT Press (2005)
24. Waterson, S.J., et al.: What Did They Do? Understanding Clickstreams with the WebQuilt Visualization System. In: Proceedings of the Working Conference on Advanced Visual Interfaces, pp. 94–102. ACM, New York (2002)

Looking for Books in Social Media: An Analysis of Complex Search Requests

Marijn Koolen[1], Toine Bogers[2], Antal van den Bosch[3], and Jaap Kamps[1]

[1] University of Amsterdam, The Netherlands
{marijn.koolen,kamps}@uva.nl
[2] Aalborg University Copenhagen, Denmark
toine@hum.aau.dk
[3] Radboud University, Nijmegen, The Netherlands
a.vandenbosch@let.ru.nl

Abstract. Real-world information needs are generally complex, yet almost all research focuses on either relatively simple search based on queries or recommendation based on profiles. It is difficult to gain insight into complex information needs from observational studies with existing systems; potentially complex needs are obscured by the systems' limitations. In this paper we study explicit information requests in social media, focusing on the rich area of social book search. We analyse a large set of annotated book requests from the LibraryThing discussion forums. We investigate 1) the comprehensiveness of book requests on the forums, 2) what relevance aspects are expressed in real-world book search requests, and 3) how different types of search topics are related to types of users, human recommendations, and results returned by retrieval and recommender systems. We find that book search requests combine search and recommendation aspects in intricate ways that require more than only traditional search or (hybrid) recommendation approaches.

Keywords: Book Search, Social Media, Evaluation, Recommendation.

1 Introduction

The rise of social media has had a major impact on how we search for and share information. For instance, it has radically changed the nature of book discovery, which has become easier than ever due to *social cataloging sites*, such as LibraryThing, GoodReads, Shelfari, BookLamp, Libib, and The Reading Room. We focus on LibraryThing[1] (LT), a popular social cataloguing site. The book collections shared on LT by its 1.8 million members cover over 8 million unique works in total. They describe not only the contents of those books, but also how the books engaged them, what their impact was, and how this related to other reading experiences. LT also offers a popular discussion forum (see Figure 1) for readers to discuss and review books, authors, and literature in general. A prominent use of the LT forum is book discovery: thousands of LT members use the forum to receive or provide recommendations for which books to read next. These book requests display a remarkable breadth, ranging from search-type requests for books on

[1] `http://librarything.com/`, last accessed January 11, 2015.

A. Hanbury et al. (Eds.): ECIR 2015, LNCS 9022, pp. 184–196, 2015.

specific topics or for certain moods, to recommendation-type requests for books similar to what a member has already read.

The general aim of this paper is to investigate whether explicit information requests in such social media, in particular related to book search, can be used to gain insight in complex information needs, i.e., those that cannot be solved by a straightforward look-up search. We study this in the context of the INEX Social Book Search Track[2] [12, 14, 15]. In recent years, this track has focused on book requests posted on the LT discussion forums. This paper provides a more detailed investigation into the nature of such requests. In the forums anyone can ask for book recommendations for a specific topic and other members reply with book suggestions. These suggestions can be seen both as relevance judgments and recommendations. The search requests go beyond topical relevance [13] and include many subjective aspects such as quality, interestingness, engagement, and familiarity. Cosijn and Ingwersen [7] and Saracevic [20] are among many that argue for the existence of different types of relevance in addition to pure topical relevance, such as situational, motivational, and affective relevance. A comprehensive survey of different interpretations of relevance is given by Borlund [4]. In this paper, we explore the relevance aspects present in the book domain by annotating and analyzing a large set of book requests from the LT forums.

We aim to address the following research questions in this paper:

RQ1. How comprehensive are book requests on the LT forum in terms of explicit information on the information need, the context of use, and the context of the user?

RQ2. What topical and non-topical relevance aspects are present in book search requests on the LT forums?

RQ3. How do different types of topics relate to user characteristics, human recommendations, and retrieval and recommender system results?

The rest of this paper is organized as follows. Section 2 presents related work, followed by an overview of the rich contextual data about book requests we can extract from the LT discussion forums in Section 3. Section 4 analyses the book requests with respect to the topical and non-topical relevance aspects expressed in them. Section 5 explores how book requests relate to the context of the user, human book recommendations, and retrieval or recommender system results. Finally, in Section 6, we discuss our results and draw conclusions.

2 Related Work

The INEX Social Book Search Track [12, 14, 15] investigates book search in collections with both professional metadata and social media content. For evaluation they use book requests on the LT discussion forums as search topics and book suggestions by members as as relevance judgments and recommendations. Koolen et al. [13] observed that these requests are complex and contain non-topical aspects, and found that the forum suggestions are different in nature than editorial relevance judgments with respect to system evaluation. In this paper we focus on the search requests themselves.

[2] All the data used in this paper are made available as part of the CLEF/INEX 2014 SBS Track.

Fig. 1. Book request on the LibraryThing forum

Ross [19] found that readers use a variety of clues to choose books. Reading a book is a substantial investment of time and energy, so readers look for recommendations from trusted sources for selection. Reuter [18] studied book selection by children and identify a list of 46 factors influencing their choices. Buchanan and McKay [5] investigated search activities of customers in bookshops. They find that enquiries often arise from cultural context—reading with others, references and reviews in media—and argue that customers' mental models may deviate from the standard bibliographic metadata. Cunningham et al. [8] studied collaborative information behaviour in bookshops. They found that groups of customers use many different ways to share information about books, e.g., talking aloud, pointing, reading, and searching together, and that they use these interactions to achieve agreement on which books to select. The gap between their mental model and the access points for online book collections may be why users turn to the LT forum for requests.

A considerable amount of related work exists on forum search, where the focus is typically on retrieving results from the collection of threads in a single forum. Examples of such approaches include work by Elsas and Carbonell [10] and Bhatia and Mitra [3]. In contrast, we analyze the initial forum posts describing a user's information need, in order to perform cross-collection search using these need descriptions. Our overall aim to use the forums to shed light on complex search requests, their context and relevance aspects, is related to a wealth of studies in information seeking. Some of the most comprehensive earlier studies predate the web and modern search systems (e.g., [21]). Our general approach is to tap into a new source of evidence for researching complex information seeking behavior.

3 Book Search Requests in LT Forums

In this section, we investigate RQ1: How comprehensive are book requests on the LT forum in terms of explicit information on the information need, the context of use, and the context of the user?

The LT discussion forums are used to discuss a broad range of topics, most of which are book-related. Many members turn to this forum asking for book suggestions and other members can reply and provide suggestions. In a random sample of 500 posts we found 67 (13.4%) containing an explicit book request. Given the massive scale of the forums with nearly 5 million messages and 3.5 million identified book mentions, this gives us access to a huge supply of real world complex search requests.[3] For the more straightforward search tasks, LT users are likely to use book search engines available at e.g. LT, Amazon, or libraries. In contrast, the forum requests contain more complex search needs that LT members have, expressed in natural language.

For instance, the request in Figure 1 is highly complex, providing requirements about the content as well as examples of books and authors that the poster is already familiar with, and contextual cues on usage. The user name links to the profile of the user, which provides additional context such as their personal book catalogue. The example books mentioned introduce a form of query-by-example that could also be seen as a recommendation task. These forum threads provide us with an unobtrusive method of investigating realistic, complex search requests that go well beyond traditional query log analysis. Members are not limited by the functionalities of a search engine or recommender system when expressing their request, but only by the concreteness of their information need and their ability to express it in natural language. As a result, they typically leave rich descriptions of their information need as well as many contextual clues to ensure others can understand its complexity.

Moreover, the LT forums allow users to mark up the names of books and authors through a simple wiki-like syntax using so-called *touchstones*. The system then automatically identifies the correct book/author and links the marked-up text to the right LT entity. These suggestions are a form of human relevance judgements.

Summarizing, from the forums we can derive rich statements of requests, including explicit statements on the context of use and the context of the user, with example books and 'ground truth' human recommendations. We find that such forum data give a unique opportunity to study complex search requests, and that the requests exhibit an amazing variation in topical and non-topical aspects. This prompt us to investigate what relevance aspects are used in the next section.

4 Relevance in Forum Book Search

In this section we study RQ2: What topical and non-topical relevance aspects are present in book search requests on the LT forums?

4.1 Relevance Aspects

Our first step is to investigate the complexity of these book search requests and the kind of relevance aspects expressed in them. Reuter [18] collected data from a user study in a children's library and identified 46 aspects, grouped into seven broad categories. We use those categories as our guide for analyzing the relevance aspects of book search requests. Due to its prominence in the LT forums, we introduce *known-item* search as an additional aspect. This resulted in the following eight relevance aspects:

[3] `https://www.librarything.com/zeitgeist`, last accessed on January 11, 2014.

Accessibility The language, length, or level of difficulty of a book.

Content Topic, plot, genre, style, or readability of a book.

Engagement Affective types of reading experiences evoked by books.

Familiarity Books similar to known books or related to a previous experience.

Known-item Descriptions of known books to identify the title and/or author.

Metadata Aspects like title, author, publication year and format.

Novelty Books that are unusual or quirky, or have novel content.

Socio-cultural Books related to the user's socio-cultural background or values, have (had) a particular cultural or social impact, or are popular or obscure.

4.2 Annotating Book Search Requests

To determine how prominent these different relevance aspects are on the LT forums, we annotated a sample of topic threads for relevance and other characteristics. We selected forum threads likely to contain requests for book recommendations using a simple regular-expression-based classifier, which filtered out all topics that did not contain one or more 'trigger' expressions, such as '*suggest*', '*looking for*' and '*which books*'. This resulted in a set of 9,403 topic threads containing touchstones. A random set of 2,646 of these topics were annotated by eight different Information Science students, three from the Royal School of Library and Information Science in Copenhagen, three from the Oslo and Akershus University of Applied Sciences, and two from Aalborg University Copenhagen. Each topic was annotated by a single annotator. We created a Web interface to help our annotators (1) identify topic threads as either *book requests* or *non-requests*; (2) annotate the requests by which relevance aspect(s) they express; and (3) annotate the suggestions provided by other LT members in the thread. This task included questions on whether the suggestion providers appeared to have read the suggested books and whether their recommendation was positive, negative, or neutral.

Of the 2,646 topics annotated by the students, 944 topics (36%) were identified as containing a book request (recall that 13.4% of a random sample contained book requests). For each identified book request, annotators could specify multiple relevance aspects. For example, for topic 99,309 on the "*politics of multiculturalism*" (partly shown in Figure 1), the topic starter asks for suggestions about a particular topic (*content* relevance), but also asks for books similar to what he has already read on the topic (*familiarity*), but written in a less annoying style (*engagement*).

4.3 Analysis

The distribution of relevance aspects in our annotated set of 944 book requests is shown in the left half of Table 1. The majority of book search information needs on the LT forums express *content* aspects (698 topics or 74%). *Familiarity* is the second most frequent aspect at 36%. These two aspects are often combined in a single book request: 267 topics (28%) express both aspects. An example of such a request is "*Can someone recommend a book that has all the joy, charm, numerous characters, pathos, adventure, love of language, etc. that the novel David Copperfield has?*" (topic 10392). The searcher wants recommendations based on the book *David Copperfield*, but also describes aspects of the book to base these recommendations on. This is querying by

Table 1. Aspect distribution and overlap in the 944 forum topics (left side) and the conditional probability *P(column | row)* (right side)

	Aspect overlap								Conditional probability							
	A	**C**	**E**	**F**	**K**	**M**	**N**	**S**	**A**	**C**	**E**	**F**	**K**	**M**	**N**	**S**
Accessibility	*152*	109	44	50	15	39	8	27	*1.00*	0.72	0.29	0.33	0.10	0.26	0.05	0.18
Content		*698*	172	267	100	176	26	99	0.16	*1.00*	0.25	0.38	0.14	0.25	0.04	0.14
Engagement			*213*	91	17	50	11	24	0.21	0.81	*1.00*	0.43	0.08	0.23	0.05	0.11
Familiarity				*338*	12	83	17	45	0.15	0.79	0.27	*1.00*	0.04	0.25	0.05	0.13
Known-item					*202*	85	0	1	0.07	0.50	0.08	0.06	*1.00*	0.42	0.00	0.00
Metadata						*264*	11	26	0.15	0.67	0.19	0.31	0.32	*1.00*	0.04	0.10
Novelty							*34*	10	0.24	0.76	0.32	0.50	0.00	0.32	*1.00*	0.29
Socio-cultural								*134*	0.20	0.74	0.18	0.34	0.01	0.19	0.07	*1.00*

example as well as description, which is a form of querying that is not supported by any current systems.

Other frequently labeled aspects include *metadata* (28%), *engagement* (23%), and *known-item* (21%). On the LT forum, *metadata* is an interesting aspect. When searching a catalog, metadata is often used to find specific books or books by a certain author, but such straightforward lookup tasks are not typically posted on the forums. Of the 264 topics labelled with *metadata*, only 22 (8%) have no other relevance aspect. These topics typically ask for recommendations on which books to read from specific authors, publishers, or series, or for the proper sequence in which to read a set of books. In most cases, *metadata* is combined with other aspects, and is used to focus the suggestions. *Engagement* is something that is hard to express through a search engine query. For instance, how can a user search for text books that are 'funny' or for books that challenge the reader's own views on a topic? Such complex relevance criteria may be a reason to ask for suggestions on the LT forum. The same holds for *known-item* topics where the user can only recall certain elements of the plot or attributes of certain characters. Most book search services are of limited use for such known-item topics, as they do not allow full-text search. Forum members, however, may be able to help out with such requests. *Accessibility*, *novelty* and *socio-cultural* aspects are less prominent in our sample.

The rest of Table 1 shows the distribution of the relevance aspects and their co-occurences. We can see a pattern emerging of relevance aspects being combined with either *content*, *familiarity*, or both, forming groups of topics clustered around these two aspects. *Known-item* requests are an exception as they seem to be a separate group. *Content* requests tend to be more typical of search tasks, as they provide a specific description of the desired books. The *familiarity* aspect seems related to recommendation-oriented tasks. The other aspects are more contextual in nature: dealing with books for certain scenarios (e.g., waiting at an airport, selecting reading material for a book club), for certain age groups or personality traits (e.g., trying to get a spouse to pick up reading), or certain moods (e.g., books that are comforting or challenge ones views). Dealing with such contextual information is an active research topic for both search [9] and recommender systems [1].

Summarizing, in a large sample of book requests annotated by their relevance aspects we find that most requests combine multiple aspects. We observed the largest clusters

Table 2. Topic groups in terms of example books and requested prose genre

Feature		KI	Cx	F	Co+F	Co	All
Example books		0.08	0.26	0.54	0.50	0.16	0.27
Genre	Fiction	0.77	0.29	0.49	0.53	0.35	0.50
	Non-fiction	0.06	0.06	0.10	0.15	0.26	0.16
	Mix	0.03	0.21	0.10	0.13	0.15	0.12
	Uncertain	0.13	0.44	0.30	0.19	0.24	0.23

around *content* and *familiarity* aspects, or both, and the *known-item* class. In the next section, we will divide the requests into different groups based on these relevance aspects and study them in more detail.

5 Impact of Content and/or Familiarity

In this section, we investigate RQ3: How do different types of topics relate to user characteristics, human recommendations, and retrieval and recommender system results?

5.1 Grouping Topics on Relevance Aspects

In the previous section we saw a prevalence of *content* and *familiarity* aspects, in isolation and in combination, suggesting a grouping of the requests based on these relevance aspects. With the *known-item* requests as a separate group and the four logical combinations of *content* and *familiarity* aspects, this results in the following five topic groups:

Known-item (KI) contains all 202 *known-item* topics. This is the most content-specific information need, but different from the rest in that the user wants a specific book.

Context (Cx) contains all 78 topics without *content*, *familiarity*, or *known-item*. There are no content-based aspects on which to base document similarity.

Familiarity (F) contains 66 topics with *familiarity*, but no *content*. Users search for books similar to a specific (set of) book(s) or genre(s). Document similarity is underspecified, i.e., the user gives no content aspects to base similarity on.

Content and Familiarity (Co+F) contains 260 topics with both *content* and *familiarity* aspects, articulating explicit and implicit topic aspects. The similarity of the desired books is expressed at the level of books as well as at the finer-grained level of specific textual aspects of the books.

Content (Co) contains 338 topics with *content*, but no *familiarity*. Users are searching for books matching specific content aspects. Here, document similarity is more explicit, corresponding to a more specific information need.

5.2 Analysis of Genre, Popularity, and Personal Catalogues

To understand how our groupings correspond to actual differences in the nature of topic groups, we compare them on characteristics of the request, the requester, and the suggested books: (1) the presence of example books in *touchstones*, (2) the genre of books

Table 3. Catalogue size in requester catalogue (median of each topic group)

Feature	KI	Cx	F	Co+F	Co	All
Pre-topic	0	38	104	100	177	84
Post-topic	4	80	81	65	108	65
Total	16	155	195	201	415	197

they target, (3) the size of the requester's book catalogues, and (4) the popularity of books in the requester's catalogue or those suggested at the forum.

Providing Example Books. For some topics, requesters add example books to their initial post using *touchstones*. These examples can serve different purposes: (1) positive examples of what they want (more of); (2) negative examples that match some relevance aspect(s), but not all; or (3) examples of what they have already read. Out of the 944 topics in total, only 256 (27%) have example books in the initial request, as shown in Table 2. We expect that examples are common among **F** topics based on previous reading experiences, and rare among **KI** and **Co** topics. These expectations are supported by the relevance aspects: the majority of the **F** topics include examples (54%), whereas only 8% of the KI topics contain examples. This lends credence to our decision to split the topics into groups based on the *content* and *familiarity* aspects.

Genre. Our annotators indicated whether requests were for fiction, non-fiction, or both. Table 2 shows that, of the 944 topics in total, 469 (50%) asked for suggestions on fiction books, 150 (16%) on non-fiction, and 113 (12%) on both fiction and non-fiction. For 212 topics (22%) the annotator could not tell. Fiction was the most common prose genre for **KI** topics at 77%, whereas only 6% of the topics were non-fiction. Fiction was also common for the **F** group at 53%. **Cx** topics have no specific content aspects, so it makes sense that mixed-genre topics and ambiguous topics are more common. In contrast, the **Co** topics are focused on non-fiction books more frequently than the other topic groups. Intuitively, this makes sense, as the topical content is arguably the main reason for reading a non-fiction book. Requests for fiction books are more likely to refer to examples, because what one is looking for in fiction may be more difficult to express and less explicitly related to the topical content of the book. This provides further evidence that the criteria for the topic groups are meaningful for analysis.

Cataloguing Behavior. Next, we count how many books the topic creator catalogued before posting the request (pre-topic), after posting it (post-topic), and in total; results are listed in Table 3. **KI** topics are often posted by LT members who have no books in their catalogue. Private profiles are an unlikely explanation for this, as these are rare. It seems these LT members use the forums mainly as a search engine and discussion board instead of as a tool for managing their book collections. Requesters of **Cx** topics tend to have small pre-topic catalogues, but add more books afterwards. These may be relatively new users with limited reading experience and have difficulty describing in detail what books they are looking for. Instead, they describe the context in which they want to read books. **F** and **Co+F** topics tend to come from more active users who have

Table 4. Median book popularity in requester catalogue, forum suggestions, system results

Feature		KI	Cx	F	Co+F	Co	All
Requester catalogue	topic group median	47	60	41	56	39	46
	topic group mean	91	98	76	92	72	86
	topic group std.dev.	101	97	78	90	79	89
Forum suggestions	topic group median	174	681	531	235	192	237
Retrieval	Top 10	55	58	107	57	42	53
	Top 1000	23	24	18	25	20	21
Recommender	Top 10	5146	5685	6022	4028	5163	5997
	Top 1000	1076	958	985	908	852	959

over 100 books pre-topic and remain active cataloguers post-topic. This suggests they know what they like and that their needs have become more specific, but are still broad enough that they only need to implicitly describe what they want by giving examples. We speculate that users with **Co** topics are typically heavy readers, who have large pre-topic catalogues and remain very active users. They can explicitly describe what they are looking for and may in fact leave out examples to avoid responders from picking up on the wrong similarity clues from those examples.

Book Popularity. Chandler [6] examined the different strategies of GoodReads users for discovering new books to read and how these relate to the popularity of discovered books. They found that the popularity distribution of books discovered through search has a long tail of less popular books, whereas for GoodReads recommendations the distribution is concentrated around the mid- to high-popularity books.

How are the five topic groups related to the popularity of books discussed on the LT forums? The popularity *Pop(d)* of a book *d* is the number of users who have *d* in their catalogues in our profile crawl. The top half of Table 4 shows the median popularity of books in searchers' catalogues and the forum suggestions. The catalogues of requesters tend to have a mix of popular and obscure books—the topic group mean is higher than the median indicating the distribution is skewed with a minority of highly popular books. There is no big difference between the popularity distributions of requesters with **F**, **Co** and **Co+F** topics. For the forum suggestions we see larger relative differences between topic groups, however. Forum members suggest more popular books for **Cx** and **F** topics than for **Co** and **Co+F** topics. The popularity of suggested books diminishes as content-specificity increases. For **KI** topics—the group with arguably the highest content specificity—suggestions are even less popular. Relating this to the findings of Chandler [6], suggestions for **Co** and **Co+F** topics are closer to search-related discoveries and **F** and **Cx** closer to recommendation-related discoveries. In terms of suggestions, book search on the LT forums seems to have a mix of search and recommendation-oriented tasks.

5.3 Retrieval and Recommendation Results

We analyse the books returned by standard retrieval and recommender systems for the forum topics, and compare them to the actual suggestions given by LT forum members.

Table 5. Performance evaluation of retrieval, recommender and best case fusion results

nDCG@10	**KI**	λ	**Cx**	λ	**F**	λ	**Co+F**	λ	**Co**	λ	**All**	λ
Retrieval	0.207		0.086		0.101		0.050		0.088		0.095	
Recommendation	0.002		0.007		0.008		0.009		0.005		0.006	
Fusion	0.215	.85	0.101	.70	0.106	.85	0.056	.70	0.090	.80	0.098	.75
MRR												
Retrieval	0.249		0.153		0.188		0.122		0.161		0.163	
Recommendation	0.003		0.037		0.033		0.038		0.018		0.025	
Fusion	0.256	.85	0.176	.70	0.219	.70	0.146	.60	0.167	.80	0.171	.75

For the retrieval system, we use the Amazon/LibraryThing collection [2] that is also used in the INEX Social Book Search Track [15]. This collection contains book metadata for 2.8 million books, including formal metadata (title, author, publisher, publication date), professional subject metadata (subject headings, Dewey Decimal System codes) and user-generated content (Amazon user reviews, LT user tags). We use Indri [11] and index all content with Krovetz stemming and stopword removal using a list of 319 stopwords. Specifically, we use a standard Language Model run with Dirichlet smoothing ($\mu = 2500$), using a combination of the thread title, a query provided by the annotators and the name of the discussion group as a query. This combination gives the best performance with standard Language Model settings.

For the recommender system we use a set of 84,210 user profiles, with information on which books the user catalogued and when, to compute nearest neighbours. We represent each user by a vector of book IDs and compute tf·idf similarity using GenSim [17]. The recommendation score for a book is the sum of the similarities of the individual neighbours who catalogued that book. We use a standard k-NN model run with recommendations from the 100 nearest neighbours based on catalogue similarity.

The lower half of Table 4 shows the median book popularity of returned results of the two runs. The rankings show a strong popularity effect for the recommender system, with the retrieval systems picking up less popular books in general than the recommender system. The popularity effect of recommender systems is also known as the "Harry Potter" problem [16]. The top of the rankings show relative differences between content and familiarity topics, with especially recommender systems returning more popular books for **F** topics than for **Co+F** and **Co** topics. The query terms of **Co** topics target less popular books than query terms of **F** topics, and their users are similar to users with a smaller fraction of highly popular books in their catalogues. In terms of book popularity, forum suggestions are roughly as popular as retrieval results. This is in line with earlier results on recommendations from friends as found on GoodReads [6]. Even though forum members do not know the requester personally, the statement of request is comprehensive enough for them to target the right types of books.

Finally, we look at system performance of the retrieval and recommender systems on the forum requests and suggestions, based on the Qrels from the INEX 2014 SBS Track for evaluation. We focus on basic retrieval and recommender models to observe relative performance of the two approaches on the various request types under well-understood conditions. In future work we will explore different models and query and user models in more detail. In addition to the baseline, we assess the potential of hybrid systems merging the results lists of both baselines using a weighted sum, $S_{fusion}(d,q) = \lambda \cdot S_{retrieval}(d,q) + (1-\lambda) \cdot S_{recommendation}(d,q)$. The performance scores are shown in Table 5. As expected, the retrieval system outperforms the recommender system, but there are differences between the topic groups. The recommender system scores relatively well on the **Cx**, **F** and **Co+F** topics, while the retrieval system performs relatively better on **KI** and **Co** than on **Cx** and **Co+F**. More importantly, on all topic groups, fusion of the results lists leads to improvements.Topics with *familiarity* or non-content aspects show the largest relative improvement. Finally, as a combination of recommendation and retrieval aspects, the **Co+F** topic set shows through its λ value of 0.6 that a more balanced fusion produces the best results. This suggests that the type of request plays an important role in the design of book discovery systems.

Summarizing, we analyzed topic groups related to known-item search and the logical combinations of content and/or familiarity. We observed varying degrees of combinations of contextual search and recommendation aspects. In terms of cataloguing behaviour, the content-specificity of requests is related to the size of the requester's catalogue. In terms of popularity, forum suggestions for topics with *content* aspects are more similar to retrieval results, and those for topics with *familiarity* aspects more similar to recommendation results. We demonstrated there is room for improvement by combining retrieval and recommendation approaches.

6 Conclusions

The aim of this paper was to investigate complex search requests in social media, in particular focusing on book search as observed on the LibraryThing discussion forums.

First, we found that the LT forums provide an unobtrusive way to study realistic, complex book search requests, which show a broad variation in topical and contextual relevance aspects. Second, we annotated the relevance aspects expressed in book requests at the LT forums. We found that the two dominating aspects are the *content* of the book and looking for *familiar* reading experiences, while other aspects are more oriented toward the reading context. The combination of content, context, and examples in a search request is a form of querying that is not supported by any current systems. Third, we found that these topic groups based on content and familiarity aspects can be differentiated by whether the requesters provide example books, what genre they are looking for (fiction or non-fiction), their cataloging activity, and the popularity of the suggested books. Retrieval systems can effectively use the content aspects of the search requests, and recommender systems can pick up signals in the requester's catalogue. We demonstrated the possibility for improvement when combining both approaches, in particular for topic groups where context and familiarity play a role. This suggest that the request type has an important role to play in the design of book discovery systems.

Our analysis was focused on the book search domain, yet similar rich profiles and contextual information is available in many modern search scenarios, in particular in mobile search and increasingly aggregated to mixed-device search scenarios. Research access to such mobile search logs and social media data is difficult due to privacy and commercial constraints, making the more constrained and less sensitive book search domain an attractive alternative to study many aspects of complex contextualized search.

We highlighted the diversity of complex search requests, and observed a mixture of content and context going beyond currently existing systems. This is an important first step toward the development of novel information access systems that blend traditional search and (hybrid) recommendation approaches into a coherent whole.

Acknowledgments. This research was supported by the Netherlands Organization for Scientific Research (README project, NWO VIDI # 639.072.601; ExPoSe project, NWO CI # 314.99.108).

References

[1] Anand, S.S., Mobasher, B.: Contextual recommendation. In: Berendt, B., Hotho, A., Mladenic, D., Semeraro, G. (eds.) WebMine 2007. LNCS (LNAI), vol. 4737, pp. 142–160. Springer, Heidelberg (2007)
[2] Beckers, T., Fuhr, N., Pharo, N., Nordlie, R., Fachry, K.N.: Overview and results of the INEX 2009 interactive track. In: Lalmas, M., Jose, J., Rauber, A., Sebastiani, F., Frommholz, I. (eds.) ECDL 2010. LNCS, vol. 6273, pp. 409–412. Springer, Heidelberg (2010)
[3] Bhatia, S., Mitra, P.: Adopting Inference Networks for Online Thread Retrieval. In: Proceedings of AAAI 2010. AAAI Press (2010)
[4] Borlund, P.: The Concept of Relevance in IR. JASIST 54(10), 913–925 (2003)
[5] Buchanan, G., McKay, D.: The bookshop: examining popular search strategies. In: Proceedings of JCDL 2011, pp. 269–278. ACM (2011)
[6] Chandler, O.: How Consumers Discover Books Online. Tools of Change for Publishing, O'Reilly (2012)
[7] Cosijn, E., Ingwersen, P.: Dimensions of Relevance. Information Processing & Management 36, 533–550 (2000)
[8] Cunningham, S.J., Vanderschantz, N., Timpany, C., Hinze, A., Buchanan, G.: Social information behaviour in bookshops: Implications for digital libraries. In: Aalberg, T., Papatheodorou, C., Dobreva, M., Tsakonas, G., Farrugia, C.J. (eds.) TPDL 2013. LNCS, vol. 8092, pp. 84–95. Springer, Heidelberg (2013)
[9] Dean-Hall, A., Clarke, C.L.A., Kamps, J., Thomas, P., Simon, N., Voorhees, E.: Overview of the TREC 2013 contextual suggestion track. In: Proceedings of TREC 2013. NIST (2013)
[10] Elsas, J.L., Carbonell, J.G.: It Pays to be Picky: An Evaluation of Thread Retrieval in Online Forums. In: Proceedings of SIGIR 2009, pp. 714–715. ACM (2009)
[11] Indri. Language modeling meets inference networks (2014), `http://sourceforge.net/projects/lemur/`
[12] Koolen, M., Kazai, G., Kamps, J., Preminger, M., Doucet, A., Landoni, M.: Overview of the INEX 2012 Social Book Search Track. In: Proceedings of INEX 2012. LNCS, Springer,
[13] Koolen, M., Kamps, J., Kazai, G.: Social Book Search: Comparing Topical Relevance Judgements and Book Suggestions for Evaluation. In: Proceedings of CIKM 2012, pp. 185–194. ACM (2012a)
[14] Koolen, M., Kazai, G., Kamps, J., Doucet, A., Landoni, M.: Overview of the INEX 2011 books and social search track. In: Geva, S., Kamps, J., Schenkel, R. (eds.) INEX 2011. LNCS, vol. 7424, pp. 1–29. Springer, Heidelberg (2012)
[15] Koolen, M., Kazai, G., Preminger, M., Doucet, A.: Overview of the INEX 2013 social book search track. In: Proceedings of the CLEF 2013 Evaluation Labs and Workshop (2013)

[16] Linden, G.: Geeking with Greg: Early Amazon: Similarities (2006), http://glinden.blogspot.nl/2006/03/early-amazon-similarities.html
[17] : Software Framework for Topic Modelling with Large Corpora. In: Proceedings of the LREC 2010 Workshop on New Challenges for NLP Frameworks, pp. 45–50. ELRA (2010)
[18] Reuter, K.: Assessing aesthetic relevance: Children's book selection in a digital library. JASIST 58(12), 1745–1763 (2007)
[19] Ross, C.S.: Finding without seeking: the information encounter in the context of reading for pleasure. Information Processing & Management 35(6), 783–799 (1999)
[20] Saracevic, T.: Relevance Reconsidered. In: Proceedings of COLIS 2, pp. 201–218 (1996)
[21] Saracevic, T., Kantor, P.B.: A Study of Information Seeking and Retrieving. I, II, III. JASIS 39(3), 161–216 (1988)

How Do Gain and Discount Functions Affect the Correlation between DCG and User Satisfaction?

Julián Urbano[1] and Mónica Marrero[2]

[1] Universitat Pompeu Fabra, Barcelona, Spain
julian.urbano@upf.edu
[2] Barcelona Supercomputing Center, Spain
monica.marrero@bsc.es

Abstract. We present an empirical analysis of the effect that the gain and discount functions have in the correlation between *DCG* and user satisfaction. Through a large user study we estimate the relationship between satisfaction and the effectiveness computed with a test collection. In particular, we estimate the probabilities that users find a system satisfactory given a *DCG* score, and that they agree with a difference in *DCG* as to which of two systems is more satisfactory. We study this relationship for 36 combinations of gain and discount, and find that a linear gain and a constant discount are best correlated with user satisfaction.

1 Introduction

Test collections are used to evaluate how well systems help users in an Information Retrieval task. In conjunction with an effectiveness measure such as Average Precision, they are an abstraction of the search process that allows us to systematically evaluate and improve systems by assessing how good a system is, and which of two systems is better. In particular, collections are an abstraction of the static component in the search process (e.g., documents, topical relevance), while effectiveness measures are an abstraction of the dynamic component (e.g., user behavior, interactions between documents). This user abstraction is advantageous because it makes evaluation experiments inexpensive, easy to run, and easy to reproduce. However, they make several assumptions about how users interact with a system and the perceived utility of the documents it retrieves.

Imagine a system that obtains an effectiveness score $\phi \in [0,1]$ for some query. The best we can interpret ϕ is to assume that $\phi \cdot 100\%$ of users will be satisfied by the system, or $P(Sat|\phi)=\phi$. If we obtain $DCG = 0.85$, we somehow interpret it as 85% probability of user satisfaction. Similarly, if the difference between two systems A and B is $\Delta\phi > 0$, we expect users to agree and prefer A over B. In fact, we expect them to do so *regardless* of how large $\Delta\phi$ is, or $P(Pref|\Delta\phi)=1$. If the test collection tells us that A is superior to B, we expect users to agree. The extent to which these interpretations are valid depends on whether the assumptions mentioned above hold or not. For instance, relevance judgments are subjective, meaning that we should expect $P(Pref|\Delta\phi) < 1$. Similarly, different effectiveness measures are based on different user models and thus result in different ϕ scores, so $P(Sat|\phi)=\phi$ is not necessarily true.

A. Hanbury et al. (Eds.): ECIR 2015, LNCS 9022, pp. 197–202, 2015.

We present a novel method to investigate these relationships, and study the specific case of *DCG* in a music recommendation task with informational queries. Through a user study where subjects told us which of two systems they preferred, we empirically map *DCG* scores onto $P(Sat)$ and $P(Pref)$. An analysis of these mappings for 6 gain and 6 discount functions suggests that the usual exponential gain underestimates satisfaction, and that all forms of discount do so too.

2 Formulations of *DCG*

Let $\mathcal{L} = \{0, 1, \ldots, n_{\mathcal{L}}-1\}$ be the set of $n_{\mathcal{L}}$ relevance levels used to make judgments, and let $r_i \in \mathcal{L}$ be the relevance given to document i. The Discounted Cumulative Gain at k documents retrieved is $DCG@k = \sum_{i=1}^{k} g(r_i) \cdot d(i)$, where $g : \mathcal{L} \rightarrow \mathbb{R}^{\geq 0}$ is a monotonically increasing *gain* function to map a relevance level onto a utility score, and $d : \mathbb{N}^{>0} \rightarrow \mathbb{R}^{>0}$ is a monotonically decreasing *discount* function to reduce utility as documents appear down the ranking. The original formulation used linear gain $g(\ell) = \ell$ and logarithmic discount $d(i) = 1/\max(1, \log_2 i)$ [4]. However, the choice of functions is open. The de facto formulation in IR uses exponential gain $g(\ell) = 2^{\ell}-1$ to emphasize the utility of highly relevant documents, and $d(i) = 1/\log_2(i+1)$ to penalize all but the first document retrieved [3].

A drawback of *DCG* is that the upper bound depends on k, $\mathcal{L}$, g and d. *nDCG* was proposed to normalize scores dividing by the *DCG* score of an ideal ranking of documents [4]. However, *nDCG* does not correlate well with user satisfaction when there are less than k highly relevant documents, because systems inevitably retrieve non-relevant documents among the top k [1]. To normalize $DCG@k$ between 0 and 1, we divide by the maximum theoretically possible with k documents. This formulation is better correlated with user satisfaction because it yields $DCG@k = 1$ only when all k documents have the highest relevance:

$$DCG@k = \frac{\sum_{i=1}^{k} g(r_i) \cdot d(i)}{\sum_{i=1}^{k} g(n_{\mathcal{L}}-1) \cdot d(i)}$$

In our experiments we study 6 different gain functions: Linear $g(\ell) = \ell$, Exponential $g(\ell) = b^{\ell}-1$ with bases $b = 2$, 3 and 5, and Binary $g(\ell) = I(\ell \geq \ell_{min})$ with minimum relevance $\ell_{min} = 1$ and 2. We also study 6 variants of discount: Zipfian $d(i) = 1/i$, Linear $d(i) = (k+1-i)/k$, Constant $d(i) = 1$ (i.e. null), and Logarithmic $d(i) = 1/\log_b(b+i-1)$ with bases $b = 2$, 3 and 5. Note that the Constant discount reduces *DCG* to Precision with Binary gains and to *CG* with the rest.

3 Methods and Data

We ran an experiment with actual users that allowed us to map system effectiveness onto user satisfaction. Similar to Sanderson et al. [7], subjects were presented with different examples, each containing a query and two ranked lists of results as if retrieved by two systems A and B. Subjects had to select one of these options: system A provided better results, system B did, they both provided *good* results, or they both returned *bad* results. Behind the scenes, we know the relevance of all documents, so the effectiveness scores ϕ_A and ϕ_B are known. Subjects indicating that both systems are *good* suggest that they are satisfied with both ranked lists, meaning that ϕ_A and ϕ_B translate into user satisfaction; if they indicate that both systems are *bad*, it means that they do not translate into satisfaction. Subjects that show preference for one of the systems suggest that there is a difference large enough to be noticed, meaning that $\Delta\phi_\mathsf{AB}$ translates into users being more satisfied with one system than with the other. Whether this preference agrees with $\Delta\phi_\mathsf{AB}$ depends on which system they prefer.

To compute reliable estimates of $P(Sat|\phi)$ and $P(Pref|\Delta\phi)$ we needed enough examples to cover the full range of ϕ and $|\Delta\phi|$ scores for all 36 *DCG* formulations under study. To do so, we split the $[0, 1]$ range in 10 equally sized bins, and randomly generate examples until we have at least 200 per bin and *DCG* formulation. We used an iterative greedy algorithm that at each iteration selects the bin and formulation with the least examples so far, generates a new example for that case, and then updates the corresponding bin in the other formulations.

As search task, we used music recommendation, where the query is the audio of a song and the result of the system is a ranked list of songs deemed as similar (relevant) to the query. This choice has several advantages over a traditional text search task for our purposes. First, it is a purely informational task where the user wants as much relevant information (similar songs) about the query as possible, which makes it a good choice to study *DCG@k*. Second, it is a task known to be enjoyable by assessors and that does not require much time per judgment, considerably reducing assessor fatigue [6]. Third, because subjects have to actively listen to the returned documents, their preferences are not confounded by other factors such as document titles and result snippets. The queries and documents are music clips 30 seconds long, taken from the corpus used in the MIREX audio music similarity and retrieval task (MIREX is a TREC-like evaluation campaign focused on Music IR tasks; see `http://www.music-ir.org/mirex/wiki/`). We used data from the 2007–2012 editions, comprising 22,074 relevance judgments across 439 queries. After running the greedy selection algorithm, we ended up with a total of 4,115 examples covering 432 unique queries and 5,636 unique documents. As per the task guidelines, all judgments are made on a scale with $n_\mathcal{L}=3$ levels, and systems retrieve $k=5$ documents (see [8] for details and the task design).

User preferences for all 4,115 examples were collected via crowdsourcing, as this has been shown to be a reliable method to gather this kind of relevance judgments [6], and it offers a large and diverse pool of subjects to help us generalize results. We used the platform Crowdflower to gather user preferences, as it provides quality control that separates good from bad workers by means of trap examples, as in [7,6,8] (some examples have known answers, provided by us, to estimate worker quality). We manually selected 20 trap questions with answers uniformly distributed. We collected only one answer per example because we are interested precisely in the user variability, not in an aggregated answer reflecting the majority preference. We paid \$0.03 per example; the total was nearly \$250.

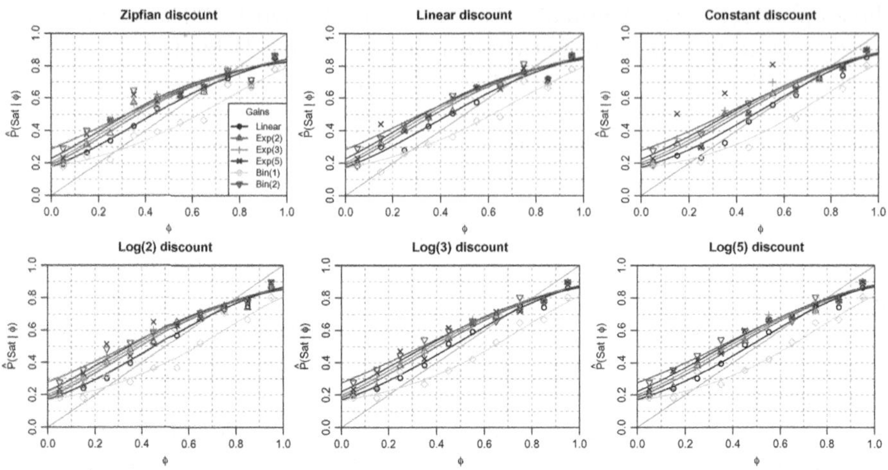

Fig. 1. $\hat{P}(Sat|\phi)$ estimated with 4,050 ranked lists judged as *good* or *bad* and all *DCG* formulations. Points show averages within bins of ϕ, lines show a quadratic logit fit.

4 Results

A total of 547 workers provided 11,042 answers in less than 24 hours. Crowdflower only trusted 175 workers (32%); their trust scores ranged from 73% to 100%, with an average of 90%. After removing answers to trap questions, 113 unique workers were responsible for the answers to our 4,115 examples.

User Satisfaction. For 2,025 of the 4,115 examples (49%) subjects judged both systems as equally good or bad, so we have 4,050 ranked lists judged as satisfactory or unsatisfactory. Fig. 1 shows the estimate $\hat{P}(Sat|\phi)$ for these examples and all *DCG* formulations. The pattern is extremely similar across discount functions: satisfaction is underestimated for low ϕ scores and overestimated beyond $\phi \approx 0.8$. This suggests that users do not discount the utility of documents based on their rank. Within discount functions, we see a subtle but clear pattern as well: gain functions that emphasize highly relevant documents tend to underestimate user satisfaction. For instance, Bin(2) is mostly above the diagonal because only documents with relevance 2 are considered useful by the gain function; those with relevance 1 are deemed as useless, though users did find them useful to some extent. Notice that the exact opposite happens with Bin(1). Similarly, we can see that exponential gains tend to underestimate proportionally to the base. Highly relevant documents are assumed to be much more useful than others (more so with larger bases), so the gain function inherently penalizes mid-relevants because they are not as relevant as they could *supposedly* be.

User Preferences. For 2,090 of the 4,115 examples (51%) subjects indicated that one system provided better results than the other one; whether those preferences agree with the sign of $\Delta\phi_{\mathsf{AB}}$ depends on the *DCG* formulation. Surprisingly, Fig. 2 shows that $P(Pref|\Delta\phi)$ is proportional to $\Delta\phi$, rather than always 1 as we expected. This means that users tend to agree with the test collection,

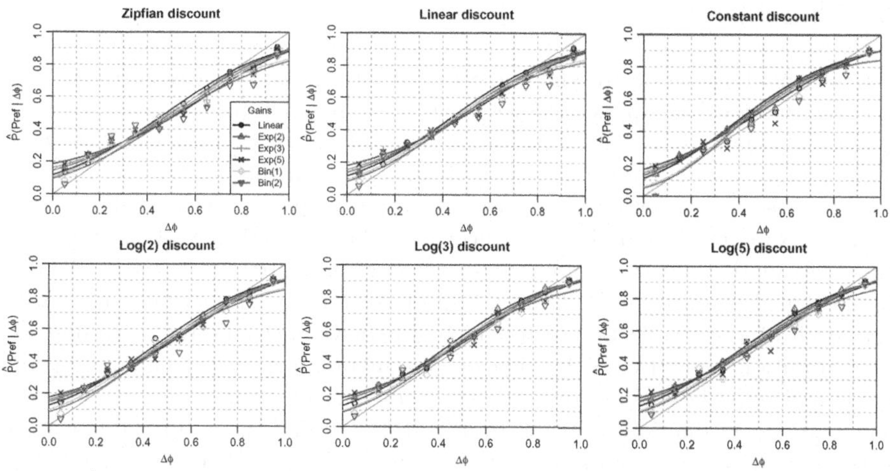

Fig. 2. $\hat{P}(Pref|\Delta\phi)$ estimated with 2,090 examples judged with a preference. Points show averages within bins of $|\Delta\phi|$, lines show a quadratic logit fit.

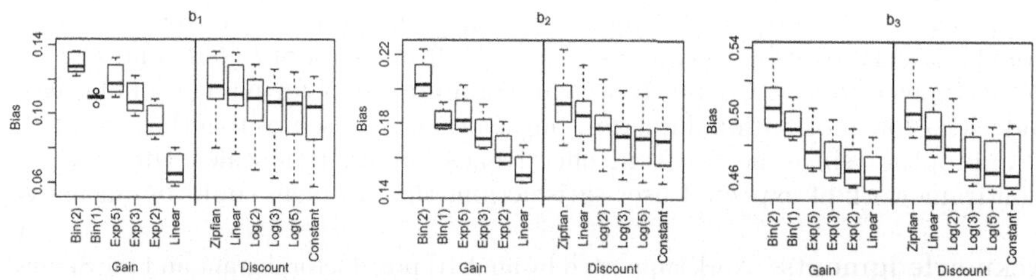

Fig. 3. Bias distributions for all 36 combinations of gain and discount functions

but differences in effectiveness need to be quite large for the majority of users to do so. On average, unless $\Delta\phi \gtrsim 0.5$ users just can not decide. A subtle but clear pattern appears again: gain functions that overemphasize highly relevant documents work better with low $\Delta\phi$ scores, as the mid-relevant documents that make that difference are found to be more useful than the gain function predicts.

To further analyze what functions correlate best with satisfaction, we computed three bias indicators. The first one, $b_1 = \int |\hat{P}(Sat|\phi) - \phi|\ d\phi$, tells how much off the ideal $P(Sat|\phi) = \phi$ we are in Fig. 1 (note that a large b_1 bias score does not necessarily mean that the *DCG* formulation is bad; it is just not as easy to interpret as expected). The second one, $b_2 = [\hat{P}(Sat|0) + 1 - \hat{P}(Sat|1)]/2$, tells how large the gaps are at the endpoints $\phi = 0$ and $\phi = 1$ in Fig. 1 (it captures user disagreement and the goodness of the *DCG* user model). The third indicator, $b_3 = \int 1 - \hat{P}(Pref|\Delta\phi)\ d\Delta\phi$, tells how far apart from the ideal $P(Pref|\Delta\phi) = 1$ we are in Fig. 2 (it measures user discriminative power). For all indicators, an ANOVA analysis shows significant differences among gain and discount functions. Fig. 3 shows that bias is proportional to the emphasis that gain functions give to highly relevant documents, and the steepest discounts are consistently more biased. The Linear gain and Constant discount are the least biased overall.

5 Conclusion

We presented a method to study how well effectiveness measures correlate with user satisfaction, and applied it for a music recommendation task with *DCG* and a range of gain and discount functions. Our results show that the usual choice of exponential gain underestimates user satisfaction, and that all types of discount tend to do so too, reflecting that users do not pay attention to the ranking. However, the apparent lack of discount effect could be due to the small cutoff used in this task, or the high level of engagement often presented by its users. We also found that differences in *DCG* need to be large for users to actually agree with the result of a test collection as to which of two systems is better. This suggests that traditional practice of looking at system rankings (e.g. Kendall's τ) and point null hypotheses in statistical significant testing (e.g. $H_0 : \Delta\phi = 0$) oversimplifies the evaluation problem. In qualitative terms, our results largely agree with previous work on both user satisfaction [2,1,7] and reliability of *DCG* [5].

Future work will investigate the relationship between user satisfaction and system effectiveness for Text IR tasks, as the results presented here do not necessarily generalize. In particular, we will study several other measures, especially for navigational queries and diversity. A similar mapping onto user satisfaction would allow us to evaluate systems within the framework of $P(Sat)$ and $P(Pref)$ for all types of query. Currently we can compute *ERR* for navigational queries and *DCG* for informational queries, but averaging all scores together might not be appropriate since they measure effectiveness on different scales. Under a common framework of expected user satisfaction, this problem could be mitigated.

Acknowledgments. Work supported by an A4U postdoctoral grant and the Spanish Government (HAR2011-27540). We thank the reviewers for their comments.

References

1. Al-Maskari, A., Sanderson, M., Clough, P.: The Relationship between IR Effectiveness Measures and User Satisfaction. In: ACM SIGIR (2007)
2. Allan, J., Carterette, B., Lewis, J.: When Will Information Retrieval Be 'Good Enough'? In: ACM SIGIR (2005)
3. Burges, C., Shaked, T., Renshaw, E., Lazier, A., Deeds, M., Hamilton, N., Hullender, G.: Learning to Rank Using Gradient Descent. In: ICML (2005)
4. Järvelin, K., Kekäläinen, J.: IR Evaluation Methods for Retrieving Highly Relevant Documents. In: ACM SIGIR (2000)
5. Kanoulas, E., Aslam, J.A.: Empirical Justification of the Gain and Discount Function for nDCG. In: ACM CIKM (2009)
6. Lee, J.H.: Crowdsourcing Music Similarity Judgments using Mechanical Turk. In: ISMIR (2010)
7. Sanderson, M., Paramita, M.L., Clough, P., Kanoulas, E.: Do User Preferences and Evaluation Measures Line Up? In: ACM SIGIR (2010)
8. Urbano, J., Downie, J.S., Mcfee, B., Schedl, M.: How Significant is Statistically Significant? The case of Audio Music Similarity and Retrieval. In: ISMIR (2012)

Different Rankers on Different Subcollections

Timothy Jones[1], Falk Scholer[1], Andrew Turpin[2], Stefano Mizzaro[3], and Mark Sanderson[1]

[1] RMIT University
[2] University of Melbourne
[3] University of Udine

Abstract. Recent work has shown that when documents in a TREC ad hoc collection are partitioned, different rankers will perform optimally on different partitions. This result suggests that choosing different highly effective rankers for each partition and merging the results, should be able to improve overall effectiveness. Analyzing results from a novel oracle merge process, we demonstrate that this is not the case: selecting the best performing ranker on each subcollection is very unlikely to outperform just using a single best ranker across the whole collection.

Keywords: Collection Partitioning, Subcollections, Retrieval effectiveness.

1 Introduction

Recent work by Sanderson et al. [7] and Jones et al. [3] showed that when TREC collections are partitioned based on the source of a document (e.g. LA Times, Financial Times, etc.) and retrieval systems are evaluated on those partitions, the ordering of the systems differs significantly across the partitions. Their results show that there are some systems that are more effective on some partitions of the TREC collections than others.

The researchers hypothesize that if a document collection can be partitioned in such a way that specific rankers work well on specific partitions, then it should be possible to merge the results from the selected rankers, producing a system that overall is more effective than a single state-of-the-art ranker retrieving from the whole collection. This hypothesis is explored in this paper.

2 Related Work

One of the first works to investigate fusing results from multiple runs to improve effectiveness was by Fox and Shaw [2]. Their work focused on combining similarity scores produced by multiple retrieval strategies. They tested six combination strategies, and showed improvement over a single run strategy. Of particular note, the researchers described the combination strategies *CombSUM* (summing the similarity scores for each document) and *CombMNZ* (multiplying the sum by the number of systems that gave the document a non-zero score), both of which were found to result in improvements over a single run.

Beitzel et al. note that while this type of data fusion is sometimes effective, it is not understood why or where it is effective [1]. Previously, it had been hypothesized

A. Hanbury et al. (Eds.): ECIR 2015, LNCS 9022, pp. 203–208, 2015.

Table 1. The sixteen Terrier rankers used in the reported experiments

A) BB2c1	B) BM25b0	C) DFR_BM25c1	D) DFRee_1
E) DLH13_9	F) DLH_8	G) DPH_0	H) Hiemstra_LM0
I) IFB2c1	J) In_expB2c1	K) In_expC2c1	L) InL2c1
M) LemurTF_IDF_12	N) LGDc1	O) PL2c1	P) TF_IDF_15

that data fusion was effective where there was a greater overlap of relevant documents than overlap of non-relevant documents between lists [5]. However, using a series of experiments where system differences (indexer, stemmer, word definition, etc) were constant, but ranking strategies varied, Beitzel et al. show that where the document lists of runs were similar, fusion was unlikely to show improvement—as scores were in effect scaled by the fusion process. Instead, fusion is more effective when new relevant documents are introduced.

More recent merging research has been conducted. LambdaMerge, was described by Sheldon et al. [8]. The technique produces a more effective merged list than simpler techniques, or any of the single strategy source lists. Wu et al. [11] examined data fusion when including evidence from anchor text in web pages. They note that data fusion can be broken into two categories: *search result fusion* (using the score or ranking of a document to produce the final ranking), and *evidence fusion* (using multiple types of evidence as input to the ranking function). They showed that the most effective fusion method from each class produced a significant improvement over baseline retrieval, but that there was no significant difference between the best fusion methods.

3 Methods and Data

To investigate the hypothesis of our paper we require: test collections, collection partitions, a set of rankers, and a rank merging strategy.

Collections: This research investigates a hypothesis about certain properties of collections that were established in past work [3, 7]. We therefore use the same ad hoc collections from TREC 4-8 here. We partition the TREC collections into subcollections based on the publication *source* of the documents (e.g. Financial Times, LA Times, Federal Register, etc.) because this was the style of partitioning used by Sanderson et al. [7]. We also examined a partitioning based on document *length*, due to the work of Wilkie and Azzopardi [10], who showed rank inconsistencies are common when document length is varied. The latter partition produces four equal-sized subcollections. Both partitions show disagreement on the best ranker, measured using the methodology of [3]. These subcollections were selected because they strongly disagreed about the best ranker–which indicates that any possible improvement due to disagreement should be large.

Rankers: As rankers, we used sixteen different parameterizations of the Terrier system [6]. Table 1 details the names of those rankers, which include variants of Language Modeling (LM); Divergence From Randomness (DFR), BM25 and TF_IDF. These are the same rankers used by Jones et al. [3].

Merging approaches: Two merging strategies are used: CombSUM [2], and a novel oracle merging scheme that preserves the order of documents between relevant documents in the lists to be merged, but optimally chooses the order of lists from which to merge. The latter scheme provides a reasonable upper bound on the effectiveness of any merging algorithm.

4 Experiments and Results

Initial approaches to improve overall effectiveness by leveraging the best performing ranker on each subcollection were not successful. Due to lack of space we do not describe this initial work here — instead, we show the results of exhaustively searching all possible combinations of rankers across the different partitioning strategies. This produces distributions of effectiveness measures for different combinations of rankers. With sixteen rankers and four partitions, for each collection splitting scheme we have 16^4 possible combinations, each merged by normalizing document scores linearly, and then ranking by the normalized scores.

We conduct an exhaustive search of ranker combinations on the TREC 8 collection, and compare effectiveness scores (MAP) with the best ranker applied across the full TREC 8 collection, which was 0.257.

Figure 1 shows the distribution of MAP scores for the combinations on the source-based partitions. The plots each show the same data, from the perspective of a different subcollection. Each plot indicates the range of scores achievable when using a particular ranker for that subcollection. The A boxplot in top right of figure 1 shows the scores of all combinations with ranker A in subcollection 1, the B all combinations with ranker B in subcollection 1, etc. In general, there is little difference between the rankers. The main result, however, is that the maximum MAP value over any combination is never higher than 0.21, substantially lower than single ranker effectiveness which is 0.257.

Figure 2 shows the results for the length-based partitioning of the collection. Although the maximum MAP is higher than for the source-based partitions, even with an exhaustive search of combinations of rankers, all MAP scores are below 0.25.

Despite evidence in past work to suggest that such partitioning might result in improvements in effectiveness, it would appear that it does not. If a particular ranker (or group of rankers) were the best for a given subcollection, one would expect it to show up as highly effective in Figures 2 or 1. While this does not appear to be the case, it can be seen that some rankers perform especially badly on particular subcollections—for example, ranker `H` in Figure 1 or ranker `O` in Figure 2.

However, these results do not conclusively show that the different ranker per subcollection strategy is failing to work, it is possible that the low scores are due to an ineffective merging strategy. This was investigated next.

4.1 An Upper Bound on Improved Retrieval

To examine an upper bound on retrieval after ranker selection has been made, we follow the best possible performance methodology introduced by Thomas and Shokouhi [9], where design choices for an element of a retrieval system are compared by assuming

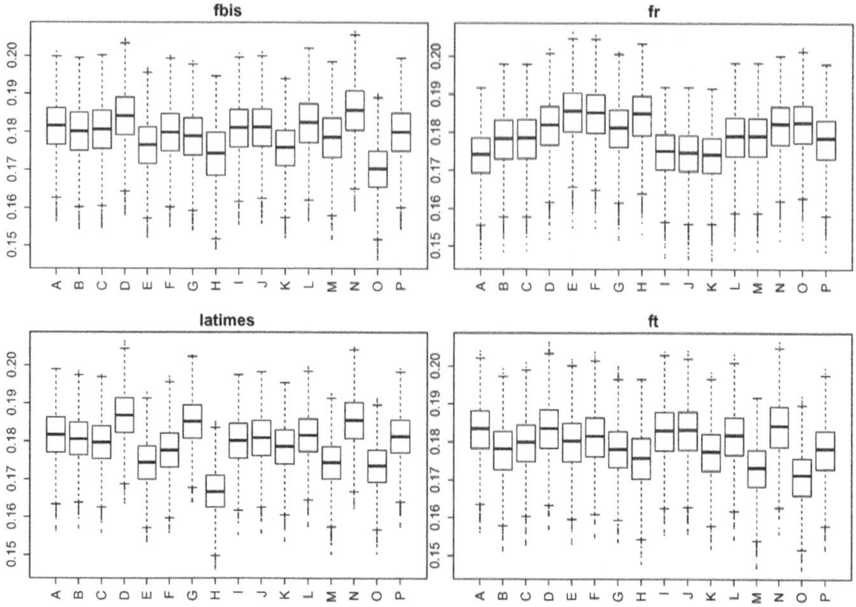

Fig. 1. MAP@1000 scores for all 16^4 ranker combinations separated according to the ranker used on each subcollection for TREC 8. Scores combined by linear normailsation. Letters A–P indicate the ranker selected for that subcollection, box plots indicate the range of scores achieved when all combinations of rankers for the other subcollections are tried.

best possible performance from all subsequent elements of the system. Here, we introduce a merging strategy called *perfect merge* (PM), which operates with the following principles (assuming rankings have no overlap). It does not violate the ordering from the ranked lists to be merged: if document a occurs below document b in one of the ranked lists, then it must occur in that order in the final list. PM does not skip any items: if document a appears at rank 1 in a ranked list, then no documents below rank 1 in that list can be selected until document a has been selected. A PM of a set of ranked lists is defined as the merge that achieves the highest score under a particular evaluation measure. In this work, we use MAP@10, since evaluating deeper in the result list produces a much larger state space that is computationally expensive to compute.

Calculating all possible merges of a set of input rankings to determine the PM would be infeasible. Fortunately, the criteria above allow us to make some assumptions that reduce the determination of a PM into a natural fit for a branch and bound solution [4]. Branch and bound is a state space search that calculates the highest possible outcome from a branch of the state tree, and ignores that branch if the outcome cannot be better than the best known solution so far.

Since we are not allowed to skip any items in the merge process, the candidate solutions must start with the document at rank one in one of the source lists. Additionally, since we know the number of relevant documents that any solution can contain, and

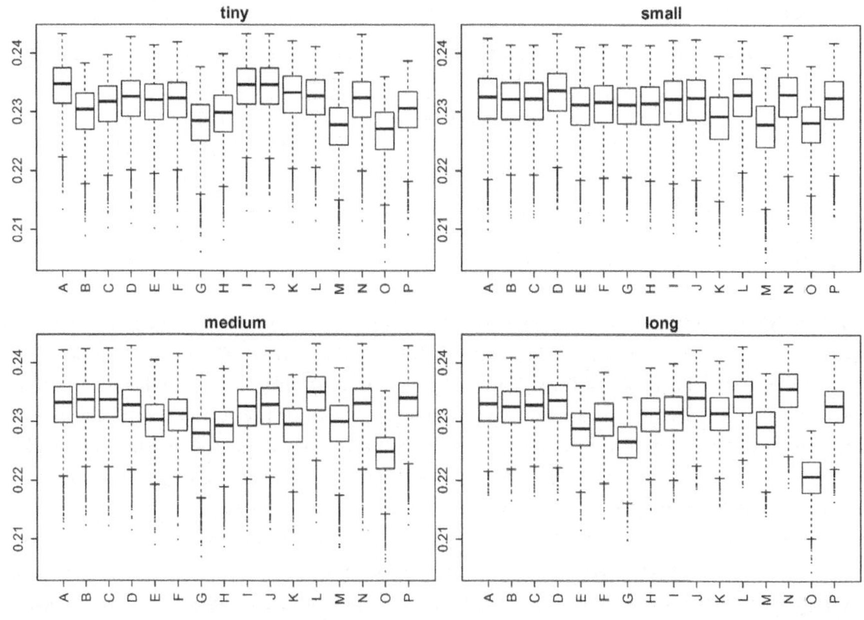

Fig. 2. MAP@1000 scores from length based subcollections displayed as Figure 1

since the best possible score for any partial ranked list can be obtained by putting all remaining relevant documents next in any partial list, we can produce an upper bound on the possible score of a partial solution. This allows us to rapidly eliminate solutions that will score poorly, and quickly converge on the optimal solution. We initialize the lower bound as the highest scoring run from any subcollection, since each input ranked list is also a candidate solution.

4.2 Ranker Selection Strategies

To simulate the best possible performance in a search system that selects different rankers for each subcollection, we assign the best performing ranker to each subcollection, measured using MAP@1000. Runs from each subcollection are then merged using PM. We call the resulting score *HybridPM*.

Since this strategy is an oracle, we need an oracle baseline to compare with. We pick one ranker that has the best mean MAP@1000 score across all subcollections. Then, we run the results for each query on each subcollection–using this one ranker–through the perfect merging process described above. We call this score *TraditionalPM*.

Table 2 shows the comparison of HybridPM and TraditionalPM on TREC 4–8 on source-based subcollections. What we find is that HybridPM does not significantly outperform TraditionalPM (paired t-test, $p > 0.05$). Under ideal merging, there is no advantage in picking different (best) rankers per subcollection.

Table 2. MAP@10 of TraditionalPM and HybridPM on source-based subcollections

	TREC 4	TREC 5	TREC 6	TREC 7	TREC 8
TraditionalPM	0.5464	0.3643	0.5205	0.4914	0.5593
HybridPM	0.5809	0.3537	0.5309	0.4926	0.5643
Percentage improvement	6.3%	2.9%	1.9%	0.2%	0.8%

5 Conclusions

Using different rankers on different subsets of a collection intuitively sounds like an approach that can improve overall search effectiveness, since previous work using the same collections and rankers had shown significant disagreement on which ranker was best. Using a simple merge, no improvements were found. Hypothesizing that this was due to the merging step, we introduced perfect merge, a strategy for producing an idealized merge of a set of input result lists. Even with perfect merge, using different rankers on each subcollection showed no improvement over using a single ranker. It can therefore be concluded that on the TREC ad hoc collections, it is very unlikely that using different rankers on each subcollection will improve retrieval. A limitation of this analysis may be that the subcollections of a TREC collection are too similar. An avenue for future work is to use subcollections that have greater variation in style and content.

References

[1] Beitzel, S.M., Jensen, E.C., Chowdhury, A., Grossman, D., Goharian, N., Frieder, O.: Recent results on fusion of effective retrieval strategies in the same information retrieval system. In: Callan, J., Crestani, F., Sanderson, M. (eds.) SIGIR 2003 Ws Distributed IR 2003. LNCS, vol. 2924, pp. 101–111. Springer, Heidelberg (2004)
[2] Fox, E.A., Shaw, J.A.: Combination of multiple searches. Proc. TREC 2, 243–252 (1994)
[3] Jones, T., Turpin, A., Mizzaro, S., Scholer, F., Sanderson, M.: Size and source matter: Understanding inconsistencies in test collection-based evaluation. In: CIKM 2014 (2014)
[4] Land, A.H., Doig, A.G.: An automatic method of solving discrete programming problems. Econometrica 28(3), 497–520 (1960)
[5] Lee, J.H.: Analyses of multiple evidence combination. In: Proc. SIGIR, pp. 267–276. ACM Press, New York (1997)
[6] Ounis, I., Amati, G., Plachouras, V., He, B., Macdonald, C., Lioma, C.: Terrier: A high performance and scalable information retrieval platform. In: Proc. OSIR Workshop, pp. 18–25 (2006)
[7] Sanderson, M., Turpin, A., Zhang, Y., Scholer, F.: Differences in effectiveness across subcollections. In: CIKM 2012, pp. 1965–1969. ACM (2012)
[8] Sheldon, D., Shokouhi, M., Szummer, M., Craswell, N.: Lambdamerge: Merging the results of query reformulations. In: WSDM 2011, pp. 795–804. ACM, New York (2011)
[9] Thomas, P., Shokouhi, M.: Evaluating server selection for federated search. In: Gurrin, C., He, Y., Kazai, G., Kruschwitz, U., Little, S., Roelleke, T., Rüger, S., van Rijsbergen, K. (eds.) ECIR 2010. LNCS, vol. 5993, pp. 607–610. Springer, Heidelberg (2010)
[10] Wilkie, C., Azzopardi, L.: Efficiently estimating retrievability bias. In: de Rijke, M., Kenter, T., de Vries, A.P., Zhai, C., de Jong, F., Radinsky, K., Hofmann, K. (eds.) ECIR 2014. LNCS, vol. 8416, pp. 720–726. Springer, Heidelberg (2014)
[11] Wu, M., Hawking, D., Turpin, A., Scholer, F.: Using anchor text for homepage and topic distillation search tasks. JASIST 63, 1235–1255 (2012)

Retrievability and Retrieval Bias: A Comparison of Inequality Measures

Colin Wilkie and Leif Azzopardi

University of Glasgow, 18 Lilybank Gardens, G12 8QQ, Glasgow, UK
{colin.wilkie,leif.azzopardi}@glasgow.ac.uk

Abstract. The disposition of a retrieval system to favour certain documents over others can be quantified using retrievability. Typically, the Gini Coefficient has been used to quantify the level of bias a system imposes across the collection with a single value. However, numerous inequality measures have been proposed that may provide different insights into retrievability bias. In this paper, we examine 8 inequality measures, and see the changes in the estimation of bias on 3 standard retrieval models across their respective parameter spaces. We find that most of the measures agree with each other, and that the parameter setting that minimise the inequality according to each measure is similar. This work suggests that the standard inequality measure, the Gini Coefficient, provides similar information regarding the bias. However, we find that Palma index and 20:20 Ratio show the greatest differences and may be useful to provide a different perspective when ranking systems according to bias.

1 Introduction

An interesting concept within the field of Information Retrieval (IR) is retrievability. Retrievability aims to estimate the likelihood that a document will be retrieved [3]. Recent work has shown that the performance of a system is linked to the retrievability of documents, often expressed as retrievability bias (or inequality) [3,4,7,10,11]. Intuitively, this is because for a document to be considered relevant, it must first be retrieved [3]. Previous work using retrievability has focussed on its applications within IR (i.e. search engine bias, reverted index, etc) and its relationship with more traditional evaluation measures (i.e. MAP, P@10, etc). In these studies the retrievability bias expressed by the system has been computed using one particular measure of inequality, the Gini Coefficient (which was first suggested by Azzopardi and Vinay [3] and used ever since). However, a range of inequality measures exist. In this short paper, we will examine a range of different inequality measures, and compare them to the Gini Coefficient to determine whether they provide a different perspective on system bias (given the retrievability scores). And whether the measures of inequality agreed on which system/model/parameter configuration exerts the least amount of bias.

A. Hanbury et al. (Eds.): ECIR 2015, LNCS 9022, pp. 209–214, 2015.

2 Background

The document-centric evaluation measure, retrievability, was first introduced by Azzopardi and Vinay [3] taking ideas from transportation planning and applying them to the domain of IR [2]. This measure evaluates how likely a document is to be retrieved by an IR system given a very large query set. The retrievability r of a document d with respect to an IR system is defined as:

$$r(d) \propto \sum_{q \in Q} O_q . f(k_{dq}, \{c, g\}) \tag{1}$$

where q is a query from the universe of queries Q, meaning O_q is the probability of a query being chosen. k_dq is the rank at which d is retrieved given q and $f(k_{dq}, \{c, g\})$ is an access function denoting how retrievable d is given q at rank cut-off c with discount factor g. To calculate retrievability, we sum the $O_q.f(k_{dq}, \{c, g\}$ across all q's in the query set Q. As it is not possible to launch all queries, a large set of queries is automatically generated from the collection. The measure essentially encodes that the more queries that retrieve d before the rank cut-off c, the more retrievable d is. The simplest model to compute the retrievability is the cumulative scoring model. In this model, an access function $f(k_{dq}, c)$ is used, such that $f(k_{dq}, c) = 1$ if d is retrieved in the top c documents given q, otherwise $f(k_{dq}, c) = 0$. Simply, if d is in the top c results, it accrues a score of 1. When done across a large cross section of queries, we get the sum of how many time d was returned above rank c

2.1 Retrievability Bias

Retrievability bias has traditionally been measured using the Gini Coefficient (Gini). However, a range of inequality measures similar to Gini exist in socio-political science. The following inequality measures have been standardised to describe how they estimate retrievability bias. In the following equations $r(d_i)$ denotes the retrievability score of the ith document while N is the number of documents in the collection. $M = \sum_{i=1}^{N} r(d_i)$ i.e. the total retrievability available for a system to distribute. D represents all the documents in the collection and d_i is the ith document in the collection. We describe the following measures in terms of this particular domain, where the population is a document collection and the wealth distributed among the documents is retrievability. Inequality refers to the level of retrievability bias present in a system.

Gini Coefficient: The Gini Coefficient is a ratio analysis method that produces values ranging from 0 to 1 (as there is no negative retrievability scores in a collection) [5]. The measure avoids statistical averages by using individual retrievability scores rather than relying on an average retrievability for large groups of documents in the collection. In a retrievability analysis, 0 denotes total equality, i.e. every document has an equal chance of being retrieved meaning the retrieval function must be random. A value of 1 represents the highest level of inequality possible where one document is retrieved for every query and no other document is ever retrieved.

$$G = \frac{1}{N}(N + 1 - 2\left(\frac{\sum_{i=1}^{N}(N + 1 - i)r(d_i)}{M}\right)) \tag{2}$$

Atkinson Index: The Atkinson index is the only parameterised measure included in this study. The parameter ϵ allows the measure to be altered to identify what areas of the collection contribute the most to the inequality [1]. For example, a low ϵ is highly sensitive to changes in the distribution of retrievability in the set of most retrievable documents. Conversely, high values of ϵ are sensitive to changes in the set of least retrievable documents. This parameter is very useful for measuring inequality in a specific area of the collection (e.g. the most and least retrievable documents).

$$A(\epsilon) = 1 - \frac{\prod_{i=1}^{N}(r(d_i)^{1/N})}{\overline{M}} \tag{3}$$

Theil Index: Two Theil Index measures were proposed, Theil T Equation 4 and Theil L Equation 5. These measures examine the entropy of the data as a way to measure inequality [9]. This is done by computing the maximum possible entropy and taking away the observed entropy from this value. Theil returns values between 0, which represents a uniform distribution of retrievability scores across the collection, and lnN, which represents when only one document is retrievable.

$$T_T = \frac{1}{N}\sum_{i=1}^{N}\left(\frac{r(d_i)}{\overline{M}}.ln\frac{r(d_i)}{\overline{M}}\right) \tag{4}$$

$$T_L = \sum_{i=1}^{N} ln\left(\frac{M}{r(d_i).N}\right) \tag{5}$$

Hoover Index: The Hoover Index is commonly referred to as the *Robin Hood Index* in economics. Applied to retrieval, it measures the portion of the total amount of retrievability that would need to be re-distributed across the collection to ensure all documents have an equal chance of retrieval [6]. Simply, how much retrievability must be taken from the highly retrievable documents and given to the poorly retrievable documents in order for them all to have an equal chance of retrieval.

$$H = \frac{1}{2}\sum_{i=1}^{N}\left|\frac{r(d_i)}{M} - \frac{i}{N}\right| \tag{6}$$

Palma Index: The Palma index examines the ratio between the top 10% most wealthy (most high retrievabile) and the poorest bottom 40% (least retrievable) [8]. It is expected that the middle 50% will posses roughly 50% of the wealth (or in this case retrievability), and so the Palma Index ignores this part of the distribution to examine the disparity between the rich and the poor.

$$P = \frac{\sum_{i=1}^{\lfloor \frac{N}{10} \rfloor} r(d_i)}{\sum_{i=\lceil \frac{6N}{10} \rceil}^{N} r(d_i)} \tag{7}$$

20:20 Ratio: The 20:20 Ratio is a simple measure of inequality like the Palma Index that excludes a large amount of the collection. As the name suggests, this measure examines the 20% most and 20% least retrievable documents. Again, the idea is to remove the statistical averages and highlight the disparity of the extremes and increase the impact of the extremes.

Table 1. Points of minimum bias for each model on AQ and DG. The actual value recorded and the correlation with the Gini Coefficient. * denotes statistical significance at $p < 0.05$.

Col.	Model	Gini	Atkinson	Hoover	Palma	20:20	Theil L	Theil T
AQ	BM25	0.52	0.24(0.99*)	0.38(1.00*)	4.29(0.97*)	32.80(0.96*)	0.05(1.00*)	0.04(0.96*)
	LM	0.56	0.27 (0.99*)	0.42(1.00*)	5.86(0.99*)	47.87(0.96*)	0.05(0.99*)	0.04(0.97*)
	PL2	0.58	0.29(1.00*)	0.43(1.00*)	6.43(0.98*)	46.03(0.97*)	0.05(0.99*)	0.05(0.98*)
DG	BM25	0.59	0.31(0.99*)	0.44(0.99*)	7.82(0.99*)	66.33(0.98*)	0.06(0.99*)	0.05(0.94*)
	LM	0.61	0.33(1.00*)	0.46(0.99*)	9.64(1.00*)	76.22(0.99*)	0.07(0.99*)	0.05(0.97*)
	PL2	0.77	0.52(0.99*)	0.61(0.99*)	35.01(0.75)	81.12(0.31)	0.11(0.98*)	0.10(0.96*)

$$R20 = \frac{\sum_{i=1}^{\lfloor \frac{2N}{10} \rfloor} r(d_i)}{\sum_{i=\lceil \frac{8N}{10} \rceil}^{N} r(d_i)} \quad (8)$$

Each of these inequality measures, provide a different take on measuring the inequality within a population. In the next section, we will empirically explore these measures in the context of parameter tuning.

3 Experimental Method

The aim of these experiments was to explore the following research questions:

1. How related are the inequality measures? i.e. do they provide different insights into how biased a system is?
2. Which inequality measure(s) should we use?

To conduct this analysis, we selected two TREC test collections that have been used in a number of previous retrievability experiments [3,11], i.e. .Gov (DG) and Aquaint (AQ). We selected three standard retrieval models, BM25, PL2 and Language Modelling with Bayes Smoothing, where we manipulated their length normalisation parameter, b (0..1), c (0..100) and β (0..10000), respectively. We leave BM25's other parameters at $k_1 = 1.2$ and $k_3 = 8$.

The query set Q used for each collection was generated by extracting the top 300,000 most frequently occurring bigrams from each collection. The $r(d)$ scores were then computed using the cumulative measures with a cut-off of 100[1]. Given the retrievability scores for the documents in each collection, for each retrieval model/parameter setting, we then computed the values for the eight inequality measures described in Section 2. To determine how similar the different inequality measures were to the typically used Gini Coefficient, we computed the Pearson's Correlation Coefficient across the parameter space for each model/collection and denote whether the correlation was significant if $p < 0.05$ (See Table 1).

4 Results and Analysis

Examining the plots of Figures 1 and 2, we see several interesting patterns across the range of inequality measures. The first pattern is that most of the inequality

[1] We also examined different cut-offs, however, our findings were similar, just different magnitudes.

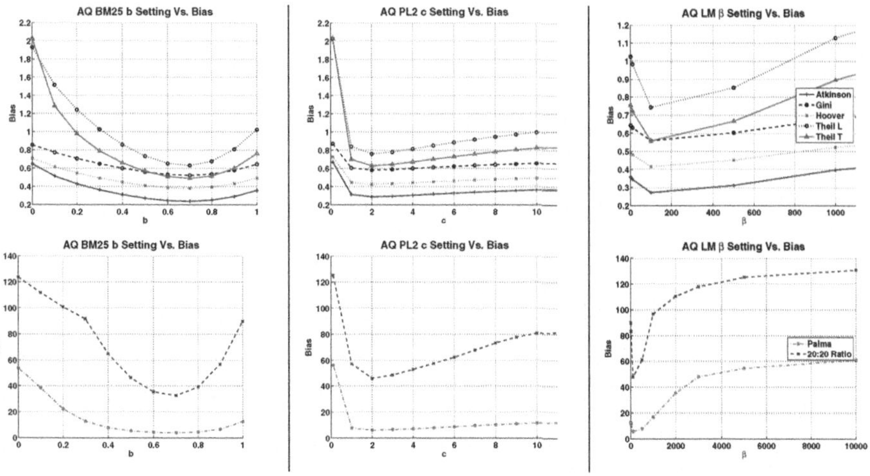

Fig. 1. Graphs for AQ for each model, BM25 (Left), PL2(Middle) and LM (right). Atikinson, Gini, Hoover, Theil T and Theil L (Top) and Palma and 20:20 Ratio (Bottom) plotted along the parameter settings against the inequality. All curves follow the same point of minimum inequality.

measures follow a similar shape, and that the parameter setting which minimises the bias (given the measure) is the same, i.e. on AQ b=0.7 for BM25, and b=0.9 on DG, across all the inequality measures. In previous work [10,11,7], the parameter setting was chosen by minimising the bias given the Gini Coefficient. This work suggests that, regardless of inequality measure, the parameter setting would have been similar.

The strength of the relationship between Gini and the other measures is also confirmed by the correlations shown in Table 1, where most are very close to 1 and statistically significant. However, the Palma and 20:20 Ratio measures show the lowest correlations with the Gini Coefficient, suggesting that these measures are the most different. Indeed, when we consider the ranking of the systems based on the inequality measures we see that on AQ, Gini ranks the systems: BM25, LM then PL2 (least biased to most biased), while the 20:20 Ratio ranks the systems, BM25, PL2, and then LM. This suggests that in terms of ranking systems, there might be differences between measures. However, we leave this for future examination.

5 Conclusions and Future Work

From the findings presented in Section 4 several conclusions can be made to answer our research questions. It is apparent that using various inequality measures does not change the point at which minimum inequality is found when using the Gini Coefficient. Therefore it is fair to continue these investigations using the Gini Coefficient. Addressing our second question, the results show thatall of these measures are applicable to quantifying retrievability bias. However, the measures used in this study have various degrees of sensitivity to changes in inequality which may make certain measures more applicable under certain constraints.

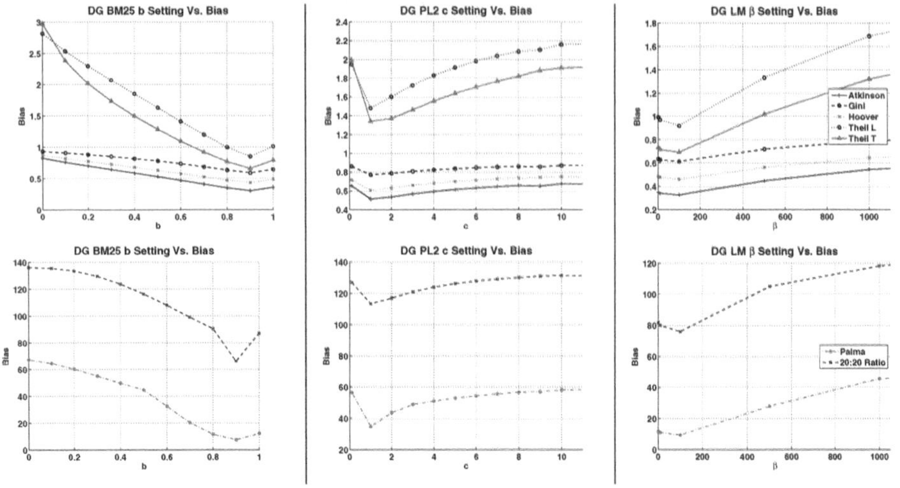

Fig. 2. Graphs for DG for each model, BM25 (Left), PL2(Middle) and LM (right). Atikinson, Gini, Hoover, Theil T and Theil L (Top) and Palma and 20:20 Ratio (Bottom) plotted along the parameter settings against the inequality. All curves follow the same point of minimum inequality. The separation is due to the different y-scales.

References

1. Atkinson, A.: On measurements of inequality. Journal of Economic Theory, 244–263 (1970)
2. Azzopardi, L., Vinay, V.: Accessibility in information retrieval. In: Macdonald, C., Ounis, I., Plachouras, V., Ruthven, I., White, R.W. (eds.) ECIR 2008. LNCS, vol. 4956, pp. 482–489. Springer, Heidelberg (2008)
3. Azzopardi, L., Vinay, V.: Retrievability: An evaluation measure for higher order information access tasks. In: Proc. of the 17th ACM CIKM, pp. 561–570 (2008)
4. Bashir,S.,Rauber,A.:Improvingretrievabilityandrecallbyautomaticcorpuspartitioning.In:Hameurlain,A.,Küng,J.,Wagner,R.,BachPedersen,T.,Tjoa,A.M.(eds.)TransactionsonLarge-ScaleData.LNCS,vol.6380,pp.122–140.Springer,Heidelberg(2010)
5. Gastwirth, J.: The estimation of the lorenz curve and gini index. The Review of Economics and Statistics 54, 306–316 (1972)
6. Hoover, E.: An Introduction to Regional Economics (1984)
7. Noor, S., Bashir, S.: Evaluating bias in retrieval systems for recall oriented documents retrieval (2013)
8. Palma, J.G.: Homogeneous middles vs. heterogeneous talls. Cambridge Working Papers in Economics (2011)
9. Theil, H.: Economics and Information Theory. North-Holland, Amsterdam (1967)
10. Wilkie, C., Azzopardi, L.: Relating retrievability, performance and length. In: Proc. of the 36th ACM SIGIR Conference, SIGIR 2013, pp. 937–940 (2013)
11. Wilkie, C., Azzopardi, L.: Best and fairest: An empirical analysis of retrieval system bias. In: de Rijke, M., Kenter, T., de Vries, A.P., Zhai, C., de Jong, F., Radinsky, K., Hofmann, K. (eds.) ECIR 2014. LNCS, vol. 8416, pp. 13–25. Springer, Heidelberg (2014)

Judging Relevance Using Magnitude Estimation

Eddy Maddalena[1], Stefano Mizzaro[1], Falk Scholer[2], and Andrew Turpin[3]

[1] University of Udine, Udine, Italy
{eddy.maddalena,mizzaro}@uniud.it
[2] RMIT University, Melbourne, Australia
falk.scholer@rmit.edu.au
[3] University of Melbourne, Melbourne, Australia
aturpin@unimelb.edu.au

Abstract. Magnitude estimation is a psychophysical scaling technique whereby numbers are assigned to stimuli to reflect the ratios of their perceived intensity. We report on a crowdsourcing experiment aimed at understanding if magnitude estimation can be used to gather reliable relevance judgements for documents, as is commonly required for test collection-based evaluation of information retrieval systems. Results on a small dataset show that: (i) magnitude estimation can produce relevance rankings that are consistent with more classical ordinal judgements; (ii) both an upper-bounded and an unbounded scale can be used effectively, though with some differences; (iii) the presentation order of the documents being judged has a limited effect, if any; and (iv) only a small number repeat judgements are required to obtain reliable magnitude estimation scores.

1 Introduction and Background

The gathering of document-level relevance judgements is a common activity in the evaluation of information retrieval (IR) systems. In evaluation campaigns such as TREC, relevance judgements were traditionally gathered on a binary (categorical) scale, although more recently ordinal scales, allowing for more fine-grained distinctions between relevance levels, have become more popular. While using ordinal scales is common practice, it seems natural to ask questions such as: how many levels should be used for an ordinal scale? Why not a continuous scale? Why not an unbounded scale?

In this paper we investigate the application of Magnitude estimation (ME) for the gathering of reliable relevance judgements. ME is a psychophysical scaling technique used to measure the intensity of stimuli. Respondents indicate the intensity of a stimulus through the assignment of a number. ME has been successfully applied for the scaling of both physical stimuli (such as the intensity of light and sound) and non-physical stimuli (including perceptions of the severity of crimes and punishments [7] or the usability of computer interfaces [3]). A key virtue of ME is that it results in a ratio scale of measurement [2], meaning that it is possible to carry out all mathematical operations (as compared to an ordinal scale, where for example the mean is not a meaningful measure of central tendency).

Applying the ME technique to measuring the relevance of documents therefore essentially involves assigning a number representing the perceived "amount of relevance"

A. Hanbury et al. (Eds.): ECIR 2015, LNCS 9022, pp. 215–220, 2015.

to a topic/document pair. In ME, usually an unbounded scale is used, such that the assigned numbers can be chosen from the range: $]0, +\infty[$. Since fractional numbers are allowed, a judge can never "run out" of values; there is always a higher number, or smaller fraction, to be used if the perception of the stimulus is higher or lower than what has been perceived previously. We also experiment with an upper-bounded version of the ME scale: $]0, +100[$, since a bounded scale may be more familiar to judges.

The application of ME to the problem of judging relevance has been limited to measuring the relevance of curated abstracts returned from a library cataloguing system [1], and for judging documents returned from a library database while carrying out a personal research project [6]. In contrast, we investigate the application of ME to the problem of judging the relevance of documents, as required for test collection-based evaluation of IR systems. In this paper we used workers recruited through the crowdsourcing platform CrowdFlower to perform a preliminary experiment aimed at understanding: (i) if ME judgements produce relevance rankings that are overall consistent with category judgements; (ii) whether there is a practical difference between using an (upper) bounded or an unbounded scale; (iii) if presentation order has an effect; and (iv) how many repeat judgements are needed to obtain stable ME scores.

2 Experimental Setup

To investigate whether ME is consistent with existing relevance judgements on an ordinal scale, we chose topics 351, 355, and 408 from the TREC-7 and 8 ad hoc tracks for our experiments. For these topics, a set of *expert judgements* made by carefully trained judges on a 4-level ordinal scale are available [5]: not relevant (`N`), marginally relevant (`M`), relevant (`R`), and highly relevant (`H`). To account for possible ordering and priming effects, we constructed four pre-defined templates of document relevance orderings based on the expert judgements: increasing (`NMRH`), decreasing (`HRMN`), non-relevant (`NNNN`), and medium (`MRMR`). To limit variability from document differences, the same documents of particular ordinal relevance levels were re-used where possible; therefore, for each topic, we selected 1 `H`, 3 `R`, 3 `M` and 4 `N` documents. The study was a between-subjects design, with each participant being asked to judge four documents for one given topic, presented using one of the four previously defined orderings. With three topics, four document relevance orderings, and two possible scales (bounded and unbounded) in total we had 3 x 4 x 2 = 24 experimental conditions. Each of these was repeated by 10 judges, for a total of 240 judges and 240 x 4 = 960 document judgements.

The experimental process involved each CrowdFlower judge being shown instructions; the TREC topic *title*, *description* and *narrative* fields; a simple initial question to test that the topic was understood correctly; and then the four documents, one at a time. The instructions were similar to those reported in the ME literature [1, 2] and are available in full at `http://www.cs.rmit.edu.au/~fscholer/ME/ECIR15`. Participants had to enter a numeric value in a text box shown under each document, and were also required to enter a short text justification. Each document was shown on a separate page, meaning that participants could not go back to revise their judgements. Each participant was paid $0.10.

3 Results and Discussion

Data Cleaning and Descriptive Statistics. The crowdsourced responses were filtered by removing the 27 judges that did not answer the initial test question correctly, and 1 judge that performed all judgements in a HRMN task in under 4 seconds. For a further 3 judges the results were not recorded due to a technical issue. This left 209 judges and 836 judgements for analysis. The breakdown of judges over the four orderings was: 55 for NMRH, 52 for HRMN, 53 for MRMR, and 49 for NNNN. The minimum, median, and maximum ME scores assigned were 1e-09, 4, and 2606203094 for the unbounded scale and 1e-73, 20, and 99 for the bounded scale.

There are several ways to analyse ME scores [4]; one option is to normalise them. Fundamentally, however, the idea of ME is that the ratios of the assigned magnitudes are meaningful. We therefore focus on the ratios of the raw scores.

Agreement with Expert Judgements: Ratios. Figure 1 shows the ratios of all ME relevance scores, categorised according to the expert-assigned ordinal relevance levels. Each column represents one ratio, in one of the document orderings (as depicted on the x-axis), and a dot in the column is the ratio value for a single judge. For example, if one judge in the NNNN task gave scores of `0.1 0.01 0.2 0.05`, this would generate points in the first column at ratios of $0.1/0.01$, $0.1/0.2$, and so on for all 12 combinations of the numbers. When the two document levels contributing to the ratio are the same, then both the ratio and its inverse are included; when the two document levels are different, then the highest ordinal relevance category forms the numerator and the lowest the denominator (so for example for NMRH not 12 but 6 ratios per judge are generated). Assuming that judges would assign a higher ME score to a document from a higher expert category, all ratios of this type should be greater than 1.

As can be seen in Figure 1, the median ratios for N/N, M/M and R/R are all one, and every other median ratio (except M/N in the HRMN condition) is above one, indicating that the median scores assigned by the judges were consistent in rank with the ordinal levels of the expert judges. There are many individuals, however, whose ratios fall below 1 when the two expert-assigned ordinal levels disagree. This indicates that either these judges were not performing the task in the same manner as the experts, or that the particular task has natural variance. One can see this by comparing the number of judges with ratios less than 1 for the R/M column in the MRMR condition, of which there are many, with the number less than 1 for the H/N condition, for which there are few. Intuitively we would expect the task of assigning a higher score to an H document than an N document to be easier than that of distinguishing R and M documents, and this is borne out in this data. (A possible further data filtering step could be to exclude judges who had a ratio of H/N less than 1, indicating that they scored the N document higher than the H document, but we have not done so in this paper.)

Agreement with Expert Judgements: Pairwise Swaps. To further investigate whether the rankings of document relevance obtained using our ME technique is consistent with the ordinal levels, we computed the proportion of pairwise judgements that agree with the order one would expect from the expert judgements, i.e., the percentage of dots in

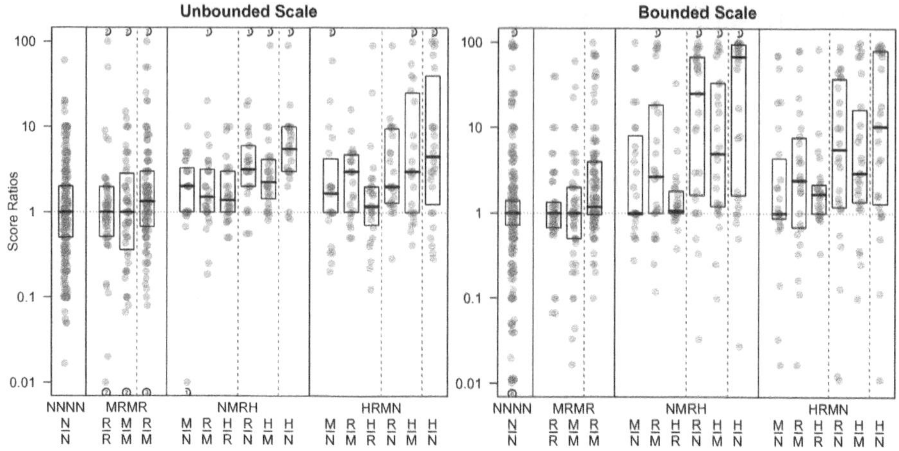

Fig. 1. Each dot represents a ratio between the ME scores assigned by a judge to documents in different expert-assigned ordinal levels. The x-axis describes the ratio depicted in each column, with the score for the higher category forming the numerator, and the lower category the denominator, thus scores higher than 1 agree with the expert ordering. Boxes show the median and inter-quartile range of each ratio. Ratios that fall outside the plot region, if any, are counted in the small circles at the upper and lower ends of each column. Vertical lines separate each condition, and dotted vertical lines separate sections of "distance" between the expert judgements underlying each ratio. Note that the y-axis shows a log scale.

each column of Figure 1 that stay above one. Table 1 shows the results. As expected from Figure 1, all are above 50%, except for `N <M` with the bounded scale, and in general all comparison rates increase as the gap between expert relevance levels widens.

Ordering and Scale Effects. The rates between the `NMRH` and `HRMN` conditions in Table 1 are largely consistent for each scale (bounded or unbounded), suggesting that there are no substantial ordering effects present. A Kolmogorov-Smirnov test between each column confirms that there is no statistically significant difference between the document ordering `NMRH` and `HRMN` ($p > 0.2$) in this data. However, it should be noted that this initial study only considered the judging of four documents, and that we cannot check for ordering effects in the `NNNN` and `MRMR` conditions. These are interesting avenues for future work.

There is no clear evidence to indicate whether using a bounded or an unbounded scale should be preferred. Perhaps from the results so far we could lean towards unbounded being more consistent with the expert judgements, as all medians are greater than one, and the inter-quartile ranges of the ratios (boxes in Figure 1) are smaller—much smaller when taking into account the log scale of the y-axis.

True Relevance Levels. Having determined that median ME scores are consistent with expert relevance assessments made using an ordinal scale, we can now investigate the actual magnitudes of these ratios. In particular, if we anchor the relevance score of

Table 1. The percentage of pairs of scores that are consistent in ordering with the expert rankings for the NMRH and HRMN categories

	NMRH		HRMN	
	Unbounded	Bounded	Unbounded	Bounded
N < M	54%	48%	56%	46%
M < R	68%	70%	70%	65%
R < H	68%	59%	59%	69%
N < R	86%	81%	67%	81%
M < H	82%	81%	85%	85%
N < H	89%	81%	70%	81%

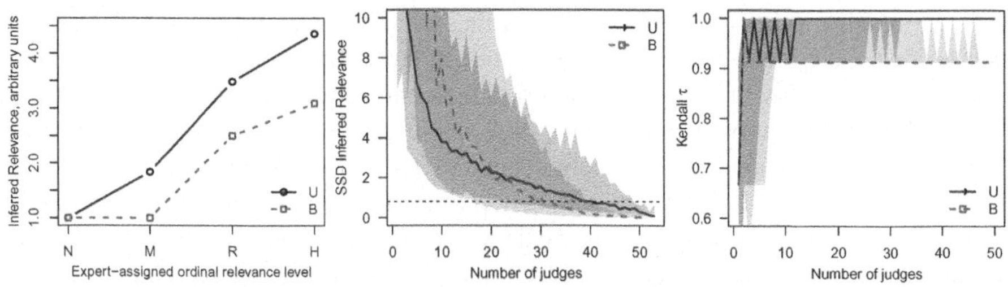

Fig. 2. Inferred relevance (left). Sum of squared differences (SSD) between the 3 inferred relevance levels M,R,H and median of judges (centre). Kendall's τ between the rank obtained by median of judges ratios and expert judgements (right).

an N document at an arbitrary level of 1, then we can infer the relative ME-assigned scores for the other ordinal levels using the median ratios from Figure 1. These are plotted in Figure 2 (left). It appears that based on unconstrained perceptions of the level of relevance, the N and M levels of the ordinal scale are much closer together than the M and R levels of the ordinal scale. This could have implications for situations where ordinal scales are folded down to binary categories (for example, to calculate effectiveness metrics such as MAP), strongly suggesting that the two lowest levels (not relevant and marginally relevant) should be combined, rather than treating not relevant as one category and folding the other three levels together as has often been done.

Number of Workers. An important consideration regarding the use of ME relevance assessments, is how many repeat measurements are required to obtain stable values. Figure 2 (centre) shows the sum of squared differences (SSD) between the median ratios of the inferred ME relevance scores for the M, R and H ordinal levels, for the number of judges shown on the x-axis and using all the judges. The two thicker lines show the median SSD, while the shading shows the 25th and 75th percentiles obtained from 1000 random permutations of the order of the judges in our data. It can be seen that the SSD falls quickly, and drops below 0.8 (the smalles gap between levels in the figure on the left for U) at about 40 for the Unbounded scale, and 30 using the Bounded scale. Since the figure aggregates data for the three topics, the median judgement appears to stabilise after about 10 to 13 judgements for each topic.

Figure 2 (right) examines the agreement (Kendall's τ) between the ranking of documents obtained using the median ratios from Figure 1 for the number of workers indicated on the x-axis and the expert judgements. That is, if the median ratio of document relevance levels x/y is greater than one, then x is ranked higher than y, and vice-versa. These stabilise after only 10 judgements, or around 3 per topic in our case: using just three judgments per topic provides a good agreement with expert judgments.

4 Conclusions and Future Work

Whether ME can be used to gather reliable relevance judgements is an interesting question that has been raised some time ago but, perhaps surprisingly, has no clear answer yet. Our results, although preliminary, hint that ME can be an effective technique for measuring the degree of relevance at the document level: ME-based judgements are reliable and consistent with categorical expert judgements; bounded and unbounded scales behave differently but both can be used; ordering effects were not found in our results; and not many crowdsourced judges are needed to get stable judgements.

This study can be considered as a first step towards understanding the effect of several parameter choices that need to be made when gathering ME relevance assessments. We are already working on a larger (more topics and more documents) and more complete study. Beyond the consideration of order effects, use of bounded or unbounded scale, and required number of judges, which we have examined here, the next study will focus on document presentation (one document per page, or all documents on one page?), learning effects (are the judges learning how to properly use a ratio scale, and therefore perhaps the first expressed judgements are less reliable than the last ones?), variations in the instructions, filtering of spam judges on the basis of their actions (such as time to express the judgements or the text-only comment). We believe that this research will also shed some light on classical ordinal relevance scales, and in particular the "true value" of each scale item.

References

[1] Eisenberg, M.: Measuring relevance judgements. Information Processing and Management 24, 373–389 (1988)
[2] Gescheider, G.: Psychophysics: The Fundamentals. Lawrence Erlbaum Associates, 3rd edn. (1997)
[3] McGee, M.: Usability magnitude estimation. Proceedings of the Human Factors and Ergonomics Society Annual Meeting 47(4), 691–695 (2003)
[4] Moskowitz, H.R.: Magnitude estimation: notes on what, how, when, and why to use it. Journal of Food Quality 1(3), 195–227 (1977)
[5] Sormunen, E.: Liberal relevance criteria of TREC: Counting on negligible documents? In: 25th SIGIR, pp. 324–330. ACM, New York (2002)
[6] Spink, A., Greisdorf, H.: Regions and levels: Measuring and mapping users' relevance judgments. JASIST 52(2), 161–173 (2001)
[7] Stevens, S.S.: A metric for the social consensus. Science 151(3710), 530–541 (1966)

Retrieving Time from Scanned Books

John Foley and James Allan

Center for Intelligent Information Retrieval
University of Massachusetts Amherst
Amherst, MA

Abstract. While millions of scanned books have become available in recent years, this vast collection of data remains under-utilized. Book search is often limited to summaries or metadata, and connecting information to primary sources can be a challenge.

Even though digital books provide rich historical information on all subjects, leveraging this data is difficult. To explore how we can access this historical information, we study the problem of identifying relevant times for a given query. That is – given a user query or a description of an event, we attempt to use historical sources to locate that event in time.

We use state-of-the-art NLP tools to identify and extract mentions of times present in our corpus, and then propose a number of models for organizing this historical information.

Since no truth data is readily available for our task, we automatically derive dated event descriptions from Wikipedia, leveraging the both the wisdom of the crowd and the wisdom of experts. Using 15,000 events from between the years 1000 and 1925 as queries, we evaluate our approach on a collection of 50,000 books from the Internet Archive. We discuss the tradeoffs between context, retrieval performance, and efficiency.

1 Introduction

With the growing number of digital libraries and the growing size of digital collections, a vast number of historical documents have been made publicly available. For example, the Internet Archive has over six million books available for free online[1]. These books are available in many languages, are from many cultures and time periods, and cover all subjects. Information retrieval in this broad historical domain is an important and interesting challenge. We believe that time is the key to success on many interesting tasks in this domain, and the first step to better use the historical information in these documents is to extract and predict times.

An example of a query with time *explicitly* specified is `lincoln april 14 1865` whereas a query for `lincoln assassination` would have that same temporal connotation, but with time *implicitly* included. In this work, we will refer to these kinds of information needs as "events" where the event here is the assassination of President Abraham Lincoln in 1865. There are many other details

[1] `https://archive.org`

A. Hanbury et al. (Eds.): ECIR 2015, LNCS 9022, pp. 221–232, 2015.

about this event, such as the name of theater, the assassin, and all of these details are part of this event which is often described in text. Given a query describing all or part of an event, we hope to retrieve a specific piece of information assigned to it: the date or set of dates on which it occurred.

In this work, we consider events to be the basis for information needs. Query log analysis from related work suggests that events, or at least temporally motivated queries are common in web search: Metzler et al. [18] report that queries with implicit temporal facets are close to 7% of web queries. Nunes et al. automatically detect temporal expressions in 1.5% of web queries and also report low recall [19].

In digital books in particular, preliminary findings suggest that temporal information is critical. The Hathi-Trust is an academic resource that allows full-text searching within books. A random sample of 600 queries from their logs shows that about 10% of these queries contain a date or time facet [25]. Although there is no analysis about what kinds of dates are included, this suggests that dates are more important for book search than for general web search, and that the percentage of queries for which there are unspecified times could be quite high.

In addition to looking at events with a single, obvious time point, we are interested in the case where a user only has a partially-specified information need, or they are interested in a class of events. Consider a user interested in the early history of Deerfield, Massachusetts, who might enter a query like `raid on deerfield`. The implicit time association of this query is ambiguous, whether the user is aware of that or not, and may refer to either of the following events:

February 29, 1704 French and Native American forces attacked the English frontier settlement at Deerfield, Massachusetts.

September 12, 1675 Deerfield was sacked and "the people as had not been butchered fled to Hatfield" [23, p. 272].

An ideal search system would be able to present the user with information to help understand and reformulate their search results with respect to time. Such a system would allow a user interested in the raids of Deerfield to consider either or both relevant dates in the collection through query reformulation.

In section 3, we discuss how we use state-of-the-art NLP tools to extract temporal information from our corpora. In section 4, we propose several models for organizing this extracted data as retrievable events, and unsupervised methods for predicting years from the highest scoring events. In section 5, we look at how to automatically derive a gold standard for this task. We discuss the results of our evaluation on 50,000 digitally-scanned books (16 million pages) in section 6. We show that our new, hybrid model is the most effective while resulting in about 14% smaller retrieval indices.

2 Related Work

This study is motivated in part by other work that addresses the task of selecting a time for queries when none is explicitly specified. Metzler et al. [18] identify the problem of recognizing queries that have implicit year qualification—for example, the name of a conference where there is often a user need for the particular year. They resolve the year ambiguity using an approach that mines query logs. We explore this issue for collections where this no query log available as well as where the range of potential years is substantially greater.

Campos et al. [3] look at date-tagging web queries by collecting the set of words that co-occur in snippets gathered from a commercial search engine containing candidate dates and choosing dates based on the most similar set.

Kanhabua and Nørvåg [12] address this challenge by using "time language models" directly with pseudo-relevance feedback (PRF), and indirectly using the publication dates of the PRF documents. Results from their intrinsic evaluation demonstrate that using publication dates significantly outperforms the language-model baseline. In contrast, we find that publication dates are not helpful in our larger and "messy" corpus.

2.1 Finding Times for Documents

Another line of research that is very similar to the task we study here is estimating dates for *documents*, typically aiming to identify the correct publication date. Kanhabua and Nørvåg [11] improve on de Jong et al.'s document date-finding techniques [6].

Kumar et al. build language models of years based on Wikipedia biography pages in order to estimate the focus time of other documents [14]. Jatowt et al. [9] use a large corpus of news documents to date a small corpus of Wikipedia, book, and web timeline events.

2.2 Other Uses of Time in IR

The value of time as a dimension for general information retrieval is well studied [1]. As one example, there is a lot of work that tries to leverage time expressions or time in order to improve retrieval [2,5,10]. These papers highlight the importance of having correct dates associated with a query or document, motivating our work to understand the best techniques for *finding* those dates when they are missing or suspect.

Much work on time expressions can be traced back to TimeML, an XML-based format for expressing explicit, implicit, and relative references to time in text. [21]. In this study, we focus on explicit references to time.

There is little work on retrieving times in archival or historical documents. Smith considers the task of detecting and browsing events, using documents with times manually annotated by historians [22]. In contrast, we explore approaches to a different task using only automatically annotated times.

Language modeling is a standard approach to general information retrieval tasks [20], as well as those in the time domain [15]. The basis of our approach is language modeling and the sequential dependence model [17] which incorporates term dependencies from adjacent query terms.

Question answering (QA) is an area of Information Retrieval that works toward constructing natural language responses for natural language questions. Often, the simplest technique applied to QA tasks is a form of passage or sentence retrieval [24], although much modern work is focused on creating and exploiting structured resources [7]. We choose to explore a similar task over unstructured documents in order to understand results based on primary sources.

3 Extracting Temporal Information

We ran the Stanford CoreNLP toolkit [16], version 3.3.1, on our book corpus, yielding sentence boundaries and date/time expressions (along with other annotations that we did not use for this study).

Since we were not investigating the task of extracting times, our processing decisions focused on precision over recall. To this end, we ignored all relative time expressions ("last year", "next Christmas") rather than introduce normalization errors. For the same reason, we kept only sentences that contained exactly one (absolute) time expression, minimizing the ambiguity by avoiding any sentences referring to multiple events.

While we wanted to use fine-grained time information, it was rare in this corpus, so we focused on years alone. While 71% of time expressions extracted had years, less than 20% of those had a day or a month included.

As an example of how the time expressions relate to the topics, consider Sylvester's work describing one of the Deerfield Massacres [23]. He mentions the year 1704 in 23 sentences, and Deerfield is mentioned in only five of them. The entire corpus mentions that year in 10,176 sentences. In all sentences with an absolute time reference, "Deerfield" is mentioned just 643 times.

4 Methods

4.1 Event Modeling for Retrieval

In this work we propose a number of ways to model events using our time-tagged documents. We describe these models, the intuition behind each, and how they were evaluated against an input query event.

Sentence-Event Model. The SENTENCE-EVENT model we propose uses the hypothesis that every sentence mentioning a time describes a unique event. To evaluate this model, we treat each sentence as a separate document and use state-of-the-art baseline retrieval methods to rank them in relation to our queries.

Document-Event Models. Another model to consider is one that assumes every *book* discusses a single event. This is our BOOK-EVENT model. Certainly, there are many books that fit this model, i.e. ones discussing a civil war battle in depth, but there many history books that cover numerous such events. As a result of this, we also considered a model that assumes every *page* of every book would describe an event, called the PAGE-EVENT model.

These models are similar to those used in systems like these that run on newswire collections. In such collections, the assumption that an article discusses a single event is more intuitively correct: since such publications are often much shorter and more focused. In related work, these models were much stronger than their counterparts in the books collection we used here.

To evaluate the BOOK-EVENT model and the PAGE-EVENT model, we treat each book or page as an independent document, and rank them.

Year-Event Models. Since our task involves predicting the year of an event as a query, it makes sense to try and model all the events within a single year directly, and simply use this aggregate model to predict the best year for each input query.

This approach was used by many others, as there has been substantial work proposing using time-based language models as a means of retrieving times or time expressions [6,11,12,14].

To evaluate our YEAR-EVENT models, we construct a language model from all the sentences mentioning a particular year, and rank the years by their similarity to the language model of our query events.

Book-Year-Event Models. The model we propose in this work unifies the intuition between the BOOK-EVENT and the YEAR-EVENT models. Since books are likely to be topically coherent, the assumption that all events corresponding to the same year will be similar is more likely to be valid within the context of a *single* book than across all books. This has the advantage of being between the too-few events of the YEAR-EVENT approach and the too-many events of the BOOK-EVENT, PAGE-EVENT, or especially SENTENCE-EVENT approach.

To evaluate this approach, we grouped our sentences containing unique, absolute time references into models by originating document and year pairs. These models were then ranked by similarity to the query events.

4.2 Year Ranking and Prediction

Regardless of the event-modeling framework, we need to rank our event models (and thus possible years) by the input event query. As we mentioned before, we use two popular, state-of-the-art baseline retrieval methods: query likelihood (QL) [20] and the sequential dependence model (SDM) [17].

QL is a unigram approach, like those that have been studied in the literature for retrieving relevant times [3,6,11,12]. The markov-random-field model of term dependencies in SDM is consistently a top performer in standard retrieval metrics across collections, so we present results using both techniques.

In evaluation of all of the models except the YEAR-EVENT models, we have the possibility for multiple event models to predict the same year. For example, with the YEAR-BOOK-EVENT models, we might retrieve two books discussing the sinking of the Titanic, with reference to the same year.

Although our initial model considers these as separate events, we can use the multiple-hypotheses generated by the highest-scoring models in order to improve our prediction.

To ensure applicability to the long-tail of history, we treated this problem of selecting years from high-scoring event models as an unsupervised re-ranking problem. We only discuss the most successful method here, briefly, for space reasons. *Reciprocal Rank Weighting* was used, which assigns every occurrence of a year a score equal to 1/rank, and sums them across all occurrences. That means that a year that occurs at ranks 1, 3 and 4 would achieve a score of $1 + 1/3 + 1/4$. This approach is based on the intuition that multiple occurrences are important, but less important as you travel down the ranked list.

4.3 Evaluation Metrics

We evaluate queries with a single relevant year with mean reciprocal rank (MRR). Since there is only one relevant year, it makes sense to use this metric as it directly measures the rank of the relevant document (year). MRR is a common evaluation for question answering [24], and fits this class of queries well.

We evaluate queries with multiple relevant years with normalized discounted cumulative gain, or NDCG [8]. In our experiments we only considered binary relevance: 1 if a year was related to that query, and 0 otherwise. The results we show are the mean NDCG across all queries. Evaluating with average precision (AP) gave us similar results, so we do not include it here.

5 Collecting Queries

This work is based on the idea that it is valuable to know the time—year or years, specifically—that are related to a query. In a typical retrieval task, a system's job is to rank documents in an order that reflects the chance that they are relevant to the information need. In contrast, in this study our task is, given a query, to rank *years* by the chance that they are relevant to the information need.

To understand which approaches are most effective at finding the correct years, we need queries and corresponding years. The question answering corpora from past community evaluations [24,4] include a small number of questions that have year as an answer. Unfortunately, there are only a handful of such questions and even fewer that overlap with the time periods of our document sets. (Most focus on modern news or Web corpora.)

To create a large number of queries we turn to Wikipedia. Nearly every year has a "year page" within Wikipedia that lists events, births, and deaths within that year, typically with references to Wikipedia pages with additional details. For example, as of Summer 2014, the year page for 1704 (`http://en.wikipedia.org/wiki/1704`) lists 13 events with a specific month or date (e.g., "September: War of the Spanish Succession" and "February 29: Raid on Deerfield (Queen Anne's War): French-Canadians and Native Americans sack Deerfield, Massachusetts, killing over 50 English colonists."), 9 events with unknown date within the year (e.g., "Isaac Newton publishes his *Opticks*"), 14 births, and 20 deaths.

For our purposes, we only consider the "Events" sections of the Wikipedia year pages. This means that we discard all dates of birth or death. We also ignored any year pages 2014 and higher, which are the future of our June 2013 english XML dump.[2].

5.1 Queries with a Single Relevant Year

We converted all Wikipedia markup into plain text and extracted all event entries. We removed any facts that were made up entirely of stop words[3] and we explicitly removed from the entry any numbers that could be year or day references. We also removed the mention of months for those entries that had them. Our goal in that processing was to remove all mentions of dates other than the entry's corresponding year, which was used only as the relevance judgment for that entry as a query.

That processing resulted in 40,356 facts with associated years, spanning 560 B.C. through A.D. 2013. Table 1 shows some example events. For the one-year task, where the goal is to select the correct single year for a query, we used the event description directly without further processing.

Based on the domain of years extracted from our test collection, we down-sampled these years to only those that were actually discussed in our books: roughly 1000-1925 A.D.

Table 1. Example queries with a single relevant year

Year	Fact
1178	The Sung Document is written, detailing the discovery of "Mu-Lan-Pi" (suggested by some to be California) by Muslim sailors.
1298	Residents of Riga and the Grand Duchy of Lithuania defeat the Livonian Order in the Battle of Turaida
1535	Manco Inca Yupanqui, nominally Sapa Inca, is imprisoned by the Spanish Conquistadors of Peru.
1704	French-Canadians and Native Americans sack Deerfield, Massachusetts.
1733	British colonist James Oglethorpe founds Savannah, Georgia.

5.2 Queries with Multiple Relevant Years

Some queries are ambiguous with respect to time, an issue we aim to also explore in this work. In order to have queries that were relevant to a *set* of years, we merged similar queries from different years as follows.

As mentioned previously, the events in the Wikipedia year pages contain links to articles the discuss the entities involved in an event. If two events from different years link to exactly the same pages, then then there is temporal ambiguity regarding that collection of pages. We therefore grouped all one-year queries by

[2] `http://dumps.wikimedia.org/enwiki`

[3] We used the Lemur 418 stopword list from `http://lemurproject.org/`

co-occurring links, creating one query from many. For example, consider the following three one-year queries from Wikipedia year pages, with links to Wikipedia articles shown in small caps:

- (1221) The MAYA of the YUCATÁN revolt against the rulers of CHICHEN ITZA.
- (1528) The MAYA peoples drive SPANISH CONQUISTADORES out of YUCATÁN.
- (1848) The Independent Republic of YUCATÁN joins Mexico in exchange for Mexican help in suppressing a revolt by the indigenous MAYA population.

Those three entries all mention "Maya" and "Yucatán" so we join them together. To generate a query, we selected just their common words — in this case, the query *maya, yucatán*. Table 2 shows several other resulting queries and their multiple relevant years. Although not all these keyword intersections will be meaningful, the large number of queries generated allows for comparison of techniques even in the presence of noise. Certainly not all user-queries would be meaningful, either. As the random sample in this table shows, the majority of these generated queries had only two relevant years, although a few had more.

Table 2. Example queries with a multiple relevant years

Years	Shared Terms
1221, 1528, 1848	yucatán, maya
1862, 1863	battle, general, ambrose, confederate, civil, war, american, union, burnside
1700, 1721	pope, xi, succeeds, innocent, clement
1380, 1382	horde, tokhtamysh, blue, khan, golden, mamai
1916, 1821	republic, colombia, venezuela
1588, 1577	spanish, plymouth, francis, drake

6 Results and Discussion

This set of experiments was run on a collection of 50,228 scanned books taken from the the Internet Archive,[4]. Rather than select books at random, we chose to use the books selected by the INEX book track [13], to simplify reproducibility. In order to generalize to millions of other books available online, we did not use any of the structured XML information provided for those challenges.

Some of the books in the collection are quite large. On average, there are 86,871 terms in a book, and about 270 terms on each page. There are 16 million pages in this collection. 10.2 million sentences with absolute year references were extracted by the tagger.

15,739 single year queries were generated and were distributed evenly by year. 3,235 queries with multiple relevant years were distributed randomly. Reranking was tuned on 1/3 of the queries, and the other 2/3 was used for evaluation.

[4] https://archive.org

6.1 Results

Our results for single-year queries on the books corpus are shown in Figure 1. We find that the YEAR-BOOK-EVENT and the SENTENCE-EVENT models were most effective. Of notice is that the YEAR-EVENT model performed poorly in comparison on queries with a single relevant year, suggesting that events occurring in the same year suffer from being being lumped together, such that all events are included and rare events may be overshadowed by the countless mentions of popular events that year.

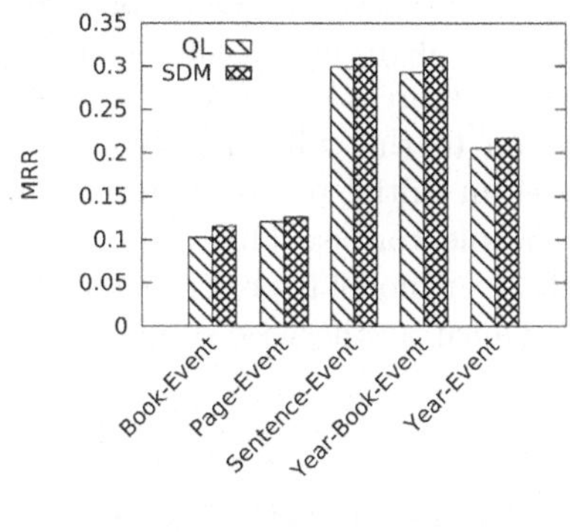

Fig. 1. MRR on queries with 1 relevant year

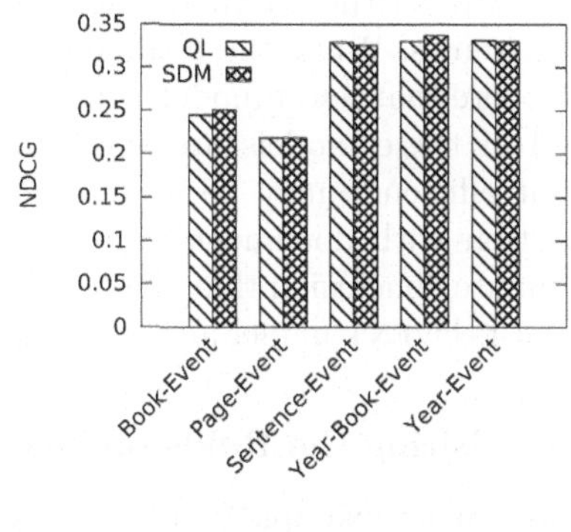

Fig. 2. NDCG on queries with two or more relevant years

While document neighbors were found to be effective in related work, both BOOK-EVENT and PAGE-EVENT performed poorly in comparison to the other models. The poor performance of BOOK-EVENT was to be expected, as there are 188 years mentioned per book on average, and so it makes sense that the intuition of one event per book is incorrect, however, the poor performance of PAGE-EVENT is more surprising, as there was only a single date on every other page on average.

We note that when there is a single mention of a year on a page, the year is represented by its containing sentence in SENTENCE-EVENT but by all text on the page in PAGE-EVENT. We hypothesize that this extra material causes spurious words to have high probability for the year, a problem avoided by the more focused SENTENCE-EVENT model.

Figure 2 shows the NDCG results for the queries with multiple relevant dates. In this case recall is slightly more important and having broad coverage of topics matters. The YEAR-BOOK-EVENT and SENTENCE-EVENT models continue to perform well on this task, but the YEAR-EVENT model roughly matches them.

In almost all cases, the proximity features included in the sequential dependence model (SDM) improved results, except in the many relevant year case with the SENTENCE-EVENT model, although those documents are so short that if two terms occur, they are already close, and the more general language of those queries might be causing problems.

Across both tasks, we find that YEAR-BOOK-EVENT and SENTENCE-EVENT perform the best overall, and that performance-wise, neither seems to have a clear advantage.

6.2 Considering Efficiency

One of the advantages of the YEAR-BOOK-EVENT is that it offers an interesting tradeoff not only in terms of effectiveness, but also in size. The inverted index for the YEAR-EVENT is the most efficient, occupying only 227 MiB on disk, whereas the postings for the SENTENCE-EVENT is over twice the size: 553 MiB. The 477 MiB of the YEAR-BOOK-EVENT postings provides a tradeoff between those extremes. While these three models were all built from the same context, we note here that fewer models with more context compress better on disk. Note that all of these models are small in comparison to the collection (90 GiB), but that efficiency may be a concern if used as an initial retrieval step or in conjunction with document retrieval over the whole corpus. This leads us to the ultimate conclusion that the YEAR-BOOK-EVENT model is preferable to the SENTENCE-EVENT model although their performance is otherwise similar.

6.3 Revisiting the Raids on Deerfield

The motivating example in the introduction was a user interested in raids on the town of Deerfield, Massachusetts, during the colonial era. We revisit this query directly on our best performing models to give us a concrete sense of the strengths and weaknesses of each. The two relevant years are 1675 and 1704.

Issuing this query against the YEAR-EVENT model, we find 1704 at rank 2, and 1675 at rank 10. The rank 1 result is a false positive created by text in the margin next to a sentence summarizing the 1704 raid. This demonstrates how even though the YEAR-EVENT might be an efficient and simple approach, it suffers when looking for specific events, as it is susceptible to more frequent mentions of Deerfield among other years.

The SENTENCE-EVENT approach is more robust, giving a topical result first: a 1713 attempt to recover prisoners, and then the 1704 raid itself in multiple appearances, with the 1675 raid buried under the several pages of results (it is only rarely mentioned in our corpus).

The results from the YEAR-BOOK-EVENT are more balanced. We still get 1713, 1704, and the false positive of 1602, but the 1675 raid appears at rank 10, after a series of mostly true positives, which suggests that of all the models, for this query, it was most able to balance the precision of avoiding the noise and breadth of the books corpus with the recall of still being able to retrieve rare events.

7 Conclusion

We present models for using historical data to predict the year of a query. Unlike past work in the newswire domain, find that that year modeling is not performant for our task. The leading techniques are a nearest sentence approach and a joint year-document model. While the nearest sentence approach is generally as efficient as the more complicated model, it is more expensive in terms of space and in terms of the number of documents to rank. Altogether, we conclude that the YEAR-BOOK-EVENT model is prefered for this task.

In addition to our experimental results, we contribute an automatically collected set of over 40,000 queries tied to a single year. We also described a mechanism for creating merged "under-specified" queries with multiple years as their target. This dataset is publicly available[5].

We believe that the promising results shown by our joint document and year event models suggest applicability for general entity models. The authors hope that work in this area will begin to unlock the possibilities of using the millions of digital books available online.

Acknowledgments. This work was supported in part by the Center for Intelligent Information Retrieval and in part by NSF grant #IIS-0910884. Any opinions, findings and conclusions or recommendations expressed in this material are those of the authors and do not necessarily reflect those of the sponsor.

References

1. Alonso, O., Strötgen, J., Baeza-Yates, R.A., Gertz, M.: Temporal information retrieval: Challenges and opportunities. TWAW 11, 1–8 (2011)
2. Brucato, M., Montesi, D.: Metric spaces for temporal information retrieval. In: de Rijke, M., Kenter, T., de Vries, A.P., Zhai, C., de Jong, F., Radinsky, K., Hofmann, K. (eds.) ECIR 2014. LNCS, vol. 8416, pp. 385–397. Springer, Heidelberg (2014)
3. Campos, R., Dias, G., Jorge, A., Nunes, C.: Gte: A distributional second-order co-occurrence approach to improve the identification of top relevant dates in web snippets. In: CIKM 2012, New York, NY, USA, pp. 2035–2039 (2012)
4. Dang, H.T., Owczarzak, K.: Overview of the TAC 2008 opinion question answering and summarization tasks. In: Proc. of the First Text Analysis Conference (2008)
5. Daoud, M., Huang, J.: Exploiting temporal term specificity into a probabilistic ranking model (2011)
6. Jong, F.d., Rode, H., Hiemstra, D.: Temporal language models for the disclosure of historical text. Royal Netherlands Academy of Arts and Sciences (2005)
7. Hoffart, J., Suchanek, F.M., Berberich, K., Weikum, G.: Yago2: A spatially and temporally enhanced knowledge base from wikipedia. Artificial Intelligence 194, 28–61 (2013)
8. Järvelin, K., Kekäläinen, J.: IR evaluation methods for retrieving highly relevant documents. In: SIGIR 2000, pp. 41–48. ACM (2000)
9. Jatowt, A., Au Yeung, C.-M., Tanaka, K.: Estimating document focus time. In: CIKM 2013, pp. 2273–2278. ACM, New York (2013)
10. Kanhabua, N., Nørvåg, K.: A comparison of time-aware ranking methods. In: SIGIR 2011, pp. 1257–1258 (2011)

[5] `http://ciir.cs.umass.edu/downloads/`

11. Kanhabua, N., Nørvåg, K.: Improving temporal language models for determining time of non-timestamped documents. In: Christensen-Dalsgaard, B., Castelli, D., Ammitzbøll Jurik, B., Lippincott, J. (eds.) ECDL 2008. LNCS, vol. 5173, pp. 358–370. Springer, Heidelberg (2008)
12. Kanhabua, N., Nørvåg, K.: Determining time of queries for re-ranking search results. In: Lalmas, M., Jose, J., Rauber, A., Sebastiani, F., Frommholz, I. (eds.) ECDL 2010. LNCS, vol. 6273, pp. 261–272. Springer, Heidelberg (2010)
13. Kazai, G., Koolen, M., Kamps, J., Doucet, A., Landoni, M.: Overview of the INEX 2010 book track: Scaling up the evaluation using crowdsourcing. In: Geva, S., Kamps, J., Schenkel, R., Trotman, A. (eds.) INEX 2010. LNCS, vol. 6932, pp. 98–117. Springer, Heidelberg (2011)
14. Kumar, A., Baldridge, J., Lease, M., Ghosh, J.: Dating texts without explicit temporal cues. arXiv preprint arXiv:1211.2290 (2012)
15. Li, X., Croft, W.B.: Time-based language models. In: CIKM 2003, pp. 469–475. ACM (2003)
16. Manning, C.D., Surdeanu, M., Bauer, J., Finkel, J., Bethard, S.J., McClosky, D.: The Stanford CoreNLP natural language processing toolkit. In: Proceedings of 52nd Annual Meeting of the Association for Computational Linguistics: System Demonstrations, pp. 55–60 (2014)
17. Metzler, D., Croft, W.B.: A markov random field model for term dependencies. In: SIGIR 2005, pp. 472–479. ACM (2005)
18. Metzler, D., Jones, R., Peng, F., Zhang, R.: Improving search relevance for implicitly temporal queries. In: SIGIR 2009, pp. 700–701. ACM (2009)
19. Nunes, S., Ribeiro, C., David, G.: Use of temporal expressions in web search. In: Macdonald, C., Ounis, I., Plachouras, V., Ruthven, I., White, R.W. (eds.) ECIR 2008. LNCS, vol. 4956, pp. 580–584. Springer, Heidelberg (2008)
20. Ponte, J.M., Croft, W.B.: A language modeling approach to information retrieval. In: SIGIR 1998, pp. 275–281. ACM (1998)
21. Pustejovsky, J., Castano, J.M., Ingria, R., Sauri, R., Gaizauskas, R.J., Setzer, A., Katz, G., Radev, D.R.: TimeML: Robust specification of event and temporal expressions in text. New Directions in Question Answering 3, 28–34 (2003)
22. Smith, D.A.: Detecting and browsing events in unstructured text. In: SIGIR 2002, pp. 73–80. ACM (2002)
23. Sylvester, H.M.: Indian Wars of New England, vol. 2 (1910), `https://archive.org/details/indianwarsneweng02sylvrich`
24. Voorhees, E.M., et al.: The TREC-8 Question Answering Track Report. TREC 99, 77–82 (1999)
25. Willis, C., Efron, M.: Finding information in books: characteristics of full-text searches in a collection of 10 million books. Proceedings of the American Society for Information Science and Technology 50(1), 1–10 (2013)

A Noise-Filtering Approach for Spatio-temporal Event Detection in Social Media

Yuan Liang, James Caverlee, and Cheng Cao

Department of Computer Science and Engineering, Texas A&M University
College Station, Texas, USA
{yliang,caverlee,chengcao}@cse.tamu.edu

Abstract. We propose an iterative spatial-temporal mining algorithm for identifying and extracting events from social media. One of the key aspects of the proposed algorithm is a signal processing-inspired approach for viewing spatial-temporal term occurrences as signals, analyzing the noise contained in the signals, and applying noise filters to improve the quality of event extraction from these signals. The iterative event mining algorithm alternately clusters terms and then generates new filters based on the results of clustering. Through experiments on ten Twitter data sets, we find improved event retrieval compared to two baselines.

1 Introduction

As users of services like Twitter and Facebook react to and report on their experiences – like political debates, earthquakes, and other real-world events – there is an opportunity for large-scale mining of these *socially sensed* events, leading to services that support intelligent emergency monitoring, finding nearby activities (e.g., rallies), and improving access to online content [5,14,20,29]. While there has been a long history of *event extraction* from traditional media like news articles, e.g., [1,25], the growth of user-contributed and often *on-the-ground* reaction by regular social media users has begun to spark new approaches.

In general, existing event detection methods can be categorized into two types: *document-pivot* approaches and *feature-pivot* approaches [7]. Document-pivot approaches identify events by clustering documents (e.g., news articles) based on semantic similarity, and then treating each cluster as an event. A series of works like [9,23] have shown the effectiveness of this method over long-form documents like news articles, which typically provide a rich source of context for event detection. Social media content, in contrast, often provides only a short description, title, or tags, (and thereby little textual narrative) limiting the effectiveness of semantic similarity based event detection techniques. As a result, many social media event detection algorithms have relied on *feature-pivot approaches*, which group similar event-related terms, for example by finding terms with a similar temporal distribution. In this way, event-related terms may be clustered together based on these common signals (treating each term as a frequency function over either time or space). These feature-pivot approaches,

A. Hanbury et al. (Eds.): ECIR 2015, LNCS 9022, pp. 233–244, 2015.

e.g., [3,4,29], have shown the potential of this approach for scaling to event detection over user-contributed social media posts.

While encouraging, these approaches may be susceptible to *noise* in both the temporal and spatial signals they use, which can hinder the quality of event detection. For example, topics not directly related to a specific event may introduce noise (e.g., discussion of a political candidate that is unrelated to a specific rally), as well as related but different events (e.g., reports of tornados in one city may pollute the signal of tornados in another city), and by data sparsity, in-correct timestamps or locations, mislabeled geo-coordinates, and so on.

Hence, we explore in this paper a *signal-processing* inspired event detection framework designed to target these sources of noise. We view spatial-temporal term occurrences as signals, analyze their noise, and apply filters to improve the quality of event extraction from these signals. We incorporate this noise-filtering approach into an iterative spatial-temporal event mining algorithm for identifying and extracting events from social media. This approach alternately clusters terms using their filtered signals, and then generates new filters based on the results of clustering. Over ten Twitter-based event datasets – we find that the noise filtering method results in a 7-10% improvement versus alternatives.

2 Related Work

Event detection refers to the discovery of a specific activity that happens at a certain time and in a certain place. Event detection is typically categorized into two types: retrospective detection and on-line detection [25]. The former is to detect events from collected historical documents [15,13], and the latter tries to extract events from real-time documents [1,24,8]. Early detection approaches usually adopted clustering methods based on document similarity, e.g., [1] used a modified version of TF/IDF to measure the distance of documents. [25] added a time window and a decay factor for the similarity measurement between documents. In this paper, we focus on retrospective detection where the collection consists of user-generated content in social media.

User-generated content in social media has different characteristics from traditional document collections, so many clustering approaches have considered event-related metadata rather than directly measuring semantic relatedness. For instance, the work in [30] detects events from click-through web data by considering each event as a set of query-page pairs. In [14], a tweet is segmented into pieces and Wikipedia is exploited for identifying events. Via co-occurrence, [18] and [23] measured closeness of tags for landmark detection and tag recommendation. [21] constructed a keyword graph where co-occurrence frequency was used to assign weights on edges and then applied a shortest path based scheme to do community detection. [2] considered graph structure to bind all associated heterogeneous metadata, and proposed a co-clustering scheme to partition them into different events.

Separately, many approaches have adopted learning-based methods, including [4,5,11] or focused on temporal and spatial features. [17] utilized temporal

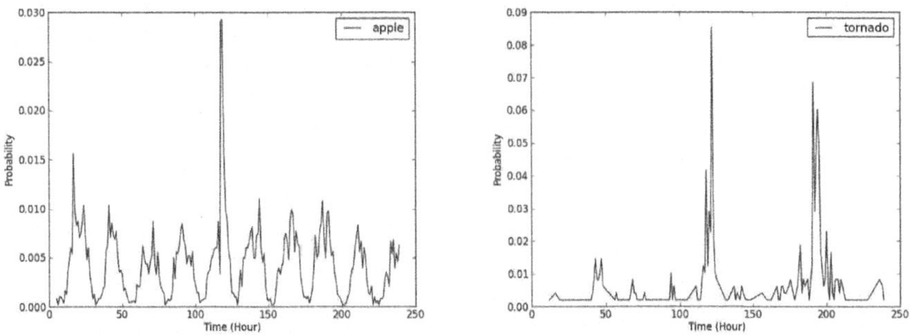

(a) Temporal Distribution for "apple" (b) Temporal Distribution for "Tornado"

Fig. 1. Examples of Noise in Term Temporal Signals

information to determine a set of bursty features in different time windows, and then detected bursty events based on the feature distributions. [7] observed the spatial-temporal patterns for tags, and adopted a wavelet transform-based method to find tags with significant peaks in spatial-temporal distribution. Similarly, [19] looked for tags with bursts in temporal and spatial patterns for event detection. [29] compared spatial-temporal distributions between terms as the measurement of the closeness of different terms, and clustered terms based on the distances to extract events. At the same time, efforts such as [12,16,22,26,27,28] integrate geo-location information, showing the potential of spatial features.

3 Noise-Aware Event Detection

Given a collection of user-contributed social media documents $D = \{d_1, d_2, ..., d_T\}$, each document d_i can be viewed as $\langle W, t, l \rangle$, where W is a list of terms from vocabulary V, t is a timestamp indicating when d_i was posted, and $l = (la, lo)$ is the associated geo-location, consisting of latitude and longitude coordinates. We assume that there are K events $\theta = \{\theta_1, ..., \theta_K\}$ hidden in D and each document belongs to one of these events. Our goal is to detect these K hidden events from the observed documents. For our purposes, an *event* refers to a specific activity that happens in a specific time and place [7]. Therefore, given a group of terms, if it represents an event, the group of terms should satisfy three constraints: 1) the terms are semantically consistent, 2) the terms should happen in the same time period, and 3) the terms should appear in similar locations. Hence, we define event detection as: given a set of terms S, to detect subsets from S so that each subset $S_k \in S$ is a set of terms satisfying these constraints.

We propose to tackle event detection from a signal-processing perspective, where terms may be viewed as signals. For example, we could view a single term as a sequence of (normalized) counts for every minute of the day, resulting in a *temporal time signal*. That is, term w_i is represented by a temporal sequence of counts: $F_{t,w_i} = \{f_{i,1}, f_{i,2}, ..., f_{i,T}\}$, where t denotes the *temporal* signal domain.

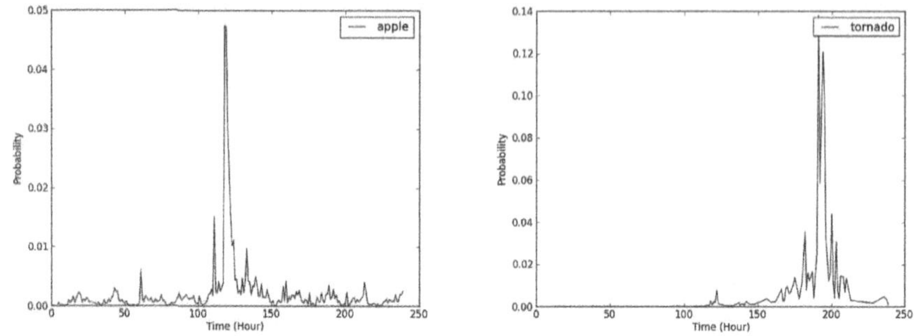

(a) Temporal Distribution for "apple" (b) Temporal Distribution for "Tornado"

Fig. 2. Temporal Signals After Filtering Using the Proposed Method

Similarly, we could view a term as a two-dimensional *spatial term signal* by bucketing terms into a grid over the latitude-longitude space (denoted as F_{l,w_i} for a term w_i in the *location* signal domain). Both perspectives can additionally be merged into a three-dimensional *spatial-temporal term signal*, denoted by F_{t,l,w_i}. Together, we view the overall event signal corresponding to event θ_k as an aggregation of the signals of the terms belong to event θ_k. Hence, given a set of terms S_k associated with event θ_k, the event signal is:

$$F_{t,l,\theta_k} = \sum_{E(w_i)=\theta_k} F_{t,l,w_i} \lambda_{w_i,\theta_k} \quad (1)$$

where $E(w_i)$ refers to the corresponding event of w_i and $\lambda_{w_i,\theta}$ is the weight of w_i. Unfortunately, these event signals are necessarily *noisy*, meaning the detection faces significant challenges. We broadly classify three prominent types of noise:
Background-topic noise refers to the signals caused by unrelated topics to the event of interest, but that may overlap with the event of interest. For example, background discussion of "apple" as in Figure 1(a), which is unrelated to a major Apple announcement (the spike of attention).
Multi-event noise refers to the burst signal caused by other unrelated events. A term w_i can belong to multiple events, so its spatial-temporal signals are actually the combination of signals from multiple events, i.e., $F_{t,l,w_i} = \sum_k F_{t,l,w_i,\theta_k}$. For example, Figure 1(b) shows two tornado events.
Random noise refers to the random signals introduced by the sparsity of data, in-correct timestamps or locations, mislabeled geo-coordinates, and so on.

3.1 An Iterative Event Extraction Method

With these challenges in mind, we propose an iterative noise-aware event extraction method that seeks to limit the impact of noise. Concretely, we view that the term signals F_{t,l,w_i} for w_i are comprised of three components: (i) the event signal of interest F_{t,l,w_i,θ_e}; (ii) random noise F_{t,l,w_i,θ_r}; and (iii) event noise $F_{t,l,w_i,\theta_{S-e}}$, where S is the set of all the events: $F_{t,l,w_i} = F_{t,l,w_i,\theta_e} + F_{t,l,w_i,\theta_{S-e}} + F_{t,l,w_i,\theta_r}$

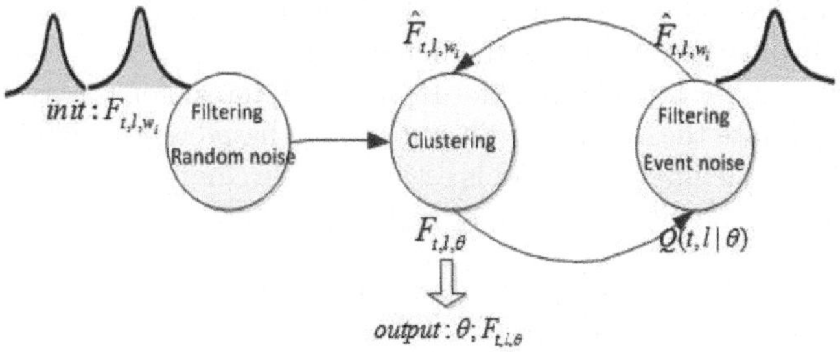

Fig. 3. Structure of Iterative Event Extraction Method

Our goal is to estimate the event signals F_{t,l,w_i,θ_e}, in effect cleaning the signal to focus primarily on the event of interest as illustrated in Figure 2. The overall approach is shown in Figure 3, where term signals are first filtered of random noise and then the signals are repeatedly clustered and filtered of event noise, until a final set of events is identified.

Filtering Random Noise. We begin with the first filter, for reducing random noise from the term signals. In speech and image processing, the *mean filter* is an effective way to smooth the signal and reduce un-correlated random noise [10]. In our context, we also assume that the random noise contained in the term signals are un-correlated, and therefore we can directly apply the mean filter to the signals. The key point of a mean filter is using the neighbors to average the signal values. For every point in the signals, the value is smoothed by:

$$F'_{t,l,w_i} = \sum_{t' \in N(t), l' \in N(l)} F_{t,l,w_i} Q(t', l') \qquad (2)$$

For the mean filter, $Q(t', l')$ is set with $1/M$, where M is the number of neighbors, $N(t)$ refers to the set of neighbor points of t. A neighbor here is the point with adjacent time unit to t and close location to $l = (la, lo)$. For example, if we define $N(t) = [t-2, t+2]$ and $N(l) = [l-2, l+2]$, then all the points within 2 time units and 2 "distance" units (which could correspond to kilometers) at (t, la, lo) are regarded as the neighbors of the unit of (t, l).

Filtering Event Noise. After filtering random noise, we alternately cluster terms using their filtered signals, and then generate new filters based on the results of clustering, toward identifying groups of event-related terms. For the initial clustering, we adopt an existing co-occurrence based method [6] to group related term signals; alternately, other clustering methods could also be applied. These clusters could be immediately viewed as events, but for the inherent multi-event and background noise in the signals. Hence, we adopt a *band-pass filter* to limit the impact of these sources of noise. The intuition of the band-pass filter is to pass the signals in a Region-of-Interest, but filter or reduce the signals in other regions. After applying the band-pass filter, the cleaned term signals are clustered again. This iterative clustering and noise filtering proceed until the clusters of terms do not change or the iteration count reaches a threshold. Finally, we output the clusters as the detected events.

The key issues are how to find the Region-of-Interest for a particular event, and how to estimate the band-pass filter $Q(t,l|\theta_k)$ based on the detected Region-of-Interest. Once the filter $Q(t,l|\theta_k)$ is estimated, we can use F_{t,l,w_i} and $Q(t,l|\theta_k)$ to retrieve the signals belonging to θ_k with Equation 3:

$$F_{t,l,w_i,\theta_k} = F_{t,l,w_i} Q(t,l|\theta_k) \tag{3}$$

where $Q(t,l|\theta_k)$ is the band-pass filter for θ_k in the spatial-temporal domain.

To detect the Region-of-Interest for a certain event θ_k, we propose to aggregate all the signals of the terms belonging to event θ_k, and then label the region which contains the strongest signals as the Region-of-Interest. The idea behind this method is to use the neighbors to filter un-correlated noises and strengthen the signals belonging to θ_k. In signal processing, mean filtering is used to sum multiple polluted signals. For example, if $s_1, s_2, ..., s_K$ are K different samples of the signal s polluted by noise, then the mean filter uses $\lambda_1 s_1 + \lambda_2 s_2 + ... + \lambda_K s_K$, $(\lambda_1 + \lambda_2 + ... + \lambda_K = 1)$ to find the un-polluted signal s. If the noise and signal are un-correlated, then by increasing K, the strength of the noise will be reduced to $1/\sqrt{K}$ [10]. Here, since individual terms can be polluted by some event noises which are usually uncorrelated, by averaging the signals of term w_i with the signals of its neighbors, the noise introduced by different events will be reduced.

Unlike the neighbors for random noise filtering which are found based on the adjacent time unit or spatial grid, the neighbors here refer to the terms belonging to the same event as determined by the clustering component. We first use a clustering method to find the neighbors for term w_i, then the signals belonging to the same cluster are averaged using Equation 1 to arrive at the estimated event signals. Regarding the clustering method, *k-means* is adopted in this paper if the number of actual clusters is already known, and *Affinity Propagation* is used if it is unknown.

We consider several different band-pass filters to explore their appropriateness for event detection from social media: a Gaussian band-pass filter, an Ideal band-pass filter, and an average band-pass filter.

Gaussian Band-Pass Filter: In the Gaussian filter, we assume that $Q(t,l|\theta_k)$ for θ_k can be represented as a single Gaussian. Then we use the event signals F_{t,l,θ_k} to train the parameters of $Q(t,l|\theta_k)$ where x is the vector of $\langle t,l\rangle$:

$$Q(t,l|\theta_k) = \frac{1}{\sigma\sqrt{2\pi}} exp\{-\frac{(x-\mu)^2}{2\sigma^2}\} \tag{4}$$

Ideal Band-Pass Filter: In the Ideal filter, we assume each point in the region (where the center is the point with strongest signal) has a weight much larger than points outside the region.

$$Q(t,l|\theta_k) = \begin{cases} \frac{\lambda}{r} & x \in [x_u, x_d] \\ \eta * \frac{1-\lambda}{R-r} & else \end{cases} \tag{5}$$

where λ is the cumulative frequency probability of the region $[x_u, x_d]$, x_u and x_d are the left-up and right-down coordinators respectively. r is the area of the region, R is the whole area of the boundary, and $\eta = 0.1$ is a penalty factor.

Average Band-Pass Filter: In the Average filter, $Q(t, l|\theta_k)$ the λ_{w_i,θ_k} in Equation 1 is set with $1/N$, where N is the number of terms belonging to θ_k.

4 Experiments

In this section, we evaluate the effectiveness of the proposed filter-based method for event extraction. We first investigate the impact of noise filtering and then compare the quality of the proposed approach versus two alternatives.

4.1 Data Collection

Our experiments are over ten different tweet datasets containing multiple events each (as shown in Table 1). We manually selected 20 events from Wikipedia between February 2011 to February 2013 and grouped them into six categories: *seasonal*, *burst*, *long-term*, *short-term*, *global area* and *local area*. An event may belong to more than one category, e.g., Christmas Eve can be in seasonal, short-term, and global. For each category, we manually select 10 hashtags that reflect the events in the category; collect all of the co-occurring hashtags; and finally rank by co-occurrence frequency. The top 10 hashtags are assumed as *relevant* for representing these events. We augment this group of six event categories with four additional datasets with a narrower geographic scope by (i) determining keywords that best describe an event; and (ii) using selected keywords to retrieve tweets for the event. We start with identifying one or two obvious keywords for an event, e.g., Irene for *Hurricane Irene*. Then we go through our tweets and find those terms that frequently appear together with our selected keyword(s). We select the top 15 terms to expand our keywords for each event, and retrieve the tweets containing the selected words.

4.2 Parameter Setup

For each selected term in the dataset, we first compute the temporal and spatial signals for them and measure the distance between each pair of terms based on the extracted signals as follows:

Temporal Distance: Given a complete time span, all the timestamps for each term w_i can be bucketed into bins: $\langle F_{t_1,w_i}, F_{t_2,w_i}, ..., F_{t_n,w_i}\rangle$. Then these temporal frequencies are normalized and used as the temporal signals. The width of each bin is set as 1 hour. The temporal distances based on F_{t,w_i} between w_i and w_j is defined as:

$$D_t(w_i, w_j) = \sum_t |F_{t,w_i} - F_{t,w_j}| \tag{6}$$

Table 1. Event Dataset

Dataset	Events	Period	Bounding[1]
SEASON	NBA, NFL, MLB, UEFA, Thanksgiving, Christmas, Halloween	02/01/2011 -02/01/2013	(0, 0) (90, 180)
BURST	Japan Tohoku earthquake 2011, Irene Hurricane 2011, Royal Wedding 2011, Sandy Hurricane 2012, London Olympics 2012, Arab Spring (2011–2012), US presidential election 2012	02/01/2011 -02/01/2013	(0, 0) (90, 180)
LONG	NBA, NFL, MLB, UEFA,Arab Spring (2011–2012), London Olympics 2012, US presidential election 2012	02/01/2011 -02/01/2013	(0, 0) (90, 180)
SHORT	Irene Hurricane 2011, Japan Tohoku Earthquake 2011 Royal Wedding 2011, Sandy Hurricane 2011, the Oscars 2013, the Cannes 2013, Steve Jobs' dearth 2011	02/01/2011 -02/01/2013	(0, 0) (90, 180)
GLOBAL	Arab Spring (2011-2012), London Olympics 2012, the Oscars 2013, the Cannes 2013, UEFA	02/01/2011 -02/01/2013	(0, 0) (90, 180)
LOCAL	Oktoberfest Beer Festival 2012, the Super bowl 2012, Memphis In May International Festival 2012	02/01/2011 -02/01/2013	(0, 0) (90, 180)
IRENE	Irene Hurricane 2011, Steve Jobs' resignation 2011, US Virginia earthquake 2011	08/20/2011 -08/30/2011	(29.6, -125.5) (49.1, -69.3)
JPEQ	Fire, Transportation, Asylum, Nuclear, General information of Tohoku Earthquake	03/11/2011 -03/20/2011	(30.4, 129.5) (45.4, 147.0)
MARCH	Japan Tohoku Earthquake 2011, Arab Spring (2011), New Zealand Christchurch earthquake 2011, Federal shutdown March 2011, background topic	03/01/2011 -03/30/2011	(29.6, -125.5) (49.1, -69.3)
AUGUST	Irene Hurricane 2011, Steve Jobs' resignation 2011, US Virginia earthquake 2011, Arab Spring (2011), background topic	08/01/2011 -08/30/2011	(29.6, -125.5) (49.1, -69.3)

[1] The geo-coordinates (latitude, longitude) of the left-up and right-down points of the rectangle bounding area.

Spatial Distance: The geographical bounding-boxes for terms are separated into $N*M$ mesh grids, and all the geo-coordinates for each term w_i are retrieved and bucketed into these grids: $\langle F_{l_1,w_i}, F_{l_2,w_i}, ..., F_{l_n,w_i}\rangle$. The N and M are set with 90 and 180 (1 degree for the width of grid). Based on the normalized spatial signals, the spatial distance between any w_i and w_j is defined as:

$$D_l(w_i, w_j) = \sum_l |F_{l,w_i} - F_{l,w_j}| \tag{7}$$

We then construct the noise filters as follows:

Average band-pass Filter: The weight λ in Equation 1 is set to $1/N$, where N is the size of the cluster.

Gaussian band-pass Filter: The μ in Equation 4 is estimated with the t with the highest term frequency (for temporal signals). σ is estimated with the d where $P((t-d) : (t+d)|\theta) = 0.68$. For spatial distributions, the μ in is estimated with the index of the grid l owning the highest term frequency, and the σ is estimated with the width of the square area, centered with μ, covering 68% percentage term frequencies.

Ideal band-pass Filter: The area $[x_u, x_d]$ in Equation 5 is computed via: 1) identify the center c by finding the bin with highest term frequency in temporal or spatial domain; 2) find the areas (1 dimension area in temporal domain, and 2 dimension square area in spatial domain) centered at c and covering 68% term frequencies. γ is set as 0.68 and λ is 0.1.

4.3 Results

To evaluate the effects of filters using our method, the first set of experiments is to separately test different filters considering both temporal features and spatial features. Concretely, we consider three filters: Average band-pass, Ideal, and Gaussian filters. K-means is used as the clustering method, and the average results of 5 times experiments are used for evaluation

Filtering Temporal Signals: To observe the effects of filters in temporal domain, the Average, Ideal and Gaussian band-pass filters are used on the temporal signals for terms, and temporal distance in Equation 6 is used to measure the similarity between terms. The clustering results using filtered signals and unfiltered signals are compared in Table 2. Table 2 indicates that generally the Event noise filters reduces the noises contained in temporal signal, resulting in better estimation of the distances, and thus achieves better clustering results. Compared with the method with un-filtered signals, the average purities on the 10 data sets using Average filter, Ideal filter and Gaussian band-pass filter are increased by 8.08%, 3.16%, and 1.95% on purity respectively. The probability-based filter – Average filter achieves the better results than the window-based filters (Gaussian and Ideal band-pass filter), most likely since the Gaussian and Ideal band-pass filters put large weights on the detected ROI region, which dramatically changes the power of the signals. If the ROI region is not detected correctly, it will incorrectly filter out the actual event signals.

Table 2. Purity Results for Filtering Temporal Signals

	Filter			
Dataset	No-filter	Average	Ideal	Gaussian
SEASON	0.662	**0.728**	0.693	0.680
BURST	0.749	0.774	0.753	**0.779**
LONG	0.722	**0.782**	0.733	0.760
SHORT	0.673	0.674	0.671	**0.678**
GLOBAL	0.683	0.648	0.693	**0.707**
LOCAL	0.604	**0.675**	0.582	0.496
IRENE	0.750	**0.813**	0.822	0.795
JPEQ	0.683	0.654	**0.706**	0.702
MARCH	0.400	**0.539**	0.426	0.427
AUGUST	0.429	**0.582**	0.477	0.455
Average	0.636	**0.687**	0.656	0.648

In addition, the improvements on March and August data sets by the noise-filters are more substantial than those on other data sets. These two datasets contain more noise corresponding to general topics due to the inclusion of common words like 'we' and 'like'. In an encouraging direction, we see that the proposed filters perform well in these cases of high noise.

Filtering Spatial Signals: In this experiment, the spatial distance in Equation 7 is used, and the Average, Ideal and Gaussian band-pass filters are compared in spatial domain. Table 3 shows the clustering results on the 10 data sets using the spatial signals of terms. Compared with the methods with un-filtered spatial signals, the Average filter improves the clustering result by 3.73%, while the window-based methods degrade the clustering performance. One possible reason is that we assume the Gaussian window and rectangle window in the Gaussian and Ideal filters have only one center. However in the spatial domain, there are usually multiple centers for some events. For example, for the Irene event, there might exist multiple topic centers due to the transition of the center of hurricane. Therefore a single Gaussian or rectangle will incorrectly filter the real event signals, and thus degrade the clustering purities.

Also we can see that the filters have better performance in the temporal domain than the spatial domain. One possible reason could be that the spatial signals are more likely to be largely affected by the population density of different regions. If the ROI regions is incorrectly detected due to the population-affected tweet density, the filter will mistakenly filter out the actual event signals.

Table 3. Purity Results for Filtering Spatial Signals

	Filter			
Dataset	No-filter	Average	Ideal	Gaussian
SEASON	0.688	0.614	**0.731**	0.728
BURST	0.724	0.782	**0.811**	0.725
LONG	0.746	0.736	**0.782**	0.754
SHORT	0.667	0.659	0.635	**0.677**
GLOBAL	0.683	0.737	**0.844**	0.730
LOCAL	0.605	0.551	0.703	**0.735**
IRENE	0.681	**0.818**	0.590	0.727
JPEQ	0.662	**0.727**	0.246	0.246
MARCH	**0.375**	0.338	0.352	0.357
AUGUST	0.378	**0.479**	0.391	0.288
Average	0.621	**0.644**	0.609	0.597

Comparison with Baselines: Based on the results in the last section, we adopt the Average band-pass filter to filter noise in temporal and spatial signals. We combine the spatial and temporal distances into a unified distance as $D_{t,l,o}(w_i, w_j) = (D_o(w_i, w_j)+1)(D_t(w_i, w_j)+D_l(w_i, w_j))$, where $D_o(w_i, w_j)$ is a co-occurrence distance defined in [6]. As baselines we consider two alternatives: a co-occurrence based method [6] and a wavelet-based spatial-temporal method [7]. From Table 4, we observe that among three methods, the co-occurrence based and wavelet-based methods achieve comparable performances. Our proposed noise filtering method performs the best overall. On average, the proposed method has an improvement of 10.60% and 7.06% over the co-occurrence based and wavelet-based methods. The results indicate the proposed method is effective in filtering event-based noise, leading to higher quality event identification.

Table 4. Average Purity Comparison

	Methods		
Dataset	Co-occur	Wavelet	Proposed Method
SEASON	0.781	0.953	**0.984**
BURST	0.869	**0.920**	0.902
LONG	0.835	0.791	**0.851**
SHORT	0.828	0.714	**1.000**
GLOBAL	0.755	0.783	**0.857**
LOCAL	0.667	**0.836**	0.744
IRENE	0.718	0.773	**0.782**
JPEQ	0.734	0.716	**0.747**
MARCH	0.444	0.438	**0.450**
AUGUST	0.454	0.395	**0.519**
Average	0.709	0.732	**0.784**

5 Conclusion

The key insight of this paper is to view spatial-temporal term occurrences as signals, and then to apply noise filters to improve the quality of event extraction from these signals. The iterative event mining algorithm alternately clusters terms using their filtered signals, and then generates new filters based on the results of clustering. Over ten Twitter-based event datasets – we find that the noise filtering method results in a 7-10% improvement versus alternatives, suggesting the viability of noise-aware event detection.

Acknowledgment. This work was supported in part by NSF grant IIS-1149383.

References

1. Allan, J., Papka, R., Lavrenko, V.: On-line new event detection and tracking. In: SIGIR (1998)
2. Bao, B.-K., Min, W., Lu, K., Xu, C.: Social event detection with robust high-order co-clustering. In: ICMR (2013)
3. Batal, I., et al.: Mining Recent Temporal Patterns for Event Detection in Multivariate Time Series Data. In: KDD (2012)
4. Becker, H., Iter, D., Naaman, M., Gravano, L.: Identifying content for planned events across social media sites. In: WSDM (2012)
5. Becker, H., Naaman, M., Gravano, L.: Beyond Trending Topics: Real-World Event Identification on Twitter. In: ICWSM (2011)
6. Begelman, G., Keller, P., Smadja, F.: Automated tag clustering: Improving search and exploration in the tag space. In: Collaborative Web Tagging Workshop (2006)
7. Chen, L., Roy, A.: Event Detection from Flickr Data through Wavelet-based Spatial Analysis. In: CIKM (2009)
8. Chen, Y., Amiri, H., Li, Z., Chua, T.-S.: Emerging topic detection for organizations from microblogs. In: SIGIR (2013)
9. Garg, N., Weber, I.: Personalized tag suggestion for Flickr. In: WWW (2008)

10. Gonzalez, R.C., Woods, R.E.: Digital Image Processing. Prentice-Hall (2007)
11. He, Q., Chang, K., Lim, E.P.: Analyzing feature trajectories for event detection. In: SIGIR (2007)
12. Hong, L., Ahmed, A., Gurumurthy, S., Smola, A.J., Tsioutsiouliklis, K.: Discovering geographical topics in the twitter stream. In: WWW (2012)
13. Kleinberg, J.: Bursty and hierarchical structure in streams. Data Min. Knowl. Discov. (2003)
14. Li, C., Sun, A., Datta, A.: Twevent: Segment-based event detection from tweets. In: CIKM (2012)
15. Li, Z., Wang, B., Li, M., Ma, W.Y.: A probabilistic model for retrospective news event detection. In: SIGIR (2005)
16. Mei, Q., Liuy, C., Suz, H., Zhaiy, C.: A Probabilistic Approach to Spatiotemporal Theme Pattern Mining on Weblogs. In: WWW (2006)
17. Moxley, E., Kleban, J., Xu, J., Manjunath, B.S.: Not all tags are created equal: Learning Flickr tag semantics for global annotation. In: ICME (2009)
18. Papadopoulos, S., Zigkolis, C., Kompatsiaris, Y.: Cluster-Based Landmark and Event Detection for Tagged Photo Collections. In: Multimedia (2010)
19. Rattenbury, T., Good, N., Naaman, M.: Towards automatic extraction of event and place semantics from Flickr tags. In: SIGIR (2007)
20. Sakaki, T., Okazaki, M., Matsuo, Y.: Earthquake shakes twitter users: real-time event detection by social sensors. In: WWW (2010)
21. Sayyadi, H., Hurst, M., Maykov, A.: Event Detection and Tracking in Social Streams. In: ICWSM (2009)
22. Sengstock, C., Gertz, M., Flatow, F., Abdelhaq, H.: A probablistic model for spatio-temporal signal extraction from social media. In: SIGSPATIAL (2013)
23. Sigurbjörnsson, B., van Zwol, R.: Flickr tag recommendation based on collective knowledge. In: WWW (2008)
24. Valkanas, G., Gunopulos, D.: How the live web feels about events. In: CIKM (2013)
25. Yang, Y., Pierce, T., Carbonell, J.: A study of retrospective and on-line event detection. In: SIGIR (1998)
26. Yin, Z., Cao, L., Han, J., Zhai, C., Huang, T.: Geographical Topic Discovery and Comparison. In: WWW (2011)
27. Yin, Z., Cao, L., Han, J., Zhai, C., Huang, T.: LPTA: A Probabilistic Model for Latent Periodic Topic Analysis. In: ICDM (2011)
28. Yuan, Q., Cong, G., Ma, Z., Sun, A., Thalmann, N.M.: Who, where, when and what: Discover spatio-temporal topics for twitter users. In: KDD (2013)
29. Zhang, H., Korayem, M., You, E., Crandall, D.J.: Beyond Co-occurrence: Discovering and Visualizing Tag Relationships from Geo-spatial and Temporal Similarities. In: WSDM (2012)
30. Zhao, Q., Liu, T.-Y., Bhowmick, S.S., Ma, W.-Y.: Event detection from evolution of click-through data. In: KDD (2006)

Timeline Summarization from Relevant Headlines

Giang Tran, Mohammad Alrifai, and Eelco Herder

L3S Research Center and Leibniz University Hannover
Appelstr. 9, 30167 Hannover, Germany
{gtran,alrifai,herder}@L3S.DE

Abstract. Timeline summaries are an effective way for helping newspaper readers to keep track of long-lasting news stories, such as the Egypt revolution. A good timeline summary provides a concise description of only the main events, while maintaining good understandability. As manual construction of timelines is very time-consuming, there is a need for automatic approaches. However, automatic selection of relevant events is challenging due to the large amount of news articles published every day. Furthermore, current state-of-the-art systems produce summaries that are suboptimal in terms of relevance and understandability. We present a new approach that exploits the headlines of online news articles instead of the articles' full text. The quantitative and qualitative results from our user studies confirm that our method outperforms state-of-the-art system in these aspects.

1 Introduction

More than two years after the Egyptian revolution of 2011, political conflicts in Egypt were back again in the breaking news headlines in 2013. While trying to relate current events to past events, newspaper readers may ask themselves several questions, such as: *How and Why did the Egyptian revolution start back in 2011? What happened in Egypt since then? Why are there many new protests again in Egypt?* A compact summary that represents the development of the story over time, highlighting its most important events - possibly with links to sources for further details - would be very beneficial for fulfilling readers' information needs.

Timeline summarization (*TS* for short) has become a widely adopted, natural way to present long news stories in a compact manner. News agencies often manually construct and maintain timelines for major events, but constructing such visual summaries often requires a considerable amount of human effort and does not scale well. Existing approaches for TS aim to tackle one of two problems: (i) select a subset of important dates as the major points of the timeline (e.g, [12], [4]) and/or (ii) generate a good daily summary for each of these dates (e.g, [6], [27], [4]). In this study, we set our focus on the second problem.

Previous work on the generation of daily summaries usually focuses on the extraction of relevant sentences from article text. The main drawback of such approaches is that it does not guarantee good *understandability* as well as high *relevance* for the daily summary. Low relevance is often caused by the nature of textual data - it is hard to select the right sentence from a large number of sentences; low understandability is

A. Hanbury et al. (Eds.): ECIR 2015, LNCS 9022, pp. 245–256, 2015.

Table 1. Examples of summaries with low understandability

(A.1) It will end as soon as the people vote on a constitution, **he** told state television... (A.2) ...**President Mohamed Mursi** hopes will help to end a crisis..."
(B.1) On **Wednesday** , two protesters were killed Aden , a southern port city..... (B.2) On **Thursday** , dozens of people were reportedly injured in clashes.....
(C.1) Anti-government protesters in Yemen have resumed demonstrations to try to force Ali Abdullah Saleh , the president , to quit , ... (C.2) The students , some of whom were also armed with batons , responded .

often caused by inconsistencies and lack of continuity between the selected sentences. The following examples in Table 1 present 3 summaries generated by a state-of-the-art system, ETS [27], showing a few understandability problems: "he" in sentence (A.1) is ambiguous and can be misunderstood as "Mohamed Mursi" in (A.2) (*daily summary A)*, time inconsistency between sentences (B.1). and (B.2.), which should not be used in the same daily summary *(daily summary B)* and content incoherence between (C.1) and (C.2) of *daily summary C.*

In addition to this, finding a good order for selected sentences to make a coherent summary is on itself already a difficult task in the NLP community (for example, see [3], [5]). This makes it even more challenging to generate a summary with good understandability by ordering selected sentences.

Headlines of online news articles have shown to be a reliable source for adequately providing a high-level overview of the news events[2]. Headlines are comprehensible to the reader without requiring too much reading time [20],[7]. The information provided in headlines is usually self-contained, timely and complete, and therefore suitable for creating coherent daily summaries. For this reason, we consider headlines as good candidates for TS generation.

There are some technical challenges that make using news headlines for TS far from being straightforward. First, one needs to distinguish informing news updates from other non-informing news headlines, which includes background information, reviews and opinions[1]. In this work, we focus on informing news headlines, which tell *what* happens in the story instead of opinions or background. Second, one needs to identify duplicates among the headlines, to minimize redundancy in the produced summary. Because headlines are often short and do not follow syntactic structures, duplicate detection among headlines is a challenging task. Third, one needs to make a selection of the most relevant headlines for making daily summaries that are as informative as possible. To our knowledge, there are no previous studies on generating TS from headlines.

The contribution of this paper is a novel approach for the generation of timeline summaries of news stories, based on the headlines of news articles. We present a *headline selection algorithm* based on a random walk model (Section 3). Further, we show the results of *quantitative and qualitative evaluations* of the proposed methods in comparison with the-state-of-the-art methods (Section 4)

[1] See Freund et al. (2011) [10] for news genre taxonomy.

2 Related Work

There is a plethora of research on the generation of timeline summaries. Typical studies in this domain include Swan and Allan [24], Allan et al. [1], Chieu at al.[6], Yan et al. [27], Tran et al. [4]. These studies share the same approach of extracting the most relevant and descriptive sentences from the full article texts. Experimental evaluations of these approaches have shown that the n-gram overlaps (*typically using ROUGE scores*) between the generated summaries and some manually created summaries for the same time period (or dates) is significant. Nonetheless, to the best of our knowledge, none of these approaches has been evaluated using qualitative analysis.

Our assumption is that the full-text extraction approach that is adopted in the aforementioned research works does not guarantee the (subjective) quality or readability of the produced summaries, as this cannot be measured using the ROUGE score. We use a different approach, directly based on the news article headlines. Our qualitative user evaluation shows that users tend to rate the summaries produced by existing solutions with lower quality scores.

Timeline summarization is a special case of *multi-document summarization* (MDS for short), which organizes events by date. Basically, TS can be generated by MDS systems by applying summarization techniques on news articles for every individual date to create a corresponding daily summary. However, because MDS techniques do not make use of the inter-date connections between news articles, they tend to be less robust than state-of-the-art methods specifically designed for TS generation (e.g., as discussed in[27]). Beside the difference in the approach (using headlines instead of the full text), our framework differs from MDS in that it takes the relations among events across dates into account. As there is already a rich body of research on multi-document summarization (for example,[22],[21], [16], [8], [17]), in this study we also investigate how good they are in producing daily summaries using only headlines, in the same setting as our approaches.

3 Problem Statement and Selection Model

The focus of this study is on generating timeline summaries that represent *what* happened in a news story. More formally, we focus on the following problem:

Problem 1 (Selection of Headlines for TS.). Let H_d be the set of headlines from published news articles of a dated, select c most relevant headlines to make daily summary of that date.

In this section, we discuss aspects of headlines that are relevant for the creation of TS: the headline's Informing value, its Spread and Influence. After that, we develop a random walk model based on personalized Pagerank on the top of these aspects. In summary, the model estimates duplicates among headlines (the *Spread*) and creates a graph in which the nodes represent the headlines and the edges are weighted by the probability that two corresponding headlines are duplicated. The model biases the random walker to prefer headlines with high *Influence* scores. Finally, we conduct a greedy algorithm based on submodularity to select a set of relevant headlines using the *Informing* aspect and the backward probabilities (i.e, rank).

3.1 Aspects of Relevant Headlines

In this section, we describe three important aspects that characterize relevant headlines: their Informing value and their Spread and Influence.

Informing. We consider a headline as an Informing news headline when it informs about a news event[2] An Informing headline typically delivers self-contained information to the readers, as it explicitly describes an event that has occurred. By contrast, non-informing news headlines often provide author opinions or reviews on the event. Although opinions or reviews are helpful in highlighting different aspects of the events, especially when they come from influential columnists, they are typically provide opinionated, subjective views of the authors and hence introduce some bias to the TS. We leave opinion-based TS for another study.

We calculate the Informing aspect by using a machine learning classification approach. For the sake of simplicity, we follow Yu and Hatzivassiloglou [28] as it performed well on our testing set. Let F(h) denote the probability of a headline *h* being an informing news headline. When a headline *h* is classified as *positive*, we assign F(h) = 1, otherwise F(h) = 0. For training purposes, we use 20K headlines as positive examples that are randomly extracted from news articles using APIs of the *WikiTimes*[3] system [25]. Those news articles are references to actual events in the Wikipedia Current Events portal [4]. In contrast, negative examples are 20K headlines of articles from the New York Time corpus that are annotated as opinion, reviews or other non-informing categories until 2007. By using these two sets of headlines for training the SVM model, instead of sentences from the full text of news articles, the machine learning model is fitted well with our headline input. Our experimental results show that the model reaches 76% accuracy by cross-validation. Due to space limitations, we do not go further into details.

Influence. An event is likely to be relevant for timeline summaries when it is influential in what will happen in the future. For example, *Mubarak resigns* will lead to a *new election event*, then lead to the *presidency of Mohamed Mursi*, and so on. We observed that influential events are those that are most often mentioned in news articles that are published in the future.

We compute Influence as follows. Let I(h) quantify the influence of headline *h*. We analyze temporal information in the content of the respective news article to heuristically locate references to this particular headline in news articles that are published after that. Let $\mathcal{E}_{V \to u}$ be the cluster of all sentences that are not published in u but refer to date u. Using the Heideltime toolkit [23] for temporal tagging, given a headline h of date u, we define its influence on future events by the similarity of its word distribution, $\theta(h)$ to the word distribution of the cluster $\theta(\mathcal{E}_{V \to u})$. The computation is done as follows:

$$I(h)_u = \sum_{w \in h} p(w|\theta(h)) * p(w|\theta(\mathcal{E}_{V \to u})) \qquad (1)$$

where $p(w|\theta)$ is probability of word w in θ.

[2] We only focus on actual news stories, not on other articles such as Photo essays, Infographics or Weather reports.

[3] http://wikitimes.l3s.de

[4] http://en.wikipedia.org/wiki/Portal:Current_events

Spread. Cluster hypothesis suggests that headlines that are similar to one another confirm the relevance of each other [26], as they are virtually members of the same clusters. We observed that a relevant event is typically spread among various headlines, as it is very often reported by different news agencies. The following example shows how the event "Mubarak resigns" is reported in different headlines:

- **Huffington Post:** *Mubarak Steps Down Tahrir Square , Egypt Erupts In Cheers.*
- **The Guardian:** *Hosni Mubarak resigns and Egypt celebrates a new dawn.*
- **CNN:** *Egypt's Mubarak resigns after 30-year rule.*
- **NBC:** *'Egypt is free,' crowds cheer after Mubarak quits.*

We quantify the *Spread* of a headline by measuring p_{ij} as *the probability that two headlines* h_i *and* h_j *are duplicated* (i.e., they report about the same event). Intuitively, more duplications and higher confidence (by mean of probability) indicate higher *Spread* value. Obviously, *Spread* is transitive: h_i and h_k may be duplicated if they both are duplicated with h_j. Due to this transitivity, the *Spread* value of a headline can be propagated through its duplicated headlines. Using this graph, a random walk model on the duplication graph of headlines is able to estimate the *Spread* value. We will present an algorithm for that estimation in a combined model with other aspects in Section 3.2.

Estimation of Duplication Probability . Now we describe how we computed the duplication probability p_{ij} using a Logistic Regression model. It is worth mentioning that even though this task is similar to sentence paraphrase detection, headlines are of shorter length and sometimes do not follow grammatical rules (but are fancy and catchy). In addition, here we only care about the core message reported in the headline, while in sentence paraphrase detection, the meaning of the entire sentence is taken into account. That makes available labeled corpora for sentence paraphrase detection not a good fit for our learning strategy. Therefore, we constructed our own training data by leveraging the wisdom of the Wikipedia crowd. Due to space limitations, we will only summarize the steps we followed: (1) extract positive examples: pairs of headlines from any pair of news articles on *an event* in Wikipedia's current events portal[5] ; (2) extract negative examples: pairs of *cross-event* headlines (i.e, each headline is from an event). In the end, we obtained a dataset of 16K with a ratio between positive and negative examples of about 50/50. Our intuition is that headlines of news articles that are references of an event are likely to be duplicated.

For training the Logistic Regression model, we use state-of-the-art semantic similarity measures that are popular in paraphrase detection: corpus-based similarity, as proposed by Mihalcea et al. [18] and Malik et al. [15], and Wordnet-based paraphrase similarity [9]. In addition, we extracted prior co-occurrence probabilities of any verb pair in the whole WikiTimes dataset as a signal for two corresponding headline pairs being duplicated. A verb pair is counted as one co-occurrence if both verbs appear in two headlines of the same event. That model results in 77% accuracy with 10% improvement gained by additionally using prior co-occurrence probability feature.

[5] To save the engineering cost, we use *WikiTimes* data: `http://wikitimes.l3s.de`

3.2 Headline Selection Model

Overview. Our target is to select headlines that maximize all three aspects Influence, Spread and Informing value. Among available propagation algorithms, personalized PageRank [11] on a graph of headlines appears to be suitable for this task, as it both takes the link graph structure (Spread aspect) into account and considers the personalized probability (Influence aspect) while performing random walks. Then, by using PageRank score as the probability of being relevant for TS, we formulate headline selection as an optimization problem that can be solved by submodular-based techniques, which we describe in the remainder of this section.

Formation of Headlines Graph From the set of headlines $H = \{h_1, h_2, ..., h_n\}$ of a day, we create an undirected event-based similarity graph G = (E, V), in which each node of V is a headline in H and each edge between 2 nodes (i, j) is weighted by the duplication probability $p_{ij} \in [0, 1]$.

Influence-based Random Walk In order to integrate the multiple aspects of the headline, we use a random walk model that follows the personalized PageRank method for ranking headlines. Headline relevance (R) is estimated by its probability of being visited by the random walker in the model, which is iteratively computed using the equation 2.

$$R(j) = d \sum_{i} \frac{p_{ij}}{\sum_k p_{ik}} * R(i) + (1 - d) * \frac{I(h_j)}{\max_{h \in H} I(h)} \quad (2)$$

where the damping factor d = 0.85 and the transitional probability is normalized from the duplication probability to satisfy the Markov property. We guide the random walker to headlines that have high influence scores *I(h)*.

Algorithm 1: Algorithm for selection of relevant headlines

$S \leftarrow \emptyset$
$Q \leftarrow H$
while $Q \neq \emptyset$ and $|S| < c$ **do**
$h_i \leftarrow \arg\max_{h \in Q} \mathrm{R}(S \bigcup h) - \mathrm{R}(S)$
$p(h_i, S) \leftarrow \max_{h_j \in S} p_{ij}$
subject to: $p(h_i, S) < \theta \wedge F(h_i) = 1$ *(#no duplication and be informing news)*
$S \leftarrow S \bigcup h_i$
$Q \leftarrow Q \setminus h_i$
end while

Submodular Method for Selecting Events. Based on the R scores of all headlines, we greedily select the top headlines as long as they do not violate any constraint: no pair of selected headlines is duplicated and selected headlines must be informing. Formally, we have to solve the following optimization problem:

$$\begin{aligned} \underset{S \subseteq H_d}{\text{maximize}} \quad & \mathrm{R}(S) \\ \text{subject to} \quad & \mathrm{R}(S) = \sum_{h_i \in S} \mathrm{R}(i) \\ & |S| = c \\ & F(h) = 1 \; \forall h \in S \\ & p_{ij} < \theta \; (i, j) \in \Omega_S \end{aligned}$$

Given our constrains: (a) no duplicated pairs in selected subset $S \subset H$, (b) budget size(S) = **c** as the number of headlines for each day summary, and (c) all selected headlines should be Informing news headlines. Our objective function is monotone and submodular [13], and therefore we may use the greedy Algorithm 1 to solve it with accuracy guarantee $1 - \frac{1}{e}$, where θ is the threshold for identifying whether one pair is duplicated, determined by the trained Logistic Regression model for duplication probability estimation.

4 Experiments

In this section, we evaluate the proposed framework by measuring the *relevance* and *understandability* of the TS output and comparing it to that of state-of-the-art systems. Our evaluation methodology is based on *human evaluation* instead of automatic n-gram based overlap metrics like Rouge scores [14], especially because headlines exhibit different characteristics than article text sentences, and n-gram based measures hardly capture paraphrases in event reporting, for instance, *'Egypt is free,' crowds cheer after Mubarak quits.* v.s *Hosni Mubarak resigns and Egypt celebrates a new dawn.*

The relevance score measures how well the selected headlines perform in reporting and summarizing important events of the news story, while the understandability score measures the readability and comprehensibility of the summary that is constructed from the selected headlines. In other words, we consider one summary better than another if it covers more relevant events and/or if users understand its description of the events better.

4.1 Dataset and Experimental Setting

We constructed a dataset that consists of news articles, which serve as input, and expert timeline summaries, which serve as ground-truth summaries (the ideal output). The articles focus on long-span stories on the *armed conflicts* Egypt Revolution, Syria War, Yemen Crisis and Libya War [6].

News articles We collected news articles by simulating users searching for articles relevant for the timelines of the aforementioned news stories - for this purpose, we used Google and targeted the same 24 news agencies that were used for creating the timelines used as the ground truth. We constructed several queries, such as "Egypt (revolution OR crisis OR uprising OR civil war)", as queries with the time filter option [Jan/2011 - July/2013] and the "site" specification. For each query, we took the top-300 answers. Using this method, we obtained 15.534 news articles, of which the distribution is summarized in the #News column of Table 2.

Expert Timeline Summaries Arguably, timeline summaries that have been published by well-known news agencies are the most trustful base for ground-truth timeline summaries, as they have been manually created by professional journalists. We manually collected 25 timeline summaries from 24 populair news agencies, including the BBC,

[6] Available at `http://www.l3s.de/~gtran/timeline/`

CNN and Reuters. These ground-truth timeline summaries are offered to the participants of our study as a baseline for deciding whether the automatically selected headlines are relevant or not. Table 2 gives an overview of these timelines.

Table 2. Overview of expert timeline summaries

Story	#TL	#Timepoint	#GT-Date	TL-Range	#a.sent	#News
Egypt Revolution	4	112	18	01/2011-07/2013	2	3869
Libya War	7	118	51	02/2011-11/2011	2	3994
Syria War	5	106	15	03/2011-08/2012	2	4071
Yemen Crisis	5	81	22	01/2011-02/2012	2	3600

Number of timelines (#TL), total number #Timepoints of all timelines, number of groundtruth dates(#GT-Date), the time ranges and rounded average sentences per date of each timelines (#a.sent.), number of news articles (#news)

4.2 Systems for comparison

We compare our approach with systems for TS generation as well as for traditional MDS. In addition, we consider two other baselines, SumSim and Longest. To make the generated summaries comparable with expert summaries in term of length, we use the same setting $c = 2$ for all systems in our evaluation, which means that each system will generate daily summaries of 2-sentence length.

Timeline Summarization. We choose two state-of-the-art methods for TS generation that focus on daily summaries: ETS and Chieu et al. Both systems have originally been designed to work with article texts. However, in addition to that, we developed one version of Chieu et al. for headlines only, named SumSim. Due to the design of the algorithm and the spare word distribution, it is not easy to adapt ETS to work with just headlines. We leave it for future investigation.

ETS is by far one of the best unsupervised TS systems in the news domain. It takes advantage of the similarity between the word distributions in a sentence and the word distribution in an entire corpus as well as within the neighboring dates. We implemented the ETS algorithm described by the authors in [27].

Chieu et al.[6] utilize the popularity of a sentence on date t_i as the sum of TF-IDF similarity scores with other sentences that are published in around t_i +- k days to estimate how important this sentence is. We select k=10, following the author's setting.

Traditional Document Summarization. Since ETS and Chieu et al. extract sentences from the full text of news articles for timeline summaries, as shown in the experiments of Yan et al. [27], we also would like to see how good (multi-)document summarization would work on the headlines dataset. We consider the following state-of-the-art methods: Centroid [22], LexRank [8], TextRank [19][7]

SumSim selects top news reports and non-duplicated headlines that maximize the sum of TF-IDF similarity with other headlines that are published in the previous and next 10 days. Conceptually, it works similarly to Chieu et al., but on the headline level.

[7] For Centroid, we used the MEAD toolkit, for LexRank and TextRank, we used the sumy toolkit `https://github.com/miso-belica/sumy`.

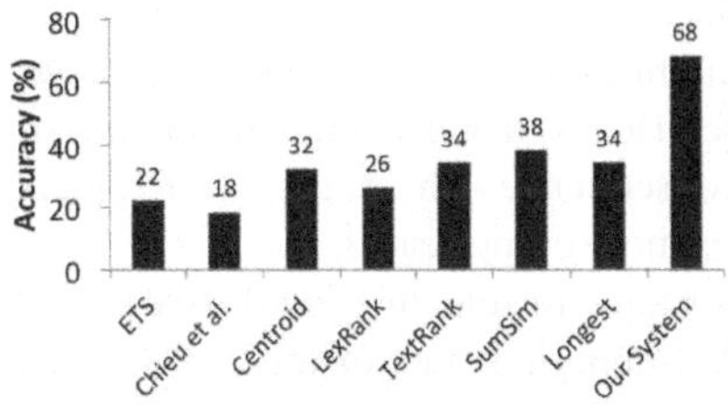

Fig. 1. Relevance evaluation of the produced summaries by the different systems in comparison with expert manual summaries

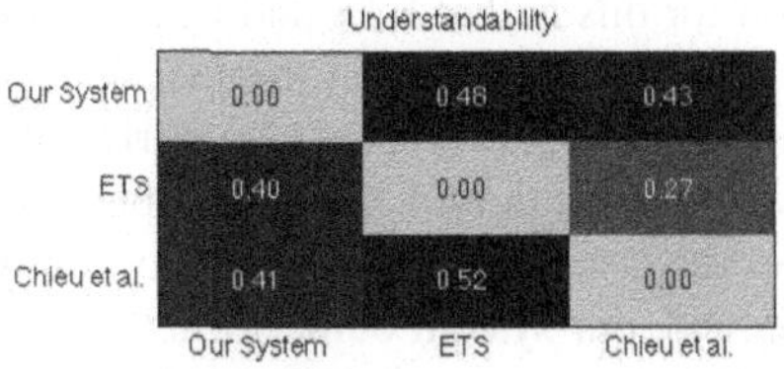

Fig. 2. Pairwise comparison of the *understandability* of summaries produced by the different systems

Longest selects top news reports and non-duplicated headlines ordered by their length. Conceptually, it assumes that the longest headlines are the most important ones.

4.3 Relevance Evaluation

We examine the performance of our approach for producing day summaries by comparing it with the aforementioned baselines. As discussed, we rely on human evaluation. We recruited 3 annotators, who confirmed to be familiar with the news stories used in our study, to annotate the relevance of the collected headlines in our dataset. We extracted all headlines of the news articles from 106 dates that appear in the ground truth TS (the expert timeline summaries of news agencies) and ignored dates on which fewer than 10 news articles were found. We asked the 3 annotators to label each headline as relevant ('1') or not ('0'), based on whether the headline reports events mentioned in the expert summary of that date. Among the annotators, one is a co-author of this study and the other two are graduate students. In total, 1319 headlines were annotated with an average agreement of $\kappa = 0.89$ between any two annotators. We kept only the dates that contain at least one relevant headline and kept the major judged answers among annotators. At the end, 1123 headlines were annotated for 47 dates.

Judging relevance for short summaries produced by the baseline systems can be a little more difficult than that of headlines. Therefore, to collect more judgments, we used CrowdFlower[8] for recruiting users to judge the relevance of the daily summaries produced by *ETS* and *Chieu et al.*. The users were requested to read the ground-truth summaries of a given date and to specify the relevance of sentences from the summaries from ETS or Chieu et al. Before working, each user was trained with at least 12 questions that we used as gold questions. During the job, they were secretly requested to answer gold questions. In total we collected 5104 judgments. Only answers from users who passed gold questions with a high agreement (≥ 0.85) were taken into account. We gathered between 5 and 10 trustful answers from separate trustful users for each question.

Results: Figure 1 shows the *Accuracy@2* of selected headlines (our system and MDS) and sentences (by Chieu et al. and ETS systems).

First, it can be seen that the results of the TS baselines ETS and Chieu et al. are not as good as those produced by our system and other headline-based baselines. The main

[8] http://www.crowdflower.com/

reason for this is that ETS and Chieu et al. select sentences from the full text of news articles and do not exploit the fact that the headlines themselves quite often serve as high-quality expert-created summaries of these articles. Their approach benefits from the rich distribution of the words, but - as a consequence - the huge list of sentences makes the task to create high-quality summaries more complicated.

Second, our system outperforms the MDS systems in selecting good headlines that reflect important events. This result implies that applying state-of-the-art MDS techniques does not ensure highly relevant events in the TS. This observation confirms the need for further investigation on TS generation using just headlines.

Third, *SumSim* perform slightly better than MDS. That is mostly because SumSim uses the information from neighbor articles (from the previous and next 10 dates) while MDS (and also Longest) do not. That is not a surprise, but confirms that the temporal aspect is crucial for TS generation, even for headline-based approaches. SumSim also outperforms its brother Chieu et al., and it shows the benefits of using headlines instead of article full-text here.

Fourth, our system outperforms all others with much higher scores for the generated timelines. The better performance can be explained by the following facts: (1) headlines are written by experts and mostly report the most important event; using the headline is therefore a better solution than selecting sentences from the full text. (2) different from the SumSim and TS baselines, our method leverages temporal information by using the influence aspect of headlines, which focuses on selective sentences with visible temporal tagging instead of all sentences; we observed that sentences with visible temporal tagging often highlight important information. (3) the combination of influence and the network structure (headline graph) produce better estimations of the importance, horizontally (Spread) and vertically (Influence). Last but not least, it is worth mentioning that the improvement is statistically significant.

4.4 Understandability Evaluation

With this experiment, we aim to evaluate the readability and understandability of the summaries from a user perspective. We compare our summaries one by one with the summaries produced by ETS and Chieu et al. More specifically, we investigate whether the selection of headlines produces summaries that are more coherent and comprehensible than extracted summaries that are composed from selected sentences from the article full-text.

Task setting: We provide CrowdFlower users with our collected ground-truth daily summaries from professional journalists, followed by 2 daily summaries, say A and B, which are alternately produced by either our system or ETS or Chieu et al. Users can answer “1” if A is more understandable than B, “-1” if A is less understandable than B, or “0” otherwise. The quality of answers is checked by the agreement with that of a small set of gold questions, secretly delivered to the users during their working sessions. In total, 141 summary pairs are presented to CrowdFlower users.

Result: We collected 2244 judgments from 122 users, of which 1552 judgments are from trusted users, who earned at least 0.85% correct on our gold questions and 0.85 *trust* gained from their work on CrowdFlower. Those 1552 judgments are used for our evaluation. The results are shown in Figure 2, where the value *m[Y][X]* in each matrix is the percentage of users who judged system *X* better than system *Y*. The higher the number, the darker its color. The rest *(1 - m[X][Y] - m[Y][X])*, which is not presented in the figure, is the percentage of users who considered X and Y to be equal.

Analysis: Generally, our headline-based approach results in better understandability than the other systems. We noticed that the confidence, the highest value among *m[Y][X], m[X][Y], and 1- (m[X][Y] + m [Y][X])*, is not very high, which indicates that the comparison of text quality is a hard task. User feedback confirmed that the task was clear (rated 4/5), but that they found it difficult to select the answer (rated 3/5).

While the relevance results showed that ETS is slightly better than the summaries provided by Chieu et al., users tend to rate the Chieu et al. summaries better in term of understandability than ETS. The reason could be that the ETS algorithm provides daily summaries that are related to summaries of the neighbor dates. Therefore, missing a piece of information from the connection between summaries between the neighbor dates can make ETS's day summaries less understandable than Chieu et al., which simply focuses on the daily events.

5 Conclusion

We presented a novel framework for automatically constructing a timeline summary for a news story from a collection of news articles. Different from previous work, where the proposed solutions extract sentences from article texts, our framework makes use of headlines. The intuition is that a careful selection of news headlines can result in summaries that are more informative and understandable than summaries that are composed of selected sentences from different parts of the news articles. Indeed, the qualitative user study showed that users prefer the timeline summaries produced by our headline-based approach over the summaries that are produced by other extractive approaches.

Unlike traditional MDS, our approach exploits temporal information to estimate the impact of an event on the future development of a new story. Therefore, it is worth mentioning that our approach best fits scenarios of retro-active summarization. Experimental evaluations have shown that the use of temporal information resulted in summaries of more relevant events than the ones selected by MDS methods.

Acknowledgements. The first author thanks Dr. Katja Markert for her valuable comments. The work was partially funded by the European Commission for the FP7 project EUMSSI (611057) and the ERC Advanced Grant ALEXANDRIA (339233).

References

1. Allan, J., Gupta, R., Khandelwal, V.: Temporal summaries of new topics. In: Proceedings of SIGIR 2001, pp. 10–18 (2001)
2. Althaus, S.L., Edy, A.J., Phalen, P.: Using substitutes for full-text news stories in content analysis: Which text is best? American Journal of Political Science, 707–723 (2001)
3. Barzilay, R., Lapata, M.: Modeling local coherence: An entity-based approach. Computational Linguistics 34(1), 1–34 (2008)

4. Binh Tran, G., Alrifai, M., Quoc Nguyen, D.: Predicting relevant news events for timeline summaries. In: Proceedings of WWW Companion (2013)
5. Branavan, S.R.K., Kushman, N., Lei, T., Barzilay, R.: Learning high-level planning from text. In: The 50th Annual Meeting of the ACL, pp. 126–135 (2012)
6. Chieu, H.L., Lee, Y.K.: Query based event extraction along a timeline. In: Proceedings of SIGIR 2004, pp. 425–432 (2004)
7. Dor, D.: On newspaper headlines as relevance optimizers. Journal of Pragmatics (2003)
8. Erkan, G., Radev, D.R.: Lexrank: graph-based lexical centrality as salience in text summarization. J. Artif. Int. Res. 22(1), 457–479 (2004)
9. Fernando, S., Stevenson, M.: A semantic similarity approach to paraphrase detection (2008)
10. Freund, L., Berzowska, J., Lee, J., Read, K., Schiller, H.: Digging into digg: Genres of online news. In: Proceedings of the iConference (2011)
11. Haveliwala, T.H.: Topic-sensitive pagerank. In: WWW, pp. 517–526 (2002)
12. Kessler, R., Tannier, X., Hagège, C., Moriceau, V., Bittar, A.: Finding salient dates for building thematic timelines. In: Proceedings of ACL 2012 (2012)
13. Khuller, S., Moss, A., Naor, J.: The budgeted maximum coverage problem (1999)
14. Lin, C.-Y., Hovy, E.: Automatic evaluation of summaries using n-gram co-occurrence statistics. In: Proceedings of NAACL 2003, vol. 1, pp. 71–78 (2003)
15. Malik, R., Subramaniam, L.V., Kaushik, S.: Automatically selecting answer templates to respond to customer emails. In: IJCAI 2007 (2007)
16. McKeown, K., Barzilay, R., Chen, J., Elson, D.K., Evans, D.K., Klavans, J., Nenkova, A., Schiffman, B., Sigelman, S.: Columbia's newsblaster: New features and future directions. In: HLT-NAACL (2003)
17. Metzler, D., Kanungo, T.: Machine learned sentence selection strategies for query-biased summarization. In: Proceedings of the 2008 ACM SIGIR LR4IR Workshop (2008)
18. Mihalcea, R., Corley, C., Strapparava, C.: Corpus-based and knowledge-based measures of text semantic similarity. In: AAAI, pp. 775–780 (2006)
19. Mihalcea, R., Tarau, P.: Textrank: Bringing order into text. In: EMNLP (2004)
20. Perfetti, C.A., Beverly, S., Bell, L., Rodgers, K., Faux, R.: Comprehending newspaper headlines. Journal of Memory and Language 26(6), 692–713 (1987)
21. Radev, D.R., Allison, T., Blair-Goldensohn, S., Blitzer, J., Ç elebi, A., Dimitrov, S., Drbek, E., Hakim, A., Lam, W., Liu, D., Otterbacher, J., Qi, H., Saggion, H., Teufel, S., Topper, M., Winkel, A., Zhang, Z.: Mead - a platform for multidocument multilingual text summarization. In: Proceedings of LREC'04 (2004)
22. Radev, D.R., Jing, H., Sty, M., Tam, D.: Centroid-based summarization of multiple documents, pp. 919–938 (2004)
23. Strötgen, J., Gertz, M.: Heideltime: High quality rule-based extraction and normalization of temporal expressions. In: Proceedings of the SemEval 2010, pp. 321–324 (2010)
24. Swan, R.C., Allan, J.: Timemine: visualizing automatically constructed timelines. In: SIGIR, p. 393 (2000)
25. Tran, G.B., Alrifai, M.: Indexing and analyzing wikipedia's current events portal, the daily news summaries by the crowd. In: WWW (Companion Volume), pp. 511–516 (2014)
26. van Rijsbergen, C.J.: Information retrieval (1979)
27. Yan, R., Wan, X., Otterbacher, J., Kong, L., Li, X., Zhang, Y.: Evolutionary timeline summarization: a balanced optimization framework via iterative substitution. In: Proceedings of SIGIR 2011, pp. 745–754 (2011)
28. Yu, H., Hatzivassiloglou, V.: Towards answering opinion questions: separating facts from opinions and identifying the polarity of opinion sentences. In: Proceedings of EMNLP 2003, pp. 129–136 (2003)

A Self-training CRF Method for Recognizing Product Model Mentions in Web Forums

Henry S. Vieira[1,2], Altigran S. da Silva[1], Marco Cristo[1], and Edleno S. de Moura[1]

[1] Institute of Computing, Federal University of Amazonas, Manaus, Brazil
{henry,alti,marco.cristo,edleno}@icomp.ufam.edu.br
[2] FPF Tech, Manaus, Brazil

Abstract. Important applications in product opinion mining such as opinion summarization and aspect extraction require the recognition of product mentions as a basic task. In the case of consumer electronic products, Web forums are important and popular sources of valuable opinions. Forum users often refer to products by means of their model numbers. In a post a user would employ model numbers, e.g., "BDP-93" and "BDP-103", to compare Blu-ray players. To properly handle opinions in such a scenario, applications need to correctly recognize products by their model numbers. Forums, however, are informal and many challenges for undertaking automatic product model recognition arise, since users mention model numbers in many different ways. In this paper we propose the use of a self-training strategy to learn a suitable CRF model for this task. Our method requires only a set of seed model numbers. Experiments in four different settings demonstrate that our method, by leveraging unlabeled sentences from the target forum, yielded an improvement of 19% in recall and 12% in F-measure over a supervised CRF model.

1 Introduction

Opinion mining is concerned with people's sentiments, opinions, attitudes and emotions towards some entity and its aspects [5]. One of the basic tasks associated with opinion mining is extracting target entities [3, 5]. Indeed, mining user opinions is more useful when the target is known. In our work, we focus on entities from a given category of a specific and relevant application domain: consumer electronic products such as Blu-ray players. Such products are the main subject of opinions posted by users in discussion forums. We observe that users very often refer to a particular product by means of its *model number*[1].

The main task we focus in this paper is recognizing model numbers of products mentioned in forum posts for a given category. We regard this task as an instance of the Named Entity Recognition (NER) [4,10] problem, for which state-of-the-art techniques use Conditional Random Fields (CRF) [7]. Although effective,

[1] In here, we adopted the same jargon of retail stores, in which "model number" refers to a code that identifies a particular product. This code is not necessarily a number.

A. Hanbury et al. (Eds.): ECIR 2015, LNCS 9022, pp. 257–264, 2015.

these models are difficult to directly apply to the problem we focus on here because they require voluminous representative labeled data for training.

In this paper we propose a novel method, called ModSpot[2], for learning a CRF[3] to undertake the task of identifying products model numbers occurring in forum posts. For enabling the learning process, our method requires only a set of seed model numbers, which means it does not require that annotated training sentences from the target forum/category. The category is implicitly determined by the provided seeds. We argue that obtaining these sets of seeds is fairly easy, since they are available in product listings from retail Web sites.

ModSpot has two main steps. In the first step, it performs a bootstrapping process, where input seeds are expanded into multiple surface forms to account for variations. Each expanded surface form is annotated in input sentences to train an initial CRF. In the second step, a self-training [9] process is carried out. ModSpot uses the output of the initial CRF to discover new model numbers in unlabeled sentences. New model numbers with high probability are added to the set of seeds and are again expanded into multiple surface forms, that are again annotated in unlabeled input sentences to train a new CRF. This process runs until no new seeds are found.

Experiments in four different settings demonstrate that ModSpot achieves similar or better results compared to using a supervised CRF. Our method converges at 9-14 iterations, where there is no growth in the seed set. All the experimented settings exhibit higher F-measures when the self-training process converges, and the number of seeds is about 40% larger by the end of the process. In particular, the expansion in seeds helps to achieve higher recall levels.

2 Method Description

The method we propose is based on the self-training framework [9]. Thus, our algorithm makes extensive use of unlabeled data for training a CRF. This simple strategy has two drawbacks: incorrect labeled instances can be included in the training set and errors are reinforced. To cope with these problems, we ensure reliable labeling by specific recognition criteria. In our self-training setting, we observe that the probabilities of the instances converge such that a final CRF is obtained after a number of iterations.

To make the process independent from the target forum, our method also includes a bootstrapping step, that takes as input a set of *seeds*, i.e., examples of product model numbers, to automatically generate an initial training set of labeled sentences. To maximize the number of sentences in this initial training set, we also detect variations, i.e., distinct surface forms, of the given seeds.

We detail our method in Algorithm 1. Let S_0 be an initial set of seeds, that is, examples of product model numbers, and U be a set of unlabeled sentences extracted from posts of a target forum. An initial training set L is automatically generated by bootstrapping from U (Lines 1-2). In this bootstrapping process, we

[2] Product **Mod**el Number **Spot**ter.

[3] Through the text, we use CRF as a synonym for CRF model.

detect surface form variations using our *SFDetection* algorithm. This detection should account for the various ways users typically mention product models. Product model mention variations are automatically annotated in sentences from U to generate training set L. Non-annotated sentences are assigned to set T. This set will be used later to enhance the seeds set with newly discovered seeds using a linear-chain CRF.

Algorithm 1. ModSpot

Require: Set of seeds S_0, set of unlabeled sentences U
1: $L \leftarrow$ SFDetection(U, S_0) ▷ *Bootstraps and detects Surface Forms*
2: $T \leftarrow U - L$
3: Build the initial CRF $\hat{\theta}_0$ from L only
4: $i \leftarrow 1$ ▷ *Self-training Process*
5: **repeat**
6: Use $\hat{\theta}_{i-1}$ to label unlabeled sentences in T ▷ *Use CRF to predict new labels*
7: $C \leftarrow$ the set of sentences labeled by $\hat{\theta}_{i-1}$
8: $M \leftarrow$ SeedExpansion(C) ▷ *Selects likely seeds*
9: $S_i \leftarrow S_{i-1} \cup M$
10: $L \leftarrow$ SFDetection(U, S_i) ▷ *Detects Surface Forms*
11: $T \leftarrow U - L$
12: Build a new CRF $\hat{\theta}_i$ from L only ▷ *Re-train CRF with new labels*
13: **until** $|S_i| = |S_{i-1}|$ ▷ *No new seeds are found*
14: **return** $\hat{\theta}_i$ ▷ *Return last CRF generated*

An initial CRF $\hat{\theta}_0$ is trained using the automatically annotated sentences in L (Line 3). CRF training is performed with stochastic gradient descent and L1 regularization. Now, with a bootstrapped CRF, our self-training iteration process (Lines 5-13) begins. The algorithm iterates until it converges to a state where output from the trained CRFs does not change from one iteration to the next. In Lines 6-9, the algorithm performs the label prediction step to discover new likely seeds. The current CRF $\hat{\theta}$ labels the unlabeled sentences in T, creating a labeled sentence set C. From the sentences in C, we run our *SeedExpansion* step that discovers new seeds into set C. Finally, set M is added to the current seeds set S_i (Line 9) expanding the initial seeds set with newly discovered product model mentions. Between the label prediction and the CRF re-training is another bootstrapping process (Lines 10-11). This process is the same that automatically generated the training set L during initialization, but uses the expanded seeds set S_i as input. Again product model mention variations are automatically annotated in sentences from U to generate the training set, and each non-annotated sentence is added to T.

In Line 12, the algorithm trains a new CRF, which is the final step in our method. This step estimates the CRF $\hat{\theta}_i$ parameters using the automatically annotated sentences in L generated from the bootstrapping process executed after the label prediction step. Our self-training algorithm convergence is determined by the difference of the seed set S_i from the current iteration and the seed set from the previous iteration S_{i-1} (Line 13).

To account for surface form variations, we devised a surface form detection algorithm. Let s be a seed from a set S containing examples of product model numbers. We model s as a sequence $x_1, x_2, \ldots, x_n$, where each x_i is a token composed of only letters or only digits. Each token x_i is called a *block*. Thus, s is a sequence of blocks. As an example, take product model number "BDP-51FD". Its sequence of blocks is "BDP", "51", "FD". We define a *surface form* f of s as being a sequence of blocks such that one of the conditions below holds: (1) f is a sub-sequence $f_1, f_2, \ldots, f_n$ of s, with $n > 1$, occurring in at least one input sentence, or; (2) f is a single block of s, composed by digits only, occurring in at least one input sentence, and the context in which f occurs in input sentences is *similar* to the context in which some known surface form of s occurs in the input sentences.

According to (1), possible surface forms of "BDP-51FD" are "BDP51FD", "BDP51", and "51FD", if they occur in at least one sentence. In the case of the second condition, "51" is a possible surface, given that the context in which it occurs in the input sentences is similar to that of another occurrence of s.

To avoid confusing any number occurring in sentences as a surface form, we use the context implied by the input sentence for disambiguation. We consider as context portions of terms occurring before and after a surface form. Then, the context similarity is computed as follows. Consider a surface form t, which satisfies our first condition, represented by a vector $\mathbf{v}$ in which v_i is the frequency of term v_i in t within a fixed size context in the same input sentence. Also consider the same vector representation $\mathbf{w}$ for a candidate surface form t_c. We define the similarity between t_c and t as:

$$sim(\mathbf{w}, \mathbf{v}) = \frac{\mathbf{w} \cdot \mathbf{v}}{|\mathbf{w}||\mathbf{v}|} = \frac{\sum_{i=1}^{m} w_i v_i}{\sqrt{\sum_{i=1}^{m} w_i^2}\sqrt{\sum_{i=1}^{m} v_i^2}} \tag{1}$$

where m is the size of a common vocabulary used by $\mathbf{v}$ and $\mathbf{w}$.

We consider the two contexts as being similar if $sim(\mathbf{w}, \mathbf{v})$ is above a predefined threshold. We arbitrarily determined a value of 0.5 for this threshold. In addition, after a few initial experiments, not reported here, we concluded that a context of size 3 is suitable for our application.

The automatic seed expansion process must be carried out without adding spurious seeds to the seeds sets. Thus, our method adopts strict criteria in order to use only high confidence seeds. Our first criterion is CRF labeling confidence. We use the so-called *posterior decoding* (Forward-Backward Algorithm) [1, 2] instead of the classical Viterbi decoding. This algorithm allows CRF to output normalized scores by evaluating all possible paths given an observation. The Forward pass is defined by:

$$\alpha_{i+1}(y_j) = \sum_{y' \in \mathbf{y}} \left[\alpha_i(y') \exp\left(\sum_{k=1}^{n} \lambda_k f_k(y', y_j, \mathbf{x}, i) \right) \right] \tag{2}$$

where α_{i+1} is the Forward-values vector used by the algorithm. the Backward pass is symmetric to the Forward pass. We determined a high threshold value of 0.9 for our probability confidence.

Our second criterion is the number and type of blocks from the terms labeled by the CRF. Consider that a labeled term is also modeled as a sequence

$x_1, x_2, \ldots, x_n$, where each x_i is a token composed of only letters or only digits, and each token x_i is a block. A valid seed has at least one block of each type, and the blocks have length greater than one.

3 Experimental Results

To evaluate our method, we used four distinct datasets[4] from three different product categories of consumer electronics. Posts are in English in all datasets but the HT dataset, which uses Portuguese. It was included to verify the resilience of our method to different languages. We randomly sampled 200 posts for manual labeling from each dataset. Table 1 gives statistics for all the datasets.

Table 1. Datasets statistics – 200 posts per dataset

Dataset	Labeled Sentences	Product Model Number Mentions	Numeric only Mentions	Posts with Mentions	Avg. Mentions per Post
AVS AVR	986	234	115 (49.1%)	99 (49.5%)	2.4
AVS BDP	1151	280	110 (39.3%)	96 (48.0%)	2.9
AVS LCD	963	135	31 (23.0%)	60 (30.0%)	2.2
HT BDP	875	148	42 (28.4%)	71 (35.5%)	2.0

The initial input seeds were collected from Amazon.com for each product category. The amount of initial seeds for each category is 747 for AVR, 323 for BDP, and 1375 for LCD.

Our experiments compare ModSpot with a supervised CRF generated for each dataset. We consider supervised CRF to be a suitable baseline for comparison with our method, since it is regarded as very effective for NER tasks [7]. In the experiments, the results obtained were evaluated against the golden set. We used the well-known precision, recall, and F-measure metrics.

We adopt features widely used in previous work [4,7,10]. These features are described in Table 2. Although CRFs are flexible enough to allow specific features for different domains, we used the same set of features and configurations in all experiments. It is important to note that our self-training procedure uses the same set of features and configurations as the baseline.

Table 2. Features used by CRF

Set	Description
0	Current token
1	Tokens in a context window of size 3
2	Part-of-speech tag of the current token and of the tokens in the context window
3	Token begins with uppercase, token is all uppercase and token has a character that is uppercase
4	Token is numeric, token is a combination of alphanumeric characters and token has punctuation

The first experimental result we report is in Table 3. ModSpot results are compared with supervised CRF. The evaluation was calculated from the final CRF generated at convergence. CRF results are the average from cross-validation.

[4] Available at `http://shine.icomp.ufam.edu.br/~henry/datasets.html`.

Table 3. ModSpot vs. CRF

Forum	Category	Method	P	R	F
AVS	AVR	CRF	**0.77**	0.55	0.63
		ModSpot	**0.77**	**0.69**	**0.73**
AVS	BDP	CRF	**0.93**	0.73	0.81
		ModSpot	0.84	**0.88**	**0.86**
AVS	LCD	CRF	**0.81**	0.34	0.47
		ModSpot	0.68	**0.72**	**0.70**
HT	BDP	CRF	0.86	0.55	0.65
		ModSpot	**0.88**	**0.63**	**0.73**

Table 4. ModSpot (no SFD) vs. SFD only

Forum	Category	Method	P	R	F
AVS	AVR	ModSpot-SFD	0.67	0.22	0.33
		SFD	0.96	0.19	0.31
AVS	BDP	ModSpot-SFD	0.96	0.25	0.40
		SFD	0.99	0.53	0.69
AVS	LCD	ModSpot-SFD	0.71	0.42	0.53
		SFD	0.95	0.41	0.57
HT	BDP	ModSpot-SFD	0.87	0.51	0.65
		SFD	1.00	0.37	0.54

We can see that ModSpot achieved higher values for recall and F-measure in all forums and categories. On average, our recall value is approximately 19% higher while the F-measure value is approximately 12% higher. This is a direct result of our Surface Form Detection algorithm. Table 4 highlights the importance of the Surface Form Detection algorithm showing the results obtained when this procedure is not used. This configuration is equivalent to the methods presented in [6,8]. These results correspond to lines labeled "ModSpot-SFD". In two forum/category pairs ModSpot achieved higher or equal precision despite higher recall. Also, in Table 4, we report the results of Surface Form Detection alone against the manually labeled golden set. These results correspond to lines labeled "SFD". The recall for all datasets is not high, since we do not have all products in the initial seeds. This demonstrates the effectiveness of using a CRF in our self-training approach to achieve higher levels of recall, and also the accuracy of our bootstrap approach.

In Figure 1, we detail the results from Table 3 by showing the results of each self-training iteration by forum and product category.

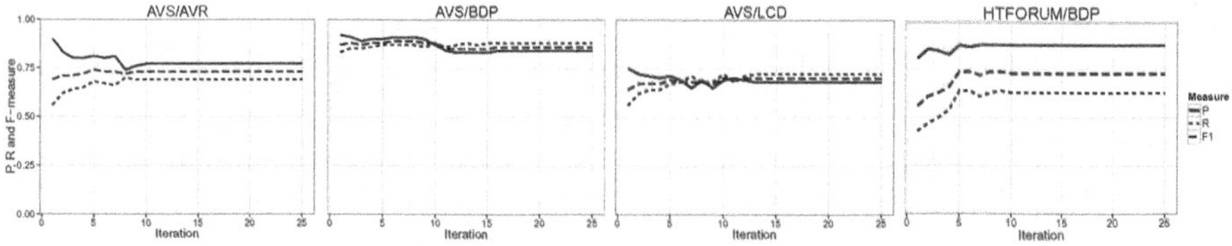

Fig. 1. Precision, recall and F-measure for different datasets per self-training iteration

We can see that our method converges at around 9-14 iterations, where there is no growth in the seed set. All the experimented datasets exhibit higher recall and F-measure when the method converges. This is caused by newly discovered seeds that are used to annotate new training sentences.

Figure 2 shows the number seeds in each iteration. The first seeds correspond to the initial input seeds manually extracted from products descriptions; further seeds were automatically expanded during the self-training process, and incorporated into our method to annotate new training sentences. The number of seeds is, on average, about 40% larger by the end of the process.

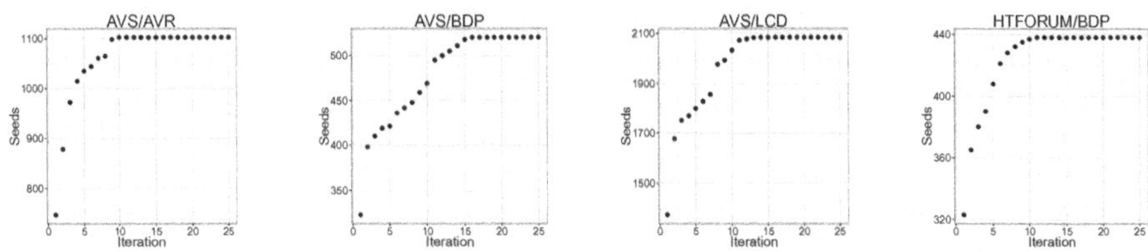

Fig. 2. Seed growth in each self-training iteration

4 Conclusions

We presented ModSpot (Product **Mod**el Number **Spot**ter), a method for learning a CRF to undertake the task of identifying model numbers of products. The method is based on a self-training process that requires only a set of initial seed model numbers from consumer products, which means it does not require annotated training sentences to be provided. Experiments in four settings demonstrated that our method achieved similar or better results when compared to a supervised CRF with the same feature set. All the experimented settings exhibited higher F-measures when our process finished, and the seed set is about 40% larger. In particular, the expansion in seeds performed by the method helped to achieve higher recall levels. Our method converged at around 9-14 iterations, when ModSpot could not identify new seeds. Finally, based on our product model mention detection, we plan to investigate product disambiguation and linking.

Acknowledgments. This work is partially supported by projects TTDSW (PRONEM/FAPEAM) and eSpot (CNPq grant 461231/2014-0), by individual CNPq fellowship grants to Altigran S. da Silva, Marco Cristo and Edleno S. de Moura, and by a FAPEAM/RHTI scholarship to Henry Vieira.

References

1. Chen, M., et al.: Crf-opt: An efficient high-quality conditional random field solver. In: Proc. of the 2008 AAAI, pp. 1018–1023 (2008)
2. Culotta, A., et al.: Confidence estimation for information extraction. In: Proc. of the 2004 HLT-NAACL, pp. 109–112 (2004)
3. Feldman, R.: Techniques and applications for sentiment analysis. Communications of the ACM 56(4), 82–89 (2013)
4. Jakob, N., et al.: Extracting opinion targets in a single-and cross-domain setting with conditional random fields. In: Proc. of the 2010 EMNLP, pp. 1035–1045 (2010)
5. Liu, B.: Sentiment analysis and opinion mining. Synthesis Lectures on Human Language Technologies 5(1), 1–167 (2012)
6. Putthividhya, D.P., et al.: Bootstrapped named entity recognition for product attribute extraction. In: Proc. of the 2011 EMNLP, pp. 1557–1567 (2011)

7. Sarawagi, S.: Information extraction. FTDB 1(3), 261–377 (2008)
8. Teixeira, J., et al.: A bootstrapping approach for training a ner with conditional random fields. In: Proc. of the 2011 EPIA, pp. 664–678 (2011)
9. Yarowsky, D.: Unsupervised word sense disambiguation rivaling supervised methods. In: Proc. of the 1995 ACL, pp. 189–196 (1995)
10. Zhang, L., et al.: Entity set expansion in opinion documents. In: Proc. of the 2011 ACM HT, pp. 281–290 (2011)

Information Extraction Grammars

Mónica Marrero[1] and Julián Urbano[2]

[1] Barcelona Supercomputing Center, Spain
monica.marrero@bsc.es
[2] Universitat Pompeu Fabra, Barcelona, Spain
julian.urbano@upf.edu

Abstract. Formal grammars are extensively used to represent patterns in Information Extraction, but they do not permit the use of several types of features. Finite-state transducers, which are based on regular grammars, solve this issue, but they have other disadvantages such as the lack of expressiveness and the rigid matching priority. As an alternative, we propose Information Extraction Grammars. This model, supported on Language Theory, does permit the use of several features, solves some of the problems of finite-state transducers, and has the same computational complexity in recognition as formal grammars, whether they describe regular or context-free languages.

1 Introduction

Information Extraction (IE) seeks to identify entities in a text and the relations among them, altogether satisfying the user search needs. Named Entity Recognition (NER) is one of the main areas in IE, needed to identify entities of interest such as persons, locations and dates. It is generally accepted that IE systems should be capable of adapting to different entities and domains [4], but patterns to recognize named entities usually respond to different lexicons and grammatical structures, and they often require high-level features (e.g., lemma, letter case, gazetteers). Besides machine learning-based models, grammars are also widely used to represent these patterns, especially regular expressions. However, they can only recognize entities that respond to one type of feature (usually characters), so their use is restricted to entities that follow a simple pattern.

Cascade grammars are used to overcome this limitation. They are built with transducers (finite-state automata that also generate an output language) concatenated such that the output alphabet from one transducer is the input alphabet to the next one [5]. This way, it is possible to use several features in the same pattern. The Common Pattern Specification Language (CPSL) [2] standardizes this representation, although it presents several drawbacks, as detailed in [3].

First, and because they are in cascade, each of the finite-state automata is independent of the others, which can lead to ambiguities and the application of incorrect rules early in the recognition process. For example, we can have a cascade grammar to recognize person names with two transducers, $P1$ and $P2$. The input alphabet to $P1$ are the tokens, and the resulting alphabet contains

A. Hanbury et al. (Eds.): ECIR 2015, LNCS 9022, pp. 265–270, 2015.

tags from two gazetteers that identify first and last person names: F and L, respectively. $P2$ identifies full names of persons from these tags by recognizing one of three rules: F, FL or FLL. However, if a text chunk can be tagged as F and L in $P1$, until $P2$ we can not decide which one is better. Consider for instance the text *"Lisa Brown Smith"*, where *"Smith"* can be both F and L. This language is ambiguous, but our domain knowledge may help us disambiguate it by assigning more priority to rule FLL than to FL and F. However, even though the language would no longer be ambiguous, the pattern still would, because $P1$ must decide between F and L *before* executing $P2$.

Second, and because they are as expressive as regular grammars, they can not be used to describe complex languages. Context-free grammars are especially useful when we face markup languages, because it is common to find nested structures that can not be recognized with regular grammars. Wrappers are often used in these cases, but they are task-specific.

To partially overcome the first problem, CPSL has been customized to different extents by different platforms. GATE (`http://gate.ac.uk`) is one of the most successful ones with its JAPE notation. In this scenario we propose Information Extraction Grammars (IEG) as an alternative to represent patterns for entity recognition. It solves the ambiguity issue of cascade grammars, has the expressiveness of context-free grammars, and at the same time it provides more flexibility than wrappers. Furthermore, the main advantage of our proposal is that it contributes to the development of pattern generation methods that can work independently of the kind of features used and the expressiveness of the language to recognize [7].

2 Information Extraction Grammars

Context-free grammars are defined with a tuple $G = (\mathcal{V}, S, \Sigma, \mathcal{P})$, where $\mathcal{V}$ is the set of non-terminal symbols, $S \in \mathcal{V}$ is the initial symbol, Σ is the set of terminal symbols making up the input alphabet, and $\mathcal{P}$ is the set of production rules which recursively define the language recognized by G. Productions are defined by a non-terminal, followed by the production symbol $\rightarrow$ and a sequence of terminals and non-terminals—the production body. The language recognized by G is the set of strings of terminal symbols that can be derived from S:

$$L(G) = \{\omega \in \Sigma^* \mid S \overset{*(G)}{\Longrightarrow} \omega\}$$

where $\overset{*(G)}{\Longrightarrow}$ represents derivations in zero or more steps, that is, replacements of the non-terminals with the body of one of their productions, sequentially from the initial symbol until we reach strings with terminal symbols alone.

These grammars do not support the recognition of more than one input alphabet at the same time. For example, it is not possible to recognize the syntax of a text and whether its tokens are included in gazetteers or not. To solve this problem we have associated conditions to the non-terminal symbols, so that for an input string $\omega \in \Sigma^*$ and a non-terminal $A \in \mathcal{V}$ such that $A \overset{*(G)}{\Longrightarrow} \omega$, ω will be

$$S \to FLL \mid FL \mid F \qquad F \to T \qquad \mathcal{C}_F = \{(FirstGaz, true)\}$$
$$T \to [\texttt{a-zA-Z0-9}]+ \qquad L \to T \qquad \mathcal{C}_L = \{(LastGaz, true)\}$$

Fig. 1. IEG for the recognition of full person names

recognized by A only if it meets all conditions associated to A. Each condition is defined with a tuple (f, y), where $f : \Sigma^* \to \mathcal{Y}_f$ is a feature function that receives a string ω and is expected to return $y \in \mathcal{Y}_f$. Note that the set $\mathcal{Y}_f$ of possible values returned by f depends on the particular type of feature. For example, if f returned the lemma of a term, we would have $\mathcal{Y}_f \subset \Sigma^*$; if f returned its length, we would have $\mathcal{Y}_f = \mathbb{N}$. We thus define an IE grammar as $IEG = (G, \mathcal{C})$, where $\mathcal{C}$ is the set of all condition sets assigned to non-terminals. All derivations must therefore meet:

$$A \overset{*(IEG)}{\Longrightarrow} \omega := A \overset{*(G)}{\Longrightarrow} \omega \text{ and } \forall (f, y) \in \mathcal{C}_A \; : \; f(\omega) = y$$

That is, each and every condition associated to A must return the expected value. Figure 1 shows an IEG to recognize person names as in our previous example. To solve the ambiguity of the pattern, we can assign priorities to the different productions of a non-terminal, as we did: $S \to FLL$ first, followed by $S \to FL$ and $S \to F$. At the same time, we could add new rules indicating that the person names have to appear inside specific HTML tags, or new conditions forcing the matching with other features such as POS tagging or font family.

We note that an IEG is similar in definition to an S-attributed grammar, widely used in compilers and language translators [1]. In the latter, any symbol of the grammar may have a set of attributes, and the attributes of the non-terminals are computed in a bottom-up fashion by semantic rules associated to their productions. In particular, these rules are applied when the production reduces an input substring or after the whole input is parsed. However, semantic rules are only used to incorporate semantics to the parse tree; they do *not* intervene in the syntactic analysis. This is precisely the purpose of our conditions: to avoid applying a production, *during* syntactic analysis, when the conditions are not met by the substring.

3 Analysis of Computational Complexity

Text recognition with regular grammars is often performed with automata; for context-free grammars, the Cocke-Younger-Kasami (CYK) is one of the most common algorithms used [6]. But the introduction of conditions to a formal grammar requires modifying these algorithms, which could have an impact in terms of efficiency. Next, we show that the time complexity does not increase as long as the conditions meet some requirements.

3.1 Regular Grammars

Regular grammars are usually represented with regular expressions. A regular expression of length r can be converted to an ε-NFA (non-deterministic finite

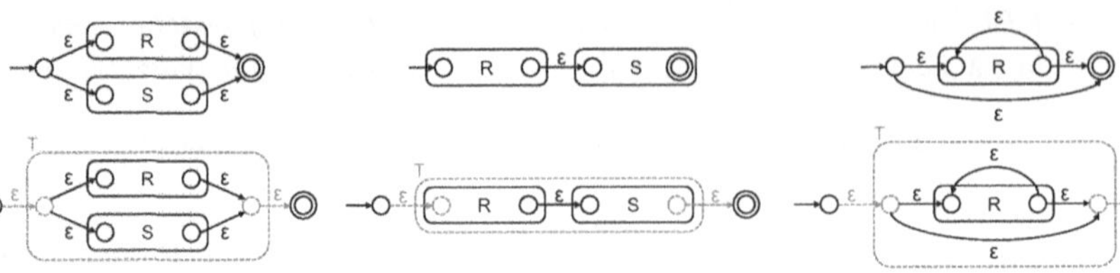

Fig. 2. ε-NFA automata for the union, concatenation and Kleene closure operations (top), and equivalent automata with additional ε-transitions (bottom)

state automaton with ε-transitions) in $\mathcal{O}(r)$ time. Given an input string of length n, an ε-NFA takes $\mathcal{O}(ns^2)$ in the worst case to recognize it, where $s \leq 2r$ is the number of states. This automaton can be transformed into a DFA (deterministic finite state automaton), which takes $\mathcal{O}(n)$ to recognize an input string [6].

We can represent every rule of the IEG with an associated condition as a T-automaton with two extra states and ε-transitions (see Figure 2, bottom). The extra ε-transition at the beginning is used to save the position in the input string; the one at the end is used to check all conditions associated to the regular expression represented by the T-automaton. It will only continue to the acceptance state if all conditions in $\mathcal{C}_T$ are met with the current substring (see Figure 3). Since the union, concatenation and Kleene closure of a regular language is also a regular language (see Figure 2, top), we can concatenate all T-automata to obtain an ε-NFA. For each symbol in the input string, we have to check at most m conditions in each of the t T-automatons, each taking d time. Therefore, the time complexity of recognition remains linear with respect to n. By using appropriate indexing mechanisms, d can be $\mathcal{O}(1)$, so the time complexity would be $\mathcal{O}(n(tm + s^2))$ in the worst case. In practice though, the conditions reduce the number of active states in the ε-NFA, reducing the s^2 factor.

A cascade grammar with $\mathcal{O}(m)$ transducers would have a recognition time of $\mathcal{O}(mns^2)$ if using ε-NFA. If each transducer is converted to a DFA, it can take $\mathcal{O}(mn)$. However, that conversion requires $\mathcal{O}(2^r)$ for each transducer, so it can be simpler and more efficient to use ε-NFA directly [6].

3.2 Context-free Grammars

For an arbitrary string $\omega = \alpha_1\alpha_2 \ldots \alpha_n$, the CYK algorithm builds a triangular table X. Each cell X_{ij} in the table contains those non-terminals capable of deriving the substring $\alpha_i\alpha_{i+1} \ldots \alpha_j$ in one or more steps. The table is filled bottom-up, so that in the bottom row we will have those non-terminals that directly derive each of the terminals in ω. In the row above we will have those

Fig. 3. T-automaton with associated conditions

$S \rightarrow \alpha A \beta$	$S \rightarrow \alpha\beta \mid \alpha A \beta$
$A \rightarrow \gamma \mid \varepsilon$	$A \rightarrow \gamma$ ~~$\mid \varepsilon$~~
$\mathcal{C}_S = \{(f_{S1}, y_{S1}), \ldots\}$	$\mathcal{C}_S = \{(f_{S1}, y_{S1}), \ldots\}$
$\mathcal{C}_A = \{(f_{A1}, y_{A1}), \ldots\}$	$\mathcal{C}_A = \{(f_{A1}, y_{A1}), \ldots\}$

Fig. 4. IEG with ε-productions (left) and its equivalent without ε-productions (right)

non-terminals whose production bodies contain the non-terminals below to form the corresponding substring, and so on. The algorithm stops if in the uppermost cell there is a non-terminal capable of generating the whole input string ω.

This algorithm has a time complexity of $\mathcal{O}(n^3)$, because it takes $\mathcal{O}(n)$ to compute any cell in the table and there are $n(n+1)/2 = \mathcal{O}(n^2)$ cells. If the non-terminals have associated conditions, then the time to check them is added to the time taken by the algorithm to fill up each cell. Again, if the time needed to check each condition is independent of the input string, the overall complexity is kept. But there are two prerequisites for the application of the CYK algorithm: the grammar can not have ε-productions, where the right side contains just the empty string ε; and it must be defined in Chomsky Normal Form (CNF).

Removal of ε-productions. Even though the use of ε-productions may facilitate the design of grammars, they are not essential for any language other than the empty string. Thus, if a language L has a grammar, then $L - \{\varepsilon\}$ has one too without ε-productions [6]. The algorithm for this transformation identifies the nullable non-terminals. A non-terminal A is nullable if $A \stackrel{*}{\Longrightarrow} \varepsilon$. Whenever A appears in a production body, we make two versions of the production: one with A and one without it, thus removing all productions whose right part is ε.

In our case we need to determine what happens when those nullable non-terminals, or the non-terminals which contain them, have associated conditions. Non-terminals whose production body is just ε can be removed altogether, because we know beforehand whether the conditions are met or not for ε. On the other hand, we have to create two versions for the non-terminals containing them in their production bodies. This does not pose any problem either for non-terminals that contain these nullable symbols, because the conditions are applied upon the resulting substring, regardless of how it is reduced (see Figure 4).

Transformation to CNF. A grammar is in Chomsky Normal Form when all its production rules have one of these forms [6]: (i) $A \rightarrow BC$, where A, B and C are non-terminals, or (ii) $A \rightarrow \alpha$, where A is a non-terminal and α is a terminal. To achieve this, after removing all ε-productions we need to (i) if there are two or more terminals in a production body, replace each of them with a new non-terminal that produces the terminal itself, and (ii) iteratively reduce production bodies so that they contain two non-terminals at most, by creating again new non-terminals and production rules for them.

In both cases we are adding sublanguages of the language recognized by a non-terminal A to the new non-terminals in its production body. Because all conditions apply to the language recognized by A, and not to its sublanguages, the result is not affected when A has conditions (see Figure 5).

$S \rightarrow \alpha\beta\gamma$	step 1: $S \rightarrow ABC$	step 2: $S \rightarrow RC$
$\mathcal{C}_S = \{(f_{S1}, y_{S1}), \ldots\}$	$A \rightarrow \alpha$	$R \rightarrow AB$
	$B \rightarrow \beta$	$A \rightarrow \alpha$
	$C \rightarrow \gamma$	$B \rightarrow \beta$
	$\mathcal{C}_S = \{(f_{S1}, y_{S1}), \ldots\}$	$C \rightarrow \gamma$
		$\mathcal{C}_S = \{(f_{S1}, y_{S1}), \ldots\}$

Fig. 5. Transformation of an IEG into CNF

4 Conclusions and Future Work

The complexity of patterns for Named Entity Recognition often requires automatic learning methods. Grammar-based models currently used to represent patterns are limited in the features they support and the expressiveness of the languages they recognize. As a consequence, different learning methods are developed for different representation models. For end users, this often requires a previous analysis of the entities to choose a model and learning method.

We propose Information Extraction Grammars as a common model to represent patterns. This model supports a custom set of features, it has the expressiveness of context-free grammars, and it avoids the ambiguity issue of cascade grammars, thus facilitating the development of portable learning methods. An analysis of the time complexity in recognition shows that it is competitive when used by standard recognition algorithms, though further research is needed to optimize them. Another extension may be the use of probabilities in the feature functions, allowing us to select the most likely parse tree for an ambiguous input.

Acknowledgments. Work partially supported by an A4U postdoctoral grant. We thank the reviewers for their comments.

References

1. Aho, A.V., Lam, M.S., Sethi, R., Ullman, J.D.: Compilers: Principles, Techniques, and Tools. Addison-Wesley, 2nd edn. (2006)
2. Appelt, D., Onyshkevych, B.: The common pattern specification language. In: TIPSTER Workshop, pp. 23–30 (1998)
3. Chiticariu, L., Krishnamurthy, R., Li, Y., et al.: Systemt: An algebraic approach to declarative information extraction. In: ACL, pp. 128–137 (2010)
4. Freitag, D.: Toward general-purpose learning for information extraction. In: ACL, pp. 404–408 (1998)
5. Hobbs, J.R., Riloff, E.: Information Extraction. In: Indurkhya, N., Damerau, F. (eds.) Handbook of Natural Language Processing, pp. 511–532. CRC Press (2010)
6. Hopcroft, J., Motwani, R., Ullman, J.: Introduction to automata theory, languages, and computation. Addison-Wesley, 3rd edn. (2006)
7. Marrero, M.: Diseño y generación semi-automática de patrones adaptables para el reconocimiento de entidades. Journal of the SEPLN 52, 87–90 (2014)

Target-Based Topic Model for Problem Phrase Extraction

Elena Tutubalina

Kazan (Volga Region) Federal University, Kazan, Russia
tutubalinaev@gmail.com

Abstract. Discovering problems from reviews can give a company a precise view on strong and weak points of products. In this paper we present a probabilistic graphical model which aims to extract problem words and product targets from online reviews. The model extends standard LDA to discover both problem words and targets. The proposed model has two conditionally independent variables and learns two distributions over targets and over text indicators, associated with both problem labels and topics. The algorithm achieves a better performance in comparison to standard LDA in terms of the likelihood of a held-out test set.

Keywords: information extraction, problem phrase extraction, Latent Dirichlet Allocation, topic modeling.

1 Introduction

Information extraction has received much attention in the past decade because of its capability to recognize entities of certain types in unstructured text. Most methods extract named entities, relations, facts (or events), and sentiment information. Problem phrase extraction allows a company to identify and fix problems, improving a product or a service. Here is an example text with a problem phrase:

Example. 1. After hours with tech support, I discovered that my amazing **iPhone cannot connect to the car**.

The "iPhone" is the target of the sentence, problems with which were discussed by users in a review, and this target can be related to an electronic product. Targets are common components of an object in a particular domain. They are presented by topics such as "performance", "value" and "feature set", "design" (e.g., a computer domain). A particular review about a phone contains users' comments about a network connection or a color of phone cover as major features and not mentions about the phone's processor. The goal is to discover a set of targets and problem words of some particular category for each review in an unsupervised manner. Problem detection and extraction of problem phrases have been studied in several papers ([1], [2], [3], [4]). Recent studies on problem phrase extraction proposed different approaches to extract targets of problem phrases:

A. Hanbury et al. (Eds.): ECIR 2015, LNCS 9022, pp. 271–277, 2015.

using a supervised classifier to select top ranking noun phrases as targets [1], and using dependency relations between the targets and the problem indicators to detect targets [3]. These approaches are limited due to dictionaries' size and lower results after shifting to another domain.

Extracting mining information from unstructured text, such as user reviews, news texts, or microblogs, has received much attention in sentiment analysis ([5], [6], [7]), event detection, and public sentiment tracking. State-of-the-art papers have implemented probabilistic topic models, such as Latent Dirichlet Allocation (LDA) [8], for multiaspect analysis tasks ([9], [10], [11], [12]). It identifies a topic structure of textual data using co-occurrence of terms.

In this study we focus is to apply probabilistic modeling techniques to extract problem targets, mentioned in user reviews, and to detect problem indicators in the text. Our contribution of this work is in organizing product targets and text indicators by integrating domain-independent knowledge about problem phrases.

2 Related Work

Problem phrase extraction, based on detection of the targets, has been studied in [1] and [3]. Gupta studied the extraction of problems with AT&T products and services from English Twitter messages [1]. He proposed that each possible problem phrase had several candidate target noun phrases. The author used a supervised method to train a maximum entropy classifier. Gupta reported the best performance F-measure of 75% for the identification of the target phrase. However, the author did not test the maximum entropy classifier in other domain. In [3] authors focused on finding domain-specific targets of problem sentences, based on user reviews of electronic and automotive products. The targets were extracted using problem phrase structure with dependency relations. WordNet categories were used to reduce targets that are not semantically related to a product domain. However, the proposed approach has not considered user interest in the particular aspect of the product (e.g., price, package design, and device sound quality). As a result, the average F1-measure was decreased from 0.84 to 0.77 after reducing non-domain-specific targets in a target set.

Probabilistic topic models have been successfully used for sentiment analysis ([9], [10], [11], [12]). In [11] user reviews were analyzed to discover sentiment aspects that depended on the authors of the reviews. They supposed that different users express similar opinions with different sentiment polarities. The proposed sLDA method, which modeled the user information, obtained a better accuracy of 0.51 in comparison with the unsupervised LDA with the SVM classifier with an accuracy of 0.39. In [13] Zhao et al. proposed a multigrain LDA-based model to jointly discover and separate aspects and opinion words. According to the paper, each word in a sentence describes aspect-specific (e.g., "friendly", which is associated with "waiter") or general opinion sentiments. They integrated a maximum entropy component to use POS tags for separating aspect opinion words. Several studies ([13], [14], [11]) showed that supervised topic models (sLDA), trained on samples with specific labels (e.g., sentiment labels), give better results. However, an issue of supervised topic models is an insufficient training data. The model we propose is unsupervised algorithm, that extract problem targets and relate text indicators with problem labels and topics, using a collection of user reviews.

3 Model Description

In this section we describe a proposed model for problem phrase extraction, called PrPh-LDA, which is an extension of LDA. The LDA model, as shown in Figure 1(a), presents the documents as a mixture of topics, where a topic is a probability distribution over words. The key difference (versus LDA) is that PrPh-LDA contains two latent distributions over words in each user review, as shown in Figure 1(b).

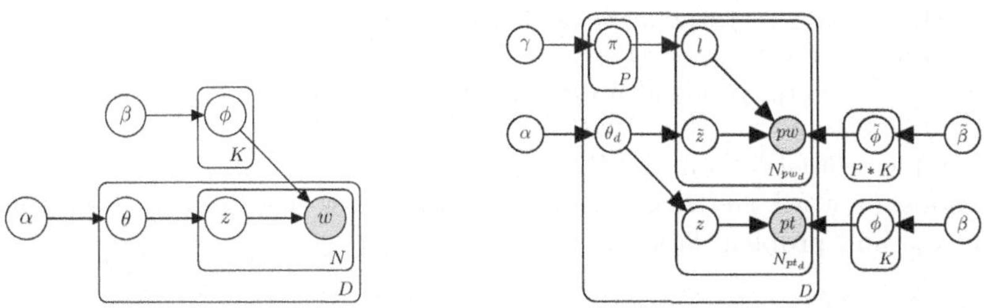

(a) LDA model **(b)** PrPh-LDA

Fig. 1. Graphical model representation for (a) LDA and (b) PrPh-LDA

In PrPh-LDA model we modify the LDA model to handle multiple classes of problem words and targets. Targets are usually nouns (i.e., products or product part of the products). A problem word is an indicator and may be represented as a verb (e,g, *replace* and *return*), a noun with a preposition (e.g., *problem with* and *error in*), an adverb (e.g., *too* and *still*), and an adjective (e.g., *bad* and *horrible*)[1]. The intuition behind the model is that every review d contains some number of problem words N_{pw_r} and targets N_{pt_r}. We assume that the problem words don't depend on the targets, because the targets may be treated as factual information of problem mentions and they do not affect the polarity of the problem words.

Assume that we have a collection of reviews $D = \{d_1, d_2, \ldots, d_D\}$, where each review is a sequence of N_d words (with $N_d = N_{pt_d} + N_{pw_d}$, the sum of all indicators plus targets). Each word in the review is denoted by item from a vocabulary $\{1, 2, \ldots, V\}$. Let K be the total number of topics, and P be the number of problem labels[2]. α, β, and $\tilde{\beta}$ are Dirichlet smoothing parameters, while θ is the topic-review distribution. The procedure of generating a problem word pw in the review d consists of several steps. The model selects a topic $\tilde{z}$

[1] Manually created dictionaries of problem words are discussed in [2] and [4].

[2] In this paper, the problem label is a binary variable $l \in \{l_{pr}, l_{no-pr}\}$, so P equals to 2.

from the document specific topic distribution θ_d. A problem label is chosen from the per-document problem distribution π. The problem word is chosen from $\tilde{\phi}_{\tilde{z},l}$, defined by both topic $\tilde{z}$ and problem label l. This part of the model is similar to other models, described in [15]. The procedure of generating noun targets is similar to standard LDAs, except we generate a sequence of N_{pw_d} targets.

Since the product targets are usually nouns in user reviews, we extract other words as text indicators to discover problem words. We use domain-independent knowledge about problem phrases by multiplying $\tilde{\beta}$ by λ_i for each topic $k \in \{1, \ldots, K\}$, where $\lambda_i = 1$ if word i is contained in the problem dictionaries ([2],[4]) and is assigned to problem label l_{pr}, otherwise $\lambda_i = 0$ for words with l_{pr}.

PrPh-LDA assumes the following generative process for a review:

1. For all d reviews sample $\theta_d \sim Dir(\alpha)$
2. For each problem label $l \in \{1, \ldots, P\}$ sample $\pi_{d,l} \sim Dir(\gamma_l)$
3. For each of the N_{pt_d} targets pt_i in the review d:
   (a) sample a topic $z_{d,pt_i} \sim Mult(\theta_d)$ and choose a target $pt_i \sim Mult(\phi_{z_{d,pt_i}})$
4. For each of the N_{pw_d} indicators pw_i in the review d:
   (a) sample a topic $\tilde{z}_{d,pw_i} \sim Mult(\theta_d)$
   (b) sample a label l_i from $Mult(\pi_{d,l})$
   (c) choose a word pw_i from the distribution over words defined by topic $\tilde{z}_{d,pw_i}$ and problem label l_i

3.1 Model Inference

In this section, we describe the inference algorithm for PrPh-LDA. The posterior distribution of the latent variables for the LDA models is difficult to compute [8]. Several approximate inference methods are applied, such as expectation propagation, variational inference, and Gibbs sampling. Following [16], we use a Gibbs sampling approach for inference because it is easy to extend. Due to the space limitation, we don't provide a detailed derivation and present the sampling parameters. The sampling methods of z and $\tilde{z}$ in PrPh-LDA are as follows:

$$P(\mathbf{z}_i = z|\mathbf{pt}_i = t, \mathbf{z}_{-i}, \mathbf{pt}_{-i}, \alpha, \beta) \propto \frac{\left\{n_d^{(z)}\right\}_{-i} + \alpha}{\{n_d\}_{-i} + K\alpha} \frac{\left\{n_z^{(t)}\right\}_{-i} + \beta}{\{n_z\}_{-i} + N_{pt}\beta} \tag{1}$$

$$P(\tilde{\mathbf{z}}_i = z, \mathbf{l}_i = l|\mathbf{pw}_i = t, \tilde{\mathbf{z}}_{-i}, \mathbf{l}_{-i}, \mathbf{pw}_{-i}, \alpha, \tilde{\beta}, \gamma) \propto$$
$$\frac{\left\{n_d^{(z)}\right\}_{-i} + \alpha}{\{n_d\}_{-i} + K\alpha} \frac{\left\{n_{z,l}^{(t)}\right\}_{-i} + \tilde{\beta}}{\{n_{z,l}\}_{-i} + N_{pw}\tilde{\beta}} \frac{\left\{n_{d,l}^{(z)}\right\}_{-i} + \gamma_p}{\{n_{d,l}\}_{-i} + \sum_{p=1}^{P} \gamma_p}, \tag{2}$$

where $n_z^{(t)}$ is the number of times a word t was assigned to topic z, $n_d^{(z)}$ is the number of words in review d assigned to topic z, $n_{z,l}^{(t)}$ is the number of times a word t were assigned to topic z and problem label l, $n_{d,l}^{(z)}$ is the number of times a word with problem label l from review d was assigned to topic z, $n_{z,l}$ is the number of times words were assigned to topic z and problem label l. The subscript $-i$ denotes a quantity excluding the current one.

4 Experiments and Evaluation

For our experiments, we collected 1,519 sentences about electronic products from the HP website[3]. We employed sentences from Amazon reviews[4] about baby and car products. We examine each review as a single sentence for analysis of short texts. These data were preprocessed with basic natural language processing techniques: we removed all the stopwords, the punctuations, and applied stemming to reduce the dimensionality of word spaces. Words with related negations are modified in conjunction with the negation.

Table 1. Summary of customer review dataset

Product domain	No. of sentences	No. of nouns	No. of other words
Electronics	1,519	7,176	8,516
Baby products	6,921	34,198	40,834
Cars	93,731	461,237	518,496

Table 2. Example words (stemmed), discovered by PrPh-LDA

Electronics				Baby products				Cars			
targets	*targets*	*pw*	*no-pw*	*targets*	*targets*	*pw*	*no-pw*	*targets*	*targets*	*pw*	*no-pw*
laptop	print	neg_work	long	babi	seat	old	love	batteri	oil	replace	perfect
batteri	cartridg	still	quick	chair	car	expen	soft	car	tire	wrong	black
comput	page	receiv	easili	bed	pad	wet	well	unit	air	old	left
price	ink	replac	consid	crip	strap	stuck	sit	power	gaug	return	bright
product	screen	neg_get	simpli	sheet	cover	neg_abl	give	time	pressur	help	highli
window	comput	help	highli	tabl	rail	disappoint	find	charger	fuel	expen	run

Table 3. Performance metric of LDA and PrPh-LDA models

Domain	LDA			PrPh-LDA	
	full review	targets	indicators	targets	indicators
Electronics	1150.92	941.71	902.96	630.21	506.78
Baby products	2146.6	1973.82	2105.80	1524.25	1612.01
Cars	3559.7	2189.51	2263.22	1776.06	1548.62

Dataset statistics are presented in Table 1. Table 2 shows the targets, the problem (*pw*) and no-problem (*no-pw*) words extracted for each dataset. The model anchor one topic to each column in Table 2. For the evaluation we hold out 10% of the reviews for testing purposes and use the remaining 90% to train the model. Topic models are evaluated using perplexity on held-out test data. We computed perplexity with different number of topics and selected k=5 topics due to better results, shown in Table 3. For all models, posterior inference was drawn using 1000 Gibbs iterations and set $\alpha = 0.5$, $\gamma = 0.05$, β=0.01, $\tilde{\beta}$=0.01.

[3] `http://reviews.shop.hp.com`

[4] The dataset is available at `https://snap.stanford.edu/data/web-Amazon.html`.

5 Conclusion and Future Work

In this paper we have proposed a probabilistic graphical model, called PrPh-LDA, that discovers a set of problem targets and problem words from user reviews of products in an unsupervised manner. PrPh-LDA contains two latent distributions over words in user review. We consider that the problem words don't depend on the targets and the targets do not affect the polarity of the problem words. PrPh-LDA has better performance for problem phrase detection than LDA, according to perplexity values. In our future work, we plan to create a labeled dataset with target category to analyze an accuracy of the model and find a correlation between model perplexity and classification metrics. We plan to add additional variables to improve target extraction with sentiment ratings.

Acknowledgement. This work was funded by the subsidy allocated to Kazan Federal University for the state assignment in the sphere of scientific activities (Project No. 3056).

References

[1] Gupta, N.: Extracting phrases describing problems with products and services from twitter messages. In: Proceedings of the Conference on Intelligent Text Processing and Computational Linguistics, CICling 2013(2013)
[2] Tutubalina, E., Ivanov, V.: Clause-based approach to extracting problem phrases from user reviews of products. In: Proceedings of the 3rd International Conference, AIST 2014 (2014)
[3] Tutubalina, E., Ivanov, V.: Unsupervised approach to extracting problem phrases from user reviews. In: Proceedings of the AHA! Workshop on Information Discovery in Text in conjunction with COLING 2014 (2014)
[4] Ivanov, V., Solovyev, V.: Dictionary-based problem phrase extraction from user reviews. In: Text, Speech and Dialogue, pp. 225–232 (2014)
[5] Liu, B.: Sentiment analysis and opinion mining. In: Synthesis Lectures on Human Language Technologies, pp. 1–167 (2012)
[6] Choi, Y., Breck, E., Cardie, C.: Identifying expressions of opinion in context. In: IJCAI, pp. 2683–2688 (2007)
[7] Etzioni, O., Popescu, A.M.: Extracting product features and opinions from reviews. In: Natural Language Processing and Text Mining, pp. 9–28 (2007)
[8] Ng, A., Blei, D., Jordan, M.: Latent dirichlet allocation. Journal of Machine Learning Research 3, 993–1022 (2003)
[9] Cardie, C., Lu, B., Ott, M., Tsou, B.K.: Multi-aspect sentiment analysis with topic models. In: 2011 IEEE 11th International Conference on Data Mining Workshops (ICDMW), pp. 81–88 (2011)
[10] Moghaddam, S., Ester, M.: The flda model for aspect-based opinion mining: Addressing the cold start problem. In: WWW, pp. 909–918 (2013)
[11] Liu, S., Li, F., Wang, S., Zhang, M.: Suit: A supervised user-item based topic model for sentiment analysis. In: Twenty-Eighth AAAI Conference on Artificial Intelligence (2014)
[12] McDonald, R., Titov, I.: Modeling online reviews with multi-grain topic models. In: Proceedings of the 17th WWW, pp. 111–120 (2008)

[13] Yan, H., Li, X., Zhao, W.X., Jiang, J.: Jointly modeling aspects and opinions with a maxent-lda hybrid. In: EMNLP (2010)
[14] Wang, C., Blei, D.: Collaborative topic modeling for recommending scientific articles. In: Proceedings of the 17th ACM SIGKDD, pp. 448–456 (2011)
[15] Zhu, X., Li, F., Huang, M.: Sentiment analysis with global topics and local dependency. In: AAAI (2010)
[16] Heinrich, G.: Parameter estimation for text analysis (2009)

On Identifying Phrases Using Collection Statistics

Simon Gog[1,2], Alistair Moffat[1], and Matthias Petri[1]

[1] Department of Computing and Information Systems,
The University of Melbourne, Australia 3010
[2] Institute of Theoretical Informatics,
Karlsruhe Institute of Technology, Germany

Abstract. The use of phrases as part of similarity computations can enhance search effectiveness. But the gain comes at a cost, either in terms of index size, if all word-tuples are treated as queryable objects; or in terms of processing time, if postings lists for phrases are constructed at query time. There is also a lack of clarity as to which phrases are "interesting", in the sense of capturing useful information. Here we explore several techniques for recognizing phrases using statistics of large-scale collections, and evaluate their quality.

1 Introduction and Related Work

Many concepts are expressed as multi-word expressions, for example, "United States of America", and "letter of condolence". But most information retrieval techniques segment both queries and source documents in to words, and compute similarity over those words as if they were independent, an apparent mismatch that suggests that improved retrieval effectiveness is possible if phrases are also employed. For example, in 1991 Croft et al. [5] wrote "there has always been the feeling that phrases, if used correctly, should improve the specificity of the indexing language". That goal has been realized recently by a range of techniques that make use of phrases and consistently – if perhaps modestly – improve search quality [2,3,6,8,9,19].

Computational cost has been a factor that has prevented the wider use of multi-word phrases in query evaluation. If index space is a dominant concern, the most economical way of handling phrases is to store a positional inverted index [20], and compute postings list intersections as queries are handled. If query time is important, then phrase-based indexing approaches can be employed, in which some or all terms' postings lists are augmented by information about following words [17]. A third possibility is to directly index certain phrases, and give them postings lists. The number of phrases admitted to the index provides a tunable tradeoff between index size and execution cost.

The question then is, which phrases should be indexed? Based on characteristics of the document collection, is it possible to generate an ordering of phrases that could then be used to decide which phrases should be granted dedicated postings lists? If the text has embedded markup, then it can be used to identify word sequences of interest, including those to be displayed as headings or with different typography, and those used as anchors for hyperlinks [7]. For plain text, a range of automatic extraction methods based on occurrence frequencies have been proposed from both linguistics [4] and computing [12,16,18]. However the sheer volume of data that must be processed in order

A. Hanbury et al. (Eds.): ECIR 2015, LNCS 9022, pp. 278–283, 2015.

to compute word and phrase statistics over large texts has been an impediment in the past. Our work in this paper is possible because of recent advances in data structures, including the development of succinct self-index technologies, see Navarro [11] for an overview and Patil et al. [13] for one implementation approach.

2 Phrase-Finding

We consider three methods for finding multi-word phrases in text. Two of them are based on previous mechanisms for identifying word bigrams of interest; we extend them to the multi-word situation.

Mutual Information. The concept of mutual information can be used to determine an *association ratio* between two words [4]. Given words w_1 and w_2, mutual information compares the probability of a co-occurrence to the probabilities of observing each word independently. If w_1 and w_2 are associated, the observed probability of the two words occurring together will be much larger than the probability of a co-occurrence by chance. Similar to Church and Hanks [4], we use the number of occurrences of word w_i normalized by the size of the corpus as an estimate of its probability $P(w_i)$. Multi-word expressions can be handled by extending the formulation (see also Van de Cruys [14]):

$$\text{MI-EXT}(w_1, w_2 \ldots w_n) = \log_2 \frac{P(w_1 w_2 \ldots w_n)}{P(w_1)P(w_2)\cdots P(w_n)} .$$

Pearson's χ^2. The χ^2 (CHI^2) metric can also be used to test the independence of an observation. The independence of a word bigram $w_1 w_2$ is evaluated by comparing its observed frequency in the collection to its expected frequency [15]. Expected frequencies require bigram statistics such as $F(w_1 w_2)$ and $F(w_1 \neg w_2)$ to be computed, both of which can be efficiently performed using a self-index, a technology that has only recently been available at the required scale.

We extend bigram scores to allow computation of n-gram scores: if $\chi^2(w_i w_{i+1})$ is the score for the word-pair w_i followed by w_{i+1}, we compute

$$\text{CHI}^2\text{-EXT} = \left(\min_{1 \leq i < n} \chi^2(w_i w_{i+1}) \right) \cdot \ln n ,$$

where the multiplication by $\ln n$ counteracts the diminishing nature of the min operator, and up-weights longer phrases that are the concatenation of shorter stronger ones.

Existence. We implemented one further mechanism, denoted EXISTENCE, defined as the ratio between the number of documents which contain all words of the candidate phrase and documents which contain the candidate phrase. In this mechanism document boundaries are used, a concept not employed in the first two approaches. For example, if there are five documents in the collection that contain all of w_1, w_2, and w_3, and the sequence $w_1 w_2 w_3$ appears as a phrase in three of them, then the (undamped) conditional probability of existence is given by $3/5 = 0.6$. In practice, to avoid every unique substring being assigned a score of 1.0, we use a dampening constant K, and compute

$$\text{EXISTENCE}(w_1 w_2 \ldots w_n) = \frac{F(w_1 w_2 \ldots w_n)}{F(w_1, w_2, \cdots, w_n) + K}$$

Table 1. Example stemmed phrases extracted from *Query Set I*. The CHI^2-EXT method produced the same top-10 results as CHI^2. Phrases corresponding to Wikipedia page titles are in bold.

Rank	MI-EXT	CHI^2/ CHI^2-EXT	EXISTENCE
1.	new mexico senator pete domenici	**punta gorda**	**sri lanka**
2.	**equine protozoal myeloencephaliti**	**puerto rico**	**punta gorda**
3.	virus hpv genital wart	**bryn mawr**	**corpus christi**
4.	methyl ether tertiary butyl	**saudi arabia**	**puerto rico**
5.	civil war 1861 1865	**corpus christi**	**st croix**
6.	1922 fordney mccumber	**sri lanka**	**pro tempore**
7.	oldsmobile ciera cutlass	**cabernet sauvignon**	**saudi arabia**
8.	bull terrier staffordshire	**monte carlo**	**los angeles**
9.	holiday inn sunspree	antirobe aquadrop	**wilke barre**
10.	pratt whitney jt8d	**chichen itza**	**bryn mawr**

where $F(s)$ is the document frequency of s in the collection, and $K = 5$ is used, to ensure that a phrase occurs at least five times if its score is greater than 0.5.

Stop words. We further apply stop word trimming. Any word for which the maximum value of the BM25 similarity computation between the word and any document is less than one when using the default parameters (see Zobel and Moffat [20]) is defined to be a stop word. Stop words at the beginning and end of candidate phrases are removed.

3 Experiments and Results

Source Data. We took the 426 GB Gov2 collection and built a self-index structure [11]. To determine potential phrases, we randomly sampled two query sets each containing 10,000 queries from the TREC Million Query Track. We selected unique queries containing two or more words such that each word appeared at least once in Gov2. Each sub-phrase in each query was then evaluated as a candidate using the index, and assigned a score by each of the mechanisms described in the previous section. For example, a four word query generates six candidate phrases.

Table 1 lists the top phrases discovered using *Query Set I*. The Mutual Information-based approach favors longer phrases, whereas the other methods rank two-word phrases higher. The first phrase of length larger than two occurs at rank 160 for CHI^2-EXT and at rank 60 for EXISTENCE.

Forming Judgments. We then sought to compare the lists of candidate phrases. The first step is to make a judgment, for each identified word sequence, as to whether it is indeed a plausible phrase. Once each algorithm's phrase ranking has been suitably annotated, a score can be derived. But generating labeled evaluation data is problematic. One option is to employ experts to create "gold standard" determinations. Another is to use non-expert judgments via a crowd-sourcing service. Both methods have their disadvantages – experts are expensive, and will not necessarily agree with each other no matter how precise their instructions; the wisdom of the crowd may generate more reliable data overall for less money, but is vulnerable to hasty workers.

To obtain preliminary results, we have employed a third alternative, and make use of Wikipedia for implicit decisions. In particular, many multi-word entities have Wiki pages associated with them, for example, `http://en.Wikipedia.org/wiki/White_House` is the page for the "*White House*".

To automate the judging process we downloaded 10,947,620 Wiki page titles[1]. The titles were filtered and normalized as follows: categorization suffixes of titles were deleted (for example, the suffix "_(film)" in the title "Personal_Best_(film)"); single term titles were removed; underscores were translated to spaces; and words lowercased and stemmed using a Krovetz stemmer. Phrases were then deemed to be valid if and only if they were in this processed list. This mechanism fails for many interesting phrases, but also works a surprising fraction of the time, including, for example, for "*standing ovation*", "*personal best*", and "*laugh out loud*".

Table 2 gives a breakdown of the set of reference phrases identified from the Wiki URLs. More than seven million Wiki pages had multi-word titles, with around half of them two words long, a quarter three words long, and so on. The "7+" category includes phrases such as "*1954 britain empire and commonwealth games medal count*". A further 3,562,553 Wiki page URLs consisted of a single word, or were explicit disambiguation pages. The phrases identified were then used as ground truth in the evaluation.

Table 2. Distribution of Wiki URLs

Length	Number	Fraction
2	3,498,885	47.4%
3	1,895,699	25.7%
4	914,401	12.4%
5	483,618	6.5%
6	264,400	3.6%
7+	328,064	4.4%
Total	7,385,067	100.0%

Applying a Metric. Once judgments have been formed, a metric can be used to compute a quality score for the ordered list of phrases generated by each of the algorithms. Any IR metric can be used, provided that it is agnostic to the total number of positive judgments. For example, the first 1,000 phrases in each list might be examined, and the fraction of them that are valid expressed as a *precision*@1,000 score. In the results reported below, we use the top-weighted arbitrary-depth RBP metric [10], with two parameters, $p = 0.99$ and $p = 0.999$, in both cases using generated rankings of 10,000 candidate phrases in decreasing score order. With these parameters, rank-biased precision (RBP) provides deep coverage in the ranked list (to an expected depth of 100 items and 1,000 items, respectively), with a relatively mild bias in favor of positions near the front of the ranking. With p values near 1.0, RBP can be expected to yield outcomes that are closely correlated with precision scores when evaluated to comparable cutoffs.

Table 3. Rank-biased precision scores for three phrase-finding mechanisms

p	Query set	MI-EXT	CHI^2	CHI^2-EXT	EXISTENCE
0.99	I	0.380	0.824	0.823	0.857
0.99	II	0.406	0.831	0.830	0.858
0.999	I	0.406	0.557	0.546	0.562
0.999	II	0.331	0.553	0.538	0.554

[1] File `enwiki-20140502-all-titles-in-ns0`, accessed 10 June 2014.

Table 4. False positives and false negatives for the EXISTENCE method and *Query Set I*

False positives		False negatives	
30.	california arnold	19944.	**social death**
31.	marriott wardman	19804.	**nation league**
32.	canton massillon	19630.	**civil movement**
39.	mountain lab	19463.	**project jersey**
50.	cmc heartland	19294.	**independence declaration**
66.	displace homemaker	19247.	**early island**
70.	paul biane	19089.	**north purchase**
71.	2006 2007	19068.	**last snow**
90.	nasa launch	19047.	**thomas plate**
104.	phs 5161	18913.	**satellite states**

Results. Table 3 shows that the methods achieved consistent scores over two query sets, and that the EXISTENCE and CHI^2 methods achieve good performance. Note that these are all lower bounds – the Wiki URLs used to provide relevance judgments are not a complete set of phrases, and are biased in favor of entities such as events, people, and places. False positives occur when a candidate phrase is scored highly by an algorithm, but does not appear in the Wiki listing; false negatives when a phrase that is a Wiki page title, is scored lowly by the algorithm. Table 4 shows the top ten false positives identified by the EXISTENCE method, and the ten lowest-scoring candidate phrases that corresponded to Wiki page titles. False positives tend to follow certain patterns: "*paul biane*", "*cmc heartland*" and "*canton massillon*" are names of people, companies or places not present in Wikipedia; and "*PHS 5161*" is the name of a form referenced often in Gov2. False negatives include ambiguous phrases such as "*last snow*", which is the name of a novel not referenced in Gov2; similarly, "*project jersey*" refers to a java framework only created after the Gov2 corpus was crawled.

4 Conclusion and Future Work

To identify phrases in collections that might warrant being explicitly indexed so as to provide fast querying, we have explored techniques for automatically extracting them using only the statistics provided by the collection itself. Using Wikipedia page titles as a reference point, we have compared those techniques, and found that the new document-aware EXISTENCE method creates the best set of phrase candidates. The benefit of the new methodology – compared, for example, to the obvious alternative of simply using the Wikipedia titles directly – is that an ordered list of phrases is created, and that they are sourced from the collection. The latter is important when technical or medical text is being stored, since Wikipedia titles would not provide useful guidance.

Our next task is to embed the phrase-finding technology into a retrieval system. That will involve the complete suffix tree traversal of the text to find candidate phrases. An index can then be constructed to fit any given space bound, taking terms in to it, plus postings lists, for how ever many phrases can best fit. It will then be possible to fully explore the complex relationships between query processing speed, index space required, and retrieval effectiveness; see, for example, Anand et al. [1].

Acknowledgments. This work was supported under the Australian Research Council's Discovery Projects funding scheme (project DP110101743), and by the Victorian Life Sciences Computation Initiative (grant VR0052) on its Peak Computing Facility at the University of Melbourne, an initiative of the Victorian State Government, Australia.

References

1. Anand, A., Mele, I., Bedathur, S., Berberich, K.: Phrase query optimization on inverted indexes. In: Proc. CIKM, pp. 1807–1810 (2014)
2. Broschart, A., Berberich, K., Schenkel, R.: Evaluating the potential of explicit phrases for retrieval quality. In: Gurrin, C., He, Y., Kazai, G., Kruschwitz, U., Little, S., Roelleke, T., Rüger, S., van Rijsbergen, K. (eds.) ECIR 2010. LNCS, vol. 5993, pp. 623–626. Springer, Heidelberg (2010)
3. Chieze, E.: Integrating phrases in precision-oriented information retrieval on the web. In: Proc. Conf. Inf. Know. Eng., pp. 54–60 (2007)
4. Church, K.W., Hanks, P.: Word association norms, mutual information, and lexicography. Comp. Ling. 16(1), 22–29 (1990)
5. Croft, W.B., Turtle, H.R., Lewis, D.D.: The use of phrases and structured queries in information retrieval. In: Proc. SIGIR, pp. 32–45 (1991)
6. Geva, S., Kamps, J., Lethonen, M., Schenkel, R., Thom, J.A., Trotman, A.: Overview of the INEX 2009 ad hoc track. In: Geva, S., Kamps, J., Trotman, A. (eds.) INEX 2009. LNCS, vol. 6203, pp. 4–25. Springer, Heidelberg (2010)
7. Lehtonen, M., Doucet, A.: Phrase detection in the Wikipedia. In: Fuhr, N., Kamps, J., Lalmas, M., Trotman, A. (eds.) INEX 2007. LNCS, vol. 4862, pp. 115–121. Springer, Heidelberg (2008)
8. Liu, S., Liu, F., Yu, C.T., Meng, W.: An effective approach to document retrieval via utilizing wordnet and recognizing phrases. In: Proc. SIGIR, pp. 266–272 (2004)
9. Metzler, D., Croft, W.B.: A Markov random field model for term dependencies. In: Proc. SIGIR, pp. 472–479 (2005)
10. Moffat, A., Zobel, J.: Rank-biased precision for measurement of retrieval effectiveness. ACM Trans. Information Systems 27(1), 2.1–2.27 (2008)
11. Navarro, G.: Spaces, trees and colors: The algorithmic landscape of document retrieval on sequences. ACM Comp. Surv. 46(4), 1–47 (2014)
12. Nevill-Manning, C.G., Witten, I.H.: Compression and explanation using hierarchical grammars. Comp. J. 40(2/3), 103–116 (1997)
13. Patil, M., Thankachan, S.V., Shah, R., Hon, W.K., Vitter, J.S., Chandrasekaran, S.: Inverted indexes for phrases and strings. In: Proc. SIGIR, pp. 555–564 (2011)
14. Van de Cruys, T.: Two multivariate generalizations of pointwise mutual information. In: Proc. Wkshp. Distr. Semantics & Compositionality, pp. 16–20 (2011)
15. Villada Moirón, M.B.: Data-driven identification of fixed expressions and their modifiability. Ph.D. thesis, University of Groningen (2005)
16. Wang, X., McCallum, A., Wei, X.: Topical n-grams: Phrase and topic discovery, with an application to information retrieval. In: Proc. ICDM, pp. 697–702 (2007)
17. Williams, H.E., Zobel, J., Bahle, D.: Fast phrase querying with combined indexes. ACM Trans. Information Systems 22(4), 573–594 (2004)
18. Witten, I.H., Paynter, G.W., Frank, E., Gutwin, C., Nevill-Manning, C.G.: KEA: Practical automatic keyphrase extraction. In: Proc. ACM Conf. Dig. Lib., pp. 254–255 (1999)
19. Zhang, W., Liu, S., Yu, C.T., Sun, C., Liu, F., Meng, W.: Recognition and classification of noun phrases in queries for effective retrieval. In: Proc. CIKM, pp. 711–720 (2007)
20. Zobel, J., Moffat, A.: Inverted files for text search engines. ACM Comp. Surv. 38(2), 6:1-6:56 (2006)

MIST: Top-k Approximate Sub-string Mining Using Triplet Statistical Significance

Sourav Dutta

Max-Planck Institute for Informatics, Germany
sdutta@mpi-inf.mpg.de

Abstract. Efficient extraction of strings or sub-strings similar to an input query string forms a necessity in applications like *instant search, record linkage*, etc., where the similarity between two strings is usually quantified by *edit* distance. This paper proposes a novel top-k approximate sub-string matching algorithm, $MIST$, for a given query, based on *Chi-squared* statistical significance of string triplets, thereby avoiding expensive edit distance computation. Experiments with real-life data validate the run-time effectiveness and accuracy of our algorithm.

Keywords: Approx. string search, Edit distance, χ^2 statistical significance, n-grams.

1 Introduction

The enormous applicability of sequence data catering to diverse domains such as search engines [3], DNA sub-sequence search [17], spell checks & text correction [12], record linkage, etc., demand robustness in tackling incorrect spellings and random noises by mapping the input query to similar records present in the corpus. This defines the *string similarity search* or *approximate sub-string matching* problem as: `given text` T_x `with a query` Q`, extract strings having matching sub-texts to the query with minimal ''deviation''`. For example, the word $cast$ might be misspelled as $cazt$, but the application should retrieve $cast$ as the approx. similar string.

Several string similarity measures such as cosine similarity, Jaccard similarity, dice score, Hamming distance, and Jaro-Winkler distance have been proposed [2] to extract approximate (sub-)strings from archives. The most widely used similarity measure is the *Levenshtein* or *edit* distance [13], which quantifies the similarity between two strings by the number of *insert*, *delete*, and *substitute* operations required to transform one string to the other. For example, the edit distance between "cat" and "coat" is 1. However, it suffers from a high run-time complexity of $O(n^2)$ for strings of length n. Hence, methods to prune the candidate search space have been proposed [8].

Contributions: This paper proposes a novel algorithm $MIST$, *Mining with Inferred Statistics on Triplets*, for top-k approximate sub-string extraction by combining *n-gram* indexing and statistical significance of grams, hence bypassing expensive edit distance computations. $MIST$ maps 3-grams to symbols, computes their probability of occurrences, and based on the *Chi-squared* statistical measure [19] reports sub-strings exhibiting high significance as approximate sub-string matches. Statistical measures handle the

A. Hanbury et al. (Eds.): ECIR 2015, LNCS 9022, pp. 284–290, 2015.

presence of noise (small amounts), inherently modeling deviations of the query from actual strings without costly merging operations as in inverted index approaches.

2 *Mining with Inferred Statistics on Triplets* (MIST) Algorithm

Assume, an alphabet set Σ of cardinality w, $\Sigma = \{a_1, a_2, \ldots, a_w\}$ and an input sequence data set $D = (s_1 s_2 \ldots s_n)$ of size n composed of alphabets $s_i \in \Sigma$. Given a query Q of length l, $Q = q_1 q_2 \ldots q_l$ where $q_i \in \Sigma$, we need to extract the top-k approximate sub-strings to Q present in D.

Initially, all triplets $t_i t_j t_k$ ($t_{i,j,k} \in \Sigma$) present in D are extracted (with 1-sliding window protocol). D is then represented by a tabular structure T, storing a mapping between positions and the number of occurrences, *count* of each triplet. There exists w^3 different triplet combinations for the alphabet set Σ; however w is usually in the order of tens (26 for an English dictionary) and hence the memory footprint of T is not expensive. As an example, let $\Sigma = \{u, x, y, z\}$ and $D = zxyxyx$ where $w = 4$ and $n = 6$, with query $Q = xyxx$ having length $l = 4$. $MIST$ initially extracts all the *four* 3-grams present in D (namely zxy, xyx, yxy and xyx) and constructs the table, T storing the occurrence counts of triplets at various positions of D. Hence, at position 3 (of D) triplet zxy has a *count* of 1, while xyx has *count* = 2 at position 6.

$MIST$ then extracts the $l - 2$ triplets present in Q. Based on the similarity with the query triplets, each of the possible w^3 triplets is classified into *similarity classes*. The similarity between two triplets is categorized into 4 hierarchical classes, each represented by a unique symbol. The similarities between a triplet t and all triplets of Q are computed, and t is classified by the highest similarity class obtained. $MIST$ represents this classification information of triplets by a many-to-one function, $f : \Sigma^3 \rightarrow Symbol$, and computes the probabilities of occurrences of the symbols.

(1) Exact Match (EM)**:** The highest category of similarity between two triplets, the *exact match* (represented by symbol σ_3), occurs when both triplets are exactly the same (i.e., comprises the same alphabets in the same order). In our above example, triplet $xyx \in D$ is an EM with $xyx \in Q$. The probability of occurrence of such a triplet (henceforth assuming uniform probability distribution over the alphabets in Σ for simplicity of analysis) is given by, $\boldsymbol{P(\sigma_3) = \frac{1}{w^3}}$

(2) Significant Match (S_gM)**:** $MIST$ considers two triplets to have *significant match* if they differ at only one position. The triplet yxx in D is S_gM w.r.t. the query triplet yxy. Mis-typing or mis-spelling leads to inadvertent swapping of adjacent alphabets, and hence to intelligently handle such scenarios, $MIST$ also marks such triplets (with swaps in only 1 adjacent position) as significant matches. Hence, the triplet $yxx \in D$ is also a significant match compared with $xyx \in Q$. We represent S_gM by symbol σ_2 with occurrence probability as, $\boldsymbol{P(\sigma_2) = \binom{3}{2} \cdot \frac{(w-1)}{w^3} + \frac{3}{w^3} = \frac{3}{w^2}}$

(3) Slight Match (S_lM)**:** A triplet is said to be in this category if it has only one position of similarity with the query triplets (except in the case of swaps as stated in S_gM), and is represented by symbol σ_1. In our example, triplets $zxy \in D$ and $yxx \in Q$ match only at their middle character and hence, zxy is labeled as S_lM. The occurrence probability of an S_lM triplet is, $\boldsymbol{P(\sigma_1) = \binom{3}{1} \cdot \frac{(w-1)^2}{w^3} - \frac{3}{w^3} = \frac{3(w-2)}{w^2}}$

(4) No Match (NM): σ_0 is used to represent triplets having *no match* at all to an input query triplet. A triplet such as uuu would be classified as an NM in our example. The probability of occurrence of an NM triplet is, $\boldsymbol{P(\sigma_0) = \left(\frac{w-1}{w}\right)^3}$
A triplet is categorized by its highest classification, hence $xyx \in D$ is considered an EM with $xyx \in Q$ (not S_gM for $yxx \in Q$) and represented by symbol σ_3. Using the mapping function f and table T, MIST transforms D into a sequence of symbols (based on triplet similarity). Our example data set D thus becomes $D' = \sigma_1\sigma_3\sigma_2\sigma_3$.

Extraction of the top-k approximate strings is performed on this modified data set, D' using the occurrence probabilities of the symbols. $MIST$ employs the linear-time *Chi-squared (χ^2) score based $AGMM$* algorithm proposed in [5] to compute (with high accuracy) the sub-strings exhibiting the highest statistically significance. The occurrence count of symbols in D' is easily retrieved using table T, and a heap of size k stores the top-k significant sub-strings obtained. Assuming $k = 1$ in our example, the sub-string $\sigma_3\sigma_2\sigma_3$ in D' provides the highest χ^2 value ($= 34$). Hence, using the position of the most significant sub-string obtained, the sequence $xyxyx \in D$ is retrieved as the top approximate sub-string match to $Q(= xyxx)$.

For each query triplet, $O(w^2)$ mapping points of f are accessed; but as w is small or constant and the $AGMM$ method is linear, the complexity of MIST becomes $O(l)$ per query, for query length l. This makes MIST efficient (for large data sets) compared to the state-of-the-art $O(l^3)$ approach [10]. Different probability distributions of alphabets in Σ require only re-conditioning of the symbol occurrence equations and other significant sequence mining approaches can also be employed within the $MIST$ framework.

3 Experimental Evaluation

We empirically evaluated the performance of $MIST$ against the naïve approach involving brute force strategy of computing edit distances of all sub-strings and finding the top-k approximate string matches (using a heap). We consider the original strings from which the queries were generated to be "gold results", for assessing the correctness of the results obtained. We benchmark the *accuracy* and *run-time* of $MIST$ on two real data sets. Only strings having an edit distance less than the *deviation* threshold, τ from the query were considered to be approximate. Preliminary experiments were also conducted comparing the run-time of $MIST$ to the edit distance based dynamic programming approach of [4], using the `Author` data set [4], and observed to be comparable (sometimes better) due to no edit-distance computation. All experiments were conducted on an Intel-i5 2.50 GHz processor with 8 GB RAM running Ubuntu 12.04.

DBLP Title Data Set. This data set contained 3.72 million DBLP publication titles extracted from `dblp.uni-trier.de/xml/`. Similar to the experimental setup in [10], we duplicated the original data 5 times to obtain around $800MB$ of data. A random edit operation (insert, delete, or substitute) was performed at each position of the strings with a probability of 0.1, and 50,000 uniformly randomly sampled titles were treated as the

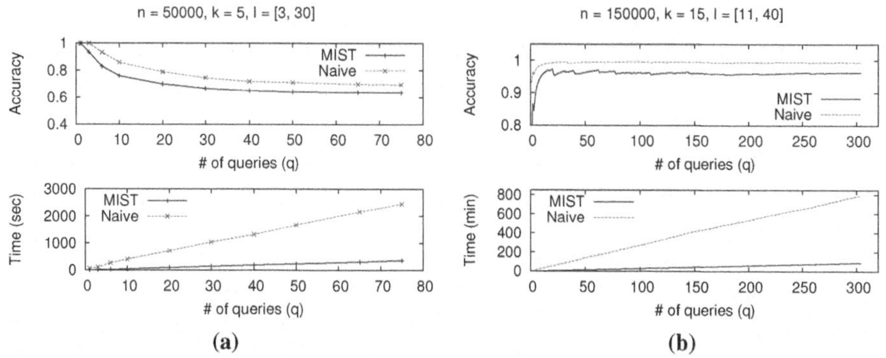

Fig. 1. Accuracy and Run-time results for (a) DBLP data set and (b) English Dictionary data set

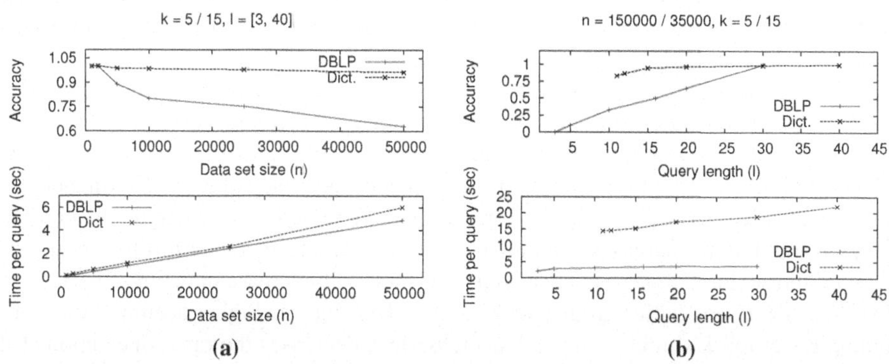

Fig. 2. Performance effect of $MIST$ due to varying (a) Data set size, (b) Query length

data set. We generated 100 queries from these titles, for random lengths between 3 to 30 with an average of 15.28, and retrieved the *top-5* approximate sub-strings. Fig. 1(a) depicts the results obtained with *deviation* parameter, $\tau = 4$. $MIST$ stabilizes at 64% accuracy while the naïve algorithm reports around 70% accuracy; however MIST obtains a run-time speed-up of nearly 7× compared to the naïve approach. The brute-force strategy also fails to achieve 100% accuracy (compared to the "gold results") due to random evictions from the heap when multiple candidate strings have the same edit distance.

English Dictionary Data Set. We next evaluated the performance of $MIST$ on an English dictionary (from `www.outpost9.com/files/WordLists.html`) containing around 320,000 words. Since the length of English words are small, we concatenated 10 consecutive words to obtain a modified data set with an average word length of 98.28. Each string was then replicated 15 times with each character position edited randomly (as in the DBLP data set) with 0.05 probability, and 150,000 such modified strings were then selected as the data set. We randomly generated around 300 queries with lengths ranging from 11 – 40 and average of 25.66, with parameter $\tau = 3$. Fig. 1(b) reports the findings for the *top-15* matching approximate sub-strings. Similar to the DBLP data set, $MIST$ attains nearly equivalent accuracy (96%) as that of the naïve approach (99%). However, $MIST$ takes around 1.5 hours as compared to 13.3 hours taken by the brute-force approach, providing $9\times$ improvement. Hence, for both the data sets, $MIST$ efficiently extracts approximate sub-strings with high accuracy and with significantly speed-up.

Parameter Variation. To explore the scalability of the $MIST$ algorithm, we performed experiments with varying parameter values as: (i) data set size, n (ii) query length, l, and (iii) top-k, k.
(1) `Data Set Size:` To simulate different input sizes, we randomly selected subsets of the data varying n from 1000 to $150,000$. The number of triplets present increases with n, leading to an increase in the probability of occurrence of the query triplets. This enables sub-strings (possibly false positives) to acquire a higher χ^2 value, leading to a decrease in the accuracy of MIST. Fig. 2(a) exhibits similar behavior for the two data sets with the accuracy of $MIST$ decreasing with increase in n, stabilizing at the average accuracy. With increase in data size, the search space for top-k approximate sub-string matches increases and we observe a linear increase in the run-time of $MIST$.
(2) `Query Length:` The number of query triplets increases with query length, l thereby enhancing the difference among dissimilar strings and decreasing the probability of random contiguous occurrence of similar triplets, leading to better pruning. Hence, in Fig. 2(a) we observe an improvement in the accuracy of $MIST$ with increase in l when varied between 3 and 40. However with increase in query length, the number of χ^2 computations increases, leading to an increase in the average run-time per query. $MIST$ thus exhibits linear increase in its run-time with increase in l (Fig. 2(b)).
(3) `Top-K:` k was varied from 1 to 15 and $MIST$ reports higher accuracy for computing fewer top-k similar strings. A decrease in k decreases the random evictions (of true positives) from the heap for strings with same statistical significance, leading to an increase in accuracy. The effect of k on run-time was observed to be insignificant.

4 Related Work

Traditional approaches to solve the *approximate sub-string match* problem involve aligning the query Q with input words represented by tries. To speed-up computations, pre-filtering techniques [1] and indexing schemes such as SSI [6], B^+ trees, inverted indices [20], and Suffix tries [21] have been proposed. Strategies involving neighborhood generation [16], fixed-length q-grams with edit distance [20] and variable-length q-grams with associated dictionary [15] were also explored. Inverted index based approaches suffer from an expensive merge step, and hence [18] proposed a combination of q-grams, filtering, and "ScanCount" merging [14]. Recently, [10] provided a dynamic programming based filtering algorithm with inverted index for theoretically bounding the edit distances between sub-strings. [11] introduces the modeling of frequent patterns with suffix array based indexing for approximate string search, while [4] proposed a pruning based dynamic-programming technique using edit distance.

Statistical modeling determines the relationship between observed experimental outcome and factors influencing the system, or to pure chance. The p-value, z-score, log-likelihood ratio (G^2) [19] and Hotelling's T^2 measure [7] are popular for capturing the significance of a pattern. Although the p-value provides a precise decision, it is computationally exponential. The Pearson's χ^2 statistic provides a good approximation [19] and is used in this paper. The χ^2 distribution is characterized by *degrees of freedom* (symbol set size minus one). The more a string deviates from the expected behavior the more significant it is, and consequently larger its χ^2 value. Mining interesting patterns in time-series databases using suffix trees was proposed in [9], while a linear-time greedy algorithm, using *blocking* procedure, with high accuracy was explored in [5].

5 Conclusions and Future Work

This paper proposed a novel algorithm, *Mining with Inferred Statistics on Triplets* ($MIST$), combining the 3-gram indexing model and χ^2 statistical significance for retrieving the top-k similar strings to a query. Experiments with real data exhibit efficiency of MIST both in accuracy and run-time. Deriving theoretical bounds on the accuracy, and extension to *similar phrase queries* provide interesting future works.

References

1. Baeza-Yates, R., Navarro, G.: New and Faster Filters for Multiple Approximate String Matching. Random Structures & Algorithms 20(1), 23–49 (2002)
2. Baeza-Yates, R.A., Ribeiro-Neto, B.A.: Modern Information Retrieval - the Concepts and Technology behind Search. Pearson Edu. Ltd. (2011)
3. Cucerzan, S., Brill, E.: Spelling Corrections as an Interactive Process that Exploits the Collective Knowledge of Web Users. In: EMNLP, pp. 293–300 (2004)
4. Deng, D., Li, G., Feng, J., Li, W.S.: Top-k string similarity search with edit-distance constraints. In: ICDE, pp. 925–936 (2013)
5. Dutta, S., Bhattacharya, A.: Most Significant Substring Mining based on Chi-Square Measure. In: Zaki, M.J., Yu, J.X., Ravindran, B., Pudi, V. (eds.) PAKDD 2010, Part I. LNCS, vol. 6118, pp. 319–327. Springer, Heidelberg (2010)
6. Fenz, D., Lange, D., Rheinländer, A., Naumann, F., Leser, U.: Efficient Similarity Search in Very Large String Sets. In: Ailamaki, A., Bowers, S. (eds.) SSDBM 2012. LNCS, vol. 7338, pp. 262–279. Springer, Heidelberg (2012)
7. Hotelling, H.: Multivariate Quality Control. Tech. of Statistical Analysis 54, 111–184 (1947)
8. Kahveci, T., Singh, A.K.: Efficient Index Structures for String Databases. In: VLDB, pp. 351–360 (2001)
9. Keogh, E., Lonardi, S., Chiu, B.: Finding Surprising Patterns in a Time Series Database in Linear Time and Space. In: SIGKDD, pp. 550–556 (2002)
10. Kim, Y., Shim, K.: Efficient Top-k Algorithms for Approximate Substring Matching. In: SIGMOD, pp. 385–396 (2013)
11. Kimura, M., Takasu, A., Adachi, J.: FPI: A Novel Indexing Method Using Frequent Patterns for Approximate String Searches. In: EDBT Workshops, pp. 397–403 (2013)
12. Kukich, K.: Techniques for Automatically Correcting Words in Texts. ACM Computing Surveys 24(4), 377–439 (1992)
13. Levenshtein, V.I.: Binary Codes capable of Correcting Deletions, Insertions, and Reversals. Soviet Physics Doklady 10(8), 707–710 (1966)
14. Li, C., Lu, J., Lu, Y.: Efficient Merging and Filtering Algorithms for Apprx. String Searches. In: ICDE, pp. 257–266 (2008)

15. Li, C., Wang, B., Yang, X.: VGRAM: Improving Performance of Approximate Queries on String Collections using Variable-length Grams. In: VLDB, pp. 303–314 (2007)
16. Myers, G.: A Sublinear Algorithm for Approximate Keyword Searching. Algorithmica 12(4), 345–374 (1994)
17. Navarro, G.: A Guided Tour to Approximate String Matching. ACM Computing Surveys 33(1), 31–88 (2001)
18. Patil, M., Cai, X., Thankachan, S.V., Shah, R., Park, S.J., Foltz, D.: Approximate String Matching by Position Restricted Alignment. In: EDBT, pp. 384–391 (2013)
19. Read, T., Cressie, N.: Goodness-of-fit Stats. for Discrete Multivariate Data. Springer (1988)
20. Yang, Z., Yu, J., Kitsuregawa, M.: Fast Algorithms for Top-k Approximate String Matching. In: AAAI, pp. 1467–1473 (2010)
21. Zhang, Z., Hadjieleftheriou, M., Ooi, B.C., Srivastava, D.: Bed-Tree: An All-purpose Index Structure for String Similarity Search based on Edit Dist. In: SIGMOD, pp. 915–926 (2010)

Active Learning Applied to Rating Elicitation for Incentive Purposes

Marden B. Pasinato[1], Carlos E. Mello[2], and Geraldo Zimbrão[1]

[1] COPPE, Federal University of Rio de Janeiro, Rio de Janeiro, Brazil
[2] DCC, Federal Rural University of Rio de Janeiro, Nova Iguaçu, Brazil
{marden,zimbrao}@cos.ufrj.br
carlos.mello@ufrrj.br

Abstract. Active Learning (AL) has been applied to Recommender Systems so as to elicit ratings from new users, namely *Rating Elicitation for Cold Start Purposes*. In most e-commerce systems, it is common to have the purchase information, but not the preference information, i.e., users rarely evaluate the items they purchased. In order to acquire these ratings, the e-commerce usually sends annoying notifications asking users to evaluate their purchases. The system assumes that every rating has the same impact on its overall performance and, therefore, every rating is worth the same effort to acquire. However, this might not be true and, in that case, some ratings are worth more effort than others. For instance, if the e-commerce knew beforehand which ratings will result in the greatest improvement of the overall system's performance, it would be probably willing to reward users in exchange for these ratings. In other words, rating elicitation can go together with incentive mechanisms, namely *Rating Elicitation for Incentive Purposes*. Like in cold start cases, AL strategies could be easily applied to *Rating Elicitation for Incentive Purposes* in order to select items for evaluation. Therefore, in this work, we conduct a extensive benchmark, concerning incentives, with the main AL strategies in the literature, comparing them with respect to the overall system's performance (MAE). Furthermore, we propose a novel AL strategy that creates a k-dimensional vector space, called *item space*, and selects items according to the density in this space. The *density-based strategy* has outperformed all others while making weak assumptions about the data set, which indicates that it can be an efficient default strategy for real applications.

Keywords: Recommender Systems, Active Learning, Rating Elicitation.

1 Introduction

The variety of options used to be the hallmark of a successful business, but now it is turning to be a drawback. It has been reported that users feel anxious in the midst of a great number of options mainly because the risk of making the wrong choice, and thus regretting, arises [17].

This gets even worst when it comes to the Web. The increase of bandwidth speed allied with social networks have reshaped the way people live and, specially,

A. Hanbury et al. (Eds.): ECIR 2015, LNCS 9022, pp. 291–302, 2015.

the way they trade. For this reason, e-commerce has flourished and also the number of options available on-line.

In the last decade, Recommender Systems (RS) have emerged with much intensity both in the industry and in the academia [3,16]. They are turning into an essential tool, not only due to their capability of increasing sales, but primarily because they help users to navigate among the myriad of available options.

By far the most widely used RS technique, Collaborative Filtering (CF) falls short when the system does not have enough ratings, particularly in the case of new users [15]. Many have tried to tackle this issue by applying Active Learning (AL) strategies so as to elicit ratings from new users and thus expanding the system's knowledge about them [5,13,14]. In order to draw a distinction between the ways of applying AL strategies to RS, we shall refer to this approach as *Rating Elicitation for Cold Start Purposes.*

Besides eliciting new users' preferences, AL strategies can be applied to RS so as to implement incentive mechanisms, namely *Rating Elicitation for Incentive Purposes.* Since evaluating is not a common habit among users, many e-commerce systems have the information about purchased items, but not their evaluations. Asking users to evaluate all the items they have ever purchased might be bothersome, especially for regular customers. An alternative and more elegant approach is to identify, among the purchased items, the top N that, if evaluated, would result in the greatest improvement of the overall system's performance. The RS could then ask users to evaluate only those items instead and, to ensure that they will contribute with these ratings, it could offer some incentive in exchange (e.g., discount on future sales, points to upgrade the account, etc.).

The goal of this work is, firstly, to point out this alternative way of applying AL strategies to RS, i.e., *Rating Elicitation for Incentive Purposes*, which seems to have passed unnoticed so far. Secondly, we propose a novel AL strategy that selects items based on their corresponding density in the item space, namely *the density-based strategy.* This space is created by applying the Singular Value Decomposition (SVD) to the purchase matrix. By selecting the items that account for the densest regions in the item space, we are actually generating the smallest sample that best estimates the Probability Density Function (PDF) of all items. Finally, we have conducted an extensive benchmark including the density-based strategy and the common strategies used in the literature. We concluded that the former has shown better performance, with respect to the overall system's performance (MAE), without making strong assumptions about the data set like the others.

This paper is divided in 6 sections, of which this is the first one. In section 2, we detail the main works in the literature that are related to ours. In section 3, we present the theory underlying the density-based strategy and how it was adapted for RS problems. In section 4, we describe our experiments including the methodology, the baseline strategies and the chosen data set. In section 5, we present and discuss our results. Finally, in section 6, we draw our final conclusions highlighting the contributions of this work and interesting aspects that should be addressed in the future.

2 Related Works

To the best of our knowledge, so far, Rating Elicitation for Incentive Purposes has never been directly addressed by the academia. However, some works have been proposed so as to motivate users to evaluate their purchased items. In [4], for instance, authors try to motivate users by developing an interface that favors user-user and user-system interactions, or, as they put it, a conversational interface. In [9], group experiments inspired by social theories were conducted in order to discover the factors that prompt human contribution to the RS. A continuation of this work was carried out in [12] where authors attempted to motivate users by displaying the value of their contribution.

In [1], an economic model was applied in order to understand the user behavior. However, the incentives considered in this modeling were subjective such as the joy from receiving a good recommendation; the excitement of searching for an item; and the fun of evaluating an item. On the other hand, [2] studies the possibility of giving objective incentives (e.g., cash payments) in exchange for ratings. Although the notion of incentives present in [2] matches with ours, it does not view ratings as feedback for the RS. In fact, authors don't even consider a RS, they propose rating elicitation just for the sake of sharing knowledge among users.

One of the pioneers in dealing with Rating Elicitation for Cold Start Purposes, [13] proposes an AL strategy based on the rating entropy. It was found that mere entropy can be misleading, hence a combined strategy was proposed that takes into account both the entropy and the logarithm of the popularity. In [14], the problems arising from the use of entropy are discussed in more depth. Besides, it also proposes other strategies that ease the undesirable effects of entropy (e.g., entropy0, HELF and IGCN).

By far the most complete work in the literature, [5] has served as the main inspiration for our research. It proposes a whole methodology for evaluating different AL strategies. Moreover, [5] has compared several AL strategies, according to different metrics, including MAE, and concluded that there is no silver bullet. In our work, we applied this methodology so as to compare the density-based strategy against the baseline strategies in the literature with respect solely to the system's overall performance (MAE).

An attempt involving incentives in exchange for ratings was carried out in [8]. In this work, authors try to leverage the performance of their RS by obtaining ratings from users of the Amazon Mechanical Turk[1] platform. These users are included in the RS and receive incentives in exchange for ratings. Since there is no record of the items they can evaluate (purchased items), this approach basically deals with Rating Elicitation for Cold Start Purposes in a larger scale. However, this work also addresses an important issue related to Rating Elicitation for Incentive Purposes which is how to attest that the ratings received are reliable. This problem will not be covered in our work.

[1] `https://www.mturk.com`

3 The Density-Based Strategy

In many modern applications, the amount of data simply overwhelms the computational capacity. Therefore, these situations require alternative approaches if one wants to make any profit at all from such data sets. [10] presents an interesting way of dealing with large data sets, i.e., it proposes a Data Reduction (DR) method that could be easily applied as an AL strategy to RS.

The data set can be viewed as a very large sample of instances belonging to $\mathbb{R}^k$, all of them drawn from a single population. From this sample P, it is assumed that the population's Probability Density Function (PDF) can be well estimated. However, P is considered to be so large that dealing directly with it is infeasible. A much smaller sample Q is considered to be a good representation of P, if the PDF estimated from Q is close enough to the PDF estimated from P.

To estimate the PDFs of both P and Q, one could propose the use of some parametric distribution that represents the densest regions in $\mathbb{R}^k$ such as the exponential distribution or the normal distribution. Since we want to make Q as small as a few instances ($|Q| \ll |P|$), one would not be able to tell which parametric distribution best fits the data. This kind of estimation requires a considerable amount of instances to be accurate.

Thus, [10] decides for a non-parametric estimation technique called Kernel Density Estimation (KDE) which is considered appropriate when data is a scarce resource (the case we want to make for Q). The PDF is given by applying a kernel function to each instance in the sample and taking the average of these functions. To be considered a kernel, a function needs to have its integral from minus infinity to infinity equal to 1. In the DR method proposed by [10], the Gaussian function is used as kernel such that the PDFs of Q and P, called $\hat{q}$ and $\hat{p}$, are given according to equations 1 and 2, respectively.

$$\hat{q}(x) = \frac{1}{|Q|} \sum_{i \in Q} G_\sigma(x, i) \tag{1}$$

$$\hat{p}(x) = \frac{1}{|P|} \sum_{i \in P} G_\sigma(x, i) \tag{2}$$

Where $G_\sigma(x, i)$ is the Gaussian function with i as mean, σ as covariance matrix and x as the domain, i.e., $\mathbb{R}^k$ (assuming that $k > 1$). After both PDFs were estimated, one needs now to compute the distance between them. There are many ways to compute this distance called *divergences*. [10] decided for the Integrated Square Error (ISE) (given in equation 3), because it shows that, when put together with the KDE, this yields significant performance improvements. Therefore, the best instance to take from P and insert into Q is the one that will most reduce the ISE between the two PDFs.

$$ISE(\hat{q}, \hat{p}) = \int_{-\infty}^{\infty} [\hat{q}(x) - \hat{p}(x)]^2 dx \tag{3}$$

Since we are interested in finding the best set of items in order to elicit user ratings, we first need to have the items mapped into a vector space, otherwise we will not be able to compute their correspondent PDF. Unfortunately, this is not a trivial task, because, in most RS, items are only represented by their demographic information, which usually comes in textual format.

Nonetheless, in many works concerning RS, a vector representation of items and users in $\mathbb{R}^k$ is achieved by applying the SVD decomposition to the rating matrix or the purchase matrix. This decomposition has emerged in the literature as one of the main tools for dealing with RS. Its success can be attributed to the fact that the SVD decomposition yields a representation of users and items in a k-dimensional space called *user space* and *item space*, respectively [19].

The SVD is a numerical method, therefore it must be applied to a full matrix. The unknown positions in the rating matrix are usually set to zero before the decomposition takes place. By doing that, one is asserting that the great majority of ratings are of the least kind, which is very unlikely and can lead to biased vector representations. In order to avoid introducing erroneous information when decomposing unfilled matrices, we applied the SVD to the purchase matrix instead. This matrix has the same dimensions as the rating matrix, but its positions are binary values (0 or 1) that indicate if the item has been purchased (or acquired) by the user. Thus, this full matrix actually accounts for the system's reality and leads to unbiased vector representations when decomposed. Besides, in real applications the purchase matrix can be less sparse than the rating matrix.

By applying the SVD decomposition to the purchase matrix $B_{n \times m}$ one gets 3 outcomes: matrix $U_{n \times k}$ that represents users in a k-dimensional space (user space); matrix $S_{k \times k}$ that comprises the singular values of B in its diagonal; and matrix $V_{m \times k}$ that represents items in a k-dimensional space (item space). The number of dimensions k is an arbitrary parameter of the SVD decomposition.

The whole schema for the density-based strategy is presented in algorithm 1. For each user u there are two sets of items: Q that comprises the items user u has purchased and evaluated and C that comprises the items user u has purchased but did not evaluate (the ones for which ratings will be elicited). The set P comprising all the items in the data set is used as reference for all users. The PDF of P, namely $\hat{p}$, is given by the KDE and, once estimated, it remains unaltered. Each item j belonging to C is temporarily inserted into Q and $\hat{q}$, the PDF of $Q \cup \{j\}$, is estimated. The ISE between $\hat{q}$ and $\hat{p}$ is assigned as score to item j. Finally, the item with the lowest score is permanently inserted into Q and removed from C. This process is repeated in order to find the second best item to be inserted into Q and so forth until we find the best N items. Once they are found, the user is asked to rate them, which, in AL terms, means that the instances are labeled and included in the training set.

In order to choose the best parameters for the density-based strategy, we have set k equal to 2, 3, 5, 10 and 50. For each of those values, we have tested σ equal $0.00005I$, $0.001I$, $0.5I$, $2I$ and $10I$, where I is the $k \times k$ identity matrix. We compared the several combinations of k and σ according to the methodology presented in section 4.1 and concluded that there is little variation when it comes to overall system's performance (MAE). Nonetheless, using $k = 2$ with

$\sigma = 0.00005I$ has yielded slightly better results, probably due to the fact that estimating PDFs with the KDE is easier in low-dimensional spaces. Since the goal of this work is just to make the case for the density-based strategy, we chose these parameters empirically and left the fine-tuning aspects for a future work.

Algorithm 1. The Density-based Strategy

$[U, S, V] \leftarrow SVD(B, k)$
$\hat{p} \leftarrow KDE(V, \sigma)$
for each user u **do**
  $list \leftarrow \emptyset$
  while $|list| < N$ **do**
    for each item $j \in C$ **do**
      $\hat{q} \leftarrow KDE(Q \cup \{j\}, \sigma)$
      $score(j) \leftarrow ISE(\hat{q}, \hat{p})$
    end for
    $j' \leftarrow \text{argmin}_j \, score(j)$
    $list \leftarrow list \cup \{j'\}$
    $Q \leftarrow Q \cup \{j'\}$
    $C \leftarrow C \setminus \{j'\}$
  end while
  $u \xleftarrow{ask} list$
end for

4 Experiments

4.1 Methodology

The methodology for comparing the performance of different strategies is the one proposed by [5] with a slight modification. Since [5] is concerned with Rating Elicitation for Cold Start Purposes, it cannot assume that users will evaluate all solicited items. In fact, one of the metrics by which [5] compares the strategies is the percentage of elicited ratings. In our scenario, as we are dealing with Rating Elicitation for Incentive Purposes, it is fair to assume that users can and will evaluate all solicited items, because they have been already purchased, plus users will receive a persuasive incentive to evaluate them.

We also assume that all given evaluations are reliable, i.e., no user gives random ratings just for the sake of receiving the incentive. Despite not being very realistic, this turns the setup of our experiments simpler. In future works, we will loosen this assumption by considering methods that identify spam users.

The rating matrix R is randomly divided into two distinct matrices T and Tr, with 20% and 80% of the ratings, respectively. Tr is also randomly divided into K and X, with 5% and 95% of the ratings, respectively. These divisions are depicted in algorithm 2 by function $M_2 = RAND(M_1, y)$, where parameters M_1 and M_2 are matrices with the same dimensions and y is the percentage of nonzero ratings that will be removed from M_1 and inserted into M_2. The ratings that were randomly chosen to be removed from M_1 will be replaced by zero values and the unfilled positions in matrix M_2 will also receive zero values.

Matrix K represents the initial knowledge of the RS. We seek to expand this knowledge by eliciting ratings in X with a strategy $S(u, N, K, C)$, which is a user-oriented function that assigns a score for each item in C. The set C comprises all items user u has purchased but not yet evaluated (items that have ratings in X given by u). Those items receive a score that is computed based only on matrix K. As result, the strategy returns a list L with the N highest (or the lowest) scored items.

We apply S to elicit the ratings of user u, i.e., to transfer them from X to K. At each iteration, S is applied for all n users and a maximum of $n \times L$ ratings are inserted into K. The recommendation model is then trained with the updated matrix K and its predictions are evaluated against the test set T. In algorithm 2, this step is depicted by function $E = MODEL(M_1, M_2)$, where M_1 is the training set, M_2 the test set and E is the value of the Mean Absolute Error (MAE) [15]. The recommendation model used is the Regularized SVD Model [11].

Likewise [5], we have applied the 5-fold cross validation so as to achieve more statistical confidence in our experiments. The results displayed in section 5 are the averaged MAE values obtained from 5 executions of algorithm 2. The ratings in R were randomly scattered into 5 distinct matrices, with the same dimensions of R, each having exactly 20% of the nonzero ratings. At each execution of the algorithm 2, one of these matrices is used as T while the others are combined into Tr.

Algorithm 2. Methodology for Evaluating a Strategy S

```
N ← 10
iter ← 15
T ← RAND(R, 20%)
Tr ← R \ T
K ← RAND(Tr, 5%)
X ← Tr \ K
i ← 1
while i ≤ iter do
    for each user u do
        L ← S(u, N, K, C)
        K ← K ∪ L
        X ← X \ L
        C ← C \ L
    end for
    error(i) ← MODEL(K, T)
    i++
end while
```

4.2 Strategies

We chose the following strategies as baseline for our experiments:

- **random** - Selecting items randomly is the simplest of all strategies and can even be considered the absence of any strategy. It has been applied as baseline to almost all works regarding Active Learning.
- **popularity** - Also a very basic strategy, popularity was proposed by [13] and selects the most evaluated items in K.
- **entropy** - The entropy measures the uncertainty that users have about an item. Therefore, this strategy tries to reduce the overall uncertainty by asking ratings for items with high entropy. It was first proposed by [13].
- **log(pop)*ent** - It has been verified that entropy alone can be misleading, because items with very few ratings can yield very high values of entropy. Therefore, this strategy tries to balance entropy with popularity [13].
- **entropy0** - This strategy also tries to balance entropy with popularity, however it considers the popularity of an item by incorporating the zero ratings in the entropy computation. The amount of zero ratings usually outnumbers the others, therefore [14] proposes the use of weighted entropy instead.
- **HELF** - This strategy is another approach in [14] that tries to balance entropy with popularity. However, instead of the simple multiplication, authors propose the use of the Harmonic Mean or F1 measure.
- **IGCN** - This strategy is based on the Information Gain (IG) and not on pure entropy. Users are hierarchically distributed into clusters according to their arrangement in the user space. The computation of IG for every item takes into account the entropy of the users' distribution. According to [14], this can be viewed as training a Decision Tree where the leaves represent clusters and the middle nodes represent a condition on a specific item.
- **variance** - Likewise the entropy, the variance also measures the uncertainty associated to an item. In essence, those strategies are the same, once they operate by the same principle [5].
- **sqrt(pop)*var** - This strategy tries to combine variance with popularity in a similar way to log(pop)*ent and HELF. It has presented the second best performance in the benchmark carried out by [6].
- **bin pred** - This strategy stands for *binary prediction* and differs significantly from the ones presented so far, since it is based on the model and not on uncertainty reduction. It uses the purchase matrix of K as training set for the model and then selects items with the highest predictions. In other words, this strategy selects items that are more likely to be consumed or acquired by users [5].
- **high pred** - This strategy stands for *highest predicted* and, unlike *bin pred*, it uses matrix K itself as training set for the model and then selects items with highest predictions. In others words, this strategy selects items that are more likely to receive high ratings [5].
- **low pred** - This strategy stands for *lowest predicted* and can be considered the exact opposite of high pred. It uses matrix K as training set for the

model and then selects items with lowest predictions. In other words, this strategy selects items that are more likely to receive low ratings [5].
- **high-low pred** - This strategy stands for *highest and lowest predicted* and can be viewed as a combination of high pred and low pred. It uses matrix K as training set for the model and then selects items whose predictions are more distant from the average rating. In other words, this strategy selects items that are more likely to receive extreme ratings [5].

4.3 Data Set

We decided to run our experiments in the data set known as MovieLens 100k, or simply MovieLens. This is a famous data set in the RS literature, made available by GroupLens [7] and used in the experiments carried out by [5]. It has 100,000 ratings, given to 1682 items (movies) from 943 users, being 6110 equal to 1; 11370 equal to 2; 27145 equal to 3; 34174 equal to 4; and 21201 equal to 5. Those ratings account only to 6.3% of all positions in the rating matrix.

5 Results

Figure 1 shows the results for all strategies described in section 4.2 plus the density-based strategy. Although the details about each strategy are not very clear, by looking at this chart one can evidently conclude that there are two major groups of strategies: those that outperform random closely and those that lag far behind.

Apart from low pred, high-low pred and density-based, random has outperformed all other strategies, which is not a very intuitive result. However, this same pattern was found in [5] and, indeed, [18] points out that beating random is not that trivial. Since all strategies make some assumption about the data, if such assumption is not verified in the data set used, the rating selection will follow an unrealistic criterion that will probably lead to a biased training set. Random, on the other hand, is unbiased by definition.

A biased training set can either improve or harm the system's performance. For instance, all strategies based on uncertainty reduction (e.g., entropy, entropy0, log(pop)*ent, HELF, IGCN, variance and sqrt(pop)*var) assume that asking ratings for items with high uncertainty will improve the model's accuracy. However, by doing so, one is actually creating a training set with only highly uncertain items. In practice, these items are the ones that pose great difficulty to users. Therefore, by favouring these instances, one is adding to the training set the most controversial items and, consequently, the model will struggle to find a rating pattern.

As for the model-based strategies, we notice that bin pred and high pred do not stand out, whereas low pred and high-low pred achieve good results. Considering the rating distribution in the data set, we see that high ratings (4 and 5) account for approximately 55% of all ratings. The initial training set (matrix K) reflects the natural unbalance of ratings in the entire data set, because it is created

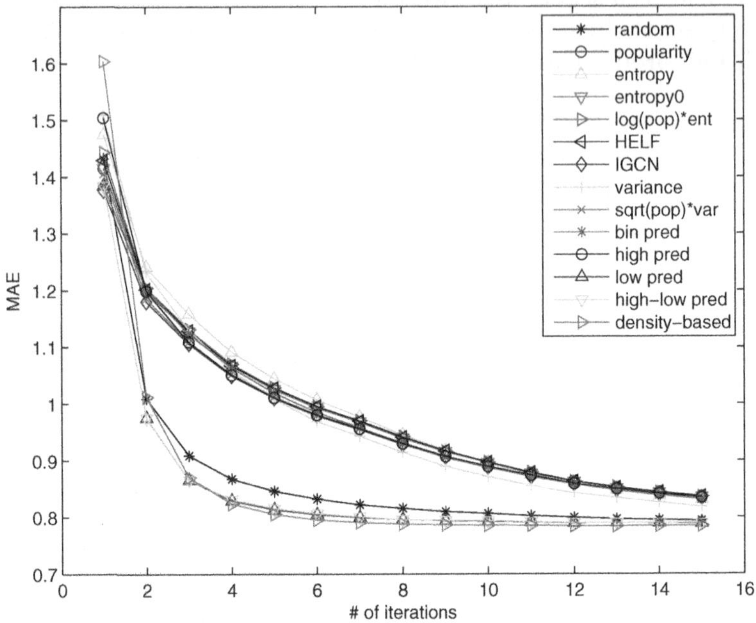

Fig. 1. Benchmark with all AL strategies

randomly. Thus, by selecting items that are more likely to receive high ratings, high pred is actually worsening the unbalance in K. As for bin pred, it is likely to just keep the unbalance in K, because it ignores the rating values. On the contrary, since low ratings (1 and 2) and extreme ratings (1 and 5) account for 17% and 27% of the data set, respectively, both low pred and high-low pred are balancing K by asking these unusual ratings. A model trained with a balanced training set can achieve good accuracy for all rating values and, consequently, a good overall accuracy.

In figure 2, we zoom in the performance of random, low pred, high-low pred and density-based taking into account the standard deviation given by the 5-fold cross validation. We see that density-based, from iteration 5 onwards, clearly outperforms the others. Since low pred and high-low pred make strong assumptions about the data set (that low and extreme ratings are minority), they might not be applicable in most cases, especially in real applications, where the data set's characteristics are constantly changing. Density-based, on the other hand, makes a weak and intuitive assumption that turned out to be very efficient in our experiments: *an unbiased training set will best favour the model.* But, unlike random, density-based is not subjected to randomness, i.e., it follows a greedy heuristic (ISE minimization) that guarantees that the most representative items, of the item space's densest regions, are selected.

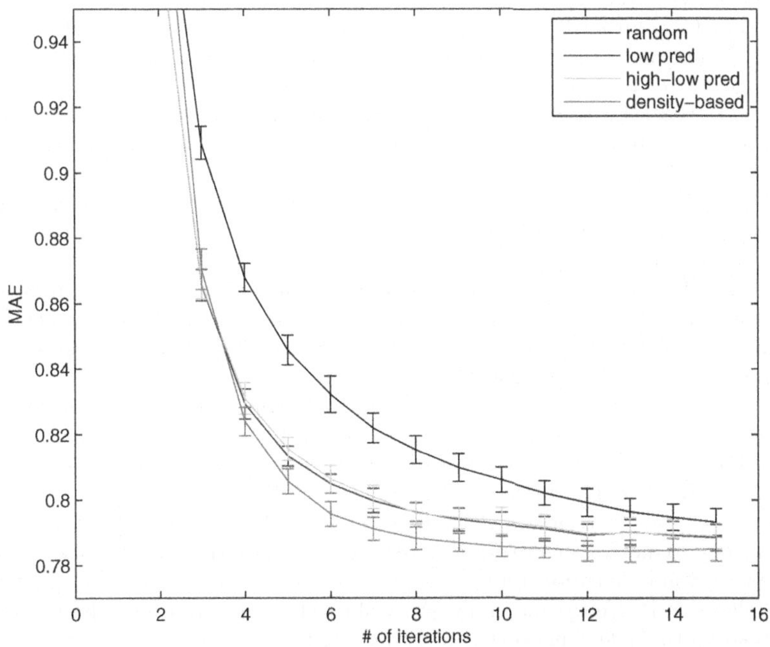

Fig. 2. Statistical confidence of random, low pred, high-low pred and density-based

6 Conclusions

In this work, we have proposed a novel AL strategy for dealing with Rating Elicitation for Incentive Purposes that has outperformed all baseline strategies concerning overall accuracy (MAE) and statistical confidence. Density-based has shown promising evidences that it can perform well regardless of the data set's characteristics, i.e., it can be an efficient default strategy for real applications. In future works, we intend to conduct experiments with larger data sets so as to confirm our findings and also consider a streaming scenario where purchases are analysed on-the-fly. Moreover, we would like to conduct a theoretical study in order to find out under which conditions it would be better to opt for a biased (balanced) training set over an unbiased (unbalanced) one, and vice-versa.

References

1. Harper, F.M., Li, X., Chen, Y., Konstan, J.A.: An economic model of user rating in an online recommender system. In: Ardissono, L., Brna, P., Mitrović, A. (eds.) UM 2005. LNCS (LNAI), vol. 3538, pp. 307–316. Springer, Heidelberg (2005)
2. Avery, C., Resnick, P., Zeckhauser, R.: The market for evaluations. The American Economic Review 89(3), 564–584

3. Bell, R.M., Koren, Y.: Lessons from the netflix prize challenge. SIGKDD Explor. Newsl. 9(2), 75–79 (2007)
4. Carenini, G., Smith, J., Poole, D.: Towards more conversational and collaborative recommender systems. In: Proceedings of the 8th International Conference on Intelligent User Interfaces, IUI 2003, pp. 12–18 (2003)
5. Elahi, M., Ricci, F., Rubens, N.: Active learning strategies for rating elicitation in collaborative filtering: A system-wide perspective. ACM Trans. Intell. Syst. Technol. 5(1), 1–33 (2014)
6. Golbandi, N., Koren, Y., Lempel, R.: On bootstrapping recommender systems. In: Proceedings of the 19th ACM International Conference on Information and Knowledge Management, CIKM 2010, pp. 1805–1808 (2010)
7. Herlocker, J.L., Konstan, J.A., Borchers, A., Riedl, J.: An algorithmic framework for performing collaborative filtering. In: Proceedings of the 22Nd Annual International ACM SIGIR Conference on Research and Development in Information Retrieval, SIGIR 1999, pp. 230–237 (1999)
8. Lee, J., Jang, M., Lee, D., Hwang, W.S., Hong, J., Kim, S.W.: Alleviating the sparsity in collaborative filtering using crowdsourcing. In: Workshop on Crowdsourcing and Human Computation for Recommender Systems (CrowdRec), p. 5 (2013)
9. Ling, K., Beenen, G., Ludford, P., Wang, X., Chang, K., Li, X., Cosley, D., Frankowski, D., Terveen, L., Rashid, A.M., Resnick, P., Kraut, R.: Using social psychology to motivate contributions to online communities. Journal of Computer-Mediated Communication 10(4)
10. de Mello, C.E.R.: Active Learning: An Unbiased Approach. Ph.D. thesis, Federal University of Rio de Janeiro (UFRJ), Brazil (2013)
11. Paterek, A.: Improving regularized singular value decomposition for collaborative filtering. In: 13th ACM Int. Conf. on Knowledge Discovery and Data Mining, Proc. KDD Cup Workshop at SIGKDD 2007, pp. 39–42 (2007)
12. Rashid, A.M., Ling, K., Tassone, R.D., Resnick, P., Kraut, R., Riedl, J.: Motivating participation by displaying the value of contribution. In: Proceedings of the SIGCHI Conference on Human Factors in Computing Systems, CHI 2006, pp. 955–958 (2006)
13. Rashid, A.M., Albert, I., Cosley, D., Lam, S.K., McNee, S.M., Konstan, J.A., Riedl, J.: Getting to know you: Learning new user preferences in recommender systems. In: Proceedings of the 7th International Conference on Intelligent User Interfaces, IUI 2002, pp. 127–134 (2002)
14. Rashid, A.M., Karypis, G., Riedl, J.: Learning preferences of new users in recommender systems: An information theoretic approach. SIGKDD Explor. Newsl. 10(2), 90–100 (2008)
15. Ricci, F., Rokach, L., Shapira, B., Kantor, P.B. (eds.): Recommender Systems Handbook. Springer, New York (2011)
16. Schafer, J.B., Konstan, J., Riedl, J.: Recommender systems in e-commerce. In: Proceedings of the 1st ACM Conference on Electronic Commerce, EC 1999, pp. 158–166 (1999)
17. Schwartz, B.: The paradox of choice. ECCO, New York (2005)
18. Settles, B.: Active learning literature survey. Computer Sciences Technical Report 1648, University of Wisconsin–Madison (2009)
19. Strang, G.: Introduction to Linear Algebra. Wellesley-Cambridge Press, Wellesley (2009)

Entity-Centric Stream Filtering and Ranking: Filtering and Unfilterable Documents

Gebrekirstos G. Gebremeskel and Arjen P. de Vries

Information Access, CWI, Amsterdam,
Science Park 123, 1098 XG Amsterdam, The Netherlands
gebre@cwi.nl, arjen@acm.org

Abstract. Cumulative Citation Recommendation (CCR) is defined as: given a stream of documents on one hand and Knowledge Base (KB) entities on the other, filter, rank and recommend citation-worthy documents. The pipeline encountered in systems that approach this problem involves four stages: filtering, classification, ranking (or scoring), and evaluation. Filtering is only an initial step that reduces the web-scale corpus into a working set of documents more manageable for the subsequent stages. Nevertheless, this step has a large impact on the recall that can be attained maximally. This study analyzes in-depth the main factors that affect recall in the filtering stage. We investigate the impact of choices for corpus cleansing, entity profile construction, entity type, document type, and relevance grade. Because failing on recall in this first step of the pipeline cannot be repaired later on, we identify and characterize the citation-worthy documents that do not pass the filtering stage by examining their contents.

1 Introduction

The maintenance of knowledge bases (KBs) has increasingly become quite a challenge for their curators, considering both the growth of the number of entities considered and the huge amount of online information that appears every day. In this context, researchers have started to create information systems that support the task of Cumulative Citation Recommendation (CCR): given a stream of documents and a set of entities from a Knowledge Base (KB), filter, rank and recommend those documents that curators would consider "citation-worthy".

KB curators will expect the input stream to cover all the (online) information sources that could contain new information about the entities in the KB, varying from mainstream news sources to forums and blogs. State-of-the-art CCR systems need to operate on web-scale information resources. Current systems therefore divide up their overall approach in multiple stages, e.g., filtering, classification, ranking (or scoring), and evaluation. This paper zooms into this first stage, filtering, an initial step that reduces the web-scale input stream into a working set of documents that is more manageable for the subsequent stages. Nevertheless, the decisions taken in this stage of the pipeline are critical for recall, and therefore impact the overall performance. The goal of our research is

A. Hanbury et al. (Eds.): ECIR 2015, LNCS 9022, pp. 303–314, 2015.

to increase our understanding how design decisions in the filtering stage affect the citation recommendation process.

We build on the resources created in the Knowledge Base Acceleration (KBA) track of the Text REtrieval Conference (TREC), introduced in 2012 with Cumulative Citation Recommendation as the main task. As pointed out in the 2013 track's overview paper [9] and confirmed by our own analysis of participants' reports, the approaches of the thirteen participating teams all suffered from a lack of recall. Could this be an effect of short-comings in the initial filtering stage?

While all TREC-KBA participants applied some form of filtering to produce a smaller working set for their subsequent experiments, the approaches taken vary widely; participants rely on different techniques and resources to represent entities, algorithms may behave differently for the different document types considered in the heterogeneous input stream, and teams use different versions of the corpus. Given these many factors at play, the task of drawing generically applicable conclusions by just comparing overall results of the evaluation campaign seems infeasible. Our paper therefore investigates systematically the impact of choices made in the filtering stage on the overall system performance, varying the methods applied for filtering while fixing the other stages of the pipeline.

The main contributions of the paper are an in-depth analysis of the factors that affect entity-based stream filtering, identifying optimal entity profiles without compromising precision, shedding light on the roles of document types, entity types and relevance grades. We also present a failure analysis, classifying the citation-worthy documents that are not amenable to filtering using the techniques investigated.

The remaining part of the paper is organized as follows. After a brief related work, Section 3 describes the dataset and approach, followed by experiments in Section 4. Sections 5 and 6 discuss their results and a failure analysis. Section 7 summarizes our conclusions.

2 Related Work

Automatic systems to assist KB curators can be seen as a variation of information filtering systems, that "sift through a stream of incoming information to find documents relevant to a set of user needs represented by profiles" [14]. In entity-centric stream filtering, user needs correspond to the KB entities to be curated. However, since the purpose of the filtering component in cumulative citation recommendation is to reduce the web-scale stream into a subset as input for further processing, the decision which documents should be considered citation-worthy is left to later stages in the pipeline.

Other related work addresses the topic of entity-linking, where the goal is to identify entity mentions in online resources and link these to their corresponding KB profiles. Relevant studies include [5,7], and evaluation resources are developed at the Knowledge Base Population (KBP) track of the Text Analysis Conference (TAC) [11]. Though related, entity linking emphasizes the problem of locating an entity's mentions in unstructured text, where the primary goal of CCR is to identify an entity's most relevant documents.

Our study is rooted in the research carried out in context of TREC KBA. The problem setup has been essentially the same for both the 2012 and 2013 KBA tracks, but the large size of the 2013 corpus had the effect that all participants resorted to reducing the data-set using an initial filtering stage. Approaches varied significantly in the way they construct entity profiles. Many participants rely on name variants taken from DBpedia, such as labels, names, redirects, birth names, alias, nicknames, same-as and alternative names [15,6,12]. Two teams considered (Wikipedia) anchor text and the bold-faced words of the first paragraph of the entity's Wikipedia page [4,13]. One participant used a Boolean *and* expression built from the tokens of canonical names [8].

Due to the large variety in the methods applied in different stages of the pipeline, it is difficult to infer which approaches are really the best. By focusing on a single component of the pipeline and analyzing the effects of its design choices in detail, we aim at more generally applicable results.

3 Approach

We use the TREC-KBA 2013 dataset[1] to compare the effectiveness of different choices for document and entity representation in the filtering stage. Cleansing refers to pre-processing noisy web text into a canonical "clean" text format. In the specific case of TREC KBA, the organisers provide two versions of the corpus: one that is already cleansed, and one that is the raw data as originally collected by the organisers. Entity profiling refers to creating a representation of the entity based on which the stream of documents is filtered, usually by straightforward matching of their textual contents.

3.1 Dataset Description

The TREC-KBA 2013 dataset consists of three main parts: a time-stamped stream corpus, a set of KB entities to be curated, and a set of relevance judgments. The stream corpus comes in two versions: raw and cleansed. The raw data is a dump of HTML pages. The cleansed version is the raw data after its HTML tags have been stripped off, considering only the documents identified as English (by the Chromium Compact Language Detector[2]). The stream corpus is organized in hourly folders, each of which contains many "chunk files". Each chunk file contains hundreds to hundreds of thousands of semi-structured documents, serialized as thrift objects (one thrift object corresponding to one document). Documents are blog articles, news articles, or social media posts (including tweets). The stream corpus has been derived from three main sources: TREC KBA 2012[3](blogs, news, and urls that were shortened at `bitly.com`), arXiv[4] (e-prints), and spinn3r[5] (blogs).

[1] `http://trec-kba.org/trec-kba-2013.shtml`
[2] `https://code.google.com/p/chromium-compact-language-detector/`
[3] `http://trec-kba.org/kba-stream-corpus-2012.shtml`
[4] `http://arxiv.org/`
[5] `http://spinn3r.com/`

The KB entities in the dataset consist of 20 Twitter and 121 Wikipedia entities. The entities selected by the organizers of the TREC KBA evaluation are "sparse" (on purpose): they occur in relatively few documents and have an underdeveloped KB entry.

TREC-KBA provides relevance judgments, which are given as document-entity pairs. Documents with citation-worthy content to a given entity are annotated as *vital*, while documents with tangentially relevant content, lacking freshliness or with content that can be useful only for initial KB-dossier creation are annotated as *relevant*. Documents with no relevant content are labeled *neutral*, spam documents are labeled as *garbage*. In total, the set of relevance judgments contains 24162 unique vital-relevant document-entity pairs (9521 vital and 17424 relevant).[6] The relevance judgments have been categorized into 8 source categories: 0.98% arXiv, 0.034% classified, 0.34% forum, 5.65% linking, 11.53% mainstream-news, 18.40% news, 12.93% social and 50.2% weblog. We have regrouped these source categories into three groups, "news", "social", and "other", for two reasons. First, mainstream-news and news are very similar, and can only be distinguished by the underlying data collection process; likewise for weblog and social. Second, some sources contain too few judged document-entity pairs to usefully distinguish between these. The majority of vital or relevant annotations are "social" (63.13%) and "news" (30%). The remaining 7% are grouped as "other".

3.2 Entity Profiling

The names of the entities that are part of the URL are referred to as their "canonical names". E.g., entity `http://en.wikipedia.org/wiki/Benjamin_Bronfman` has canonical name "Benjamin Bronfman", and `https://twitter.com/RonFunchesFor` has canonical name "RonFunchesFor". For the Wikipedia entities, we derive additional name variants from DBpedia: name, label, birth name, alternative names, redirects, nickname, or alias. For the Twitter entities, we copied the display names manually from their respective Twitter pages. On average, we extract approximately four different name variants for each entity.

For each entity, we create four entity profiles: canonical (cano), canonical partial (cano-part), all name variants combined (all) and their partial names (all-part). Throughout the paper, we refer to the last two profiles as name-variant and name-variant partial, using the terms in parentheses in the Table captions.

3.3 Evaluation Measures

Our main measure of interest is the recall, as documents missed in this stage cannot be recovered during further processing. We also report the overall performance of a standard high performing setup for the subsequent stages of the

[6] The numbers of vital and relevant do not add up to 24162 because some documents are judged as both vital and relevant, by different assessors.

pipeline, that we keep constant. Here, we compute the track's standard evaluation metric, max-F, using the scripts provided [9]. Max-F corresponds to the maximally attained F-measure over different cutoffs, averaged over all entities. The default setting takes the vital rating if a document-entity pair has both vital and relevant judgments.

4 Experiments and Results

4.1 Cleansing: Raw or Cleansed

Tables 1 and 2 show that recall (on retrieving each relevance judgment) is higher in the raw version than in the cleansed one. Recall increases on Wikipedia entities vary from 13% to 16.4%, and on Twitter entities from 62.8% to 357.9%. At an aggregate level, recall improvement ranges from 15% to 20.5%. The recall increases are substantial. To put it into perspective, an 15% increase in recall on all entities is a retrieval of 2864 more unique document-entity pairs.

4.2 Entity Profiles

The aggregate recall increase from canonical partial to name-variant partial is 25% and from canonical names to name variants is 35% (see Table 2). This means that a quarter of the documents mentioned the entities by partial names of non-canonical name variants and more than one-third of the documents mention the entities by non-canonical names, respectively. Generally, recall increases as we move from canonicals to canonical partial, to name-variant, and to name-variant partial. The only exception is that using canonical partial leads to a better recall for Wikipedia entities than using the name-variants.

4.3 Relevance Rating: Vital and Relevant

The primary objective of cumulative citation recommendation is to identify the citation-worthy documents. We would like to know if there is a difference between filtering vital and relevant documents (as measured by recall). This could be helpful to make choices that improve the retrieval of citation-worthy documents selectively. In Table 3, we observe that recall performances considering vital documents only are in general higher than those that consider relevant documents as well. Especially for Wikipedia entities, the vital documents tend

Table 1. Vital recall for cleansed

	cano	cano-part	all	all-part
Wikipedia	61.8	74.8	71.5	77.9
Twitter	1.9	1.9	41.7	80.4
Aggregate	51.0	61.7	66.2	78.4

Table 2. Vital recall for raw

	cano	cano-part	all	all-part
Wikipedia	70.0	86.1	82.4	90.7
Twitter	8.7	8.7	67.9	88.2
Aggregate	59.0	72.2	79.8	90.2

Table 3. Breakdown of recall performances by document source category

		Aggregate			Wikipedia			Twitter		
		other	news	social	other	news	social	other	news	social
Vital	cano	82.2	65.6	70.9	90.9	80.1	76.8	8.1	6.3	30.5
	cano part	90.4	80.6	83.1	100.0	98.7	90.9	8.1	6.3	30.5
	all	94.8	85.4	83.1	96.4	95.9	85.2	81.1	42.2	68.8
	all part	100	99.2	95.9	100.0	99.2	96.0	100	99.3	94.9
Relevant	cano	84.2	53.4	55.6	88.4	75.6	63.2	10.6	2.2	6.0
	cano part	94.7	68.5	67.8	99.6	97.3	77.3	10.6	2.2	6.0
	all	95.8	90.1	72.9	97.6	95.1	73.1	65.2	78.4	72.0
	all part	98.8	95.5	83.7	99.7	98.0	84.1	83.3	89.7	81.0
All	cano	81.1	56.5	58.2	87.7	76.4	65.7	9.8	3.6	13.5
	cano part	92.0	72.0	70.6	99.6	97.7	80.1	9.8	3.6	13.5
	all	94.8	87.1	75.2	96.8	95.3	75.8	73.5	65.4	71.1
	all part	99.2	96.8	86.6	99.8	98.4	86.8	92.4	92.7	84.9

to mention the entities by their canonical name. This observation can be explained by the intuition that a highly relevant document usually will mention the entity multiple times, using different forms to refer to it. Those documents are therefore likely to pass the filtering stage.

4.4 Document Categories and Entity Types

The study of recall across document categories (news, social, other) helps us understand how types of documents behave with respect to filtering. Our documents are divided mainly between social and news. Table 3 shows that for Wikipedia entities recall for news documents is higher than for social. In Twitter entities, however, the recall for social documents is higher than for news, except in name-variant partial. Regarding the two types of entities (Wikipedia and Twitter), we see that Wikipedia entities achieve higher recall than Twitter entities (see Tables 1, 2 and 3).

4.5 Impact on Classification

We now will conduct experiments to see how the different choices we made at the filtering stage impact the subsequent steps of the pipeline. Based on the findings of previous work [1,2,10], we use a standard pipeline, where the documents passing the filtering stage are classified into their relevance grades. We take the state of the art WEKA's[7] Classification Random Forest and the set of features used in [10], for they are small in number, and the resulting classifier is known to be effective for the CCR problem. We follow the official TREC KBA training and testing setting, that is, we train on the number of documents that our filtering system retrieves from the training data and test on those documents

[7] `http://www.cs.waikato.ac.nz/~ml/weka/`

retrieved from the test set. For example, when we use cleansed data and canonical profile, we train on training relevance judgments that we retrieve from the cleansed corpus, using the canonical profile, and test on the corresponding test relevance judgments that we retrieve from the cleansed corpus. The same applies for other combinations of choices. In here, we present results showing how the cleansing, entity type, document category, and entity profile impact classification performance.

Table 4. Cleansed: vital max-F

	cano	cano-part	all	all-part
all-entities	0.241	0.261	0.259	0.265
Wikipedia	0.252	0.274	0.265	0.271
twitter	0.105	0.105	0.218	0.228

Table 5. Raw: vital max-F

	cano	cano-part	all	all-part
all-entities	0.240	0.272	0.250	0.251
Wikipedia	0.257	0.257	0.257	0.255
twitter	0.188	0.188	0.208	0.231

Table 6. Cleansed: vital-relevant max-F

	cano	cano-part	all	all-part
all-entities	0.497	0.560	0.579	0.607
Wikipedia	0.546	0.618	0.599	0.617
twitter	0.142	0.142	0.458	0.542

Table 7. Raw: vital-relevant max-F

	cano	cano-part	all	all-part
all-entities	0.509	0.594	0.590	0.612
Wikipedia	0.550	0.617	0.605	0.618
twitter	0.210	0.210	0.499	0.580

Tables 4 and 5 show the max-F performance for vital relevance ranking. On Wikipedia entities, with the exception of canonical entity profiles, the max-F performance using the cleansed version of the corpus is better than that using the raw one. On Twitter entities however, the performance obtained using the raw corpus is better on all entity profiles, with the exception of name-variant partial. This result is interesting, because we saw in previous sections that *recall* when using the raw corpus is substantially higher than using cleansed one. This gain in recall for the raw corpus does however not translate into a gain in max-F for recommending vital documents. In fact, in most cases overall CCR performance decreased. Canonical partial for Wikipedia entities and name-variant partial for Twitter entities achieve the best results. Considering the vital-relevant category (Tables 6 and 7), the results are different. The raw corpus achieves better results in all cases (except in canonical partial of Wikipedia). Summarizing, we find that using the raw corpus has more effect on relevant documents and Twitter entities.

5 Analysis and Discussion

There are 3 interesting observations: 1) cleansing impacts relevant documents and Twitter entities negatively. This is validated by the observation that recall gains in Twitter entities and the relevant categories in the raw corpus also translate into overall performance gains. Cleansing removes more relevant documents than it does vital, which can be explained by the fact that it removes related links and adverts which may contain a mention of the entities. One example we saw was that cleansing removed an image with a text of an entity name which was actually relevant. Cleansing also removes more social documents than news, as can be seen by the fact that most of the missing documents from cleansed are social documents. Twitter entities are affected because of their relation to relevant documents and social documents. Examination of the relevance judgments show that about 70% of relevance judgments for Twitter entities are relevant.

2) Taking both performance (recall at filtering and overall F-score) into account, the trade-off between using a richer entity-profile and retrieval of irrelevant documents results in Wikipedia's canonical partial and Twitter's name variant partial as the two best profiles for Wikipedia and Twitter respectively. This is interesting because TREC KBA participants did not consider Wikipedia's canonical partial as a viable entity profile. Experiments with richer profiles for Wikipedia entities increase recall, but not overall performance.

3) The analysis of entity profiles, relevance ratings, and document categories reveal three differences between Wikipedia and Twitter entities. a) Wikipedia entities achieve higher recall and higher overall performance. b) The best profiles for Wikipedia entities are canonical partial and for Twitter entities name-variant partial. c) The fact that Twitter canonical names achieve very low recall means that documents (specially news and others) almost never use Twitter user names to refer to Twitter entities. However, comparatively speaking, social documents refer to Twitter entities by their user names than news and others suggesting a difference in adherence to standard in names and naming.

The high recall and subsequent higher overall performance of Wikipedia entities can be due to two reasons. First, Wikipedia entities are relatively better described than Twitter entities. The fact that we can retrieve different name variants from DBpedia is an indication of rich description. On the contrary, the fact that the Twitter's richest profile achieves both the highest recall and the highest max-F scores indicates that there is still room for enriching the Twitter entity profiles. Rich description plays a role in both filtering and computation of features such as similarity measures in later stages of the pipeline. By contrast, we have only two names for Twitter entities: their user names and their display names. Second, unfortunately, no standard DBpedia-like resource exists for Twitter entities, from which alternative names can be collected.

In the experimental results, we also observed that recall scores in the vital category are higher than in the relevant category. Based on this result, we can say that the more relevant a document is to an entity, the higher the chance that it will be retrieved with alternative name matching. Across document categories, we observe a pattern in recall of others, followed by news, and then by social. Social documents are the hardest to retrieve, a consequence of the fact that social documents (tweets and blogs) are more likely to point to a resource where the entity is mentioned, mention the entity with short abbreviation, or talk without mentioning the entities but with some context in mind. By contrast news documents mention the entities they talk about using the common name variants more than social documents do. However, the greater difference in percentage recall between the different entity profiles in the news category indicates news refer to a given entity with different names, rather than by one standard name.

6 Failure Analysis: Vital or Relevant, but Missing

The use of name-variant partial for filtering is an exhaustive attempt to retrieve as many relevant documents as possible, at the cost of bringing in many irrelevant documents. However, we still miss about 2363 (10%) of the vital-relevant documents. If these are not even mentioned by their partial name variants, what type of expressions were they mentioned by?

Table 8 shows the documents that we miss with respect to cleansed and raw corpus. The upper part shows the number of documents missing from cleansed and raw versions of the corpus. The lower part of the table shows the intersections and exclusions in each corpus.

Table 8. The number of documents missing from raw and cleansed extractions (upper part cleansed, lower part raw)

category	Vital	Relevant	Total
Cleansed	1284	1079	2363
Raw	276	4951	5227
missing only from cleansed	1065	2016	3081
missing only from raw	57	160	217
Missing from both	219	1927	2146

One would naturally assume that the set of document-entity pairs retrieved from the cleansed corpus would be a sub-set of those that are retrieved from the raw corpus. We find that this is however not the case; we even find that we retrieve documents from the cleansed corpus that we miss from the raw corpus. Examining the content of the documents reveals that this can be attributed to missing text in the corresponding document representations. Apparently, a (part of) the document content has been lost in the cleansing process, where the removal of HTML tags and non-English content resulted in a loss of partial or entire content. Documents missing from the raw corpus are all social ones (tweets, blogs, posts from other social media), where the conversion to the raw data format (a binary byte array) may have faulted. In both cases, the entity mention happens to be on the part of the text cut out in the transformation.

The most surprising failures correspond to judged documents that do not pass the filtering stage, neither from the raw nor from the cleansed version of the corpus. These may indicate a fundamental shortcoming of filtering the stream using string-matching, requiring potentially more advanced techniques.

Our failure analysis identifies 2146 unique document-entity pairs, the majority (86.7%) of which are social documents, 219 of these judged as vital, and related to 35 entities (28 Wikipedia and 7 Twitter).

We observed that among the missing documents, different document ids can have the same content, and be judged multiple times for a given entity.[8] Avoiding duplicates, we randomly selected 35 distinct documents, 13 news and 22 social, one for each entity. Based on this subset of the judgements, we categorized situations under which documents can be vital, without mentioning the entity in ways captured by the entity profiling techniques investigated.

Outgoing link mentions: posts with outgoing links mentioning the entity.

Event place - event: A document that talks about an event is vital to the location entity where it takes place. For example Maha Music Festival takes place in Lewis and Clark_Landing, and a document talking about the festival is vital for the park. There are also cases where an event's address places the event in a park and due to that the document becomes vital to the park. This is basically being mentioned by address which belongs to a larger space.

Entity - related entity: A document about an important figure such as artist, athlete can be vital to another. This is specially true if the two are contending for the same title, one has snatched a title, or award from the other.

Organization - main activity: A document that talks about an area on which the company is active is vital for the organization. For example, Atacocha is a mining company and a news item on mining waste was annotated vital.

Entity - group: If an entity belongs to a certain group (class), a news item about the group can be vital for the individual members. FrankandOak is named innovative company and a news item that talks about the group of innovative companies is relevant for it.

Artist - work: Documents that discuss the work of artists can be relevant to the artists. Such cases include books or films being vital for the book author or the director (actor) of the film. Robocop is film whose screenplay is by Joshua Zetumer. A blog that talks about the film was judged vital for Joshua Zetumer.

Politician - constituency: A major political event in a certain constituency is vital for their politicians. Take e.g. a weblog that talks about two north Dakota counties being drought disasters. The news is considered vital for Joshua Boschee, a politician, a member of North Dakota democratic party.

Head - organization: A document that talks about an entity's organization can be vital: Jasper_Schneider is USDA Rural Development state director for North Dakota and an article about problems of primary health centers in North Dakota is judged vital for him.

World knowledge, missing content, and disagreement: Some judgments require world knowledge. For example "refreshments, treats, gift shop specials, ... free and open to the public" is judged relevant to Hjemkomst_Center. Here, the person posting this on social media establishes the relation, not the text itself. Similarly "learn about the gray wolf's hunting and feeding ... 15 for

[8] For a more detailed analysis of the effect of duplicate documents on evaluation using the KBA stream corpus, refer to [3].

members, 20 for nonmembers" is judged vital to Red_River_Zoo. For a small remaining number of documents, the authors found no content or could otherwise not reconstruct why the assessors judged them vital.

7 Conclusions

In this paper, we examined the effect of the chain of interactions of cleansing, entity profiles, the effect of the type of entities (Wikipedia or Twitter), categories of documents (news, social, or others) and the relevance ratings (vital or relevant) on recall and overall performance. There is a difference between vital and relevant rankings with respect to filtering: it is easy to achieve higher recall for vital documents only than vital or relevant ones. Given the importance of vital documents (those are the ones we definitely do not want to miss), this is good news for the development of high performing CCR systems.

Cleansing may remove (partial) document content, thereby reducing recall up to 21%. But, this affects the performance of retrieving the relevant documents more than that of vital ones. Looking beyond recall, the overall performance on ranking vital documents improves for Wikipedia entities. Considering also the relevant documents, cleansing affects overall performance negatively. If one is interested in vital documents, then we recommend cleansing, but if one is interested in relevant documents too, then cleansing seems disadvantageous. For KB curation, the emphasis is likely on vital documents, but other tasks (such as filtering information for journalists) may require a high performance on both relevance grades.

Regarding entity profiles, the most effective profiles of Wikipedia entities rely on their canonical partial representation, while the partial name variants perform best for Twitter entities. Because entity type and relevance grade both exhibit differences regarding filtering, they should be dealt with differently to maximize performance. Similarly, social posts and news should be treated differently.

Despite an exhaustive attempt to retrieve as many vital documents as possible, we observe that there are still documents that defy retrieval. About 10% of the vital or relevant documents cannot be identified using our entity profiling techniques, establishing a 90% recall as an upper bound for the full pipeline. The circumstances under which this happens are many. We found that some judged documents are not fully represented in the collection, and in a few cases it is simply not clear why assessors deemed those documents vital. However, the main circumstances under which vital documents can defy filtering can be summarized as outgoing link mentions, venue-event, entity - related entity, organization - main area of operation, entity - group, artist - artist's work, party - politician, and world knowledge. More advanced entity profiling techniques will be necessary to resolve these situations in the future.

Acknowledgments. This study is financed by the COMMIT/ program, as part of the Infiniti project.

References

1. Balog, K., Ramampiaro, H.: Cumulative Citation Recommendation: Classification vs. Ranking. In: Proceedings of the 36th International ACM SIGIR Conference on Research and Development in Information Retrieval, pp. 941–944 (2013)
2. Balog, K., Ramampiaro, H., Takhirov, N., Nørvåg, K.: Multi-step Classification Approaches to Cumulative Citation Recommendation. In: Proceedings of the 10th Conference on Open Research Areas in Information Retrieval, pp. 121–128 (2013)
3. Baruah, G., Roegiest, A., Smucker, M.D.: The Effect of Expanding Relevance Judgements with Duplicates. In: SIGIR 2014 Proceedings of the 37th International ACM SIGIR Conference on Research & Development in Information Retrieval, pp. 1159–1162 (2014)
4. Bouvier, V., Bellot, P.: Filtering Entity Centric Documents Using Numerics and Temporals Features within RF Classifier. In: TREC 2013 (2013)
5. Dalton, J., Dietz, L.: A Neighborhood Relevance Model for Entity Linking. In: Proceedings of the 10th Conference on Open Research Areas in Information Retrieval, pp. 149–156 (2013)
6. Dietz, L., Dalton, J.: Umass at TREC 2013 Knowledge Base Acceleration Track. In: TREC 2013 (2013)
7. Dredze, M., McNamee, P., Rao, D., Gerber, A., Finin, T.: Entity disambiguation for knowledge base population. In: Proceedings of the 23rd International Conference on Computational Linguistics, pp. 277–285 (2010)
8. Efron, M., Willis, C., Organisciak, P., Balsamo, B., Lucic, A.: The University of Illinois' Graduate School of LIS at TREC 2013. In: TREC 2013 (2013)
9. Frank, J.R., Bauer, J., Kleiman-Weiner, M., Roberts, D.A., Tripuraneni, N., Zhang, C., Ré, C., Voohees, E., Soboroff, I.: Evaluating Stream Filtering for Entity Profile Updates for TREC 2013. In: TREC 2013 (2013)
10. Gebremeskel, G.G., He, J., De Vries, A.P., Lin, J.: Cumulative Citation Recommendation: A Feature-aware Comparisons of Approaches. In: Database and Expert Systems Applications (DEXA), pp. 193–197. IEEE (2014)
11. Ji, H., Grishman, R.: Knowledge Base Bopulation: Successful Approaches and Challenges. In: Proceedings of the 49th Annual Meeting of ACL: Human Language Technologies, pp. 1148–1158 (2011)
12. Liu, X., Fang, H.: A Related Entity Based Approach for Knowledge Base Acceleration. In: TREC 2013 (2013)
13. Nia, M.S., Grant, C., Peng, Y., Wang, D.Z., Petrovic, M.: University of Florida Knowledge Base Acceleration. In: TREC 2013 (2013)
14. Robertson, S.E., Soboroff, I.: The TREC 2002 Filtering Track Report. In: TREC 2012 (2002)
15. Wang, J., Song, D., Lin, C.Y., Liao, L.: BIT and MSRA at TREC KBA Track 2013. In: TREC 2013 (2013)

Generating Music Playlists with Hierarchical Clustering and Q-Learning

James King and Vaiva Imbrasaitė

Computer Laboratory
University of Cambridge
JamesK@cantab.net, Vaiva.Imbrasaite@cl.cam.ac.uk

Abstract. Automatically generating playlists of music is an interesting area of research at present, with many online services now offering "radio channels" which attempt to play through sets of tracks a user is likely to enjoy. However, these tend to act as recommendation services, introducing a user to new music they might wish to listen to. Far less effort has gone into researching tools which learn an individual user's tastes across their existing library of music and attempt to produce playlists fitting to their current mood. This paper describes a system that uses reinforcement learning over hierarchically-clustered sets of songs to learn a user's listening preferences. Features extracted from the audio are also used as part of this process, allowing the software to create cohesive lists of tracks on demand or to simply play continuously from a given starting track. This new system is shown to perform well in a small user study, greatly reducing the relative number of songs that a user skips.

Keywords: Music playlist generation, reinforcement learning, hierarchical clustering, user study.

1 Introduction

We listen to music in a variety of ways. Many people listen to individual albums one at a time, some prefer to listen to tracks in a random order, whilst others opt to create playlists by hand. Each of these options suits different individuals better than others, but all of them come with drawbacks. With the growing popularity of digital music, the motivation for more intelligent *automatic playlist generation* is also growing, as people's music collections become unmanageable.

Many existing solutions to this problem make use of large online databases, or rely heavily on tagged audio files. In this paper we show a solution that removes any dependence on these types of data sources or any external services to achieve a completely personal and independent music player.

The player monitors the user's actions continuously using reinforcement learning and updates its matrices based on user behaviour. Using implicit user behaviour only, our player is able to learn user preferences providing users with a better music listening experience. We show this by executing both a quantitative user study as well as some more qualitative tests.

A. Hanbury et al. (Eds.): ECIR 2015, LNCS 9022, pp. 315–326, 2015.

2 Background

The increasing use of online music streaming services in recent years has seen a surge in research investigating music recommendation and playlisting. Here we focus solely on Automatic Playlist Generation (APG), and while this shares many traits with recommendation, there are some important differences. Music recommendation systems assume access to external data and new songs, while APG systems should be expected to run with only local data. Irregularities in local music collections, along with the lack of external data, make this a hard problem. Furthermore, recommendation algorithms focus on discovery, whereas APG tends to be more concerned with coherence within a set of ordered tracks.

2.1 Commercial Services

Vast online meta-data resources are being increasingly utilised by a wide range of services [2]. Of particular note is The Echo Nest[1], which beyond simple song meta-data provides an array of similarity, personalisation and learning tools for its extensive database. Some interesting 'acoustic attributes' are extracted from audio, such as 'danceability', 'energy', 'speechiness' and 'liveness'. Unfortunately, as it is a commercial project, most of the implementation details used are hidden.

Spotify[2] is a well known audio player that provides a radio feature which creates automatic playlists. However, Spotify is an online service, and currently only learns song similarities across its database of users. It can generate random playlists, but only based on a user's favourites. The primary technique Spotify uses is *collaborative filtering* [8].

iTunes Genius[3], another popular playback tool, is known to use latent factor analysis [10] to extract song recommendations from huge data sets of listening patterns from other users.

2.2 Existing Research

Existing solutions to APG tend to focus on two techniques: Collaborative Filtering (CF) [7,17], and Content-Based (CB) approaches [16], as well as hybrids of the two [21,3]. CF has seen extensive use of online marketplaces, and follows the reasoning that if a user likes A (in our case, a music track), and many other users like both A and B, then we can recommend B to the user. Well known issues with this are dense grouping of tracks by the same artist, and the cold-start problem in which new tracks cannot easily be introduced. CB methods instead look directly at audio data to recommend similar tracks to a user, but this clearly relies on the assumption that audio similarity is a key factor in what a user likes, when many other factors are also involved.

To overcome the problems above, these approaches have often been fused with other data, such as social media information and sensory data (spatiotemporal)

[1] `http://the.echonest.com`
[2] `http://www.spotify.com`
[3] `http://www.apple.com/uk/itunes/features`

on mobile devices [5,19]. These context-aware methods seek to choose tracks which fit with a current situation or mood. Closely tied with this, there has also been a more limited investigation into what a 'good' playlist actually is [9,14], and how we might evaluate whether a given APG tool creates good playlists.

The approach we have taken is to use CB methods to inform an unsupervised machine learning algorithm (Q-learning), which over time can learn from implicit user feedback to generate playlists which are personalised to different listeners. We do not use CF or other methods which require external data. A number of other machine learning approaches have been applied to this problem [6,4], and implicit user feedback methods have been tried before [15], but to our best knowledge the Q-learning with clustering approach taken here is novel.

One final related work has been the creation of automatic 'DJ's [11]. Perhaps most notable of these is the Microsoft Research AutoDJ project. This uses Gaussian Process Regression [18] to learn priors for selecting new tracks. It also looks carefully at the issues surrounding similarity measures, which are a related issue we do not focus on in this paper.

3 Audio Analysis and Clustering

3.1 Feature Extraction

Given the requirement for the generator to not use meta-data, it is necessary to gather information about files from the audio data itself. Much work has been done into extracting meaningful statistics about signals, and many of these apply directly to audio signals in the context of music.

We divide the 16KHz signal into windows of 512 samples and extract 14 features using the jAudio [12] library: spectral centroid, spectral roll-off, spectral flux, compactness, spectral variability, root mean square of the power, fraction of low energy windows, zero crossings rate, strongest beat, beat sum, strength of the strongest beat, Mel-Frequency Cepstrum Coefficients (MFCC), Linear Predictive Coding (LPC) and the statistical method of moments. All values are mean averaged across all windows, forming an array of 39 values in total.

3.2 Clustering

It may not be obvious why we need to cluster the songs before going further—after all, the aim is to create playlists from individual songs and learn the probabilities of transitioning between them. As we are aiming to learn the relationship between individual songs, the number of values we would have to learn for all the transitions would be impractically large for all but the smallest of music libraries (n^2 transition probabilities for a library with n songs). So, the solution is to first cluster the songs, and then learn transitions between these clusters. Provided the clusters are small enough, this will be accurate enough that users cannot tell the difference, and it will reduce the number of learning values to K^2, where K is the number of clusters.

Hierarchical Clustering. Although k-means clustering of songs is a good starting point for tackling the playlist learning problem, it has one obvious deficiency. Namely, there is no universal method for choosing K without generating an infeasible number of clusters, or clusters that are simply too big. If K is too small, we tend towards the K = 1 case in which the clusters contain so many songs that knowing which cluster to choose provides no useful information. As K grows, the number of transition probabilities grows, and so we go back to the original problem where the matrices are too large to learn.

The solution we propose is therefore to use *hierarchical* clusters, in which a tree of clusters is constructed using k-means clustering, and Q-learning (explained below) is performed on nodes at multiple levels. This keeps the benefits of clustering without introducing large transition matrices or large groups of songs to choose between. In fact, this reduces the space complexity from $\mathcal{O}(n^2)$ to just $\mathcal{O}(n)$.

Space Complexity. Consider an arbitrary cluster tree. Let n be the number of songs clustered, and m be the limit set on the number of clusters per node, so that $m = \mathrm{K}$ in terms of the k-means clustering algorithm. In an ideal tree m will also be the branching factor making the tree balanced. Finally, let h be the height of the tree so that a 1-node tree has height 0, the $1 + m$ node tree has height 1, and so on.

Now, we have that $m^h = n$ for a balanced tree, since the branching factor is m. Each node has a transition matrix in this model, so we are interested in the total number of nodes. At level 0 there is $m^0 = 1$ node, at level 1—$m^1 = m$ nodes, at level 2—m^2 nodes, and at the bottom level $m^h = n$ nodes. However, the bottom level nodes are the individual tracks, so we only need to consider up to level $h - 1$ which has m^{h-1} nodes.

If we sum the total number of nodes, we get $S = \sum_{x=0}^{h-1} m^x$, and this is a simple geometric series $S = \frac{m^h - 1}{m-1} = \frac{n-1}{m-1}$. As each node's transition matrix has m^2 entries, the total number of values is therefore the product, $m^2 \frac{n-1}{m-1}$, which is $\mathcal{O}(mn)$. Since m is a small constant, this tends to $\mathcal{O}(n)$, compared with the $\mathcal{O}(n^2)$ complexity we get without clustering.

4 Learning from User Behaviour

4.1 Reinforcement Learning

Markov Decision Processes [1] (MDPs), which are an extension of Markov Chains, are used in situations where a decision maker has some control over choices with random outcomes. An MDP is a discrete-time stochastic control process, in which the agent is in some state s at time t, and must choose an action a which will move the process to a new state s' and give the agent a reward $\mathcal{R}(s, a)$. Since all previous states are ignored under these conditions, MDPs possess the *Markov property*. In addition to the set of states S, an initial state s_0 and a set of actions A, the agent also requires:

- A transition function $\mathcal{S}\colon S \times A \rightarrow S$ which gives the probability that action a causes a state transition $s \rightarrow s'$
- A reward function $\mathcal{R}\colon S \times A \rightarrow \mathbb{R}$ giving the reward (a real number) of choosing action a in state s

In our case the set of states S corresponds to the set of clusters, and not the individual songs. Furthermore, since the clusters are hierarchical, we will need one model per node in the tree, with S being only the node's child clusters. In order to make the choice on the next track, or to update policies based on feedback, we will need to walk the tree and update multiple nodes along the path (the exact solution is described below). Note that this is not the same as hierarchical reinforcement learning, where a problem is solved by abstracting it to multiple levels.

4.2 Q-Learning

Q-learning [20] is a model-free reinforcement learning technique which learns a Q-function that gives the expected utility of taking a particular action a. It is model-free because the agent has no knowledge of the transition function $\mathcal{S}$ or the reward function $\mathcal{R}$. In the music player, the agent does actually know $\mathcal{S}$, since this is just a deterministic function in which a_1 means "transition to s_1", a_2—"transition to s_2", and so on. However, $\mathcal{R}$ is clearly unknown because the rewards are assigned by the user. This is known as active learning, since the agent is learning what to do in each state rather than simply looking for some goal state. Of course, in the music player there is no goal state since it is the combination and path through the states which matters, and playback may continue indefinitely.

Q-learning works with a value iteration update formula:

$$Q_{t+1}(s_t, a_t) = Q_t(s_t, a_t) + \alpha_t(s_t, a_t) \times \left[R_{t+1} + \gamma \max_a Q_t(s_{t+1}, a) - Q_t(s_t, a_t)\right]$$

In our case it can be simplified as follows, with transition matrix P:

$$P'_{r,c} = P_{r,c} + \alpha_t \times \left[R_t + \gamma \max_a P_{c,a} - P_{r,c}\right] \tag{1}$$

where r is the matrix row, c the column, and t is the time of the update. The α and R in this formula, though indexed by t for clarity, are functions which may depend on other parameters including the current song choice.

The intuition behind this formula is that when a transition from state r to c receives some reward R_t, we essentially want to change the respective entry in Q by the amount R_t. However, since the goal is really to optimise the long-term gain, there is also a factor based on the maximum attainable reward in the next state c. This is weighted by the *discount factor* γ, and then the whole update is weighted by the *learning rate* factor denoted α_t. The purpose of α_t is to prevent the Q value being set immediately equal to the reward, but instead smooth updates so that both the reward and the current value are taken into account.

So the reward is multiplied by α_t and the old value by $1-\alpha_t$, which explains why α_t is typically fairly low (less than 0.1), since with updates occurring frequently it is desirable to place more weight on the accumulated past updates than on a single new reward.

The value of γ, such that $0 \leq \gamma \leq 1$, is a constant which sets a trade-off for how much immediate rewards are valued compared to future rewards. A value of 0 makes the agent *myopic* and leads to a greedy algorithm since it only considers current rewards, whereas $\gamma = 1$ will make it aim for a higher long-term utility.

Q-learning algorithms often iterate within 'episodes' until local convergence is reached. This method doesn't apply well to the music player scenario, so instead there is just one update per user action. This reflects that rather than having a single clear goal for which an optimum path must be found, we are continually trying to find good states and may continue indefinitely. So there are no absorbing states in the MDP, and its transition graph may contain loops.

Optimal Policies. The Q-learning algorithm is designed to converge to an optimal policy, where a policy is simply the assignment of weights to possible actions for each state (taken from P). The optimal policy here is the 'discounted cumulative reward' which maximises $r_{tot} = \sum_{i=0}^{\infty} \gamma r_{t+i}$, where t is the start time, and we use policy π for infinite time after starting in some state s_t. The r_t are rewards at time t. It is important to notice that this is only defined for a given γ, since clearly a greedy algorithm would have a different sense of optimal reward to a long-term optimiser. However, the definition is unconditional on α, which only affects how quickly and effectively it can learn this optimal policy.

4.3 Calculating Rewards

Since the learning agent can only adapt itself based on the rewards obtained from the user, the design of these rewards is vital to the success of the agent. We want the user feedback to be implicit—avoiding features such as voting buttons whose usage inevitably drops over time and leads to inconsistencies. However, the user's normal behaviour provides plenty of clues about suitable rewards, providing they are reasonably active whilst listening. The design chosen for the reward system for music playback is outlined below. All values are chosen so that $-1 < r < 1$, and generally linear functions are sufficient to interpolate between the two extremes.

Track Skipped or Finished. The basic measure of implicit reward is listening time, an assertion which Chi et al. [6] have provided evidence to support. So, when a track plays all the way through without interruption from the user, a positive reward to it from the prior track is established. In our system we take it to be the maximum reward $r = 1.0$.

The converse of this is how soon a track was skipped, which allows a negative reward to be assigned. It has been shown that even a fairly simple heuristic based on this principle can be effective [15]. With this reward, the earlier a track is skipped, the greater the negative weight on the reward assigned. This leads us to $r = -1.0$ as the greatest negative reward for a skip after 0 seconds, and $r = 1.0$ as the greatest positive reward for not skipping at all.

So a track finishing is just a special case of this general rule, and we interpolate linearly between the two reward extremes based on how far through the track we got to before the skip.

There is another issue raised by skipping a track early (here defined as within 10s): we should ignore it in all the updates regarding the previous state. So, if the following track then completes, the associated positive reward will 'jump' over the skipped track and relate the song before and the song after the skipped track. In other words, whenever a track is skipped quickly, the previous song should be treated as the current track when future rewards are calculated.

Playlist Rewards. This is a separate class of reward from those mentioned previously, as it does not result from specific user behaviours, and can never be triggered by an action on behalf of the agent. However, a user's existing playlists offer a huge insight into their listening preferences. A small reward should link every song in the playlist (as the songs are meant to be listened to at roughly the same time), with a larger reward for consecutive songs to emphasise the importance of a specific ordering. In practice, the small rewards should only be applied for smaller playlists since the $\mathcal{O}(n^2)$ time requirement can be constraining, and the reward itself becomes negligible.

4.4 Learning with Hierarchical Clusters

As we have mentioned above, to perform Q-learning across a hierarchical tree with many levels, a single update may need to trigger Q-learning updates at several different nodes. Each node is represented by a Q-matrix except for the bottom level, where nodes are leaves holding individual songs.

Firstly, we need to find the lowest common ancestor (LCA) between two nodes. We then define the level multiplier as an exponential function based around the maximum cluster count K of any node, $multiplier = K^{-(LCA-level)}$. This takes the value 1 at the LCA, and rapidly approaches 0 elsewhere with speed dependent on K. Note that this multiplies the learning rate, and not the reward, which would lead to inconsistencies with the updates. The intuition behind this multiplier is that a node shouldn't learn as fast if it is updated more frequently. We want the net rate of learning to be roughly constant across all nodes. The root cluster is involved in every update, so it will be weighted with the smallest multiplier.

After every action all the clusters starting from the root and down to the LCA get an update that is weighted by the *multiplier*. The updates also involve re-normalising the affected matrices, since it is required that every row sums to 1 when we come to base probabilities off these values.

4.5 Track Choices from Hierarchical Clusters

Track choices are made using the HEURISTIC function, and a method for choosing the next cluster to explore from a particular point in the tree. These vital two components will be described separately in the following two sections. The basic idea is to traverse down the tree choosing the next cluster from each node, until a point is reached where using the heuristic is either required or sensible. If no choice can be found, the algorithm repeats itself one level higher in the tree, until a choice of song is made.

In order to achieve a good playlist generator we need to shift gradually from exploration to exploitation. We want to explore the different clusters in the early stages of training to learn what the user's preferences are. We define the overall randomness probability λ and initialise it with a large value. We then gradually lower it as the agent learns more about the user. We can also give the user some control over the randomness, and define the actual degree of randomness used in the system as a combination of these two.

Cluster Choice. The algorithm for choosing the cluster traces the probabilities down the cluster tree, picking a cluster at each level. The choice of a cluster depends on the randomness setting used—if the randomness is set to 0, then the cluster with the highest probability is chosen; if the randomness is set to 1, then all the clusters have an equal chance of being selected.

Heuristic Function. Once a cluster is chosen (which happens when a node with leaves as children is reached, or the cluster probability matrix is no longer applicable at this depth in the tree), the heuristic function picks a song to play. The algorithm works by establishing several features which vote for the viability of choosing different songs: the distance from the current track (the closer the better), a list of the immediate listening history (the further the better), and the overall 'preferred' tracks. All three of these are scaled to avoid any one from causing another voter to be ignored, and combined to provide the final vote. An extra setting prevents the current track from being immediately repeated by the heuristic. Then the song with the highest vote gets picked as the next song.

4.6 System Parameters

One of the most important parameters in our system is K, the number of clusters enforced by the k-means algorithm. We chose K to be 6, as this means a hierarchy for a typical library will have around 4–5 levels, and each matrix will have only 36 elements. Furthermore, 6 is roughly the right magnitude to represent the different number of genres at the top of the hierarchy.

The parameters for the Q-learning agent are also vital. The learning rate α was chosen as a balance between a high, oscillation-prone value, and a lower value which may never converge over the course of a user study. Generally smaller values are a wiser choice, so we used $\alpha = 0.05$ for the user study.

We used a very low discount factor, γ, since each state is equally important in such a system—there is no golden end state like we might find with a robot exploring to reach a goal, so it shouldn't heavily optimise for the future.

The initial condition used in the learning may also be quite important. We decided to initialise all matrices to the identity matrix, with a utility of 1 for returning back to the same cluster (itself) again. This was a safe option because the NEXT–CLUSTER algorithm incorporates a degree of randomness irrespective of the user setting.

5 Evaluation

5.1 Methodology

We chose to evaluate our playlist generator in two distinct ways to try to demonstrate its merits. Firstly, a user study was conducted with a small sample of active users to gather both qualitative and quantitative data about the generator's performance. Secondly, a series of set experiments were performed to demonstrate that the player can learn user's preferences and can be trained to select or avoid certain types of song depending on the user's listening history.

Some related work has gone into other methods for evaluating music playlist generators [13]—a surprisingly difficult problem. People do not agree on what defines a 'good playlist', beyond a set of fairly basic assumptions. We cannot just enumerate all good playlists, because there are an intractable number of possible song selection and ordering options for all but the most trivial of music collections. Many APG papers nonetheless attempt to devise metrics which will assign fair scores to attempts by different playlist generators. But without a standard method used across all research in the field, these inevitably self-optimise for the solution in question.

5.2 User Study — Data

For our evaluation study, we recruited 20 participants which we separated into two groups: the control group (5 participants) and the experimental group (15 participants). The participants were asked to use the player for 28 days (the average listening time was 41 minutes per day). The player had 3 modes: manual playlist creation, shuffle mode and the Smart Playlist mode. For the experimental group, the Smart Playlist was generated using our algorithm (and was the most frequently used mode by the participants), and all three modes were used for learning. For the control group, the Smart Playlist's behaviour was identical to shuffle mode. The participants were not aware of which group they belonged to.

The user behaviour was tracked throughout the user study and the data was collected and reported in the form of the following features: total listening time and the number of tracks played, the time and the number of tracks listened to in the three modes we provided, the total number of skips and the number of early skips, the number of jumps and the number of queued songs, the total number of song searches, and, finally, the size of the stored library.

Whilst many interesting conclusions can be drawn from the data we collected, the most important one for the evaluation of our method is the skip count. Tracking the total number of counts, one the other hand, can lead to incorrect conclusions. The absolute skip count values would be higher on more active days, without the *proportion* of songs skipped necessarily changing. We are therefore interested in the relative number of skips—normalising the total number of skips by the total time the user has spent listening to music. Figure 1 plots the relative skip counts in both the experimental and control groups.

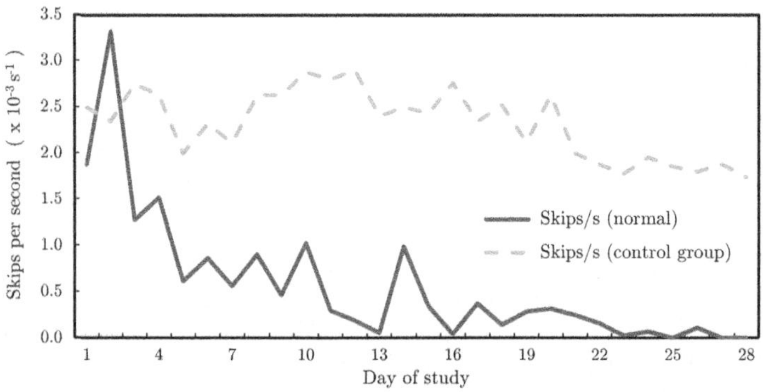

Fig. 1. Skip count results from the user study, normalised by listening time

This graph shows a very convincing trend—one we were hoping for, but did not expect to be as clear. The drop in the relative number of skips is very sharp even after only a couple of days of use, while the relative number of skips for the control group remains stable throughout the duration of the study.

5.3 User Study — Survey

The qualitative side of the user study data was provided by a survey sent out at the end of the study to the participants in the experimental group. The responses were generally very positive. Users seemed to be either fairly active or somewhat dormant in using the player, with no middle ground. Most users had libraries of 100–1000 songs, which aligns well with the cluster size setting we had chosen. The player was described as very easy to use, and users spent most of the time in Smart Play mode. Static playlists were not heavily used, although a number of users claimed this as one of their usual listening practices. The participants were also pleased with the player's utility in rediscovering old music.

5.4 Testing Learning Capabilities

In order to test the learning capabilities of the playlist generator we artificially devised a library with 3 clusters corresponding to 3 different genres. The agent could then be trained by repeatedly creating playlists with tracks chosen from the different clusters as required. For instance, selecting one track from each cluster in order was shown to result in a Q-matrix with high transition probabilities between those adjacent tracks in the playlist. Throughout this process, the randomness was set near to 0 to enable the exploitation phase of the algorithm.

A similar method was used to demonstrate that the agent can be trained to choose particular types of songs more often, such as a specific musical style or a set of the user's favourite tracks. In this case the learning process was simulated by the skipping of certain tracks during playback with a high learning rate set. The converse process was used to demonstrate that tracks can also be learned to be chosen less frequently, such as for a style the user dislikes.

6 Conclusions

In this paper we have described a novel method for generating music playlists that relies entirely on the analysis of music and can be used offline. It requires no explicit user input, yet still learns user's preferences and is able to generate a playlist that is personalised and adapted to user needs. The playlist generator is also especially suited for large libraries as the use of hierarchical clustering enables it to scale extremely well, while still being able to learn user's preferences.

In addition to that, we have executed a user study to evaluate our algorithm and using the objective skip count metric showed that the algorithm behaves as intended and that it outperforms a baseline shuffle mode. We are making the player publicly available for download at `www.james.eu.org/musicplayer`.

7 Discussion and Future Work

While the system we have proposed shows potential to improve the users' experience when listening to generated playlists, a perfect solution that achieves a zero skip count is obviously impossible—people are always unpredictable to some degree when deciding what music they would like to listen to.

A system that requires no explicit user input is even more difficult to achieve. The reward system for the Q-learning assumes that whenever a song finishes playing, the user must have enjoyed that song. Of course, if a user happens to leave the room or becomes distracted, then this assumption breaks down entirely, so an active listener is required. Possible future extensions include:

- Choosing songs by considering the feature difference between the end of one song and start of the next one would ease the playlist into less similar clusters during exploration more gently and improve coherence.
- Experimentation with different feature weightings. At present, all audio features are given equal weight, but it has not been shown that this is the ideal approach. For instance, the speed of a track may well be a more salient feature than its spectral flux.
- Weighting of user actions by 'user activeness'. So if a user very rarely intervenes in playback, then when they do skip a track this is more significant.

References

1. Bellman, R.: A Markovian Decision Process. Journal of Mathematics and Mechanics 6 (1957)
2. Bertin-Mahieux, T., Ellis, D.P., Whitman, B., Lamere, P.: The Million Song Dataset. ISMIR 12 (2011)
3. Bu, J., Tan, S., Chen, C., Wang, C., Wu, H., Zhang, L., He, X.: Music Recommendation by Unified Hypergraph: Combining Social Media Information and Music Content. In: Proc. ICMR, pp. 391–400, New York, USA (2010)
4. Chen, S., Moore, J.L., Turnbull, D., Joachims, T.: Playlist Prediction via Metric Embedding. In: Proc. 18th ACM SIGKDD International Conference on Knowledge Discovery and Data Mining, pp. 714–722 (2012)
5. Cheng, Z., Shen, J.: Just-for-Me: An Adaptive Personalization System for Location-Aware Social Music Recommendation. In: Proc. ICMR (2014)
6. Chi, C.-Y., Lai, J.-Y., Tsai, R.T.-H., Jen Hsu, J.Y.: A Reinforcement Learning Approach to Emotion-based Automatic Playlist Generation. In: International Conference on Technologies and Applications of Artificial Intelligence, vol. 12, pp. 60–65 (2010)
7. Schafer, J., et al.: Collaborative filtering recommender systems. The Adaptive Web, 291–324 (2007)
8. Hu, Y., Koren, Y., Volinsky, C.: Collaborative Filtering for Implicit Feedback Datasets. In: ICDM, pp. 263–272 (2008)
9. Jannach, D., Kamehkhosh, I., Bonnin, G.: Analyzing the Characteristics of Shared Playlists for Music Recommendation. In: Proceedings of the 6th Workshop on Recommender Systems and the Social Web (2014)
10. Koren, Y.: Factorization Meets the Neighborhood: A Multifaceted Collaborative Filtering Model. In: KDD (2008)
11. Liebman, E., Stone, P.: DJ-MC: A Reinforcement-Learning Agent for Music Playlist Recommendation. AAMAS 13 (2014)
12. McEnnis, D., Fujinaga, I., McKay, C., DePalle, P.: JAudio: A feature extraction library. In: ISMIR (2005)
13. McFee, B., Lanckriet, G.: The Natural Language of Playlists. In: Proc. ISMIR, Miami, FL, USA (October 2011)
14. McFee, B., Lanckriet, G.: Hypergraph Models of Playlist Dialects. In: Proc. ISMIR, Porto, Portugal (October 2012)
15. Pampalk, E., Pohle, T., Widmer, G.: Dynamic Playlist Generation Based on Skipping Behavior. In: Proc. ICMIR, pp. 634–637 (2005)
16. Pazzani, M.: Content-based recommendation systems. The Adaptive Web, 325–341 (2007)
17. Platt, J.C.: Fast embedding of sparse music similarity graphs. In: NIPS (2003)
18. Platt, J.C., Burges, C.J.C., Swenson, S., Weare, C., Zheng, A.: Learning a Gaussian Process Prior for Automatically Generating Music Playlists. Microsoft Corporation (2001)
19. Schedl, M., Breitschopf, G., Ionescu, B.: Mobile Music Genius: Reggae at the Beach, Metal on a Friday Night? In: Proc. ICMR (2014)
20. Watkins, C.: Learning from Delayed Rewards. PhD thesis, King's College, University of Cambridge (1989)
21. Yoshii, K., Goto, M., Komatani, K., Ogata, T., Okuno, H.G.: An Efficient Hybrid Music Recommender System Using an Incrementally Trainable Probabilistic Generative Model. Trans. Audio, Speech and Lang. Proc. 16(2), 435–447 (2008)

Time-Sensitive Collaborative Filtering through Adaptive Matrix Completion

Julien Gaillard[1] and Jean-Michel Renders[2]

[1] University of Avignon, Avignon, France
[2] Xerox Research Center Europe, Meylan, France

Abstract. Real-world Recommender Systems are often facing drifts in users' preferences and shifts in items' perception or use. Traditional state-of-the-art methods based on matrix factorization are not originally designed to cope with these dynamic and time-varying effects and, indeed, could perform rather poorly if there is no "reactive", on-line model update. In this paper, we propose a new incremental matrix completion method, that automatically allows the factors related to both users and items to adapt "on-line" to such drifts. Model updates are based on a temporal regularization, ensuring smoothness and consistency over time, while leading to very efficient, easily scalable algebraic computations. Several experiments on real-world data sets show that these adaptation mechanisms significantly improve the quality of recommendations compared to the static setting and other standard on-line adaptive algorithms.

1 Introduction

The fact that item perception and user tastes and moods vary over time is well known. Still, most recommender systems fail to offer the right level of reactivity that users are expecting, i.e. the ability to detect and to integrate changes in needs, preferences, popularity, etc. Suggesting a movie a week after its release might be too late [1]. In the same vein, it could take only a few ratings to make an item go from *not advisable* to *advisable*, or the other way around.

One of the motivations of this work was based on the observation of the dramatic drop in performance when going from random train/test splits as in a standard cross-validation setting towards a strict temporal split. For instance, the difference in rating prediction accuracy as measured by the RMSE (Root Mean Squared Error) exceeds 5% (absolute) when using the famous MovieLens (1M ratings) data set. Another motivation was to ensure the efficiency and the scalability of the algorithms, to respect the real-time constraints on very large recommendation platforms, so that we excluded from our scope some approaches based on Bayesian probabilistic inference methods (e.g. those based on probabilistic matrix factorizations and non-linear Kalman filters).

In this paper, we propose an Adaptive Matrix Completion method that makes the system very flexible with respect to dynamic behaviors. The factor matrices are dynamically and continuously updated, in order to provide recommendations in phase with the very recent past. It should be noted that the method

A. Hanbury et al. (Eds.): ECIR 2015, LNCS 9022, pp. 327–332, 2015.

is truly adaptive and not only incremental, in the sense that it could give more weight to recent data – and not uniform weights to all observations – if this is needed. We are considering the case where no other information than the set of <user, item, ratings> tuples is given and, consequently, we are not addressing the "(strictly) cold start" problem, where a completely new user or a new item is appearing, with no associated information. The method's principle is that, when receiving a new observation (<user, item, rating> tuple), we update the corresponding entries (rows and columns) of the factor matrices, controlling the trade-off between fitting as close as possible to the new observation and being smooth and consistent with respect to the previous entries. This gives raise to a least-squares problem with temporal regularization, coupling the update of both users- and items-related factors. We will show that the problem could be solved by a simple iterative algorithm (requiring the inversion of a $K \times K$ matrix, where K is the reduced rank in the matrix factorization), converging in a few iterations (typically 2 or 3), so that it could easily update the models even with an arrival rate of several thousands ratings per second.

2 Related Work

One of the first works to stress the importance of temporal effects in Recommender Systems and to cope with it was the *timeSVD++* algorithm [4]. The approach is to explicitly model the temporal patterns on historical rating data, in order to remove the "temporal drift" biases. This means that the time dependencies are modelled parametrically as time-series, typically in the form of linear trends, with a lot of parameters to be identified. Other approaches (see [5,2,9]) rely on a Bayesian framework and on probabilistic matrix factorization, where a state-space model is introduced to model the temporal dynamics. One of their main advantages is that they could easily be extended to include additional user- or item-related features (addressing in this way the cold-start problem). But, in order to remain computationally tractable, they update only either the user factors, or the items factors, but never both factors simultaneously; otherwise, they should rely on rather complex non-linear Kalman filter methods. An earlier work ([8]) also proposed to incrementally update the item- or user-related factor corresponding to a new observation by performing a (stochastic) gradient step of a quadratic loss function, but allowing only one factor to be updated; the updating decision is taken based on the current number of observations associated to a user or to an item (for instance, a user with a high number of ratings will no longer be updated).

Interestingly, tensor factorization approaches have also been adopted to model the temporal effects of the dynamic rating behavior ([10]): user, item and time constitute the 3 dimensions of the tensors. Tensor factorization is useful for analyzing the temporal evolution of user and item-related factors, but it could hardly extrapolate rating behavior in the future. More recently, [3] introduced a "reactivity" mechanism in the similarity-based approach to Collaborative Filtering, which updates the similarity measures between users and between items with some form of forgetting factor, allowing to decrease the importance of old ratings.

3 Adaptive Matrix Completion

Starting from one of the standard static settings of matrix completion for Collaborative Filtering (CF), we will extend it to the time-varying case, by adopting an incremental, on-line approach based on temporal regularization.

Let X be a $n \times m$ sparse rating matrix (n users, m items). One of the standard state-of-the-art CF approaches amounts to approximate X by a low-rank matrix $\tilde{X}$ that optimizes a criterion mixing:

- the approximation quality over observed ratings, typically the sum of squared errors;
- a complexity penalty, typically the nuclear norm of $\tilde{X}$, as a way to recover a low-rank matrix.

Assuming the decomposition $\tilde{X} = L.R^T$ (with L and R having K columns if $\tilde{X}$ is rank K at most), and introducing user- and item-specific biases (often called user subjective bias and item popularity) noted as $\mathbf{a}$ and $\mathbf{b}$, the nuclear norm problem can be approximated by the following minimization problem [6]:

$$\min \sum_{(i,j)\in\omega} (X_{i,j} - m - a_i - b_j - \sum_{k=1}^{K} L_{i,k} R_{j,k})^2 + \mu_a \|\mathbf{a}\|^2 + \mu_b \|\mathbf{b}\|^2 + \mu_L \|L\|_F^2 + \mu_R \|R\|_F^2 \quad (1)$$

where ω designates the set of available rating tuples, m is the average rating over ω, a_i, b_j, $X_{i,j}$, $L_{i,k}$ and $R_{j,k}$ are respectively the elements of $\mathbf{a}$, $\mathbf{b}$, X, L and R, corresponding to user i, item j and latent factor k. $\|M\|_F^2$ is the squared Frobenius norm of matrix M.

It should be noted that the regularization terms, including the ones related to $\mathbf{a}$ and $\mathbf{b}$, are particularly critical in our case. Indeed, in real world cases, the test sets are chronologically posterior to the training and development sets so that, in practice, the standard *iid* (independent and identically distributed) assumption between the training and the test sets is far from correct and a strong regularization is needed. One usual way of solving this optimisation problem is to use Alternating (Regularized) Least Squares or Stochastic Gradient Descent (see [7] for instance). Typically, the choice of the μ_a, μ_b, μ_L, μ_R, K parameters are done by grid search on a development set.

3.1 Adaptation of a_i and b_j

Let us first consider the simple model including only the item and user biases, before describing the extension to the complete model based on matrix factorization (MF). When observing a new tuple $< i, j, X_{i,j} >$, we update a_i and b_j by minimizing the following criterion:

$$\min \quad (X_{i,j} - m - a_i - b_j)^2 + \alpha_1 (a_i - \tilde{a_i})^2 + \beta_1 (b_j - \tilde{b_j})^2 \quad (2)$$

where $\tilde{a_i}$ and $\tilde{b_j}$ are the values before the adaptation. This criterion is a trade-off between approximation quality with respect to the new observation and smoothness in the evolution of the biases. For new users and items, $\tilde{a_i}$ and $\tilde{b_j}$ are set to 0.

The values of α_1 and β_1 are obtained by a grid search on a development set, which is chronologically posterior to the training set.

Solving this optimization problem leads to the following simple update equations:

$$a_i = \frac{(\alpha_1 + \alpha_1/\beta_1)\tilde{a_i} + X_{i,j} - m - \tilde{b_j}}{1 + \alpha_1 + \alpha_1/\beta_1} \tag{3}$$

$$b_j = \frac{(\beta_1 + \beta_1/\alpha_1)\tilde{b_j} + X_{i,j} - m - \tilde{a_i}}{1 + \beta_1 + \beta_1/\alpha_1} \tag{4}$$

3.2 Adaptation of L_i and R_j

Latent factor matrices L and R are adapted too, according to the same idea: observe $< i, j, X_{i,j} >$ then update L_i and R_j (respectively the i-th row of L and the j-th row of R), such that:

$$\min \quad (\widehat{X_{i,j}} - \sum_k L_{i,k} R_{j,k})^2 + \alpha_2 \|L_i - \widetilde{L_i}\|_F^2 + \beta_2 \|R_j - \widetilde{R_j}\|_F^2 \tag{5}$$

where $\widehat{X_{i,j}}$ is equal to $X_{i,j} - m - a_i - b_j$ (i.e. the residual rating), while $\widetilde{L_i}$ and $\widetilde{R_j}$ are the values of the corresponding rows before adaptation. For new users and items, the entries of $\tilde{L_i}$ and $\tilde{R_j}$ are set to 0. The values of α_2 and β_2 are obtained by a grid search on the development set.

Unfortunately, there is no closed-form solution to this problem, due to the coupling between L_i and R_j. However, this could be solved iteratively by applying recursively the following equations:

$$L_i^{(t)} = (\alpha_2 I + (R_j^{(t-1)})^T.R_j^{(t-1)})^{-1}.(\alpha_2 \widetilde{L_i} + \widehat{X_{i,j}}.R_j^{(t-1)}) \tag{6}$$

$$R_j^{(t)} = (\beta_2 I + (L_i^{(t)})^T.L_i^{(t)})^{-1}.(\beta_2 \widetilde{R_j} + \widehat{X_{i,j}}.L_i^{(t)}) \tag{7}$$

with $L_i^{(0)} = \widetilde{L_i}$ and $R_j^{(0)} = \widetilde{R_j}$. Experimentally, for all datasets we used and the corresponding values of α_2 and β_2, two or three iterations were sufficient to converge.

4 Results

Experiments have been performed on 3 datasets: MovieLens (1M ratings), Vodkaster (2M), Netflix (2M), divided into 3 temporal (chronologically ordered) splits: Train, Dev (20k), Test (20k); the development set is used to tune the different parameters of the algorithm. Vodkaster is a recently-born Movie Recommendation website, dedicated to rather movie-educated people. These datasets show very different characteristics: Netflix has a high number of users and is spread over a short time period (less than 10 months, Dev and Test sets represent each 1 week). MovieLens has a high number of users and is spread over a long time period. Vodkaster has a low number of users and is spread over a short time period (one year), but users are very "loyal" and active. Note that methods such as *TimeSVD++* or Temporal Tensor Factorization could not be applied here as, in their original version, they do not allow extrapolation to future time periods, because they need to identify time-specific parameters based on data from the same time-period.

The results show that Adaptive methods improve the performances according to RMSE, MAE and MAPE metrics (Table 1). Figure 1 displays explicitly the effect of dynamic adaptation over time. Each point of coordinates $x = n$ corresponds to the average RMSE after observing n ratings from the user (starting from the beginning of the test set), the average being computed over the

Table 1. Results with Matrix Factorization on Vodkaster, Netflix and MovieLens Test sets. RegLS corresponds to the simple model with biases identified by regularized least squares, SGD corresponds to minimizing (5) using an adaptive Stochastic Descent Gradient method with constant learning rates (tuned on the Dev set), while MF designates the prediction model based on Matrix Factorization

		RMSE	MAE	MAPE
Vodkaster	RegLS	0.8465	0.6603	0.3477
	MF(on residuals)	0.8177	0.631	0.3294
	On-line SGD with fixed learning rate	0.7976	0.6184	0.3141
	Adapting a_i and b_j - L_i and R_j	**0.7805**	**0.5993**	**0.3031**
Netflix	RegLS	0.9344	0.7322	0.2755
	MF(on residuals)	0.9161	0.7118	0.27
	On-line SGD with fixed learning rate	0.8824	0.6819	0.2594
	Adapting a_i and b_j - L_i and R_j	**0.8685**	**0.6678**	**0.2506**
MovieLens	RegLS	0.9194	0.713	0.3011
	MF(on residuals)	0.9047	0.7012	0.2943
	On-line SGD with fixed learning rate	0.8544	0.6632	0.2625
	Adapting a_i and b_j - L_i and R_j	**0.8435**	**0.6528**	**0.2576**

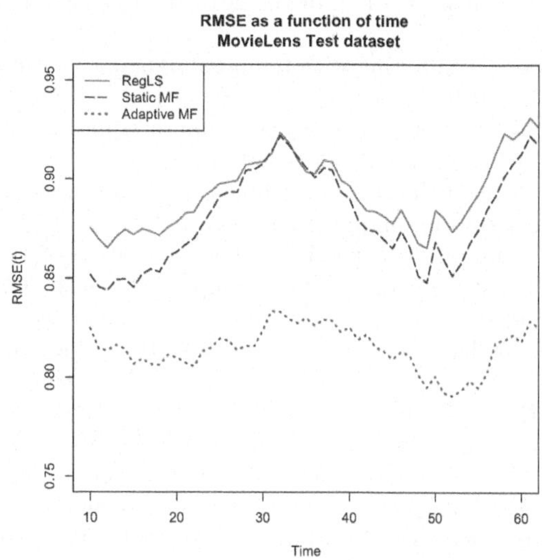

Fig. 1. RMSE as a function of relative time MovieLens Test set

users who have rated at least n items in the test set. This corresponds to a relative user-centric timescale and shows that, without adaptation, prediction errors increase, while it is stabilizing to a much lower value with adaptation.

5 Conclusion

We have proposed an Adaptive Matrix Completion method that allows Recommender Systems to be highly dynamic. Experimental results showed that this method improves significantly the accuracy of predicted ratings, even if there is still a residual noise which seems unavoidable when using only rating data and no other features. Future works should now focus on extending this scheme to time-varying user- and item- features, but also investigate other matrix regularizers to automatically determine the optimal reduced rank K. Ideally, we should also introduce some meta-adaptation that allows the adaptation rates $(\alpha_1, \beta_1, \alpha_2, \beta_2)$ to vary over time,

Acknowledgment. This work was partially funded by the French Governement under the grant <ANR-13-CORD-0020> (ALICIA Project).

References

1. Adomavicius, G., Tuzhilin, A.: Toward the next generation of recommender systems: A survey of the state of the art and possible extensions. IEEE Trans. on Knowledge and Data Engineering, 734–749 (2005)
2. Agarwal, D., Chen, B., Elango, P.: Fast online learning through offline initialization for time-sensitive recommendation. In: Proc. of KDD 2010, pp. 703–712 (2010)
3. Gaillard, J., El-Beze, M., Altman, E., Ethis, E.: Flash reactivity: adaptive models in recommender systems. In: Proc. of the 2013 International Conference on Data Mining (2013)
4. Koren, Y.: Collaborative filtering with temporal dynamics. Communications of the ACM 53(4), 89–97 (2010)
5. Lu, Z., Agarwal, D., Dhillon, I.: A spatio-temporal approach to collaborative filtering. In: Proc. of RecSys 2009, pp. 13–20 (2009)
6. Recht, B., Fazel, M., Parrilo, P.: Guaranteed minimum rank solutions of matrix equations via nuclear norm minimization. SIAM Review (2010)
7. Recht, B., Ré, C.: Parallel stochastic gradient algorithms for large-scale matrix completion. Mathematical Programming Computation 5(2), 201–226 (2013)
8. Rendle, S., Schmidt-thieme, L.: Online-updating regularized kernel matrix factorization models for large-scale recommender systems. In: Proc. of RecSys 2008 (2008)
9. Stern, D., Herbrich, R., Graepel, T.: Matchbox: large scale online bayesian recommendations. In: Proc. of WWW 2009, pp. 111–120 (2009)
10. Xiong, L., Chen, X., Huang, T.-K., Schneider, J., Carbonel, J.: Temporal collaborative filtering with bayesian probabilistic tensor factorization. In: Proc. of the SIAM International Conference on Data Mining (SDM), vol. 10, pp. 211–222 (2010)

Toward the New Item Problem: Context-Enhanced Event Recommendation in Event-Based Social Networks

Zhenhua Wang[1,2], Ping He[2], Lidan Shou[2],
Ke Chen[2], Sai Wu[2], and Gang Chen[2]

[1] Huawei Technologies, Hangzhou, China
[2] College of Computer Science and Technology, Zhejiang University, Hangzhou, China
{wzh-cs,andyepeacebow,should,chenk,wusai,cg}@zju.edu.cn

Abstract. Increasing popularity of event-based social networks (EBSNs) calls for the developments in event recommendation techniques. However, events are uniquely different from conventional recommended items because every event to be recommended is a new item. Traditional recommendation methods such as collaborative filtering techniques, which rely on users' rating histories, are not suitable for this problem. In this paper, we propose a novel context-enhanced event recommendation method, which exploits the rich context in EBSNs by unifying content, social and geographical information. Experiments on a real-world dataset show promising results of the proposed method.

Keywords: Event recommendation, event-based social network, new item problem, learning to rank.

1 Introduction

Event-based social networks such as Meetup, Plancast and Douban Events, have been experiencing rapid growth in recent years. These services allow people to organize, distribute and attend social events (e.g., movie nights, technical conferences and out-door recreation) by linking individuals in both online interactions and offline vis-a-vis communications. Event recommendation plays a significant role in developing EBSN services, since good recommendation results can greatly improve online experience and promote offline participations.

However, traditional recommendation techniques are not as effective in the context of event recommendation in EBSNs: 1) Different from conventional recommended items like movies, books and POIs, each event is held at a specific location and specific time in the future, thus for a newly proposed event, its user participation cannot be known in advance. That is, there will be no *rating histories* until the event happens. On the other hand, recommending previous events makes no sense as users cannot participate in those expired events. Therefore, every event to be recommended is a new item. Traditional recommendation methods such as collaborative filtering techniques [1], which rely on the user-item rating matrix (rating histories), are not suitable for this problem. 2) EBSN

A. Hanbury et al. (Eds.): ECIR 2015, LNCS 9022, pp. 333–338, 2015.

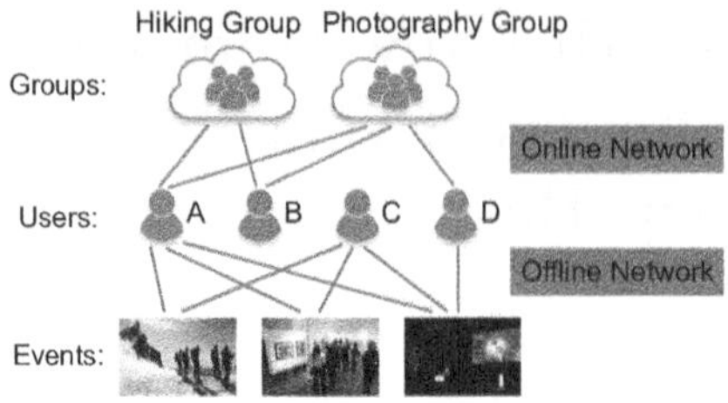

Fig. 1. An example of social networks in Meetup

is a new type of social network [3], in the sense that it consists of both online and offline social connections. Figure 1 depicts an example of user interactions in Meetup, where users' group co-memberships form the online network and their event co-participations form the offline network. For instance, user A and B are *online friends* because they are members of a hiking/photography group, so that they can share comments and photos online; user A and C have attended events together (e.g., a mountain hike, a photography exhibition and a product conference), so they are *offline friends* due to their common behaviors. Since social aspects such as people attending the same events have strong priority and influence on decision making [5], offline friends can play an important role in EBSNs. Existing methods [6,7] utilize online social information but ignore offline interactions. Qiao et al. [4] employ heterogenous relationships for recommendation, but their method is also based on CF approaches, thus cannot deal with the new item problem. 3) Moreover, existing methods still suffer from the cold-start problem, when facing new users or users who have few ratings or friends.

In this paper, we propose a novel context-enhanced event recommendation method, which addresses the above issues by exploiting the rich context in EBSNs, including content, social and geographical information. To deal with the new item problem, we utilize content information (i.e., textual description of users and events), and capture users' personal preference through topic relatedness between users and events. The *user preference* indicates the extent to which an event matches a user's interest. Therefore, it can be considered as the user's *pseudo rating*, which allows us to extract CF-based features of *social influence* from both online and offline interactions. Furthermore, to alleviate the cold-start problem, we propose the *local popularity* to measure the similarity between an event and a user's local interest. The local interest is estimated based on all the events held in the neighborhood of the user. Finally, we aggregate the features of user preference, online/offline social influence and local popularity, and formulate event recommendation as a learning to rank problem. Experiments on a real-world Meetup dataset show the effectiveness of the proposed approach.

2 Methodology

This section describes our method in detail. It consists of two phases: first, we exploit the rich context information to extract descriptive features of individual preference, social influence and local popularity; second, we fuse these features into a learning to rank model, and train a ranking model to recommend events.

2.1 Contextual Feature Extraction

User Preference. In EBSNs, every event is a new item in the recommender system. Hence, collaborative filtering techniques are not suitable. Fortunately, with the rich content information in EBSNs, content-based features can play a key role in dealing with the new item problem. Therefore, we try to capture users' preference by content similarity between users and events.

Specifically, we build an event e_j's profile by combining its name and description, which is in a textual form. A user u_i's profile consists of the tags chosen by herself and the profiles of the events that she has attended before, so that it captures both her self-description and previous behaviors. Then, we compare the profiles of u_i and e_j to obtain their similarity. However, raw textual profiles are difficult to compare because they are high-dimensional and sparse. Thus, we convert the textual profiles to small-sized vectors by topic modeling. We employ the Latent Dirichlet Allocation (LDA) as the generative model to generate topic distributions θ_{u_i} and θ_{e_j}, for u_i and e_j respectively. Now, the user preference $Pref(u_i, e_j)$ is defined as follows:

$$\begin{aligned} Pref(u_i, e_j) &= 1 - D_{js}(\theta_{u_i}, \theta_{e_j}) \\ &= 1 - \frac{1}{2}(D_{kl}(\theta_{u_i}||M) + D_{kl}(\theta_{e_j}||M)), \end{aligned} \tag{1}$$

where D_{js} is the *Jensen-Shannon Divergence* (JSD), which is a symmetrical and smoothed metric of measuring the similarity between two probability distributions. M is the average of the two distributions, and $D_{kl}(\theta||M)$ is the *Kullback-Leibler Divergence* which defines the divergence of distribution M from θ:

$$D_{kl}(\theta||M) = \sum_z \theta(z) log \frac{\theta(z)}{M(z)}, \tag{2}$$

where $\theta(z)$ is the weight of the z-th topic in the user/event profile.

Social Influence. The user preference indicates the extent to which the event matches the user's personal interest, i.e., the probability that the user will participate in the event. Thus, we propose to use it as a *pseudo rating*, which enables us to extract CF-based features of social influence.

As illustrated in Figure 1, the characteristic that distinguishes EBSNs from conventional social networks is the co-existence of online and offline social connections. Therefore, both the influences of online and offline friends should be taken into consideration. The social influence features are defined as follows:

$$Inf^{on}(u_i, e_j) = \frac{\sum_{v \in F_i^{on}} \omega_{u_i,v} \cdot Pref(v, e_j)}{\sum_{v \in F_i^{on}} \omega_{u_i,v}}, \tag{3}$$

$$Inf^{off}(u_i, e_j) = \frac{\sum_{v' \in F_i^{off}} \omega_{u_i,v'} \cdot Pref(v', e_j)}{\sum_{v' \in F_i^{off}} \omega_{u_i,v'}}, \tag{4}$$

where F_i^{on}/F_i^{off} is the online/offline friend set of u_i, $Pref(v, e_j)/Pref(v', e_j)$ is the pseudo rating of friend v/v' on e_j. $\omega_{u_i,v}/\omega_{u_i,v'}$ is the profile similarity between user u_i and v/v', which prefers online/offline friends with more similar interests to u_i, since such social ties are more likely to cause participation.

Local Popularity. For cold-start users, features extracted from user preference and social influence may not work well, due to lack of information of user profiles or friends. In this case, popularity can be a promising factor for recommendation. We measure the local popularity of event e_j as the topical similarity between e_j and the local interest around the neighborhood of user u_i. The local interest can vary across geographical regions, e.g., events in Silicon Valley are mainly about IT technologies, while most events in Hawaii are about outdoor recreation. Assume events' topics within a region follow a Gaussian distribution, we evaluate the local interest θ_{R_i} of that region using maximum likelihood estimation.

$$\theta_{R_i} = \frac{\sum_{e' \in R_i} \theta_{e'} \cdot N_{e'}}{\sum_{e' \in R_i} N_{e'}}, \tag{5}$$

where R_i is the region around the neighborhood of u_i, e' is an event held in R_i, and $N_{e'}$ is the number of participants of e'. To partition the space into regions, any geographical granularities or user-defined shapes are applicable. In our experiments, we use the city-level granularity.

Then, the local popularity of e_j is calculated as the similarity between θ_{R_i} and θ_{e_j}:

$$Pop(R_i, e_j) = 1 - D_{js}(\theta_{R_i}, \theta_{e_j}). \tag{6}$$

2.2 Recommendation by Learning to Rank

After the feature extraction phase, we obtain contextual features derived from a user-event pair. To better understand the roles of these features in recommendation, we aggregate them into a ranking function, and formulate the recommendation task as a learning to rank problem. Since the user-event participation relationship is binary, i.e., a user either attends an event or not, we use the pairwise learning to rank method.

Specifically, features derived from a user-event pair is represented as a feature vector x. We assume that the ranking function f is a linear function $f(x) = \langle w, x \rangle$ where w is the weight vector and $\langle \cdot, \cdot \rangle$ denotes an inner product. Training data is given as $((x_i^1, x_i^2), y_i), i = 1, ..., m$. Each instance consists of two feature vectors (x_i^1, x_i^2) and a label y_i. The two vectors are derived from two user-event pairs $(u(i), e^1)$ and $(u(i), e^2)$, and should correspond to the same user $u(i)$. y_i denotes which vector should be ranked ahead: if the user attends event e^1 but not e^2, $y_i = +1$; conversely, $y_i = -1$.

Then, we learn a SVM for classifying the order of pairs of feature vectors and apply the SVM in ranking. We employ the IR SVM model [2] and the loss function is defined as follows:

$$\min_{w} \sum_{i=1}^{m} \frac{1}{N_{u(i)}} [1 - y_i \langle w, x_i^1 - x_i^2 \rangle]_+ + \lambda ||w||^2, \tag{7}$$

where $[z]_+$ indicates function $max(z, 0)$. The first term is the hinge loss and the second term is a regularizer of w. $u(i)$ is the user with whom the ith instance is associated, and $N_{u(i)}$ is the number of instances associated with $u(i)$. The weight $\frac{1}{N_{u(i)}}$ is used to balance losses from different users based on their activeness, i.e., it avoids training a model biased toward users having more training instances.

3 Experiments

Dataset. A public Meetup dataset is available in [3], but it does not contain content information of events. We collect event information using Meetup's API[1]. Since a part of events has been unavailable at the time of collecting, we use those remaining events to extract users who have attended at least one of them. The resulting dataset has 104,927 users, 86,643 events, and 367,878 participations.

Methods in Comparison. We compare four methods with the proposed method. Observe that each of the features extracted in Section 2.1 can be used as a stand-alone recommendation method by ranking events according to the feature values. We denote these methods as *Pref*, *Inf-on*, *Inf-off*, and *Pop* respectively. Note that *Pref* is a content-based method, *Inf-on* and *Inf-off* are social-based CF methods using pseudo ratings, and *Pop* is based on popularity. Hence, these methods are representative competitors.

In our method (denoted as *Context*), the number of topics in LDA is set at 20. To learn the IR SVM model, we randomly select 70% users to derive training instances. The remaining users are used as the test set. Since about 96% user participations take place within 100 km from users' home locations, we perform geographical searches to reduce the candidate event space for both training and testing. Because the number of training instances is still very large, we use stochastic gradient descent for optimizing the loss function in Equation (7), in order to accelerate the training process.

Experimental Results. We use three metrics to evaluate the performance of these methods: HitRate@k, Precision@k and Recall@k. The metrics are averaged over the test set. Table 1 reports the results (shown as percentages). Recall that the dataset has a very low density, which usually results in low precision and recall values [7,8]. In this paper, we focus on the relative improvements we achieve, instead of the absolute metric values. We can observe that our method outperforms all the four baseline methods on three metrics for all k values (we only show four of them to save space). The results show the effectiveness of our context-enhanced method, which is a step towards solving the new item problem.

[1] `http://www.meetup.com/meetup_api/`

Table 1. Recommendation performance (shown as percentages), $k = \{1, 5, 25, 50\}$

Metric	HitRate@k				Precision@k				Recall@k			
k	1	5	25	50	1	5	25	50	1	5	25	50
Pref	0.768	3.500	14.70	28.51	0.768	0.854	0.965	1.139	0.247	1.318	6.336	14.15
Inf-on	0.780	3.338	10.91	17.22	0.780	0.857	0.815	0.812	0.257	1.350	5.071	8.441
Inf-off	0.777	3.166	11.09	17.43	0.777	0.824	0.825	0.819	0.260	1.273	5.128	8.494
Pop	0.701	3.274	11.54	18.72	0.701	0.801	0.843	0.864	0.230	1.228	5.227	9.220
Context	**0.854**	**3.796**	**14.99**	**30.18**	**0.854**	**0.923**	**0.980**	**1.181**	**0.389**	**1.477**	**6.407**	**14.86**

4 Conclusions

In this paper, we propose a context-enhanced event recommendation method which takes advantage of the rich context in EBSNs, by integrating content, social and geographical information. We extract the features of user preference, online/offline social influence and local popularity. Using the learning to rank model, we aggregate these features to rank events. Experiments show promising results of the proposed method. In the future, more features can be extracted and incorporated into our method, e.g., users' relationships with event organizers, time-aware features like day of week or time of day, etc.

Acknowledgments. The work is supported by National Science Foundation of China (Grant No. 61170034 and 61472348), National High Technology Research and Development Program of China (Grant No. SS2013AA040601), National Key Basic Research Program of China (973 Grant No. 2015CB352400) and the Fundamental Research Funds for the Central Universities.

References

1. Adomavicius, G., Tuzhilin, A.: Toward the next generation of recommender systems: A survey of the state-of-the-art and possible extensions. TKDE 17(6), 734–749 (2005)
2. Cao, Y., Xu, J., Liu, T.-Y., Li, H., Huang, Y., Hon, H.-W.: Adapting ranking svm to document retrieval. In: SIGIR, pp. 186–193 (2006)
3. Liu, X., He, Q., Tian, Y., Lee, W.-C., McPherson, J., Han, J.: Event-based social networks: Linking the online and offline social worlds. In: KDD, pp. 1032–1040 (2012)
4. Qiao, Z., Zhang, P., Cao, Y., Zhou, C., Guo, L., Fang, B.: Combining heterogenous social and geographical information for event recommendation. In: AAAI, pp. 145–151 (2014)
5. Troncy, R., Fialho, A.T.S., Hardman, L., Saathoff, C.: Experiencing events through user-generated media. In: Proceedings of the First International Workshop on Consuming Linked Data (2010)
6. Yang, X., Steck, H., Guo, Y., Liu, Y.: On top-k recommendation using social networks. In: RecSys, pp. 67–74 (2012)
7. Ye, M., Yin, P., Lee, W.-C., Lee, D.-L.: Exploiting geographical influence for collaborative point-of-interest recommendation. In: SIGIR, pp. 325–334 (2011)
8. Yuan, Q., Cong, G., Ma, Z., Sun, A., Magnenat-Thalmann, N.: Time-aware point-of-interest recommendation. In: SIGIR, pp. 363–372 (2013)

On the Influence of User Characteristics on Music Recommendation Algorithms

Markus Schedl[1], David Hauger[1], Katayoun Farrahi[2], and Marko Tkalčič[1]

[1] Department of Computational Perception, Johannes Kepler University, Linz, Austria
[2] Department of Computing, Goldsmith's University of London, UK

Abstract. We investigate a range of *music recommendation algorithm* combinations, *score aggregation functions*, *normalization techniques*, and *late fusion techniques* on approximately 200 million listening events collected through *Last.fm*. The overall goal is to identify superior combinations for the task of artist recommendation. Hypothesizing that user characteristics influence performance on these algorithmic combinations, we consider specific user groups determined by age, gender, country, and preferred genre. Overall, we find that the performance of music recommendation algorithms highly depends on user characteristics.

1 Introduction

Music recommendation within the field of recommender systems is becoming increasingly important since the advent of music streaming platforms that provide access to tens of millions of tracks. At the same time, listeners reveal a lot of personal information in social media, which might play an important role on the quality of music recommendations. However, the relationship between user characteristics and quality of music recommendations has not been thoroughly explored. In this paper, we provide an analysis of various combinations of *recommendation algorithms*, *score aggregation functions*, *normalization techniques*, and *late fusion techniques* on a dataset of almost 200 million listening events from *Last.fm*. Hypothesizing that age, gender, country, and preferred genre influence the quality of recommendations, we further group users according to these aspects and evaluate performance on the resulting user groups.

This work is organized as follows. Section 2 overviews related work. In Sections 3 and 4, we present the aspects we categorize users into and the recommendation models and settings we investigate, respectively. We introduce the dataset used for the experiments, explain the experimental setup, and analyze results in Section 5, before concluding in Section 6.

2 Related Work

Current work on music recommender systems typically employs the same recommendation algorithm to serve different user groups. While it can be argued that matrix factorization techniques may take into account various user aspects,

A. Hanbury et al. (Eds.): ECIR 2015, LNCS 9022, pp. 339–345, 2015.

such as temporal dynamics, they still use a single algorithm [4]. In contrast, in this work, we assess how different algorithmic variants of music recommenders perform for different groups of users. In the same vein, Farrahi et al. [5] analyze how aspects of listening frequency, diversity, and mainstreaminess influence recommendation models, but they use a relatively small and sparse dataset mined from microblogs.

Work that integrates user aspects into music recommendation algorithms includes Kaminskas et al. [7], who propose a hybrid matching method to recommend music for places of interest. Baltrunas et al. [1] target music recommendation in a car, taking into account driver and traffic conditions. Zangerle et al. [10] propose an approach that exploits user-based co-occurrences of music items mined from *Twitter* data. Chen and Shen [2] propose a recommendation approach that integrates user location, listening history, music descriptors, and global music popularity trends inferred from microblogs.

In this work, we chose the music platform *Last.fm* to gather a real-world dataset, since it has been shown to attract users of a wide variety of music tastes [9]. In contrast, existing work commonly makes use of rather small and noisy datasets, typically gathered from *Twitter* and including a maximum of a few million listening events [6].

3 User Characteristics

To investigate which aspects of the listener influence the performance of music recommendation algorithms, we categorize each user according to the following attributes. Typewriter font is used to indicate the abbreviations for categories used to indicate user sets for the results.

Age: Listeners from 8 possible age groups are considered. These ranges are [6-17], [18-21], [22-25], [26-30], [31-40], [41-50], [51-60], [61-100]: `US_age_[Start-End]`.

Gender: A listener's gender is considered (i.e. male or female): `US_gender_[male|female]`.

Country: *Last.fm* provides the user with a choice of 240 countries to select from. For reasons of computational complexity and significance of results, we focus on users from the top 6 countries (USA, UK, Brazil, Russia, Germany, and Poland). Each of these has more than 500 users and all other countries are assigned less than half of the number of users of any top 6 country: `US_country_[US|UK|BR|RU|DE|PL]`.

Genre: We categorize listeners according to their preferred genre(s). Assuming that people are typically highly affine to at most 3 different genres, we compute the share of a user's listening events for each genre among all her listening events. Each user is then categorized into all genre classes for which her listening share exceeds 30% of her total listening events. This way a user is assigned none, one, two, or three genre classes. We finally create user sets for 5 representative genres: `US_genre_[jazz|rap|folk|blues|classical]`.

4 Recommendation Methods

We assess several recommendation algorithms for the task of music artist recommendation, in particular, standard *user-based collaborative filtering* (CF), a *popularity-based* algorithm (PB), and an algorithm based on user distance with respect to political or *cultural regions* (CULT). The PB algorithm recommends the most popular artists (i.e. most frequently played) in the dataset. The CULT method defines the target user's nearest neighbors as those that reside in the same country, and recommends their preferred music. As baseline, we include a recommender that proposes items of randomly picked users (RB). For the CF and CULT algorithms, we define two *aggregation functions* (arithmetic mean and maximum) which are used to create an overall ranking of artists to recommend, as an aggregation of similarity scores of the target user's nearest neighbors.

In addition to single methods (PB, $\mathsf{CF}_{[\mathrm{mean,max}]}$, $\mathsf{CULT}_{[\mathrm{mean,max}]}$, and RB), we analyze combinations of two and three algorithms. More precisely, we look into all possible variants: PB+CF, PB+CULT, CF+CULT, and PB+CF+CULT. For these combined variants, a variety of *normalization functions* (n) and *fusion functions* (f) are defined. We consider four methods to normalize the scores of different recommendation methods before fusing their results: n_{none} indicates no normalization is performed; n_{gauss} refers to Gaussian normalization; n_{sumto1} and n_{maxto1} linearly stretches the scores so that their sum equals 1 or their maximum value equals 1, respectively. After scores have been normalized, the results of individual recommenders can be fused. Five fusion functions are investigated: f_{max}, f_{mean}, f_{sum}, $f_{multiply}$, and f_{borda}. While the former four fuse the scores of the individual recommenders directly, by computing their maximum, arithmetic mean, sum, or product, the latter performs rank aggregation based on Borda count [3]. To facilitate perception of individual experiments, we define a standardized scheme. We use sans-serif font for denominations of experiments. For instance, $\mathsf{PB}+\mathsf{CF}_{mean}+\mathsf{CULT}_{max}$ $(n_{gauss}, f_{multiply})$ refers to an experiment in which three algorithms (PB, CF, and CULT) are combined. While the CF recommender employs the mean as aggregation function, CULT employs the maximum. Before fusing the results of the three recommenders by multiplying the item scores, Gaussian normalization is performed.

5 Evaluation

5.1 Dataset

In order to conduct experiments on a large scale, a dataset of almost 200 million listening events has been fetched through the *Last.fm* API.[1] To this end, we select a random subset of 16,429 active users and obtain their listening histories of up to 20,000 listening events. After data cleansing, this eventually yields 191,108,462 listening events to 1,140,014 unique artists. The average number of listening events per user is $11{,}603 \pm 7{,}130$.

[1] `http://www.last.fm/api`

Table 1. Average and maximum precision, recall, and F-measure for best performing methods and algorithmic combinations, on categories `US_age` (upper part) and `US_gender` (lower part)

Method	Precision		Recall		F-score	
	avg	max	avg	max	avg	max
US_age_06-17						
$RB(n_{none})$	1.44	2.06	7.43	19.81	1.63	2.05
$PB + CF_{mean}(n_{none}, f_{max})$	**4.26**	**8.20**	16.44	**34.87**	4.37	5.61
$PB + CF_{mean}(n_{none}, f_{borda})$	4.21	7.41	**16.93**	34.76	**4.42**	**5.69**
US_age_18-21						
$RB(n_{none})$	1.51	1.94	5.45	14.37	1.64	2.28
$CF_{mean}(n_{none})$	5.15	**9.66**	14.47	33.02	4.94	6.04
$PB + CF_{mean}(n_{none}, f_{borda})$	**5.37**	8.92	**15.36**	**33.41**	**5.27**	**6.46**
US_age_22-25						
$RB(n_{none})$	1.61	1.98	4.60	11.98	1.65	2.37
$PB(n_{none})$	***5.25***	9.02	***11.61***	**25.20**	***4.74***	**5.98**
$CF_{mean}(n_{none})$	4.93	**9.85**	8.34	21.90	4.10	5.74
US_age_26-30						
$RB(n_{none})$	1.62	1.95	3.85	10.23	1.59	2.36
$PB(n_{none})$	***5.46***	8.77	***10.24***	**22.41**	***4.73***	5.95
$CF_{mean}(n_{none})$	5.09	**8.97**	9.57	22.30	4.32	**5.27**
US_age_31-40						
$RB(n_{none})$	1.71	1.85	3.35	8.92	1.59	2.51
$PB(n_{none})$	***5.90***	**9.93**	***9.18***	**20.20**	***4.72***	**5.88**
US_age_41-50						
$RB(n_{none})$	1.79	2.30	3.48	9.38	1.62	2.65
$CF_{mean}(n_{none})$	**6.07**	**9.68**	**9.53**	**20.61**	***4.84***	**6.18**
US_age_51-60						
$RB(n_{none})$	1.85	2.36	3.69	9.40	1.68	2.56
$CF_{mean}(n_{none})$	**6.02**	**10.78**	**9.64**	**20.12**	**4.74**	**6.14**
US_age_61-						
$RB(n_{none})$	1.45	1.67	3.65	8.75	1.42	2.30
$CF_{mean}(n_{none})$	4.23	**7.51**	8.47	18.74	3.55	**4.43**
$PB + CF_{mean}(n_{none}, f_{max})$	**4.24**	7.51	8.52	18.88	**3.56**	4.43
$PB + CF_{mean}(n_{none}, f_{borda})$	3.87	5.63	**8.59**	**19.37**	3.47	4.27

US_gender_male						
$RB(n_{none})$	0.74	1.54	1.70	8.22	0.77	2.10
$PB(n_{none})$	**2.47**	**6.64**	**4.92**	**19.88**	**2.45**	**5.51**
$PB + CF_{mean}(n_{sumto1}, f_{max})$	0.79	6.34	0.79	6.34	0.79	**6.34**
US_gender_female						
$RB(n_{none})$	1.78	2.13	5.31	13.87	1.85	2.70
$PB(n_{none})$	**5.63**	9.28	***12.88***	**27.72**	***5.18***	6.47
$PB + CF_{mean}(n_{sumto1}, f_{max})$	3.03	**9.86**	1.52	6.62	1.66	**6.62**

5.2 Experimental Setup

We perform 5-fold cross-validation on a per-user basis, i.e. using 80% of each user's listening history for training and 20% for testing. Given the components of one recommendation experiment, there is a total of 1,640 different algorithmic combinations per user set (recommendation model, number of recommended items, aggregation function, normalization function, and fusion technique). The investigated 4 user categories with a total of 21 attributes thus require 34,440 individual runs.

We measure performance in terms of precision, recall, and F-measure, for various numbers [10–1000] of recommended artists. Please note that there exists a natural upper limit for achievable recall, because several artists in the dataset are listened to by only a single user, can hence never be recommended. This upper limit is 38.63% for the entire dataset, not grouping by any user set.

Table 2. Average and maximum precision, recall, and F-measure for best performing methods and algorithmic combinations, on categories `US_country` (upper part) and `US_genre` (lower part)

Method	Precision		Recall		F-score	
	avg	max	avg	max	avg	max
`US_country_US`						
$RB(n_{none})$	2.00	2.58	4.63	11.93	1.94	2.81
$CF_{mean}(n_{none})$	6.12	**10.93**	11.50	26.57	5.21	6.31
$PB + CF_{max}(n_{sumtol}, f_{sum})$	3.12	6.72	10.94	**27.34**	4.11	**6.96**
$PB + CF_{max}(n_{none}, f_{borda})$	**6.34**	10.46	**12.21**	27.24	**5.52**	6.85
`US_country_UK`						
$RB(n_{none})$	2.11	2.47	4.89	12.61	2.07	3.00
$CF_{mean}(n_{none})$	***6.79***	**12.07**	**12.20**	**26.92**	***5.67***	**7.10**
`US_country_BR`						
$RB(n_{none})$	1.93	2.75	7.37	18.18	2.07	2.74
$CF_{mean}(n_{none})$	***6.44***	**12.35**	**19.41**	**42.59**	***6.30***	**7.87**
`US_country_RU`						
$RB(n_{none})$	1.28	1.65	3.44	9.08	1.26	1.83
$PB(n_{none})$	***4.79***	**8.25**	***10.18***	**21.97**	***4.16***	**5.13**
`US_country_DE`						
$RB(n_{none})$	1.58	1.79	4.06	10.63	1.54	2.29
$CF_{mean}(n_{none})$	***5.73***	**10.16**	**11.94**	**26.79**	***4.96***	**6.18**
`US_country_PL`						
$RB(n_{none})$	1.64	1.96	5.81	14.97	1.77	2.46
$CF_{mean}(n_{none})$	***5.68***	**10.70**	**15.16**	**34.17**	***5.34***	**6.63**
`US_genre_jazz`						
$RB(n_{none})$	1.21	1.49	9.56	26.64	1.47	1.82
$PB(n_{none})$	**2.78**	**5.01**	**16.74**	35.44	**3.13**	**3.88**
$PB + CF_{mean}(n_{none}, f_{multiply})$	2.71	4.92	16.10	**35.63**	3.01	3.76
`US_genre_rap`						
$RB(n_{none})$	0.88	1.00	9.03	25.77	1.17	1.57
$CF_{mean}(n_{none})$	2.58	**5.24**	16.20	33.28	2.90	**3.85**
$PB + CF_{mean}(n_{none}, f_{max})$	**2.59**	5.24	16.42	34.73	**2.90**	3.85
$PB + CF_{mean}(n_{none}, f_{multiply})$	2.22	3.73	16.78	**36.67**	2.66	3.48
$PB + CF_{mean}(n_{none}, f_{borda})$	2.48	4.77	**17.21**	35.94	2.87	3.60
`US_genre_folk`						
$RB(n_{none})$	1.15	1.50	9.12	25.53	1.46	1.94
$CF_{mean}(n_{none})$	3.57	**7.42**	18.41	38.10	3.86	5.10
$PB + CF_{mean}(n_{none}, f_{multiply})$	3.18	5.74	18.44	**39.55**	3.59	4.58
$PB + CF_{mean}(n_{none}, f_{borda})$	3.46	7.05	**18.99**	39.19	3.82	4.96
$PB + CF_{mean} + CULT_{mean}(n_{none}, f_{max})$	**3.57**	7.42	18.56	38.86	**3.87**	**5.10**
`US_genre_blues`						
$RB(n_{none})$	1.59	2.88	6.66	23.99	1.77	3.18
$PB + CF_{mean}(n_{maxtol}, f_{mean})$	2.73	4.73	**24.20**	**53.88**	3.30	4.10
$PB + CF_{max}(n_{none}, f_{multiply})$	**2.85**	**5.93**	23.11	52.68	**3.32**	3.96
$PB + CF_{mean} + CULT_{mean}(n_{maxtol}, f_{mean})$	2.72	4.67	24.20	53.88	3.30	**4.11**
`US_genre_classical`						
$RB(n_{none})$	1.28	2.35	3.74	11.65	1.27	2.35
$CF_{mean}(n_{none})$	2.29	**7.08**	6.58	16.49	2.18	4.38
$PB + CF_{mean}(n_{maxtol}, f_{mean})$	2.77	5.54	17.31	37.77	**3.13**	4.42
$PB + CF_{mean}(n_{none}, f_{borda})$	2.79	6.23	16.85	37.10	3.08	**4.56**
$PB + CF_{max}(n_{sumtol}, f_{mean})$	**2.81**	6.23	17.07	37.35	3.11	4.52
$PB + CF_{max}(n_{maxtol}, f_{mean})$	2.76	5.38	**17.32**	**37.85**	3.13	4.50

5.3 Discussion

Due to space limitations, we cannot provide here the entire set of results for each algorithmic combination. We hence only show the results of the best performing variants (in terms of average and maximum precision, recall, and F-measure) for each user category and attribute. Results for categories age and gender are shown in Table 1; results for country and genre are depicted in Table 2.

Main general findings from these results are that (i) fusing scores of different recommenders frequently outperforms single variants, (ii) using the mean as aggregation function for CF almost always outperforms the maximum,[2] and (iii) recommendations are overall better when categorizing users according to age and country than according to gender or genre. Analyzing the results per category in detail, we make other interesting observations:

- Younger people seem to be easier to satisfy by recommending overall popular music, whereas mid-aged and elder listeners (41-100) should be offered collaborative filtering recommendations (or combinations that include CF).
- By recommending music using the PB approach it seems slightly easier to satisfy women than men; otherwise no substantial differences between genders can be made out.
- While listeners in most investigated countries are served well by CF approaches, Russian listeners seem to prefer highly popular mainstream music (PB). For US citizens, the right mixture of popular music and music listened to by like-minded people yields best results.
- Including cultural aspects (CULT) most strongly contributes to increased performance for lovers of folk and blues. The surprisingly good performance of PB for jazz aficionados indicates that they may prefer overall popular jazz music, whereas combinations of PB and CF provide most accurate recommendations for fans of rap and classical music.

In order to see if there is a significant winning method within the user groups, we perform pairwise significance tests between the best method within each group and the other methods within that group. As the distributions of the performance metrics (precision, recall, and F-score) are not normal, we employ the Mann-Whitney U test for equal medians of two samples [8]. We mark significant results in italics in the results tables.

6 Conclusion and Outlook

The overall finding of our study is that music recommendation can be improved by tailoring to different listener categories. There is no single method that fits everyone; rather a combination of individual recommendation models and variants should be considered for each user group.

We are currently conducting experiments using a larger variety of user-specific factors, including categories related to listening frequency, temporal aspects of music consumption, and openness to unknown music. In the future, we would also like to investigate the influence of personality traits on music recommendation and music taste in general. Song-level recommendation experiments and the related topic of addressing data sparsity, as well as looking into content-based algorithms, constitute other research directions.

[2] This is not the case for currently investigated content-based recommenders.

Acknowledgments. This research is supported by the EU-FP7 project no. 601166 and by the Austrian Science Fund (FWF): P25655. The authors are furthermore most thankful to Johann Messner for providing his expertise in scientific computing.

References

1. Baltrunas, L., Kaminskas, M., Ludwig, B., Moling, O., Ricci, F., Lüke, K.-H., Schwaiger, R.: InCarMusic: Context-Aware Music Recommendations in a Car. In: Proc. EC-Web, Toulouse, France (August–September 2011)
2. Cheng, Z., Shen, J.: Just-for-Me: An Adaptive Personalization System for Location-Aware Social Music Recommendation. In: Proc. ICMR, Glasgow, UK (April 2014)
3. de Borda, J.-C.: Mémoire sur les élections au scrutin. Histoire de l'Académie Royale des Sciences (1781)
4. Dror, G., Koenigstein, N., Koren, Y., Weimer, M.: The Yahoo! Music Dataset and KDD-Cup 2011. In: JMLR: Proceedings of KDD-Cup 2011 Competition, pp. 18:3–18 (October 2012)
5. Farrahi, K., Schedl, M., Vall, A., Hauger, D., Tkalčič, M.: Impact of Listening Behavior on Music Recommendation. In: Proc. ISMIR, Taipei, Taiwan (October 2014)
6. Hauger, D., Schedl, M., Košir, A., Tkalčič, M.: The Million Musical Tweets Dataset: What Can We Learn From Microblogs. In: Proc. ISMIR, Curitiba, Brazil (November 2013)
7. Kaminskas, M., Ricci, F., Schedl, M.: Location-aware Music Recommendation Using Auto-Tagging and Hybrid Matching. In: Proc. RecSys 2013, Hong Kong, China (October 2013)
8. Prajapati, B., Dunne, M., Armstrong, R.: Sample Size Estimation and Statistical Power Analyses. Optometry Today 16(7) (2010)
9. Schedl, M., Tkalčič, M.: Genre-based Analysis of Social Media Data on Music Listening Behavior. In: Proc. ACM Multimedia Workshop: ISMM, Orlando, FL, USA (November 2014)
10. Zangerle, E., Gassler, W., Specht, G.: Exploiting Twitter's Collective Knowledge for Music Recommendations. In: Proc. WWW Workshop: #MSM (April 2012)

A Study of Smoothing Methods for Relevance-Based Language Modelling of Recommender Systems

Daniel Valcarce, Javier Parapar, and Álvaro Barreiro

Information Retrieval Lab.,
Computer Science Department,
University of A Coruña, Spain
{daniel.valcarce,javierparapar,barreiro}@udc.es

Abstract. Language Models have been traditionally used in several fields like speech recognition or document retrieval. It was only recently when their use was extended to collaborative Recommender Systems. In this field, a Language Model is estimated for each user based on the probabilities of the items. A central issue in the estimation of such Language Model is smoothing, i.e., how to adjust the maximum likelihood estimator to compensate for rating sparsity. This work is devoted to explore how the classical smoothing approaches (Absolute Discounting, Jelinek-Mercer and Dirichlet priors) perform in the recommender task. We tested the different methods under the recently presented Relevance-Based Language Models for collaborative filtering, and compared how the smoothing techniques behave in terms of precision and stability. We found that Absolute Discounting is practically insensitive to the parameter value being an almost parameter-free method and, at the same time, its performance is similar to Jelinek-Mercer and Dirichlet priors.

Keywords: Recommender systems, Collaborative filtering, Smoothing, Relevance Models.

1 Introduction

In a world with a growing amount of available information, Recommender Systems are key in satisfying the increasing demands of the users. These systems generate personalised suggestions saving the customers the time of searching for relevant information. Many approaches to recommendation have been proposed being collaborative filtering one of the most successful techniques. This family of methods exploits the past ratings of the users in order to generate recommendations. Parapar et al. recently proposed the use of Relevance-Based Language Models for collaborative filtering [5] obtaining superior figures in precision w.r.t. the state of the art methods. Following previous results in using Language Models for the recommendation task [6], the authors decided to use Jelinek-Mercer for smoothing the different probabilities arguing that Dirichlet priors can demote the weight of those items recently introduced in the systems. In this paper

A. Hanbury et al. (Eds.): ECIR 2015, LNCS 9022, pp. 346–351, 2015.

we tested those intuitions and analysed the performance of different smoothing techniques in the recommendation task. For doing so, we followed the Zhai and Lafferty study of smoothing techniques for document retrieval [7] and adapted their methodology to collaborative filtering. We analysed Jelinek-Mercer, Dirichlet priors and Absolute Discounting smoothing in the context of Relevance-Based Language Models for collaborative filtering in different collections. We somehow obtained similar trends as in document retrieval but, in contrast, the better behaviour in terms of stability of Absolute Discounting makes it more suitable for the recommendation task than the other approaches. In the following sections, first, we briefly introduce Relevance Models for recommendation, then we present the smoothing techniques, the experimental conditions, the results and finally we conclude with some remarks about our findings and future work.

2 Relevance Models for Recommendation

Recommender Systems help users with the finding of relevant items. We denote the set of users by $\mathcal{U}$ and the set of items by $\mathcal{I}$. We refer to the rating that the user u expressed to the item i by the notation $\mathrm{r}_{u,i}$. Also, $\mathcal{I}_u$ is used to indicate the set of items that were rated by the user u.

In the context of Statistical Language Models (LM), Relevance-Based Language Models (usually referred as Relevance Models or RM) [3], are a pseudo relevance feedback technique for text retrieval. Given a query and a set of pseudo relevant documents, RM suggest terms to expand the original query and, thus, improve the text retrieval performance. Recently, RM has been applied as a collaborative filtering technique achieving high accuracy figures. Users play the role of both documents and queries whilst items are equivalent to the terms. In this way, we can expand users with new items as we expanded queries with new terms. To perform query expansion via RM, we need a set of pseudo relevant documents that, in this case, is the neighbourhood of the target user. Parapar et al. proposed the use of Posterior Probability Clustering (PPC [2]), a matrix factorization clustering algorithm, for calculating the neighbourhoods.

Two approaches of RM were proposed for recommendation: RM1 and RM2. Recommendations are generated by computing the Relevance Model of every user, R_u, and estimating the relevance of each item under it as shown in (1) and (2) for methods RM1 and RM2, respectively.

$$p(i|R_u) \propto \sum_{v \in V_u} p(v) p(i|v) \prod_{j \in \mathcal{I}_u} p(j|v) \quad (1) \qquad p(i|R_u) \propto p(i) \prod_{j \in \mathcal{I}_u} \sum_{v \in V_u} \frac{p(i|v)p(v)}{p(i)} p(j|v) \quad (2)$$

where V_u is the set of neighbours of the user u. Also, $p(i)$ and $p(v)$ are considered uniform. Finally, the probability of an item given a user $p(i|u)$ can be computed by smoothing the maximum likelihood estimate $p_{ml}(i|u)$:

$$p_{ml}(i|u) = \frac{\mathrm{r}_{u,i}}{\sum_{j \in \mathcal{I}_u} \mathrm{r}_{u,j}} \quad (3)$$

3 Smoothing Methods in Recommendation

Smoothing is a well studied aspect of Language Models for text retrieval [4, 7]. The maximum likelihood estimator suffers from data sparsity, i.e., in the recommendation task each item is only rated by some users. Therefore, it is necessary to apply smoothing to adjust the estimator to prevent the apparition of zeros in (3). Furthermore, smoothing also plays a similar role to the *idf* (inverse document frequency).

In this paper, we studied the effect of applying three different smoothing methods for recommendation. All these techniques employ a background model which is the following collection model.

$$p(i|\mathcal{C}) = \frac{\sum_{v \in \mathcal{U}} \mathrm{r}_{v,i}}{\sum_{j \in \mathcal{I},\, v \in \mathcal{U}} \mathrm{r}_{v,j}} \tag{4}$$

Jelinek-Mercer. JM performs a linear interpolation between the maximum likelihood estimator and the collection model controlled by the parameter λ.

$$p_\lambda(i|u) = (1-\lambda)\, p_{ml}(i|u) + \lambda\, p(i|\mathcal{C}) \tag{5}$$

Bayesian Smoothing with Dirichlet Priors. DP uses Dirichlet priors for Bayesian analysis which results in the following expression with parameter μ.

$$p_\mu(i|u) = \frac{\mathrm{r}_{u,i} + \mu\, p(i|\mathcal{C})}{\mu + \sum_{j \in \mathcal{I}_u} \mathrm{r}_{u,j}} \tag{6}$$

Absolute Discounting. AD subtracts a constant, δ, from the count of the seen words.

$$p_\delta(i|u) = \frac{\max(\mathrm{r}_{u,i} - \delta, 0) + \delta\, |\mathcal{I}_u|\, p(i|\mathcal{C})}{\sum_{j \in \mathcal{I}_u} \mathrm{r}_{u,j}} \tag{7}$$

4 Experiments

4.1 Evaluation

We conducted our experiments in the *Movielens 100k*[1], the *R3 Yahoo! Music*[2] and the *Movielens 1M*[1] collections. The statistics of these datasets are presented in the Table 1.

In this work, we considered the precision of the recommendations which is the fraction of items included in the recommendation list that are relevant to the user. We evaluated this metric at a cut-off rank of five, following the *TestItems* methodology described in [1].

[1] `http://grouplens.org/datasets/movielens`

[2] `http://webscope.sandbox.yahoo.com/catalog.php?datatype=r`

Table 1. Datasets statistics

Dataset	#users	#items	#ratings	Sparsity
Movielens 100k	943	1682	100,000	6.30%
R3 Yahoo! Music	15400	1000	365,703	2.37%
Movielens 1M	6040	3952	1,000,209	4.19%

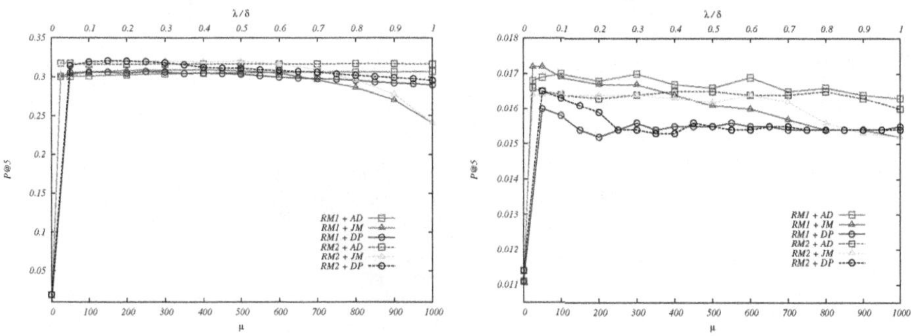

Fig. 1. Precision at 5 of the RM1 and the RM2 algorithms using Absolute Discounting (AD), Jelinek-Mercer (JM) and Dirichlet priors (DP) smoothing methods in the Movielens 100k (left) and the R3 Yahoo! Music (right) collections

4.2 Results and Discussion

First, we studied the P@5 of the RM1 and the RM2 algorithms with the different smoothing techniques in the three collections. The results of the Movielens 100k and Yahoo datasets are illustrated in Fig. 1 (in this experiment, Movielens 1M presented the same trends as 100k). We must remark that the precision values for the Yahoo dataset are low because of the very few of available testing ratings (only ten per user) which makes the recommendation a very hard task.

We notice that smoothing plays a key role in accuracy and a small amount of smoothing is sufficient to achieve good results. We can appreciate that Jelinek-Mercer performance deteriorates with a high value of λ. The same behaviour is observed when Dirichlet priors are applied, although on a lesser scale. It is very interesting that only AD does not present statistically significant differences (Wilcoxon test, $p < 0.05$) in precision when varying the smoothing parameter, in contrast to JM and DP. This points out that the performance of the system when using JM or DP will be dependent on choosing the optimal smoothing value, which unfortunately depends on the data in collection, as can be observed in Fig. 1. In fact, in the R3 Yahoo! Music dataset, the demotion of precision when using DP and increasing the smoothing is more visible. Moreover, in this collection, RM1 works better than RM2. It seems that RM1 may be better for dealing with very sparse datasets, although further work is required to establish this. These trends are similar to the reported by Zhai and Lafferty for text

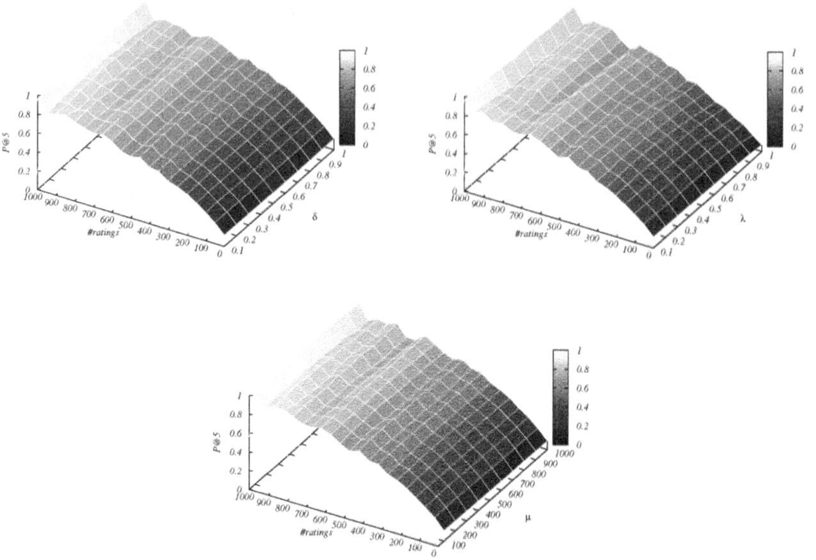

Fig. 2. Precision at 5 of RM2 algorithm using Absolute Discounting (top left), Jelinek-Mercer (top right) and Dirichlet priors (bottom) smoothing methods when varying the smoothing intensity and considering different $|\mathcal{I}_u|$, i.e., the number of ratings in the user profiles (they have been binned in steps of 50 and the average precision for each bin is plotted).

retrieval [7], except for the fact that AD is more stable w.r.t. the smoothing parameter and that DP does not outperform the other methods.

The second experiment analyses the effect of the smoothing, in terms of precision, when recommending to users with different amount of rated items. Losada and Azzopardi have extensively studied the effects of the document length in text retrieval [4]. We aim to determine if there is such parallelism with the length of the user profiles. In the Fig. 2, we show the average precision achieved by the RM2 algorithm with each method when varying the intensity of smoothing and the size of the user profiles. The precision of the system improves with the number of rated items, achieving near perfect recommendations for users with a long rating history. The performance of DP and AD is very similar, although DP slightly degrades with high values of smoothing. The same effect, more intense, is observed in the case of the JM method. Additionally, JM does not seem to be a good technique for recommending to users with many ratings.

In the light of the results, we can recommend the use of AD because parameter optimization is not critical as long as a small amount of smoothing is applied. Furthermore, it obtains a good performance for each size of user profile.

5 Conclusions and Future Work

We studied three techniques of LM smoothing in the context of Relevance Modelling for Recommender Systems. Through empirical analysis, we get insights of the behaviour of smoothing for the recommendation task. The evidence indicates that smoothing methods are crucial for achieving high precision: tiny values of the smoothing parameters produce notably superior results.

The current findings suggest that there is no big difference in terms of optimal precision among the studied smoothing techniques. However, Dirichlet priors and, specially, Jelinek-Mercer suffer a significant decrease in precision when a high amount of smoothing is applied, in contrast to Absolute Discounting. Thus, AD is the best election for a recommender system that makes use of Relevance Modelling because it saves the developers the time to tune properly the smoothing parameter for each domain and collection. An almost parameter-free smoothing method is very useful when no training data is available.

As a future work, it would be interesting to study how these smoothing methods behave w.r.t. different aspects such as novelty and diversity.

Acknowledgments. This work was funded by grants TIN2012-33867 and GPC2013/070 from the *Ministerio de Economía y Competitividad* and the Galician Government.

References

1. Bellogín, A., Castells, P., Cantador, I.: Precision-oriented evaluation of recommender systems. In: RecSys 2011, p. 333. ACM Press, New York (2011)
2. Ding, C., Li, T., Luo, D., Peng, W.: Posterior probabilistic clustering using NMF. In: SIGIR 2008, pp. 831–832. ACM, New York (2008)
3. Lavrenko, V., Croft, W.B.: Relevance based language models. In: SIGIR 2001, pp. 120–127. ACM Press, New York (2001)
4. Losada, D.E., Azzopardi, L.: An analysis on document length retrieval trends in language modeling smoothing. Information Retrieval 11(2), 109–138 (2008)
5. Parapar, J., Bellogín, A., Castells, P., Barreiro, Á.: Relevance-based language modelling for recommender systems. Information Processing & Management 49(4), 966–980 (2013)
6. Wang, J.: Language Models of Collaborative Filtering. In: Lee, G.G., Song, D., Lin, C.-Y., Aizawa, A., Kuriyama, K., Yoshioka, M., Sakai, T. (eds.) AIRS 2009. LNCS, vol. 5839, pp. 218–229. Springer, Heidelberg (2009)
7. Zhai, C., Lafferty, J.: A study of smoothing methods for language models applied to information retrieval. ACM Transactions on Information Systems 22(2), 179–214 (2004)

The Power of Contextual Suggestion

Adriel Dean-Hall and Charles L.A. Clarke

University of Waterloo, Waterloo, Canada

Abstract. The evaluation process for the TREC Contextual Suggestion Track consumes substantial time and resources, taking place over several weeks and costing thousands of dollars in assessor remuneration. The track evaluates a point-of-interest recommendation task, using crowdsourced workers as a source of user profiles and judgments. Given the cost of assessment, we examine track data to provide guidance for future experiments on this task, particularly with respect to the number of assessors required. To provide insight, we first consider the potential impact of fewer assessors on the TREC 2013 experiments. We then provide recommendations for future experiments. Our goal is to minimize costs, while still meeting the requirements of those experiments.

1 Introduction

The TREC Contextual Suggestion Track [6] envisions a traveller visiting an unfamiliar city and seeking recommendations for appropriate points-of-interest. The Contextual Suggestion Track provides a framework for evaluating point-of-interest recommendation systems. Given information about a traveller's preferences, systems aim to make recommendations tailored to that traveller.

In molding a contextual suggestion task to fit the typical TREC task format, the delay between data release and solution submission precludes the involvement of real travellers, who are unlikely to tolerate a delay of weeks or months to receive their recommendations. Instead, the Contextual Suggestion Track substitutes crowdsourced assessors for the travellers. Each of these assessors rate selected attractions in a designated home city according to their own personal preferences. For TREC 2013, Philadelphia was designated as the home city, with 562 assessors rating 50 attractions on a five-point scale. The rating from each assessor form a *profile* for that assessor. These profiles were provided to participating groups, who are posed with the problem of returning a ranked list of attractions for each assessor in target cities. For TREC 2013 [6], groups were given six weeks to submit experimental runs comprised of suggestions for 50 target cites, with 19 groups submitting a total of 34 experimental runs.

For each assessor+city pair, the top-five suggestions from each group are combined into a pool, with each pool containing suggestions for one assessor from one city. Assessors are invited back to rate these suggestions, again according to their own personal preferences. For TREC 2013, 136 assessors agreed to return, with 39 assessors rating attractions for one city only (one pool) and 97 assessors rating attractions for two cities (two pools). These ratings form the basis

A. Hanbury et al. (Eds.): ECIR 2015, LNCS 9022, pp. 352–357, 2015.

for computing evaluation measures to quantify the performance of the various systems, including precision@5 (P@5) [5].

To support future experiments in this area, we analyze data from the track to minimize these costs while still maintaining an ability to detect significant differences between systems. For example, if we view a difference of 0.1 in precision@5 to be of practical significance, how many assessors should be recruited to achieve statistical significance at the 95% level at least 80% of the time? In other words, with these requirements, how many assessors would be needed to achieve a statistical power of 80%. To provide some insight, we first consider the impact a reduction in the number of assessors would have had on the results of the TREC 2013 track. Extending this analysis, we make recommendations for future efforts, depending on experimental requirements.

2 Background and Related Work

Point-of-interest recommendation is a topic of active research. Braunhofer et al. [4] developed and evaluated a mobile application for making recommendations within particular cities. Their application asks the user several personality questions and then makes suggestions that take their responses into account. Adomavicius et al. [1] adopted a collaborative filtering approach, but also consider contextual features such as the day of the week and the weather. Baltrunas et al. [2]. consider contextual information such as the user's budget or familiarity with a city. Despite this ongoing research, comparing approaches to point-of-interest recommendation remains a challenging problem, with no standard test collections or other methods for conducting robust and repeatable experiments.

The goal of the TREC Contextual Suggestion Track is to facilitate the comparison of different approaches to the problem of point-of-interest recommendation. Several different strategies have been explored by track participants. The general strategy used compares candidate attractions from the target city against ratings for attractions in the profile in order to rank them. A variety of methods were used to make this comparison, some systems used textual similarity between attractions rated positively in the profile and the candidate suggestions [7]. Other systems based their rankings on ratings, reviews [13], attraction categories, and other features.

The problem of designing evaluation tasks in order to reliably recognize differences between systems is a longstanding research issue. Voorhees and Buckley [11] examine the problem of selecting the number of topics to use in TREC-style adhoc retrieval experiments. They note that if too few topics are used, this will cause large enough errors such that the evaluation cannot be viewed as reliable. They indicate that at least 50 topics are needed in order to reliably order systems, and this has become a common choice for many experiments.

In order to examine the utility of significance testing with respect to information retrieval evaluation tasks, Zobel [14] split the topic set in half and examined how often statistical significance in one half translated into statistical significance in the other half. Zobel used ANOVA, Wilconox and the t-test, finding all three

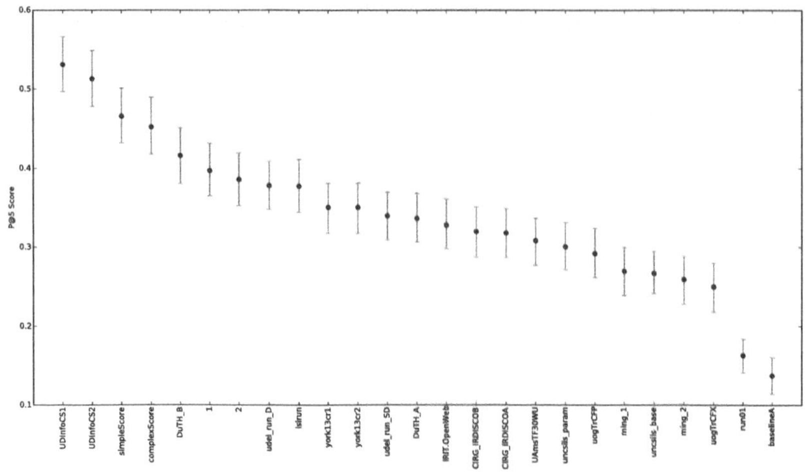

Fig. 1. Mean P@5 scores of all open web runs with 95% confidence intervals

to be reliable at detecting the order of systems. Sanderson and Zobel [9] revisited the reliability of statistical tests, as well as the number of topics needed to reliably order systems. Here again, they conclude that small topic sets (less than 25) do not allow researchers to reliably order systems. In other work, Webber et al. [12] employed statistical power to determine, given a specific number of topics, how large of a difference in scores is needed before we can reliably determine the true ordering of systems.

Finally, in order to determine the number of topics Sakai developed a model that accounts for the depth of the pool size and the desired size of the confidence in order to have statistically reliable evaluation results [8].

3 The Impact of Fewer Assessors

Once the judgements were made a P@5 score was calculated for each ranked list of suggestions. The score for each run is the mean of these scores calculated across all the runs. Figure 1 shows all the open web runs from TREC 2013 ordered by P@5 score, as well as the 95% confidence interval around each P@5 score, calculated using the bootstrap method [10]. This work does not mix open web and ClueWeb12 runs addressing concerners raised by Bellogín et al. [3].

While confidence intervals usually overlap for adjacent runs, we clearly see that many runs outperform others. From the standpoint of the goals of track, this was a very positive outcome, implying we are able to detect differences between many of the systems. For the purposes of our analysis, we define two systems as being "different" in performance if their confidence intervals do not overlap. Additionally, we view each of our $n = 223$ assessor+city pairs as independent, even through some assessors judged two pools, while some judged only one.

We would expect that reducing the number of pairs would reduce the number of differences, but how much degradation occurs? In this section, we consider

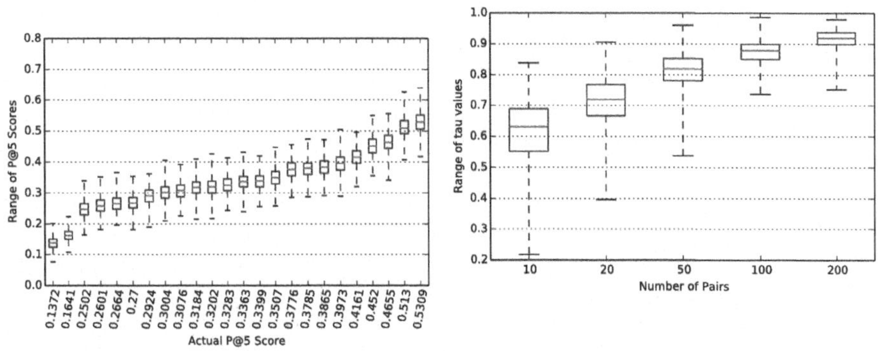

(a) Actual P@5 score vs. P@5 for a random set of 100 pairs

(b) Kendall's τ of experiment iteration ranking vs total ranking for random set sizes

Fig. 2. Results with min., max., mean, and 25th/75th percentiles marked

the hypothetical outcome if we had had a smaller number of assessor+city pairs. From the set of all pairs, we randomly selected, with replacement, a subset of $m < n$ pairs, for various m, and calculated the mean P@5 scores for each system. A total of 1000 iterations of this process was completed for each value of m equal to 10, 20, 50, 100, and 200.

Figure 2a shows the ranges of P@5 scores for $m = 100$ vs. the P@5 score across all 223 pairs. While, with 100 pairs many differences would still be apparent. However, the range for most systems is approximately 0.2, which is as large as the difference between many systems. We also consider how the number of assessor+city pairs impacts the overall ranking of systems. To compare rankings, we compute Kendall's τ between the actual system ranking using all $n = 223$ pairs vs. using a random selection of m pairs. Figure 2b shows the range of scores for different values of m. Clearly 10 pairs does a poor job of ranking systems, but rankings with 200 pairs are highly correlated with the actual ranking.

4 Statistical Power

In the previous section we examined the potential impact of fewer assessor+city on TREC 2013. This analysis illustrates the level of degradation in the evaluation we may have seen with a smaller group of assessors. It would also be interesting to see if larger groups of assessors would have provided more accurate evaluation results that are of practical significance.

Our goal here is to determine the number of assessors which provides us with a practical, useful ordering of systems. However, beyond a certain point there is no practical significance between systems. A user may notice if the difference between the mean scores of the two systems is 0.2, but they would probably not notice if the difference was 0.01, even if the user visits hundreds of cities.

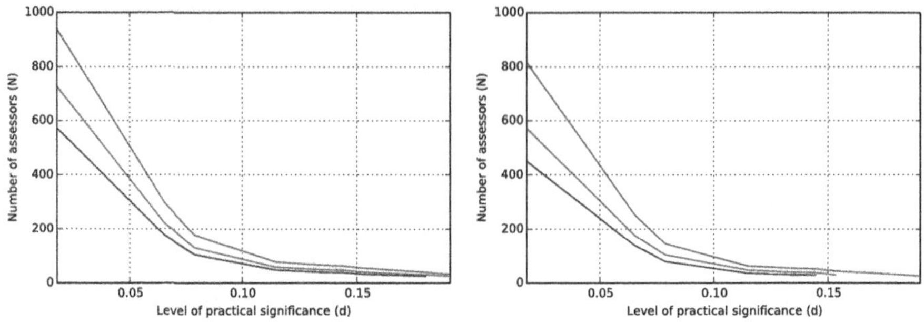

(a) $\alpha = 0.05$ (b) $\alpha = 0.1$

Fig. 3. Level of practical significance vs number of assessors with $\alpha = .1$ or .05. The three lines, from top to bottom, indicate a desired statistical power of .7, .8, and .9.

In order to determine how many assessors are needed for future tasks, we turn to statistical power, which tells us the probability of rejecting the null hypothesis given that it is false. Here our null hypothesis is that the two systems being compared provide equivalent suggestions, producing the same mean effectiveness value. Besides the level of practical significance (the desired difference between the P@5 means) d, to compute statistical power we need two other parameters: the desired level of statistical significance (α) and the desired level of power $(1-\beta)$. α is the probability of a Type I error, and β is the probability of a Type II error. It is desirable to have a higher power and a lower β.

In order to calculate our statistical power we employ a simple nonparametric approach, simulating assessors by sampling from the pool of real assessors. For each round we pick m of assessors with replacement. We then pick two systems where the difference between the means is d, this allows us to see if we are able to detect the difference between systems with this d or greater (i.e., we set our practical level of difference to be d). We also set α to be either 0.05 or 0.1, providing two reasonable values of statistical significance. We then determine power values of .7, .8, or .9, providing three reasonable levels of statistical power.

For a given m, d, and α we then randomly select different sets of assessors multiple times (10,000 times). Over these samples, we compute how often we are able to recognize a difference between two systems. For this computation, we recognize a difference using statistical significance calculated by the Wilcoxon signed-rank test, with α as given. This results in a single power estimate, for the given combination of m, d, and α.

We varied m from 25 to 1000, and the level of practical significance from 0.02 to 0.18. Here, it is reasonable to assume that we would not want to detect smaller differences than 0.02 between systems. On the other side 0.18 is such a large difference that we may not have a useful evaluation if this was the smallest difference that we could detect.

Given a choice of α, β, and level of practical significance, we can now estimate how many assessors we need in order to reasonably rank systems. The analysis so far has focused on P@5, however it can easily be extended to other metrics 3.

From Figure 3a we see with a reasonable set of parameter values $\alpha = .05$, power $= .8$, and level of practical difference $= .05$, the number of assessors needed is about 300, which is slightly greater than the number actually used in the task.

5 Conclusion

One of the major decisions in the evaluation of the Contextual Suggestion Track is the determination of how many assessors are needed. For TREC 2013, 223 assessor+city pools were obtained. In this paper, we have examined the level of degradation which might occur if only a subset of these assessors had judged suggestions. Looking forward to future experiments, we have also considered the number of assessors needed, given a desired level of practical significance, statistical power, and statistical significance. As it turns out, while 223 assessors provide for reasonable experimental requirements increasing the number of assessors to 300 would have been a good investment of resources.

References

1. Adomavicius, G., Tuzhilin, A.: Context-aware recommender systems. In: Recommender Systems Handbook, pp. 217–253. Springer (2011)
2. Baltrunas, L., Ludwig, B., Peer, S., Ricci, F.: Context relevance assessment and exploitation in mobile recommender systems. Personal Ubiquitous Comput. 16(5), 507–526 (2012)
3. Bellogín, A., Samar, T., de Vries, A.P., Said, A.: Challenges on combining open web and dataset evaluation results: The case of the contextual suggestion track. In: de Rijke, M., Kenter, T., de Vries, A.P., Zhai, C., de Jong, F., Radinsky, K., Hofmann, K. (eds.) ECIR 2014. LNCS, vol. 8416, pp. 430–436. Springer, Heidelberg (2014)
4. Braunhofer, M., Elahi, M., Ricci, F.: Usability assessment of a context-aware and personality-based mobile recommender system. In: Hepp, M., Hoffner, Y. (eds.) E-Commerce and Web Technologies, vol. 188, pp. 77–88. Springer, Heidelberg (2014)
5. Büttcher, S., Clarke, C.L.A., Cormack, G.V.: Information retrieval: Implementing and evaluating search engines. MIT Press (2010)
6. Dean-Hall, A., Clarke, C.L.A., Kamps, J., Thomas, P., Simone, N., Voorhees, E.: Overview of the TREC 2013 contextual suggestion track. In: Proc. of TREC (2013)
7. Milne, D., Thomas, P., Paris, C.: Finding, weighting and describing venues: Csiro at the 2012 trec contextual suggestion track. In: Proc. of TREC (2012)
8. Sakai, T.: Designing test collections for comparing many systems. In: Proc. CIKM, pp. 61–70 (2014)
9. Sanderson, M., Zobel, J.: Information retrieval system evaluation: Effort, sensitivity, and reliability. In: Proc. ACM SIGIR, pp. 162–169 (2005)
10. Soboroff, I.: Computing confidence intervals for common ir measures. In: Proc. of EVIA (2014)
11. Voorhees, E.M., Buckley, C.: The effect of topic set size on retrieval experiment error. In: Proc. of ACM SIGIR, pp. 316–323 (2002)
12. Webber, W., Moffat, A., Zobel, J.: Statistical power in retrieval experimentation. In: Proc. of ACM CIKM, pp. 571–580 (2008)
13. Yang, P., Fang, H.: An opinion-aware approach to contextual suggestion. In: Proc. of TREC (2013)
14. Zobel, J.: How reliable are the results of large-scale information retrieval experiments? In: Proc. of ACM SIGIR, pp. 307–314 (1998)

Exploiting Semantic Annotations for Domain-Specific Entity Search*

Tuukka Ruotsalo[1] and Eero Hyvönen[2]

[1] Helsinki Institute for Information Technology HIIT, Aalto University, Finland
[2] Semantic Computing Research Group (SeCo), Aalto University, Finland
first.last@aalto.fi

Abstract. Searches on the Web of Data go beyond the retrieval of textual Web sites, and shifts the focus of search engines towards domain-specific entity data, for which the units of retrieval are domain-specific entities instead of textual documents. We study the effect of using semantic annotation in combination with a knowledge graph for domain-specific entity search. Different reasoning, indexing and query-expansion strategies are compared to study their effect in improving the effectiveness of entity search. The results show that the use of semantic annotation and background knowledge can significantly improve the retrieval effectiveness, but require graph structures to be exploited beyond standard reasoning. Our findings can help to develop more effective information and data retrieval methods that can enhance the performance of semantic search engines that operate with structured domain-specific Web data.

1 Introduction

Recent studies have shown that a large portion of Web search queries are targeted to find information about entities [18,10]. As a consequence, a variety of methods and systems have been proposed to retrieve entity data that can provide the end users of Web of Data more structured information than possible with the conventional Web search [2,4,14]. As the end users of the Web of Data are mostly unaware of the underlying formalisms that are used to represent the structured data, such as RDF or micro formats, the key affordance of semantic entity search is the utilization of information retrieval methods that can operate on the structured knowledge and can work without explicit assumptions about the underlying schema or expressive query languages. For example, a user issuing a topical query "astronomy", could benefit from not only results with an exact match to "astronomy", but also to other information available via a knowledge graph, such as famous astronomers "Newton", "Copernicus", and "Galileo", and tools and techniques used by astronomers, such as "sundial" or "law of gravitation".

This paper contributes by systematically investigating the effectiveness of using knowledge graph constructed from a set of semantic annotations and ontologies as a source of data and semantics for domain-specific entity search. Recent research shows that despite the proliferation of structured knowledge on the Web, most of this knowledge is weakly interlinked across domains [8]. Consequently, we concentrate on

* The work was partially supported by the Academy of Finland (278090).

A. Hanbury et al. (Eds.): ECIR 2015, LNCS 9022, pp. 358–369, 2015.

domain-specific search for which high quality background knowledge is available and can be directly utilized in the indexing and retrieval process.

We consider two entity retrieval tasks: ad-hoc entity search and search-by-entity [4,14]. In the ad-hoc entity search task, a query is issued to a system and the system responds with a ranked list of entities that best match the query. In the search-by-entity task, one of the entities in the collection is used as a query to find a ranked list of related entities. We compare structured indexing with and without RDF(S)[1] reasoning, knowledge graph based query-expansion, and random-walk based query and document expansion for domain-specific semantic search. The findings can be summarized as follows:

1. Background knowledge can significantly improve the effectiveness of domain-specific entity search.
2. Employing only standard RDF(S) reasoning is only marginally useful in domain-specific entity search.
3. Background knowledge -based query-expansion and random-walk -based query and document expansion can improve the precision of the top-ranked entities over 100% compared to the standard RDF(S) reasoning.

2 Background

We follow the definition by Pound et al. [18], which is also considered by recent related works [14,10], and define entity search as the task of answering arbitrary information needs related to particular aspects of entities, expressed as unconstrained natural language queries or structured queries, and resolved using a collection of structured data. Similar approaches to semantic search have also been referred to as object search [18] or semantic data retrieval [19].

While a wide range of semantic entity search solutions have been proposed in the past, they make limited use of background knowledge to resolve the semantic connections that might be useful in answering the query. Conversely, many experiments rely on either structured indexing and using standard information retrieval methods on top of the resulting index. For example, this strategy was used by most of the systems that participated in the SemSearch campaigns [3,18], and they use at best only the structure of the data to compute query-independent features [16,17]. Furthermore, many experiments and evaluation campaigns limit their analysis only in queries expressed in some formal query language, such as SPARQL [13], or focus more on a recommendation scenario [6,21]. Another line of research focuses on extending text retrieval with ontological query expansion [7,5,24,12,15,1]. However, these studies do not consider searching structured knowledge: they only utilize ontologies as a source for query expansion in text retrieval.

Semantic entity search, sometimes referred to as object search, has recently been studied extensively [18,4,14,10,22]. Much of the recent work has used the SemSearch dataset [9] to evaluate the effectiveness of semantic entity search. A well-known project that provides access methods to RDF resources is Sindice [17]. Sindice initially started

[1] `http://www.w3.org/TR/rdf-schema/`

as a look-up index for RDF objects, but has advanced to a search engine and now includes ranking and interactive search functionality. However, Sindice makes use of only query-independent features computed using the RDF graph. Similar approaches have been used in the RSS system [16] as well as in the Semantic Search Engine project [13], which enable ranked search on semantically annotated data through ranking algorithms that measure the global importance of resources in the data graph.

As a consequence, the question of whether a knowledge graph constructed from semantic annotations, domain ontologies, and vocabularies can improve the effectiveness of entity search and how it should be best utilized in the entity search process has remained fairly unstudied. This advocates the importance of information retrieval methods that can shed light on the value of different types of approaches making use of such knowledge.

3 Using Semantic Annotations for Entity Search

The use of semantic markup for entity search has an intuitive appeal: the structured entity descriptions contain domain-specific knowledge about the entities and their relations to background knowledge, which can be used to enhance information retrieval. The hypothesis is that increased knowledge about the entities, either for indexing or query representation, can improve the effectiveness of entity search even when queries are vague, underspecified, or expressed using a different vocabulary than the entity descriptions.

We focus on Web data for which the RDF model has been proposed as a W3C standard for data representation and interchange. As in many previous works [18,4,14,10], we omit RDF specific features, such as blank nodes, and employ a general graph-structured data model. This allows us to make more general contribution while still connecting our work to existing research lines. Formally, the graph-structured data model is an RDF graph represented as a directed and labeled graph $G = (N, P)$. The set of nodes N is a disjoint union of resources and literals and the set of P edges is a set of RDF properties. As we are not focusing on retrieving RDF resources (i.e. any node in the graph), but meaningful entity instances described using the graph, we consider a separate set of nodes E that are resource identifiers for entities. In essence, each entity in $e \in E$ can be seen to be annotated with G. Note, that each entity itself is a part of the complete graph. Intuitively, this means that the entities are present in the same graph as any other node of the knowledge graph that is used in the annotation, but we focus the retrieval to a specific set of entities E. We investigate several document and query expansion strategies which are summarized in the following subsections.

3.1 Direct Triples (DT)

Direct triples is a baseline indexing strategy in which the nodes in G are treated as literals, i.e. indexed only as terms without any reasoning based on the knowledge graph. This approach is also used in the SemSearch benchmark campaigns [4,14]. In essence, we start from the entity being indexed $e \in G$ and follow the edges (i.e. the properties) from e to the distance of one, i.e. the index for an entity e is the union of all property triples and nodes directly reachable from the entity e. In case the resulting node of the graph is a literal, we use the Porter stemmed value directly, and in the case of resources we use the Porter stemmed label of the resource.

3.2 Subsumption Reasoning (SR)

As our target is to also study the effect of a knowledge graph, we consider subsumption properties as a special case and use them to infer additional information to the graph-structured data model, as specified in RDF(S) semantics. The knowledge graph is used to compute the transitive closures of the RDF(S) subsumption relations, and the resulting triples are added to the RDF graph G. The indexing is then performed similarly as in the direct triples strategy. Now, the union of all property triples and nodes that are directly reachable from the entity e consists of present or inferred properties and nodes.

3.3 Subsumption Reasoning and Query Expansion (SR-QE)

Query expansion allows also resources that do not belong to the transitive closure of subsumption relations to be used for expansion at query time. The query is augmented with more general resources (and more specific resources via the subsumption reasoning) by using the subsumption hierarchies present in the graph G. We dynamically adjust the expansion with respect to the position of a resource in the subsumption hierarchy by using the Wu-Palmer measure [26,23]. The deeper in the subsumption hierarchy the resource is, the more expansion is allowed. We set a cut-off value of 0.8 for the Wu-Palmer measure.

3.4 Random Walks (RW)

Random walks is another approach to expand the index or query with resources from the neighborhood of the query or entity description by performing a random walk in the graph G. We use a modified version of the personalized PageRank [11] method. This allows to perform restarts of the random walk from the preference nodes instead of random teleportation. The RDF graph G is considered to be undirected (i.e. although RDF graphs are directed, we allow undirected walks to expand the query and indexing).

The entity $e \in G$ is modeled as a preference node q in the graph. Computing the resulting weight vector v for a given preference node can then be formalized as follows. Let an entity e be a resource that is described with a set of other resources in the graph G, such that $e$is connected to the resources in one of the triples that form the graph. A resource identifier is denoted as r, and $I(r)$ and $O(r)$ denote the set of in-neighbors and out-neighbors of r in G, respectively. Let A be the matrix corresponding to the RDF graph G describing what resources are connected to each other, where $A_{ij} = \frac{1}{|O_{ij} \cup I_{ij}|}$ if resource i links to resource j or vice versa, and $A_{ij} = 0$ otherwise. For a given q, the random walk equation can be written as

$$v = (1 - c)Av + cq,$$

where $c = 0.85$. The solution v is a steady-state distribution of random surfers, where a surfer teleports at each step to a resource r with probability $c \cdot q(r)$. In this way, the personalized PageRank allows restart behavior instead of random teleportation. We compute the steady distribution by using the power iteration method with 100 iterations.

The solution v is the personalized PageRank vector for the preference node q, i.e. in our case it is computed directly for the entity. The weights of the v can now be directly used as a vector that expands the original representation of e. The random walks were applied at indexing time and no expansion was performed at query time as the indexing time expansion is computed over the entire RDF graph.

3.5 Random Walks Directed by Subsumption (RW-S)

Random walks directed by subsumption use the same random walk approach, but restrict the walk to subsumption hierarchies. However, this approach still allows walks to the upper nodes of the source node (i.e. we consider the subsumption graph as undirected). In essence, the scoring can be formalized as follows. Let $A_{ij} = \frac{1}{|O_{ij} \cup I_{ij}|}$ if resource i links to resource j via a subsumption relation, and $A_{ij} = 0$ if i links to j with any other relation. The other parameters and indexing strategy are the same as in the basic random walk approach.

3.6 Ranking Model and Indexing

We use the vector space model with cosine similarity to index and rank entity descriptions. The data are indexed using field-based indexing. Each field is based on property-based splitting of the data in which each triple is indexed in a separate vector space based on the property in the triple [19,4]. Index construction for an exemplar triple with one inference step (traversal of one step in the subsumption hierarchy) is shown in Table 1. The original triple is first expanded using the chosen strategy to a set of inferred triples. The property and the object of the triple are both expanded, but the subject of the triple is not as it is the identifier of the entity. In the example, the property `dc:subject` is expanded to `rdf:Property` as this is the super property of `dc:subject`. The concept `aat:astronomy` is expanded to its super concepts `aat:physical_sciences` along with the Porter stemmed literal values. The terms of the vector spaces are the weights w assigned using the $tf - idf$ weighting that are computed separately for each vector space. Intuitively, the same resource can have different weights depending on the property context. Resources that have more references in a specific property context will have lower weights due to the weighting.

4 Experiments

In order to measure the effectiveness of using the knowledge graph in entity search, we conducted an experiment with two retrieval tasks: an ad-hoc task and a search-by-entity task. This section describes the data, the relevance assessments obtained to produce a ground truth for the data, the evaluation measures, and the analysis methods.

Table 1. An exemplar index and query representation for a triple describing an entity

Original triple	**Vector space**	**Vector space dimensions**
<imss:402015, dc:subject, aat:astronomy>	dc:subject	aat:astronomy
	dc:subject	"astronomi"
Inferred triples	**Vector space**	**Vector space dimensions**
<imss:402015, rdf:Property, aat:astronomy>	rdf:Property	aat:astronomy
	rdf:Property	"astronomi"
<imss:402015, dc:subject, aat:physical_sciences>	dc:subject	aat:physical_sciences
	dc:subject	"physic"
	dc:subject	"scienc"
<imss:402015, rdf:Property, aat:physical_sciences>	rdf:Property	aat:physical_sciences
	rdf:Property	"physic"
	rdf:Property	"scienc"

4.1 Entity Data, Queries and Tasks

We used a dataset comprised of four individual datasets. Two of the datasets describe museum collections and two point-of-interests for two cities. The museum datasets describe museum items, including artwork, fine arts, and scientific instruments. The points-of-interest datasets describe locations to visit, as well as statues, sights, and museums. The data were obtained from the Museo Galileo (museum dataset and point-of-interest dataset) in Florence, Italy, the Fine Arts Museum of Malta (museum dataset), and the Heritage Malta (point-of-interest dataset) as a part of the SmartMuseum project [20]. The datasets were selected because they represent a single domain, but originate from different data providers, thus ensuring heterogeneity in vocabulary, and the level of annotation describing the entities.

Altogether, the dataset consists of one thousand entities and 13,761 raw triples indexing these entities. The triples refer to vocabularies that contain over two million concepts (and after the inference step lead to a dataset of several million triples). The entities are described using Dublin Core properties with extensions for the cultural heritage domain, such as material, object type, and place of creation for the item described. An example annotation of a document describing a scientific instrument from the Museo Galileo is presented in Figure 1. Although the dataset is compact in size, it contains all of the entities exhibited in the museums and a representative sample of the points-of-interests present in the collections at the time the data were obtained. This makes the dataset realistic and allows the experiments to provide insights into real retrieval scenarios of domain-specific data, which is fairly typical in the Web of Data cloud.

4.2 Knowledge Graph

The data are indexed with a knowledge graph constructed by combining the annotations with the following vocabularies and ontologies [2]:

[2] http://www.getty.edu/research/conducting_research/vocabularies/

```
<dc:identifier> <urn:imss:instrument:402015> .
<physicalLocation> <http://www.imss.fi.it/> .
<dc:title> "Horizontal dial" .
<dc:subject> "Measuring time" .
<dc:description> "Sundial, complete with gnomon..." .
<dc:subject> <aat:300054534> . #Astronomy
<dateOfCreation> <time_1501_1600> .  #16th Century
<material> <aat:300010946> . #Gilt Brass
<objectType> <aat:300041614> . #Sundial
<placeOfCreation>  <tgn:7000084> #Germany
<processesAndTechniques> <aat:300053789> . #Gilding
<dc:terms/isPartOf> "Medici collections" .
<rdf:type> <Instrument> .
```

Fig. 1. A partial example of an entity description from the Museo Galileo in Turtle syntax. The dc:description has been shortened due to the shortage of space. The subject of the triples is the same in each triple (URI for the entity) and thus omitted.

The Art and Architecture Thesaurus (AAT) is a structured vocabulary of around 34,000 concepts, including 131,000 terms, descriptions, and other information relating to fine art, architecture, decorative arts, archival materials, and material culture.

The Getty Thesaurus of Geographic Names (TGN) is a structured vocabulary containing around 912,000 records, including 1.1 million names, place types, coordinates, and descriptive notes, which focus on important places for the study of art and architecture.

The Union List of Artist Names (ULAN) is a structured vocabulary containing around 120,000 records, including 293,000 names and biographical and bibliographic information about artists and architects, including a wealth of variant names and pseudonyms.

These vocabularies were transformed into RDF(S) from the original vocabularies. The taxonomies were transformed to class and property subsumption hierarchies both for concepts and relations respectively [25]. Different types of related term relations were transformed to custom properties using the RDF(S) definitions. Geographical instances that are structured in meronymical hierarchies that represent geographical inclusion were transformed to subsumption hierarchies. Temporal data were described using a separate structured format that has concepts for each year, decade, century, and millennium organized in a subsumption hierarchy.

4.3 Ground Truth, Tasks and Queries

We used two sets of inputs for entity search: expert written queries for the ad-hoc retrieval task and a subset of selected entities for the search-by-entity task. In total, the experts created 40 queries and selected 40 entities as sources for the search-by-entity task.

The ground truth was constructed by the same domain experts who created the queries. Two domain experts from each data provider assessed each query against all entities, i.e. a full recall assessment was conducted and pooling or top-k assessments were not used. The experts were instructed to have in mind a search scenario, in which a tourist would be searching entities (ad-hoc task) or in which entities would be recommended for the tourist based on a single entity that the tourist had selected (search-by-entity task). The assessments were conducted one query at a time and a binary assessment (relevant or not relevant) was provided for each entity against each query. In the search-by-entity task, the experts assessed whether each other entity was relevant given the source entity.

Notably, our assessment procedure is very precise as all queries were assessed against all other entities in the collection. The approach is different from the existing benchmark datasets, such as the Semantic Search Workshop data [4,9], in which the relevance assessments were determined for top-ranked entities by pooling and assessed by Mechanical Turk workers, and where the queries were very short and sampled from search engine logs. Our approach ensures that the ground truth can also be used to evaluate matches that are non-trivial and require deep understanding of the domain. This is important because semantic search techniques usually have the advantage of improving the quality of the results due to the explicit use of non-trivial and domain-specific deep semantic connections as opposed to the simple textual query matching used in conventional search. In our case, the queries do not necessarily have trivial connections to the entities that are expected as answers. For example, in our ground truth, a query including the entity "seascapes" is not only expected to return entities that are typed as "seascapes", but also entities typed with related concepts, such as "landscapes" and entities that have related subject-matters, such as "harbors" or "docks", and even person entities that have illustrated such subjects.

4.4 Evaluation Measures

The effectiveness of the retrieval methods were measured using Mean Average Precision (MAP) and precision at recall points of 10 (P@10), 15 (P@15), and 20 (P@20) entities. In addition, we plotted precision-recall curves to gain an understanding of the overall performance differences between the methods. The statistical significance of the differences in the results obtained using different combinations of methods were ensured using the the Friedman test, which is a non-parametric test based on ranks and is suitable for comparing more than two related samples. The statistical significance between method pairs was then analyzed using a paired Wilcoxon Signed-Rank test with Bonferroni correction as a post-hoc test. The differences between the method variants were found to be statistically significant ($p<0.001$) unless reported otherwise.

5 Results

The results of the experiments are given in Table 2. The results show that approaches with extensive query or document expansion beyond the standard subsumption reasoning achieve significantly higher MAP and precision for top-ranked documents. This result holds for both the ad-hoc and search-by-entity tasks. In fact, subsumption reasoning seems to hurt the precision among the top-ranked documents and shows decreased performance, even in MAP. This indicates that, while knowledge graphs are highly useful

Table 2. Results of the retrieval experiments. The highest values are bold (several in case no statistically significant differences between the methods could be found). DT = Direct triples, SR = Subsumption reasoning, SR-QE = Subsumption reasoning and using the best performing query expansion, RW-S = Random walks directed by subsumption, and RW = Random walks.

Measure	Ad-hoc task				
	DT	SR	SR-QE	RW-S	RW
MAP	0.22	0.21 (-4.5%)	**0.39 (+77.3%)**	0.29 (+31.8%)	**0.37 (+68.2%)**
P@10	**0.61**	0.43 (-29.5%)	0.53 (-13.1%)	0.53 (-13.1%)	0.53 (-13.1%)
P@15	0.39	0.4 (+2.6%)	**0.46** (+18.0%)	0.34 (-12.8%)	**0.43 (+10.3%)**
P@20	0.21	0.19 (-9.5%)	**0.44 (+119.0%)**	0.29 (+62.0%)	**0.41 (+104.8%)**
	Search-by-entity task				
MAP	0.63	0.52 (-17.5%)	0.58 (-8.0%)	0.60 (-5.0%)	**0.68 (+8.0%)**
P@10	0.66	0.81 (+23.0%)	**0.85 (+29.0%)**	0.70 (+6.0%)	0.78 (+18.0%)
P@15	0.63	0.69 (+9.5%)	**0.77 (+22.2%)**	0.58 (-8.0%)	0.69 (+10.0%)
P@20	0.59	0.62 (+5.1%)	**0.64 (+8.5%)**	0.55 (-6.8%)	**0.65 (+10.2%)**

sources for background knowledge and seem to significantly increase the effectiveness of entity search, the subsumption reasoning alone may not be useful. A possible explanation is that the retrieval method has to employ deep semantic connections to achieve improvements over indexing with direct triples.

5.1 Ad-Hoc Task

In the ad-hoc task (Table 2), the best MAP was achieved using subsumption reasoning with query expansion (0.39) and the random walk approach (0.37). No statistically significant difference between the two best performing approaches could be shown. Interestingly, indexing directly with triples (i.e. with no usage of the knowledge graph) shows the best performance for precision at 10. This is in line with the assumption that, in the ad-hoc task, background knowledge is mainly useful for achieving better recall (and therefore show better performance at higher recall levels) than simple indexing.

Detailed analyses show that precision at 20 entities already shows over 100% improvement for query expansion and random walk approaches, and even the random walk approach that only makes use of the subsumption hierarchies achieves 62% improvement, compared to simple indexing with triples. Using only subsumption reasoning hurts the precision by 29.5% at 10 entities and by 9.5% at 20 entities compared to indexing with triples, and shows the worst overall performance.

The performance of the retrieval strategies in the ad-hoc task is also illustrated in Figure 2a, which shows the precision-recall curve for the compared strategies. The SR and RW-S are omitted because they constantly perform worse than SR-QE and RW and are derivatives of these strategies. The precision-recall curve illustrates that the RW-S strategy is only better among very few top documents and SR-QE and RW strategy outperform other strategies at a recall level of 0.1. Compared to indexing with triples, the SR-QE and RW are significantly better throughout the precision recall curve, particularly at recall levels 0.1 and 0.2. However, the RW and SR-QE strategies, which both make use of the knowledge graph beyond standard reasoning, clearly outperform DT, SR and RW-S (Table 2). No statistically significant difference could be found between the the top two methods. A possible explanation is that in the ad-hoc task the queries

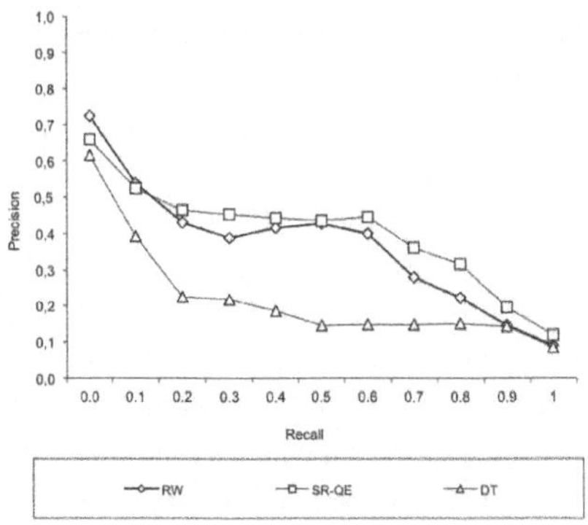

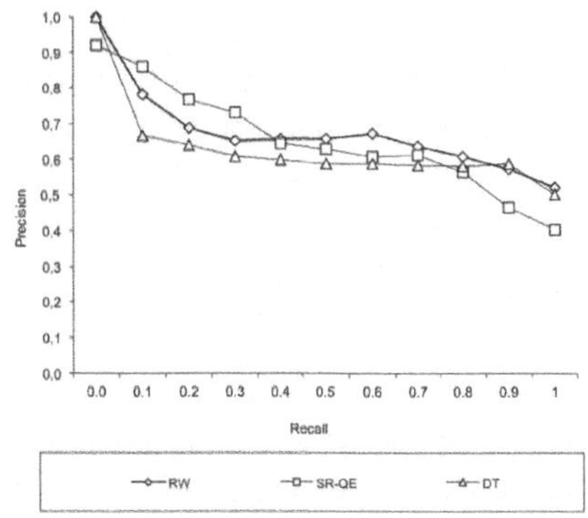

Fig. 2. Precision plotted on 11 recall levels for different reasoning and indexing strategies for the ad-hoc entity search task (a) and for the search-by-entity task (b). DT = Direct triples, SR-QE = Subsumption reasoning and using the best performing query expansion, and RW = Random walks.

contain very little evidence about the user's search intent and the use of background knowledge leads to larger gains in retrieval effectiveness.

5.2 Search-by-Entity Task

The results for the search-by-entity task are shown in Table 2. A similar trend as in the ad-hoc task can be observed also in the search-by-entity task. The best MAP is achieved by using RW (0.68) and can be mainly attributed to improved recall. Both random walks and reasoning combined with query expansion are found to be more effective than SR in the search-by-entity task. Surprisingly, indexing with direct triples has relatively good performance—in particular for MAP (0.52) and is competitive against indexing with SR, which achieves a MAP of 0.52. No statistically significant difference between the two best performing approaches could be shown. The corresponding precision-recall curve for the search-by-entity task is shown in Figure 2b. The SR and RW-S are also omitted in this curve as they constantly perform worse than SR-QE and RW. The precision-recall curve shows that all methods behave similarly. SR-QE shows slightly better performance at high-recall levels than RW. A possible explanation is that random walks expand the queries and entity descriptions more and favor recall instead of precision at low-recall levels. In general, the results show that background knowledge has less effect on the search-by-entity task. The highest improvement was 29% for precision at 10 compared to the over 100% improvement in precision at 20 that was achieved in the ad-hoc task.

6 Discussion and Conclusions

In this paper, we report experiments that evaluate the usefulness of background knowledge in the form of a knowledge graph for domain-specific entity retrieval. We compared the effectiveness of reasoning and query expansion, both using RDF(S) semantics and random walks -based approaches and studied two commonly occurring semantic search tasks: ad-hoc entity search and search-by entity.

In our experiments, knowledge graph approaches were found to be more effective for entity search and they consistently outperformed trivial structured indexing with triples. The best improvements—over 100% in precision at 20 and over 77% in MAP—were found using approaches that combine reasoning with query expansion or utilize random walks. The only exception was that standard RDF(S) reasoning was found to have no or very little effect for retrieval performance, even when compared to simple indexing with triples. The effectiveness was only improved when RDF(S) reasoning was combined with query expansion, or when the random walks approach was used to determine the local conceptual neighborhood of the entity. The effects were particularly prominent in the ad-hoc search task, in which the conceptual representation of queries and indexing vocabulary can vary more than in the search-by-entity task. The results suggest that background knowledge can be effective for entity search, but it's use should not be restricted to standard deductive reasoning. Our results also suggest that, in addition to the current success of entity retrieval research and existing evaluation campaigns, the evaluation of entity search should consider the nature of the domain-specificity of many of the data collections currently available in the Linked Open Data Cloud and recognize the role of methods that go beyond the current, rather simple semantic matching and relevance assessment strategies. We conclude that the information encoded in knowledge graphs of the Web of Data should be more carefully exploited in semantic search systems to reveal the true power of the structured Web of Data.

References

1. Bhogal, J., Macfarlane, A., Smith, P.: A review of ontology based query expansion. Inf. Proc. & Man. 43(4), 866–886 (2007)
2. Bizer, C., Heath, T., Berners-Lee, T.: Linked Data - the story so far. Int. J. Semantic Web Inf. Syst. 5(3), 1–22 (2009)
3. Blanco, R., Halpin, H., Herzig, D.M., Mika, P., Pound, J., Thompson, H.S., Tran, T.: Repeatable and reliable semantic search evaluation. Web Semantics: Science, Services and Agents on the World Wide Web 21 (2013)
4. Blanco, R., Mika, P., Vigna, S.: Effective and efficient entity search in RDF data. In: Aroyo, L., Welty, C., Alani, H., Taylor, J., Bernstein, A., Kagal, L., Noy, N., Blomqvist, E. (eds.) ISWC 2011, Part I. LNCS, vol. 7031, pp. 83–97. Springer, Heidelberg (2011)
5. Castells, P., Fernandez, M., Vallet, D.: An adaptation of the vector-space model for ontology-based information retrieval. IEEE TKDE 19(2), 261–272 (2007)
6. Di Noia, T., Mirizzi, R., Ostuni, V.C., Romito, D., Zanker, M.: Linked open data to support content-based recommender systems. In: I-SEMANTICS 2012, pp. 1–8. ACM, New York (2012)
7. Frnandez, M., Cantador, I., Lpez, V., Vallet, D., Castells, P., Motta, E.: Semantically enhanced information retrieval: An ontology-based approach. Web Semantics: Science, Services and Agents on the World Wide Web 9(4), 434–452 (2011)
8. Guéret, C., Groth, P., van Harmelen, F., Schlobach, S.: Finding the achilles heel of the web of data: Using network analysis for link-recommendation. In: Patel-Schneider, P.F., Pan, Y., Hitzler, P., Mika, P., Zhang, L., Pan, J.Z., Horrocks, I., Glimm, B. (eds.) ISWC 2010, Part I. LNCS, vol. 6496, pp. 289–304. Springer, Heidelberg (2010)

9. Halpin, H., Herzig, D., Mika, P., Blanco, R., Pound, J., Thompon, H., Duc, T.T.: Evaluating ad-hoc object retrieval. In: Proc. Works. Eval. of Sem.Tech., vol. 666, Shanghai, China, CEUR (November 2010)
10. Herzig, D.M., Mika, P., Blanco, R., Tran, T.: Federated entity search using on-the-fly consolidation. In: Alani, H., Kagal, L., Fokoue, A., Groth, P., Biemann, C., Parreira, J.X., Aroyo, L., Noy, N., Welty, C., Janowicz, K. (eds.) ISWC 2013, Part I. LNCS, vol. 8218, pp. 167–183. Springer, Heidelberg (2013)
11. Jeh, G., Widom, J.: Scaling personalized web search. In: Proc. WWW 2003, pp. 271–279. ACM, New York (2003)
12. Kiryakov, A., Popov, B., Ognyanoff, D., Manov, D., Kirilov, A., Goranov, M.: Semantic annotation, indexing, and retrieval. In: Fensel, D., Sycara, K., Mylopoulos, J. (eds.) ISWC 2003. LNCS, vol. 2870, pp. 484–499. Springer, Heidelberg (2003)
13. Lei, Y., Uren, V.S., Motta, E.: Semsearch: A search engine for the semantic web. In: Staab, S., Svátek, V. (eds.) EKAW 2006. LNCS (LNAI), vol. 4248, pp. 238–245. Springer, Heidelberg (2006)
14. Neumayer, R., Balog, K., Nørvåg, K.: On the modeling of entities for ad-hoc entity search in the web of data. In: Baeza-Yates, R., de Vries, A.P., Zaragoza, H., Cambazoglu, B.B., Murdock, V., Lempel, R., Silvestri, F. (eds.) ECIR 2012. LNCS, vol. 7224, pp. 133–145. Springer, Heidelberg (2012)
15. Ning, X., Jin, H., Jia, W., Yuan, P.: Practical and effective ir-style keyword search over semantic web. Inf. Proc. & Man. 45(2), 263–271 (2009)
16. Ning, X., Jin, H., Rw, H.: Rss: A framework enabling ranked search on the semantic web. Inf. Proc. & Man. 44(2), 893–909 (2008); Evaluating Exploratory Search Systems; Digital Libraries in the Context of Users' Broader Activities
17. Oren, E., Delbru, R., Catasta, M., Cyganiak, R., Stenzhorn, H., Tummarello, G.: Sindice.com; a document oriented lookup index for open linked data. Int. J. Metadata Semant. Ontologies 3(1), 37–52 (2008)
18. Pound, J., Mika, P., Zaragoza, H.: Ad-hoc object retrieval in the web of data. In: Proc. WWW 2010, pp. 771–780. ACM, New York (2010)
19. Ruotsalo, T.: Domain specific data retrieval on the Semantic Web. In: Simperl, E., Cimiano, P., Polleres, A., Corcho, O., Presutti, V. (eds.) ESWC 2012. LNCS, vol. 7295, pp. 422–436. Springer, Heidelberg (2012)
20. Ruotsalo, T., Haav, K., Stoyanov, A., Roche, S., Fani, E., Deliai, R., Mäkelä, E., Kauppinen, T., Hyvönen, E.: SMARTMUSEUM: A mobile recommender system for the web of data. Web Semantics: Science, Services and Agents on the World Wide Web 20, 50–67 (2013)
21. Ruotsalo, T., Hyvönen, E.: A method for determining ontology-based semantic relevance. In: Wagner, R., Revell, N., Pernul, G. (eds.) DEXA 2007. LNCS, vol. 4653, pp. 680–688. Springer, Heidelberg (2007)
22. Ruotsalo, T., Jacucci, G., Myllymäki, P., Kaski, S.: Interactive intent modeling: Information discovery beyond search. Commun. ACM 58(1) (January 2015)
23. Ruotsalo, T., Mäkelä, E.: A comparison of corpus-based and structural methods on approximation of semantic relatedness in ontologies. Int. J. Sem. Web and Inf. Syst. 5(4), 39–56 (2009)
24. Vallet, D., Fernández, M., Castells, P.: An ontology-based information retrieval model. In: Gómez-Pérez, A., Euzenat, J. (eds.) ESWC 2005. LNCS, vol. 3532, pp. 455–470. Springer, Heidelberg (2005)
25. van Assem, M.: Converting and Integrating Vocabularies for the Semantic Web. PhD thesis, VU University Amsterdam (2010)
26. Wu, Z., Palmer, M.: Verbs semantics and lexical selection. In: Proc. ACL 1994, pp. 133–138. ACL (1994)

Reachability Analysis of Graph Modelled Collections

Serwah Sabetghadam, Mihai Lupu, Ralf Bierig, and Andreas Rauber

Institute of Software Technology and Interactive Systems
Vienna University of Technology
Vienna, Austria
{sabetghadam,lupu,bierig,rauber}@ifs.tuwien.ac.at

Abstract. This paper is concerned with potential recall in multimodal information retrieval in graph-based models. We provide a framework to leverage individuality and combination of features of different modalities through our formulation of faceted search. We employ a potential recall analysis on a test collection to gain insight on the corpus and further highlight the role of multiple facets, relations between the objects, and semantic links in recall improvement. We conduct the experiments on a multimodal dataset containing approximately 400,000 documents and images. We demonstrate that leveraging multiple facets increases most notably the recall for very hard topics by up to 316%.

1 Introduction

There is rapid growth of online multimodal content as well as personal data generation in our daily life. This trend creates severe challenges in multimodal information retrieval. Multimodal retrieval is defined as searching for the relevant modality with textual queries (keywords, phrases, or sentences) and/or image examples, music files or video clips. Many approaches have been tested in recent years, ranging from associating image with text search scores to sophisticated fusion of multiple modalities [7,5,12].

In addition to the observation that data consumption today is highly multimodal, it is also clear that data is now heavily semantically interlinked. This can be through social networks (text, images, videos of users on LinkedIn, Facebook or the like), or through the nature of the data itself (e.g. patent documents connected by their metadata - inventors, companies). Structured data is naturally represented by a graph, where nodes denote entities and directed/indirected edges represent the relations between them. Such graphs are heterogeneous, describing different types of objects and links. Connected data poses structured IR as an option for retrieving more relevant data objects.

Previous works [9,4,6] introduced models to leverage both structured and unstructured IR. There, a question arises: Is the graph model conducive to retrieval performance? In this work, we propose an analysis on reachability of relevant objects in a graph modelled data. In our previous works [15,16], we introduced a model that enriches the available data by extracting inherent information of

A. Hanbury et al. (Eds.): ECIR 2015, LNCS 9022, pp. 370–381, 2015.

objects in the form of facets. This has support in the principles of Information Retrieval, most notably in the theory of poly-representation [10]. The aim is to leverage cognitive and functional representations of information objects to improve IR results, but there is currently no understanding of how using different representations of the same objects (what we call here facets) affects the reachability of relevant items.

We showed previously that our model matches the efficiency of non-graph based indexes, while having the potential to exploit different facets for better retrieval [16]. In this work, we illustrate the effect of multiple facets on reachability of relevant nodes in a collection. Further, we enrich the relations in the collection by adding corresponding semantic links from DBpedia. We demonstrate how it helps improving recall for hard and very hard topics. We provide extensive experimental evidence for our conclusions, based on the ImageCLEF 2011 Wikipedia dataset [19].

The paper is structured as follows: in the next section, we address the related work, followed in Section 3 by the basic definition of our model, graph traversal and weighting. The experiment design is shown in Section 4. Results are discussed in Section 5, and finally, conclusions and future work are presented in Section 6.

2 Related Work

2.1 Content-Based Retrieval

There are many efforts in multimodal retrieval, e.g. by mining the visual information of images to improve text-based search. Martinent et al. [12] propose to generate automatic document annotations from inter-modal analysis. They consider visual feature vectors and annotation keywords as binary random variables. In combination of text and images, given the massive web data, relevant web images can be readily obtained by using keyword based search [7,5].

I-Search, as a multimodal search engine [11], defines relations between different modalities of an information object, e.g. a lion's image, its sound and its 3D representation. They define neighbourhood relation between two multimodal objects which are similar in at least one of their modalities. However, in I-Search, the semantic relation between objects (e.g. a dog and a cat object) is not considered. They do not consider explicit links between information objects. We take advantage of the context through links in the context graph whose nodes represent different modalities in the search set.

2.2 Graph-Based Retrieval

Srinivasan and Slaney [18] add content based information to image characteristics as visual information to improve their performance. Their model is based on random walks on bipartite graphs of joint model of images and textual content. Jing et al. [8] employ the PageRank to rerank image search. The hyperlinks between images are based on visual similarity of search results. Yao et al. [20] make a similarity graph of images and aim to find authority nodes as result for image queries. Through this model, both visual content and textual information of the images is explored. The structured search engine NAGA [9], provides the results of a structured (not keyword) query by using subgraph pattern on an Entity-Relationship graph. Rocha et al. [13] use spreading activation for relevance propagation applied to a semantic model of a given domain. select subgraphs to match the query and do the ranking by means of statistical language models. We build upon these works and complement them with the concept of faceted search.

In our model, in addition to similarity links between facets of the same type, we have other types of links like semantic or part-of, which enables the framework to model a collection with diverse relation types between information objects. Further, by extracting inherent information of objects in the form of facets, we provide a framework with higher flexibility to prioritize a specific feature. We will show that our model can effectively integrate multiple facets of different modalities to improve performance.

3 Model Representation

We define a model to represent information objects and their relationships, together with a general framework for computing similarity. We see the information objects as a graph $G = (V, E)$, in which V is the set of vertices (including data objects and their facets) and E is the set of edges. By facet we mean inherent information of an object, otherwise referred to as a representation of the object. For instance, an image object may have several facets (e.g. color histogram, texture representation). Each of these is a node linked to the original image object. Each object in this graph may have a number of facets. We define four types of relations between the objects in the graph. The relations and their characteristics and weightings are discussed in detail in [14]. We briefly repeat them here for completeness of the presentation:

- **Semantic** (α): any semantic relation between two objects in the collection (e.g. the link between lyrics and a music file). The edge weight w_{uv} is inversely proportional the number of outgoing α links from u.
- **Part-of** (β): a specific type of semantic relation, indicating an object as part of another object, e.g. an image in a document. This is a containment relation, and therefore has default weight to 1.
- **Similarity** (γ): relation between objects with the same modality, e.g. between the same facets of two objects. The weight is the similarity value.
- **Facet** (δ): linking an object to its representation(s). It is a directed edge from facet to the object.
  Weights are given by perceived information content of features, with respect to the query type. For instance, with a query like "blue flowers", the color histogram is a determining facet that should be weighted higher. These weights should be learned for a specific domain, and even for a specific query if we were to consider relevance feedback.

3.1 Traversal Method - Spreading Activation

There are different methods to traverse a graph of which random walks and spreading activation are two well-known methods. We proved that these two methods are principally the same [17]. However, spreading activation provides more options to customize the graph traversal. The SA procedure, always starts with an initial set of activated nodes, usually the result of a first stage processing of the query. During propagation, surrounding nodes are activated and ultimately, a set of nodes with respective activation are obtained. After t steps, we use the method provided by Berthold et al. [2], to compute the nodes' activation value: $a^{(t)} = a^{(0)} \cdot W^t$ where $a^{(0)}$ is the initial activation vector, W is the weight matrix—containing different edge type weights—, and $a^{(t)}$ is the final nodes' activation value used for ranking.

Memory Spreading Activation algorithm. In this variation of spreading activation, we propose an input function on received energy to manage the amount of energy spreading in the network. The amount of energy a node receives in each step t, is the sum of the energy of its neighbours. Part of this received energy has been sent two steps before from the same node to its neighbours. We subtract this part from the whole received energy to prevent energy bias near activated nodes. We define the *energy capacity* of nodes as vector sm, which contains the sum of the edge weights for each node. We define the energy capacity of node i as $sm_i = \sum_{j=1}^{n} W_{ij}$ where j goes over the columns for row i. This is the energy it is able to carry to its neighbours. It may be less or more than the energy it has at any point in time, as a function of the weights of its outgoing edges. We denote $M = diag(sm)$ which converts vector sm to the diagonal matrix with the vector values on the diagonal. Here, we define the energy received in each step of t as: $a^{(t)} = a^{(0)}.W^t - a^{(t-2)}.M$. In each step, we deduct the self-energy received by subtracting the multiplication of activation value of two steps before to the energy capacity of this node $(a^{(t-2)}.M)$. In the expanded form it is:

$$a^{(t)} = a^{(0)} \sum_{k=0}^{t-1} (-1)^k.W^{t-2k}.M^k \qquad (1)$$

3.2 Hybrid Search

We proposed to leverage the combination of faceted search with graph search to find relevant objects [15]. The use of results from independent modality indexing neglect a) that data objects are interlinked through different relations and b) that many relevant images can be retrieved from a given node by following semantic or 'part-of' relations. Our hybrid ranking method consists of two steps: 1) In the first step, we perform an initial search with Lucene and/or LIRE to obtain a set of activation nodes, which is based on specific facet indexed results. . 2) In the second step, using the initial result set of data objects (with normalized scores) as seeds, we exploit the graph structure and traverse it.

The number of transitions is determined by imposing different stop rules: distance constraint [3], fan-out constraint [3] or type constraint[13]. In this version of our model, we use the distance constraint to stop the traversal.

4 Experiment Design

4.1 Data Collection

We applied the ImageCLEF 2011 test collection as a benchmark. ImageCLEF 2011 is based on Wikipedia pages and their associated images. It is a multimodal collection (consisting of 125,828 documents and 237,434 images), and an appropriate choice for testing the rich and diverse set of relations in our model.

Each image in this collection has metadata providing name, location, one or more associated parent documents in up to three languages (English, German and French), and textual image annotations (i.e. caption, description and comment). We parsed the image metadata and created nodes for all parent documents, images and corresponding facets. We created different relation types: the β relation between parent documents and images (as part of the document), δ relation between information objects and their facets.

4.2 Adding Semantic links

We connect the ImageCLEF 2011 Wikipedia collection to DBPedia through the equivalent pages in DBpedia for each wiki page in the collection. The ImageClef2011 Wikipedia collection uses the ImageCLEF 2010 Wikipedia collection, which is based on the September 2009 Wikipedia dumps. Therefore we downloaded DBPedia version 3.4 which is based on Wiki dump September 2009.

Among all DBPedia RDF, we only consider those linking two existing documents in our collection. We add α relations between semantically related documents. The result is a more connected, large scale graph. This way, after visiting a document, we follow its neighbours that may be images or other documents connected through semantic links. For instance, document named *Battle of Leyte Gulf*, contained 6 images as neighbours. After adding semantic links, this document connects to 13 other documents in the collection (e.g. *Pacific War* and *World War II*).

In total 55,544 links are added, which is considerable with respect to the number of documents in the collection (125,828). These links are valuable in the sense that they provide a more connected graph of objects.

4.3 Standard Text and Image Search

In the indexed search approach, as first phase of our hybrid search, we use standard indexing results both for documents and images. The computed scores in both modalities are normalized per topic between (0,1). Different indexings based on different facets are:

- **Text tf.idf facet:** We utilize default Lucene indexer, based on tf.idf, as text facet. We refer the result set of this facet as **R1**.
- **CEDD facet:** For image facets, we selected the Color and Edge Directivity Descriptor (CEDD) feature since it is considered the best method to extract purely visual results [1]. We refer to the image results of this facet as **R2**.
- **Image textual annotation tf.idf facet (Tags):** We use metadata information of the images (provided by the collection), as image textual facets (Tags). Meta-data XML files of ImageCLEf 2011, includes textual information (caption, comment and description) of images. Using Lucene we can index them as separate fields, and search based on a multi-field indexing. Tags search result make **R3** result set.

In the second phase, starting from standard indexed results, we conduct the graph search based on spreading activation to the number of t steps.

4.4 Evaluation Method

The aim of these experiments is to obtain an understanding of the collection of how the relevant images are distributed in the graph. We conduct experiments starting from different indexed facets (Text, CEDD, and Tags).

Through these investigations, we want to see how far and up to how much recall we are able to reach in the graph. There are 50 topics in ImageCLEF 2011 Wikipedia collection. We conduct the traversal up to 50 steps for each of these topics. In each step, we check if we visit new related images for that specific topic. Different topics show different recall behaviour as we go further in the graph. In order to interpret these behaviours, we partitioned the results based on the topic categorization done by Tsikrika et .al [19]. They divide the topics to four categories of easy (17 topics), medium (10 topics), hard (16 topics) and very hard (7 topics). They show 10 topics in easy and hard categories in their report which we use in this work.

5 Results and Discussion

In the first part of the experiments we provide an exploratory data analysis over the collection. In the second part we show the effectiveness of our graph model, leveraging different facets on the collection. In the last part, we perform the same experiments on the semantic enhanced collection.

5.1 Relevant Objects Distribution

Figure 1 shows the distribution of relevant nodes in the collection as we start from all three facets (R1,R2 and R3). The x axis is the number of steps we traverse the graph, and y axis is the Id of the query topics we have. In each step we count the number of new related images we visit. Existence of a shape (circle/square/star/triangle) indicates visiting at least a true positive. The size of a shape is the ratio of number-of-related-seen-nodes-in-this-step/number-of-total-related-ones for the specific query topic Id.

We observe the large number of large shapes, in the first steps. It indicates visiting more related images initiating from different facet results.

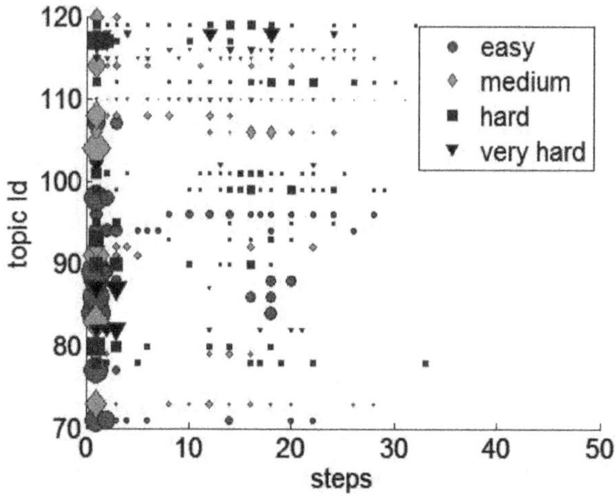

Fig. 1. Relevant node distribution for different categories of topics: easy, medium, hard and very hard

Distribution Per Topic Categories. The distribution of relevant objects for different categories of topics is shown by different shapes in Figure 1. We observe that easy topics points (circles) are mostly at the very beginning steps. For hard and very hard topics (squares and triangles) there are more distributed related nodes as we continue the traversal. They show almost constant increase as we traverse the graph. This observation demonstrates that the distribution of related results for hard and very hard topics is in about 30 steps from the beginning.

5.2 Potential Recall

Here we observe the behaviour of potential recall leveraging different facets.

Different Facet Combinations. We performed the experiment for different combination of facets: R1, R1-R2, R1-R3, and R1-R2-R3 (Figure 2). We observe the changes in the recall values using each combination. The diagram demonstrates that when adding more text (R1-R3) or more image features (R1-R2) we are visiting different objects. In fact R1-R3 results are near to those of R1, while R1-R2 obtains higher recall values, closer to those obtained when using all features (R1-R2-R3). This highlights the importance of different, diverse representations to the data in order to cover all aspects of the relevant objects. The

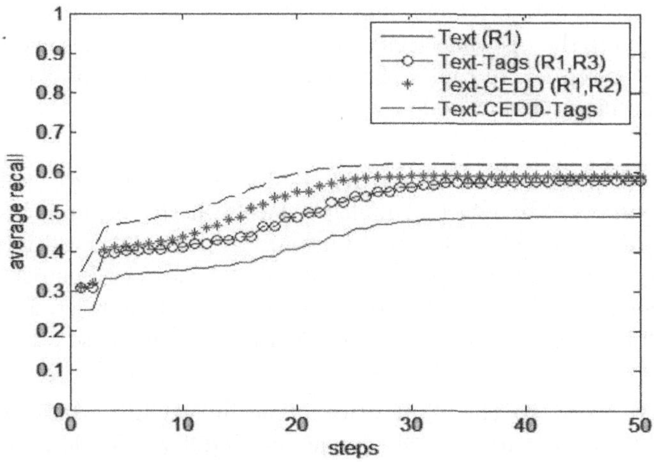

Fig. 2. Average recall for different facets

addition of more textual features, as represented by the meta-data fields (Tags), has produced a lower increase in recall than the addition of CEDD.

We investigate the effect of R1 and R1-R2-R3 facets individually on recall for different categories of topics in next experiments.

Text Facet. In this experiment, we include only R1 results to start search in the graph. Figure 3a shows the average recall for different categories. We observe that easy topics meet 0.66 recall after 27 steps and keeps this value to the 50th step. For medium topics is the same after 25th step with maximum value of 0.51. Hard and very hard topics continue increasing the recall value until 30th step and up to the values of 0.37 and 0.43 respectively. An interesting observation is the behaviour of very hard topics after 3rd step which outpaces hard topics. This demonstrates that as we go farther in the graph we cover higher percentage of recall for very hard topics rather than hard topics. Although we used only Text facet, with the graph modelled collection, we can reach these recall values.

Another observation is the increase rate of average recall in each category. Easy topics show the increase rate of 37.5% (from 0.48 to 0.66), where it is 18.6% for medium topics (from 0.43 to 0.51), 131% for hard topics (from 0.16 to 0.37) and 258% for very hard topics (from 0.12 to 0.43). The values show that hard and very hard topics benefit more than easy topics from the graph structure. While easy and medium topics are apparently answerable by direct query, it is in the hard and very hard topics that the graph model shows most promise.

Further, we observe that recall is increasing up to 30th step and then goes to a plateau for all categories. Two results are obtained from this observation: first is that by conducting the traversal, we can expect increase in recall in the graph to about 30 steps. Because we are still visiting related nodes as we go farther every one or two steps. Second is that after the 30th step we are not visiting relevant images any more, and recall is still less than 0.7 even for easy topics. This shows the disconnectivity of the graph. Our log files show no more node after 40th step for all topics. Therefore, the probability of continuing the traversal and seeing relevant node is zero.

All Facets. We use the R1, R2 and R3 results to start the propagation (Figure 3b). We observe the effect of multiple facets in the beginning steps (1st to 5th) with higher recall values. In addition, the potential recall level can be reached earlier with all facets. We have the same values between 5th and 15th steps here, compared to 15th to 25th step with only Text facet. Further, the average recall has increased to 0.5 for very hard topics (increase rate of 316%).

Still Limited View to the Collection. The ImageCLEF 2011 has 363,262 nodes. We counted the number of all seen nodes for different topics. We obtained the average of 93,232 nodes seen starting from all three facets. This illustrates our limited view to this particular collection, by traversing one fourth of the collection size. In addition, the convergence of traversal performance at about 25th-30th step for all topic categories (despite of their different magnitude) is another confirmation to this limited perspective. To tackle this challenge we added semantic links to the collection towards more connectivity.

5.3 Potential Recall - Semantically Enhanced Collection

We perform the same experiments for the collection enhanced with semantic links.

Text Facet. In this experiment, we conduct the test on the enhanced version of the collection including semantic links. It is apparent that we obtain a more connected graph and consequently expect higher recall. We show the reachability result starting from Text facet in Figure 3c. We observe that recall in all categories reaches a plateau in 11th step compared to the graph version without semantic links which was 30 steps. Further, the diagram shows that all categories have a shift in their final value of average recall in comparison to the collection without semantic links: easy topics from %66 to %88, medium from %51 to %75, hard from %37 to %64, and very hard from %43 to %73. In this experiment, hard and very hard topics with 300% and 508%, outpaced other categories.

All Facets. By starting from R1,R2 and R3 results, we reach approximately the same with R1 experiment for different categories (Figure 3d) after 11 steps. The reason is that we have a highly connected graph, of which where to start to search through does not differ after many steps. However, starting from different facets, affects in the initiating steps (1st to 5th step) leading to steeper slope at the beginning. It is considerable since in steps 6, 7 and 8, we are visiting about 30,000 new node in each step. Therefore, for few steps it is worth leveraging different facets, even in highly connected collection.

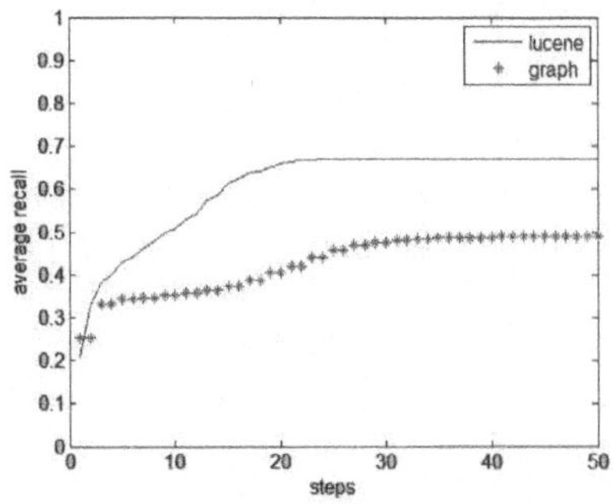

(a) Average recall using Text facet (R1)

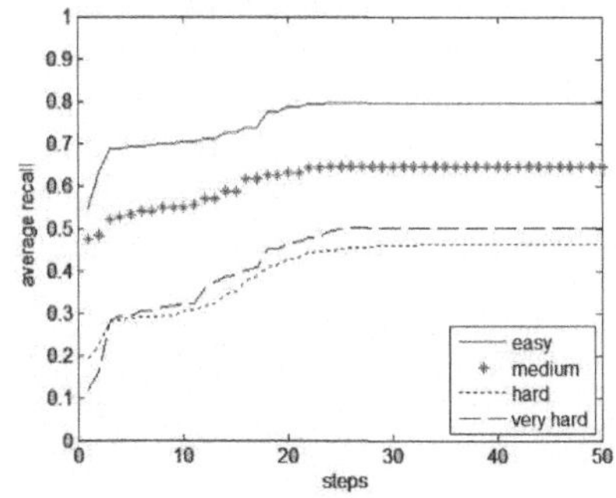

(b) Average recall using all facets (R1, R2, R3)

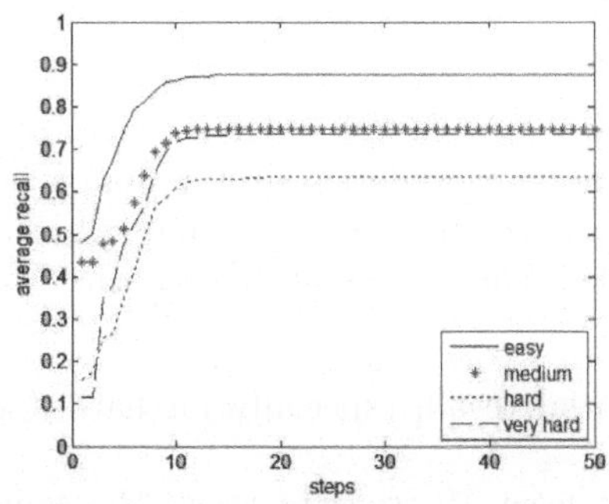

(c) Semantic links added, average recall using Text facet (R1)

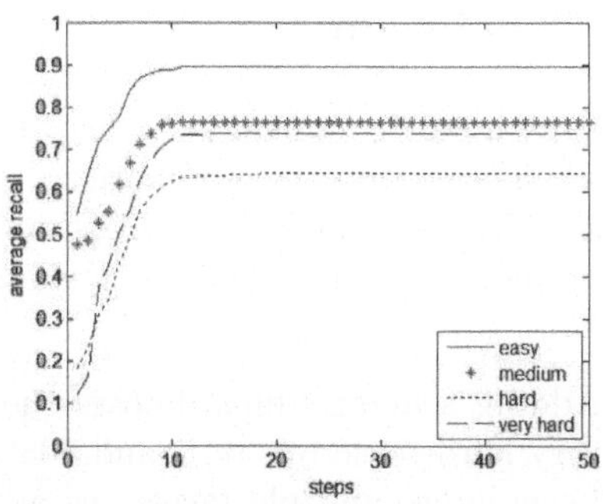

(d) Semantic links added, average recall using all facets (R1, R2, R3)

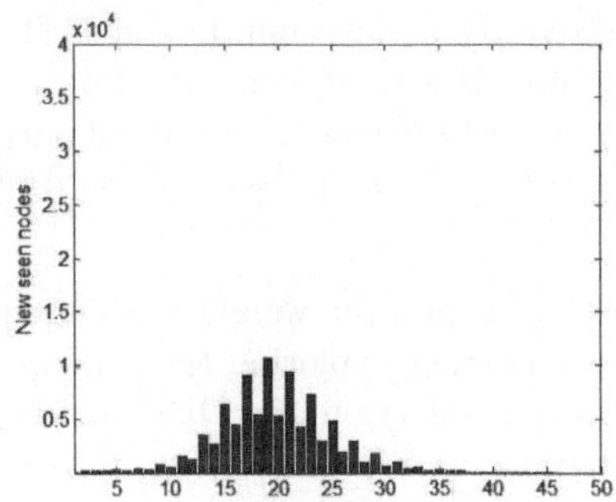

(e) Number of new seen nodes per step in the collection

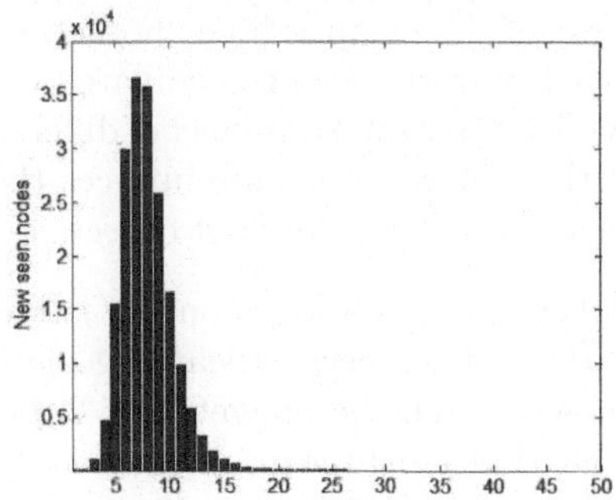

(f) Number of new seen nodes per step, semantic links added

Fig. 3. 3a, 3b, 3c, and 3d show recall under different conditions. 3e and 3f show the number of new nodes visited in each step of traversal.

Number of Nodes Seen in Steps. Figure 3e shows the average number of new seen nodes for all topics in each step. We observe that it starts to increase after 11th step up to 30th step to the total size of 93,330 nodes (about 25% of the collection size). The oscillation of the seen nodes in even steps is because of seeing documents in even steps and seeing images in odd steps. The number of images are more than twice of documents in the collection.

The same analysis on the collection containing semantic links demonstrates that the number of nodes are mainly increasing in the first steps up to 11th steps (Figure 3f), to the total size of 188,830 node (about 50% of the collection size). This observation indicates lower number of steps needed to traverse the reachable nodes with semantic links. Further, we touch half of the collection due to more connected collection, leading to visiting more relevant nodes. However, it challenges the precision. Since we visit new nodes in the scale of thousands including few related nodes about 0,001 of the nodes.

6 Conclusion

We presented experiments on the reachability of relevant objects in a graph modelled collection. We compared a graph model where data objects had a set of facets based on their inherent features with a graph model where data objects are additionally connected by semantic links. The results are summarized as below:

- Adding semantic links boosts the potential recall, especially for hard and very hard topic by 300% and 508%.
- Leveraging multiple facets, we saved at least 10 steps to reach the same potential recall compared to using only one facet. Further it increased recall for very hard topics by up to 258%.
- Leveraging semantic links, potential recall reached a plateau in 11 steps. This saved at least 19 steps compared to the traversal without semantic links.
- We demonstrated the effect of different facets leading to visiting different parts of the collection. This reinforces the importance of the poly-representation idea to touch the relevant objects.

Our future work will focus on the following: 1) Learning the weight of different facets through supervised learning methods. 2) Further exploring the semantic relations between the ImageCLEF 2011 Wikipedia collection and DBPedia. For example, traversing the graph starting from the collection and spreading through DBPedia until returning to the collection, considering the effect of semantic links. 3) Using concept extraction to create additional, more meaningful semantic links between query topics and image textual annotations(caption, comment and description of the image)

Acknowledgments. This research was partly funded by the Austrian Science Fund (FWF) project numbers P25905-N23 (ADmIRE) and I1094-N23 (MUCKE, under the CHIST-ERA Program for Transnational Research Projects).

References

1. Berber, T., Vahid, A.H., Ozturkmenoglu, O., Hamed, R.G., Alpkocak, A.: Demir at imageclefwiki 2011: Evaluating different weighting schemes in information retrieval. In: CLEF (2011)
2. Berthold, M.R., Brandes, U., Kotter, T., Mader, M., Nagel, U., Thiel, K.: Pure spreading activation is pointless. In: CIKM 2009 (2009)
3. Crestani, F.: Application of spreading activation techniques in information retrieval. Artificial Intelligence Review 11 (1997)
4. Delbru, R., Toupikov, N., Catasta, M., Tummarello, G.: A node indexing scheme for web entity retrieval. In: Aroyo, L., Antoniou, G., Hyvönen, E., ten Teije, A., Stuckenschmidt, H., Cabral, L., Tudorache, T. (eds.) ESWC 2010, Part II. LNCS, vol. 6089, pp. 240–256. Springer, Heidelberg (2010)
5. Duan, L., Li, W., Tsang, I.W., Xu, D.: Improving web image search by bag-based reranking. IEEE Transactions on Image Processing 20(11) (2011)
6. Elbassuoni, S., Blanco, R.: Keyword search over RDF graphs. In: CIKM (2011)
7. Fergus, R., Fei-Fei, L., Perona, P., Zisserman, A.: Learning object categories from google's image search. In: Proc. of Intl. Conf. on Computer Vision (2005)
8. Jing, Y., Baluja, S.: Visualrank: Applying pagerank to large-scale image search. IEEE Trans. Pattern Anal. Mach. Intell. (2008)
9. Kasneci, G., Suchanek, F., Ifrim, G., Ramanath, M., Weikum, G.: Naga: Searching and ranking knowledge. In: ICDE (2008)
10. Larsen, B., Ingwersen, P., Kekäläinen, J.: The polyrepresentation continuum in ir. In: Proc. of IIiX (2006)
11. Lazaridis, M., Axenopoulos, A., Rafailidis, D., Daras, P.: Multimedia search and retrieval using multimodal annotation propagation and indexing techniques. Signal Processing: Image Comm. (2012)
12. Martinet, J., Satoh, S.: An information theoretic approach for automatic document annotation from intermodal analysis. In: Workshop on Multimodal Information Retrieval (2007)
13. Rocha, C., Schwabe, D., Aragao, M.P.: A hybrid approach for searching in the semantic web. In: Proc. of WWW (2004)
14. Sabetghadam, S., Lupu, M., Rauber, A.: Astera - a generic model for multimodal information retrieval. In: Integrating IR Tech. for Prof. Search Workshop (2013)
15. Sabetghadam, S., Lupu, M., Rauber, A.: A combined approach of structured and non-structured IR in multimodal domain. In: ICMR (2014)
16. Sabetghadam, S., Bierig, R., Rauber, A.: A hybrid approach for multi-faceted IR in multimodal domain. In: Kanoulas, E., Lupu, M., Clough, P., Sanderson, M., Hall, M., Hanbury, A., Toms, E. (eds.) CLEF 2014. LNCS, vol. 8685, pp. 86–97. Springer, Heidelberg (2014)
17. Sabetghadam, S., Lupu, M., Rauber, A.: Which one to choose: Random walk or spreading activation. In: Lamas, D., Buitelaar, P. (eds.) IRFC 2014. LNCS, vol. 8849, pp. 112–119. Springer, Heidelberg (2014)
18. Srinivasan, S., Slaney, M.: A bipartite graph model for associating images and text. In: Workshop on Multimodal Information Retrieval (2007)
19. Tsikrika, T., Popescu, A., Kludas, J.: Overview of the wikipedia image retrieval task at imageclef 2011. In: CLEF (2011)
20. Yao, T., Mei, T., Ngo, C.-W.: Co-reranking by mutual reinforcement for image search. In: Proc. of CIVR (2010)

Main Core Retention on Graph-of-Words for Single-Document Keyword Extraction

François Rousseau and Michalis Vazirgiannis

LIX, École Polytechnique, France

Abstract. In this paper, we apply the concept of *k-core* on the *graph-of-words* representation of text for single-document keyword extraction, retaining only the nodes from the main core as representative terms. This approach takes better into account proximity between keywords and variability in the number of extracted keywords through the selection of more *cohesive* subsets of nodes than with existing graph-based approaches solely based on *centrality*. Experiments on two standard datasets show statistically significant improvements in F1-score and AUC of precision/recall curve compared to baseline results, in particular when weighting the edges of the graph with the number of co-occurrences. To the best of our knowledge, this is the first application of graph degeneracy to natural language processing and information retrieval.

Keywords: single-document keyword extraction, graph representation of text, weighted graph-of-words, k-core decomposition, degeneracy.

1 Introduction

Keywords have become ubiquitous in our everyday life, from looking up information on the Web via a search engine bar to online ads matching the content we are currently browsing. Researchers use them when they write a paper for better indexing as well as when they consult or review one to get a gist of its content before reading it. Traditionally, keywords have been manually chosen by the authors but the explosion of the number of available textual contents made the process too time-consuming and costly. Keyword extraction as an automated process then naturally emerged as a research issue to satisfy that need.

A graph-of-words is a syntactic graph that encodes co-occurrences of terms as opposed to the traditional bag-of-words and state-of-the-art approaches in keyword extraction proposed to apply PageRank and HITS on it to extract its most salient nodes. In our work, we capitalize on the k-core concept to propose a novel approach that takes better into account proximity between keywords and variability in the number of extracted keywords through the selection of more cohesive subsets of vertices. The proposed approach presents some significant advantages: (1) it is totally *unsupervised* as it does not need any training corpus; (2) it is *corpus-independent* as it does not rely on any collection-wide statistics such as IDF and thus can be applied on any text out of the box; (3) it *scales* to any document length since the algorithm is linearithmic in the number of unique

A. Hanbury et al. (Eds.): ECIR 2015, LNCS 9022, pp. 382–393, 2015.

terms as opposed to more complex community detection techniques and (4) the method in itself is *parameter-free* as the number of extracted keywords adapts to the structure of each graph through the k-core principle.

The rest of the paper is organized as follows. Section 2 provides a review of the related work. Section 3 defines the preliminary concepts upon which our work is built. Section 4 introduces the proposed approach and compares it with existing graph-based methods. Section 5 describes the experimental settings and presents the results we obtained on two standard datasets. Finally, Section 6 concludes our paper and mentions future work directions.

2 Related Work

In this section, we present the related work published in the areas of *keyword extraction* and *graph representation of text*. Mihalcea and Tarau in [19] and Litvak and Last in [16] are perhaps the closest works to ours since they also represent a text as a graph-of-words and extract the most salient keywords using a graph mining technique, solely based on centrality unlike k-core though.

2.1 Keyword Extraction

In the relevant literature, keyword extraction is closely related to text summarization. Indeed, Luhn in [17], which is one of the earliest works in automatic summarization, capitalizes on the *term frequency* to first extract the most salient keywords before using them to detect sentences. Later, the research community turned the task into a *supervised learning problem* with the seminal works of Turney in [24] based on genetic algorithms and of Witten *et al.* in [26] based on Naive Bayes. We refer to the survey of Nenkova and McKeown in [20] for an in-depth review on automatic summarization and by extension on keyword extraction. Briefly, the published works make several distinctions for the general task of keyword extraction: (a) *single-* [11,16,19] vs. *multi-document* [18] depending on whether the input is from a single document or multiple ones (e.g., a stream of news), (b) *extractive* [11,16,19] vs. *abstractive* [4] depending on whether the extracted content is restricted to the original text or not (e.g., use of a thesaurus to enrich the keywords), (c) *generic* [11,16,19,24] vs. *query-based* [25] vs. *update* [12] depending on whether the extracted keywords are generic or biased towards a specific need (e.g., expressed through a query) or dependent of already-known information (e.g., browsing history) and finally (d) *unsupervised* [7,16,19] vs. *supervised* [11,16,24] depending on whether the extraction process involves a training part on some labeled inputs. Our work falls within the case of *unsupervised generic extractive single-document keyword extraction.*

2.2 Graph Representation of Text

Graph representations of textual documents have been around for a decade or so and we refer to the work of Blanco and Lioma in [3] for an in-depth review. These representations have been mainly investigated as a way of taking into account *term dependence* and *term order* compared to earlier approaches that did not. In NLP, text has historically been represented as a *bag-of-words*, i.e. a multiset of terms that assumes independence between terms and the task of keyword extraction was no exception to that representation. But graph structures allow to challenge that *term independence assumption* and somewhat recently Rousseau and Vazirgiannis introduced in [22] the denomination of *graph-of-words* to encompass that idea of using a graph whose vertices represent unique term and whose edges represent some meaningful relation between pairs of terms. This relation can either be based solely on statistics or use deeper linguistic analysis, leading respectively to *syntactic* [16,19] and *semantic* [15] graphs. More generally, what a vertex of the graph represents depends entirely on the level of granularity needed. This can be a sentence [7,19], a word [3,16,19,22] or even a character [9]. Most of the existing works based on graph-of-words were explored for automatic summarization, in particular the seminal works of Erkan and Radev in [7] and Mihalcea and Tarau in [19]. More recent papers [3,22] proposed applications in ad hoc IR to challenge the well-established tf-based retrieval models.

3 Preliminary Concepts

In this section, we define the preliminary concepts upon which our work is built: the notions of graph, k-core and graph-of-words.

3.1 Graph

Let $\mathcal{G} = (\mathcal{V}, \mathcal{E})$ be a *graph* (also known as a network), $\mathcal{V}$ its set of *vertices* (also known as nodes) and $\mathcal{E}$ its set of *edges* (also known as arcs or links). We denote by n the number of vertices ($n = |\mathcal{V}|$) and m the number of edges ($m = |\mathcal{E}|$). A graph can represent anything, from a protein-interaction network to a power grid or in our case a textual document. This is the natural representation to model interactions between entities and we believe text makes no exception.

Depending on the nature of these interactions, an edge and by extension a graph can be weighted and/or directed. This impacts the definition of the *degree* of a vertex, which measures the interactions a node has with its neighbors and somewhat its importance or influence in the network. We denote by $deg_{\mathcal{G}}(v)$ the *degree* of a vertex $v \in \mathcal{G}$ in $\mathcal{G}$. In the undirected case, this corresponds to the sum of the weights of the adjacent edges (unit weight in the unweighted case). In the directed case, the notion of degree is usually split in two: *indegree* and *outdegree* corresponding to the (weighted) number of inlinks and outgoing links.

3.2 K-core

The idea of a *k-degenerate* graph comes from the work of Bollobás in [5, page 222] that was further extended by Seidman in [23] into the notion of a *k-core*, which explains the use of *degeneracy* as an alternative denomination in the literature. Henceforth, we will be using the two terms interchangeably. Let k be an integer. A subgraph $\mathcal{H}_k = (\mathcal{V}', \mathcal{E}')$, induced by the subset of vertices $\mathcal{V}' \subseteq \mathcal{V}$ (and a fortiori by the subset of edges $\mathcal{E}' \subseteq \mathcal{E}$), is called a *k-core* or a *core of order k* iff $\forall v \in \mathcal{V}'$, $deg_{\mathcal{H}_k}(v) \geq k$ and $\mathcal{H}_k$ is the maximal subgraph with this property, i.e. it cannot be augmented without losing this property. In other words, the k-core of a graph corresponds to the maximal connected subgraph whose vertices are at least of degree k within the subgraph.

The *core number* core(v) of a vertex v is the highest order of a core that contains this vertex. The core of maximum order is called the *main core* and the set of all the k-cores of a graph (from the 0-core to the main core) forms what is called the *k-core decomposition* of a graph.

Thanks to Batagelj and Zaveršnik in [2], the k-core decomposition of a weighted graph can be computed in linearithmic time ($\mathcal{O}(n + m \log n)$) and linear space ($\mathcal{O}(n)$) using a *min-oriented binary heap* to retrieve the vertex of lowest degree at each iteration (n in total). We implemented their algorithm in our experiments. Note that in the unweighted case, there exists a linear version in time that uses *bin sort* since there are at most $\Delta(\mathcal{G}) + 1$ distinct values for the degrees where $\Delta(\mathcal{G}) = max_{v \in \mathcal{V}}(deg_{\mathcal{G}}(v)) = \mathcal{O}(n)$. For the directed case, Giatsidis *et al.* in [10] proposed a two-dimensional k-core decomposition that is beyond the scope of this paper. We still explored the effects of degeneracy on directed graphs in our experiments, considering either the indegree or the outdegree instead of the degree but not both of them at the same time.

3.3 Graph-of-words

We model a textual document as a *graph-of-words*, which corresponds to a graph whose vertices represent unique terms of the document and whose edges represent co-occurrences between the terms within a fixed-size sliding window. The underlying assumption is that all the words present in a document have some relationships with the others, modulo a window size outside of which the relationship is not taken into consideration. This is a statistical approach as it links all co-occurring terms without considering their meaning or function in the text. The graph can be weighted to take into account the number of co-occurrences of two terms. Similarly, the graph can be directed to encode the *term order*, forward edges indicating the natural flow of the text.

Regarding the preprocessing steps, we applied the following on the input text: (1) tokenization; (2) part-of-speech[1] annotation and selection (nouns and adjectives like in [19]); (3) stopwords[2] removal; and (4) stemming[3]. All the remaining terms constitute the vertices of the graph-of-words. The edges were drawn between terms co-occurring within a fixed-size sliding window W of size 4 over the processed text, value consistently reported as working well [3,7,16,19,22]. For the whole process, the complexity is $\mathcal{O}(nW)$ in time and $\mathcal{O}(n + m)$ in space.

Figure 1a illustrates the weighted graph-of-words representation of one of the documents (id 1938) from the dataset introduced by Hulth in [11]. Edge weight

[1] http://opennlp.apache.org/documentation/1.5.3/manual/opennlp.html
[2] http://jmlr.org/papers/volume5/lewis04a/a11-smart-stop-list/english.stop
[3] http://tartarus.org/~martin/PorterStemmer

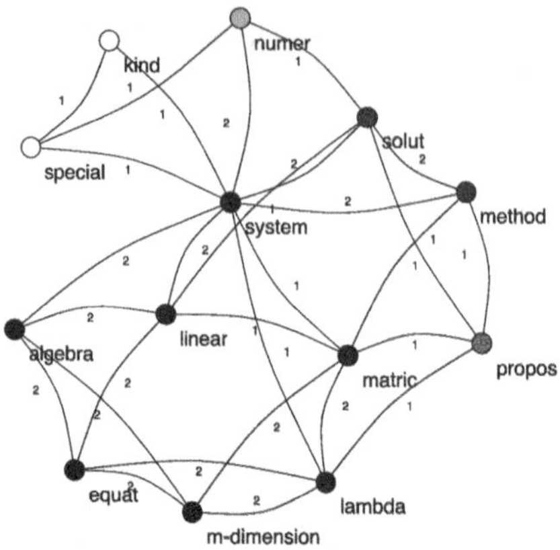

(a) Graph-of-words representation

WK-core		PageRank		HITS	
system	6	**system**	1.93	**system**	0.45
matric	6	**matric**	1.27	**matric**	0.38
lambda	6	solut	1.10	**linear**	0.32
linear	6	**lambda**	1.08	**lambda**	0.31
equat	6	**linear**	1.08	solut	0.30
algebra	6	**equat**	0.90	method	0.28
m-dim...	6	**algebra**	0.90	propos	0.25
method	5	**m-dim...**	0.90	**algebra**	0.25
solut	5	propos	0.89	**m-dim...**	0.23
propos	4	method	0.88	**equat**	0.22
numer	3	special	0.78	**numer**	0.18
specia	2	**numer**	0.74	special	0.15
kind	2	kind	0.55	kind	0.12

(b) Ranked list of scored keywords

Fig. 1. Subfigure (a) illustrates a graph-of-words representation of a textual document. Edge weight corresponds to the number of co-occurrence, node color to the node core number (gray scale). Table (b) shows the ranked lists of scored keywords extracted with weighted k-core, PageRank and HITS. Bold font indicates the golden keywords and the dashed lines the cutoff for each method.

corresponds to the number of co-occurrences. Node color indicates the highest core a vertex belongs to, from 2-core (white) to 6-core (black).

4 Graph-based Keyword Extraction

In this section, we present the existing state-of-the-art graph-based methods for keyword extraction as well as our proposed approach.

4.1 Existing Approaches: PageRank and HITS on Graph-of-words

There exist two algorithms that have been successfully used for graph-based keyword extraction: PageRank and HITS, considered in this context first in [19] and [16] respectively. Both methods are based on eigenvector centrality and define recursively the weight of a vertex as a measure of its *influence* inside the network, regardless of how cohesive its neighborhood is. For PageRank, it is defined as the sum of the weights of its incoming neighbors (the ones giving it support) and the vertex itself gives a weighted portion of its own weight to each of its outgoing neighbors. For HITS, it is slightly different as it defines two types of influential nodes: the *authorities* that are being pointed at by a lot of nodes and the *hubs* that are pointing to a lot of nodes (these are the same in the undirected case).

4.2 Our Contribution: k-core Decomposition on Graph-of-words

Our idea was to consider the vertices of the main core as the set of keywords to extract from the document. Indeed, it corresponds to the most cohesive connected component(s) of the graph and thus its vertices are intuitively good candidates for representing the entire graph-of-words. Additionally, assuming a set of golden keywords to compare with, we considered more and more cores in the decomposition and expected an increase in the recall of the extracted keywords without a too important decrease in precision.

We illustrate in Table 1b the process by presenting the list of keywords extracted using all three algorithms on the graph-of-words presented in Figure 1a. The horizontal dashed lines indicate where the cutoff is applied for each method and in bold the golden keywords according to human annotators. For our approach, it is the main core (of order 6 in this example). For PageRank and HITS, Mihalcea and Tarau suggested extracting the top third of the vertices (top 33%), relative numbers helping accounting for documents of varying length.

Overall, we notice that "numer" is never extracted by any method. The term appears only once in the original text and around terms of lesser importance (except "system"). Extracting it can be considered as a form of *overfitting* if we assume that the pattern to extract is related to term repetition, standard assumption in NLP and IR. Both PageRank and HITS retrieve "solut", which is not a golden keyword, because of its centrality but not k-core because of its neighbors. Again, the main core corresponds to *cohesive set(s) of vertices* in which they all contribute equally to the subgraph they belong to – removing any node would collapse the entire subgraph through the *cascading effect* implied by the k-core condition. PageRank and HITS, on the other hand, provide scores for each vertex based on its centrality yet somewhat independently of its neighborhood, therefore not capturing the proximity between keywords.

4.3 Keywords are Bigrams

For the 500 abstracts from the *Inspec* database that we used in our experiments, only 662 out of the 4,913 keywords manually assigned by human annotators are unigrams (13%). The rest of them range from bigrams (2,587 – 52%) to 7-grams (5). Similar statistics were observed on the other dataset. Because higher order n-grams can be considered as multiple bigrams, we make the general claim that human annotators tend to select *keywords that are bigrams*. Thus, to improve the performances of an automated system, one needs to capture the interactions between keywords in the first place – hence, the explored graph-of-words representation to challenge the traditional bag-of-words.

Even if both the existing models and our approach extract unigrams because of the way the graph-of-words is constructed, the edges do represent co-occurrences within a sliding window. And for small-enough sizes (which is typically the case in practice), we can consider that two linked vertices represent a long-distance bigram [1], if not a bigram. Hence, by considering cohesive subgraphs, we make sure to extract unigrams that co-occur together and thus are bigrams, if not higher order n-grams. On the contrary, PageRank and HITS may extract unigrams that are central because they co-occur with a lot of other words but these words may not be extracted as well because of a lower weight. Extracting salient bigrams would require to include bigrams as vertices but the number of nodes increases exponentially with the order of the n-gram.

4.4 K-cores are Adaptive

Most current techniques in keyword extraction assign a score to each term of the document and then take the top ones. For a given collection of homogeneous documents in size or because of specific constraints, an *absolute* number may make sense. For example, Turney in [24] limited to the top 5 keywords while Witten *et al.* in [26] to the top 15. Mihalcea and Tarau argued in [19] that a *relative* number should be used for documents of varying length or when no prior is known. We claim that the numbers of retrieved keywords should be decided at the document level and not at the collection level. For instance, for two documents, even of equal size, one of them might require more keywords to express the gist of its content (because it deals with more topics for example).

The size of each core, i.e. the number of vertices in the subgraph, depends on the structure of the graph. In the unweighted case, it is lower-bounded by $k+1$ since each vertex has at least k neighbors but can potentially be up to n in the case of a complete graph. Hence, we think that degeneracy can capture this variability in the number of extracted keywords, in particular for a fixed document length (PageRank and HITS still extract more and more keywords as the document length increases when using relative numbers). In Section 5, we will show distributions of extracted keywords per document length for all models and from human annotators to support our claim.

5 Experiments

In this section, we describe the experiments we conducted to test and validate our approach along with the results we obtained.

5.1 Datasets

We used two standard datasets publicly available[4]: (1) *Hulth2003* – 500 abstracts from the *Inspec* database introduced by Hulth in [11] and also used by Mihalcea and Tarau in [19] with PageRank; and (2) *Krapi2009* – 2,304 ACM full papers (references and captions excluded) introduced by Krapivin *et al.* in [14]. For Hulth2003, we used the "uncontrolled" golden keywords since we do not want to restrict the keywords to a given thesaurus and for Krapi2009, we used the ones chosen by the authors of each ACM paper. Since all approaches are unsupervised and single-document, the scalability of the methods are measured with regards to the document length, not the collection size (that only needs to be large enough to measure the statistical significance of the improvements).

[4] https://github.com/snkim/AutomaticKeyphraseExtraction

5.2 Models

For the graph-of-words representation, we experimented with undirected, forward edges (natural flow of the text – an edge $term_1 \rightarrow term_2$ meaning that $term_1$ precedes $term_2$ in a sliding window) and backward edges (the opposite). In terms of keyword-extracting methods, we considered (1) PageRank, (2) HITS, (3) k-core on an unweighted graph-of-words and (4) k-core on a weighted one (the edge weight being the number of co-occurrence). We extracted the top third keywords (top 33%) on Hulth2003 and the top 15 keywords on Krapi2009 for PageRank and HITS and the main core for our approaches (the k values differs from document to document). The choice between relative (top X%) and absolute numbers (top X) comes from the fact that for relatively short documents such as abstracts, the longer the document, the more keyword human annotators tend to select while past a certain length (10-page long for ACM full papers), the numbers vary far less. Hence, in all fairness to the baselines, we selected the top 33% on the abstracts like in the original papers and the top 15 for the full papers (15 being the average number of unigrams selected as keywords by the papers' authors). Note that for HITS, we only display the results for the authority scores since the hub scores are the same in the undirected case and symmetric in the directed case (the hub score for forward edges corresponds to the authority score for backward edges).

5.3 Evaluation

For each document, we have a set of golden keywords manually assigned by human annotators and a set of extracted keywords, leading to *precision*, *recall* and *F1-score* per document and per method that are then *macro-averaged* at the collection level. The statistical significance of improvement over the PageRank baseline for each metric was assessed using the Student's paired t-test, considering two-sided p-values less than 0.05 to reject the null hypothesis.

Note that we convert the golden keywords into unigrams to easily compute precision and recall between this golden set and the set of extracted unigrams. Mihalcea and Tarau in [19] suggested to "reconcile" the n-grams as a post-processing step by looking in the original text for adjacent unigrams but then questions arise such as whether you keep the original unigrams in the final set, impacting the precision and recall – hence an evaluation based on unigrams. Indeed, it is not clear how to penalize a method that, given a golden bigram to extract, would return part of it (unigram) or more than it (trigram).

5.4 Macro-averaged Results

We present in Table 1 the macro-averaged precision, recall and F1-score (in %) for PageRank, HITS, k-core and weighted k-core (columns) for the different variants of graph-of-words considered (rows) on each dataset. Overall, PageRank and HITS have similar results, with a precision higher than the recall as reported in previous works. It is the opposite for k-core, which tends to extract a main core

Table 1. Macro-averaged precision, recall and F1-score for PageRank, HITS, K-core and Weighted K-core (WK-core). Bold font marks the best performance in a block of a row. * indicates statistical significance at $p < 0.05$ using the Student's t-test w.r.t. the PageRank baseline of the same block of the same row.

Graph	Dataset	Macro-averaged precision (%)				Macro-averaged recall (%)				Macro-averaged F1-score (%)			
		PageRank	HITS	K-core	WK-core	PageRank	HITS	K-core	WK-core	PageRank	HITS	K-core	WK-core
undirected edges	Hulth2003	58.94	57.86	46.52	**61.24***	42.19	41.80	**62.51***	50.32*	47.32	46.62	49.06*	**51.92***
	Krapi2009	50.23	49.47	40.46	**53.47***	48.78	47.85	**78.36***	50.21	49.59	47.96	46.61	**50.77***
forward edges	Hulth2003	55.80	54.75	42.45	**56.99***	41.98	40.43	**72.87***	46.93*	45.70	45.03	**51.65***	50.59*
	Krapi2009	47.78	47.03	39.82	**52.19***	44.91	44.19	**79.06***	45.67	45.72	44.95	46.03	**47.01***
backward edges	Hulth2003	59.27	56.41	40.89	**60.24***	42.67	40.66	**70.57***	49.91*	47.57	45.37	45.20	**50.03***
	Krapi2009	51.43	49.11	39.17	**52.14***	49.96	47.00	**77.60***	50.16	**50.51**	47.38	46.93	50.42

with a lot of vertices since the k-core condition can be interpreted as a set of keywords that co-occur with at least k other keywords. For the weighted case, it corresponds to a set of keywords that co-occur at least k times in total with other keywords leading to cores with fewer vertices but with stronger links and the extraction of important bigrams, hence the increase in precision (at the cost of a decrease in recall) and an overall better F1-score.

Edge direction has an impact but not necessarily a significant one and is different across methods. This disparity in the results and the lack of a dominant choice for edge direction is consistent with the relevant literature. Mihalcea and Tarau in [19] and Blanco and Lioma in [3] used undirected edges, Litvak and Last in [16] backward edges and Rousseau and Vazirgiannis in [22] forward edges. Hence, we recommend the use of undirected edges for ease of implementation but other techniques that would try to extract paths from the graph-of-words for instance might need the edge direction to follow the natural flow of the text like for multi-sentence compression [8].

5.5 Precision/Recall Curves

Additionally, instead of just considering the main core or the top X% vertices, we computed precision and recall at each core and at each percent of the total number of terms to get *precision/recall curves*. We used relative numbers because the documents are of varying length so the top 10 keywords for a document of size 10 and 100 do not mean the same while the top 10% might.

We show on Figure 2a the resulting curves on the Hulth2003 dataset, one for each model (100 points per curve, no linear interpolation following Davis and Goadrich in [6]). The final recall for all models is not 100% because human annotators used keywords that do not appear in the original texts. We observe that the curve for the weighted k-core (green, solid circle) is systematically above the others, thus showing improvements in *Area Under the Curve* (AUC) and not just in point estimates such as the F1-score. The curve for k-core (orange, diamond) is overall below the other curves since it tends to only find a few cores with a lot of vertices, lowering the precision but insuring some minimum recall (its lowest value of recall is greater than for the other curves).

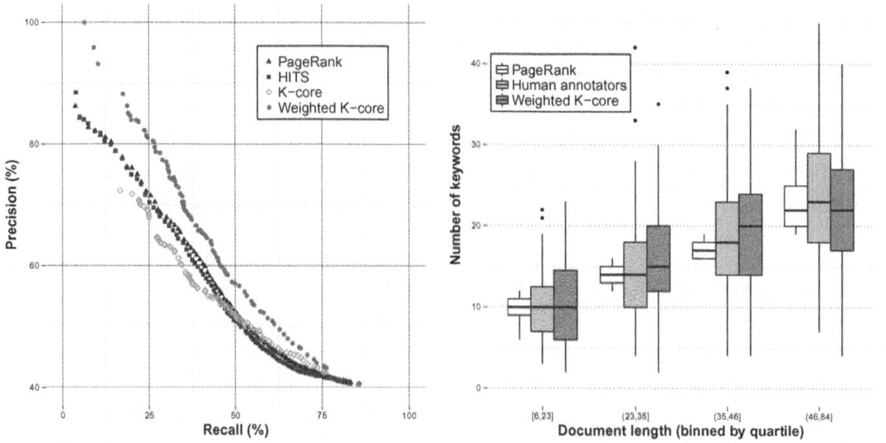

(a) Precision/recall curves (b) Distribution of the number of keywords

Fig. 2. Weighted k-core on graph-of-words consistently captures more golden keywords than PageRank and HITS (subfigure a) and provides a variability in the number of extracted keywords closer to the human one (subfigure b)

5.6 Distribution of the Number of Keywords

Human annotators do not assign the same number of keywords to all documents. There is a variability that is partially due to varying document length (the number increases with the length) but not only. Indeed, when computing a distribution per document length, we can still observe some dispersion. With PageRank, by extracting a relative number of unigrams (top 33%), one accounts for varying length but does not fully capture the variability introduced by human annotators while k-core does better. Similarly, for Krapi2009, where documents are of the same length (10-page), some authors may have chosen more keywords for their paper than others because for instance there are alternative denominations for the concept(s) developed in their work and as a result more keywords.

We present in Figure 2b above groups of three box plots computed for Hulth2003. In each group, the left one corresponds to PageRank (in white, similar results for HITS), the middle one to human annotators (light gray) and the right one to weighted k-core (dark gray). We do not show any box plot for unweighted k-core since the method tends to overestimate the number of keywords (higher recall, lower precision). For space constraints and also for sparsity reasons, we binned the document lengths by quartile (i.e. the bins are not of equal range but contains the same number of documents – 25% each).

As expected, the number of keywords for a given bin varies little for PageRank – the increase in median value across the bins being due to the baseline taking the top third unigrams, relative number that increases with the document length. For weighted k-core, we observe that the variability is taken much more into account: the first, second and third quartiles' values for the number of keywords are much closer to the golden ones. For PageRank and HITS, it would be much harder to learn the number of keywords to extract for each document while it is inherent to graph degeneracy. Alone, these results would not mean much but because higher accuracy has already been established through consistent and significant higher macro-averaged F1-scores, they support our claim that k-core is better suited for the task of keyword extraction because of its adaptability to the graph structure and therefore to the document structure.

6 Conclusions and Future Work

In this paper, we explored the effects of k-core on graph-of-words for single-document keyword extraction. Similarly to previous approaches, we capitalized on syntactic graph representations of text to extract central terms. However, by retaining only the main core of the graph, we were able to capture more cohesive subgraphs of vertices that are not only central but also densely connected. Hence, the extracted keywords are more likely to form bigrams and their number adapts to the graph structure, as human annotators tend to do when assigning keywords to the corresponding document.

As a final example, here are the stemmed unigrams belonging to the main core of the graph-of-words corresponding to this paper (references, captions and this paragraph excluded): {*keyword, extract, graph, represent, text, weight, graph-of-word, k-core, degeneraci, edg, vertic, number, document*}. Using PageRank, "work" appears in the top 5, "term" and "pagerank" in the top 10, and "case" and "order" in the top 15. Existing methods tend to extract central keywords that are not necessarily part of a cohesive subgraph as opposed to our proposed approach, which provides closer results to what humans do on several aspects.

Possible extension of this work would be the exploration of the clusters of keywords in the top cores to elect representatives per cluster for topic modeling.

References

1. Bassiou, N., Kotropoulos, C.: Word clustering using PLSA enhanced with long distance bigrams. In: Proceedings of ICPR 2010, pp. 4226–4229 (2010)
2. Batagelj, V., Zaversnik, M.: Fast algorithms for determining core groups in social networks. Advances in Data Analysis and Classification 5(2), 129–145 (2011)
3. Blanco, R., Lioma, C.: Graph-based term weighting for information retrieval. Information Retrieval 15(1), 54–92 (2012)
4. Blank, I., Rokach, L., Shani, G.: Leveraging the citation graph to recommend keywords. In: Proceedings of RecSys 2013, pp. 359–362 (2013)
5. Bollobs, B.: Extremal Graph Theory. Academic Press, London (1978)
6. Davis, J., Goadrich, M.: The relationship between precision-recall and ROC curves. In: Proceedings of ICML 2006, pp. 233–240 (2006)
7. Erkan, G., Radev, D.R.: LexRank: graph-based lexical centrality as salience in text summarization. Journal of Artificial Intelligence Research 22(1), 457–479 (2004)
8. Filippova, K.: Multi-sentence compression: finding shortest paths in word graph. In: Proceedings of COLING 2010, pp. 322–330 (2010)

9. Giannakopoulos, G., Karkaletsis, V., Vouros, G., Stamatopoulos, P.: Summarization system evaluation revisited: N-gram graphs. ACM Transactions on Speech and Language Processing 5(3), 1–39 (2008)
10. Giatsidis, C., Thilikos, D.M., Vazirgiannis, M.: D-cores: Measuring collaboration of directed graphs based on degeneracy. In: Proceedings of ICDM 2011, pp. 201–210 (2011)
11. Hulth, A.: Improved automatic keyword extraction given more linguistic knowledge. In: Proceedings of EMNLP 2003, pp. 216–223 (2003)
12. Karkali, M., Plachouras, V., Stefanatos, C., Vazirgiannis, M.: Keeping keywords fresh: A BM25 variation for personalized keyword extraction. In: Proceedings of TempWeb 2012, pp. 17–24 (2012)
13. Kleinberg, J.M.: Authoritative sources in a hyperlinked environment. Journal of the ACM 46(5), 604–632 (1999)
14. Krapivin, M., Autaeu, A., Marchese, M.: Large dataset for keyphrases extraction. Technical Report DISI-09-055, University of Trento (May 2009)
15. Leskovec, J., Grobelnik, M., Milic-Frayling, N.: Learning semantic graph mapping for document summarization. In: Proceedings of KDO 2004 (2004)
16. Litvak, M., Last, M.: Graph-based keyword extraction for single-document summarization. In: Proceedings of MMIES 2008, pp. 17–24 (2008)
17. Luhn, H.P.: The automatic creation of literature abstracts. IBM Journal of Research and Development 2(2), 159–165 (1958)
18. McKeown, K., Passonneau, R.J., Elson, D.K., Nenkova, A., Hirschberg, J.: Do summaries help. In: Proceedings of SIGIR 2005, pp. 210–217 (2005)
19. Mihalcea, R., Tarau, P.: TextRank: bringing order into texts. In: Proceedings of EMNLP 2004, pp. 404–411 (2004)
20. Nenkova, A., McKeown, K.R.: Automatic summarization. Foundations and Trends in Information Retrieval 5(2-3), 103–233 (2011)
21. Page, L., Brin, S., Motwani, R., Winograd, T.: The PageRank citation ranking: Bringing order to the web. Technical Report 1999-0120, Stanford University (1999)
22. Rousseau, F., Vazirgiannis, M.: Graph-of-word and TW-IDF: new approach to ad hoc IR. In: Proceedings of CIKM 2013, pp. 59–68 (2013)
23. Seidman, S.B.: Network structure and minimum degree. Social Networks 5, 269–287 (1983)
24. Turney, P.D.: Learning to extract keyphrases from text. Technical report, National Research Council of Canada, Institute for Information Technology (1999)
25. Turpin, A., Tsegay, Y., Hawking, D., Williams, H.E.: Fast generation of result snippets in web search. In: Proceedings of SIGIR 2007, pp. 127–134 (2007)
26. Witten, I.H., Paynter, G.W., Frank, E., Gutwin, C., Nevill-Manning, C.G.: KEA: practical automatic keyphrase extraction. In: Proceedings of DL 1999, pp. 254–255 (1999)

Entity Linking for Web Search Queries

Deepak P[1], Sayan Ranu[2], Prithu Banerjee[1], and Sameep Mehta[3]

[1] IBM Research - India, Bangalore, India
[2] Dept. of Comp. Sc. and Engg., Indian Institute of Technology Madras, India
[3] IBM Research - India, New Delhi, India
{deepak.s.p,prithuba,sameepmehta}@in.ibm.com,
sayan@cse.iitm.ac.in

Abstract. We consider the problem of linking web search queries to entities from a knowledge base such as Wikipedia. Such linking enables converting a user's web search session to a footprint in the knowledge base that could be used to enrich the user profile. Traditional methods for entity linking have been directed towards finding entity mentions in text documents such as news reports, each of which are possibly linked to multiple entities enabling the usage of measures like entity set coherence. Since web search queries are very small text fragments, such criteria that rely on existence of a multitude of mentions do not work too well on them. We propose a three-phase method for linking web search queries to wikipedia entities. The first phase does IR-style scoring of entities against the search query to narrow down to a subset of entities that are expanded using hyperlink information in the second phase to a larger set. Lastly, we use a graph traversal approach to identify the top entities to link the query to. Through an empirical evaluation on real-world web search queries, we illustrate that our methods significantly enhance the linking accuracy over state-of-the-art methods.

1 Introduction

Web search queries issued by users provide very powerful insights into the interests of the user. Information from query logs have been shown to be useful for scenarios such as interest-based ad-targeting[1] and improving query recommendations [1]. We study the problem of linking web search queries to entities in a knowledge base such as Wikipedia. Such linkages enrich the queries with semantic information that may be leveraged by downstream processes utilizing search queries for a variety of applications. Linking phrases in text documents to entities from a knowledge base such as Wikipedia has been a subject of much research [10,5]. Though entity linking may be seen as a specialized version of information extraction (IE) using flat entity data, the rich graph-text structure among entities in Wikipedia and YagoDB [2] has led to entity linking techniques to become fairly specialized to be able to exploit such structures. However, entity linking techniques have had limited success in processing web search

[1] http://www.nytimes.com/2009/03/11/technology/internet/11google.html?_r=0
[2] http://www.mpi-inf.mpg.de/departments/databases-and-information-systems/research/yago-naga/yago/

A. Hanbury et al. (Eds.): ECIR 2015, LNCS 9022, pp. 394–399, 2015.

Table 1. Example Queries

Query	Method	Linked Entities
honda pilot 2014	AIDA	Honda
	Manual Tagging	Honda Pilot
		2014
russian submarine kursk fleet	AIDA	Russian Language
		Kursk
	Manual Tagging	Northern Fleet
		Russian submarine --K-141 Kursk

Table 2. Summary of Entity Linking Approaches

	Supervised	**Unsupervised**
Documents	[14,11,7]	[10,9,5,2,8,13]
Short Texts	[3,4]	Our Method

queries. We will reason as to why some traditional entity linking considerations almost break down in the case of super-short texts such as web queries, leading to our argument that entity linking on them requires profoundly different considerations.

Entity Linking Framework: Conventional entity linking algorithms for text start by identifying *mentions*, i.e., phrases/words that are candidates to be linked to entities. For a document, the starting point is the set of mentions extracted from the document; entities are then linked to mentions using the following criteria:

1. Similarity: Entities that are highly similar to the mention are prioritized for linking. Similarity could be textual [2], or quantified using keyphrase overlaps [9].

2. Entity Popularity: Among similar entities, popular entities [9,3] may be chosen to link to a mention, since they are statistically more probable.

3. Coherence: Coherence prefers a choice of candidate entities so that the set of entities linked to a text document lead to a compact footprint in the entity space. Coherence between entities is quantified by number of common hyperlinks [9], similarity to other mentions [2], graph propagation [6] or ensured by choosing a dense subgraph [15].

Motivation and Contribution: We observe that the usage of the phases of *mention detection* followed by *entity assignment* often leads to discovery of *shallow* linkages. Table 1 illustrates the issue using a couple of entity linking examples performed by the AIDA system (on queries from a real dataset) along with corresponding manually tagged entities. In the query *honda pilot 2014*, it is presumable that AIDA chose to link the entity *Honda* in preference to *Honda Pilot* due to the higher popularity of the former, since there is not much to choose between these on other criteria. However, it may be seen that *Honda Pilot* is intuitively a better choice. The second example illustrates a query that could be seen as having many legitimate entity mentions (e.g., Russia, Submarine, Kursk, Fleet); AIDA's choice of *Russian Language* and *Kursk* is a coherent set since the latter is a Russian town, where Russian is presumably spoken. However, the correct referent for this *Kursk* turns out to be a Russian submarine of the same name, which is more evident if one considers the query in entirety. Since web search queries are seldom more than a few words in length, the mention identification oriented approach is forced to work with very limited context. We develop techniques that prefer entities related to the entire query; for this, we blend techniques from IR where queries are always considered in toto and those from the entity linking that exploit relatedness between entities, to develop techniques for linking web search queries to entities.

2 Related Work

Existing work in entity linking can be grouped on multiple facets: supervised vs. unsupervised and document-targeted vs. text snippet targeted. Table 2 presents a view of the current state of the art. As can be seen, no work exists on entity disambiguation in an unsupervised manner in short text snippets; we address this void.

As in Table 2, majority of the existing work focuses on linking entities to a document. The challenges of the same task in the short-text scenario, and more specifically search queries, are different. Methods for documents fail to perform well. Further, many techniques use natural language processing such as topic modeling [5] to link entities to documents. These strategies do not transfer to search queries since they often do not have the redundancy to yield good topic models. TagMe[3] and [4] are the only two techniques that target the short text scenario. However, both of them employ supervised approaches, whereas we focus on an unsupervised scenario. An unsupervised approach is more robust since it can scale to a large entity corpus and can automatically adapt to the inevitable evolution of the entity database. Also, TagMe and [4] focus on tweets and web page snippets that have tens of words; we focus on much shorter search queries.

3 Our Method

Our method has three phases. The first uses IR to narrow down to a seed set of entities; the second phase then expands it using the hyperlink structure among entities. Lastly, the entities are ranked using Random Walk with Restarts (RWR). Thus, the first and second phases are text and graph based respectively; the third expoits both.

Phase 1: Seed Set Construction:

We use Lucene's[3] default scoring method[4] to collect result entities for each query. We use two indexes with different entity representations:

$$Article(e) = text\ in\ wikipedia\ page\ for\ e$$

$$Anchor(e) = \underset{l \in Links}{CONCAT} \begin{cases} l.anchortext & \text{if } l.target = e \\ \phi & \text{otherwise} \end{cases}$$

where $Links$ denotes the set of all hyperlinks internal to Wikipedia, and $l.target$ and $l.anchortext$ denote the target entity and anchor text of the link respectively. The $CONCAT$ operator simply appends the text fragments in the input set to create a larger text sequence. We collect the top-k results from each index and merge.

$$Seed(q) = Index_{Article}(q, k) \cup Index_{Anchor}(q, k)$$

where $Index(q, k)$ denotes the top-k documents (i.e., entities) returned from $Index$ in response for q. We uniformly use $k = 3$ for our method.

Phase 2: Expansion:

Having retrieved lexically similar entities, we use the hyperlink structure to select all entities linked to from the seed set, to create an expanded set. Consider the example in

[3] `http://lucene.apache.org/`

[4] `http://ipl.cs.aueb.gr/stougiannis/default.html`

Table 1; the entity *Northern Fleet* is not highly similar to the query *russian submarine kursk fleet* based on either the article text or the anchor text (we observed that the anchor text mostly contains the phrase *northern fleet*). However, it is linked to from most entities related to Russian Navy as well as from *K-141 Kursk.*

$$Exp(q) = \{e|e \in Seed(q) \vee (\exists e' \in Seed(q) \wedge (e' \rightarrow e) \in Links)\}$$

$Exp(q)$ denotes the expanded set of entities to be processed in the third phase.

Phase 3: Scoring and Selection:

We now rank the entities based on the twin criteria of graph proximity and textual similarity within a single model using RWR. The entities in $Exp(q)$ form nodes in the graph in addition to q itself which also forms a node; edges from q to entity nodes are weighted wrt unigram language model probabilities:

$$\omega(q \rightarrow e) = \frac{max\{L^q_{Article(e)}, L^q_{Anchor(e)}\}}{\sum_{e' \in Exp(q)} max\{L^q_{Article(e')}, L^q_{Anchor(e')}\}}$$

where $L^q_{Article(e)}$ and $L^q_{Anchor(e)}$ denote the unigram language probability [12] of q from models constructed using $Article(e)$ and $Anchor(e)$ respectively. The weights for entity-entity links are defined with the help of a boolean function $SOL(e, e')$:

$$SOL(e, e') = ((e \rightarrow e') \in Links) \vee (e = e'))$$

Thus, $SOL(.,.)$ is turned on for such pairs where there is a link from the first entity to the second or if both refer to the same entity (i.e., an implicit self-link). Based on $SOL(.,.)$ we assign the weight of edges between entities as follows:

$$\omega(e \rightarrow e') = \begin{cases} \frac{\tau}{\sum_{e'' \in Exp(e)} I(SOL(e,e''))} & \text{if } SOL(e, e') \\ \frac{1.0-\tau}{\sum_{e'' \in Exp(e)} I(\neg SOL(e,e''))} & \text{otherwise} \end{cases}$$

where $I(.)$ maps the boolean values to 1 and 0. Informally, for an entity e, τ fraction of the weight is assigned uniformly to nodes to which it is connected using $SOL(.,.)$. The remaining (1.0-τ) weight is used to create links uniformly to other entities. Keeping aside a mass for unlinked entities ensures that the Random Walk does not get stuck at nodes that do not have outward links. The normalization in the construction of the edge weights ensures that for every node n, $\sum_{n'} \omega(n \rightarrow n') = 1.0$. We set τ to 0.8 so that most of the mass is assigned to entities connected through hyperlinks.

RWR: An RWR is run starting from the query node for several iterations. At any iteration, the RWR either resets to the query node, or hops from the current node to a neighbor in accordance with the link weights. We set the restart probability to 0.4 consistently. As the distribution of the number of visits among nodes in $Exp(q)$ stabilizes, the nodes are scored based on the decreasing order of frequency of RWR visits.

Selection: It is intuitive to assume that longer queries may be linked to more entities than shorter ones. Accordingly, we take the top $\frac{l}{t}$ entities where l denotes the query length and t is a parameter that we set to 2.0 for all experiments.

4 Experimental Study

Setup: We used the only public dataset with entity-labeled queries, the Yahoo! WebScope Query to Entity Dataset[5]. We excluded queries with fewer than three words, and evaluated the techniques on the remaining 1777 queries. We compare our system with the state of the art unsupervised entity linking system, AIDA [9,13]. Additionally, we also compare with Lucene-based IR since our method uses that as the first phase in our approach; we use the same two lucene indexes, one that indexes article text and another than uses anchor texts. We use the labeled entities for each query (from the dataset) as true labels, and use Precision, Recall and F-Measure to evaluate each technique.

Table 3. Experimental Results

Method	Precision	Recall	F-Score
AIDA-CockTailParty	5.41%	3.21%	3.84%
AIDA-PriorOnly	5.73%	3.34%	4.00%
IR-Article	9.59%	7.41%	7.88%
IR-Anchor	25.81%	22.17%	22.65%
Our Method	**36.46%**	**30.71%**	**31.60%**

Experimental Results:
The summary of the results appear in Table 3; the best-performing variants of AIDA were seen to reach only up to 4% on the F-measure. This confirms our initial hunch that usual entity linking considerations do not work well on short texts. The IR baseline on anchor texts is seen to score 22.65 on F-Measure. The 9 point improvement that our method achieves over this is illustrative of the value that graph considerations (in the second and third phases) bring in, to our technique. Our study confirms that our method would be the preferred method for entity linking on web search queries. Further, randomization tests show that the improved performance of our method is statistically significant at a p-value of < 0.01. The example result in Table 4 is illustrative of the difference in flavor between our method and the best baseline; the differ only in the third result where our method correctly prioritizes *Northern Fleet*. This is enabled due to the graph proximity consideration in our approach (IR-Anchor only considers text similarity). It may also be noted that both these methods are being able identify deeper semantic linkages as compared to AIDA (Table 1) that does mention detection upfront. Though standard IR evaluation metrics such as MAP, MRR and NDCG [6] are precision-oriented and not very popular to evaluate entity linking, we analyzed the top-10 results from our approach on these measures. Our approach scored **0.49** (IR-A:0.38), **0.50** (IR-A:0.39) and **0.55** (IR-A:0.44) on **MAP**, **MRR** and **NDCG** respectively on an average across queries; all these metrics are in the [0-1] range. Our parameter uses a parameter t that is inversely related to the number of entities to be retrieved; Table 5 plots the trends across varying values of t. Our method is robust as it produces stable F-score.

[5] http://webscope.sandbox.yahoo.com/catalog.php?datatype=l
[6] http://nlp.stanford.edu/IR-book/html/htmledition/evaluation-of-ranked-retrieval-results-1.html

Table 4. Example Results

Query: russian submarine kursk fleet	
IR-Anchor	**Our Method**
Russian Submarine K-141 Kursk	Russian Submarine K-141 Kursk
Russian Submarine Kursk explosion	Russian Submarine Kursk explosion
Fleet Submarine	Northern Fleet

Table 5. Varying t

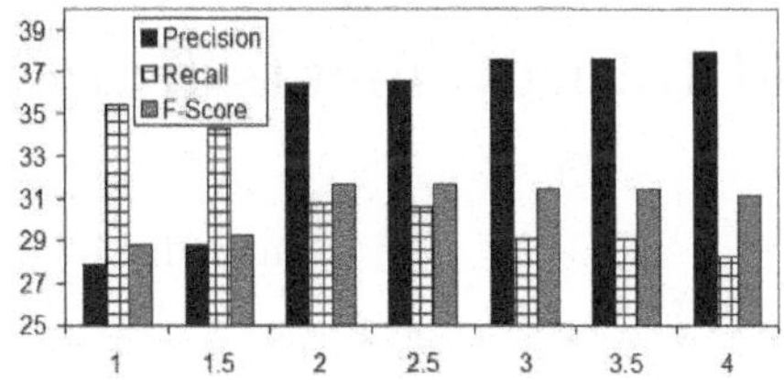

5 Conclusions

In this paper, we considered the problem of entity linking on web search queries. We outlined reasons as to why traditional entity linking algorithms designed for documents do not perform well on very short text fragments such as web search queries, and argued that considering queries in entirety would be beneficial for entity linking on them. We then proposed a method that scores entities based on both textual similarity and graph proximity within a single framework. Our empirical analysis on a real-world search query dataset, outperforms the IR-based techniques as well as the state-of-the-art entity linking method by large and statistically significant margins.

References

1. Baeza-Yates, R., Hurtado, C., Mendoza, M.: Query recommendation using query logs in search engines. In: Lindner, W., Fischer, F., Türker, C., Tzitzikas, Y., Vakali, A.I. (eds.) EDBT 2004. LNCS, vol. 3268, pp. 588–596. Springer, Heidelberg (2004)
2. Dalton, J., Dietz, L.: A neighborhood relevance model for entity linking. In: OAIR (2013)
3. Ferragina, P., Scaiella, U.: Tagme: on-the-fly annotation of short text fragments (by wikipedia entities). In: CIKM, pp. 1625–1628 (2010)
4. Habib, M.B., van Keulen, M.: A generic open world named entity disambiguation approach for tweets. In: KDIR. SciTePress, Portugal (2013)
5. Han, X., Sun, L.: An entity-topic model for entity linking. In: EMNLP-CoNLL (2012)
6. Han, X., Sun, L., Zhao, J.: Collective entity linking in web text: A graph-based method. In: SIGIR 2011, pp. 765–774. ACM, New York (2011)
7. He, Z., Liu, S., Li, M., Zhou, M., Zhang, L., Wang, H.: Learning entity representation for entity disambiguation. In: ACL (2), pp. 30–34 (2013)
8. Hoffart, J., Seufert, S., Nguyen, D.B., Theobald, M., Weikum, G.: Kore: Keyphrase overlap relatedness for entity disambiguation. In: CIKM 2012, pp. 545–554 (2012)
9. Hoffart, J., Yosef, M.A., Bordino, I., Fürstenau, H., Pinkal, M., Spaniol, M., Taneva, B., Thater, S., Weikum, G.: Robust disambiguation of named entities in text. In: EMNLP 2011 (2011)
10. Kulkarni, S., Singh, A., Ramakrishnan, G., Chakrabarti, S.: Collective annotation of wikipedia entities in web text. In: KDD, pp. 457–466 (2009)
11. Li, Y., Wang, C., Han, F., Han, J., Roth, D., Yan, X.: Mining evidences for named entity disambiguation. In: KDD 2013, pp. 1070–1078. ACM, New York (2013)
12. Liu, X., Croft, W.B.: Statistical language modeling for information retrieval. Technical report, DTIC Document (2005)
13. Nguyen, D.B., Hoffart, J., Theobald, M., Weikum, G.: Aida-light: High-throughput named-entity disambiguation. In: Linked Data on the Web, WWW (2014)
14. Pilz, A., Paa, G.: Collective search for concept disambiguation. In: COLING 2012 (2012)
15. Yosef, M.A., Hoffart, J., Bordino, I., Spaniol, M., Weikum, G.: Aida: An online tool for accurate disambiguation of named entities in text and tables. PVLDB 4(12) (2011)

Beyond Sentiment Analysis: Mining Defects and Improvements from Customer Feedback

Samaneh Moghaddam

eBay Inc.
2065 Hamilton Ave.
San Jose, California, US 95125
samoghaddam@ebay.com

Abstract. Customer satisfaction is considered as one the key performance indicators within businesses. In the current competitive marketplace where businesses compete for customers, managing customer satisfaction is very essential. One of the important sources of customer feedback is product reviews. Sentiment analysis on customer reviews has been a very hot topic in the last decade. While early works were mainly focused on identifying the positiveness and negativeness of reviews, later research tries to extract more detailed information by estimating the sentiment score of each product aspect/feature. In this work, we go beyond sentiment analysis by extracting actionable information from customer feedback. We call a piece of information actionable (in the sense of customer satisfaction) if the business can use it to improve its product. We propose a technique to automatically extract defects (problem/issue/bug reports) and improvements (modification/upgrade/enhancement requests) from customer feedback. We also propose a method for summarizing extracted defects and improvements. Experimental results showed that without any manual annotation cost, the proposed semi-supervised technique can achieve comparable accuracy to a fully supervised model in identifying defects and improvements.

Keywords: Customer Feedback Analysis, Defect Identification, Improvement Request Extraction, Opinion Mining.

1 Introduction

Customer satisfaction is a measure of how products and/or services supplied by a company meet or surpass customer expectation [17]. In fact, customer satisfaction may be the best indicator of how likely it is that the customers will do further businesses in the future [3]. Therefore, managing and monitoring customer satisfaction is essential for businesses in the current competitive marketplace. Technology has made it easier for companies to obtain feedback from their customers to manage their satisfaction levels. Customers usually share their experience with a product/service provided by a company in review sites, community blogs and forums. These are some the important sources of customer feedback.

A. Hanbury et al. (Eds.): ECIR 2015, LNCS 9022, pp. 400–410, 2015.

In the last decade many researchers have been working on mining opinion and analyzing sentimental text. In most of the early works the main goal was classifying text as positive and negative. While identifying the positiveness and negativeness of feedback provides more insight into the customer experience, this level of information is not useful to assist customers or companies in making business decisions. To answer this need, recent sentiment analysis research was focused on extracting aspect (also called features, e.g., 'zoom', 'battery life', etc. for a digital camera) and estimating their rating from the feedback. Extracted aspects and their estimated ratings provide more detailed information for the customers to make business decisions (whether to buy an item or use a service). However, this level of information is mainly useful for customers. Companies need more refined level of information, e.g., why customers dislike a specific aspect? or how can we improve that? While sentiment analysis provides a good indicator of user satisfaction, companies need more actionable information to improve it. Actionable information can be defined as the data that can be used to make specific business decisions. In this work, we propose a technique based on distant learning [10] to extract actionable information from customer feedback. Our method automatically extracts defects and improvements from customer feedback. Improvement request is a verbatim that explicitly suggests company to add/change/improve/stop specific aspects. Defect report is a verbatim that explicitly points out a difficulty/error/bug/inability in the product.

Our proposed method extracts and summarizes defects and improvements from customer feedback in order to assist companies to improve their customer satisfaction. Experimental results on a large real-life dataset showed the proposed semi-supervised technique can provide the same accuracy level as the fully supervised SVM model in both tasks with no manual annotation cost.

The remainder of the paper is organized as follows. The next section is devoted to related work. Section 3 introduces the problem statement and discusses our contributions. Section 4 presents our proposed method for the considered problem. In Section 5 we report the results of our experimental evaluation. Section 6 concludes the paper with a summary and the discussion of future work.

2 Related Work

While there is a lot of literature regarding opinion mining, we could not find many works related to extraction of detailed information from feedback to benefit the companies. In the following we review research works focusing on similar problems to ours.

Aspect-based opinion mining performs fine-grained analysis to discover sentiments on aspects of items (e.g., 'battery life', 'zoom', etc. for a digital camera). Most of the early works on aspect-based opinion mining are frequency-based approaches where some filters are applied on frequent noun phrases to identify aspects [6,7,13]. Later works are mainly model-based techniques that automatically learn model parameters from the data. Some of the proposed models are based on supervised learning techniques (HMM and CRF), however most of the current models are unsupervised topic models and based on Latent Dirichlet Allocation (LDA) [9,11,12,16]. Extracting aspects and ratings are mainly beneficial for the customers to make business decisions (whether to buy a product or use a service).

The works presented in [2,5,14] have the most connection with the problem considered in this work. The authors of [2] defined a set of general semantic patterns (e.g., "a manufacturer entity which is a subject of a modal verb used in the past tense and perfective aspect") to identify suggestions. This work is further extend in [15] to use the extracted opinions and suggestions to improve item recommendation. In [14] a set of Part-of-Speech (PoS) patterns (e.g., "modal verb + auxiliary verb + <positive opinion>") are defined to identify suggestions and also purchase wishes (wish-list items). Finally, in [5] an SVM classifier is trained for identifying wishes. In addition to Bag-of-Words the classifier also used some predefined binary wish template features (e.g., "I wish ...") to classify text as wish/not-wish. In this work, we propose a semi-supervised technique to identify not only improvement requests (also called suggestions and wishes), but also defects from customer feedback and we use the methods proposed in these related works as our comparison partners.

As our proposed method is based on Distant Learning we briefly discuss some of the related works in this area. Distant learning is first proposed in [10] to extract relational facts from text (e.g., learning that a person is employed by a particular organization). Whereas the supervised learning that needs a labeled corpus, distant supervision uses noisy signals in text as positive labels to train classifiers. This approach is further used in other problem spaces such as sentiment classification [4] and topic identification [8]. In [4] a list of emotion icons, for example :) and :(are used as noisy signals to train the sentiment classifier. In [8] a set of keywords are used for each class to train the classifier, e.g., obama and biden for politician class. In our work, we propose to use distant learning to identify defects and improvements.

3 Problem Statement and Contribution

Let $P = \{P_1, P_2, ..., P_M\}$ be a set of items (products/services) provided by the company. For each item P_i there is a set of feedback comments $R_i = \{d_1, d_2, ..., d_N\}$ provided by the customers. In some of the feedbacks customers proactively report a defect or request an improvement. In the following we define defect, improvement and the problem addressed more formally:

Improvement: Feedback that explicitly suggests/requests company to add/change/improve/stop specific aspects. In other words, the customer proactively proposes a product change/improvement to the company. For example,

- "The only thing I would like to see on this mobile app is the option to send an invoice and print a shipping label."
- "Needs ability to add an item to a specific watch list and a way to organize watch lists."

Defect: Feedback that explicitly points out a difficulty/error/bug/inability in the product. In other words, the customer reports one or more aspects of the product that does not work properly or need to be fixed. For example,

- "It lacks the ability to move a saved item from your backer back into your basket again for purchase."
- "You cannot send invoice through this App."

Problem Definition: Given a set of reviews about item P_i, the task is to identify the major defects reported about P_i and also to extract the major improvements requested by customers from the set of feedback comments R_i.

In general, this problem consists of two main tasks: 1) Extracting defect reports and improvement requests from feedback, 2) Grouping and summarizing the extracted defects and improvements.

4 Proposed Method

In this section, we first describe the proposed method for identifying defects and improvements and then discuss the summarization technique.

4.1 Mining Defects and Improvements

A simple approach for identifying defect and improvement feedback is traditional classification where the classifier is trained using the labeled feedback. Preparing a labeled dataset for classification is usually cost expensive, time consuming and labor intensive. Manual annotation cost for training a reliable classifier on imbalanced datasets, where the class distribution is not uniform among the classes, is even higher. To this end, we propose to apply distant learning to identify defects and improvements.

Distant learning uses noisy signal in text as positive labels to train classifiers [10]. The intuition of distant supervision is that any instance with a noisy class label is likely to belong to that class in some way. Since there may be many instances labeling with a given class, we can extract very large numbers of (potentially noisy) features that are combined in a classifier. In our problem, we define a set of trivial patterns for identifying defects/improvements and consider feedback extracted by these patterns as positive cases in training the distant learning model. Although using the results of patterns as positive cases can result in false positive, it provides supervision from a distance.

To prepare noisy labels for training the distant learning model, we first manually analyzed a set of user feedback for eBay App reviews with the goal of identifying various ways in which users report a defect or request an improvement. Our finding revealed that around 20% of user feedback contain some forms of defect report and/or improvement request. Based on this investigation, we defined a set of lexical-PoS (Part-of-Speech) patterns for each task. We came up with eight patterns to extract improvement requests and five patterns to find defect reports. Tables 2 and 1 show the found patterns and a sample sentence segment.

Table 1. Extracted patterns to identifying defects

Pattern	Example (selected sentence segment)
NEG (allow|let) USER	The shipping options would not allow me to put in the exact weight and dimensions of the package.
NEG (option|ability)	I have no ability to directly access my pay pal account from this App.
I NEG like	I do not like how I have to reset my search settings each and every time.
I cannot VB	I cannot pay for anything from my phone.
(bug|crash|error)	Only one thing and it happens very little when you list something and hit continue it starts over not saving anything so you gave to start over you guys need to fix that bug.

Table 2. Extracted patterns to identifying improvements

Pattern	Example (selected sentence segment)
there should be (DT)	There should be a reply to all button so I can do 100 items in just 2 minutes.
(allow|let) USER to	Allow us to open and pay the invoices.
VB (DT) option	Give me the option to search for auctions in Europe.
I wish COMPANY	I wish eBay would make my eBay emails open in the app instead of safari.
MD be (ADV) ADJ	From the standpoint of store owner would be very helpful to be able to create sale put on and off vacation mode and edit categories.
MD (like|prefer|love) to	i would love to have one button to remind all my buyer to leave feedback with my message and link to the feedback page.
stop VBG	Please stop sending me an email every time a bid happens.
ability (to|of)	Only thing I would change is to have the ability to do searches by years

In these tables DT, VB, MD, ADV, ADJ, VBG and NEG indicate determiner, verb, modal, adverb, adjective, gerund verb a negation term, respectively.

One can easily find more accurate patterns by applying a pattern mining technique on a set of labeled feedback. However, our goal here is not to find accurate patterns, but to find a trivial way to label feedback as defect or improvement. In Section 5, we will compare the result of the distant learning method with the methods proposed in the literature (unsupervised patterns and supervised SVM).

4.2 Summarizing Defects and Improvements

While categorizing customer feedback as 'reporting a defect' or 'requesting an improvement' is very useful for the business owner to improve their products, it is still hard to read through every extracted feedback. Therefore, in this section we propose a technique for summarizing the extracted defects and improvements.

We first classify sentences in each feedback as containing defect report and/or improvement request. We train SVM classifiers using manually labeled data to classify each sentence in the extracted feedback as positive or negative case. As our baseline, we use the defined patterns to identify target sentences (sentences containing defect report or improvement request). For each task one classifier is trained on all feedback independent from their feedback-level labels and another is trained on only feedback positively labeled for that task (i.e., defect sentence classifier trained only on feedback labeled as defect).

To summarize the identified sentences, we propose to apply Latent Dirichlet Allocation (LDA). This approach not only clusters similar feedback, but also identifies the top n topics of that cluster. We applied LDA on simple bag-of-words (BoW) as our baseline and compare the results with extracted topics from bag of noun phrases, verb phrases, and bi-terms.

5 Experiments

In the sections, we first briefly describe our dataset and then present the evaluation of the proposed technique.

5.1 Dataset

To evaluate the proposed method, we performed experiments on a large real-life dataset of customer feedback from "eBay App Reviews". This dataset contains 50,000 reviews written by eBay App customers on App store and Google Play. Each app review is annotated by 5 human judges with the following four categories: improvement request, defect report, both. other.

In this annotation, if a review is labeled as defect report or improvement request, the sentence(s) containing defect/improvement is also identified. In this dataset, 15% of feedback are labeled as improvement request, 8% are labeled as defect reports, and 1% are labeled as both. The judges' agreement for defect and improvement labels were 100% and 96%, respectively. In our experimental evaluation we only used cases with 100% agreements for training and testing. In the following we evaluate the accuracy of the proposed method on this dataset.

5.2 Evaluation

We held out 20% of the reviews for testing purposes and used the remaining 80% to train the model. To evaluate how well a method works, we computed precision, recall and F-measure of the held-out test set. We compare the result

Table 3. Precision, recall and F-measure of different methods for defect and improvement extraction

Task	Defect Extraction			Improvement Extraction		
Method	Precision	Recall	F-measure	Precision	Recall	F-measure
SVM	0.44	0.87	0.58	0.38	0.78	0.51
Patterns	0.61	0.29	0.40	0.64	0.21	0.32
Distant Learning	0.4	0.91	0.56	0.32	0.74	0.46

of our method with the method proposed in [5] (SVM classification) and also the one proposed in [2] (Patterns). We train an SVM classifier for each problem using the bag-of-word features. For Patterns, we use the patterns we discussed in Section 4.1.

Table 3 shows the results of the proposed method and the comparison partners. We observe that the precision, recall and F-measure of all methods (with an exception in precision of Patterns) are higher in extracting defects than those of improvement extraction. We believe this is related to the balance of the data sets, as the training data set for defect reports is more balanced than the other one (in the App review data, 15% of feedback reporting a defect, while only 8% requesting an improvement).

Comparing the results of SVM and Patterns, we can see in both tasks, patterns could achieved higher precision than SVM. The recall of patterns, however, in both tasks is lower than SVM as patterns cannot find defects/improvements in complex sentences. We also observe that distant learning improved the recall in both tasks. An interesting finding here is the closeness of precision and recall of distant learning and those of SVM classifier. This finding is very important as SVM is a fully supervised technique while the distant learning method only used noisy labels assigned by the patterns. In other words, in training the distant learning method no manually labeled data is used. Comparing the F-measure values, SVM and distant learning are performed quite similar in defect extraction and very close in improvement extraction task. Considering the cost of manual annotation for training a reliable classifier, especially in highly imbalance datasets such as ours, the proposed method is the clear winner.

To evaluate the summarization technique, we compare the precision and recall of the trained sentence classifier with those of the patterns. We also report quantitative and qualitative evaluations of the generated topic summary. Table 4 shows the accuracy of the trained SVM classifier and the patterns in identifying sentences reporting a defect or requesting an improvement. In the first two rows (SVM and Patterns), we report the results of the classifiers trained using all sentences without considering their feedback-level labels. For example, the sentence classifier for improvement used sentences from all feedback in the training data (whether the feedback is labeled as improvement or not) to train the model. However, in practice it makes more sense to apply the sentence classifier on positively labeled feedback (e.g., apply improvement sentence extraction on

Table 4. Precision, recall and F-measure of different methods for target sentence identification

Task	Defect Extraction			Improvement Extraction		
Method	Precision	Recall	F-measure	Precision	Recall	F-measure
SVM	0.21	0.89	0.34	0.20	0.75	0.32
Patterns	0.60	0.24	0.34	0.70	0.25	0.37
SVM-Labeled	0.68	0.47	0.55	0.71	0.61	0.65
Patterns-Labeled	0.73	0.27	0.40	0.88	0.24	0.38

only improvement feedback). To this end, we further evaluate the performance of both methods on positively labeled feedback (SVM-Labeled and Patterns-Labeled), e.g., the sentence classifier for improvement only used sentences from feedback labeled as improvement.

Comparing the results of SVM and patterns show that patterns are more precise than the trained SVM in identifying defect/improvement sentences, however they miss many positive cases (lower recall). Comparing SVM-labeled and Patterns-labeled with SVM and Patterns, we can see an increase in precision for both tasks. This was predictable as we only feed positive cases (feedback labeled as defect/improvement) to the classifier. While Patterns-labeled provide a higher precision than SVM-labeled, its recall is much lower (again since patterns cannot find defects/improvements in complex sentences). Comparing the F-measure values, we observe that in the absence of feedback-level labels, SVM and patterns perform the same (patterns slightly outperforms in improvement extraction). Utilizing feedback-level labels could only improve the performance of the patterns slightly (Patterns-labeled). SVM classifier, on the other hand, achieved the best results using feedback labels. To summarize, same as feedback-level classification, patterns can provide a very close performance to SVM with no manual cost. However, in the case of having labels for feedback, the trained SVM (SVM-labeled) outperforms patterns.

In the last step we summarize the extracted sentences by applying the Latent Dirichlet Allocation (LDA) [1] on different feature sets. In our experiments we trained LDA models using the following features: bag-of-words, nouns, verbs, noun phrases, verb phrases and bi-grams. We evaluate the models qualitatively by providing the top generated topics for each feature set. Tables 5 shows the top extracted defect reports and improvement requests using different models. Comparing topics, we observe that the extracted topics from bi-grams and phrase features are more informative than single-word features (i.e., bag-of-words, nouns and verbs). For example, 'slow load[ing]', '[cannot] send invoice' and 'freez[ing]' are vital defects that need to be fixed promptly. On the other hand, the ability to 'delete unsold [item]' or the option to 'pay [for] multiple [items]' are valuable improvement suggestions that product owner can consider in the next version.

We also compute precision, recall and F-measure of the extracted topics using different feature sets. Table 6 reports the value of these measures for $k = 20$. We

Table 5. Qualitative evaluation of sentence summarizations ($k = 20$)

Features	Top extracted defect topics
Bag-of-Words	item, option, app, load, update, search, slow, work, crash
Nouns	fix, error, list, watch, item, invoice, search, crash, battery, app, issue
Verbs	update, search, drop, load, hate, freeze, find, save, list, send, sell, leave
Noun Phrases	previous version, listing, watch item, crash, great app, payment option
Verb Phrases	load, leave feedback, purchase item, stop working, freeze, fix
bigrams	slow load, send invoice, app crash, save search, sell item, shipping label
	Top extracted improvement topics
Bag-of-Words	seller, invoice, app, ship, sort, unsold, multiple, option,
Nouns	app, design, item ,view, account, tracking, watch, invoice
Verbs	enjoy, work, edit, send, update, ship, attach, search, add, improve
Noun Phrases	good app, multiple item, search preference, combine invoice
Verb Phrases	search seller, send invoice to buyer, delete sold item, buy item
bigrams	delete unsold, nice app, pay multiple, contact seller, combine invoice

Table 6. Precision, recall and F-measure of the extracted topics using different feature sets for defects and improvement extraction ($k = 20$)

Task	Defect Extraction			Improvement Extraction		
Feature	Precision	Recall	F-measure	Precision	Recall	F-measure
Bag-of-Words	0.63	0.69	0.67	0.61	0.55	0.58
Nouns	0.54	0.69	0.61	0.45	0.51	0.48
Verbs	0.59	0.79	0.68	0.51	0.59	0.54
Noun Phrases	0.47	0.55	0.51	0.53	0.58	0.55
Verb Phrases	0.63	0.77	0.70	0.65	0.68	0.67
bigrams	0.72	0.89	0.70	0.65	0.77	0.71

used a gold set of actual aspects for our test set to evaluate different models. Note that, each aspect in the gold set is represented by a set of synonym words or phrases (e.g., ‘shipping label’, ‘label’, ‘print shipping label’ and ’print label’). Comparing the results, we observe that topics extracted from bi-grams and verb phrases are closer to the gold set (as we saw in Table 5 too). Among these two, bi-grams could achieve the highest precision and recall in both tasks, meaning bi-gram feature are most suitable for identifying summary topics from feedback.

6 Summary and Future Work

Most recent works on mining customer feedback were mainly focused on extracting product aspects and estimating their rating from the feedback. While extracted aspects and ratings provide more detailed information for the customers to make purchase/usage decisions, businesses usually need more detailed information to make business decision. In this work we proposed a technique to extract actionable information from customer feedback. Our method automatically extracts improvement requests and defect reports from customer feedback. Our method also summarizes extracted defect and improvement in order to assist companies to improve their customer satisfaction.

We proposed to use a set of trivial lexical-PoS patterns to prepare positive cases for training a distant learning method. Experimental results on a large real-life dataset from eBay App reviews showed the proposed semi-supervised technique can achieve a comparable accuracy to the fully supervised SVM technique with no manual annotation cost.

The results of this work suggest several direction for future research. Our proposed approach works best for feedback that contains explicit defect reports or improvement requests. However, in many feedback customers implicitly report a defect or request an improvement, e.g., "In addition to mm, you need to show if a small, regular or large band (length) [is available.] I need a small. Thank You" implicitly suggests to add a filter for size to the search engine. Identifying implicit defects and improvements is a hard but practical problem. Another future direction is identifying the correlation between of the accuracy of noisy signals (patterns) and that of the distant learning method. Finally, investigating the accuracy of the proposed approach in other problem spaces (e.g., spam detection) would be very interesting.

References

1. Blei, D.M., Ng, A.Y., Jordan, M.I.: Latent dirichlet allocation. J. Mach. Learn. Res. 3, 993–1022 (2003)
2. Hagege, C., Brun, C.: Suggestion mining: Detecting suggestions for improvement in users' comments. Research in Computing Science 70, 199–209 (2013)
3. Farris, P.W., Bendle, N.T., Pfeifer, P.E., Reibstein, D.J.: Marketing Metrics: The Definitive Guide to Measuring Marketing Performance, 2nd edn. Wharton School Publishing (2010)
4. Go, A., Bhayani, R., Huang, L.: Twitter sentiment classification using distant supervision. Processing, 1–6 (2009)
5. Goldberg, A.B., Fillmore, N., Andrzejewski, D., Xu, Z., Gibson, B., Zhu, X.: May all your wishes come true: A study of wishes and how to recognize them. In: Proceedings of Human Language Technologies: The 2009 Annual Conference of the North American Chapter of the Association for Computational Linguistics, NAACL 2009, pp. 263–271. Association for Computational Linguistics, Stroudsburg (2009)
6. Hu, M., Liu, B.: Mining and summarizing customer reviews. In: KDD (2004)

7. Liu, B., Hu, M., Cheng, J.: Opinion observer: Analyzing and comparing opinions on the web. In: WWW 2005 (2005)
8. Marchetti-Bowick, M., Chambers, N.: Learning for microblogs with distant supervision: Political forecasting with twitter. In: Proceedings of the 13th Conference of the European Chapter of the Association for Computational Linguistics, EACL 2012, pp. 603–612. Association for Computational Linguistics, Stroudsburg (2012)
9. Mei, Q., Ling, X., Wondra, M., Su, H., Zhai, C.X.: Topic sentiment mixture: Modeling facets and opinions in weblogs. In: WWW 2007 (2007)
10. Mintz, M., Bills, S., Snow, R., Jurafsky, D.: Distant supervision for relation extraction without labeled data. In: Proceedings of the Joint Conference of the 47th Annual Meeting of the ACL and the 4th International Joint Conference on Natural Language Processing of the AFNLP: Volume 2, ACL 2009, pp. 1003–1011. Association for Computational Linguistics, Stroudsburg (2009)
11. Moghaddam, S., Ester, M.: The flda model for aspect-based opinion mining: Addressing the cold start problem. In: WWW 2013 (2013)
12. Moghaddam, S., Ester, M.: ILDA: interdependent LDA model for learning latent aspects and their ratings from online product reviews. In: SIGIR 2011 (2011)
13. Moghaddam, S., Ester, M.: Opinion digger: An unsupervised opinion miner from unstructured product reviews. In: CIKM 2010 (2010)
14. Ramanand, J., Bhavsar, K., Pedanekar, N.: Wishful thinking: Finding suggestions and 'buy' wishes from product reviews. In: Proceedings of the NAACL HLT 2010 Workshop on Computational Approaches to Analysis and Generation of Emotion in Text, CAAGET 2010, pp. 54–61. Association for Computational Linguistics, Stroudsburg (2010)
15. Stavrianou, A., Brun, C.: Opinion and suggestion analysis for expert recommendations. In: Proceedings of the Workshop on Semantic Analysis in Social Media, pp. 61–69. Association for Computational Linguistics, Stroudsburg (2012)
16. Titov, I., McDonald, R.: Modeling online reviews with multi-grain topic models. In: WWW 2008 (2008)
17. Wikipedia. Customer satisfaction, `http://en.wikipedia.org/wiki/Customer_satisfaction`

Measuring User Influence, Susceptibility and Cynicalness in Sentiment Diffusion*

Roy Ka-Wei Lee and Ee-Peng Lim

Living Analytics Research Centre,
Singapore Management University, Singapore
{roylee.2013,eplim}@smu.edu.sg

Abstract. Diffusion in social networks is an important research topic lately due to massive amount of information shared on social media and Web. As information diffuses, users express sentiments which can affect the sentiments of others. In this paper, we analyze how users reinforce or modify sentiment of one another based on a set of inter-dependent latent user factors as they are engaged in diffusion of event information. We introduce these sentiment-based latent user factors, namely *influence*, *susceptibility* and *cynicalness*. We also propose the *ISC model* to relate the three factors together and develop an iterative computation approach to derive them simultaneously. We evaluate the ISC model by conducting experiments on two separate sets of Twitter data collected from two real world events. The experiments show the top influential users tend to stay consistently influential while susceptibility and cynicalness of users could changed significantly across events.

Keywords: Twitter network, sentiment diffusion.

1 Introduction

Motivation. Psychological research had shown that emotion induces and boosts social transmission of information [1,9,10]. In the context of online social networks, social transmission occurs mainly in the form of information diffusion. As social media becomes pervasive and users spend much time using them, it is now both important and feasible to study sentiment and user behavioral characteristics in information diffusion.

People generally believe that content with negative sentiment diffuse more readily than content with positive sentiment. Thelwall et al found that negative sentiment strength is more prevalent for popular events mentioned in Twitter [13]. Stieglitz and Linh conducted a study of political tweets and found that sentiment-charged tweets are more likely to be retweeted than neutral ones [12]. Tumasjan et al performed research on predicting the German Federal election outcome using sentiment charged tweets from Twitter users. They found that

* This research is supported by the Singapore National Research Foundation under its International Research Centre @ Singapore Funding Initiative and administered by the IDM Programme Office.

A. Hanbury et al. (Eds.): ECIR 2015, LNCS 9022, pp. 411–422, 2015.

the online sentiments closely follow the political landscape of Germany during the period of time [14].

Most of the above studies, however, only considered the effects of the sentiment in tweet content, while neglecting the effects of user characteristics on driving the outcome of diffusion with sentiments. Consider the following scenario. When a user v expressed a sentiment towards a piece of content introduced to him by a friend, there are two possibilities. The first is that *there is no sentiment in the original content.* The sentiment from v is new suggesting that v has intrinsic sentiment towards the content. The second possibility is that *sentiment is found in the original content.* In this case, the sentiment from v can be affected by the sentiment-charged content from his friend. How likely v expresses sentiment and what sentiment polarity v would adopt for the incoming content would depend on the characteristics of both him and his friend.

Research objectives and contributions. In this paper, we aim to identify and model latent user characteristics that contribute to sentiment-charged content diffusion in a social network. In other words, we focus on the cases whereby users express sentiments after receiving content that carries sentiment. The adoption of same sentiment polarity by v from his friend (say u) may be due to: (i) the influential personality of u, (ii) the susceptibility of v to follow the sentiment polarities from others, or (iii) v's intrinsic sentiment polarity towards the diffused content. If v adopts a sentiment polarity opposite to that of the content diffused from u, this again may be due to: (i) the influential power of u, (ii) the cynicalness of v, and (iii) v's intrinsic sentiment polarity towards the diffused content. The three latent user characteristics, influence, susceptibility and cynicalness, are the focuses of this research. The intrinsic sentiment of v towards diffused content is a user-topic specific characteristics. In this research which focuses on user characteristics only, we assume that u is intrinsically neutral on any diffused content, and leave the user-topic characteristics to our future work.

We will focus on quantifying the three user characteristics, influence, susceptibility and cynicalness. We define the influence of a user to be how easy he or she could swing the sentiments of other users towards his, the susceptibility to be how easy the user adopts the same sentiments diffused by other users, the cynicalness to be how easy the user adopts opposing sentiments diffused by other users. The inter-dependency among the three user characteristics suggests that we need a model that derives them altogether. The scenario here is similar to HITS model where both authority and hub characteristics of web pages are to be measured together[7]. Our problem context is relatively more complex as there are three quantities to be measured. The involvement of content and sentiment polarity further complicates the model definition.

This work improves the state-of-the-art of user modeling in sentiment diffusion. To the best of our knowledge, there has not been any other work addressing the same research, i.e., considering sentiment diffusion in user characteristics modeling. Our main contributions in this work are as follows:

- We introduce user influence, susceptibility and cynicalness as the latent user characteristics affecting sentiment diffusion. These user characteristics are quantifiable and they together help to explain sentiment diffusion.
- We propose a novel model called ISC that utilizes the inter-dependency between the three characteristics to measure their corresponding values simultaneously.
- We develop an iterative computation algorithm to compute the model. The algorithm is simple and be easily implemented.

We also applied the proposed model and conducted a series of experiments on two separate Twitter datasets from two highly discussed real world events. Some of the interesting findings from the experiments include:

- Vast majority of users are non-influential, non-susceptible, and non-cynical.
- The top influential users, which are mainly news media and celebrities, tend to remain consistantly influential across the two real world events.
- The susceptibility and cynicalness of users could change significantly across events.

Paper outline. The rest of the paper is organized as follows. Section 2 reviews the literature related to our study. Section 3 presents our proposed ISC model for user characteristics relevant to sentiment diffusion. The experiments on the Twitter datasets gathered for two real world social events are described in Section 4. Section 5 highlights the experiment results and analysis before the conclusion in Section 6.

2 Related Work

The effects of emotions on information diffusion has been examined in both the psychology and information systems fields. Berger, in his psychological research, showed that emotions characterized by high arousal such as anxiety or amusement are likely to boost social transmission of information more than emotions characterized by low arousal such as sadness or contentment [1]. Other psychological researchers had also conducted similar experiments and obtained similar findings [9,10].

In computer science, a number of research projects studied the effects of emotion in information diffusion for social networks such as Twitter. Stieglitz and Linh conducted a research study on political tweets in Twitter and found that sentiment-charged tweets are more likely to be retweeted than neutral ones [12]. Hansen et al, in their work, shown that negative news contents and positive non-news content are more likely to be retweeted by users in Twitter network [4]. Other research works had also shown that popular life events tend to generate more sentiment-charged tweets [13,2]. These studies, though extensive, did not cover the latent user characteristics that contribute to sentiment diffusion.

There are some recent studies on latent user characteristics in information diffusion for social networks. Hoang et al proposed to measure the virality of Twitter users by their efforts in tweeting and retweeting viral tweets [6]. Janghyuk et al also conducted a study to measure the virality of a user in a marketing campaign by the amount of time taken by the user's friends to respond to the user's recommendation, and the number of unique friends that the user sends his recommendation after adopting an item [8]. Besides measuring virality of users, there are also research works on the susceptibility of users adopting an item in information diffusion [5]. Unlike these works, our paper considers sentiment in defining user characteristics. Hence, these user characteristics are unique. In particular, we introduce susceptibility and cynicalness according to the user's tendency to change of sentiment polarity.

3 User Model for Sentiment Diffusion

In this section, we introduce our proposed user model for sentiment diffusion. We first define sentiment diffusion as an instance of sentiment diffusing from one user to another. Based on a collection of sentiment diffusions, our proposed user model is then defined.

3.1 Sentiment Diffusion Representation

Tracking sentiment diffusion in the midst of many tweets received and generated by users is non-trivial. An approach to this is to focus on diffusion via retweeting whereby a user is said to be diffused when he retweets an incoming tweet. This approach is however very restrictive in the context of sentiment diffusion as it does not account for the case whereby the user generates a new "relevant" sentiment-charged tweet (instead of a retweet) after receiving an incoming sentiment-charged tweet. To identify the tweets (which also include retweets) relevant to sentiment diffusion, we have chosen to define sentiment diffusion for an event accordingly. In our experiments, we determine event tweets by a combination of event relevant keywords and user community. Similar and more sophisticated techniques [11,3,15] to find event tweets are available but are outside the scope of this paper.

We represent a set of users $i \in U$ and their *follower-followee* relationships by a directed graph $G = (U, E)$. A directed edge $(v, u) \in E$ represents v follows u. Here, an item refers to a tweet and the item sentiment x refers to the sentiment of a sentiment-charged tweet. Sentiment charged tweets, in the context of this study, are tweets that reveal the polarity, i.e. positive, negative or neutral, of the publishing user's sentiment on a certain event. We let $X(u)$ to denote the set of item sentiments that user u adopts. We give more notations and their definitions in Table 1.

Figure 1 illustrates an example of sentiment diffusion. User u adopts a positive $(+)$ item sentiment while users $v1$ and $v2$, who are followers of u, had previously adopted neutral (0) item sentiment. Subsequently, $v1$ follows u's sentiment polarity and adopts a positive item sentiment while $v2$ adopts a negative $(-)$ item

Table 1. Notation

$x(v)$	Item sentiment x adopted by user v before diffusion
$x'(v)$	Item sentiment x adopted by user v after diffusion
$X(u)$	Set of item sentiments adopted by user u
$X_d^{\rightarrow}(u)$	Set of item sentiments diffused by user u
$X_{ds}^{\leftarrow}(v)$	Set of item sentiments diffused to user v and v adopts the same item sentiment
$X_{do}^{\leftarrow}(v)$	Set of item sentiments diffused to user v and v adopts the opposite item sentiment
$X_i^{\leftarrow}(v)$	Set of item sentiments introduced to user v
$F_d^{\rightarrow}(u,x)$	Set of followers whom user u diffuses item sentiment x to
$F_{ds}^{\leftarrow}(v,x)$	Set of followees who diffuse item sentiment x to user v and v adopts the same item sentiment
$F_{do}^{\leftarrow}(v,x)$	Set of followees who diffuse item sentiment x to user v and v adopts the opposite item sentiment
$F_r(u)$	Number of followers of user u
$F_e(u)$	Number of followees of user u
$d^{\rightarrow}(u)$	Number of times user u diffused sentiment to his followers
$d^{\leftarrow}(u)$	Number of times user u is diffused sentiment by his followees

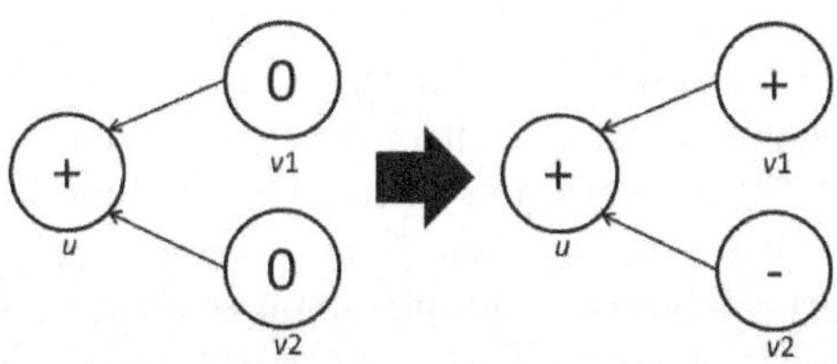

Fig. 1. Sentiment Item Diffusion

sentiment which is opposite to u's. At the point of $v1$ and $v2$'s sentiment adoption, the positive (+) item sentiment adopted by user u was the latest received tweet on $v1$ and $v2$'s Twitter timelines.

We say that u diffuses item sentiment x to v, if all the following conditions hold:(1) u adopts x before v adopts the same or opposing item sentiment x'. (2) v is a follower of u when v adopts x'. (3) u's tweet with sentiment x is the latest received tweet on v's Twitter timeline.

We assume that each user may receive and generate multiple item sentiments relevant to the same event. As a user adopts a sentiment item, he also introduces the item sentiment to his followers. A user can therefore diffuse item sentiments from multiple followees to multiple followers. We denote the set of item sentiments diffused by u to his followers by $X_d^{\rightarrow}(u)$. We also use $X_{ds}^{\leftarrow}(v)$ to denote the set of item sentiments diffused to v by his followees and v adopts the same sentiment polarities, and $X_{do}^{\leftarrow}(v)$ to denote set of item sentiments diffused to v and v adopts the opposite sentiment polarities. Every item sentiment $x(u)$ from user u has a value of 1 if x is positive, -1 if x is negative and 0 if x is neutral.

3.2 Proposed User Model

Sentiment diffusion in a network is an outcome of interactions among users. Depending on the characteristics of the diffusing and diffused users, the sentiment diffused may change accordingly. Thus, we propose a **Influence-Susceptibility-Cynical (ISC) Model** that measures user influence, susceptibility and cynicalness simultaneously based on a set of principles that help to distinguish each latent user characteristics from others. The three principles are:

- An influential user can get others, particularly the non-susceptible users and cynical users, to change and adopt the same item sentiment diffused by him.
- A susceptible user adopts same sentiments with sentiment-charged items diffused to him by non-influential users.
- A cynical user adopts opposite sentiments with sentiment-charged items diffused to him by non-influential users.

We denote the influence, susceptibility and cynicalness of a user u by $I(u)$, $S(u)$ and $C(u)$ respectively. $I(u)$ is assigned a value between 0 (denoting non-influential user) and 1 (denoting most influential user). The same applies to $S(u)$ and $C(u)$.

One of the important components of this study is the definition of change in sentiment of a follower v as a result of user u's influence. We represent this change in sentiment as $\triangle x(u, v)$ and introduce two functions to capture this change in sentiment. The first function, $f_s(x(u), x(v), x'(v))$, returns the change in sentiment when the follower v adopts same sentiment diffused by user u. Another function, $f_o(x(u), x(v), x'(v))$, returns the change in sentiment when follower v adopts the opposite sentiment diffused by user u. Both of the functions take in three parameters; $x(u)$ is the item sentiment value diffused by user u, $x(v)$ is initial item sentiment value adopted by follower v before the sentiment diffusion and $x'(v)$ is the item sentiment value adopted by v after the sentiment diffusion. Tables 2 and 3 show the definitions of $f_s(x(u), x(v), x'(v))$ and $f_o(x(u), x(v), x'(v))$ respectively. Unless specified in the tables, the function values are zero by default.

As shown in Table 2, the maximum change in sentiment (+2) is observed when v reverses his initial sentiment and adopted the same sentiment diffused by u. For example, v changes from an initial negative sentiment (−1) and adopts the same positive sentiment (1) diffused by u. In this example, the maximum change in sentiment is $|x'(v) - x(v)| = 2$. Likewise, If v changes from an initial neutral sentiment (0) and adopts the same positive sentiment (1) diffused by u, the change in sentiment is $|x'(v) - x(v)| = 1$. As a neutral sentiment diffused by u is not considered a strong sentiment, we assign a small value 0.5 to the change in sentiment when v changes from positive or negative to neutral sentiment due to the neutral sentiment diffusion by u. In contrast, Table 3 shows that the maximum change in sentiment is observed when v reverses his initial sentiment and adopts the opposite sentiment diffused by u. For example, v changes from an initial negative sentiment (−1) and adopts the opposite positive sentiment (1) diffused by u. i.e. maximum change in sentiment is $|x'(v) - x(v)| = 2$.

Table 2. Definition of $f_s(x(u), x(v), x'(v))$

$x(u)$	$x(v)$		
	1 (+ve)	0 (neutral)	−1 (−ve)
1 (+ve)	0	1 if $x'(v) = 1$	2 if $x'(v) = 1$; 1 if $x'(v)=0$
0 (neutral)	0.5 if $x'(v) = 0$	0	0.5 if $x'(v) = 0$
−1 (−ve)	2 if $x'(v) = -1$; 1 if $x'(v) = 0$	1 if $x'(v) = -1$	0

Table 3. Definition of $f_o(x(u), x(v), x'(v))$

$x(u)$	$x(v)$		
	1 (+ve)	0 (neutral)	−1 (−ve)
1 (+ve)	2 if $x'(v) = -1$; 1 if $x'(v) = 0$	1 if $x'(v) = -1$	0
0 (neutral)	0	0	0
−1 (−ve)	0	1 if $x'(v) = 1$	2 if $x'(v) = 1$; 1 if $x'(v) = 0$

We will use $f_s(x(u), x(v), x'(v))$ for $\triangle x(u, v)$ in influence and susceptibility score computation and $f_s(x(u), x(v), x'(v))$ for cynicalness computation.

In Equation 1, the *influence* of a user u is defined by the proportion of adopted items, $X(u)$, that are diffused, weighted by the proportion of diffused users having their sentiment influenced by u, $F_d^{\rightarrow}(u, x)$. Each diffused user v is further weighted by the change in sentiment of v due to u, $\triangle x(u, v)$, and the average of v's inverse *susceptibility*, $1 - S(v)$, and *cynicalness*, $C(v)$. To avoid giving high influence scores to users with very few followers diffusing sentiment well to the latter, we further weigh the influence score with $W_1(u)$, representing the amount of diffusing items from u as shown in Equation 4. N and M are large numbers to keep $W_1(u)$ within the range of $[0, 1]$. In our experiments, N and M are set to be 1000 (as 5% of users having at least 1000 followers) and 500 (as 5% of users have at least diffused sentiment to their followers for more than 500 times) respectively.

In Equation 2, the *susceptibility* of a user v is defined by the proportion of sentiment-charged items introduced to v , $X_i^{\leftarrow}(v)$, that are adopted with the same item sentiments by the set of users introducing the items, $F_{ds}^{\leftarrow}(v)$. Each user u who diffuses the sentiment-charged item to v is further weighted by the change in sentiment, $\triangle x(u, v)$, and his inverse *influence*, $1 - I(v)$. Finally, to avoid giving high susceptibility score to users with very few followees and getting diffused with sentiment, we introduce the weight $W_2(v)$ (see Equation 5) representing the amount of diffused items to. N and W are large numbers to keep $W_2(u)$ within the range of $[0, 1]$. In our experiments, P and Q are set to be 1000 (as 5% of users having at least 1000 followees) and 20 (as 5% of users have at least been diffused sentiment by their followees for more than 20 times) respectively.

In Equation 3, the *cynicalness* of a user v is defined by the proportion of sentiment-charged items introduced to v, $X_i^{\leftarrow}(v)$, that are adopted with the opposite item sentiments by the set of users introducing the items, $F_{do}^{\leftarrow}(v)$. Each user u who diffuses the sentiment-charged item to v is further weighted by the change in sentiment, $\Delta x(u,v)$, and his inverse *influence*, $1 - I(v)$. Similar to susceptibility, we finally include the weight $W_2(v)$ (see Equation 5).

$$I(u) = \frac{W_1(u)}{|X(u)|} \cdot \sum_{x \in X_d^{\rightarrow}(u)} Avg_{v \in F_d^{\rightarrow}(u,x)} \Big(\triangle\, x(u,v) \cdot \frac{(1 - S(v)) + C(v)}{2} \Big) \tag{1}$$

$$S(v) = \frac{W_2(v)}{|X_i^{\leftarrow}(v)|} \cdot \sum_{x \in X_{ds}^{\leftarrow}(v)} Avg_{u \in F_{ds}^{\leftarrow}(v,x)} \big(\triangle\, x(u,v) \cdot (1 - I(u)) \big) \tag{2}$$

$$C(v) = \frac{W_2(v)}{|X_i^{\leftarrow}(v)|} \cdot \sum_{x \in X_{do}^{\leftarrow}(v)} Avg_{u \in F_{do}^{\leftarrow}(v,x)} \big(\triangle\, x(u,v) \cdot (1 - I(u)) \big) \tag{3}$$

$$W_1(u) = \frac{F_r(u)}{F_r(u) + N} \cdot \frac{d^{\rightarrow}(u)}{d^{\rightarrow}(u) + M} \tag{4}$$

$$W_2(v) = \frac{F_e(v)}{F_e(v) + P} \cdot \frac{d^{\leftarrow}(v)}{d^{\leftarrow}(v) + Q} \tag{5}$$

3.3 Model Computation

To compute the ISC model, we employ an iterative computation method. The algorithm first initializes $I(u)$, $S(u)$ and $C(u)$ for all users u's with 0.5. It then computes $I(u)$'s using the initial scores of $S(u)$'s and $C(u)$'s. The computed $I(u)$ values are then used to compute new set of values for $S(u)$ and $C(u)$. This process repeats until the values converge.

We found that the iterative computation method works well for our dataset and could achieve convergence in less than 50 iterations. The proof of convergence for this method is however difficult and we shall leave to the future research.

4 Datasets

In this section, we describe two Twitter datasets that were used to evaluate the ISC model. The first Twitter dataset contains tweets published by users from an asian city in a day within June 2013 where the city experienced the worse haze in its history. As the haze severely affected the livelihood of the local people and the local news media covered it widely, we expect strong sentiments and sentiment diffusion among the local Twitter users. The second Twitter dataset contains tweets published by the same set of users for an riot event which took place on in a day within December 2013. As riots in are rare in this city, the event attracted much attention and aroused strong sentiments within the local social media community. We again expect sentiment diffusion in the data which can be used in our experiments.

We first crawled tweet messages from about 150,000 Twitter users from the city on the events dates; on one day in June 2013 for the haze event and another day in December 2013 for the riot event. We selected tweets that contain keywords and hastags related to the events. These include "haze" and "worsehaze", etc., for the haze event and "riot", "police", etc., for the riot event. A total of 16,190 tweets generated by 5,570 users were collected for the haze event, while 18,933 tweets generated by the same set of 5,570 users were collected for the riot event. We also collected the follower-followee relationship among these users.

Next, we assign sentiment values to tweets using the sentiment classifier C_STANFORD, the Stanford's sentiment scoring API[1], which is widely used sentiment classifier based on maximum entropy. The training of the classifier makes use of tweets that are labeled based on positive and negative keywords and emoticons. The API returns a score of -1, 0, or $+1$ for a tweet detected to have positive, negative, or neutral sentiment respectively. We also assume that a user's previous published tweet was neutral when he published his first tweet.

5 Experiment Results

In this section, we discuss the results of the experiment by first examining the overall distribution statistics of *Influential, Susceptibility* and *Cynicalness* measures of users in both haze and riot events. Next, we compare the ISC model measures results of the two events using Pearson correlation and Jaccard similarity coefficient. Lastly we examine the characteristics of the influential users in greater detail and compare the ISC model influence measure with other traditional influence measures such as *In-Degree* and *PageRank*.

5.1 Distribution Statistics

Examining into the distribution of *influence*, *susceptibility* and *cynicalness* scores of users for both events, there are very few users have very high influence scores while majority of the users have very low or zero influence scores. The same can be said for susceptibility and cynicalness scores. This suggests that there are only few users who are highly influential, susceptible and cynical.

5.2 Comparision of Haze and Riot ISC Results

The pearson correlations of *influence*, *susceptibility* and *cynicalness* scores of users in the Haze and Riot events are 0.395, 0.045 and 0.034 respectively. The *influence* scores of users in the two events are more similar with each other than the other two measures. This suggests that influential users are consistently ranked in both events while the susceptibility and cynicalness of users changed significantly across events.

The same observation can be made in Table 4, which shows the Jaccard similarity coefficient between top $k\%$ for *influence*,*susceptibility* and *cynicalness* score

[1] `http://help.sentiment140.com/api`.

Table 4. Jaccard similarity between top k% users in Haze and Riot events

k	Influence	Susceptibility	Cynicalness
1%	0.327	0.055	0.018
2%	0.227	0.073	0.036
3%	0.23	0.085	0.042
5%	0.182	0.116	-
10%	0.445	0.149	-
20%	0.785	0.212	-

Table 5. Comparison of top 1% influential users with average users

	Avg # followers	Avg # tweets	Avg # sentiment-charged tweets
All users	690	3	1
Top 1% users (haze)	22406	9	3
Top 1% users (riot)	22859	13	4

users for both events. The Jaccard similarity coefficient for top 20% *influence* score users is 0.786, which suggest that most of the top 20% influential users remain highly influential between the two events. We observed some anomaly in the Jaccard similarity coefficient for top 2-5% *influence* score users. Examining into the data, we found that the top 1% influence score users tweeted intensively for both haze and riot event which resulted in some of them ranked highly for both events contributing to significantly higher Jaccard similarity coefficient. Whereas the top 2-5% users only tweeted intensively for only one of the two events, resulting in disparity for a user's ranking in two events and eventually a low Jaccard similarity coefficient. We did not compare the *cynicalness* score beyond top 3% because only 3% and 6% of the users have a non-zero *cynicalness* score for haze and riot event respectively.

5.3 Characteristics of Influential Users

As the influential users remain consistent across both events, we examine the characteristics of these users in greater detail. Table 5 shows the comparison between top 1% influential users with an average user. We observed that the top 1% influential users in both haze and riot events have an average of more than 20,000 followers, which is almost 30 times more than that of an average user. The top influential users also generate significantly more tweets than average users.

Table 6 shows the comparison of number of sentiment sent and diffused for top 1% influential users for In-Deg and ISC model. The number of sentiment sent by a user refers to the number of time the user's sentiment-charged tweets remain the first tweet on his follower's Twitter timeline at the point when his followers make a tweet. The number of sentiment diffused refers to the number of time the followers adopt the sentiments diffused by the user. We observed that although top 1% influential user under both measures have high average of

Table 6. Comparison of top 1% influential user for In-Deg and ISC model

	Avg Sentiment Sent (Haze)	Avg Sentiment Diffused (Haze)	Avg Sentiment Sent (Riot)	Avg Sentiment Diffused (Riot)
All users	5.52	0.045	6.585	0.034
Top 1% In-Deg	353.436	4.491	362.964	3.247
Top 1% ISC	334.566	5.438	368.127	3.833

Table 7. Pearson Correlation between Influence Measures

	INF_Riot	INF_Haze	In_Deg	PageRank
INF_Riot	1	0.395	0.475	0.597
INF_Haze	-	1	0.409	0.561
In_Deg	-	-	1	0.763
PageRank	-	-	-	1

sending sentiments to their followers, the influential users under the ISC model have a slightly higher average number of sentiment diffused to their followers.

5.4 Comparison of Influence Measures

Finally, we compare the influence measure of ISC model with other popular user influence measures, namely, *In-Degree* and *PageRank*. We define the *In-Degree* of a user by the number of his followers. *PageRank* defines the stationary probability of each user by performing a random walk from every user to his followees with equal transition probability.

Table 7 shows the Pearson Correlation between the different influence measures. The table shows that the *In-Degree* and *PageRank* are more similar with each other than the *Influence* measure in our proposed ISC model. Both the *In-Degree* and *PageRank* measures focus on the user's follower-followee relationships for computing user's influence. Although the ISC model's *Influence* measure considers the follower-followee relationships as well, it also considers the magnitude sentiment change when a user diffuses a sentiment item to his followers. This makes it more different from other influence measures.

6 Conclusion

In this paper, we propose a novel framework to model latent user characteristics that contribute to sentiment diffusion in a social network. We develop the ISC model to measure user influence, susceptibility and cynicalness simultaneously. The model determines how a user influences (or is influenced by) others by diffusing (or is diffused by) sentiment-charged tweets. We also propose the algorithm for implementing the model. We extract event relevant Twitter data for our experiment evaluation. Our experiment results have shown that different latent user characteristics can be derived from the observed sentiment diffusion. The ISC model however requires accuracy in sentiment analysis. In the future work, we can improve the accuracy of sentiment mining further to enhance the ISC model. We will also study more detailed emotions from users such as fear, anger, etc., in determining latent user characteristics.

References

1. Berger, J.: Arousal increases social transmission of information. Psychological Science 22(7), 891–893 (2013)
2. Bollen, J., Pepe, A., Mao, H.: Modeling public mood and emotion: Twitter sentiment and socioeconomic phenomena. In: ICWSM (2011)
3. Chakrabarti, D., Punera, K.: Event summarization using tweets. In: ICWSM (2011)
4. Hansen, L.K., Arvidsson, A., Nielsen, F.A., Colleoni, E., Etter, M.: Good friends, bad news—affect and virality in twitter. In: Park, J.J., Yang, L.T., Lee, C. (eds.) FutureTech 2011, Part II. CCIS, vol. 185, pp. 34–43. Springer, Heidelberg (2011)
5. Hoang, T.-A., Lim, E.-P.: Virality and susceptibility in information diffusions. In: ICWSM (2012)
6. Hoang, T.-A., Lim, E.-P., Achananuparp, P., Jiang, J., Zhu, F.: On modeling virality of twitter content. In: Xing, C., Crestani, F., Rauber, A. (eds.) ICADL 2011. LNCS, vol. 7008, pp. 212–221. Springer, Heidelberg (2011)
7. Kleinberg, J.M.: Authoritative sources in a hyperlinked environment. JACM 46(5), 604–632 (1999)
8. Lee, J., Lee, J.-H., Lee, D.: Impacts of tie characteristics on online viral diffusion. JAIS 24(1) (2009)
9. Luminet, O., Bouts, P., Delie, F., Manstead, A.S.R., Rime, B.: Social sharing of emotion following exposure to a negatively valenced situation. Cognition and Emotion 14(5), 661–688 (2000)
10. Peters, K., Kashima, Y., Clark, A.: Talking about others: Emotionality and the dissemination of social information. European Journal of Social Psychology 39(2), 207–222 (2009)
11. Sakaki, T., Okazaki, M., Matsuo, Y.: Earthquake shakes twitter users: Real-time event detection by social sensors. In: WWW (2010)
12. Stieglitz, S., Linh, D.X.: Emotions and information diffusion in social media—sentiment of microblogs and sharing behavior. JMIS 29(4), 217–247 (2013)
13. Thelwall, M., Buckley, K., Paltoglou, G.: Sentiment in twitter events. JASIST 62(2), 406–418 (2011)
14. Tumasjan, A., Sprenger, T.O., Sandner, P.G., Welpe, I.M.: Predicting elections with twitter: What 140 characters reveal about political sentiment. In: ICWSM (2010)
15. Xie, W., Zhu, F., Jiang, J., Lim, E.-P., Wang, K.: Topicsketch: Real-time bursty topic detection from twitter. In: ICDM (2013)

Automated Controversy Detection on the Web

Shiri Dori-Hacohen and James Allan

Center for Intelligent Information Retrieval
School of Computer Science
University of Massachusetts, Amherst, Amherst MA 01002, USA
{shiri,allan}@cs.umass.edu

Abstract. Alerting users about controversial search results can encourage critical literacy, promote healthy civic discourse and counteract the "filter bubble" effect, and therefore would be a useful feature in a search engine or browser extension. In order to implement such a feature, however, the binary classification task of determining which topics or webpages are controversial must be solved. Earlier work described a proof of concept using a supervised nearest neighbor classifier with access to an oracle of manually annotated Wikipedia articles. This paper generalizes and extends that concept by taking the human out of the loop, leveraging the rich metadata available in Wikipedia articles in a weakly-supervised classification approach. The new technique we present allows the nearest neighbor approach to be extended on a much larger scale and to other datasets. The results improve substantially over naive baselines and are nearly identical to the oracle-reliant approach by standard measures of F_1, $F_{0.5}$, and accuracy. Finally, we discuss implications of solving this problem as part of a broader subject of interest to the IR community, and suggest several avenues for further exploration in this exciting new space.

1 Introduction

On the web today, alternative medicine sites appear alongside pediatrician advice websites, the phrase "global warming is a hoax" is in wide circulation, and political debates rage in many nations over economic issues, same-sex marriage and healthcare. Access does not translate into trustworthy information: e.g., parents seeking information about vaccines will find plenty of "proof" that they cause autism, and may not even realize the depth of the controversy involved [1]; ads for helplines displayed to users searching for "abortion" are discreetly funded by pro-life (anti-abortion) religious groups [10]. The underlying thread connecting all these examples is that users searching for these topics may not even be aware that a controversy exists; indeed, without the aid of a search engine feature or browser extension to warn them, they may never find out. We believe that informing users about controversial topics would be a valuable addition to the end-user experience; this requires detecting such topics as a prerequisite.

In prior work, we analyzed whether the structural properties of the problem allow for a solution by proxy via Wikipedia, and demonstrated that there is a correlation between controversiality of Wikipedia pages and that of the webpages related to them [7]. We performed a proof-of-concept upper-bound

A. Hanbury et al. (Eds.): ECIR 2015, LNCS 9022, pp. 423–434, 2015.

analysis, using human-in-the-system judgments as an oracle for the controversy level of related Wikipedia articles. This naturally raises the question of whether an actual controversy detection system for the web can be constructed, making use of these properties.

In this work, we are putting these insights to use by introducing a novel, fully-automated system for predicting that arbitrary webpages discuss controversial topics. Our contribution is a weakly-supervised approach to detect controversial topics on arbitrary web pages. We consider our system as distantly-supervised [16] since we use heuristic labels for neighboring Wikipedia articles, which act as a bridge between the rich metadata available in Wikipedia and the sparse data on the web. One might hypothesize that using an automated system to scoring Wikipedia articles (instead of an oracle of human annotations) would degrade the results. In fact, however, our approach achieves comparable results to the prior art, which represented an upper-bound on this approach [7], while at the same time making it applicable to any large-scale web dataset.

2 Related Work

Several strands of related work inform our work: controversy detection in Wikipedia, controversy on the web and in search, fact disputes and trustworthiness, as well as sentiment analysis. We describe each area in turn.

Controversy detection in Wikipedia. Several papers focused on detecting controversy in Wikipedia [12,17,21], largely using metadata features such as length of the talk page, proportion of anonymous editors, and certain types of edits such as reverts. We describe a few of these in more detail in Section 3.2. Wikipedia is a valuable resource, but often "hides" the existence of debate by presenting even controversial topics in deliberately neutral tones [20], which may be misleading to people unfamiliar with the debate.

While detecting controversy in Wikipedia automatically can be seen as an end in itself, these detection methods have wider reach and can be used as a step for solving other problems. Recently, Das et al. used controversy detection as a step to study manipulation by Wikipedia administrators [6]. Additionally, Wikipedia has been used in the past as a valuable resource assisting in controversy detection elsewhere, whether as a lexicon or as a hierarchy for controversial words and topics [3,15]. Likewise, we use a few of the Wikipedia-specific controversy measures described above as a step in our approach (see Section 3.2).

As described above, prior work showed an upper-bound analysis demonstration using related Wikipedia articles as a proxy for controversy on the web, by using human annotations as an oracle rating the controversy of the articles [7]. In contrast, we use automatically-generated values for the Wikipedia articles.

Controversy on the web and in search. Outside of Wikipedia, other targeted domains such as news [3,5] and Twitter [15] have been mined for controversial topics, mostly focusing on politics and politicians. Some work relies on domain-specified sources such as Debatepedia[1] [3,11] that are likewise politics-heavy. We

[1] `http://dbp.idebate.org/`

consider controversy to be wider in scope; medical and religious controversies are equally interesting. A query completion approach might be useful in detecting controversial queries [9]; assuming one knows that a query is controversial, diversifying search results based on opinions is a useful feature [11].

Fact disputes and trustworthiness are often related to controversial topics [8,19]. Similar to our goal, the Dispute Finder tool focused on finding and exposing disputed claims on the web to users as they browse [8]. However, Dispute Finder was focused on manually added or bootstrapped fact disputes, whereas we are interested in scalably detecting controversies that may stem from fact disputes, but also from disagreement on values or from moral debates.

Sentiment analysis can naturally be seen as a useful tool as a step towards detecting varying opinions, and potentially controversy [5,15,18]. However, as mentioned elsewhere [3,7], sentiment alone may not suffice for detecting controversy, though it may be useful as a feature.

3 Nearest Neighbor Approach

Our approach to detecting controversy on the web is a nearest neighbor classifier that maps webpages to the Wikipedia articles related to them. We start from a webpage and find Wikipedia articles that discuss the same topic; if the Wikipedia articles are controversial, it is reasonable to assume the webpage is controversial as well. Prior work demonstrated that this approach worked using human judgment [7], leaving open the question of whether a fully-automated approach can succeed.

The choice to map specifically to Wikipedia rather than to any webpages was driven by the availability of the rich metadata and edit history on Wikipedia [12,17,21]. We consider our approach as a distantly-supervised classifier in the relaxed sense (c.f. [16]), since we are using automatically-generated labels, rather than truth labels, for an external dataset (Wikipedia) rather than the one we are training on (web). While some of these labels were learned using a supervised classifier on Wikipedia, none of them were trained for the task at hand, namely classifying webpages' controversy.

To implement our nearest neighbor classifier, we use several modules: matching via query generation, scoring the Wikipedia articles, aggregation, thresholding and voting. We describe each in turn.

3.1 Matching via Query Generation

We use a query generation approach to map from webpages to the related Wikipedia articles. The top ten most frequent terms on the webpage, excluding stop words, are extracted from the webpage, and then used as a keyword query restricted to the Wikipedia domain and run on a commercial search engine. We use one of two different stop sets, a 418 word set (which we refer to as "Full" Stopping [4]) or a 35 word set ("Light" Stopping [13]). Wikipedia redirects were followed wherever applicable in order to ensure we reached the full Wikipedia article with its associated metadata; any talk or user pages were ignored.

We considered the articles returned from the query as the webpage's "neighbors", which will be evaluated for their controversy level. Based on the assumption that higher ranked articles might be more relevant, but provide less coverage, we varied the number of neighbors in our experiments from 1 to 20, or used all articles containing all ten terms. A brief evaluation of the query generation approach is presented in Section 5.1.

3.2 Automatically-Generated Wikipedia Labels

The Wikipedia articles, found as neighbors to webpages, were labeled with several scores measuring their controversy level. We use three different types of automated scores for controversy in Wikipedia, which we refer to as **D**, **C**, and **M** scores. All three scores are automatically generated based on information available in the Wikipedia page and its associated metadata, talk page and revision history. While we use a supervised threshold on the scores, the resulting score and prediction can be generated with no human involvement.

The D score tests for the presence of **D**ispute tags that are added to the talk pages of Wikipedia articles by its contributors [12,17]. These tags are sparse and therefore difficult to rely on [17], though potentially valuable when they are present. We test for the presence of such tags, and use the results as a binary score (1 if the tag exists or -1 if it doesn't). Unfortunately, the number of dispute tags available is very low: in a recent Wikipedia dump, only 0.03% of the articles had a dispute tag on their talk page. This is an even smaller dataset than the human annotations provided in prior work [7]; the overlap between these articles and the 8,755 articles in the dataset is a mere 165 articles.

The C score is a metadata-based regression that predicts the controversy level of the Wikipedia article using a variety of metadata features (e.g. length of the page and its associated talk page, number of editors and of anonymous editors). This regression is based on the approach first described by Kittur et al. [12]. We use the version of this regression as implemented and trained recently by Das et al. [6], generating a floating point score in the range (0,1).

The M score, as defined by Yasseri et al., is a different way of estimating the controversy level of a Wikipedia article, based on the concept of mutual reverts and edit wars in Wikipedia [21]. Their approach is based on the number and reputation of the users involved in reverting each others' edits, and assumes that "the larger the armies, the larger the war" [21]. The score is a positive real number, theoretically unbounded (in practice it ranges from 0 to several billion).

3.3 Aggregation and Thresholding

The score for a webpage is computed by taking either the maximum or the average of all its Wikipedia neighbors' scores, a parameter we vary in our experiments. After aggregation, each webpage has 3 "controversy" scores from the three scoring methods (**D**, **C** and **M**). We trained various thresholds for both **C** and **M** (see Section 4.1), depending on target measures.

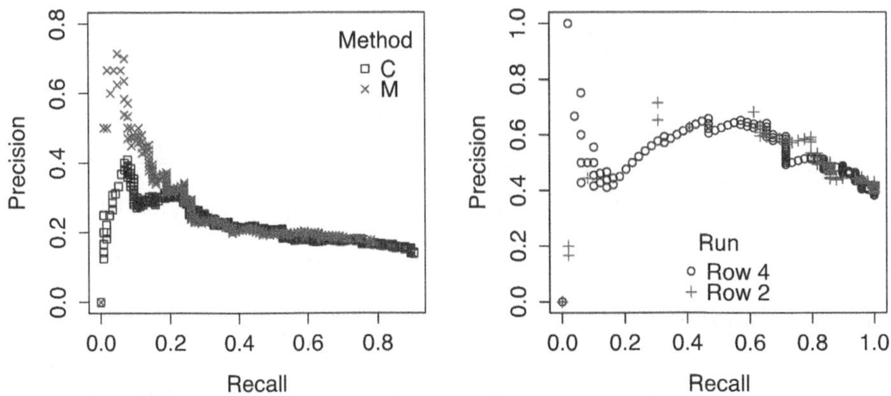

Fig. 1. Precision-Recall curves (uninterpolated). Left: PR curve for C and M thresholds on the Wikipedia NNT set. Right: PR curve for select runs on the Test set. Row numbers refer to Table 2.

3.4 Voting

In addition to using each of the three labels in isolation, we can also combine them by voting. We apply one of several voting schemes to the binary classification labels, after the thresholds have been applied. The schemes we use are:

- Majority vote: consider the webpage controversial if at least two out of the three labels are "controversial".
- Logical *Or*: consider the webpage controversial if any of the three labels is "controversial".
- Logical *And*: consider the webpage controversial only if all the three labels are "controversial".
- Other logical combinations: we consider results for the combination $(Dispute \vee (C \wedge M))$, based on the premise that if the dispute tag happens to be present, it would be valuable[2].

4 Experimental Setup and Data Set

To compare to prior work, we use the dataset used in previous experiments [7], consisting of webpages and Wikipedia articles annotated as controversial or non-controversial. This publicly-released dataset includes 377 webpages, and 8,755 Wikipedia articles. Of the Wikipedia articles annotated in the set, 4,060 were the Nearest Neighbors associated with the Training set ("NNT" in Table 1), which we use later (see Section 4.1). For evaluation, we use Precision, Recall, Accuracy, F_1 and $F_{0.5}$ using the classic IR sense of these metrics, with "controversial" and "non-controversial" standing in for "relevant" and "non relevant", respectively.

[2] D's coverage was so low that other voting combinations were essentially identical to the majority voting; we therefore omit them.

Table 1. Data set size and annotations. "NNT" denotes the subset of Wikipedia articles that are Nearest Neighbors of the webpages Training set.

Webpages		
Set	Pages	Controversial
All	377	123 (32.6%)
Training	248	74 (29.8%)
Testing	129	49 (38.0%)

Wikipedia articles			
Set	Articles	Annotated	Controversial
All	8,755	1,761	282 (16.0%)
NNT	4,060	853	115 (13.5%)

4.1 Threshold Training

C and **M** are both real-valued numbers; in order to generate a binary classification, we must select a threshold above which the page will be considered controversial. (**D** score is already binary.) Since the public corpus has annotations on some of the Wikipedia articles [7], we trained the thresholds for **C** and **M** for the subset of articles associated with the training set (labeled "NNT" in Table 1). The Precision-Recall curve for both scores is displayed in Figure 1. We select five thresholds for the two scoring methods, based on the best results achieved on this subset for our measures.

For comparison, we also present single-class acceptor baselines on this task of labeling the Wikipedia articles, one which labels all pages as non-controversial and one which labels all pages as controversial. Finally, two random baselines which label every article as either controversial or non-controversial based on a coin flip, are presented for comparison (average of three random runs). One

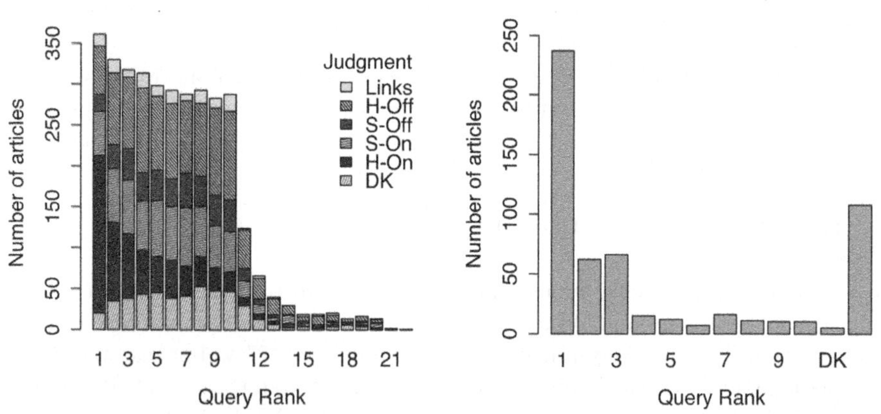

Fig. 2. Evaluation of Matching scheme. Left: Judgments on Wikipedia articles returned by the automatically-generated queries, by rank. Annotators could choose one of the following options: H-On="Highly on [webpage's] topic", S-On="Slightly on topic", S-Off="Slightly off topic", H-Off="Highly off topic", Links="Links to this topic, but doesn't discuss it directly", DK="Don't Know". Right: Frequency of page selected as best, by rank. DK="Don't Know", N="None of the above".

of these baselines flips a coin with 50% probability, and the other flips it with 29.8% probability (the incidence of controversy in the training set).

5 Evaluation

We treat the controversy detection problem as a binary classification problem of assigning labels of "controversial" and "non-controversial" to webpages. We present a brief evaluation for the query generation approach before turning to describe our results for the controversy detection problem.

5.1 Judgments from Matching

A key step in our approach is selecting which Wikipedia articles to use as nearest neighbors. In order to evaluate how well our query generation approach is mapping webpages to Wikipedia articles, we evaluated the automated queries and the relevance of their results to the original webpage. This allows an intrinsic measure of the effectiveness of this step - independent of its effect on the extrinsic task, which is evaluated using the existing dataset's judgments on the webpages' controversy level[3]. We annotated 3,430 of the query-article combinations (out of 7,630 combinations total) that were returned from the search engine; the combinations represented 2,454 unique Wikipedia articles. Our annotators were presented with the webpage and the titles of up to 10 Wikipedia articles in alphabetical order (not ranked); they were not shown the automatically-generated query. The annotators were asked to name the single article that best matched the webpage, and were also asked to judge, for each article, whether it was relevant to the original page. Figure 2 shows how the ranked list of Wikipedia articles were judged. In the figure, it is clear that the top-ranking article was viewed as highly on topic but then the quality dropped rapidly. However, if both "on-topic" judgments are combined, a large number of highly or slightly relevant articles are being selected. Considering the rank of the best article as the single relevant result, the Mean Reciprocal Rank for the dataset was 0.54 (if the best article was "don't know" or "none of the above", its score was zero).

5.2 Our Results Compared to Baseline Runs

We compare our approach to several baselines, a sentiment analysis approach based on a logistic regression classifier [2] trained to detect presence of sentiment on the webpage, whether positive or negative; sentiment is used as a proxy for controversy. We add single-class and random baselines (average of three runs). Finally, the best results from our prior work [7] are reported. As described in Section 3, we varied several parameters in our nearest neighbor approach:

1. **Stopping set** (Light or Full)
2. **Number of neighbors** (k=1..20, or no limit)

[3] Both sets are publicly released - see `http://ciir.cs.umass.edu/downloads`

3. Aggregation method (average or max)
4. Scoring or voting method (C, M, D; Majority, Or, And, $D \vee (C \wedge M)$)
5. Thresholds for C and M (one of five values, as described in Section4.1).

These parameters were evaluated on the training set and the best runs were selected, optimizing for F_1, $F_{0.5}$ and Accuracy. The parameters that performed best, for each of the scoring/voting methods, were then run on the test set.

The results of our approach on the test set are displayed in Table 2. For ease of discussion, we will refer to row numbers in the table. For space considerations, highly similar runs are omitted.

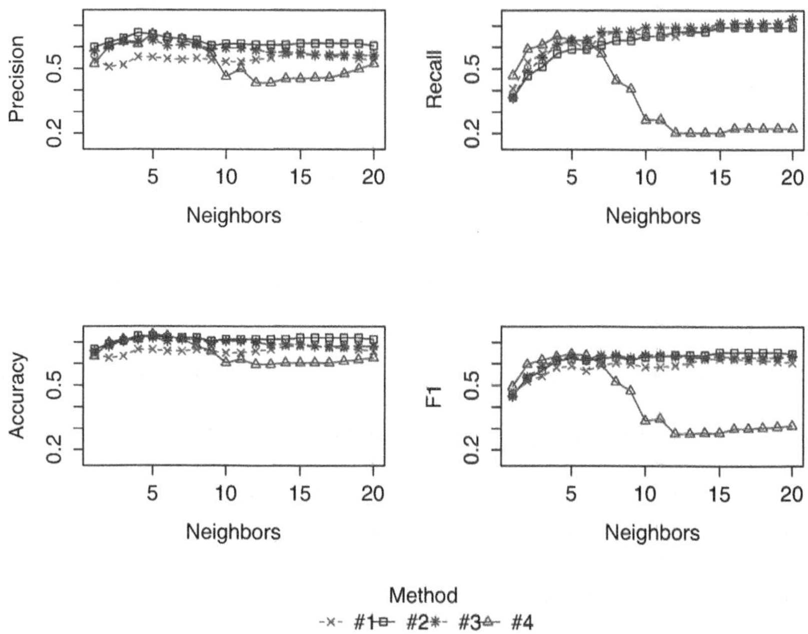

Fig. 3. Evaluation metrics vary with number of neighbors (k). Refer to rows 1-4 in Table 2: rows 1 and 4 use avg aggregator with low thresholds, while 2 and 3 use max with high thresholds.

6 Discussion

As the results in Table 2 show, our fully-automated approach (rows 1-11) achieves results higher than all baselines (rows 15-19), in all metrics except recall (which is trivially 100% in row 19). The method that optimized for $F_{0.5}$ on the training set among all the single score approaches was the run using Light Stopping, M with a rather high (discriminative) threshold, and aggregating over the maximal value of all the result neighbors (row 2 in Table 2). Using k values of 8 through 12 achieved identical results on the training set. These runs ended up achieving

Table 2. Results on Testing Set. Results are displayed for the best parameters on the training set, using each scoring method, optimized for F_1, Accuracy and $F_{0.5}$. The overall best results of our runs, in each metric, are displayed in bold; the best prior results (rows 12-14 [7]) and baseline results (rows 15-19) are also displayed in bold. See text for discussion.

	Parameters						Target	Test Metric				
#	Stop	Score	k	agg	Thres C	Thres M	Target	P	R	F_1	Acc	$F_{0.5}$
1	Full	M	8	avg	–	*84930*	F_1, Acc	0.55	0.67	0.61	0.67	0.57
2	Light	M	8	max	–	2.85×10^6	$F_{0.5}$	0.63	0.63	0.63	**0.72**	**0.63**
3	Light	C	15	max	*0.17*	–	F_1	0.57	0.71	0.64	0.69	0.60
4	Light	C	7	avg	4.18×10^{-2}	–	Acc, $F_{0.5}$	**0.64**	0.57	0.60	0.71	0.62
5	Light	D	19	max	–	–	F_1	0.43	0.57	0.49	0.55	0.45
6	Full	D	5	max	–	–	Acc	0.53	0.37	0.43	0.64	0.49
7	Light	D	6	max	–	–	Acc, $F_{0.5}$	0.44	0.35	0.39	0.58	0.41
8	Light	Maj.	15	max	*0.17*	2.85×10^6	F_1	0.59	0.73	**0.65**	0.70	0.61
9	Full	Maj.	5	max	4.18×10^{-2}	2.85×10^6	Acc, $F_{0.5}$	0.59	0.61	0.60	0.69	0.59
10	Light	And	no	max	*0.17*	*84930*	F_1, Acc, $F_{0.5}$	0.52	0.51	0.51	0.64	0.52
11	Light	D|CM	7	avg	4.18×10^{-2}	*84930*	Acc, $F_{0.5}$	0.63	0.55	0.59	0.70	0.61
12	Oracle-based [7], best run for P, Acc and $F_{0.5}$							**0.69**	0.51	0.59	**0.73**	**0.65**
13	Oracle-based [7], best run for R							0.51	**0.84**	0.64	0.64	0.56
14	Oracle-based [7], best run for F_1							0.60	0.69	**0.64**	0.70	0.61
15	Sentiment [7]							0.38	0.90	0.53	0.40	0.43
16	Random_{50}							**0.42**	0.53	0.47	0.54	**0.44**
17	$\text{Random}_{29.8}$							0.23	0.19	0.21	0.61	0.22
18	All non-controversial							0	0	0	**0.62**	0
19	All Controversial							0.38	**1.00**	**0.55**	0.38	0.43

some of the best results on the test set; with value k=8 the results were the best for $F_{0.5}$ as well as Accuracy (row 2), with 10.1% absolute gain in accuracy (16.3% relative gain) over the non-controversial class baseline, which had the best accuracy score among the baselines. For $F_{0.5}$ this run showed 19.5% absolute gain (44.5% relative gain) over the best $F_{0.5}$ score, which was achieved by the Random_{50} baseline. Even though none of the results displayed in the table were optimized for precision, they still had higher precision than the baselines across the board (compare rows 1-11 to rows 15-19). Among the voting methods, the method that optimized for F_1 on the training set was the Majority voting, using Light Stopping, aggregating over the maximal value of 15 neighbors, with discriminative thresholds for both M and C (row 12). This run showed a 10.4% (18.9% relative gain) absolute gain on the test set over the best baseline for F_1.

The results of the sentiment baseline (row 15) were surprisingly similar to a trivial acceptor of "all controversial" baseline (row 19); at closer look, the sentiment classifier only returns about 10% of the webpages as lacking sentiment, and thus its results are close to the baseline. We tried applying higher confidence thresholds to the sentiment classifier, but this resulted in lower recall without improvement in precision. We note that the sentiment classifier was not trained to detect controversy; it's clear from these results, as others have noted, that sentiment alone is too simplistic to predict controversy [3,7].

When comparing our results (rows 1-11) to the best oracle-reliant runs from prior work (rows 12-14, see [7]), the results are quite comparable. Recall that this prior work represents a proof-of-concept upper-bound analysis, with a human-in-the-loop providing judgments for the relevant Wikipedia pages, rather than an automatic system that can be applied to arbitrary pages[4]. When comparing the best prior work result (row 12) to our best run (row 2) using a zero-one loss function, the results were not statistically different. This demonstrates that our novel, fully-automated system for detecting controversy on the web is as effective as upper-bound, human-mediated predictions [7].

We observe that when using a max aggregator, results were generally better with more discriminative thresholds and a large number of neighbors (k); when average was used, a lower threshold with smaller k was more effective. To understanding this phenomenon, we fixed all the parameters from rows 1-4 above except for k, and plotted system results as a function of k (see Figure 3). Consider that the max function is more sensitive to noise than the average function - a higher threshold can reduce the sensitivity to such noise while extending coverage by considering more neighbors. In most runs depicted, precision drops a little but remains fairly consistent with k, while recall increases steadily. However, in the parameters from row 4, there is a penalty to both precision and recall as k increases, demonstrating the noise sensitivity of the max function.

7 Conclusions and Future Work

We presented the first fully automated approach to solving the recently proposed binary classification task of web controversy detection [7]. We showed that such detection can be performed by automatic labeling of exemplars in a nearest neighbor classifier. Our approach improves upon previous work by creating a scalable distantly-supervised classification system, that leverages the rich metadata available in Wikipedia, using it to classify webpages for which such information is not available. We reported results that represent 20% absolute gains in F measures and 10% absolute gains in accuracy over several baselines, and are comparable to prior work that used human annotations as an oracle [7].

Our approach is modular and therefore agnostic to the method chosen to score Wikipedia articles; like Das et al. [6], we can leverage future improvements in this domain. For example, scores based on a network collaboration approach [17] could be substituted in place of the $\mathbf{M}$ and $\mathbf{C}$ values, or added to them as another feature. The nearest neighbor method we described is also agnostic to the choice of target collection we query; other rich web collections which afford controversy inference, such as Debate.org, Debatabase or procon.org, could also be used to improve precision.

[4] Note that this is not a strict upper-bound limit in the theoretical sense, but in principle it's reasonable to assume that a human annotator would perform as well as an automated system. In fact, in a few cases the automated system performed better than the oracle-reliant approach, see e.g. F1 on row 8 vs. row 14.

Future work could improve on our method: better query generation methods could be employed to match neighbors, using entity linking for Wikification could create the links directly, or else language models could compare candidate neighbors directly. Standard machine learning approaches can be used to combine our method with other features such as sentiment analysis.

The nearest neighbor approach we presented is limited in nature by the collection it targets; it will not detect controversial topics that are not covered by Wikipedia. Entirely different approaches would need to be employed to detect such smaller controversies. Nonetheless, it's possible that some metric of sentiment variance across multiple websites could provide useful clues. Another approach could use language models or topic models to automatically detect the fact that strongly opposing, biased points of view exist on a topic, and thus it is controversial. This would flip the directionality of some recent work that presupposes subjectivity and bias to detect points of view [6,22].

We see the controversy detection problem as a prerequisite to several other interesting applications and larger problems such as: user studies on the effects of informing users when the webpage they are looking at is controversial; the evolution and incidence of controversial topics over time; and diversifying controversial search results according to the stances on them, are a few such problems.

With the growing trend towards personalization in search comes a risk of fragmenting the web into separate worlds, with search engines creating a self-fulfilling prophecy of users' bias confirmation. Informing users about fact disputes and controversies in their queries can improve trustworthiness in search; explicitly exposing bias and polarization may partially counteract the "filter bubble" or "echo chamber" effects, wherein click feedback further reenforce users' predispositions. Further development and refinement of controversy detection techniques can foster healthy debates on the web, encourage civic discourse, and promote critical literacy for end-users of search.

Acknowledgments. Our thanks go to Allen Lavoie, Hoda Sepehri Rad, Taha Yasseri and Elad Yom-Tov for valuable resources and fruitful discussions. Thanks to Sandeep Kalra, Nada Naji, Ravali Pochampally, Emma Tosch, David Wemhoener, Celal Ziftci, and anonymous reviewers for comments on various drafts. Special thanks go to Gonen Dori-Hacohen, without whose valuable and timely assistance, this paper would not have been possible. This work was supported in part by the Center for Intelligent Information Retrieval and in part by NSF grant #IIS-1217281. Any opinions, findings and conclusions or recommendations expressed in this material are the authors, and do not necessarily reflect those of the sponsor.

References

1. Activist Post. 22 Medical Studies That Show Vaccines Can Cause Autism, `http://www.activistpost.com/2013/09/22-medical-studies-that-show-vaccines.html` (September 24, 2014) (accessed)

2. Aktolga, E., Allan, J.: Sentiment Diversification With Different Biases. In: Proc. of SIGIR 2013, pp. 593–602 (2013)
3. Awadallah, R., Ramanath, M., Weikum, G.: Harmony and Dissonance: Organizing the People's Voices on Political Controversies. In: Proc. of WSDM 2012, pp. 523–532 (February 2012)
4. Callan, J.P., Croft, W.B., Harding, S.M.: The INQUERY retrieval system. In: Database and Expert Systems Applications, pp. 78–83. Springer, Vienna (1992)
5. Choi, Y., Jung, Y., Myaeng, S.-H.: Identifying Controversial Issues and Their Subtopics in News Articles. Intelligence and Security Informatics 6122, 140–153 (2010)
6. Das, S., Lavoie, A., Magdon-Ismail, M.: Manipulation Among the Arbiters of Collective Intelligence: How Wikipedia Administrators Mold Public Opinion. In: Proc. of CIKM 2013, pp. 1097–1106 (2013)
7. Dori-Hacohen, S., Allan, J.: Detecting controversy on the web. In: Proc. of CIKM 2013, pp. 1845–1848 (2013)
8. Ennals, R., Trushkowsky, B., Agosta, J.M.: Highlighting disputed claims on the web. In: Proc. of WWW 2010, p. 341 (2010)
9. Gyllstrom, K., Moens, M.-F.: Clash of the typings: finding controversies and children's topics within queries. In: Clough, P., Foley, C., Gurrin, C., Jones, G.J.F., Kraaij, W., Lee, H., Mudoch, V. (eds.) ECIR 2011. LNCS, vol. 6611, pp. 80–91. Springer, Heidelberg (2011)
10. Heroic Media. Free Abortion Help website (January 2014), `http://freeabortionhelp.com/us/` (September 24, 2014) (accessed)
11. Kacimi, M., Gamper, J.: MOUNA: Mining Opinions to Unveil Neglected Arguments. In: Proc. of CIKM 2012, pp. 2722–2724 (2012)
12. Kittur, A., Suh, B., Pendleton, B.A., Chi, E.H., Angeles, L., Alto, P.: He Says, She Says: Conflict and Coordination in Wikipedia. In: Proc. of CHI 2007, pp. 453–462 (2007)
13. Manning, C.D., Raghavan, P., Schütze, H.: Introduction to information retrieval, vol. 1. Cambridge University Press, Cambridge (2008)
14. Pariser, E.: The Filter Bubble: What the Internet is hiding from you. Penguin Press HC (2011)
15. Popescu, A.A.-M., Pennacchiotti, M.: Detecting controversial events from twitter. In: Proc. CIKM 2010, pp. 1873–1876 (2010)
16. Riedel, S., Yao, L., McCallum, A.: Modeling Relations and Their Mentions Without Labeled Text. In: Balcázar, J.L., Bonchi, F., Gionis, A., Sebag, M. (eds.) ECML PKDD 2010, Part III. LNCS, vol. 6323, pp. 148–163. Springer, Heidelberg (2010)
17. Sepehri Rad, H., Barbosa, D.: Identifying controversial articles in Wikipedia: A comparative study. In: Proc. WikiSym (2012)
18. Tsytsarau, M., Palpanas, T., Denecke, K.: Scalable detection of sentiment-based contradictions. In: DiversiWeb 2011 (2011)
19. Vydiswaran, V.G.V., Zhai, C., Roth, D., Pirolli, P.: BiasTrust: Teaching Biased Users About Controversial Topics. In: Proc. CIKM 2012, pp. 1905–1909 (2012)
20. Wikipedia. Wikipedia: Neutral Point of View Policy (January 2014)
21. Yasseri, T., Sumi, R., Rung, A., Kornai, A., Kertész, J.: Dynamics of conflicts in Wikipedia. PloS One 7(6), e38869 (2012)
22. Yom-Tov, E., Dumais, S.T., Guo, Q.: Promoting civil discourse through search engine diversity. Social Science Computer Review (2013)

Learning Sentiment Based Ranked-Lexicons for Opinion Retrieval

Filipa Peleja and João Magalhães

CITI, Departamento de Informática, Faculdade de Ciências e Tecnologia, Universidade Nova de Lisboa, 2829-516 Caparica, Portugal
{filipapeleja}@gmail.com, {jm.magalhaes}@fct.unl.pt

Abstract. In contrast to classic search where users look for factual information, opinion retrieval aims at finding and ranking subjective information. A major challenge of opinion retrieval is the informal nature of user reviews and the domain specific jargon used to describe the targeted item. In this paper, we present an automatic method to learn a space model for opinion retrieval. Our approach is a generative model that learns sentiment word distributions by embedding multi-level relevance judgments in the estimation of the model parameters. In addition to sentiment word distributions, we also infer domain specific named entities that due to their popularity become a sentiment reference in their domain (e.g. name of a movie, "Batman" or specific hotel items, "carpet"). This contrasts with previous approaches that learn a word's polarity or aspect-based polarity. Opinion retrieval experiments were done in two large datasets with over 703.000 movie reviews and 189.000 hotel reviews. The proposed method achieved better, or equal, performance than the benchmark baselines.

Keywords: Opinion retrieval, sentiment analysis, sentiment space model.

1 Introduction

The Web's increasing popularity led to changes in people's habits. In this new context, sentiment expressions or opinion expressions became important pieces of information, specially, in the context of online commerce. Therefore, modelling text to find the lexicon that is meaningful upon expressing a sentiment has emerged as an important research direction. There has been a considerable amount of research in the area of opinion retrieval [5]. Most work in this area has focused on sentiment classification as positive, negative and neutral, or joint aspect-sentiment features, only a few approaches focused on the task of automatically defining specific sentiment vocabularies [6]. However, a major challenge in opinion retrieval is the detection of the words that express a subjective preference, or, more importantly, domain related idiosyncrasies where specific sentiment words are common (jargon). Domain entities, e.g., "Batman", can also become sentiment anchors due to their popularity. Also, domain dependencies are constantly changing and opinions are not binary, hence, capturing sentiment words for opinion ranking can be particularly challenging.

A. Hanbury et al. (Eds.): ECIR 2015, LNCS 9022, pp. 435–440, 2015.

Typically available lexicons are too generic and are not designed for ranking tasks which are at the core of IR: they do not consider domain words, have fixed sentiment word weights (sometimes are simply positive/negative or have more than one sentiment weight) and do not capture word interactions [6]. We aim at providing IR tasks with a sentiment resource that is specifically designed for rank-by-sentiment tasks. The two main steps in building such resource, concerns the identification of the lexicon words and words sentiment weighting (which we argue that a simple weight is not enough).

2 Related Work

Sentiment bearing words are known as opinion words, polar words, opinion-bearing words or sentiment words. A widely used approach to extract sentiment words is to extend an initial seed of words. Turney et al. [12] extracts sentiment phrases containing an adjective or a verb. In another approach, Bethard et al. [1] devised a supervised statistical classification task where opinionated and factual documents are used to compute the relative frequency and build a sentiment lexicon. In a second approach, by Bethard et al., a sentiment word lexicon is built by using a modified log-likelihood ratio of the words relative frequency and sentiment words from a pre-built lexicon – a seed list of 1,336 manually annotated adjectives. Recently, a few studies identified sentiment words by exploring the usage of slang or domain-specific sentiment words [2]. In particular, a concept-based resource for sentiment analysis SenticNet [9]. SenticNet is concept-based resource that contains concepts along with a polarity score. Urban Dictionary[1] (UD) and Twittrat's[2] are dictionaries that aim at capturing sentiment words that traditional dictionaries fail to capture [9] (e.g. Multi-perspective Question Answering (MPQA), General Inquirer and SentiWordNet). Chen et al. [2] apply the UD in the process of identifying sentiment words by exploring the issue of slang and domain-specific sentiment words. Chen et al. applied a target-dependent strategy to extract sentiment expressions from unlabelled tweets and compares the results with gold standard sentiment lexicons – MPQA, General Inquirer and SentiWordNet. Peng et al. [8] proposes to learn a sentiment word lexicon that captures informal and domain-specific sentiment words. Peng et al. designed a matrix factorization method where each entry is the edge weight between two sentiment words. The weight is calculated from the synonyms/antonyms relations.

Yohan et al. [5] propose to apply the LDA generative model to uncover the pairs {aspect, sentiment} in which aspect refers to product aspects. To evaluate the obtained pairs, Yohan et al. follow a binary supervised sentiment classification task. In contrast to previous approaches, we go beyond simple word weights and infer sentiment word distributions over the entire range of sentiment relevance levels. Our approach is related to the Labeled LDA algorithm [10] and LDA for re-ranking [11].

---

[1] `http://www.urbandictionary.com/`

[2] `https://twitter.com/twitrratr/`

3 Learning Ranked Lexicons

The problem aims at creating a sentiment lexicon based on user reviews without human supervision. To identify the sentiment words we propose a multilevel generative model of users' reviews. We propose a generative probabilistic model that ties words to different sentiment relevance levels, creating a sentiment rank over the entire sentiment lexicon. The main contribution of the proposed approach is that the model infers a sentiment lexicon by analysing user reviews as sentiment ranked sets of documents.

3.1 Rank-LDA Sentiment Lexicon

The LDA method models co-occurrences at document level through a set of latent topics z and their associated words. However, our goal here is somewhat different: we wish to extract words associated to a sentiment level. Figure 1 presents the graphical model of the proposed Rank-LDA method. At its core, the Rank-LDA links the hidden structure (latent topics) to the document sentiment relevance level. In this hidden structure a set of hidden topics are activated for each sentiment level. Hence, while LDA defines a topic as a distribution over a fixed vocabulary Rank-LDA computes the distribution of words over topics that best describe an association to a sentiment. Notice that this sentiment-topic association is different from previous work [7] where LDA was used to capture the topic distributions over the words that best describes product aspects.

Rank-LDA is structured as follows: β is the per-corpus topic $\text{Dirichlet}(\cdot\,|\eta)$ distribution, θ is the per-document topic $\text{Dirichlet}(\cdot\,|\alpha)$ distribution, z is the per-word topic assignment following a $\text{Multinomial}(\cdot\,|\theta^{(d)})$ distribution, and w correspond to the set of words observed on each document. Finally, $s_i \in \{1, \ldots, R\}$ is the per-document sentiment relevance level and sw is the per-word random variable corresponding to its sentiment distributions across the different sentiment levels of relevance. The random variables α, η and π are the distribution priors.

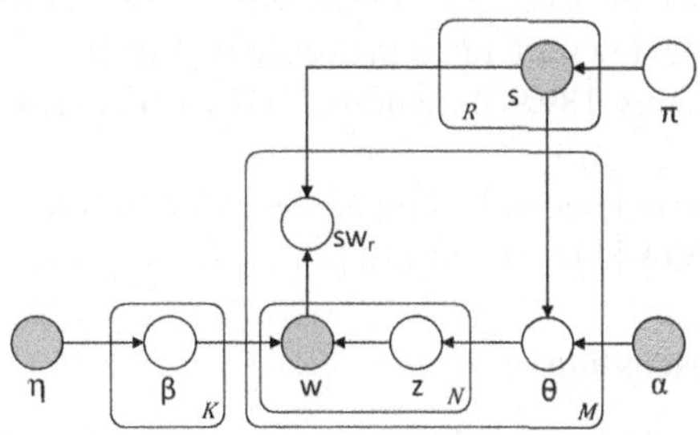

Fig. 1. The Rank-LDA graphical model

The proposed method is related to Labeled-LDA [10] but with significant differences. As mentioned, in our method, we tie the latent topics to sentiment relevance levels. The hidden topics will encode the words ranked by sentiment level and then by topic relevance. While the sLDA assumes that labels are generated by the topics, in the Labeled-LDA the labels activate and de-activate the topics.

3.2 Sentiment Word Distributions

A key characteristic of sentiment words is that reviews from different sentiment levels share sentiment words, but each review exhibits those sentiment words in different proportion. Word distributions over topics, associate a word relevance to the different sentiment levels. The sentiment word distributions are given by the density distribution

$$p(sw_i \mid s) = \int p(\theta) \cdot \prod_{n=1}^{N} p(z_n \mid \theta, s) p(sw_i \mid z_n)\, d\theta + \tau \qquad (1)$$

where we compute the marginal distribution of a word given a sentiment level, over the N latent topics of the Rank-LDA model. The variable τ is a smoothing parameter that we set to 0.01.

The sentiment word distribution function can also be used to rank words by its positive/negative weight and to calculate a word's relevance in different sentiment levels. A straightforward way of achieving this conversion is through the function

$$RLDA(sw_{i,j}) = \frac{p(sw|s=i) - p(sw|s=j)}{min(p(sw|s=i), p(sw|s=j))} \qquad (2)$$

where $p(sw|s=i)$ and $p(sw|s=j)$ denote the sentiment word sw relevance level values in rating i and j. The obtained lexicon with Rank-LDA is denoted as RLDA.

4 Evaluation

The experiment concerns opinion retrieval by rating level in which we use the evaluating metrics: P@5, P@30, NDCG and MAP.

4.1 Datasets and Methods

IMDb-Extracted: Contains over 703,000 movie reviews, corresponding to a total of 7,443,722 sentences. Reviews are rated in a scale of 1 to 10.

TripAdvisor: Contains 189,921 reviews, and each review is rated in a scale of 1 to 5 [13].

The obtained sentiment lexicon is compared to three well-known sentiment lexicons: SentiWordNet [3], MPQA [14] and Hu-Liu [4].

4.2 Results and Discussion

Opinion Retrieval. Table 1 and Table 2 show the retrieval performances of the proposed lexicons. In this task a user review is represented as a query and the relevance judgment is the rating level. RLDA lexicon is consistently effective across the four retrieval metrics (P@5, P@30, MAP and NDCG).

RLDA is the most consistent as it produces a clear improvement in relation to other lexicons, results underlined in Table 1 and Table 2. In Table 2, MPQA outperformed RLDA (P@5). However, the following aspects should be kept in mind: First MPQA provides a list of annotated words and the words provide no weight intensity – e.g. *excellent* and *good* are both labelled as positive. Secondly, the lexicon is limited to approximately 6,886 words and it is context independent, as a consequence, some users' reviews are represented with a low volume of sentiment words.

Table 1. Opinion retrieval (IMDb)

Method	P@5	P@30	MAP	NDCG
Rank-LDA	92.0%	90.7%	78.2%	56.3%
SWN	88.0%	89.0%	76.8%	53.5%
Hu-Liu	82.0%	76.7%	72.4%	43.8%
MPQA	82.0%	81.7%	73.6%	46.2%

Table 2. Opinion retrieval (TripAdvisor)

Method	P@5	P@30	MAP	NDCG
Rank-LDA	92.0%	98.7%	65.3%	81.3%
SWN	92.0%	96.0%	63.7%	80.9%
Hu-Liu	92.0%	90.0%	55.8%	78.1%
MPQA	100.0%	87.3%	58.0%	79.0%

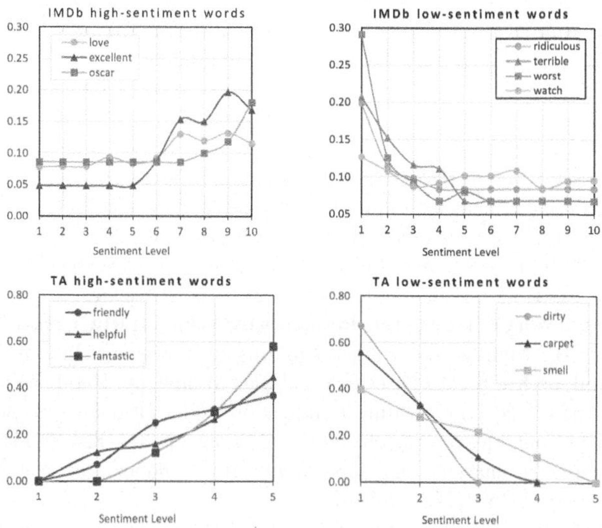

Fig. 2. Sentiment word distributions

Sentiment Word Distributions. One of the key properties of the proposed method is the sentiment word distributions for specific domains. Rank-LDA leverages on the rating scale assigned to reviews to learn a structured and generative model that represents the entire domain. Figure 2 depicts examples of sentiment word distributions. In these figures the conditional probability density functions for each word is presented. The words *love* and *excellent* (first graph) are general sentiment words that are used from a mid-range to a top-level sentiment value. It is interesting to note that in this domain the domain specific sentiment word *oscar* is only used to express a highly positive sentiment. The second graph illustrates words that are mostly used to express negative sentiment. The sentiment word *watch* is used across the entire range of sentiment expressivity. Hence, this is an important feature, because the Rank-LDA does not categorize word as neutral (or positive/negative), instead it creates a fine-grain model of how likely this word occurs at different sentiment levels. In the third and fourth graphs we observed an interesting phenomena: the most positive words were quite general and not highly domain-specific. This was not true for the most negative sentiment word distributions where the word *dirty* is highly relevant in this domain (for obvious reason), but the words *carpet* and *smell* are highly relevant because they are key for the domain in particular.

5 Conclusion

In this paper we have proposed the Rank-LDA method, a generative model, to learn a highly structured sentiment space model that learns a domain specific sentiment lexicon, characterizes words in terms of sentiment distributions and its latent structure, and also, captures word interactions creating joint distributions of sentiment words. We examined the impact of the dimensionality of the hidden structure in the sentiment word lexicon. In the experiments, the improvements of the proposed method over the baselines, were as good as, or better than, existing methods. It is interesting to note that these improvements are related to domain specific words and the sentiment word distributions inferred by the Rank-LDA method.

References

[1] Bethard, S., et al.: Automatic Extraction of Opinion Propositions and their Holders. In: Proc. AAAI Spring Symposium Exploring Attitude and Affect in Text, pp. 22–24 (2004)
[2] Chen, L., et al.: Extracting Diverse Sentiment Expressions with Target-Dependent Polarity from Twitter. In: Proc. 6th AAAI Conf. Weblogs and Social Media (ICWSM) (2012)
[3] Esuli, A., Sebastiani, F.: Sentiwordnet: A publicly available lexical resource for opinion mining. In: Proc. 5th Conf. Language Resources Evaluation (LREC), pp. 417–422 (2006)
[4] Hu, M., Liu, B.: Mining and summarizing customer reviews. In: Proc. 10th ACM SIGKDD Conf. on Knowledge Discovery and Data Mining, pp. 168–177 (2004)
[5] Jo, Y., Oh, A.H.: Aspect and sentiment unification model for online review analysis. In: Proc. 4th ACM Conf. on Web Search and Data Mining, pp. 815–824 (2011)
[6] Liu, B.: Sentiment analysis and subjectivity. In: Handbook of Natural Language Processing (2010) ISBN 978-1420085921
[7] Moghaddam, S., Ester, M.: ILDA: Interdependent LDA model for learning latent aspects and their ratings from online product reviews. In: Proc. of the 34th ACM SIGIR Conf. on Research and Development in Information Retrieval, pp. 665–674 (2011)
[8] Peng, W., Park, D.H.: Generate Adjective Sentiment Dictionary for Social Media Sentiment Analysis Using Constrained Nonnegative Matrix Factorization. In: Proc. 6th AAAI Conf. on Weblogs and Social Media (ICWSM) (2011)
[9] Poria, S., et al.: Enhanced SenticNet with Affective Labels for Concept-Based Opinion Mining. IEEE Intelligent Systems 28, 31–38 (2013)
[10] Ramage, D., et al.: Labeled LDA: A Supervised Topic Model for Credit Attribution in Multi-labeled Corpora. In: Proc. Conf. on EMNLP, pp. 248–256 (2009)
[11] Song, Y., et al.: Topic and keyword re-ranking for LDA-based topic modeling. In: Proc. 18th Conf. on Information and Knowledge Management, pp. 1757–1760 (2009)
[12] Turney, P.D., Littman, M.L.: Unsupervised Learning of Semantic Orientation from a Hundred-Billion-Word Corpus. Information Retrieval. ERB-1094, 11 (2002)
[13] Wang, H., et al.: Latent aspect rating analysis on review text data. In: Proc. 16th SIGKDD Conf. on Knowledge Discovery and Data Mining (KDD), p. 783 (2010)
[14] Wilson, T., et al.: Recognizing contextual polarity in phrase-level sentiment analysis. In: Proc. of the Conference HLT/EMNLP, pp. 347–354 (2005)

Topic-Dependent Sentiment Classification on Twitter

Steven Van Canneyt, Nathan Claeys, and Bart Dhoedt

Department of Information Technology, Ghent University - iMinds, Belgium
{steven.vancanneyt,nathan.claeys,bart.dhoedt}@ugent.be

Abstract. In this paper, we investigate how discovering the topic dicussed in a tweet can be used to improve its sentiment classification. In particular, a classifier is introduced consisting of a topic-specific classifier, which is only trained on tweets of the same topic of the given tweet, and a generic classifier, which is trained on all the tweets in the training set. The set of considered topics is obtained by clustering the hashtags that occur in the training set. A classifier is then used to estimate the topic of a previously unseen tweet. Experimental results based on a public Twitter dataset show that considering topic-specific sentiment classifiers indeed leads to an improvement.

1 Introduction

Twitter is an excellent source of opinions, as it gives us access to the unprompted views of a broad set of users on particular products or events. The opinions or expressions of sentiment about organizations, products, events and people has proven extremely useful for marketing [8] and social studies [13]. Often it is especially important to quickly detect negative opinions, so a company can respond to any criticism in a timely manner. Therefore, we will focus on detecting the tweets expressing negative sentiments.

The sentiment of words used in a tweet are often dependent on the topic of that tweet. For example the tweet 'So I juuuust started the first amazing 15 minutes of The Last of Us, when my ps3 shuts off and the red light started blinking' with a negative sentiment label contains the word 'amazing' which in general indicates a positive sentiment. However as this tweet is situated in the 'Game console' topic, 'red' is associated with the crash of the ps3 which always show the infamous red light blinking. Therefore, we propose a methodology that directly uses the topics of tweets to improve the sentiment classification. We consider a cluster of similar hashtags as a topic. For each cluster we train two classifiers: one classifier aimed at recognising tweets that talk about the corresponding topic, and one classifier aimed at detecting negative opinions in tweets talking about this topic. Given a previously unseen tweet, we use the classifiers of the former type to determine the most likely topic. Then we use the corresponding topic-specific sentiment classifier to estimate the sentiment of the tweet.

A. Hanbury et al. (Eds.): ECIR 2015, LNCS 9022, pp. 441–446, 2015.

The remainder of this paper is structured as follows. We start with a review of related work in Section 2. Next, in Section 3, we describe our topic-dependent classifier. Section 4 explains how the topics of the tweets are estimated. Details on the considered training data, test data and preprocessing steps are provided in Section 5. Subsequently, Section 6 presents the experimental results. Finally, we conclude our work and discuss future work in Section 7.

2 Related Work

Early work on sentiment analysis focused largely on blogs and reviews. Das et al. [4], for instance, used lexical resources to decide whether a post on a stock message board expresses a positive or negative sentiment by the presence of sentiment words. In addition, linguistic rules were used to deal with e.g. negation in sentences. The authors of [11] researched the performance of various machine learning based classifiers for sentiment classification of movie reviews. A more comprehensive survey about sentiment analysis on documents such as reviews can be found in [10].

In recent years, sentiment classification in Twitter has gained a lot of attention. This introduced additional challenges as tweets tend to be very short and noisy compared to reviews and blogs. The methodology described in this paper is based on the machine learning technique introduced by Go et al. [6]. They tested the suitability for sentiment classification of a number of standard classifiers, including Naive Bayes, SVM and Maximum Entropy classifiers. These classifiers were trained using emoticons in the tweets as labels, together with different types of features for the text of the tweets such as unigrams, bigrams and part-of-speech (POS) features. As using bigrams and POS features in addition to unigrams did not increase the performance of the classifiers, we only consider unigrams in this paper. The research of Bifet et al. [2] notes that the accuracy of the sentiment classifiers needs to be nuanced as it is shown that these classification algorithms often favour the most common class. This typically results in good classification performance for this class at the cost of the smaller classes. By focussing on detecting negative tweets we avoid this problem.

The hashtags used in Twitter have been used by several in the context of sentiment analysis. In addition to using hashtags as unigram features [6], they have been used as sentiment labels [5]. In the paper of Davidov et al. [5], the hashtags #happy, #sad, #crazy and #bored were used to label the training data of a classifier. Similar to our approach, Wang et al. [14] considered hashtags as topics. However, their objective is to estimate the sentiment related to a hashtag. In contrast, we consider hashtags clusters as topics and use topic-specific classifiers to improve the quality of the sentiment detection for individual tweets.

3 Sentiment Classification

The sentiment classifier estimates the probability that a tweet is negative, which allows us to sort the tweets according to the likelihood that a tweet is negative.

The sentiment classifier consists of one generic classifier C^K and a topic-specific classifier C^d. The generic classifier C^K is trained on all the tweets in training set K and estimates the generic probability that a tweet t contains negative sentiment, i.e. $P_t(\text{neg}|K)$. The topic-specific classifier C^d is only trained on the tweets in K of topic d. The classifier C^{d_t} estimates the topic-specific probability that a tweet t of topic d_t is negative, i.e. $P_t(\text{neg}|d_t)$.

For the generic and topic-specific classifiers, the Naive Bayes Multinomial classifier [9] implementation of MOA [3] is used. The feature vector V_t of the tweet t, which is used as input for these classifiers, is constructed using a bag-of-words approach. The components of vector V_t are associated with a word that appears in dictionary W. This dictionary W is the set of all words occuring in the tweets of training set K. For feature vector V_t of tweet t, the component comp_w associated with word $w \in W$ is given by:

$$\text{comp}_w = \begin{cases} \frac{\max(p_w, n_w)}{p_w + n_w} \times \frac{|K|}{p_w + n_w} & \text{if } w \in t \\ 0 & \text{otherwise} \end{cases} \tag{1}$$

with p_w and n_w being the absolute frequency of occurrences of word w in respectively positive or negative tweets in K. The first part of this equation ensures that words which occur often in only one sentiment category (positive or negative) have higher associated component values. The second part ensures that words which occur in a lot of tweets of K have lower associated component values. We also experimented with binary features and term frequency features, but as initial experiments showed that these alternatives yield worse results, we will not consider them in the remainder of the paper.

We finally define the probability $P_t(\text{neg})$ that a tweet t is negative as follows:

$$P_t(\text{neg}) = \lambda \cdot P_t(\text{neg}|d_t) + (1 - \lambda) \cdot P_t(\text{neg}|K) \tag{2}$$

with d_t the topic of tweet t, and $\lambda \in [0, 1]$.

4 Topic Classification

The definition of $P_t(\text{neg})$ in (2) assumes that we already know the topic of a tweet. Therefore, a topic classification algorithm is used to classify each tweet into a fitting topic. In this paper, topics are defined by the hashtag clusters that are present in the collection of tweets K. First, the hashtags are clustered into topics D using the Spectral Clustering algorithm with the cut-off threshold of τ [7,12]. The co-occurence distance between two hashtags $h1$ and $h2$ is used as distance measure:

$$\text{distance}(h1, h2) = 1 - \left(\frac{n_{h1,h2}}{\sum_{i=1}^{|H|} n_{h1,hi}} + \frac{n_{h1,h2}}{\sum_{i=1}^{|H|} n_{h2,hi}}\right) \times \frac{1}{2} \tag{3}$$

with H the set of hashtags that occur in the tweets of training set K, and $n_{h1,h2}$ the number of times hashtag $h1$ and $h2$ occur together in the tweets of K. The idea of using this distance measure is that hashtags which co-occur in the same tweets are associated with a similar or even the same topic such as '#cod' and '#callofduty'. As a result of this step, we have a number of clusters of hashtags. We interpret each of these clusters as a topic. The set K_D contains all tweets of K that have at least one hashtag associated with a cluster. Second, tweets in K_D are associated to their corresponding topic. Third, the binary bag-of-words feature vectors of the tweets in K_D are used to train a Naive Bayes Multinomial classifier [3,9], whereby the topics of the tweets are used as labels. Finally, this classifier is used to estimate the topics of the tweets in $K \setminus K_D$ and U. This topic classification approach is based on the methodology described in [1].

5 Data Collection and Preprocessing

We use the public available Stanford Twitter Sentiment corpus[1] introduced by Go et al. [6]. They obtained training set K by automatically labeling tweets based on their emoticons. The use of emoticons as noisy labels makes it easy to extract a large set of training data. In particular, the Twitter API was first queried between April 6, 2009 and June 25, 2009 using query ':(' and ':)' to extract tweets with respectively negative and positive sentiment. Second, the emoticons in the tweets were stripped off and retweets were removed. Finally, the first 800 000 tweets with positive emoticons and the first 800 000 tweets with negative emoticons were considered as training set K. The test set U constructed by [6] contains tweets collected by querying the Twitter API with queries indicating products, companies and people. The obtained tweets were manually annotated resulting in 177 negative, 182 positive and 139 neutral tweets.

Similar as described in [6], all collected tweets were preprocessed to reduce the feature space. In particular, the words of the tweets were converted to lower case and Porter stemmed, and user mentions and URLs were replaced by respectively 'USER_TOKEN' and 'URL_TOKEN'.

6 Results

To evaluate the advantage of using topic-specific classifiers, we compare the result of the proposed classifier with the result of using the generic classifier alone, i.e. $P_t(\text{neg}|K)$. We also evaluate the performance of using the topic-specific classifiers without the generic classifier, i.e. $P_t(\text{neg}|d_t)$. The classifiers are used to estimate the probability that the tweets in the test set are negative, and to rank them based on their associated probability. To evaluate the quality of the ranking, the average precision metric (AP) is used. We empirically set the cut-off threshold τ for the Spectral Clustering algorithm to 0.98.

The average precision for different λ values for equation (2) are shown in Figure 1(a) (for test set U). Note that only the generic classifier $P_t(\text{neg}|K)$ is used when $\lambda = 0$, and only the topic-specific classifier $P_t(\text{neg}|d_t)$ is used when

[1] `http://help.sentiment140.com/`

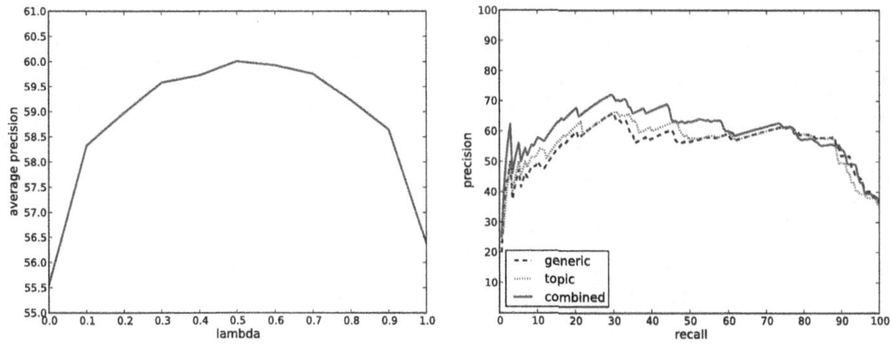

(a) (b)

Fig. 1. (a) Average precisions for different λ values. (b) Precision-recall curves of the generic classifier (baseline), the topic-specific classifier and the combined classifier.

$\lambda = 1$. As can be seen, the curve is more or less symmetric with an maximum average precision when $\lambda = 0.5$ is used. The average precision of the combined classifier with optimal λ (AP = 60.01%) is 4.5 percentage points higher than when the generic classifier (AP = 55.55%) is used. To determine if the difference in quality of the classifications are statistically significant when the different approaches are used, we consider the sign test on the classification accuracy metric. In particular, we obtained a classification accuracy of 82.3% when the combined classifier with $\lambda = 0.5$ is used, which is statistically significant better then when the generic classifier (accuracy of 79.9%) is used (sign test, $p < 0.01$). Finally, the precision-recall curves of the combined classifier with $\lambda = 0.5$, the generic classifier and topic-specific classifier are shown in Figure 1(b).

The following is an example tweet where the topic classifier shows a better probability than the generic classifier: 'I still love my Kindle2 but reading The New York Times on it does not feel natural'. This tweet contains a negative label, however the generic classifier classifies this as positive. This is most likely because the word 'love' is the only generic word which gives a real idea about the sentiment. The topic classifier however sees the word 'natural' as negative, while the generic classifier does not. This can be explained because in the cluster '...#amazon #book #kindle...' the word 'natural' refers to the problem that some users did not find reading on the Kindle2 as natural as reading a book or a newspaper. This is an example of a topic-specific feature that has a strong meaning in the topic that is non-existent in the general tweet corpus because the feature is widely used in general tweets. This sort of features allow the topic-specific classifier to make corrections to the negative probability of the generic classifier.

7 Conclusions and Future Work

We proposed a methodology to rank tweets based on the probability that they express negative sentiment. To this end, we have interpolated a generic language model for negative sentiment and a topic-specific model. In this way we can take advantage of the robustness of a generic classifier, which can be trained on a much larger training set, with the ability of topic-specific classifiers to pick up on context-specific expressions of sentiment. We used a fixed set of topics based on the hashtags from the tweets in the training set. As the topics that are discussed in tweets change over time, in future work we will consider a topic detection approach which evolves over time.

Acknowledgments. Steven Van Canneyt is funded by a Ph.D. grant of the Agency for Innovation by Science and Technology (IWT).

References

1. Antenucci, D., Handy, G., Modi, A., Tinkerhess, M.: Classification of tweets via clustering of hashtags. In: EECS 545 Project, pp. 1–11 (2011)
2. Bifet, A., Frank, E.: Sentiment knowledge discovery in Twitter streaming data. In: Pfahringer, B., Holmes, G., Hoffmann, A. (eds.) DS 2010. LNCS (LNAI), vol. 6332, pp. 1–15. Springer, Heidelberg (2010)
3. Bifet, A., Holmes, G., Kirkby, R., Pfahringer, B.: MOA: Massive Online Analysis. Journal of Machine Learning Research 11, 1601–1604 (2010)
4. Das, S., Chen, M.: Yahoo! for Amazon: Extracting market sentiment from stock message boards. Management Science 53(9), 1375–1388 (2007)
5. Davidov, D., Tsur, O., Rappoport, A.: Enhanced sentiment learning using Twitter hashtags and smileys. In: Proc. of the 23rd Int. Conf. on Computational Linguistics (2010)
6. Go, A., Bhayani, R., Huang, L.: Twitter sentiment classification using distant supervision. In: CS224N Project Report, Stanford (2009)
7. Hall, M., National, H., Frank, E., Holmes, G., Pfahringer, B., Reutemann, P., Witten, I.H.: The WEKA data mining software: An update. SIGKDD Explorations 11(1) (2009)
8. Jiang, L., Yu, M., Zhou, M., Liu, X., Zhao, T.: Target-dependent Twitter sentiment classification. In: Proc. of the 49th Annual Meeting of the Association for Computational Linguistics: Human Language Technologies, pp. 151–160 (2011)
9. Mccallum, A., Nigam, K.: A comparison of event models for naive bayes text classification. In: Proc. of the AAAI-98 Workshop on Learning for Text Categorization, pp. 41–48 (1998)
10. Pang, B., Lee, L.: Opinion mining and sentiment analysis. Foundations and Trends in Inf. Retrieval 2(1-2), 1–135 (2008)
11. Pang, B., Lee, L., Vaithyanathan, S.: Thumbs up? Sentiment classification using machine learning techniques. In: Proc. of the Conf. on Empirical Methods in Natural Language Processing, pp. 79–86 (May 2002)
12. Shi, J., Malik, J.: Normalized cuts and image segmentation. IEEE Pattern Analysis and Machine Intelligence 22(8), 888–905 (2000)
13. Thelwall, M., Buckley, K., Paltoglou, G.: Sentiment in Twitter events. Journal of the American Society for Inf. Science and Technology 62(2), 406–418 (2011)
14. Wang, X., Wei, F., Liu, X., Zhou, M., Zhang, M.: Topic sentiment analysis in Twitter: A graph-based hashtag sentiment classification approach. In: Proc. of the 20th ACM Int. Conf. on Inf., pp. 1031–1040 (2011)

Learning Higher-Level Features with Convolutional Restricted Boltzmann Machines for Sentiment Analysis

Trung Huynh[1], Yulan He[2], and Stefan Rüger[1]

[1] Knowledge Media Institute, The Open University, UK
[2] School of Engineering and Applied Science, Aston University, UK
{trung.huynh,stefan.rueger}@open.ac.uk,
y.he9@aston.ac.uk

Abstract. In recent years, learning word vector representations has attracted much interest in Natural Language Processing. Word representations or embeddings learned using unsupervised methods help addressing the problem of traditional bag-of-word approaches which fail to capture contextual semantics. In this paper we go beyond the vector representations at the word level and propose a novel framework that learns higher-level feature representations of n-grams, phrases and sentences using a deep neural network built from stacked Convolutional Restricted Boltzmann Machines (CRBMs). These representations have been shown to map syntactically and semantically related n-grams to closeby locations in the hidden feature space. We have experimented to additionally incorporate these higher-level features into supervised classifier training for two sentiment analysis tasks: subjectivity classification and sentiment classification. Our results have demonstrated the success of our proposed framework with 4% improvement in accuracy observed for subjectivity classification and improved the results achieved for sentiment classification over models trained without our higher level features.

Keywords: Sentiment analysis, Convolutional Restricted Boltzmann Machines, Stacked Restricted Boltzmann Machine, Word embeddings.

1 Introduction

Word representations have been a key element to many Natural Language Processing (NLP) tasks. Typically, a word can be represented by a vector that captures semantic and syntactic information of the word. Word representations can be induced in many ways including neural language models [1,6,2], which in this case, are often called word embeddings where each dimension of the embedding represents a latent feature of the word. Recent research has shown that using word embeddings has resulted in improved performance in a number of NLP tasks such as word sense disambiguation, named entity recognition, chunking, part-of-speech tagging and sentiment classification [1,9,2].

Most approaches typically induce word embeddings at the individual word level. In some NLP tasks, learning higher-level feature representations such as at the n-gram, phrase or even sentence level could be potentially useful. For example, in sentiment

A. Hanbury et al. (Eds.): ECIR 2015, LNCS 9022, pp. 447–452, 2015.

analysis, it has previously been shown that overall sentiment changes depend on word compositions and discourse structures [8]. Socher et al. [9] introduced semi-supervised recursive autoencoders (RAEs) that learned vector space representation of phrases and sentences. It requires a parse tree to gradually combine the contexts at the left and right children of each tree node. The model was extended to learn a distribution over sentiment labels at each node of the hierarchy constructed by RAEs. Their approach achieved the state-of-the-art performance on sentence-level sentiment classification on the Movie Review (MR) dataset[1]. Kalchbrenner et al. [3] described a Dynamic Convolutional Neural Network approach for modelling sentences. It does not rely on a parse tree but requires labelled data for training.

Different from the aforementioned work, we propose a novel unsupervised framework which uses the pre-trained word embeddings and stacked Convolutional Restricted Boltzmann Machines (CRBMs) to learn interactions amongst consecutive words and induce higher-level feature representations from n-grams and sentences. We demonstrate that these representations are meaningful, i.e., map syntactically and semantically related n-grams to closeby locations in the hidden feature space. We evaluate our proposed framework on two sentiment analysis tasks, subjectivity classification on the MPQA (v1.2) corpus[2] and sentiment classification on the MR dataset. Our experimental results show that the learned higher-level features when combined with existing bag-of-words features help improve the performance of both tasks, with 4% improvement in accuracy observed for subjectivity classification and outperforming the RAE approach on sentence-level sentiment classification.

2 Convolutional Restricted Boltzmann Machine (CRBM)

A Restricted Boltzmann Machine (RBM) is an undirected bipartite network with a set of hidden units $\mathbf{h}$, a set of visible units $\mathbf{v}$, and symmetric connection weights between these two layers represented by a weight matrix W. With an energy function $E(\mathbf{v}, \mathbf{h})$, the generative probability of the network is given by $P(\mathbf{v}, \mathbf{h}) = \frac{1}{Z} \exp(-E(\mathbf{v}, \mathbf{h}))$, where Z is the partition function, which normalises the generative probability. The hidden units are binary-valued and the visible units can be binary-valued or real-valued.

A Convolutional RBM [4] is similar to a normal RBM but weights between visible units and hidden units are shared among all locations in the hidden layer. In a two-dimensional setting, the input layer is an array with dimension of $N_V \times N_V$[3]. The hidden layer consists of K "groups" with each group an $N_H \times N_H$ array of binary units. Each of the K groups is associated with $N_W \times N_W$ filter weights, which are shared across all the hidden units within the group. This results in each hidden layer having a dimension of $N_H = N_V - N_W + 1$. To prevent overfitting, Lee et al., [4] also suggested adding a regularisation term that penalises a deviation of the expected activation of the hidden units from a fixed level p, which is called the target sparsity, a constant controlling the sparseness of the hidden units.

[1] http://www.cs.cornell.edu/People/pabo/movie-review-data/

[2] http://mpqa.cs.pitt.edu/corpora/mpqa_corpus/

[3] For notation convenience, we assume a square matrix. Note that there is no requirement that the inputs must be equal-sized or even two-dimensional in CRBM.

3 Proposed Framework

In our proposed framework, words are first represented by their pre-learned word embeddings. Higher-level features are then learned using stacked CRBMs from sentences where words of the same sentence are stacked into a two-dimensional matrix with their embeddings.

We first chunked each sentence in our datasets into separate words using NLTK's Treebank Work Tokenizer[4]. Words that were not present in the pre-learned word embeddings were replaced by "UNKNOWN". Additionally we tagged words using the NLTK POS tagger and replaced proper nouns (NNP, NNPS) with "ENTITY". We did not perform stemming and kept punctuations since some punctuations such as "!" might be indicative of sentiment. Each word is then represented by its corresponding word embedding, which has a form of a vector of length N_V.

For a sentence containing L words, when we stack its word embeddings, we construct a matrix of size $L \times N_V$ shown as visible layer in Figure 1 (left). Therefore, each sentence is represented by a matrix with the same column size (N_V) although its row size (sentence length L) differs from each other. We use the first CRBM to learn hidden features from n-grams. These first level hidden features are then fed into the second CRBM to learn another higher-level sets of features. Combining all these features improves the sentiment analysis results as will be shown in Section 4. Figure 1 illustrates how we learn higher-level features from sentences using two hidden CRBM layers. All layers are trained using contrastive divergence with 1-step Gibbs sampling.

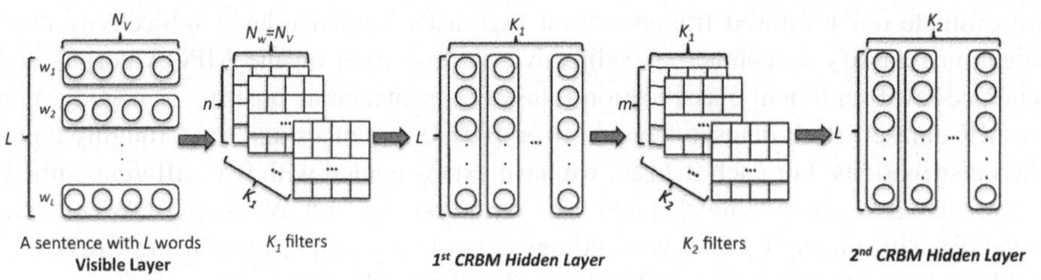

Fig. 1. A CRBM network with two hidden layers. In all layers the width of filters is always equal to the width of the input layer. This helps the network learn interactions among all input features rather than just local features.

For the first CRBM layer, we use K_1 Gaussian-binary unit filters[5]. Each filter has a size of $n \times N_w$. With the row size n, the CRBM can learn higher-level features from n-gram word sequences. In our work here, we set $n = 5$ as this is the typical word sequence length used for inducing word embeddings [1]. We add two padding words at the beginning and the end of each sentence so that convolution can be done with sentences with less than 5 words. Each convolutional filter produces a $L \times 1$ vector

[4] `http://nltk.org/_modules/nltk/tokenize.html#word_tokenize`

[5] We have real-valued visible units and binary-valued hidden units. As such, filters connecting the visible layer and the hidden layer have Gaussian-binary units.

where L is the length of an input sentence. The sum of these vectors forms a set of features that we call CRBM-layer1 features.

Stacking sequentially K_1 vectors produces a matrix of $L \times K_1$ binomial probabilistic units. We then applied another CRBM with K_2 binary-binary unit filters each of which has a size of $m \times K_1$. The parameter m is empirically set to 9. Sums of vectors produced from this CRBM form another set of features that we call CRBM-layer2 features.

Table 1 demonstrates that the hidden features learned from the CRBM network co-locate syntactically related phrases. In addition, the learned hidden higher-level features can capture semantic similarities between phrases, for example, "films provide some great insight" and "film delivers a solid mixture" conveys a similar meaning.

Table 1. Nearest neighbours of some example phrases in the CRBM first-layer hidden space with CBOW embedding on the MR dataset

films provide some great insight	**film well worth seeing**	**of the greatest family-oriented**
abilities offers a solid build-up	is well worth seeing .	of the greatest natural sportsmen
ship makes a fine backdrop	is something worth seeing	of this italian freakshow .
still offers a great deal	is certainly worth hearing .	about the best straight-up
It makes a wonderful subject	this movie worth seeing .	to the greatest generation .
howard demonstrates a great eye	still quite worth seeing .	's very best pictures .
film delivers a solid mixture	it 's worth seeing .	of the finest kind ,

4 Experiments

We evaluate our proposed framework on two tasks, sentence-level subjectivity classification (classify a sentence as subjective or objective) on the MPQA corpus and sentence-level sentiment classification (classify a sentence as positive or negative) on the MR dataset. Both datasets contain over 10,000 sentences and have roughly equal class distributions. For each dataset, we have experimented with two different embeddings, the C&W embeddings [1] and the Continuous Bag-Of-Words (CBOW) embeddings [6]. Since using CBOW embeddings consistently outperforms using C&W embeddings, we only report the results obtained with CBOW embeddings.

4.1 Experimental Setup

We trained two different embeddings for two different tasks here using publicly available code[6]. For subjectivity classification on the MPQA corpus, we trained 200-dimensional word embeddings from the first one billion characters from Wikipedia[7]. For sentiment classification on the MR dataset, we trained 100-dimensional word embeddings from 50,000 movie reviews collected from IMDB[8]. The first layer of the CRBM network has $K_1 = 200$ filters with a target sparsity of $p = 0.01$ while the second CRBM layer has $K_2 = 50$ filters with a target sparsity of $p = 0.5$. The choices of N_V, K_1, K_2 are based on emperical results.

[6] https://code.google.com/p/word2vec/
[7] http://mattmahoney.net/dc/enwik9.zip
[8] http://ai.stanford.edu/~amaas/data/sentiment/

4.2 Results

Subjectivity Classification. For sentence-level subjectivity classification, we combined word embeddings in sentences with higher level features learned from stacked CRBMs and train a linear SVM model[9] with default parameters. All models are cross-validated with 10 folds.

We compared our proposed approach with four baselines. Lexicon labelling uses the MPQA subjectivity lexicon[10] to label a sentence as subjective or objective depending on the occurrence of polarity words in the sentence. SubjLDA is a weakly-supervised Bayesian modelling approach [5] that incorporates word polarity priors from the MPQA subjectivity lexicon into a variant of the Latent Dirichlet Allocation (LDA) model for subjectivity classification. The result of Naive Bayes (NB) was previously reported in [10] where a supervised NB classifier is trained from the MPQA corpus. We also trained a linear SVM with bag-of-words features as another baseline.

It can be observed from Table 2(a) simply training SVM from CBOW word embeddings already outperforms all the baselines. Additionally incorporating higher level features learned by CRBM further improves the accuracy. In particular, adding the CRBM-layer1 features seems to be quite effective. Further adding the CRBM-layer2 features only results in marginal improvements. Overall, with our proposed approach we observed 4% improvement in accuracy upon the best baseline model.

Table 2. Experimental results

(a) Subjectivity classification on MPQA

Model	Accuracy (%)
Lexicon labelling	63.1
subjLDA [5]	71.2
Naive Bayes [10]	73.8
SVM	74.3
Sum of Embeddings (SoE)	77.3*
SoE+CRBM-layer1	78.1* †
SoE+CRBM-layer1 & 2	**78.3***

(b) Sentiment classification on MR

Model	Accuracy(%)
BoF+Reversal	76.4
Tree-CRF [7]	77.3
Greedy RAE [9]	77.7
BoF+Reversal+SoE	78.1*
BoF+Reversal+SoE+CRBM-layer1	78.5* †
BoF+Reversal+SoE+CRBM-layer1 & 2	**78.7*** †

* statistical significance ($p < 0.05$) with respect to the baselines
† statistical significance with respect to its next best model

Sentiment Classification. For sentence-level sentiment classification, we compare our results with three baselines, combining Bag-of-Features with Polarity Reversal (BoF+Reversal), a dependency tree based classification method employing Conditional Random Fields (Tree-CRF) [7], and the greedy Recursive Autoencoder (RAE) network [9]. Here, Bag-of-Features refer to the surface forms, base forms, and POS tags of word unigrams. Polarity reversal indicates polarity reversing caused by content-word negators. These features are trained using linear SVMs with default parameters and and validated by 10-fold cross validation.

[9] http://www.csie.ntu.edu.tw/~cjlin/liblinear/
[10] http://mpqa.cs.pitt.edu/lexicons/

It can be observed from Table 2(b) that using CBOW word embeddings gives similar performance as Greedy RAE. With additional features from a two-layer CRBMs, the model outperforms all the baselines. Although the improvement may appear modest, they are very notable in comparison to the scale of improvements reported in similar literatures [7,9].

5 Conclusions

In this paper we have proposed a novel framework, which uses Convolutional Restricted Boltzmann Machines (CRBMs) to learn useful higher-level features of sentences with pre-trained word embeddings. These features have been shown to boost the performance of simple shallow linear models (linear SVM) outperforming existing approaches in subjectivity classification and has proven to improve sentiment classification accuracy.

In future development, we are going to explore the differences in our model and other unsupervised architectures, e.g. autoencoders. Another potential application is to use this architecture as a pre-training setting for Deep Convolutional Network [3]. It is also interesting to investigate the effect of training with other algorithms such as Dropout and Maxout on the current architecture. In addition, we intend to investigate training an embedding that can distinguish between words with different sentiment.

References

1. Collobert, R., Weston, J.: A Unified Architecture for Natural Language Processing: Deep Neural Networks with Multitask Learning. In: ICML (2008)
2. Dahl, G.E., Adams, R.P., Larochelle, H.: Training Restricted Boltzmann Machines on Word Observations. In: ICML (2012)
3. Kalchbrenner, N., Grefenstette, E., Blunsom, P.: A convolutional Neural Network for Modelling Sentences. In: ACL (2014)
4. Lee, H., Grosse, R., Ranganath, R., Ng, A.Y.: Convolutional Deep Belief Networks for Scalable Unsupervised Learning of Hierarchical Representations. In: ICML (2009)
5. Lin, C., He, Y., Everson, R.: Sentence Subjectivity Detection with Weakly-Supervised Learning. In: IJCNLP (2011)
6. Mikolov, T., Zweig, G.: Context Dependent Recurrent Neural Networkl Language Model. Tech. rep., Microsoft Research Technical Report (2012)
7. Nakagawa, T., Inui, K., Kurohashi, S.: Dependency Tree-based Sentiment Classification using CRFs with Hidden Variables. In: NAACL (2010)
8. Pang, B., Lee, L.: Opinion Mining and Sentiment Analysis. Foundations and Trends in Information Retrieval 2(12), 1–135 (2008)
9. Socher, R., Pennington, J., Huang, E.H., Ng, A.Y., Manning, C.D.: Semi-Supervised Recursive Autoencoders for Predicting Sentiment Distributions. In: EMNLP (2011)
10. Wiebe, J., Riloff, E.: Creating Subjective and Objective Sentence Classifiers from Unannotated Texts. In: Gelbukh, A. (ed.) CICLing 2005. LNCS, vol. 3406, pp. 486–497. Springer, Heidelberg (2005)

Towards Deep Semantic Analysis of Hashtags

Piyush Bansal, Romil Bansal, and Vasudeva Varma

International Institute of Information Technology
Hyderabad, Telangana, India
{piyush.bansal,romil.bansal}@research.iiit.ac.in,
vv@iiit.ac.in

Abstract. Hashtags are semantico-syntactic constructs used across various social networking and microblogging platforms to enable users to start a topic specific discussion or classify a post into a desired category. Segmenting and linking the entities present within the hashtags could therefore help in better understanding and extraction of information shared across the social media. However, due to lack of space delimiters in the hashtags (e.g *#nsavssnowden*), the segmentation of hashtags into constituent entities (*"NSA"* and *"Edward Snowden"* in this case) is not a trivial task. Most of the current state-of-the-art social media analytics systems like Sentiment Analysis and Entity Linking tend to either ignore hashtags, or treat them as a single word. In this paper, we present a context aware approach to segment and link entities in the hashtags to a knowledge base (KB) entry, based on the context within the tweet. Our approach segments and links the entities in hashtags such that the coherence between hashtag semantics and the tweet is maximized. To the best of our knowledge, no existing study addresses the issue of linking entities in hashtags for extracting semantic information. We evaluate our method on two different datasets, and demonstrate the effectiveness of our technique in improving the overall entity linking in tweets via additional semantic information provided by segmenting and linking entities in a hashtag.

Keywords: Hashtag Segmentation, Entity Linking, Entity Disambiguation, Information Extraction.

1 Introduction

Microblogging and Social Networking websites like *Twitter, Google+, Facebook* and *Instagram* are becoming increasingly popular with more than 400 million posts each day. This huge collection of posts on the social media makes it an important source for gathering real-time news and event information. Microblog posts are often tagged with an unspaced phrase, prefixed with the sign "#" known as a hashtag. 14% of English tweets are tagged with at least 1 hashtag with an average of 1.4 hashtags per tweet [1]. Hashtags make it possible to categorize and track a microblog post among millions of other posts. Semantic analysis of hashtags could therefore help us in understanding and extracting important information from microblog posts.

A. Hanbury et al. (Eds.): ECIR 2015, LNCS 9022, pp. 453–464, 2015.

In English, and many other Latin alphabet based languages, the inherent structure of the language imposes an assumption, under which the space character is a good approximation of word delimiter. However, hashtags violate such an assumption making it difficult to analyse them. In this paper, we analyse the problem of extracting semantics in hashtags by segmenting and linking entities within hashtags. For example, given a hashtag like "#NSAvsSnowden" occurring inside a tweet, we develop a system that not only segments the hashtag into "NSA vs Snowden", but also tells that "NSA" refers to "National Security Agency" and "Snowden" refers to "Edward Snowden". Such a system has numerous applications in the areas of Sentiment Analysis, Opinion Mining, Event Detection and improving quality of search results on Social Networks, as these systems can leverage additional semantic information provided by the hashtags present within the tweets. Our system takes a hashtag and the corresponding tweet text as input and returns the segmented hashtag along with Wikipedia pages corresponding to the entities in the hashtag. To the best of our knowledge, the proposed system is the first to focus on extracting semantic knowledge from hashtags by segmenting them into constituent entities.

2 Related Work

The problem of word segmentation has been studied in various contexts in the past. A lot of work has been done on Chinese word segmentation. Huang et al. [3] showed that character based tagging approach outperforms other word based segmentation approaches for Chinese word segmentation. English URL segmentation has also been explored by various researchers in the past [2][5][6]. All such systems explored length specific features to segment the URLs into constituent chunks[1]. Although a given hashtag can be segmented into various possible segments, all of which are plausible, the "correct" segmentation depends on the tweet context. For example, consider a hashtag '*notacon*'. It can be segmented into chunks 'not, a, con' or 'nota, con' based on the tweet context. The proposed system focuses on hashtag segmentation while being context aware. Along with unigram, bigram and domain specific features, content in the tweet text is also considered for segmenting and linking the entities within a hashtag.

Entity linking in microposts has also been studied by various researchers recently. Various features like commonness, relatedness, popularity and recentness have been used for detecting and linking the entities in the microposts [11][12][13]. Although semantic analysis of microposts has been studied vastly, hashtags are either ignored or treated as a single word. In this work, we analyse hashtags by linking entities in the hashtags to the corresponding Wikipedia pages.

[1] The term "chunk" here and henceforth refers to each of the segments s_i in a segmentation $S = s_1, s_2, ...s_i, ...s_n$. For example, in case of the hashtag #NSAvsSnowden, one of the possible segmentations (S) is NSA, vs, Snowden. Here, the *words* - "NSA", "vs" and "Snowden" are being referred to as chunks.

3 System Architecture

In this section, we present an overview of our system. We also describe the features extracted, followed by a discussion on training and learning procedures in Section 4.

As illustrated in Fig. 1, the proposed system has 3 major components - 1) *Hashtag Segmentations Seeder*, 2) *Feature Extraction and Entity Linking module*, and 3) *Segmentation Ranker*. In the following sections, we describe each component in detail.

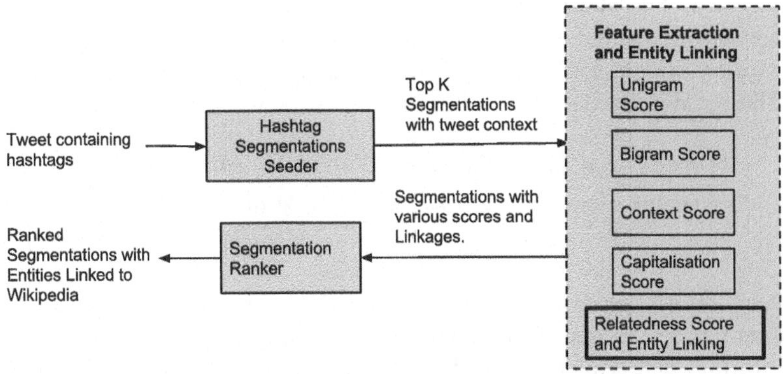

Fig. 1. Schematic Diagram of the Overall System

3.1 Hashtag Segmentations Seeder

Hashtag Segmentations Seeder is responsible for generating a list of possible segmentations of a given hashtag. We propose *Variable Length Sliding Window technique* for generating a set of highly probable hashtag segmentations for the given hashtag in the first step.

The *Variable Length Sliding Window technique* is based on an assumption that for a given hashtag "*#AXB*", if A and B are valid semantic units (a single word or a collection of words concatenated together without a space), it is reasonable to hypothesize that X is also a valid semantic unit. For example, in the hashtag "*#followUCBleague*", since, '*follow*' and '*league*' are well known dictionary words, and collectively this hashtag has some semantic meaning associated with it as it has occurred in a tweet, it is reasonable to assume that '*UCB*' is also a valid semantic unit with some meaning associated with it. The length of the sliding window(X) is varied from `MIN_LEN` to `MAX_LEN` with each iteration, and the window is slid over the hashtag. $O(n^2)$ triplets of the form (A, X, B) are generated using the sliding window technique, where n is the length of the hashtag, X is the part of the hashtag lying within the window and A and B are the parts of the hashtag (of length ≥ 0) that lie on the left and right of the window respectively.

Each segment A and B of the triplet (A, X, B) is assigned a score according to the classically known Dynamic Programming based algorithm for Word Segmentation [7], hereby referred to as $ViterbiWordSeg$.

$ViterbiWordSeg$ takes a string as input and returns the best possible segmentation $BestSeg$ (ordered collection of chunks) for that string. The score assigned to the segmentation by $ViterbiWordSeg$ is the sum of log of probability scores of the segmented chunks based on the unigram language model.

$$ViterbiWordSegScore(S) = \sum_{s_i \in BestSeg(S)} log(P_{Unigram}(s_i)) \qquad (1)$$

We used Microsoft Web N-Gram Services[2] for computing the unigram probability scores. The aforementioned corpus contains data from the web, and hence various acronyms and slang words occur in it. This holds critical importance in the context of our task. Next, for each triplet of the form (A, X, B), we compute the Sliding Window score as follows.

$$\begin{aligned} Score_{SlidingWindow}(A, X, B) &= ViterbiWordSegScore(A) + \\ constant * \log_{10}(UnigramProb(X)) &* WordLenProb(len(X)) + \\ &ViterbiWordSegScore(B) \end{aligned} \qquad (2)$$

where $WordLenProb(x)$ is the *Ordinate* value at x in Figure 2 and the `constant` is set by experimentation.

Also, for each triplet (A, X, B), the final segmentation, $Seg(A, X, B)$ is the ordered collection of chunks ($BestSeg(A)$, X, $BestSeg(B)$), where $BestSeg(A)$ and $BestSeg(B)$ refer to the best segmentation (ordered collection of chunks) returned by $ViterbiWordSeg(A)$ and $ViterbiWordSeg(B)$ respectively.

To find the suitable value of `MIN_LEN` and `MAX_LEN`, we plot the percentage of frequency vs. word length graph using 50 million tweets[3]. Figure 2 shows the plot obtained.

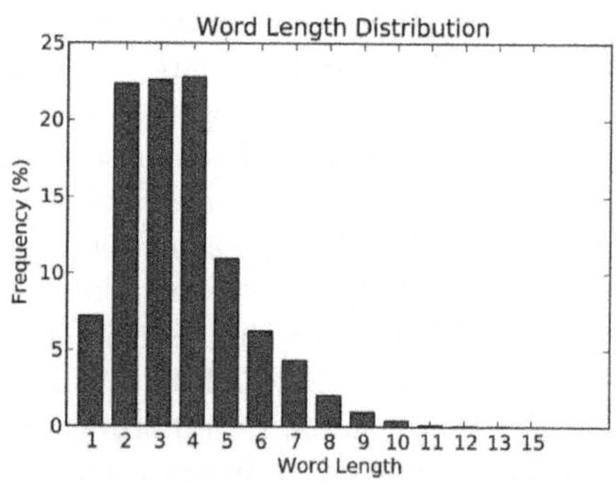

Fig. 2. Word Length vs. Frequency Percentage graph for 50M tweets

[2] Microsoft Web N-Gram Services `http://research.microsoft.com/en-us/collaboration/focus/cs/web-ngram.aspx`

[3] The dataset is available at `http://demeter.inf.ed.ac.uk/cross/docs/fsd_corpus.tar.gz`

It is observed that 79% of the tweet words are between length 2 to 6. Hence, we set `MIN_LEN` and `MAX_LEN` as 2 and 6 respectively.

The major benefit of this technique is that we are able to handle named entities and out of vocabulary (OOV) words. This is achieved by assigning score as a function of $WordLenProb$ and smoothed backoff unigram probability (Equation 2) for words within the window.

Now that we have a list of $O(n^2)$ segmentations and their corresponding $Score_{SlidingWindow}$, we pick the top k segmentations for each hashtag on the basis of this score. We set $k = 20$, as precision at 20 (P@20) comes out to be around 98%. This establishes that the subset of segmentations we seed, which is of $O(n^2)$, indeed contains highly probable segmentations out of a total possible 2^{n-1} segmentations[4].

3.2 Feature Extraction and Entity Linking

This component of the system is responsible for two major tasks, feature extraction from each of the seeded segmentations, and entity linking on the segmentations. The features, as also shown in the System Diagram, are 1) Unigram Score, 2) Bigram Score, 3) Context Score, 4) Capitalisation Score, and 5) Relatedness Score. The first feature, Unigram Score, is essentially the $ViterbiWordSegScore$ computed in the previous step. In the following sections, we describe the rest of the features.

Bigram Score: For each of the segmentations seeded by the *Variable Length Sliding Window Technique*, a bigram based score using the Microsoft Web N-Gram Services is computed. It is possible for a hashtag to have two perfectly valid segmentations. Consider the hashtag ***#Homesandgardens***. Now this hashtag can be split as "Homes and gardens" which seems more probable to occur in a given context than "Home sand gardens". Bigram based scoring helps to rank such segmentations, so that higher scores are awarded to the more semantically "appealing" segmentations. The bigram language model would score one of the above segmentations - "Homes and gardens" as

$$\begin{aligned} &P(Homes, and, gardens) \approx \\ &\quad P(Homes| < s >) * P(and|Homes) * \\ &\quad P(gardens|and) * P(< /s > |gardens) \end{aligned} \tag{3}$$

Context Score: Context based score is an important feature. This is responsible for bubbling up of the segmentations with maximum contextual similarity with the tweet content. Using the CMU TweetNLP toolkit [8], words having

[4] For a string made up of n characters, we need to decide where to put the spaces so that we can get a sequence of valid words. There are $n-1$ positions where a space can be placed, and each position may or may not have a space. Hence there are 2^{n-1} segmentations.

POS tags like verb, noun and adjective are extracted both from the candidate segmentation of the hashtag and the tweet context, i.e. the text of the tweet other than the hashtag. Next, Wu Palmer similarity from Wordnet [9] is used on these two sets of words to find how similar a candidate segmentation is to the tweet context. These scores are normalized from 0 to 1.

Capitalisation Score: Hashtags are of varied nature. Some hashtags have a camelcase-like capitalisation pattern as in ***#HomesAndGardens***, while others have everything in lowercase or uppercase characters like ***#homesandgardens***. However, we can easily see that camelcase conveys more information as it helps segment the hashtag into "Homes and gardens" and not "Home sAnd Gardens". Capitalisation score helps us to capture the information conveyed by capitalisation patterns within the hashtags. We use the following two rules. For a hashtag,

- If a set of characters occuring together are in capitals as in ***#followUCB league***, they are considered to be a part of an "assumed cluster" ("UCB" in this case).
- If it has a few capital letters separated by a group of lower case letters as in ***#SomethingGood***, we assume the capital letters are delimiters and hence derive a few assumed clusters from the input hashtag.

We calculate the capitalisation score for a given segmentation S containing chunks $s_1, s_2 ... s_i .. s_n$ as

$$Score_{Cap} = \sum_{i=1}^{n} assumedClusterNotIntact(s_i) \tag{4}$$

where $assumedClusterNotIntact(s_i)$ returns 1, if s_i fails to keep an assumed cluster intact, and 0 otherwise.

Relatedness Score: Relatedness score measures the coherence between the tweet context and the hashtag segmentation. This score is computed on the basis of semantic relatedness between the entities present within the segmented hashtag and the tweet context.

We calculated the relatedness between all the possible mentions in the segmented hashtag (M_H) to all other possible mentions in the tweet context (M_T). For computing relatedness between the two entities, we used the Wikipedia-based relatedness function as proposed by Milne and Witten [4].

Relatedness between two Wikipedia pages p_a and p_b is defined as follows:

$$rel(p_a, p_b) = 1 - \delta \tag{5}$$

where,

$$\delta = \frac{log(max(|in(p_a), in(p_b)|)) - log(|in(p_a) \cap in(p_b)|)}{log(W) - log(min(|in(p_a), in(p_b)|))} \tag{6}$$

$in(p_a)$ is the set of Wikipedia pages pointing to page p_a and W is the total number of pages in Wikipedia.

The overall vote given to a candidate page p_a for a given mention a by a mention b is defined as

$$vote_b(p_a) = \frac{\sum_{p_b \in Pg(b)} rel(p_b, p_a).Pr(p_b|b)}{|Pg(b)|} \tag{7}$$

where $Pg(b)$ are all possible candidate pages for the mention b and $Pr(p_b|b)$ is the prior probability of b linking to a page p_b.
The total relatedness score given to a candidate page p_a for a given mention a is the sum of votes from all other mentions in the tweet context (M_T).

$$rel_a(p_a) = \sum_{b \in M_T} vote_b(p_a) \tag{8}$$

Now the overall relatedness score for a given hashtag segmentation, h is

$$score_h = \frac{\sum_{m \in M_H} rel_m(p_a).Pr(p_a|m)}{|M_H|} \tag{9}$$

The detected page p_a for a given mention in the segmented hashtag is the Wikipedia page with the highest $rel_a(p_a)$. Since not all the entities are meaningful, we prune the entities with very low $rel_a(p_a)$ scores. In our case, the threshold is set to 0.1. This disambiguation function is considered as state-of-the-art and has also been adopted by various other systems [12][16]. The relatedness score, $score_h$ is used as a feature for hashtag segmentation. The entities in the segmented hashtag are returned along with the score for further improving the hashtag semantics.

3.3 Segmentation Ranker

This component of the system is responsible for ranking the various probable segmentations seeded by the *Hashtag Segmentations Seeder Module.* We generated five features for each segmentation using *Feature Extraction and Entity Linking Module* in the previous step. These scores are combined by modelling the problem as a regression problem, and the combined score is referred to as $Score_{Regression}$. The segmentations are ranked using $Score_{Regression}$. In the end, the *Segmentation Ranker* outputs a ranked list of segmentations along with the entity linkings.

In the next section, we discuss the regression and training procedures in greater detail.

4 Training Procedure

For the task of training the model, we consider the $Score_{Regression}$ of all correct segmentations to be 1 and all incorrect segmentations as 0. Our feature vector comprises of five different scores calculated in Section 3.2. We use linear regression with elastic net regularisation [14]. This allows us to learn a model that is trained with both L1 and L2 prior as regularizer. It also helps us take care of the situation when some of the features might be correlated to one another. Here, ρ controls the convex combination of L1 and L2.

The Objective Function we try to minimize is

$$\min_{w} \frac{1}{2n_{samples}} ||Xw - y||_2^2 + \alpha\rho||w||_1 + \frac{\alpha(1-\rho)}{2} ||w||_2^2 \tag{10}$$

where X, y and w are Model Matrix, Response Vector, and Coefficient Matrix respectively. The parameters $alpha(\alpha)$ and $rho(\rho)$ are set by cross validation.

5 Experiments and Results

In this section we describe the datasets used for evaluation, and establish the effectiveness of our technique by comparing our results to a well known end-to-end Entity Linking system, TAGME [12], which works on short texts, including tweets.

5.1 Evaluation Metrics and Datasets

This section is divided into two parts. First, we explain the evaluation metrics in the context of our experiments, and later, we discuss the datasets used for evaluation.

Evaluation Metrics. We evaluated our system on two different metrics. Firstly, the system is evaluated based on its performance in the segmentation task. As the system returns a list of top-k hashtag segmentations for a given hashtag, we evaluated the precision at n (P@n) scores for the hashtag segmentation task. We also compared our P@1 score with Word Breaker[5], which does the task of word segmentation. Secondly, the system is also evaluated on the basis of its entity linking performance on the hashtags. We computed Precision, Recall and F-Measure scores for the entities linked in the top ranked hashtag. For Entity Linking task, we used the same notions of Precision, Recall, and F-Measure as proposed by Marco et al. [10]. We compared our system with the state-of-the-art TAGME system.

We show that adding semantic information extracted from the hashtags leads to an improvement in the overall tweet entity linking. For this, we performed a comparative study on the output of the TAGME system when a tweet is given with un-segmented hashtag vs. when it is given with segmented and entity-linked hashtag[6]. The case when un-segmented hashtag is fed to TAGME is considered as a baseline to show how much improvement can be attributed to our method of enriching the tweet with additional semantic information mined by segmenting and linking entities in a hashtag.

---

[5] `http://web-ngram.research.microsoft.com/info/break.html`

[6] For the segmented and entity-linked case, the linked entities in a hashtag were replaced with the corresponding Wikipedia page titles.

Table 1. Comparative Accuracies on the Microposts NEEL Dataset[7]

(a) Comparative Accuracies for Hashtags Entity Linking task.

	Precision	Recall	F Score
TAGME (Baseline)	0.441	0.383	0.410
Our System	0.711	0.841	0.771

(b) Comparative Accuracies for Overall Tweet Entity Linking task.

	Precision	Recall	F Score
TAGME (Baseline)	0.63	0.69	0.658
Our System + TAGME	0.732	0.91	0.811

(c) Various $P@n$ for Hashtag Segmentation task.

n	1	2	3	5	10	20
$P@n$	0.914	0.952	0.962	0.970	0.974	0.978

Datasets. The lack of availability of a public dataset that suits our task has been a major challenge. To the best of our knowledge, no publicly available dataset contains tweets along with hashtags, and the segmentation of hashtag into constituent entities appropriately linked to a Knowledge Base. So, we approached this problem from two angles - 1) Manually Annotated Dataset Generation (where dataset is made public), 2) Synthetically generated Dataset. The datasets are described in detail below.

1. Microposts NEEL Dataset: The Microposts NEEL Dataset [15] contains over 3.5k tweets collected over a period from 15th July 2011 to 15th August 2011, and is rich in event-annotated tweets. This dataset contains Entities, and the corresponding linkages to DBPedia. The problem however, is that this dataset does not contain the segmentation of hashtags. We generate synthetic hashtags by taking tweets, and combining random number of consecutive words with each entity present within them. The remaining portion of the tweet that does not get combined is considered to be the tweet context. If no entity is present within the tweet, random words are combined to form the hashtag. This solves the problem of requiring human intervention to segment and link hashtags, since now we already know the segmentation as well as the entities present within the hashtag.

Our system achieved an accuracy ($P@1$) of 91.4% in segmenting the hashtag correctly. The accuracy of Word Breaker in this case was 80.2%. This, however, can be attributed to a major difference between our system and Word Breaker. Word Breaker is not context aware. It just takes an unspaced string, and tries to break it into words. Our method takes into account the relatedness between the

[7] "TAGME (Baseline)" refers to the baseline evaluation where we give an unsegmented hashtag to TAGME to annotate. "Our System + TAGME" refers to the evaluation, where we first do segmentation and entity linking on hashtags using our system, and then feed them to TAGME to annotate either just the hashtag (Table a) or the full tweet (Table b). This is also discussed under "Evaluation Metrics" in subsection 5.1.

Table 2. Comparative Accuracies on the Manually Annotated Stanford Sentiment Analysis Dataset

(a) Comparative Accuracies for the Hashtag Entity Linking task.

	Precision	Recall	F Score
TAGME (Baseline)	0.398	0.465	0.429
Our System	0.731	0.921	0.815

(b) Comparative Accuracies for the Overall Tweet Entity Linking task.

	Precision	Recall	F Score
TAGME (Baseline)	0.647	0.732	0.687
Our System + TAGME	0.748	0.943	0.834

(c) Various $P@n$ for Hashtag Segmentation task.

n	1	2	3	5	10	20
$P@n$	0.873	0.917	0.943	0.958	0.965	0.967

entities in a hashtag and the rest of the tweet content. Also, various other hashtag specific features like Capitalisation Score play an important part in improving the accuracy.

The comparative results of Entity Linking (in hashtags and overall), as well as $P@n$ at various values of n for segmentation task are contained in Table 1. All the values are calculated by k-fold Cross-validation with k=5.

2. Manually Annotated Stanford Sentiment Analysis Dataset: To overcome the limitation that a synthetically generated hashtag might not actually be equivalent to a real world hashtag, we sampled around 1.2k tweets randomly from the Stanford Sentiment Analysis Dataset[8], all of which contained one or more hashtags in them. After this, we generated around 20 possible segmentations for each hashtag by passing the hashtag and tweet from *Segmentations Seeder Module*. In the end we had around 21k rows which were given to 3 human annotators to annotate as 0 or 1 depending on whether or not a given segmentation is correct (for a given hashtag) according to their judgement.

Determining the "correct" segmentation for a given hashtag is particularly challenging, as there may be many answers that are equally plausible. It has been long established that there exist style disagreements among various editorial content ("Homepage" vs "Home page"). There are also various new words that come into existence like "TweetDeck" which are brand or product names. So, our annotation guidelines in case of Stanford Sentiment Analysis Dataset allow for annotators to mark multiple segmentations as correct.

The rows were labelled 0, if at least 2 annotators out of 3 agreed on the label 0, similarly the rows were labelled 1, if at least 2 out of 3 annotators agreed on the label 1. The labels are essentially $Score_{Regression}$ as described in Section 4. The value of *Fleiss' Kappa* (κ), which is a measure of inter annotator

[8] `http://cs.stanford.edu/people/alecmgo/trainingandtestdata.zip`

Table 3. Importance of each feature

Added Feature	P@1	Δ
Unigram	0.834	NA
+ Bigram	0.846	+1.2%
+ Context	0.855	+0.9%
+ Capitalisation	0.862	+0.7%
+ Relatedness	0.873	+1.1%

agreement, comes out to be 0.89, showing a good agreement between annotators. This dataset is made public to ease future research in this area[9].

Our system achieved a precision (P@1) of 87.3% in segmenting the hashtags correctly. The P@1 score of Word Breaker in this case was 78.9%. The difference in performance can again be attributed to same reasons as in the case of NEEL Dataset.

The comparative results of Entity Linking (in hashtags and overall), as well as P@n at various values of n for the task of segmentation are contained in Table 2. All the values are calculated by k-fold Cross-validation with k=5.

Results. We demonstrate the effectiveness of our technique by evaluating on two different datasets. We also show how overall Entity Linking in tweets was improved, when our system was used to segment the hashtag and link the entities in the hashtag. We achieved an improvement of 36.1% F-Measure in extracting semantics from hashtags over the baseline in case of NEEL Dataset. We further show that extracting semantics led to overall increase in Entity Linking of tweet. In case of NEEL Dataset, we achieved an improvement of 15.3% F-Measure over baseline in overall tweet Entity Linking task as can be seen in Table 1. Similar results were obtained for the Annotated Stanford Sentiment Analysis Dataset as well, as shown in Table 2. Further, we measured the effectiveness of each feature in ranking the hashtag segmentations. The results are summarized in Table 3.

6 Conclusions

We have presented a context aware method to segment a hashtag, and link its constituent entities to a Knowledge Base (KB). An ensemble of various syntactic, as well as semantic features is used to learn a regression model that returns a ranked list of probable segmentations. This allows us to handle cases where multiple segmentations are acceptable (due to lack of context in cases, where tweets are extremely short) for the same hashtag, e.g. ***#Homesandgardens***.

The proposed method of extracting more semantic information from hashtags can be beneficial to numerous tasks including, but not limited to sentiment analysis, improving search on social networks and microblogs, topic detection etc.

[9] Dataset: `http://bit.ly/HashtagData`

References

1. Weerkamp, W., Carter, S., Tsagkias, M.: How People use Twitter in Different Languages. In: Proceedings of Web Science (2011)
2. Wang, K., Thrasher, C., Hsu, B.-J.P.: Web scale NLP: a case study on url word breaking. In: Proceedings of the 20th International Conference on World Wide Web, WWW 2011 (2011)
3. Huang, C., Zhao, H.: Chinese word segmentation: A decade review. Journal of Chinese Information Processing 21(3), 8–20 (2007)
4. Milne, D., Witten, I.H.: An effective, low-cost measure of semantic relatedness obtained from wikipedia links. In: Proceedings of AAAI (2008)
5. Kan, M.-Y., Hoang Oanh Nguyen, T.: Fast webpage classification using URL features. In: Proceedings of the 14th ACM International Conference on Information and Knowledge Management. ACM (2005)
6. Srinivasan, S., Bhattacharya, S., Chakraborty, R.: Segmenting web-domains and hashtags using length specific models. In: Proceedings of the 21st ACM International Conference on Information and Knowledge Management. ACM (2012)
7. Russell, S., Norvig, P.: Articial Intelligence: A Modern Approach. Prentice-Hall (2003)
8. Gimpel, K., et al.: Part-of-speech tagging for twitter: Annotation, features, and experiments. In: Proceedings of the 49th Annual Meeting of the Association for Computational Linguistics: Human Language Technologies: Short Papers-Volume 2. Association for Computational Linguistics (2011)
9. Miller, G.A.: WordNet: A Lexical Database for English. Communications of the ACM 38(11), 39–41 (1998)
10. Cornolti, M., Ferragina, P., Ciaramita, M.: A framework for benchmarking entity-annotation systems. In: Proceedings of the 22nd International Conference on World Wide Web (2013)
11. Meij, E., Weerkamp, W., de Rijke, M.: Adding Semantics to Microblog Posts. In: Proceedings of the 5th ACM International Conference on Web Search and Data Mining (2012)
12. Ferragina, P., Scaiella, U.: Tagme: on-the-fly annotation of short text fragments (by Wikipedia entities). In: Proceedings of 19th ACM Conference on Knowledge Management (2010)
13. Bansal, R., Panem, S., Gupta, M., Varma, V.: EDIUM: Improving Entity Disambiguation via User Modeling. In: Proceedings of the 36th European Conference on Information Retrieval (2014)
14. Zou, H., Hastie, T.: Regularization and variable selection via the elastic net. Journal of the Royal Statistical Society: Series B (Statistical Methodology) (2005)
15. Cano Basave, A.E., Rizzo, G., Varga, A., Rowe, M., Stankovic, M., Dadzie, A.-S.: Making Sense of Microposts (#Microposts2014) Named Entity Extraction & Linking Challenge. In: 4th Workshop on Making Sense of Microposts (#Microposts2014) (2014)
16. Han, X., Sun, L., Zhao, J.: Collective entity linking in web text: A graph-based method. In: Proceedings of 34th International ACM SIGIR Conference on Research and Development in Information Retrieval. ACM (2011)

Chalk and Cheese in Twitter: Discriminating Personal and Organization Accounts

Richard Jayadi Oentaryo, Jia-Wei Low, and Ee-Peng Lim

Living Analytics Research Centre, Singapore Management University
80 Stamford Road, Singapore 178902
{roentaryo,jwlow,eplim}@smu.edu.sg

Abstract. Social media have been popular not only for individuals to share contents, but also for organizations to engage users and spread information. Given the trait differences between personal and organization accounts, the ability to distinguish between the two account types is important for developing better search/recommendation engines, marketing strategies, and information dissemination platforms. However, such task is non-trivial and has not been well studied thus far. In this paper, we present a new generic framework for classifying personal and organization accounts, based upon which comprehensive and systematic investigation on a rich variety of content, social, and temporal features can be carried out. In addition to generic feature transformation pipelines, the framework features a gradient boosting classifier that is accurate/robust and facilitates good data understanding such as the importance of different features. We demonstrate the efficacy of our approach through extensive experiments on Twitter data from Singapore, by which we discover several discriminative content, social, and temporal features.

Keywords: account type classification, gradient boosting, social media.

1 Introduction

Social media provide a platform not only for social interaction among users, but also for businesses, government agencies, and other interest groups to engage users with news and campaign events. As such, social media see the strong presence of both ordinary users and organizations. Unfortunately, these two kinds of social media accounts are not clearly differentiated, as the account types are not specified when accounts are created. In some cases, one could manually judge the account type by examining the account name, description, and content postings. However, such kind of intelligent judgment is a non-trivial task for machines.

We define *organization account* as a social media account that represents an institution, corporation, agencies, news media, or common interest group, whereas *personal account* is of non-organizational nature and usually managed by an individual. An accurate labeling of these account types will bring about many benefits. Firstly, organization and oersonal accounts exist for different purposes and thus demand for different types of support and services. For example,

A. Hanbury et al. (Eds.): ECIR 2015, LNCS 9022, pp. 465–476, 2015.

organization accounts may require templates to standardize the format of their content postings, and dashboard to track their social media performance, say the amount of positive and negative sentiments on their product brands. Personal accounts, in contrast, would likely benefit from friend and content recommendation. Such differentiation of services is presently not possible until the account type can be made known or accurately predicted, which is the focus of this paper.

From the information retrieval perspective, the ability to distinguish personal and organization accounts is useful for enriching and providing context to search or recommendation engines. For example, when searching for a certain trending topic, one may be interested to separate/categorize between official information coming from a credible institution or news source and subjective opinions/views from individuals. From the social science standpoint, much work on social media, such as friend recommendation, community discovery, topic modeling, etc., has been done often assuming that social media accounts are owned by ordinary users. The presence of organization accounts clearly introduces biases to the results analysis and should be treated differently from personal accounts.

In this work, we attempt to address the account type classification problem, which involves assigning social media accounts into the *personal* and *organization* categories. This problem has not been well studied in the past. Nevertheless, recently there is a surge of interest in developing methods for differentiating the two account types, such as the works in [7,12,14,15]. However, most of these works focused on a limited set of social, temporal, or content features [7,12], or relied on assumptions that may impose significant biases in their evaluation (e.g., using only geotagged tweets [15] or small data samples [12,14]).

Contributions. Deviating from the previous works, we approach the account classification task by developing a generic framework that facilitates systematic studies on a rich set of content, temporal, and social features, and that offers accurate/robust prediction method. Specifically, our key contributions include:

- We develop a generic framework for account type classification that can cater for various features using generic set of feature transformation pipelines. At its core is the *gradient boosting* classification method [8], which provides not only accurate and robust prediction but also facilitates data understanding.
- We present a new empirical study on Twitter data involving a large (unbalanced) pool of personal and organization accounts. We conduct exploratory analyses on a variety of content, social, and temporal factors associated with personal and organization accounts, based on which we systematically devise a comprehensive set of predictive features for account type classifcation.
- Extensive experiments have also been carried out to evaluate the impacts of different features, and to compare the performance of our gradient boosting approach with other classification methods. We also identify several key features important for the distinction of personal and organization accounts.

2 Related Work

The abundance of user-generated data in social media has recently attracted great interest in inferring the latent attributes of users (e.g., gender [3], political stand [6], ethnicity [5]). Most of these works, however, have treated organization and personal accounts equally. Yet, the ability to distinguish the two is practically important for marketing and information dissemination. Nonetheless, several efforts have been recently made to this end. Tavares and Faisal [12] distinguished between personal, managed, and bot accounts in Twitter, using only the temporal features of the tweets. De Choudhury *et al.* [7] classified Twitter accounts as organization, journalist/blogger, or individual. They utilized structural features, textual features, and binary features indicating the presence of named entitites and associations with news topics. Yan *et al.* [14] called the personal and organization accounts closed and open accounts respectively, and used the diversity of the follower distribution as features. Recently, Yin *et al.* [15] devised a probabilistic method that utilizes temporal, spatial and textual features to classify personal communication and public dissemination accounts.

Proposed approach. Our work differs from the abovementioned approaches in several ways. For instance, Tavares and Faisal [12] focused only on temporal features without considering other feature types. Meanwhile, Yin *et al.* [15] used only geotagged tweets, which may yield significant bias against non-mobile (e.g., desktop) users who do not share their location. In contrast, we use a comprehensive set of content, social, and temporal features, and we consider all tweets with or without geotag information. In [7], De Choudhury *et al.* utilized only simple social features based on in-degree and out-degree centrality metrics. By comparison, our work involves more sophisticated social features that go beyond simple degree centrality. Moreover, we utilize temporal features (e.g., tweet distribution per hour or weekday) in our classification model. Compared to [14], our approach takes into account a more comprehensive set of temporal and social features (encompassing many node centrality and diversity measures). We further elaborate our classification method and feature set in Sections 4 and 5.

3 Data Exploration

In this study, we use the Twitter data of users from Singapore collected from March to May 2014. Starting from a set of popular seed users (having many followers) based in Singapore, we crawled their network based on the follow, retweet, and user mention links. In turn, we added into our user base those followers/followees, retweet sources, and mentioned users who declare Singapore in their profile location. This led to a total of 160,143 public Twitter accounts whose profiles can be accessed. To establish the ground truth, we took accounts whose "urlEntities" field ends with ".com.sg", ".gov.sg", or ".edu.sg". This choice allows us to clearly identify organization accounts for deriving high-quality ground truths, though it may impose labeling bias and miss other, less common types of organization. Nevertheless, we show later that our prediction method can work well for organization accounts that are not from these domains (cf. Section 5.4).

Using this procedure, we were able to identify 885 organization accounts. Through random sampling of the Twitter data, we also obtained 1,135 personal accounts. All labels have been manually inspected by humans. In total, we have

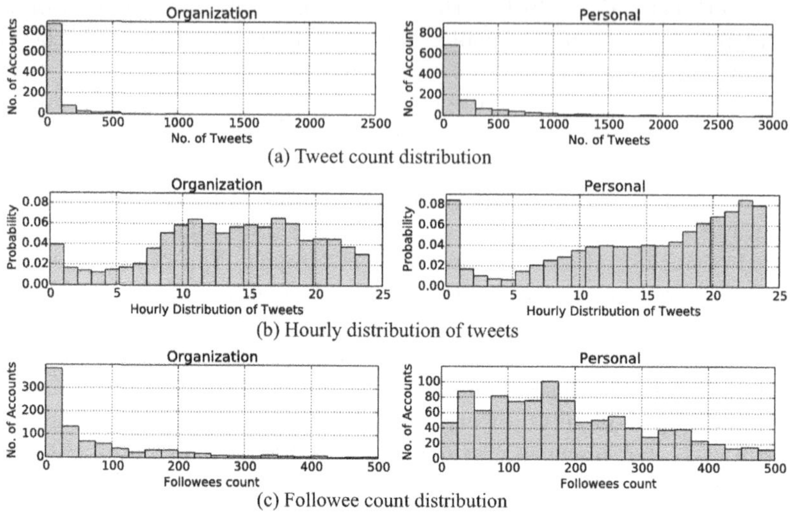

Fig. 1. Data distributions for personal and organization accounts

2,020 labeled accounts, involving 1.18 million tweets. One may argue that it is better to balance the label distribution, e.g., by using the same number (885) of personal accounts. However, we can expect that the full Twitter population would naturally have more personal accounts than organization accounts [7,14]. Hence, we maintain the current label distribution (i.e., 885 vs. 1,135), and let our classification algorithm internally take care of the skewed distribution.

Content analysis. We first conducted analysis on the number of tweets for personal and organization accounts. Fig. 1(a) shows the distribution of tweet counts for the two accounts. From the figure, we can see that the tweet counts generally follow a long-tail distribution. It is also shown that personal accounts tend to tweet more than organization accounts. We then conducted *Kolmogorov-Smirnov* (K-S) test [11] to check whether the two distributions are significantly different[1]. In this case, we obtained a p-value of 2.8779×10^{-36}, which is smaller than typical significance level (e.g., 0.01 or 0.05). Hence, we can reject the null hypothesis that the two distributions are identical. That is, the distributions of personal and organization accounts are significantly different.

Temporal analysis. Next, we conducted a temporal data analysis to check whether the tweet dynamics of the personal and organization accounts are different. Fig. 1(b) shows the hourly distribution of the tweet counts. As the purpose of setting organization accounts is chiefly about information dissemination, we can see that their tweet activities tend to be more aligned with business operation/working hours. On the other hand, we observe that personal accounts tend to tweet more towards the end of the day, peaking around midnight. Using the

[1] We use two-sample K-S test, which is a nonparametric statistical test to quantify the distance between the empirical distribution functions of two samples.

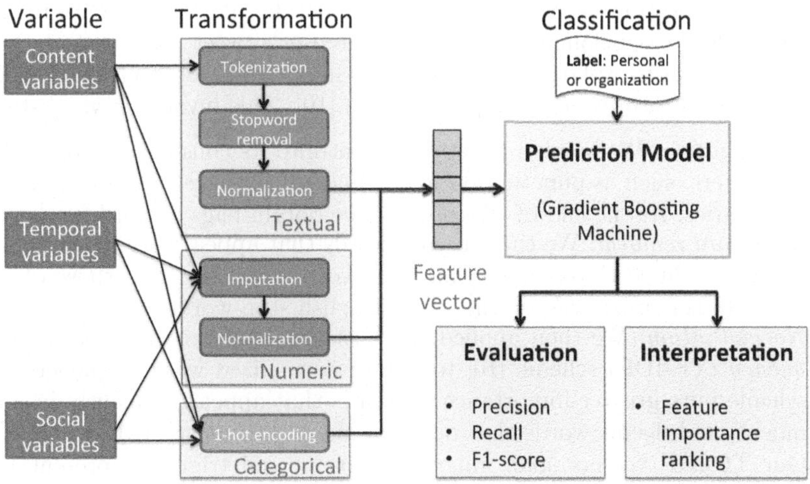

Fig. 2. Proposed framework for account type classification

K-S test, we again obtained p-value $< 10^{-100}$ and concluded a significant difference between the two. This suggests that the temporal distribution of tweets could be a useful feature for distinguishing the two account types.

Social analysis. We also analyzed the interaction patterns among accounts. Fig. 1(c) shows the distributions of the followee counts for personal and organization accounts. We can see that in general personal accounts have more followees than organization accounts. Again, this may be attributed to the fact that organization accounts are set up mainly for dissemination purposes, and so unlikely to be interested in other accounts. The significant difference between the two distributions is evident in our K-S test, with p-value of 8.07×10^{-61}.

4 Proposed Framework

Our proposed account classification framework is outlined in Fig. 2. It takes three types of (raw) input variables: *content*, *temporal*, and *social*. Each variable type goes through a specific transformation pipeline (cf. "Transformation" in Fig. 2) to derive feature vector representation suitable for our classification model. The choice of pipeline for a given variable depends on the semantics of the variable. Using the feature vector and the class label, we build the model (cf. "Predictive Model" in Fig. 2). We then evaluate the model performance based on several metrics (cf. "Evaluation" in Fig. 2). The framework also has a specialized module to extract knowledge structure from the model (cf. "Interpretation" in Fig. 2).

4.1 Feature Transformation Module

Our framework has three types of transformation pipeline, which can be generically used to transform any content, temporal, and social variables into feature vector representation for our classification model. For convenience, we refer the collection of tweets belonging to a user as the user's tweet *document*.

Textual pipeline. For text variables such as tweet documents, we convert them into *bag-of-words* vector representation [10]. This involves several steps:

- *Tokenization*: We break a tweet document into its constituent word tokens. Delimiters, such as punctuation marks and white spaces, were used as word boundaries. At the end of this process, we obtain bags of word frequencies.
- *Stop-word removal*: We then discard words that appear very frequently and contribute little to discriminating the tweets of a user from those of other users. In this work, we use the list of English stop-words in [9].
- *Normalization*: We then applied the *term frequency–inverse document frequency* (TF-IDF) scheme [10] to obtain normalized word frequencies. The scheme puts greater importance on words that appear frequently in a document, and deems words that occur in many documents as less important. Our TF-IDF vectors span unigram, bigram, and trigram representations. More advanced methods such as BM25 and part-of-speech tagging [10] can be included, but for simplicity we use only the TF-IDF method in this work.

Numerical pipeline. The transformation steps of numerical variables (such as count or ratio variables; cf. Table 1) include:

- *Imputation.* We first impute the missing feature values by replacing them with some constant value, or else the average of the other, existing feature values. In this work, we impute missing values with a constant value of zero.
- *Normalization*: This step performs feature normalization by (re)scaling each feature to a unit range [0, 1]. This normalization serves to address the feature scaling issues in classification methods that rely on some distance metric.

Categorical pipeline: In our framework, all categorical variables are binary-encoded. For example, a categorical variable with four possible values: "A", "B", "C", and "D" is encoded using four binary features: "1 0 0 0", "0 1 0 0", "0 0 1 0", and "0 0 0 1", respectively. This is also known as *one-hot encoding* scheme.

4.2 Prediction Model

For our classification task, we employ an ensemble model called *gradient boosting machine* (GBM) [8]. The learning procedure in GBM involves consecutively fitting new models to provide more accurate estimate of the response variable (i.e., class label). The centerpiece of GBM is to construct new base-learners so that they are maximally correlated with the negative gradient of the specified loss function, associated with the entire ensemble [8].

It is worth noting that the loss function used in GBM can be arbitrary, thus providing practitioners with the flexibility to select the most appropriate loss function to the task requirements. GBM is also relatively easy to implement, allowing practitioners to experiment with different model designs. In this work, we focus on using the *binomial loss* function in GBM, which is suitable for our (binary) classification task [8]. As the base learners in GBM, we choose decision tree [2] for both computational efficiency and interpretability reasons.

4.3 Evaluation Module

To evaluate our approach, we use a *stratified* 10-fold cross-validation (CV) procedure, whereby we split the Twitter data into 10 folds of training and testing data, each retaining the class label proportion as per the original data. We then report the average performance as well as its variation (i.e., standard deviation). The stratification is needed to ensure that each fold is a good representative of the whole, i.e., retains the (unbalanced) label distribution in the original data.

In this work, we consider several evaluation metrics popularly used in information retrieval, namely *Precision*, *Recall*, and *F1-score* [10]:

$$Precision = \frac{TP}{TP + FP}; Recall = \frac{TP}{TP + FN}; F1 = \frac{2 Precision Recall}{Precision + Recall} \quad (1)$$

where TP, FP and FN are the true positives, false positives, and false negatives respectively. Here we treat the organization account as the positive class.

4.4 Interpretation Module

The ability to describe and interpret the derived predictive model is important for many applications. A useful interpretation of our GBM classifier involves understanding those particular features that are most influential in contributing to the classification performance as well as its variance. To this end, we utilize the feature importance metric derived based on the decision tree influences [2]. Specifically, the feature importance corresponds to the expected fraction of the samples that each decision tree contributes within the ensemble models [8].

5 Experiment

This section presents our empirical studies on the Twitter data we have collected. All evaluations were based on the stratified 10-fold CV method (cf. Section 4.3).

5.1 Features Extracted

Based on findings in Section 3, we devised numerous content, social and temporal features for our account classification task. Table 1 lists all features used in our work, as well as their corresponding types and feature transformation pipelines. For convenience, we shall use the term "user" and "account" interchangably. We do not use categorical features in this work for now, although the implementation of the categorical pipeline is readily available in our framework.

For the textual contents of tweet documents, we use the TF-IDF representation for tweet documents, as described in Section 4.1. We also construct a number of numerical features from the content and social variables. These include count and ratio features, such as the total counts of entities (e.g., "MentionCount"), the counts of unique entities (e.g., "MentionUnique"), and the ratio of unique over total counts (e.g., "MentionUniqueRatio").

Table 1. List of features used

Feature	Type	Pipeline	Description
TweetContent	C	X	TF-IDF vector of a user's tweet document
TweetCount	C	N	No. of tweets of a user (March–May 2014)
SourceUnique	C	N	No. of unique applications a user tweets from
SourceUniqueRatio	C	N	No. of unique applications / total no. of tweets
HashtagUnique	C	N	No. of unique hashtags
HashtagCount	C	N	Total no. of hashtags
HashtagUniqueRatio	C	N	No. of unique hashtags / total no. of hashtags
HashtagCountRatio	C	N	No. of hashtags / total no. of tweets
ListedCount	S	N	No. of Twitter lists at which a user appears
FavouritesCount	S	N	No. of tweets a user has marked as favourite
MentionUnique	S	N	No. of unique (user) mentions
MentionCount	S	N	Total no. of (user) mentions
MentionUniqueRatio	S	N	No. of unique mentions / total no. of mentions
MentionCountRatio	S	N	No. of mentions / total no. of tweets
MentionClusterCoeff	S	N	Clustering coefficient for mention graph
MentionMentionedRatio	S	N	No. of mentions / no. of mentioneds
FollowersCount	S	N	No. of followers of a user
FolloweesCount	S	N	No. of followees of a user
FolloweeClusterCoeff	S	N	Clustering coefficient for followee graph
FollowerFolloweeRatio	S	N	No. of followers / no. of followees
FolloweeFollowerMean	S	N	Mean of the no. of followers of a user's followees
FolloweeFollowerMedian	S	N	Median of the no. of followers of a user's followees
FolloweeFollowerStdDev	S	N	Deviation of the no. of followers of a user's followees
FolloweeFollowerEntropy	S	N	Entropy of the no. of followers of a user's followees
FolloweeFolloweeMean	S	N	Mean of the no. of followees of a user's followees
FolloweeFolloweeMedian	S	N	Median of the no. of followees of a user's followees
FolloweeFolloweeStdDev	S	N	Deviation of the no. of followees of a user's followees
FolloweeFolloweeEntropy	S	N	Entropy of the no. of followees of a user's followees
FolloweeTraceMean	S	N	Mean of the trace of no. of followees over time
FolloweeTraceMedian	S	N	Median of the trace of no. of followees over time
FolloweeTraceStdDev	S	N	Deviation of the trace of no. of followees over time
FolloweeTraceEntropy	S	N	Entropy of the trace of no. of followees over time
FollowerTraceMean	S	N	Mean of the trace of no. of followers over time
FollowerTraceMedian	S	N	Median of the trace of no. of followers over time
FollowerTraceStdDev	S	N	Deviation of the trace of no. of followers over time
FollowerTraceEntropy	S	N	Entropy of the trace of no. of followers over time
AccountAge	T	N	Total duration from since account created till now
AverageTweetCount	T	N	No. of tweets / account age
ProbWeekend	T	N	Probability of a user tweeting on the weekend
ProbMorning	T	N	Probability of a user tweeting in the morning
ProbAfternoon	T	N	Probability of a user tweeting in the afternoon
ProbEvening	T	N	Probability of a user tweeting in the evening
ProbNight	T	N	Probability of a user tweeting at night
Hour-x	T	N	Probability of a user tweeting at hour x
Weekday-x	T	N	Probability of a user tweeting at day x

Type – C: content, S: social, T: temporal; Pipeline – N: numeric, X: textual

For social features, we also consider two-hop centrality features such as clustering coefficient (CC) and some first- and second-order statistics of the followees' followees (or followees' followers) of a user. The purpose of including two-hop features is to allow us to account for a sufficiently large community of users. The CC metric measures the extent to which a user's neighborhood form a clique. For a user i, CC is the number of edges between the user's neighbors N_i divided by the total number of possible edges between them, i.e., $|N_i| \times (|N_i| - 1)$.

As for the statistics of the followees/followers, we use first-order statistics such as mean and median, as well as second-order statistics such as standard deviation and entropy. The second-order metrics are used to quantify the *diversity* of the entities associated with a user's neighborhood. To obtain the entropy, we first

Table 2. Impacts of different features for account type classification (10-fold CV)

(a) Classification results

Features	Precision	Recall	F1-Score
C	0.841 ± 0.030	0.782 ± 0.046	0.809 ± 0.025
S	0.865 ± 0.031	0.858 ± 0.040	0.860 ± 0.025
T	0.758 ± 0.029	0.777 ± 0.054	0.766 ± 0.028
C, S	0.879 ± 0.036	0.886 ± 0.034	0.882 ± 0.024
C, T	0.854 ± 0.019	0.861 ± 0.058	0.856 ± 0.033
S, T	0.890 ± 0.032	0.889 ± 0.047	0.889 ± 0.024
C, S, T	$\mathbf{0.909 \pm 0.023}$	$\mathbf{0.904 \pm 0.041}$	$\mathbf{0.906 \pm 0.016}$

C: content, S: social, T: temporal

(b) Statistical significance (p-value)

	C, S	C, T	S, T	C, S, T
C	0.005**	0.005**	0.005**	0.005**
S	0.007**	0.333	0.013*	0.005**
T	0.005**	0.005**	0.005**	0.005**
C, S	-	0.021*	0.241	0.009**
C, T	-	-	0.009**	0.007**
S, T	-	-	-	0.008**

C: content, S: social, T: temporal
* / **: significant at 95% / 99%

Table 3. Benchmarking results of different algorithms (10-fold CV)

Algorithm	Precision	Recall	F1-Score	p-value
Support vector machine	0.859 ± 0.045	0.843 ± 0.033	0.850 ± 0.029	0.005**
Logistic regression	0.863 ± 0.037	0.842 ± 0.039	0.852 ± 0.030	0.005**
Decision tree	0.808 ± 0.023	0.827 ± 0.043	0.817 ± 0.023	0.005**
Random forest	0.878 ± 0.028	0.899 ± 0.032	0.888 ± 0.029	0.008**
Gradient boosting	$\mathbf{0.909 \pm 0.023}$	$\mathbf{0.904 \pm 0.041}$	$\mathbf{0.906 \pm 0.016}$	-

**: significant at 99%

take the normalized count (i.e., probability density) $p_{i,j}$ for each neighbor $j \in N_i$ of user i, and then compute the entropy $-\sum_{j=1}^{|N_i|} p_{i,j} \log p_{i,j}$.

We also devise more advanced social features dubbed *trace*, describing the dynamics of social entities over time. For instance, the "FolloweeTraceMean" feature in Table 1 means the average of the trace vector of followee counts over time. Here each element in the trace vector is the followee count observed for time period t . In this work, we set the observation period as $t = 3$ days.

Finally, we devise a number of temporal features based on the periodicity of the tweet counts observed at different time spans. In particular, we bin the tweets by time and compute the probability of tweeting in the morning (4:00-11:59am), afternoon (12:00pm-4:59pm), evening (5:00-7:59pm), and night (8:00pm-3:59am). We also compute the probability of the tweets occurring in the weekend. To capture daily and hourly distribution of tweets (cf. Fig. 1(b)), we also compute the probability of tweeting at Weekday-x (where $x \in \{0, 1, \ldots, 6\}$ for Monday to Sunday), and Hour-x (where $x \in \{0, 1, \ldots, 23\}$ for 24 hours).

5.2 Performance Assessment

We first evaluated the impact of different features to the overall classification performance of GBM, and then compared the GBM results using all features to the results of several other popular classification algorithms. Table 2 illustrates the impact of different features. Looking at the results of individual content (C), social (S), and temporal (T) features, we can see that the social features alone gave the highest F1-score, followed by the content features and temporal features. The performance of combination of content and social features is higher than either of the individual baseline. The same conclusion applies for the combination of social and temporal features. Lastly, the GBM model that uses all content, social, and temporal features was able to achieve the highest F1 score.

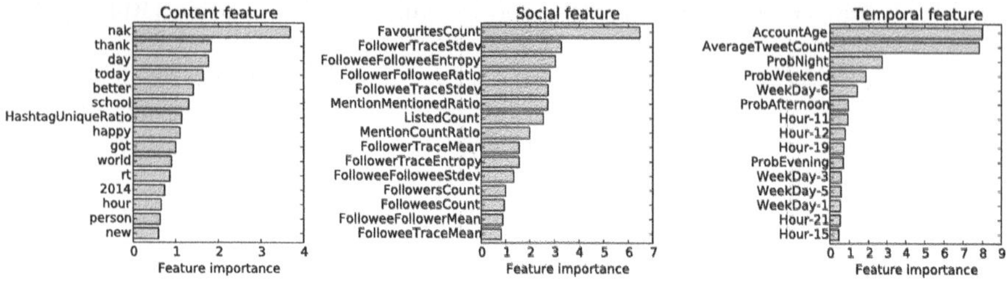

Fig. 3. Top 15 features for Twitter account type classification

Table 4. Prediction results on unseen data

(a) Confusion matrix

		Predicted									
		Top 20		Top 40		Top 60		Top 80		Top 100	
		Per	*Org*	*Per*	*Org*	*Per*	*Org*	*Per*	*Org*	*Per*	*Org*
Actual	*Per*	20	0	40	0	60	0	79	1	99	1
	Org	1	19	3	37	5	55	7	73	12	88

Per: personal, *Org*: organization

(b) Organization accounts

Domain	**No. of accounts**
.com	66
.sg	14
.org	5
None	3

To evaluate the contributions of different feature combinations, we conducted the *Wilcoxon signed-rank* statistical test [13][2]. From the p-values in Table 2(b), we can see that the overall pairwise differences of $F1$ are statistically significant, except for two cases. Nevertheless, it is clear that combining all feature types (content, social, and temporal) gave substantially better results than using the constituent features (cf. Table 2(b), last column), which is our primary interest.

We further benchmarked the results of our approach against those of other classification algorithms. These include *support vector machine* and *logistic regression* [4], which are linear models widely used in information retrieval. We also used decision tree baseline [2], as well as *random forest* [1]—a popular bootstrap aggregating method to create an ensemble of decision trees. For fairness, we used all three feature types in this benchmark. As evident from Table 3, our GBM method consistently outperforms the other algorithms across all evaluation metrics. We also found that the improvements are statistically significant according to the Wilcoxon signed-rank test, as per the last column of Table 3. This in turn justifies the accuracy and robustness traits of our approach.

5.3 Feature Importance

Using the trained GBM model, we can now evaluate the importance of different features, as described in Section 4.4. Fig. 3 shows the top 15 most important features produced by GBM for each feature type. Several interesting insights are observed. For example, the top textual feature "nak" is the short form of

[2] The Wilcoxon test provides a non-parametric alternative to the t-test for matched pairs, when the pairs cannot be assumed to be normally distributed.

"want" in Malay language, which is often used for informal communication. From Fig. 3 and our manual inspections, we also found that special word such as "rt" (which stands for retweet) is indicative of the account type (in this case, personal accounts tend to retweet more). Among the non-textual features, "HashtagUniqueRatio" is ranked among the top. A closer look at the data shows that organization accounts often have more unique hashtags than personal accounts, suggesting that the former have a more focused topic of interest.

As for social features, it is shown that "FavouritesCount" emerges as the top feature. Indeed, our internal inspections reveals that personal accounts tend to have larger favourite counts. Despite this observation, using "FavouritesCount" alone is not sufficient to obtain good classification results, and the collective contribution of the other social features remains important. We also found the diversities of the no. of followees/followers over time (e.g., "FollowerTraceStdDev", "FolloweeTraceStdDev", "FollowerFolloweeEntropy") to be discriminative of the account types. From our inspections, we found that the deviations of followee trace for organization accounts are moderate in general. This is likely due to the fact that most organizations utilize Twitter as a dissemination platform.

With regard to temporal features, we discovered that organization accounts have gained traction on Twitter only in the recent 2-3 years, whereas many personal accounts were created 4-5 years ago. This explains why the account age is one of the top features. We also noticed that personal accounts have higher "AverageTweetCount" than organization accounts. In addition, we conclude from "ProbWeekend" and "Weekday-6" that personal accounts tend to tweet more than organization accounts during the weekend. The results also suggest that the probability of tweeting in the afternoon ("ProbAfternoon") or evening ("ProbEvening") is discriminative. Lastly, there are several critical hours (e.g., "Hour-11", "Hour-12", "Hour-19"–possibly related to lunch/dinner time) as well as critical days (e.g., "Weekday-1" (Tuesday), "Weekday-3" (Thursday), "Weekday-5" (Saturday)) that are useful for the account type classification.

5.4 Out-of-Sample Generalization

To assess the ability of our model to generalize, we used our trained GBM model to predict for all unlabeled data. We then picked the top K organization accounts and top K personal accounts based on the prediction scores. We varied K from 20 to 100 and examined the prediction results for all the top accounts, so as to see how the GBM predictions match with our manually-examined labels. Table 4(a) summarizes the results. It is shown that, under varied K, our approach produced good performance on unseen data, achieving robust accuracies of $98.75 - 100\%$ for personal accounts and $88 - 95\%$ for organization accounts.

Table 4(b) shows the domain type breakdown of the 88 correct predictions for the top 100 organization accounts. We can see that our approach can correctly predict for organization accounts with domain types other than those of the labeled (training) data. Note here that the domain extensions ".com" and ".sg" in the unlabeled data are different from the ".com.sg" extension in the labeled data. In sum, these results justify the generalization ability of our approach.

6 Conclusion

We put forward a generic framework for discriminating personal and organizational accounts in social media. Our framework provides a generic set of feature transformation pipelines that supports integration of rich content, social, and temporal features. With gradient boosting as its core, our approach achieves accurate/robust performance and provides useful insights on the data. We have empirically demonstrated the effectiveness and interpretability of our approach using Singapore Twitter data. Moving forward, we wish to apply our method to Twitter data from a larger region. We also plan to build a multi-attribute prediction method that can integrate information from heterogeneous social networks.

Acknowledgments. This research is supported by the Singapore National Research Foundation under its International Research Centre@Singapore Funding Initiative and administered by the IDM Programme Office, Media Development Authority (MDA).

References

1. Breiman, L.: Random forests. Machine Learning 45(1), 5–32 (2001)
2. Breiman, L., Friedman, J.H., Olshen, R.A., Stone, C.J.: Classification and regression trees (1984)
3. Burger, J.D., Henderson, J., Kim, G., Zarrella, G.: Discriminating gender on Twitter. In: EMNLP, pp. 1301–1309 (2011)
4. Chang, C.-C., Lin, C.-J.: LIBSVM: A library for support vector machines. ACM-TIST 2(27), 1–27 (2011)
5. Chang, J., Rosenn, I., Backstrom, L., Marlow, C.: ePluribus: Ethnicity on social networks. In: ICWSM, pp. 18–25 (2010)
6. Cohen, R., Ruths, D.: Classifying political orientation on Twiter: It's not easy? In: ICWSM, pp. 91–99 (2013)
7. De Choudhury, M., Diakopoulos, N., Naaman, M.: Unfolding the event landscape on Twitter: Classification and exploration of user categories. In: CSCW (2012)
8. Friedman, J.H.: Greedy function approximation: A gradient boosting machine. The Annals of Statistics 29(5), 1189–1232 (2001)
9. Lewis, D.D., Yang, Y., Rose, T.G., Li, F.: RCV1: A new benchmark collection for text categorization research. JMLR 5, 361–397 (2004)
10. Manning, C.D., Raghavan, P., Schütze, H.: Introduction to Information Retrieval. Cambridge University Press (2008)
11. Smirnov, N.: Table for estimating the goodness of fit of empirical distributions. The Annals of Mathematical Statistics 19(2), 279–281 (1948)
12. Tavares, G., Faisal, A.A.: Scaling-laws of human broadcast communication enable distinction between human, corporate and robot Twitter users. PloS One 8(7), 1–11 (2013)
13. Wilcoxon, F.: Individual comparisons by ranking methods. Biometrics Bulletin 1, 80–88 (1945)
14. Yan, L., Ma, Q., Yoshikawa, M.: Classifying Twitter users based on user profile and followers distribution. In: Decker, H., Lhotská, L., Link, S., Basl, J., Tjoa, A.M. (eds.) DEXA 2013, Part I. LNCS, vol. 8055, pp. 396–403. Springer, Heidelberg (2013)
15. Yin, P., Ram, N., Lee, W.-C., Tucker, C., Khandelwal, S., Salathé, M.: Two sides of a coin: Separating personal communication and public dissemination accounts in Twitter. In: Tseng, V.S., Ho, T.B., Zhou, Z.-H., Chen, A.L.P., Kao, H.-Y. (eds.) PAKDD 2014, Part I. LNCS(LNAI), vol. 8443, pp. 163–175. Springer, Heidelberg (2014)

Handling Topic Drift for Topic Tracking in Microblogs

Yue Fei, Yihong Hong, and Jianwu Yang*

Institute of Computer Science and Technology Peking University, China
{feiyue,hongyihong,yangjw}@pku.edu.cn

Abstract. Microblogs such as Twitter have become an increasingly popular source of real-time information, where users may demand tracking the development of the topics they are interested in. We approach the problem by adapting an effective classifier based on Binomial Logistic Regression, which has shown to be state-of-art in traditional news filtering. In our adaptation, we utilize the link information to enrich tweets' content and the social symbols to help estimate tweets' quality. Moreover, we find that topics are very likely to drift in microblogs as a result of the information redundancy and topic divergence of tweets. To handle the topic drift over time, we adopt a cluster-based subtopic detection algorithm to help identify whether drift occurs and the detected subtopic is regarded as the current focus of the general topic to adjust topic drift. Experimental results on the corpus of TREC2012 Microblog Track show that our approach achieves remarkable performance in both T11SU and F-0.5 metrics.

1 Introduction

The boom of various online social media has successfully facilitated the way of information creation, sharing, and diffusion among web users. As a popular form of social media, microblogging services such as Twitter have raised much attention. Twitter's real-time nature enables users broadcast and share information about their opinions, statuses and activities ranging from daily life to current events, news stories, and other interests anytime and anywhere [7]. Receiving more than 500 million tweets per day[1], information provided by Twitter is not only invaluable but also overwhelming, which makes users more difficult to track with the evolution of topics they are interested in.

In this paper, we study the problem of topic tracking in Twitter, which aims at filtering information relevant to a given topic from real-time tweet streams. Topic drift over time, which emerges when the topic-related contents are enriched and different aspects of the general topic are derived along with the development of event, is one of the most challenging problems in topic tracking. To be more comprehensible, we take the topic "boxing competition" as an example. Before

* Corresponding Author.
[1] https://blog.twitter.com/2013/new-tweets-per-second-record-and-how

A. Hanbury et al. (Eds.): ECIR 2015, LNCS 9022, pp. 477–488, 2015.

the competition starts, users' interests may focus on the physical and competitive condition of the participant. When the competition begins, the topic's focus may change to the development of the match such as who has an advantage over his opponent and who performs beyond audiences' expectation. After the match, the focus of this event may drift again to the discussion about the winner.

Several approaches [2,15,23,21,3] have been proposed to resolve the drift problem in the filtering of newswire, which can be referred in the topic tracking of tweet streams. However, microblog differs from newswire in several respects. Due to the real-time nature in Twitter, when an event arises, users respond to the event quickly by posting or reposting large volume of related tweets. As a result, event diffuses and develops much faster than that in newswire. In addition, the length limitation (140 characters) and informality of tweet content, makes it tricky to obtain the topic of a tweet accurately than a news article. Specially, we have observed that users' short-term interests may focus on different aspects of a specific event in Twitter, which means tweets of different subtopics may intermingle. In this case, previous works may not work well as they are mostly designed to detect gradual changes. All of these distinctions lead to the fact that handling drift over the real-time tweet streams will be a challenging problem.

In this paper, we propose an approach to explore topic tracking in continuous tweet streams by employing an effective classifier based on the Binomial Logistic Regression. A cluster-based subtopic detection algorithm is introduced to deal with the topic drift over time, which groups tweets obtained from pseudo-relevance feedback into subtopics dynamically. We regard the detected subtopic as the current focus of the general topic, which can help adjust the topic drift over time. In addition, drift is detected once a new subtopic emerges and event's development can be observed from the subtopic set. We conduct several experiments to evaluate our approach using the corpus of TREC2012 Microblog Track [18]. The results show that our approach achieves remarkable performance in both T11SU and F-0.5 metrics.

Our contribution in this work are as follows:

1. Adapting the traditional filtering approach in newswire with several microblog characteristics to deal with the short text and spotty quality of tweets.
2. Proposing a novel approach to handle topic drift with a cluster-based subtopic detection algorithm.

2 Related Work

Adaptive filtering with the purpose of handling drift has been studied in the topic tracking of newswire. Some adaptive algorithms focused on how to maintain training examples of classifier. For example, to recognize the concept changes, Klinkenberg et al. [8] adopted a window to choose the training data, whose size was either fixed or automatically adapted to the current extent of concept. Some adaptive filtering approaches treated the problem as a retrieval task and used an adaptive threshold model on the retrieval score to make a binary decision.

For example, Bayesian inference [3] and Okapi probabilistic model [15] were proposed to determine the threshold of score. Incremental Rocchio Algorithm [2] and Logistic Regression (LR) Model [20] are two effective approaches in the information filtering of news. Several works [21,23] analyzed the robustness of Incremental Rocchio Algorithm and Logistic Regression Model in real-time filtering of news and found that Logistic Regression is more effective than Incremental Rocchio Algorithm. The aforementioned methods all depend on the pseudo positive examples from the filtering results, which are also popularly adopted in the topic tracking on Twitter.

In 2011, Lin et al. [10] first defined the topic tracking problem in Twitter, which can be summarized as "Given a continuous stream of incoming tweets, we are interested in filtering and retaining only those that are relevant to a particular topic". In their work, they explored the smoothing techniques integrating foreground models captured recently with background models, as well as different techniques for retaining history both intrinsically and extrinsically. However, they did not consider Twitter's social characteristic and the topic drift. In 2013, Albakour et al. [1] proposed an effective approach to deal with the sparsity and drift for real-time filtering in Twitter. In their approach, query expansion based on pseudo-relevance feedback was used to enrich the representation of user profile which improved the filtering performance a lot. Furthermore, a set of recent relevant tweets were utilized to represent the users' short term interests and tackle the drift issue. Three strategies were introduced to decide the size of the tweets set: arbitrary adjustments, daily adjustments and event detection strategy based on CombSUM voting technique [11] and Grubb's test [5]. The event detection significantly improved recall at the cost of a marginal decrease in the overall filtering performance. In [6], Hong et al. presented an effective real-time approach, which consists of a content model and Pseudo Relevance Feedback model, to exploit the topic tracking in Twitter. More specifically, document expansion and tweet's quality were taken into consideration in the content model, and a fixed-width window aiming at keeping the recent relevant tweets was applied in order to make their filtering system adapt to the drift. However, the size of the window is fixed which may not portray the drift properly as the emerging of topic drift is unexpected and irregular. Magdy et al. [12] proposed an unsupervised approach for tracking short messages from Twitter that are relevant to broad and dynamic topics, which initially gets a set of user-defined fixed accurate (Boolean) queries that cover the most static part of the topic and updates a binary classifier to adapt to dynamic nature of the tracked topic automatically. However, it's not easy to find such user-defined queries to capture emerging subtopics of a broad topic.

Differed from those existing algorithms, in this paper, we consider that multiple subtopics of the general event can have time overlap, in other words, the general topic may have more than one subtopics during a same period, while previous works are mostly designed to detect gradual changes and assume the focus of the topic will remain unchanged in a certain period.

3 Topic Tracking in Microblogs

3.1 Problem Definition

In the definition of topic tracking in Twitter, we first assume that the tweets in set $D = \langle d_1, d_2, d_3, ...\rangle$ arrive in a strictly chronological order, and suppose a user has seen a trigger tweet T_0 posted at event start time t_Q and becomes interested in new relevant tweets about the same topic. Then we treat tweets that arrive before time t_Q as background corpus and tweets that arrive after time t_Q as foreground corpus respectively. For each new tweet in foreground corpus, the tracking approach should make a decision about whether to show the tweet to users or not without utilizing future information. If the approach decides to show the tweet, it could access the tweet's relevance judgement (if any) as immediate relevance feedback, but not otherwise [10,18].

3.2 Tracking with Logistic Regression

We regard the topic tracking problem as a classification problem and choose Logistic Regression as our basic classifier. Actually, any effective classifier would be qualified here, still we prefer Logistic Regression for two reasons. First, Logistic Regression has shown to be state-of-art in adaptive news filtering [23]. Second, Logistic Regression is robust and has been applied successfully to a lot of real-time tasks for its efficiency in both training and inference.

We adopt two types of features to describe characteristics of tweets, namely semantic feature and quality feature.

Semantic Feature is used to measure the similarity between tweets and the general topic. We can utilize different IR models such as Vector Space Model, Language Model and Boolean Model to generate semantic features. Due to the 140 characters limitation, feature sparsity is a challenging problem in topic tracking in microblog. In order to enrich the semantic information of tweets, we take advantage of document expansion approach by collecting all the external URLs contained in our corpus and extracting their topic information. An efficient and effective method proposed by Liang et al. [9] is adopted to obtain the topics of webpages. Noting that web pages might be deleted as time elapsed, we only crawl a portion of the external URL set.

Quality Feature is used to estimate content quality of tweets. When users want to obtain the information about a certain event, they prefer reading tweets that are relevant but informative. We state the informative tweets as high-quality tweets. We believe the quality of a tweet can be inferred from entities in it such as hashtags, URLs, mentions and retweets. In most cases, tweets containing such symbols tend to be informative.

To combat the topic drift, we specially introduce a drift feedback feature which estimates the similarity between a tweet and its corresponding subtopic. This feature helps adapt the general topic to the current subtopic, thus diminishing the influence of concept change over time.

Table 1. Top 5 frequent words in each subtopic of topic "BBC World Service staff cuts"

Subtopic	Top 5 frequent words
1	online, new, budget, 25, job
2	job, new, 650, language, media
3	new, outline, office, radio, quarter

3.3 Tracking with Subtopic Detection

Topic drift is one of the most challenging problems in topic tracking, which occurs when user's focus on the general topic changes over time. In microblog, topic drift is more prevalent due to the highly dynamic environment. In this section, we first discuss the drift problem in Twitter with an intuitive example, then we describe our cluster-based subtopic detection algorithm.

3.3.1 An Intuitive Example of Drift

We adopt the training topics in the real-time filtering task of the Microblog track TREC2012 to observe the characteristics of drift in Twitter. Taking the topic named "BBC World Service staff cuts" as example, we group the topic's relevant tweets into three clusters. Each cluster is regarded as a subtopic of the general topic. The top 5 frequent words (words in query are excluded) of each subtopic are shown in Table 1. Besides, we calculate the number of tweets posted every day in each subtopic and present the distribution in Figure 1. Here we only focus on the distribution from Jan. 24, 2011 to Jan. 29, 2011, since most of the tweets were posted during this period.

From Figure 1 and Table 1, we can observe that on Jan. 24, 2011, it was announced that BBC would cut its online budget, and tweets about this topic are all relevant to subtopic 1. On the next day, users transferred their focus and became interested in the quantity of to-be-cut services and jobs, since BBC announced they would slash 650 jobs and cut five BBC language services. With the emerging of subtopic 2, the percentage of tweets relevant to subtopic 1 decreased. And on the following days, subtopic 3 concerning which office and why this office would be closed was derived.

Figure 1 shows that there is time overlap among different subtopics, which means that at the same time, the focus of the event may be more than one. In the previous approaches retaining latest relevant tweets, it is widely assumed that there is only one focus at a time and the selected tweets will be cleared up when new subtopics evolved. These approaches fail to describe the phenomenon we observed. In the next section, we describe a cluster-based drift detection algorithm which can detect the topic drift in Twitter and well model the coexistence of several subtopics.

3.3.2 Detecting Subtopics

we employ an incremental clustering algorithm to detect subtopics over time. As discussed above, the focus of an event can be more than one during its

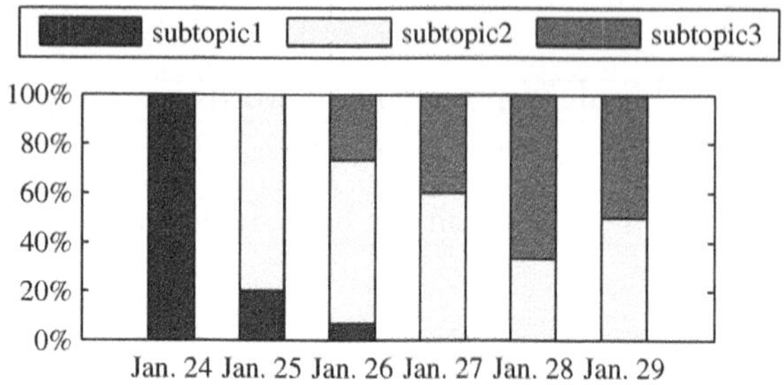

Fig. 1. Distribution of each subtopic's tweet percentage over time of Topic "BBC World Service staff cuts"

development. By clustering relevant tweets of the event into subtopics, we can well keep track of the dynamic focuses. What's more, clusters that aggregate a couple of tweets into one subtopic also compensate for the brevity of tweet content, which enriches the topical information of the certain subtopic.

Stream data clustering has been studied and applied in a variety of web services such as query expansion in retrieval, news filtering, topic detection, text summary and etc [17,22,4]. Algorithm 1 shows how our clustering procedure contributes to the tracking task. A set of clusters C is defined to represent different subtopics of an event. At the event start time t_Q, we use the trigger tweet T_0 to form an initial cluster, and after that, the cluster set contains the initial cluster only.

Suppose a tweet T arrives at time t_T, and there are m active clusters at that time. At first, we calculate the similarities between tweet and every cluster in the cluster set. Then, we get the cluster c^* whose centroid is closest to T according to Eq. 1 and obtain the similarity $score^*$ based on Eq. 2.

$$c^* = \underset{c_i, i \in [1,m]}{\arg\max}\, Sim(c_i, T) \tag{1}$$

$$score^* = Sim(c^*, T) \tag{2}$$

Next, $score^*$ is regarded as the drift feedback feature used in the classifier we defined in Section 3.2 and the classifier will give out the decision whether tweet T is relevant to the topic. If tweet T is judged relevant to the topic, it will be added to the tweet set of a certain cluster. Note that although c^* is closest to T, it does not mean T belongs to c^*. A new cluster will be created if T is still very distant from c^*. In order to determine whether to create a new cluster or not, we set a clustering threshold β. If $score^*$ is smaller than β, then T is upgraded to a new cluster and a new subtopic is detected. Otherwise, T is added to the tweet set of the closest cluster c^*. With the detected subtopic set, we can easily summarize the event's development at any moment by referring to the proportion of each subtopics among the related tweets.

Algorithm 1. Topic Tracking with Subtopic Detection

```
Input:  tweet stream stream,
        event query q,
        trigger tweet T_0,
        model parameters ω, b,
        clustering threshold β.
c_0 ← {T_0}
clusterSet ← {c_0}
while stream.hasNext() do
   T ← stream.next()
   score* ← 0
   c* ← null
   for c ∈ clusterSet do
      s = Sim(T, c)
      if s > score* then
         score* = s
         c* = c
      end if
   end for
   x = getFeauture(T, q)
   x ← x ∪ {score*}
   res = LRClassfier(ω, b, x)
   if res = relevant then
      Display T
      if score* > β then
         c* ← c* ∪ {T}
      else
         newC ← {T}
         clusterSet ← clusterSet ∪ {newC}
      end if
   end if
end while
```

4 Experiments

In this section, we first introduce the dataset and the evaluation metrics applied in our experiments. Then we conduct several experiments to estimate the performance of our approach. Our experimental results will also be discussed in this section.

4.1 Dataset

Tweets11 Corpus. We adopt the standard dataset of TREC2012 real-time filtering pilot task in our experiments. Tweets11 corpus is obtained using a donation of the unique identifiers of a sample of tweets from Twitter [14]. It is created by sampling 16 million tweets from January 24, 2011 to February 8, 2011, covering big events all around world during the period. We crawl the HTML version copy of the corpus with the provided tools[2]. The simple re-tweeted tweets beginning with RT in the Tweets11 corpus are removed based on the assumption that such tweets have no extra information beyond the original ones.

The dataset contains 49 topics. Ten of which are for training while others are for testing. Our classifier and the clustering threshold β are trained based on the 10 topics in training set.

4.2 Evaluation Method

For the TREC2012 real-time filtering pilot task, approaches are expected to make a binary decision to accept or reject a tweet for each topic. Therefore, the result set consists of an unranked list of tweets. The main measurement is utility (i.e. T11U), which assigns a reward of two points to every relevant tweet retrieved and a penalty of one point to every irrelevant tweet retrieved [18,16]. Another measurement used in the TREC2012 real-time filtering pilot task is F_γ. And γ is set as 0.5 which gives an emphasis on precision. Average T11SU score and F-0.5 score among the 39 testing topics are regarded as the final evaluation metrics.

4.3 Experimental Results

In this section, we will discuss our experimental results and verify the effectiveness of our approach.

4.3.1 Evaluation of LR model

In this section, we compare experimental results of LR model with different settings. The first run is **LROrg**, which uses LR model with basic similarity between tweets and general topic as semantic feature. The second run is **LRDE**, which enriches semantic feature by document expansion. The third run is **LRDE+URL**, which is based on **LRDE**, including a quality feature that indicates whether a tweet contains URLs. The KL-divergence score [13] is applied to calculate similarities between topic and tweets, which has been proved effective in microblog retrieval [9].

The experimental results are shown in Table 2. Both **LRDE** and **LRDE+URL** outperform **LROrg** significantly according to a paired t-test ($p < 0.05$) in F-0.5, indicating that document expansion is effective in handling feature sparsity. Especially, **LRDE+URL**, which we utilise both document expansion and quality information, achieves significant improvements against **LRDE** in F-0.5, demonstrating the effectiveness of quality feature. However, we have also experimented other quality features that not listed here like hashtags, mentions and retweets, which turn out to harm the overall performance. As a result, our following experiments are all based on **LRDE+URL**.

[2] `https://github.com/lintool/twitter-corpus-tools`

Table 2. Experimental results with different corpora. † denotes a statistically significant increase over **LROrg**. Statistical significance is estimated with a paired t-test at ($p < 0.05$).

Run	T11SU	F-0.5	Precision	Recall
LROrg	0.3363	0.0480	0.1662	0.0144
LRDE	0.3670	0.1780 †	0.5394 †	0.0653
LRDE+URL	**0.3976**	**0.2896** †	0.5267 †	**0.1813** †

Table 3. Experimental results with subtopic detection algorithm and other approaches. † denotes a statistically significant increase over **LRDE+URL**. Statistical significance is estimated with a paired t-test at ($p < 0.05$).

Run	T11SU	F-0.5	Precision	Recall
LRDE+URL	0.3976	0.2896	0.5267	0.1813
LRDE+URL+simCls	**0.4341(+9.18%)**	**0.4005(+38.29%)** †	0.5427(+3.04%)	0.3733(+105.9%) †
M-Dyaa1	0.3771	0.3573	0.3256	0.3415
LMDynDEAllPRF	0.4336	0.3691	-	-
hitUWT	0.4117	0.3338	**0.6219**	0.1740

4.3.2 Evaluation of Subtopic Detection Algorithm

Table 3 shows the experimental results with subtopic detection algorithm. Here **LRDE+URL+simCls** uses similarity between tweets and current subtopic as drift feedback feature. **M-Dyaa1** is the experimental result which adopts CombSUM voting technique and Grubb's test to detect the drift in tweet streams [1]. **LMDynDEAllPRF** combines a content model with Pseudo Relevance Feedback model and employs a fixed-width window to adapt to drift [6]. The Precision and Recall results are not presented in their paper. Both **M-Dyaa1** and **LMDynDEAllPRF** are evaluated upon **Tweets11** corpus. **hitUWT** is the best run in TREC2012. Compared with **LRDE+URL**, **LRDE+URL+simCls** achieve significant improvements in F-0.5 measure according to a paired t-test ($p < 0.05$). The improvements in Recall are extremely remarkable, which all increased over 100% compared to **LRDE+URL**. The big gain in Recall can be explained by the subtopic detection algorithm which tends to consider a tweet relevant to the general topic if the tweet relates to any subtopics. Meanwhile, the clustering threshold β helps avoid a sharp drop in Precision. This experimental results strongly prove the effectiveness of our subtopic detection approach.

LRDE+URL+simCls achieves substantial improvements in F-0.5 measure over all these four runs below and the increase in T11SU metric is also notable over all these four runs except **LMDynDEAllPRF**. In a word, our cluster-based subtopic detection approach achieves a better balance of precision and recall by improving recall significantly as well as maintaining a comparative precision. The experimental performance are remarkable in both T11SU and F-0.5 metrics.

4.4 Discussion

4.4.1 Sensitivity Analysis of Clustering Threshold β

The threshold β which decides the creation of new subtopic is important in our approach. In this section, we study the robustness of the threshold β. We let β

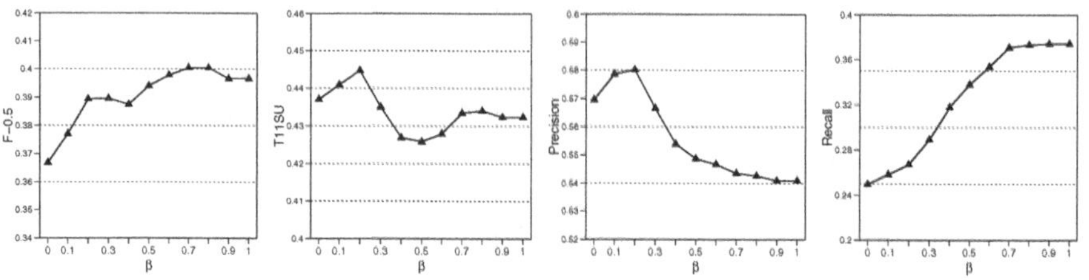

Fig. 2. Sensitivity analysis of the threshold β, which determines whether to create a new cluster

vary from 0 to 1. When β is set as 0, every relevant tweet will be put into one cluster since the similarity can never be less than 0, thus no new subtopic will be derived and the event will consist of only one subtopic. In this scenario, the drift detection algorithm can be regarded as a pseudo relevance feedback approach. When β is set as 1, it means that every relevant tweet will be regarded as a single subtopic. If we fix the number of subtopics, this will be identical to the window-based approach, a most common concept drift handling technique, which is based on instance selection and consists in generalizing from a window that moves over recently arrived instances and uses the learnt concepts for prediction only in the immediate future [19]. The performances of **LRDE+URL+simCls** in F-0.5 and T11SU metrics with different β are summarized in Figure 2.

We can observe that the optimal result is achieved when the value of β is around 0.7 in F-0.5, while the value is around 0.3 in T11SU. This is cause by the fact that T11SU tilts towards precision more heavily than F-0.5. As threshold β increases, more subtopics will be detected since the criteria to aggregate tweets into one cluster becomes more strict. The raise in the number of subtopics results in more relevant tweets, which can well explain why there is a sharp increase in recall the curve. At first, more subtopics helps to filter relevant tweets more precisely, but there comes a sudden drop in the precision curve when β is above 0.3. The fall in precision results from the over-dose of subtopics, which produces massive noise.

5 Conclusions and Future Works

In this paper, we propose an approach to explore topic tracking in continuous tweet streams by employing an effective classifier on the basis of Binomial Logistic Regression. Since topic drift over time is one of the most challenging problems in the tweets real-time filtering, we first analyze the drift phenomenon in Twitter, and then integrate a cluster-based subtopic detection algorithm into our classifier to handle the topic drift over time. In our approach, we dynamically group tweets obtained from pseudo-relevance feedback into subtopics, which contribute to the relevance prediction of new tweets. Furthermore, drift emerges when a new subtopic is detected, thus the event's development can be observed from the subtopic set generated. Experimental results using the corpus of TREC2012 Microblog Track show that our approach achieves good performance in both T11SU and F-0.5 metrics.

There still remain plenty of studies for future works. For example, deciding the number of emerging subtopics dynamically is worth exploring in the future. And we assume each tweet belongs to one subtopic in our approach, while it's not the case in the real world. How to assign a tweet to multiple subtopics also deserves researching in the future.

Acknowledgments. The work reported in this paper was supported by the National Natural Science Foundation of China Grant 61370116.

References

1. Albakour, M.D., Macdonald, C., Ounis, I.: On sparsity and drift for effective real-time filtering in microblogs. In: Proceedings of the 22nd ACM International Conference on Information and Knowledge Management, CIKM 2013, pp. 419–428. ACM, New York (2013), http://doi.acm.org/10.1145/2505515.2505709
2. Allan, J.: Incremental relevance feedback for information filtering. In: Proceedings of the 19th Annual International ACM SIGIR Conference on Research and Development in Information Retrieval, SIGIR 1996, pp. 270–278. ACM, New York (1996), http://doi.acm.org/10.1145/243199.243274
3. Callan, J.: Learning while filtering documents. In: Proceedings of the 21st Annual International ACM SIGIR Conference on Research and Development in Information Retrieval, SIGIR 1998, pp. 224–231. ACM, New York (1998), http://doi.acm.org/10.1145/290941.290998
4. Chen, Y., Amiri, H., Li, Z., Chua, T.S.: Emerging topic detection for organizations from microblogs. In: Proceedings of the 36th International ACM SIGIR Conference on Research and Development in Information Retrieval, SIGIR 2013, pp. 43–52. ACM, New York (2013), http://doi.acm.org/10.1145/2484028.2484057
5. Grubbs, F.E.: Procedures for detecting outlying observations in samples. Technometrics 11(1), 1–21 (1969)
6. Hong, Y., Fei, Y., Yang, J.: Exploiting topic tracking in real-time tweet streams. In: Proceedings of the 2013 International Workshop on Mining Unstructured Big Data Using Natural Language Processing, UnstructureNLP 2013, pp. 31–38. ACM, New York (2013), http://doi.acm.org/10.1145/2513549.2513555
7. Java, A., Song, X., Finin, T., Tseng, B.: Why we twitter: understanding microblogging usage and communities. In: Proceedings of the 9th WebKDD and 1st SNA-KDD 2007 Workshop on Web Mining and Social Network Analysis, pp. 56–65. ACM, New York (2007), http://doi.acm.org/10.1145/1348549.1348556

8. Klinkenberg, R., Renz, I.: Adaptive information filtering: Learning in the presence of concept drifts. In: Workshop Notes of the ICML/AAAI-98 Workshop Learning for Text Categorization, pp. 33–40. AAAI Press (1998)
9. Liang, F., Qiang, R., Yang, J.: Exploiting real-time information retrieval in the microblogosphere. In: Proceedings of the 12th ACM/IEEE-CS Joint Conference on Digital Libraries, pp. 267–276. ACM (2012), `http://doi.acm.org/10.1145/2232817.2232867`
10. Lin, J., Snow, R., Morgan, W.: Smoothing techniques for adaptive online language models: Topic tracking in tweet streams. In: Proceedings of the 17th ACM SIGKDD International Conference on Knowledge Discovery and Data Mining, SIGIR 2011, pp. 422–429. ACM (2011), `http://doi.acm.org/10.1145/2020408.2020476`
11. Macdonald, C., Ounis, I.: Voting for candidates: Adapting data fusion techniques for an expert search task. In: Proceedings of the 15th ACM International Conference on Information and Knowledge Management, CIKM 2006, pp. 387–396. ACM, New York (2006), `http://doi.acm.org/10.1145/1183614.1183671`
12. Magdy, W., Elsayed, T.: Adaptive method for following dynamic topics on twitter (2013)
13. Manning, C.D., Raghavan, P., Schütze, H.: Introduction to information retrieval, vol. 1. Cambridge University Press, Cambridge (2008)
14. Ounis, I., Macdonald, C., Lin, J., Soboroff, I.: Overview of the trec-2011 microblog track. In: Proceedings of TREC 2011 (2011)
15. Robertson, S.: Threshold setting and performance optimization in adaptive filtering. Inf. Retr. 5(2-3), 239–256 (2002), `http://dx.doi.org/10.1023/A:1015702129514`
16. Robertson, S., Soboroff, I.: The trec 2002 filtering track report. In: Proceedings of TREC 2002 (2002)
17. Shou, L., Wang, Z., Chen, K., Chen, G.: Sumblr: Continuous summarization of evolving tweet streams. In: Proceedings of the 36th International ACM SIGIR Conference on Research and Development in Information Retrieval, SIGIR 2013, pp. 533–542. ACM, New York (2013), `http://doi.acm.org/10.1145/2484028.2484045`
18. Soboroff, I., Ounis, I., Lin, J.: Overview of the trec-2012 microblog track. In: Proceedings of TREC 2012 (2012)
19. Tsymbal, A.: The problem of concept drift: Definitions and related work. Tech. rep. (2004)
20. Weisberg, S.: Applied linear regression, vol. 528. Wiley (2005)
21. Yang, Y., Yoo, S., Zhang, J., Kisiel, B.: Robustness of adaptive filtering methods in a cross-benchmark evaluation. In: Proceedings of the 28th Annual International ACM SIGIR Conference on Research and Development in Information Retrieval, SIGIR 2005, pp. 98–105. ACM, New York (2005), `http://doi.acm.org/10.1145/1076034.1076054`
22. Zhang, X., Li, Z.: Automatic topic detection with an incremental clustering algorithm. In: Wang, F.L., Gong, Z., Luo, X., Lei, J. (eds.) WISM 2010. LNCS, vol. 6318, pp. 344–351. Springer, Heidelberg (2010), `http://dl.acm.org/citation.cfm?id=1927661.1927714`
23. Zhang, Y.: Using bayesian priors to combine classifiers for adaptive filtering. In: Proceedings of the 27th Annual International ACM SIGIR Conference on Research and Development in Information Retrieval, SIGIR 2004, pp. 345–352. ACM, New York (2004), `http://doi.acm.org/10.1145/1008992.1009052`

Detecting Location-Centric Communities Using Social-Spatial Links with Temporal Constraints

Kwan Hui Lim[1,2], Jeffrey Chan[1], Christopher Leckie[1,2], and Shanika Karunasekera[1]

[1] Department of Computing and Information Systems, The University of Melbourne, Parkville, VIC 3010, Australia
[2] Victoria Research Laboratory, National ICT Australia
{jeffrey.chan,caleckie,karus}@unimelb.edu.au, limk2@student.unimelb.edu.au

Abstract. Community detection on social networks typically aims to cluster users into different communities based on their social links. The increasing popularity of Location-based Social Networks offers the opportunity to augment these social links with spatial information, for detecting location-centric communities that frequently visit similar places. Such location-centric communities are important to companies for their location-based and mobile advertising efforts. We propose an approach to detect location-centric communities by augmenting social links with both spatial and temporal information, and demonstrate its effectiveness using two Foursquare datasets. In addition, we study the effects of social, spatial and temporal information on communities and observe the following: (i) augmenting social links with spatial and temporal information results in location-centric communities with high levels of check-in and locality similarity; (ii) using spatial and temporal information without social links however leads to communities that are less location-centric.

Keywords: Community Detection, Clustering Algorithms, Foursquare, Location-based Social Networks, Social Networks.

1 Introduction

The study of communities on social networks typically involves using community detection algorithms to cluster users into different communities based on their friendships on the social network (i.e., social links). With the rising popularity of Location-based Social Networks (LBSN), it is now possible to add a spatial aspect to these traditional social links for the purpose of community detection. Many researchers have used such social-spatial links to detect location-centric communities on LBSNs [2,3]. The detection of these location-centric communities is especially important for companies embarking on location-based and mobile advertising, which are increasingly crucial to any company's marketing efforts [5]. We posit that the detection of such location-centric communities can be further improved by adding a temporal constraint to such social-spatial links,

A. Hanbury et al. (Eds.): ECIR 2015, LNCS 9022, pp. 489–494, 2015.

Table 1. Types of Links

Link Type	Description
Social (SOC)	Links based on explicitly declared *friendships* (i.e., topological links)
Social-Spatial-Temporal (SST)	*Social links* where two users share a *common check-in*, on the *same day*
Social-Spatial (SS)	*Social links* where two users share a *common check-in*, regardless of time
Spatial-Temporal (ST)	Links based on two users sharing a *common check-in*, on the *same day*

and demonstrate the effectiveness of this approach using two LBSN datasets. In addition, we study the effects of social, spatial and temporal links on the resulting communities, in terms of various location-based measures.

Related Work. The spatial aspects of LBSNs have been used in applications ranging from friendship prediction to detecting location-centric communities. For example, [4] used spatial-temporal links (photos taken at the same place and time) to infer friendships on Flickr, while [13] used spatial links (tweets sent from the same location) and tweet content similarity to predict friendships on Twitter. Similarly, [2] used social-spatial links (friends with common check-ins) to detect location-centric communities on Twitter and Gowalla. Brown et al. [3] also used social-spatial links to study the topological and spatial characteristics of city-based social networks, and [9] found that communities with common interest tend to comprise users who are geographically located in the same city.

Most of these earlier works consider the spatial aspect of check-ins and co-location without the temporal aspect (e.g., visiting the same place over any span of time), while [4] considers this temporal aspect for the purpose of friendship prediction. Our research extends these earlier works by adding a temporal constraint to social and spatial links, for the purpose of detecting location-centric communities. Using two LBSN datasets, we demonstrate the effectiveness of our proposed approach in detecting location-centric communities that display high levels of check-in and locality similarity.

Contributions. We make a two-fold contribution in this paper by: (i) enhancing existing community detection algorithms by augmenting traditional social links with both a spatial aspect and temporal constraint; (ii) demonstrating how these links result in location-centric communities comprising users that are more similar in terms of both their visited locations and residential hometown.

2 Methodology

Our proposed approach to detecting location-centric communities involves first building a social network graph $G = (N, E_t)$, where N refers to the set of users and E_t refers to the set of links of type t (as defined in Table 1). SOC links are essentially topological links that are used in traditional community detection tasks, while SS links were used in [2] to detect location-focused communities with great success. Our work extends [2] by adding a temporal constraint to these links, resulting in our SST links.[1] Furthermore, we also use ST links to determine the effects of adding this temporal constraint solely to the spatial aspects of links (i.e., without considering social information). While there are many definitions of links, these four types of links allow us to best investigate the effects of social, spatial and temporal information on location-centric communities.

Then, we apply a standard community detection algorithm on graph G, resulting in a set of communities. Thus, the different types of links (SOC, SST, SS and ST) used to construct the graph G will result in the different types of communities that we evaluate in this paper. We denote the detected communities as Com_{SOC}, Com_{SST}, Com_{SS} and Com_{ST}, corresponding to the types of links used. In this experiment, Com_{SST} are the communities detected by our proposed approach, while Com_{SOC}, Com_{SS} and Com_{ST} serve as baselines.

For the choice of community detection algorithms, we choose the Louvain [1], Infomap [12] and LabelProp [11] algorithms. Louvain is a greedy approach that aims to iteratively optimize modularity and results in a hierarchical community structure, while Infomap is a compression-based approach that uses random walkers to identify the key structures (i.e., communities) in the network. LabelProp first assigns labels to individual nodes and iteratively re-assigns these labels according to the most frequent label of neighbouring nodes, until reaching a consensus where the propagated labels denote the different communities. In principle, any other community detection algorithms can be utilized but we chose these community detection algorithms for their superior performance [6], and also to show that our obtained results are independent of any particular community detection algorithm.

3 Experiments and Results

Datasets. Our experiments were conducted on two Foursquare datasets, which are publicly available at [8] and [7]. Foursquare dataset 1 comprises 2.29M check-ins and 47k friendship links among 11k users, while dataset 2 comprises 2.07M check-ins and 115k friendship links among 18k users. Each check-in is tagged with a timestamp and latitude/longitude coordinates, which is associated with a specific location. In addition, dataset 1 provides the hometown locations that are explicitly provided by the users. We split these datasets into training and validation sets, using 70% and 30% of the check-in data respectively. The training set is used to construct the set of SST, SS and ST links, which will subsequently be used for community detection as described in Section 2.

Evaluation Metrics. Using the validation set, we evaluate the check-in activities and locality similarity of users within each Com_{SOC}, Com_{SST}, Com_{SS} and Com_{ST} community. Specifically, we use the following evaluation metrics:

[1] While SST links can also be defined as two friends who share a common check-in within D days, our experiments show that a value of D=1 offers the best results, hence the current definition of SST links. More importantly, using higher values of D days converges SST links towards SS links, which we also investigate in this work.

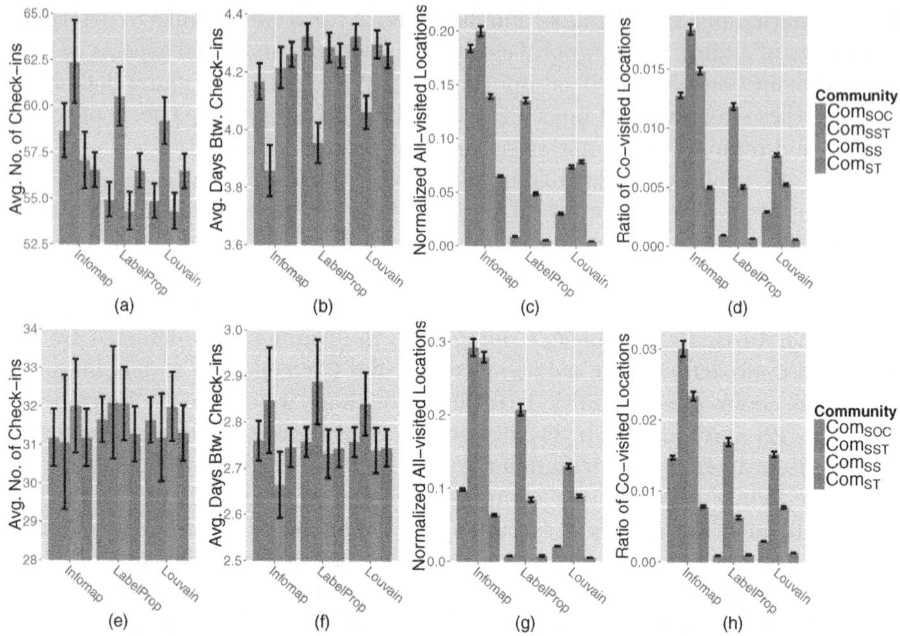

Fig. 1. Average number of check-ins, average days between check-ins, normalized number of all-visited locations and ratio of co-visited locations for Foursquare dataset 1 (top row) and dataset 2 (bottom row). For better readability, the y-axis for Fig. 1a/b/e/f do not start from zero. Error bars indicate one standard deviation. Best viewed in colour.

1. **Average check-ins**: The mean number of check-ins to all locations, performed by all users in a community.
2. **Average unique check-ins**: The mean number of check-ins to unique locations, performed by all users in a community.
3. **Average days between check-ins**: The mean number of days between consecutive check-ins, performed by all users in a community.
4. **Normalized all-visited locations**: The number of times when all users of a community visited a unique location, normalized by the community size.
5. **Ratio of co-visited locations**: Defined as $\frac{1}{|C|}\sum_{i \in C}\frac{|L_i \cap L_C|}{|L_C|}$, where L_i is the set of unique locations visited by user i, and L_C is the set of unique locations visited by all users in a community C.
6. **Ratio of common hometown**: The largest proportion of users within a community that share the same hometown location.

Evaluation metrics 1 to 3 measure the level of user check-in activity, while metrics 4 to 6 measure the user locality (check-in and hometown) similarity within each community. Ideally, we want to detect communities with high levels of check-in activity and locality similarity. As Metrics 1 to 3 are self-explanatory, we elaborate on Metrics 4 to 6. Metric 4 (normalized all-visits) determines how location-centric the entire community is based on how often the entire community visits the same locations. We normalize this metric by the number of users in a community to remove the effect of community sizes (i.e., it is more likely for a community of 50 users to visit the same location than for a community of 500 users). Metric 5 (co-visit ratio) measures the similarity of users in a community (in terms of check-in locations) and a value of 1 indicates that all users visit the exact set of locations, while a value closer to 0 indicates otherwise. Similarly, a value of 1 for Metric 6 (hometown ratio) indicates that all users in a community reside in the same location, while a value of 0 indicates otherwise.

Results. We focus on communities with >30 users as larger communities are more useful for a company's location-based and mobile advertising efforts. Furthermore, there has been various research that investigated the geographic properties of communities with ≤30 users [2,10]. In particular, [10] found that communities with >30 users tend to be more geographically distributed than smaller communities. Instead of repeating these early studies, we investigate the check-in activities and locality similarity of communities with >30 users.

In terms of the average number of check-ins (Fig. 1a/e), unique check-ins (not shown due to space constraints) and days between check-ins (Fig. 1b/f), Com_{SST} outperforms Com_{SOC}, Com_{SS} and Com_{ST} on dataset 1, regardless of which community detection algorithm used. However for dataset 2, the performance of Com_{SST} is largely indistinguishable from that of Com_{SOC}, Com_{SS} and Com_{ST}.[2] For both datasets, there is no clear difference among Com_{SOC}, Com_{SS} and Com_{ST} in terms of the average number of check-ins, unique check-ins or days between check-ins. These results show that our proposed SST links can be used to effectively detect communities that are more active in terms of check-in activity (for dataset 1), and such communities serve as a good target audience for a company's location-based and mobile advertising efforts. There is no clear difference among using SOC, SS and ST links (for both datasets). For the detection of location-centric communities, the locality similarity of these communities is a more important consideration, which we investigate next.

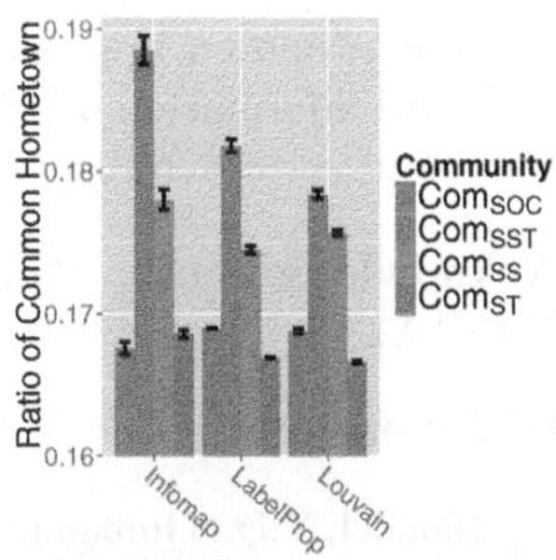

Fig. 2. Common hometown ratio for dataset 1

We examine locality similarity of the four communities in terms of the normalized number of all-visited locations (Fig. 1c/g), ratio of co-visited locations (Fig. 1d/h) and ratio of common hometown (Fig. 2). We only compare the ratio of common hometown for dataset 1 as this information is not available for dataset 2. For both datasets, Com_{SST} offers the best overall performance in terms of these three locality similarity metrics, while Com_{SS} offers the second best overall performance.[3] On the other hand, Com_{ST} resulted in the worst performance for both datasets. These results show that using our proposed SST links results in

[2] With an exception in Fig. 1f where Com_{SST} marginally underperforms Com_{SOC}, Com_{SS} and Com_{ST}.

[3] With exceptions in Fig. 1c where Com_{SS} (using Louvain) outperforms Com_{SST}, and Com_{SOC} (using Infomap) outperforms Com_{SS}.

communities comprising users who tend to frequently visit similar locations and reside in the same geographic area. Such location-centric communities are useful for the purposes of providing meaningful location-relevant recommendations and to better understand LBSN user behavior.

4 Discussions and Conclusion

We demonstrate how standard community detection algorithms can be used to detect location-centric communities by augmenting traditional social links with spatial information and a temporal constraint. Our evaluations on two Foursquare LBSN datasets show that: (i) augmenting social links with spatial information allows us to detect location-centric communities (ii) however, using spatial/temporal information (without considering social links) results in communities that are less location-centric than communities based solely on social links, thus spatial/temporal information should not be used independently; and (iii) our proposed approach of augmenting social links with both spatial and temporal information offers the best performance and results in location-centric communities, which display high levels of check-in and locality similarity.

Acknowledgments. This work was supported by National ICT Australia (NICTA).

References

1. Blondel, V.D., Guillaume, J.L., Lambiotte, R., Lefebvre, E.: Fast unfolding of communities in large networks. J. of Statistical Mechanics 2008(10), P10008 (2008)
2. Brown, C., Nicosia, V., et al.: The importance of being placefriends: Discovering location-focused online communities. In: Proc. of WOSN, pp. 31–36 (2012)
3. Brown, C., Noulas, A., Mascolo, C., Blondel, V.: A place-focused model for social networks in cities. In: Proc. of SocialCom, pp. 75–80 (2013)
4. Crandall, D.J., Backstrom, L., Cosley, D., Suri, S., Huttenlocher, D., Kleinberg, J.: Inferring social ties from geographic coincidences. PNAS 107(52) (2010)
5. Dhar, S., Varshney, U.: Challenges and business models for mobile location-based services and advertising. Communications of the ACM 54(5), 121–128 (2011)
6. Fortunato, S.: Community detection in graphs. Physics Reports 486(3) (2010)
7. Gao, H., Tang, J., Liu, H.: Exploring social-historical ties on location-based social networks. In: Proc. of ICWSM, pp. 114–121 (2012)
8. Gao, H., Tang, J., Liu, H.: gSCorr: modeling geo-social correlations for new check-ins on location-based social networks. In: Proc. of CIKM, pp. 1582–1586 (2012)
9. Lim, K.H., Datta, A.: Tweets beget propinquity: Detecting highly interactive communities on twitter using tweeting links. In: Proc. of WI-IAT, pp. 214–221 (2012)
10. Onnela, J.P., Arbesman, S., González, M.C., Barabási, A.L., Christakis, N.A.: Geographic constraints on social network groups. PLoS One 6(4), e16939 (2011)
11. Raghavan, U.N., Albert, R., Kumara, S.: Near linear time algorithm to detect community structures in large-scale networks. Phy. Review E 76(3), 36106 (2007)
12. Rosvall, M., Bergstrom, C.T.: Maps of random walks on complex networks reveal community structure. PNAS 105(4), 1118–1123 (2008)
13. Sadilek, A., Kautz, H., Bigham, J.P.: Finding your friends and following them to where you are. In: Proc. of WSDM, pp. 723–732 (2012)

Using Subjectivity Analysis to Improve Thread Retrieval in Online Forums*

Prakhar Biyani[1], Sumit Bhatia[2], Cornelia Caragea[3], and Prasenjit Mitra[4]

[1] Yahoo Labs, Sunnyvale, CA, USA
[2] IBM Almaden Research Center, San Jose, CA, USA
[3] Computer Science, University of North Texas, Denton, TX, USA
[4] Qatar Computing Research Institute, Doha, Qatar
pxb5080@yahoo-inc.com, sumit.bhatia@us.ibm.com,
ccaragea@unt.edu, pmitra@qf.org.qa

Abstract. Finding relevant threads in online forums is challenging for internet users due to a large number of threads discussing lexically similar topics but differing in the type of information they contain (e.g., opinions, facts, emotions). Search facilities need to take into account the match between users' *intent* and the type of information contained in threads in addition to the lexical match between user queries and threads. We use intent match by incorporating subjectivity match between user queries and threads into a state-of-the-art forum thread retrieval model. Experimental results show that subjectivity match improves retrieval performance by over 10% as measured by different metrics.

1 Introduction

Apart from asking questions and holding discussions, internet users search archives of online forums for threads discussing topics that are relevant to their information needs. Often, threads sharing common keywords discuss different topics and in such cases, finding relevant threads becomes challenging for users. Consider the following query issued by an internet user to a travel forum such as Trip Advisor–New York: *"best thanksgiving turkey"*. The query is subjective seeking opinions and viewpoints of different users on the quality of thanksgiving turkey served/found at different places in New York. A thread simply containing keywords *"thanksgiving"* and *"turkey"* and not having opinions of users would not satisfy the searcher. Similarly, for queries seeking factual information, threads having long discussions and opinions are likely to be not relevant. Hence, in addition to the *lexical* dimension (i.e., keyword match), search facilities in online forums need to take into account the ***intent*** dimension i.e., the type of information (e.g., opinions, facts) a searcher wants, to improve search. The current work addresses precisely this problem.

We improve an ad-hoc thread retrieval model for an online forum by combining lexical match between query and thread content with the match between searchers' intent and the type of information contained in threads. We focus on an important dimension of searchers' intent which is his preference for *subjective* and *non-subjective* information. Subjective information contains opinions, viewpoints, emotions and other private

* Work performed when Prakhar Biyani was at Pennsylvania State University.

A. Hanbury et al. (Eds.): ECIR 2015, LNCS 9022, pp. 495–500, 2015.

states whereas non-subjective information contains factual material [1]. Specifically, we identify the subjectivity of thread topics and user queries and incorporate the subjectivity match in a state-of-the-art retrieval model for online forums [2] to improve the retrieval performance. To predict thread subjectivity, we use our subjectivity classifier, specifically developed for online forum threads (Biyani et al. [3]). The classifier uses features derived from thread structure, sentiment and dialogue acts [4]. For determining query subjectivity, we use manual subjectivity tags (subjective/non-subjective).

2 Related Work

Subjectivity analysis has been actively researched in opinion mining [5], question-answering [6,7,8,9,10], and finding opinionated threads in online forums [11,12,13]. Stoyanov et al., [6] used subjectivity filter on answers, separating factual sentences from opinion sentences, to improve answering of opinion questions. Li et al., [8] used graphical models to rank answers based on their topical and sentiment relevance to opinion questions. Gurevych et.al., [7] used a rule-based lexicon based approach to classify questions as subjective or factoid. Moghaddam et al., [9] performed aspect-based question answering in product reviews and showed that taking into account the match between opinion polarities of questions and answers improved answer retrieval. Oh et al., [10] improved answering of non-factoid why-questions by using supervised classification for re-ranking answers based on their sentiment and other properties. All these previous works focused on improving question-answering of non-factoid (i.e., opinion) questions in product reviews and community QA sites. In contrast, the current work employs subjectivity analysis to improve an ad-hoc vertical retrieval model for an online forum. We show that using the subjectivity match, retrieval performance can be improved for both subjective and non-subjective queries.

3 Retrieval Model

Here, we discuss how information about subjectivity of threads can be utilized in thread retrieval systems. We use a state-of-the-art probabilistic model for forum thread retrieval [2] as a strong baseline (explained below) and incorporate subjectivity match between queries and threads in the model to see if it helps improve the retrieval performance.

3.1 Probabilistic Retrieval

Bhatia and Mitra [2] used a probabilistic model based on inference networks that utilizes the structural properties of forum threads. Given a query Q, the model computes $P(T|Q)$, the probability of thread T being relevant to Q, as follows:

$$P(T|Q) \overset{rank}{=} P(T) \prod_{i=1}^{n} \left\{ \sum_{j=1}^{m} \alpha_j P(Q_i|S_{j_T}) \right\} \quad (1)$$

where: $P(T)$ is the prior probability of a thread being relevant, Q_i is the ith term in query Q, S_{jT} is the jth structural unit in the thread T, α_j determines the weight given to component j and $\sum_{j=1}^{m} \alpha_j = 1$.

Note that the term $\prod_{i=1}^{n} \left\{ \sum_{j=1}^{m} \alpha_j P(Q_i|S_{j_T}) \right\}$ models lexical match between query and thread content. In order to estimate the likelihoods $P(Q_i|S_{j,T})$, we use the standard language modeling approach in information retrieval [14] with *Dirichlet Smoothing* as follows:

$$P(Q_i|S_{j_T}) = \frac{f_{Q_i,j_T} + \mu \frac{f_{Q_i,j_C}}{|j|}}{|j_T| + \mu} \tag{2}$$

Here,
f_{Q_i,j_T} = frequency of term Q_i in j^{th} structural component of thread T,
f_{Q_i,j_C} = frequency of term Q_i in j^{th} structural component of all the threads in the collection C.
$|j_T|$ is the length of jth structural component of thread T,
$|j|$ is the total length of jth structural component of all the threads in the collection C,
μ is the Dirichlet smoothing parameter.

In this work, we set μ to be equal to 2000, a value that has been found to perform well empirically [15]. Thus, the model computes the overall probability of a thread being relevant to the query by combining evidences from different structural units of the thread (title, initial post and reply posts).

3.2 Incorporating Subjectivity Information in the Retrieval Model

In absence of any information about thread's content, subjective threads are more likely to be relevant to subjective queries and vice versa for non-subjective threads. We conceptualize this idea by taking into account the match between subjectivities of threads and queries in addition to the lexical match between them. Specifically, we incorporate the subjectivity match using the term $P(T)$ (in Equation 1) which represents the prior probability of a thread being relevant to a query. We use the following two settings to incorporate subjectivity match between threads and queries into the retrieval model:

1. Subjectivity probability of a thread as its prior relevance probability: For subjective (or non-subjective) queries, a thread's prior probability of being relevant is taken to be its probability of being subjective (or non-subjective). More precisely, for a subjective query, Q_s, relevance score of a thread T is calculated as follows:

$$P(T|Q_s) \overset{rank}{=} P(Subj|T) \prod_{i=1}^{n} \left\{ \sum_{j=1}^{m} \alpha_j P(Q_{si}|S_{j_T}) \right\} \tag{3}$$

Here, $P(Subj|T)$ is the probability of thread T being subjective as outputted by the subjectivity classifier. Likewise, for a non-subjective query, the term $P(Subj|T)$ is replaced by $P(NSubj|T)$ which is the probability of thread T being non-subjective. For a thread T, $P(Subj|T) + P(NSubj|T) = 1$.

2. Re-ranking using subjectivity probabilities: A two-step ranking model is used. First, threads are ranked according to their lexical similarity with the query where $P(T)$

is taken as constant for all the threads and then re-ranking of threads (at various ranks) is performed based on their subjectivity probabilities. Basically, for a subjective query, re-ranking is sorting (in descending order) the ranked list of threads based on their subjectivity probabilities. Re-ranking for a non-subjective query is done similarly.

3.3 Getting Subjectivity Information for Threads and Queries

To obtain the subjectivity probability for a thread ($P(Subj|T)$), we used our subjectivity classifier developed previously (Biyani et al. [3]). We used the classifier to get confidence scores for all the threads (of belonging to the subjective class) and used the scores as the subjectivity probabilities. For determining query subjectivity, we took help of human annotators (discussed in Section 4.1).

4 Experiments and Results

4.1 Data Preparation

For our experiments, we used the dataset as used by Bhatia et al. [2]. It consists of threads crawled from a popular online forum: **Trip Advisor–New York** that contains travel related discussions mainly for New York city. It has 83072 crawled threads from the forum, a set of 25 queries and associated relevance judgments. For a query, the dataset has graded relevance judgments: 0 for totally irrelevant, 1 for partially relevant and 2 for highly relevant threads.

For annotating queries as subjective or non-subjective, we took help of three human annotators. First, two annotators tagged all the 25 queries with a percentage agreement and Kappa value of 88% and 0.743 respectively. The third annotator was then asked to disambiguate the tags of the queries on which the two annotators disagreed. Finally, we get 10 subjective and 15 non-subjective queries. Table 1 lists some of the subjective and non-subjective queries.

Table 1. Examples of subjective and non-subjective queries

Type	Example queries
Subjective	best mode of transportation from brooklyn to manhattan; safety in manhattan; best thanksgiving turkey; how safe is new york; how much to tip people
Non-subjective	new york to niagara falls; educational trips in new york; beaches in new york city; winter temperature in new york city; penn station to JFK

4.2 Experimental Setting

To conduct retrieval experiments, we used the Indri language modeling toolkit[1]. While indexing, stemming was performed using Porter's stemmerand stopwords were removed

[1] http://lemurproject.org

using a general stop word list of 429 words used in the Onix Test Retrieval Toolkit[2]. The queries and relevance judgments available with the dataset as discussed in Section 4.1 were used for retrieval experiments. For the baseline retrieval model, we used the optimal parameter settings as used in the original work [2]. In order to compare the performance of various retrieval models, we report precision, Normalized Discounted Cumulative Gain (NDCG) and Mean Average Precision (MAP) at ranks 5, 10 and 15.

Table 2. Retrieval results

Model	P@5	P@10	NDCG@5	NDCG@10	MAP@5	MAP@10
Subjective queries						
Baseline	0.56	0.52	0.7745	0.7180	0.7264	0.6662
Top 5 Re-rank	0.56	0.52	**0.8672**	0.7322	**0.8792**	**0.7355**
Top 10 Re-rank	0.58	0.52	0.7433	0.6964	0.7504	0.6873
Top 15 Re-rank	0.6	0.55	0.7370	0.7018	0.7361	0.6956
Subjectivity Prior Model	0.56	0.54	0.8010	**0.7433**	0.7880	0.6882
Non-subjective queries						
Baseline	0.546	0.546	0.6838	0.6988	0.7	0.651
Top 5 Re-rank	0.546	0.546	0.7056	0.7263	0.6688	0.6499
Top 10 Re-rank	0.56	0.546	0.8148	0.7644	0.7475	**0.7078**
Top 15 Re-rank	0.546	0.533	**0.8220**	**0.7658**	**0.7938**	0.6761
Subjectivity Prior Model	0.546	0.546	0.7827	0.7597	0.7518	0.7045
Average						
Baseline	0.552	0.536	0.7201	0.7065	0.7105	0.6572
Top 5 Re-rank	0.552	0.536	0.7703	0.7286	0.7530	0.6842
Top 10 Re-rank	0.568	0.536	0.7862	0.7372	0.7486	**0.6996**
Top 15 Re-rank	0.568	0.54	0.7880	0.7402	**0.7707**	0.6840
Subjectivity Prior Model	0.552	0.544	**0.7900**	**0.7532**	0.7663	0.6980

4.3 Results

Table 2 presents retrieval results for subjective and non-subjective queries, and the overall average result. **Subjectivity Prior Model** denotes the setting where thread's subjectivity probability is used as its prior relevance probability (as explained in Section 3.2). We see that using subjectivity information of threads improves MAP and NDCG values for both subjective and non-subjective queries against the baseline model. We also note that precision values remain almost unchanged (across all the settings). This is an interesting observation as it suggests that subjectivity match does not help much in finding more relevant threads. Instead, it improves ranking of threads by changing relative ordering of ranked threads. MAP takes into account ordering of ranked results and NDCG takes into account ordering and graded relevance (0, 1, 2) of the ranked results. For the re-ranking setting, we see that re-ranking at rank 5 outperforms re-ranking at ranks 10

[2] http://www.lextek.com/manuals/onix/stopwords1.html

and 15 for subjective queries. In contrast, for non-subjective queries, re-ranking at rank 15 outperforms the other two re-ranking settings.

5 Conclusion and Future Work

We combined the two dimensions of *lexical similarity* and *intent match* in a forum thread retrieval model and showed that the combination performs better than the model based only on lexical similarity. In future, we plan to explore automatic subjectivity classification of user queries, investigate other dimensions of user intent, and build fully automated thread retrieval systems.

References

1. Bruce, R., Wiebe, J.: Recognizing subjectivity: A case study in manual tagging. Natural Language Engineering 5(2), 187–205 (1999)
2. Bhatia, S., Mitra, P.: Adopting inference networks for online thread retrieval. In: AAAI, pp. 1300–1305 (2010)
3. Biyani, P., Bhatia, S., Caragea, C., Mitra, P.: Thread specific features are helpful for identifying subjectivity orientation of online forum threads. In: COLING, pp. 295–310 (2012)
4. Bhatia, S., Biyani, P., Mitra, P.: Classifying user messages for managing web forum data. In: Proceedings of the 15th International Workshop on the Web and Databases, pp. 13–18 (2012)
5. Liu, B.: Sentiment analysis and subjectivity. In: Handbook of Natural Language Processing, 2nd edn., pp. 627–666 (2010)
6. Stoyanov, V., Cardie, C., Wiebe, J.: Multi-perspective question answering using the opqa corpus. In: EMNLP-HLT, pp. 923–930 (2005)
7. Gurevych, I., Bernhard, D., Ignatova, K., Toprak, C.: Educational question answering based on social media content. In: AIE, pp. 133–140 (2009)
8. Li, F., Tang, Y., Huang, M., Zhu, X.: Answering opinion questions with random walks on graphs. In: ACL, pp. 737–745 (2009)
9. Moghaddam, S., Ester, M.: Aqa: Aspect-based opinion question answering. In: ICDMW, pp. 89–96. IEEE (2011)
10. Oh, J.H., Torisawa, K., Hashimoto, C., Kawada, T., De Saeger, S., Kazama, J., Wang, Y.: Why question answering using sentiment analysis and word classes. In: EMNLP-CoNLL, pp. 368–378 (2012)
11. Biyani, P., Caragea, C., Singh, A., Mitra, P.: I want what i need!: Analyzing subjectivity of online forum threads. In: Proceedings of the 21st ACM International Conference on Information and Knowledge Management, pp. 2495–2498 (2012)
12. Biyani, P., Bhatia, S., Caragea, C., Mitra, P.: Using non-lexical features for identifying factual and opinionative threads in online forums. Knowledge-Based Systems 69(0), 170–178 (2014)
13. Biyani, P., Caragea, C., Mitra, P.: Predicting subjectivity orientation of online forum threads. In: Proceedings of the 14th International Conference on Intelligent Text Processing and Computational Linguistics, pp. 109–120 (2013)
14. Ponte, J.M., Croft, W.B.: A language modeling approach to information retrieval. In: SIGIR, pp. 275–281 (1998)
15. Zhai, C., Lafferty, J.: A study of smoothing methods for language models applied to ad hoc information retrieval. In: SIGIR, pp. 334–342 (2001)

Selecting Training Data for Learning-Based Twitter Search

Dongxing Li, Ben He, Tiejian Luo, and Xin Zhang

University of Chinese Academy of Sciences, Beijing, P.R. China
{lidongxing12,zhangxin510}@mails.ucas.ac.cn, {benhe,tjluo}@ucas.ac.cn

Abstract. Learning to rank is widely applied as an effective weighting scheme for Twitter search. As most learning to rank approaches are based on supervised learning, their effectiveness can be affected by the inclusion of low-quality training data. In this paper, we propose a simple and effective approach that learns a query quality classifier, which automatically selects the training data on a per-query basis. Experimental results on the TREC Tweets13 collection show that our proposed approach outperforms the conventional application of learning to rank that learns the ranking model on all training queries available.

Keywords: Microblog search, Learning to rank, Social networks.

1 Introduction

With the rapid development of social networks on the World Wide Web, the microblogging services such as Twitter [1] and Sina Weibo [2] have gained notable popularity in the past few years. Consequently, how to efficiently and effectively retrieve the user statuses, namely tweets, has become a trendy research topic. For instance, the Text REtrieval Conference (TREC) has been running the Microblog track since 2011, where several dozens research groups and organizations around the world actively participate in the experimentation on the retrieval from a sample of the Twitter website. As illustrated in recent TREC Microblog track real-time Twitter search tasks, most of the top runs employ learning to rank algorithms to improve the retrieval effectiveness by integrating multiple features [6]. In those methods, it is often assumed that the training data are reliable and sufficient to reflect the characteristics of the relevant and non-relevant documents, so that a robust and effective ranking model can be learned. However, as suggested in [2], this assumption does not always hold. For example, some of the queries in the TREC Microblog track have only very few relevant tweets, which not only contribute little to the training data, but also possibly bias the learning process [10].

To this end, in this paper, we propose a simple but effective approach that aims to improve the learning to rank-based approaches for Twitter search. Specifically,

[1] http://twitter.com
[2] http://weibo.com

A. Hanbury et al. (Eds.): ECIR 2015, LNCS 9022, pp. 501–506, 2015.

we propose to learn a query quality classifier using a number of query features, including the content-based relevance scores, Normalized Query Commitment (NQC) [8], and several Twitter specific features. Then, the training data for the learning to rank algorithms is classified on a per-query basis. In our experiments on the standard Tweets13 dataset, our proposed approach, called RankSVM+, markedly outperforms the baseline.

2 Selecting Training Data for Learning to Rank

In this section, we propose a training data selection approach for learning to rank based on the estimation of the retrieval performance gain brought by a given training query. The basic idea of our approach is to directly estimate the retrieval performance gain by learning weak ranking models using individual training queries. A linear relationship of a set of query features and the estimated retrieval performance gain is established to learn a query quality classifier that selects high-quality training queries out of many.

Table 1. Pre-defined query features for the query quality estimation

Feature Type	Feature ID	Description
Content relevance	PL2QE	Mean PL2 [1] score of the top-3 tweets with query expansion
	BM25QE	Mean BM25 [7] score of the top-3 tweets with query expansion
	WBcdf2QE	Mean WBcdf2 [3] score of the top-3 tweets with query expansion
	LMDir	Mean score of top 3 tweet given by the KL-divergence language model with Dirichlet smoothing [9]
NQC	PL2NQC	NQC of relevance score given by PL2 with query expansion
	BM25NQC	NQC of relevance score given by BM25 with query expansion
	WBcdf2NQC	NQC of relevance score given by WBcdf2 with query expansion
	LMDirNQC	NQC of relevance score given by KL-divergence language model with Dirichlet smoothing
Twitter-specific	URL COUNT	The percentage of the top 10 tweets' with URLs in their content
	Hashtags COUNT	The percentage of the top 10 tweets' with Hashtags in their content
	Followers COUNT	Average number of followers of the top 10 tweets' authors

Figure 1 outlines the general framework of our proposed approach. In particular, our proposed approach consists of a training phase and a test phase as follows. First, the aim of the *training phase* is to learn a linear relationship of a set of query features with the quality of a training query for learning to rank. A set of training queries with human labels[3] are required to learn such a linear function using logistic regression. Next, this linear function is used as a query quality classifier that determines if a given query should be included in the set of training queries for learning to rank. In the *test phase*, the training queries for learning to rank are classified using the linear function obtained by logistic regression in the training phase. The ranking model is then learned over the selected training queries, which are classified as being high-quality, instead of over

[3] To differentiate between the a training query used by the learning to rank algorithms and a query used in the training phase of our proposed approach, the former notion is called a *training query for learning to rank* in the rest of the paper.

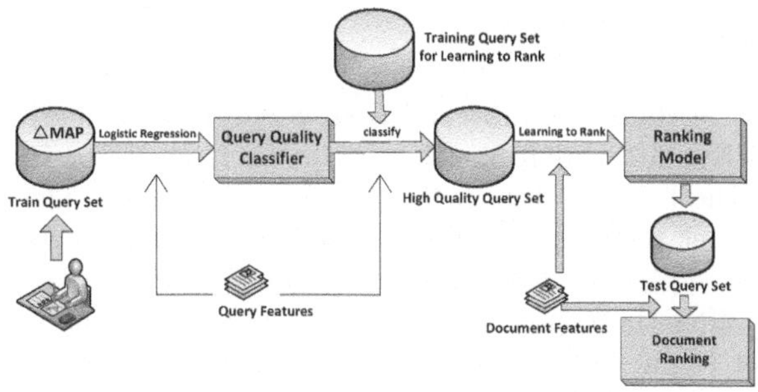

Fig. 1. General framework of the training query selection

all the training queries available, as in the conventional application of learning to rank algorithms.

Three types of query features are used for inferring the benefit brought by a given training query for learning to rank, namely content-based relevance scores, the NQC, and the Twitter specific features. Table 1 presents all the query features exploited in this paper. In particular, the NQC is based on the standard deviation of the relevance score distribution, which is originally proposed for query performance prediction in [8].

To learn a query quality classifier, a set of training queries Q_t and a separate set of validation queries Q_v are required. Both query sets should come with relevance assessments. Our method for learning such a linear function involves the steps as follows.

1) Produce an initial document ranking for the validation queries Q_v using a baseline model M_b. The mean average precision (MAP) obtained by the baseline model is denoted as MAP_{M_b}.

2) Using each individual query $q_i \in Q_t$ as a training data set, a learning to rank algorithm, e.g. RankSVM, is applied to learning a weak ranking model M_i.

3) Re-rank the documents returned for Q_v in 1) using M_i. The MAP obtained after the re-ranking is denoted as MAP_{M_i}.

4) The quality of q_i is defined as the change in MAP as follows:

$$\Delta MAP = MAP_{M_i} - MAP_{M_b}$$

Thus, ΔMAP indicates the retrieval performance gain brought by a training query q_i for learning to rank.

5) Repeat steps 2) - 4) for all $q_i \in Q_t$.

6) Extract the pre-defined query features (as in Table 1) for $q_i \in Q_t$, and learn a linear function using logistic regression.

In the test phase, the learned linear function is used as a query quality classifier. The queries with a positive predicted ΔMAP are regarded as being high-quality and therefore selected. For a given set of training queries for learning to rank, only the queries selected by the query quality classifier are used for learning the ranking model, in contrast to the conventional application of learning to rank, which uses all training queries available for the learning.

3 Experimental Setup

We experiment on the Tweets13 collection, which is a standard test collection used in the TREC 2013 Microblog track [4]. We fetch up to 10,000 tweets for each query via the track API with the official access token. The document features for the tweet representation are organized around the basic entities for the query-tweet tuples to distinguish between the relevant and non-relevant messages, including the content-based relevance scores, content richness, user authority, tweet recency etc., which were also exploited by the TREC 2013 Microblog track participants [4].

The pairwise RankSVM algorithm [5] is applied in our experiments. We do not experiment with listwise learning to rank algorithms since they did not show a clear advantage over RankSVM according to the results obtained by participants in the TREC 2013 Microblog track [4]. In the evaluation, we compare the conventional application of RankSVM that learns a ranking model on all training queries available with our approach, for which the ranking model is learned only on the automatically selected training queries. The latter is denoted as *RankSVM+* in the rest of the paper. Another viable baseline is the OptPPC approach proposed by Geng et al. in [2]. However, as their approach results in more than 10% decrease in MAP compared to RankSVM, the related results are not presented in this paper. Our guess is that their approach was developed on the datasets such as LETOR that comes with content-based scores features, and may not be suitable for Tweets13 which has various sources of features. On the 60 test queries associated to the Tweets13 collection, 3-fold cross-validation was conducted to evaluate both RankSVM and RankSVM+.

4 Experimental Results

Figure 2 plots the predicted ΔMAP against the actual ΔMAP on the test queries. From Figure 2, we can see that the ΔMAP predicted by the logistic regression has a moderate linear correlation with the actual ΔMAP. The correlation coefficient is R=0.5715, and the P-value is 0.008953 which indicates a statistically significant linear correlation at the 0.05 level.

In addition, Table 2 presents the precision and recall measures of the training query selection. A training query is selected if its predicted ΔMAP value is larger than 0, and is regarded as being high-quality if its actual ΔMAP value is larger than 0. According to Table 2, the query quality classifier learned by the logistic regression results in a decent performance in the selection of high-quality training queries. In particular, our approach successfully selects 85.71% of the high-quality queries with a precision of 66.67%. Also, the query quality classifier appears to be better in recognizing "bad" queries as it achieves a 90.91% precision in the unselected set with a recall of 76.92%. The overall accuracy of the query quality classification is 70.00%, which is not strikingly high, but still leads to markedly improved retrieval performance as shown in the comparison to the baseline.

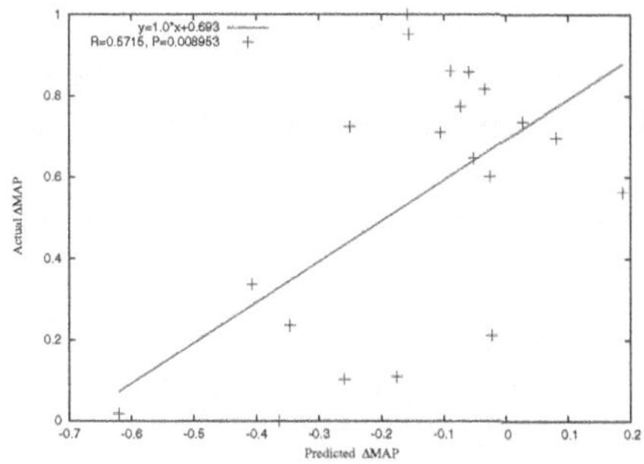

Fig. 2. The predicted ΔMAP against the actual ΔMAP on the test queries

Table 2. Precision, Recall and overall Accuracy of the training query selection on the test queries

	Precision	Recall
Selected	0.6667	0.8571
Unselected	0.9091	0.7692
Accuracy	0.7000	

Table 3. Evaluation results on test queries. A $*$ indicates a statistically significant difference according to the Wilcoxon matched-pairs signed-rank test at the 0.05 level.

Method	MAP	P30	R-prec
RankSVM	0.3115	0.5197	0.3531
RankSVM+	0.3533	0.5551	0.3872
Improvement	13.42%*	6.81%*	9.66%*

Finally, Table 3 compares the retrieval effectiveness of our approach RankSVM+ with the baseline, namely the classical RankSVM algorithm [5]. From the table, we can see that using RankSVM, learning a ranking model on the selected high-quality training queries leads to statistically significant improvement over the baseline in all three evaluation measures. In particular, RankSVM+ outperforms RankSVM by 13.42% in MAP, the official evaluation measure of the TREC 2013 Microblog track.

5 Conclusions and Future Work

We have proposed a simple and effective approach that automatically selects training queries for learning a ranking model for retrieval of the tweets. Our approach utilizes various query features to learn a query quality classifier through logistic regression, and selects the training queries for learning to rank based on their predicted benefit in the retrieval effectiveness. The experiments on the standard TREC Tweets13 dataset show that our proposed approach can indeed pick up most of the high-quality training queries for learning the ranking model, and consequently, leads to improved effectiveness in comparison to the baseline.

Moreover, an encouraging observation of this study is that it is possible to successfully select high-quality training queries for learning to rank by the direct estimation the retrieval performance gain. We plan to consider the role of query features within learning to rank techniques, as discussed by Macdonald et al. [11], which can allow the learned models to customise itself to different types of queries.

Acknowledgements. This work is supported in part by the National Natural Science Foundation of China (61103131/61472391), Beijing Natural Science Foundation (4142050) and SRF for ROCS, SEM.

References

1. Amati, G., Van Rijsbergen, C.J.: Probabilistic models of information retrieval based on measuring the divergence from randomness. ACM Trans. Inf. Syst. 20(4), 357–389 (2002)
2. Geng, X., Qin, T., Liu, T., Cheng, X., Li, H.: Selecting optimal training data for learning to rank 47, 730–741 (2011)
3. Hui, K., He, B., Luo, T., Wang, B.: Relevance weighting using within-document term statistics. In: CIKM, Glasgow, UK, pp. 99–104 (2011)
4. Lin, J., Efron, M.: Overview of the trec 2013 microblog track. In: TREC (2013)
5. Liu, T.-Y.: Learning to rank for information retrieval. Foundations and Trends in Information Retrieval 3(3), 225–331 (2009)
6. Ounis, I., Lin, J., Soboroff, I.: Overview of the trec 2011 microblog track. In: TREC (2011)
7. Robertson, S., Zaragoza, H.: The probabilistic relevance framework: Bm25 and beyond. Found. Trends Inf. Retr. 3(4), 333–389 (2009)
8. Shtok, A., Kurland, O., Carmel, D., Raiber, F., Markovits, G.: Predicting query performance by query-drift estimation. ACM Trans. Inf. Syst. 30(2), 11:1–11:35 (2012)
9. Zhai, C., Lafferty, J.: A study of smoothing methods for language models applied to ad hoc information retrieval. In: ACM SIGIR, pp. 334–342. ACM, New York (2001)
10. Zhang, X., He, B., Luo, T., Li, D., Xu, J.: Clustering-based transduction for learning a ranking model with limited human labels. In: ACM CIKM, pp. 1777–1782. ACM, New York (2013)
11. Macdonald, C., Santos, R.L.T., Ounis, I.: On the usefulness of query features for learning to rank. In: ACM CIKM, Maui Hawaii, USA, pp. 2559–2562. ACM (2012)

Content-Based Similarity of Twitter Users

Stefano Mizzaro, Marco Pavan, and Ivan Scagnetto

Dept. of Mathematics and Computer Science - University of Udine
via delle Scienze, 206. Udine, Italy
{mizzaro,marco.pavan,ivan.scagnetto}@uniud.it

Abstract. We propose a method for computing user similarity based on a network representing the semantic relationships between the words occurring in the same tweet and the related topics. We use such specially crafted network to define several user profiles to be compared with cosine similarity. We also describe an initial experimental activity to study the effectiveness on a limited dataset.

1 Introduction

There is a growing interest in analysing Social Networks (SNs) content to produce user models and to measure user similarity. The latter has been traditionally exploited in filtering and recommendation systems, and more recently in web search. A common approach is to define user similarity by exploiting the graph of the social relationships between users (e.g., friendship, sharing, liking, and commenting on Facebook, following, retweeting, and favoriting on Twitter). However, this approach has some drawbacks: the resulting system could be too strictly tailored against a peculiar SN and it may not easily adapt to other cases; it may fail in representing and comparing "lone" users, i.e., people not liking to follow other people or being followed; there might be a cold start problem. An approach based on social relationships may not be successful where these are weak or absent (e.g., messaging systems not relying on a SN).

Our approach is content based: we try to predict user similarity by relying on contents only. While we focus on Twitter, our approach is independent from the specific SN: we do not rely neither on the *following/being followed* social relationships nor on the peculiar structure of tweets (e.g., links, hashtags etc.). In our model each user is represented by a network linking the words most often posted, and other words from text enrichment procedures, with the tweets they occur in, and the latter with the topics the user is interested in. Topics are taken from the category hierarchy of Wikipedia, but, again, we are free to switch to other equivalent knowledge sources. The network allows us to evaluate several distinct approaches to users profiling and similarity.

2 Related Work

TUMS (Twitter-based User Modelling Service) is a web application building semantic profiles in RDF format starting from tweets [11]. Like our approach, TUMS features topic detection and text-enrichment, linking tweets to news articles about their context. The inferred profiles can be based on entities, topics or hashtags. It uses the Friend-Of-A-Friend (FOAF) vocabulary and the Weighted Interest vocabulary for inferring user interests (while we use the category hierarchy of Wikipedia).

A. Hanbury et al. (Eds.): ECIR 2015, LNCS 9022, pp. 507–512, 2015.

The authors of [14] exploit both textual data (tweets, with URLs and hashtags) and social structure (following and retweeting relationships), to discover communities of users in Twitter. User profiling and modelling is often a research activity tailored to a specific platform (e.g., Twitter/Facebook) and the resulting profiles are not interchangeable nor interoperable. In [10] the authors propose a framework for automatically creating and aggregating distinct profiles by means of semantic technologies. In [7] a new user similarity model is presented, combining the local context information of user ratings with the global preference of user behaviour, to help the system when only a few ratings are available. In [6] the authors develop a complex framework with a non-linear multiple kernel learning algorithm, to combine several notions of user similarities from different SNs. Experiments on a movie review data set show that the system provides more accurate recommendations than trust-based and collaborative filtering approaches. In [8] the author revisits the Page Rank algorithm and the related notion of random walks on a network, to improve ranking and recommendation systems based on the analysis of users' interactions carried out in the WWW (e.g, records of friendship relations in SNs, e-commerce transactions, messages exchanged in online communities, etc.). User similarity is also exploited in [5] as a criterion for sharing training data, and in [2] to predict evaluation outcomes in social media applications. A comprehensive survey of user modelling techniques in social media websites is available in [1].

Models and techniques borrowed from graph analytics (e.g., centrality analysis, path analysis, community detection and sub-graph isomorphism) have proven to be very effective tools in understanding and mining SNs [3]. An interesting SN analysis on a subset of tweets generated by a group of 1082 users is carried out in [13], where the in/out degree of nodes is based on the *following/being followed* relationships.

3 Proposed Approach

Our approach has the following overall steps (more details in the following):

1. The words in user's tweets are collected.
2. Text enrichment is used to add more words to each tweet and to associate *topics* to the tweets, obtained from the Wikipedia categories. We distinguish the most specific from the most generic ones (called macro-topics in the following) on the basis of Wikipedia category hierarchy.
3. Words, tweets, and topics are used to build the network.
4. Vector-based user profiles are defined, whose components are words weighted by network centralities.
5. User similarity is computed by using cosine similarity function.

3.1 The Network-Based User Model

We build a network (see a toy example in Fig. 1 (a)) with three layers of nodes, similar to those in [4,12]. The first layer represents the original words posted on Twitter and the additional words obtained from the enrichment process presented in [9]. Each node contains the string representing the word and an *ID* to assess if it was part of the original ones or those added. The second layer nodes represent *tweets*, with an *ID*, and a *timestamp* for future temporal network analysis. Each word is connected by an undirected arc to the corresponding tweet where it was published, and added to the network only once: if a word is already present, a new edge will be added to the new tweet, to avoid

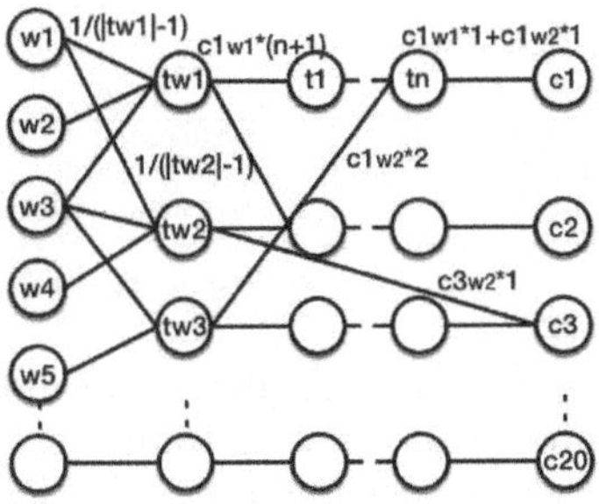

	words		words+enrich.		enrich.		topics		macro-topics	
1	war	1	august	1	august	1	Politics	1	Politics	1
2	renzi	.750	Italy	.571	Italy	.571	History	.977	History	.962
3	retirement	.750	politics	.500	politics	.357	Chronology	.675	Places	.633
4	reformation	.750	renzi	.429	sole (it. newspaper)	.357	Aux. Sciences of History	.639	Sciences	.352
5	politics	.500	retirement	.429	government	.286	Places	.613	Economics	.321

Fig. 1. Example of (a) Network-based user model and of (b) User Profile

duplicates and to emphasize the weight of that specific word. The third layer is composed of the labels, extracted during the categorisation process of [9], which represent the topics discussed in a specific tweet. We keep the relations among Wikipedia categories; the network features paths from the words, through the tweets, and the topics, up to the macro-topics, which represent the most general user interests.

The edge e_{w_i,tw_j} between the word w_i and the tweet tw_j has weight $1/(|tw_j| - 1)$, where $|tw_j|$ is the number of words in the tweet tw_j: we emphasize words contained in shorter texts, so as to give higher scores if a word strongly represents the semantics of the user's tweet, considering also the new words added by the enrichment process.

The edges connecting tweets with topics are denoted as e_{tw_i,t_j}, and their weight is computed using the relevance scores obtained by the labelling process for the macro-topics. We propagate their values along the network, emphasising topics more distant from the macro-topic, to give higher weight to nodes representing a more specific topic rather than generic ones. The weight is computed as $c_{t_{wi}} \cdot (steps(t, c_t) + 1)$, where $steps(t, c_t)$ is the number of steps necessary from the topic t to reach its related macro-topic c_t, and $c_{t_{wi}}$ is the relevance score got by the Wikipedia macro-topic c_t related to that specific topic t and the tweet tw_i, during the labelling process. Thus, the user model highlights the specific user interests, without losing information about the macro-topics.

For each tweet we have the path from the related topics to a macro-topic identifying the generic interest of the user. Therefore, we can again propagate the same relevance score to the path with the same rule previously described. Each time a tweet propagates the score, we sum the values to raise up the weight of related edges denoted as e_{t_i,t_j}. In this way we can highlight the relationships between the specific topics dealt with and the "super"-topics in the path, and define a network structure allowing the similarity computation between users at multiple levels. The final weight for these edges is:

$$\sum_{tw_k \in T} c_{t_{wi}} \cdot (steps(t_j, c_t) + 1), \tag{3.1}$$

where T is the set of all tweets in a path leading to the topic t_i and consequently t_j, related to the current edge. We say that there exists $tw_k \in T \Leftrightarrow tw_i$ *isConnectedTo* t_i.

3.2 User Profiling and User Similarity

Before comparing users, we need to define on which data (extracted from the whole user model) to carry out such comparison, i.e., we must define *user profiles*. In the following we aim at evaluating several kinds of profiles, considering multiple elements, either separately or together, in order to experiment different points of view.

Table 1. User profiling based on centralities (user "tweetpolitica"), legend: s→strength, e→eigenvector, b→betweenness, w→original words, e→enriched words

	s (w+e)	e (w+e)	b (w+e)	s (topics)	e (topics)	b (topics)
1	August 1	August 1	Italy 1	Pol. parties in Italy 1	Politics of Italy 1	Pol. parties in Italy 1
2	Italy .660	Italy .520	August .957	Politics of Italy .897	Pol. parties in Italy .798	Politics of Italy .917
3	politics .451	sole .483	politics .449	Politics by country .598	Politics by country .654	Chronology .500
4	sole .380	politics .466	euro .369	Politics .458	Politics .196	Politics by country .479
5	euro .346	gov. .448	sole .279	Chronology .306	Italian gov. .098	Sport .438

First, we set a baseline for our remaining approaches: we count only the set of words originally posted by the user, simply by considering how many times a word is connected to a tweet, and we repeat the process for the words added by the enrichment process, and finally for all the words in the network (both originally posted and added). The score list is normalised to have a final rank list. This step allows us to compare the sets of words, to evaluate if the enrichment process has led to improvements, i.e., if the added words with related scores better represent the analysed user.

Then, we build a profile based on the part of the model related to topics and macro-topics, to focus on the main interests of the user, leveraging on the scores obtained by the text enrichment and categorisation (for the details see [9]). If we want a coarse-grained profile, we restrict to macro-topics (identifying only the main interests of the user). Otherwise, we can resort to a fine-grained profile, considering the entire path of topics with related scores computed according to Formula (3.1). Fig. 1 (b) shows a profile built for a popular Italian Twitter account about politics called "tweetpolitica". We selected the first 5 words, with their scores, for each approach described above.

As second step we use network centrality measures, to assign scores to nodes taking advantage of our network structure and weights. We extract a subnetwork to make a first computation based only on the relationships between words and added words, and test the centrality-based profiling. We use both types of words since the enrichment process added a useful set of new terms to add information about user, while the original words preserve his/her original style of expression. In particular, we exploit: the strength centrality to see which are the nodes with higher degree, by analysing the edges weights (as defined in Section 3.1); the eigenvector centrality to emphasize the words often used in conjunction with the most used; and the betweenness centrality, to have information about words often present in tweets (i.e., to highlight the user style of expression). We adopt the same approach for the subnetwork composed of topics, to build a user profile based on the user major interests. Table 1 shows the final rank lists of terms with score computation based on network centralities for both words and topics.

Given two user profiles, we compute their similarity score from multiple points of view, like in the profile building process. For instance, if we consider only the macro-topics, we can say if two users have in common some general interests. Then, by considering all topics, we can get more detailed information. The similarity may be also computed by analysing just the words, to understand how users express their opinions and how they discuss their topics. Someone can use peculiar terms or grammar constructs to deal with the same topics. Users may satisfy different similarity notions.

On this basis, after focusing on a certain set of data, we build a list of terms with just those the two users have in common, with related scores. Hence, we build a geometric space defined by the features represented by the terms in common, and users are represented as vectors into this space. We compare them by using the *cosine* similarity function, to compute how "close" the users are.

Table 2. User similarity comparisons, legend: s→strength, e→eigenvector, b→betweenness, w→original words, e→enriched words, macro-t.→macro-topics

	words	w+e	enrich.	topics	macro-t.	s (w+e)	e (w+e)	b (w+e)	s (topics)	e (topics)	b (topics)
matteorenzi - beppe_grillo	.028	.054	.131	.874	.904	.510	.602	.002	.868	.998	.757
matteorenzi - tweetpolitica	.068	.089	.148	.602	.790	.714	.766	.385	.447	.003	.654
matteorenzi - Pontifex_it	.029	.024	.082	.371	.514	0	.978	0	.308	.003	.610
matteorenzi - SerieA_TIM	.019	.023	.062	.175	.147	0	.786	0	.271	0	.685

4 Evaluation and Results

Being in a prototypical phase, we carried out an expert evaluation to assess pros and cons of our method. With a dataset of at least 30 tweets (carefully processed in their right one-month long temporal context) for each of 17 selected accounts, we built their user profiles, as described in Section 3.2. Then, we computed the cosine similarity over several couples of accounts to test if our approach properly assigns scores to similar users, and how the network-based user model provides information at multiple levels (e.g., to understand if two users have in common just the main topics, if they match deeper, or if they have a similar style of expression). We compared the account "matteorenzi", the Prime Minister of Italy, due to his well defined political focus, to four accounts with different similarity w.r.t. him: "beppe_grillo", founder of the Italian political party Five Star Movement; "tweetpolitica", the account used in Section 3.2 focused on Italian political news; "Pontifex_it", the account of Pope Francis; and "SerieA_TIM", the top Italian football competition. Table 2 lists such pairs (ordered from the most to least similar, based on expert evaluation) with the computed similarity scores.

Scores based only on counting the words posted by users are very low, although the enrichment process has improved the computation. The combined solution (words + enrich.) seems to be the most reliable due to its mixed composition: enriched words make users more similar if they talk about the same topic, and the originally posted ones keep the users' style of expression. The labelling process provided a set of terms that well identify the trend of posts. The resulting scores are very high for the first pair, as we expected, still high for the second one, and lower for the remaining ones. We notice how "matteorenzi" and "Pontefix_it", apparently not related, have a considerable score. This fact is due to the nature of texts extracted during the test period; indeed, both users have talked about topics related to war and Iraq. The topics scores provide further information: users similar on macro-topics not necessarily are related also on more specific topics. For instance, the score is lower for the second and third pairs.

As to the network centralities, it is possible to see how the strength on words can give more semantics to what users post. The first pair, with high similarity got a good value also for the strength centrality: this fact indicates that the links in the network of words lead to high scores for both, representing a high correlation on expression. The users "matteorenzi" and "Pontefix_it" have a very low similarity if we consider strength on word and topics, but an high value on eigenvector for the same reason previously described. With our approach based on centralities we are able to grasp this kind of correlation, when users talk about related topics by using different modes of expression or with different purposes. The high scores for all pairs on betweenness with topics indicate a high presence of common sub topics. This fact is probably due to the extraction of "Locations" or "Geographic regions" as topics whenever texts contain names of states, regardless of their use. This is an issue to take in care for future improvements.

5 Conclusions and Future Work

In this paper we exploited the short text categorisation of [9], we designed a network-based user model to extract profiles, and we computed several user similarities. In general the enrichment process improved the profiling yielding higher scores in similarity computation, but using a simple count on the words posted was not sufficient to highlight similarities where we expected. The categorisation process provided topics and macro-topics very relevant and allowed us to have better scores. The similarity values obtained from the centralities led to a further step: the scores based on strength are strongly related to the affinities of analysed pairs, and the ones based on eigenvector and betweenness provide additional information to better understand what the users have in common. Our proposed method represents a new approach to user similarity that does not need URLs inside the text, or hash-tags, or other social media features, as it is usually done in other related works. Thus, we can analyse users also with general short texts, such as text messages, or vocal messages, on mobile phones. On this basis, we plan to run other experiments to test our approach on larger datasets and also to adapt the method to multilanguage environments to test cross-language similarities.

References

1. Abdel-Hafez, A., Xu, Y.: A survey of user modelling in social media websites. Computer and Information Science 6(4), 59 (2013)
2. Anderson, A., Huttenlocher, D., Kleinberg, J., Leskovec, J.: Effects of user similarity in social media. In: WSDM, pp. 703–712. ACM (2012)
3. Campbell, W.M., Dagli, C.K., Weinstein, C.J.: Social Network Analysis with Content and Graphs. Lincoln Laboratory Journal 20(1) (2013)
4. Kwok, K.L.: A neural network for probabilistic information retrieval. In: SIGIR 1989, pp. 21–30 (1989)
5. Lane, N.D., et al.: Exploiting social networks for large-scale human behavior modeling. IEEE Pervasive Computing 10(4), 45–53 (2011)
6. Li, X., Wang, M., Liang, T.-P.: A multi-theoretical kernel-based approach to social network-based recommendation. Decision Support Systems (2014)
7. Liu, H., Hu, Z., Mian, A., Tian, H., Zhu, X.: A new user similarity model to improve the accuracy of collaborative filtering. Knowledge-Based Systems 56, 156–166 (2014)
8. Medo, M.: Network-based information filtering algorithms: Ranking and recommendation. In: Dynamics on and of Complex Networks, vol. 2, pp. 315–334. Springer, New York (2013)
9. Mizzaro, S., Pavan, M., Scagnetto, I., Valenti, M.: Short text categorization exploiting contextual enrichment and external knowledge. In: Proc. of SoMeRA 2014, SIGIR 2014 (2014)
10. Orlandi, F., Breslin, J., Passant, A.: Aggregated, interoperable and multi-domain user profiles for the social web. In: 8th Int. Conf. on Semantic Systems, pp. 41–48. ACM (2012)
11. Tao, K., Abel, F., Gao, Q., Houben, G.-J.: TUMS: Twitter-based user modeling service. In: García-Castro, R., Fensel, D., Antoniou, G. (eds.) ESWC 2011. LNCS, vol. 7117, pp. 269–283. Springer, Heidelberg (2012)
12. Wilkinson, R., Hingston, P.: Using the cosine measure in a neural network for document retrieval. In: ACM SIGIR Conf. on Research and Development in IR, pp. 202–210. ACM (1991)
13. Yan, Q., Wu, L., Zheng, L.: Social network based microblog user behavior analysis. Physica A: Statistical Mechanics and its Applications 392(7), 1712–1723 (2013)
14. Zhang, Y., Wu, Y., Yang, Q.: Community discovery in twitter based on user interests. Journal of Computational Information Systems 8(3), 991–1000 (2012)

A Corpus of Realistic Known-Item Topics with Associated Web Pages in the ClueWeb09

Matthias Hagen, Daniel Wägner, and Benno Stein

Bauhaus-Universität Weimar
<first name>.<last name>@uni-weimar.de

Abstract. Known-item finding is the task of finding a previously seen item. Such items may range from visited websites to received emails but also read books or seen movies. Most of the research done on known-item finding focuses on web or email retrieval and is done on proprietary corpora not publically available. Public corpora usually are rather artificial as they contain automatically generated known-item queries or queries formulated by humans actually seeing the known-item.

In this paper, we study original known-item information needs mined from questions at the popular Yahoo! Answers Q&A service. By carefully sampling only questions with a related known-item web page in the ClueWeb09 corpus, we provide an environment for repeatable realistic studies of known-item information needs and how a retrieval system could react. In particular, our own study sheds some first light on false memories within the known-item questions articulated by the users. Our main finding shows that false memories often relate to mixed up names. This indicates that search engines not retrieving any result on a known-item query could try to avoid returning a zero-result list by ignoring or replacing names in respective query situations.

Our publically available corpus of 2,755 known-item questions mapped to web pages in the ClueWeb09 includes 240 questions with annotated and corrected false memories.

1 Introduction

In the field of information retrieval, *known-item search* is the common task of re-finding a previously accessed item. Types of known items include visited web sites, received or written emails, stored personal documents, but also read books, seen movies, or songs heard on the radio.

In contrast to informational or transactional searches, which can have a multitude of viable results, the goal of a known-item search usually is to retrieve a single, specific item (or syntactic/semantic aliases of it) [6]. In some cases a hub that is "one step away from the target [item]" can also be a less desirable, but still acceptable result [6]. An example for such a hub could be a web page clearly linking to the page a user is looking for or the track listing of a music album, with one of the songs being the desired known item.

Consequently, the number of relevant or useful results tends to be much smaller for known-item queries than for other query types. On the other hand,

A. Hanbury et al. (Eds.): ECIR 2015, LNCS 9022, pp. 513–525, 2015.

the user often has a larger amount of information which can be used to narrow down the results of a known-item query. These two points, the number of acceptable results and the available knowledge, are two main factors that separate known-item searches from other search tasks.

While a large amount of available information can make it easier to re-find a known item, particular attention needs to be paid to incomplete or false memories. Studies have shown that humans remember some kinds of details better than others [4,11,16]. For example, a user looking for a movie might misremember details about the setting (by thinking that it took place in Ireland, rather than Scotland), the cast (by confusing Danny Glover with Morgan Freeman) or misquote a specific line (Darth Vader never says the exact phrase "Luke, I am your father" in "The Empire Strikes Back"). False memories are problematic in that they can lead to the desired item being excluded from the results of a formulated query containing the false memory. A search engine taking the query as is (i.e., including the false memory) might not find any matching result. Presenting an empty result list should be avoided since they harm user experience. Thus, taking care of false memories on search engine side helps to avoid such situations (e.g., the search engine could try to correct the false memory or remove it from the query in a "did you mean"-way [13]). Our study will focus on identifying and characterizing typical false memories. One of our main results shows that searchers often mix up person names when looking for movies or songs.

Current research on the topic of known-item retrieval relies heavily on corpora of known-item queries and their respective known items. Unfortunately, many of those corpora (1) are proprietary and not publicly available, (2) consist of automatically generated queries [2,17,10], or (3) consist of queries generated manually from a known item itself, in a human computation game [18].

Hauff et al. [14] characterized proprietary corpora as problematic since they do not allow for repeatable experiments. Hauff et al. also stated that queries generated from the known item itself, whether automatically or manually, are rather artificial and not representative of real-world user queries since they make unrealistic assumptions: randomly failing memory in automatic query generation or almost perfect memory in human computation games where the known item actually is displayed during or shortly before query formulation. To provide an alternative to these existing corpora, Hauff et al. proposed the creation of a known-item topic set built from questions posted by users of the Yahoo! Answers platform,[1] with the aim to address the lack of public data and the unrealistic approaches to query generation they identified in prior work [14]. As a proof of concept, 103 questions by Yahoo! Answers users were crawled. Among those, 64 information needs were manually assessed, consisting of 32 website and 32 movie known items. Interestingly, even a handful of false memories could be identified.

In the paper at hand, we significantly expand on the ideas of Hauff et al. and build a large-scale corpus with a wider coverage of different information needs, suitable for use in further research. Studying known-item information needs from Yahoo! Answers, we analyze false memories in realistic situations. To ensure the

[1] `http://answers.yahoo.com`

usability of our experiments in a broader context, we only examine known items with a related web page in the ClueWeb09 corpus. For non-website items, like movies or books, this is usually their corresponding entry in the English Wikipedia. The corpus consisting of 2,755 known-item questions mapped to web pages in the ClueWeb09 corpus (including 240 questions with annotated and corrected false memories) is publically available.[2]

The paper is organized as follows. Section 2 describes related work on known-item finding. We present our methodology of corpus construction in more detail in Section 3. First analysis results are reported in Section 4, followed by conclusions and ideas for future work in Section 5.

2 Related Work

We first describe studies that investigated the process of re-finding in different contexts and then focus on studies of known-item querying in particular.

Re-finding Behavior. Blanc-Brude and Scapin [4] conducted a study investigating the ability to recall attributes of a user's own documents (both paper and digital ones) and whether the users could re-find those documents in their work place. The documents were classified as *old* (last access six or more months ago), *recent* (last access within the last six months) and *recurrent* (regularly accessed). The findings show that the study participants were most often mixing true and false memories when being asked to recall the title and keywords of a document in question. For 32% of the documents the recalled keywords were correct, while for 68% they were only partially correct. Recalling the title was even more difficult: 33% correctly recalled document titles versus 47% partially correct and 20% completely false recollections. Location, format, time, keywords, and associated events were remembered most frequently; still, many of these attributes, particularly keywords, time, and location were often only partially remembered or the recollections were incorrect.

Elsweiler et al. [8,9] performed user studies to investigate what users remember about their email messages and how they re-find them. The most frequently remembered attributes of emails were the topic, the reason for sending the email, the sender of the email and other temporal information. In the evaluation, no indication was given if the memories were (partially) false or correct but another finding, in line with research in psychology, was that memory recall declines over time. Emails that had not been accessed for a long time were less likely to have attributes remembered than recently read emails. That users are indeed accessing old documents was shown by Dumais et al. [7]: up to eight years old documents were sought by users in a work environment.

In case of re-finding behavior on the web, people also often do re-find and revisit pages they have accessed a couple of days ago [1]. The last visited documents of a previous session are typically pages to be re-found at the beginning of a later session and people tend to formulate better (i.e., shorter) queries over time, when they access the same item several times [20].

[2] http://www.webis.de/research/corpora

A range of studies [3,5] showed that users in general prefer to browse and to visually inspect items in order to re-find a target document instead of relying on provided text-based search tools. It is then argued that the current personal information management search tools are not sophisticated enough to deal with what and how users remember aspects of the target documents. This is probably also true for the web where the typical interface for re-finding also is a simple keyword-based search box—that still is highly effective for many tasks.

In our scenario, we also consider known-items that have a corresponding web document but we will mostly focus on known-items that have been seen more than just a couple of days ago. We study known-item information needs submitted to a popular question answering platform. Similar to most of the cited studies, also in our study users face the problem of false memories and problems in articulating their need as a query or question when the item was accessed a longer time ago. In contrast to many other search related studies, our corpus of 2,755 known-item information needs connected to ClueWeb09 documents is publically available in order to support further research.

Known-Item Query Generation. Since no large-scale query logs with known-item queries are available, different approaches to generate known-item queries have been proposed ranging from automatic generation to human computation games. For instance, the automatic known-item topic generation approach by Azzopardi et al. [2] works as follows: a known-item / query pair is generated by first selecting a document from the corpus in the role of the known item and by then deriving a corresponding query. The query terms are drawn from the selected document according to particular probability distributions (e.g., the most discriminative terms are selected with a higher probability) while adding some random noise models memory problems. This process was also adapted for the case of personal information management and emails [17,10]. Since such documents usually consist of different fields—emails for instance have a sender, a title, a sending date and a body—, the query terms are drawn from the fields with different probabilities to mimic human memory.

Rather than using automatic query generation, Kim and Croft [18] employ a human computation game to create more "natural" queries. Study participants were shown the known item in question and shortly thereafter they were asked to create a query that retrieves the known item as high as possible in the ranking of a standard retrieval engine. However, even though showing the known item to a user may entail natural queries (i.e., queries created by humans), it does not fully include the concept of false memories.

Hauff et al. [15,14] emphasize the importance of realistic query generation scenarios including false memories when studying search behavior in the known-item setting. They conclude that none of the existing query generation approaches are really realistic as the studied corpora are either proprietary and not publicly available, or consist of automatically generated queries, or consist of queries generated manually from a known item itself. Following Hauff et al.'s suggestions [14] our proposed methodology addresses these problems: we collect a set of 2,755 known-item topics from a popular question answering platform. The known-item topics are based on real information needs by users having problems remembering the known item fully or correctly. Our first results will show what the main issues are with false memories in these cases.

3 Corpus Construction

As discussed in the related work section, the existing approaches to constructing publically available known-item corpora tend to yield rather artificial results. We propose our new Webis Known-Item Question Corpus 2013 (Webis-KIQC-13) as an alternative to those corpora, with the goal of providing a freely available known-item corpus based on real information needs expressed by real humans and with linked items in the popular ClueWeb09 corpus. In principle, our corpus construction follows the suggestions of Hauff et al. [14]. We select questions and answers from a question answering platform where the desired known-item has a corresponding web page in the ClueWeb09 corpus. For the sampled questions and answers a manual annotation identifies the known-item intent and whether a false memory is contained (with manually annotated corrections). This section provides the details on the process of corpus construction.

3.1 Crawling Known-Item Topics from Yahoo! Answers

Web-based community question-answering (cQA) services allow users to pose questions to other users, rate answers by others and receive rewards for providing good answers to open questions. We chose the Yahoo! Answers platform for our purpose of retrieving known-item topics since it provides a public API and a broad range of information needs submitted by many different users. Users are able to submit questions expressed in natural language. These are then opened for other users to propose answers or vote for the best answer to a question. If no best answer gets selected by the asker during the open period, the community votes given by other users potentially determine the chosen answer. In both cases, the question is marked as *resolved*. If no best answer can be chosen through either method, the question is labeled as *undecided*.

For building our known-item topic set, we use the public Yahoo! Answers API, which for example allows retrieving up to 1,050 question entries matching a given keyword query. Our primary focus is on retrieving questions on three types of known items that are often searched for: websites, movies, and musical works (songs and music albums). Nine separate API queries were formulated for each of the three types; to provide a broader range of topics, ten additional queries for other types of known-item information needs were formulated, such as re-finding a book or TV series. Examples of the used API queries are shown in Table 1. To avoid the effect of low quality answers, we only sampled resolved questions from the Yahoo! Answers API. On January 21, 2013, the 37 distinct search queries were submitted to the Yahoo! Answers API, which resulted in a combined set of 24,765 unique questions.

In a second step, the comments and information about who voted for a best answer (community or asker) were scraped from each question's HTML version

Table 1. Examples of search queries used to retrieve from Yahoo! Answers

`(remember) AND (title) AND (movie)`
`(forgot) AND (name) AND (film)`
`(forgot) AND (title) AND (song)`
`(forgot) AND (url) AND (website OR (web site))`
`(remember OR forgot) AND (name OR title) AND (book)`

on the Yahoo! Answers website since they were not contained in the API results. The comments that the asker added to an answer can sometimes be a valuable indication of whether an answer actually contained the searched item and best answers selected by the original asker are a better indication of a correctly found known item than are community votes. Note that also the Yahoo! Answers point system promotes that the asker should select a best answer if there is one. In this case, 3 points are gained while a community vote (that is likely when the desired item is in an answer) does not yield any points. Six questions returned by the API were no longer accessible; among the remaining 24,759 questions only 8,825 questions had their best answer chosen by the original asker. These were kept for manual assessment.

3.2 Assessment of the Crawled Questions

The crawled questions and answers were manually assessed to ensure that they represent satisfied known-item information needs and that they correspond to some website in the ClueWeb09 corpus. The assessors were presented with a form that contains the data fields retrieved by the API query and HTML scraper and additional fields that are to be filled out manually. An external window provides a web view, which allows the assessor to view questions as they are presented to Yahoo! Answers users, to follow hyperlinks and to perform web searches. We had two assessors who checked each of the crawled questions independently. The assessors discussed their decisions afterwards for the few questions where they did not agree initially to reach a consensus.

Assessment of Question Intent. For each of the 8,825 questions with a best answer chosen by the asker, it was first judged whether the intent was to re-find a previously known item, and whether the answer was the desired known item.

For example, questions like "What is the weirdest movie you remember from your childhood?" or "What songs are similar to 'Remember The Name' by Fort Minor?" match our API queries but are posed to initiate a discussion or to receive a recommendation, rather than to satisfy a known-item information need.

For some known-item questions, the asker commented that an answer did not contain the known item, but still chose it as the best answer. This would happen if the answer was still useful to the asker (e.g., recommending a similar item), or merely so the asker would gain some points. In both cases, the questions are omitted from our corpus, as the desired known item could not be determined.

In total, 5,419 questions were discarded in this step, further narrowing down the topic set to 3,406 known-item information needs. Although similar search terms were chosen for all types of items, the proportion of discarded questions varied widely. While only about 35% of movie questions had to be discarded, for websites it were more than 95%. Possible explanations are the following.

- The default behavior of the API, to search in both the question and the answer, led to a large number of unwanted results. For instance, one of the website API queries returned almost one-hundred site support questions answered by the same user, with the same or similar stock answers containing every part of the search term. All of these had to be discarded.
- Askers may be less interested in re-finding a specific website than they are for other item types. Frequently, users are also content with an alternative website offering the same functionality, even if it is not the known item.
- The search terms in our API queries may be ill-suited for finding known-item website questions. The analysis of other cue phrases could be an interesting path for investigation in future research.
- Website re-finding questions in general may be less often submitted to Yahoo! Answers, compared to those for movies, music, or books.

Website re-finding information needs were originally supposed to form a major part of our Webis-KIQC-13. However, only 82 out of 1,706 website known-item questions remain after the intent assessment step.

Mapping of Known Items to their ClueWeb09 ID. In the next step, the assessors checked whether a known item's URL is included in the ClueWeb09. For website questions, this would be the website's URL itself. For most other types of items, we decided that an appropriate URL would be the corresponding article in the English Wikipedia, if there is one. It should be noted that a known item may have multiple semantically or syntactically equivalent aliases [6]. For example, a movie can have both a Wikipedia article and a corresponding IMDb entry, or a notable website may in turn have a Wikipedia article. In these cases, the more appropriate known-item URL in the ClueWeb09 was preferred (e.g., the URL containing more content on the known item). Also, as noted by Broder et al. [6], a so-called *hub*-type result, which is one step away from the target, can be an acceptable, although less desirable result. Examples where hub-type results were deemed acceptable by our assessors include songs not represented through a Wikipedia article of their own, but through the album they were released on, or specific pages on a website where only the main page is in the ClueWeb09.

We used the publically available ChatNoir API [19] that easily maps an item's URL to the corresponding ClueWeb09 ID. Still, the mapping of URLs to ClueWeb09 IDs often had to be done manually by the assessors as a movie or song title often could not directly be translated to a Wikipedia-URL and also the decision of whether a hub-result is contained in the ClueWeb09 had to be determined manually. For 651 out of the 3406 known items, no ClueWeb09 entry could be identified; only the 2,755 known-item questions with matching ClueWeb09 entries form our Webis-KIQC-13. Most of the discarded questions were posed for

Table 2. Examples of tagged false memories in Yahoo! Answers questions

Known item	False memory / Correction
Shooter (film)	[...] Morgan freeman offers him a job to kill a person [...] wrong actor: Danny Glover, not Morgan Freeman
Tokio Hotel	What's the english emo rock band [...] They are american [...] origin: German band, not English or American
An American Tail	[...] a Disney cartoon about a little mouse [...] company: Amblin Entertainment, not Disney
theforgottenlair.net	[...] it went somethin like the underground lair [...] URL: "forgotten", not "underground"

known items more recent than the 2009 crawl date of the ClueWeb09. Given the age of the ClueWeb09 corpus, we expected such an outcome. The differences in coverage over time will be further analyzed in Section 4.

Annotation of False Memories. Finally, the assessors determined whether a known-item question contained false memories. In these cases, the assessors tagged the question as such and added a short annotation documenting the type of error, a correction, and the misremembered property. Some examples of false memories in Yahoo! Answers questions and their annotated corrections are shown in Table 2. Of the 2,755 known-item questions in the Webis-KIQC-13 corpus, 240 (8.7%) contain at least one false memory.

Summary. Although we started from a base of 24,759 unique questions retrieved from the Yahoo! Answers API, the final topic set consists of only 2,755 suitable known-item information needs (11.1% of the original crawl). This is mostly due to the decision to exclude questions decided by community vote, which account for about two in three questions across all crawled categories. A summary of the items removed in the assessment steps is given in Table 3. The large amount of non-known-item questions that we had to discard for some topics is a little surprising. Possible explanations for the case of website information needs have already been hypothesized above. These explanations might, to a lesser degree, be applicable to other categories as well.

The amount of false memory effects identified in the corpus met our initial expectations to be in the range of 5–10% that was also found in the small-scale study by Hauff et al. [14]. The actual number of false memories may be even higher. As the annotators mostly had to rely on the answer text and the known item's corresponding ClueWeb09 document, it is likely that they missed false memories that were not explicitly mentioned therein.

As argued by Azzopardi et al. [2], the manual construction of a known-item corpus on the scale of our Webis-KIQC-13 is a laborious and time-consuming process. Our two assessors together spent approximately 400 hours on the evaluation of the 8,825 questions that had an answer chosen by the asker which translates to an average of about 80 seconds per question.

Table 3. Summary of the removed/remaining items during assessment. Note that the column "Total" also includes additional categories like books etc.

	Movies	Music	Websites	Total
Retrieved questions	5,896	6,481	5,343	24,759
Best answer chosen by voters	-3,718	-4,112	-3,637	-15,934
Best answer chosen by asker	2,178	2,369	1,706	8,825
Not known-item questions	-768	-1,451	-1,624	-5,419
Known-item questions	1,410	918	82	3,406
Not in ClueWeb09	-250	-219	-20	-651
In ClueWeb09	1,160	699	62	2,755
Containing false memories	81	74	4	240

4 Corpus Analysis

We provide a first analysis of the known-item information needs contained in our Webis Known-Item Question Corpus 2013 (Webis-KIQC-13) and their associated properties. We briefly analyze the coverage of the ClueWeb09 corpus and then focus on the types of false memories exhibited. These false memory analyses and the release of our corpus are meant as an enabler for research on the influence of false memories on retrieval processes. By no means, our first analyses can be conclusive but will shed some light on very interesting directions for future work.

4.1 ClueWeb09 Coverage

The ClueWeb09 has been crawled from the live web in January and February 2009. We examine the coverage of the known-item questions by the time of their submission to Yahoo! Answers. Note that, although the newer corpus ClueWeb12 is much younger with a crawling period between February 10, 2012 and March 10, 2012, unfortunately it does not contain Wikipedia and thus lacks the main source of known-item URLs we are aiming for.

The left part of Table 4 presents the relative ClueWeb09 coverage of the retrieved known item queries per year. The steep increase in the number of retrieved known item questions in 2008 can probably be related to an increase in Yahoo! Answers usage. Beginning from 2009, the ClueWeb09 coverage predictably decreases due to the occurrence of known items that did not exist at the time of the ClueWeb09 crawl (e.g., newer movies). While in 2007 a record high of 92.2% could be achieved, the known-item coverage fell to only 71.9% for 2012. By a closer analysis of the known-item questions, we noticed that there were two major groups of re-finding needs that are influenced differently by the ClueWeb09 crawling date. We have (1) questions for items that have not been accessed for a long time (e.g., users searching for the favorite movie of their childhood), and (2) questions for items that have only been incompletely accessed more recently (e.g., by watching the trailer of a movie the other day). Obviously, the web corpus crawling data has a much higher impact on the latter type.

Table 4. ClueWeb09 coverage of the originally crawled 3,406 known-item questions by year and domain type

	2006	2007	2008	2009	2010	2011	2012	Wikipedia	IMDb	Others	No link
Webis-KIQC-13	68	176	369	701	578	477	364	2,618	3	134	-
Not in ClueWeb09	8	15	60	112	148	140	142	405	66	94	86
Total	76	191	429	813	726	617	506	3,023	69	228	86
Coverage	89.5%	92.2%	86.0%	86.2%	79.6%	77.3%	71.9%	86.6%	4.3%	58.8%	0%

Further, we also examine the domains of the ClueWeb09 documents used to represent the known items. The right part of Table 4 shows the frequency with which websites were chosen by the assessors. As can be seen, Wikipedia is the first source the assessors checked when searching for a known item's URL, and the majority of known items were matched to their article there. This decision was made since the ClueWeb09 corpus contains a nearly complete dump of the English Wikipedia at the time of its crawl. At the time of our assessment, 3023 known items either had a Wikipedia article of their own or, as per Broder et al.'s definition [6], a *hub*-type result on the live web. However, for 405 out of them, the Wikipedia article is not part of the ClueWeb09. These 405 were then checked against IMDb or other domains. However, only three out of 69 IMDb entries found on the live web were actually part of the ClueWeb09. Note that in 86 cases, the assessors could not even find a suitable document representing the known item on the live web. These were usually items like poems or songs not released on some album with a Wikipedia entry.

4.2 False Memories

At least 240 of the 2,755 known items in the Webis-KIQC-13 contain some kind of false memory. Categories of false memories were defined ad-hoc by the assessors and were unified in a second pass over the information needs with false memories. Given the search terms used to retrieve our topic set, most of the information needs relate to works of art and entertainment. The most common types of memory errors are shown in Table 5, with an explanation and their number of occurrences. Note that especially the categories relating to persons (character, artist, and actor) with their total amount of 67 false memories form the biggest problem users had in articulating their information need. These categories mostly relate to movie and music questions. Especially for music questions, the lyrics category is another big source of problems. Some text might be mixed up or only remembered in a misheard form and thus can not lead to a good retrieval result.

Our first, and still very basic, analyses reveal two important findings for retrieval systems when taking false memories into account. First, when a query or question including person names does not yield any search result, it is not unlikely that the name is a false memory. A retrieval system could then support the user by leaving out the name for retrieval or suggesting related names (e.g., other actors) that would yield results. Second, queries or questions including lyrics tend to contain false memories. Incorporating sophisticated phonetic

Table 5. Common types of false memories in the Webis-KIQC-13

Category	False memories relating to ...	#
character	attributes of character in a work of fiction	34
lyrics	lyrics of song or poem	29
title	title of work	27
format	way work was released	21
artist	wrong attribution of artist to musical work	22
time	time a work has been produced or released	18
origin	geographical background of a work or artist	15
actor	wrong attribution of actor to movie or series	11
plot	key elements of a work's plot	9
setting	time or place a work is set in	9
company	company involved in production of item	6
scene	single scene in movie or series	5
prop	object in movie or theater play	5
mix-up	confusing attributes of two items	5
URL	URL of website	4

similarities at retrieval system side might be a research direction to support the frequent case of false memories in form of misheard lyrics (e.g., "Stayin' Alive" by the Bee Gees is often misheard as "Steak and a Knife").

5 Conclusions

Our Webis Known-Item Question Corpus 2013 (Webis-KIQC-13) enables a new approach to the evaluation of known-item retrieval tasks, based on using real information needs with a clearly stated intent of known-item re-finding. We believe that by constraining the topic set to answers selected as correct by their asker, we could minimize the error in our known-item mappings. In connection with the ClueWeb09 corpus, this topic set allows for repeatable and realistic testing of known-item information needs. The corpus is freely available.[3]

One direction we envision as particularly promising besides general known-item question analyses is the false memories we annotate in the corpus. They often relate to important details of the known item being sought. The investigation of these false memories is an interesting path for future research. Based on the false memories contained in queries, search engines might not find any reasonable result. To avoid such zero-result lists, the false memories could be identified by to-be-developed techniques and then replaced or removed in a did-you-mean manner [13].

The annotated false memories could also be used to examine the recall of different kinds of information in audiovisual media since most of the search terms we used to crawl questions from the Yahoo! Answers API acquired known-items

[3] `http://www.webis.de/research/corpora`

from the categories Arts & Humanities as well as Entertainment & Music. This places a large number of information needs in our Webis-KIQC-13 close to the field of media or video retrieval, although from a different vantage point.

Incorporating other types of known items that users might search for, such as geographical landmarks or electronic devices, is an interesting direction for future corpus enrichment to provide a representative sample of all potential known-item intents. Especially interesting in that respect would be the inclusion of many more website items. For that category, our search terms that yielded acceptable results on other categories hardly returned usable known-item information needs.

Although our corpus was originally developed as a testbed for known-item search tasks, other uses could be considered as well. Since our Webis-KIQC-13 is publically available and is linked to the widely-used ClueWeb09 corpus, repeatable research on web requests in the known-item domain is possible.

References

1. Adar, E., Teevan, J., Dumais, S.T.: Large scale analysis of web revisitation patterns. In: CHI 2008, pp. 1197–1206 (2008)
2. Azzopardi, L., de Rijke, M., Balog, K.: Building simulated queries for known-item topics: An analysis using six european languages. In: SIGIR 2007, pp. 455–462 (2007)
3. Barreau, D., Nardi, B.: Finding and reminding: File organization from the desktop. ACM SIGCHI Bulletin 27(3), 39–43 (1995)
4. Blanc-Brude, T., Scapin, D.L.: What do people recall about their documents?: Implications for desktop search tools. In: IUI (2007)
5. Boardman, R., Sasse, M.: Stuff goes into the computer and doesn't come out: A cross-tool study of personal information management. In: CHI 2004, pp. 583–590 (2004)
6. Broder, A.: A taxonomy of web search. SIGIR Forum 36(2), 3–10 (2002)
7. Dumais, S.T., Cutrell, E., Cadiz, J.J., Jancke, G., Sarin, R., Robbins, D.C.: Stuff I've seen: A system for personal information retrieval and re-use. In: SIGIR 2003, pp. 72–79 (2003)
8. Elsweiler, D., Baillie, M., Ruthven, I.: Exploring memory in email refinding. ACM Trans. Inf. Syst. 26(4), 1–36 (2008)
9. Elsweiler, D., Baillie, M., Ruthven, I.: What makes re-finding information difficult? A study of email re-finding. In: Clough, P., Foley, C., Gurrin, C., Jones, G.J.F., Kraaij, W., Lee, H., Mudoch, V. (eds.) ECIR 2011. LNCS, vol. 6611, pp. 568–579. Springer, Heidelberg (2011)
10. Elsweiler, D., Losada, D.E., Toucedo, J.C., Fernández, R.T.: Seeding simulated queries with user-study data forpersonal search evaluation. In: SIGIR 2011, pp. 25–34 (2011)
11. Elsweiler, D., Ruthven, I., Jones, C.: Towards memory supporting personal information management tools. JASIST 58(7), 924–946 (2007)
12. Gunning, R.: The technique of clear writing. McGraw-Hill (1952)
13. Hagen, M., Stein, B.: Applying the user-over-ranking hypothesis to query formulation. In: Amati, G., Crestani, F. (eds.) ICTIR 2011. LNCS, vol. 6931, pp. 225–237. Springer, Heidelberg (2011)
14. Hauff, C., Hagen, M., Beyer, A., Stein, B.: Towards realistic known-item topics for the ClueWeb. In: IIiX 2012, pp. 274–277 (2012)

15. Hauff, C., Houben, G.-J.: Cognitive processes in query generation. In: Amati, G., Crestani, F. (eds.) ICTIR 2011. LNCS, vol. 6931, pp. 176–187. Springer, Heidelberg (2011)
16. Kelly, L., Chen, Y., Fuller, M., Jones, G.J.F.: A study of remembered context for information access from personal digital archives. In: IIiX 2008, pp. 44–50 (2008)
17. Kim, J., Croft, W.B.: Retrieval experiments using pseudo-desktop collections. In: CIKM 2009, pp. 1297–1306 (2009)
18. Kim, J., Croft, W.B.: Ranking using multiple document types in desktop search. In: SIGIR 2010, pp. 50–57 (2010)
19. Potthast, M., Hagen, M., Stein, B., Graßegger, J., Michel, M., Tippmann, M., Welsch, C.: ChatNoir: A search engine for the ClueWeb09 corpus. In: SIGIR 2012, p. 1004 (2012)
20. Tyler, S.K., Teevan, J.: Large scale query log analysis of re-finding. In: WSDM 2010, pp. 191–200 (2010)

Designing States, Actions, and Rewards for Using POMDP in Session Search

Jiyun Luo, Sicong Zhang, Xuchu Dong, and Hui Yang

Department of Computer Science, Georgetown University
37th and O Street NW, Washington DC, 20057, USA
{jl1749,sz303,xd47}@georgetown.edu, huiyang@cs.georgetown.edu

Abstract. Session search is an information retrieval task that involves a sequence of queries for a complex information need. It is characterized by rich user-system interactions and temporal dependency between queries and between consecutive user behaviors. Recent efforts have been made in modeling session search using the Partially Observable Markov Decision Process (POMDP). To best utilize the POMDP model, it is crucial to find suitable definitions for its fundamental elements – *States*, *Actions* and *Rewards*. This paper investigates the best ways to design the states, actions, and rewards within a POMDP framework. We lay out available design options of these major components based on a variety of related work and experiment on combinations of these options over the TREC 2012 & 2013 Session datasets. We report our findings based on two evaluation aspects, retrieval accuracy and efficiency, and recommend practical design choices for using POMDP in session search.

Keywords: Session Search, POMDP, State, Action, Reward.

1 Introduction

Information Retrieval (IR) tasks are concerned with finding relevant documents to fulfill user's information needs. Session search, as defined in the TREC (Text REtrieval Conference) Session tracks is an information retrieval task that involves multiple queries and multiple search iterations to achieve a complex information need [11,12]. In a session, a user keeps formulating queries until he or she gets satisfied with the information need [12], bored, or frustrated [2]. Session search is a challenging research area that is characterized by rich user-system interactions, complex information needs, and temporal dependency between queries and between user behaviors.

In a session, a user interacts with the search engine to explore the information space: the user continuously reformulates queries, clicks on documents, and examines documents. This is a trial-and-error setting. Classic ad-hoc retrieval models emphasize on handling one-shot query and treating each queries in a session independently [16]. Classic relevance feedback models, such as Rocchio [9], although modeling feedbacks from the user, also treat each query in a session independently: the user feedbacks are for a particular query. The continuity

A. Hanbury et al. (Eds.): ECIR 2015, LNCS 9022, pp. 526–537, 2015.

of queries in a sequence during a session has not yet been studied much. This places unique challenge on session search for new statistical retrieval models that is able to handle the dynamics present in the task.

The family of Reinforcement Learning (RL) algorithms [6] matches well with the trial-and-error setting present in session search: the algorithm learns from repeated, varied attempts which are continued until success. The learner (also known as agent) learns from its dynamic interactions with the world, rather than from a labeled dataset as in supervised learning. In such a setting, a stochastic model assumes that the system's current state depend on the previous state and action in a non-deterministic manner [15]. Among various models in the RL family, Partially Observable Markov Decision Processes (POMDP) [19] has been applied recently on IR problems including session search [14], document re-ranking [8,22], and advertisement bidding [20]. In a POMDP, hidden information can be modeled as hidden states, while visible signals in the process can be modeled as observations or actions.

States, actions, and *reward functions* are the fundamental elements in a POMDP framework. The following principles are usually referred to when defining these elements in a POMDP framework:

- *States*: What changes with each time step?
- *Actions*: How does our system change the state?
- *Rewards*: How can we measure feedback or effectiveness?

Given the recent work on applying POMDP to session search, what is missing is a study that evaluates the design for *States, Actions,* and *Rewards.* In this paper, we strive to answer the research question – *what are the best design options to model session search using POMDP.* We use search effectiveness and search efficiency as two evaluation aspects to help select the best design under different circumstances.

However, there are only a few existing approaches that use POMDP to study IR problems. We hence expand the targeted group of approaches to a wider range of methods, including MDP [5], exploratory online learning [6] and decision theories [18], to study how they define the three major components for session search. Therefore, not all methods studied in this paper are based on POMDP, but they all share the idea of using states, actions, and rewards. We would like to find out the promising designs of those elements for our task.

In the remainder of this paper, after briefly presenting the POMDP framework (Section 2), we lay out available options for *states, actions,* and *rewards* in using POMDP for session search (Section 3). We then experiment on combinations of various options over the TREC Session 2012 & 2013 datasets [11,12] and report our findings on the impacts of various settings in terms of search accuracy and efficiency (Section 4). Finally, we recommend design choices for using POMDP in session search (Section 5) and conclude the paper (Section 6).

2 Using a POMDP Framework

Partially Observable Markov Decision Process (POMDP) can be represented as a tuple of $(S, M, A, R, \gamma, O, \Theta, B)$, , which consists of states S, state transition function M, actions A, reward function R, discount factor γ (usually between 0 and 1), observations O, observation function Θ, and belief states B. In a POMDP model, the states are hidden from the agent. The agent can only observe symbols (observations) emitted according to hidden states. At the same time, the agent forms its beliefs on the hidden states, which is an estimated probability distribution over the state space. Once the agent obtains a new observation, its belief will be updated accordingly. A detailed version of using POMDP in session search can be found in [14].

The goal of a POMDP is to find an optimal policy which maximizes the expected reward value, also known as the value function. Let $R(b, a)$ be the reward for an action a based on the current belief b. The value function can be expressed by the Bellman equation [1,10].

$$V^*(b) = \max_a \left[R(b, a) + \gamma \sum_{o'} P(o'|a, b) V^*(b') \right] \tag{1}$$

The notation $P(o'|a, b)$ represents the probability of observing o after taking action a with belief b. Let $b(s)$ denote the belief on being in state s. The new belief b' on the next state is calculated by updating b as follows:

$$b'(s') = \eta \times \Theta(o|s', a) \sum_{s \in S} P(s'|s, a) b(s) \tag{2}$$

In Eq. 2, probability function P is the transition function, and the notation $\Theta(o|s', a)$ stands for the probability to observe o given state s' and action a. Here, we use η as the normalizing constant.

There are standard algorithms, including QMDP and MC-POMDP, to solve problems formalized by POMDPs [10]. Value iteration is used in QMDP by treating the value function as a mapping from beliefs to real numbers. MC-POMDP algorithm is applicable to continuous POMDP problems. However, many approaches can only be applied to problems of very small scales. Littman et al's Witness Algorithm is a more practical approach to obtain solutions to POMDP problems [13].

Solutions to the POMDP framework for session search can be obtained by using these approaches. Our aim in this paper is not how to get a solution. When applying POMDP to session search, the definitions of the states, actions, and rewards are flexible but critical to search accuracy and efficiency. In the following sections, we focus on studying the design choices of these elements.

3 Design Choices: States, Actions, and Rewards

In this section, we summarize the existing research work to enumerate the available design choices for a POMDP model in the context of session search. These choices are discussed in three categories: states, actions and rewards. Some of the existing work mentioned in this section are not based on POMDP. However, they all share the idea of using states, actions, and rewards. Hence they are still valuable to our study.

3.1 States

State definition is essential in modeling session search by a POMDP. As we can see, related research in similar tasks have proposed a variety of state definitions. They include *queries* [5,6], *document relevance* [8,22], and *relevance vs. exploration decision making states* [14]. We group them into two design options:

(S1) Fixed number of states. Using a predefined fixed number of states can easily characterize certain properties of the session based on the current state. For instance, Zhang et al. used two binary relevance states, "Relevant" and "Irrelevant" to represent the decision-making states that the user considers the previously returned documents are relevant or not [22]. A more complete formulation of the decision-making states was presented in Luo et al. [14], where a cross-product of two decision-making dimensions – "whether the previously retrieved documents are relevant" and "whether the user desires to explore" – forms four hidden states which reflect the current status of the search process.

(S2) Varying number of states. Some approaches choose to model session search using a varying or even infinite number of states. A popular approach is to model queries in a session as states (Hofmann et al. [6] and Guan et al. [5]). In this design, the number of states changes according to session length, i.e., the number of queries in a session. There are also abstract definitions of states. For instance, Jin et al. used relevance score distribution as the states[8], which leads to an infinite number of real valued states.

As discussed above, all state definitions are used to characterize the current status of the search process. Using fixed number of states tends to reflect more specific features while using varying number of states may have more abstract characterization of the search process. Hence, we would like to point out that *state definition is an art*, which depends on the needs of the actual IR task.

3.2 Actions

It is worth noting that, as Luo et al. [14] pointed out, the user and the search engine are two autonomous agents in a session. For session search, typical user actions include: *Add query terms; Remove query terms; Keep query terms; Click on documents;* and *SAT click on documents* (click and read the documents for a long period of time). Typical search engine actions include: *increase/decrease/keep term weights; switch on or switch off query expansion; adjust the number of top documents used in Pseudo Relevance Feedback (PRF)* and *consider the ranked list itself as actions.* Here we focus on the search engine actions. Existing search engine actions in related work are grouped into:

(A1) Technology Selection. Some approaches use a meta-level modeling of actions. They don't focus on details in a single search method but on implementing multiple search methods (termed as *search technologies*), and selecting the best search technology to use. An action using technology selection can be *switching on or switching off the technology*, or *adjusting parameters in the technology*. Example technologies include query expansion and pseudo relevance feedback (PRF). To illustrate, Luo et al. selected the number of top retrieved documents to be included in PRF [14].

(A2) Term Weight Adjustment. Another idea to model search engine actions focuses on term weight adjustments. This group of actions enables the search engine to directly adjust individual terms' weights. Typical weighting schemes include *increasing term weights, decreasing term weights*, or *keeping term weights unchanged*. Guan et al. proposed four types of term weighting scheme (*theme terms, novel added terms, previously-retrieved added terms*, and *removed terms*) as actions according to the query changes detected between adjacent search iterations [5].

(A3) Portfolio A more straightforward type of search engine actions is using the document lists. We follow the naming used in [8] and call this type of actions *portfolio*. Here a ranked list of documents is a *portfolio* and is treated as a single action. The space of the document permutation is the action space, where each document ranking permutation is a different action.

These actions are in fact what a search engine can do for document retrieval. Hence, we say that *actions are essentially options in your search algorithm*.

3.3 Rewards

A clear goal is key to any success. In order to estimate the benefits from an action, we need to evaluate the reward R of taking the action at state s. Similar to the loss (risk) function in supervised learning, a reward function can guide the search engine throughout the entire dynamic process of session search. Since session search is a document retrieval task, it's natural that *the reward function is about document relevance*. Notably, the difference between session search and one-shot query search lies in that session search aims to optimize a *long term reward*, which is an expectation over the overall rewards in the whole session, while one-shot query search doesn't have to do that. We group reward functions in related work into:

(R1) Explicit Feedback. Rewards directly generated from user's relevance assessments are considered as explicit feedback. Both Jin et al. [8] and Luo et al. [14] calculated the rewards using nDCG [7], which measures the document relevance for an entire ranked list of documents with ground truth judgments.

(R2) Implicit Feedback. Other approaches used implicit feedback obtained from user behavior as rewards. For instance, Hofmann et al. used user click information as the reward function in their online ranking algorithm [6] and Zhang et al. used clicks and dwell time as reward for document re-ranking [22].

4 Experiments

In this section, we aim to examine the design choices for using POMDP in session search. As we lay out in the previous section, there are two options for states, three for actions, and two for rewards, which result in a total of $2\times3\times2 = 12$ combinations. For example, the search system proposed by [14] used a combination of $S_1A_1R_1$, which means "Fixed number of states", "Technology Selection" as the actions, and "Explicit Feedback" as the reward. We report our findings on the search accuracy and search efficiency for those design options.

4.1 Task and Datasets

We evaluate a number of systems, each of which represents a combination of design choices as mentioned in Section 3. The session search task is the same as in the recent TREC 2012 and 2013 Session Tracks [11,12]: to retrieve 2000 relevant documents for the last query in a session. Session logs, including queries, retrieved URLs, Web page titles, snippets, clicks, and dwell time, were generated by the following process. Search topics were provided to the user. The user was then asked to create queries and perform search using a standard search engine provided by TREC. TREC 2012 contains 297 queries in 98 sessions, while TREC 2013 contains 442 queries in 87 sessions. An example search topic is "You just learned about the existence of long-term care insurance. You want to know about it: costs/premiums, companies that offer it, types of policies, ..." (TREC 2013 Session 6).

We use the evaluation scripts and ground truth provided by TREC for evaluation. The metrics are mainly about search accuracy, including nDCG@10, nERR@10, nDCG, and MAP [12]. We also report the retrieval efficiency in Wall Clock Time, CPU cycles and the Big O notation. The dataset used for TREC 2012 is ClueWeb09 CatB, containing 50 million English Web pages crawled in 2009. The dataset used for TREC 2013 is ClueWeb12 CatB, containing 50 million English Web pages crawled in 2012. Spam documents are removed according to the Waterloo spam scores [3]. Duplicated documents are also removed.

4.2 Systems

Among the 12 combinations mentioned in Section 3, $S_1A_2R_2$, $S_1A_3R_1$, $S_2A_1R_2$, $S_2A_2R_2$ and $S_2A_3R_2$ are not discussed in this paper because we have not yet found a realistic way to implement them. We evaluate the remaining seven choices. For $S_2A_1R_1$, we implement two versions of it. The first is UCAIR, a re-implementation of Shen et al.'s work [18]. However, this system has only one action. To have a fair comparison with other systems, we create another $S_2A_1R_1$ system to include more actions. In total, we implement eight POMDP systems:

$\mathbf{S_1A_1R_1}$**(win-win).** This is a re-implementation of Luo et al.'s system [14]. Its configuration is "S_1 Fixed number of states" + "A_1 Technology Selection" + "R_1 Explicit Feedback". Its search engine actions include six retrieval technologies:

(1) increasing weights of the added query terms; (2) decreasing weights of the added query terms; (3) QCM [5]; (4) PRF (Pseudo Relevance Feedback) [17]; (5) Only use the last query in a session; and (6) Equally weights and combines all unique query terms in a session. The system employs 20 search engine actions in total and uses nDCG@10 as the reward.

$\mathbf{S_1A_1R_2}$. This is a variation of $S_1A_1R_1$(win-win). Its configuration is "S_1 Fixed number of states" + "A_1 Technology Selection" + "R_2 Implicit Feedback". This system also uses 20 actions. Unlike win-win, its rewards are SAT Clicks (documents that receive user clicks and the time of user dwelling on is greater than 30 seconds [4]).

$\mathbf{S_1A_2R_1}$. This system's configuration is "S_1 Fixed number of states" + "A_2 Term Weight Adjustment" + "R_1 Explicit Feedback". Specifically, the states in this approach are"Exploitation" and "Exploration". The term weights are adjusted similarly to Guan et al. [5] based on query changes. For example, if the user is currently under "Exploitation" and adds terms to the current query, we let the search engine take an action to increase the weights for the added terms.

$\mathbf{S_1A_3R_2}$. This system's configuration is "S_1 Fixed number of states" + "A_3 Portfolio" + "R_2 Implicit Feedback". It contains a single state, which is the current query. It uses the last query in a session to retrieve the top X documents as in [21] and then re-ranks them to boost the ranks of the SAT Clicked documents. The actions are portfolios, i.e., all possible rankings for the X documents. For each ranked list D_i, the system calculates a reward and selects the ranked list with the highest reward.

$\mathbf{S_2A_1R_1}$(UCAIR). This is a re-implementation of Shen et al.'s work [18]. Its configuration is "S_2 Varying number of states" + "A_1 Technology Selection" + "R_1 Explicit Feedback". Every query is a state. Query expansion and re-ranking are the two search technologies. In UCAIR, if a previous query term occurs frequently in the current query's search results, the term is added to the current query. The expanded query is then used for retrieval. After that, the re-ranking phase is performed based on the combination of each SAT Click's snippet the expanded query.

$\mathbf{S_2A_2R_1}$(QCM). This is a re-implementation of Guan et al.'s system in [5]. Its configuration is "S_2 Varying number of states"+ "A_2 Term Weight Adjustment" + "R_1 Explicit Feedback". In QCM, every query is a state. The search engine actions are term weight adjustments. QCM's actions include increasing theme terms' weights, decreasing added terms' weights, and decreasing removed terms' weights. The term weights of each query is also discounted according to an reinforcement learning framework in [5].

$\mathbf{S_2A_1R_1}$. This system's configuration is "S_2 Varying number of states"+ "A_1 Technology Selection" + "R_1 Explicit Feedback". It is built on the basis of $S_2A_2R_1$(QCM). Its search engine actions are two: QCM with or without spam detection. The spam detection is done by using Waterloo's spam scores. The rest settings are the same as in QCM.

Table 1. Search accuracy on TREC 2012 and TREC 2013 Session Tracks

Approach (2012)	nDCG@10	nDCG	MAP	nERR@10
$S_1A_1R_1$(win-win)	0.2916	0.2875	0.1424	0.3368
$S_2A_1R_1$	0.2658	0.2772	0.1307	0.3105
$S_1A_1R_2$	0.2222	0.2733	0.1251	0.2464
$S_2A_2R_1$(QCM)	0.2121	0.2713	0.1244	0.2302
$S_2A_1R_1$(UCAIR)	0.2089	0.2734	0.1225	0.2368
$S_1A_3R_2$	0.1901	0.2528	0.1087	0.2310
$S_1A_2R_1$	0.1738	0.2465	0.1063	0.1877
$S_2A_3R_1$(IES)	0.1705	0.2626	0.1184	0.1890
Approach (2013)	nDCG@10	nDCG	MAP	nERR@10
$S_1A_1R_1$(win-win)	0.2026	0.2609	0.1290	0.2328
$S_2A_1R_1$	0.1676	0.2434	0.1132	0.1914
$S_2A_2R_1$(QCM)	0.1316	0.1929	0.1060	0.1547
$S_2A_1R_1$(UCAIR)	0.1182	0.1798	0.0927	0.1360
$S_2A_3R_1$(IES)	0.1076	0.1851	0.0966	0.1133
$S_1A_3R_2$	0.0987	0.1538	0.0761	0.1064
$S_1A_1R_2$	0.0964	0.2159	0.0689	0.1041
$S_1A_2R_1$	0.0936	0.1499	0.0740	0.0995

$\mathbf{S_2A_3R_1}$(IES). This is a re-implementation of Jin et al.'s work [8]. Its configuration is "S_2 Varying number of states" + "A_3 Portfolio" + "R_1 Explicit Feedback". This system uses the top K documents as pseudo relevance feedback to re-rank the retrieved documents. It assumes each document's true relevance score is a random variable following a multi-variable normal distribution $\mathcal{N}(\boldsymbol{\theta}, \Sigma)$. $\boldsymbol{\theta}$ is the mean vector and is set as the relevance score calculated directly by [21]. The Σ is approximated using document cosine similarity. IES also uses Monte Carlo Sampling and a greedy algorithm called "Sequential Ranking Decision" to reduce the action space.

4.3 Search Accuracy

Table 1 shows the search accuracy of the above systems using TREC's effectiveness metrics for both datasets. The systems are decreasingly sorted by nDCG@10 in the table.

As we can see, $S_1A_1R_1$ (win-win) outperforms all other systems in both datasets. For example, in TREC 2012, $S_1A_1R_1$ (win-win) shows 37.5% improvement in nDCG@10 and 46.3% in nERR@10 over $S_2A_2R_1$ (QCM), a strong state-of-the-art session search system which uses a single search technology [5]. The improvements are statistically significant ($p < 0.05$, t-test, one-sided). It also shows 6.0% nDCG and 14.5% MAP improvements over QCM, however they are not statistically significant. Another system $S_2A_1R_1$, which also uses technology selection, improves 25.3% in nDCG@10 and 34.9% in nERR@10 over QCM, too. The improvements are statistically significant ($p < 0.05$, t-test, one-sided).

Table 2. Efficiency on TREC 2012 and 2013 Session Track. O(L) is the time complexity of conducting a Language Modeling retrieval. l is the number of alternative actions. K is the top K ranks. O(X) is the time complexity of re-ranking X documents. Z is the sample size of feedback documents.

Approach	TREC 2012		TREC 2013		BigO
	Wall Clock	CPU cycle	Wall Clock	CPU cycle	
$S_2A_3R_1$(IES)	9.7E4s	2.6E14	8.0E4s	2.2E14	$O(L+KZX^3)$
$S_1A_1R_2$	3.2E4s	8.6E13	1.8E4s	4.8E13	$O(lL)$
$S_1A_1R_1$(win-win)	3.1E4s	8.4E13	1.3E4s	3.5E13	$O(lL)$
$S_2A_1R_1$	6.6E3s	1.8E13	8.6E3s	2.3E13	$O(lL)$
$S_2A_2R_1$(QCM)	2.2E3s	5.8E12	1.9E3s	5.2E12	$O(L)$
$S_2A_1R_1$(UCAIR)	1.8E3s	4.8E12	0.8E3s	2.0E12	$O(L)$
$S_1A_2R_1$	1.1E3s	3.0E12	0.4E3s	1.0E12	$O(L)$
$S_1A_3R_2$	0.8E3s	2.2E12	0.3E3s	0.8E12	$O(L+X)$

It suggests that "A_1 Technology Selection", the meta-level search engine action, is superior to a single search technology, for example, term weight adjustment in QCM. Moreover, $S_1A_1R_1$(win-win) performs even better than $S_2A_1R_1$, where the former uses more search technologies than the latter. We therefore suggest that using more alternative search technologies can be very beneficial to session search.

4.4 Search Efficiency

In this section, we report the efficiency of these systems using a hardware support of 4 CPU cores (2.70 GHz), 32 GB Memory, and 22 TB NAS. Table 2 presents the wall clock running time, cpu cycles, as well as the Big O notation for each system. The systems are decreasingly ordered by wall clock time , which is measured in seconds.

All approaches, except $S_2A_3R_1$(IES), are able to finish within 1 day. Moreover, the experiment shows that $S_1A_3R_2$, $S_1A_2R_1$, $S_2A_1R_1$ (UCAIR), $S_2A_2R_1$(QCM) and $S_2A_1R_1$ are quite efficient and finished within 2.5 hours. $S_1A_1R_1$(win-win) and $S_1A_1R_2$ also show moderate efficiency and finished within 9 hours.

$S_2A_3R_2$(IES) is the slowest system, which took 27 hours to finish. We investigate the reasons behind its slowness. Based on Algorithm 1 in IES [8], the system first retrieves X documents using a standard document retrieval algorithm [21], then the algorithm has three nested loops to generate top K results by re-ranking. The first loop enumerates each rank position and its time complexity is $O(K)$. The second loop iterates over each retrieved document, thus its time complexity is $O(X)$. Inside the second loop, it first samples Z documents from the top K documents, then runs the third loop. The third loop enumerates each sample and has a time complexity of $O(Z)$. Inside the third loop, there is a matrix multiplication calculation for every retrieved document, which alone

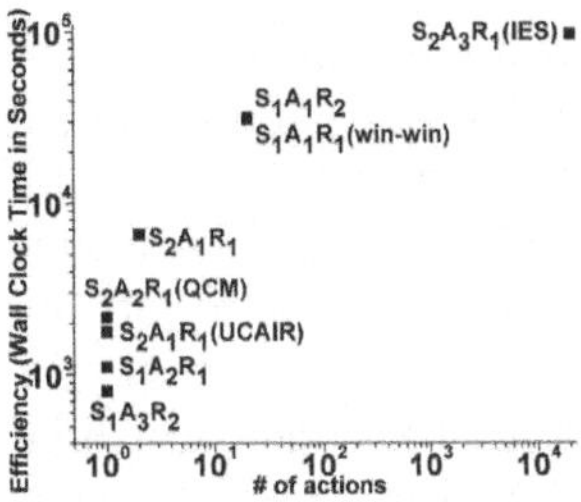

Fig. 1. Efficiency vs. # of Actions on TREC 2012

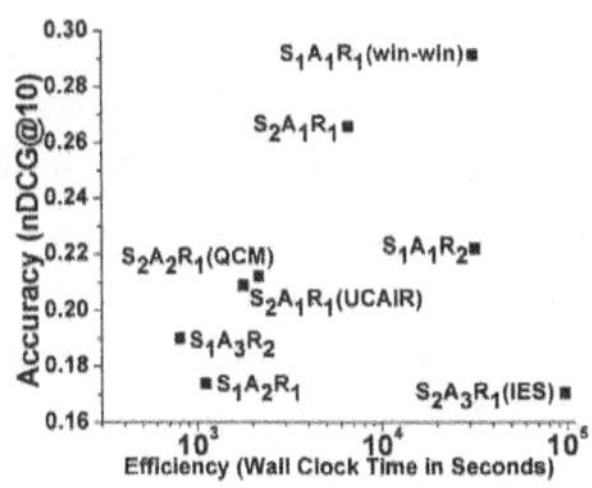

Fig. 2. Accuracy vs. Efficiency on TREC 2012

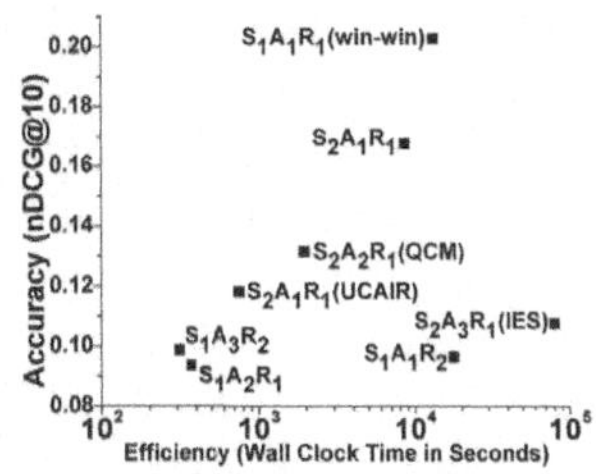

Fig. 3. Accuracy vs. Efficiency on TREC 2013

attributes to a time complexity of $O(X^2)$. Therefore, IES's total time complexity is $O(KZX^3)$, which makes IES computationally demanding.

We also look into the time complexity of other systems and present their Big O notations in Table 2. We notice that $S_2A_2R_1$(QCM), $S_2A_1R_1$(UCAIR) and $S_1A_2R_1$ only perform one document retrieval, hence their time complexity is $O(L)$. $S_1A_1R_2$, $S_1A_1R_1$(win-win) and $S_2A_1R_1$ conduct l document retrievals, hence their time complexity is $O(lL)$. $S_1A_3R_2$ performs one document retrieval and one document re-ranking, hence its time complexity is $O(L + X)$. Their time complexities range from linear, e.g. $O(L)$ or $O(X)$, to quadratic, e.g. $O(lL)$, which suggests that these systems are efficient.

We see an interesting association between efficiency and the number of actions used in a system. Figure 1 shows that in TREC 2012, the systems' running time increases monotonically as the number of actions increases. It suggests that besides time complexity, the number of actions used in POMDP is another important factor in deciding its running time. We do not observe similar association between actions and accuracy for the systems under evaluation.

4.5 Tradeoff between Accuracy and Efficiency

Based on the search accuracy and efficiency results, we observe a trade-off between them, which is presented in Figures 2 and 3. They show that accuracy tends to increase when efficiency decreases. This is because systems with higher accuracy tend to be more computationally demanding. For instance, $S_1A_1R_1$(win-win) could achieve better accuracy but worse efficiency than $S_2A_1R_1$. We also find that $S_2A_1R_1$ (UCAIR) strikes a good balance between search accuracy and efficiency. With a simple feedback mechanism based on the vector space model, this system reaches high efficiency while can still achieve quite good nDCG@10. Overall, $S_1A_1R_1$(win-win) gives impressive accuracy with a fair degree of efficiency.

5 Our Recommendations

Giving the TREC Session task and typical computational resource as described in Section 4.4, our recommendation is the following. If more emphasis is put on accuracy rather than efficiency, we recommend $S_1A_1R_1$ (win-win) [14], whose settings are "Fixed number of states", "Technology Selection", and "Explicit Feedback" as the reward, for its highest search accuracy (Tables 1 and 2). If more emphasis is put on efficiency, e.g. with a limit of finishing the experiments within 1 hour, our recommendation will be $S_2A_2R_1$(QCM) [5], whose settings are "Varying number of states", "Term Weight Adjustment" as actions, and "Explicit Feedback" as the reward, for its high accuracy within the time constraint. In addition, we also recommend $S_2A_1R_1$ (UCAIR) [18], which is the runner-up in search accuracy among runs finishing within 1 hour, while only taking half as much time as QCM.

We have noticed that the number of actions heavily influences the search efficiency. Specifically, using more actions may benefit the search accuracy, while hurts the efficiency. For instance, with a lot of action candidates, $S_1A_1R_1$ (win-win) outperforms other runs in accuracy. However, the cost of having more actions in the model is that it requires more calculations and longer retrieval time. Therefore, we recommend a careful design of the number of total actions, when creating a new POMDP model, to balance between accuracy and efficiency.

6 Conclusion

This paper aims to provide guidelines for using POMDP models to tackle session search. Based on an extended set of IR algorithms that share the use of state, action and reward, we evaluate the various design options in designing suitable states, actions and reward functions for session search. The design options are evaluated against two major factors, search accuracy and search efficiency. We experiment and report our findings on the TREC 2012 and 2013 Session Track datasets. Finally, we make recommendations for a typical session search task for IR researchers and practitioners to use POMDP in session search.

From our experiments, we have learned that a model with more action options tends to have better accuracy but worse efficiency. It once again proves the importance of managing a good balance between accuracy and efficiency. We hope our work can motivate the use of POMDP and other reinforcement learning models in session search and provide a general guideline for designing *States*, *Actions*, and *Rewards* in session search.

Acknowledgments. The research is supported by NSF grant CNS-1223825, DARPA grant FA8750-14-2-0226, and a sponsorship from the China Scholarship Council. The U.S. Government is authorized to reproduce and distribute reprints for Governmental purposes notwithstanding any copyright notation thereon.

References

1. Bellman, R.: Dynamic Programming. Princeton University Press (1957)
2. Chilton, L.B., Teevan, J.: Addressing people's information needs directly in a web search result page. In: WWW 2011, pp. 27–36

3. Cormack, G.V., Smucker, M.D., Clarke, C.L.: Efficient and effective spam filtering and re-ranking for large web datasets. Inf. Retr. 14(5), 441–465 (2011)
4. Fox, S., Karnawat, K., Mydland, M., Dumais, S., White, T.: Evaluating implicit measures to improve web search. ACM Trans. Inf. Syst. 23(2), 147–168
5. Guan, D., Zhang, S., Yang, H.: Utilizing query change for session search. In: SIGIR 2013, pp. 453–462 (2013)
6. Hofmann, K., Whiteson, S., de Rijke, M.: Balancing exploration and exploitation in learning to rank online. In: Clough, P., Foley, C., Gurrin, C., Jones, G.J.F., Kraaij, W., Lee, H., Mudoch, V. (eds.) ECIR 2011. LNCS, vol. 6611, pp. 251–263. Springer, Heidelberg (2011)
7. Järvelin, K., Kekäläinen, J.: Cumulated gain-based evaluation of IR techniques. ACM Trans. Inf. Syst. 20(4) (October 2002)
8. Jin, X., Sloan, M., Wang, J.: Interactive exploratory search for multi page search results. In: WWW 2013, pp. 655–666 (2013)
9. Joachims, T.: A probabilistic analysis of the Rocchio algorithm with TFIDF for text categorization. In: ICML 1997, pp. 143–151 (1997)
10. Kaelbling, L., Littman, M., Cassandra, A.: Planning and acting in partially observable stochastic domains. Artificial Intelligence 101(1-2), 99–134 (1998)
11. Kanoulas, E., Carterette, B., Hall, M., Clough, P., Sanderson, M.: Overview of the trec 2012 session track. In: TREC 2012 (2012)
12. Kanoulas, E., Carterette, B., Hall, M., Clough, P., Sanderson, M.: Overview of the trec, session track. In: TREC 2013 (2013)
13. Littman, M.L.: The witness algorithm: Solving partially observable Markov decision processes. Technical report, Providence, RI, USA (1994)
14. Luo, J., Zhang, S., Yang, H.: Win-win search: Dual-agent stochastic game in session search. In: SIGIR 2014 (2014)
15. Norris, J.R.: Markov Chains. Cambridge University Press (1998)
16. Robertson, S., Zaragoza, H.: The probabilistic relevance framework: BM25 and beyond. Found. Trends Inf. Retr. 3(4), 333–389 (2009)
17. Salton, G., Buckley, C.: Improving retrieval performance by relevance feedback. Readings in Information Retrieval 24, 5 (1997)
18. Shen, X., Tan, B., Zhai, C.: Implicit user modeling for personalized search. In: CIKM 2005, pp. 824–831 (2005)
19. Sondik, E.: The optimal control of partially observable markov processes over the infinite horizon: Discounted cost. Operations Research 26(2), 282–304 (1978)
20. Yuan, S., Wang, J.: Sequential selection of correlated ads by POMDPs. In: CIKM 2012, pp. 515–524 (2012)
21. Zhai, C., Lafferty, J.: Two-stage language models for information retrieval. In: SIGIR 2002, pp. 49–56 (2002)
22. Zhang, S., Luo, J., Yang, H.: A POMDP model for content-free document re-ranking. In: SIGIR 2014 (2014)

Retrieving Medical Literature for Clinical Decision Support

Luca Soldaini, Arman Cohan, Andrew Yates,
Nazli Goharian, and Ophir Frieder

Information Retrieval Lab, Georgetown University
{luca,arman,andrew,nazli,ophir}@ir.cs.georgetown.edu

Abstract. Keeping current given the vast volume of medical literature published yearly poses a serious challenge for medical professionals. Thus, interest in systems that aid physicians in making clinical decisions is intensifying. A task of Clinical Decision Support (CDS) systems is retrieving highly relevant medical literature that could help healthcare professionals in formulating diagnoses or determining treatments. This search task is atypical as the queries are medical case reports, which differs in terms of size and structure from queries in other, more common search tasks. We apply query reformulation techniques to address literature search based on case reports. The proposed system achieves a statistically significant improvement over the baseline (29% – 32%) and the state-of-the-art (12% – 59%).

Keywords: medical literature search, medical query reformulation, query expansion, query reduction.

1 Introduction

A Clinical Decision Support (CDS) system is a system designed to assist clinicians in providing patient care by offering timely and actionable health knowledge. One of tasks a CDS system could be designed to solve is the retrieval of key medical literature that can assist the practice of healthcare professionals given a medical case report (an example is shown in Fig. 1). We propose a system that addresses this need, which we refer to as CDS search.

CDS search presents some unique challenges: (*i*) compared to queries in traditional search domains, clinical case reports are substantially longer; (*ii*) although retrieval techniques for long queries have been widely studied in other domains (e.g., legal/patent search), case reports, unlike queries in those instances, have a narrative structure instead of being keyword based; (*iii*) most importantly, CDS search highly favors precision over recall, since healthcare professionals can only afford to spend limited time reading medical literature while practicing [4,16].

Biomedical literature retrieval has been studied in the TREC genomics track[1]. CDS search, while sharing some aspects with it – descriptive queries, domain

[1] http://ir.ohsu.edu/genomics/

A. Hanbury et al. (Eds.): ECIR 2015, LNCS 9022, pp. 538–549, 2015.

A 19-year-old African American student reports that he can "feel his heartbeat". It happens with exercise and is associated with some lightheadedness and shortness of breath. On examination, his heart has a regular rate and rhythm, but you hear a holosystolic murmur along his left sternal border. It increases with Valsalva maneuver.

Fig. 1. Example of a medical case report

specific lexicon – is not limited to the genomics domain, but spans across multiple fields in medicine. Consequently, CDS search systems must process a variety of literature styles written with a wide domain specific vocabulary. Therefore, it is necessary to re-evaluate the effect of known IR techniques for this domain.

In this work we study the impact of query expansion and reduction methods that take advantage of medical domain knowledge, as well as general purpose IR techniques. Finally, we propose an approach that combines such methods, achieving a statistically significant improvement over the baseline (29%-32%) and an over all other approaches (12%-59%), including state-of-the-art.

Currently, no benchmark dataset containing case reports or medical publications can be used to evaluate a CDS search system. Clinical reports from last years' ShARe/CLEF eHealth Evaluation Lab [15,10] are designed to test information extraction systems. OHSUMED [7] provides relevance annotations on medical literature, but its queries are considerably shorter than a case report (6 vs 67.6 terms on average) and are keyword based. NIST's TREC has added a CDS search track to the TREC 2014[2]; however, the system we propose was conceived and tested before the ground truth (q-rels) was publicly released. Thus, we developed an alternative, fully automated experimental framework for evaluating CDS search system based on the practice material for the United States Medical Licensing Examination (USMLE). Such dataset is publicly available[3] to other researchers; the performance obtained by our system on it were found to be comparable to TREC's [14].

In summary, our contributions are: (*i*) a system for retrieving highly relevant, and thus actionable, medical literature in support of clinical practice, (*ii*) an adaptation and evaluation of query reformulation techniques for CDS search, and (*iii*) publicly available experimental framework and benchmark for CDS search.

2 Related Work

Historically, search systems in the medical domain have focused on short and/or keyword-heavy queries. In PubMed, for example, the query is expanded by mapping each term to MeSH terms and then considered as a boolean conjunctive query. Such an approach is ill-suited when considering long, narrative case reports as queries. We approach CDS search as a reformulation problem. Many reduction and expansion approaches have been introduced over the years; here, we give an overview of domain-specific and domain-independent methodologies.

[2] `http://www.trec-cds.org/2014.html`

[3] `https://github.com/Georgetown-IR-Lab/CDS-search-dataset`

Query reduction algorithms have been extensively studied as a way to remove noisy terms from the original query. Their impact has mostly been tested in the web search domain. For example, Kumaran and Carvalho [11] used SVM^{rank} [9] to find the best sub-query using a series of clarity predictors and similarity measures as features. Balasubramanian et al. [3] also studied how to improve performance by reducing queries using quality predictors; however, their system only removes up to one term from the query. This approach is not viable when dealing with long, descriptive case reports. To the best of our knowledge, the only work that has adopted query reduction in the medical domain is by Luo et al. [12]. They built a search engine that performs query reduction by filtering non-important terms based on their tf-idf score. Unlike CDS search, their system is designed for lay people performing health search on the Web and does not focus on medical literature retrieval.

Over the past years, query expansion techniques were successfully employed in medical literature retrieval. Hersh et al. [8] expanded queries with terms manually selected from UMLS Metathesaurus relationships to enhance retrieval performance. Experimental results showed that thesaurus based query expansion did not necessarily improve search efficiency. Yu et al. [17] experimented with relevance feedback in PubMed; their system used RankSVM to re-arrange retrieved results based on explicit users' feedback. Abdou and Savoy [1] used pseudo relevance feedback methods to improve the retrieval of MEDLINE abstracts; their system was tested on manually crafted, keyword based queries substantially shorter than the case reports in our dataset (14 vs. 67.6 terms). In a preliminary version of this work, Cohan et al. [5] explored the use of pseudo relevance feedback for CDS search.

Another line of research related to CDS search is clinical question answering, given the shared goal of improving medical understanding. Demner-Fushman and Lin [6] focused on extracting medical concepts from MEDLINE abstracts that match the information need of the question. Sneiderman et al. [13] examined three knowledge-based methods to evaluate their efficiency in helping clinicians retrieve answers from MEDLINE. In contrast to our work, question answering search systems are designed to handle queries that are much shorter than a case report and are strictly formulated as query. Furthermore, they usually generate an answer rather than returning relevant resources.

3 Methodology

We approached CDS as a query reformulation problem. As such, we capitalized on query reduction (section 3.1) and expansion (section 3.2) techniques. For query reduction, we used a domain specific tool, MetaMap (*MMselect*), and Wikipedia (*HT*), to prune non-medical terms from the query. We also implemented one of the state-of-the-art techniques for domain-agnostic query reduction (*QQP*). Finally, we introduced a refined version of *QQP* that takes advantage of domain specific resources (*Fast QQP*). We then evaluated several query expansion techniques: one (*MMexpand*) takes advantage of a medical thesaurus, another (PRF) uses pseudo relevance feedback to incorporate key terms in the original query. Finally, we introduced a new method (*HT-PRF*) that combines a domain specific approach with pseudo relevance feedback. As shown in section 5, this method outperforms all others, including *QQP* and *Fast QQP* (state-of-the-art and its derivative).

As a baseline, we considered an algorithm that submits the unmodified case report (after removing stopwords) to the search engine.

3.1 Query Reduction Techniques

UMLS Concepts Selection (*MMselect*). We extract concepts from queries based on concepts defined in the Unified Medical Language System[4] (UMLS) to perform query reduction. For this extraction we utilize MetaMap[5], a tool designed for UMLS concept extraction. We reformulated the query by removing all the terms that did not have a mapping to any UMLS concepts.

Health-related Terms Selection (*HT*). Rather than selecting health-related words based on a medical thesaurus, we leverage Wikipedia as an external resource. Specifically, for each word candidate c_l in the original query, we estimate its likelihood of being associated with a health-related Wikipedia entry by computing the odds ratio between the probability of a Wikipedia page P being health-related when $c_l \in P$ over the probability of P not being health-related over all the Wikipedia pages.

$$\mathrm{OR}(c_l) = \frac{\Pr\{P \text{ is health-related} \mid c_l \in P\}}{\Pr\{P \text{ is not health-related} \mid c_l \in P\}} \tag{1}$$

A word $c_l \in \{c_1, \ldots, c_m\}$ is kept as part of the reduced query if $\mathrm{OR}(c_l) \geqslant \delta$, where δ is a tuning parameter.

We used a Wikipedia dump from November 4, 2013 (2,794,145 unique entries). Those pages whose infobox[6] contain one or more of the following medically-related code entries were determined to be health-related: OMIM, eMedicine, MedlinePlus, DiseasesDB and MeSH (24,654 pages); the rest were considered to be not health-related. The optimal value for δ was empirically found to be 2.

Query Quality Predictors for Optimal Sub-query Identification (*QQP*). We implemented the system suggested by Kumaran and Carvalho [11]. Their method uses quality predictors as features to rank sub-queries of the original query using SVM^{rank}. The following predictors are considered as features:

- *Mutual information*: each sub-query is represented as a fully connected weighted graph, where each vertex represents a term in the sub-query. Edges are weighted by mutual information. For each graph, the heaviest spanning tree is extracted; the average weight of the edge is used as query predictor.

[4] http://www.nlm.nih.gov/research/umls/

[5] http://metamap.nlm.nih.gov/

[6] An infobox is template containing structured information that appear on the right of Wikipedia pages to improve concepts representation.

- *Query clarity*: estimation of the divergence of the query model from the collection model using the top 500 documents retrieved per sub-query.
- *Simplified clarity score*: simplified version of clarity score that estimates the probability of a term in the language model by considering the likelihood of it appearing in the query.
- *Query scope*: measure of the size of the retrieved set of documents relative to the size of the collection. Sub-queries showing high query scope are expected to perform poorly since they contain terms that are too broad.
- *Similarity to orginal query*: *tf-idf* similarity is considered as one of the quality predictors under the hypothesis that the closer a sub-query is to the original query, the less likely it is to cause intent drift.

In addition to the previously listed features, *QQP* considers, for each sub-query, statistical measures[7] over the term frequency, document frequency and collection frequency of the terms in the sub-query as features for SVM^{rank}. The length of each sub-query is also considered as a feature. We refer the reader to the original paper for more details.

Since most of the query predictors are query dependent, they cannot be computed ahead of time, thus slowing the sub-query selection process. Therefore, as suggested by the authors, we implemented a set of heuristics to reduce the number of candidate sub-queries, which, prior to pruning, is exponential to the size of the original query: (*i*) select queries with length between three and six terms; (*ii*) select only the top twenty five sub-queries ranked by MI; (*iii*) select only the sub-queries containing name entities. The parameters for SVM^{rank} were set as suggested in [11].

Faster Query Quality Predictors with Medical Features (*Fast QQP*). Since *QQP* was not designed specifically for CDS search, its performance is negatively affected by the greatly reduced length of the generated sub-queries and by the lack of domain-specific features. Because of the unique formulation of case reports, we implemented a set of sub-query candidates pruning heuristics that resulted in statistically significant improvements over the original formulation while reducing the processing time.

First, we increased the maximum length M_{subq} of a sub-query candidate from 6 to 16 terms (empirically determined). This is motivated by the fact that case reports are, on average, much longer than the queries in [11] (16.2 vs. 67.6 terms). The minimum length of a sub-query was not altered (i.e., $m_{\text{sub-q}} = 3$).

As the size of the candidates set grows exponentially when the maximum number of tokens increases linearly, *Fast QQP* prunes the list of candidates after each increase in length of candidate sub-queries. In other words, for each $i \in \{m_{\text{subq}}, \ldots, M_{\text{subq}}\}$, the set of candidates C_i is ranked by MI; the top-k sub-queries are then extracted (set $C_{i,k}$) and used to build the set C_{i+1} accordingly with the following formula:

[7] Maximum and minimum value; arithmetic, harmonic, and geometric mean; standard deviation and coefficient of variation.

$$C_{i+1} = \{s_l \cup \{q_h\} \mid s_l \in C_{i+1} \wedge q_h \in Q\} \cup C_{i,k} \tag{2}$$

where Q is the original query. After empirical evaluation, we set $k = 50$.

We further improved *Fast QQP* by including some domain-specific features:

- number of UMLS concepts in the candidate sub-query,
- semantic type of the UMLS concepts in the candidate sub-query,
- statistical features[7] over the likelihood of each term in the candidate sub-query of being health related, as estimated by equation (1), and
- number of MeSH terms in the candidate sub-query.

3.2 Query Expansion Techniques

UMLS Concepts Extraction (*MMexpand*). Similar to MM Select method, this method identifies UMLS Metathesaurus concepts that exist in the query using MetaMap. However, rather than filtering out terms, this method expands the query using new terms associated with the concepts identified. After detecting the concepts in the query, expansion terms were chosen by querying UMLS for new terms that were synonyms of the concepts in the query and were marked as preferred terms by UMLS; the query was expanded with all these terms. Given the extensive coverage of UMLS, we limited concept expansion to concepts containing drugs, diseases, and findings to prevent query drift.

Pseudo Relevance Feedback (*PRF*). Pseudo relevance feedback was modeled after the "IDF Query Expansion" method proposed in [1]. We modified the algorithm to adapt it to our experimental setup: instead of directly altering term weights, our system determines a boosting coefficient for each term in the reformulated query. The query Q is expanded as follows: it tokenizes the top k retrieved documents retrieved for Q; it then builds the root set $\mathcal{R}_Q$, which consists of the union of the set containing all the terms in Q with the set of all the terms in the retrieved documents for Q. The boost coefficient b_j for each term $t_j \in \mathcal{R}_Q$ is calculated as:

$$\begin{aligned} b_j &= \log_{10}(10 + w_j) \\ w_j &= \alpha \cdot \mathrm{I}_Q(t_j) \cdot tf_j + \beta/k \sum_{i=1}^{k} \mathrm{I}_{D_i}(t_j) \cdot idf_j \end{aligned} \tag{3}$$

where t_j is the j-th term in the top Q documents, $\mathrm{I}_Q(t_j)$ is an indicator of the presence of term t_j in Q, $\mathrm{I}_{D_i}(t_j)$ is an indicator of the presence of term t_j in the document D_i, idf_j is the inverse document frequency of the j-th term in the top k documents. Finally, α and β are smoothing factors.

Once all the weights have been determined, the terms in $\mathcal{R}_Q$ are ranked by their boost coefficient; the top m terms not in the original query are added to Q; each term in the reformulated query is boosted by its boosting factor. Tuning parameters were set as suggested in [1]: $\alpha = 2$, $\beta = .75$, $k = 10$, $m = 20$.

Health Terms Pseudo Relevance Feedback (*HT-PRF*). We explored the effect of combining a pure IR approach – pseudo relevance feedback – with domain specific knowledge (health terms). *HT-PRF* operates similarly to *PRF*– it retrieves the top k documents, builds the root set $\mathcal{R}_Q$ of the query, scores each term in the root set using the equation (3) – but instead of always expanding with top m candidates, it calculates, for each term, the odds of it being health related using equation (1), retaining only those whose odds ratio is greater or equal to δ', where δ' is a tuning parameter of the system. Because of this, the number of terms m'_q added to each query varies.

Finally, we would like to stress the fact that, despite taking advantage of *HT*, *HT-PRF* is not a reduction method: non-health specific terms are only pruned off the list of candidates for query expansion; the original query is left untouched.

4 Experimental Setup

As stated in the introduction, the lack of datasets designed to evaluate a CDS search system required us to create our own. To create a benchmark for evaluation, we developed an approach to automatically identify relevant documents to case reports by making use of external information about each case report (the correct diagnosis, treatment or test associated with each one as well as explanations about the correctness of such relations). Our dataset contains two components: medical papers and medical case reports. The medical literature was obtained from Open Access Subset of PubMed central[8], a free full-text archive of health journals (728,455 documents retrieved January 1, 2014).

495 medical case reports were obtained from three USMLE preparation books[9] Each case report contains a description of a patient followed by a question asking for the correct diagnosis, treatment, or test that should be executed. Case reports from USMLE are modeled after real clinical situations with goal of assessing the ability of future physicians in applying clinical knowledge, concepts and principles for effective patient care[10].

Given a case report, our goal is to retrieve documents (medical publications) that can help a physician diagnose the patient, treat the patient's condition, or request a test relevant to the case; the content of three USMLE prep books were used to determine which documents in our collection were relevant. In detail, we took advantage of the multiple answer choices associated with the case reports as well as the explanation of why an answer is correct. To determine relevant documents for each case report, we separately issued as queries the explanation paragraph (q_{E}) and each answer choice individually ($q_{a_0}, \ldots, q_{a_3}$). Documents retrieved by the correct answer $q_{a_{\mathrm{corr}}}$ and q_{E} received a relevance score of two, while documents retrieved by q_{E} and any incorrect answer choice were given a score of one. By using this approach, we were able to take into account that not

[8] `http://www.ncbi.nlm.nih.gov/pmc/tools/openftlist`

[9] `https://github.com/Georgetown-IR-Lab/CDS-search-dataset`

[10] Bulletin of Information, `http://www.usmle.org/pdfs/bulletin/2012bulletin.pdf`

only the correct documents retrieved by querying the correct answer contribute to determine the right treatment/test/diagnosis, but also those related to the incorrect options. Any answer choice query ($q_{a_i}, \in \{0, \ldots, 3\}$) that contained more than 200 documents was discarded under the assumption that the query was too broad. A case report was discarded if its correct answer choice query was discarded. This process left us with 195 valid queries (i.e., case reports).

Three human assessors were then instructed to read each of these case reports and determine their validity. Specifically, they were asked to categorize each one as invalid or as asking for a diagnosis, treatment, or test. Invalid queries were those that were primarily quantitative (i.e., contained only numeric values about some tests or vital signs e.g. blood pressure, heart rate, body temperature, etc). The three assessors' inter-rater agreement was 0.56 as measured by Fleiss' kappa[11]. Any query deemed invalid by at least two assessors was discarded. This left us with 85 case reports; of those, 17 were reserved for parameters tuning, while the remaining 68 were used for testing.

We used ElasticSearch v1.2.1, a search server built on top of Lucene v4, to index the medical documents in our dataset and to retrieve results. The default tokenizer and the divergence from randomness retrieval model [2] were used.

5 Results and Discussion

We validate our query reformulation approach for CDS search by running two experiments. First, we compare the performance of each method introduced in section 3; second, we describe the tuning process for the best performing method. In both experiments, we retrieve 1000 documents for each test query.

5.1 Comparison of Reformulation Methods

As previously mentioned, CDS search is a precision oriented task; it is meant to support healthcare professionals who are looking for findings that could help them determine the next action in the care of a patient. For this reason, performance at the first ten points of precision (Fig. 2) is key to assert the quality of a reformulation method. We focus our analysis on precision at five documents retrieved (P@5), as the performance of each method is consistent throughout the first ten points (Fig. 2, left) of precision and show no significant difference up to P@100 (Fig. 2, right). Recall and nDCG are also reported (Table 1); these metrics, albeit less key to the task, are still useful indicators to assert the overall quality of each method. We use a paired Student's t-test to measure whether the difference between any two methods is statistically significant ($p < 0.01$).

MMselect performed significantly worse than the baseline. We attribute such difference to the fact that, while it successfully identifies most medical concepts in the query, it often discards terms that have a key role connecting domain

[11] The moderate level of agreement between assessors is attributable to the hardness of the task. The evaluators reported that many reports laid in the spectrum between fully quantitative and fully qualitative, thus representing a noteworthy challenge.

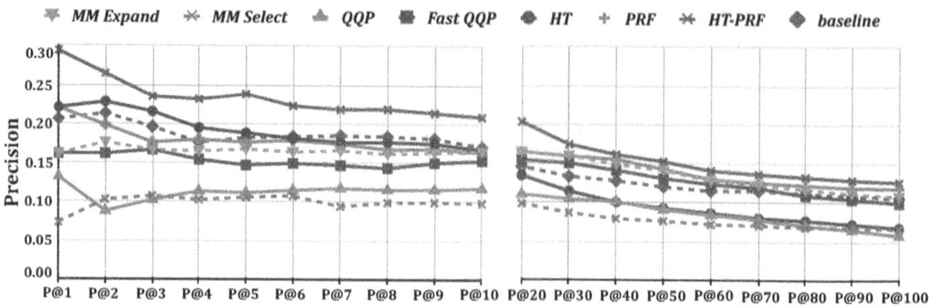

Fig. 2. Points of precision for each method. The best performing method, *HT-PRF*, achieves a 43% increase over the baseline for P@1.

specific expression. For example, for the case report in Fig. 1, *MMselect* fails to identify "increases" as relevant term (last sentence), which is key in understanding the outcome of the "Valsalva maneuver" on the patient. *MMexpand* showed a minor but significant gain in terms of nDCG and recall over the baseline, but it performed worse (although not significantly) than the baseline in terms of P@5. We attribute the modest difference to the limited coverage of the portion of the synonym map in UMLS *MMexpand* uses with respect to the size of our dataset. This tradeoff was necessary to prevent query drift.

QQP performed very poorly. Its limited performance is due to its aggressive reduction algorithm, which reduces the original query to at most six terms. As result, the reduced query loses most of the information content of the case report.

Fast QQP showed substantially better nDCG and recall results, but fell short in terms of P@5. We attribute the improvement to the fact that the inclusion of domain specific features and a more conservative approach lead to a more effective reduction. On the other hand, the worsening in terms of P@5 is likely due to the insufficient coverage of medical terms in the query: in medical literature, the same concept is often expressed using different terms and expression; thus a method that only performs reduction is likely to miss documents that are relevant to the case report, but differ from it in terms of vocabulary.

Table 1. Each method's performance (∘ for query reduction, • for expansion). A △/▽ indicate a significant improvement/worsening ($p < 0.01$) over the baseline. ▲ indicates a significant improvement over Simple and methods marked with △.

	nDCG		Recall		P@5	
baseline	0.2855	–	0.2741	–	0.1824	–
MMselect ∘	0.1622▽	(−43.2%)	0.1486▽	(−45.8%)	0.1059▽	(−41.9%)
MMexpand •	0.3020△	(+5.8%)	0.2958△	(+7.9%)	0.1676	(−8.1%)
QQP ∘	0.2557▽	(−10.4%)	0.2494▽	(−9.0%)	0.1118▽	(−38.7%)
Fast QQP ∘	0.3177△	(+11.3%)	0.3129△	(+14.2%)	0.1471▽	(−19.4%)
HT ∘	0.3328△	(+16.5%)	0.3262△	(+19.0%)	0.1882	(+3.2%)
PRF •	0.3390△	(+16.5%)	0.3263△	(+19.0%)	0.1765	(−3.4%)
HT-PRF •	**0.3768▲**	**(+32.0%)**	**0.3520▲**	**(+28.9%)**	**0.2382▲**	**(+30.5%)**

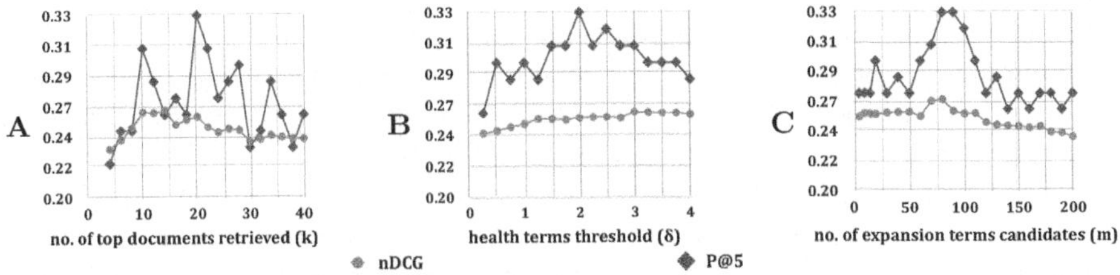

Fig. 3. Effect of different parameter values for *HT-PRF* in terms of nDCG and P@5 (other precision levels exhibit similar behavior). The best precision performances are achieved when $k = 20$, $\delta' = 2$, $m = 90$.

Both *HT* and *PRF* methods showed a statistically significant improvement over the baseline in terms of nDCG and recall; *HT* removes common non-health-related terms, whereas *PRF* reweights the entire query, increasing the importance of health-related terms, which naturally have a high IDFQE coefficient given the domain of the dataset. In *HT* some improvement is expected, as it keeps more generalized medical concepts in comparison with the UMLS concept selection method. Neither *HT* nor *PRF* showed significant improvement in terms of P@5. *HT* is likely to suffer from the same limitation in terms of vocabulary coverage *Fast QQP* has, while *PRF* is partially affected by query drift.

We achieved the most noteworthy results by using the *HT-PRF*. The nDCG and recall values shown in Table 1 are statistically significant not only with respect to the baseline but also over simple *PRF* and *HT*. Moreover, *HT-PRF* consistently improves over the baseline for each precision level shown in Fig. 2 ($p < 0.01$). The substantial increase in performances of *HT-PRF* is due to the fact that it combines two very effective techniques: by expanding the query using the most relevant document, it is able to broad its vocabulary; on the other side, filtering the list of candidate terms for expansion prevents query drifting.

5.2 Parameter Tuning for *HT-PRF*

In this section we detail the tuning process for *HT-PRF*. We studied the outcome of varying the number k of the top ranking documents used by pseudo relevance feedback to build the list of candidate terms for query expansion (Fig. 3A), the value δ' of the conditional probability threshold used to select expansion terms from the list of candidate terms (Fig. 3B), as well as the number m of candidate terms for query expansion (Fig. 3C).

The results we present were obtained on a subset of 17 separate case reports we reserved for tuning purposes. For all three tuning parameters, we preferred those values that yielded better performance in terms of P@5. As in section 5.1, we chosen to report the performances in terms of P@5, as we observed comparable behavior at all the other precision levels between one and ten (the differences between methods are not statistically significant after ten results).

Fig. 3A shows that the highest performance in terms of P@5 is obtained when the number of top documents k is equal to 20. However, we also noticed an ample variation in terms of P@5 for small differences in the number of retrieved documents. This variation clearly depends on which terms are used to expand the original query. Since the terms picked for expansion are the most representative terms of the top k documents retrieved, their effectiveness in improving the retrieval performance depends on whether the top k documents are relevant or not. Given the fact that the top document set is small, each time a new document is added (i.e. k increases) the set of terms picked for expansion varies substantially. In other words, when a non-relevant document is added to the set of top documents, the relevance of the terms selected for expansion decreases, thus causing query drift. Similarly, when a relevant document is included in the top k documents, the relevance of terms selected for expansion increases, leading to better performance. Nevertheless, we observed that the retrieval performance decreases as the number of top documents increases past 20. This outcome is expected, since the more documents the system considers, the more likely it is to suffer from query drift, as less relevant terms are picked for expansion.

With health terms' threshold (Fig. 3B) we noticed a much more defined trend: the best precision is achieved when $\delta' = 2$. The bigger δ' is, the more aggressive the filter is. And for higher values of δ', precision starts to decrease. That is, because bigger values of δ' result in selection of more focused and specific medical terms, many more general key terms for optimal retrieval are being discarded. In fact, the lower performance of thesaurus based methods further reveals the fact that considering only highly focused medical terms decreases P@5. On the other side, when δ' is smaller the method is more likely to consider all sorts of terms for query expansion, which eventually results in query drift.

Finally, we recorded the best retrieval performance when the number of candidates for expansion m is set to 90 (Fig. 3C). Different values of m tend to cause query drift when they are larger than the optimal and cause key terms to be removed from the when they are smaller than the optimal.

6 Conclusions

We described CDS search based on medical case reports, which is a search task intended to help medical practitioners retrieve relevant publications to clinical case reports. We used query reformulation to perform CDS search, and found that the best methods for this task are a query reduction method retaining only health-related terms and a pseudo relevance feedback query expansion method. Both methods independently improved performance significantly (as measured by nDCG and recall), yet showed limited improvements in terms of precision. However, when combined, the resulting method outperformed each individual method and greatly improved precision. We conclude that while this method decisively improved retrieval performance, there is still room for improvement; this stresses that CDS search is significantly different than other types of health-related search, making it a novel search task worthy of further study.

Acknowledgments. This work was partially supported by the US National Science Foundation through grant CNS-1204347.

References

1. Abdou, S., Savoy, J.: Searching in medline: Query expansion and manual indexing evaluation. Information Processing & Management (2008)
2. Amati, G., Van Rijsbergen, C.J.: Probabilistic models of information retrieval based on measuring the divergence from randomness. In: ACM Transactions on Information Systems (TOIS) (2002)
3. Balasubramanian, N., Kumaran, G., Carvalho, V.R.: Exploring reductions for long web queries. In: Proceedings of the 33rd International ACM SIGIR Conference on Research and Development in Information Retrieval. ACM (2010)
4. Burke, D.T., DeVito, M.C., Schneider, J.C., Julien, S., Judelson, A.L.: Reading habits of physical medicine and rehabilitation resident physicians. American Journal of Physical Medicine & Rehabilitation (2004)
5. Cohan, A., Soldaini, L., Yates, A., Goharian, N., Frieder, O.: On clinical decision support. In: Proceedings of the 5th ACM Conference on Bioinformatics, Computational Biology, and Health Informatics, pp. 651–652. ACM (2014)
6. Demner-Fushman, D., Lin, J.: Answering clinical questions with knowledge-based and statistical techniques. Computational Linguistics (2007)
7. Hersh, W., Buckley, C., Leone, T., Hickam, D.: Ohsumed: An interactive retrieval evaluation and new large test collection for research. In: Clough, P., Foley, C., Gurrin, C., Jones, G.J.F., Kraaij, W., Lee, H., Mudoch, V. (eds.) ECIR 2011. LNCS, vol. 6611, pp. 251–263. Springer, Heidelberg (2011)
8. Hersh, W., Price, S., Donohoe, L.: Assessing thesaurus-based query expansion using the umls metathesaurus. In: Proceedings of the AMIA Symposium. American Medical Informatics Association (2000)
9. Joachims, T.: Training linear svms in linear time. In: Proceedings of the 12th SIGKDD International Conference on Knowledge Discovery and Data Mining. ACM (2006)
10. Kelly, L., et al.: Overview of the shARe/CLEF eHealth evaluation lab 2014. In: Kanoulas, E., Lupu, M., Clough, P., Sanderson, M., Hall, M., Hanbury, A., Toms, E. (eds.) CLEF 2014. LNCS, vol. 8685, pp. 172–191. Springer, Heidelberg (2014)
11. Kumaran, G., Carvalho, V.R.: Reducing long queries using query quality predictors. In: Proceedings of the 32nd International ACM SIGIR Conference on Research and Development in Information Retrieval. ACM (2009)
12. Luo, G., Tang, C., Yang, H., Wei, X.: Medsearch: A specialized search engine for medical information retrieval. In: Proceedings of the 17th ACM Conference on Information and Knowledge Management. ACM (2008)
13. Sneiderman, C.A., Demner-Fushman, D., Fiszman, M., Ide, N.C., Rindflesch, T.C.: Knowledge-based methods to help clinicians find answers in medline. Journal of the American Medical Informatics Association (2007)
14. Soldaini, L., Cohan, A., Yates, A., Goharian, N., Frieder, O.: Query reformulation for clinical decision support search. In: The Twenty-Third Text REtrieval Conference Proceedings (TREC 2014) (2015)
15. Suominen, H., et al.: Overview of the shARe/CLEF eHealth evaluation lab 2013. In: Forner, P., Müller, H., Paredes, R., Rosso, P., Stein, B. (eds.) CLEF 2013. LNCS, vol. 8138, pp. 212–231. Springer, Heidelberg (2013)
16. Tenopir, C., King, D.W., Clarke, M.T., Na, K., Zhou, X.: Reading patterns and preferences of pediatricians. Journal of the Medical Library Association (2007)
17. Yu, H., Kim, T., Oh, J., Ko, I., Kim, S.: Refmed: relevance feedback retrieval system fo pubmed. In: Proceedings of the 18th ACM Conference on Information and Knowledge Management. ACM (2009)

PatNet: A Lexical Database for the Patent Domain

Wolfgang Tannebaum and Andreas Rauber

Institute of Software Technology and Interactive Systems,
Vienna University of Technology, Austria
{tannebaum,rauber}@ifs.tuwien.ac.at
http://www.ifs.tuwien.ac.at

Abstract. In the patent domain Boolean retrieval is particularly common. But despite the importance of Boolean retrieval, there is not much work in current research assisting patent experts in formulating such queries. Currently, these approaches are mostly limited to the usage of standard dictionaries, such as *WordNet*, to provide synonymous expansion terms. In this paper we present a new approach to support patent searchers in the query generation process. We extract a lexical database, which we call *PatNet*, from real query sessions of patent examiners of the United Patent and Trademark Office (USPTO). *PatNet* provides several types of synonym relations. Further, we apply several query term expansion strategies to improve the precision measures of *PatNet* in suggesting expansion terms. Experiments based on real query sessions of patent examiners show a drastic increase in precision, when considering support of the synonym relations, US patent classes, and word senses.

Keywords: Patent searching, Query term expansion, Query log analysis.

1 Introduction

In the patent domain Boolean retrieval is particularly common. Virtually all search systems of the patent offices and commercial operators process Boolean queries. This is not because this kind of retrieval is the most effective one. Rather, Boolean queries are easy for patent experts to manipulate and they provide a record of what documents were searched [3]. But despite the importance of Boolean retrieval in patent searching, as shown in [8], there is not much work in current research assisting patent experts in formulating such queries, preferable via automatic query term expansion.

In this paper we present a new approach to support patent searchers in the query generation process. We extract a lexical database, which we call *PatNet*, from real query sessions of patent examiners of the USPTO. First, we review related work on automatic query term expansion in patent searching. We then describe the approaches to detect several types of synonym relations in the query logs. Following we present the lexical database *PatNet*. Finally, we provide the experiments to improve the precision measures of *PatNet* followed by conclusions and an outlook on future work.

Hanbury et al. (Eds.): ECIR 2015, LNCS 9022, pp. 550–555, 2015.

2 Related Work

Related approaches to enhance query term expansion in patent searching are mostly limited to computing co-occurring terms in a patent corpus for query expansion, while patent searchers predominately use synonyms and equivalents for query term expansion [1,5]. An analysis of real query sessions of patent examiners has shown that about 60% of the used expansion terms *(ETs)* are synonyms and equivalents [8]. Further, [9] shows that the highly specific vocabulary used in the patent domain is not included in standard dictionaries, such as *WordNet.* Patent examiners use the terms created by the patent applicants, such as "*pocketpc*" for "*notebook*", "*watergas*" for "*steam*", or "*passcode*" for "*password*" for synonym expansion. Hence, the challenge is to learn the synonyms directly from the patent domain to assist patent searchers in formulating Boolean queries. An approach to extract synonyms directly from patent documents is presented in [5]. Claim sections of granted patent documents from the European Patent Office including the claims in English, German and French are aligned to extract translation relations for each language pair. Based on the language pairs having the same translation terms, synonyms are learned in English, French and German. Contrary to the extraction of the synonyms from patents, as indicated in [5], we propose to extract them from query logs as presented in [7] and in particular from query logs of patent examiners as suggested in [9]. This allows us to extract specific terms, in particular the query and expansion terms to the patent applications.

3 Extracting Synonyms from Query Logs of Patent Examiners

For our experiments we downloaded and preprocessed 103,896 query log files of USPTO patent examiners from Google as mentioned in [9].[1] We kept 7,500 log files as a hold-out set for evaluation and used 96,396 files for the following experiments.

In [9] the Boolean Operator "OR", which indicates that two query terms are synonyms or can at least be considered as equivalents, was used for detecting synonyms (single term relations) in the text queries. Expanding the approach, we now use the proximity operator "ADJ" to detect keyword phrases and the Boolean operator "OR" to learn synonyms thereto. Table 1 shows several types of synonym relations provided by the search operators "OR" and "ADJ" and for each type of relation an example.

Table 1. Synonym Relations provided by the Search Operators "OR" and "ADJ"

Type	Definition	Example
single term	term OR term	drill OR burr
single term to phrase	(term ADJ term) OR term	(digital ADJ assistant) OR blackberry
	term OR (term ADJ term)	transponder OR (data ADJ carrier)
phrase to phrase	term ADJ (term OR term)	force ADJ (sensor OR detector)
	(term OR term) ADJ term	(control OR instrument) ADJ panel
	(term ADJ term) OR (term ADJ term)	(duty ADJ cycle) OR (band ADJ width)

[1] http://www.google.com/googlebooks/uspto-patents.html

The process to detect single term relations works as follows: We filter all 3-grams generated from the text queries in the form "X *b* Y", where *b* is the Boolean operator "OR" and X and Y are query terms. To exclude mismatches and misspellings, we consider those 3-grams that were encountered at least three times. To detect single term to phrase and phrase to phrase relations, we filter all 5-grams generated from the text queries in the form "X *b* Y *p* Z" and " X *p* Y *b* Z", and all 7-grams in the form "X *p* Y *b* Z *p* W", where X, Y, Z and W are query terms, *p* the proximity operator "ADJ" and *b* the Boolean operator "OR". To exclude mismatches, we consider the correctly set parentheses. Table 2 shows the detected synonym relation frequencies.

Table 2. Detected Synonyms based on the Search Operators

Type of Relation	Code	#Relations	#Terms
single term	STR	27,798	17,105
single term to phrase	STPR	628	928
phrase to phrase	PPR	409	701
Σ	-	**28,835**	**17,643**

In addition, we learned that patent examiners may also rely on a default operator, which can be set to "OR" or "AND". This is indicated by the default operator element in the query logs. To detect these synonyms, we use all text queries where the default operator is set on "OR" and the approach to detect synonyms as mentioned above, but we excluded the "OR" operator in the 3-, 5- and 7-grams. We obtained 1,871 single term relations, 394 single term to phrase, and 165 phrase to phrase relations.

4 *PatNet*: A Lexical Database

Based on the detected synonym relations, we learn in this section a lexical database for the patent domain, which we call *PatNet*. The lexical database resembles a thesaurus of English concepts that can be used for semi-automatic query term expansion. To query the lexical database we use the open source thesaurus management software *TheW32* [2].

Table 3. Synonym Relations provided by *PatNet*

Type of Relation	Code	#Relations	#Terms
single term	STR	29,477	18,804
single term to phrase	STPR	920	1,523
phrase to phrase	PPR	530	984
Σ	-	**30,927**	**19,040**

As shown in Table 3, *PatNet* provides 30,927 unique synonym relations and 19,040 unique query terms in total. *PatNet* suggests to a single query term: (1) single synonym terms, (2) synonym phrases, and (3) single terms, which in combination with the query term constitute a keyword phrase and finally suggests a synonym phrase.

Table 4. Suggested *STR*, *STPR* and *PPR* for the query term "*voice*"

Term	Type of Relation			
	STR	***STPR***		***PPR***
voice	acoustic	voice exchange	voice mail	machine mail
	audio	voice mail	voice print	speech recognition
	sound	voice message	voice sample	speech sample
	speak	voice print	-	-
	speech	voice response	-	-
	telephony	voice sample	-	-
	verbal	-	-	-

Table 4 shows the provided *ETs* for the term "*voice*". *PatNet* suggests single terms (*STR*), keyword phrases (*STPR*), and single terms, which in combination with the query term constitute a keyword phrase and finally suggests synonym phrases (*PPR*).

5 Experiments

In this section we apply several query term expansion strategies to suggest *ETs* in a useful order to avoid time-consuming term selection. For the single terms *PatNet* provides, on average, 11 *ETs*. But the maximum number rise up to 92 terms, for common terms, such as "*sensor*". For the experiments we use the test set from Subsection 3.1. and measure the performance of *PatNet* based on real query sessions of patent examiners (gold standard), because (1) benchmark data sets with synonym relations are not available for the patent domain and (2) the performance of thesauri in *IR* depends on contextual factors, as shown [4].

At first, we rank the synonym relations of *PatNet* according to their support in the training set and carry out five expansion steps ($Step_1$ *to* $Step_5$) which is a realistic value in real query sessions. We start with the top-5 *ETs* (having the highest ranking r_1) in $Step_1$ followed by additional *ETs* based on the rankings r_2 to r_5 in $Step_2$ *to* $Step_5$. For each expansion step we calculate recall (we compare the suggested *ETs* from *PatNet* with the synonyms used by the examiners in the test set) and precision (we compare the synonyms used by the examiners with all *ETs* suggested by *PatNet*). For recall we consider the obtained scores of the previous expansion steps.

Table 5. Recall and Precision achieved when successively suggesting the highest ranked *ETs*

Expansion Step	*Ranking*	*Positions*	**Recall**	**Precision**
$Step_1$	r_1	*1 – 5*	**38.46**	23.10
$Step_2$	r_2	*6 – 10*	48.72	**24.81**
$Step_3$	r_3	*11 – 15*	55.38	22.31
$Step_4$	r_4	*16 – 20*	58.38	20.45
$Step_5$	r_5	*21 - 25*	**62.54**	**20.00**

As shown in Table 5, in $Step_1$ *to* $Step_5$, on average, 1 out of 5 terms that are suggested by *PatNet* as synonyms were used by the examiners for query expansion (on average

22% precision). Further, after $Step_2$ *PatNet* already provides almost half of the *ETs* used (49% recall). Compared to suggesting all possible *ETs* in one single step (on average 70% recall and 5% precision), there is a drastic increase in precision (up to 25%) and only a minor decrease in recall (63%).

Next, we consider specific and related US patent classes, as presented in [9], to suggest *ETs* in a certain context (patent class). In addition, we use the idea behind Relevance Feedback *RF* to take the *ETs* that are initially suggested for a *QT* and to use information about whether or not those are relevant to perform a new expansion step. At first, we consider the US patent classes of the *QTs* and expand the terms with class-specific *ETs* ($Step_1$). Then, we expand the relevant *ETs* from $Step_1$ with further *ETs* appearing in related classes ($Step_2$). Finally, we expand the relevant *ETs* from $Step_2$ with additional *ETs* from all other classes ($Step_3$).

Table 6. Recall and Precision achieved when using intersections between US patent classes

Expansion Step	*Expansion Terms*	**Recall**	**Precision**
$Step_1$	*class-specific*	**49.38**	**18.50**
$Step_2$	*class-related*	50.86	17.37
$Step_3$	*class-independent*	**54.99**	**12.21**

Table 6 shows that after $Step_1$ almost half of the used *ETs* are provided by the class-specific *ETs* with best precision (19%). In $Step_2$, the recall measure could be further improved, while we notice only a minor decrease in precision (17%). In $Step_3$ precision fall to 12% and recall rises to 55%. In light of suggesting all possible *ETs* in one step, there is a significant increase in precision, but also a major decrease in recall.

Finally, we perform word sense disambiguation (*WSD)* to suggest the most suitable *ETs*. We determine the sense of an *ET* based on the overlap of the sense definitions of the target word, as mentioned in [6]. We consider the *QTs*, which appear before the *STR* in the training and test set (reflecting real query expansion scenarios, where information from past queries can be used). We use a context size of n = 20 words. We rank the *ETs* according the number of common words (highest overlap) and initially suggest the highest ranked *ETs* followed by additional ones.

Table 7. Recall and Precision achieved when using *WSD*

Expansion Step	*Ranking*	*Overlap*	**Recall**	**Precision**
$Step_1$	r_1	≥ 5	**6.06**	**44.44**
$Step_2$	r_2	4	9.09	37.50
$Step_3$	r_3	3	12.12	36.36
$Step_4$	r_4	2	18.18	19.35
$Step_5$	r_5	1	**30.30**	**11.24**

As shown in Table 7, compared to the expansion strategies applied before, there is a further increase in precision (up to 44% in $Step_1$). But now also a decrease in recall has to be noticed. Recall measures already decrease from 70% to 30%, when considering only one common term in the context words. Further experiments show that also a considerable decrease in recall has to be noticed (from 70% to 56%), when using a context size of 50 terms, while now the precision scores, on average, rise up to 20%.

6 Conclusions and Future Work

In this paper we presented a new approach to support patent experts in formulating Boolean queries. We used real query expansion sessions of patent examiners to learn the lexical database *PatNet*. We have shown that *PatNet* can be used to support patent searchers in the time-consuming query generation process. Experiments showed that the achieved precision scores significantly exceed the scores achieved in related work for patent searching and are comparable to numbers reported for professional academic search [3,9,10]. Specifically, we notice only a minor decrease in recall, when considering support of the extracted relations and successively suggesting the highest ranked *ETs* (while precision increases). In future work we want to evaluate *PatNet* based on the relevant documents cited by the patent examiners in their search reports to measure the performance of our query expansion approach in document retrieval.

References

1. Andersson, L., Mahdabi, P., Hanbury, A., Rauber, A.: Exploring patent passage retrieval using nouns phrases. In: Serdyukov, P., Braslavski, P., Kuznetsov, S.O., Kamps, J., Rüger, S., Agichtein, E., Segalovich, I., Yilmaz, E. (eds.) ECIR 2013. LNCS, vol. 7814, pp. 676–679. Springer, Heidelberg (2013)
2. De Vorsey, K., Elson, C., Gregorev, N., Hansen, J.: The Development of a local thesaurus to improve access to the anthropological collections of the American Museum of Natural History. D-Lib Magazine 12(4) (2006)
3. Kim, Y., Seo, J., Croft, W.B.: Automatic Boolean query suggestion for professional search. In: Proceedings of the 34th International ACM SIGIR Conference on Research and Development in Information Retrieval (SIGIR 2011), Beijing, China, pp. 825–834 (2011)
4. Kless, D., Milton, S.: Towards Quality Measures for Evaluating Thesauri. In: Sánchez-Alonso, S., Athanasiadis, I.N. (eds.) MTSR 2010. CCIS, vol. 108, pp. 312–319. Springer, Heidelberg (2010)
5. Magdy, W., Jones, G.J.F.: A Study of Query Expansion Methods for Patent Retrieval. In: Proceedings of PaIR 2011, Glasgow, Scotland, pp. 19–24 (2011)
6. Navigli, R.: Word sense disambiguation: A survey. ACM Computing Surveys (CSUR) 41(2), Article 10 (2009)
7. Silvestri, F.: Mining Query Logs: Turning Search Usage Data into Knowledge. Foundations and Trends in Information Retrieval 4(1-2), 1–174 (2010)
8. Tannebaum, W., Rauber, A.: Mining Query Logs of USPTO Patent Examiners. In: Forner, P., Müller, H., Paredes, R., Rosso, P., Stein, B. (eds.) CLEF 2013. LNCS, vol. 8138, pp. 136–142. Springer, Heidelberg (2013)
9. Tannebaum, W., Rauber, A.: Using Query Logs of USPTO Patent examiners for automatic Query Expansion in Patent Searching. Information Retrieval 17(5-6), 452–470 (2014)
10. Verberne, S., Sappelli, M., Kraaij, W.: Query Term Suggestion in Academic Search. In: de Rijke, M., Kenter, T., de Vries, A.P., Zhai, C., de Jong, F., Radinsky, K., Hofmann, K. (eds.) ECIR 2014. LNCS, vol. 8416, pp. 560–566. Springer, Heidelberg (2014)

Learning to Rank Aggregated Answers for Crossword Puzzles

Massimo Nicosia[1,2], Gianni Barlacchi[2], and Alessandro Moschitti[1,2]

[1] Qatar Computing Research Institute
[2] University of Trento
{m.nicosia,gianni.barlacchi}@gmail.com, amoschitti@qf.org.qa

Abstract. In this paper, we study methods for improving the quality of automatic extraction of answer candidates for automatic resolution of crossword puzzles (CPs), which we set as a new IR task. Since automatic systems use databases containing previously solved CPs, we define a new effective approach consisting in querying the database (DB) with a search engine for clues that are similar to the target one. We rerank the obtained clue list using state-of-the-art methods and go beyond them by defining new learning to rank approaches for aggregating similar clues associated with the same answer.

1 Introduction

CPs are among the most popular language games. Automatic solvers mainly use AI techniques for filling the puzzle grid with candidate answers. The basic approach is to optimize the overall probability of correctly filling the grid by exploiting the likelihood of each candidate answer, fulfilling the grid constraints. Previous work [4] clearly suggests that providing the solver with an accurate list of answer candidates is vital. These can be (i) partially retrieved from the Web and (ii) most importantly they can be recuperated from a DB of previously solved CPs (CPDB). The latter contains clues from previous CPs, which are often reused: querying CPDB with the target clue may allow for recuperating the same (or similar) clues. It is interesting to note that all previous automatic CP solvers use standard DB techniques, e.g., SQL Full-Text query, for querying CPDBs. In [2], we showed that IR techniques can improve clue retrieval but our approach was limited on providing better ranking of clues whereas CP solvers require the extraction of the answer. In other words, given the list of similar clues retrieved by an IR system, a clue aggregation step and a further reranking process is needed to provide the list of answer candidates to the solver. More specifically, each clue c_i in the rank is associated with an answer a_{c_i}. A typical approach to select or rerank answers is to consider c_i as a vote for a_{c_i}. However, this is subject to the important problem that clues are relevant to the query with different probability. Trivially, a clue very low in the rank is less reliable than clues in the first position. One solution may use the score provided by the learning to rank algorithm (LTR) as a vote weight but as we show, its value is not uniformly distributed with respect to the probability of the correctness of c_i: this makes voting strategies less effective. In this paper, we study and propose different techniques for answer aggregation and reranking with the aim of solving the

A. Hanbury et al. (Eds.): ECIR 2015, LNCS 9022, pp. 556–561, 2015.

problem above. First of all, we apply logistic regression (LGR) to the scores produced by LTR algorithms for transforming them into probabilities. This way, we can apply a voting approach with calibrated probabilities, which improves on previous work. Secondly, we propose an innovative machine learning model for learning to combine the information that each c_i bring to their a_{c_i}: we define a representation of each a_{c_i} based on aggregate features extracted from c_i, e.g., their average, maximum and minimum reranking score. We experiment with this new answer representation with both LGR as well as SVM^{rank} [7]. Thirdly, another important contribution is the construction of the dataset for clue retrieval: it is constituted by 2,131,034 of clues and associated answers. This dataset is an interesting resource that we made available to the research community. Eventually, using the above dataset, we carried out two sets of experiments on two main tasks: (i) clue reranking, which focuses on improving the rank of clues c_i retrieved for a query; and (ii) answer reranking, which targets the list of a_{c_i}, i.e., their aggregated clues. The results of our experiments with the above dataset demonstrate that (i) standard IR greatly improves on DB methods for clue reranking, i.e., BM25 improves on SQL query by 6 absolute percent points; (ii) kernel-based rerankers using several feature sets, improves SQL by more than 15 absolute percent points; and (iii) using our answer aggregation reranking methods, the improvement on Recall (Precision) at rank 1, increases by additional 2 points absolute over the best results.

2 Related Work

There have been many attempts to build automatic CP solving systems. Their goal is to outperform human players in solving crosswords, more accurately and in less time. Knowledge about previous CPs is essential for solving new ones as clues often repeat in different CPs. Thus, all systems contain at least a module for clue retrieval from CPDBs. Proverb [8] was the first system for automatic resolution of CPs. It includes several modules for generating lists of candidate answers. These lists are merged and used to solve a Probabilistic-Constraint Satisfaction Problem. Proverb relies on a very large crossword database as well as several domain-specific expert modules. WebCrow [4] extends Proverb by applying basic linguistic analysis such as POS tagging and lemmatization. It uses semantic relations contained in WordNet, dictionaries and gazetteers. To exploit the database of clue-answer pairs, WebCrow applies MySQL match and Full-Text search functions. We used WebCrow as baseline as its CPDB module is one of the most accurate among CP resolution systems. This makes it one of the best system for Automatic CP resolution. The authors kindly made it available to us. It should be noted that, to the best of our knowledge, the state-of-the-art system is Dr. Fill [6], which targets the crossword filling task with a Weighted-Constraint Satisfaction Problem. However, its CPDB module is comparable to the one of WebCrow.

3 Advanced Learning to Rank Algorithms

We used the reranking framework applied to CPs described in [2]. This uses a preference reranking approach [7] exploiting structural kernels [10] and feature vectors.

Structural Kernels. The model described in [11] are fed with a textual query and the list of related candidates, retrieved by a search engine (used to index a DB) according to some similarity criteria. Then, the query and the candidates are processed by an NLP pipeline, which contains many text analysis components: the tokenizer[1], sentence detector[1], lemmatizer[1], part-of-speech (POS) tagger[1], chunker[2] and stopword marker[3]. The output of these processors are used for building tree representations of clues. We use kernels applied to syntactic trees and feature vectors to encode pairs of clues in SVMs, which reorder the candidate lists. Since the syntactic parsing accuracy can impact the quality of our trees, and thus the accuracy of SVMs, we used shallow syntactic trees.

3.1 Feature Vectors

In addition to structural representations, we also used features for capturing the degrees of similarity between clues.

iKernels features (iK). these are a set of similarity features taking into account syntactic information captured by n-grams, and using kernels:
– *Syntactic similarities.* Several cosine similarity measures are computed on n-grams (with $n = 1, 2, 3, 4$) of word lemmas and part-of-speech tags.
– *Kernel similarities.* These are computed using (i) string kernels applied to clues, and tree kernels applied to structural representations

DKPro Similarity (DKP). We used similarity features used in Semantic Textual Similarity (STS) tasks, namely features in DKPro from the UKP Lab [1]. These features were effective in predicting the degree of similarity between two sentences:
– *Longest common substring measure* and *Longest common subsequence measure*. They determine the length of the longest substring shared by two text segments.
– *Running-Karp-Rabin Greedy String Tiling*. It provides a similarity between two sentences by counting the number of shuffles in their subparts.
– *Resnik similarity*. The WordNet hypernymy hierarchy is used to compute a measure of semantic relatedness between concepts expressed in the text.
– *Explicit Semantic Analysis* (ESA) similarity [5]. It represents documents as weighted vectors of concepts learned from Wikipedia, WordNet and Wiktionary.
– *Lexical Substitution* [3]. A supervised word sense disambiguation system is used to substitute a wide selection of high-frequency English nouns with generalizations. Resnik and ESA features are computed on the transformed text.

WebCrow features (WC). We included the similarity measures computed on the clue pairs by WebCrow and the Search Engine as features:
– *Lucene Score*. BM25 score of the target candidate.
– *Clue distance*. It quantifies how dissimilar the input clue and the retrieved clue are. This formula is mainly based on the well known Levenshtein distance.

[1] `http://nlp.stanford.edu/software/corenlp.shtml`
[2] `http://cogcomp.cs.illinois.edu/page/software_view/13`
[3] Stopwords: `https://github.com/mimno/Mallet/blob/master/stoplists/en.txt`

4 Aggregation Models for Answer Reranking

CP resolution is a sort of question answering task: it requires extracting the answer rather than a set of ranked clues. Groups of similar clues retrieved from the search engine can be associated with the same answers. Since each clue receives a score from the reranker, a strategy to combine the scores is needed. We aim at aggregating clues associated with the same answer and building meaningful features for such groups. We designed two different strategies: (i) apply LGR to the scores of our reranker to obtain probabilities and then sum together those referring to the same answer candidates; and (ii) represent each answer candidate with features derived from all the clues associated with it, i.e., their aggregation using standard operators such average, min. and max.

Logistic Regression Model. The search engine or the reranker associate clues with scores that are not probabilities and have their own distributions. In contrast, LGR assigns probabilities to answer candidates. Such probabilities, learned using also additional features, are more effective for aggregation. We apply the following formula: $Score(G) = \frac{1}{n} \sum_{c \in G} \frac{P^{LR}(y=1|\boldsymbol{x}_c)}{rank_c}$ to obtain a single final score for each different answer candidate, where c is the answer candidate, G is the set of clue answers equal to c, and n is the size of the answer candidate list. $\boldsymbol{x}_c$ is the feature vector associated with $c \in G$, $y \in \{0, 1\}$ is the binary class label ($y = 1$ when c is the correct answer). $rank_c$ is the rank assigned from the reranker to the word c. Eventually, we divide the probability by the rank of the answer candidate to reduce the contribution of bottom candidates. The conditional probability computed by the linear model is the following: $P^{LR}(y = 1|c) = \frac{1}{1+e^{-y\boldsymbol{w}^T \boldsymbol{x}_c}}$, where $\boldsymbol{w} \in \mathbb{R}^n$ is a weight vector [12].

Learning to Rank Aggregated Answers. We apply SVM^{rank} to rerank each set of clues having the same answer candidate. To build the feature vectors associated with such groups, we average the features used for each clue by the first reranker, i.e., those described in Sec. 3.1. We call these features **FV**. Additionally, we compute the sum and the average of the scores, the maximum score, the minimum score and the term frequency of the word in the CPDB Dataset. We call them (**AVG**). Eventually, we model the occurrences of the answer instance in the list by means of positional features: we use n features, where n is the size of our candidate list (i.e., 10). Each feature corresponds to the positions of the answer instance in the list. We call them (**POS**).

5 Experiments

The experiments compare different ranking models. i.e., WebCrow, BM25 and several rerankers, for the task of clue retrieval. Most importantly, we show innovative models for aggregating and reranking answers based on LGR and SVM^{rank}.

5.1 Database of Previously Resolved CPs (CPDB)

We compiled a crosswords corpus combining (i) the downloaded CPs from the Web[4] and (ii) the clues database provided by Otsys[5]. We removed duplicates, fill-in-the-blank

[4] http://www.crosswordgiant.com
[5] http://www.otsys.com/clue

Table 1. Similar Clue Reranking

Model	MRR	SUC@1	SUC@5
WebCrow (WC)	64.65	57.14	74.98
BM25	75.17	63.78	90.40
RR (iK)	78.01	67.34	92.32
RR (iK+DKP)	80.89	71.62	93.14
RR (iK+DKP+WC)	**81.70**	**72.50**	**94.02**

Table 2. Answer reranking

Model	MRR	SUC@1	SUC@5
Raw voting	41.33	17.44	78.48
LGR voting	83.16	73.18	96.68
SVM (AVG+POS)	83.49	73.82	96.78
SVM (AVG+POS+FV)	**83.95**	**74.60**	**96.78**
LGR (AVG+POS+FV)	81.70	73.54	96.74

clues (which are better solved by using other strategies) and clues representing anagrams or linguistic games. The resulting compressed dataset, called CPDB, contains 2,131,034 unique and standard clues, with associated answers.

5.2 Experimental Setup

We used SVM-light-TK[6] to train our models, with default parameters. It enables the use of structural kernels [10] in SVM-light [7]. We applied a polynomial kernel of degree 3 to the explicit feature vectors. To measure the impact of the rerankers as well as the baselines, we use: success at rank 1 (SUC@1), which is the percentage of questions with a correct answer in the first position; Mean Reciprocal Rank (MRR), which is computed by $\frac{1}{|Q|}\sum_{q=1}^{|Q|}\frac{1}{rank(q)}$, where *rank(q)* is the position of the first correct answer in the candidate list; and success at rank 5 (SUC@5), which is the percentage of questions with at least one correct answer in the first 5 reranked clues.

5.3 Ranking Results

To build the reranking training and test set, we used the clues contained in CPDB for querying the search engine, which retrieves a list of candidates from the indexed clues excluding the input clue. For each input clue, similar candidate clues are retrieved and used to form a first list for the reranker. The training set is composed by 8,000 unique pairs of clue/answer that have at least one correct answer in the first 10 candidates retrieved by the search engine. We also created a test set containing 5,000 clues that are not contained in the training set. We tested two different models: (i) BM25 and (ii) reranking models (RR). Since WebCrow includes a database module, we also report its accuracy. We used the BM25 implementation of Lucene [9] as the IR Baseline: lists are ordered using Lucene scores. To rerank these lists, we tried different combinations of features for the rerankers, described in Section 3.1. The results in Tab. 1 show that: (i) BM25 produces an MRR of 75.17%, which improves on WebCrow by more than 6.5 absolute percent points, demonstrating the superiority of an IR approach over DB methods; (ii) RR (iK) achieves a higher MRR, up to 4 percent absolute of improvement over BM25 and thus about 10.5 points more than WebCrow. With respect to this model, the improvement on MRR of (iii) RR (iK+DKPro) is up to 1.2 percent points and finally, (iv) RR (iK+DKP+WC) improves the best results of another full percent point. Tab. 2 shows the results for answer reranking: (i) voting the answer using the raw score of the reranker is not effective; (ii) voting, after transforming scores into probabilities with LGR, improves on the best clue reranking model in terms of SUC@1 and MRR; (iii)

[6] `http://disi.unitn.it/moschitti/Tree-Kernel.htm`

the SVM^{rank} aggregation model using AVG and POS feature sets improves on the LGR voting model; (iv) when FV are added we notice a further increase in MRR and SUC@1; (v) LGR on the same best model AVG+POS+FV is not effective, showing that ranking methods are able to refine answer aggregation better than regression methods.

6 Conclusions

In this paper, we improve the answer extraction from DBs for automatic CP resolution. We design innovative learning to rank aggregation methods based on SVMs on top of state-of-the-art rerankers designed for clue reordering. Our approach first retrieves clues using BM25, then applies SVMs based on several features and tree kernels and eventually, collapses clues with the same answers, thus modeling answer reranking. The latter uses innovative aggregation features and positional features. The comparisons with state-of-the-art CP solvers, i.e., WebCrow, show that our model relatively improves it by about 30% (16.4 absolute percent points) in SUC@1 and even more on MRR. For our study, we collected over 6 millions of English clues and we created a dataset for clue similarity with over 2 millions of English clues. This is an important resource for IR research that we make available to the community.

References

1. Bär, D., Zesch, T., Gurevych, I.: Dkpro similarity: An open source framework for text similarity. In: Proceedings of ACL (System Demonstrations) (2013)
2. Barlacchi, G., Nicosia, M., Moschitti, A.: Learning to rank answer candidates for automatic resolution of crossword puzzles. In: Proceedings of the Eighteenth Conference on Computational Natural Language Learning. Association for Computational Linguistics (June 2014)
3. Biemann, C.: Creating a system for lexical substitutions from scratch using crowdsourcing. Lang. Resour. Eval. 47(1), 97–122 (2013)
4. Ernandes, M., Angelini, G., Gori, M.: Webcrow: A web-based system for crossword solving. In: Proc. of AAAI 2005, pp. 1412–1417. AAAI Press, Menlo Park (2005)
5. Gabrilovich, E., Markovitch, S.: Computing semantic relatedness using wikipedia-based explicit semantic analysis. In: Proceedings of the 20th International Joint Conference on Artifical Intelligence, IJCAI 2007, pp. 1606–1611 (2007)
6. Ginsberg, M.L.: Dr.fill: Crosswords and an implemented solver for singly weighted csp. J. Artif. Int. Res. 42(1), 851–886 (2011)
7. Joachims, T.: Optimizing search engines using clickthrough data. In: Proceedings of the Eighth ACM SIGKDD International Conference on Knowledge Discovery and Data Mining, KDD 2002, pp. 133–142. ACM, New York (2002)
8. Littman, M.L., Keim, G.A., Shazeer, N.: A probabilistic approach to solving crossword puzzles. Artificial Intelligence 134(1-2), 23–55 (2002)
9. McCandless, M., Hatcher, E., Gospodnetic, O.: Lucene in Action, Second Edition: Covers Apache Lucene 3.0. Manning Publications Co., Greenwich (2010)
10. Moschitti, A.: Efficient convolution kernels for dependency and constituent syntactic trees. In: ECML, pp. 318–329 (2006)
11. Severyn, A., Moschitti, A.: Structural relationships for large-scale learning of answer reranking. In: Proceedings of ACM SIGIR, New York, NY, USA (2012)
12. Yu, H.F., Huang, F.L., Lin, C.J.: Dual coordinate descent methods for logistic regression and maximum entropy models. Mach. Learn. 85(1-2), 41–75 (2011)

Diagnose This If You Can
On the Effectiveness of Search Engines in Finding Medical Self-diagnosis Information

Guido Zuccon[1], Bevan Koopman[2], and João Palotti[3]

[1] Queensland University of Technology, Brisbane, Australia
g.zuccon@qut.edu.au
[2] Australian e-Health Research Centre, CSIRO, Brisbane, Australia
bevan.koopman@csiro.au
[3] Vienna University of Technology, Vienna, Austria
palotti@ifs.tuwien.ac.at

Abstract. An increasing amount of people seek health advice on the web using search engines; this poses challenging problems for current search technologies. In this paper we report an initial study of the effectiveness of current search engines in retrieving relevant information for diagnostic medical circumlocutory queries, i.e., queries that are issued by people seeking information about their health condition using a description of the symptoms they observes (e.g. hives all over body) rather than the medical term (e.g. urticaria). This type of queries frequently happens when people are unfamiliar with a domain or language and they are common among health information seekers attempting to self-diagnose or self-treat themselves. Our analysis reveals that current search engines are not equipped to effectively satisfy such information needs; this can have potential harmful outcomes on people's health. Our results advocate for more research in developing information retrieval methods to support such complex information needs.

Keywords: Medical Information Retrieval, Self-Diagnosis, Evaluation, Medical Circumlocution.

1 Introduction and Motivations

The use of the Web as source of health-related information is a wide-spread phenomena. Qualitative research carried out by the Pew Research Center has found that 80% of the interviewed U.S.-based population uses the Web to acquire health information [2]. Health-related websites available on the Internet range from those providing information and support for people with diagnosed conditions, to those (developed both from private companies and recognised healthcare providers) suggesting diagnoses for particular symptoms, and those providing self-treatment options and cures [6].

Search engines are commonly used as a means to access health information available online. An analysis of query logs obtained a dozen of years ago from

A. Hanbury et al. (Eds.): ECIR 2015, LNCS 9022, pp. 562–567, 2015.

three commercial search engines revealed that health-related queries amounted to about 10% of the total number of queries issued to web search engines [7]. This trend has grown enormously in recent years [2]. A survey from the Pew Research Center reports that nearly 70% of search engine users in the U.S. have performed health-related searches; many of these searches were for self-diagnosis purposes, and of these about half lead to users seeking professional medical attention [2].

Previous research has, however, shown that exposing people with no or scarce medical knowledge to complex medical language may lead to erroneous self-diagnosis and self-treatment [1]. White and Horvitz have shown that access to medical information on the Web can lead to the escalation of concerns about common symptoms (e.g., cyberchondria) [9].

It is therefore important to develop and evaluate search methodologies that effectively support users in finding topical, high-quality, and accessible health information on the web. The ShARe/CLEF eHealth Evaluation Labs 2013 and 2014 (Task 3) have focused on evaluating information retrieval systems aimed at health consumers to improve how they access medical information on the Web [3,4]. The tasks focused on queries used by health consumers to find information about their diseases or disorders as reported in a discharge summary they were given upon discharge from a hospital admission. The results from the 2014 campaign showed that effective systems can be created using statistical language modelling techniques along with sophisticated query expansion mechanisms based on structured domain knowledge and the exploitation of information from discharge summaries.

The queries investigated by the CLEF evaluation labs so far were seeking information about a medical term (usually the name of a medical condition) users encountered in their discharge summaries. As mentioned above, these are only one part of the health-related queries issued to search engines, with queries aimed at self-diagnosis purposes being another important type of health-related information needs [2,9,10,8]. A recent study by Stanton et al. [8] has suggested that self-diagnosis queries observed from search engines query logs tend to be in a *circumlocutory* form, where the information seeker describes the symptoms they are observing in a colloquial way and using a "talking around" style, instead of the actual medical expression, e.g., [white part of the eye turned green] in place of [jaundice]. Answering such circumlocutory self-diagnosis queries correctly is of critical importance to avoid the risk of harm from incorrect self-diagnosis or self-treatment.

Our Contribution. In this paper, we perform an initial investigation of the effectiveness of current commercial search engines in retrieving information that helps the information seekers to correctly self-diagnose themselves. We investigate 8 main symptoms and for each of these we consider 3 to 4 queries (26 queries in total) obtained from the work of Stanton and colleagues [8], who have proposed a method to generate medical circumlocution diagnostic queries that resemble what users may issue to search for self-diagnosis information. Queries are issued to two commercial search engines (Google and Bing), their search results recorded and assessed to evaluate whether users may find relevant information

Table 1. Crowdsourced queries with associated symptoms obtained from [8] and used in this work to evaluate the effectiveness of state-of-the-art search engines

Symptom Group	Crowdsourced Circumlocutory Queries
alopecia	baldness in multiple spots, circular bald spots, loss of hair on scalp in an inch width round
angular cheilitis	broken lips, dry cracked lips, lip sores, sores around mouth
edema	fluid in leg, puffy sore calf, swollen legs
exophthalmos	bulging eye, eye balls coming out, swollen eye, swollen eye balls
hematoma	hand turned dark blue, neck hematoma, large purple bruise on arm
jaundice	yellow eyes, eye illness, white part of the eye turned green
psoriasis	red dry skin, dry irritated skin on scalp, silvery-white scalp + inner ear
urticaria	hives all over body, skin rash on chest, extreme red rash on arm

that helps self-diagnoses their conditions (the 8 main symptoms). The results reveal that only half of the top 10 results retrieved by the considered search engines provide information that is somewhat relevant to the self-diagnosis of the medical condition; only about 3 out 10 results on average are highly useful for self-diagnosis purposes.

2 Methodology

We use the 26 crowdsourced queries from the work of Stanton and colleagues [8]. Along with the queries, we extracted the name of the symptoms each queries referred to: queries can be divided in 8 groups which correspond to the 8 different symptoms. We used this symptom information for relevance assessment. The considered queries and symptoms are reported in Table 1.

Two large, commercial search engines (Google and Bing) were used as representative of current state-of-the-art search engines; these search engines were used to retrieve the top-10 results in answer to each of the 26 queries. Queries were issued against the (deprecated) Google Ajax API and the Microsoft Azure Marketplace API from Australia on the same day. The URL of the returned top 10 results were recorded.

A purposely customised version of the Relevation! assessment tool [5] was used to carry out the relevance assessment exercise. Eight higher degree students and researchers from Queensland University of Technology were employed to assess the relevance of the retrieved results. The assessors were not medical experts: this was deliberate to realistically simulate the situation of people with little or no medical knowledge searching for health information on the Web, similar to the actual task we investigate. Web pages returned for queries belonging to the same symptom were shown to a single assessor. Assessors were instructed to evaluate whether each webpage provided relevant information that

Table 2. Retrieval effectiveness achieved by two widely used commercial search engines when prompted with circumlocutory medical queries aimed at self-diagnosis purposes. Results are averaged over 26 queries.

	ndcg@1		ndcg@5		ndcg@10		P@5		P@10	
System	Rel	Hrel	Rel	Hrel	Rel	Hrel	Rel	Hrel	Rel	Hrel
Bing	.3846	.2308	.3812	.2654	.3802	.2764	.4385	.2769	.4308	.2769
Google	.3846	.3077	.4242	.3142	.4252	.3138	.5000	.3154	.4923	.3115

would allow the information seeker to self-diagnose, i.e., individuate the correct medical term of the symptom they are experiencing. Assessors could assign one of the following relevance label to each result: Not relevant (assigned to 226 documents), On topic but unreliable (assigned to 54 documents), Somewhat relevant (assigned to 87 documents) and Highly relevant (assigned to 153 documents). Queries, webpage URLs and relevance assessments are made available at `https://github.com/ielab/ecir2015-DignoseThisIfYouCan`.

To evaluate the effectiveness of two search engines we consider precision at ranks 5 and 10 (P@5, P@10), which indicates the proportion of relevant documents among the top 5 (10) search results, and nDCG at 1, 5 and 10 (ndcg@1, ndcg@5, ndcg@10), which indicates the usefulness, or gain, of the document ranking based on the position of relevant documents in the result list.

3 Results and Analysis

Table 2 reports the effectiveness of the two commercial search engines. We distinguish between Somewhat relevant (Rel) and Highly relevant only documents (Hrel only) (see below for an analysis of these two relevance categories). The results reveal differences in effectiveness between the two search engines (in particular beyond rank 1). Similarly, Figure 1 reports the effectiveness of the systems at a query level, showing that differences are not due to the contribution of outliers, e.g., a single query where one system was particular good or bad. More importantly though, the results highlight that, on average, only about 4 to 5 out of the first 10 results provide information that can help people self-diagnose themselves. This reduces to 3 out of the first 10 documents if highly relevant information is sought.

An analysis of documents assessed as "Somewhat relevant" reveals that a prototypic somewhat relevant document contained information that was not focused on only the relevant symptom, e.g., it provided a list of symptoms with corresponding definition that included the relevant symptom. A similar analysis revealed that documents assessed as highly relevant instead contained information that was mostly solely focused on the relevant symptom, providing descriptions and causes of the symptoms, often aided by photographic material showing visual examples of symptoms occurrences. Pages that were deemed as on topic but unreliable were considered irrelevant for the purpose of this evaluation. These pages

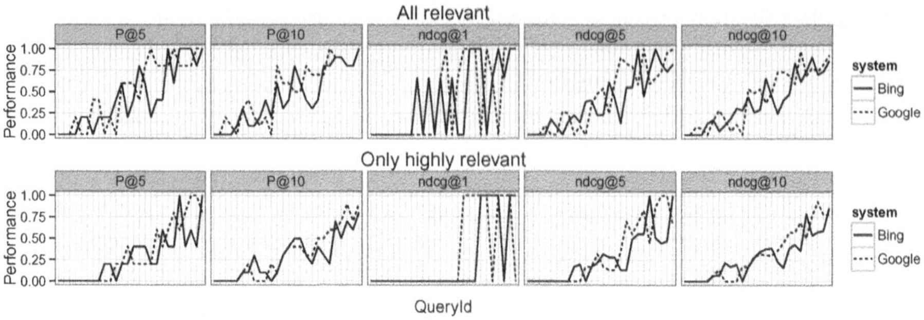

Fig. 1. Retrieval effectiveness of the two studied search engines for each individual query; results are reported for different level of relevance

contained information that was somewhat related to the sought symptoms, but it was of suspicious origin and often involved the purchase of a service or a product (for example, selling anti hair loss shampoos for alopecia or glasses for jaundice).

Both search engines retrieved documents that were judged irrelevant by the assessors. A large number of irrelevant documents did contain the query terms but were suggesting a different medical symptom than that underlying the issued query. Other irrelevant documents instead did not related to the medical intent of the query (for example the Amazon page selling copies of "Yellow Eyes" by R. G. Montgomery for the query [yellow eyes] but referring to the jaundice symptom) or related to health problems not in human beings (for example a page about cat bald spot diagnosis for the query [baldness in multiple spots]).

The results obtained in this initial investigation suggest that people searching the Web for information for self-diagnosis is likely to encounter misleading advice that could confuse them or, ultimately, cause harm.

4 Conclusion

Previous research has considered the development and evaluation of techniques to support health information seeking; recent efforts have mostly focused on the problem of searching for information that describes or explains a specialistic medical term and effective information retrieval methods have been developed for this task [3,4].

In this paper we have investigated the effectiveness of current state-of-the-art commercial web search engines for retrieving diagnostic information in answer to a different type of health queries: those that describe symptoms in a circumlocutory, colloquial manner, similar to those observed in query logs and likely be issued by people seeking to self-diagnose themselves. The empirical results suggest that current retrieval techniques may be poorly suited to such queries. We advocate for more research be directed towards improving search systems to support such type of queries, as previous research has highlighted that the access to not relevant information can lead to erroneous self-diagnosis and self-treatment and ultimately to possible harm [6,9].

The evaluation reported in this study presents a number of limitations. Firstly, only a small amount of queries were considered in the empirical experiments; nevertheless, the queries refer to common symptoms and are thus likely to appear in search activities. Secondly, the evaluation considered an ad hoc scenario, where only one query was considered while it is likely that health-related queries are part of more complex search sessions [7] and thus the effectiveness of the sessions, rather than the single queries, should also be accounted for. Finally, we did not *fully* consider the factors that come into play when information seekers consider the relevance of the documents: for health information seeking in particular, it has been shown how the reliability and understandability of the retrieved information is critical to determine its utility and these should be accounted for in the evaluation [11].

Acknowledgements. Guido Zuccon is supported by a QUT ECARD grant, and João Palotti is supported by the EU Project FP7/2007-2013 under grant agreement n°257528 (KHRESMOI) and by the FWF project I1094-N23 (MUCKE). The experiments were ran on hardware funded through QUT SEF Large Equipment Grant 94. The authors would like to thank the assessors that took part to this study for their time.

References

1. Benigeri, M., Pluye, P.: Shortcomings of health information on the internet. Health Promotion International 18(4), 381–386 (2003)
2. Fox, S.: Health topics: 80% of internet users look for health information online. Pew Internet & American Life Project (2011)
3. Suominen, H., et al.: Overview of the ShARe/CLEF eHealth evaluation Lab 2013. In: Forner, P., Müller, H., Paredes, R., Rosso, P., Stein, B. (eds.) CLEF 2013. LNCS, vol. 8138, pp. 212–231. Springer, Heidelberg (2013)
4. Kelly, L., et al.: Overview of the shARe/CLEF eHealth evaluation lab 2014. In: Kanoulas, E., Lupu, M., Clough, P., Sanderson, M., Hall, M., Hanbury, A., Toms, E. (eds.) CLEF 2014. LNCS, vol. 8685, pp. 172–191. Springer, Heidelberg (2014)
5. Koopman, B., Zuccon, G.: Relevation!: An open source system for information retrieval relevance assessment. In: Proc. of SIGIR 2014 (2014)
6. Ryan, A., Wilson, S.: Internet healthcare: Do self-diagnosis sites do more harm than good? Expert Opinion on Drug Safety 7(3), 227–229 (2008)
7. Spink, A., Yang, Y., Jansen, J., Nykanen, P., Lorence, D.P., Ozmutlu, S., Ozmutlu, C.: A study of medical and health queries to web search engines. Health Information & Libraries Journal 21(1), 44–51 (2004)
8. Stanton, I., Ieong, S., Mishra, N.: Circumlocution in diagnostic medical queries. In: Proc. of SIGIR 2014, pp. 133–142 (2014)
9. White, R.W., Horvitz, E.: Cyberchondria: studies of the escalation of medical concerns in web search. ACM TOIS 27(4), 23 (2009)
10. White, R.W., Horvitz, E.: Experiences with web search on medical concerns and self diagnosis. In: Proc. of AMIA, vol. 2009, p. 696 (2009)
11. Zuccon, G., Koopman, B.: Integrating understandability in the evaluation of consumer health search engines. In: Proc. of MedIR 2014, vol. 29 (2014)

Sources of Evidence for Automatic Indexing of Political Texts

Mostafa Dehghani[1], Hosein Azarbonyad[2], Maarten Marx[2], and Jaap Kamps[1]

[1] Institute for Logic, Language and Computation, University of Amsterdam
[2] Informatics Institute, University of Amsterdam
{dehghani,h.azarbonyad,maartenmarx,kamps}@uva.nl

Abstract. Political texts on the Web, documenting laws and policies and the process leading to them, are of key importance to government, industry, and every individual citizen. Yet access to such texts is difficult due to the ever increasing volume and complexity of the content, prompting the need for indexing or annotating them with a common controlled vocabulary or ontology. In this paper, we investigate the effectiveness of different sources of evidence—such as the labeled training data, textual glosses of descriptor terms, and the thesaurus structure—for automatically indexing political texts. Our main findings are the following. First, using a learning to rank (LTR) approach integrating all features, we observe significantly better performance than previous systems. Second, the analysis of feature weights reveals the relative importance of various sources of evidence, also giving insight in the underlying classification problem. Third, a lean-and-mean system using only four features (text, title, descriptor glosses, descriptor term popularity) is able to perform at 97% of the large LTR model.

Keywords: Automatical Indexing, Political Texts, Learning to Rank

1 Introduction

Political texts are pervasive on the Web, with a multitude of laws and policies in national and supranational jurisdictions, and the law making process as captured in debate notes of national and local governments. Access to this data is crucial for government transparency and accountability to the population, yet notoriously hard due to the intricate relations between these documents. Indexing documents with a controlled vocabulary is a proven approach to facilitate access to these special data sources [9]. There are serious challenges in the increased production of the political text, making human indexing very costly and error-prone.[1] Thus, technology-assisted indexing is needed which scale and can automatically index any volume of texts.

There are different sources of evidence for the selection of appropriate indexing terms for political documents, including variant document and descriptor term representations. For example, descriptor terms can be expanded by their textual descriptions or glosses, which is useful for calculating the similarity of a descriptor term with the content of documents [7]. Also the structure of thesauri, if existing, could be another useful source for finding the semantic relations between descriptor terms and taking

[1] Iivonen [2] focuses on search (with the same information mediators that do subject cataloguing), and lists 32.1% pairwise agreement on the chosen terms, but 87.6% agreement when taking into account terms that are close in terms of the thesauri relations.

A. Hanbury et al. (Eds.): ECIR 2015, LNCS 9022, pp. 568–573, 2015.

these relations into account [6, 7]. One of the main sources of evidence is to use a set of annotated documents, with the descriptor terms assigned. These documents are considered as train data in supervised methods [4, 5].

The main research problem of this paper is: How effective are different sources of evidence—such as the labeled training data, textual glosses of descriptor terms, and the thesaurus structure—for automatically indexing political texts? Our approach is based on learning to rank (LTR) as a means to take advantage of all sources of evidence, similar to [11], considering each document to be annotated as a query, and using all text associated with a descriptor term as documents. We evaluate the performance of the proposed LTR approach on the English version of JRC-Acquis [8] and compare our results with JEX [9] which is one of the state of the art systems developed for annotating political text documents. JEX treats the problem of indexing document as a profile-based category ranking task and uses textual features of documents as well as description of categories to index documents.

Our first research question is: How effective is a learning to rank approach integrating a variety of sources of information as features? We use LTR also as an analytic tool, leading to our second research question: What is the relative importance of each of these sources of information for indexing political text? Finally, based on the analysis of feature importance, we study our third research question: Can we select a small number of features that approximate the effectiveness of the large LTR system?

2 Sources of Evidence

In this section, we briefly introduce the sources of evidence used: 1) labeled documents, 2) textual glosses of descriptor terms,and 3) the thesaurus structure. We construct formal models of documents and descriptor terms, and use them to extract features.

Models are based on both title and body text of documents, which are available in all political document collections. The constructed model of documents is as follows:

$$Model_D = \langle M(title_D), M(text_D)\rangle, \tag{1}$$

where $Model_D$ is the model generated for the document D. This model is composed of different submodels: $M(title_D)$ based on only the title and $M(text_D)$ based on all text in the document (including titles). To construct these models, title and text of the document are considered as bag of words with stopword removal and stemming.

Similarly, the model of a descriptor terms is defined as:

$$Model_{DT} = \langle M(title_{DT}), M(text_{DT}), M(gloss_{DT}), M(anc\_gloss_{DT})\rangle, \tag{2}$$

where $M(title_{DT})$ and $M(text_{DT})$ are the union of the title models and text models of all documents annotated by descriptor term DT. $M(gloss_{DT})$ is the descriptor model of DT and defined as the bag of words representation of glossary text of DT. $M(anc\_gloss_{DT})$ is considers all descriptor terms that are ancestors of the descriptor term DT in the thesaurus hierarchy, and takes the union of their descriptor models.

These models lead to eight possible combinations of a document and descriptor term submodel (2 times 4, respectively). For each combination, we employ three IR measures: a) language modeling similarity based on KL-divergence using Dirichlet smoothing, b) the same run using Jelinek-Mercer smoothing, and c) Okapi-BM25.

In addition, we define a number of features for reflecting the characteristics of descriptor terms independent of documents. First, the statistics of the descriptor terms in train data is considered as the prior knowledge for determining what is the likelihood of selecting a descriptor term for annotating documents. That is, we define the number of times that a descriptor term has been selected for annotating documents in training data as its *popularity*. Second, in automatic indexing of documents, the degree of ambiguity of a descriptor term implicitly affects its chance for being assigned to the documents. We have modeled *ambiguity* with two different features, the number of parents of a descriptor term in thesaurus hierarchy graph and the number of its children. Another factor for determining the chance of a descriptor term for being an annotation of a given document is its *generality*. We quantify the generality of a descriptor term as its level in the thesaurus hierarchy. We consider the level of a descriptor term as the length of its shortest path to the root of thesaurus hierarchy.

Exploiting LTR enables us to learn an effective way to combine features and generate a final ranking list using all features. Finally the top-k (typically 5) descriptor terms in the ranking list are selected as the labels of a document.

3 Experiments

In this section, we detail the experimental settings (data, parameters and pre-processing), followed by the experimental results and analysis.

3.1 Experimental Settings

We use JRC-Acquis dataset [8], a widely used collection for automatic indexing of political texts. The documents of this corpus have been manually labeled with EuroVoc concepts [1]. EuroVoc contains 6,796 hierarchically structured concepts, used to annotate political documents and news within the EU and in national governments. Since the structure of documents has changed over the years, we only use the documents of the last five years: from 2002 to 2006. We use the English version of JRC-Acquis, which contains 16,824 documents, each labeled with 5.4 concepts on average.

In order to evaluate the proposed methods, we divide the collection respecting its chronological order. The first part which contains the 70% oldest of documents is used to construct the models of descriptor terms (as documents in LTR). The remaining 30% of the collection is used to construct the test and train data (train and test query in LTR). To avoid missing information, in the second part we have removed descriptor terms that do not exist in the first part as annotation. This leads to 1,639 different descriptor terms in our dataset. We do 5-fold cross validation on the second part. To have a comparable evaluation, for 5-fold cross validation on JEX, we added the first 70% part of the collection to the training data used in each fold, to train its model. We have trained the ranking model using different LTR algorithms. Among them, AdaRank [10] has a slightly better performance and we report the results of this method.

We compare our results with JEX [9]. The pre-processing done in this paper is same as in JEX. We employ Porter stemmer and consider the 100 top frequent words in the collection as stopwords. We use different parameters for similarity functions according to the type of queries and documents. Based on pilot experiments, for short queries (considering titles of documents as queries) we use these parameters: $\mu = 1,000$ for

Table 1. Performance of JEX, best single feature, and LTR methods. We report incremental improvement and significance (▲ indicates t-test, one-tailed, p-value < 0.05)

Method	P@5 (%Diff.)	Recall@5 (%Diff.)
JEX	0.4353	0.4863
BM25-TITLES	0.4798 (10%)▲	0.5064 (4%)▲
LTR-ALL	0.5206 (20%)▲	0.5467 (12%)▲

LM-Dirichlet, $\lambda = 0.2$ for LM-JM, and $b = 0.65$ and $k_1 = 1.2$ for Okapi BM25. For long queries (the text of documents) we use these parameters: $\mu = 2,000$ for LM-Dirichlet, $\lambda = 0.6$ for LM-JM, and $b = 0.75$ and $k_1 = 1.2$ for Okapi BM25.

3.2 Experimental Results

We now discuss our results, following the three research questions.

Effectiveness of LTR. Starting with our first research question: How effective is a learning to rank approach integrating a variety of sources of information as features? Table 1 shows the evaluation results of the proposed method compared to the baseline system and JEX in terms of P@5, Recall@5. We use P@5 as the main measure to evaluate different methods, since the average number of descriptor terms per document in our dataset is about 5. Therefore, P@5 approximately could be considered as R-Precision as well. BM25-TITLES ranks the descriptor terms based on the similarities of them with title submodels of documents. This is the best performing single feature, and significantly better than JEX. The proposed LTR-ALL method significantly outperforms both BM25-TITLES and JEX. This demonstrates that the additional sources of evidence are effective for the indexing task.

Importance of Different Information Sources. Next, we continue with our second research question: What is the relative importance of each of these sources of information for indexing political text? We use the trained model of SVM-Rank [3] as well as the *P*@5 of employing each individual feature. SVM-Rank tries to learn weights of features and combines them linearly based on their weights. For feature analysis, we assume the weight of each feature is a reflection of its importance. Figure 1 illustrates the importance of a selected set of exploited features. We pick only one of the similarity methods (BM25) from each feature type since the other two get very similar scores.

Similarity of titles of documents and descriptor terms is the most efficient feature. The performance is statistically better than the performance of the feature defined using text models of both descriptor terms and documents. Similarity of text of the given document and titles of the descriptor terms is also efficient. Therefore titles can be considered as a succinct predictor of classes. Titles of political documents tend to be directly descriptive of the content, making the title the most informative part of the document. In addition, to human annotators will pay considerable attention to the titles.

Among the query-independent features, generality and ambiguity do not help a lot while popularity stands out. Investigating the hierarchy graph of the concepts, we see that there is little variation in generality: the average number of levels in the hierarchy is 3.85 and its standard deviation is 1.29. There is considerable difference in ambiguity: the average number of children is 4.94 (standard deviation is 4.96) and the average

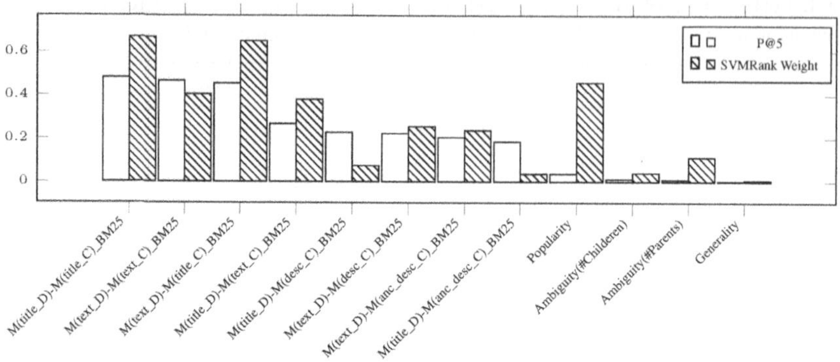

Fig. 1. Feature importance: 1) $P@5$ of individual features, 2) weights in SVM-Rank model

Table 2. Performance of LTR on all feature, and on four selected features

Method	P@5 (%Diff.)	Recall@5 (%Diff.)
LTR-ALL	0.5206 (-)	0.5467 (-)
LTR-TTGP	0.5058 (-3%)	0.5301 (-3%)

number of parents is 1.08 (standard deviation is 0.25). Ambiguity may have low importance because it is not discriminative on this data. Although popularity of classes cannot achieve a high performance by itself, it gets a high weight in SVM-Rank model. It means that considering the fact that a descriptor term is frequently assigned in general, increases the quality along with other features. This feature is important due to skewness of assigned descriptor term frequency in JRC-Acquis [1].

Lean and Mean Approach. Based on the feature analysis, we now continue with our third research question: Can we select a small number of features that approximate the effectiveness of the large LTR system? The designed LTR-ALL uses a large set of features that is very complex, hence we try to carve out a lean-and-mean system which has a better efficiency/effectiveness trade-off.

Our lean-and-mean system is an LTR trained system on four selected features: the BM25 similarities of text submodel of documents with all text, titles only, and textual glosses of descriptor terms, and popularity of descriptor terms. Table 2 indicates the performance of this LTR-TTGP approach using only four features. The LTR-TTGP approach is significantly better than JEX and BM25-TITLES before. Although the performance of LTR-ALL is significantly better than the LTR-TTGP method, the performance of LTR-TTGP is 97% of the large LTR-ALL system. Therefore, making the selective LTR approach a computationally attractive alternative to the full LTR-ALL approach.

4 Conclusion and Future Work

Our broad motivation is to build connections between political data from different national and international jurisdictions (such as EU versus national laws and parliamentary debates, or between different national parliaments). Such connections are essential for researchers, both at the level of whole documents and individual document parts. This paper addresses an important initial step, trying to replicate the human indexing of EU laws and policies based on the EuroVoc vocabulary functioning as pivot language.

Our main findings are the following. First, using a learning to rank (LTR) approach integrating all features, we observe significantly better performance than previous systems. Second, the analysis of feature weights reveals the relative importance of various sources of evidence, also giving insight in the underlying classification problem. Third, a lean-and-mean system using only four features (text, title, descriptor glosses, descriptor term popularity) is able to perform at 97% of the large LTR model.

Are the proposed systems "good enough" for the motivating task at hand. Clearly we are far from exactly replicating the choices of the human indexer. However, considering the inter-indexer agreement and the (soft) upperbound of the full LTR approach, the room for improvement seems limited. However, as Iivonen [2] observes, indexers that disagree pick terms that are near to each other in the concept hierarchy. Anecdotal inspection of our automatic indexing reveals the same: wrong descriptors tend to be conceptually close to the gold standard indexing term. Hence, this give support to the utility of the current systems for discovering conceptual cross-connections in political texts, as well as suggests ways to improve the current approaches by clustering and propagating descriptors to similar terms.

Acknowledgements. This research was supported by the Netherlands Organization for Scientific Research (ExPoSe project, NWO CI # 314.99.108; DiLiPaD project, NWO Digging into Data # 600.006.014) and by the European Community's Seventh Framework Program (FP7/2007-2013) under grant agreement ENVRI, number 283465.

References

[1] EuroVoc. Multilingual thesaurus of the european union, http://eurovoc.europa.eu/
[2] Iivonen, M.: Consistency in the selection of search concepts and search terms. IPM 31, 173–190 (1995)
[3] Joachims, T.: Training linear svms in linear time. In: SIGKDD, pp. 217–226 (2006)
[4] Nam, J., Kim, J., Loza Mencía, E., Gurevych, I., Fürnkranz, J.: Large-scale multi-label text classification - revisiting neural networks. In: Calders, T., Esposito, F., Hüllermeier, E., Meo, R. (eds.) ECML PKDD 2014, Part II. LNCS, vol. 8725, pp. 437–452. Springer, Heidelberg (2014)
[5] Pouliquen, B., Steinberger, R., Ignat, C.: Automatic annotation of multilingual text collections with a conceptual thesaurus. In: EUROLAN, pp. 9–28 (2003)
[6] Ren, Z., Peetz, M.-H., Liang, S., van Dolen, W., de Rijke, M.: Hierarchical multi-label classification of social text streams. In: SIGIR, pp. 213–222 (2014)
[7] Rousu, J., Saunders, C., Szedmak, S., Shawe-Taylor, J.: Kernel-based learning of hierarchical multilabel classification models. J. Mach. Learn. Res. 7, 1601–1626 (2006)
[8] Steinberger, R., Pouliquen, B., Widiger, A., Ignat, C., Erjavec, T., Tufis, D.: The JRC-Acquis: A multilingual aligned parallel corpus with 20+ languages. In: LREC, pp. 2142–2147 (2006)
[9] Steinberger, R., Ebrahim, M., Turchi, M.: JRC EuroVoc indexer JEX-A freely available multi-label categorisation tool. In: LREC, pp. 798–805 (2012)
[10] Xu, J., Li, H.: Adarank: A boosting algorithm for information retrieval. In: SIGIR, pp. 391–398 (2007)
[11] Yang, Y., Gopal, S.: Multilabel classification with meta-level features in a learning-to-rank framework. Machine Learning 88(1-2), 47–68 (2012)

Automatically Assessing Wikipedia Article Quality by Exploiting Article–Editor Networks

Xinyi Li[1,2], Jintao Tang[1], Ting Wang[1], Zhunchen Luo[3], and Maarten de Rijke[2]

[1] National University of Defense Technology, Changsha, China
{tangjintao,tingwang}@nudt.edu.cn
[2] University of Amsterdam, Amsterdam, The Netherlands
{x.li,derijke}@uva.nl
[3] China Defense Science and Technology Information Center, Beijing, China
zhunchenluo@gmail.com

Abstract. We consider the problem of automatically assessing Wikipedia article quality. We develop several models to rank articles by using the editing relations between articles and editors. First, we create a basic model by modeling the article-editor network. Then we design measures of an editor's contribution and build weighted models that improve the ranking performance. Finally, we use a combination of featured article information and the weighted models to obtain the best performance. We find that using manual evaluation to assist automatic evaluation is a viable solution for the article quality assessment task on Wikipedia.

1 Introduction

Wikipedia is the largest online encyclopedia built by crowdsourcing, on which everyone is able to create and edit the contents. Its articles vary in quality and only a minority of them are manually evaluated high quality articles.[1] Since manually labeling articles is inefficient, it is essential to automatically assess article quality. Content quality criteria are known to help retrieval; in a web setting they are often based on link structure [7, 8] but in the setting of social media and collaboratively created content, content-based features are often used [11]. Here, we study the quality assessment of Wikipedia articles by exploiting the article-editor network. We view this task as a ranking problem. Our task is motivated by the assumption that automatic procedures for assessing Wikipedia article quality can help information retrieval that utilizes Wikipedia resources [2] and information extraction on Wikipedia [13] to obtain high quality information.

There have been different approaches to the content quality assessment problem. One branch of research uses simple metrics, such as article length, number of links and citations etc. [1, 6, 9, 12]. These authors do not consider the interactions between editors and articles, which differentiates Wikipedia from traditional encyclopedias. Other work takes into account the network of articles and editors. Hu et al. [4] proposes what they call a probabilistic review model to rank articles. The model is tested on a dataset of only 242 articles. Suzuki and Yoshikawa [10] uses a combination of survival ratio method and link analysis to score articles. They use relative evaluation metrics to measure the performance of models. It remains to be seen to which degree they can achieve satisfactory ranking results in more realistic settings.

[1] Only 0.1% of all Wikipedia articles are featured articles.

A. Hanbury et al. (Eds.): ECIR 2015, LNCS 9022, pp. 574–580, 2015.

We examine the editing actions of editors and find that the majority of them are field-specific, i.e., they specialize in a certain category of articles. These field-specific editors outnumber all-around editors to a great extent. Since the editor-article networks of different categories only share very few nodes, ranking articles should be done in separate categories. As featured articles are manually-tagged high quality articles, we select them as the ground truth for our task. We develop several models to rank articles by quality. Our first motivation is to see if the importance of a node in the network can indicate quality. So we develop a basic PageRank-based model. Additionally, instead of treating links as equal in the basic model, we tweak the model by putting weights on the links to reflect the difference of editor contributions. Finally, we utilize existing manual evaluation results to improve automatic evaluation. So we incorporate manual evaluation results into our model. We use articles of different quality levels to measure the levels of editors, and then assist ranking.

The experiments carried out on multiple datasets covering different fields show that ranking performance is related to the number of high quality articles we utilize. In particular, the higher the percentage of high quality articles used, the better the ranking performance. We also find that the basic model does not yield satisfactory ranking results, but that using weights boosts performance.

2 Models

We introduce the models and explain how each model is computed, including a baseline model, weighted models, and weighted models with probabilistic initial value.

2.1 Baseline Model

First, we develop a basic quality model based on Pagerank. PageRank is widely applied for ranking web pages, where pages are seen as nodes and hyperlinks as edges [7]. The node value represents its importance in the network. In our basic model we treat both articles and editors as nodes connected by edges that represent editing relations. For instance, if article A is edited by B then there is a bidirectional edge that connects A and B. The value of the nodes are distributed through the edges during each iteration of the PageRank computation. As shown in (1), the value of node v is determined by nodes in the set $U(v)$ that connect to it, where $N(u)$ is the number of edges that point out of node u.

$$PR(v) = (1-d) + d\sum\nolimits_{u \in U(v)} \tfrac{PR(u)}{N(u)}. \quad (1)$$

In this basic model, we give all nodes the same initial value and iteratively compute the node value until they converge. The articles will then be ranked by node value.

2.2 Weighted Models

The baseline model treats edges as equal. However, consider an article that has multiple editors, which is quite common. When the value of the article node is distributed toward its editors during computation, editors that make a higher contribution should get more.

There should be a weight to address this difference. It is therefore necessary to measure how users contribute to article quality and how articles contribute to user authority in return. While it is hard to precisely quantify the contribution, we can use editing actions during an article's history as an approximation. An intuitive measure is to use the edit counts between article and editor as a measure, defined in (2):

$$Contribution1 = \#edits. \tag{2}$$

We define the weighted model based on this equation as the *simple weighted* (SW) model. By further parsing the editing actions, we can obtain a more complex measure that takes different editing behaviors into account, which is defined in (3):

$$Contribution2 = \#insertions + \#deletions + \#replacements. \tag{3}$$

An editor's contribution to an article is the sum of words affected by their editing actions. The editing actions are insertion (insert new content), deletions (delete content) and replacements (insert new content right after deletion), which are shown to have a strong correlation with article quality[5]. As Wikipedia only provides history versions of articles, we obtain the editing actions by comparing adjacent article revisions with a diff-algorithm [3]. We define this model as the *complex weighted* (CW) model. After defining the contribution, we put the contribution value on each edge as the weight. The value of nodes is defined in (4).

$$PR(v) = (1-d) + d\sum_{u \in U(v)} PR(u)\frac{C_{uv}}{\sum C_u}. \tag{4}$$

In this equation the value of node u will be multiplied by the proportion of the weight value C_{uv} against the weight sum $\sum C_u$.

2.3 Weighted Models with Probabilistic Initial Value

To further improve ranking, we incorporate manual evaluation results into our weighted models. Our hypothesis is that featured articles and other articles have different levels of editors. Using articles of different quality to differentiate editors' levels may improve article ranking.

To do so, we simply give articles different initial values before computation. Their value will then be distributed to editors through editing relations. An article's initial value is determined by its probability of being high quality. We assign an initial value of 1.0 to featured articles because they have a probability of 100% to be high quality articles. Likewise, we set the initial value of other articles as the proportion of featured articles to all articles in that particular category. We set the initial value of editors as 0. We define the models as the *simple weighted probabilistic* (SWP) model and the *complex weighted probabilistic* (CWP) model based on different contribution measures.

3 Experimental Setup

3.1 Datasets

We select three categories from an English Wikipedia dump[2] as a case study. These categories cover different fields and contain both high quality articles and articles of

[2] Data dump of March 15, 2013, fetched from `https://dumps.wikimedia.org/`.

Table 1. Statistics of datasets

Category	#articles	#editors	#featured articles
Chemistry	7,796	392,055	36
Meteorology	4,218	187,637	138
Geography	38,543	1,360,508	180

unknown quality. The statistical information of the articles in these categories is shown in Table 1. We find that most editors specialize in one field, and only a minority of them are all-around editors. Therefore the article-editor networks of different categories only share a tiny proportion of common nodes. Based on this structure of the article-editor network, we will apply ranking by category.

3.2 Metrics

We assess article quality by ranking. Since featured articles are the best quality articles on Wikipedia, they are frequently used as the gold standard to measure ranking performance. However, common metrics such as RMSE are not suitable for this task as Wikipedia does not give a specific ranking for featured articles. We consider recall scores at the first N items in the result set, as well as precision-recall curves.

3.3 Parameter Settings

In the baseline model and weighted models, we initially assign 1.0 to all nodes and iteratively compute their values. The iterations can be halted for any desired mean error of the ranking being less than 0.01. For the SWP and CWP models, we will initialize them using probabilistic values as explained earlier.

4 Experimental Evaluation

We address two main research questions. We contrast our four methods, i.e., the Baseline method, the simple weighted model (SW), the complex weighted (CW) model, as well as two variants with probabilistic initial values (SWP, CWP). But first we examine the impact of the number of featured articles used for initialization in SWP and CWP. We want to find out how this number affects ranking performance.

Table 2 shows that in most cases, the more featured articles used for initialization in SWP or CWP, the better the ranking performance. We notice a few exceptions to this finding, especially in categories that have more featured articles. This is because many of the featured articles used in initialization are ranked atop, reducing the chance for other articles in the ground truth to rank high. Still, by using all featured articles for initialization we achieve the best recall performance.

Next, we compare SWP and CWP in this best case with the previous models in Figure 1. To determine whether the observed differences between two models are statistically significant, we use Student's t-test, and look for significant improvements

Table 2. Recall (N) of SWP and CWP in different categories

featured%	r@100		r@200		r@300		r@400	
	SWP	CWP	SWP	CWP	SWP	CWP	SWP	CWP
chemistry								
25%	.556	.363	.767	.667	.867	.793	.440	.874
50%	.644	.378	.778	.694	.861	.833	.972	.883
75%	.756	.400	.911	.744	.956	.911	1.000	.944
meteorology								
25%	.111	.092	.246	.175	.365	.317	.498	.421
50%	.101	.103	.274	.165	.438	.346	.607	.486
75%	.140	.114	.346	.200	.517	.357	.703	.514
geography								
25%	.173	.086	.342	.168	.426	.283	.496	.369
50%	.163	.069	.357	.182	.497	.317	.562	.422
75%	.149	.051	.376	.162	.518	.327	.596	.407

(two-tailed) at a significance level of 0.99. We find that both SWP and CWP statistically significantly outperform other models in all categories. We also note that the SWP model performs better than the CWP model in most cases, which is contrary to the previous experiments where the complex contribution measure yields better results. The best ranking performance is achieved by the SWP model when using all available high quality articles in initialization. And the recall levels are up to an applicable value. E.g., the recall value at $N = 200$ is 0.756 in geography, meaning that the 180 featured articles in that category have a probability of 75.6% to appear in the top-200 list.

We also notice that ranking performance is related to the number of featured articles in each category. E.g., chemistry, which has the fewest featured articles, is higher in precision than other categories at a given recall level. Meanwhile, the curves of meteorology and geography both experience a rise and then gradually descend. This is

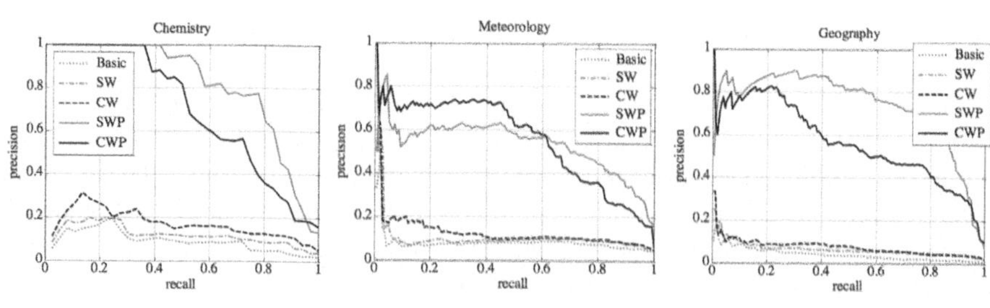

Fig. 1. Precision-recall curves for the baseline (Basic), simple weighted (SW), complex weighted (CW), simple weighted probabilistic (SWP), complex weighted probabilistic (CWP) model

because at the top of the result list are mostly featured articles, and then false positives are appearing at an increasing speed in the list, causing the curve to go downwards.

Our SWP model has achieved the best precision/recall performance and is far better than using content-based features. For instance applying our SWP model in chemistry category gives a recall score of 0.889 out of the top 100 items, while using content-based features in Blumenstock [1] only yields 0.306. Our evaluation metrics are also more applicable for ranking purpose than the relative measures used in Suzuki and Yoshikawa [10], so that we can apply our model in a practical setting.

5 Conclusion

We have developed several models for estimating Wikipedia article quality based on the article-editor network. They include a basic model, a weighted model, which addresses the difference of editors' contributions, and probabilistic weighted models incorporating manual evaluation results. The experimental results show that by using featured articles, we are able to differentiate editor levels and then improve ranking performance.

Additionally, the baseline model we considered (based on PageRank) does not yield satisfactory ranking results, but when we put weights on the links, the ranking results receive a boost. The improvements are not as significant as using featured articles.

Summarizing, the combination of existing manual evaluation results (featured articles) with the article-editor network yields a state-of-the-art solution for assessing article quality. For future work, we will improve our model by adding features of editors, and also conduct a systematic comparison with the methods presented in [4, 10].

Acknowledgments. This research was partially supported by the National Natural Science Foundation of China (Grant No. 61472436 and 61202337), the European Community's Seventh Framework Programme (FP7/2007-2013) under grant agreement nr 312827 (VOX-Pol), the Netherlands Organisation for Scientific Research (NWO) under project nrs 727.011.005, 612.001.116, HOR-11-10, 640.006.013, 612.066.930, the Center for Creation, Content and Technology (CCCT), the Dutch national program COMMIT, the ESF Research Network Program ELIAS, the Elite Network Shifts project funded by the Royal Dutch Academy of Sciences (KNAW), the Netherlands eScience Center under project number 027.012.105, the Yahoo! Faculty Research and Engagement Program, the Microsoft Research PhD program, and the HPC Fund.

References

[1] Blumenstock, J.E.: Size matters: Word count as a measure of quality on Wikipedia. In: WWW (2008)
[2] Cassidy, T., Ji, H., Ratinov, L.-A., Zubiaga, A., Huang, H.: Analysis and enhancement of wikification for microblogs with context expansion. In: COLING (2012)
[3] Ferschke, O., Zesch, T., Gurevych, I.: Wikipedia revision toolkit: Efficiently accessing Wikipedia's edit history. In: ACL (2011)
[4] Hu, M., Lim, E.-P., Sun, A., Lauw, H.W., Vuong, B.-Q.: Measuring article quality in Wikipedia: models and evaluation. In: CIKM (2007)

[5] Li, X., Luo, Z., Pang, K., Wang, T.: A lifecycle analysis of the revision behavior of featured articles on Wikipedia. In: ISCC (2013)
[6] Lih, A.: Wikipedia as participatory journalism: Reliable sources? metrics for evaluating collaborative media as a news resource. In: ISOJ (2004)
[7] Liu, B.: Web Data Mining. Springer, Heidelberg (2007)
[8] Page, L., Brin, S., Motwani, R., Winograd, T.: The PageRank citation ranking: Bringing order to the web. Technical report, Stanford University (1999)
[9] Stvilia, B., Twidale, M.B., Smith, L.C., Gasser, L.: Assessing information quality of a community-based encyclopedia. In: IQ (2005)
[10] Suzuki, Y., Yoshikawa, M.: Assessing quality score of Wikipedia article using mutual evaluation of editors and texts. In: CIKM (2013)
[11] Weerkamp, W., de Rijke, M.: Credibility-based reranking for blog post retrieval. Information Retrieval Journal 15(3-4), 243–277 (2012)
[12] Wilkinson, D., Huberman, B.: Cooperation and quality in Wikipedia. In: WikiSym (2007)
[13] Wu, F., Weld, D.S.: Open information extraction using Wikipedia. In: ACL (2010)

Long Time, No Tweets! Time-aware Personalised Hashtag Suggestion

Morgan Harvey[1] and Fabio Crestani[2]

[1] Dept. of Maths and Info. Sciences, Northumbria University at Newcastle, UK
[2] Faculty of Informatics, University of Lugano, Switzerland
morgan.harvey@northumbria.ac.uk, fabio.crestani@usi.ch

Abstract. Microblogging systems, such as the popular service Twitter, are an important real-time source of information however due to the amount of new information constantly appearing on such services, it is difficult for users to organise, search and re-find posts. Hashtags, short keywords prefixed by a # symbol, can assist users in performing these tasks, however despite their utility, they are quite infrequently used. This work considers the problem of hashtag recommendation where we wish to suggest appropriate tags which the user could assign to a new post. By identifying temporal patterns in the use of hashtags and employing personalisation techniques we construct novel prediction models which build on the best features of existing methods. Using a large sample of data from the Twitter API we test our novel approaches against a number of competitive baselines and are able to demonstrate significant performance improvements, particularly for hashtags that have large amounts of historical data available.

1 Introduction

Social-media update streams are fast becoming a key mode of information access on the web, with many services basing their offerings on this paradigm. One of the most popular of these is Twitter, which has become remarkably successful in recent years (10% of online Americans use the service on a typical day [7]). Twitter is a *microblogging* platform which allows users to post short messages (of up to 140 characters) to share thoughts, opinions, useful links, and insights from their personal experiences. Users are encouraged to "follow" others on the service whose posts (or *tweets*) may be of interest to them. Doing so results in all of the posts created by that user appearing on the follower's stream.

Although Twitter represents a highly valuable, user-driven and up-to-date source of information of unprecedented volume, evidence suggests that high volumes of tweets can become overwhelming for users. Nearly half of all Twitter search tasks involve re-finding previously seen tweets from the stream, a task which was reported to be amongst the most difficult [7]. A feature called *hashtags*, short keywords prefixed by a # symbol, a practise which emerged organically through use of the system, allows the topic(s) of each tweet to be specifically defined by the author. Hashtags provide users with a means to more easily search,

A. Hanbury et al. (Eds.): ECIR 2015, LNCS 9022, pp. 581–592, 2015.

browse and re-find tweets, form ad-hoc communities based around a hashtag's topic and follow the evolution of discussions or breaking news stories [10].

Despite the clear utility of hashtags and their ability to promote the tweets to which they are assigned [10], only a relatively small number of tweets - as few as 8% [11] - contain them. As there is no pre-defined set of hashtags to choose from when writing a tweet, users can choose any terms they wish, leading to vocabulary mismatch problems. Given the benefits of appropriate hashtag usage and the reluctance many users have in employing them (perhaps because they find selecting the best terms difficult), the problem of hashtag recommendation is important. By recommending hashtags during the tweeting process we aim to support users in allocating terms to their posts and increase the homogeneity of hashtag usage on Twitter as a whole. Since hashtag usage has been shown to be heavily dependent on time, user interests and of course the topics of the parent tweet, we attempt to incorporate these three sources of information into our recommendation models. We test several novel approaches on real Twitter data collected over a period of one month and compare the performance of our models against competitive baselines from the literature.

2 Related Work

Twitter's popularity and the existence of a public API has led to it becoming a common topic of research interest. A large amount of early work focused on understanding how networks and communities of users on such services grow and what kind of content is posted [1] which led to studies on how and why people actually use Twitter [24]. Analysis of search behaviour showed that while users often express the desire to re-find tweets, this is usually extremely difficult [7]. Twitter content has been used for various purposes: to identify and locate events as they are occurring [4], to replace tags as information sources for URLs [8] and to predict and track natural disasters [16] or the outcome of elections [18].

Two key interactive features of Twitter have been investigated in detail: the @syntax (which allows tweets to reference a particular user) [9] and hashtags. Hashtags have been used for many applications such as tweet and topic recommendation/filtering [2], to augment existing tags on other social media sites [3] and to detect communities of users [23]. Cunah et al. [5] found that hashtag popularity follows a power-law distribution and that they are used to classify tweets, propagate ideas and to promote specific topics. Elsweiler et al. [7] state that "hashtags can be helpful ... [searching and re-finding] become noticeably more difficult for users when they are not present." Hashtags encourage convergence in query terminology, are used to promote content and to find other tweets about a given topic or other users who are interested in the same topic(s) and popular queries are much more likely to contain a hashtag than unpopular ones [17,10].

The distribution of hashtags in Twitter changes rapidly and as such the most frequent terms in one hour may look very different from those in the next ("churn") [14]. Analysis of how hashtag popularity evolves over time shows several types of distribution with many being "bursty" and short-lived [12]. Huang et al. [10] used the standard deviation of hashtag ages (relative to some fixed time point) to measure the spread of hashtag usage over time, asserting that many short-lived hashtags can be explained by the appearance of "micro-memes" - time-sensitive, ad hoc discussions around a topic - and breaking news stories. They showed that a hashtag's temporal spread (as determined by standard deviation) can indicate whether or not it has been triggered by a micro-meme.

Despite the utility of hashtags and the clear advantage in promoting their use, the problem of recommending them has received little attention thus far [13]. An early approach [21] used similarity metrics to compare the vectors of terms to rank tweets in terms of their closeness to the one being written. The method then took the union of hashtags from a number of top-ranked tweets as candidate hashtags to present as suggestions. Three weighting methods for the candidate hashtags were tested with one based on the score of the most similar tweet proving to be most effective. Later work [11] improved on this by using the previous hashtags chosen by the target user to introduce some personalisation, however only raw frequencies of hashtags within the top candidate tweets were used as weights. The authors found that including the user's own hashtag choices improved performance slightly, particularly in cases where the number of top tweets chosen to draw hashtags from was small.

An alternative formulation of the problem instead tried to predict which hashtags will be reused in the future [15,20]. Yang at al. [20] considered methods for prediction of hashtag adoption and tested the hypothesis that hashtags serve as a tag of content and a symbol of community membership. They found evidence for this and built models to predict whether a user will adopt each potential hashtag within the next 10 days. However, they do not predict which tags will be assigned to a given tweet and therefore their methods are not applicable to hashtag suggestion. In this work we aim to bring together the insights from previous work together with features to exploit the strong temporal trends in the usage of hashtags by users in order to improve recommendations.

3 Recommending Hashtags

We wish to recommend hashtags to Twitter users after they have finished writing a new *target tweet* and therefore have information about the *target user* (i.e. the one who is writing the tweet), the content of the new tweet and the current time. We also have a collection of tweets which were publicly made available prior to the user beginning to write the target tweet - some of which may also have been written by the target user. The content of each tweet can be separated into two groups of terms: *hashtags* (prefixed with a # symbol) and *content terms*. To increase the likelihood of them being useful, suggested hashtags should be: (a) topically appropriate to the content of the target tweet, (b) related to the interests and vocabulary choices of the target user, and (c) temporally relevant.

We have access to a sample of tweets D with a combined vocabulary W, a combined hashtag vocabulary H, written by a set of users U. Each individual tweet i is composed of a number of content terms from T and hashtags from

H (both potentially of length 0). The counts of the wth content term and hth hashtag in the ith tweet are denoted $Cw_{i,w}$ and $Ch_{i,h}$, the author and posting time of the ith tweet are denoted $u(i)$ and $t(i)$. The summation of term counts for term w over all tweets in D is Cw_w. Each user u can also be represented by the set of all of the content terms and hashtags of their tweets (their term and hashtag profiles) using similar notation: $Cw_{u,w}$ being the count of the wth term in the uth user's profile.

Identifying Candidate Hashtags. Given a new candidate tweet j written by user $u(j)$ at time $t(j)$, we first identify similar tweets in D from which to draw candidate hashtags. This can be achieved (with some success) by using the content terms and ranking tweets by their similarity to j [22,11,20,13] using the cosine similarity between vectors of TFIDF-weighted content terms. Any similarity metric could be used, but we take this approach as it reported to be the best performing [22] and calculate the similarity thus: $Sim(i,j) = \frac{\mathbf{i} \cdot \mathbf{j}}{||\mathbf{i}|| \cdot ||\mathbf{j}||}$ Where $\mathbf{i}$ (and $\mathbf{j}$) are vectors of TFIDF weights over all content terms in W such that: $\mathbf{i}_w = Cw_{i,w} \cdot IDF(w)$ and $||\mathbf{i}||$ is the magnitude (or length) of vector $\mathbf{i}$ as computed by the euclidian norm. The Inverse Document Frequency (IDF), defined as $IDF(w) = log\left(\frac{|D|}{\sum^{D} I\{Cw_{i,w}>0\}}\right)$, reduces the importance of terms which occur too frequently in the collection (in this case, in too many tweets) and therefore have little discriminative power.

Now that we have a similarity score of each tweet i and the target tweet j we can rank these in descending order and, after choosing the top k most similar, we can extract the union of all hashtags within these tweets.

Personalisation To personalise the suggestions we can also look for candidate hashtags which are related to the interests and vocabulary use of u. We could follow the same approach as above but instead of looking for tweets similar to the target tweet, we look for those similar to the target user. While this approach may work well in some cases, many Twitter user have only a small number of prior tweets and as such the amount of term frequency information available will be very small. Instead we can employ a collaborative filtering-like method where we take advantage of Twitter's following mechanism and make the assumption that the hashtags used by those people who u follows are likely correlated with the interests of u. Studies have found strong evidence of homophily between users and those they follow [19] meaning that they share similar topics of interest.

For each user in U we construct a vector of TFIDF values over all hashtags in H such that $\mathbf{u}_h = Ch_{u,h}$ and the new IDF is as follows: $IDF(h) = log\left(\frac{|U|}{\sum^{U} I\{Ch_{u,h}>0\}}\right)$. Using the same similarity measure as before we identify users who share interests with u and rank these in descending order of similarity. We again choose the top k most similar and extract the union of all hashtags within tweets posted by these users. We will refer to the set of top k tweets as $\hat{D}$, the set of top k users as $\hat{U}$ and the combined set of candidate hashtags as $\hat{H}$.

Weighting Candidate Hashtags. We now address the problem of weighting the candidate hashtags such that their likelihood of being relevant to the new tweet is maximised. Previous work has investigated methods for doing this [11,21], proposing the following simple approaches:

1. *OverallPopularity* - frequency over entire collection.
2. *SamplePopularity* - frequency over the sub-set of k tweets most similar to the target.
3. *MaxSimilarity* - the greatest similarity score over the k most similar tweets.

Of these the *MaxSimilarity* method was found to be most effective [21], although some work [11] used the *SamplePopularity* method considering tweets similar to the target tweet and to the target user. Despite the Zipf-like distribution of hashtag popularity in Twitter, the *OverallPopularity* method does not seem to return particularly good rankings.

We would like a method which includes candidate hashtags from both similar tweets and similar users such that the similarity scores from the selection step are included in the score and the influence the two sets of scores have on the final candidate weighting can be varied. Our approach computes the sum of scores from the two sets and linearly combines them into an interpolated sum:

$$score(h) = \lambda \left(\sum_{i=1}^{|\hat{D}|} I\{Ch_{i,h} > 0\} Sim(i,j) \right) + (1-\lambda) \left(\sum_{i=1}^{\hat{U}} I\{Ch_{u,h} > 0\} Sim(i,u) \right)$$

where λ is a free parameter which allows us to vary the relative influence of the scores from similar tweets and similar users.

Considering Temporal Relevance. As discussed in the related work section, analyses of hashtag usage have uncovered evidence of strong temporal patterns [12,10]. By looking at the timestamps of tweets to which a given hashtag had been assigned Huang et al. [10] identified two categories of hashtags: those used for "organisational" means (used over long periods of time, have high variance); and "conversational" ones (short lifespan, low variance).

Figure 1 shows how two hashtags were used over the first 20 days of January 2014 with the lines representing 2-period moving averages calculated over time bins of 6 hours (4 per day). Although both hashtags are used with approximately the same frequency (266 and 280 instances respectively), they have very different temporal characteristics. The first, #happykanginday, is an example of a conversational tag and refers to the birthday of South Korean celebrity Kangin - which falls on the 17th of January - while #marketing is clearly much more general in nature. Note that the popularity of #happykanginday on the 17th is so great that it exceeds the y-axis, having a count for this bin of 246.

Imagine that we want to re-weight candidate hashtags based on this temporal information. If the target tweet is being written on the 17th and one of the candidate tags happens to be #happykanginday then an increase in the weight of this tag would be sensible. If instead the tweet was being written on another day then it is much less likely to be relevant and therefore should be assigned a negative temporal weight. However, for the #marketing tag the likelihood of relevance is uniform over time and therefore we would not want to assign it such an extreme temporal weight (neither negative nor positive).

We need a way to measure, in a single point statistic, how spread out the distribution of the ages of previous tweets is. An obvious candidate is the standard deviation, which was used by Huang et al. [10] and is easy to calculate. Another is the entropy of the relative frequencies over evenly-spaced time windows, likely a better measure as it uses more information about the distribution and does not assume that is symmetrical [6]. If we know the frequencies of occurrence of the hashtag over a continuous set of time windows index by i, $Ch(i)$, we can calculate the normalised entropy as follows:

$$\mathbb{H}(X) = -\frac{\sum^{X} P(x_i) log_b(P(x_i))}{log_b(|X|)}, \quad where\ P(x_i) = \frac{Ch(i) + 0.01}{\sum^{X} Ch(i) + 0.01|X|}$$

Note that the probability calculations are smoothed to ensure that the entropy is always finite. In our Twitter data (described later) high-entropy examples are general topical terms or long-running entertainment phenomenon (such as the TV series *The Walking Dead* and the Chicago Bears) are appear with uniform frequency over time. The low-entropy one are instead more specific and usually related to mercurial Internet memes or short-term news events.

To understand how to model the temporal patterns in the hashtags we analysed how the probability of a hashtag being relevant at a given time is related to its age. We split a data set of tweets obtained in January 2014 into two parts with an 80:20 ratio. For each tweet in the 20% part we try to predict which hashtags were actually assigned to it using the method described earlier. For each one we output the top 10 candidate hashtags and the following statistics: entropy, standard deviation, minimum age, maximum age, mean age and median age as well as whether or not each candidate was relevant (i.e. was actually assigned to the target tweet).

The hashtags are separated into two categories - those with entropy less than 0.5 and those with entropy equal to or greater than 0.5 - and then divided into 100 equal-sized bins. For each bin we calculate the probability of relevance as the number of relevant hashtags divided by the total number of hashtags within that bin. Logistic regression models predicting the relevance of a hashtag using each of the measures of location determined that the minimum age has the greatest predictive power. To understand why this is so, and to see how age affects relevance differently for the high- and low-entropy tags, in figure 2 we plot the probability of relevance over the range of minimum ages.

The figure shows that for both sets there is a clear trend of decay in the probability of relevance as minimum age increases, however the rate is much steeper for the low-entropy queries. This confirms the intuition that low-entropy tags relate to short-lived topics or are merely conversational in nature. If the hashtag has a low entropy and the minimum age (i.e. the time since it was last used) is high then it is unlikely that it will be used again and its score in the ranking function should be heavily penalised. However, if it has a high entropy then although there will still be some decay of interest over time, we should not penalise it so aggressively. Note that in the case of hashtags which have general relevance (such as the #marketing example) the entropy will be high and the

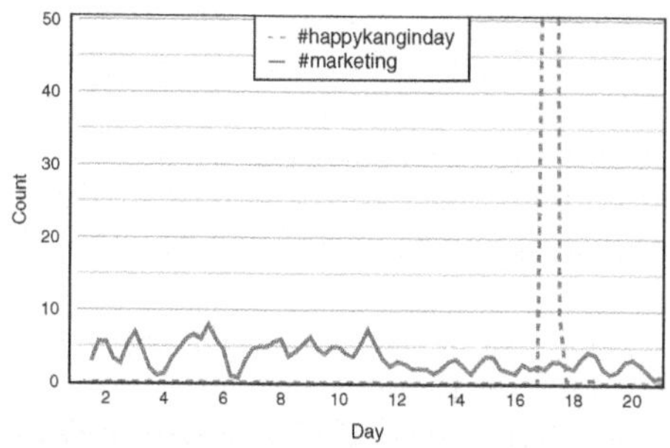

Fig. 1. Trend lies for temporal activity of two hashtags #happykanginday and #marketing

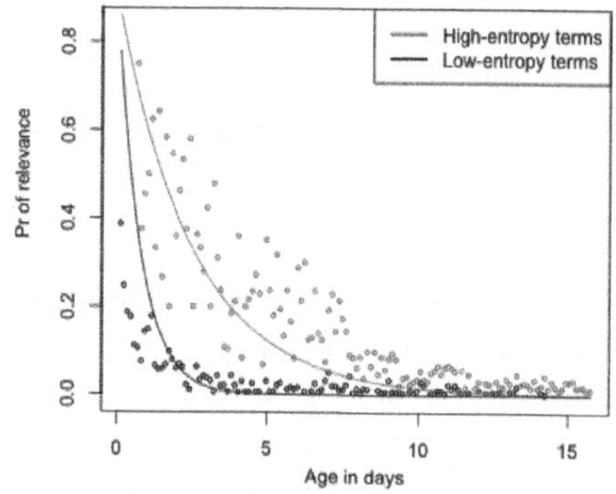

Fig. 2. Pr. of relevance for low- and high-entropy hashtags by minimum age

min age will be low meaning that it should receive a positive temporal weighting as the likelihood of being used is always quite high.

Figure 2 also shows lines fitted to each of the two sets calculated using an exponential decay function: $N(t) = e^{-\eta t}$. $N(t)$ is the expected value at time t and η is the rate of decay, which can be learned from the data as in the example in the figure. The output of this function is between 0 and 1 however since the weight should have a positive effect where $N(t)$ is high and a negative effect when it is low we add a constant of 0.5. Multiplying the original similarity-based scores with the output of this function gives an increased weight where $N(t) > 0.5$ (as the output will be between 1 and 1.5) and a decrease when $N(t) < 0.5$. To model the two categories of hashtag we use two different values of η in the function: one for the low- and one for the high-entropy hashtags (η_l and η_h). When weighting the candidate hashtags we calculate their entropies over previous tweets in the data set and if the entropy is < 0.5 we use η_l, otherwise we use η_h.

We have devised sensible functions for identifying candidate hashtags and then ranking those tags based on their similarity to the target tweet and the target author's expanded interest profile weighted by their temporal relevance. We now detail how we collected a suitable data set for testing our methods and describe the results achieved by them. We conclude by discussing the results and commenting on potential avenues for future work.

4 Experiments

Data Set. A sample of 5,016 Twitter users was collected from the Twitter API [1] by first downloading tweets from the Twitter streaming API - which we assume to be random - and then listing all users who posted any sampled tweets. The account details of these users were obtained and the list filtered by

[1] Twitter REST API version 1.1:
`https://dev.twitter.com/docs/api/1.1`

removing: verified users (usually celebrities or news organisations), those with unusually high numbers of friends (spammers), those who had joined within the past week and had more than 1000 tweets (spammers) and users with no followers (potential spammers), resulting in a list of 2,576 users. From this list we randomly sampled 300 users and collected all tweets written by users they follow - 379,919 - between the 1st and 20th of January 2014, yielding 3,303,016 tweets, appearing a total of 3,528,564 times (a single tweet can appear on more than one user's timeline).

Since this data will be used to suggest hashtags we restricted our dataset to those tweets that have at least 1 hashtag and, in keeping with literature, we do not use retweets in our data set as our similarity search would return an identical retweet, clearly distorting the results. The final data set consisted of 333,784 tweets (10.1% of the original tweets) from 23,476 unique authors with a hashtag vocabulary of 51,899 unique tags.

Models and Baselines. Here the models used for hashtag suggestion are briefly described. An * indicates that the method was newly developed for this work.

1. *TweetMax* - hashtags drawn from similar tweets only, weights each candidate by max. similarity score. Best-performing method of Zangerle et al. [21].
2. *UserMean** - uses only similar users to draw hashtags from, weights each candidate by the mean similarity score over all similar users.
3. *CombCount* - uses union of hashtags from similar users and tweets, weighted by total count of hashtag over all similar tweets and users. Slightly more sophisticated version of best-performing method used by Kywe et al. [11].
4. *CombInt** - uses union of hashtags from similar users and tweets, candidates weighted by linearly interpolated scores from similar tweets and users.
5. *TemporalTweetMax** - hashtags drawn from similar tweets only, weighted by the maximum similarity score multiplied by temporal relevance score.
6. *TemporalCombInt** - uses union of hashtags from similar users and tweets, candidates weighted by linearly-interpolated scores from similar tweets and users multiplied by temporal relevance score.

Splitting the Data Set and Optimising Parameters. The data was sorted by time in ascending order and split into two sections in the ratio 80:20. Although it is normal to use split-fold testing with multiple splits, this is not possible as we are interested in the specific temporal aspects of the data and therefore cannot test on data generated before the training data. The last 20% of the largest split was used to optimise any model parameter values: the η values for the low- and high-entropy exponential decay functions (η_l and η_h respectively) and the λ parameter controlling the linear interpolation of hashtag ranking scores from similar tweets and users. All parameters were optimised via an exhaustive search resulting in the following optimised values: $\eta_l = 1.2$, $\eta_h = 0.6$ and $\lambda = 0.4$.

The smaller split of the data (66,757 tweets) was used to test the models. For each model we wish to predict which hashtags were actually chosen by the author. To do so all data which existed prior to each tweet in time was used to train the similarity models and learn the entropies and minimum ages of the hashtags. The content terms of each test tweet as well as the user ID of the author were then input into each model which returned a ranked list of 5 candidate hashtags. These suggestions were then compared with the hashtags actually assigned to the tweet (which we take to be relevant, with all other hashtags being non-relevant). The standard IR metrics of precision and recall were calculated for ranks 1 through 5, where precision is the number of relevant returned over the number returned and recall is the number of relevant returned over the total number relevant. Note that often there is only one relevant tag.

4.1 Results

Table 1 summarises the performance of the 6 hashtag suggestion models. P@1 indicates a model's ability to return a relevant tag at position one in the ranking and P@5 indicates the ability to return at least one relevant tag within the top 5 candidates. R@5 describes, on average, what ratio of all relevant tag the model is able to suggest. We can see that the additions made to the basic models in this work served to increase both the accuracy and coverage of the suggested hashtags. Figures in parentheses indicate the percentage difference of each model relative to the most competitive baseline (*TweetMax*).

The worst-performing model is *CombinedCount*, probably because of its lack of sophisticated weighting, relying as it does on combined frequencies of each candidate hashtag over the similar tweets and similar users. In terms of P@1, *TweetMax* is able to achieve better performance than *UserMean*, which is expected as it relies on information about the tweet itself and not just about the user, however surprisingly *UserMean* is able to out-perform it over the next 4 rank positions. Linearly interpolating candidate hashtags and scores (*CombInt*) performs significantly better than either of the single components on their own.

The addition of the temporal weighting seems to have a very positive impact on suggestion performance as the two models which include this weighting returned better performance than the equivalent models without it. The most sophisticated method (*TemporalCombInt*) yields the best performance figures for all metrics and does particularly well in terms of recall, being able to predict 42.9% of all hashtags correctly within the top 5 rank positions.

Changes in Rank Position. To examine the performance improvements resulting from including temporal information in the ranking we look in more detail

Table 1. Results table for all methods compared. * indicates a statistically significant improvement over *TweetMax*, 2-sample t at 95% confidence

Method	P@1	P@5	R@5
TweetMax	0.256	0.086	0.311
CombinedCount	0.153	0.069	0.267
UserMean	0.238 (-8.2%)	0.089 (3.5%)	0.348* (11.9%)
CombInt	0.292* (14.1%)	0.106* (23.3%)	0.416* (33.8%)
TemporalTweetMax	0.310* (21.9%)	0.102* (18.6%)	0.359* (15.4%)
TemporalCombInt	0.314* (22.7%)	0.109* (26.7%)	0.429* (37.9%)

at the relative performance between *TweetMax* and *TemporalTweetMax* (which are otherwise identical). The variation in performance can be better understood by considering the difference in the ranks of the relevant hashtags. Figure 3 shows the distribution of the difference in the ranking of relevant hashtag for single-hashtag tweets. Red bars show the number of tweets where the ranking was improved by using temporal information, while the green ones indicate a deterioration and "other" refers to all rank changes greater than 5. The chart shows - as one would expect from table 1 - that the temporal information results in a better ranking far more often than a worse one (74% of cases are better). However it also shows that in the majority of negative cases, the ranking is only deteriorated by a couple of rank positions - 48.1% of deteriorated rankings are only by one or two positions. On the other hand, in 37.8% of cases where the temporal information has a positive effect the improvement is dramatic (i.e. an improvement of more than 5 rank positions).

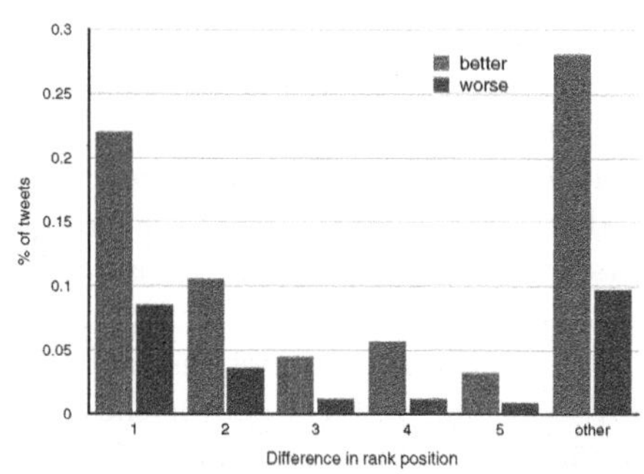

Fig. 3. Δ in rank position of relevant hashtag between the rankings from *TweetMax* and *TemporalTweetMax*

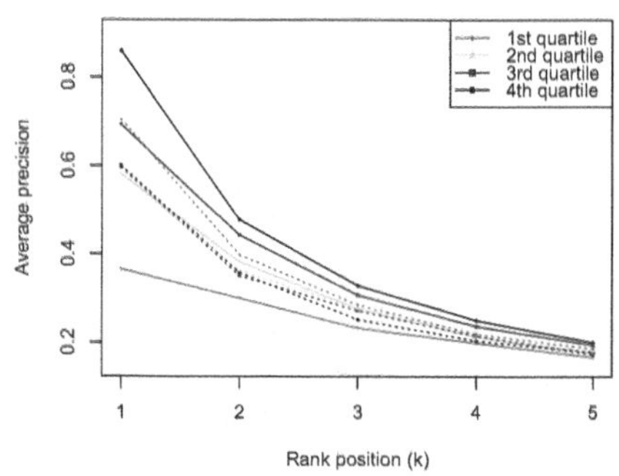

Fig. 4. Average precision by amount of historical data. Solid lines = *TemporalTweetMax*, dotted = *TweetMax*.

Do We Have Enough Data?. Given that the temporal part of our models is based on historical information about each hashtag and our data set represents only a small sample of all tweets posted on Twitter between the crawling dates, we now investigate how the quantity of information available about a hashtag affects performance. We again compare the performance of *TweetMax* and *TemporalTweetMax* and only consider instances where the target tweet has a single hashtag. However, here we sample to ensure that both models were able to return the single relevant hashtag somewhere within the first 20 rank positions.

Figure 4 shows how the performance of the two models changes (over the first 5 rank positions) as we vary the amount of historical data available in the training set about the relevant hashtag (in 4 equal quartiles). The *TemporalTweetMax* (solid lines) returns poorer performance when we have less information about the relevant hashtag but much better performance when we have more information. This pattern is, however, not evident for the *TweetMax* model (dotted lines) which returns similar performance regardless of the amount of data available about the relevant hashtag. This indicates that the performance improvements given by the inclusion of temporal information could be even greater if we had more training data to base our entropy and minimum age statistics on.

5 Conclusions and Future Work

In this paper we proposed new methods for hashtag suggestion which could lead to more frequent, accurate and useful assignment of hashtags to tweets. We began by identifying the most effective measures for basic hashtag recommendation in the literature and proceeded to investigate ways to improve performance by including more information in the model and using existing information in a more intelligent fashion. We analysed temporal patterns in hashtags from the perspective of relevance and identified trends which we hypothesised could be exploited to make suggestions more temporally relevant. By analysing the ages of tweets containing candidate hashtags, relative to when a new tweet was posted, we developed a method to re-weight candidate scores by their temporal relevance.

Using a sample of real-world Twitter data from January 2014, we tested the performance of our novel methods against two competitive baselines from the literature, demonstrating significant performance improvements, although these were restricted by the amount of training data available and therefore have the potential to be better still. We showed that these improvements came from both the temporal information and the more sophisticated use of user interest data, augmented by a collaborative filtering approach. Further analysis showed that the improvements in rank position of relevant hashtags brought by including temporal information in rankings are often quite large. Perhaps more importantly, in the few instances where the temporal weighting is not successful, it rarely results in a large detrimental change to the ranking.

In future work we would like to first investigate more nuanced ways of identifying similar tweets and similar users, perhaps using some form of dimensionality reduction to mitigate the issue of vocabulary mismatch. Similar approaches to addressing this problem could also consider term expansion of the initial list of candidate hashtags. We also intend to investigate how the temporal information could be more subtly utilised in the models. Instead of grouping hashtags into two categories (low- and high-entropy) with tuned η values, it may be possible to learn a smooth mapping between a hashtag's entropy and the appropriate value of η in the temporal weighting function.

References

1. Boyd, D., Golder, S., Lotan, G.: Tweet, tweet, retweet: Conversational aspects of retweeting on twitter. In: HICSS, pp. 1–10 (2010)
2. Chen, K., Chen, T., Zheng, G., Jin, O., Yao, E., Yu, Y.: Collaborative personalized tweet recommendation. In: SIGIR, pp. 661–670 (2012)
3. Correa, D., Sureka, A.: Mining tweets for tag recommendation on social media. In: SMUC 2011, pp. 69–76. ACM, New York (2011)

4. Cui, A., Zhang, M., Liu, Y., Ma, S., Zhang, K.: Discover breaking events with popular hashtags in twitter. In: CIKM, pp. 1794–1798 (2012)
5. Cunha, E., Magno, G., et al.: Analyzing the dynamic evolution of hashtags on twitter: A language-based approach. In: LSM, pp. 58–65 (2011)
6. Ebrahimi, N., Maasoumi, E., Soofi, E.S.: Ordering univariate distributions by entropy and variance. J. of Econometrics 90(2), 317–336 (1999)
7. Elsweiler, D., Harvey, M.: Engaging and maintaining a sense of being informed: Understanding the tasks motivating twitter search. JASIST (2014)
8. Harvey, M., Carman, M., Elsweiler, D.: Comparing tweets and tags for URLs. In: Baeza-Yates, R., de Vries, A.P., Zaragoza, H., Cambazoglu, B.B., Murdock, V., Lempel, R., Silvestri, F. (eds.) ECIR 2012. LNCS, vol. 7224, pp. 73–84. Springer, Heidelberg (2012)
9. Honeycutt, C., Herring, S.C.: Beyond microblogging: Conversation and collaboration via twitter. In: HICSS, pp. 1–10 (2009)
10. Huang, J., Thornton, K.M., Efthimiadis, E.N.: Conversational tagging in twitter. In: HT, pp. 173–178 (2010)
11. Kywe, S.M., Hoang, T.-A., Lim, E.-P., Zhu, F.: On recommending hashtags in twitter networks. In: Aberer, K., Flache, A., Jager, W., Liu, L., Tang, J., Guéret, C. (eds.) SocInfo 2012. LNCS, vol. 7710, pp. 337–350. Springer, Heidelberg (2012)
12. Lehmann, J., Gonçalves, B., Ramasco, J.J., Cattuto, C.: Dynamical classes of collective attention in twitter. In: WWW, pp. 251–260 (2012)
13. Li, T., Yu Wu, Y.Z.: Twitter hashtag prediction algorithm. In: WORLDCOMP (2011)
14. Lin, J., Mishne, G.: A study of "churn" in tweets and real-time search queries. In: ICWSM (2012)
15. Ma, Z., Sun, A., Cong, G.: Will this #hashtag be popular tomorrow? In: SIGIR, pp. 1173–1174 (2012)
16. Sakaki, T., Okazaki, M., Matsuo, Y.: Earthquake shakes twitter users: Real-time event detection by social sensors. In: WWW, pp. 851–860 (2010)
17. Teevan, J., Ramage, D., Morris, M.R.: #twittersearch: A comparison of microblog search and web search. In: WSDM, pp. 35–44 (2011)
18. Tumasjan, A., Sprenger, T.O., Sandner, P.G., Welpe, I.M.: Predicting elections with twitter: What 140 characters reveal about political sentiment. In: ICWSM, pp. 178–185 (2010)
19. Weng, J., Lim, E.-P., Jiang, J., He, Q.: Twitterrank: Finding topic-sensitive influential twitterers. In: WSDM, pp. 261–270 (2010)
20. Yang, L., Sun, T., Zhang, M., Mei, Q.: We know what @you #tag: Does the dual role affect hashtag adoption? In: WWW, pp. 261–270 (2012)
21. Zangerle, E., Gassler, W.: Recommending #-tags in twitter. In: CEUR Workshop (2011)
22. Zangerle, E., Gassler, W., Specht, G.: On the impact of text similarity functions on hashtag recommendations in microblogging environments. Social Network Analysis and Mining 3(4), 889–898 (2013)
23. Zhang, Y., Wu, Y., Yang, Q.: Community discovery in twitter based on user interests. Journal of Computational Information 3, 991–1000 (2012)
24. Zhao, D., Rosson, M.B.: How and why people twitter: The role micro-blogging plays in informal comms at work. In: GROUP, pp. 243–252 (2009)

Temporal Multinomial Mixture for Instance-Oriented Evolutionary Clustering

Young-Min Kim[1], Julien Velcin[2], Stéphane Bonnevay[2], and Marian-Andrei Rizoiu[2]

[1] Korea Institute of Science and Technology Information, South Korea
[2] ERIC Lab., University of Lyon 2, France
ymkim@kisti.re.kr,
{julien.velcin,stephane.bonnevay,marian-andrei.rizoiu}@univ-lyon2.fr

Abstract. Evolutionary clustering aims at capturing the temporal evolution of clusters. This issue is particularly important in the context of social media data that are naturally temporally driven. In this paper, we propose a new probabilistic model-based evolutionary clustering technique. The Temporal Multinomial Mixture (TMM) is an extension of classical mixture model that optimizes feature co-occurrences in the trade-off with temporal smoothness. Our model is evaluated for two recent case studies on opinion aggregation over time. We compare four different probabilistic clustering models and we show the superiority of our proposal in the task of instance-oriented clustering.

Keywords: Evolutionary clustering, mixture model, temporal analysis.

1 Introduction

Clustering is a popular way to preprocess large amount of unstructured data. It can be used in several ways, such as data summarization for decision making or representation learning for classification purpose. Recently, evolutionary clustering aims at capturing temporal evolution of clusters in data streams. This is different from traditional incremental clustering, for evolutionary clustering methods optimize another measure that builds the clustering model at time $t+1$ by taking into account of the model at time t in a retrospective manner [1,2,3]. Applications range from clustering photo tags in `flickr.com` to document clustering in textual corpora.

The existing methods fall into two different categories. *Instance-oriented* evolutionary clustering mostly aims at primarily regrouping objects and *topic-oriented* evolutionary clustering aims at estimating distributions over components (*e.g.*, words). While the former extracts tightest clusters in the feature space, the latter improves the smoothness of temporally consecutive clusters. In this work, we focus on developing a new temporal-driven model of the first category, motivated by two case studies.

We propose a new probabilistic evolutionary clustering method that aims at finding dynamic instance clusters. Our model, Temporal Mixture Model (TMM),

A. Hanbury et al. (Eds.): ECIR 2015, LNCS 9022, pp. 593–604, 2015.

is an extension of the classical mixture model to categorical data streams. The main novelty is not to use Dirichlet prior in order to relax smoothness constraint. While our model can further be improved in terms of more advanced properties, such as learning the number of clusters as in non-parametric models [4,5], in this work we mainly focus on realizing our basic idea and studying the performance of the model. Using internal evaluation measures, we demonstrate that TMM outperforms a typical *topic-oriented* dynamic model and achieves similar compactness results with two static models. This result is achieved at the slight expense of cluster smoothing ability through temporal epochs.

In the following sections, we first motivate and present in detail the proposed TMM model. Then we present the experimental results of TMM as well as three other methods of the literature, showing the superiority of our method with new type of datasets in opinion mining. Finally we conclude with some perspectives and future works.

2 Motivation and Related Work

2.1 Motivation

Document clustering and topic extraction are sometimes considered as equivalent problems, and the methods desired to address each problem are used interchangeably [6]. However, there is a fundamental difference in terms of clustering objective between them and this draws a clear algorithmic difference. Even though this issue has not been actively mentioned in the clustering literature, it is indirectly confirmed by the fact that topic modeling is not recommended to be used directly for document clustering in general. [7] have empirically shown that even simple mixture models outperform Dirichlet distribution-based topic models for document clustering, when directly using model parameters. A recent work [8] is dealing with this issue by proposing an integrated graphical model for both document clustering and topic modeling. However, the great success of topic models in unsupervised learning has often led researchers to use them as instance clustering in practice. This observation remains valid for evolutionary clustering, for which one hardly finds an alternative to topic models using Dirichlet smoothing. The situation is identical when dealing with more classical categorical data, which is the case of our work. This paper starts from this significant issue in evolutionary clustering.

To the best of our knowledge, this is the first attempt to use a non-Dirichlet mixture model for temporal analysis of data steams. The reason why we abandon Dirichlet prior reflects our (maybe peculiar) point of view towards the Dirichlet distribution. That is, the power of topic models mainly comes from their ability to smoothen distributions via the Dirichlet prior. It is effective for extracting representative topics or for making inference on new data. However, in case of clustering instances, a hasty smoothing of the distributions risks to mix data samples with no common feature. In this paper, target datasets are not necessarily textual; therefore the clustering process can be more sensitive to this effect than when dealing with a large feature space (such as a vocabulary of words).

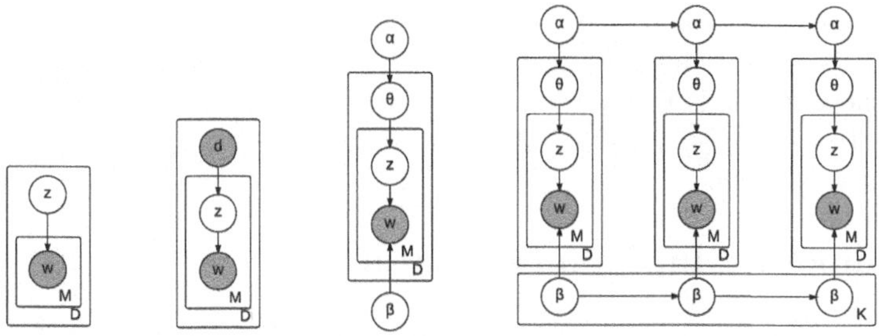

(a) (b) (c) (d)

Fig. 1. Graphical representation of (a) MM, (b) PLSA, (c) LDA, and (d) DTM

In our case, each feature becomes more important, thus special attention must be given to the actual matching between the cluster distribution and the observed feature co-occurrences. This is the reason why we decide to build our method on top of a simple mixture model expecting to minimize the discussed risk.

2.2 Related Work

Our new evolutionary clustering model, *Temporal Multinomial Mixture* (TMM), has been designed with the assumption that regrouping non co-occurring features is highly prejudicial. TMM is a temporal extension of the *Multinomial Mixture* (MM), a simple probabilistic generative model for clustering. More complex mixture models such as *Probabilistic Latent Semantic Analysis* (PLSA) [9] or *Latent Dirichlet Allocation* (LDA) [10] seem less suitable for clustering non-textual data as mentioned in Section 2.1. Non co-occurring features are often mixed together in the same cluster because of additional hidden layers added to these models, either for instance-topic distributions (PLSA) or as Dirichlet prior (LDA). The graphical representation of these models are given in Fig. 1(a)-(c).

Despite the obvious difference between our purpose and dynamic topic models, since the temporal approaches in unsupervised learning usually stand on the basis of topic models, it is inevitable to introduce the state-of-the arts of topic models. Most of the current techniques in clustering introducing a temporal dimension are topic models taking Dirichlet distribution [11,12] since the development of *Dynamic Topic Model* (DTM, Fig. 1(d)) [13], a simple extension of LDA. This kind of dynamic topic analysis has been the object of numerous studies over recent years and more complex models such as DMM [11] or MDTM [12] have been developed. In comparison, TMM is much simpler and we experimentally show the power of simple modeling by comparing three clustering methods, MM, PLSA and DTM with ours.

On the other hand, some pioneer works were designed for data points that basically last during more than two time periods. These stand on various theoretical bases such as k-means, agglomerative hierarchical method, spectral clustering, and even generative model [1,2,14]. However, the underlined property of data points is contrary to the case of data stream, which is our concern here.

Whatsoever, several applications in temporal analysis are intended for dealing with text corpora. Being designed for text hinders the "out-of-the-box" application of these methods to unfamiliar data such as image, gene, market, network data etc. In comparison, TMM is an evolutionary clustering dedicated to general categorical datasets.

3 Temporal Multinomial Mixture

We propose Temporal Multinomial Mixture (TMM) for instance-oriented clustering over time. TMM is a temporal extension of MM and the relation between TMM and MM is analogous to that between DTM and LDA. While the majority of existing temporal topic analysis tend to complicate the modeling process, TMM rather goes against this trend. We assume that complicated distributional structures confuse the instance-oriented clustering. Therefore our method assumes the form of a simple mixture model. As in many other evolutionary clusterings and temporal topic analysis, data instances are associated with a time epoch. A time epoch indicates a time period between two adjacent moments. Dataset is generally divided into subsets by epoch. Instances are assumed to be described by features weighted with a frequency[1].

3.1 Generative Process

The graphical representation of TMM is given in Fig. 2. The extension from MM is realized by encoding the temporal dependency into the relation between data components w of the current epoch and the clusters z of the previous epoch. The generation process of an instance $d^t = i$ at the epoch t is as follows:

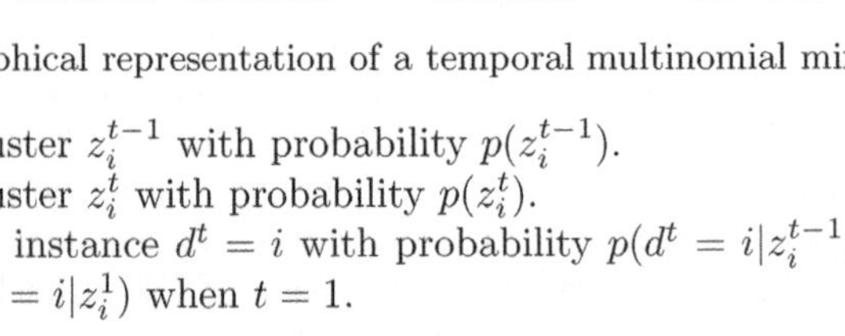

Fig. 2. Graphical representation of a temporal multinomial mixture model

- Choose a cluster z_i^{t-1} with probability $p(z_i^{t-1})$.
- Choose a cluster z_i^t with probability $p(z_i^t)$.
- Generate an instance $d^t = i$ with probability $p(d^t = i|z_i^{t-1}, z_i^t)$ when $t > 1$ or with $p(d^1 = i|z_i^1)$ when $t = 1$.

The last step is realized by repeatedly generating the components $w_{im}^t, \forall m$, sequential features in the instance $d^t = i$, as illustrated in the graphical representation. Unlike most temporal graphical models, it is a connected network

[1] For the sake of understanding, the reader can see a feature as a unique word over a vocabulary and a data component as a word occurrence in a document even if an instance is not a document here.

Table 1. Notations

Symbol	Description
d^t	instance d at epoch t
w^t_{im}	mth component in the instance $d^t{=}i$ at epoch t
z^t_i	assigned cluster for instance $d^t{=}i$ at epoch t
D^t	sequence of instances at epoch t
Z^t	sequence of cluster assignments for D^t
$\mathbf{D}$	sequence of all instances, $\mathbf{D} = (D^1, D^2, ..., D^T)$
$\mathbf{Z}$	sequence of cluster assignments for $\mathbf{D}$, $\mathbf{Z} = (Z^1, Z^2, ..., Z^T)$
T	number of epochs
$\|D^t\|$	number of instances at epoch t
M^t_d	number of components in the instance d at epoch t
V	number of unique components (number of features)
K	number of clusters
$\boldsymbol{\phi}^t_k$	multinomial distribution of cluster k over components at epoch t
π^t_k	prior probability of cluster k at epoch t
α	weight for the component generation from the clusters of previous epoch, $0 < \alpha < 1$

considering the correlation of all topics of t and $t-1$. The notations used in TMM are shown in Table 1. We mostly referred the notations in [15] and [16]. Because of the variable dependency between different time epochs, we need sequential expression of features. This is the reason why we cannot use the simple notation of MM.

3.2 Parameter Estimation via Approximate Development

The objective function to be maximized is the expectation of log-likelihood [17]:

$$\mathbb{E}(\tilde{\mathcal{L}}) = \sum_{\mathbf{Z}} p(\mathbf{Z}|\mathbf{D}, \Theta_{old}) \cdot \log\left(p(\mathbf{D}, \mathbf{Z}|\Theta)\right) \tag{1}$$

Because of the dependency between the variables z^t and z^{t-1}, the log-likelihood cannot be simplified using marginalized latent variables as in MM or PLSA. Instead, we start with the joint distribution of instances and assigned clusters (latent variables):

$$p(\mathbf{D}, \mathbf{Z}) = \left\{ \prod_{d=1}^{|D^1|} p(z^1_d) \cdot p(d^1|z^1_d) \right\} \left\{ \prod_{t=2}^{T} \prod_{d=1}^{|D^t|} p(z^t_d) \cdot p(d^t|z^t_d, z^{t-1}_d) \right\} \tag{2}$$

Eq. 1 can be simplified by taking only the valid latent variables per term:

$$\mathbb{E}(\tilde{\mathcal{L}}) = \sum_{i=1}^{|D^1|} \sum_{k=1}^{K} p(z^1_i = k|d^1 = i) \log\{p(z^1_i = k) p(d^1 = i|z^1_i = k)\}$$
$$+ \sum_{t=2}^{T} \sum_{i=1}^{|D^t|} \sum_{k=1}^{K} \sum_{k'=1}^{K} p(z^t_i{=}k, z^{t-1}_i{=}k'|d^t{=}i) \log\{p(z^t_i{=}k) p(d^t{=}i|z^t_i{=}k, z^{t-1}_i{=}k')\} \tag{3}$$

At epoch 1, $p(d^1{=}i|z^1_i{=}k)$ can be rewritten using ϕ^1_k and $n^1_{i,j}$, the frequency of unique component j included in the instance i, such as $\prod_{j=1}^{V} (\phi^1_{k,j})^{n^1_{i,j}}$. On the other hand, the instance generation at epoch t, $\forall t \geq 2$ is dependent also on the clusters of the previous epoch. Thus the conditional probability of an instance i given current and previous clusters k and k', is inferred as follows with Bayes Rule:

$$p(d^t{=}i|z_i^t{=}k,z_i^{t-1}{=}k'){=}\prod_{m=1}^{M_i^t}\frac{p(z_i^t{=}k|w_{im}^t,z_i^{t-1}{=}k')p(z_i^{t-1}{=}k'|w_{im}^t)p(w_{im}^t)}{p(z_i^t{=}k,z_i^{t-1}{=}k')} \quad (4)$$

Under the assumptions of graphical model, the analytical calculation of $p(z_i^t|w_{im}^t, z_i^{t-1})$ is so complicated because the latent variables are related by the explaining away effect. To tackle this issue, we make an important hypothesis that $p(z_i^t|w_{im}^t, z_i^{t-1})$ can be *approximated* by $p(z_i^t|w_{im}^t)$. Consequently, Eq. 4 is rewritten using $p(w_{im}^t{=}j|z_i^t{=}k)$ as well as $p(w_{im}^t{=}j|z_i^{t-1}{=}k')$, which is equivalent to the previous epoch's parameter $\phi_{k',j}^{t-1}$. Penalizing the influence rate of the previous cluster with α, a weighted parameter value $(\phi_{k',j}^{t-1})^\alpha, 0 < \alpha < 1$ is used instead of $\phi_{k',j}^{t-1}$. Letting the constant $\prod_{m=1}^{M_i^t} 1/p(w_{im}^t)$ be C_i^t, we obtain the following equation.

$$p(d^t = i|z_i^t = k, z_i^{t-1} = k') = C_i^t \cdot \prod_{j=1}^{V} (\phi_{k,j}^t)^{n_{i,j}^t} (\phi_{k',j}^{t-1})^{\alpha \cdot n_{i,j}^t} \quad (5)$$

Using the parameters Θ, the $\mathbb{E}(\tilde{\mathcal{L}})$ becomes:

$$\mathbb{E}(\tilde{\mathcal{L}}){=}\sum_{i=1}^{|D^1|}\sum_{k=1}^{K} p(z_i^1{=}k|d^1{=}i)\cdot\left\{\log\pi_k^1 + \sum_{j=1}^{V} n_{i,j}^1\cdot\log\phi_{k,j}^1\right\}$$
$$+\sum_{t=2}^{T}\sum_{i=1}^{|D^t|}\sum_{k=1}^{K}\sum_{k'=1}^{K} p(z_i^t{=}k,z_i^{t-1}{=}k'|d^t{=}i)\cdot\left\{\log\pi_k^t{+}\log C_i^t{+}\sum_{j=1}^{V} n_{i,j}^t\cdot(\log\phi_{k,j}^t{+}\alpha\cdot\log\phi_{k',j}^{t-1})\right\}$$

3.3 EM Algorithm

We solve the following optimization problem to obtain the parameter values.

$$\arg\max_{\Theta}\ \mathbb{E}(\tilde{\mathcal{L}}),\ \text{ subject to} \sum_{j=1}^{V}\phi_{k,j}^t = 1,\ \forall t,k \quad \text{and} \quad \sum_{k=1}^{K}\pi_k^t = 1,\ \forall t.$$

The EM algorithm is updated as follows.

Initialization
Randomly initialize parameters $\Theta = \{\phi_k^t, \pi_k^t \mid \forall t, k\}$

$$\text{subject to} \sum_{j=1}^{V}\phi_{k,j}^t = 1,\quad \forall t,k \quad \text{and} \quad \sum_{k=1}^{K}\pi_k^t = 1,\quad \forall t.$$

E-step
Compute the expectation of posteriors as follows.

$$p(z_i^t{=}k, z_i^{t-1}{=}k'|d^t{=}i){=}\frac{\prod_{j=1}^{V}(\phi_{k,j}^t)^{n_{i,j}^t}(\phi_{k',j}^{t-1})^{\alpha\cdot n_{i,j}^t}\cdot\pi_k^t\cdot\pi_{k'}^{t-1}}{\sum_{a=1}^{K}\sum_{a'=1}^{K}\prod_{j=1}^{V}(\phi_{a,j}^t)^{n_{i,j}^t}(\phi_{a',j}^{t-1})^{\alpha\cdot n_{i,j}^t}\cdot\pi_a^t\cdot\pi_{a'}^{t-1}}, 2{\le}t{\le}T, \forall k, k', i. \quad (6)$$

$p(z_i^1 = k|d^1 = i)$ is similarly calculated by eliminating the variables of $t-1$.

M-step
Update the parameters maximizing the objective function.

$$\phi_{k,j}^t = \frac{\sum_{i=1}^{|D^t|}\sum_{k'=1}^{K} n_{i,j}^t \cdot p(z^t{=}k, z^{t-1}{=}k'|d^t{=}i) + \sum_{i=1}^{|D^{t+1}|}\sum_{k'=1}^{K} \alpha \cdot n_{i,j}^{t+1} \cdot p(z_i^{t+1}{=}k', z_i^t{=}k|d^{t+1}{=}i)}{\sum_{i=1}^{|D^t|}\sum_{j'=1}^{V}\sum_{k'=1}^{K} n_{i,j'}^t \cdot p(z^t{=}k, z^{t-1}{=}k'|d^t{=}i) + \sum_{i=1}^{|D^{t+1}|}\sum_{j'=1}^{V}\sum_{k'=1}^{K} \alpha \cdot n_{i,j'}^{t+1} \cdot p(z_i^{t+1}{=}k', z_i^t{=}k|d^{t+1}{=}i)},$$

$$2 \le t \le T-1, \quad \forall j, k. \tag{7}$$

$\phi_{k,j}^1$ is calculated by eliminating the variables of $t-1$ from the above formula and $\phi_{k,j}^T$ is done by eliminating both variables and terms of $t+1$.

$$\pi_k^t = \frac{\sum_{i=1}^{|D^t|}\sum_{k'=1}^{K} p(z^t{=}k, z^{t-1}{=}k'|d^t{=}i) + \sum_{i=1}^{|D^{t+1}|}\sum_{k'=1}^{K} p(z_i^{t+1}{=}k', z_i^t{=}k|d^{t+1}{=}i)}{\sum_{i=1}^{|D^t|}\sum_{a=1}^{K}\sum_{k'=1}^{K} p(z^t{=}a, z^{t-1}{=}k'|d^t{=}i) + \sum_{i=1}^{|D^{t+1}|}\sum_{k'=1}^{K}\sum_{a=1}^{K} p(z_i^{t+1}{=}k', z_i^t{=}a|d^{t+1}{=}i)}, \tag{8}$$

$$2 \le t \le T-1, \quad \forall k$$

π_k^1 and π_k^T are calculated as in $\phi_{k,j}^1$ and $\phi_{k,j}^T$.

3.4 Instance Assignment and Cluster Evolution

The assignment of each instance is eventually obtained from the estimated distributions. For $t = 1$, we assign to the instance i the cluster that maximizes the posterior probability $p(z_i^1{=}k|d^1{=}i)$. For the instances in the other epochs, we integrate out z_i^{t-1} to obtain the instance cluster such that $p(z_i^t{=}k|d^t{=}i) = \sum_{k'=1}^{K} p(z_i^t{=}k, z_i^{t-1}{=}k'|d^t{=}i)$.

TMM being a connected network, all the clusters in the epoch $t-1$ can contribute to the clusters in the epoch t. Please note that the same cluster index in different epochs does not mean that the corresponding clusters are identical over time. That is why we need to find which cluster of the previous epoch contributes most to the specific cluster k of the current epoch. The dynamic correlation between clusters of the adjacent epochs is fully encoded in the distribution $p(z_i^t{=}k, z_i^{t-1}{=}k'|d^t{=}i)$. By integrating out z_i^t instead of z_i^{t-1} from $p(z_i^t{=}k, z_i^{t-1}{=}k'|d^t{=}i)$, we can deduce the most likely cluster at the previous epoch for the instance $d^t{=}i$. We call it the origin of the instance. Given the specific cluster $z^t = k$, we have the classified instances and their origins. By counting we find the most frequent origin and we can eventually relate the most influential cluster of the previous epoch to $z^t = k$. Since this is a surjective function from t to $t-1$, the division of a cluster over time is traceable. Conversely, the merge of multiple clusters can also be caught if we choose not only the most likely cluster but also the second or the third likely one.

Table 2. Statistics of datasets and features we define

	ImagiWeb opinion dataset	RepLab 2013
source	Political opinion tweets	English & Spanish opinion tweets
annotation size	11527 tweets (7283 unique)	26709 tweets (all unique)
subsets	Entity (politician P, politician Q)	Domain (automative, music)
feature space	Aspect-polarity pairs	Entity-polarity pairs
	9 aspects, 6 polarities	20 entities per domain, 3 polarities

4 Experiments

We compare four different generative models in order to evaluate the performance of TMM. DTM is selected as a Dirichlet-based model; MM and PLSA are used as static baselines for highlighting the effect of introducing a temporal dimension. Finally, we show that TMM outperforms the other models on two datasets of opinion mining, by finding a trade-off between compactness and temporal smoothing.

4.1 Datasets

ImagiWeb political opinion dataset.[2] The first dataset is comprised of a set of about 7000 unique tweets related to two politicians (each politician is analyzed separately). The manual annotation process has been supervised by domain experts of public opinion analysis and it has followed a detailed procedure with the design of 9 aspects (*e.g.*, project, ethic or political line) targeted by 6 possible opinion polarities (-2=very negative, -1=negative, 0=neutral, +1=positive, +2=very positive, NULL=ambiguous). For instance, the tweet "RT @anonym: P's project is just hot air" can be described by the pair `(project,-2)` attached to the politician P. Each pair corresponds to a feature w whose value is the occurrence of the corresponding opinion for describing the studied entity. The full procedure and dataset are described in [18]. Because of the length limit of a tweet as well as for clustering purpose, we decide to combine the annotations by author for each time epoch.

RepLab 2013 Corpus. This corpus has been used for the RepLab 2013, second evaluation campaign on Online Reputation Management. It consists of a collection of tweets referring to 61 entities from four domains. We select two dominant domains out of four, automative and music, where the number of entities is 20 respectively. The clustering is done for each *domain* separately this time instead of entity. Tweets are annotated with three polarities: positive, negative and neutral. We let the features be the *entity-polarity* pairs instead of aspect-polarity pairs, so that the opinion aggregation is based on co-occurring entities. It means that the opinion groups are constructed by users, who are interested in same entities with similar polarities. Tab. 2 sums up basic statistics on the two datasets.

[2] It will be distributed to the public in Spring 2015 on the ImagiWeb official website, `http://mediamining.univ-lyon2.fr/velcin/imagiweb`

4.2 Evaluation Measures

The ground truth is hardly available when evaluating clustering output for evolutionary clustering. We instead develop the following three quantitative measures with the object of well detecting clustering quality.

Co-occurrence level. Our main interest lies in detecting compact clusters, which means that the number of observed co-occurring features actually match the estimated distribution. This measure counts the real number of co-occurring feature couples in each sample among the non-zero features grouped in a cluster.

Unsmoothness. This catches the dissimilarity between corresponding clusters through different time epochs using Kullback-Leibler (KL) divergence. If a temporal clustering method well detects the evolution of clusters, the cluster signatures having same identity would be similar to each other. Therefore we develop 'unsmoothness' to measure how suddenly a cluster changes over time.

Homogeneity. This measures the degree of unanimity of grouped tweets in a cluster in terms of polarity. Opposite opinions hardly co-occur because an author usually keep his opinion stance in a sufficiently short time. By ignoring the degree of polarity, the homogeneity of a cluster is simply defined as follows[3]:

$$\text{Homogeneity} = (|\#(\text{positive}) - \#(\text{negative})|)/(\#(\text{positive}) + \#(\text{negative}))$$

This is intuitive and easy to be visually represented but is an indirect evaluation.

4.3 Result

Clustering is conducted at subset level. For a given clustering method and subset, experiments are repeated 10 times by changing initialization to get the statistical significance. Since MM and PLSA are time-independent, temporal clusters are obtained via two stages: normal clustering per epoch and heuristic matching between clusters of two adjacent epochs judged by their distributional form.

The first sub-table of Table 3 shows the experimental results of four methods on the ImagiWeb dataset. Once clustering is done per subset, we merge the results to analyze together the reputation of two competitors. The number of epochs is fixed at two by splitting data by an actual important political event date. Each value is the averaged result of 10 experiments as well as the standard deviation in brackets. The bold number indicates the best result among four methods and the underlined one is the second best. The gray background of bold number means the result statistically outperforms the second best and the light-gray means it does not outperform the second best, but does the third one. The value of α in TMM has been set to 0.7 after several pre-experiments judged by visual representation of clusters (as shown in Fig. 3) as well as balance among cluster sizes. We manually choose the value by varying α from 0.5 to 1. Larger value increases distributional similarity whereas decreases separation of opposite opinions. The hyper parameters of DTM have also been set to the best ones after several experiments.

Globally, TMM outperforms the others in terms of two measures except unsmoothness. Then DTM and MM are in the second place. PLSA produces the

[3] #(polarity) is the number of tweets annotated with this polarity.

Table 3. Evaluation of temporal clustering for four methods on ImagiWeb opinion dataset(left) and RepLab 2013 for automative(middle) and music(right)

	ImagiWeb opinion dataset				RepLab(Auto)				RepLab(Music)			
	TMM	DTM	MM	PLSA	TMM	DTM	MM	PLSA	TMM	DTM	MM	PLSA
Avg. Homogen.	**0.86**	0.70	**0.86**	0.67	**0.76**	0.67	0.73	0.70	**0.77**	0.75	0.75	0.76
(stand. deviation)	(0.02)	(0.06)	(0.02)	(0.05)	(0.02)	(0.05)	(0.03)	(0.04)	(0.03)	(0.05)	(0.02)	(0.03)
Co-occurr. level	**123**	113	122	111	**40**	34	**40**	33	**26**	22	25	22
(stand. deviation)	(1.98)	(1.02)	(0.88)	(1.48)	(1.21)	(1.18)	(0.58)	(1.52)	(0.74)	(0.80)	(0.40)	(0.35)
Avg. Unsmooth.	2.27	**1.57**	3.16	3.61	4.30	**1.37**	6.35	6.91	4.5	**2.54**	6.12	7.75
(stand. deviation)	(0.23)	(0.10)	(0.33)	(0.21)	(0.90)	(0.12)	(0.82)	(0.69)	(0.90)	(0.51)	(0.87)	(1.11)

worst result for all measures. Since homogeneity is a direct basis to evaluate if the tested method well detects the difference between negative and positive opinion groups, it becomes more important when the mix of opposite opinions is a crucial error. Co-occurrence level also directly shows if the captured clusters are really based on the co-occurring features. Given that both measures evaluate cluster quality of a specific time epoch, it is encouraging that TMM provides identical or even slightly better result than MM because TMM can be thought of as a relaxed version of MM in the point of view of data adjustment over time. The result therefore demonstrates that TMM successfully makes use of the generative advantage of MM. For homogeneity, TMM and MM both obtain 0.86, which perfectly outperform the second best DTM in terms of Mann-Whitney test with the p-value of 0.00001. Meanwhile, for unsmoothness the best one is DTM with a clearly better result, 1.57 than the others. DTM concentrates on the distribution adjustment over time at the expense of well grouping opinions that is the principal objective in the task. The second best TMM also perfectly outperforms MM with the p-value of 0.0002. It proves the time dependency encoded in TMM successfully enhances MM for capturing cluster evolution.

In addition to the quantitative evaluation, we visualize a TMM clustering result in Fig. 3. It is the evolution of two clusters with five different time epochs on politician P subset. The zoomed figure shows a negative group about P at epoch 1 especially on the aspects "political line" and "project". TMM captures the dynamics of the cluster over time as shown in the figure. As time goes by, opinions about "project" disappear (at t=5) but the other negative opinions about "ethic" appear in the cluster. The cluster in the second line groups mainly positive and neutral opinions about various aspects at epoch 1, but some aspects gradually disappear with time.

The experimental results on RepLab 2013 corpus are given in the middle and right sub-tables in Table 3. Number of epochs is also fixed at two and the data is split by the median date. This corpus is not originally constructed for opinion aggregation, therefore we do not have sufficient feature co-occurrences. The proportion of instances having at least two components is only 5.2% for automative and 2.9% for music. Despite the handicap, we rather expect that we would emphasize the characteristics of each model via experiments with this restrictive dataset. The α value has been set to 1 to make maximum use of the effect of previous clusters regarding lack of co-occurrences.

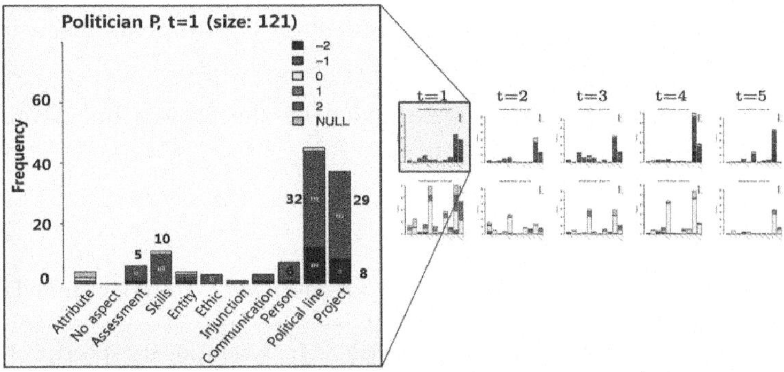

Fig. 3. Visualization of the evolution of two clusters extracted from a TMM clustering result with five different time epochs on politician P subset

Two outstanding methods are TMM and DTM but there is an obvious difference between their results. TMM gives a better performance in terms of local clustering quality such as homogeneity and co-occurrence whereas DTM outperforms the others in temporal view. Homogeneity does not seem really meaningful here because the opposite opinions about different entities can be naturally mixed in an opinion group. However, from the fact that co-occurring features are rarely observed and, moreover, only 10% of total opinions are negative in the corpus, negative and positive opinions seldom co-occur. Therefore, the high homogeneity can be a significant measure here also. As in the ImagiWeb dataset, the co-occurrence level of TMM is clearly better than that of DTM. On the other hand, even though DTM gives a perfectly better result for unsmoothness, the captured distributions are not really based on the real co-occurrences when we manually verify the result. Nevertheless, when the dataset is extremely sparse as in this case, smoothing distribution would anyway provide the opportunity not to ignore rarely co-occurring features.

5 Conclusions

The proposed TMM model succeeds in effectively extending MM, by taking into consideration the temporal factor for clustering. Our method captures the dynamics of clusters much better than the heuristic matching of single clustering results using MM or PLSA, without losing clustering quality at local time epoch. TMM clearly outperforms DTM in terms of local cluster quality. DTM tends to produce well-smoothed distributions over time, but as shown through its low performance with the other measures, high smoothness does not always signify that the cluster evolution is well detected.

An inherent hypothesis in TMM is that clusters evolve progressively over time and it has enabled the modeling of direct dependency between two adjacent epochs. However if abrupt changes arrive, the distributions found for each cluster can be incoherent. A future developmental direction is taking such changes into account. A possible way could be to establish an automatic adjustment of the dependency rate α. Another interesting direction is to develop means to infer more exactly the conditional probability $p(z_i^t|w_{im}^t, z_i^{t-1})$.

Acknowledgments. This work was funded by the project ImagiWeb ANR-2012-CORD-002-01.

References

1. Chakrabarti, D., Kumar, R., Tomkins, A.: Evolutionary clustering. In: KDD 2006, pp. 554–560. ACM (2006)
2. Chi, Y., Song, X., Zhou, D., Hino, K., Tseng, B.L.: Evolutionary spectral clustering by incorporating temporal smoothness. In: KDD 2007, pp. 153–162. ACM (2007)
3. Xu, T., Zhang, Z.M., Yu, P.S., Long, B.: Dirichlet process based evolutionary clustering. In: ICDM 2008, pp. 648–657. IEEE Computer Society (2008)
4. Teh, Y.M., Jordan, M.I., Blei, D.M.: Hierarchical Dirichlet processes. Journal of the American Statistical Association 101(476), 1566–1581 (2006)
5. Ahmed, A., Xing, E.: Dynamic non-parametric mixture models and the recurrent chinese restaurant process: With applications to evolutionary clustering. In: SIAM International Conference on Data Mining (2008)
6. Zhang, J., Song, Y., Zhang, C., Liu, S.: Evolutionary hierarchical Dirichlet processes for multiple correlated time-varying corpora. In: KDD 2010, pp. 1079–1088. ACM (2010)
7. Pessiot, J.F., Kim, Y.M., Amini, M.R., Gallinari, P.: Improving document clustering in a learned concept space. Inform. Process. & Manag. 46(2), 180–192 (2010)
8. Xie, P., Xing, E.P.: Integrating document clustering and topic modeling. In: UAI (2013)
9. Hofmann, T.: Probabilistic latent semantic analysis. In: UAI 1999, pp. 289–296 (1999)
10. Blei, D.M., Ng, A.Y., Jordan, M.I.: Latent Dirichlet allocation. J. Mach. Learn. Res. 3, 993–1022 (2003)
11. Wei, X., Sun, J., Wang, X.: Dynamic mixture models for multiple time series. In: IJCAI 2007, pp. 2909–2914. Morgan Kaufmann Publishers Inc. (2007)
12. Iwata, T., Yamada, T., Sakurai, Y., Ueda, N.: Online multiscale dynamic topic models. In: KDD 2010, pp. 663–672. ACM (2010)
13. Blei, D.M., Lafferty, J.D.: Dynamic topic models. In: International conference on Machine learning, ICML 2006, pp. 113–120. ACM (2006)
14. Lin, Y.R., Chi, Y., Zhu, S., Sundaram, H., Tseng, B.L.: Facetnet: a framework for analyzing communities and their evolutions in dynamic networks. In: WWW 2008, pp. 685–694. ACM (2008)
15. AlSumait, L., Barbará, D., Domeniconi, C.: On-line LDA: Adaptive topic models for mining text streams with applications to topic detection and tracking. In: ICDM 2008, pp. 3–12. IEEE Computer Society (2008)
16. He, Y., Lin, C., Gao, W., Wong, K.F.: Dynamic joint sentiment-topic model. ACM Transactions on Intelligent Systems and Technology (2013)
17. Bishop, C.M.: Pattern Recognition and Machine Learning (Information Science and Statistics). Springer-Verlag New York, Inc., Secaucus (2006)
18. Velcin, J., et al.: Investigating the image of entities in social media: Dataset design and first results. In: LREC 2014 (2014)

Temporal Latent Topic User Profiles for Search Personalisation

Thanh Vu[1], Alistair Willis[1], Son N. Tran[2], and Dawei Song[1,3]

[1] The Open University, Milton Keynes, United Kingdom
[2] City University London, London, United Kingdom
[3] Tianjin University, Tianjin, P.R. China
{thanh.vu,alistair.willis,dawei.song}@open.ac.uk, son.tran.1@city.ac.uk

Abstract. The performance of search personalisation largely depends on how to build *user profiles* effectively. Many approaches have been developed to build user profiles using topics discussed in relevant documents, where the topics are usually obtained from human-generated online ontology such as Open Directory Project. The limitation of these approaches is that many documents may not contain the topics covered in the ontology. Moreover, the human-generated topics require expensive manual effort to determine the correct categories for each document. This paper addresses these problems by using Latent Dirichlet Allocation for unsupervised extraction of the topics from documents. With the learned topics, we observe that the search intent and user interests are dynamic, i.e., they change from time to time. In order to evaluate the effectiveness of temporal aspects in personalisation, we apply three typical time scales for building *a long-term profile*, *a daily profile* and *a session profile*. In the experiments, we utilise the profiles to re-rank search results returned by a commercial web search engine. Our experimental results demonstrate that our temporal profiles can significantly improve the ranking quality. The results further show a promising effect of temporal features in correlation with click entropy and query position in a search session.

Keywords: User Profiles, Temporal Aspects, Latent Topics, Search Personalisation, Re-ranking.

1 Introduction

As one of the key components in advanced search engines (e.g., Google and Bing), *Search Personalisation* has attracted increasing attention [1,9,12,15,16,19]. The personalisation is expected to improve the usefulness of search algorithms. Unlike the search methods which don't use personalisation, personalised search engines utilise the personal data of each user to tailor search results, which depend not only on the input query but also on the user's interest (as context of the query). Such personal data can be used to construct a *user profile* which is crucial to effective personalisation.

Normally, one of the most common approaches is to represent the profile with the main topics discussed in documents which the user has previously clicked

A. Hanbury et al. (Eds.): ECIR 2015, LNCS 9022, pp. 605–616, 2015.

on [1,8,11,16,19]. The topics of a document are often obtained from a human-generated online ontology, such as the Open Directory Project (ODP) [1,11,19]. This approach has a limitation that many topics may not appear in the ontology. Moreover, it requires expensive manual effort to determine the correct categories for each document, as mentioned in [8]. In order to solve this problem, recent approaches [8,16] focus on learning latent topics from the relevant documents, using unsupervised models (i.e., Latent Dirichlet Allocation (LDA) [2]).

Latent topics have been successfully used to build user profiles, but little attention has been paid to the *temporal* aspects in the latent topic profiles, which reflect an important type of context. In this paper, we propose a novel temporal modelling approach for building user profiles from latent topics. We then carry out a comprehensive study on the effectiveness of temporal features in learning the topical interest of a user, with application to search results re-ranking. Our main goal is to address the following research questions: (1) Can temporal profiles help to improve search performance? and (2) How do temporal aspects affect the re-ranking quality?

To this end, we construct three temporal latent topic profiles for each user using the relevant documents with different time scales in the user's search history. We name the profiles as *session profile*, *daily profile* and *long-term profile*, as they are built from the topics learned from the documents within a session, a day and a whole history respectively. We note that the three profiles represent the user interest in different time scales (from short-term to long-term). In order to extract topics from the relevant documents, we employ the same approach proposed in [16] that utilises a topic modelling method (i.e., LDA [2]) to automatically derive *latent topics* instead of using a human-generated ontology as in [1,11,19].

The rest of this paper is structured as follows. In Section 2, we present the related work on user modelling for search personalisation. Section 3 describes our personalisation framework for building the temporal profiles and using the profiles to re-rank the returned result list. In Section 4, we describe our experiment setting. We then report the results in Section 5 and conclude the paper in Section 6.

2 Related Work

The user profile maintains the user's information on an individual level, typically based on the terms that represent user's search interests. To represent a user profile, Bennett *et al.* [1] mapped the user's interest onto a set of topics, which are extracted from large online ontologies of web sites, namely the ODP. This approach suffers from a limitation that many documents may not appear in the online categorisation scheme. Moreover, it requires expensive manual effort to determine the correct categories for each document. Harvey *et al.* [8] and Vu *et al.* [16] applied a latent topic model (i.e., LDA) to determine these topics. This means that the topic space is determined based purely on relevant documents extracted from query logs and does not require human involvement to define the topics. However, in their researches, the authors used all relevant documents extracted from the user's whole search history to construct the user profile (i.e., long-term profile). Moreover, they treated the relevant documents equally without considering temporal features (i.e., the time of documents being clicked and viewed).

The user interests could be long-term [6,8,14,16] or short-term [18,19]. Long-term interests, in the context of *IR* systems, are stable interests that can be exhibited for a long time in the user's search history. The long-term interests have been shown helpful for improving the search results [6,8,16]. Typically, the interests are represented as frequent terms or topics which have been extracted from the text of user's queries and clicked results. Alternatively, they can be also extracted from other personal data such as computer files and emails etc. [14]. In the application of re-ranking, [8,14,16], these terms/topics that represent long-term interests are used to re-rank relevant documents with the future queries.

Short-term interests, on the other hand, are temporary interests of a searcher during a relatively short time (e.g. in one or some continuous search sessions). The short-term interests are usually obtained from the submitted queries and the clicked documents in a search session and used to personalise the search within the session [18,19]. Bennett *et al.* [1] studied the interaction between long-term and short-term and found that the long-term behaviour provided advantages at the start of a search session while short-term session played a very important role in the extended search session. Furthermore, the combination of short-term and long-term interactions outperformed using either alone.

In this paper, in constrast to Bennett *et al.* [1] and White *et al.* [19], we apply LDA to automatically derive the latent topics from the user's relevant documents. Furthermore, in contrast to [8,16] as building a single user profile statically, we propose three temporal user profiles (i.e., long-term, daily and session profiles) which can represent both long-term and short-term user interests. It is worth noting that our long-term profile is different from Vu *et al.* [16] in term of considering the view-time of the relevant document (Section 3.2). We then thoroughly investigate the effectiveness of the proposed profiles in search personalisation.

3 Personalisation Framework

3.1 Extracting Topics from Relevant Documents

We briefly describe the method to extract topics from relevant documents, which was initially proposed in [16]. We first extract the relevant data of each user from the query logs. A log entry consists of an anonymous user-identifier, a submitted query, top-10 returned URLs, and clicked results along with the user's dwell time. We use the SAT criteria detailed in [7] to identify satisfied (SAT) clicks (as relevant data) from the query logs as either a click with a dwell time of at least 30 seconds or the last result click in a search session. To identify a *session*, we use the common approach of demarcating session boundaries by 30 minutes of user inactivity [11].

After that, we employ LDA [2] to extract latent topics (Z) from the SAT clicked documents (D) of all users. LDA represents each topic as a multinomial distribution over the entire vocabulary. Furthermore, each document is also described as a multinomial distribution over topics.

3.2 Constructing User Profiles

Modelling a User Profile. Formally, the user variable is denoted as U. Let u denote an instance of U. We build a user profile based on the topics of the user's relevant documents. Let $D_u = \{d_1, d_2, .., d_n\}$ be a relevant document set of the user u. We define the user profile of u (given D_u) as a distribution over the topic Z. The probability of a topic z given u is defined as a mixture of probabilities of z given relevant document $d_i \in D_u$ as follows

$$p(z|u) = \sum_{d_i \in D_u} \lambda_i p(z|d_i) \tag{1}$$

Here, $\sum_i \lambda_i = 1$ to guarantee that $\sum_z p(z|u) = 1$. The simple approach as used in Vu *et al.* [16] is to treat relevant documents equally when calculating $p(z|u)$. It means that $\lambda_1 = \lambda_2 = ... = \lambda_n = \frac{1}{|D_u|}$. Therefore, we have

$$p(z|u) = \frac{1}{|D_u|} \sum_{d_i \in D_u} p(z|d_i) \tag{2}$$

Temporal weighting. Since the search intent and user interest change over time, the more recent relevant documents could express more about the user interest than the distant one. This characteristic can be captured by introducing a decay function [18,1]. In this paper, instead of treating all the relevant documents equally (e.g. [16]), we model λ_i as the exponential decay function of t_{d_i}, which is the time the user u clicked on the document d_i, as follows

$$\lambda_i = \frac{1}{K} \alpha^{t_{d_i} - 1} \tag{3}$$

where $K = \sum_{d_i} \alpha^{t_{d_i} - 1}$ is a normalisation factor; $t_{d_i} = 1$ indicates that d_i is the most recent relevant (SAT click) document. By applying Eq. 3 to Eq. 1, we have

$$p(z|u) = \frac{1}{K} \sum_{d_i \in D_u} \alpha^{t_{d_i} - 1} p(z|d_i) \tag{4}$$

Motivating example. Previous work [8,16] on latent topic-based user profiles only used a single user profile (i.e., long-term profile). This work treated all the relevant documents equally and used the user's whole search history to construct the profile. In this paper, however, we treat the relevant documents temporally based on the viewing time of the user on the document. Furthermore, a single long-term profile cannot quickly represent the short-term interest of a user in a search session or in a specific day. For example, with a user having a strong law background, the long-term profile of the user has been constructed from thousands of law-related documents. On the first day of the World Cup (WC) 2014, even though she submitted WC-related queries and clicked on WC-related documents, the updated long-term profile cannot change promptly to express the football interest and does not seem to help personalising the WC-related queries. Therefore, apart from the long-term profile, we model two other profiles, namely *daily* and *session* profiles using the user's relevant documents in the current searching day and current search session respectively. It is worth clarifying that the long-term profile represents the permanent/long-term interest of the user. Otherwise, the session profile describes the provisional interest of the current user. The daily profile indicates the user interest over a searching day. Finally, we construct the three user profiles using different relevant datasets which change overtime as follows:

Long-term Profile. We build the long-term user profile of u using relevant documents D_w extracted from the user's whole search history as follows

$$p_w(z|u) = \frac{1}{K} \sum\nolimits_{d_i \in D_w} \alpha^{t_{d_i}-1} p(z|d_i) \tag{5}$$

Daily Profile. We build the daily user profile of u using relevant documents D_d extracted from the search history of u in the current day as follows

$$p_d(z|u) = \frac{1}{K} \sum\nolimits_{d_i \in D_d} \alpha^{t_{d_i}-1} p(z|d_i) \tag{6}$$

Session Profile. We build the session user profile of u using relevant documents D_s extracted from the current search session of u as follows

$$p_s(z|u) = \frac{1}{K} \sum\nolimits_{d_i \in D_s} \alpha^{t_{d_i}-1} p(z|d_i) \tag{7}$$

3.3 Re-ranking Search Results Using User Profiles

We utilise the user profiles to re-rank the original list of documents returned by a search engine. The detailed steps are as follows

(1) We download the top n ranked search results (as recorded in a data set of query logs) from the search engine for a query. We denote a downloaded web page as d and its rank in the search result list as $r(d)$.

(2) We then compute a similarity measure, $Sim(d|p)$, between each web page d and user profile p. Because both d and p are models as D, P distributions over topic Z, respectively, we use Jensen-Shannon divergence ($D_{JS}\lfloor.||.\rfloor$) to measure the similarity between the two probability distributions as follows

$$Sim(d|p) = D_{JS}\lfloor D||P\rfloor = \frac{1}{2} D_{KL}\lfloor D||M\rfloor + \frac{1}{2} D_{KL}\lfloor P||M\rfloor \tag{8}$$

Here $D_{KL}\lfloor.||.\rfloor$ is the Kullback-Leiber divergence and $M = \frac{1}{2}(D+P)$. After this step, we get three personalised scores, denoted as $f_w = Sim(d|p_w)$, $f_d = Sim(d|p_d)$, and $f_s = Sim(d|p_s)$, with respect to long-term, daily, and session profiles respectively. We consider the three scores as the personalised features of the document d.

(3) The personalised features only represent the user interest on a returned document. Therefore, apart from these features, we also extract other non-personalised features of input query q and the search result d. The full description of these features is presented in Table 1.

Table 1. Summary of the document features

Feature	Description
Personalised Features	
LongTermScore	The similarity score between the document and the long-term profile
DailyScore	The similarity score between the document and the daily profile
SessionScore	The similarity score between the document and the session profile
Non-personalised Features	
DocRank	Rank of the document on the original returned list
QuerySim	The cosine similarity score between the current query and the previous query
QueryNo	Total number of queries that have been submitted to the Search Engine

(4) After extracting the document features, to re-rank the top n returned URLs instead of using a simple ranking function [16], we employ a learning to rank algorithm (LambdaMART [3]) to train ranking models. Among many learning to rank algorithms, LambdaMART has been regarded as one of the best performing algorithms [4], and has been chosen as the base learning algorithm in various state of the art approaches to search personalisation[1] [1,12,13,17]. However, it is worth noting that our proposed features are insensitive to ranking algorithm, thus any reasonable learning-to-rank algorithm would likely provide similar results.

4 Experimental Methodology

4.1 Dataset and Evaluation Methodology

Dataset In the experiment, we evaluate the approaches using the search results produced by a commercial search engine. The data used in our experiments is the query logs of 1166 anonymous users in four weeks, from 01st July 2012 to 28th July 2012. Each sample in the query logs consists of: an anonymous user identifier, an input query, the query time, top 10 returned URLs and clicked results along with the user's dwell time. We also download the content of these URLs for the learning of the topics.

We then partition the whole dataset into profiling, training and test sets. The profiling set is used to build the long-term user profile, the training set is for training the ranking model using LambdaMART and the test set is used for evaluation of the approaches. In particular, the profiling set contains the log

[1] Indeed, an ensemble of LambdaMART rankers won Track 1 of the 2010 Yahoo! *Learning to Rank Challenge* [5].

data in the first 13 days; the training set contains the query logs in next 2 days; and the test set contains the log data in the remaining 13 days. Table 2 shows the basic statistics on the three datasets.

Table 2. Basic statistics of the evaluation search log set

Item	ALL	Profiling	Training	Test
#days	28	13	2	13
#queries	520010	240066	29834	236615
#distinct queries	176029	85641	12112	89445
#search session	94972	43462	5655	45886
#clicks	433277	200119	25805	207353
#SAT clicks	334227	154753	19513	159961
#SAT clicks/#queries	0.6427	0.6446	0.6541	0.6760

Evaluation Methodology. For evaluation, we use the SAT criteria [7] to identify the satisfied clicks (SAT click) from the query logs. We assign a positive (relevant) label to a returned URL if it is a SAT click. Furthermore, similar to [1], we also assign a positive label to a URL if it is a SAT click in one of the repeated/modified queries in the same search session[2]. The remainder of the top-10 URLs are assigned negative (irrelevant) labels. We use the rank positions of the positive labelled URLs as the ground truth to evaluate the search performance before and after re-ranking. We also apply a simple pre-processing on these data sets as follows. At first, we remove the queries whose positive label set is empty from the dataset. After that we discard the domain-related queries (e.g. Facebook, Youtube). Finally, we normalise the relevance features (both personalised and non-personalised features) to zero mean and standard deviation (i.e., z-score) from the training set.

4.2 Experimental Settings

Personalisation Methods and Baselines. We empirically investigate the effect of different temporal aspects in latent topic-based personalisation by using the three proposed profiles and their combination to generate the following features:

1. LongTermScore from long-term profile (LON)
2. DailyScore from daily profile (DAI)
3. SessionScore from session profile (SES)
4. AllScore from combination of three profiles (ALL)

We further combine these features with the non-personalised features to enrich the personalisation with relevant information from all users. As mentioned earlier, our first baseline, named as *Default*, is the search results (ranking of

[2] A query q' is a modification of query q if the returned URLs (top 10) of q' contains at least one SAT click of q.

URLs) returned by the commercial search engine, where we obtain the log data. The second baseline we would like to compare with is the combination of non-personalised features and the topic features proposed by Vu *et al.* [16], which does not take the temporal features into account. We named the second baseline as *Static*.

In the following we present the setting of LDA and LambdaMART for learning the topics and for learning the ranking function respectively. Note that in order to make a fair comparison we use the same topic distributions for all personalisation approaches and baselines.

LDA & LambdaMART. We train the LDA model on the relevant documents extracted from the query logs, as mentioned in Section 3.1. The number of topics is decided by using a held-out validation set which consists of 10% of all the relevant documents. The selected number of topics is the one that gives the lowest perplexity value. We also use the validation set to select the temporal weighting parameter α.

The ranking function is learned using LambdaMART. After getting the features from the approaches, we randomly extract 10% of the training set for validation. We used the default setting for LambdaMART's prior parameters[3]. We follow the same model selection process as in [1,12].

Evaluation Metrics. The evaluation is based on the comparison between our personalised approaches and the baselines. For completeness, we use four evaluation metrics which are: Mean Average Precision (MAP), Precision (P@k), Mean Reciprocal Rank (MRR) and Normalized Discounted Cumulative Gain (nDCG@k). These are standard metrics which have been widely used for performance evaluation in document ranking [10]. For each evaluation metric, the higher value indicates the better ranking.

5 Experimental Results

5.1 Overall Performance

In this experiment, we analyse the effect of temporal aspects on latent topic profiles as proposed in Section 3 using six metrics: MAP, P@1, P@3, MMR, nDCG@5 and nDCG@10. Table 3 shows promising results when the temporal features are used to build user profiles. One can see that all three temporal profiles (i.e., session, daily, long-term profiles) have led to improvements over the original ranking and the use of non-temporal profile. Especially, the combination of all features (ALL) achieves the highest performance. This interesting result shows that a comprehensive user profile should capture different temporal aspects of the user's history. It should be noted that the improvements over the baselines reported in Table 3 are all significant with paired t-test of $p < 0.001$.

In the comparison between the temporal profiles, Table 3 shows that the session profile (SES) achieves better performance than the daily profile (DAI). It also shows that the daily profile (DAI) gains advantage over the long-term

[3] Specifically, number of leaves = 10, minimum documents per leaf = 200, number of trees = 100 and learning rate = 0.15.

Table 3. Overall performance of the methods

Models	MAP	P@1	P@3	MMR	nDCG@5	nDCG@10
Default	*0.7494*	*0.6471*	*0.3320*	*0.7699*	*0.7805*	*0.8197*
Static	0.7460	0.6464	0.3289	0.7683	0.7751	0.8175
LON	0.7577	0.6601	0.3377	0.7813	0.7911	0.8267
DAI	0.7760	0.6897	0.3473	0.8016	0.8080	0.8406
SES	0.7936	0.7207	0.3537	0.8214	0.8238	0.8540
ALL	**0.7964**	**0.7283**	**0.3543**	**0.8254**	**0.8251**	**0.8563**

profile (LON). This indicates that the short-term profiles capture more details of user interest than the longer ones. The results are also consistent with what has been found in [1]. The difference is that our profiles are based on the learned latent topics while they use the ODP.

5.2 Click Entropies

In search personalisation, click entropy plays an important role in deciding the search performance. In [6], Dou et al. have argued that a small click entropy may deteriorate the quality of the search results. The click entropy of a query is defined as:

$$ClickEntropy(q) = \sum\nolimits_{d \in D_q} -p(d|q) \log_2 p(d|q) \tag{9}$$

Here D_q is a collection of web pages which are clicked for the distinct query q, and $p(d|q)$ is the percentage of the clicks on document d among all the clicks for q. A smaller query click entropy value indicates more agreement between users on clicking a small number of web pages. In this paper, we are also interested in investigating the effect of the click entropy on the performance of the temporal latent topic profiles. In the experimental data, about 67.25% and 16.34% queries have a low click entropy from 0 to 0.5 and from 0.5 to 1 respectively; 10.05% and 3.95% queries have a click entropy from 1 to 1.5 and from 1.5 to 2 respectively; and only 2.41% queries have a high click entropy (≥ 2).

In Figure 1, we show the improvement of the temporal profiles over the *Default* ranking from the search engine in term of MAP metric for different magnitudes of click entropy. Here the statistical significance is also guaranteed with the use of paired t-test ($p < 0.001$). The results show that when users have more agreement over clicked documents, with respect to smaller value of click entropy, the re-ranking performance is only slightly improved. For example, with click entropy between 0 and 0.5, the improvement of the MAP metric from long-term profile is of only 0.39%, in comparison with the original search engine. One may see that the effectiveness of the temporal profiles is increasing proportionally according to the value of click entropy. In particular, the improvement of personalised search performance increases significantly when the click entropy becomes larger, especially with click entropies ≥ 0.5, and the highest improvements are achieved when click entropies are ≥ 2. This result contributes a case study on temporal latent topic profiles to the study of click entropy for personalisation besides the static latent topic profile [16].

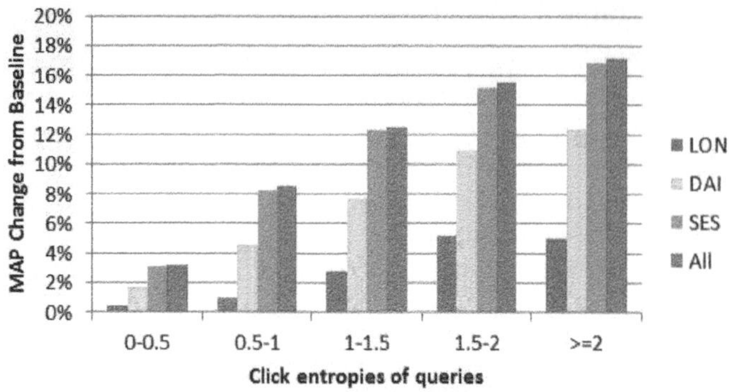

Fig. 1. Search performance improvements over *Default* with different click entropies

5.3 Query Positions

A query usually has a broader influence in a search session than only returning a list of URLs. The position of a query in a search session is also important because it may be fine-tuned by a user after the unsatisfactory results from previous queries. Therefore, in order to get into the insights of the user's information need, a search engine should take into account the position of an input query in a search session. In this experiment we aim to study whether the position of a query has any effect on the performance of the temporal latent topic profiles. For each session, we label the queries by their positions during the search. The first five queries are numbered from one to five according to the order of the time that they have been entered to the search engine, the remaining queries are labelled as ≥ 6, similarly as in [1].

We show the MAP performances of the temporal latent topic profiles for different query positions in Figure 2. From the MAP values, we can see that the first query always received higher satisfaction than the others. It shows that the advanced search engine where we extracted the logs has managed to produce reasonably relevant results at the first query. The higher query positions achieve smaller value of MAP in a search session, which can be explained as users tend to search for supplementary information after the first query, and that the latter queries are so similar to the previous one that the search results contains many URLs which have already appeared in the previous search result. Our result is consistent to what has been mentioned in [19].

Note that we cannot build a session profile for the first query because there is no previously observed relevant document for the query. For long-term and daily profiles, we found that their search performances are similar to the search engine performance of the first query. This can be explained by the fact that the single long-term and daily profiles are diverse and cannot sufficiently represent the user recent interests for the first query. Furthermore, as shown in Figure 2, the search engine satisfies most the user's information need for the first query (MAP value of 0.8353 out of 1). However, for the next queries in the search session, the temporal latent topic profiles show a significant improvement. It

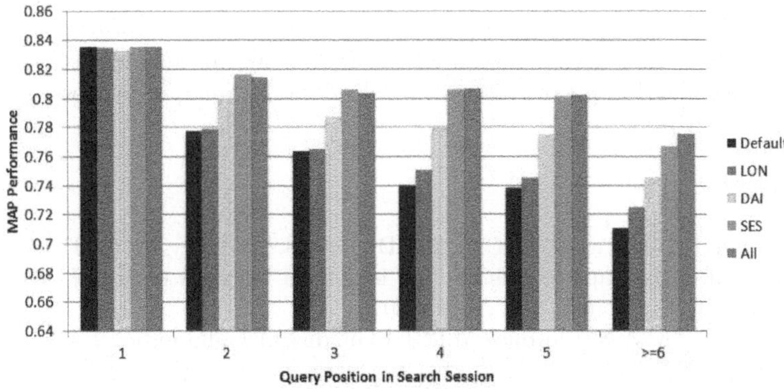

Fig. 2. Performances of the methods by position of query in search session

shows that temporal profiles can quickly adapt to represent the user interest. For example, the session profile achieves the highest performance on the second and the third queries in a session whilst the combination of profiles outperforms the other models on the queries from the fourth positions. This new result is interesting because it shows that the temporal features can help tuning the search performance in further queries which has not been done successfully by the original search engine.

6 Conclusions

We have presented a study on the temporal aspects for building user profiles with latent topics learned from the documents. For each user, we used relevant documents at different time scales to build long-term, daily, and session profiles. Each user profile is represented as a distribution over latent topics from which we extract the features and combine them with non-personalised features to learn a ranking function using LambdaMART. We performed a set of experiments to study the effectiveness of the temporal latent topic-based profiles.

The results showed that the temporal features help improve search performance over the competitive ranker of the original search engine and over the static latent topic profile. We also found that the session profile captures the most interests of a user and is able to generate helpful features for learning the re-ranking function. The best performance was achieved by the combination of all three temporal profiles, indicating that a good personalisation should take into account all temporal aspects from user's search history. Other experimental results confirmed that the impact of the query's click entropy on temporal latent topic profile is similar to that on the static latent topic profile. Finally, another interesting finding is the usefulness of the temporal profile in tuning the search results for the next queries in a search session.

Acknowledgements. This work is partially funded by the Chinese National Program on Key Basic Research Projects (973 Program, grant no. 2013CB329304,

2014CB744604) and the National Science Foundation of Chinese (grant no. 61272265). We thank the reviewers for their valuable comments. We would also like to thank Jingfei Li for providing the data & his help in pre-processing it.

References

1. Bennett, P.N., White, R.W., Chu, W., Dumais, S.T., Bailey, P., Borisyuk, F., Cui, X.: Modeling the impact of short- and long-term behavior on search personalization. In: SIGIR, pp. 185–194. ACM (2012)
2. Blei, D.M., Ng, A.Y., Jordan, M.I.: Latent dirichlet allocation. J. Mach. Learn. Res., 993–1022 (2003)
3. Burges, C.J., Ragno, R., Le, Q.V.: Learning to rank with nonsmooth cost functions. In: NIPS, pp. 193–200. MIT Press (2007)
4. Burges, C.J.C.: From ranknet to lambdarank to lambdamart: An overview. Technical Report MSR-TR-2010-82, Microsoft Research (July 2010)
5. Chapelle, O., Chang, Y., Liu, T.: Yahoo! learning to rank challenge overview. In: JMLR, pp. 1–24 (2011)
6. Dou, Z., Song, R., Wen, J.-R.: A large-scale evaluation and analysis of personalized search strategies. In: WWW, pp. 581–590. ACM (2007)
7. Fox, S., Karnawat, K., Mydland, M., Dumais, S., White, T.: Evaluating implicit measures to improve web search. ACM Trans. Inf. Syst., 147–168 (2005)
8. Harvey, M., Crestani, F., Carman, M.J.: Building user profiles from topic models for personalised search. In: CIKM, pp. 2309–2314. ACM (2013)
9. Hassan, A., White, R.W.: Personalized models of search satisfaction. In: CIKM, pp. 2009–2018. ACM (2013)
10. Manning, C.D., Raghavan, P., Schütze, H.: Introduction to Information Retrieval. Cambridge University Press (2008)
11. Raman, K., Bennett, P.N., Collins-Thompson, K.: Toward whole-session relevance: Exploring intrinsic diversity in web search. In: SIGIR, pp. 463–472 (2013)
12. Shokouhi, M., White, R.W., Bennett, P., Radlinski, F.: Fighting search engine amnesia: Reranking repeated results. In: SIGIR, pp. 273–282. ACM (2013)
13. Song, Y., Shi, X., White, R., Awadallah, A.H.: Context-aware web search abandonment prediction. In: SIGIR, pp. 93–102. ACM (2014)
14. Teevan, J., Dumais, S.T., Horvitz, E.: Personalizing search via automated analysis of interests and activities. In: SIGIR, pp. 449–456. ACM (2005)
15. Teevan, J., Morris, M.R., Bush, S.: Discovering and using groups to improve personalized search. In: WSDM, pp. 15–24. ACM (2009)
16. Vu, T.T., Song, D., Willis, A., Tran, S.N., Li, J.: Improving search personalisation with dynamic group formation. In: SIGIR, pp. 951–954. ACM (2014)
17. Wang, H., Song, Y., Chang, M.-W., He, X., Hassan, A., White, R.W.: Modeling action-level satisfaction for search task satisfaction prediction. In: SIGIR, pp. 123–132. ACM (2014)
18. White, R.W., Bennett, P.N., Dumais, S.T.: Predicting short-term interests using activity-based search context. In: CIKM, pp. 1009–1018. ACM (2010)
19. White, R.W., Chu, W., Hassan, A., He, X., Song, Y., Wang, H.: Enhancing Personalized Search by Mining and Modeling Task Behavior. In: WWW, pp. 1411–1420. ACM (2013)

Document Priors Based On Time-Sensitive Social Signals

Ismail Badache and Mohand Boughanem

IRIT - Paul Sabatier University, Toulouse, France
{Badache,Boughanem}@irit.fr

Abstract. Relevance estimation of a Web resource (document) can benefit from using social signals. In this paper, we propose a language model document prior exploiting temporal characteristics of social signals. We assume that a priori significance of a document depends on the date of users actions (social signals) and on the publication date (first occurrence) of the document. Particularly, rather than estimating the priors by simply counting signals related to the document, we bias this counting by taking into account the dates of the resource and the action. We evaluate our approach on IMDb dataset containing 167438 resources and their social data collected from several social networks. The experiments show the interest of temporally-aware signals at capturing relevant resources.

Keywords: Social Information Retrieval, Social Signals, Signal Time, Resource Publication Date, Social Ranking, Language Models.

1 Introduction

Web search engines are expected to return relevant search results for a query. Classic notions of relevance focus on textual relevance. Recently, majority of search engines include social signals (e.g. +1, like) as non-textual features to relevance. However, in the existing works signals are considered time-independent. They are taken into account by only counting the signal frequency on a resource.

In this paper, we hypothesise that signals are time-dependent, the date when the user action has happened is important to distinguish between recent and old signals. Therefore, we assume tha the recency of signals may indicate some recent interests to the resource, which may improve the a priori relevance of document. Secondly, number of signals of a resource depends on the resource age. Generally, an old resource may have much more signals than a recent one.

We introduce the time-aware social approach that incorporates temporal characteristics of users' actions as prior in the retrieval model. Precisely, instead of assuming uniform document priors in this retrieval model, we assign document priors based on the signals associated to that document biased by both the creation date of the signals and the age of the document. Research questions addressed in this paper are the following:

1. How to take into account signals and their date to estimate the priors?
2. What is the impact of temporally-aware signals on IR system performance?

A. Hanbury et al. (Eds.): ECIR 2015, LNCS 9022, pp. 617–622, 2015.

The remainder of this paper is organized as follows. Section 2 reviews some related work. Section 3 presents details of our approach. In section 4, we describe our experiments. Finally, we conclude the paper and announce some future work.

2 Related Work

While considerable work has been done in the context of temporal query classification there is still lack of user studies that would analyze users' actions in temporal search from diverse viewpoints. Major existing works [1, 2] focus on how to improve IR effectiveness by exploiting users' actions and their underlying social network. For instance, Chelaru et al. [2] study the impact of social signals (like, dislike, comment) on the effectiveness of search on YouTube. Badache and Boughanem [1] show the impact of different signals individually and grouped.

The works that are most related to our approach include [3, 4], which attempt to improve ranking in Web search. Inagaki et al. [3] propose a set of temporal click features, called ClickBuzz, to improve machine learning recency ranking by favoring URLs that have been of recent interest for users. Khodaei and Alonso [4] propose incorporating time as aspect when investigating social search. They categorized user social interests into five classes: recent, ongoing, seasonal, past and random, and then analyzed Twitter and Facebook data on users activities.

Our work has a similar motivation as those previous efforts, i.e., harnessing any temporal features around a resource to improve relevance ranking of conventional text search. However, our approach is based on novel characteristics which are incorporated into language model. Our goal is to estimate the significance of a resource by taking into account the signal recency and the age of the resource.

3 Time-Aware Social Signals

Our approach focuses on the temporal dimension of users' actions. We rely on language model to model temporally-aware signals as a prior probability.

3.1 Preliminaries and Context

Social information that we exploit within the framework of our model can be represented by 5-tuple $< U, R, A, T, SN >$ where U, R, A, T, SN are finite sets of instances: *Users*, *Resources*, *Actions*, *Times* and *Social networks.*

Resources. We consider a collection $C=\{D_1, D_2, ...D_n\}$ of n documents. Each document D can be a Web page, video or other type of Web resources. We assume that resource D can be represented both by a set of textual keywords $D_w=\{w_1, w_2, ...w_z\}$ and a set of social actions A performed on D, $D_a=\{a_1, a_2, ...a_m\}$.

Actions. We consider a set $A=\{a_1, a_2, ...a_m\}$ of m actions (signals) that users can perform on resources. These actions (e.g. *like*, *share*, *comment* on Facebook) represent the relation between users $U=\{u_1, u_2, ...u_h\}$ and resources C.

Time. Time T represents two types of temporal dimensions:

1. The history of each social action, let $T_{a_i}=\{t_{1,a_i}, t_{2,a_i},..t_{k,a_i}\}$ a set of k moments (datetime format) at which action a_i was produced, noted t_{k,a_i}.
2. Age of resource, let $T_d=\{t_{D_1}, t_{D_2},..t_{D_n}\}$ a set of n dates (datetime format) at which each resource D was published, noted t_D.

3.2 Query Likelihood and Document Priors

We exploit language models [5] to estimate the relevance of document to a query. The language modelling approach computes the probability $P(D|Q)$ of a document D being generated by query Q as follows:

$$P(D|Q) \stackrel{\text{rank}}{=} P(D) \cdot P(Q|D) = P(D) \cdot \prod_{w_i \in Q} P(w_i|D) \tag{1}$$

$P(D)$ is a document prior i.e. query-independent feature representing the probability of seeing the document. The document prior is useful for representing and incorporating other sources of evidence to the retrieval process. w_i represents words of query Q. Estimating of $P(w_i|D)$ can be performed using different models (Jelineck Mercer, Dirichlet) [5]. The main contribution in this paper is how to estimate $P(D)$ by exploiting social signals.

3.3 Estimating Time-Aware Priors

According to our previous approach [1], the priors are estimated by a simply counting of actions performed on the resource. We assume that signals are independent, the general formula is the following:

$$P(D) = \prod_{a_i \in A} P(a_i) \tag{2}$$

$P(a_i)$ is estimated using maximum-likelihood: $P(a_i) = \dfrac{Count(a_i, D)}{Count(a_\bullet, D)}$ (3)

To avoid Zero probability, we smooth $P(a_i)$ by collection C using Dirichlet. The formula becomes as follows:

$$P(D) = \prod_{a_i \in A} \left(\frac{Count(a_i, D) + \mu \cdot P(a_i|C)}{Count(a_\bullet, D) + \mu} \right) \tag{4}$$

$P(a_i|C)$ is estimated using maximum-likelihood: $P(a_i|C) = \dfrac{Count(a_i, C)}{Count(a_\bullet, C)}$ (5)

Where: $P(D)$ represents the a priori probability of D. $Count(a_i, D)$ represents number of occurrence of action a_i on resource D. $a_\bullet$ is the total number of social signals in document D or in collection C.

We assume that this simple counting of signals may boost old resources compared to recent ones, because resources with long life in the Web has much more chance to get more signals than recent ones. In addition, we assume that resources that have recent signals are more likely to interest user. We propose to consider the dates associated with a signal and the creation of a resource. To estimate priors, we distinguish two ways to handle it:

a. By considering time of signal: we assume that a resource associated with fresh (recent) signals should be promoted comparing to those associated with old signals. Each time a given signal appears, it is associated with its occurrence time. Therefore, instead of counting each occurrence of a given signal, we bias the counting, noted $Count_{t_a}$, by the date of the occurrence of the signal.

$$Count_{t_a}(t_{j,a_i}, D) = \sum_{j=1}^{k} f(t_{j,a_i}, D) = \sum_{j=1}^{k} \exp\left(-\frac{\| t_{current} - t_{j,a_i} \|^2}{2\sigma^2}\right) \quad (6)$$

Where: $f(t_{j,a_i}, D)$ represents signal-time function, we use Gaussian Kernel [6] to estimate a distance between current time $t_{current}$ and t_{j,a_i} with $\sigma \in R_+$.

The prior $P(D)$ is estimated using formula 4 but by replacing $Count()$ by $Count_{t_a}()$. Notice that if the signal time is not considered $f(t_{j,a_i}, D) = 1 \ \forall t_{j,a_i}$.

b. By considering the age of resource: the resource publication date plays an important role on the social life of this resource, i.e. an old resource has a greater chance to have a large number of interactions compared to a recently published resource. So to cope with this issue we propose to normalize the distribution of signals associated with a resource through resource publication date. We divide the number of signals by the current lifespan of the resource.

$$Count_{t_D}(a_i, D) = \frac{Count(a_i, D)}{Age(D)} = \frac{Count(a_i, D)}{\exp\left(-\frac{\|t_{current} - t_D\|^2}{2\sigma^2}\right)} \quad (7)$$

The prior $P(D)$ is estimated using formula 4 but by replacing $Count()$ by $Count_{t_D}()$ for document and $Count_{t_C}()$ for collection.

4 Experimental Evaluation

To evaluate our approach, we conducted a series of experiments on IMDb dataset. The baseline is a retrieval process without using document priors. Our main goal in these experiments is to evaluate the impact of temporally-aware signals on IR.

4.1 Description of Test Dataset

We used a collection IMDb documents provided by INEX[1]. Each document describes a movie, and is represented by a set of metadata, and has been indexed

[1] `https://inex.mmci.uni-saarland.de/tracks/dc/2011/`

according to keywords extracted from fields [1]. For each document, we collected specific social data via their corresponding API of 5 social networks listed in table 1. The nature of these social signals is a counting of each social actions on the resource. We chose 30 topics with their relevance judgments provided by INEX IMDb 2011[2]. In our study, we focused on the effectiveness of the top 1000 results. Table 1 shows an example of a document with their social data.

Table 1. Instance of document with social data

		Facebook			Google+	Delicious	Twitter	LinkedIn
Film Title	Id	Like	Share	Comment	+1	Bookmark	Tweet	Share
Sinister	*tt1922777*	14763	13881	22914	341	12	2859	14
		Facebook						
Film Title	Id	Last Share		Last Comment		Publication Date		
Sinister	*tt1922777*	2014-09-29T02:49:01		2014-09-28T00:41:01		2011-05-07T19:00:57		

Unfortunately, the date of the different actions are not available except the last date of Facebook actions (*comment* and *share*). Therefore, we represent results using formula 6 biased only by the last date of *comment* and *share*.

4.2 Result and Discussion

We conducted experiments with models based only on documents (Lucene Solr model and Hiemstra language model without prior [7]), as well as approaches combining textual content and social features with temporal aspects as prior of document. We note that the best value of $\mu \in [90, 100]$.

Tables 2 summarizes the results of precision@k for $k \in \{10, 20\}$, nDCG (Normalized Discounted Cumulative Gain) and MAP. We evaluated different configurations, by taking into account social actions, actions time (labeled signal^{T_a}) and resource age (labeled signal^{T_D}). We have already shown that exploiting time-independent signals as prior improve search. In order to check the significance of the results, we performed the Student test and attached * (significance against baselines) to the performance number of each row in the table 2 when the p-value is 0.05 confidence level, compared to the corresponding baselines results.

First, we investigate the retrieval performance attainable by considering the *action time*, in our case date of last *comment* and *share*. Table 2 (With Considering Action Time) shows that the nDCG and precisions are in general slightly better than the nDCG and precision scores where *action time* is ignored, but remain very comparable. Second, we investigate the retrieval performance attainable by considering the *publication date of resource.*Table 2 (With Considering Age of Resource) shows that the nDCG and precisions are in general better than the nDCG and precision scores where *publication date* is ignored (Without Considering Time). Finally, the best results are obtained by (All Criteria^{T_D}) run with considering the *publication date*. Therefore, *publication date* factor is the most effective temporal aspect to enhance a search. Concerning date of signals, we did not really evaluated the real impact of the proposal because of the lack

[2] `https://inex.mmci.uni-saarland.de/tracks/dc/2011/`

of suitable data (dates of different actions). We exploited only the date of the last action which is not enough to draw effective conclusion.

Table 2. Results of P@k for $k \in \{10, 20\}$, nDCG and MAP

IR Models	P@10	P@20	nDCG	MAP	IR Models	P@10	P@20	nDCG	MAP
Baselines: Without Priors					**With Considering Action Time T_a**				
Lucene Solr	0.3411	0.3122	0.3919	0.1782	Share^{T_a}	0.4148*	0.3681*	0.5472*	0.2970*
ML.Hiemstra	0.3700	0.3403	0.4325	0.2402	Comment^{T_a}	0.3861*	0.3601*	0.5207*	0.2844*
Baselines: Without Considering Time [1]					**With Considering Publication Date T_D**				
Like	0.3938	0.3620	0.5130	0.2832	Like^{T_D}	0.4091*	0.3620*	0.5308*	0.2907*
Share	0.4061	0.3649	0.5262	0.2905	Share^{T_D}	0.4177*	0.3721*	0.5544*	0.2989*
Comment	0.3857	0.3551	0.5121	0.2813	Comment^{T_D}	0.3912*	0.3683*	0.5285*	0.2874*
Tweet	0.3879	0.3512	0.4769	0.2735	Tweet^{T_D}	0.3918*	0.3579*	0.4903*	0.2779*
+1	0.3826	0.3468	0.5017	0.2704	+1^{T_D}	0.3900	0.3511	0.5246	0.2748
Bookmark	0.3730	0.3414	0.4621	0.2600	Bookmark^{T_D}	0.3732	0.3427	0.4671	0.2618
Share (LIn)	0.3739	0.3432	0.4566	0.2515	Share^{T_D} (LIn)	0.3762	0.3449	0.4606	0.2542
All Criteria	0.4408	0.4262	0.5974	0.3300	All Criteria^{T_D}	0.4484*	0.4305*	0.6200*	0.3366*

5 Conclusion

In this paper, we studied the impact of time related to users' actions and resource on IR. We proposed to estimate a social priors of a document by considering the time of the action and the publication date of the resource. Experiments conducted on IMDb dataset show that taking into account social features and temporal aspects in a textual model improves the quality of returned search results. The main contribution of this work is to show that time of user's action and the ratio of signals are fruitful for IR systems. An important issue that we did not address is the exploitation of times associated for each action. Unfortunately, currently social networks APIs do not allow extraction of these informations. For future work, we plan to estimate the impact of signals diversity with respect of their ages. Further experiments on another dataset are also needed.

References

[1] Badache, I., Boughanem, M.: Social priors to estimate relevance of a resource. In: IIiX Conference. IIiX 2014, pp. 106–114. ACM, NY (2014)
[2] Chelaru, S., Orellana-Rodriguez, C., Altingovde, I.S.: How useful is social feedback for learning to rank youtube videos? In: World Wide Web, pp. 1–29 (2013)
[3] Inagaki, Y., Sadagopan, N., Dupret, G., Dong, A., Liao, C., Chang, Y., Zheng, Z.: Session based click features for recency ranking. In: AAAI Press (2010)
[4] Khodaei, A., Alonso, O.: Temporally-aware signals for social search. In: SIGIR 2012 Workshop on Time-aware Information Access (2012)
[5] Ponte, J.M., Croft, W.B.: A language modeling approach to information retrieval. In: SIGIR Conference, pp. 275–281. ACM, NY (1998)
[6] Phillips, J.M., Venkatasubramanian, S.: A gentle introduction to the kernel distance. CoRR abs/1103.1625 (2011)
[7] Hiemstra, D.: A linguistically motivated probabilistic model of information retrieval. In: Nikolaou, C., Stephanidis, C. (eds.) ECDL 1998. LNCS, vol. 1513, pp. 569–584. Springer, Heidelberg (1998)

Prediction of Venues in Foursquare Using Flipped Topic Models

Wen-Haw Chong, Bing-Tian Dai, and Ee-Peng Lim

Singapore Management University 80 Stamford Road, Singapore 178902
whchong.2013@phdis.smu.edu.sg, {btdai,eplim}@smu.edu.sg

Abstract. Foursquare is a highly popular location-based social platform, where users indicate their presence at venues via check-ins and/or provide venue-related tips. On Foursquare, we explore Latent Dirichlet Allocation (LDA) topic models for venue prediction: predict venues that a user is likely to visit, given his history of *other* visited venues. However we depart from prior works which regard the users as documents and their visited venues as terms. Instead we 'flip' LDA models such that we regard venues as documents that attract users, which are now the terms. Flipping is simple and requires no changes to the LDA mechanism. Yet it improves prediction accuracy significantly as shown in our experiments. Furthermore, flipped models are superior when we model tips and check-ins as separate modes. This enables us to use tips to improve prediction accuracy, which is previously unexplored. Lastly, we observed the largest accuracy improvement for venues with fewer visitors, implying that the flipped models cope with sparse venue data more effectively.

Keywords: Foursquare, venue prediction, topic models.

1 Introduction

The prevalence and growing popularity of social media in recent years have led to an explosive grow in observable user behavior data. In particular, location-based platforms such as Foursquare and Gowalla provide rich context and user-visitation data. For example, Foursquare users can indicate their presence at venues via check-ins. They can optionally write reviews about visited venues, referred to as tips. These data are fast growing, fine-grained and vast in volume. Currently Foursquare[1] reports a user base of over 50 million, with more than 6 billion check-ins generated. Thus it is not surprising that check-ins has been especially well studied for user profiling and modeling [2,4,5,7].

In this work, we focus on Foursquare due to its market dominance and the ease of accessing related data. Our problem of interest is to predict venues that a user will visit. This translates easily to applications of commercial value, such as user profiling, venue analysis and targeted advertising. For example, venue owners may want to direct their advertisements or promotions at selected new users based on their propensity of visitations.

[1] https://foursquare.com/about

A. Hanbury et al. (Eds.): ECIR 2015, LNCS 9022, pp. 623–634, 2015.

We explore several topic models. Although our work is carried out on Foursquare, the models are easily applicable on venue visitation logs from other platforms. In addition, we also proposed models to handle user generated reviews/tips that are tied to venues. We discussed the targeted problem next.

1.1 Problem Definition

Our prediction task is straightforward: predict venues that a user will likely visit, given historical information of his *other* visited venues. We cast this as a ranking problem. Given a list of candidate venues for each user, we seek to rank venues such that high ranking venues are more likely to be visited by the user.

Our defined problem serves a different purpose and differs from *next* venue prediction [6,8,13] and *time-aware* venue prediction [9,7]. Next venue prediction aims to predict the next venue a user will visit, given additional factors such as a user's current location, time of the day, location of friends etc. Time-aware venue prediction is highly similar, but prediction is for a certain time slot and the user's current location may not be known. In contrast, for our venue prediction task, we do not assume that additional information or contextual constraints such as time are available. The task can also be understood as inferring the overall propensity of a user to visit a venue.

In many cases, the lack of additional information makes venue prediction task harder than next or time-aware venue prediction. For example, consider next venue prediction. With spatial constraints, a user's next venue is likely to be geographically near his current venue [13,6]. Time constraints help as well, e.g. food venues are obviously more likely to be visited during meal times [7]. In addition, for both next and time-aware venue prediction, a venue may be repeatedly visited [13,8] in a user's visitation history, e.g. his home or workplace. All these help to rank or narrow the list of candidate venues. In contrast for our problem, we consider candidate venues that are not visited by the user according to the observed visitation data. Hence, many methods for next venue and time-aware venue prediction tasks are less appropriate to solve the proposed problem.

1.2 Proposed Research Idea

Approach. Our approach is based on Latent Dirichlet Allocation (LDA) [1]. LDA was first introduced for modeling topics in text corpus. Since then, topic models have been widely applied in various domains, including social media platforms. Recent works [4,5] had applied LDA on Foursquare check-ins. Both works model the users as high level documents containing venues as terms. For discussion, we denote this as the base model: **LDA-Udoc**.

Our research idea originates from the key observation that in Foursquare [2], there are many more users than venues. There are many users with little visitation data. On the other hand, venues are often visited by many users who leave traces of check-in's and tips. Hence if we regard venues as documents containing users as terms, we obtain fewer, but longer documents over a larger term dictionary. The question is how these changes affect venue prediction. Based on this insight, we define the LDA-Vdoc model which is essentially a flipped version of LDA-Udoc, while retaining all the underlying LDA mechanisms. Remarkably, LDA-Vdoc easily outperforms LDA-Udoc in venue prediction.

We consider further LDA extensions, whereby we model check-ins and tips as two separate modes of user behavior. Again, we compare the two design choices. **Vdoc** uses venues as high level documents while **Udoc** does so with users. Our experiments indicate that Vdoc performs better. In fact, the Vdoc model enables us to exploit tips to improve prediction accuracy. To the best of our knowledge, the venue as document idea and multi-modal extension were unexplored in prior works [4,5,6,7] which focused on check-ins (or location logs) only. Our research findings further reveal that accuracy improvement is largest for unpopular venues where there are fewer users, and hence sparser data. Obviously, venues may also have fewer users if they are newly added, thus there are parallels with the *cold-start* problem for new items in recommendation tasks. In such cases, Vdoc outperforms other models significantly.

Contributions. Flipping and the inclusion of tips constitute the novel aspects of our work. In summary, we present two flipped models, Vdoc-LDA and Vdoc for venue prediction in Foursquare. Vdoc-LDA models a single mode. If tips are available as well, we propose to apply Vdoc. Vdoc also copes with sparse venue data more effectively for prediction. This is important since new venues are continuously being added to Foursquare.

2 Models

We shall describe explored models, starting with the vanilla LDA models. Let the number of users, venues and topics be U, V and K respectively. Also let tip words be from a vocabulary of size W. We represent symmetric Dirichlet distributions with hyperparameters α as $\mathrm{Dir}(\alpha)$; and multinomials with parameter vector θ as $\mathrm{Mult}(\theta)$. Other notations are introduced in an inline manner for ease of reading.

2.1 LDA Models

We begin with the base model: **LDA-Udoc**. Traditionally, LDA assumes a text document is generated by sampling a topic for each word, followed by sampling the word conditional on the topic. Let us now regard a document as a user and a word as a check-in/tip venue. Each user u has a latent vector θ_u with a Dirichlet prior $\mathrm{Dir}(\alpha)$. θ_u specifies his distribution over topics z which in turn specifies distributions over venues. The model assumes a single venue mode without differentiating whether users have chosen to check-in and/or write tips at venues. Note that prior work [4,5] had simply used check-ins. However we include venues from tips[2] such that prediction accuracies of all uni-modal and multi-modal models can be fairly compared on a common venue set. Tip words are ignored in LDA-Udoc. Formally, LDA-Udoc has the generative process:

[2] Some users write tips about a venue without generating check-ins.

1. For each user u, sample $\theta_u \sim \text{Dir}(\alpha)$
2. For each topic k, sample $\phi_k \sim \text{Dir}(\beta)$
3. For venue v_i in check-in/tip i of user u, sample:
   (a) Topic $z_i \sim \text{Mult}(\theta_u)$, Venue $v_i \sim \text{Mult}(\phi_{z_i})$

Now we flip the model and propose the **LDA-Vdoc** model, whereby venues *attract* users to check-in and/or write tips. Hence venues play a more active generative role and generate the users. Note that topics are now defined over users instead and denoted by y. LDA-Vdoc also does not differentiate between users from check-ins or tips. Tip words are ignored. The generative process is:

1. For each venue v, sample $\theta_v \sim \text{Dir}(\alpha)$
2. For each topic k, sample $\phi_k \sim \text{Dir}(\beta)$
3. For user u_i in check-in/tip i of venue v, sample:
   (a) Topic $y_i \sim \text{Mult}(\theta_v)$, User $u_i \sim \text{Mult}(\phi_{y_i})$

2.2 Multi-modal Models

We now propose models Udoc and Vdoc which generate check-ins and tips in distinct weakly coupled modes, unlike previous LDA models. With Udoc, venues from check-ins and tips are treated as distinct entity modes generated by check-in and tip topics respectively. However we also tie the mentioned two modes of topics with a common topic indicator. This accounts for the weak coupling and can be viewed as a form of regularization between the two modes. Vdoc is defined in a similar way.

Udoc generates venues, tip content and is a direct, non-flipped extension of the base model Udoc-LDA. It seeks to exploit all information from tips, including the tip words. Since tips are short with a character limit of 200 imposed by Foursquare, we assume each to cover only a single topic. We also attribute each tip word to either the venue or topic with a Bernoulli switch Bern(η), with a prior from a beta distribution Beta(λ). The intuition is that certain venues may have a large influence on tip content.

For each user, venues are now differentiated as check-in venues $\tilde{v}$ and tip venues $\hat{v}$, generated via check-in topics $\tilde{z}$ and tip topics $\hat{z}$. Let each tip contains N_w words w. Udoc's generative process is listed below (best understood with the plate diagram in Figure 1).

1. For each user u, sample $\theta_u \sim \text{Dir}(\alpha)$
2. For each topic indicator k, sample distributions for tip topics: $\phi_k \sim \text{Dir}(\beta)$, $\gamma_k \sim \text{Dir}(\omega)$, and check-in topics: $\tilde{\phi}_k \sim \text{Dir}(\tilde{\beta})$
3. For each venue v, sample $\hat{\gamma}_v \sim \text{Dir}(\hat{\omega})$
4. Sample a global Bernoulli vector for flags: $\eta \sim \text{Beta}(\lambda)$
5. For tip i of user u, sample tip topics, tip venues and words:
   (a) Topic $\hat{z}_i \sim \text{Mult}(\theta_u)$, Venue $\hat{v}_i \sim \text{Mult}(\phi_{\hat{z}_i})$
   (b) For the j-th word $w_{i,j}$
      i. Sample a flag $x_{i,j} \sim \text{Bern}(\eta)$
      ii. Sample $w_{i,j} \sim \text{Multi}(\gamma_{\hat{z}_i})$ if $x_{i,j}=0$, else sample $w_{i,j} \sim \text{Multi}(\hat{\gamma}_{\hat{v}_i})$

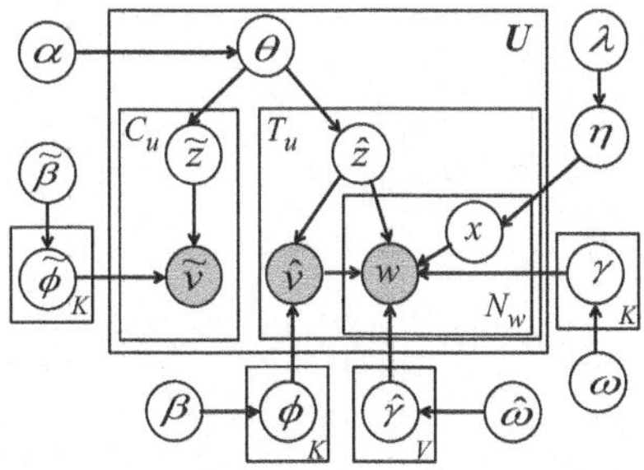

Fig. 1. Udoc model. Each user u has C_u check-ins and T_u tips.

6. For check-in i of user u, sample check-in topics and check-in venues:
   (a) Topic $\tilde{z}_i \sim \text{Mult}(\theta_u)$, Venue $\tilde{v}_i \sim \text{Mult}(\tilde{\phi}_{\tilde{z}_i})$

Vdoc is a flipped version of Udoc and regards each venue as a document unit. Intuitively, each venue attracts users to either check-in, write tips or do both. In addition, we observed in our Foursquare dataset of an Asian city, (refer Section 3.1) that 76% of venues have both check-ins and tips. In contrast, only 21% of users both check-in and write tips, with the rest being biased towards only one behavior mode. In this sense, more venue documents have both modes and can be regarded as more 'complete' than user documents. This will impact prediction accuracy as shown in our experiments (refer section 3.2).

For each venue, users are now differentiated as check-in/tip users ($\tilde{u}/\hat{u}$), generated via check-in/tip topics ($\tilde{y}/\hat{y}$). We also let tip words to be attributable to either tip topics or tip users. We now define Vdoc's generative process with the corresponding plate diagram shown in Figure 2.

1. For each venue v, sample $\theta_v \sim \text{Dir}(\alpha)$
2. For each topic indicator k, sample distributions for the tip mode: $\phi_k \sim \text{Dir}(\beta)$, $\gamma_k \sim \text{Dir}(\omega)$, and check-in mode: $\tilde{\phi}_k \sim \text{Dir}(\tilde{\beta})$
3. For each user u, sample $\hat{\gamma}_u \sim \text{Dir}(\hat{\omega})$
4. Sample a global Bernoulli vector for flags: $\eta \sim \text{Beta}(\lambda)$
5. For tip i at venue v, sample tip topics, tip users and words:
   (a) Topic $\hat{y}_i \sim \text{Mult}(\theta_v)$, User $\hat{u}_i \sim \text{Mult}(\phi_{\hat{y}_i})$
   (b) For the j-th word $w_{i,j}$
      i. Sample a flag $x_{i,j} \sim \text{Bern}(\eta)$
      ii. Sample $w_{i,j} \sim \text{Multi}(\gamma_{\hat{y}_i})$ if $x_{i,j}{=}0$, else sample $w_{i,j} \sim \text{Multi}(\hat{\gamma}_{\hat{u}_i})$
6. For check-in i at venue v, sample check-in topics and check-in users:
   (a) Topic $\tilde{y}_i \sim \text{Mult}(\theta_v)$, User $\tilde{u}_i \sim \text{Mult}(\tilde{\phi}_{\tilde{y}_i})$

2.3 Inference

We use Collapsed Gibbs Sampling (CGS) to infer parameters for all the models. CGS draws a sequence of samples to approximate joint distributions. It has been widely used for inference [3] in LDA-based models. For the multi-modal models,

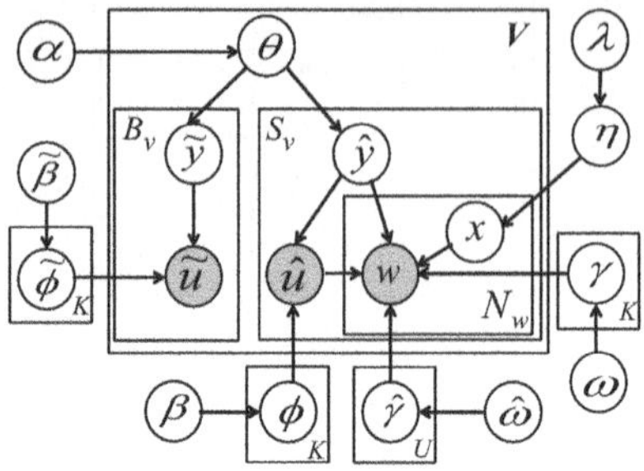

Fig. 2. Vdoc model. Each venue v has B_v check-ins and S_v tips.

Udoc and Vdoc's sampling equations are highly similar in form. Due to space constraints, we only present sampling equations for Vdoc topics.

The topic inference task is to sample for tip and check-in topics. For notation simplicity, we also omit hyperparameters which are implicitly conditioned upon during sampling. Recall that in Vdoc, venues $\boldsymbol{v}$ are not differentiated while users are differentiated as check-in users $\widetilde{\boldsymbol{u}}$ and tip users $\hat{\boldsymbol{u}}$. Given a tip i with bag of words $\boldsymbol{w}_i$, we sample its topic as follows:

$$p(\hat{y}_i = k|, \hat{\boldsymbol{y}}_{-i}, \hat{\boldsymbol{u}}, \boldsymbol{v}, \boldsymbol{w}, \boldsymbol{x}) \propto \frac{N^{TV}_{kv_i,-i} + \alpha}{\sum_{k'} N^{TV}_{k'v,-i} + K\alpha} \frac{N^{\hat{U}T}_{\hat{u}_i k,-i} + \beta}{\sum_{\hat{u}'} N^{\hat{U}T}_{\hat{u}'k,-i} + U\beta} \prod_{\substack{w \in \boldsymbol{w}_i,\\ x_w = 0}} \frac{N^{WT}_{wk,-i} + \omega}{\sum_{w'} N^{WT}_{w'k,-i} + W\omega} \quad (1)$$

where subscript $-i$ means contributions from tip i are excluded. N^{TV}, $N^{\hat{U}T}$ and N^{WT} are respective count matrices for assignments of topics to venues, tip users to topics and tip words to topics. Subscripts reference the matrix elements. For a check-in i, we sample its topic as:

$$p(\widetilde{y}_i = k|, \widetilde{\boldsymbol{y}}_{-i}, \widetilde{\boldsymbol{u}}, \boldsymbol{v}) \propto \frac{N^{TV}_{kv_i,-i} + \alpha}{\sum_{k'} N^{TV}_{k'v,-i} + K\alpha} \frac{N^{\widetilde{U}T}_{\widetilde{u}_i k,-i} + \widetilde{\beta}}{\sum_{\widetilde{u}'} N^{\widetilde{U}T}_{\widetilde{u}'k,-i} + U\widetilde{\beta}} \quad (2)$$

where $N^{\widetilde{U}T}$ counts assignments of check-in users to topics and N^{TV} is previously defined. Similarly, the sampling equations for flag assignments per tip word can be readily derived. We omit their discussion here for brevity.

2.4 Prediction

Our goal is to predict the venues that a user is likely to visit. We do not differentiate between check-in/tip venues in prediction, hence the targeted quantity is $p(v|u)$. This is used to rank candidate venues. While in practice, a user can tip without having actually visited a venue, extensive inspections of sample tips indicate that it is reasonable to assume most tips are generated post-visits.

For Udoc-LDA, $p(v|u)$ is computed via topic marginalization: $\sum_z p(v|z)p(z|u)$. To obtain $p(v|u)$ for Udoc, topic marginalization is done for each mode and then combined with the two observed empirical probabilities of u performing a check-in and tip. For Vdoc-LDA, we marginalized out topics over users and then apply Bayes theorem $p(v|u) \propto p(u|v)p(v)$. The same formula applies to Vdoc as well, however we first need to compute $p(u|v)$. Assume that a venue v generates check-ins and tips with conditional probabilities $p(c|v)$ and $p(t|v)$. We compute $p(u|v)$ by marginalizing over modes: $m = \{c, t\}$ and applying the chain rule:

$$p(u|v) = p(u, m = c|v) + p(u, m = t|v) = p(\widetilde{u}|v)p(c|v) + p(\hat{u}|v)p(t|v) \quad (3)$$

Note that $p(\widetilde{u}|v)$ and $p(\hat{u}|v)$ in (3) are obtained via marginalizing out the topics:

$$p(\widetilde{u}|v) = \sum_{\widetilde{y}} p(\widetilde{u}|\widetilde{y})p(\widetilde{y}|v), \quad p(\hat{u}|v) = \sum_{\hat{y}} p(\hat{u}|\hat{y})p(\hat{y}|v) \quad (4)$$

where $p(\widetilde{y}|v)$, $p(\hat{y}|v)$, $p(\hat{u}|\hat{y})$ and $p(\widetilde{u}|\widetilde{y})$ are estimated with count matrices from CGS in a similar fashion as proposed in [3].

3 Experiments

3.1 Data and Setup

In our experiments, we use two Foursquare datasets: United States (US) check-ins from [2] and check-ins plus tips which we extract from users in Singapore (SG), spanning Mar 2012 to Dec 2013. The latter comprises of check-ins posted as tweets on the user's Twitter timeline[3] and tips crawled directly using the Foursquare API. Following standard noise filtering practices [4,7,10,6], we exclude inactive users with too few venues and inactive venues with too few users. We used a common threshold of 6 for both user and venue filtering, i.e. ≥ 6.

For each user, we randomly select one of his venues as the test venue. We then hide *all* his tips and check-ins from the test venue. His remaining tips/check-ins are then included in the training set for model building. This process is repeated for all users. We generate 10 trials of training/test sets whereby trials differ due to random sampling of test venue per user. Also note that *prediction here is in terms of retrieving hidden venues*, and that to support multiple trials, we have not restricted hidden venues to be necessarily the most recent visited venues.

On average, the US training set contains 48,900+ users, 14,900+ venues and 252,000+ check-ins. The SG training set contains 24,400+ users, 17,600+ venues, 62,900+ tips and 1,062,200+ check-ins. Comparing both datasets, the US dataset has more users and fewer venues than the SG dataset.

Note that for the US dataset, we only apply LDA-Vdoc and LDA-Udoc since tips are not available. For the SG dataset, we ignore tip content when applying uni-modal models, such that there are no differentiation between entities (users or venues) from check-ins/tips. With each model, we rank candidate venues for

[3] Check-ins are visible only if posted as tweets, otherwise they are hidden.

each user (excluding those in his training set). Hence for each user, the number of candidates is slightly less than the number of venues per dataset. We then extract the rank of the hidden test venue and compute the Mean Reciprocal Rank (*MRR*), a standard information retrieval measure defined as:

$$MRR = \frac{1}{Q}\sum_{i}^{Q} 1/rank_i \tag{5}$$

where $rank_i$ is the rank of the hidden test venue i predicted by the model and Q is the total number of test cases. (Each test case consists of a user and his hidden test venue.) MRR lies between 0 and 1 with the latter implying perfect ranking accuracy. We compute the average MRR across the 10 trials.

All models are fitted using 500 iterations of CGS with a burn-in of 200 iterations. For estimating distributions required for prediction, we collect samples with a lag of 20 iterations in between. We have experimented with various number of topics and observed that relative prediction performance of models are fairly consistent, e.g. Vdoc being consistently the best performer. In subsequent discussion, we present results involving 20 topics.

3.2 Prediction Results

In this section, we compare the models quantitatively. We regard LDA-Udoc as the baseline and focus on how other models perform relative to it. Table 1 presents the prediction results. Also recall that our notion of documents depends on the models. For LDA-Vdoc and Vdoc, documents are venues while for LDA-Udoc and Udoc, documents are users.

Table 1. Average MRR with standard deviations (bracketed). Gain is % improvement over LDA-Udoc. (US: United States check-ins, SG: check-ins & tips in Singapore).

Dataset	Model	Ave. MRR	Gain (%)
US	LDA-Vdoc	**0.1302 (2.05E-3)**	**22.35**
	LDA-Udoc	0.1064 (1.77E-3)	-
SG	Vdoc	**0.0575 (1.34E-3)**	**7.06**
	Udoc	0.0532 (1.21E-3)	-0.89
	LDA-Vdoc	0.0564 (0.93E-3)	4.92
	LDA-Udoc	0.0537 (1.25E-3)	-

On both datasets, LDA-Vdoc easily outperforms the previously proposed LDA-Udoc model [4,5]. This supports the argument of flipping. Accuracy gain is especially large at over 20% on the US dataset. As described in section 3.1, the US dataset has more users and yet, fewer venues than the SG dataset. This means that in the former, LDA-Vdoc's characteristics are even more pronounced, i.e. modeling fewer and longer documents. Hence we expect a larger accuracy gain over LDA-Udoc, compared to the SG dataset.

On the SG dataset, Vdoc is the best performer with more than 7% improvement over the baseline. The difference is consistent across different runs and statistically significant (using the Wilcoxon signed rank test) with a p-value of less than 0.01. In addition, LDA-Vdoc consistently emerges as the second best performer (p-value < 0.01) when compared with LDA-Udoc). Hence models using venues as documents (as in Vdoc, LDA-Vdoc) consistently perform better than models with users as documents.

Vdoc's superiority over LDA-Vdoc indicates that tips contain useful information, which the former had exploited. However we note that while Udoc considers tips as well, its performance is essentially the same as LDA-Udoc. We attribute this to overly sparse co-occurrence information. Obviously, in breaking up entities into different modes, some co-occurrence information is lost. (To see this, imagine treating every entity as a unique mode. This leads to a total loss of co-occurrence information.) Thus additional information from tips may have been cancelled off in Udoc. Vdoc is however more robust to this effects.

We attribute Vdoc's robustness to previously discussed characteristics such as having fewer, but longer documents. In addition, Vdoc's documents are more complete than Udoc's documents in containing entities from both modes. As mentioned in Section 2.2, 76% of venues (Vdoc's documents) from the SG dataset contain users from both tips and check-ins. In contrast, with users as documents (as in Udoc), only 21% contains both tip and check-in venues. This is a direct consequence of how users utilize Foursquare, i.e. leaning towards either generating check-ins or writing tips rather than doing both in a more balanced manner.

3.3 Prediction Results by Venue Popularity

For a more in-depth analysis, we bin test cases for the SG dataset by test venue popularity. This allows us to examine how various models perform on venues of different popularities. We quantify venue popularity by two measures: combined tip/check-in count and number of unique users per venue. We divide test cases into three bins of equal size, corresponding to venues of *low*, *medium* and *high* popularities. Figure 3 shows the MRR of venues with different popularities.

Figures 3(a) and 3(d) show that Vdoc's accuracy improvement over other models is biggest for the least popular venues. The improvement decreases as we consider more popular venues. For low popularity venues, Vdoc outperforms the baseline LDA-Udoc by around 200% for both popularity measures, hence indicating that Vdoc makes better use of sparse venue data. This takes on an even greater importance if we consider a common scenario in Foursquare: newly created venues will usually belong to the unpopular bins simply by virtue of having little or no previous data. Predicting for them is analogous to recommending for new items in recommender systems, which relates to the *cold start* problem. In such cases, prediction/recommendation difficulty increases due to data sparsity. Compared with other models, Vdoc is more accurate in such scenarios.

For highly popular venues, Figures 3(c) and 3(f) show that Vdoc's improvement over LDA-Udoc is smaller at 4-5% for both popularity measures. Hence, even though popular test venues are easier to predict for, Vdoc still manages

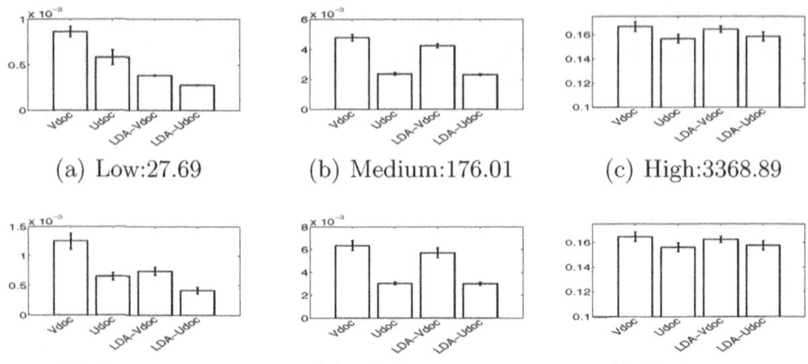

(a) Low:27.69 (b) Medium:176.01 (c) High:3368.89

(d) Low:19.85 (e) Medium:117.9 (f) High:1654.62

Fig. 3. MRR binned by combined tip/check-in count (a,b,c) and unique user count (d,e,f). Each sub-figure plots Vdoc, Udoc, LDA-Vdoc and LDA-Udoc (left to right). Numbers are mean tip/check-in count for (a,b,c), and mean user count for (d,e,f).

some improvement over Udoc-LDA. We also compare Vdoc with LDA-Vdoc. Their performance differs more for low popularity venues, and less with increased venue popularity. Since unpopular venues have much less data for models to exploit, content information in the few related tips will be relatively more important. Vdoc is able to exploit this additional information in contrast to LDA-Vdoc which totally ignores content.

4 Sample Topics

We illustrate some Vdoc's topics over tip words on the SG dataset. By inspecting the topics, one easily gets a understanding of user interests and the aspects that they care about enough to write tips. Table 2 shows the top 12 words of 6 sample topics (out of 20) from Vdoc. As can be seen, the topics are easily interpretable.

Table 2. Top 12 words of sample Vdoc topics. We manually annotate the displayed topics (labels in bold) for ease of understanding.

Service: service food staff slow bad time order long wait good don waiting
Transport: bus service time interchange train long will queue wait station mins morning
Pastry: ice cream chocolate cake tea nice good love best sweet caramel awesome
Tea/Coffee: tea milk ice coffee nice best good jelly drink love sugar green
Western Food: good chicken cheese beef pasta fries sauce great nice awesome fish pizza
Opening hours: hours closed open public till sat sun mon fri daily place opens

5 Related Work

As mentioned, [4,5] had applied LDA to model Foursquare check-ins. They presented qualitative analysis of the topics instead of quantitative results. In [6],

Kurashima et. al modeled venues conditional on both topics and each user's movement history. The model was used to predict the last visited venue of the user. Note that all the above mentioned works treated users as documents, venues as terms and topics as distributions over venues.

Some works [11,12,10] had explored topic models of geo-located tweets. Tweet contents and originating locations are used in [11,12] while [10] included time information as well. The aim is to predict geographic coordinates that tweets are sent from. This problem is less applicable on Foursquare since tips can possibly be generated post-visits by users when they may not be physically present at the venue locations. Nonetheless, we note that all the proposed models [11,12,10] had utilized users as documents, instead of spatial regions or locations as documents. Potentially model flipping can be investigated for accuracy gains.

Other researchers had explored non topic modeling approaches in next venue [8,13] and time-aware venue prediction [7,9]. In [8], dynamic Bayesian networks were used to model a user's locations as hidden states. Each state is conditional on the last state and emit observations such as time information and the locations of friends. Noulas et al. [13] trained M5 model trees with mobility and temporal features to predict a user's next check-in venue. Yuan et al. [7] constructed a time-aware collaborative filtering model to predict user locations conditional on time. Cho et al. [9] conducted a similar task with a mixture of Gaussians.

Lastly we mention works more applicable to our prediction task, but with models in the continuous space [9,15,14]. (We mentioned [9] earlier for time-aware venue prediction, but it can be adapted for this). Typically continuous distributions such as Gaussian mixtures [9,15] or kernel estimated densities [14], are fitted to model the spatial coordinates of venues. In contrast, we model venues in the discrete space and do not require spatial coordinates. Both continuous and discrete modeling have their strengths and weaknesses. For example, different venues can occur at the same coordinates, by being at different levels of the same building. Predicting between these venues is tricky with continuous modeling, which by far, had mainly utilized two dimensional distributions [9,14,15]. Nonetheless, in our further work we will be interested in fusing the models presented here with continuous techniques such that the strengths of both can be leveraged on. We describe a possible research direction in our conclusion.

6 Conclusion

We have explored several LDA based models for venue prediction in Foursquare. In particular, we consider flipped models such that venues are treated as documents and users as terms. Flipping is extremely easy to apply, and yet leads to significant accuracy gains in venue prediction. It also has the additional benefit of allowing us to exploit tips. Without flipping, it is uncertain that including tips can increase accuracy, e.g. Udoc does not improve over Udoc-LDA.

In ongoing research, we are exploring the fusion of the models here with continuous models [15,14,9]. Instead of designing ever more complex generative models, one possible approach is to combine different models, either linearly or otherwise. This allows information from various diverse aspects, e.g. tips, spatial and social influence to contribute to the prediction task. In addition, the inferred combination weights serve to indicate the relative importance of various aspects.

Lastly, given the huge variety of topic models out there in different applications, many can potentially be flipped and the performance investigated. Researchers can also consider flipped/non-flipped versions in the design of any new models. Hence our works here has served as a motivating example.

Acknowledgements. This research is partially supported by DSO National Laboratories, Singapore; and the Singapore National Research Foundation under its International Research Centre@Singapore Funding Initiative and administered by the IDM Programme Office, Media Development Authority (MDA).

References

1. Blei, D.M., Ng, A.Y., Jordan, M.I.: Latent Dirichlet Allocation. Journal of Machine Learning Research 3, 993–1022 (2003)
2. Levandoski, J.J., Sarwat, M., Eldawy, A., Mokbel, M.F.: LARS: A Location-Aware Recommender System. In: ICDE (2012)
3. Griffiths, T.L., Steyvers, M.: Finding Scientific Topics. Proceedings of the National Academy of Sciences 101, 5228–5235 (2004)
4. Joseph, K., Tan, C.H., Carley, K.M.: Beyond "Local", "Categories" and "Friends": Clustering Foursquare Users with Latent "Topics". In: UbiComp (2012)
5. Long, X., Jin, L., Joshi, J.: Exploring Trajectory-driven Local Geographic Topics in Foursquare. In: UbiComp (2012)
6. Kurashima, T., Iwata, T., Hoshide, T., Takaya, N., Fujimura, K.: Geo topic Model: Joint Modeling of User's Activity area and Interests for Location Recommendation. In: WSDM (2013)
7. Yuan, Q., Cong, G., Ma, Z., Sun, A., Magnenat-Thalmann, N.: Time-aware Point-of-Interest Recommendation. In: SIGIR (2013)
8. Sadilek, A., Kautz, H.A., Bigham, J.P.: Finding your Friends and Following them to Where You Are. In: WSDM (2012)
9. Cho, E., Myers, S.A., Leskovec, J.: Friendship and Mobility: User Movement in Location-based Social Networks. In: KDD (2011)
10. Yuan, Q., Cong, G., Ma, Z., Sun, A., Magnenat-Thalmann, N.: Who, Where, When and What: Discover Spatio-temporal Topics for Twitter Users. In: KDD (2013)
11. Hong, L., Ahmed, A., Gurumurthy, S., Smola, A., Tsioutsiouliklis, K.: Discovering Geographical Topics in the Twitter Stream. In: WWW (2012)
12. Hu, B., Ester, M.: Spatial Topic Modeling in Online Social Media for Location Recommendation. In: ACM Conference on Recommender Systems (2013)
13. Noulas, A., Scellato, S., Lathia, N., Mascolo, C.: Mining User Mobility Features for Next Place Prediction in Location-Based Services. In: ICDM (2012)
14. Lichman, M., Smyth, P.: Modeling Human Location Data with Mixtures of Kernel Densities. In: KDD (2014)
15. Zhao, S., King, I., Lyu, M.R.: Capturing Geographical Influence in POI Recommendations. In: Lee, M., Hirose, A., Hou, Z.-G., Kil, R.M. (eds.) ICONIP 2013, Part II. LNCS, vol. 8227, pp. 530–537. Springer, Heidelberg (2013)

Geographical Latent Variable Models for Microblog Retrieval

Alexander Kotov[1], Vineeth Rakesh[1], Eugene Agichtein[2], and Chandan K. Reddy[1]

[1] Department of Computer Science, Wayne State University, Detroit MI 48226, USA
{kotov,ed3424}@wayne.edu, reddy@cs.wayne.edu
[2] Department of Mathematics and Computer Science, Emory University, Atlanta GA, 30322, USA
eugene@mathcs.emory.edu

Abstract. Although topic models designed for textual collections annotated with geographical meta-data have been previously shown to be effective at capturing vocabulary preferences of people living in different geographical regions, little is known about their utility for information retrieval in general or microblog retrieval in particular. In this work, we propose simple and scalable geographical latent variable generative models and a method to improve the accuracy of retrieval from collections of geo-tagged documents through document expansion that is based on the topics identified by the proposed models. In particular, we experimentally compare the retrieval effectiveness of four geographical latent variable models: two geographical variants of post-hoc LDA, latent variable model without hidden topics and a topic model that can separate background from geographically-specific topics. The experiments conducted on TREC microblog datasets demonstrate significant improvement in search accuracy of the proposed method over both the traditional probabilistic retrieval model and retrieval models utilizing geographical post-hoc variants of LDA.

Keywords: Microblog Retrieval, Latent Variable Models.

1 Introduction

Collections of microblog documents pose difficult challenges and offer unique opportunities to retrieval systems at the same time. On one hand, microblog retrieval systems need to overcome severe vocabulary mismatch problem (i.e. how to retrieve very short documents, which might be conceptually relevant, but do not explicitly contain some or all of the query terms), while having to deal only with scarce relevance signals that can be derived from the text of the tweets alone. Furthermore, relevance in the context of microblog retrieval (MBR) is a multi-faceted phenomenon and involves many other factors besides content matching, such as recency, content quality, and geographical focus. On the other hand, social media documents in general and microblogs in particular naturally combine many different types of data besides textual content: timestamps, manually assigned topical tags (hashtags), geographical location of the users who

A. Hanbury et al. (Eds.): ECIR 2015, LNCS 9022, pp. 635–647, 2015.

created the tweets and their social networks (followers and followees), which can be leveraged in retrieval models as additional non-textual dimensions and indicators of relevance. As a result, combining lexical with non-lexical relevance signals, such as re-tweets [4] and timestamps [7] [6], has become a dominant theme across most of the recent developments in microblog retrieval.

While most such extrinsic dimensions of relevance (particularly, temporal) have recently received some degree of attention, geographical locations in textual form associated with Twitter user accounts is one important additional dimension and type of meta-data provided by Twitter, which remains relatively overlooked. The importance of accounting for geographical context can be illustrated by using the topic MB04 "Mexico drug war" from the 2011 TREC Microblog track query set as an example. The distribution of geographical locations of the authors of relevant tweets for this topic is shown in Figure 1.

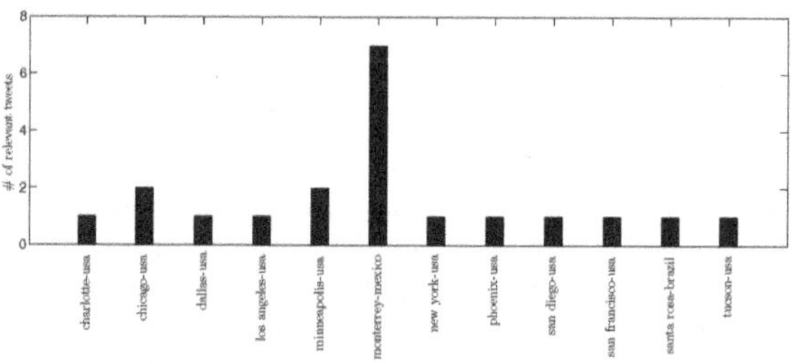

Fig. 1. Distribution of geographical locations of the authors of relevant tweets for the topic MB04 "Mexico drug war"

It is clear from Figure 1 that the majority of relevant tweets were authored by the users in a major city in Mexico as well as the cities in the United States, which are close to the Mexican border. Furthermore, from Table 1 it follows that the query terms "mexico" and "drug" individually occur in only about half of 111 relevant tweets for this topic, while the tweets in the other half include different, but conceptually related terms reflecting other aspect of this topic ("border", "catapult", "pot", "fire", "violence", "smuggler"). Only 16 relevant tweets contain include the query terms "mexico" and "drug" together and just 8 (less than 10%) relevant tweets contain all three query terms. Some relevant tweets, such as *"El Paso, Juarez Citizens Unite to Protest Border Violence: 'No Mas Sangre' http://amplify.com/u/..."*, do not include any query terms at all.

This example illustrates the fact that queries are often geographically contextualized (i.e. in regions close to the border between the United States and Mexico, "war" is associated with different concepts than in its traditional definition). While many terms can be related to "war" in general, only a subset of these terms are relevant, given additional geographical ("mexico") and lexical ("drug") contexts. Therefore, the key intuition behind the methods proposed in this work is that accurately addressing vocabulary mismatch problem in MBR through expansion of microblog posts with conceptually related terms requires taking into account their geographical context.

Table 1. Top 10 most frequently occurring terms in the relevant tweets for the query "Mexico drug war"

term	# rel. tweets
mexico	58
drug	58
border	46
catapult	40
mexican	29
u.s.	22
pot	16
fire	15
found	15
smuggler	14

Table 2. Top 10 geographical locations associated with the most number of tweets in 2011 TREC Microblog track corpus

location	# tweets
new york-usa	83,479
monterrey-mexico	65,473
sao paulo-brazil	62,243
rio branco-brazil	54,411
london-uk	48,496
los angeles-usa	36,096
caracas-venezuela	33,860
chicago-usa	33,394
jackarta-indonesia	29,795
san francisco-usa	23,642

In this work, geographical context is determined by projecting the tweets into lower-dimensional semantic space by leveraging the probabilistic machinery of latent variable generative models. In particular, we propose latent variable models (LVMs), which incorporate geographical locations as observed textual labels. Our work extends the line of information retrieval research, which addresses the problem of vocabulary mismatch through dimensionality reduction, by converting sparse and potentially noisy representation of documents as distributions over terms in the collection vocabulary into more compact representation as distributions over hidden topics, or clusters of semantically related terms. Although state-of-the-art methods to perform dimensionality reduction of document collections, such as Latent Dirichlet Allocation (LDA) [2] topic model, have been previously successfully applied to ad hoc information retrieval [18] [20], little is known about the utility of geographical topic models for information retrieval in general or MBR in particular. In this work, we propose latent variable models (LVMs) that utilize textual geographical locations in the profiles of Twitter users and document expansion methods that leverage the output of the proposed LVMs to address the vocabulary mismatch problem in MBR through *geographically-focused document expansion.*

The rest of this paper is organized as follows. In Section 2, we provide a brief overview of the previous work in closely related areas. Proposed geographical LVMs are discussed in detail in Section 4 and the method to derive document expansion LMs from the output of the proposed LVMs is presented in Section 4.5. Results of an experimental evaluation of the proposed methods are presented in Section 5 and our key contributions are summarized in Section 5.3.

2 Related Work

Microblog Retrieval. The main challenges in MBR, such as defining units of retrieval and relevance, factoring in quality, authority and timeliness of tweets as well as addressing the vocabulary mismatch problem are discussed in detail in [5], while [17] highlights the key differences between web search and microblog search. Most previously proposed methods for microblog IR focused on incorporating specific types of meta-data (e.g., temporal [7], [15], [3], [10], [1], [14] or social [11]) into retrieval models to address the issue of vocabulary mismatch.

Leveraging timestamps of tweets to model temporal relevance in pseudo-relevance feedback is one of the most well-explored directions in MBR up to date. In particular, Efron et al. proposed a document expansion method [7], in which each tweet is first submitted as a pseudo-query and then the retrieved tweets that are the closest to the timestamp of the original tweet are selected as expansion documents. Efron et al. also proposed a method [6] for re-ranking initial results by estimating the temporal density of relevant documents. A query expansion method proposed in [15] first obtains the initial results for a given query to construct its temporal profile as well as the temporal profiles for each top retrieved document. It then selects those expansion documents to construct the relevance model, for which the temporal profile is the closest to the query temporal profile as measured by Bhattacharyya distance. Amati [1] experimented with exponential, log-normal, log-logistic and Zipf-Mandelbrot distributions to model the freshness aspect of temporal relevance. Miyanishi et al. [14] proposed two methods to select query expansion terms based on analyzing temporal properties of queries and documents. The first method selects the expansion terms one by one from the top retrieved documents by constructing and comparing the temporal profiles for the original and expanded queries. The second method favors recency and selects the expansion terms for which the sample mean of the timestamps in the profile of the expanded query is close to the sample mean of the timestamps in the temporal profile of the original query.

Geographical Topic Models. A series of recent studies [8] [9] [21] [13] have demonstrated that geography-aware topic models can capture lexical preferences and nuances of language use by people in different geographical locations. Unfortunately, these models are not usable for microblog IR, since they are computationally complex, only work with geographical coordinates and can only handle very small vocabularies. While previous studies [18] [20] have shown the effectiveness of basic topic models in improving retrieval accuracy in traditional ad-hoc IR scenario, with the exception of the preliminary work of Kotov et al. [12], who applied basic post-hoc geographical variant of LDA to MBR and reported promising results, no other work studied the utility of geographical topic models for MBR. In this work, we propose several new geography-aware topic models that use textual geographical locations rather than coordinates and compare their effectiveness for MBR.

3 Retrieval Model

Our proposed methods are based on the query likelihood retrieval model, in which a document d is scored and ranked against a query q according to the likelihood of generating q from the language model (LM) of d:

$$P(q|d) = \prod_{w \in q} p(w|\Theta_d) \tag{1}$$

The maximum likelihood estimate $p_{ml}(w|\Theta_d) = \frac{c(w,d}{|d|}$ of document LM is normally smoothed to avoid zero probabilities for query terms that don't occur in d, for example using the Dirichlet prior smoothing:

$$p(w|\Theta_d) = \frac{|d|}{|d| + \mu} p_{ml}(w|\Theta_d) + \frac{\mu}{|d| + \mu} p(w|\mathbb{C}) \quad (2)$$

where $p_{ml}(w|\Theta_d)$ and $p(w|\mathbb{C})$ are the probabilities of w in the maximum-likelihood estimates of document LM and collection LM respectively, and $\mu \geq 0$ is the Dirichlet prior. A combination of query likelihood retrieval method with Dirichlet prior smoothing is used as one of the baselines in our experiments (denoted as **QL-DIR**).

Within the language modeling retrieval framework, the issue of vocabulary mismatch is typically addressed through expansion of either query or document LMs. We adopt the latter approach, in which a document expansion LM $\hat{\Theta}_d$ is first derived for each document by leveraging semantic terms associations mined either from external resources or the collection itself. Then the expanded document LM $p(w|\tilde{\Theta}_d)$ is obtained from the original document LM Θ_d through linear interpolation with a document expansion LM $\hat{\Theta}_d$ with the coefficient α:

$$p(w|\tilde{\Theta}_d) = \alpha p(w|\Theta_d) + (1 - \alpha) p(w|\hat{\Theta}_d) \quad (3)$$

The key idea behind document expansion is to add more terms into the document LM that are conceptually relevant to the terms in the original document. In this work, we leverage geography-aware LVMs to identify clusters of semantically related terms within particular geographical regions. In the following sections, we present and discuss the details of the proposed LVMs.

4 Geographical Latent Variable Models

4.1 Post-hoc Geographical Variants of LDA

We use retrieval methods based on two post-hoc geographical variants of Latent Dirichlet Allocation (LDA) [2], a popular topic model, as baselines. LDA considers each document d in the collection as a mixture of K multinomials (topics) ϕ_z drawn from a symmetric Dirichlet prior β.

Geo-specific topics can be mined from a geo-tagged document collection $\mathbb{C} = \{(d_1, l_{d_1}), \ldots, (d_M, l_{d_M})\}$, in which each document d is associated with textual location l_d from a set of L distinct locations, using standard LDA in a post-hoc way by grouping the documents labeled with each distinct geo-tag $l \in \mathcal{L}$ into sub-collections and running a separate instance of LDA on each sub-collection. The following two variants of this method are used as baselines in our experimental evaluation:

PH-GLDA: this variant uses the same number of geo-specific topics K^{loc} for each location sub-collection. The optimal number of local topics is determined by fitting LDAs with the same number of topics (starting with 2) for each location sub-collection to determine the setting that minimizes perplexity. Therefore, this method finds the optimal *global configuration* of post-hoc LDA and was used to obtain geo-specific topics in [12].

OPT-GLDA: this variant uses different number of geo-specific topics $K^{loc,l}$ for each location sub-collection. The optimal setting is determined by exhaustively trying different numbers of topics for each sub-collection LDA to find the setting that minimizes perplexity on the testing portion of each location sub-collection. This method finds the optimal *local configuration* of post-hoc LDA.

4.2 GLTA

Geographic Latent Term Allocation (**GLTA**) associates a latent variable with each word, which determines its type (whether a word is generated from a background or geo-specific LM) instead of topical assignment. It considers each document d labeled with geo-tag l_d as a mixture of the *background LM* ϕ^{bg}, which is drawn from β^{bg} (all Dirichlet priors in this work are symmetric and have a single hyper-parameter), and *location-specific LM* ϕ^{loc,l_d}, which is drawn from β^{loc}. GLTA models document generation according to the following probabilistic process:

1. draw $\lambda_d \sim Beta(\gamma)$, a binomial distribution controlling the mixture of local and a background LMs for d
2. for each word position i of N_d in d:
   (a) draw Bernoulli switching variable $m_{d,i} \sim \lambda_d$
   (b) if $m_{d,i} = bg$:
      i. draw a word $w_{d,i} \sim \phi^{bg}$
   (c) if $m_{d,i} = loc$:
      i. draw a word $w_{d,i} \sim \phi^{loc,l_d}$

Figure 2a shows the graphical model of GLTA in plate notation. GLTA is a probabilistic extension of the geography-aware Naïve Bayes method proposed in [19].

4.3 GLDA

Geographical LDA (**GLDA**) considers each document d labeled with geo-tag l_d as a mixture of the *background topic* ϕ^{bg} drawn from β^{bg} and K^{loc} *location-specific* topics ϕ^{loc,l_d} drawn from β^{loc} and models document generation according to the following probabilistic process:

1. draw $\lambda_d \sim Beta(\gamma)$, a binomial distribution controlling the mixture of local topics and a background topic for d
2. draw $\Theta_d^{loc,l_d} \sim Dir(\alpha^{loc})$
3. for each word position i of N_d in d:
   (a) draw Bernoulli switching variable $m_{d,i} \sim \lambda_d$
   (b) if $m_{d,i} = bg$:
      i. draw a word $w_{d,i} \sim \phi^{bg}$
   (c) if $m_{d,i} = loc$:
      i. draw a topic $z_{d,i} \sim \Theta_d^{loc,l_d}$
      ii. draw a word $w_{d,i} \sim \phi_{z_{d,i}}^{loc,l_d}$

The graphical model of GLDA in plate notation is presented in Figure 2b.

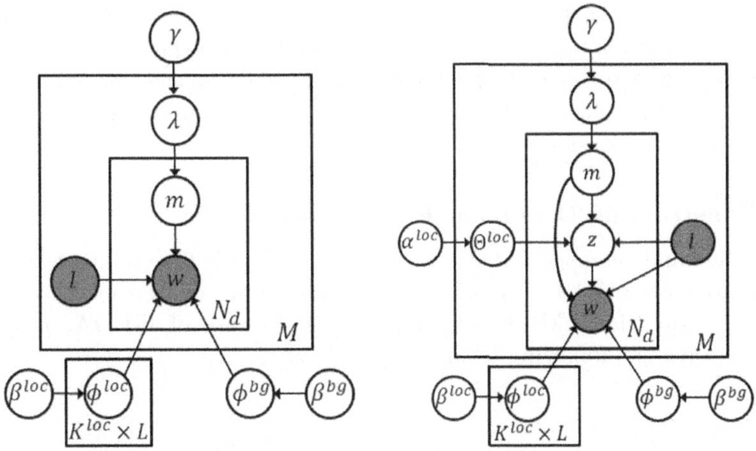

(a) a) **GLTA** (b) b) **GLDA**

Fig. 2. Graphical models of the proposed LVMs in plate notation

4.4 Posterior Inference

Posterior inference for **GLTA** is done using Gibbs sampler, which at each iteration selects the topic type $m_{d,i}$ for a word at every position i in each document in $\mathbb{C}$ based on the following formulas:

$$p(m_{d,i} = bg|\boldsymbol{m}_{\neg i}) \propto \frac{n(d, bg)_{\neg i} + \gamma}{n_d + 2\gamma - 1} \times \frac{n(w_{d,i}, bg)_{\neg i} + \beta^{bg}}{\sum_{j=1}^{N} n(w_j, bg) + N\beta^{bg} - 1} \tag{4}$$

$$p(m_{d,i} = loc|\boldsymbol{m}_{\neg i}) \propto \frac{n(d, loc)_{\neg i} + \gamma}{n_d + 2\gamma - 1} \times \frac{n(w_{d,i}, loc)_{\neg i} + \beta^{loc}}{\sum_{j=1}^{N} n(w_j, loc) + N\beta^{loc} - 1} \tag{5}$$

The Gibbs sampler for **GLDA** at each iteration selects both the topic type $m_{d,i}$ and topical assignment $z_{d,i}$ for each word based on the following formulas:

$$p(m_{d,i} = bg|\boldsymbol{z}_{\neg i}, \boldsymbol{m}_{\neg i}) \propto \frac{n(d, bg)_{\neg i} + \gamma}{n_d + 2\gamma - 1} \times \frac{n(w_{d,i}, bg)_{\neg i} + \beta^{bg}}{\sum_{j=1}^{N} n(w_j, bg) + N\beta^{bg} - 1} \tag{6}$$

$$p(z_{d,i}^{loc,l_d}, m_{d,i} = loc|\boldsymbol{z}_{\neg i}, \boldsymbol{m}_{\neg i}) \propto \frac{n(d, loc)_{\neg i} + \gamma}{n_d + 2\gamma - 1} \times \frac{n(w_{d,i}, z_{d,i}^{loc,l_d})_{\neg i} + \beta^{loc}}{\sum_{j=1}^{N} n(w_j, z_{d,i}^{loc,l_d}) + N\beta^{loc} - 1} \times \frac{n(d, z_{d,i}^{loc,l_d})_{\neg i} + \alpha^{loc}}{\sum_{k=1}^{K^{loc}} n(d, z_k^{loc,l_d}) + K^{loc}\alpha^{loc} - 1} \tag{7}$$

where $n(w,z)_{\neg i}$ is the number of times a term w is assigned to a topic z in the entire collection and $n(d,z)_{\neg i}$ $n(d,bg)$, $n(d,loc)$ are the number of terms in document d that are assigned to topic z, background or geo-specific topics (all counts exclude the current assignments of topic category m and topic z to the word at position i in document d).

4.5 Constructing Document Expansion LMs

Background and geo-specific topics, per document topic type mixtures and topic distributions obtained by the LVMs presented above can be used to derive a document expansion LM $p(w|\hat{\Theta}_d)$ for each d. In case of **GLDA**, $p(w|\hat{\Theta}_d)$ is obtained using the following formula:

$$p(w|\hat{\Theta}_d) = p(bg|\lambda_d)p(w|\phi^{bg}) + p(loc|\lambda_d)\sum_{k=1}^{K^{loc}} p(w|\phi_k^{loc,l_d}) \times p(z_k^{loc,l_d}|\Theta_d^{loc}) \quad (8)$$

5 Experiments

We used the 2011 TREC Microblog track [16] corpus, which is a 1% sample of Twitter over a period of 2 weeks, as the base dataset for all experiments in this work. The query set for 2011 TREC Microblog track was used to tune the parameters of LVMs and retrieval model, while the 2012 query set was used for the final comparison of retrieval performance. To avoid the sparsity issue (only $1 \sim 2\%$ of microblog posts have geographic coordinates [9]), all microblog posts were labeled with the location of their authors extracted from their Twitter account. Potential noise that may be introduced by a fraction of tweets that are not about the user's primary location can be tolerated by the proposed LVMs, since they identify *major* topical patterns in large volumes of textual data created by many users (there are about 600,000 unique users in TREC dataset). Since the proposed retrieval method requires geographical meta-data, which is not available in the original TREC corpus, we performed additional data collection and pre-processing steps. Firstly, we post-processed the corpus by filtering out all non-English tweets (tweets that do not include any words from the English dictionary of the spell-checking program *aspell*). Secondly, we determined all unique users, who authored the tweets in TREC dataset, extracted their locations from their Twitter profiles and normalized those location to the common "city-country" format using a manually compiled dictionary of suburbs and popular name variants of major cities (e.g. ny, nyc, brooklyn, bronx were all converted to "new york-usa") and Google Geocoding API [1]. Then we selected the top 150 locations (top 10 of which are in Table 2) and used only the documents labeled with those locations to train the proposed LVMs.

Although all retrieval runs are based on using the original TREC corpus, for the purpose of unbiased evaluation of all retrieval models, we only considered the relevant tweets which are covered by the geo-coded subset of the original dataset.

[1] Data is available at `http://www.cs.wayne.edu/kotov/code.html#geombr`

5.1 Optimization of Topic Models

In order to determine the optimal number of local topics for **PH-GLDA** and **GLDA**, we trained both models on 90% of the documents in each location sub-collection and estimated the perplexity on the remaining 10% of documents. During both training and testing the Gibbs sampler was run for 1000 iterations for all models. We found out that **GLDA** achieves significantly lower perplexity than both post-hoc baselines (**OPT-GLDA** and **PH-GLDA**) and **GLTA**, which we attribute to the inclusion of an additional hidden variable, which determines the topic type. Our experiments indicated that the optimal number of geo-specific topics per location for **GLDA** is 30. Examples of the topics discovered by **PH-GLDA**, **OPT-GLDA**, **GLTA** and **GLDA** are provided in Table 3.

Table 3. Sample geographically-specific topics and LMs extracted by the proposed LVMs

PH-GLDA						OPT-GLDA					
chicago-usa			cairo-egypt			chicago-usa			cairo-egypt		
topic 1	topic 2	topic 3	topic 1	topic 2	topic 3	topic 1	topic 2	topic 3	topic 1	topic 2	topic 3
snow	tax	new	protest	tahrir	sunni	come	chicago	get	egypt	tahrir	will
take	nice	day	night	old	regime	snomg	mayor	bulls	revolut	protest	people
close	idea	game	fun	light	problem	blizzard	story	beat	police	egypt	mubarak
weather	rahm	race	intern	square	bandar	snow	rahm	point	please	cairo	peace
inch	chi	bears	airport	govt	mobile	stuck	today	rose	thug	freedom	kill
GLTA						GLDA					
bkg	chicago-usa		cairo-egypt			bkg	chicago-usa		cairo-egypt		
	geo LM		geo LM				topic 1	topic 2	topic 1	topic 2	topic 3
new	chicago	blizzard	egypt	cairo	regime	rt	snow	court	egypt	muslim	amin
rt	snow	home	thug	people	arab	go	storm	rahm	tahrir	silent	shahira
love	day	good	tahrir	square	arrest	new	blizzard	mayor	square	brotherhood	mustafa
time	bears	snow	mubarak	police	army	love	inch	emanuel	protest	aljazeera	heenim
video	game	work	protest	revolut	afp	time	shovel	ballot	mubarak	moham	sabah

5.2 Optimization of Parameters and Training Performance Summary

We used 2011 TREC Microblog track query set to optimize the parameters of different document expansion-based retrieval models proposed in this work with respect to precision at 20 (P@20). First, we optimized the value of Dirichlet prior μ in **QL-DIR** and achieved the best performance when $\mu = 50$. After that we optimized the value of interpolation coefficient α. Sensitivity of retrieval performance of the document expansion methods in terms of mean average precision (MAP) and P@20 on different settings of interpolation coefficient α is shown in Figures 3a and 3b, respectively. Retrieval performance of the proposed methods and the baselines on the training query set is summarized in Table 4.

As follows from Table 4, document expansion methods leveraging the output of geography-aware LVMs all improve over the baseline retrieval model (**QL-DIR**), while **GLDA** consistently achieves the best performance across all metrics and outperforms a state-of-the-art baseline (**PH-GLDA**).

5.3 Testing Performance Summary

Table 5 summarizes and compares with the baselines the retrieval effectiveness of document expansion methods that use the output of the proposed latent variable

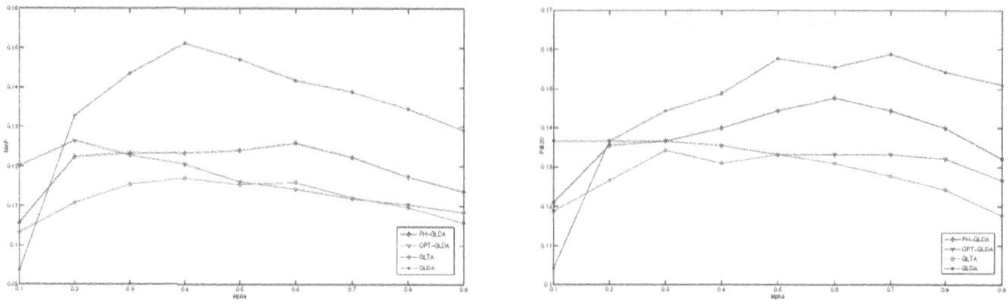

(a) MAP (b) P@20

Fig. 3. Performance of document expansion methods based on different LVMs by varying the interpolation coefficient α

Table 4. Comparison of the best performance of different document expansion-based methods optimized with respect to P@20 on the training query set

method	MAP	GMAP	P@20	Bpref
QL-DIR	0.1015	0.0333	0.1189	0.6223
PH-GLDA	0.1259	0.0464	0.1478	0.6264
OPT-GLDA	0.1266	0.0397	0.1367	0.6170
GLTA	0.1156	0.0356	0.1344	0.6225
GLDA	**0.1390**	**0.0503**	**0.1589**	**0.6445**

models on 2012 TREC Microblog query set, which we use as a testing set in this work. Both the proposed methods and the baselines are used with the optimal parameters determined on the training set as described in Section 5.2.

Table 5. Summary of retrieval performance of document expansion methods based on the output of the proposed latent variable models on testing query set using the optimal parameters determined on the training query set. The magnitude of improvement (↑) or degradation (↓) in percentage relative to QL-DIR baseline is shown in parenthesis. ▲ indicates statistically significant improvement according to the paired *t*-test ($p < 0.05$).

method	MAP	GMAP	P@20	Bpref
QL-DIR	0.0849	0.0469	0.106	0.6135
PH-GLDA	0.1167 (↑37.45%)	0.0662 (↑41.15%)	0.1664 (↑56.98%)	0.6053 (↓1.34%)
OPT-GLDA	0.1123 (↑32.27%)	0.0493 (↑5.11%)	0.1603 (↑51.23%)	0.571 (↓6.93%)
GLTA	0.1123 (↑32.27%)	0.0575 (↑22.6%)	0.1466 (↑38.3%)	0.5976 (↓2.59%)
GLDA	**0.1289 (↑51.83%▲)**	**0.0745 (↑58.85%▲)**	**0.1698 (↑60.19%▲)**	**0.6323 (↑3.06%)**

The results in Table 5 indicate that document expansion based on the post-hoc geographical variants of LDA (**PH-GLDA** and **OPT-GLDA**) and our proposed latent variable models (**GLTA** and **GLDA**) all result in significant improvement over **QL-DIR** baseline according to MAP, GMAP and P@20. Remarkably, retrieval accuracy in terms of Bpref measure is improved only when document expansion based on **GLDA** is used and gets worse in case of both post-hoc LDA variants and **GLTA**. Hence, **GLDA**-based document expansion is able to not only retrieve more relevant documents at higher ranks, but also consistently ranks relevant documents above non-relevant ones. Furthermore, **GLDA**-based document expansion results in the highest improvement of retrieval accuracy relative to **QL-DIR** baseline across all metrics.

Table 6 compares the retrieval accuracy of the proposed latent variable models and post-hoc LDA variants relative to the state-of-the-art baseline (**PH-GLDA**). As follows from Table 6, only **GLDA** was able to consistently outperform **PH-GLDA**. The improvement is substantial (over 10%) in terms of MAP and GMAP and statistically significant across most metrics.

Table 6. Improvement (↑) or degradation (↓) of retrieval performance of document expansion LMs derived from the proposed latent variable models relative to PH-GLDA baseline. • indicates statistically significant improvement ($p < 0.05$).

method	MAP	GMAP	P@20	Bpref
OPT-GLDA	↓3.77%	↓25.53%	↓3.67%	↓5.67%
GLTA	↓3.77%	↓13.14%	↓11.9%	↓1.27%
GLDA	↑10.45%•	↑12.54%•	↑2.04%	↑4.46%•

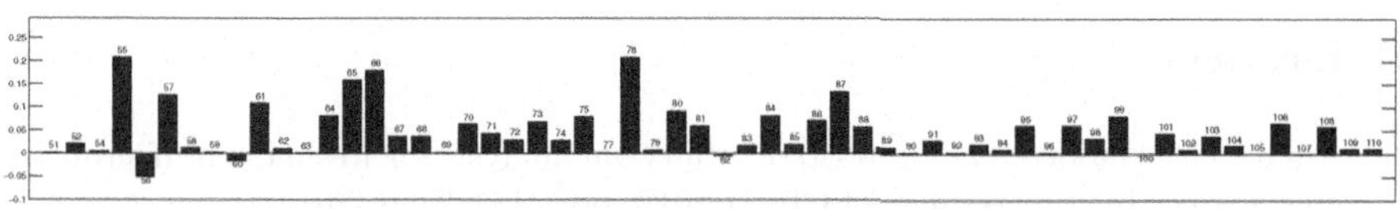

(a) Between GLDA and QL-DIR

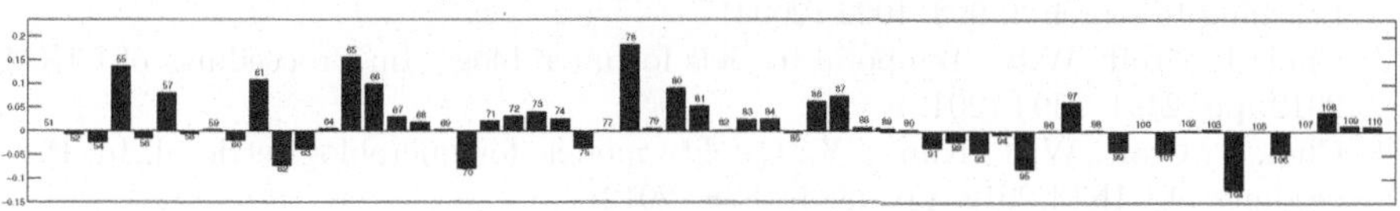

(b) Between GLDA and PH-GLDA

Fig. 4. Per-topic difference in average precision between GLDA and the baselines

Figure 4 shows per-topic differences in average precision between the best preforming document expansion method (based on **GLDA** model) and **QL-DIR** and **PH-GLDA** baselines. As follows from both Figure 4a and 4b, improvement in retrieval accuracy varies by the topic. The highest improving queries are shared between both baselines, which indicates that they are both benefiting from accounting for the same phenomena in retrieval, and include: *MB57 "Chicago blizzard"*, *MB61 "Hu Jintao visit to the United States"*, *MB65 "Michelle Obama's obesity campaign"*, *MB66 "Journalists' treatment in Egypt"*, *MB78 "McDonalds food"*, *MB86 "Joanna Yeates murder"*.

In contrast, the topics, for which applying geographically focused document expansion results in decreased retrieval performance (e.g. *MB62 "Starbucks Trenta cup"*, *MB70 "farmers markets opinions"*, *MB70 "texting and driving"*) are broad queries that are not tied to any particular geographical location.

Summary and Conclusions

The main contribution of the present work are as follows:

- we proposed new geography-aware LVMs that work with *textual* geographical labels;
- we proposed a method to derive document expansion LMs that leverages the output of the proposed LVMs and compared the retrieval effectiveness of the proposed LVMs on standard TREC datasets for MBR evaluation. Unlike most of the previously proposed methods for microblog IR, our approach does not rely on pseudo-relevance feedback, and hence is more robust and efficient.

Our work has implications beyond microblog retrieval. In particular, the proposed methods can be applied to any geo-tagged document collections other than microblogs. We believe that an interesting direction for future research would be to consider an interplay between different dimensions of relevance in microblog retrieval, such as geographic and temporal.

References

1. Amati, G., Amodeo, G., Gaibisso, C.: Survival analysis for freshness in microblogging search. In: Proceedings of CIKM 2012, pp. 2483–2486 (2012)
2. Blei, D.M., Ng, A.Y., Jordan, M.I.: Latent dirichlet allocation. Journal of Machine Learning Research 3, 993–1022 (2003)
3. Choi, J., Croft, W.B.: Temporal models for microblogs. In: Proceedings of CIKM 2012, pp. 2491–2494 (2012)
4. Choi, J., Croft, W.B., Kim, J.Y.: Quality models for microblog retrieval. In: Proceedings of CIKM 2012, pp. 1834–1838 (2012)
5. Efron, M.: Information search and retrieval in microblogs. ASIS&T 62(6), 996–1008 (2011)
6. Efron, M., Lin, J., He, J., de Vries, A.: Temporal feedback for tweet search with non-parametric density estimation. In: Proceedings of SIGIR 2014, pp. 33–42 (2014)
7. Efron, M., Organisciak, P., Fenlon, K.: Improving retrieval of short texts through document expansion. In: Proceedings of SIGIR 2012, pp. 911–920 (2012)
8. Eisenstein, J., O'Connor, B., Smith, N.A., Xing, E.P.: A latent variable model for geographic lexical variation. In: Proceedings of EMNLP 2010, pp. 1277–1287 (2010)
9. Hong, L., Ahmed, A., Gurumurthy, S., Smola, A., Tsioutsiouliklis, K.: Discovering geographical topics in the twitter stream. In: Proceedings of WWW 2012, pp. 769–778 (2012)
10. Keikha, M., Gerani, S., Crestani, F.: Time-based relevance models. In: Proceedings of SIGIR 2011, pp. 1087–1088 (2011)
11. Kotov, A., Agichtein, E.: The importance of being socially-savvy: Quantifying the influence of social networks on microblog retrieval. In: Proceedings of CIKM 2013, pp. 1905–1908 (2013)
12. Kotov, A., Wang, Y., Agichtein, E.: Leveraging geographical metadata to improve search over social media. In: Proceedings of WWW 2013, pp. 151–152 (2013)

13. Mei, Q., Liu, C., Su, H., Zhai, C.: A probabilistic approach to spatiotemporal theme pattern mining on weblogs. In: Proceedings of WWW 2006, pp. 533–542 (2006)
14. Miyanishi, T., Seki, K., Uehara, K.: Combining recency and topic-dependent temporal variation for microblog search. In: Serdyukov, P., Braslavski, P., Kuznetsov, S.O., Kamps, J., Rüger, S., Agichtein, E., Segalovich, I., Yilmaz, E. (eds.) ECIR 2013. LNCS, vol. 7814, pp. 331–343. Springer, Heidelberg (2013)
15. Miyanishi, T., Seki, K., Uehara, K.: Improving pseudo-relevance feedback via tweet selection. In: Proceedings of CIKM 2013, pp. 439–448 (2013)
16. Ounis, I., Macdonald, C., Lin, J., Soboroff, I.: Overview of the trec-2011 microblog track. In: Proceedings of TREC 2011 (2011)
17. Teevan, J., Ramage, D., Morris, M.R.: #twittersearch: A comparison of microblog search and web search. In: Proceedings of ACM WSDM 2011, pp. 35–44 (2011)
18. Wei, X., Croft, W.B.: Lda-based document models for ad-hoc retrieval. In: Proceedings of ACM SIGIR 2006, pp. 178–185 (2006)
19. Wing, B.P., Baldridge, J.: Simple supervised document geolocation with geodesic grids. In: Proceedings of the ACL 2011, pp. 955–964 (2011)
20. Yi, X., Allan, J.: A comparative study of utilizing topic models for information retrieval. In: Boughanem, M., Berrut, C., Mothe, J., Soule-Dupuy, C. (eds.) ECIR 2009. LNCS, vol. 5478, pp. 29–41. Springer, Heidelberg (2009)
21. Yin, Z., Cao, L., Han, J., Zhai, C., Huang, T.: Geographical topics dsicovery and comparison. In: Proceedings of WWW 2011, pp. 247–256 (2011)

Nonparametric Topic Modeling Using Chinese Restaurant Franchise with Buddy Customers*

Shoaib Jameel, Wai Lam, and Lidong Bing

Key Lab of High Confidence Software Technologies,
Ministry of Education (CUHK Sub-Lab)
Department of Systems Engineering and Engineering Management,
The Chinese University of Hong Kong, Hong Kong
{msjameel,wlam,ldbing}@se.cuhk.edu.hk

Abstract. Many popular latent topic models for text documents generally make two assumptions. The first assumption relates to a finite-dimensional parameter space. The second assumption is the bag-of-words assumption, restricting such models to capture the interdependence between the words. While existing nonparametric admixture models relax the first assumption, they still impose the second assumption mentioned above about bag-of-words representation. We investigate a nonparametric admixture model by relaxing both assumptions in one unified model. One challenge is that the state-of-the-art posterior inference cannot be applied directly. To tackle this problem, we propose a new metaphor in Bayesian nonparametrics known as the "Chinese Restaurant Franchise with Buddy Customers". We conduct experiments on different datasets, and show an improvement over existing comparative models.

1 Introduction

Assuming the bag-of-words representation in documents has been the holy-grail in probabilistic topic modeling such as Latent Dirichlet Allocation (LDA) [1]. The bag-of-words assumption simplifies the modeling [1], and has an advantage for computational efficiency [2]. However, this assumption has some disadvantages. One major disadvantage is that many unigram words discovered in the latent topics are not very insightful to a reader [3]. Another disadvantage is that the model is not able to consider semantic information that is conveyed by the order of the words in the document [2]. This results in an inferior performance in some text mining tasks as shown by different topic models [4,5,6,7]. These models may discover many general words in latent topics with high probability instead of relevant content words [8]. In order to tackle this problem, general words are commonly removed from the corpus during text pre-processing [9], but this

* The work described in this paper is substantially supported by grants from the Research Grant Council of the Hong Kong Special Administrative Region, China (Project Code: CUHK413510) and the Microsoft Research Asia Urban Informatics Grant FY14-RES-Sponsor-057. This work is also affiliated with the CUHK MoE-Microsoft Key Laboratory of Human-centric Computing and Interface Technologies.

A. Hanbury et al. (Eds.): ECIR 2015, LNCS 9022, pp. 648–659, 2015.

leads to further problems especially when processing natural language or speech data [8]. Wallach [8] has described that incorporating word order removes many general words dominating the latent topics. McCallum et al., in [9] have shown that using asymmetric priors in the LDA model can also help reduce the problem, but still the topic interpretability problem remains [3].

In order to address the limitations inherent in the unigram based topic models, some parametric topic models have been proposed which maintain the order of the words in the document. Such models are able to not only discover phrasal terms in topics [3], but also demonstrate a superior performance on several text mining tasks such as document classification [2] and document modeling [8]. It is intuitive that generating a phrasal term such as "air conditioner" is more insightful than just discovering "air" and "conditioner" independently [3,10,11]. These models have a fixed parameter space and some parameters, such as the number of topics, need to be pre-defined by the user. This might be impractical because the user may not always know the true number of latent topics inherent in the data.

One way to address the model selection issue is to train several models with different number of topics, and choose the one that has the best performance [12]. But this is not a principled approach and it is very time consuming [12]. A desirable way to deal with the problem is to automatically infer the number of latent topics based on the characteristic of document collection. Such models are known as nonparametric probabilistic topic models which are characterized by an infinite-dimensional parameter space. Most importantly, these nonparametric latent topic models impose as few assumptions as possible [13] making them more powerful than parametric latent topic models. Parametric models might face over-fitting and under-fitting issues when there is a mis-match between the model complexity and the data. In contrast, nonparametric models are less prone to this problem [14]. Models such as Hierarchical Dirichlet Processes (HDP) [15] when used as a topic model [16,17] can automatically infer the number of latent topics based on the data characteristic, but it imposes the bag-of-words assumption in documents. The "Chinese Restaurant Franchise" (CRF) metaphor has been proposed to compute the posterior distribution of HDP, which generates data from an exchangeable distribution. It thus inherits some of the limitations of the unigram based topic models.

To tackle the above issues, we propose a new metaphor in Bayesian nonparametrics called "Chinese Restaurant Franchise with Buddy[1] Customers" (CRF-BC) that not only maintains the word order, but also infers automatically the number of latent topics based on the data characteristic. Our metaphor falls in the class of non-exchangeable distributions for Bayesian nonparametric models [18]. Using the buddy assignment scheme, our model can discover n-gram words in topics. By n-gram we mean that we can discover a unigram or a bigram or even a higher-order-gram depending upon the buddy assignments. One challenge is that the state-of-the-art posterior inference cannot be applied directly. We refine the traditional Gibbs sampling algorithm for nonparametric topic modeling for

[1] Buddy is an informal term meaning a close friend. -Source: Wikipedia.

our metaphor. We conduct experiments on document modeling and show that our framework can outperform state-of-the-art topic models.

2 Related Work

Much work has been done in the parametric topic modeling literature where the order of words in documents is maintained. There are some models which use the LDA model to discover n-gram words, for example, [4]. Wallach [8] proposed the Bigram Topic Model (BTM) for text data that maintains the order of the words. Griffiths et al., [19] extended the BTM model and proposed the LDA-Collocation Model (LDACOL). In the Topical N-gram model (TNG) [10] the topic assignments for the two words in a bigram may not be alike. Lindsey et al., [3] proposed a topic model that incorporates the Hierarchical Pitman-Yor Processes (HPYP) in the LDA model. But the main concern is that the model cannot scale to accommodate large text collections due to the HPYP model [20]. Lau et al., [21] presented a study investigating whether word collocations can help improve topic models. In Johri et al., [22], the authors introduced a multi-word enhanced author-topic model for text data. In [23], the authors proposed some improvements to the n-gram topic models. Their method uses Chinese Restaurant Process (CRP) for sampling, with a fixed dimensional parameter space.

The seminal nonparametric topic model is the Hierarchical Dirichlet Processes (HDP) model proposed by Teh et al., [15]. This model assumes that words in a document are exchangeable, and thus cannot capture short-range word dependencies. CRF metaphor is also used to describe this model [17]. Considering the order of words in Bayesian nonparametrics[2] has attracted some attention recently. Goldwater et al., [25] presented two nonparametric models for word segmentation. Observing that the ordering of words could play a dominant role, Goldwater et al., extended the unigram based model to a bigram based model called the "Bigram HDP" model. The model closely resembles the HPYP model and cannot generate latent topics. It is well suited for the word segmentation task. Johnson [26] incorporated nonparametric adaptor grammars to discover word collocations instead of just unigrams. However, one disadvantage is that it adopts a two-stage approach towards collocation discovery whereas our approach can tackle it in a single model. In [27], the author introduced a nonparametric model that can extract phrasal terms based on the mutual rank relation. It employs a heuristic measure for the identification of phrasal terms. In [28], the authors introduced the notion of an extension pattern, which is a formalization of the idea of extending lexical association measures defined for bigrams. In [29], the authors presented a Bayesian nonparametric model for symbolic chord sequences. Their model is designed to handle n-grams in chord sequences for music information retrieval. Recently, we have proposed a nonparametric topic model to discover more interpretable latent topics in [6]. One main weakness of the model is that only the first term in the bigram has a topic assignment

[2] Due to space limit, we do not present a detailed background of Bayesian nonparametrics. We request inquisitive readers to consult some excellent resources [24,13].

whereas the second term does not. The model uses existing posterior inference schemes to discover collocations. Our model proposed in this paper bears some theoretical resemblance with the Distance Dependent Chinese Restaurant Process (ddCRP) [30] in which customers are first assigned to each other and this customer-customer assignment can directly be related to a clustering property. In our model, customers are first assigned to each other using the buddy assignment scheme and then the customers are assigned to tables. A franchise based model based on the ddCRP has been proposed in [31], but this model does not consider the order of words in the document. Some interesting extensions have been proposed in the past with slight modifications to the basic CRF metaphor. For example, Fox et al., [32] proposed the "Chinese Restaurant Franchise with Loyal Customers". "Chinese Restaurant Franchise with Preferred Seating" has been proposed in [33].

Our proposed model is different from the above models. In contrast with [6,10], our framework gives the same topic assignment to all the words in an n-gram. We derive a posterior inference scheme which is different from the one employed in existing models.

3 Our Proposed Model

3.1 Chinese Restaurant Franchise (CRF) Background

One perspective associated with the HDP mechanism can be expressed by the Chinese Restaurant Franchise (CRF) [15] which is an extension of the Chinese Restaurant Process (CRP). The HDP model makes use of this metaphor to generate samples from the posterior distribution given the observations. In order to describe the sharing among the groups, the notion of "franchise" has been introduced that serves the same set of dishes globally. When applied to text data, each restaurant corresponds to a document. Each customer corresponds to a word. Each dish corresponds to a latent topic. A customer sits at a table, one dish is ordered for that table and all subsequent customers who sit at that table share that dish. The dishes are sampled from the base distribution which corresponds to discrete topic distributions. Multiple tables in multiple restaurants can serve the same dish. A table can be regarded as the topic assignment of the words in documents.

3.2 Our Proposed CRF-BC Model

We propose a new class of non-exchangeable metaphor which considers the order of words in the document. In this metaphor, customers are first assigned to each other outside the restaurant, and subsequently, individual customers enter the restaurant and sit at tables just as in the CRF metaphor. However in order to capture n-grams words, we need to refine the existing HDP model and its inference framework which uses CRF because the existing framework does not consider word order. Our new metaphor known as "Chinese Restaurant Franchise with Buddy Customers" (CRF-BC) can capture friendship associations between customers in the entire customer-franchise setup. Our model follows a Markovian assumption on the order of words and also imposes a transitive property on that order in sequence to discover n-grams. It means that if w_i^d (w_i^d is a word at position i in the document d) is a buddy of w_{i-1}^d, and w_{i-1}^d is a buddy of w_{i-2}^d, then w_i^d is also a buddy of w_{i-2}^d. Similarly, if w_{i-1}^d is a buddy of w_{i-2}^d, and w_{i-2}^d is a buddy of w_{i-3}^d, then w_{i-3}^d and w_i^d are also buddies. Following this rule, we can obtain higher order n-grams. One can certainly impose higher order Markovian assumptions, but it would impose problems with data sparsity and high computational complexity. The idea of employing first order Markovian assumption on word order has also been used in other parametric topic models such as [2].

The general idea behind this metaphor can be described in this way. Consider a Chinese franchise with a shared menu which is shared across the restaurants. Each restaurant has an infinite set of tables as in the original CRF scheme and each restaurant corresponds to a document. Consider a set of customers, which are mainly words in the document. Some of the customers have pre-planned their visit so that they can spend time together with their "good old buddies" and eat the same food in the table. These buddies have already reserved their tables beforehand. In this scheme, we assume that the customers are waiting in the queue outside the restaurant in the same order as that of the words in a document. This assumption is different from the CRF metaphor. There might be "loners" too in the same queue who may have no buddies. They too can sit and eat in the same restaurant in any of the other unreserved tables or share the table with other lonely customers. Just as in the CRF metaphor, we assume that the loners share the same dish with other customers in that table. Note that inside the restaurant, exchangeability is still valid i.e. tables are exchangeable and so are customers who are sitting at those tables as buddies can sit in any seat at the reserved table. As every customer carries with herself a table, a buddy and word order assignments, we can easily get n-gram words in topics from these three information. We present a detailed generative mechanism of our probabilistic CRF-BC in the "restaurant-franchise representation" below.

1. Draw ϕ from $\mathbf{Dirichlet}(\beta\tau)$, where β is the concentration parameter, and τ is the corpus-wide distribution over vocabulary. ϕ is the word-topic distribution matrix. We place a $\mathbf{Dirichlet}(\kappa_\tau)$ prior over τ. We also place a $\mathbf{Gamma}(\kappa_\beta^1, \kappa_\beta^2)$ over β. $\kappa_\beta^1, \kappa_\beta^2$ are the shape and scale parameters respectively. One can notice that we infer the priors by placing priors over those priors to find their posteriors. Thus the resulting inferences are less influenced by these "hyper-hyperparameters" than they are by fixing the original hyperparameters to specific values [13].
2. Draw μ from $\mathbf{GEM}(\eta)$.
   We place a $\mathbf{Gamma}(\kappa_\eta^1, \kappa_\eta^2)$ prior over η to compute its posterior. $(\kappa_\eta^1, \kappa_\eta^2)$ are the shape and scale parameters of the Gamma distribution respectively. Readers can consult [13] for description about GEM distribution. μ actually supplies the corpus-wide distribution over topics information which follows the stick-breaking representation.
3. Draw $\mathbf{Discrete}(\sigma)$ from $\mathbf{Dirichlet}(\delta)$.
   σ is the distribution over "buddies", and δ is its conjugate prior. We place a $\mathbf{Gamma}(\kappa_\delta^1, \kappa_\delta^2)$ prior over this prior to compute the posterior of this prior.

4. Draw **Bernoulli**(ω) from **Beta**(γ_0, γ_1), where γ_0 and γ_1 are the shape parameters of the Beta distribution.
   ω is the distribution over "buddy assignment variables".
5. For each document d,
   (a) Draw **Multinomial**($\widetilde{\theta}^d$) from **Dirichlet**(α).
   The variable $\widetilde{\theta}^d$ will contain the per-document topic distribution, α is the prior or concentration parameter, and we determine the value of this prior by placing another prior, for example, **Gamma**($\kappa_\alpha^1, \kappa_\alpha^2$), where $\kappa_\alpha^1, \kappa_\alpha^2$ are the shape and the scale parameters of the Gamma distribution respectively.
   (b) Draw k_t^d from μ, where k_t^d is the topic index variable for each table t in d. μ comes from the stick breaking process.
   (c) For each word w_i^d at the position i in the document d (we are considering the word order here),
      i. Draw b_i^d from **Bernoulli**($\omega_{t_{i-1}^d w_{i-1}^d}$).
      This is where we conduct buddy assignments. The underlying meaning is that, if $b_i^d = 0$, where b_i^d is a buddy assignment variable, then the customer (word) is a "loner" and is not a buddy with the previous customer standing in that queue, and if $b_i^d = 1$, then customer who is waiting outside the restaurant is a "buddy" with the previous customer (word) standing in the same queue. Previous customer means a customer standing in front of the current customer in the queue. This partitioning of customers or buddy assignments outside the restaurant is done based on corpus wide statistics. The first customer in the queue assumes $b_i^d = 0$. Buddy assignments not only consider the co-occurrence information, but also consider the latent topic of the previous word. In the initial run of the algorithm, this assignment is done randomly which may change by the sampler during future iterations.
      ii. Draw t_i^d from $\widetilde{\theta}^d$ if $b_i^d = 0$, otherwise $t_i^d = t_{i-1}^d$.
      This process says that if the current customer is not a buddy with the previous customer then the current customer draws a new table assignment for herself. Otherwise, if the new customer is a buddy and sits at the same table as its previous buddy and shares the same dish. t is a table or an indication of a cluster for the word i in the document d.
      iii. Draw w_i^d from $\phi_{k_{t_i^d}^d}$ if $b_i^d = 0$ else draw $\sigma_{w_{i-1}^d}$. $\phi_{k_{t_i^d}^d}$ refers to a specific value in the matrix ϕ by following the path of the table and dish assignments if the customers are not buddies. Otherwise, buddies are drawn from a distribution of the previous buddy (word). Another way to describe the process is that the customer w_i^d in the restaurant d, sat at table t_i^d while the table t in the restaurant d serves the dish k_t^d.

3.3 Posterior Inference in CRF-BC

To find the latent variables that best explain the observed data, we use Gibbs sampling. One of the main advantages of using this sampling is that it samples from a true posterior. It requires some resources on book-keeping leading to a more effective algorithm [15]. Note that in our model, we have to make significant changes at the restaurant level, and little at the franchise level of the CRF metaphor as the buddy allocation happens outside the restaurant. Due to space constraint, we present an outline of our algorithm.

We will sample t_i^d which is the table index for each word w at the position i in the document d. Let K be the total number of topics, which can either increase or decrease as the number of iterations of the sampler increases. Let $\hat{k}$ denote the new topic being sampled. We will then sample k_t^d which is the topic (dish) index variable for each table t in d. Let n_{tk}^d be the number of customers at restaurant d, sitting at table t eating dish k. We define $\mathbf{w}$ as $(w_i^d : \forall d, i)$ and $\mathbf{w}_t^d$ as $(w_i^d : \forall i \text{ with } t_i^d = t)$, $\mathbf{t}$ as $(t_i^d : \forall d, i)$ and $\mathbf{k}$ as $(k_t^d : \forall d, t)$. Let $m_{.k}$ denote the number of tables belonging to the topic k in the corpus. Let $m_{..}$ denote the total number of tables in the corpus. $f_{\hat{k}}^{\neg w_i^d}(w_i^d)$ is the prior density of w_i^d. When a sign $\neg$ in the superscript is attached to a set of variables or count, for example, $(\mathbf{k}^{\neg dt}, \mathbf{t}^{\neg di})$, it means that the variables corresponding to the superscripted index is removed from the set or from the calculation of the count. Let $f_k^{\neg w_i^d}(.)$ denote the conditional likelihood density for some previously used table, which can be derived based on the type of the problem we are solving. In [15], the authors only presented HDP in general and not for topic modeling in particular. In case of topic modeling, we can follow a widely used Dirichlet-Multinomial paradigm, where the base measure is a Dirichlet, and the density F (same F as used in [15]) as Multinomial. We also introduce a notion of reserved tables using r. We use v to denote an unreserved table. We use the symbol $\hat{t}$ or $\hat{k}$ to denote a new table and dish, respectively. Also, note that buddies will be in their own buddy circles (commonly known as friendship circle) waiting outside the restaurant in queue, so different buddy groups take their own reserved tables. The likelihood of w_i^d who is a loner for $t_i^d = \hat{t}$, where $\hat{t}$ is the new table being sampled, is written as:

$$P(w_i^d = \texttt{Loner} | t_i^d = \hat{t} = v, \mathbf{t}^{\neg di}, \mathbf{k}, b_i^d = 0, w_{i-1}^d, t_{i-1}^d) = \sum_{k=1}^{K} \frac{m_{.k}}{m_{..} + \eta} f_k^{\neg w_i^d}(w_i^d) + \frac{\eta}{m_{..} + \eta} f_{\hat{k}}^{\neg w_i^d}(w_i^d) \quad (1)$$

The above equation lays a restriction on the "loner" not to occupy the reserved table. This is because $b_i^d = 0$ associated with the loner will disallow this loner to occupy any of the reserved tables. But the loner can request a new table of the same topic (by ordering the same dish k as those of the reserved tables) as that of the reserved table or a different dish $\hat{k}$, with probability value proportional to α. The loner can also share an unreserved table with other loners with a value proportional to n_{tk}^d. The mechanism for buddies choosing a table is different. b_i^d indicates whether a customer is a buddy with the previous customer. The first buddy, w_i^d, who enters the restaurant carries with herself $b_i^d = 0$ because this customer is not a buddy with the previous customer who has just entered the restaurant. This customer is certainly not a loner, but will follow Equation 1 due to the buddy assignment variable. Therefore, this customer can either share an unreserved table with other loners, or requests a new table and sits alone. But when the second customer w_{i+1}^d in that buddy group enters the restaurant, this customer knows that the previous customer is her buddy. So this customer requests new table serving the same dish if the previous customer sat at an unreserved shared table, or shares the table with the previous buddy in case that buddy had requested a new table for herself and happens to be the first customer to sit there. The table is then set to reserved. The changes made by w_i^d using Equation 1 (if used) have to be reset to the previous state. This is where we make slight changes at the franchise level where we decrement the count from the existing unreserved table where w_i^d sat. The previous buddy then joins the buddy in that table. The scheme at the restaurant level can be expressed as:

$$P(w_i^d = \texttt{First}|t_i^d, \mathbf{t}^{\neg di}, \mathbf{k}, b_i^d = 1, w_{i-1}^d, t_{i-1}^d) = \begin{cases} \frac{\eta}{m_{..}+\eta} f_{\hat{k}}^{\neg w_i^d}(w_i^d) \ \& \ k_{i-1}^d = k_i^d \text{ if } t_i^d = \hat{t}, b_{i-1}^d = 0, b_i^d = 1 \\ \sum_{k=1}^{K} \frac{m_{.k}}{m_{..}+\eta} f_k^{\neg w_i^d}(w_i^d) \ \& \ t_i^d = t_{i-1}^d, \hat{t} = r \text{ if } b_{i-1}^d = 0, t_{i-1}^d = \hat{t}, b_i^d = 1 \end{cases} \quad (2)$$

Others, in the buddy group sit in the same table one by one requested by the "First Buddy" (denoted by `First` in Equation 2) i.e. ($t_i^d = t_{i-1}^d$) and share the same dish k.

$$P(w_i^d = \texttt{Other}|t_i^d = r, \mathbf{t}^{\neg di}, \mathbf{k}, b_i^d = 1, w_{i-1}^d, t_{i-1}^d) = \sum_{k=1}^{K} \frac{m_{.k}}{m_{..}+\eta} f_k^{\neg w_i^d}(w_i^d) \ \& \ t_i^d = t_{i-1}^d, k_i^d = k_{i-1}^d \quad (3)$$

We present the buddy assignment scheme below which is based on global statistics. The idea is to compute the probabilities of how often two customers (words) consecutively come in sequence. Then based on the probability value, the buddy indicator variable is set to either 0 or 1. Let $p_{t_{i-1}^d w_{i-1}^d b_i^d}$ be the number of times the buddy indicator variable b_i^d has been set to 0 or 1 given the previous word and the table of the previous word. $n_{w_i^d}^{w_{i-1}^d}$ is the number of times the word w_i^d comes after the word w_{i-1}^d in the entire corpus. Let V be the total number of words in the vocabulary. n_{kw} is the number of times a word has appeared in topic k.

$$P(b_i^d = 0|\mathbf{b}^{\neg di}, \mathbf{w}, \mathbf{t}) = \frac{p_{t_{i-1}^d w_{i-1}^d 0} + \omega_0}{\sum_{c=0}^{1} p_{t_{i-1}^d w_{i-1}^d c} + \omega_0 + \omega_1} \times \frac{(\beta \tau_{w_i^d} + n_{k w_i^d} - 1)}{\sum_{v=1}^{V} (\beta \tau_v + n_{kv}) - 1} \quad (4)$$

$$P(b_i^d = 1|\mathbf{b}^{\neg di}, \mathbf{w}, \mathbf{t}) = \frac{p_{t_{i-1}^d w_{i-1}^d 1} + \omega_1}{\sum_{c=0}^{1} p_{t_{i-1}^d w_{i-1}^d c} + \omega_0 + \omega_1} \times \frac{n_{w_i^d}^{w_{i-1}^d} + \delta_{w_i^d} - 1}{\sum_{v=1}^{V} (n_v^{w_{i-1}^d} + \delta_v) - 1} \text{ and } t_i^d = t_{i-1}^d \quad (5)$$

Using the above equations at the restaurant level and the franchise level of the CRF, one can compute the posterior estimates to get the topic distributions for a corpus.

4 Experiments and Results

In our experiments, we evaluate different aspects of our model in terms of its generalization ability on unseen data and the words generated in the topics. In all experiments, the Gibbs sampler was run for 1000 iterations. We found that this number of iterations is sufficient because the joint likelihood of the sampled hidden variables and the words indicated convergence in the Markov chain. The topic models were run for five times, and the average of those five runs was taken.

4.1 Document Modeling

Document modeling using perplexity has been widely used in topic modeling. We use the same formula for perplexity as used in [15]. We use both small and large scale datasets for this experiment. The datasets that we use are: 1) AQUAINT-1 that comes with TREC HARD track (1,033,461 documents), 2) NIPS dataset (1,830 documents) commonly used for topic models 3) OHSUMED, a popular dataset used in the information retrieval community (233,448 documents), 4) Reuters collection (806,791 documents). We used the same text pre-processing strategy as used in [2], which also maintains order of words. We create five folds for each of these datasets and conduct five-fold cross validation. Each fold is created by randomly sampling 75% of the entire documents into the training set, and the rest into the test set.

The comparative methods that we use in experiments consist of both parametric and nonparametric topic models. The parametric topic models are: `LDA` [1], `BTM` [8], `LDACOL` [19], `TNG` [10], and a recently proposed method `NTSeg` [2]. The nonparametric topic models are `HDP` [15], and a recently proposed model `NHDP` [6]. We use the best experimental settings including hyperparameter sampling for these models as described in their respective works. `HDP` and `NHDP` both use CRF to sample from the posterior.

We use a tuning method to determine the number of latent topics in the parametric models. In the tuning process, in each fold, we first divide the training set into the development set which is 75% of the total number of documents in the training set, and the rest goes into the tuning set. We train the model using the development set and vary the number of topics. Then we compute the perplexity for each number of topics using the tuning set in each fold. Note that we also run the Gibbs sampler with 1000 iterations in each fold. Then we choose the best performing model through this procedure i.e. the model with the lowest average perplexity. We repeat five times and take the average. The number of topics with the lowest average perplexity is chosen as the output of the tuning process. We then merge the development and the tuning sets together to get the

Table 1. Document modeling results

Model	Perplexity			
	AQUAINT-1	NIPS	OHSUMED	Reuters
LDA	4599.48	834.45	2305.32	3490.12
BTM	4578.57	833.75	2229.96	3411.98
LDACOL	4501.44	831.45	2398.22	3298.76
TNG	4423.76	828.32	2315.72	3108.43
NTSeg	4400.76	811.32	2295.72	3112.43
HDP	4322.32	825.43	2240.23	3192.54
NHDP	4495.32	820.56	2299.45	3102.53
Our	4107.75	766.90	2192.44	3089.44

training set where we train the model using the number of topics obtained from the tuning process. We test the model using the same number of topics on the test set in each fold, by running five times and compute the average.

Table 1 depicts the result of document modeling. In all the four datasets, we see that our model, labeled as "Our" is the best performing one. The improvements are statistically significant based on two-tailed test with $p < 0.05$ against each of the comparative methods. The reason why our model performs better in generalizability is mainly due to its ability to determine the number of topics based on the data characteristic. In addition, considering word order is another advantage. Our model also performs better than the n-gram parametric models. For parametric models, despite using the tuning step, the data fitting might be an issue in the test set. Unigram models cannot capture word order information.

4.2 Qualitative Results

We present some high probability words in decreasing order obtained from the nonparametric topic models in Table 2. Following the result illustration technique from [10], we present unigrams and n-grams separately as we are comparing with the HDP model. We show the results obtained from AQUAINT-1 (presented left) and Reuters (presented right). The topics shown in the tables have been selected randomly from these two collections. Although qualitative comparison in topic models is not a strong predictor for measuring the robustness of a model, we can see from the results that our model has discovered better topical words than the comparative models. Bigrams such as "january february" do not convey much meaning in a topic in the NHDP model in Reuters. Similarly, in the same collection, the word "report" discovered by the HDP model is not very insightful. In AQUAINT-1, n-gram such as "talk real person" by the NHDP model is also not very insightful, and same goes for word "new" discovered by the HDP model.

Table 2. High probability words in descending order obtained from a topic in two different collections. The table on the left shows results from AQUAINT-1 collection, and on the right, we depict results from Reuters collections.

HDP	NDHP		Our	
	Unigrams	N-grams	Unigrams	N-grams
year	test	internet sale	phone	web site
game	computer	search engine	digit	cell phone
music	year	create search engine	computers	high technology
computer	project	internet user	technology	microsoft windows
train	modern	index html	information	computer technology
new	service	state department	web	computer device
team	software	computer software	mail	laptop equipment
church	internet	computer bulletin	user	recognition software
transit	editor	latin america	online	large comfortable keyboard
time	technology	talk real person	network	speech technology

HDP	NDHP		Our	
	Unigrams	N-grams	Unigrams	N-grams
report	year	oil product	oil	oil price
bank	japan	crude oil	trade	gulf war
win	iraq	new oil product	cargo	oil stock
pakistan	oil	january february	high	crude oil
oil	crude	saudi arabia	market	domestic crude
rate	demand	total product	price	iraq ambassador
net	gasoline	crude export	fuel	oil product
french	saudi	gasoline distillation	tonne	indian oil
launch	arabia	thousand barrel	crude	run oil company
qatar	uae	oil import	week	world price

5 Conclusions

We have proposed a new metaphor in Bayesian nonparametrics called the Chinese Restaurant Franchise with Buddy Customers that takes into account the order of words in documents. Our model is able to discover n-gram words in latent topics. We have introduced a notion of buddy assignments in the basic CRF metaphor where we find out whether customers standing in order are friends with each other. All buddies occupy their reserved table in the restaurant which is not shared by other customers who do not belong to their friendship circle. We have tested our model on some text collections, and have shown that improvements are achieved in both quantitative performance and quality of topical words.

References

1. Blei, D.M., Ng, A.Y., Jordan, M.I.: Latent Dirichlet allocation. JMLR 3, 993–1022 (2003)
2. Jameel, S., Lam, W.: An unsupervised topic segmentation model incorporating word order. In: SIGIR, pp. 472–479 (2013)
3. Lindsey, R.V., Headden III, W.P., Stipicevic, M.J.: A phrase-discovering topic model using hierarchical Pitman-Yor processes. In: EMNLP, pp. 214–222 (2012)
4. Kim, H.D., Park, D.H., Lu, Y., Zhai, C.: Enriching text representation with frequent pattern mining for probabilistic topic modeling. ASIST 49, 1–10 (2012)
5. Barbieri, N., Manco, G., Ritacco, E., Carnuccio, M., Bevacqua, A.: Probabilistic topic models for sequence data. Machine Learning 93, 5–29 (2013)
6. Jameel, S., Lam, W.: A nonparametric N-gram topic model with interpretable latent topics. In: Banchs, R.E., Silvestri, F., Liu, T.-Y., Zhang, M., Gao, S., Lang, J. (eds.) AIRS 2013. LNCS, vol. 8281, pp. 74–85. Springer, Heidelberg (2013)
7. Kawamae, N.: Supervised N-gram topic model. In: WSDM, pp. 473–482 (2014)
8. Wallach, H.M.: Topic modeling: beyond bag-of-words. In: ICML, pp. 977–984 (2006)
9. McCallum, A., Mimno, D.M., Wallach, H.M.: Rethinking LDA: Why priors matter. In: NIPS, pp. 1973–1981 (2009)
10. Wang, X., McCallum, A., Wei, X.: Topical N-grams: Phrase and topic discovery, with an application to Information Retrieval. In: ICDM, pp. 697–702 (2007)
11. Fei, G., Chen, Z., Liu, B.: Review topic discovery with phrases using the Pólya urn model. In: COLING, pp. 667–676 (2014)

12. Darling, W.: Generalized Probabilistic Topic and Syntax Models for Natural Language Processing. PhD thesis (2012)
13. Blei, D.M., Griffiths, T.L., Jordan, M.I.: The nested Chinese restaurant process and Bayesian nonparametric inference of topic hierarchies. JACM 57, 7 (2010)
14. Teh, Y.W.: Dirichlet process. In: Encyclopedia of Machine Learning, pp. 280–287 (2010)
15. Teh, Y.W., Jordan, M.I., Beal, M.J., Blei, D.M.: Hierarchical Dirichlet processes. JASA 101, 1566–1581 (2006)
16. Teh, Y.W., Kurihara, K., Welling, M.: Collapsed variational inference for HDP. In: NIPS, pp. 1481–1488 (2007)
17. Sudderth, E.B.: Graphical models for visual object recognition and tracking. PhD thesis, Massachusetts Institute of Technology (2006)
18. Foti, N., Williamson, S.: A survey of non-exchangeable priors for Bayesian nonparametric models (2013)
19. Griffiths, T.L., Steyvers, M., Tenenbaum, J.B.: Topics in semantic representation. Psychological Review 114, 211 (2007)
20. Bartlett, N., Pfau, D., Wood, F.: Forgetting counts: Constant memory inference for a dependent hierarchical Pitman-Yor process. In: ICML, pp. 63–70 (2010)
21. Lau, J.H., Baldwin, T., Newman, D.: On collocations and topic models. TSLP 10, 10:1–10:14 (2013)
22. Johri, N., Roth, D., Tu, Y.: Experts' retrieval with multiword-enhanced author topic model. In: NAACL. SS 2010, pp. 10–18 (2010)
23. Noji, H., Mochihashi, D., Miyao, Y.: Improvements to the Bayesian topic n-gram models. In: EMNLP, pp. 1180–1190 (2013)
24. Teh, Y.W., Jordan, M.I.: Hierarchical Bayesian nonparametric models with applications. Bayesian Nonparametrics, 158–207 (2010)
25. Goldwater, S., Griffiths, T., Johnson, M.: A Bayesian framework for word segmentation: Exploring the effects of context. Cognition 112, 21–54 (2009)
26. Johnson, M.: PCFGs, topic models, adaptor grammars and learning topical collocations and the structure of proper names. In: ACL, pp. 1148–1157 (2010)
27. Deane, P.: A nonparametric method for extraction of candidate phrasal terms. In: ACL, pp. 605–613 (2005)
28. Petrovic, S., Snajder, J., Basic, B.D.: Extending lexical association measures for collocation extraction. Computer Speech and Language 24, 383–394 (2010)
29. Yoshii, K., Goto, M.: A vocabulary-free infinity-gram model for nonparametric Bayesian chord progression analysis. In: ICMIR, pp. 645–650 (2011)
30. Blei, D.M., Frazier, P.I.: Distance dependent Chinese restaurant processes. JMLR 12, 2461–2488 (2011)
31. Kim, D., Oh, A.: Accounting for data dependencies within a hierarchical Dirichlet process mixture model. In: CIKM, pp. 873–878 (2011)
32. Fox, E., Sudderth, E., Jordan, M., Willsky, A.: A sticky HDP-HMM with application to speaker diarization. APS 5, 1020–1056 (2011)
33. Tayal, A., Poupart, P., Li, Y.: Hierarchical double Dirichlet process mixture of Gaussian processes. In: AAAI (2012)

A Hierarchical Tree Model for Update Summarization

Rumeng Li[1] and Hiroyuki Shindo[2]

[1] Peking University, Beijing 100871, China
alicerumeng@foxmail.com
[2] Nara Institute of Science and Technology, Nara 630-0192, Japan
shindo@is.naist.jp

Abstract. Update summarization is a new challenge which combines salience ranking with novelty detection. This paper presents a generative hierarchical tree model (HTM for short) based on Hierarchical Latent Dirichlet Allocation (hLDA) to discover the topic structure within history dataset and update dataset. From the tree structure, we can clearly identify the diversity and commonality between history dataset and update dataset. A summary ranking approach is proposed based on such structure by considering different aspects such as focus, novelty and non-redundancy. Experimental results show the effectiveness of our model.

1 Introduction

Update summarization, put forward in the DUC (Document Understanding Conference) and TAC (Text Analysis Conference), aims to generate a concise and fluent "update" summary for a collection of documents of the same topic(update documents for short), under the assumption that users have already read the earlier documents (history documents for short) about the same topic. The purpose of the update summary is to inform readers of new and different information about the topic. It differs from generic summarization in that besides salience ranking, it also emphasizes novelty detection.

Multiple approaches have been developed to extract novel information in update documents based on traditional summarization techniques. These techniques include Maximal Marginal Relevance (MMR) [3] algorithm which tries to exclude sentences similar to the history documents according to tf-idf [2], or TextRank [5, 10, 14–16] which re-ranks the salience scores of sentences by considering both salience and novelty. However, these approaches tend to view the update summarization task more as a redundancy removal problem rather than a novel topic detection problem.

Recently, topic models have been widely used in NLP tasks due to its probabilistic robustness in discovering correlated word clusters from the corpus. Topic models based approaches have been applied in summarization task [4,8]. Topic models offer clear and rigorous probabilistic interpretations of documents which many other techniques lack. Delort and Alfonseca introduced *DualSum* based on Latent Dirichlet Allocation [1] that tries to distinguish update topic-word distribution from earlier one. Another work close to this task is the approach developed by [9–11] which uses a Hierarchical Dirichlet Process model [13] to discover the novel topic described in datasets and proposes a sentence ranking method that penalizes topics already addressed in history datasets. However, even though these approaches can indeed detect the new topics in update

A. Hanbury et al. (Eds.): ECIR 2015, LNCS 9022, pp. 660–665, 2015.

datasets, they can hardly capture the evolvement of the old topics, which should also be described in update summary.

In this paper, to detect not only new topics in update datasets, but also the evolvement of old ones, we present HTM: a sentence-level probabilistic topic model building on Hierarchical Latent Dirichelt Allocation (hLDA) [1] for update summarization. In the tree structure got from hLDA, each sentence is assigned to a path to the node in the tree with a vector of topics where each node is associated with a topic distribution over words. The children of each tree node denote its subtopics. From the hierarchical tree structure of topics, we can easily identify both the novel topics discussed and the evolving pattern of old topics in update datasets. In addition, we propose a novel ranking function according to such tree structure by considering different aspects such as focus, novelty and non-redundancy. We experiment our model on update summarization datasets in TAC2010 and TAC2011. Experimental results demonstrate the effectiveness of our approach.

2 HTM for Update Summarization

In this section, we firstly give a standard formulation of our problem and then get down to the details of hLDA model in HTM and our sentence selection strategies.

2.1 Problem Formulation

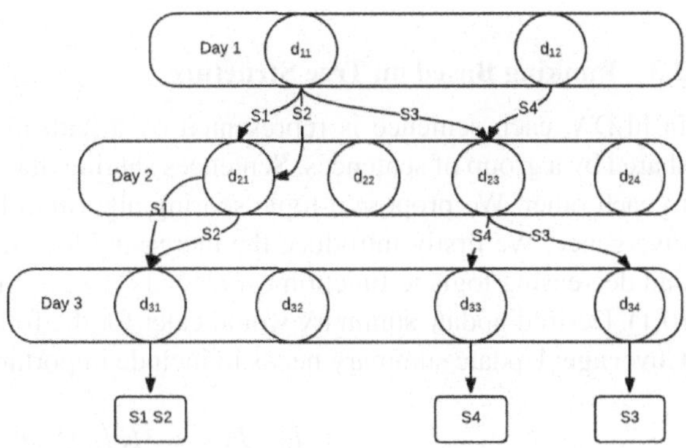

Fig. 1. A Graphical Illustration of nCRP. Each node represents a restaurant in which customers choose tables.

We are presented with two dataset: history dataset D^H and update dataset D^U. Each dataset is comprised of a collection of documents $D^H = \{d_i\}_{i=1}^{i=N}$, where N denotes the number of documents. Each document is comprised of a collection of sentences $d_i = \{s_{ij}\}_{i=1}^{i=N_i}$ where N_i denotes the number of sentences in current document and each sentence is comprised of a collection of words w. V denotes the vocabulary size. The input for the algorithm is two datasets D^H and D^U and the output is the update summarization I where $I \subset D^U$.

2.2 HTM

Our system, HTM, is based on a hierarchical topic model to extract sentences for update summarization generation.We discover the hidden topic distributions of sentences based on a revised version of hLDA [1].

1. for each topic $k \in [1, K]$:
   draw topic-word distribution $\beta_k \sim Dir(\eta)$
2. for each document m in D^H and D^U:
   draw an L-dimensional topic proportion
   $\theta_m \sim Dir(\alpha)$
3. for each sentence s:
   3.1 draw a path from root to leaf
   $c_s \sim nCRP(\gamma_0)$
   3.2 draw $z_l, l \in [1, L]$ from $Mult(\theta)$
   3.3 for each word w in s:
   draw w from β according to z

Fig. 2. Generative story for HTM

hLDA represents distribution of topics by organizing topics into a tree model. Each candidate sentence s is assigned to a path from the root to the leaf in the tree and each node is associated with a topic distribution over words. The algorithm can be illustrated by a metaphor as nested Chinese Restaurant Process (nCRP) [7]. Fig. 1 gives a graphic illustration of nCRP. The Generative story for HTM is shown in Fig. 2.

In this paper, we use Gibbs sampling to fit hLDA model, which iteratively sample the latent parameters c and z. In our experiments, we set the value of α to 0.1, η to 0.001 and γ_0 to 10^{-4}. After each iteration, we can calculate β_{kw}, which is the posterior distribution that word w generated from topic k and ϕ_{mk} which denotes the posterior probability of topic k in document m as follows:

$$\beta_{kw} = \frac{N^{w}_{z=k,C=c} + \eta}{\sum_{w'} N^{w'}_{z=k,C=c} + V\eta}; \ \phi_{mk} = \frac{E^k_m + \alpha}{\sum_k E^k_m + K\alpha} \tag{1}$$

where $N^{w}_{z=k,C=c}$ denotes the number of replicates of word w at topic k and E^k_m denotes the number of sentences in document m that have been assigned to topic k.

2.3 Ranking Based on Tree Structure

In hLDA, each sentence is represented by a path in the tree, and each path can be shared by a group of sentences. Sentences sharing the same path would be more similar to each other. We propose a topic scoring algorithm based on Kullback-Leibler (KL) divergence. We firstly introduce the increasing logistic function$\zeta_1(x) = e^x/(1+e^x)$ and decreasing logistic function $\zeta_2(x) = 1/(1+e^x)$ to map the distance into interval (0,1). Desired update summary would cater for the following properties:

Coverage: Update summary needs to include important contents described in D^U.

$$F_C(I) = \zeta_2[KL(I||D^U)]$$

Novelty: Update summary should avoid describing information already mentioned in D^H. We first define the score for each sentence in D^H. For each $s \in D^U$, $F(s) = \{s'|s' \in D^H, s' \in Min(s)\}$, which denotes the collection of history sentences that are located closest to s at tree structure. Clearly, we prefer that there is large KL divergence between sentence in update summarization and sentences from its nearest history sentences.

$$F_N(s) = \frac{1}{|F(s)|} \sum_{s' \in F(s)} \zeta_1(KL(s|s')); F_N(I) = \frac{1}{|I|} \sum_{s \in I} F_N(s)$$

Non-redundancy: We prefer that update summary covers multiple aspects and that sentences in that summary has larger KL divergence with each other.

$$F_{NR}(I) = \frac{1}{|I| \cdot (|I|-1)} \sum_{s \in I} \sum_{s' \in I, s' \neq s} \zeta_1[KL(s||s')]$$

The final score of summary I is the combination of three parts discussed above. Let w_i denotes the factor controls the influence of three parts proposed above, $\sum_i w_i = 1$.

$$F(I) = w_1 F_C(I) + w_2 F_N(I) + w_3 F_{NR}(I); I = \arg\max_{I^\star} F(I^\star) \quad (2)$$

During each iteration, we select the sentence which largest increases the score of $F(I)$. The sentence selection process is as follows:
for sentence set D^U and the summary set I, (1) we initialize by let $I = \phi$, $X = \{s_i | s_i \in D^U\}$. (2) While words($I$) less than L: $s = \arg max_{s \in X}(F(I + \{s\}) - F(I))$, $I = I + \{s\}$, $X = X - \{s\}$. (3)Repeat (2) until words(I) no longer less than L and output the summary set I.

3 Experiments

In our experiments, we use four years of TAC(2008-2011) data. For each topic, two docsets, D^H and D^U are given[1]. As for the automatic evaluation, we use the widely used ROUGE (Recall Oriented Understudy for Gisting Evaluation) (Lin and Hovy, 2003) measures, including ROUGE-2 and ROUGE-SU4.

We firstly remove stop words using a stopword list of 598 words. Words are then stemmed using Porter Stemmer[2]. Since sentence compression would largely improve linguistic quality of summaries [6, 12, 17], we use the sentence compression technique described in [12].

We tune parameters w_1, w_2 and w_3 at TAC2008 and TAC2009 and test the model on TAC2010 and TAC2011. We use a gradient research strategy which changes one parameter at one time with other parameters fixed. Due to the space limit, we just report the results that we achieve the highest ROUGE scores when w_1 is set to 0.4, w_2 is set to 0.45 and w_3 is set to 0.15 correspondingly.

3.1 Comparison

We compare our model with multiple approaches. For fair comparison, we adopt the same processing techniques for all baselines. We use the following baselines:
DualSum: The Bayesian approach proposed by [4] that considers 4 topics: background, document-specific, history and update topic.
HDP: The aprroach proposed by [10] that uses a Hierarchical Dirichlet Process (HDP) to discover the novel topics latent in update datasets.
MMR: A sentence scoring algorithm derived from MMR proposed by [2] which prefers to select those sentences dissimilar to history dataset.

---

[1] These two docsets are named docset A and docset B in TAC. For convenience in describing the model, we name the two docsets as docset H and docset U

[2] `http://tartarus.org/martin/PorterStemmer/.`

PNR2: A negative reinforcement between sentences which turned historical sentences to sink points proposed by [16].
Results are presented in Fig. 3 and Fig. 4. Our approach achieves relatively comparative results with, if not better than HDP. Bayesian approaches (HTM,HDP and DualSum) achieve better results than traditional document summarization approaches such as MMR and PNR^2.

	ROUGE-2	ROUGE-SU4
HTM	**0.0886** (0.0848-0.0924)	**0.1288** (0.1237-0.1339)
DualSum	0.0768 (0.0721-0.0815)	0.1140 (0.1082-0.1198)
HDP	0.0840 (0.0799-0.0881)	0.1232 (0.1198-0.1276)
MMR	0.0718 (0.0672-0.062)	0.1032 (0.0980-0.1084)
PNR	0.0724 (0.0690-0.0758)	0.1019 (0.0974-0.1064)

Fig. 3. Baseline Comparison in TAC2011

	ROUGE-2	ROUGE-SU4
HTM	**0.1052** (0.1012-0.1092)	**0.1386** (0.1344-0.1428)
DualSum	0.0928 (0.0892-0.0964)	0.1288 (0.1242-0.1234)
HDP	0.1017 (0.0980-0.1054)	01362 (0.1310-0.1414)
MMR	0.0854 (0.0812-0.0896)	0.1222 (0.1178-0.1266)
PNR	0.0848 (0.0810-0.0886)	0.1230 (0.1194-0.1266)

Fig. 4. Baseline Comparison in TAC2012

3.2 Manual Evaluation

We also conduct manual evaluation and in this subsection, we compare HTM with HDP. We ask 4 professional annotators (who are not the authors and are highly experienced in annotating for various NLP tasks and fluent in English) to assign a score to each summary with respect to each of the following four criteria: 1) Overall Responsiveness. 2) Focus (containing less irrelevant details). 3) Novelty (containing novel information beyond D^H). 4) Non-redundancy (repeating less similar information). The score is an integer between 1 (very poor) and 5 (very good). We randomly select 20 topics from TAC2011 data and assign each of them to two annotators.

Fig. 5 reports the average score and standard deviation from manual evaluation results which indicate HTM is significantly better than HDP. According to the results, besides Overall, Focus and Novelty, HTM outperforms HDP in Non-redundancy by a large margin. This demonstrates that the tree structure from hLDA helps to get a less-redundant summary.

4 Conclusion

In this paper, we propose HTM, a novel approach based on hLDA for update summarization. The performance of our model outperforms multiple update summarization systems, which illustrates the effectiveness of our model.

System	HTM	HDP
Overall	**3.84** $\pm$ 0.54	3.40 $\pm$ 0.46
Focus	3.50 $\pm$ 0.43	**3.56** $\pm$ 0.56
Novelty	**3.92** $\pm$ 0.41	3.42 $\pm$ 0.56
Non-redundancy	**4.01** $\pm$ 0.39	3.52 $\pm$ 0.42

Fig. 5. Results of manual evaluation

References

1. Blei, D.M., Ng, A.Y., Jordan, M.I.: Latent dirichlet allocation. Journal of Machine Learning Research 3, 993–1022 (2003)
2. Boudin, F., El-Bèze, M.: A scalable mmr approach to sentence scoring for multi-document update summarization. In: Proceedings of COLING 2008: Poster, pp. 23–26. The COLING 2008 Organizing Committee (2008)
3. Carbonell, J., Goldstein, J.: The use of mmr, diversity-based reranking for reordering documents and producing summaries. In: Proceedings of the 21st Annual International ACM SIGIR Conference on Research and Development in Information Retrieval, pp. 335–336. ACM (1998)
4. Delort, J.Y., Alfonseca, E.: Dualsum: A topic-model based approach for update summarization. In: Proceedings of the 13th Conference of the European Chapter of the ACL, pp. 214–223. ACL (2012)
5. Du, P., Guo, J., Zhang, J., Cheng, X.: Manifold ranking with sink points for update summarization. In: Proceedings of the 19th ACM International Conference on Information and Knowledge Management, pp. 1757–1760. ACM (2010)
6. Gillick, D., Favre, B.: A scalable global model for summarization. In: Proceedings of the Workshop on Integer Linear Programming for Natural Langauge Processing, pp. 10–18. ACL (2009)
7. Griffiths, D., Tenenbaum, M.: Hierarchical topic models and the nested chinese restaurant process. Advances in Neural Information Processing Systems 16, 17 (2004)
8. Haghighi, A., Vanderwende, L.: Exploring content models for multi-document summarization. In: Proceedings of Human Language Technologies: The 2009 Annual Conference of the North American Chapter of the ACL, pp. 362–370. ACL (2009)
9. Li, J., Cardie, C.: Timeline generation: tracking individuals on twitter. In: Proceedings of the 23rd International Conference on World Wide Web, pp. 643–652. International World Wide Web Conferences Steering Committee (2014)
10. Li, J., Li, S., Wang, X., Tian, Y., Chang, B.: Update summarization using a multi-level hierarchical dirichlet process model. In: Proceedings of COLING 2012, pp. 1603–1618. The COLING 2012 Organizing Committee, Mumbai (2012)
11. Li, J., Ott, M., Cardie, C.: Identifying manipulated offerings on review portals. In: The 2013 Conference on Empirical Methods on Natural Language Processing, pp. 1933–1942 (2013)
12. Li, P., Wang, Y., Gao, W., Jiang, J.: Generating aspect-oriented multi-document summarization with event-aspect model. In: Proceedings of the Conference on Empirical Methods in Natural Language Processing, pp. 1137–1146. ACL (2011)
13. Teh, Y.W., Jordan, M.I., Beal, M.J., Blei, D.M.: Hierarchical dirichlet processes. Journal of the American Statistical Association 101(476) (2006)
14. Wan, X.: Timedtextrank: adding the temporal dimension to multi-document summarization. In: Proceedings of the 30th Annual International ACM SIGIR Conference on Research and Development in Information Retrieval, pp. 867–868. ACM (2007)
15. Wang, X., Wang, L., Li, J., Li, S.: Exploring simultaneous keyword and key sentence extraction: improve graph-based ranking using wikipedia. In: Proceedings of the 21st ACM International Conference on Information and Knowledge Management, pp. 2619–2622. ACM (2012)
16. Wenjie, L., Furu, W., Qin, L., Yanxiang, H.: Pnr 2: ranking sentences with positive and negative reinforcement for query-oriented update summarization. In: Proceedings of COLING 2008, vol. 1, pp. 489–496. The COLING 2008 Organizing Committee (2008)
17. Zajic, D., Dorr, B.J., Lin, J., Schwartz, R.: Multi-candidate reduction: Sentence compression as a tool for document summarization tasks. Information Processing & Management 43(6), 1549–1570 (2007)

Document Boltzmann Machines for Information Retrieval

Qian Yu[1], Peng Zhang[1,*], Yuexian Hou[1], Dawei Song[1,2], and Jun Wang[3]

[1] Tianjin Key Laboratory of Cognitive Computing and Application, Tianjin University, China

[2] The Computing Department, The Open University, United Kingdom

[3] Department of Computer Science, University College London, United Kingdom

{yqcloud,darcyzzj,krete1941,dawei.song2010,seawan}@gmail.com

Abstract. Probabilistic language modelling has been widely used in information retrieval. It estimates document models under the multinomial distribution assumption, and uses query likelihood to rank documents. In this paper, we aim to generalize this distribution assumption by exploring the use of fully-observable Boltzmann Machines (BMs) for document modelling. BM is a stochastic recurrent network and is able to model the distribution of multi-dimensional variables. It yields a kind of Boltzmann distribution which is more general than multinomial distribution. We propose a Document Boltzmann Machine (DBM) that can naturally capture the intrinsic connections among terms and estimate query likelihood efficiently. We formally prove that under certain conditions (with 1-order parameters learnt only), DBM subsumes the traditional document language model. Its relations to other graphical models in IR, e.g., MRF model, are also discussed. Our experiments on the document reranking demonstrate the potential of the proposed DBM.

1 Introduction

Probabilistic models for information retrieval (IR) can be divided into two categories: document-generation and query-generation [3]. Query-generation based models assume that a query is generated from a document model and rank documents based on the query likelihood. To calculate the query likelihood, one should first decide what kind of term distribution is adopted for document modelling. Language model [5] is a representative query-generation model with a multinomial term distribution assumption.

In this paper, our aim is to generalize the distribution assumption used in the traditional language modelling approach, based on fully-observable Boltzmann Machine (BM) [1] that can yield a kind of Boltzmann distribution. In statistical mechanics, the Boltzmann distribution (also known as Gibbs distribution) is used as probability distribution over possible states of a system. Analogously, a document can also be treated as a dynamic system, and each segment sampled from

* Corresponding Author.

A. Hanbury et al. (Eds.): ECIR 2015, LNCS 9022, pp. 666–671, 2015.

it can be seen as a possible state of this system. Therefore, it is theoretically feasible to model a document with a Boltzmann distribution. This is appealing, as Boltzmann distributions yielded by BMs do not necessarily assume the independence of terms as in multinomial distribution. In other words, term dependence information, which has been proven important in IR modelling, can be naturally captured with the BM mechanisms, without the need of explicitly involving any external term dependence detection method. With this motivation, we propose a document model called Document Boltzmann Machine (DBM), using BM as a tool to learn automatically from segments sampled from a document.

2 Document Boltzmann Machines Model

2.1 A Brief Introduction to Boltzmann Machines

Boltzmann machine [1] is a stochastic version of the deterministic network, and is widely used as a generative model in statistical mechanics and machine learning. The general Boltzmann machine has an energy function formulated as:

$$\begin{aligned} E(\mathbf{x}; W) &= E(\mathbf{x}; \mathbf{w}^1, \mathbf{w}^2, \mathbf{w}^3, \cdots) \\ &= -\sum_i w_i^1 x_i - \sum_{i<j} w_{ij}^2 x_i x_j - \sum_{i<j<k} w_{ijk}^3 x_i x_j x_k - \cdots \end{aligned} \tag{1}$$

In the above equation, $\mathbf{x}$ is the variable vector, and $\mathbf{w}^1$, $\mathbf{w}^2$, $\mathbf{w}^3$, etc. are different subsets of parameter set W. 1-order parameter w_i^1 models the state of node i, 2-order parameter w_{ij}^2 models the connection strength between node i and node j, and higher order parameters are able to model connection strength among more than two nodes. Thus, BM naturally models the term dependencies when each node represents one term's state. The log likelihood of the sample $\mathbf{x}$ is:

$$\log p(\mathbf{x}; W) = -E(\mathbf{x}; W) - \log[Z(W)] \tag{2}$$

where $Z(W) = \sum_{\mathbf{x}} \exp[-E(\mathbf{x}; W)]$ is the partition function.

2.2 Procedure of Document Modelling with Boltzmann Machines

The procedure of Document Boltzmann Machine (DBM) modelling is displayed in Figure 1. At first, we define a structure of BM where each node corresponds to one term, and we call this structure as the BM template. With this BM template, we sample from each document d_i and model it with a learnt BM_i.

For any document d_i, we sample segments from it. We use a sliding window to get overlapped document segments with a fixed window size σ and we set step length of the sliding to 1. Then, based on these segments, we prepare samples for BM using a simple method. Specifically, each dimension of the resultant samples represents whether or not a term exist in this segment. In other words, we assign 1 to dimensions whose corresponding terms appear in the current document segment, and 0 otherwise. The sample set (X) is used for learning a model for

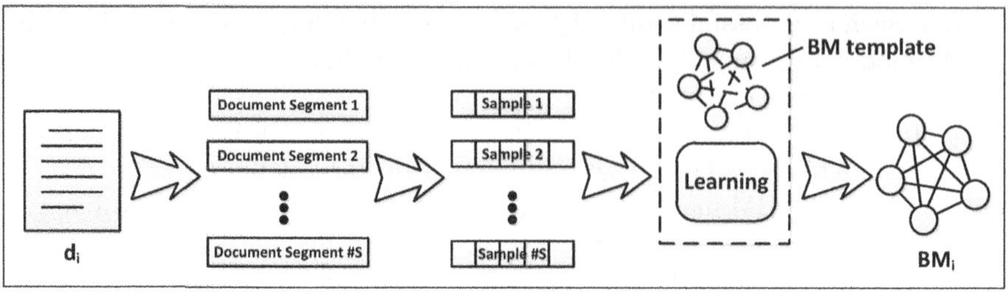

Fig. 1. Procedure of Document Modelling with Boltzmann Machines

document i using maximum likelihood (ML) method. According to Equation 2, we can get the learning objective, i.e., the samples log likelihood $L(W;X) = -\frac{1}{N}\sum_{n=1}^{N} E(\mathbf{x}_n;W) - \log Z$ and its gradient for parameters updating:

$$\partial L(W;X)/\partial W = -\langle \partial E(\mathbf{x};W)/\partial W \rangle_0 + \langle \partial E(\mathbf{x};W)/\partial W \rangle_\infty \quad (3)$$

where $\langle \cdot \rangle_0$ and $\langle \cdot \rangle_\infty$ denote averages in respect to samples distribution and the current model distribution, respectively [2]. For more details of the learning algorithm, please refer to [2].

The learning algorithm will assign parameters with a set of values W_i, i.e., a learnt Boltzmann machine model BM_i for each document d_i. We call it Document Boltzmann Machine (DBM) model. A DBM model represents the probability distribution of segments in one document, and the probability of a segment $\mathbf{x}$ of document d_i given BM_i can be written as:

$$\log p(\mathbf{x}|d_i) = \log p(\mathbf{x}|BM_i) = \log p(\mathbf{x};W_i) \quad (4)$$

where $\log p(\mathbf{x};W_i)$ can be calculated as in Equation 2.

2.3 Query Generation via Document Boltzmann Machines

If DBM is regarded as the probability distribution of dynamic system states, then the query stands for a state, and the system appears in this state with a certain probability. In other words, since a query can be a value of Boltzmann machine input, it is straightforward to get its probability, i.e., query likelihood.

Ideally, we can incorporate each terms in vocabulary into the structure of Boltzmann machine, but it will cause computational burden for the model learning and probability calculation. In our implementation, given a query $Q = (q_1, q_2, ..., q_i, ...)$, we assign one node for each query term q_i, which is enough for calculation of query likelihood. In this way, once we get a Boltzmann machine model BM with parameter value W for a document d, we can obtain the query likelihood by

$$\log p(\mathbf{x}_Q|BM) = \log p(\mathbf{x}_Q;W) = \sum_i w_i^1 + \sum_{i<j} w_{ij}^2 + \cdots - \log Z_d \quad (5)$$

where $\mathbf{x}_Q$ is the all-one vector since each query term appears in Q.

2.4 Smoothing with Collection Statistics

Similar to LM's Dirichlet smoothing method [8], we now present the smoothing method for Boltzmann machine based on its conjugate prior function. The objective function for BM learning is then changed to the posterior function:

$$L(W; X, \chi, \nu) = \log[p(W|X, \chi, \nu)] \propto -\sum_{n}^{N} E(\mathbf{x}_n; W) + \nu W^T \chi - \log Z^{N+\nu} \quad (6)$$

where χ and ν are hyperparameters in prior function, $f(\chi, \nu)$ is the normalization constant, and Z is the function of W as in Equation 2. χ has the same dimension number as the features (i.e., x_i, $x_i x_j$, $\cdots$) in energy function. We assign χ with frequencies of these features in collection, and ν is a single value to be tuned ($\nu = 0$ for no smoothing). ML method is available for smoothed DBMs learning.

3 Analysis of DBM

3.1 Relation to Language Model

We use BM^1 to denote a 1-order BM, i.e., only the first-order, namely the bias parameters $\mathbf{w}^1$, are learnt. The probability of active state for one dimension, namely $x_i = 1$ can be proved to be $p(x_i = 1; BM^1) = [\exp(w_i^1)]/[1+\exp(w_i^1)]$. If a BM^1 is completely learnt, the probability for a vector is almost the same as its proportion in samples. Formally, $p_i = p(x_i = 1; BM^1)$, where p_i is the proportion of samples whose i^{th} dimension is 1. If we sample segments uniformly from a document (e.g., the sampling method described in Section 2.2), we approximately assign each term a probability which is the corresponding term frequency in the document. In this case, the query likelihood $\log p(\mathbf{x}_Q|D_{BM})$ becomes:

$$\log \frac{\exp(\mathbf{w}^1 \cdot \mathbf{1})}{\sum_{\mathbf{x}} \exp(\mathbf{w}^1 \mathbf{x})} = \log \frac{\prod_i \exp(w_i^1)}{\prod_i [1 + \exp(w_i^1)]} = \sum_i \log p(x_i = 1; BM^1) = \sum_i \log p_i$$

which has exactly the same form as LM. This equivalence reveals that without higher order parameters, BM^1 still results in a multinomial distribution.

3.2 Comparison with MRF Model

The relation and difference between DBM model and the popular Markov random field model [4] need to be clarified. At first, MRF model is used in [4] to model the joint distribution $P(Q, D)$ over queries Q and documents D, whilst the DBM focuses on the document modelling. Thus, it is possible to exploit DBM model for tasks where no query is involved, for example, calculation of document similarity. Secondly, DBM, making use of a stochastic generative model, needs the principled learning & smoothing methods as described in Section 2. Due to this reason, new progress in machine learning area may be applied in DBM, e.g., parameter selection principle. Besides, MRF is quite a general framework with the form of $P(D|Q) = \sum_{c \in C(G)} \lambda_c f(c)$, and our DBM can serve as one of the potential functions: $\phi_B(c) = \lambda_B \log p(Q|B_D)$. This provides a principled way to incorporate DBM with other query generation probabilistic models, e.g., LM.

3.3 Distinction between Our Model and Restricted/Deep BM

Restricted BMs (also known as harmoniums) [7] and the recent deep BMs [6] are applied in IR. Essentially, both the models with hidden layers are trained as feature extractors for document representation. Hence, these two models are different from ours in motivations, structures, and specific application tasks.

4 Experiments

Experiments are conducted on four standard TREC collections: AP8889 (query 151-200), WSJ8792 (query 151-200), ROBUST2004 (query 601-700) and WT10G (query 501-550). Collections are indexed by *Indri 5.3* with Porter stemming and stopwords removed. Reranking is performed on top 50 documents, with initial ranking by LM. We fix window size σ (see Section 2.2) to 16, and we also tested other settings (10 - 20) and observed similar results. Statistically significant improvements according to Wilcoxon test at level 0.05 are marked by * in tables.

Firstly, we compare different document models without smoothing. LM with Dirichlet smoothing parameter μ assigned to 0 (LM(μ=0)) and a small positive value ϵ (LM(μ=+ϵ)) act as baseline. DBM with 1-order (1-DBM) and 1&2-order (2-DBM) parameters are tested, and $\nu = 0$ for both models. We can tell from Table 1 that DBM without smoothing outperforms LM on most collections, and 2-DBM performs always better than 1-DBM, which indicates the usefulness of higher order parameters. Next, we evaluate the smoothed DBM model. Table 2 shows the performance of different query generation approaches. DBM uses

Table 1. Evaluation of DBM without smoothing on reranking. Percentages of change are calculated based on LM with μ=+ϵ.

	AP8889		WSJ8792		ROBUST2004		WT10G	
Metric	MAP	P@10	MAP	P@10	MAP	P@10	MAP	P@10
LM(μ=0)	0.0744	0.2960	0.0870	0.2600	0.1339	0.3273	0.0452	0.1490
LM(μ=+ϵ)	0.1542	0.4440	0.1562	0.3680	**0.1800**	**0.3697**	0.0600	0.1755
1-DBM ($\nu = 0$)	0.1761 +14.20%*	0.4540 +2.25%*	0.1752 +12.16%*	0.3660 -0.54%	0.1603 -10.04%	0.3121 -15.58%	0.0741 +23.50%*	0.2327 +32.59%*
2-DBM ($\nu = 0$)	**0.1782** +15.56%*	**0.4560** +2.70%*	**0.1835** +17.48%*	**0.3880** +5.43%*	0.1713 -4.83%	0.3333 -9.85%	**0.0806** +34.33%*	**0.2449** +37.97%*

Table 2. Evaluations of DBM with smoothing on reranking. Smoothing parameters for LM and DBM, i.e., μ and ν are selected by maximizing MAP.

	AP8889		WSJ8792		ROBUST2004		WT10G	
Metric	MAP	P@10	MAP	P@10	MAP	P@10	MAP	P@10
LM	0.1814	0.4900	0.2181	0.4640	0.2214	0.4212	0.1085	0.2776
DBM	0.1875 +3.36%*	0.5020 +2.45%*	0.2204 +1.05%	**0.5020** +8.19%*	0.2250 +1.63%	**0.4313** +2.40%*	**0.1099** +1.29%	**0.2878** +3.67%*
LM-BM	**0.1881** +3.69%*	**0.5040** +2.86%*	**0.2206** +1.15%	0.4920 +6.03%*	**0.2261** +2.12%*	0.4303 +2.16%*	**0.1099** +1.29%	**0.2878** +3.67%*

parameters with order up to 2. LM-BM is the combination of LM and DBM (see Section 3.2), and the parameters for potential functions are selected as in MRF framework. It is demonstrated that DBM model helps improving the performance of LM. The performance similarity between DBM and LM-BM supports the argument that DBM can hardly benefit from its special case: LM.

5 Conclusion and Future Work

This work is the first step towards exploiting the full potential of applying the fully-observable Boltzmann machine (BM) in IR. We have proposed a Document BM (DBM) model that generalizes the multinomial distribution used in LM. This generalization is appealing in that, it facilitates the extraction and utilization of connection strengths among terms. We have formally proved that DBM subsumes the LM as a special case. Equipped with principled learning & smoothing methods, DBM can be practical and efficient. Experimental results on reranking task reveal that DBM helps obtaining a more reasonable query likelihood estimation. In the future, we would like to develop more robust smoothing methods, examine higher order parameters and long queries for DBM.

Acknowledgments. This work is supported in part by the Chinese National Program on Key Basic Research Project (973 Program, grant No.2013CB329304, 2014CB744604), the Natural Science Foundation of China (grant No. 61402324, 61272265, 61105072), and Research Fund for the Doctoral Program of Higher Education of China (grant no. 20130032120044).

References

1. Aarts, E.H., Korst, J.H.: Boltzmann machines and their applications. In: Treleaven, P.C., Nijman, A.J., de Bakker, J.W. (eds.) PARLE 1987. LNCS, vol. 258, pp. 34–50. Springer, Heidelberg (1987)
2. Carreira-Perpinan, M.A., Hinton, G.E.: On contrastive divergence learning. In: Proceedings of the 10th International Workshop on AISTATS, pp. 33–40. Citeseer (2005)
3. Lafferty, J., Zhai, C.: Probabilistic relevance models based on document and query generation. In: Language Modeling for IR, pp. 1–10. Springer (2003)
4. Metzler, D., Croft, W.B.: A markov random field model for term dependencies. In: Proceedings of the 28th ACM SIGIR Conference, pp. 472–479. ACM (2005)
5. Ponte, J.M., Croft, W.B.: A language modeling approach to information retrieval. In: Proceedings of the 21st ACM SIGIR Conference, pp. 275–281. ACM (1998)
6. Srivastava, N., Salakhutdinov, R.R., Hinton, G.E.: Modeling documents with deep boltzmann machines. arXiv preprint arXiv:1309.6865 (2013)
7. Welling, M., Rosen-Zvi, M., Hinton, G.E.: Exponential family harmoniums with an application to information retrieval. In: Advances in NIPS, pp. 1481–1488 (2004)
8. Zhai, C., Lafferty, J.: A study of smoothing methods for language models applied to ad hoc information retrieval. In: Proceedings of the 24th ACM SIGIR Conference, pp. 334–342. ACM (2001)

Effective Healthcare Advertising Using Latent Dirichlet Allocation and Inference Engine

Yen-Chiu Li and Chien Chin Chen

Department of Information Management, National Taiwan University
No.1, Sec. 4, Roosevelt Rd., Taipei City 10617, Taiwan (R.O.C.)
{r01725001,patonchen}@ntu.edu.tw

Abstract. The growing access to healthcare websites has aroused the interest of designing a specific advertising system focusing on healthcare products. In this paper, we develop an advertising method which analyzes the messages posted by users on a healthcare website. The method integrates semantic analysis with an inference engine for effective healthcare advertising. Based on our experiment results, healthcare advertising systems could be enhanced by using the domain-specific knowledge to augment the content of user messages and ads.

Keywords: semantic analysis, knowledge base, advertising.

1 Introduction

Nowadays, people have become interested in utilizing the resources on the Internet to investigate and help manage their health problems. For instance, WebMD, the most popular healthcare website worldwide, has exceeded 100 million unique visitors per month[1]. Many business models have been designed to make a profit from the healthcare website users. Among them, online advertising is perhaps most promising. The essence of online advertising is to present ads relevant to a user's interests. This is because the more relevant an ad is to a user's interests, the higher click-through rate the ad will have [3].

In this paper, we propose an intelligent healthcare advertising method. We especially focus on shared patient experience (PEx) forums where illness messages are posted to exchange healthcare information. Due to the complexity in medical science, PEx users rarely are able to find helpful healthcare products like over-the-counter (OTC) drugs by themselves [5], and for this reason, PEx forums are excellent venues for online healthcare advertising. However, the illness messages posted by users are often very short. To resolve the information sparseness of user messages, the method we propose first segments an illness message into a set of symptoms which are applied to a rule-based knowledge base to infer the potential diseases regarding the message. The descriptions of the diseases together with the identified symptoms are analyzed by the

[1] http://investor.shareholder.com/wbmd/releasedetail.cfm?releaseid=801330&CompanyID=WBMD

A. Hanbury et al. (Eds.): ECIR 2015, LNCS 9022, pp. 672–677, 2015.

latent Dirichlet allocation (LDA) [1], a state-of-the-art semantic analysis method, to discover the latent topics of the illness message. Finally, the relevance of the illness message and ads is computed in terms of the latent topics and the relevant ads are presented to the user.

2 Method

Figure 1 depicts our healthcare advertising method which consists of two stages: the offline stage and online stage. In the offline stage, a considerable number of authoritative healthcare articles were gathered. The articles and ads are applied to LDA to discover latent topics of a healthcare domain and to represent ads in the latent topic space. To clarify the illness of a user message, in the online stage, we extract important symptoms from the message. The symptoms then are applied to an inference engine to reason potential diseases regarding the message. Finally, the message, together with the descriptions of the potential diseases, is folded in the latent topic space. The method then computes the relevance of ads and the user message by means of the latent topic space and relevant ads are advertised.

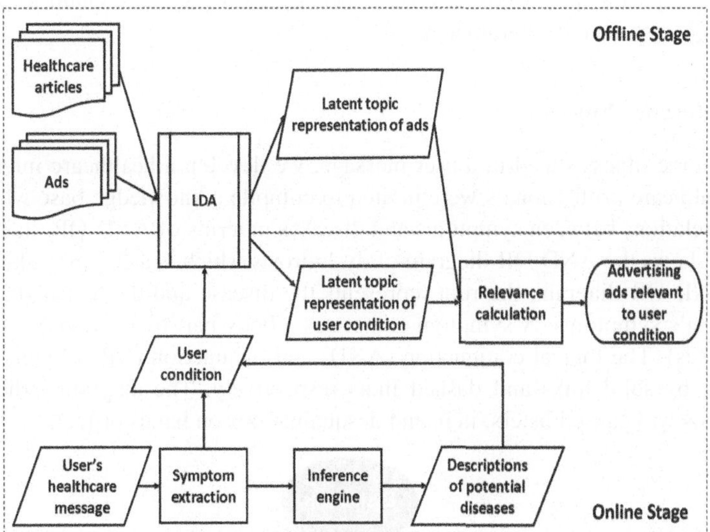

Fig. 1. The proposed healthcare advertising system

2.1 Semantic Analysis

We employ LDA to learn important topics of a healthcare domain. LDA is a generative probability model. The document generation process is as follows. Let $D = \{d_1, d_2, ..., d_M\}$ be a document corpus and it consists of a set of healthcare articles and ads. Let $Z = \{z_1, z_2, ..., z_K\}$ be a set of latent topics and V be the number of unique terms in D. Variables $\vec{\alpha}$ and $\vec{\beta}$ are a K-dimensional vector and V-dimensional vector respectively. They are hyperparameters used to generate the model's Dirichlet priors. For simplicity, the entries in $\vec{\alpha}$ and $\vec{\beta}$ are all identical respectively. LDA first picks a multinomial topic-term distribution vector $\overrightarrow{\phi_k}$ from a Dirichlet distribution with hyperparameter $\vec{\beta}$ for each latent topic z_k. All the $\overrightarrow{\phi_k}$'s form a $K \times V$ matrix $\boldsymbol{\phi}$. To generate a document d_m, LDA picks a multinomial document-topic distribution vector $\overrightarrow{\theta_m}$ from a Dirichlet distribution with hyperparameter $\vec{\alpha}$, and the $\overrightarrow{\theta_m}$'s of all documents in D form an $M \times K$ matrix $\boldsymbol{\theta}$. To generate a word in the document d_m, a topic z_k is chosen in accordance with the multinomial distribution $\overrightarrow{\theta_m}$, and afterward the word is selected based on the topic-term distribution $\overrightarrow{\phi_k}$.

In D, only the words in d_m's are observed and the distributions in $\boldsymbol{\Theta}$ and $\boldsymbol{\phi}$ are unknown. The goal of LDA is to find $\boldsymbol{\phi}$ and $\boldsymbol{\theta}$ that maximize the following likelihood function.

$$\mathrm{p}(d_1, d_2, \dots, d_M | \boldsymbol{\Theta}, \boldsymbol{\phi}) = \prod_{m=1}^{M} p(d_m | \overrightarrow{\theta_m}, \boldsymbol{\phi}) = \prod_{m=1}^{M} \prod_{i=1}^{N_m} p\left(d_{m,i} | \overrightarrow{\theta_m}, \boldsymbol{\phi}\right), \tag{1}$$

where N_m is the number of words in d_m and $d_{m,i}$ is the ith word in d_m In other words, LDA aims at inferring $\boldsymbol{\Theta}$ and $\boldsymbol{\phi}$ that best fit the observed words for the above generative process. Inferring the parameters $\boldsymbol{\theta}$ and $\boldsymbol{\phi}$ from D is an intractable problem. Here, we employ Gibbs sampling [4], a special case of Markov chain Monte Carlo (MCMC), to estimate the parameters.

2.2 Inference Engine

To reason the illness stated in a user message, we develop a healthcare inference engine. Healthcare professionals were invited to compile a knowledge base which modeled the relations between symptoms and diseases in terms of AND-OR diagrams [9]. Figure 2 shows the AND-OR diagram of dyshidrosis which is a common skin disease. In the AND-OR diagram, the root represents the disease and the leaf nodes indicate the disease's symptoms. A symptom comprises a body part (e.g., hands) and a lesion (e.g., rashes). The logical conjunction (AND) and disjunction (OR) of symptoms are presented by solid links and dashed links respectively. The diagram indicates that dyshidrosis will cause blisters, itch, and desquamations on hands or feet.

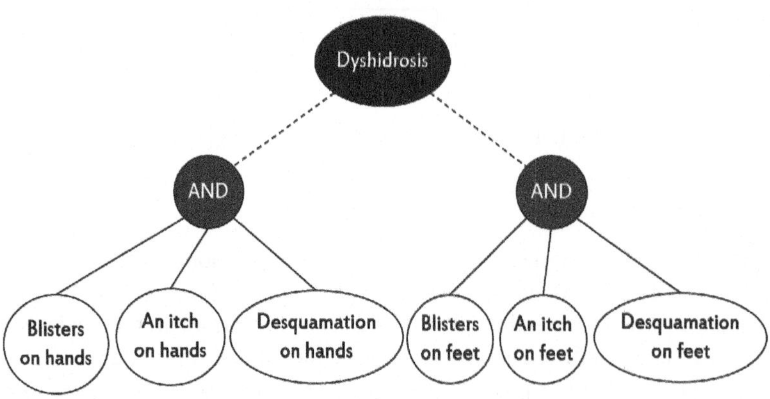

Fig. 2. An example of the AND-OR diagram

We implement the well-known backward reasoning algorithm [9] which employs the depth-first search strategy to infer potential diseases regarding a user message. The algorithm starts from the root of a disease's AND-OR diagram and recursively explores all the root's children. Every node has a certainty score ranging from 0% to 100%. When a leaf node is explored and the corresponding symptom matches a symptom in $S = \{s_1, \ldots, s_L\}$ which represents a set of symptoms extracted from the user message, the certainty score of the leaf node is 100%; otherwise it is zero. Logically, if any child of an AND node is not 100% certain, the certainty score of the AND node is 0%. However, user messages are often so short that the extracted symptoms hardly cover all necessary symptoms of a disease. We adopt the following sum-up approach to compute the certainty score of an AND node *n*.

$$n.certainty = \frac{\sum_{c \in n's\, children} c.certainty}{\#\, of\, n's\, children}. \quad (2)$$

In other words, the certainty of an AND node is the normalized sum of the children's certainties. For an OR node, we adopt the following independent-or formula to derive its certainty score.

$$n.certainty = 1 - \prod_{c \in n's\, children}(1 - c.certainty). \quad (3)$$

The algorithm returns the certainty score of the root (i.e., the disease). The higher the score, the more relevant the disease to the user message. We rank diseases according to their certainty scores and combine the descriptions of the top-ranked diseases with the user message to advertise appropriate ads.

2.3 Relevance Calculation

In the offline stage, LDA is employed to learn the topic distribution vector $\overrightarrow{\theta_a}$ for each ad *a*. In the online stage, we treat a user condition c (i.e., a user message together with the text descriptions of the potential diseases) as a new document and apply a folding-in approach to learn the *c*'s topic distribution vector. The folding-in approach is also based on Gibbs sampling which assigns each word in *c* a topic. However, it keeps the word-topic assignments in *D* unchanged to accelerate the learning process. Once *c*'s topic distribution vector is obtained, we measure the relevance between it and ads. Here, we employ the Kullback Leibler (KL) divergence, which is a popular measure used to calculate the difference between probability distributions. The KL divergence computes the difference between the topic vectors of *a* and *c* as follows.

$$\mathrm{KL}(a, c) = \sum_{k=1}^{K} \theta_a^k log_2 \frac{\theta_a^k}{\theta_c^k}. \quad (4)$$

Due to the asymmetric property of the KL divergence, we average KL(*a*,*c*) and KL(*c*,*a*) instead.

$$\mathrm{Diff}(a, c) = \frac{1}{2}[KL(a, c) + KL(c, a)]. \quad (5)$$

The lower the Diff (a,c) is, the higher the relevance is between a and c. Then, all ads are ranked according to their relevance degrees, and relevant ads are advertised.

3 Experiments

We evaluate our healthcare advertising performance on dermatology. We crawled many authoritative articles about disease treatment and nursing from the consulting forum in KingNet[2], which is the largest healthcare website in Taiwan. The articles were applied to LDA to discover latent topics of dermatology. Furthermore, a knowledge base which covers 44 common skin diseases and 202 symptoms was compiled and revised by dermatologists. We also collected 860 healthcare ads from an online retailer and four websites of pharmaceutical companies. To test our method, 200 cases edited by dermatologists were used for performance evaluation. Each case contains an illness message stated by a user and a disease name concluded by a dermatologist. When testing, we examine whether the top three ads advertised by our method are related to the concluded disease and measure the advertising performance in terms of the precision and coverage defined as follows.

$$\text{Precision} = |relevant\ ads\ in\ examined\ ads| / |examined\ ads| . \quad (6)$$

$$\text{Coverage} = |cases\ that\ have\ at\ least\ one\ relevant\ ads| / |examined\ cases| . \quad (7)$$

We compare our method with five advertising methods: the bag-of-word method (BOW) method, the language model (LM) method [8], the web relevance feedback (WRF) method [2], the keyword-topic (KT) method [7] and the hidden topic (HT) method [6]. Note that the last two methods are also based on LDA. To ensure a fair comparison, the system parameters of the compared methods are set as suggested by the original papers.

Table 1. The advertising performance of the compared methods

	Precision	Coverage
Our method (*K*=80)	0.360	0.588
Our method (*K*=20, no inference engine)	0.164**	0.336**
BOW(TF-IDF)	0.077**	0.140**
LM	0.067**	0.175**
WRF	0.206**	0.365**
KT(*K*=100)	0.107**	0.215**
HT(*K*=200)	0.131**	0.276**

The results marked with ** show the improvements achieved by our method over the compared methods with 99% confidence levels based on the z-statistic for two propositions.

As shown in Table 1, the results of BOW, LM, KT, HT and our method (no inference engine) are poor. Nevertheless, KT, HT and our method (no inference engine) is better than BOW and LM. The result indicates that the latent topics learned by LDA

[2] `http://www.kingnet.com.tw/`

are able to enhance the relevance estimation between user messages and ads. Our method achieves the best advertising performance and it significantly outperforms WRF which considers the Web as an external knowledge base to enrich the sparse content of user messages. This is because the knowledge base we used (i.e., the AND-OR diagrams of diseases) is compiled by domain experts and is therefore more reliable than the returned pages used by WRF.

4 Conclusions

In this paper, we have proposed a method which integrates techniques of semantic analysis and inference engine for effective healthcare advertising. In future work, we will extend our method to different healthcare domains. In addition, effective symptom weighting scheme will be developed to polish the corresponding inference results.

Acknowledgements. This research was supported in part by NSC 100-2628-E-002-037-MY3 from the National Science Council, Republic of China and MOST 103-2221-E-002-106-MY2 from the Ministry of Science and Technology, Republic of China.

References

1. Blei, D.M., Ng, A.Y., Jordan, M.I.: Latent dirichlet allocation. Journal of Machine Learning Research 3, 993–1022 (2003)
2. Broder, A.Z., Ciccolo, P., Fontoura, M., Gabrilovich, E., Josifovski, V., Riedel, L.: Search advertising using web relevance feedback. In: Proceedings of the 17th ACM Conference on Information and Knowledge Management, pp. 1013–1022. ACM, New York (2008)
3. Chatterjee, P., Hoffman, D.L., Novak, T.P.: Modeling the clickstream: Implications for web-based advertising efforts. Marketing Science 22(4), 520–541 (2003)
4. Griffiths, T.L., Steyvers, M.: Finding scientific topics. Proceedings of the National academy of Sciences of the United States of America 101(suppl. 1), 5228–5235 (2004)
5. Luo, G., Thomas, S.B., Tang, C.: Automatic home medical product recommendation. Journal of Medical Systems 36(2), 383–398 (2012)
6. Phan, X.H., Nguyen, C.T., Le, D.T., Nguyen, L.M., Horiguchi, S., Ha, Q.T.: A hidden topic-based framework toward building applications with short Web documents. IEEE Transactions on Knowledge and Data Engineering 23(7), 961–976 (2011)
7. Phuong, D.V., Phuong, T.M.: A keyword-topic model for contextual advertising. In: Proceedings of the Third Symposium on Information and Communication Technology, pp. 63–70. ACM, New York (2012)
8. Raghavan, H., Iyer, R.: Probabilistic first pass retrieval for search advertising: from theory to practice. In: Proceedings of the 19th ACM International Conference on Information and Knowledge Management, pp. 1019–1028. ACM, New York (2010)
9. Russell, S., Norvig, P.: Artificial Intelligence: A Modern Approach, 3rd edn. Prentice-Hall, Inc., Upper Saddle River (2010)

User Simulations for Interactive Search: Evaluating Personalized Query Suggestion

Suzan Verberne[1], Maya Sappelli[1,2], Kalervo Järvelin[3], and Wessel Kraaij[2,1]

[1] Institute for Computing and Information Sciences, Radboud University Nijmegen
[2] TNO, Delft
[3] School of Information Sciences, University of Tampere

Abstract. In this paper, we address the question "what is the influence of user search behaviour on the effectiveness of personalized query suggestion?". We implemented a method for query suggestion that generates candidate follow-up queries from the documents clicked by the user. This is a potentially effective method for query suggestion, but it heavily depends on user behaviour. We set up a series of experiments in which we simulate a large range of user session behaviour to investigate its influence. We found that query suggestion is not profitable for all user types. We identified a number of significant effects of user behaviour on session effectiveness. In general, it appears that there is extensive interplay between the examination behaviour, the term selection behaviour, the clicking behaviour and the query modification strategy. The results suggest that query suggestion strategies need to be adapted to specific user behaviours.

Keywords: interactive search, academic search, user simulations, user interaction, query suggestion.

1 Introduction

Effective search for information often needs more than one iteration: An initial query is modified multiple times to increase precision or recall. Query suggestion is a functionality of a search engine that suggests the user a list of queries to proceed the search session with. If the query suggestion algorithm works well, it reduces the cognitive load of users and makes them more efficient in their search for information [2]. For web search, query logs are a good source for query suggestion [11]. However, for search tasks addressing highly specialized topics, where there are no relevant queries from other users available, the only realistic option we have is to fall back to the user's own data (previous queries, clicked documents) [18]. In this paper, we evaluate query suggestion for interactive search tasks in a scientific domain. After the initial (user-formulated) query, query suggestion can assist the user in entering effective follow-up queries. Our system generates candidate follow-up queries from the documents clicked by the user and presents these candidate queries (extensions or adaptations of the previous query) in a ranked list. This is a potentially effective method for personalized query suggestion, as shown in previous work [20], but we hypothesize that its

A. Hanbury et al. (Eds.): ECIR 2015, LNCS 9022, pp. 678–690, 2015.

effectiveness heavily depends on user behaviour. We therefore address the following research question in this paper: what is the influence of user search behaviour on the effectiveness of personalized query suggestion?

We set up a series of simulation experiments in which we explore a range of possible user behaviours in order to find out what the important variables are. We are especially interested in what happens to the effectiveness of personalized query suggestion when user behaviour is not perfect. For example, a non-perfect (realistic) user formulates underspecified queries, clicks on irrelevant documents, selects suboptimal queries from the query suggester, and ends his search before he has reached full recall of relevant results. We investigate the following aspects of user behaviour: (1) Query modification strategies, as proposed by [3]; (2) Examination and click behaviour, using an adapted version of the Click Chain Model by [9]; (3) Query selection strategies (model proposed in this paper); and (4) Time-driven session stopping behaviour [3].

Simulation of user behaviour is a powerful tool to evaluate systems for a large range of user behaviours without bringing in hundreds of real users. We use simulations as *what-if* experiments: we observe how the effectiveness of our system changes with varying user behaviours [1]. It should be noted that even if a model cannot be fully validated with user data (because of the lack of sufficient suitable data), the model can still be very useful to see the relation between user behaviour and system effectiveness.

The contributions of this paper are: (a) Session simulations of interactive search, based on the combination of four user models: a click model, a model for time-based stopping behaviour, a model for query formulation strategies and a new model for query selection strategies; (b) An adaptation of the Click Chain Model that accounts for lower examination probabilities for lower ranked results; (c) Large-scale simulations to measure the effectiveness of query suggestion under influence of diverse user behaviours.

2 Related Work

Query Suggestion. The most used source for query suggestion are query logs. These are especially useful when the queries of other users can be reused by the current user, for example because the queries occur in similar sessions [11]. For personalization purposes, the user's own previous queries are sometimes used as a source for query suggestion, but this data is sparse and topic-dependent [7]. When there are no relevant query logs available, documents in the retrieval collection can be used as an alterative source for query suggestion. The idea is to extract query terms from the documents in the collection that are most relevant to the user's current query. Relevance can either be defined by the search engine itself, using the top-n highest ranked documents ('pseudo-relevance feedback'), or by the user's clicks, using the documents that are clicked by the user ('relevance feedback') [5]. One advantage of using clicked documents as source for query suggestion, is that the suggested queries are geared towards the user's current information need, since he will click more often on documents that seem relevant to him [18]. This aspect makes the use of clicked documents suitable for search tasks addressing highly specialized topics. For personalized query suggestion in a scientific topic domain, we therefore implemented the recent successful approach by [20], which extracts terms from documents clicked in the current session and uses these terms as suggestions for follow-up queries.

User Simulations. Most previous work on user simulations for information retrieval focuses on models for result examination (snippet scanning) and clicking behaviour. In the current paper, we use the Dependent click model by [10] and the Click chain model by [9] for simulating examination and clicking behaviour. Examination and click models describe the user behaviour for one query; less attention has been paid to simulation of *session behaviour*. For simulating complete sessions, query modification strategies [4] need to be defined, as well as session stopping behaviour. For both, we use the models proposed by [3]. For the evaluation of query suggestion methods, we also need a model for query selection behaviour. Previous works on query suggestion either assume that the user always selects the first-ranked query (fully trusting the query suggester) [16] or uses expert assessments to determine which queries are selected [20,6] . The main drawback of the latter approach is that each newly implemented query suggestion method will generate new terms that need to be judged. Therefore, we propose a model for query selection behaviour in this paper that allows query selection to be part of user simulations.

3 Methodology

3.1 Data

The iSearch collection of academic information seeking behaviour [17] consists of 65 natural search tasks (topics) from 23 researchers and students from university physics departments. The topic owners were given a task description form with five fields: (a) What are you looking for? (information need); (b) Why are you looking for this? (work task context); (c) What is your background knowledge of this topic? (knowledge state) (d) What should an ideal answer contain to solve your problem or task? (ideal answer); (e) Which central search terms would you use to express your situation and information need? (search terms). A collection of 18K book records, 144K full text articles and 291K metadata records from the physics field is distributed together with the topics. For each topic, 200 documents were manually assessed on their relevance using a 4-point scale.

3.2 Retrieval Set-Up

We indexed the iSearch collection with the Indri search engine[1]. We used the Indri API to set up a query interface to the combined index of Metadata, Book and Article records. All characters that are not alphanumeric, no hyphen or whitespace are removed from the query terms. Multiple query terms are concatenated and combined using the `combine` function in the Indri query language. For example, the two terms ‘ZNO’ and ‘Transparent conductive oxides’ together form the Indri query `#combine(zno transparent conductive oxides)`. Thus, we

[1] `http://www.lemurproject.org/indri/`

convert all queries to bag-of-words representations. As ranking model, we use the Indri LM with default Dirichlet smoothing ($\mu = 50$). Per query, we retrieve 100 results from the combined index.

3.3 Simulation of Query Modification Strategies

We implemented query modification strategies S1–S5 from [3], based on physicians' information seeking behaviour [15]: S1 creates queries of one term[2] where each follow-up query is a different term; S2 creates queries of two terms of which the first term is kept and the last term is varied; S3 creates queries of three terms of which the first two are kept and the last one is varied; S4 creates incrementally growing queries starting with one term and adding one term to each follow-up query; S5 creates incrementally growing queries starting with two terms. For a given topic, the first query of the session is always the first term (or first and second term) from field e 'search terms' in the iSearch data. When adding more terms from the iSearch data, we maintain the original order as created by the topic owner. For example, consider the search terms field "ZnO, transparent conductive oxides, magnetron sputtering, doping". With query modification strategy S4 , the initial query is 'zno'; the second query (without query suggestion) is 'zno transparent conductive oxides'; the third query 'zno transparent conductive oxides magnetron sputtering' and the fourth query 'zno transparent conductive oxides magnetron sputtering doping'.

A search session is defined by a pre-defined time limit; all actions in the session (query formulation, result examination) are associated with *costs* in terms of number of seconds. The user continues formulating queries as long as he has time left in the session and search terms left in his task description in the iSearch data. For query modification, we adopt the costs from [3]: Formulating the initial query costs 3 seconds in S1 and S4, 6 seconds in S2 and S5, and 9 seconds in S3. Each subsequent query costs 3 seconds.

3.4 Simulation of Examination Behaviour and Clicks

We use the Click Chain Model (CCM) by [9] to simulate examination and clicking behaviour on the result list. Like all cascade models, CCM assumes that the user examines the result list from top to bottom.

Click Probabilities. The conditional probability that a document is clicked, given that its snipped is scanned/examined ($P(C_i = 1|E_i = 1)$, where E_i means: "the snipped of the ith result is examined"), is determined by R_i, the perceived relevance of the examined document. For estimating R_i, we use a model that gives the probability that a document is *perceived* relevant given the *actual* relevance of the document (which is given by the relevance assessments in the iSearch data). This probabilistic model is adopted from the dependent click model by [10], who defined probabilities for three different user/query types: perfect (the user never clicks an irrelevant document), informational and navigational (see Table 1). Furthermore, in order to make evaluation straightforward, we assume in the simulation that the user remembers his/her clicks for the short

[2] A term can consist of more than one word.

Table 1. The click probabilities that we use for user simulation. The model has been adapted from [10], converting a 3-level relevance scale to a 4-level relevance scale.

relevance grade	0	1	2	3
perfect	0.00	0.33	0.67	1.00
informational	0.40	0.60	0.75	0.90
navigational	0.05	0.33	0.67	0.95

duration of a session and therefore never clicks on a document he has clicked on before in the same session.

Examination Probabilities. In CCM, the probability of examining the next result ($E_{i+1} = 1$) is zero if the current result is not examined (cascade assumption: if the user does not scan the ith snippet, he will also not scan the $i+1$th snippet). If the current result is examined ($E_i = 1$) and not clicked ($C_i = 0$), the probability of examining the next result is a constant α_1 . The higher α_1, the more persevering the examination behaviour. We make one adaptation to the model: We argue that the examination probability should not only depend on user perseverance but also on the rank of the current result i: the further down in the result list a result is (the higher i), the lower the probability that the user examines the next result [14,8]. Even a highly persevering user will at some point stop examination of a (long) result list. Therefore, we adapt the examination probability as follows, using a sigmoid function to model the decreasing examination probability with higher ranks:

$$P(E_{i+1} = 1 | E_i = 1, C_i = 0) = \frac{1}{1 + e^{k(i-\gamma)}} \tag{1}$$

where i is the rank of the current result, k is a parameter representing the steepness of the slope (a higher k makes the sigmoid less linear and more threshold-like) and γ is a parameter that defines the center of the sigmoid; the rank at which $P(E_{i+1} = 1) = 0.5$. We use a sigmoid function because the examination probabilities that are reported in the literature (based on eye-tracking fixations) can be fitted using a sigmoid: With the parameters $k = 0.5$ and $\gamma = 5$, we can fit equation (1) to the distribution of fixations reported by [8] with a Mean Squared Error of 1.7% and the distribution of fixations reported by [14] with a Mean Squared Error of 3.0%. Both distributions are for web search.

In the situation where the current result was clicked ($C_i = 1$), the probability that the next result is clicked ($E_{i+1} = 1$) depends on the perceived relevance of the current result R_i, and two parameters α_2 and α_3 . In order to make the examination probability for $C_i = 1$ also rank-dependent, we use the same sigmoid function as eq. (1), but we set k to: $k = \alpha_2 * (1 - R_i) + \alpha_3 * R_i$, using α_2 and α_3 as in the original CCM. A larger difference between α_2 and α_3 leads to a larger influence of R_i.

Like query formulation, the examination of the result list is associated with costs. We also adopt these from [3]: the scanning of a snippet costs 3 seconds.

3.5 Query Suggestion Method

We implemented the method for personalized query suggestion from [20]. The simulated user gets 10 suggestions for query terms to be added to the next query. These terms have automatically been extracted from all the documents that the user clicked on in the current search session, including the clicked documents from earlier queries.[3] All word n-grams with $n = \{1, 2, 3\}$ in these documents are considered candidate terms. The terms are ranked by scoring them with Kullback-Leibler divergence [19] between the probability distributions for a term in two collections: the collection of documents clicked by the user and a background collection of general English (the Corpus of Contemporary American English, COCA). The output score denotes the informativeness of the term for the collection of clicked documents. The terms are ranked by this score and presented to the (simulated) user as suggested query terms, either to expand the previous query or replace the final term of the previous query, depending on the query modification strategy.

3.6 Simulation of Query Selection Behaviour

We propose the following model for simulating the selection of a query in a query suggestion scenario. The input for the model is the output of the query suggestion method: an ordered list of suggested query terms $L = t_1, t_2, ...t_k$. We simulate the user's decision with four variables S_{ts}, S_{rel}, S_{in}, S_{st}. Each term in L takes a value for each of these four variables:

- S_{ts}: The term suggester score. This is the output of the query suggestion method (See Section 3.5).
- S_{rel}: The output of a term scorer that determines the informativeness of the term for the subcollection of documents that are relevant for the current topic, using Kullback-Leibler divergence between this subcollection and a background corpus of general English (COCA). If a term from L does not occur in the subcollection, $S_{rel} = 0$
- S_{in}: The output of a term scorer that determines the informativeness of the term for the user's explicit information need (a concatenation of the fields a, b and d for the current topic in the iSearch data), using Kullback-Leibler divergence between the collection and a background corpus of general English (COCA). If a term from L does not occur in the information need, $S_{in} = 0$.
- S_{st}: The output of a binary scorer that determines whether or not the term is in the set of the user-formulated search terms (from the iSearch data). If the term (normalized for case, whitespace and hyphenation) is in the list of search terms then $S_{st} = 1$, otherwise $S_{st} = 0$.

These four variables are justified as follows: S_{in} and S_{st} were given in the iSearch data by the searchers whose query selection behaviour we aim to simulate. S_{ts} is the score given by the term suggestion algorithm, the evaluation of which is central to our simulations. And S_{rel} is higher for terms that come from documents

[3] In the case of metadata and book records, the terms are extracted from the fields 'title' and 'description'; in the case of articles in PDF, for which no metadata is available, the terms are extracted from the first 200 words of the document.

that are judged as relevant by the searcher; a competent searcher will be more likely to select one of these terms than a term from an irrelevant document. The term selection simulator is a tuple of integer weights $W = (W_{ts}, W_{rel}, W_{in}, W_{st})$. The simulated user selects a term by solving:

$$\arg\max_{t \in L} \frac{W_{ts} * S_{ts} + W_{rel} * S_{rel} + W_{in} * S_{in} + W_{st} * S_{st}}{\sum_{x \in \{ts, rel, in, st\}} W_x} \quad (2)$$

The higher the weight for S_{in}, S_{rel} and S_{st}, the more informed the simulated user is. A higher weight for S_{ts} implies more trust in the query suggester; a simulated user with a 0-weight for S_{in}, S_{rel} and S_{st} fully trusts the query suggester and will always take the top-ranked term. A simulated user with a 0-weight for S_{ts} and S_{rel} is very critical and will only select terms that are in his explicit information need or list of search terms.

It is yet unknown what a realistic time cost is for selecting a query from a drop-down list. We assume that it takes less time to select a query than to fomulate one: we set the time-cost of query term selection to 1 second.

3.7 Evaluation

We evaluate the effectiveness of the session using Cumulated Gain (CG) [13], following the arguments by [3] for no discounting and no normalization: discounting has value in a one-query evaluation setting but is not sensible over a complete session, and normalization may leed to counterintuitive results when user behaviour is based on time costs instead of result ranks. CG is the sum of the relevance scores of all seen documents in the session. Thus, the goal for the simulated user was to collect as much gain as possible in a 5 minute session. For each query in the session, we evaluate the result list up to the last examined result, keeping track of the relevance of the seen documents. Documents that have been seen by the user previously in the same session are disregarded. In all experiments, we set the session time limit (a bit arbitrarily) to 300 seconds (5 minutes). We leave it for later papers to examine the effect of session length.

4 Experiments and Results

4.1 The Effect of Examination Behaviour

We measured the effect of the examination parameters (k, γ, $\alpha_{2,3}$) in two settings: the setting where the user selects his own queries using terms from field e in the iSearch data, and the setting where the user gets query suggestions. In both settings, the query modification strategy was S4 (see Section 3.3).

In [9], the following values for the examination parameters are suggested for informational queries: $\alpha_1 = 1$ (the user inspects all results; queries without clicks were disregarded), $\alpha_2 = 0.40$ and $\alpha_2/\alpha_3 = 1.5$. Since we aim to investigate the influence of user aspects on the effectiveness of personalized query suggestion, we inspect a *range* of parameter values instead of fixing them for a given user type. We experimented with the following grid of parameter values around the values suggested in [9]: α_2 in the range 0.1–1.0 with steps of 0.1 and the proportion

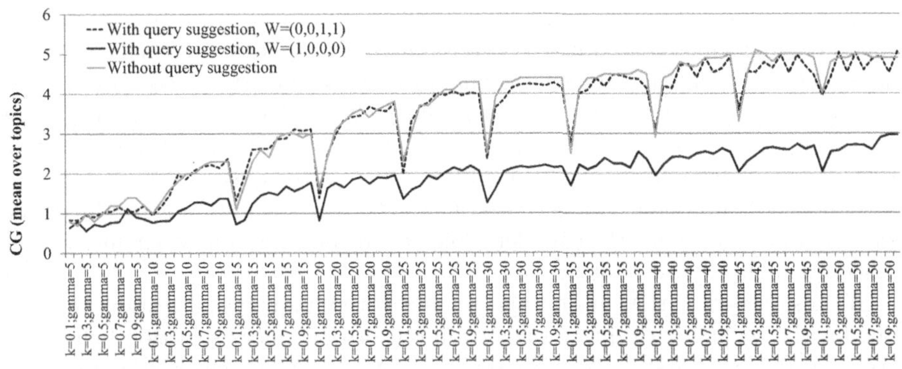

Fig. 1. Mean CG ($N = 65$ topics) for the grid of inspection parameter values (k and γ), with and without query suggestion. $W = (1, 0, 0, 0)$ represents 'lazy' query suggestion behaviour; $W = (0, 0, 1, 1)$ represents critical query selection behaviour.

α_2/α_3 in the range 1.0–3.0 with steps of 0.5. We fix $k = \alpha_2$. For γ (the center of the sigmoid), we had found that $\gamma = 5$ gave the best fit when fitting the sigmoid to examination probabilities in web search (see section 3.4). For interactive search on scientific topics, we extend the range of this parameter to higher values, to represent a more recall-oriented user [16]: we use a grid in the range of 5–50 with steps of 5.

For the setting without query suggestions, we obtained CG values (averaged over queries) ranging from 0 to 5.1. The parameter that has the largest influence on the effectiveness of the session is γ: there is a strong positive relationship between γ and CG (Kendall's $\tau = 0.78$, $P < 0.0001$), while the relationship between α_2 (or k) and CG is weak (Kendall's $\tau = 0.20$, $P < 0.001$). The effect of γ on the effectiveness of the session is a consequence of the number of documents examined per query: in the sessions with $\gamma = 5$, the average number of results examined per query is 3.7, while in the sessions with $\gamma = 50$, the average number of results examined per query is 33.8. In other words, a higher γ represents more persevering user behaviour.

In the setting *with* query suggestions, we used the perfect click model (see Table 1) and we evaluated two extreme configurations of the term selection model W (See Section 3.6):

- $W = (1, 0, 0, 0)$: the user fully relies on the term suggester.
- $W = (0, 0, 1, 1)$: the user only selects terms that are in his explicit information need or his list of search terms. If none of the terms is, the user formulates his own query using the terms in the field e from the iSearch data (like in the setting without query suggestion).

Figure 1 shows the results. We see that a user who fully trusts the query suggester ($W = (1, 0, 0, 0)$) ends up with lower CG than the user who formulates his own query. The difference between the two is the smallest for the lowest values of γ and k. This suggests that the more persevering the user is in examining results (higher k and γ), the larger the importance of formulating or selecting the right query. On average, the lazy user who fully trusts the query suggester is faster: he can enter more follow-up queries because he spends less time formulating each query. For $\gamma = 10$, the user who formulates his own queries enters 5.1 queries per 300 second-session, while the user who picks the first suggestion from the query suggester enters 11.8 queries per session on average. In addition, the user who does not get query suggestions often runs out of query terms before the session time is up, while the user who picks the first suggestion from the query suggester mostly spends the complete 300 second-session formulating queries and stops when the 300 seconds are up.

We also see that the line for the setting without query suggestion and the line for the setting with query suggestion but critical query selection behaviour ($W = (0, 0, 1, 1)$) are strongly related to each other. This is because the user with critical query selection behaviour only selects a query term from the suggester if it is in his own list of query terms, or his explicit information need. Analysis of the query selection behaviour shows that for $k = 1$ and $\gamma = 50$, the user with $W = (0, 0, 1, 1)$ selects a suggested query for only 11.6% of his queries; in all other cases, he formulates his own query using a term from the iSearch data. For lower values of k and γ this percentage is even lower. In the next subsection, we more precisely investigate the effect of the term selection model W.

4.2 The Effect of Query Selection Behaviour (W)

For measuring the effect of the term selection model W, we experiment with a grid of weight integer values $\{0, 1, 10\}$ for all four weights. Thus, we get configurations such as $(0, 10, 0, 1)$, $(1, 0, 10, 1)$, etc. We compared two values of the most important examination parameter: $\gamma = \{10, 50\}$. We fix k to 0.5, $\alpha_2 = k$ and $\alpha_2/\alpha_3 = 1.5$ because their effect on the effectiveness of the session is much smaller than of γ and having too many variables makes analysis of the results complex. In all cases, the query modification strategy was S4 and we used the perfect click model.

We found that the only W-parameter that has a significant relationship with CG is W_{ts}; this relationship is moderately negative (Kendall's $\tau = -0.37$, $P < 0.0001$ and $\tau = -0.34$, $P < 0.001$ respectively for $\gamma = 10$ and $\gamma = 50$). This means that the higher the weight for the term suggester score (thus the more the user trusts the term suggester), the lower the CG. Overall, the best-scoring parameter settings for query selection behaviour are $W = \{0, 0, 1, 0\}$ (CG for $\gamma = 50$ is 5.2, CG for $\gamma = 10$ is 2.3) and $W = \{0, 0, 1, 1\}$.

4.3 The Effect of Clicking Behaviour

For investigating the effect of the click behaviour on the effectiveness of query suggestion, we evaluated the perfect, informational and navigational click models (see Section 3.4). We again compared two values of the most important examination parameter: $\gamma = \{10, 50\}$ and we fix k to 0.5, $\alpha_2 = k$ and $\alpha_2/\alpha_3 = 1.5$. For the query selection behaviour, we compared two parameter settings: $W = \{(1, 0, 0, 0), (0, 0, 1, 1)\}$. In all cases, the query modification strategy was S4. The results are in Figure 2.

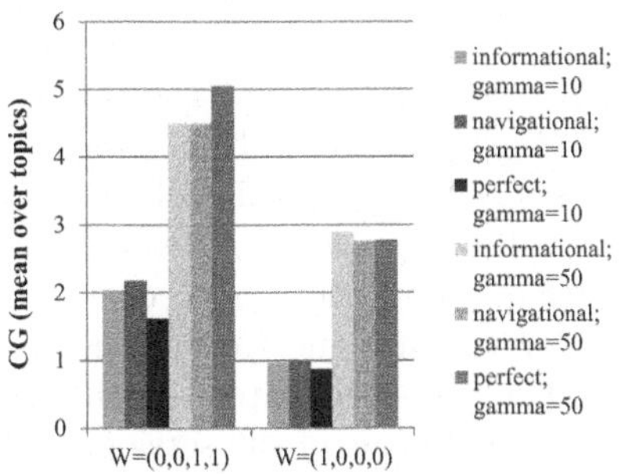

Fig. 2. Mean CG ($N = 65$ topics) for the 3 click models, 2 different examination behaviour types γ, and two different query selection behaviours W. In all cases, the query modification strategy is S4.

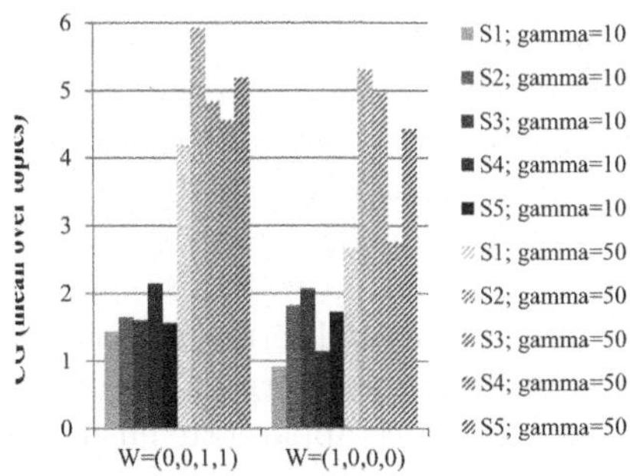

Fig. 3. Mean CG ($N = 65$ topics) for the five query strategies, two different examination behaviour types γ, and two different query selection behaviours W. In all cases, we used the perfect click model.

The results show that a bigger effect comes from the examination parameter γ than from the click model and the query selection model W. Besides that, we see that with critical query selection behaviour ($W = (0,0,1,1)$), the perfect click model seems to give the highest results for the persevering user ($\gamma = 50$), while the navigational model gives the highest results for the 'lazy' examination behaviour ($\gamma = 10$).[4] This suggests that for lazy examination behaviour, it can be profitable to click on more documents, even if not all of them are relevant. This is not the case when the user always selects the highest ranked suggested query ($W = (1,0,0,0)$); then the three models perform almost equally.

4.4 The Effect of the Query Modification Strategy

We evaluated the five query modification strategies (see Section 3.3). We again compared two values of the most important examination parameter: $\gamma = \{10, 50\}$ and we fix k to 0.5, $\alpha_2 = k$ and $\alpha_2/\alpha_3 = 1.5$. We used the perfect click model and for the query selection behaviour, we compared two parameter settings: $W = \{(1,0,0,0), (0,0,1,1)\}$. The results are in Figure 3.

Again, the biggest effect comes from the examination parameter γ. However, query strategy does have a big influence. The effect of query strategy is bigger for the user who trusts the query suggester ($W = (1,0,0,0)$) than for the user with critical query selection behaviour ($W = (0,0,1,1)$). Surprisingly, the best performing query strategy is S2 (subsequent queries of two terms of which the first term is kept and the last term is varied). The poorest performing query strategy is S1 (issueing one term at the time), followed by S4 (adding each new query term to the previous query) in most combinations of W and γ. An

[4] A paired t-test (with the CG scores for individual topics paired) shows that the difference between the navigational and the perfect click model for $\gamma = 10$ is significant with $P < 0.01$. For $\gamma = 50$, the difference is not significant.

exception is the setting where $W = (0,0,1,1)$ and $\gamma = 10$ (lazy examination behaviour with critical query selection behaviour); in that case S4 is the most effective strategy. The differences between S1,4 and S2,3,5 are bigger for $W = (1,0,0,0)$ than for $W = (0,0,1,1)$. We think this is because picking the highest-ranked term from the query suggester for each follow-up query can lead to topic drift in the session. In S2, S3 and S5, the first query of the session consists of two user-formulated terms, whereas in S1 and S4, the first query only consists of one user-formulated term. The combination of two user-formulated search terms in the first query apparently ensures a better topical focus of the session.

5 Conclusions

We addressed the following research question in this paper: what is the influence of user behaviour on the effectiveness of personalized query suggestion? Query suggestion can make the user more efficient, because it takes less time to select a query than to formulate one. But query suggestion is not profitable for all user types. We found the following significant effects of user behaviour on the effectiveness of query suggestion: (1) The more persevering the user is in examining result lists, the larger the importance of selecting the right query from the query suggester. This might be because lower in the result list the documents are less relevant and therefore the suggested terms are of lower quality. The persevering user should therefore be more critical in where he clicks or more critical in selecting suggested queries; (2) The less critical the user is in selecting terms from the query suggester (the more trust he has in the suggestions), the lower the cumulated gain of the session; (3) If the user examines few results ('lazy examination behaviour'), clicking on more documents results in higher cumulated gain, even if not all of the clicked documents are relevant. This is in line with previous findings for pseudo-relevance feedback [12]. (4) Because of the risk of topic drift when the user adds suggested terms from clicked documents, it is profitable to start the session with a query consisting of more than one term.

It appears that there is extensive interplay between the examination behaviour and the term selection behaviour on the one hand, and the clicking behaviour or the query modification strategy on the other hand: both the choice of the most effective query modification strategy and the most effective click model depend on how persevering the examination behaviour and how critical the query selection is. This suggests that query suggestion strategies need to be adapted to specific user behaviours.

In future work, we plan to collect real user data for search tasks in a scientific domain. With these data, we will (1) model the query modification strategies that are typical for academic search sessions, including the associated time costs; (2) validate (and improve) our examination model with rank-dependent examination probabilities; and (3) optimize our query suggestion method and compare it to other methods.

References

1. Azzopardi, L., Järvelin, K., Kamps, J., Smucker, M.D.: Report on the sigir 2010 workshop on the simulation of interaction. SIGIR Forum 44(2), 35–47 (2011)
2. Azzopardi, L., Kelly, D., Brennan, K.: How query cost affects search behavior. In: Proceedings of the 36th International ACM SIGIR Conference on Research and Development in Information Retrieval, pp. 23–32. ACM (2013)
3. Baskaya, F., Keskustalo, H., Järvelin, K.: Time drives interaction: simulating sessions in diverse searching environments. In: Proceedings of the 35th International ACM SIGIR Conference on Research and Development in Information Retrieval, pp. 105–114. ACM (2012)
4. Bates, M.J.: Information search tactics. Journal of the American Society for information Science 30(4), 205–214 (1979)
5. Belkin, N.J., Cool, C., Kelly, D., Lin, S.J., Park, S., Perez-Carballo, J., Sikora, C.: Iterative exploration, design and evaluation of support for query reformulation in interactive information retrieval. Information Processing & Management 37(3), 403–434 (2001)
6. Bhatia, S., Majumdar, D., Mitra, P.: Query suggestions in the absence of query logs. In: Proceedings of the 34th International ACM SIGIR Conference on Research and Development in Information Retrieval, pp. 795–804. ACM (2011)
7. Feild, H., Allan, J.: Task-aware query recommendation. In: Proceedings of the 36th International ACM SIGIR Conference on Research and Development in Information Retrieval, SIGIR 2013, pp. 83–92. ACM, New York (2013)
8. Guan, Z., Cutrell, E.: An eye tracking study of the effect of target rank on web search. In: Proceedings of the SIGCHI Conference on Human Factors in Computing Systems, pp. 417–420. ACM (2007)
9. Guo, F., Liu, C., Kannan, A., Minka, T., Taylor, M., Wang, Y.M., Faloutsos, C.: Click chain model in web search. In: Proceedings of the 18th International Conference on World wide Web, pp. 11–20. ACM (2009)
10. Hofmann, K., Schuth, A., Whiteson, S., de Rijke, M.: Reusing historical interaction data for faster online learning to rank for ir. In: Proceedings of the Sixth ACM International Conference on Web Search and Data Mining, WSDM 2013, pp. 183–192. ACM, New York (2013)
11. Huang, C.K., Chien, L.F., Oyang, Y.J.: Relevant term suggestion in interactive web search based on contextual information in query session logs. Journal of the American Society for Information Science and Technology 54(7), 638–649 (2003)
12. Järvelin, K.: Interactive relevance feedback with graded relevance and sentence extraction: simulated user experiments. In: Proceedings of the 18th ACM Conference on Information and Knowledge Management, pp. 2053–2056. ACM (2009)
13. Järvelin, K., Kekäläinen, J.: IR evaluation methods for retrieving highly relevant documents. In: Proceedings of the 23rd Annual International ACM SIGIR Conference on Research and Development in Information Retrieval, pp. 41–48. ACM (2000)
14. Joachims, T., Granka, L., Pan, B., Hembrooke, H., Gay, G.: Accurately interpreting clickthrough data as implicit feedback. In: Proceedings of the 28th Annual International ACM SIGIR Conference on Research and Development in Information Retrieval, pp. 154–161. ACM (2005)

15. Keskustalo, H., Järvelin, K., Pirkola, A., Sharma, T., Lykke, M.: Test collection-based IR evaluation needs extension toward sessions – A case of extremely short queries. In: Lee, G.G., Song, D., Lin, C.-Y., Aizawa, A., Kuriyama, K., Yoshioka, M., Sakai, T. (eds.) AIRS 2009. LNCS, vol. 5839, pp. 63–74. Springer, Heidelberg (2009)
16. Kim, Y., Seo, J., Croft, W.B.: Automatic boolean query suggestion for professional search. In: Proceedings of the 34th International ACM SIGIR Conference on Research and Development in Information Retrieval, SIGIR 2011, pp. 825–834. ACM, New York (2011)
17. Lykke, M., Larsen, B., Lund, H., Ingwersen, P.: Developing a test collection for the evaluation of integrated search. In: Gurrin, C., He, Y., Kazai, G., Kruschwitz, U., Little, S., Roelleke, T., Rüger, S., van Rijsbergen, K. (eds.) ECIR 2010. LNCS, vol. 5993, pp. 627–630. Springer, Heidelberg (2010)
18. Shen, X., Tan, B., Zhai, C.: Implicit user modeling for personalized search. In: Proceedings of the 14th ACM International Conference on Information and Knowledge Management, pp. 824–831. ACM (2005)
19. Tomokiyo, T., Hurst, M.: A language model approach to keyphrase extraction. In: Proceedings of the ACL 2003 Workshop on Multiword Expressions: Analysis, Acquisition and Treatment, vol. 18, pp. 33–40. Association for Computational Linguistics (2003)
20. Verberne, S., Sappelli, M., Kraaij, W.: Query term suggestion in academic search. In: de Rijke, M., Kenter, T., de Vries, A.P., Zhai, C., de Jong, F., Radinsky, K., Hofmann, K. (eds.) ECIR 2014. LNCS, vol. 8416, pp. 560–566. Springer, Heidelberg (2014)

The Impact of Query Interface Design on Stress, Workload and Performance

Ashlee Edwards[1], Diane Kelly[1], and Leif Azzopardi[2]

[1] School of Information and Library Science, University of North Carolina, USA
[2] School of Computing Science, University of Glasgow, UK
{aedwards,diane.kelly}@unc.edu, leif.azzopardi@glasgow.ac.uk

Abstract. We investigated how the design of the query interface impacts stress, workload and performance during information search. Two query interfaces were used: a standard interface which looks similar to contemporary, general purpose search engines with a standard query box, and an experimental (structured) interface that was designed to slow people down when querying by presenting a series of boxes for query terms. We conducted a between subjects laboratory experiment where participants were randomly assigned to use one of the query interfaces to complete two assigned search tasks. Stress was measured by recording physiological signals and with the Short Stress State Questionnaire. Workload was measured with the NASA-TLX and log data was used to characterize search behavior. The differences in stress and search behaviors were not significant, but participants who used the structured interface rated their success significantly higher than those who used the standard interface, and reported significantly less workload.

1 Introduction

Search user interfaces are comprised of many features to assist searchers with the information search process such as specifying initial queries, deciding which results to examine and reformulating queries. The mechanism searchers use to communicate their information needs to the system is perhaps the most essential element of the search user interface. Past research has shown the design of the query mechanism influences how people construct queries and engage in search, and small changes to this mechanism can lead to more positive search outcomes, such as the construction of more effective queries [1,5,6,9,13,24]. Qvarfordt et al. [24], for example, presented searchers with a query preview tool to help them evaluate the extent to which their queries were retrieving new documents and found when using this tool, searchers spent more time formulating queries, went deeper in the search results list and retrieved more diverse documents. Azzopardi et al. [5] found similar results when examining a query interface that required people to slow down when querying by entering each of their query terms into unique boxes; searchers using this interface spent more time constructing initial queries, issued fewer queries, went to greater depths in the search results list and viewed more documents. This study also found evidence that the design

A. Hanbury et al. (Eds.): ECIR 2015, LNCS 9022, pp. 691–702, 2015.

of the interface impacted how searchers evaluated their experiences. Specifically, people who used the slow query interface reported less mental demand, temporal demand, and frustration, and greater success with their queries.

In this work, we explore the hypothesis that the type of search behaviors supported by the standard query interface - issuing more queries, evaluating fewer documents per query and shallowly evaluating search results lists - has negative affective consequences in terms of stress and workload. Related work suggests that this is likely the case [c.f., [8]]. However, the stress and workload imposed by different query interfaces has not be investigated in depth. In this paper, we systematically investigate the relationship between the query interface and participants' experiences of stress and workload using log data, physiological sensors and psychometric scales.

2 Related Work

The query mechanism is one of the most important parts of the search user interface since it is the facility searchers use to express their information needs. Today, query interface design for general-purpose search engines has converged on a standard: a small, rectangular box. Most contemporary research is focused on accelerating the query process through query auto-completion and suggestions. However, a small body of research has shown small changes to the query interface can lead to more positive outcomes, both with respect to retrieval effectiveness and user experience [1,5,6,9,13,24]. Franzen and Karlgren [9] found increasing the size of the query box increased the length of searchers' queries. Belkin et al. [6] found searchers provided longer queries when shown a prompt next to the query box. Agapie et al. [1] placed an interactive halo around the query box that changed colors as searchers typed their queries and found this halo increased query length. Qvarfordt et al. [24] and Azzopardi et al. [5] found searchers spent more time formulating queries and went deeper in the search results list when introducing a query preview tool and structured query interface, respectively. These studies demonstrate how changes to the query interface can potentially impact search behaviors and retrieval performance. However, they do not provide evidence about how the query interface impacts searchers' affective experiences, including experiences of workload.

Researchers have just recently started studying affect and emotion during information search. Lopatovska and Arapakis [16] provide an overview of theories and methods used to study emotions, and of a small number of studies that have been conducted about emotions in the context of search. Nahl and Bilal [21] provide an overview of information and emotion in information behavior research. To date, most research has examined the relationship between search behaviors and experiences of affect and emotion during information search. For example, Moshfeghi and Jose [19] studied affect and search success. They found anxiety and anger were present at the end of the search task, even in cases of successful search. Gwizdka and Lopatovska [10] focused on a more broadly defined set of self-report variables, such as happiness, interest, and satisfaction and found participants' emotions changed as a result of search. For example, participants who reported feeling unhappy before starting a search task reported greater happiness after successfully completing the task. Lopatovska [15] investigated the relationship between emotions and micro-search behaviors, such as clicks and scrolls, and found people experienced negative emotions when querying and positive emotions when examining search results. Feild et al. [8] found searchers experienced the most frustration when querying and were frustrated with about half of their queries. These latter two works in particular, suggest the query mechanism and interface design could have an impact on both search behaviors and emotional experiences.

Several studies have examined the relationship between emotions and search tasks [3,10,20,23]. Arapakis et al. [3], who measured affective responses by analyzing participants' facial expressions as they searched, found as task difficulty increased, people's emotions moved from positive to negative valence. Poddar and Ruthven [23] found positive emotions were highly correlated with interest in the task, and tasks participants rated as less difficult before searching were associated with more positive emotions after searching. Of particular note, they found people's own search tasks were associated with more positive emotions than assigned search tasks.

In addition to self-report measures, a small number of studies have used physiological measures [4,18,19,22,25]. For example, Trimmel et al. [25] found both heart rate and mental effort increased as system response times increased. Arapakis et al. [4] combined physiological signals, such as heart rate, galvanic skin response, skin temperature, with facial expression analysis, to predict the relevance of both textual and audio-visual content and found participants experienced more emotions when viewing audio-visual content. Moshfeghi and Jose [19] used a similar set of physiological signals and facial expressions to create predictive models of relevance, and found when combined with traditional implicit feedback signals such as dwell time, the signals could distinguish between relevant and non-relevant documents.

To our knowledge, physiological measures have not been used to understand user experience in the context of alternative search interfaces. Furthermore, most past studies have focused on emotions like frustration, or negative emotions, more generally; in this work we focus on experiences of stress, a psychological and physical reaction that occurs when a person encounters a situation that taxes or exceeds his or her resources [14]. Specifically, we seek to understand how the design of the query interface affects people's search behaviors and experiences of stress and workload during search.

3 Method

We conducted a between subjects experiment where participants were randomly assigned to use one of two search interfaces. Each participant completed one practice task and two experimental tasks, whose orders were counterbalanced across interface condition. The interfaces, tasks, collection and system settings

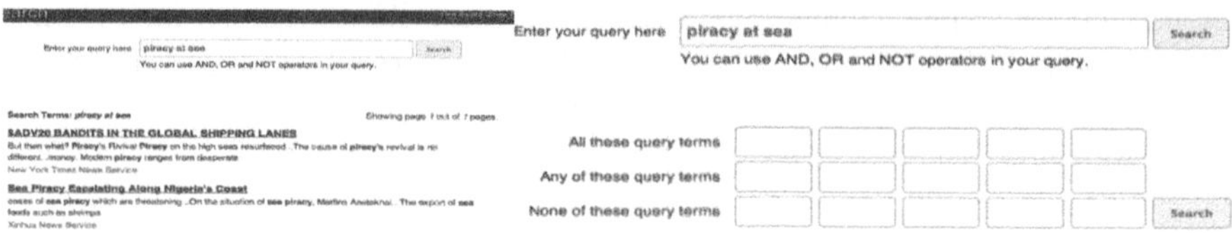

Fig. 1. Left: Standard Interface. Right: Standard and Structured Query Boxes.

were the same used in [5] as we wanted to try to replicate and extend these findings. In addition, we collected physiological data from participants, as well as self-report data regarding stress and workload.

Interfaces, System, Collection and Search Topics. One baseline interface was used (standard interface) which looked similar to contemporary, general purpose search engines with a standard query box (Figure 1). The experimental interface (structured interface) presented a series of boxes to participants. Participants entered one term per box and had to click in each box to type This interface more closely resembles the advanced search interfaces in Google, the ACM Digital Library and those provided by commercial database vendors.

The Whoosh IR Toolkit was used as the core retrieval system. Query terms entered via the standard interface were implicitly AND-ed, unless operators were included in the query. A tip below the query box indicated to participants they could use operators when formulating queries to ensure they had the same functionality as those who used the structured interface. With the structured interface, a Boolean query was automatically constructed from the terms entered into the ALL, ANY or NOT query boxes, which translated to AND, OR and NOT, respectively. BM25, with standard parameters, was used to rank results.

The TREC Robust Track collection [26] was used, along with two test topics from [5]: Topic 347 (Wildlife Extinction) and Topic 435 (Curbing Population Growth). Participants were required to find documents relevant to the topics and were told to imagine they were a newspaper reporter and needed to write a detailed report about the given topics. System tutorials were not provided.

Measures of Stress. BioPac (MP35) was used to collect physiological data. We measured electrodermal activity (EDA) (skin conductance) and heart rate (HR). We chose these two measures because studies have shown that these are the most prominent signals of stress aside from saliva samples and electroencephalography (EEG) signals [2,7]. To measure skin conductance, we attached electrodes to the thenar and hypothenar eminences of participants' palms. To measure heart rate, we attached electrodes to participants' collarbones and ribs. Participants performed a practice search task to insure their EDA and HR were at normal levels before beginning the actual search tasks.

The 24-item Short State Stress Questionnaire (SSSQ) [12] was used to measure three aspects of stress: engagement, distress, and worry. This scale is a shortened version of the Dundee Stress State Questionnaire [17]. Participants responded to items using a 5-point scale (1= never and 5=very often).

Pre-Task and Post-Task Questionnaires. Participants completed pre- and post-task questionnaires before and after each search (Table 1). The pre-task

Table 1. Pre-Task and Post-Task Questionnaire Items

Pre-Task Questionnaire	
1. How much do you know about this topic?	[Nothing...I know details]
2. How interested are you to learn more about this topic?	[Not at all...Very Much]
3. How relevant is this topic to your life?	[Not at all...Very Much]
4. Have you ever searched for information related to this topic?	[Never...Very Often]
5. How difficult do you think it will be to search for info. about this topic?	[Very Easy...Very Difficult]
Post-Task Questionnaire	
1. How difficult was it to find relevant documents?	[Very Easy...Very Difficult]
2. How would you rate your skill and ability at finding relevant documents?	[Not Good...Very Good]
3. How would you rate the system's ability at retrieving relevant documents?	[Not Good...Very Good]
4. How successful was your search?	[Unsuccessful...Successful]
5. How many of the relevant documents do you think you found?	[A few of them...All of them]

questionnaire assessed participants' prior knowledge and experiences searching the topic, interest in the topic and estimated difficulty. The post-task questionnaire assessed participants' experienced search difficulty and search success. Items were evaluated on 5-point scales.

Workload. The NASA-TLX [11] was used to measure workload. This scale contains six items assessing mental demand, physical demand, temporal demand, performance, effort and frustration. Participants responded on a 7-point scale (1=very low; 7=very high). This instrument also consists of paired-comparisons of factors, where participants compare all factors and select the one that contributed the most to their workload. The comparisons are used to derive personalized weights for each factor.

Search Behaviors and Performance. The search behavior measures are divided into actions and time (in seconds) spent performing different actions. Actions included: number of queries issued, number of documents and SERPs examined, deepest click on SERP, deepest SERP page examined, and hover depth. Time-based measures included: total time, time spent per query, and time spent querying, viewing documents and SERPs. Performance was evaluated by number of documents saved, number of TREC relevant documents saved and P@10.

Procedure. After an introduction to the study, electrodes were attached to participants. A 3-minute resting period occurred before the practice task. Then, participants were shown the first task description and completed the pre-task questionnaire. After completing the task, participants were given the post-task questionnaire. Once both tasks were completed, participants had the electrodes removed. They then took the SSSQ and NASA-TLX. Finally, participants were given a $20USD honorarium.

Participants: Participants were recruited via email solicitation to undergraduate students at the University of North Carolina. The initial sample size was 34 participants. However, we discovered problems with the BioPac equipment after running 14 participants, which compromised the accuracy of these participants' physiological data, so these participants were excluded. The sample described in this paper included 20 undergraduates (15 women; 5 men) with a mean age of 20 years (SD=1.18). There were 15 humanities majors, 2 business majors, 2 in the natural sciences and 1 undecided.

4 Results

4.1 Stress: Physiological Data

Table 2 reports the physiological data according to practice and temporal tasks, and interface (practice task values are excluded from interface totals). Task numbers are defined temporally (i.e., Temporal Task 1: first task participants conducted) to allow us to examine how these signals changed during the course of the session. Data from the first 60 seconds of each task is also presented, given that a physiological reaction might occur when participants first viewed the interfaces. This also corresponds to when participants would be issuing their initial queries. Recordings were taken at 1000 samples per second.

The initial row of values for each task report participants' normalized EDA in microsiemens and HR in beats per minutes (bpm). To aid with interpretation of these data, research has shown EDA can range from 1-30 microsiemens, with baselines that typically hover around 2 microsiemens, but can vary amongst individuals [7]. Studies have shown that under stress, EDA can spike to levels of 1-2 microsiemens or higher above the individual baseline [7]. Similarly, average baseline resting HR is roughly 60-80 bpm [2], but baselines can vary according to individual age, health, and physical activity, and can be affected by changes in emotion. The second and third rows for each task present the amount of change participants' experienced in these measures. For example, for the practice task, the EDA values for those who used the structured interface changed more than for those who used the standard interface. The fourth and fifth rows for each task present the average direction of these changes. For example, during the practice task, on average, participants' EDA decreased for both interfaces, while their HRs increased. Thus, while the change values describe the absolute amount of fluctuation of EDA and HR, the directional values describe the average valence of these fluctuations.

Participants experienced similar levels of EDA when conducting the practice and first tasks regardless of interface, and the greatest amount when conducting the second task, with those participants in the structured group experiencing the greatest EDA. The change values show those in the structured group experienced more variations in EDA. When looking at the direction of these changes, those in the structured group usually experienced decreases in EDA, except during the first 60 seconds. This suggests those in the structured group experienced an initial increase in EDA at the start of their searches, which then gradually returned to normal rates during the course of the session. The standard deviations for these change measures are much larger for those in the structured group suggesting more variability in how the interface affected the EDA of these participants. When viewing the data at the interface level, participants in the structured group experienced higher EDA and more change; the overall direction of the change reveals nearly equal amounts of increases and decreases. Statistical tests found no significant differences in any of these measures.

There was little difference in average HR or amount of change experienced according to interface or task. Except for the practice task, participants in the standard group more often experienced decreases in HR during their searches, most notably for the second task they conducted, which potentially indicates a slight relaxation during the search. The direction of change for the first 60 seconds shows those in the standard group also experienced greater decreases in HR at the start of their searches. While those in the structured group did not experience as much decrease in HR during their searches, their normalized HR values were similar to those in the standard group, which suggests that they did not experience more elevated HRs. None of these differences were significant.

Table 2. Average (SD) normalized electrodermal activity (EDA) (microsiemens) and heart rate (HR) (bpm), amount of change experienced and direction of change

	Electrodermal (EDA)		Heart Rate (HR)	
	Standard	Structured	Standard	Structured
Practice Task	2.60 (1.46)	2.95 (0.92)	77.52 (11.54)	75.57 (7.75)
Amount of Change Experienced	0.97 (1.12)	2.87 (5.70)	6.14 (3.08)	5.82 (2.99)
Amount of Change (first 60 seconds)	1.07 (2.03)	2.41 (5.72)	5.76 (4.31)	4.74 (2.69)
Direction of Change	-0.52 (1.42)	-1.29 (6.31)	-3.88 (5.88)	-4.45 (4.97)
Direction of Change (first 60 seconds)	+0.94 (2.10)	+1.78 (5.97)	-2.82 (6.82)	-2.21 (5.18)
Temporal Task 1	2.58 (1.44)	2.96 (0.94)	75.94 (9.46)	74.80 (7.41)
Amount of Change Experienced	1.64 (0.95)	4.04 (5.09)	6.01 (4.52)	6.62 (3.02)
Amount of Change (first 60 seconds)	1.27 (0.80)	3.23 (4.70)	6.63 (3.80)	5.32 (2.49)
Direction of Change	-0.03 (1.97)	-0.85 (6.57)	-5.37 (5.35)	-4.99 (5.49)
Direction of Change (first 60 seconds)	+0.66 (1.40)	+2.40 (5.23)	-4.21 (6.60)	-2.88 (5.33)
Temporal Task 2	3.68 (3.49)	4.42 (3.02)	73.33 (8.58)	76.43 (5.49)
Amount of Change Experienced	1.81 (0.96)	4.29 (5.39)	7.54 (5.64)	7.20 (3.82)
Amount of Change (first 60 seconds)	2.57 (2.51)	3.62 (5.81)	6.49 (5.72)	7.25 (6.59)
Direction of Change	+0.01 (2.14)	-1.01 (6.94)	-7.54 (5.64)	-3.38 (7.72)
Direction of Change (first 60 seconds)	-0.35 (3.67)	+2.37 (6.48)	-5.77 (6.53)	-1.12 (10.02)
Interface Totals	3.13 (2.13)	3.69 (1.98)	74.63 (9.02)	75.61 (6.45)
Amount of Change Experienced	1.72 (0.12)	4.16 (0.18)	6.78 (1.07)	6.91 (0.41)
Direction of Change	-0.02 (2.05)	-0.93 (6.75)	-6.45 (5.85)	-4.18 (6.60)

4.2 Short Stress State Questionnaire (SSQ)

Overall, participants in both groups reported similar levels of stress (structured: M=2.65, SD=0.43; standard: M=2.56; SD=0.45; ns). When examined at the component level, the structured group reported higher engagement (structured: M=3.64, SD=0.49; standard: M=3.18, SD=0.62) and lower distress (structured: M=1.69, SD=0.42; standard: M=2.03, SD=0.52), but these differences were not significant. The amount of worry reported by groups was similar (structured: M=2.47, SD=1.09; standard: M=2.37, SD=0.90; ns).

4.3 Workload

Table 3 shows participants' ratings of the NASA-TLX items and the mean weight of each item (i.e., the average number of times the item was picked during the paired-comparisons). Because of an error, weights were not recorded for two participants in the structured group. Those in the standard interface group reported greater mental demand, temporal demand and effort, lower performance (a higher number reflects lower performance) and slightly less frustration. These participants' overall unweighted scores were also higher. However, none of these differences were significant except for performance [$t(18) = 2.16, p < 0.05$]. When examining the weights, participants who used the standard interface indicated performance contributed most to their workload, while those in the structured group selected mental demand more often. When combining weights and ratings, participants in the standard group reported significantly greater workload than those in the structured group [$t(16) = 2.05, p < 0.05$].

Table 3. Mean (SD) ratings of the NASA-TLX items and weights. Overall ratings are computed as the sum of each participant's factor ratings, while overall weights are a combination of factor weights and ratings. * indicates $p < 0.05$

	Ratings		Weights	
	Standard	Structured	Standard	Structured
Mental Demand	4.20 (1.14)	3.60 (1.17)	2.63 (1.06)	3.80 (1.13)
Physical Demand	1.40 (0.52)	1.50 (0.53)	0.13 (0.35)	0.90 (0.99)
Temporal Demand	4.00 (1.25)	3.10 (1.80)	2.88 (1.13)	2.00 (1.41)
Performance (Success)	4.60* (1.27)	3.20* (1.62)	3.25 (0.89)	3.50 (0.85)
Effort	4.20 (1.40)	3.60 (1.35)	2.75 (1.91)	2.40 (1.27)
Frustration	3.00 (1.67)	3.90 (2.03)	2.38 (1.69)	1.50 (1.90)
Overall	21.45 (4.13)	18.90 (5.78)	4.03* (0.72)	3.20* (0.95)

4.4 Pre-task and Post-task Questionnaires

Data describing participants' responses to the pre- and post-task items are not included because of space limitations. We conducted two ANOVAs using temporal task order (and unique task) as a repeated measures variable and interface as a between subjects variable. There were no significant main effects for temporal task order or for interface. There was one significant main effect for unique task, with participants indicating greater prior knowledge of Task 435 (M=2.30; SD=1.03) than Task 347 (M=1.45; SD=0.61), $F(1, 20) = 7.81, p < 0.05$. There were no significant interaction effects.

With respect to participants' responses to the post-task items, no significant main effects were found for temporal task order or unique task. Participants' who used the structured interface rated their own skill at finding relevant documents significantly better than those using the standard interface [Structured: M=3.40, SD=0.99; Standard: M=2.60, SD=0.42; $F(1, 20) = 4.20, p < 0.05$]. These participants also rated the system's performance significantly better [Structured: M=3.65, SD=0.85; Standard: M=2.90, SD=0.52; $F(1, 20) = 5.68, p < 0.05$].

4.5 Search Behaviors and Performance

An ANOVA was conducted using interface as a between subjects variable and unique task (and temporal task order) as a repeated measures variable to examine the effects of interface and task on search behaviors and performance. No significant effects were found for unique task or temporal task order, so we only present descriptive data for interface (Table 4). Generally, participants who used the standard interface exerted more effort and saved fewer documents, but no main effects were found, except for time spent per query $[F(1,20) = 4.48, p < 0.05]$, where those in the structured group spent longer formulating their queries.

Table 4. Means (SDs) for search behaviors and performance according to interface

	Interface	
	Standard	Structured
#queries	4.45 (3.14)	3.40 (2.04)
#docsViewed	10.30 (8.72)	10.90 (6.37)
docsperq	3.71 (3.27)	3.81 (2.26)
depth	9.27 (9.13)	7.12 (3.86)
#SERPs	6.40 (4.60)	4.10 (2.02)
SERPdepth	2.40 (1.85)	2.00 (1.21)
hoverdepth	13.21 (9.91)	10.14 (5.13)
docssaved	5.00 (2.22)	6.00 (3.99)
docsrel	2.05 (1.28)	3.00 (1.99)
P@10	0.29 (0.18)	0.27 (0.13)
timetotal	527.90 (324.51)	552.85 (207.65)
timequery	64.50 (58.56)	111.65 (90.17)
timeperq	15.30 (10.81)	38.56 (37.64)
timedocs	246.55 (221.72)	239.85 (137.89)
timeperd	24.15 (16.37)	22.91 (7.85)
timeSERPs	134.00 (103.90)	130.05 (58.03)

5 Discussion and Conclusion

Overall, participants in both interface groups experienced similar levels of electrodermal activity (EDA), with the greatest amount experienced when completing the second search task. The structured group experienced more changes in EDA with the most occurring during the first 60 seconds of each task where there were increases. Given that this corresponded to the time when participants would have submitted their initial queries, it is likely that the novel query interface induced some stress. However, no significant differences were found. This potentially is a result of the large variances observed in the structured group. It is likely the case that the novel query interface generated greater increases in EDA for some participants, but not others. There were no significant differences in participants' heart rates (HR) according to interface or task. While the normalized HR values were similar, those in the standard group experienced greater changes in HR during the first 60 seconds of each task, especially for the first and second search tasks, but these differences were not significant.

Participants' responses to the SSSQ corroborated the findings of the physiological measures in that there were no significant differences between groups. This might be a result of the artificial tasks used in this study. Since these were not participants' genuine search tasks, participants might not have much emotional energy invested in the outcome; for example, in Poddar and Ruthven [23] there were differences in emotions reported by participants when they were doing their own tasks versus assigned tasks. Another explanation is the instrument might be too coarse to capture people's experiences during search since it was given at the conclusion of both search tasks and represents participants' cumulative assessments. Any negative feelings might have been forgotten or minimized since participants were able to complete both search tasks.

The workload and post-task questionnaire results provide evidence the standard interface imposed a greater workload on participants, and participants were less satisfied with their overall performance, with their skill at finding relevant documents, as well as the system's ability to find relevant documents. The weighted NASA-TLX scores showed a significant difference in workload with those in the structured group reporting less workload than those in the standard group. Although most behavioral measures were not significant, participants who used the standard interface submitted more queries, went to greater depths and hovered and viewed more SERPs, which likely contributed to their experiences of workload. Compared to the NASA-TLX scores for system load in [5], our participants reported less workload overall regardless of group; also consistent participants in the structured group reported less overall workload than those in the standard group. Our participants completed two experimental search tasks, while in [5] they completed three, so this might explain their higher scores.

There were no significant differences in the types of search behaviors engaged in by the two groups, except participants in the structured group spent significantly longer formulating their first queries. Based on Azzopardi et al.'s [5] findings, our hypothesis was that the interfaces would cause participants to engage in different search behaviors, which would in turn impact the amount of stress participants experienced while searching. Essentially the premise was rapid-fire querying and shallow result list examination are behaviors that are associated with more stress because of the accelerated pace. However, since participants who used the different query interfaces did not exhibit significantly different search behaviors, we were unable to fully explore this hypothesis.

We are cautious in interpreting our results, especially our inability to replicate the results of Azzopardi et al. [5] as our study was underpowered and the possibility of a Type II error exists. Although we originally planned for a sample of 34, we discovered problems with the equipment, which compromised the accuracy of the physiological data of the first 14 participants, so these participants were excluded. Because of budgetary constraints, we could not enroll additional participants. Aside from the power of the study, our inability to replicate the results of Azzopardi et al. might also be because we did not offer participants bonus money to motivate them to complete the tasks. It might also be the case that the sensors caused participants to behave differently. Participants in [5] were given a time limit of ten minutes per topic. While our participants were not given a time limit, the average time they spent completing tasks was about 10 minutes, so the sensors did not seem to impact time spent searching. We note while not significant, many of the behavioral measures were in the same direction as that reported in [5].

This was one of the first studies to systematically examine how different query interfaces impact people's affective experiences, especially with respect to stress. Most other studies in the field have manipulated some aspects of task (e.g., type or difficulty), focused on general emotions, or been correlational in nature. This was one of the first studies in IR to use BioPac to collect these data; most other studies have used less expensive equipment designed for consumer use (e.g., heart rate monitors for exercise) which provides less precise data. While our results with respect to the physiological data were not prodigious, we believe the contributions are still useful given the recent call for more work in this area [16]. To our knowledge, this was the first study in IR to use the SSSQ to measure stress. We hope our introduction of it will increase researchers' awareness of its existence and potential usefulness for measuring the user experience.

Our experiences collecting physiological data allow us to make several observations about use of this data in IR studies. First, collecting this data comes with its own challenges including applying the sensors correctly, determining an appropriate sampling rate, dealing with large datasets and interpreting the results. Second, it might be the case that the usefulness of this type of data varies based on the IR research situation being investigated. In studies from other fields that use physiological data, the more common research design is to study stimuli that can be introduced quickly and at multiple times during an activity; the isolated onset of the stimulus allows researchers to more precisely see how it impacts physiological signals at any given point in time. When studying interfaces, where the interface itself is the stimulus and the person will use it continuously for some period of time (e.g., 10 minutes), it is more difficult to associate changes in physiological signals to specific aspects of the interface since people likely habituate to the interface. It might be the case that physiological data are more useful for understanding IR interactions when the researcher can control the onset of the stimulus.

In conclusion, we found participants who used the structured query interface rated their success and skills significantly higher than those who used the standard interface, reported significantly less workload, but did not experience greater stress. At the very least, this result questions accepted wisdom that the standard query interface provides the best user experience and calls for more investigations of alternative query interfaces to fine tune the user experience.

Acknowledgements. Our thanks to Dr. Moshfeghi for his helpful comments.

References

1. Agapie, E., Golovchinsky, G., Qvardordt, P.: Encouraging behavior: A foray into persuasive computing. In: Proceedings of HCIR (2012)
2. Appelhans, B.M., Luecken, L.J.: Heart rate variability as an index of regulated etional responding. Review of General Psychology 10(3), 229 (2006)
3. Arapakis, I., Jose, J.M., Gray, P.D.: Affective feedback: an investigation into the role of emotions in the information seeking process. In: Proceedings of the 31st ACM SIGIR, pp. 395–402. ACM (2008)
4. Arapakis, I., Konstas, I., Jose, J.M.: Using facial expressions and peripheral physiological signals as implicit indicators of topical relevance. In: Proceedings of the 17th ACM International Conference on Multimedia, pp. 461–470. ACM (2009)

5. Azzopardi, L., Kelly, D., Brennan, K.: How query cost affects search behavior. In: Proc. of the 36th ACM SIGIR Conference, SIGIR 2013, pp. 23–32 (2013)
6. Belkin, N.J., Kelly, D., Kim, G., Kim, J.Y., Lee, H.J., Muresan, G., Tang, M.C., Yuan, X.J., Cool, C.: Query length in interactive information retrieval. In: Proc. of the 26th ACM SIGIR Conference, pp. 205–212 (2003)
7. Boucsein, W.: Electrodermal Activity. Springer (2012)
8. Feild, H.A., Allan, J., Jones, R.: Predicting searcher frustration. In: Proc. of the 33rd International ACM SIGIR Conference, pp. 34–41 (2010)
9. Franzen, K., Karlgren, J.: Verbosity and interface design. In: SICS Research Report. Swedish Institute of Computer Science (2000)
10. Gwizdka, J., Lopatovska, I.: The role of subjective factors in the information search process. JASIST 60, 2452–2464 (2009)
11. Hart, S.G., Staveland, L.E.: Development of nasa-tlx: Results of empirical and theoretical research. Advances in Psychology 52, 139–183 (1988)
12. Helton, W.S.: Validation of a short stress state questionnaire. Proc. of the Human Factors & Ergonomics Soc. Annual Meeting 48, 1238–1242 (2004)
13. Kelly, D., Dollu, V.D., Fu, X.: The loquacious user: a document-independent source of terms for query expansion. In: Proc. of the 28th ACM SIGIR, pp. 457–464 (2005)
14. Lazarus, R.S., Folkman, S.: Stress. Appraisal, and Coping, p. 456 (1984)
15. Lopatovska, I.: Emotional correlates of information retrieval behaviors. In: IEEE Workshop on Affective Computational Intelligence (WACI), pp. 1–7. IEEE (2011)
16. Lopatovska, I., Arapakis, I.: Theories, methods and current research on emotions in library and information science, information retrieval and human–computer interaction. Information Processing & Management 47, 575–592 (2011)
17. Matthews, G., Joyner, L., Gilliland, K., Campbell, S., Falconer, S., Huggins, J.: Validation of a comprehensive stress state questionnaire: Towards a state big three. Personality Psychology in Europe 7, 335–350 (1999)
18. Mooney, C., Scully, M., Jones, G.J., Smeaton, A.F.: Investigating biometric response for information retrieval applications. In: Lalmas, M., MacFarlane, A., Rüger, S.M., Tombros, A., Tsikrika, T., Yavlinsky, A. (eds.) ECIR 2006. LNCS, vol. 3936, pp. 570–574. Springer, Heidelberg (2006)
19. Moshfeghi, Y., Jose, J.M.: An effective implicit relevance feedback technique using affective, physiological and behavioural features. In: Proc. of the 36th ACM SIGIR Conference on Research and Development in IR, pp. 133–142 (2013)
20. Moshfeghi, Y., Jose, J.M.: On cognition, emotion, and interaction aspects of search tasks with different search intentions. In: WWW, pp. 931–942 (2013)
21. Nahl, D., Bilal, D.: Information and emotion: The emergent affective paradigm in information behavior research and theory. Information Today, Inc. (2007)
22. O'Brien, H.L., Lebow, M.: Mixed-methods approach to measuring user experience in online news interactions. Journal of the American Society for Information Science and Technology 64, 1543–1556 (2013)
23. Poddar, A., Ruthven, I.: The emotional impact of search tasks. In: Proc. of the 3rd Symposium on Information Interaction in Vontext, pp. 35–44 (2010)
24. Qvarfordt, P., Golovchinsky, G., Dunnigan, T., Agapie, E.: Looking ahead: query preview in exploratory search. In: Proc. of the 36th ACM SIGIR Conference, pp. 243–252 (2013)
25. Trimmel, M., Meixner-Pendleton, M., Haring, S.: Stress response caused by system response time when searching for information on the internet. Human Factors: The J. of the Human Factors and Ergonomics Society 45, 615–622 (2003)
26. Voorhees, E.M.: Overview of the trec 2003 robust retrieval track. In: TREC, pp. 69–77 (2003)

Detecting Spam URLs in Social Media via Behavioral Analysis

Cheng Cao and James Caverlee

Department of Computer Science and Engineering, Texas A&M University
College Station, Texas, USA
{chengcao,caverlee}@cse.tamu.edu

Abstract. This paper addresses the challenge of detecting spam URLs in social media, which is an important task for shielding users from links associated with phishing, malware, and other low-quality, suspicious content. Rather than rely on traditional blacklist-based filters or content analysis of the landing page for Web URLs, we examine the behavioral factors of both who is posting the URL and who is clicking on the URL. The core intuition is that these behavioral signals may be more difficult to manipulate than traditional signals. Concretely, we propose and evaluate fifteen click and posting-based features. Through extensive experimental evaluation, we find that this purely behavioral approach can achieve high precision (0.86), recall (0.86), and area-under-the-curve (0.92), suggesting the potential for robust behavior-based spam detection.

1 Introduction

URL sharing is a core attraction of existing social media systems like Twitter and Facebook. Recent studies find that around 25% of all status messages in these systems contain URLs [7,17], amounting to millions of URLs shared per day. With this opportunity comes challenges, however, from malicious users who seek to promote phishing, malware, and other low-quality content. Indeed, several recent efforts have identified the problem of spam URLs in social media [1,5,9,16], ultimately degrading the quality of information available in these systems.

Our goal in this paper is to investigate the potential of *behavioral analysis* for uncovering which URLs are spam and which are not. By behavioral signals, we are interested both in the aggregate behavior of *who is posting* these URLs in social systems and *who is clicking* on these URLs once they have been posted. These behavioral signals offer the potential of rich contextual evidence about each URL that goes beyond traditional spam detection methods that rely on blacklists, the content of the URL, its in-links, or other link-related metadata. Unfortunately, it has historically been difficult to investigate behavioral patterns of posts and clicks. First, many social systems provide restricted (or even no) access to posts, like Facebook. Second, even for those systems that do provide research access to a sample of its posts (like Twitter), it has been difficult to assess how these links are actually received by the users of the system via clicks.

A. Hanbury et al. (Eds.): ECIR 2015, LNCS 9022, pp. 703–714, 2015.

As a result, much insight into behavioral patterns of URL sharing has been limited to proprietary and non-repeatable studies.

Hence, in this paper, we begin a behavioral examination of spam URL detection through two distinct perspectives (see Figure 1): (i) the first is via a study of how these links are posted through publicly-accessible Twitter data; (ii) the second is via a study of how these links are received by measuring their click patterns through the publicly-accessible Bitly click API. Concretely, we propose and evaluate fifteen click and posting-based behavioral features, including: for postings – how often the link is posted, the frequency dispersion of when the link is posted (e.g., is it posted only on a single day in a burst? or is it diffusely posted over a long period?), and the social network of the posters themselves; and for clicks – we model the click dynamics of each URL (e.g., does it rapidly rise in popularity?) and consider several click-related statistics about each URL, including the total number of clicks accumulated and the average clicks per day that a URL was actually clicked. Through extensive experimental study over a dataset of 7 million Bitly-shortened URLs posted to Twitter, we find that these behavioral signals provide overlapping but fundamentally different perspectives on URLs. Through this purely behavioral approach for spam URL detection, we can achieve high precision (0.86), recall (0.86), and area-under-the-curve (0.92). Compared to many existing methods that focus on either the content of social media posts or the destination page – which may be easily manipulated by spammers to evade detection – this behavior-based approach suggests the potential of leveraging these newly-available behavioral cues for robust, on-going spam detection.

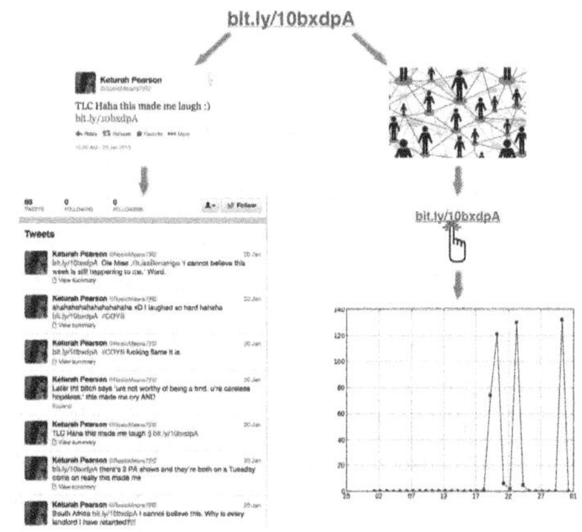

Fig. 1. Studying spam URL detection in social media from two perspectives: (i) Posting behavior (left); (ii) Click behavior (right)

2 Related Work

URLs (and in particular, shortened URLs) have been widely shared on social media systems in recent years. Antoniades et al. [1] conducted the first comprehensive analysis of short URLs in which they investigated usage-related properties such as life span. With the rising concern of short URLs as a way to conceal untrustworthy web destinations, there have been a series of studies focused on security concerns of these URLs, including: a study of phishing attacks through short URLs [5], geographical analysis of spam short URLs via usage logs [9], an examination of security and privacy risks introduced in shortening services [16], and a long-term observation of shortening services on security threats [14].

Separately, Twitter spam detection has been widely studied in recent years. In general, three types of approaches have been proposed: user profile based, content based, and network relation based. User profile based methods [10,21,19] build classifiers using features extracted from account profiles, e.g., profile longevity. Content-based features [8,19] focus on the posting text. Network-based features [4,18,27] are those extracted from the social graph such as clustering coefficient. Many detection systems of suspicious Web URLs have been developed. Some of these [11,12,13,15] directly use URL lexical features, URL redirecting patterns, and URL metadata such as IP and DNS information. Some [3,6] consider features extracted from the HTML content of the landing page. Additionally, several dynamic spam URL filtering systems have also been developed [20,24,26].

Several recent works have used clicks extracted from the Bitly API, typically to study the properties of known spam links. For example, Grier et al. [8] recovered clicking statistics of blacklisted Bitly links, with the aim of measuring the success of those spam links on Twitter. Maggi et al. [14] submitted malicious long URLs to the Bitly API in order to examine the performance in terms of spam pre-filtering. Chhabra et al. [5] shortened a set of known phishing long URLs and analyzed factors like the referrer and location. There recently has been some research on using proprietary server-side click log data to defend against some types of spam (e.g., [23,25]). In contrast, our aim is to investigate how large-scale publicly-available click-based information may be used as behavioral signals in the context of spam URL detection on social media.

3 Behavior-Based Spam URL Detection

In this section, we investigate a series of behavioral-based features for determining whether a URL shared in social media is spam or not. Hence, for both the posting-based and click-based perspectives, we are interested to explore questions like: What meaningful patterns can we extract from these publicly-available resources? Are posting or click-based features more helpful for spam URL detection? And which specific features are most informative?

3.1 Problem Statement and Setup

Given a URL v that has been shared on a social media platform, the *behavior-based spam URL detection problem* is to predict whether v is a spam URL through a classifier $c : v \rightarrow \{\text{spam, benign}\}$, based only on behavioral features. In this paper, we consider two types of behavioral features associated with each URL – a set of posting-related behavioral features F_p and a set of click-based behavioral features F_c. Such a behavior-based approach requires both a collection of URLs that have been shared, as well as the clicks associated with each URL. Since many social media platforms (like Facebook) place fairly stringent limits on crawling, we targeted Bitly-shortened URLs.

URL Postings. Concretely, we first used the Twitter public streaming API to sample tweets during January 2013. We collected only tweets containing at least one Bitly URL (that is, a URL that had been shortened using the Bitly link shortening service). In total, we collected 13.7 million tweets containing 7.29 million unique Bitly-shortened URLs. We observed the typical "long tail" distribution: only a few URLs have been posted upwards of 100,000 times, whereas most have been posted once or twice.

URL Clicks. We accessed the Bitly API to gather fine-grained click data about each of the 7.29 million URLs. For example, we can extract the number of clicks per time unit (e.g., minute, hour, day, month) and by country of origin. In total, we find that nearly all – 7.27 million out of 7.29 million – of the URLs have valid click information, and that 3.6 million (49.5%) of the URLs were clicked at least once during our study focus (January 2013). As in the case of postings, we find a "long tail" distribution in clicks.

3.2 Posting-Based Features

In the first perspective, we aim to study the URLs through the posting behaviors associated with them. For example, some URLs are posted by a single account and at a single time. Others may be posted frequently by a single account, or by many accounts. Similarly, URLs may be temporally bursty in their posting times are spread more evenly across time. Our goal in this section is to highlight several features that may describe each URL based on its posting behavior.

Posting Count. The first feature of posting behavior is the total number of times a URL has been posted on Twitter during our study window. Our intuition is that this count can provide an implicit signal of the topic of the link destination as well as the intent of the sharer: e.g., URLs that are posted only a few times may indicate more personal, or localized interest. We formulate this feature as *posting count*, denoted as $PostCount(u)$ given a short URL u.

Posting Standard Deviation. A Weather Channel URL and a CNN breaking news URL may have a similar *posting count* on Twitter. However, the Weather Channel URL may be posted every day of the month (linking to a routine daily forecast), whereas a breaking news URL may be posted in a burst of activity in a single day. To capture this posting concentration, we consider the standard deviation of the days in which a URL is posted. Concretely, for each URL u we have a list of days when u was posted. We refer to this list as u's *posting days*, denoted by $PostDays(u)$. We define the *posting standard deviation* of a URL u as the standard deviation of all elements in $PostDays(u)$, denoted as $std(u)$. For example, if a URL u was posted 10 times on January 22nd and not tweeted on any other day, we have $std(u) = 0$. On the contrary, a URL u shared once per day will have a much larger $std(u)$.

Posting Intensity. The posting standard deviation gives insight into how concentrated a URL has been posted, but it does not capture the total intensity of the posting. For example, two URLs both of which have only one single posting day will have the same posting standard deviation, even if one was posted thousands of times while the other appeared only once. To capture this difference, we introduce *posting intensity* to capture how intense the posting behaviors of a URL are. Given a URL u, we calculate u's "intensity score" via the following:

$$intensity(u) = \frac{PostCount(u)}{(std(u) * |set(PostDays(u))|) + 1}$$

where $|set(PostDays(u))|$ is the number of distinct posting days of u. For those URLs whose scores are the highest, they have high posting frequency, but also a low intensity of posting days. To illustrate, we find in our dataset that the URL with the highest intensity score was posted nearly 30,000 times on a single day.

Posting User Network. The sharer's personal network and reputation have certain connection with what and why she posts. A typical example is the comparison between celebrities and spammers. Spammers whose networks commonly are sparse tend to post spam links to advertise, whereas a celebrity may not share such low-quality links. Thus, for each URL, we consider features capturing the poster's personal network. We use the counts of followers and friends as simple proxies for user popularity, and take the *median* among all posters.

3.3 Click-Based Features

Now we turn our attention to how URLs are received in social media by considering the clicks that are associated with each URL in our dataset. We consider two kinds of clicking patterns: *clicking timeline* features that consider the temporal series of daily received clicks, and *clicking statistics* features that capture overall statistics of the clicks. For the first kind of clicking pattern, we have every short URL's fine-grained daily clicking data – which we can plot as its *clicking timeline*. We adopt three features extracted from this clicking timeline curve:

Rises + Falls. The first question we are interested is: how to capture the overall shape of a URL's clicks – do some go up continuously? Or do some periodically go up and down? To measure these changes, let n_i denote the number of clicks on the ith day. We define a *rise* if there exists an i such that $n_{i+1} - n_i > \alpha * n_i$ where α is a threshold and we set it to be 0.1, ensuring the change is non-trivial. Based on this criteria, we observe eight rises in Figure 2b (some are quite small). Similarly, let n_i denote the number of clicks on the ith day. We define a *fall* if there exists an i such that $n_i - n_{i-1} > \beta * n_{i-1}$ where β is a threshold value (set to 0.1 in our experiments). We observe eleven falls in Figure 2b.

Spikes + Troughs. In Figure 2b, we observe that while there are 8 rises, there are only 5 spikes of interest. So rather than capturing consecutive monotonic changes (as in the rises and falls), we additionally measure the degree of fluctuation of a URL through its *spikes* and *troughs*. That is, if there is an i such that $n_{i-1} < n_i > n_{i+1}$ we call it a *spike*. If there exists an i satisfying $n_{i-1} > n_i < n_{i+1}$, then it is a *trough*. Figure 2b has 5 spikes and 3 troughs.

Peak Difference. Naturally, there is a relationship between how and when a URL is posted and the clicks the URL receives. For example, Figure 2a illustrates a close relationship between posting and clicking for a URL. In contrast, Figure 2b demonstrates a much looser connection, indicating some external interest in the

URL beyond just its Twitter postings (in this case, the URL refers to a university website which attracts attention from many sources beyond Bitly-shortened links on Twitter). To capture the extent to which posting behaviors influence clicks, we define the *peak difference*. For each URL, we identify its *clicking peak* as the day it received the most clicks. Similarly, we identify its *posting peak* as the day it was posted the most. Note that a URL may have more than one posting peak and clicking peak. Here we define the *peak difference* as the minimum difference between two peaks among all pairs. The range of peak difference is from 0 to 30. In this way, peak difference can represent the level of tightness between clicking and posting.

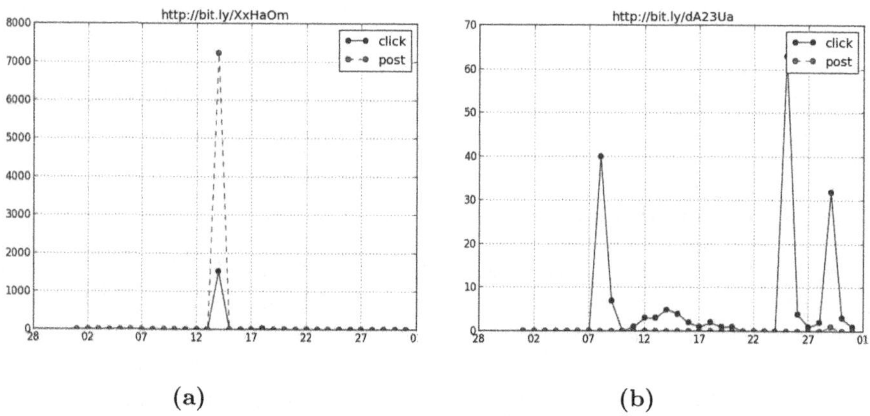

Fig. 2. The click and post timelines for two URLs. In (a), post and click behaviors are tightly coupled. In (b), the relationship is more relaxed.

We augment these timeline-based features with several click statistics:

Total Clicks. The first statistic is the *total clicks* a URL received in the period of study, which is a clear indicator of the popularity of a URL.

Average Clicks. Given a URL's total clicks and posting count, we can measure its *average clicks* per posting. By intuition more exposures bring more clicking traffic, but the average clicks is not necessarily large. Compared to total clicks, average clicks has a starker representation of popularity: many clicks via few postings suggest highly popular.

Clicking Days. We measure the number of *clicking days* in which a URL received clicks. This feature captures the consistency of attention on a URL.

Max Clicks. *Max clicks* is the maximum daily clicks. Unlike total clicks, this statistic can distinguish URLs that receive a burst of attention.

Effective Average Clicks. For those URLs with great total clicks, we observe some have a large number of clicking days while some have only one clicking day but thousands of clicks. Since average clicks considers only the relationship between total clicks and posting count, here we introduce *effective average clicks* defined as the following: Effective average clicks $= \frac{\text{total clicks}}{\text{clicking days}}$.

Click Standard Deviation. We already have features representing the fluctuation of timelines, now we consider a feature for the fluctuation of daily clicks given that we have specific sequence of daily clicks. We can calculate the standard deviation of daily clicks, defined as *click standard deviation*. Note that we fix a month as our time window of study. So, for each short URL we have a sequence of 31 daily clicks and we can compute the standard deviation.

Mean Median Ratio. Finally, given 31 daily clicks of a URL u, we can calculate its mean and median daily clicks, denoted as $mean(u)$ and $median(u)$ respectively. Now suppose we have a URL obtaining thousands of clicks on a day but very few on other days. It may have a considerable mean value but a low median. To build a connection between mean and median, we define *mean median ratio* of u as the following: Mean median ratio (u) $= \frac{mean(u)}{median(u)+1}$.

4 Experiments

In this section, we report a series of experiments designed to investigate the capacity of these two behavioral perspectives – posting-based and click-based – on the effectiveness of spam URL detection. Recall that our goal here is to examine the effectiveness of *behavioral signals alone* on spam detection. The core intuition is that these signals are more difficult to manipulate than signals such as the content of a social media post or the content of the underlying destination page. Of course, by integrating additional features such as those studied in previous works – e.g., lexical features of tweet texts, features of user profiles, and so forth – we could enhance the classification performance. Since these traditional features may be more easily degraded by spammers, it is important to examine the capability of a behavioral detector alone.

4.1 Experimental Setup

We consider two different sources of spam labels:

Spam Set 1: List Labeled. For the first set of spam labels, we use a community-maintained URL-category website *URLBlacklist* (`http://urlblacklist.com`) that provides a list of millions of domains and their corresponding high-level category (e.g., "News", "Sports"). Among these high-level categories are two that are clearly malicious: "Malware" and "Phishing", and so we assign all URLs in our dataset that belong to one of these two categories as *spam*. We assign all URLs that belong to the category "Whitelist" as *benign*. It is important to note that many URLs belong to potentially dangerous categories like "Adult", "Ads", "Porn", and "Hacking"; for this list-based method we make the conservative assumption that all of these URLs belong to the *unknown* class. For all remaining URLs, we assume they are *unknown*. This labeling approach results in 8,851 spam URLs, 223 benign, and 1,009,238 unknown. Of these URLs, we identify all with at least 100 total clicks, resulting in 1,049 spam, 21 benign, and 60,012 unknown. To balance the datasets, we randomly select 1,028 URLs from the unknowns (but avoid those above-mentioned dangerous categories), and consider them as *benign*, leaving us with 1,049 spam and 1,049 benign URLs.

Spam Set 2: Manually Labeled. We augment the first spam set with this second collection. We randomly pick and manually label 500 short URLs, each of which has been posted at least 30 times along with at least 5 original tweets (i.e., not a retweet, nor a reply tweet). We label a URL as "spam" if its landing page satisfies one of the following conditions: (1) The browser client (Google Chrome in our work) or Bitly warns visitors that the final page is potentially dangerous before redirecting; (2) The page is judged as a typical phishing site; (3) After several redirectings, the final page is judged to be a typical "spam page"; (4) Apparent Crowdturfing Web sites such as what were introduced in [22]. Finally, we end up with 124 manually-labeled malicious URLs: 79 spam ones, 30 irrelevant ads ones, and 15 pornographic ones. We also collect 214 benign URLs: 85 news ones, 70 blog ones, 49 video-audio ones, and 10 celebrity-related ones.

For each dataset, we construct the five posting-based features and the ten click-based features for all of the URLs. Then, we adopt the Random Forest classification algorithm (which has shown strong results in a number of spam detection tasks, e.g., [2,4,19]), using 10-fold cross-validation. The output of the classifier is a label for each URL, either *spam* or *benign*. We evaluate the quality of the classifier using several standard metrics, equally-weighted for both classes.

4.2 Experimental Results

Classification on the List-labeled Dataset. For the first dataset, we report the evaluation results in Table 1. We find that using all features – both posting-based and click-based – leads to a 0.74 precision, recall, and F-Measure, and a ROC area of 0.802. These results are quite compelling, in that with no access to the content of the tweet nor the underlying web destination, spam URLs may be identified with good success using only behavioral patterns.

Next, we ask whether posting-based features or click-based features provide more power in detecting spam URLs. We first exclude the five posting-based features and report the *Click-based only* result in the table. We see even in this case we find a nearly 0.65 precision, recall, and F-Measure. When we drop click-based features in favor of a *Posting-based only*, we see a similar result. These results show that individually the two feature sets have reasonable distinguishing power, but that in combination the two reveal complementary views of URLs leading to even better classification success. We additionally consider the very restricted case of *clicking statistics only* (recall that our click-based features include both clicking statistics and clicking timeline features). Using only the seven click statistics, we observe only a slight degradation in quality relative to all click-based features.

To provide more insights into the impact of each feature, we use the Chi-square filter to evaluate the importance of features to the classification result. The top 6 features are shown in Table 2. Median friends and average clicks are the most two important features. Generally speaking, click-based features tend to play more important roles than posting-based features. Recall that our list-labeled dataset are those URLs with abundant clicks received, but it is not guaranteed that they have adequate posting counts, which may explain the ranking. For

Table 1. Evaluation results for the list-based dataset

Set of features	Precision	Recall	F-Measure	ROC area
All 15 features	0.742	0.737	0.736	0.802
Click-based only	0.647	0.647	0.647	0.705
Posting-based only	0.648	0.695	0.694	0.756
Clicking statistics only	0.622	0.622	0.622	0.679

instance, if most URLs, either malicious or benign, have only one or two posting days and posting counts is less than 5, their posting counts and posting standard deviations will tend to be similar.

Table 2. Top-6 features for list-labeled dataset (Chi-square)

Rank	Features	Score	Category
1	Median friends	277.43	Posting
2	Average clicks	199.11	Clicking
3	Median followers	159.53	Posting
4	Effective average clicks	150.72	Clicking
5	Click standard deviation	141.62	Clicking
6	Mean median ratio	141.49	Clicking

Classification on the Manually-labeled Dataset. We repeat our experimental setup over the second dataset and report the results here in Table 3. When we use the complete 15 features, the precision, recall, and F-Measure are all even higher than in the list-labeled dataset case, around 0.86, with a ROC area of around 0.92. These results are encouraging. We attribute the increase in performance relative to the first dataset to the more expansive labeling procedure for the second dataset. In the list-labeled dataset, we only considered extremely "bad" URLs since we considered only the "Malware" and "Phishing" categories. This conservative assumption may lead to many spam-like URLs lurking in the set of benign URLs. In contrast, the manually-labeled dataset considers more broadly the context of what makes a spam URL.

Continuing our experiments, we again consider subsets of features in the classification experiment. Again, we find that using only a single feature type – either *Click patterns only* or *Posting patterns only* – leads to fairly strong classification performance. But that in combination, the two provide complementary views on URLs that can be used for more successful spam URL detection

Again, we use Chi-square filter to rank features, as shown in Table 4. Interestingly, the ranking is quite different from what we found in Table 2, though again we observe a mix of both posting and click-based features. We attribute some of this difference to the click data's availableness in the manually-labeled dataset; most of the URLs in the manually-labeled dataset have abundant posting information and we can see that the posting behavior features play important roles in classification. On the contrary, most of the URLs in the manually-labeled

Table 3. Evaluation results for the manually-labeled dataset

Set of features	Precision	Recall	F-Measure	ROC area
All 15 features	0.860	0.861	0.859	0.921
Click-based only	0.828	0.828	0.828	0.888
Posting-based only	0.839	0.84	0.837	0.904
Clicking statistics only	0.842	0.843	0.841	0.875

dataset do not have very large clicking traffic to support clicking-based features. However, these two results – on the two disparate ground truth datasets – demonstrate the viability of integrating click-based features into spam URL detection in social media, and the importance of integrating complementary perspectives (both posting-based and click-based) into such tasks.

Table 4. Top-6 features for manually-labeled dataset (Chi-square)

Rank	Features	Score	Category
1	Average clicks	149.41	Clicking
2	Posting count	144.23	Posting
3	Median followers	123.24	Posting
4	Median friends	118.19	Posting
5	Score function	87.00	Posting
6	Posting standard deviation	63.66	Posting

To further illustrate the significance of click and posting-based features, we consider two of the top-ranked features in both datasets (recall Table 2 and Table 4): median friends and average clicks. We compare the distributions of these two strongly correlated features for all spam URLs and benign URLs, in Figure 3. For URLs in the list-based dataset, as in Figure 3a, around 20% spam URLs are posted by users with a median friends count of 0, and yet around 20% have a median friends count that exceeds 1,000. These two types of posters could correspond to newly-registered accounts (0 friend) and "high-quality" accounts

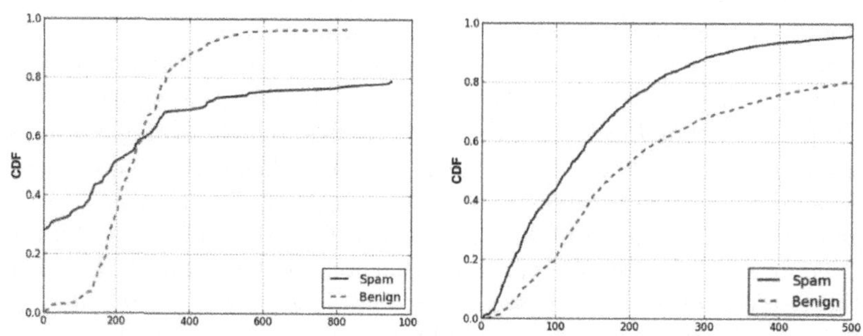

(a) Median Friends (b) Average clicks

Fig. 3. Example feature comparison for spam and benign URLs like those in a for-pay campaign. In contrast, legitimate accounts who posted benign URLs have relatively "normal" distribution of median friends, that is, most have median friends less than 300 and almost none has a zero median. For URLs in manually-labeled dataset, as in Figure 3b, we see that spam URLs tend to have a lower average clicks. A potential reason is that malicious URLs require more exposure or other "abnormal means" to support consistent clicks, while legitimate URLs can survive longer due to their appealing contents. We find similar distributions for other click-based statistics, including the click standard deviation and the effective average clicks.

5 Conclusions

In summary, this paper investigated the potential of behavioral analysis aiding in uncovering spam URLs in social media. Purely by behavioral signals, we have considered two perspectives – (i) how links are posted through publicly-accessible Twitter data; and (ii) how links are received by measuring their click patterns through the publicly-accessible Bitly click API. The core intuition is that these signals are more difficult to manipulate than signals such as the content of a social media post or the content of the underlying destination page. Through an extensive experimental study over a dataset of 7 million Bitly-shortened URLs posted to Twitter, we find accuracy of up to 86% purely based on these behavioral signals. These results demonstrate the viability of integrating these publicly-available behavioral cues into URL spam detection in social media.

Acknowledgment. This work was supported in part by AFOSR Grant FA9550-12-1-0363.

References

1. Antoniades, D., et al.: we.b: the web of short urls. In: WWW (2011)
2. Benevenuto, F., Magno, G., Rodrigues, T., Almeida, V.: Detecting spammers on twitter. In: CEAS (2010)
3. Canali, D., Cova, M., Vigna, G., Kruegel, C.: Prophiler: a fast filter for the large-scale detection of malicious web pages. In: WWW (2011)
4. Castillo, C., Donato, D., Gionis, A., Murdock, V., Silvestri, F.: Know your neighbors: web spam detection using the web topology. In: SIGIR (2007)
5. Chhabra, S., Aggarwal, A., Benevenuto, F., Kumaraguru, P.: Phi.sh/$ocial: the phishing landscape through short urls. In: CEAS (2011)
6. Cova, M., Kruegel, C., Vigna, G.: Detection and analysis of drive-by-download attacks and malicious javascript code. In: WWW (2010)
7. Cui, A., Zhang, M., Liu, Y., Ma, S.: Are the urls really popular in microblog messages? In: CCIS (2011)
8. Grier, C., Thomas, K., Paxson, V., Zhang, M.: @spam: the underground on 140 characters or less. In: CCS (2010)

9. Klien, F., Strohmaier, M.: Short links under attack: geographical analysis of spam in a url shortener network. In: HT (2012)
10. Lee, K., Caverlee, J., Webb, S.: Uncovering social spammers: social honeypots + machine learning. In: SIGIR (2010)
11. Lee, S., Kim, J.: WarningBird: Detecting suspicious URLs in Twitter stream. In: NDSS (2012)
12. Ma, J., Saul, L.K., Savage, S., Voelker, G.M.: Beyond blacklists: learning to detect malicious web sites from suspicious urls. In: KDD (2009)
13. Ma, J., Saul, L.K., Savage, S., Voelker, G.M.: Identifying suspicious urls: an application of large-scale online learning. In: ICML (2009)
14. Maggi, F., et al.: Two years of short urls internet measurement: security threats and countermeasures. In: WWW (2013)
15. McGrath, D.K., Gupta, M.: Behind phishing: an examination of phisher modi operandi. In: LEET (2008)
16. Neumann, A., Barnickel, J., Meyer, U.: Security and privacy implications of url shortening services. In: W2SP (2010)
17. Rodrigues, T., Benevenuto, F., Cha, M., Gummadi, K., Almeida, V.: On word-of-mouth based discovery of the web. In: SIGCOMM (2011)
18. Song, J., Lee, S., Kim, J.: Spam filtering in twitter using sender-receiver relationship. In: Sommer, R., Balzarotti, D., Maier, G. (eds.) RAID 2011. LNCS, vol. 6961, pp. 301–317. Springer, Heidelberg (2011)
19. Stringhini, G., Kruegel, C., Vigna, G.: Detecting spammers on social networks. In: ACSAC (2010)
20. Thomas, K., Grier, C., Ma, J., Paxson, V., Song, D.: Design and evaluation of a real-time url spam filtering service. In: SP (2011)
21. Thomas, K., Grier, C., Song, D., Paxson, V.: Suspended accounts in retrospect: an analysis of twitter spam. In: IMC (2011)
22. Wang, G., et al.: Serf and turf: crowdturfing for fun and profit. In: WWW (2012)
23. Wang, G., et al.: You are how you click: Clickstream analysis for sybil detection. In: USENIX (2013)
24. Wang, Y., et al.: Automated web patrol with strider honeymonkeys: Finding web sites that exploit browser vulnerabilities. In: NDSS (2006)
25. Wei, C., et al.: Fighting against web spam: A novel propagation method based on click-through data. In: SIGIR (2012)
26. Whittaker, C., Ryner, B., Nazif, M.: Large-Scale automatic classification of phishing pages. In: NDSS (2010)
27. Yang, C., Harkreader, R.C., Gu, G.: Die free or live hard? Empirical evaluation and new design for fighting evolving twitter spammers. In: Sommer, R., Balzarotti, D., Maier, G. (eds.) RAID 2011. LNCS, vol. 6961, pp. 318–337. Springer, Heidelberg (2011)

Predicting Re-finding Activity and Difficulty

Sargol Sadeghi[1], Roi Blanco[2], Peter Mika[2],
Mark Sanderson[1], Falk Scholer[1], and David Vallet[3]

[1] RMIT University, Melbourne, Australia
[2] Yahoo! Research , Barcelona, Spain
[3] Google, Sydney, Australia
{seyedeh.sadeghi,mark.sanderson,falk.scholer}@rmit.edu.com,
{roi,pmika}@yahoo-inc.com,
dvallet@google.com

Abstract. In this study, we address the problem of identifying if users are attempting to re-find information and estimating the level of difficulty of the re-finding task. We propose to consider the task information (e.g. multiple queries and click information) rather than only queries. Our resultant prediction models are shown to be significantly more accurate (by 2%) than the current state of the art. While past research assumes that previous search history of the user is available to the prediction model, we examine if re-finding detection is possible without access to this information. Our evaluation indicates that such detection is possible, but more challenging. We further describe the first predictive model in detecting re-finding difficulty, showing it to be significantly better than existing approaches for detecting general search difficulty.

Keywords: Re-finding Identification, Difficulty Detection, Behavioral Features.

1 Introduction

Re-finding is a task where people seek information they have previously encountered. Examining a year of web search logs, Teevan et al. [18] determined that 40% of queries are attempts to *re-find*. While many such tasks are simple, such as searching for a home page, there are re-finding tasks that are more difficult, such as when only the broad sense of what was previously encountered can be recalled [17]. Current search engines are not optimised for re-finding [4,16]. Being able to detect and estimate how *difficult* a re-finding task is proving to be, would enable a search engine to employ services to help the user, such as biasing results towards a searcher's history, or customizing snippets to include texts and images that might be more memorable.

Research on re-finding difficulty has focused on users coping with changes to web sites and search results [2,16]. Difficulties have been studied for specific application areas, such as email search [3,4]. Beyond re-finding, identifying user difficulties has been explored for different task types. For example, Liu et al. [12,13] have shown that it is useful for IR systems to predict when a user is struggling, where systems could consequently adapt search results.

Current re-finding prediction is limited to the level of queries [18]. Because a re-finding user will likely engage in multiple searches, prediction of re-finding beyond a

A. Hanbury et al. (Eds.): ECIR 2015, LNCS 9022, pp. 715–727, 2015.

single query is crucial. Past research has also emphasized the importance of *tasks* either in identifying re-finding behavior [2], or generally detecting difficulties [12]. Although task-level re-finding has been examined [2,19], the work is limited in using behavioural features to predict re-finding tasks. As users can easily encounter information items through browsing, or receiving information via a social network, it is also important to examine how the identification of re-finding can be performed independent of the search history of the user using behavioural features.

Two research questions are explored: (1) Re-finding identification: How can re-finding tasks be differentiated from general web search tasks? (2) Re-finding difficulty: What features characterize user difficulties in completing a re-finding task?

We first describe past work, followed by a description of the experimental methodology. Next, we explain the features used in the predictive model. We then detail the setup of the prediction models, along with results from a range of experiments exploring different types of re-finding and feature sets.

2 Related Work

Re-finding Identification. In one of the first studies on web-based re-finding, Teevan et al. [18] used query log features to predict if the same result would be clicked on by a user given that they had re-submitted a previously entered query. Tyler and Teevan [19] studied re-finding at the level of sessions, finding that queries change more across sessions than within. Later, Tyler et al. [20] examined query features and the rank of the clicks to identify re-finding. Capra [2], studying 18 search tasks of users, found it difficult to distinguish between generic web search engine use and re-finding. From a diary study by Elsweiler and Ruthven [5], re-finding tasks were classified using the granularity of the information to be re-found (lookup, one-item, and multi-item).

Many search features were studied in the related area of predicting task continuation and cross-session tasks [11,21]. In a study by Kotov et al. [11], session-based features (e.g. "number of queries since the beginning of the session"), history-based features (e.g. "whether the same query appeared in the user's search history"), and pair-wise features (e.g. "number of overlapping terms between two queries") were examined.

Overall, current studied behavioural features for the re-finding context are limited and dependent on the search history of the user. However, for identifying particularly difficult re-finding tasks, it is required to examine a broader range of features.

Re-finding Difficulty. Capra [2] explored features to detect user difficulty including the number of search URLs, task completion time, and the elapsed time between search tasks. The best features included task frequency, topic familiarity, and determining that target information had been moved from the page where it was originally found. Teevan highlighted [16] information being moved, as well as changes in target document rank position, as causes of re-finding difficulty. She found that changes in the path to reach target information was a stronger indicator of user difficulty than temporal features. Elsweiler and Ruthven [5] studied the granularity of information and found no significant influence of granularity on difficulty. However, they reported that longer time gaps could indicate that users were having difficulties for some re-finding.

In general web search, large-scale query log features have been used to predict search difficulty [12,13], as well as user frustration, dissatisfaction, or success/failure [1,7,8,9]. Features ranged from temporal to user behavioral, and search result ranks. Examples of studied features include time interval between queries, number of clicks with high dwell time, and mean reciprocal ranks of clicks for each query.

Overall, current examined features for detecting difficulties in re-finding are mainly limited to user's self assessed features (e.g. topic familiarity) or target information (e.g. moved web page), and the construction of predictive models using behavioural features has not been considered.

3 Experimental Methodology

Our prediction model is based on the analysis of query logs. In this section, we describe the explored data sets and the methodology for evaluation.

3.1 Dataset

Our data consists of a sample of logs taken from 30 days of interactions with the Yahoo search engine gathered from the $1^{st} - 30^{th}$ of October 2012. The interactions of 2,847,028 unique anonymised users were logged including submitted queries, the URL, the rank position of clicked search results, and a timestamp for each event. The terms of service and privacy policies of Yahoo were followed.

To identify task boundaries, the logs were segmented into *goals*, which is defined as a group of related queries and corresponding clicks submitted by a user to perform a task with an atomic search need. Goal segmentation was performed using the technique described by Jones and Klinkner [10], where classifiers are used to predict goal boundaries based on features indicative of relatedness between queries (e.g. number of words in common) with an accuracy of 92. Note that other log segmentation approaches are either less accurate (e.g. sessions), or consist of more than one information need (e.g. missions) [10], and therefore we considered the goal segmentation. All goals from the same user were extracted and ordered by their timestamp, and all possible goals were *paired*. As we were not interested in short-term re-finding, paired goals that occurred less than thirty minutes apart were not considered. In total, 39,683,301 paired search goals were extracted.

3.2 Potential Re-finding Goals

Teevan et al. [18] classified pairs of queries and clicks into different types of re-finding. They examined whether the paired queries were equal or not, and explored result click overlap. We extend the approach to the level of pairs of goals across multiple queries and clicks.

We measure queries and clicks equivalence using a 5-point scale, resulting in a total of 25 combined classes. For queries, this includes sharing a term, term stem, or term corrections (simple edits for the purpose of spelling correction). For clicks, equivalence levels include overlapping URLs as well as at what point in the goal the overlapping clicks

Query Overlap	URL Overlap
Query	Last URL + URL
Query Term	Last URL + URL Root
Term Correction	Last URL
Term Stem	URL
No Query Overlap	URL Root

Original Goal
Q: bleacher report college football T: 2
C(3): `www.cbssports.com/collegefootball` T: 15
C(10): `bleacherreport.com/college-football`
Re-finding Goal
Q: college fottball T: 2 *Query term overlap*
Q: college fottball T: 9 *Query term correction*
C(1): `espn.go.com/college-football/` T: 16
C(39): `www.cbssports.com/collegefootball` T: 20 *URL overlap*
C(43): `bleacherreport.com/college-football` *Last URL overlap*
Classification: Query term overlap, Last URL + URL overlap

Fig. 1. Left: Definitions of query and click overlaps used across paired goals. Right: An example paired goal from the logs, with its classification.

occurred. For example, common clicks that occurred at the end of a goal (*last URL*) are distinguished. We also considered whether two URLs matched fully or only partially (based on the server name or *URL root*). As an example the overlap between these two URLs is considered as the URL root overlap: `en.wikipedia.org/wiki/Doc_Martin` and `en.wikipedia.org/wiki/Dr._Martin`. The query and click levels with some examples are illustrated in Figure 1. If a paired goal could belong to more than one class, the most restrictive class was selected. Paired goals where there was no URL overlapping were eliminated, as some minimum level of click commonality was required [18,19]. From the overlapping classes, 4,968,243 paired goals were extracted for our dataset. Note that the proposed classes are means to identify potential re-finding cases through the overlapping between parts of a paired goal; however, this does not mean that each overlapping is certainly a re-finding case. For example, users might repeat the same query but with a different search need, or clicks might have overlapping in their root URL, while referring to two different documents.

On the other hand, we note that there are other potential types of re-finding as shown in Figure 2. The paired goals might not always have overlapping in clicked URLs, such as cases where the URL has changed by the time that re-finding is attempted, but the corresponding web document is the same; or where the user failed to reach the same target document, thus having the same task but not resulting on overlapping URLs. We refer to this type of re-finding as *paired* but with *no URL overlapped*. Moreover, we made an assumption that there is a corresponding original search for each identified re-finding task (*paired* goals); whereas in some cases re-finding could occur in an *isolated* form. An example of the isolated re-finding is when the searcher cannot be identified (e.g. no login information, accessing from a different location), or the information being re-found may originally have been found by means other than searching (e.g. browsing, or social links). While, these cases might be more likely to include difficult re-finding, the identification of such cases is challenging from a query log study and is left for future work. However, we focused on those *URL overlapped paired* goals that are *non-navigational* and more likely include difficult cases.

Teevan et al. noted that much re-finding, such as navigational searches, are easy to detect. The navigational searches were identified based on equal query and single identical clicks. As the focus of our work was detecting more challenging forms of re-finding, we created a set of filters to remove easy cases. Paired goals where the queries contained

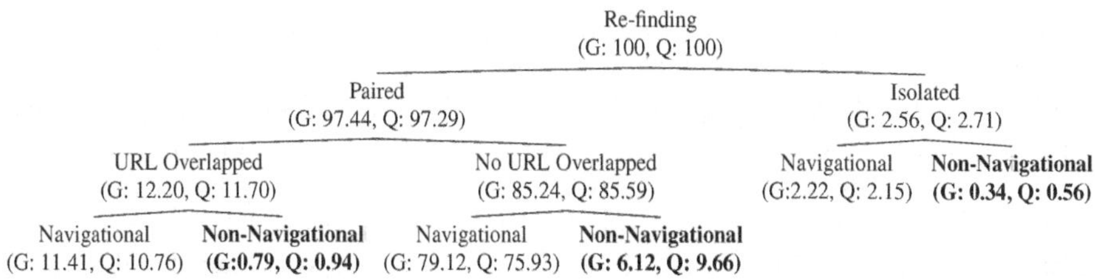

Fig. 2. The landscape of re-finding tasks. G: The percentage of goals, Q: The percentage of queries.

only popular domain names[1] or terms such as "www" and ".com" were removed. If the domain name of the clicked URL matched the corresponding submitted query, or was a spell-corrected version of the query, the paired goals were also removed. Only paired goals where each element of the pair contained more than one query or one click were considered. Filter accuracy was checked by manual investigation of a sample of paired goals. Our analysis of removed pairs showed that at worst only 1.6% were incorrectly removed.

After removing the easy paired goals, 322,639 pairs remained. The large reduction in size of data does not necessarily reflect that the re-finding problem we study is small; rather, applying our filtering rules is a way of giving us a dataset where we are confident we will find a concentration of challenging re-finding problems. Once a re-finding classification is constructed from this data set, other examples of re-finding can be explored in the full query logs. The summed percentages of non-navigational goals in Figure 2 is 7.25% (i.e. 0.79% + 6.12%+ 0.34%). Detecting and eventually helping users with their re-finding goals from this notable fraction of the query log has the potential to provide help on the most difficult re-finding tasks.

3.3 Ground Truth Dataset

We manually label re-finding activity and re-finding difficulty. To the best of our knowledge, this study is the first to include labeling of re-finding difficulty in web search logs.

We designed a labeling interface where paired goals were presented showing queries, clicked URLs plus their rank, the time gap between queries and clicks, as well as the gap between the paired search goals. Each assessor was asked to answer two questions: 1) *"Do you think that in the second search the user is re-finding document(s) that were found in the first search?"* (Responses: "yes", "no", and "not sure")? [2] 2) *"In terms of search difficulty, would you say the second search is..."* (Responses: "easy", "difficult", "not sure")? The notion of "difficulty" was defined for assessors in a broad sense of whether it seems that the user is struggling to find the target document. Specifically assessors were instructed to consider the effort of the user in a) reformulating queries,

[1] Identified through top 50 ranked websites from `Alexa.com` (e.g. "youtube").

[2] Note, we ask about re-finding the same *document(s)* not the same *information*, as there could be cases where it might not be possible to infer whether the user was searching for the same information (due to the dynamic content of web documents, for example news pages).

b) clicking relevant documents, and c) recognizing the target document. Examples of easy and difficult cases were shown to assessors.

All 25 combined classes of query overlap/URL overlap levels in Figure 1 were uniformly sampled (75 pairs from each class). However, eight were low in frequency (fewer than 25), and so were not considered in our sampling. In total, 1,275 paired goals were labeled by an *experienced* assessor, who had conducted the same labeling exercise on a separate dataset. The fraction of "not sure" labels was 8%, which reduced the size of our data to 1,167.

Examining ground-truth data reliability, we randomly sampled sixty instances and asked three other experienced assessors to assign labels once more. Mean pairwise Cohen's kappa (κ) for inter-assessor agreement was 0.89 and 0.47 for identifying re-finding and difficulty assignments respectively. By comparison, κ agreement scores in the range of 0.23-0.71 were achieved for relevance judgments for the TREC Legal Track [22]. This gave us confidence that the ground-truth data is sufficiently consistent.

We noticed a low frequency of goals labeled "difficult" making the data set imbalanced, which could be due to the limitation in the identification of re-finding based on query/click overlapping as discussed in Section 3.2; whereas in more difficult cases a fewer number of overlapping could occur, as the user might not be able to repeat queries and clicks from the original search. Consequently, we employed a form of active learning to increase the frequency of difficult instances in our training set. A classification model was learnt on our original labeled data and applied to unlabeled goals taken from the unlabelled data. The goals were ranked based on the estimated probability of them belonging to the "difficult" class. The top fifty, along with ten random instances from the rest of the predictions, were manually labeled and added to our data set. The procedure was repeated for ten iterations; at this point, a balanced number of "difficult" labels (48.3% of the identified paired goals) were obtained, and the procedure was stopped. After removing the "not sure" labels, the size of our final training set was 1,706 (with 74.4% re-finding case). This data was used for training and evaluating our classification models.

4 Features

This section explains the set of features that were used to construct predictive models for the identification and difficulty classification of re-finding.

4.1 Feature Categories

Features in three main groups are considered: (1) baseline query-level features from past research [18] ; (2) features from general web search related studies on detecting search difficulty and failure ; and (3) new features extended in our study for the re-finding context. All features considered are listed in Table 1. Most features are numerical, except for some Boolean features such as *"ended with query"*, *"exist advanced query syntax"*, *"all common clicks skipped"*, *"exist jumped common clicks"*, *"exist non-sequential clicks"*, and *"exist common clicks in different ranks between original and re-finding"* goals.[3]

[3] A detailed description of features: `http://tinyurl.com/feature-description`

Some features, indicated by '*' in the table, can be measured across the paired goals, in addition to being measured on each goal independently. For example, for the feature *"goal length in no. of queries"*, the pair-wise version of this feature would measure the relative difference of the goal length between the original and re-finding paired goal. For starred numerical features, we measure the difference between the paired goals; for Boolean features, we apply logical 'and' between the corresponding values of each goal. Given the defined notions in Table 1, the total number of features that could be calculated for a paired goal is 124.

We further separate the features into two broader groups: those requiring access to the original goal (*history-dependent*) and those that do not (*history-independent*, i.e. current goal only). This could be particularly useful for identifying *no URL overlapped* and *isolated* re-finding tasks illustrated in Figure 2.

4.2 Feature Discussion

The two features *"all common clicks skipped"* and *"exist jumped common clicks"* were inspired by a related study [15], which re-ranks repeated search results based on the behavior of users in clicking, skipping, or missing results. As our log data did not contain viewed results, we implemented a similar idea for clicked results in relation to their ranks. The first feature indicates whether there is a click at a lower rank, followed by the common clicks at higher ranks. The second feature indicates whether there is a common click, followed by a click at a higher rank. These assumptions are based on the fact that the user is likely to browse the result page from top to bottom. The feature *"exist common click in different ranks within pairs"* was inspired by Teevan's study [16], where changes in the rank of the clicks make re-finding difficult. Moreover, we added a condition that common clicks in following result pages could increase the difficulty of the re-finding task (*"no. of non-first-page ranked clicks"*). Some features considered the position of common clicks. For example, *"common click in relation to the last click"* examines whether a common click occurred in the last click of the original and re-finding paired goal. In terms of the importance of engaged clicks, we developed the feature, *"missed engaged later clicks in original"*, which is true if, after a common click, there are engaged clicks in the original goal that have not been clicked in the re-finding goal.

A dwell time of greater than 30 seconds has been highlighted as an indication of *engaged* and relevant clicks [9]. We added "relative dwell time", which is computed in terms of the fraction of click dwell time to the total time-span of the goal. Dwell time after clicks might not be entirely reflective of search time, as the user might spend time on acquiring knowledge, or inspecting a document. Therefore, we define *"effective search time"*: the total dwell time after queries and those clicks that have low dwell time (less than 30 seconds).

The feature *"query overlap/URL overlap"* is defined in terms of the classification between query and click commonalities of paired goals (see Figure 1). More commonality could increase the chance of re-finding. On the other hand, differences could be indicative of greater difficulties. As an example, *"first query transformation type within pairs"* measures the differences between the initial queries of the original and re-finding

Table 1. Features used to detect re-finding and difficulties. Each feature could be related to either original goal: †, or re-finding goal: ‡, or a relative difference between both goals: ∗. Features signed by † and ∗ are *history-dependent*; whereas, ‡ features are *history-independent*.

Baseline query level features (from past re-finding work)
equal query class ∗
equal query elapsed time ∗
equal query length ∗
equal query no. of original clicks †
equal query no. of common clicks ∗
equal query no. of original uncommon clicks †
General web search (related) difficulty features
goal length in no. of both queries and clicks ‡
goal length in no. of unique/all queries ‡
goal length in no. of unique/all clicks ‡
mean no. of clicks across all queries ‡
time to the first click ‡
min/max/mean time to the first click of all queries ‡
min/max/mean inter-query time ‡
min/max/mean inter-click time ‡
no. of engaged clicks (dwell time >30 seconds) ‡
no. of clicks on next page ‡
ended with query ‡
exist advanced query syntax (e.g. quotes) ‡
queries per second ‡
clicks per query ‡
fraction of queries for which no click ‡
time span of goal ‡
Extended re-finding features
query overlap/URL overlap ∗
no. of common/uncommon/all clicks † ‡
mean query length of common/all clicks † ‡
mean no. of query common/all clicks † ‡
mean no. of uncommon clicks of all queries † ‡
mean no. of uncommon clicks of common click queries † ‡
days between paired goals ∗
effective search time † ‡ ∗
total dwell time after all queries † ‡
total dwell time after all clicks † ‡
total time to reach to the first common click † ‡
rank of the first reached common click † ‡
mean reciprocal rank of common clicks † ‡
rank of the last click † ‡
no. of non-first-page ranked clicks in common/all clicks † ‡
all common clicks skipped † ‡
exist jumped common clicks † ‡
exist non-sequential clicks † ‡
mean dwell time/relative dwell time of common clicks † ‡
no. of repetitions of common clicks † ‡
fraction of queries with no common clicks † ‡
re-finding is longer than original in length ∗
re-finding is longer than original in no. of queries ∗
re-finding is longer than original in no. of clicks ∗
re-finding missed engaged later clicks in original ∗
first query transformation type within pairs ∗
exist common click in different ranks within pairs ∗
common click in relation to the last click ∗
mean relative goal position of common clicks † ‡
min/max goal position of common clicks † ‡
mean relative common clicks goal position (early, middle, late) † ‡
goal length in no. of both queries and clicks † ∗
goal length in no. of unique/all queries † ∗
goal length in no. of unique/all clicks † ∗
mean no. of clicks across all queries †
time to the first click †
min/max/mean time to the first click of all queries †
min/max/mean inter-query time †
min/max/mean inter-click time †
no. of engaged clicks (dwell time >30 seconds) †
no. of clicks on next page †
ended with query † ∗
exist advanced query syntax (e.g. quotes) † ∗
queries per second † ∗
clicks per query † ∗
fraction of queries for which no click † ∗
time span of goal †

goals (based on query reformulation types: "exactly the same", "error correction", "specialization", "generalization", and non-trivial transitions considered as "other").

5 Prediction Models

We used Support Vector Machines as our classification model, trained with a Sequential Minimal Optimization (SMO) algorithm, as this has been shown to work well in similar classification scenarios [18]. We trained a binary classifier to classify a goal as re-finding or not; and the second one to predict re-finding difficulty (easy or difficult).

We employed a ten times ten-fold cross-validation approach, which repeats ten-fold cross-validation and measures the average of classification results [14]. We report precision, recall, and F-measure scores. A paired two-tailed t-test was used to test for statistically significant differences in effectiveness.

Table 2 reports the accuracy when using different groups of features (see Table 1). Considering the columns *all features* in Table 2 (using all features in Table 1), our SMO classifier achieves an F-measure of 91.6 on the identification problem (left table), and 82.7 on the difficulty prediction problem (right table). We replicated a model proposed

Table 2. Re-finding classification performance of feature sets measured using P: Precision, R: Recall, and F: F-measure

	All features	History-dependent	History-independent
Baseline query level identification	P: 89.8 [1] R: 89.8 F: 89.8	P: 89.8 R: 89.8 F: 89.8	-
Re-finding identification	P: 91.6 R: 91.7 F: 91.6	P: 91.6 R: 91.7 F: 91.6	P: 67.6 R: 74.0 F: 70.7

[1] The same as history-dependent.

	All features	History-dependent	History-independent
General web search difficulty	P: 79.2 [2] R: 78.9 F: 79.0	-	P: 79.2 R: 78.9 F: 79.0
Re-finding difficulty	P: 82.8 R: 82.7 F: 82.7	P: 81.0 R: 80.9 F: 80.9	P: 79.3 R: 79.0 F: 79.1

[2] The same as history-independent.

by Teevan et al. [18] as a state of the art baseline, which used the "Baseline query level features" introduced in Table 1. It can be seen that re-finding identification improves from 89.8 to 91.6, a relative increase of 2.0%. Examining re-finding difficulty, we obtain 4.7% relative improvements compared to the best found baseline, which was trained on "General web search difficulty features" in Table 1. The changes in F-measure scores are all statistically significant ($p < 0.05$) with the Cohen's effect size of 1.4 and 1.2 for re-finding identification and difficulty detection using all features.

The vast majority of re-finding research has focussed on re-finding where the information was originally found with a search engine and that finding activity was logged. We also consider the detection of re-finding without the information from the original (historical) goal. Using only history-independent features reduces re-finding accuracy (F-measure of 70.7); past work has not considered this type of identification, so there is no baseline to compare to (and the scores of the baseline using all features and history-dependent features are the same). We plan to improve the performance of this classification by studying history-independent features in future work, which enables the identification of more challenging re-finding tasks. Examining the history-independent column for the difficulty problem, similar accuracy was obtained for both re-finding and general search. However, features from the history-dependent group improve the performance of the classifier (i.e. 80.9).

6 Feature Importance Analysis

We calculated the information gain of each individual feature in order to assess their importance for the two prediction tasks. This measure estimates the amount of information that can be obtained about the class prediction from each feature [6]. The ten with the highest information gain are shown in Table 3. Some features are related to commonalities between paired goals (e.g. *"min goal position of common clicks"*), whilst others record measurements across a goal (e.g. *"effective search time"*). We start by analysing all features from paired goals.

All Features. Perhaps unsurprisingly, the most important feature was (*"query overlap/URL overlap"*) measuring the level of query and clicked URL overlap between the paired goals. This categorization ranks higher than all features used in past work [18]. Contextual features appeared to be important for identifying re-finding such as *"com-*

mon click in relation to the last click", *"no. of common clicks"*, or *"mean query length of common clicks"*.

The first ranked feature for difficulty detection was *"effective search time"*, which was a stronger indicator than the length of the search measured in queries and clicks (e.g. *"goal length in no. of both queries and clicks"*). The *"total dwell time after all queries"* was second. The corresponding feature for clicks (i.e. *"total dwell time after all clicks"*) did not appear in the top ten, suggesting that time spent after submitting queries is more likely to be representative of task difficulty than the time allocated after clicks.

Among other features in Section 4.2 that were not ranked in the top ten, but still ranked relatively strongly, *"all common clicks skipped"* and *"missed engaged later clicks in original"* appeared to be more effective in the identification of re-finding rather than difficulty detection. These features could provide signals that the user is not interested in previously seen documents, and therefore the underlying task is not re-finding. The information gain of *"no. of non-first-page ranked common clicks"* indicated that when the user navigates to the next result page, it is more indicative of search difficulty than re-finding. Similarly for the *"exist jumped common click"*, jumping to the previously seen document could be more indicative of an easy task in recognizing a target document rather than a particular re-finding behavior.

History-Independent Features. We also ranked history-independent features as shown in Table 3. It appeared that time-based features are important in identifying re-finding tasks independent of the search history of the user. As an example, *"max inter-click time"* acquired the highest information gain. Here, the time spent between clicks seems to be more important than the time between queries (i.e. *"max inter-query time"*). Other features indicative of the goal length in terms of number of queries/clicks and also the length of the queries obtained the top ranks in the identification of re-finding.

Table 3. Top 10 features for re-finding identification and difficulty detection ranked by information gain. A †, ‡, or ∗ indicate feature related to original, re-finding or both, respectively.

	All features	History-independent
Re-finding identification	1. query overlap/ URL overlap ∗	1. max inter-click time ‡
	2. common click in relation to the last click ∗	2. goal no. of all queries ‡
	3. no. of common clicks ∗	3. max inter-query time ‡
	4. equal query class ∗	4. total dwell time after clicks ‡
	5. mean no. of clicks for common click queries ‡	5. mean inter-click time ‡
	6. max goal position of common clicks ‡	6. mean inter-query time ‡
	7. min goal position of common clicks †	7. total dwell time ‡
	8. mean relative goal position of common clicks ‡	8. clicks per query ‡
	9. mean no. of clicks for common click queries †	9. mean no. of clicks across all queries ‡
	10. mean query length of common clicks ‡	10. mean query length of all clicks ‡
Re-finding difficulty	1. effective search time ‡	1. effective search time ‡
	2. total dwell time after all queries ‡	2. total dwell time after all queries ‡
	3. max goal position of common clicks ‡	3. goal no. of all clicks ‡
	4. goal length in no. of all clicks ‡	4. goal length in no. of both queries and clicks ‡
	5. goal length in no. of both queries and clicks ‡	5. max time to first clicks ‡
	6. max time to the first click of all queries ‡	6. mean time to first query clicks ‡
	7. mean time to the first click of all queries ‡	7. goal no. of unique clicks ‡
	8. goal length in no. of unique clicks ‡	8. no. of engaged clicks ‡
	9. no. of engaged clicks ‡	9. goal no. of all queries ‡
	10. goal length in no. of all queries ‡	10. no. of clicks on next page ‡

The top features indicative of difficulty in re-finding (discussed above) are history-independent (*"effective search time"* and *"total dwell time after queries"*). Apart from the proposed features in this study, there are other features from past research, which are also indicative of difficulty in re-finding, such as time to the first click and the number of engaged clicks [9].

In comparing re-finding identification features with difficulty indications, it can be seen that "goal no. of all queries" is a stronger signal for the identification of re-finding; whereas, "goal no. of all clicks" is more important in detecting the difficulty of the task. Some features particularly indicative of re-finding difficulty were history-independent and some could be computed during the search (e.g. *"mean time to first clicks"*). The latter features are referred to as *real-time* in the literature [13], and search engines that make use of them could provide real-time predictions. Using all the developed features in this study, we measured the accuracy of predictions given partial information from the beginning of re-finding tasks (after 2, 4, 8, 16, 32, and 64 seconds). The average F-score of 83.7 and 74.3 were obtained for re-finding identification and difficulty detection respectively, which could indicate the predictability of these two tasks at real-time for an online user support that can be further explored in future work.

7 Conclusions and Future Work

This paper focuses on better understanding re-finding behavior by answering two questions: a) how can re-finding tasks be differentiated from general web search tasks; and b) what features characterize user difficulties in completing a re-finding task.

We proposed a set of features and constructed prediction models for both re-finding identification and difficulty detection. Classifiers built using our feature sets achieved an F-measure of 91.6 for identifying re-finding, and 82.7 for predicting re-finding difficulty. Our model significantly outperformed existing state of the art re-finding identification approaches, which are based on query repetitions and dependent on the search history of the user, with a 2.0% improvement in accuracy. To the best of our knowledge, our work is the first to investigate the re-finding difficulty classification problem; we therefore compared our results against an adaptation of general web task difficulty detection approaches, resulting in a significant improvement of 4.7% for difficulty detection. We examined the effectiveness of predictors based on features, which can be computed without identifying the user and their search history. In this case, we obtained F-measure scores of 70.7 and 79.1 for detecting re-finding and difficulty respectively. The history-independent analysis could enable the identification of more complex re-finding tasks, which was not addressed in past research.

An analysis of the effectiveness of individual features for the two re-finding classification problems demonstrated that our proposed features, such as *"query overlap/URL overlap"* and the use of the *"effective search time"*, are ranked highly in terms of their information gain impact. Our analysis showed that some top ranked features can be calculated as the search task progresses (e.g. *"time to first click"*), which means that search engines can potentially take advantage of real-time prediction, even if there is no access to the search history of the user.

In future work, we plan to investigate further improvements to our predictive models by incorporating more real-time and fewer history-dependent features, and identify more distinctive behavioural features from a general search task. Moreover, some basic hypotheses in this study can be extended and further examined. For instance, instead of pairing sequential goals from the same user, we could also take into consideration chains of goals (due to the repeated nature of re-finding tasks). Furthermore, it would be interesting to carry out controlled user experiments to identify and incorporate user-side factors that cannot be derived from query log analysis.

References

1. Ageev, M., Guo, Q., Lagun, D., Agichtein, E.: Find it if you can: A game for modeling different types of web search success using interaction data. In: Proc. SIGIR, pp. 345–354. ACM (2011)
2. Capra III, R.G.: An investigation of finding and refinding information on the web. Ph.D. thesis, Virginia Polytechnic Institute and State University (2006)
3. Elsweiler, D., Baillie, M., Ruthven, I.: What makes re-finding information difficult? A study of email re-finding. In: Clough, P., Foley, C., Gurrin, C., Jones, G.J.F., Kraaij, W., Lee, H., Mudoch, V. (eds.) ECIR 2011. LNCS, vol. 6611, pp. 568–579. Springer, Heidelberg (2011)
4. eLSWEILER, D., Harvey, M., Hacker, M.: Understanding re-finding behavior in naturalistic email interaction logs. In: Proc. SIGIR, pp. 35–44. ACM (2011)
5. Elsweiler, D., Ruthven, I.: Towards task-based personal information management evaluations. In: Proc. SIGIR, pp. 23–30 (2007)
6. Hall, M.A.: Correlation-based feature selection for machine learning. Ph.D. thesis, The University of Waikato (1999)
7. Hassan, A., Jones, R., Klinkner, K.L.: Beyond DCG: User behavior as a predictor of a successful search. In: Proc. WSDM, pp. 221–230. ACM (2010)
8. Hassan, A., Shi, X., Craswell, N., Ramsey, B.: Beyond clicks: Query reformulation as a predictor of search satisfaction. In: Proc. CIKM, pp. 2019–2028. ACM (2013)
9. Hassan, A., Song, Y., He, L.W.: A task level metric for measuring web search satisfaction and its application on improving relevance estimation. In: Proc. CIKM, pp. 125–134. ACM (2011)
10. Jones, R., Klinkner, K.L.: Beyond the session timeout: Automatic hierarchical segmentation of search topics in query logs. In: Proc. CIKM, pp. 699–708. ACM (2008)
11. Kotov, A., Bennett, P.N., White, R.W., Dumais, S.T., Teevan, J.: Modeling and analysis of cross-session search tasks. In: Proc. SIGIR, pp. 5–14. ACM (2011)
12. Liu, J., Gwizdka, J., Liu, C., Belkin, N.J.: Predicting task difficulty for different task types. Proc. ASIST 47(1), 1–10 (2010)
13. Liu, J., Liu, C., Cole, M., Belkin, N.J., Zhang, X.: Exploring and predicting search task difficulty. In: Proc. CIKM, pp. 1313–1322. ACM (2012)
14. Nadeau, C., Bengio, Y.: Inference for the generalization error. Machine Learning 52(3), 239–281 (2003)
15. Shokouhi, M., White, R.W., Bennett, P., Radlinski, F.: Fighting search engine amnesia: Reranking repeated results. In: Proc. SIGIR, pp. 273–282. ACM (2013)
16. Teevan, J.: Supporting finding and re-finding through personalization. Ph.D. thesis, Massachusetts Institute of Technology (2006)
17. Teevan, J.: How people recall, recognize, and reuse search results. TOIS 26(4), 19 (2008)
18. Teevan, J., Adar, E., Jones, R., Potts, M.A.: Information re-retrieval: Repeat queries in yahoo's logs. In: Proc. SIGIR, pp. 151–158. ACM (2007)
19. Tyler, S.K., Teevan, J.: Large scale query log analysis of re-finding. In: Proc. WSDM, pp. 191–200. ACM (2010)

20. Tyler, S.K., Wang, J., Zhang, Y.: Utilizing re-finding for personalized information retrieval. In: Proc. CIKM, pp. 1469–1472. ACM (2010)
21. Wang, H., Song, Y., Chang, M.W., He, X., White, R.W., Chu, W.: Learning to extract cross-session search tasks. In: Proc. WWW, pp. 1353–1364. International World Wide Web Conferences Steering Committee (2013)
22. Webber, W., Toth, B., Desamito, M.: Effect of written instructions on assessor agreement. In: Proc. SIGIR, pp. 1053–1054. ACM (2012)

User Behavior in Location Search on Mobile Devices

Yaser Norouzzadeh Ravari, Ilya Markov, Artem Grotov,
Maarten Clements, and Maarten de Rijke

University of Amsterdam, Amsterdam, The Netherlands
{y.norouzzadehravari,i.markov,a.grotov,
m.clements,derijke}@uva.nl

Abstract. Location search engines are an important part of GPS-enabled devices such as mobile phones and tablet computers. In this paper, we study how users behave when they interact with a location search engine by analyzing logs from a popular GPS-navigation service to find out whether mobile users' location search characteristics differ from those of regular web search. In particular, we analyze query- and session-based characteristics and the temporal distribution of location searches performed on smart phones and tablet computers. Our findings may be used to improve the design of search interfaces in order to help users perform location search more effectively and improve the overall experience on GPS-enabled mobile devices.

1 Introduction

Location search engines (LSEs) are widely used to search for *points of interest* (POIs) such as restaurants, shops, filling stations, etc. and to navigate to them. Despite their importance, they have not yet been studied extensively, while most of research in the past has focused on local search and location-related queries submitted to regular web search engines. There are important differences between location search on the one hand and local search and location-related queries on the other. First, LSEs are aimed at finding POIs, rather than local information as in local search. Second, LSEs differ from location search in other systems (e.g., maps) with regard to user intents. In many cases, people use LSEs to navigate to a POI (in our logs, more than 70% of the sessions and more than 50% of the queries result in actual navigation), while in other systems users aim at locating relevant places and getting information about them.

The differences just noted make the design of an LSE a unique problem, which needs to be studied in order to improve user satisfaction. In this paper, we make the first step in this direction and study user search interaction with LSE. The main research questions that guide our work are the following: (1) Does user search interaction with LSE differ from that in web search? (2) Does user search interaction with LSE depend on the type of device?

To answer these questions, we analyze a recent log from the LSE of a popular GPS-navigation system. The studied LSE receives a keyword query and then finds relevant locations. Users can locate results on a map, check location-related information and navigate to selected results. The studied LSE is primarily focussed on car navigation and, therefore, is mostly used in the car or for pre-trip planning from home.

A. Hanbury et al. (Eds.): ECIR 2015, LNCS 9022, pp. 728–733, 2015.

We analyze user search interactions with the LSE installed on *tablets* (more specifically, iPads) and *mobile phones* (more specifically, iPhones). We are interested in studying query- and session-related characteristics of user interaction as well as its temporal aspect. The results of our study can be used in search personalization, user modeling, interface design, query refinement and query suggestion.

2 Related Work

Local search and mobile search have been important research topics in recent years. In order to better understand user behavior in local search, researchers performed user studies and analyzed logs of web search engines. Berberich et al. [1] analyze logs of business web sites, customer ratings, GPS-traces, and logs with driving-direction requests. They measured the geographic distance between a user and a search result to infer relevance and to improve search. Zheng et al. [8] work with logs of GPS-enabled devices to find interesting locations and common travel sequences in a region.

Recently, several studies have focused on the device type and analyzed its effect on user behavior in desktop and mobile web search. Kamvar et al. [3] analyze mobile, tablet and desktop users and suggested that no single interface can fit all user needs and search experience should change based on the type of device. Song et al. [5] also compare the above devices and conclude that a single ranker cannot be used for all of them. They propose to use the characteristics of user behavior on tablet/mobile to improve rankers.

Researchers have also used context, such as location and temporal information, to improve local search results. Lane et al. [4] propose the Hapori framework that utilizes location, time and weather for local search. Teevan et al. [6] conduct a user study, asking participants about their location when searching, desired destination, plans about visiting a place, etc. The authors report that participants mostly search on the go and plan to visit destinations soon after querying.

Also, location related queries have been analyzed in web search engines. Gan et al. [2] study geographic searches using queries from AOL. The authors classify queries into geo and non-geo queries and report that non-geo queries are related to geo ones. In [7], the authors study web search logs to explore the relation between mobile queries and their locations. The authors propose a statistical model to predict whether a user is soon observed at the searched location.

The above studies are mostly concerned with user behavior in local search and are based on logs of a general web search engine on desktop, mobile or tablet. Our work differs as we study user interaction with a *LSE* within a GPS-navigation system. We first compare user behavior in location search to that in general web search. Then we compare user search behavior across different devices, namely tablet and mobile.

3 Dataset

For this study, we sampled the log of LSE of a popular navigation application during the period from February to June 2014. We considered search sessions from the USA and UK and filtered out non-English queries. Sessions were logged on the following

Table 1. User search behavior statistics for the LSE in a GPS-navigation system on tablet and mobile devices, compared to that in standard web search on desktop, tablet and mobile [5]. All statistics for the tablet LSE are significantly different from those for the mobile LSE ($p < 0.01$).

	#sessions (%)	#queries (%)	avg. queries per session	avg. session length in mins	avg. query length
Desktop [5]	N/A	13,928,038	1.89	8.61	2.73
Tablet [5]	N/A	8,423,111	1.94	9.32	2.88
Mobile [5]	N/A	9,732,938	1.48	7.62	3.05
		Tablet LSE			
All	21,936	38,129	1.74	2.69	1.93
Click	15,770 (72%)	21,208 (56%)	1.82	3.22	1.84
No click	6,166 (28%)	16,921 (44%)	1.53	1.34	2.05
Route	15,277 (70%)	19,580 (51%)	1.79	3.16	1.83
		Mobile LSE			
All	423,509	632,288	1.49	1.86	1.87
Click	305,104 (72%)	360,343 (57%)	1.49	2.22	1.78
No click	118,405 (28%)	271,945 (43%)	1.49	0.94	1.99
Route	296,568 (70%)	340,953 (54%)	1.47	2.18	1.78

devices: iPhone ("mobile") and iPad ("tablet"). Each session may consist of multiple queries. Sessions are separated by a period of inactivity of more than 30 minutes or based on closing the application. Overall, we collected 445,446 search sessions consisting of 670,417 queries: 21,936 sessions and 38,129 queries for tablet, 423,509 sessions and 632,288 queries for mobile. The uneven distribution of the number of sessions and queries between tablet and mobile is due to the difference in device usage frequency in the sampled part of our log.

In a typical scenario of user interaction, the session starts when a user opens the navigation application. After submitting a query, the user is presented with a list of location results and can click on them to see the map centered on the result, its phone number and web site address, sharing buttons and the route planning button. Then, the user can contact the chosen location, check more information about it, share the location and plan a route to it.

4 Analysis

In this section we answer our research questions by analyzing the our logs described in the previous section. First, we compare user interaction with an LSE to that with general web search. Then we compare user interaction with an LSE on tablet vs. mobile.

Table 1 shows user search statistics: the number of sessions, number of queries, average number of queries per session, average session length in minutes and average query length in words. The first block of Table 1 shows the statistics for general web search on desktop, tablet and mobile devices, as reported by Song et al. [5].

The second block reports the statistics of user search sessions in tablet and mobile LSEs. The first row for each device type shows the overall user search statistics. The second row presents the statistics for sessions and queries in which a user clicked on one or more results. The third row shows the statistics for sessions and queries in which a user did not click on any result. Since the goal of LSEs is to help users plan a route to a desired POI, the last row shows the statistics for sessions and queries that contain the "route to" action. Absence of the route action does not mean that a user is not satisfied with the search results—in many cases users are interested in checking the results without navigating to them (e.g., pre-trip planning). The differences between the corresponding tablet and mobile LSE statistics in Table 1 are statistically significant according to the Mann-Whitney U-test at the 0.01 level.

Note that the number of sessions in the mobile LSE is much larger than the number of sessions on tablet. This is due to the fact, that LSEs are mostly used on the go and, therefore, users tend to prefer mobile to tablet. Also, the form factor of mobile phones makes them much more popular for in-car navigation, which is further stimulated by the availability of phone docking stations.

In the following, we first compare tablet/mobile LSEs with general web search, and then compare tablet LSE with mobile LSE.

Location Search in LSE vs. Web Search. According to Table 1, users submit more queries per session while performing web search on tablet compared to LSE for the same device type. The opposite is true when users interact with mobile devices but the difference is much smaller. This suggests that the way users interact with LSEs is more similar to how they interact with mobiles rather than with tablets.

Users spend less time interacting with LSEs than performing web search: three times less on tablet and four times less on mobile, even though the average number of queries per session is roughly the same. This observation can be interpreted as saying that users of an LSE are mostly on the move and have less time for searching compared to the web search scenario. Also, users can easily understand if a location is relevant or not, while in web search users spend more time on examining results.

In general, queries in location search are shorter than in web search. This can be explained by the fact that queries in location search are limited to places as opposed to web search queries, which can be about anything. This suggests that LSE would greatly benefit from custom NLP techniques different from those of general web search.

Tablet vs. Mobile LSE. The number of sessions and queries indicate that the mobile LSE is used much more often than the tablet LSE. On the other hand, the average number of queries per session, average session length and average query length for the tablet LSE are all larger than those for the mobile LSE, which means that users spend more time when using tablet devices. These observations can be explained as follows. Tablets are more often used for pre-trip planning, while mobile phones are used on the go. In trip planning, people spend more time and use more queries because they want to explore all possible results (e.g., finding appropriate hotels, restaurants, etc.). Instead, people on the move execute more targeted searches and are mainly looking for the nearest available POI that solves their direct needs (e.g., petrol station, parking, fast-food, etc.).

It is interesting to note that the above behavior is similar to that in web search (see the first block of Table 1). This means that the different form factor between tablet and mobile devices has a similar effect on how people use them for location and web search.

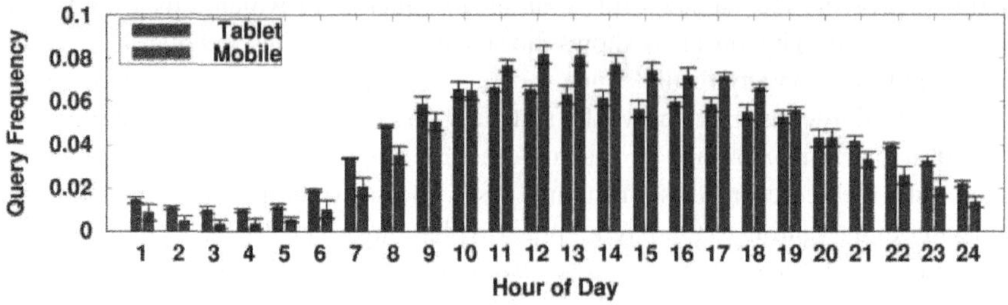

Fig. 1. Query frequency distribution in a GPS-navigation system on tablet and mobile devices

When we consider the percentages of session that have at least one click, both LSEs are similar. In user interactions with LSEs, the routing action is a strong signal of user satisfaction. The percentage of routing in tablet and mobile LSEs reaches 70% of sessions (97% of sessions with clicks), therefore if a user clicks on a result, it is almost certain that her intent is to plan a route somewhere. In the remainder of the sessions the user was either unable to locate relevant POIs or did not want to plan a route. This can mean that a click is a reliable indicator of user intent while interacting with an LSE.

In sessions with both click and route actions (which we assume to be successful), the average number of queries per session and the average session length are usually larger than the average for all sessions. This can be explained as follows: users who do not click anywhere give up fast and submit few queries; users who are more persistent in finding relevant POIs have to click on returned results and submit more queries.

Temporal Characteristics. Here, we compare user behavior in tablet and mobile LSEs along the temporal dimension. The query frequency distribution during the day is shown in Figure 1. The graph shows that users prefer to interact with LSEs using mobile during working hours (from 11am till 7pm) and prefer to use tablet while mostly at home (from 9pm to 10am). This observation is not surprising, because users usually carry their mobiles with them, but may keep their tablets at home. Moreover, tablets are used more for pre-trip planning, usually done during non-working hours, while mobiles are used for actual navigation. We also analyzed the query frequencies for different days of the week and found that the relative number of queries in mobile LSE is lower than on tablet during weekdays, but larger during weekends. The smaller size of mobile devices may explain this difference: during weekends people are on the go and tend to use mobile devices more than tablets.

5 Conclusions and Future Work

In this paper we analyzed LSE logs of a popular GPS-navigation system and compared user interaction with an LSE to that of general web search. We also checked if user interaction with an LSE depends on the type of device, i.e., tablet and mobile.

We showed that user search interaction with an LSE and web search has certain similarities and differences. The similarities include the number of queries per session and the relative session length on tablets compared to mobile. On the other hand, due to specific usage scenarios of LSEs (e.g., on the go), sessions and queries are shorter in location search compared to web search. Our observations on LSEs vs. web search have implications for the interaction design and underlying technology for LSEs.

Our statistical observations also showed similarities and differences between tablet and mobile LSEs. People use the mobile LSE more, especially in working hours. In addition, mobile LSE sessions and queries are shorter than on tablets. This is because tablets are more often used for pre-trip planning, while mobile phones are used on the go. These observations suggest that the interface of the mobile LSE should be adapted to be used in movement, so should be simple and provide basic functionality, while the interface of the tablet LSE can contain more details and support more complex interactions.

In future, we are interested to investigate more characteristics of user interaction with LSE to find how much users are satisfied with results and how we can improve location search. We would like to find common sequences of user activities and determine which sequences are successful and which are not. Moreover, the combined analysis of queries and destinations is a promising direction for future research.

Acknowledgments. This research was partially supported by grant P2T1P2_152269 of the Swiss National Science Foundation (SNF), the European Community's Seventh Framework Programme (FP7/2007-2013) under grant agreement nr 312827 (VOX-Pol), the Netherlands Organisation for Scientific Research (NWO) under project nrs 727.-011.005, 612.001.116, HOR-11-10, 640.006.013, 612.066.930, the Center for Creation, Content and Technology (CCCT), the Dutch national program COMMIT, the ESF Research Network Program ELIAS, the Elite Network Shifts project funded by the Royal Dutch Academy of Sciences (KNAW), the Netherlands eScience Center under project number 027.012.105, the Yahoo! Faculty Research and Engagement Program, the Microsoft Research PhD program, and the HPC Fund.

References

[1] Berberich, K., König, A.C., Lymberopoulos, D., Zhao, P.: Improving local search ranking through external logs. In: SIGIR 2011, pp. 785–794. ACM (2011)
[2] Gan, Q., Attenberg, J., Markowetz, A., Suel, T.: Analysis of geographic queries in a search engine log. In: LOCWEB 2008, pp. 49–56. ACM (2008)
[3] Kamvar, M., Kellar, M., Patel, R., Xu, Y.: Computers and iphones and mobile phones, oh my!: A logs-based comparison of search users on different devices. In: WWW 2009, pp. 801–810. ACM (2009)
[4] Lane, N.D., Lymberopoulos, D., Zhao, F., Campbell, A.T.: Hapori: Context-based local search for mobile phones using community behavioral modeling and similarity. In: UbiComp 2010, pp. 109–118. ACM (2010)
[5] Song, Y., Ma, H., Wang, H., Wang, K.: Exploring and exploiting user search behavior on mobile and tablet devices to improve search relevance. In: WWW 2013, pp. 1201–1212. ACM (2013)
[6] Teevan, J., Karlson, A., Amini, S., Brush, A.J.B., Krumm, J.: Understanding the importance of location, time, and people in mobile local search behavior. In: MobileHCI 2011, pp. 77–80. ACM (2011)
[7] West, R., White, R.W., Horvitz, E.: Here and there: Goals, activities, and predictions about location from geotagged queries. In: SIGIR 2013, pp. 817–820. ACM (2013)
[8] Zheng, Y., Zhang, L., Xie, X., Ma, W.-Y.: Mining interesting locations and travel sequences from gps trajectories. In: WWW 2009, pp. 791–800. ACM (2009)

Detecting the Eureka Effect in Complex Search

Hui Yang, Jiyun Luo, and Christopher Wing

Department of Computer Science, Georgetown University
37th and O Street NW, Washington DC, 20057, USA
huiyang@cs.georgetown.edu, {jl1749,cpw26}@georgetown.edu

Abstract. In search tasks that show a high complexity, users with zero or little background knowledge usually need to go through a learning curve to accomplish the tasks. In the context of patent prior art finding, we introduce a novel notion of Eureka effect in complex search tasks that leverages the sudden change of user's perceived relevance observable in the log data. *Eureka effect* refers to the common experience of sudden understanding a previously incomprehensible problem or concept. We employ non-parametric regression to model the learning curve that exists in learning-intensive search tasks and report our preliminary findings in observing the Eureka effect in patent prior art finding.

Keywords: Complex Search, Prior Art Retrieval, Learning Curve, Eureka Effect.

1 Introduction

State-of-the-art Information Retrieval (IR) research is extremely valuable for a wide range of applications but they are subject to a limited number of search task types. Most search tasks attracting lots of current research efforts are one-shot query tasks. Although those search tasks account for a large portion of online Web search activities, a great deal of other complex search tasks remain understudied. These tasks lie along the spectrum of tasks that require plenty of professional expertise such as patent prior art finding [11] and e-discovery [5], to tasks that are complex but do not require much expertise such as travel search that [4]. Typically, these search tasks require multiple queries in a search session and involves rich user and system interactions.

One fundamental type of challenge in these complex search tasks is represented by the users' changing perception of document relevance. This problem can be understood by using the notion of learning curve, that is, the rate of a user's progress in gaining knowledge, experience or new skills. By modeling the learning curve, user's changing understanding of document relevance can be reflected in search algorithms and evaluation metrics. Commonly used learning curve formulas are observed from industrial production lines and are usually used to determine expected labor and materials costs. Most of them are linear formulations in the form of $Y_x = aX^b$, where Y is the cumulated average time required to produce X units, a is the time required to produce the first output,

A. Hanbury et al. (Eds.): ECIR 2015, LNCS 9022, pp. 734–740, 2015.

Table 1. A prior art finding session

query id	# returned docs	query	timestamp
1	109	(SAKAKI near1 YUZO).in	2012/06/07 13:27
2	855	428/827, 828, 829, 830.ccls	2012/06/07 13:27
3	195	428/836.2.ccls	2012/06/07 13:28
4	0	S1 and S2 and S3	2012/06/07 13:28
5	74829	CoCrPtRu((“Co.sub.”$2) same(Ru ruthenium))	2012/06/07 13:29
6	31	S2 and S3 and S5	2012/06/07 13:30
7	2	(“20040184176” — “20050181237”).PN	2012/06/07 13:45
8	402914	samung kikitsu.in.	2012/06/07 14:02
9	8	(S2 S3) and S8 and bernatz	2012/06/07 14:02
10	3456	lee.in and (Ku anisotropy)	2012/06/07 14:05
11	22	428/826-827.ccls and S10	2012/06/07 14:06
12	2	jp adj “2008090913”	2012/06/07 14:10
13	2	“20020012816”	2012/06/07 14:11
14	1	cn adj “1870145”	2012/06/07 14:12
15	2	“2006024791”.pn.	2012/06/07 14:15

and b is the learning rate. Another popular formulation is the “S-shape” learning curve using the Sigmoid functions [9,6]. However, it is unclear whether existing learning curve formulas are suitable in the context of complex search.

This paper uses patent prior art search as a motivating example of complex search tasks constrained by time. As an illustration, we give an example taken from a query log from the U.S. Patent and Trademark Office (USPTO). We extracted and analyzed the prior art finding sessions that U.S Patent Examiners conducted for a patent application on “light controlling”. There are more than 15 distinct queries in the session. Most of these are structural queries, where Boolean operators (AND, OR) and proximity operators (within 2 words) are used to pose constraints on the query. Another common operator is browsing, where from a seed document, more documents are browsed from its references or from the document class it belongs to. The search lasted for around 2 hours. We noted that at the moment that the patent examiner came across the passage *“a control device for controlling hue of light emitted by a light source, device comprises: a body with a surface containing a visible representation of a plurality of selectable combinations of hue available for said light source”*, the time spent on examining a single document is suddenly decreased from 15 minutes per document to less than 1 minute per document. It is illustrated in Table 1 at query $S9$.

This sudden change of the reading time in general indicates a change of user's status of mind of understanding the related topics; which we call the “Eureka effect”. “Eureka!” is the word shouted out by Archimedes, the Greek mathematician, when he suddenly discovered how to calculate the volume of an irregular object and leap out of a public bath. Here we use “Eureka effect” to refer to the common experience of suddenly understanding a previously incomprehensible problem or concept.

On the other hand, this example suggests that if we are able to recognize the sudden drop of reading time per document, we could obtain a novel learning curve formulation specifically designed for complex search tasks, which will allow search engines to create better search algorithms and better evaluation mechanisms.

Based on these observations, we propose the following definition of Eureka effect in complex search tasks:

- *Eureka effect is the phenomenon that in complex search process where we detect a sharp increase of users' understanding of the domain and the related documents.*

There are two main issues involved in this definition, namely the computation of the gap between a document's user received relevance (URR) and user perceived relevance (UPR), and the modeling of the learning curve to detect the Eureka effect. We show that detection of Eureka effect can be tackled by a non-parametric regression algorithm. The solution consists of two steps. First, automatically extracting relevance judgments from office action documents submitted by patent examiners. Second, fitting the difference between user perceived relevance and user received relevance to a non-parametric regression model, in which the model parameters define the learning curve and the Eureka effect. Particularly, we are particularly interested in situations where users start the search with zero or little background and study the Eureka effect for them.

2 Related Work

In complex search tasks, retrieval results usually have different reading difficulties and users also show various reading proficiencies. Borlund [1] pointed out that relevance is a dynamic concept that depends on a user's judgment at a certain point of time. Heilman et al. [7] and Kidwell et al. [8] provided two statistical approaches to estimate a passage's reading difficulty by utilizing lexical and grammatical features. Collins-Thompson et al. [3] provided a Language Modeling Approach to estimate reading difficulties. In their further work, Collins-Thompson et al. [2] pointed out that users' satisfaction are enhanced when they are shown with materials that match with their reading proficiency. Scholer et al. [10] conducted a user study on eighty-two users and discovered that the relevance of documents viewed early impacts the assessment of subsequent documents.They also observed that the more difficult the search topics are, the more significant the difference between the two user groups. In this work, we conduct user study with students who has little background in searching patent documents, which increases the difficulty of the task and fits well with our purpose – detecting the Euraka effect.

3 Method

Our proposed method include the following main steps. Automatic extraction of human relevance judgments: (1) extract subtopics (claims) from the patent documents, (2) extract passage-level relevance from Office Actions as the truth data (user received relevance), (3) extract the user perceived relevance from query logs. Then, we (4) fit the difference between URR and UPR into a local polynomial regression model, and (5) based on the model parameters, determine the existence of the Eureka Effect.

Table 2. Final rejection data statistics

#docs txt	#docs XML	avg # total claims	avg # claims rej.	avg # prior art cited	avg docs / claim rej.
1.6M	3.0M	12.74	8.94	1.55	2.20

3.1 Automatic Extraction of Relevance Judgments

We propose an automatic approach to generate ground truth (URR) from the official action (OA) documents that are available on USPTO PAIR.[1] An OA is written by patent examiners and explains which prior art they used as evidence to *reject* various claims in the patent application. We extract this information from their descriptions and transfer them to the input format of our metric scripts. For a patent application, its corresponding office actions $O = \{O1, O2, O3, ...\}$, including non-final office actions, final office actions and the examiners' answers, are processed to extract a set of evidence, including reference documents, reference passages and reasoning paragraphs. The evidence is then used to extract the actual passages and documents from the references. Most OAs used in this process issue rejections to patent applications.

We confine our data collection to Final Rejection office action documents. The dataset is constituted by a series of official actions from the year 2012. All image information and cover sheet have been removed. We assessed and appended the relevance score of each prior art cited within the Final Rejection to the patent application. Each Final Rejection typically cites between 1 to 4 prior arts with an average number of 1.87 citations. On average, 1.29 documents are used to reject a given claim. Each Final Rejection had an average number of 12.19 claims rejected. More dataset statistics can be found in Table 2.

3.2 Detecting the Eureka Effect

The Eureka effect can be formulated as a function about the difference between UPR and URR. We propose to use non-parametric regression to model the learning curve. To reduce the model bias, local polynomial regression estimator is chosen for our task instead of the most commonly used kernel regression.

For a polynomial $P_x(u; a(x)) = a_0(x) + a_1(x)(u - x) + \frac{a_2(x)}{2!}(u - x)^2 + ... + \frac{a_p(x)}{p!}(u - x)^p$, its coefficients $a(x)$ can be estimated by minimizing the weighted sums of squares

$$\sum_{i=1}^{n} w_i(x)(Y_i - P_x(X_i))^2,$$

where $w_i(x) = \frac{K(X_i - x)}{h}$.

The local polynomial regression estimation can be solved by minimizing the least squared error and we get:

$$\hat{m}_n(x) = \sigma_{i=1}^{n} l_i(x) Y_i$$

[1] http://portal.uspto.gov/pair/PublicPair.

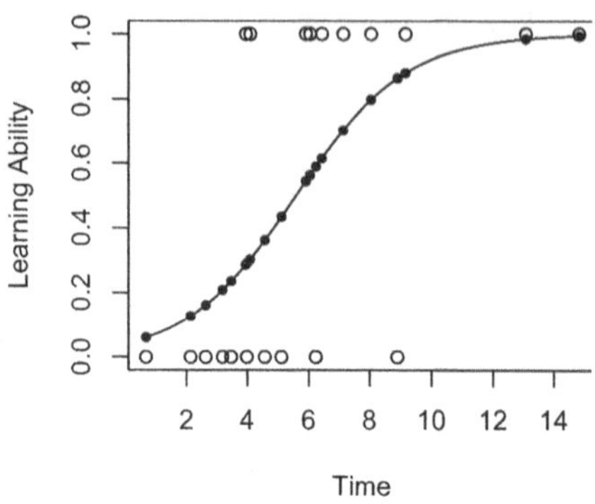

Fig. 1. User's learning ability

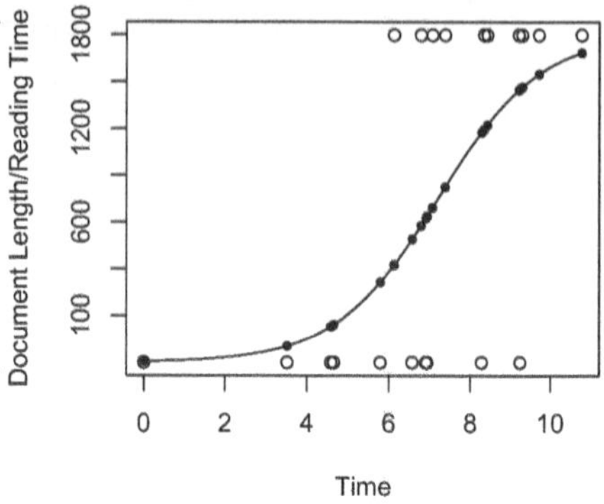

Fig. 2. Reading speed

where $l(x)^T = e_1^T(X_x^T W_x X_x)^{-1} X_x^T W_x$, $e_1 = (1, 0, ..., 0)^T$, X_x is a vector representation of the coefficients, and W_x is a diagonal matrix whose (i, i) component is $w_i(x)$. We then measure the gap between the adjacent learned coefficient $w_i(x)$. When a large gap is detected, we say an Eureka effect is found.

4 Preliminary Experimental Results

In this section, we report findings in our preliminary experiments. We conducted a user study to evaluate the relationship between UPR and URR. Twelve graduate students from various majors participated in the study. They are proficient with the use of computers, highly proficient in English, and have little knowledge about the topic described in the patent documents. This experiment setting makes sure that the user information needs are highly complex and the topics of search tasks are unfamiliar to the users. The dataset is the publicly available patent dataset[2] from the U.S. Patent and Trademark Office (USPTO). We automatically extract ground truth relevant documents from the official search reports published at PublicPAIR as described earlier.

One quantity to measure users' ability of making correct judgment is the difference between the relevance grade given by ground truth (URR) and by the users (UPR). We consider it is a measure of user's learning ability. Fig. 1 plots the curve for learning ability fitted by local polynomial regression (LPR). LPR suggests an Eureka effect happens in the middle of the S-shaped learning curve.

Fig. 2 plots the curve for average reading speed per document fitted by local polynomial regression. The plot suggests an S-shaped learning curve too. We can see that user's learning speed is low at the beginning for a relatively long time, and it accelerates steeply after the user spends more time learning and has accumulated enough background knowledge. As a user continues learning and the accumulated knowledge reaches a high plateau, the learning speed tapers off. An Eureka effect happens in the middle of the S-shaped learning curve.

[2] http://www.google.com/googlebooks/uspto-patents-applications-text.html.

5 Discussion and Conclusion

In search tasks that show a high complexity, users with zero or little background knowledge usually have the common experience of sudden understanding a previously incomprehensible problem or concept. In the context of patent prior art search, this paper introduces a novel notion of Eureka effect in complex search tasks that leverages the sudden change of user's perceived relevance observable in the search log data. An initial set of preliminary experiments are done using non-parametric regression to model the learning curve. The preliminary experimental results are encouraging – we are able to observe the S-shape learning curve in the search process. It suggests that in patent prior art search, at the beginning a user could not easily distinguish relevant documents from non-relevant ones since the terms used in patent documents are often very abstract, rare, and difficult. As the user learns more about the search topic from the retrieved documents, it is possible that he can suddenly understand quite a lot of related materials, which are not previously comprehensible, all at once.

As part of attempts to model the learning curve, our work focuses on detecting the Eureka effect in complex search. Learning curve is an important concept in learning-intensive search tasks, which will potentially enable search engines to improve on providing users with the right documents at the right time.

Acknowledgments. The research is supported by NSF grant CNS-1223825 and DARPA grant FA8750-14-2-0226. The U.S. Government is authorized to reproduce and distribute reprints for Governmental purposes notwithstanding any copyright notation thereon. Portions of this work were conducted under the umbrella of a larger project at the USPTO.

References

1. Borlund, P.: The concept of relevance in ir. Wiley Online Library 2003. Journal of the American Society for information Science and Technology 54(10), 913–925 (2003)
2. Collins-Thompson, K., Bennett, P.N., White, R.W., de la Chica, S., Sontag, D.: Personalizing web search results by reading level. In: CIKM 2011 (2011)
3. Collins-Thompson, K., Callan, J.: Predicting reading difficulty with statistical language models. Wiley Subscription Services, Inc., A Wiley Company 2005. Journal of the American Society for Information Science and Technology 56(13), 1448–1462 (2005)
4. Dean-Hall, A., Clarke, C.L.A., Hall, M., Kamps, J., Thomas, P., Voorhees, E.: Overview of the trec 2012 contextual suggestion track. In: TREC 2012 (2012)
5. Grossman, M.R., Cormack, G.V., Hedin, B., Oard, D.W.: Overview of the trec 2011 legal track. In: TREC 2011 (2011)
6. Hartz, S., Ben-Shahar, Y., Tyler, M.: Logistic growth curve analysis in associative learning data. Animal Cognition (2001)
7. Heilman, M., Collins-Thompson, K., Eskenazi, M.: An analysis of statistical models and features for reading difficulty prediction. In: EANL 2008 (2008)
8. Kidwell, P., Lebanon, G., Collins-Thompson, K.: Statistical estimation of word acquisition with application to readability prediction. In: EMNLP 2009 (2009)

9. Murre, J.M.: S-shaped learning curves. Psychonomic Bulletin & Review (2013)
10. Scholer, F., Kelly, D., Wu, W.C., Lee, H.S., Webber, W.: The effect of threshold priming and need for cognition on relevance calibration and assessment
11. Zhao, L., Callan, J.: How to make manual conjunctive normal form queries work in patents search. In: TREC (2011)

Twitter Sentiment Detection via Ensemble Classification Using Averaged Confidence Scores

Matthias Hagen, Martin Potthast, Michel Büchner, and Benno Stein

Bauhaus-Universität Weimar
<first name>.<last name>@uni-weimar.de

Abstract. We reproduce three classification approaches with diverse feature sets for the task of classifying the sentiment expressed in a given tweet as either positive, neutral, or negative. The reproduced approaches are also combined in an ensemble, averaging the individual classifiers' confidence scores for the three classes and deciding sentiment polarity based on these averages. Our experimental evaluation on SemEval data shows our re-implementations to slightly outperform their respective originals. Moreover, in the SemEval Twitter sentiment detection tasks of 2013 and 2014, the ensemble of reproduced approaches would have been ranked in the top-5 among 50 participants. An error analysis shows that the ensemble classifier makes few severe misclassifications, such as identifying a positive sentiment in a negative tweet or vice versa. Instead, it tends to misclassify tweets as neutral that are not, which can be viewed as the safest option.

1 Introduction

We reproduce three state-of-the-art approaches to classifying the sentiment expressed in a given tweet as either positive, neutral, or negative, and combine the three approaches to an ensemble based on the individual classifiers' confidence scores.

With about 271 million active users per month and about 350,000 tweets per minute, Twitter is one of the biggest social networks that can be mined for opinions. It is used as a communication platform by individuals, but also companies and organizations. The short text messages, or tweets, shared on Twitter cover a range of topics like information and comments on ongoing events but also opinions on products, brands, etc. Making the latter piece of information accessible to automatic analysis is not straightforward. A central tool required for this is sentiment detection, which determines whether a given tweet is rather positive or rather negative. However, sentiment detection for tweets is a challenge in itself. Unlike Amazon reviews, for instance, that typically have a length of several sentences or even paragraphs, the tweets come with a length limit of just 140 characters, which forces people to use abbreviations, slang, and genre-typical expressions. The language used in tweets often differs significantly from what is observed in other large text collections.

To facilitate the development of effective approaches for the analysis of tweets, corresponding shared tasks have been organized at SemEval from 2013 onwards. The goal of these tasks is to grasp the opinions expressed in microblogging forums like Twitter and to thus gain a better understanding of what matters to the users, e.g., what they like and what they do not like. Platforms such as Twitter, with their wealth and diversity

A. Hanbury et al. (Eds.): ECIR 2015, LNCS 9022, pp. 741–754, 2015.

of users, have often been found to be an accurate source for tracking opinions in societies that increasingly express themselves online. People share their reviews of books or movies, express pros and cons on political topics, or just give feedback to companies or restaurants. Such expressions are meaningful to the respective reviewed subjects, for instance, to design new products, but they are also meaningful to the general public to get a better idea whether a specific product is useful. Especially for information retrieval, incorporating the sentiment of a piece of text is an important signal for diversification of retrieval results.

In particular, we focus on subtask B in SemEval 2013's task 2 and SemEval 2014's task 9 "Sentiment Analysis in Twitter," where the goal is to classify the whole tweet as either positive, neutral, or negative. Note that this is a slightly different task than classifying the sentiment of a tweet expressed for or against a given target topic (e.g., a product or a brand). In the setting we address, the tweet as a whole could express several sentiments (positive for topic A but negative against topic B) and the goal is to identify the sentiment that dominates. Besides the aforementioned SemEval tasks, this variant of the Twitter sentiment detection problem has attracted quite some research interest.

Since notebook descriptions accompanying submissions to shared tasks are understandably very terse, it is often a challenge to reproduce the results reported. Therefore, we attempt to reproduce three state-of-the-art Twitter sentiment detection algorithms that have been submitted to the aforementioned tasks. Furthermore, we combine them in an ensemble classifier. Since the individual approaches employ diverse feature sets, the goal of the ensemble is to combine their individual strengths. Our experimental evaluation shows that our re-implementations of the three selected approaches outperform their originals in two cases, whereas one achieves the same results as its original. Furthermore, our ensemble approach outperforms its components. The ensemble would have been ranked in the top-5 ranks among all the 50 participants of SemEval 2013 and 2014. An error analysis of our approach reveals that there are hardly any misclassifications of a positive tweet as negative or vice versa. Instead, misclassifications of positive or negative tweets usually result in a neutral classification. This could be viewed as the safest option when the classifier is in doubt.

The remainder of the paper is organized as follows. In Section 2 we briefly review related work on sentiment detection with a focus on Twitter. The detailed description of the three individual approaches as well as our ensemble approach follows in Section 3. Our experimental evaluation within the SemEval task's setting is described in Section 4. Some concluding remarks and an outlook on future work close the paper in Section 5.

2 Related Work

Sentiment detection in general is a classic problem of text classification. Unlike other text classification tasks, the goal is not to identify topics, entities, or authors of a text but to rate the expressed sentiment typically as positive, negative, or neutral. Most approaches used for sentiment detection have also been useful for other text classification tasks and usually involve methods from machine learning, computational linguistics, and statistics. Typically, several approaches from these fields are combined for sentiment detection [32, 41, 14]. Linguistic considerations range from tokenizing the to-be-classified texts to other syntactic analyses. Statistical considerations typically involve frequencies of tokens or phrases, e.g., the occurrence of many "positive" words in a text, or similar statistics. The respective features then usually are combined by machine learning algorithms to classify the sentiment of arbitrary texts.

Machine learning methods are applied to sentiment detection as a matter of course, both supervised or unsupervised. Without training data available, one of the earliest unsupervised sentiment detection methods is based on word polarity dictionaries and pointwise mutual information (PMI) of part-of-speech (POS) sequences [41]. PMI is a measure of the statistical dependency of two terms. First, the PMI scores of POS-tagged noun phrases like "long movie" with positive and negative words from the polarity dictionary like "excellent" or "poor" are computed. The two PMI scores are then subtracted and, based on the result the original noun phrase, tagged as either positive or negative. For a given text, the polarity scores of all phrases are added and the sum's algebraic sign "detects" the text's overall sentiment. Originally, the PMI scores were determined via search engine requests but also large text corpora such as the ClueWeb or similar can be applied. The accuracy of the PMI method is not too impressive and can usually be improved when labeled data is available for training.

Supervised methods are trained on labeled data (i.e., texts with known polarity). In case of reviews, the actual assessment like "5 stars" or "1 star" can easily be translated to a sentiment. However, in case of sentiment detection in tweets, acquiring training data typically is a laborious and costly manual process. Typical methods of supervised learning for sentiment detection involve features like unigrams, bigrams, trigrams, polarity dictionaries, etc. Standard learning approaches range from Naive Bayes or Maximum Entropy to Support Vector Machines that learn the actual classifier from labeled training data [32]. Our approach is based on reproducing three supervised approaches, which are trained on data obtained from the SemEval Twitter sentiment detection task.

Several state-of-the-art methods for sentiment detection in texts exist but the important question then is for what scenarios the sentiment detection is actually useful [22]—besides determining the polarity of a text. Since we deal with Twitter data, the question is about use cases of detecting the sentiment of tweets: corresponding papers apply state-of-the-art sentiment detection on Twitter data to identify the general public's mood on given events from media, politics, culture, or economics [6]. This way, sentiment detection enables sociological studies at scale and almost in real time. Another paper studies the evaluation of politicians' TV debate performance based on sentiments expressed on Twitter [12]. This gives direct feedback to political election campaigns on what specific topics the voters are interested in and how the candidate's perceived performance is on that topic. Similarly, the general sentiment, or rather opinion, on products or events can be extracted from the Twitter stream [5], and aid in economic or sociological studies. As for companies, besides detecting sentiment for products or for marketing campaigns, also identifying the employees' mood can be beneficial for employee development programs and the like [27]. Of course, in this case the work force should be rather big and Twitter-savvy to get meaningful results.

As for more retrieval-oriented tasks, the ranking of products and reviews benefits from sentiment detection [10]: by identifying categories important to the users from sentiments expressed on Twitter, products ca be re-ranked accordingly. Moreover cross-language retrieval and ranking can incorporate sentiments and their respective translations [19].

Finally, annotating search results with the expressed general sentiment can be helpful as a facet in result presentation [11].

Due to the different applications in mining and retrieval, and since Twitter is one of the richest sources of opinion, a lot of different approaches to sentiment detection in tweets have been proposed. Different approaches use different feature sets ranging from standard word polarity expressions or unigram features also applied in general sentiment detection [17, 23], to the usage of emoticons and uppercases [4], word lengthening [8], phonetic features [13], multi-lingual machine translation [3], or word embeddings [40]. The task usually is to detect the sentiment expressed in a tweet as a whole (also focus of this paper). But it can also be to identify the sentiment in a tweet with respect to a given target concept expressed in a query [21]. The difference is that a generally negative tweet might not say anything about the target concept and must thus be considered neutral with respect to the target concept.

Both tasks, namely sentiment detection in a tweet, and sentiment detection with respect to a specific target concept, are part of the SemEval sentiment analysis tasks since 2013 [28, 38]. SemEval thus fosters research on sentiment detection for short texts in particular, and gathers the best-performing approaches in a friendly competition. The problem we are dealing with is formulated as subtask B in the SemEval 2013 task 2 and in the SemEval 2014 task 9: given a tweet, decide whether its message is positive, negative, or neutral. A few examples from the annotated SemEval 2013 training set give the gist of the task:

Positive: `Gas by my house hit $3.39!!!! I'm going to Chapel Hill on Sat. :)`
Negative: `Dream High 2 sucks compared to the 1st one.`
Neutral: `Battle for the 17th banner: Royal Rumble basketball edition`

State-of-the-art approaches have been submitted to the SemEval tasks. However, the organizers never trained a meta classifier based on the submitted approaches to determine what can be achieved when combining them, whereas each participating team only trains their individual classifier using respective individual feature set. Our idea is to combine three of the best-performing approaches with different feature sets, and to form an ensemble classifier that leverages the individual classifiers' strengths.

Ensemble learning is a classic approach of combining several weak classifiers to a more powerful ensemble [30, 33, 36]. The classic approaches of Bagging [7] and Boosting [39, 15] try to either combine the outputs of different classifiers trained on different random instances of the training set or on training the classifiers on instances that were misclassified by the other classifiers. Both rather work on the final predictions of the classifiers just as for instance averaging or majority voting on the predictions [1] would do. In our case, we employ the confidence scores of the participating classifiers. Several papers describe different ways of working with the classifiers' confidence scores, such as learning a dynamic confidence weighting scheme [16], or deriving a set cover with averaging confidences [37]. Instead, we simply average the three confidence scores of the three classifiers for each individual class. This straightforward approach performs superior to its individual parts and performs competitive in the SemEval competitions. Thus, its sentiment detection results can be directly used in any of the above use cases for Twitter sentiment detection.

3 Approaches

We select three state-of-the-art approaches for sentiment detection among the 38 participants of subtask B of the SemEval 2013 sentiment detection task. To identify worthy candidates—and to satisfy the claim "state of the art"—we picked the top-ranked approach by team NRC-Canada [26]. However, instead of simply picking the approaches on ranks two and three to complete our set, we first analyzed the notebooks of the top-ranked teams in order to identify approaches that are significantly dissimilar from the top-ranked approach. We decided to handpick approaches this way so they complement each other in an ensemble. As a second candidate, we picked team GU-MLT-LT [18] since it uses some other features and a different sentiment lexicon. Incidentally, it was ranked second. As a third candidate, we picked team KLUE [35], which was ranked fifth. We discarded the third-ranked approach as it is using a large set of not publicly available rules, whereas the fourth-ranked system seemed too similar to NRC and GU-MLT-LT to add something new to the planned ensemble.

This way, reproducing three approaches does not deteriorate into reimplementing the feature set of one approach and reusing it for the other two. Moreover, combining the three approaches into an ensemble classifier actually makes sense, since, due to the feature set diversity, they tap sufficiently different information sources. In what follows, we first briefly recap the features used by the individual classifiers and then explain our ensemble strategy.

3.1 NRC-Canada

Team NRC-Canada [26] used a classifier with a wide range of features. A tweet is first preprocessed by replacing URLs and user names by some placeholder. The tweets are then tokenized and POS-tagged. An SVM with linear kernel is trained using the following feature set:

N-grams. The occurrences of word 1-grams up to word 4-grams are used as features as well as occurrences of pairs of non-consecutive words where the intermediate words are replaced by a placeholder. No term-weighting like *tf·idf* is used. Similarly, character 3-grams up to character 5-grams are used as features.

ALLCAPS. The number of words written all capitalized is used as a feature.

Parts of speech. The occurrences of part-of-speech tags is a feature.

Polarity dictionaries. In total, five polarity dictionaries are used. Three of these were manually created: the NRC Emotion Lexicon [24, 25] with 14,000 words, the MPQA Lexicon [42] with 8,000 words, and the Bing Liu Lexicon [20] with 6,800 words. Two other dictionaries were created automatically. For the first one, the idea is that several hash tags can express sentiment (e.g., #good). Team NRC crawled 775,000 tweets from April to December 2012 that contain at least one of 32 positive or 38 negative hash tags that were manually created (e.g., #good and #bad). For word 1-grams and word 2-grams in the tweets, PMI-scores were calculated for each of the 70 hash tags to yield a score for the n-grams (i.e., the ones with higher positive hash tag PMI are positive, the others negative). The resulting dictionary contains 54,129 unigrams, 316,531 bigrams, and 308,808 pairs of non-consecutive words. The second automatically created dictionary is not based on PMI for hash tags but for emoticons. It was created similarly to the hash tag dictionary and contains 62,468 unigrams, 677,698 bigrams, and 480,010 pairs of non-consecutive words.

For each entry of the five dictionaries, the dictionary score is either positive, negative, or zero. For a tweet and each individual dictionary, several features are computed: the number of dictionary entries with a positive score and the number of entries with a negative score, the sum of the positive scores and the sum of the negative scores of the tweet's dictionary entries, the maximum positive score and minimum negative score of the tweet's dictionary entries, and the last positive score and negative score.

Punctuation marks. The number of non-single punctuation marks (e.g., !! or ?!) is used as a feature and whether the last one is an exclamation or a question mark.

Emoticons. The emoticons contained in a tweet, their polarity, and whether the last token of a tweet is an emoticon are employed features.

Word lengthening. The number of words that are lengthened by repeating a letter more than twice (e.g., cooooolll) is a feature.

Clustering. Via unsupervised Brown clustering [9] a set of 56,345,753 tweets by Owoputi [31] clustered into 1,000 clusters. The IDs of the clusters in which the terms of a tweet occur are also used as features.

Negation. The number of negated segments is another feature. According to Pang et al. [32] a negated segment starts with a negation (e.g., shouldn't) and ends with a punctuation mark. Further, every token in a negated segment (words, emoticons) gets a suffix NEG attached (e.g., perfect_NEG).

3.2 GU-MLT-LT

Team GU-MLT-LT [18] was ranked second in the SemEval 2013 ranking and trains a stochastic gradient decent classifier on a much smaller feature set compared to NRC. For feature computation, they use the original raw tweet, a lowercased normalized version of the tweet, and a version of the lowercased tweet where consecutive identical letters are collapsed (e.g., helllo gets hello). All three versions are tokenized. The following feature set is used:

Normalized unigrams. The occurrence of the normalized word unigrams is one feature set. Note that no term weighting like for instance $tf \cdot idf$ is used.

Stems. Porter stemming [34] is used to identify the occurrence of the stems of the collapsed word unigrams as another feature set. Again, no term weighting is applied.

Clustering. Similar to the NRC approach, the cluster IDs of the raw, normalized, and collapsed tokens is a feature set.

Polarity dictionary. The SentiWordNet assessments [2] of the individual collapsed tokens and the sum of all tokens' scores in a tweet are further features.

Negation. Normalized tokens and stems were added as negated features similarly to the NRC approach.

3.3 KLUE

Team KLUE [35] was ranked fifth in the SemEval 2013 ranking. Similarly to NRC, team KLUE first replaces URLs and user names by some placeholder and tokenizes the lowercased tweets. A maximum entropy-based classifier is trained on the following features.

N-grams. Word unigrams and bigrams are used as features but in contrast to NRC and GU-MLT-LT not just by occurrence but frequency-weighted. Due to the short tweet length this however often boils down to a simple occurrence feature. To be part of the feature set, an n-gram has to be contained in at least five tweets. This excludes some rather obscure and rare terms or misspellings.

Length. The number of tokens in a tweet (i.e., its length) is used as a feature. Interestingly, NRC and GU-MLT-LT do not explicitly use this feature.

Polarity dictionary. The employed dictionary is the AFINN-111 lexicon [29] containing 2,447 words with assessments from -5 (very negative) to $+5$ (very positive). Team KLUE added another 343 words. Employed features are the number of positive tokens in a tweet, the number of negative tokens, the number of tokens with a dictionary score, and the arithmetic mean of the scores in a tweet.

Emoticons and abbreviations. A list of 212 emoticons and 95 colloquial abbreviations from Wikipedia was manually scored as positive, negative, or neutral. For a tweet, again the number of positive and negative tokens from this list, the total number of scored tokens, and the arithmetic mean are used as features.

Negation. Negation is not treated for the whole segment as NRC and GU-MLT-LT do but only on the next three tokens except the case that the punctuation comes earlier. Only negated word unigrams are used as an additional feature set. The polarity scores from the above dictionary are multiplied by -1 for terms up to 4 tokens after the negation.

3.4 Remarks on Reimplementing the Original Approaches

As was to be expected, it turned out to be impossible to re-implement all features precisely as the original authors did. Either not all data was publicly available, or the features themselves were not sufficiently explained in the notebooks. We deliberated to contact the original authors to give them a chance to supply missing data as well as to elaborate on missing information. However, we ultimately opted against doing so for the following reason: our goal was to reproduce their results, not to repeat them. The difference between reproducibility and repeatability is subtle, yet important. If an approach can be re-implemented with incomplete information and if it then achieves a performance within the ballpark of the original, it can be considered much more robust than an approach that must be precisely the same as the original to achieve its expected performance. The former hints reproducibility, the latter only repeatability. This is why we have partly re-invented the approaches on our own, wherever information or data was missing. In doing so, we sometimes found ourselves in a situation where departing from the original approach would yield better performance. In such cases, we decided to maximize performance rather than sticking to the original, since in an evaluation setting, it is unfair to not maximize performance wherever one can.

Table 1. F1-scores of the original and reimplemented classifiers on the SemEval 2013 test data

Classifier	Original SemEval 2013	Reimplemented Version
NRC	69.02	69.44
GU-MLT-LT	65.27	67.27
KLUE	63.06	67.05

In particular, the emoticons and abbreviations added by the KLUE team were not available, such that we only choose the AFINN-111 polarity dictionary and re-implemented an emoticon detection and manual polarity scoring ourselves. We also chose not to use the frequency information in the KLUE system but only Boolean occurrence like NRC and GU-MLT-LT, since pilot studies on the SemEval 2013 training and development sets showed that to perform much better. For all three approaches, we unified tweet normalization regarding lowercasing and completely removing URLs and user names instead of adding a placeholder. As for the classifier itself, we did not use the learning algorithms used originally but L2-regularized logistic regression from the LIBLINEAR SVM library for all three approaches. In our pilot experiments on the SemEval 2013 training and development set this showed a very good trade-off between training time and accuracy. We set the cost parameter to 0.05 for NRC and to 0.15 for GU-MLT-LT and KLUE.

Note that neither of our design decisions hurt the individual performances but instead improve the accuracy for GU-MLT-LT and KLUE on the SemEval 2013 test set. Table 1 shows the performance of the original SemEval 2013 ranking and that of our re-implementations. Corresponding to the SemEval scoring, we report the averaged F1-score for the positive and negative class only. As can be seen, the NRC performance stays the same while GU-MLT-LT and KLUE are improved.

Altogether, we conclude that reproducing the SemEval approaches was generally possible but involved some subtleties that lead to difficult design decisions. As outlined, our resolution is to maximize performance rather than to dogmatically stick to the original approach. Our code for the three reproduced approaches as well as that of the ensemble described in the following section is publicly available.[1]

3.5 Ensemble Combination

In our pilot studies on the SemEval 2013 training and development sets, we tested several ways of combining the above three classifiers to an ensemble method. One of the main observations was that each individual approach classifies some tweets correctly that others do fail for. This is not too surprising given the different feature sets but also supports the idea of using an ensemble to combine the individual strengths. Although we briefly tried different ways of bagging and boosting the three classifiers, it soon turned out that some simpler combination performs better. A problem, for instance, was that some misclassified tweets are very difficult (e.g., the positive `Cant wait for`

[1] `http://www.uni-weimar.de/medien/webis/publications/by-year/#stein_2015b`

`the UCLA midnight madness tomorrow night`). Since often at least two classifiers fail on a hard tweet, this rules out some basic combination schemes, such as the majority vote among the three systems (the majority vote turned out to perform worse on the SemEval 2013 development set than NRC alone).

The solution that we finally came up with is motivated by observing how the three classifiers trained on the SemEval 2013 training set behave for tweets in the development set. Typically, not the three final decisions but the respective confidences or probabilities of the individual classifiers give a good hint on uncertainties. If two are not really sure about the final classification, sometimes the remaining third one favors another class with high confidence. Thus, instead of looking at the classifications, we decided to use the confidence scores or probabilities to build the ensemble. This approach is also motivated by old and also more recent papers on ensemble learning [1, 16, 37]. But instead of computing a weighting scheme for the different individual classifiers or learning the weights, we decided to simply compute the average probability of the three classifiers for each of the three classes (positive, negative, neutral).

Our ensemble thus works as follows. The three individual re-implementations of the NRC, the GU-MLT-LT, and the KLUE classifier are individually trained on the SemEval 2013 training set as if being applied individually—without boosting or bagging. As for the classification of a tweet, the ensemble ignores the individual classifiers' classification decisions but requests the classifiers' probabilities (or confidences) for each class. The ensemble decision then chooses the class with the highest average probability—again, no sophisticated techniques like dynamic confidence weighting [16] or set covering schemes [37] are involved. Thus, our final ensemble method is a rather straightforward system based on averaging confidences instead of voting schemes on the actual classifications of the individual classifiers. It can be easily implemented on top of the three classifiers and thus incurs no additional overhead. It also proves very effective in the following experimental evaluation.

4 Evaluation

To evaluate our ensemble approach, we employ the data sets provided for the SemEval 2013 and 2014 Twitter sentiment analysis tasks. More precisely, our setting is that of the subtask B (detect the sentiment of a whole tweet) while subtask A asks to detect the sentiment in a specific part of a tweet.

4.1 Evaluation Setup

The datasets used for the SemEval Twitter sentiment detection subtask B consist of a training set of 9,728 tweets (3,662 positive, 1,466 negative, 4,600 neutral), a developer set of 1,654 tweets (575 positive, 340 negative, 739 neutral), and two test sets. The test set from 2013 contains 3,813 tweets (1,572 positive, 601 negative, 1,640 neutral) while the smaller test set from 2014 contains 1,853 tweets (982 positive, 202 negative, 669 neutral). The tweets were crawled by the task organizers with a focus on topics relevant in the crawling period of January 2012 to January 2013 (the test set of 2014 was added later), including entities (e.g., Gaddafi, Steve Jobs), products (e.g., Kindle, Android phone), or events (e.g., Japan earthquake, NHL playoffs) [28, 38].

Table 2. Ranking and classification results on the SemEval Twitter data based on the F1-scores. The left part shows the original top ranks and the average score of 38 participants of SemEval 2013 with our ensemble included. The right part shows the results for SemEval 2014 (F1-scores for the 2014 and 2013 test data): the top ranks and average of 50 participants with our ensemble included according to its weaker rank on the 2014 data.

Ranking SemEval 2013		**Ranking SemEval 2014**		
Team	F1-score	Team	F1 on 2014	F1 on 2013
Our ensemble	**71.09**	TeamX	70.96	72.12
NRC-Canada	69.02	coooolll	70.14	70.40
GU-MLT-LT	65.27	RTRGO	69.95	69.10
teragram	64.86	NRC-Canada	69.85	70.75
BOUNCE	63.53	**Our ensemble**	**69.79**	**71.09**
KLUE	63.06	TUGAS	69.00	65.64
AMI&ERIC	62.55	CISUC KIS	67.95	67.56
FBM	61.17	SAIL	67.77	66.80
AVAYA	60.84	SWISS-CHOCOLATE	67.54	64.81
SAIL	60.14	Synalp-Empathic	67.43	63.65
Average	53.70	Average	60.41	59.78

The evaluation and ranking at SemEval 2013 and SemEval 2014 is based on the F1-score for the positive and negative tweets only. To compute the positive precision $prec_{\text{pos}}$, the number of tweets that were correctly classified as positive by the system is divided by the total number of tweets classified as positive by the system. Likewise, positive recall rec_{pos} is computed by dividing the number of tweets that were correctly classified as positive by the number of positive tweets in the gold standard test data. The F1-score for the positive class is computed as usual: $F_{\text{pos}} = \frac{2(prec_{\text{pos}}+rec_{\text{pos}})}{prec_{\text{pos}} \cdot rec_{\text{pos}}}$. Similarly, F_{neg} is computed from the negative precision and recall and the final overall score is the average F1-score $F = (F_{\text{pos}} + F_{\text{neg}})/2$.

4.2 General Performance

We use the SemEval 2013 training set to train the individual classifiers of our system. The SemEval 2013 development set is used to obtain the ensemble combination as described in Section 3.5. The ensemble is then tested on the 2013 test set against the participants of SemEval 2013 and on the 2013 and 2014 test sets against the participants of SemEval 2014. The results are shown in Table 2.

As can be seen on the 2013 participants (left part of Table 2), our ensemble method outperforms all 2013 participants and thus also the individual classifiers forming the ensemble. On the 2013 test data, our ensemble still takes the second place among the participants of SemEval 2014 while on the 2014 test data, our ensemble is ranked fifth (right part of Table 2). This places our ensemble system among the top-5 approaches in the two years of Twitter sentiment detection at SemEval. It would be an interesting direction for future research to try including new top-performing systems in our ensemble and to identify approaches among the top-performing 2014 participants that implement different paradigms to complement our ensemble. In a further analysis of the evaluation results, we examine the influence of the different classifiers in the ensemble and the characteristics of tweets with classification errors.

Table 3. F1-scores of the ensemble and without each individual classifier on the 2013 test data

Ensemble	F1-score	$prec_{pos}$	rec_{pos}	$prec_{neg}$	rec_{neg}
All	71.09	72.61	79.60	65.73	66.72
All - GU-MLT-LT	70.67 (-0.42)	72.83	78.78	67.31	64.06
All - KLUE	70.56 (-0.53)	73.39	78.59	65.22	65.22
All - NRC	68.80 (-2.29)	69.15	78.71	57.97	71.38

4.3 Component Influence

To check the influence of each individual classifier in our ensemble, we compare the ensemble of three classifiers to the ensembles with just two approaches—again, classification always is done by averaging the confidence scores. As can be seen from Table 3, each component is important for the overall score of the system. Leaving out the individually best classifier NRC reduced the performance the most but still results in an ensemble better than the two individual approaches. This is not too surprising since the observations on the development set showed that each individual classifier's confidence scores can sometimes help to avoid misclassifications. Hence, none of the individual and different classifiers should be removed from the ensemble. Adding appropriate other approaches is an interesting direction for future work.

4.4 Error Analysis

Analyzing the misclassifications of our ensemble sheds light on its robustness. The confusion matrices in Table 4 show a particularly nice feature of our ensemble. There are hardly severe misclassifications like classifying a positive tweet to be negative and vice versa. Most of the misclassifications put a positive or negative tweet in the neutral class which can be viewed as a rather safe option: in doubt it is often better to leave something without clear classification.

Altogether, the experimental results show our ensemble system to be able to robustly classify sentiment in tweets across different data sets and to rank among the top-performing approaches. The system itself is build straightforward from the individual approaches, each having their share in the achieved detection scores. Most of the errors observed concerns the misclassification of a tweet as neutral which is not the worst misclassification possible.

Table 4. Confusion matrices of gold and computed classes for SemEval 2013 and 2014 test data

		Computed classification of our ensemble					
		SemEval 2013			SemEval 2014		
		positive	neutral	negative	positive	neutral	negative
Gold label	positive	1,249	234	86	768	184	30
	neutral	394	1,123	123	177	436	56
	negative	77	123	401	41	33	128

5 Conclusion and Outlook

We have reproduced three state-of-the-art approaches to sentiment detection for Twitter tweets. Our findings include that not all aspects of the approaches could be reproduced precisely, but that missing data, missing information, as well as opportunities to improve the approaches' performances lead us to re-invent them and to depart to some extent from the original descriptions. All of our changes have improved the performances of the original approaches. Moreover, we have demonstrated that the approaches can be reproduced even with incomplete information about them, which is a much stronger property than being merely repeatable.

In addition, we investigated a combination of confidence scores of the three approaches within an ensemble that altogether yields a top-performing Twitter sentiment detection system. The ensemble computation is as efficient as its components, and its effectiveness was shown on the SemEval Twitter sentiment evaluation data where our system ranks among the top-5 approaches for various data sets.

Promising directions for future research are the analysis of further potential classifiers useful as ensemble components. Following our philosophy of selecting approaches that are significantly different from each other, it will be interesting to observe how much new approaches can improve our existing ensemble. A further error analysis from a retrieval perspective might also be promising. We found that our ensemble classifier makes few severe errors, such as misclassifying a negative tweet as positive or vice versa. From a retrieval perspective, it would be nice to examine the influence of such misclassifications on the perceived usefulness of a system that incorporates sentiment detection results into search results, e.g., for diversification.

References

[1] Asker, L., Maclin, R.: Ensembles as a sequence of classifiers. In: Proc. of IJCAI, pp. 860–865 (1997)
[2] Baccianella, S., Esuli, A., Sebastiani, F.: SentiWordNet 3.0: An enhanced lexical resource for sentiment analysis and opinion mining. In: Proc. of LREC (2010)
[3] Balahur, A., Turchi, M.: Improving sentiment analysis in twitter using multilingual machine translated data. In: Proc. of RANLP 2013, pp. 49–55 (2013)
[4] Barbosa, L., Feng, J.: Robust sentiment detection on twitter from biased and noisy data. In: Proc. of COLING 2010, pp. 36–44 (2010)
[5] Bifet, A., Frank, E.: Sentiment knowledge discovery in twitter streaming data. In: Proc. of DS 2010, pp. 1–15 (2010)
[6] Bollen, J., Mao, H., Pepe, A.: Modeling public mood and emotion: Twitter sentiment and socio-economic phenomena. In: Proc. of ICWSM 2011 (2011)
[7] Breiman, L.: Bagging predictors. Machine Learning 24(2), 123–140 (1996)
[8] Brody, S., Diakopoulos, N.: Cooooooooooooooollllllllllllll!!!!!!!!!!!!!! Using word lengthening to detect sentiment in microblogs. In: Proc. of EMNLP 2011, pp. 562–570 (2011)
[9] Brown, P.F., Della Pietra, V.J., de Souza, P.V., Lai, J.C., Mercer, R.L.: Class-based n-gram models of natural language. Computational Linguistics 18(4), 467–479 (1992)
[10] de Albornoz, J.C., Plaza, L., Gervás, P., Díaz, A.: A joint model of feature mining and sentiment analysis for product review rating. In: Clough, P., Foley, C., Gurrin, C., Jones, G.J.F., Kraaij, W., Lee, H., Mudoch, V. (eds.) ECIR 2011. LNCS, vol. 6611, pp. 55–66. Springer, Heidelberg (2011)

[11] Demartini, G.: ARES: A Retrieval Engine Based on Sentiments. In: Clough, P., Foley, C., Gurrin, C., Jones, G.J.F., Kraaij, W., Lee, H., Mudoch, V. (eds.) ECIR 2011. LNCS, vol. 6611, pp. 772–775. Springer, Heidelberg (2011)

[12] Diakopoulos, N., Shamma, D.A.: Characterizing debate performance via aggregated twitter sentiment. In: Proc. of CHI 2010, pp. 1195–1198 (2010)

[13] Ermakov, S., Ermakova, L.: Sentiment classification based on phonetic characteristics. In: Proc. of ECIR 2013, pp. 706–709 (2013)

[14] Feldman, R.: Techniques and applications for sentiment analysis. CACM 56(4), 82–89 (2013)

[15] Freund, Y., Schapire, R.E.: Experiments with a new boosting algorithm. In: Proc. of ICML 1996, pp. 148–156 (1996)

[16] Fung, G.P.C., Yu, J.X., Wang, H., Cheung, D.W., Liu, H.: A balanced ensemble approach to weighting classifiers for text classification. In: Proc. of ICDM 2006, pp. 869–873 (2006)

[17] Go, A., Bhayani, R., Huang, L.: Twitter sentiment classification using distant supervision. Project Report CS224N, Stanford University (2009)

[18] Günther, T., Furrer, L.: GU-MLT-LT: Sentiment analysis of short messages using linguistic features and stochastic gradient descent. In: Proc. of SemEval 2013, pp. 328–332 (2013)

[19] He, Y.: Latent sentiment model for weakly-supervised cross-lingual sentiment classification. In: Clough, P., Foley, C., Gurrin, C., Jones, G.J.F., Kraaij, W., Lee, H., Mudoch, V. (eds.) ECIR 2011. LNCS, vol. 6611, pp. 214–225. Springer, Heidelberg (2011)

[20] Hu, M., Liu, B.: Mining and summarizing customer reviews. In: Proc. of KDD 2004, pp. 168–177 (2004)

[21] Jiang, L., Yu, M., Zhou, M., Liu, X., Zhao, T.: Target-dependent twitter sentiment classification. In: Proc. of HLT 2011, pp. 151–160 (2011)

[22] Karlgren, J., Sahlgren, M., Olsson, F., Espinoza, F., Hamfors, O.: Usefulness of sentiment analysis. In: Baeza-Yates, R., de Vries, A.P., Zaragoza, H., Cambazoglu, B.B., Murdock, V., Lempel, R., Silvestri, F. (eds.) ECIR 2012. LNCS, vol. 7224, pp. 426–435. Springer, Heidelberg (2012)

[23] Kouloumpis, E., Wilson, T., Moore, J.D.: Twitter sentiment analysis: The good the bad and the OMG! In: Proc. of ICWSM (2011)

[24] Mohammad, S.M., Turney, P.D.: Emotions evoked by common words and phrases: Using mechanical turk to create an emotion lexicon. In: Proc. of HLT 2010 Workshop CAAGET 2010, pp. 26–34 (2010)

[25] Mohammad, S.M., Turney, P.D.: Crowdsourcing a word-emotion association lexicon. Computational Intelligence 29(3), 436–465 (2013)

[26] Mohammad, S.M., Kiritchenko, S., Zhu, X.: NRC-Canada: Building the state-of-the-art in sentiment analysis of tweets. In: Proc. of SemEval 2013, pp. 321–327 (2013)

[27] Moniz, A., de Jong, F.: Sentiment analysis and the impact of employee satisfaction on firm earnings. In: de Rijke, M., Kenter, T., de Vries, A.P., Zhai, C., de Jong, F., Radinsky, K., Hofmann, K. (eds.) ECIR 2014. LNCS, vol. 8416, pp. 519–527. Springer, Heidelberg (2014)

[28] Nakov, P., Kozareva, Z., Ritter, A., Rosenthal, S., Stoyanov, V., Wilson, T.: Semeval-2013 task 2: Sentiment analysis in Twitter. In: Proc. of SemEval 2013, pp. 312–320 (2013)

[29] Nielsen, F.Å.: A new ANEW: evaluation of a word list for sentiment analysis in microblogs. In: Proc. of ESWC 2011 Workshop MSM 2011, pp. 93–98 (2011)

[30] Opitz, D.W., Maclin, R.: Popular ensemble methods: An empirical study. Journal of Artificial Intelligence Research 11, 169–198 (1999)

[31] Owoputi, O., O'Connor, B., Dyer, C., Gimpel, K., Schneider, N., Smith, N.A.: Improved part-of-speech tagging for online conversational text with word clusters. In: Proc. of HLT 2013, pp. 380–390 (2013)

[32] Pang, B., Lee, L., Vaithyanathan, S.: Thumbs up?: Sentiment classification using machine learning techniques. In: Proc. of EMNLP 2002, pp. 79–86 (2002)
[33] Polikar, R.: Ensemble based systems in decision making. IEEE CASS Mag 6(3), 21–45 (2006)
[34] Porter, M.: An algorithm for suffix stripping. Program 14(3), 130–137 (1980)
[35] Proisl, T., Greiner, P., Evert, S., Kabashi, B.: Klue: Simple and robust methods for polarity classification. In: Proc. of SemEval 2013, pp. 395–401 (2013)
[36] Rokach, L.: Ensemble-based classifiers. Artificial Intelligence Review 33(1-2), 1–39 (2010)
[37] Rokach, L., Schclar, A., Itach, E.: Ensemble methods for multi-label classification. Expert Systems with Applications 41(16), 7507–7523 (2014)
[38] Rosenthal, S., Ritter, A., Nakov, P., Stoyanov, V.: Semeval-2014 task 9: Sentiment analysis in twitter. In: Proc. of SemEval 2014, pp. 73–80 (2014)
[39] Schapire, R.E.: The strength of weak learnability. Machine Learning 5, 197–227 (1990)
[40] Tang, D., Wei, F., Yang, N., Zhou, M., Liu, T., Qin, B.: Learning sentiment-specific word embedding for twitter sentiment classification. In: Proc. of ACL 2014, pp. 1555–1565 (2014)
[41] Turney, P.D.: Thumbs up or thumbs down? Semantic orientation applied to unsupervised classification of reviews. In: Proc. of ACL 2002, pp. 417–424 (2002)
[42] Wilson, T., Wiebe, J., Hoffmann, P.: Recognizing contextual polarity in phrase-level sentiment analysis. In: Proc. of EMNLP 2005, pp. 347–354 (2005)

Reproducible Experiments on Lexical and Temporal Feedback for Tweet Search

Jinfeng Rao[1], Jimmy Lin[1], and Miles Efron[2]

[1] University of Maryland, College Park
jinfeng@cs.umd.edu, jimmylin@umd.edu
[2] University of Illinois, Urbana-Champaign
mefron@illinois.edu

Abstract. "Evaluation as a service" (EaaS) is a new methodology for community-wide evaluations where an API provides the only point of access to the collection for completing the evaluation task. Two important advantages of this model are that it enables reproducible IR experiments and encourages sharing of pluggable open-source components. In this paper, we illustrate both advantages by providing open-source implementations of lexical and temporal feedback techniques for tweet search built on the TREC Microblog API. For the most part, we are able to reproduce results reported in previous papers and confirm their general findings. However, experiments on new test collections and additional analyses provide a more nuanced look at the results and highlight issues not discussed in previous studies, particularly the large variances in effectiveness associated with training/test splits.

Keywords: TREC Microblog, evaluation as a service, search API.

1 Introduction

"Evaluation as a service" (EaaS) [11] is a new methodology that enables community-wide evaluations and the construction of test collections on documents that cannot be distributed. The basic idea is that instead of providing the document collection in a downloadable form, as is standard in most TREC, NTCIR, CLEF, and other evaluations, the organizers provide a service API through which the evaluation task can be completed [12]. Typically, the API would provide keyword search capabilities, but it can be augmented with additional features customized to the evaluation task at hand. The key point is, however, that the API provides the sole access point to the document collection, and thus it can be engineered to respect restrictions on the dissemination of content.

One important advantage of the evaluation-as-a-service model is that it enables reproducible IR experiments. Modern search systems have become complex collections of components for document ingestion, inverted indexing, query evaluation, document ranking, and machine learning. As a result, it can be difficult to isolate and attribute differences in effectiveness to specific components, algorithms, or techniques. Consider a baseline retrieval model such as BM25 or

A. Hanbury et al. (Eds.): ECIR 2015, LNCS 9022, pp. 755–767, 2015.

query-likelihood within the language modeling framework—alternative implementations may produce substantially different results due to small but consequential decisions such as the tokenization strategy, stemming algorithm, method for pruning the term space (e.g., discarding long or rare terms), minor scoring variations, and other engineering issues [16,22]. In some cases, the effects that we are hoping to study are masked by differences we are not interested in. The evaluation-as-a-service model addresses many of these issues by deploying a common API that is used by all participants. This means that everything "below" the API (e.g., indexing, tokenization, etc.) is *exactly* the same for everyone. Thus, we can be confident that differences in effectiveness can be attributed to retrieval techniques on top of the API, rather than "uninteresting" issues.

Additionally, we believe that this model is conducive to an open culture of sharing pluggable system components. There is broad recognition that open-source software advances the state of the art; a common API increases the likelihood that code components inter-operate, thus increasing the likelihood of adoption. Although there is already widespread availability of open-source search engines, nearly all systems are monolithic in that they were not designed for service decomposition along functional boundaries. This means that a particular algorithm developed for one system cannot be easily used by researchers who have written their code on another system due to interface incompatibilities. A common API begins to address these issues.

In this paper, we illustrate both advantages of the evaluation-as-a-service model by reproducing lexical and temporal feedback techniques for searching tweets in the context of the TREC Microblog tracks, which was the first to operationalize this evaluation model. By lexical feedback we mean pseudo-relevance feedback where an initial set of retrieved documents is exploited to refine the query model. Since tweets are very short, a number of researchers have suggested that the query expansion effects of pseudo-relevance feedback are beneficial to search effectiveness. To rigorously test this insight, we have reimplemented the popular RM3 approach [8] using the TREC Microblog API. We use the term *lexical* feedback to distinguish RM3 (and related models) from techniques that take advantage of the *temporal* information of documents, which is the focus of our second set of experiments. We attempt to reproduce the techniques proposed in a recent SIGIR paper by Efron et al. [5], reimplementing their proposed algorithms based on kernel density estimation (KDE) as well as two other temporal-ranking techniques. Finally, we combine both lexical and temporal feedback to explore the question of whether the effectiveness gains are cumulative. All of the source code for experiments conducted in this paper can be found in our open-source code repository.[1]

Our reproducibility efforts were largely successful and our experimental results are consistent with previous studies for the most part. However, through more extensive experiments on new test collections and additional analyses, we provide a more nuanced look at previous results. In particular, we note the large variances in effectiveness associated with training/test splits of test collections.

[1] `http://twittertools.cc/`

2 Background

The context for our study is the recent Microblog tracks at TREC [17,21,11], which have been running since 2011. Although the task has remained essentially the same, the evaluation methodology has changed over the years, and so it is worth providing an overview.

The TREC Microblog tracks in 2013 and 2014 used the evaluation-as-a-service model described in the introduction. The API served the Tweets2013 collection, which consists of 243 million tweets crawled from Twitter's public sample stream between February 1 and March 31, 2013 (inclusive). Although the "official" collection is not available for download, participants could acquire substantively similar data by also crawling the public stream during the same time (which was coordinated on the track mailing list and indeed, many participants did acquire tweets in this manner). In contrast, the 2011 and 2012 evaluations used the Tweets2011 corpus, which consists of an approximately 1% sample (after some spam removal) of tweets from January 23, 2011 to February 7, 2011 (inclusive), totaling approximately 16 million tweets. For those evaluations, the TREC organizers made the list of tweet ids that comprise the collection available, and together with a distributed crawler (also supplied by the organizers), participants could download the actual tweets from Twitter itself (this approach does not scale to the much larger Tweets2013 collection). Note, however, that the collection acquired by each participant might be slightly different due to transient network glitches, message deletions, removal of spam accounts, and a whole host of other factors. Nevertheless, a study in 2012 [13] found that these artifacts did not impact the stability of the test collection. These methodological differences add an extra dimension of interest in our studies, as we would like to examine the impact of having a local collection vs. using the service API.

The formulation of the tweet search problem for the TREC Microblog track is as follows: at time t, a user expresses an information need in the form of a query Q. The system's task is to return topically-relevant documents (tweets) posted before the query time. Thus, each topic consists of a query and an associated timestamp, which indicates when the query was issued. There are 50 topics for TREC 2011, 60 topics for TREC 2012, 60 topics for TREC 2013, and 55 topics for TREC 2014. NIST assessors used a standard pooling strategy for evaluation, assigning one of three judgments to each tweet in the pool: "not relevant", "relevant", and "highly relevant". For the purpose of our experiments, we considered both "relevant" and "highly relevant" tweets to be relevant.

In addition to the official API used for TREC 2013 and 2014 (which served the Tweets2013 collection), the organizers also provided an API that serves the smaller Tweets2011 collection so that participants could runs experiments using topics from TREC 2011 and 2012. Both APIs were identical except for the underlying document collection, and were implemented in Java using service definitions provided by Thrift[2] and Lucene[3] as the underlying search engine.

[2] http://thrift.apache.org/

[3] http://lucene.apache.org/

Ranking was provided using Lucene's implementation of query-likelihood in the language modeling framework [18]. The API returned up to 10000 hits, and each hit contained the full text of the tweet and associated metadata (statistics about the user, the source tweet if the tweet was a retweet or a reply, etc.). There is one implementation detail worth mentioning—for efficiency reasons, Lucene implements a rank-equivalent scoring model to query-likelihood,[4] which cannot be used in more complex ranking models that depend on valid log probabilities. To get around this issue, a patch was made to the service API whereby the client could (optionally) request that the system compute valid query-likelihood probabilities in a second pass after the initial retrieval. In all our experiments, we enabled this option.

Code for all experiments reported in this paper, implemented in Java, has been open sourced and integrates directly with the TREC Microblog API. Reproducing our results is as simple as executing the command-line invocations included in our documentation—the evaluation-as-a-service model obviates the need to download the document collection, build inverted indexes, etc.

3 Lexical Feedback with Relevance Models

A longstanding challenge in information retrieval is the issue of vocabulary mismatch, where queries are expressed using terms not present in relevant documents. Query expansion techniques, particularly those based on pseudo-relevance feedback, are often used to address this problem; there is a long history of research in this area dating back many decades [19]. The brevity of tweets exacerbates vocabulary mismatch, and thus query expansion techniques are likely to improve the effectiveness of tweet search. This is an insight shared by many researchers—for example, many of the most effective runs from TREC 2011 take advantage of such techniques in various guises [2,10,14]. In our first set of experiments, we wished to verify the effectiveness of pseudo-relevance feedback for tweet search, and to that end, we implemented relevance models [8] using the TREC Microblog API. Relevance models have specifically been explored for tweet search in a few previous studies [3,15], which provides a point of reference for our reproducibility efforts.

Given a query Q consisting of n query terms $\{q_1, q_2, ...q_n\}$, its relevance model $P(w|R_Q)$ is simply a weighted average of the terms in all documents, where the weights are the query likelihood scores:

$$P(w|R_Q) = \sum_{D \in \mathcal{D}} P(D)P(w|D) \prod_{i=1}^{n} P(q_i|D). \quad (1)$$

In the RM3 variant [1], the above model is interpolated with the observed query model according to a mixing parameter γ. In our experiments, we set $\gamma = 0.5$, which is the default value in the Indri implementation. In practice, RM3 is typically implemented using query expansion (e.g., augmenting the original query

[4] See equation (4) in [24] for more details.

Table 1. Results comparing query-likelihood (QL) against RM3, where the relevance models are estimated with retweets included (+retweets) or discarded (−retweets)

	2011/12		2013/14	
Method	MAP	P30	MAP	P30
QL	0.2692	0.3552	0.3266	0.5156
RM3 (+retweets)	0.3005*	0.3778*	0.3629*	0.5351*
RM3 (−retweets)	0.3003*	0.3787*	0.3597*	0.5357*

using Indri query operators); our implementation follows this approach as well. Following common parameter settings, we estimated the relevance models from $k = 50$ pseudo-relevant documents and selected $n = 20$ feedback terms.

Experimental results are shown in Table 1 for TREC 2011/12 and 2013/14, reporting mean average precision (MAP) to 1000 hits and precision at rank 30 (P30) computed with `trec_eval`. There is no training/test split because we are not tuning parameters, but simply using "best practice" defaults from the literature. Query-likelihood provides a baseline for comparison. The symbol * indicates that the difference with respect to the baseline is statistically significant ($p < 0.01$) based on Fisher's two-sided, paired randomization test [20]. Our experiments also examined the impact of one detail we have not seen much discussion about in the literature: the effect of retweets. According to the assessment guidelines, retweets that provide no additional information are considered not relevant. Thus, it makes sense to remove all retweets from the final results (which we did here and for all subsequent runs). However, it is unclear if the retweets should be included or discarded when estimating relevance models—thus, we tried both conditions.

Results show that RM3 yields significant and consistent improvements over the query-likelihood baseline in terms of MAP and P30. Furthermore, it does not appear to matter whether or not retweets are included in the estimation of the relevance model. To summarize, we have successfully reproduced previous results and confirmed that the benefits of pseudo-relevance feedback are robust.

4 Temporal Feedback with Kernel Density Estimation

4.1 Overview

Relevance models represent a popular approach to lexical feedback, taking advantage of an initial set of search results to refine the system's estimate of the term distribution of relevant documents. We can extend this idea to temporal feedback by estimating the temporal density of relevance—which characterizes where along a timeline we would expect relevant documents to appear. For information needs where temporality plays an important factor (as is common in tweet search), we would expect a non-uniform distribution of documents over time, and hence there might be a temporal relevance signal that can be exploited. In the same way that an initial set of search results can be used to estimate relevance models, we can estimate the temporal density of relevance from an initial list of retrieved documents. These are the ideas behind the work of Efron et al. [5], which we reproduce here. Below, we briefly summarize the relevant techniques.

As a starting point, consider the query-likelihood approach in the language modeling framework [18]. Documents are ranked by $P(D|Q) \propto P(Q|D)P(D)$, where $P(Q|D)$ is the likelihood that the language model that generated document D would also generate query Q, and $P(D)$ is the prior distribution.

Recency Priors. Li and Croft [9] incorporate temporal information using a prior that favors recent documents, modeling $P(D)$ with an exponential $P(D) = \lambda e^{-\lambda T_D}$, where T_D is the timestamp of document D and $\lambda \geq 0$ is the rate parameter. We refer to this as a *recency prior*, or "Recency" for short.

Moving Window Approach. Recency priors are query-independent and unable to account for information needs with different temporal profiles [7]. Dakka et al. [4] proposed a query-specific way to combine lexical and temporal evidence in the language modeling framework by separating the lexical and temporal signals into two components: W_D, the words in the document and T_D, the document's timestamp. This leads to the following derivation:

$$P(D|Q) = P(W_D, T_D|Q) \tag{2}$$
$$= P(T_D|W_D, Q)P(W_D|Q) \tag{3}$$
$$\sim P(W_D|Q)P(T_D|Q) \tag{4}$$

where the last step follows if we assume independence between lexical and temporal evidence. The result is similar to standard query-likelihood, but with the addition of the probability of observing a time T_D given the query Q.

Dakka et al. proposed several ways to estimate $P(T_D|Q)$. In the moving window approach (WIN for short), initial documents retrieved for Q are allocated among b bins according to their timestamps. For each bin b_t, we count $n(b_t)$, the number of retrieved documents in b_t. Next, bin counts are smoothed by averaging x bins into the past and x bins into the future. Let $n(b_{tx})$ be the average number of documents in the $2x$ bins surrounding b_t and b_t itself. Finally, bins are arranged in decreasing order of $n(b_{tx})$. The quantity $P(T_D|Q)$ depends on the bin associated with T_D. If T_D is in the nth ordered bin, then $P(T_D|Q) = \phi(n, \lambda)$ where ϕ is an exponential distribution with rate parameter λ.

Kernel Density Estimates. To estimate the temporal distribution of relevant documents, Efron et al. proposed using kernel density estimation (KDE) [6]. Let $\{x_1, x_2, \ldots, x_n\}$ be i.i.d. samples drawn from some distribution with an unknown density f. We are interested in estimating the shape of this function f. Its kernel density estimator is:

$$\hat{f}_\omega(x) = \frac{1}{nh}\sum_{i=0}^{n} \omega_i K\left(\frac{x - x_i}{h}\right) \tag{5}$$

where $K(\cdot)$ is the kernel—a symmetric function that integrates to one (in our case, a Gaussian)—and $h > 0$ is a smoothing parameter called the bandwidth.

For bandwidth selection we use what is known as the "robust rule of thumb" [23], which yields a bandwidth automatically. KDE additionally has the ability to handle weighted observations, given non-negative weights $\{\omega_1, \omega_2, \ldots, \omega_n\}$ such that $\sum \omega_i = 1$. Consider four different weighting schemes:

- *Uniform weights.* The simplest approach is to give all documents in the initial results equal weights.
- *Score-based weights.* We can weight each document based on its query-likelihood, i.e.,

$$\omega_i^s = \frac{P(Q|D_i)}{\sum_{j=1}^{n} P(Q|D_j)}. \tag{6}$$

- *Rank-based weights.* We can adopt a rank-based scheme that preserves ordering in the initial results, but not the actual score differences, via an exponential distribution:

$$\omega_i^r = \frac{\lambda e^{-\lambda r_i}}{\sum_{j=1}^{n} \lambda e^{-\lambda r_i}} \tag{7}$$

  where $\lambda > 0$ is the rate parameter of the exponential and r_i is the rank of document D_i in R. Though we could leave λ as a tuneable parameter, a simpler approach is to use the maximum likelihood estimate. If R contains n documents, the MLE of λ is simply $\frac{1}{\bar{r}}$, where $\bar{r}$ is the mean of the ranks $1, 2, \ldots, n$.
- *Oracle.* An upper bound can be characterized by an oracle where the density estimates are derived from documents marked relevant by human assessors.

To combine the temporal and lexical evidence, Efron et al. proposed a simple log-linear model. For a parameter $\alpha \in [0, 1]$, we have

$$\log P_\alpha(R|D, Q) = Z_\alpha + (1 - \alpha) \log P(R|W_D, Q) + \alpha \log P(R|T_D, Q) \tag{8}$$

where Z_α is a normalization constant. Since Z_α does not depend on D for ranking, we can ignore it; α is a free parameter learned from data.

4.2 Experimental Results

Experiments in Efron et al. were conducted on topics from TREC 2011 and 2012 over a local copy of the Tweets2011 corpus. During corpus preparation, all retweets were eliminated. Thus, the collection used in those experiments is substantively different from the corpus behind the official TREC Microblog API, which does include retweets. This is an additional factor that might affect the reproducibility of their results. To be clear, however, retweets are removed in all cases in the final ranked list prior to evaluation.

In our first set of experiments, we attempted to reproduce the experimental conditions in Efron et al. as closely as possible. Even-numbered topics from TREC 2011 and 2012 were used for training, and odd-numbered topics for

Table 2. Results from attempting to reproduce experiments in Efron et al. [5] as closely as possible. Metrics computed over odd-numbered topics from TREC 2011/12, training on even-numbered topics. Columns marked "original" contain results copied from the previous SIGIR paper; columns marked "reproduced" show reproduced results.

	MAP		**P30**	
Condition	**original**	**reproduced**	**original**	**reproduced**
QL	0.2363	0.2705	0.3473	0.3582
Recency	0.2467°	0.2766	0.3642°	0.3607
WIN	0.2407	0.2548	0.3515	0.3449
KDE (uniform)	0.2457°	0.2685	0.3618°	0.3534
KDE (score-based)	$0.2505^{\bullet\dagger}$	0.2719	0.3606°	0.3582
KDE (rank-based)	$0.2546^{\bullet\triangle\dagger}$	0.2724	$0.3709^{\bullet\ddagger}$	0.3649
KDE (oracle)	$0.2843^{\bullet\blacktriangle\ddagger}$	$0.3045^{\bullet\blacktriangle\ddagger}$	$0.4024^{\bullet\blacktriangle\ddagger}$	$0.3922^{\bullet\blacktriangle\ddagger}$

Table 3. Symbols indicating statistically significant change for data reporting

Symbol	Description
$\circ$, $\bullet$	improvements over the QL baseline ($p < 0.05$, $p < 0.01$)
$\triangle$, $\blacktriangle$	improvements over the recency prior ($p < 0.05$, $p < 0.01$)
$\dagger$, $\ddagger$	improvements over the WIN method ($p < 0.05$, $p < 0.01$)

testing. These results are shown in Table 2, with QL representing the query-likelihood baseline; we characterize effectiveness in terms of MAP (to rank 1000) and precision at rank 30 (P30). The free parameters for each technique were tuned via grid search to optimize MAP and P30 (separately).[5] Results are annotated with symbols indicating the statistical significance of improvements as shown in Table 3. Following Efron et al., we applied one-sided paired t-tests for significance testing.

These results are somewhat different from those reported in Efron et al. We immediately noticed the differences between the two versions of the QL baseline (Indri for the SIGIR paper and Lucene + QL recomputation for the TREC Microblog API). Although both are putatively implementing the same ranking function (Dirichlet scores), there is a fairly large difference in MAP. There are many possible sources for these differences, including the fact that the two experiments are actually on *different* collections, as well as issues related to corpus preparation such as removal of retweets, stemming, tokenization, etc. This further affirms the arguments behind the evaluation-as-a-service model in providing a common starting point for everyone—otherwise, relatively uninteresting differences could easily mask the effects of the techniques we are studying.

Consistent with the original work, we find that the KDE oracle condition is highly effective, which indicates that there is a strong temporal relevance signal. We observe improvements for rank-based KDE in terms of MAP and P30, albeit

[5] Note that in these experiments we did not compute metrics using `trec_eval` to facilitate tighter training/test coupling, and thus there are small differences between our values and those reported using `trec_eval` due to how scoring ties are handled.

Table 4. Results from TREC 2011/12. "Cross" represents training using all TREC 2013/14 topics; "Even-Odd" represents training on even-numbered topics and testing on odd-numbered topics; "Odd-Even" represents switching train/test.

	Cross		**Even-Odd**		**Odd-Even**	
Metric	**MAP**	**P30**	**MAP**	**P30**	**MAP**	**P30**
QL	0.2689	0.3562	0.2705	0.3582	0.2673	0.3541
Recency	0.2748	0.3578	0.2766	0.3607	0.2721	0.3509
WIN	0.2689	0.3578	0.2548	0.3449	0.2673	0.3541
KDE (uniform)	0.2699	0.3568	0.2685	0.3534	0.2702	0.3553
KDE (score-based)	0.2711	0.3673	0.2719	0.3582	0.2697	0.3698
KDE (rank-based)	0.2707	0.3655	0.2724	0.3649	0.2716	0.3616
KDE (oracle)	$0.3032^{\bullet\blacktriangle\ddagger}$	$0.3988^{\bullet\blacktriangle\ddagger}$	$0.3045^{\bullet\blacktriangle\ddagger}$	$0.3922^{\bullet\blacktriangle\ddagger}$	$0.3001^{\bullet\blacktriangle\ddagger}$	$0.4069^{\bullet\blacktriangle\dagger}$

not statistically significant. Furthermore, the WIN approach does not seem to be effective. We suspect that these findings may be attributed to the much improved QL baseline in our experiments.

We extended the experiments of Efron et al. in two ways. First, we evaluated the techniques on test collections from TREC 2013 and 2014. This not only provides (roughly) double the number of topics, but also allows us to examine the effects of a much larger corpus. An open question is whether the proposed techniques would remain effective for a collection spanning a much longer duration (about two weeks for Tweets2011 compared to two months for Tweets2013); we now have an opportunity to answer this question. Second, during our experiments we noticed large effectiveness differences that stemmed from different training/test splits; we wanted to explore these effects in more detail.

In terms of training regimes, one simple approach is to arbitrarily divide the test collection into halves; train on one half and test on the other half. Splitting topics by topic number is a perfectly acceptable arbitrary division: we can train on even-numbered topics and test on odd-numbered topics (as before), and also flip the two halves (i.e., train on odd, test on even). Another reasonable strategy might be to consider the TREC 2011/12 topics to be a unit, train on all those topics, and test on TREC 2013/14 topics; and also the other way around, i.e., train on TREC 2013/14 topics and test on all TREC 2011/12 topics. This condition assesses whether it may be possible to generalize parameters across different collections, i.e., a simple form of transfer learning.

Results from these experiments are shown in Table 4 for TREC 2011/12 and Table 5 for TREC 2013/14. Note that the figures reported in Table 2 are the same as those in the "Even-Odd" column in Table 4. In these experiments we used Fisher's two-sided, paired randomization test [20], reflecting better practice than the one-sided paired t-tests used in the SIGIR experiments. Results appear to show that the KDE techniques are more effective on the TREC 2013/2014 test collection. For rank-based weights, the differences are statistically significant in most cases.

To further explore the train/test split issue, we conducted a series of trials where we randomly divided the TREC 2011/12 and TREC 2013/14 topics in

Table 5. Results from TREC 2013/14. "Cross" represents training using all TREC 2011/12 topics; other conditions have the same meaning as in Table 4.

	Cross		**Even-Odd**		**Odd-Even**	
Metric	**MAP**	**P30**	**MAP**	**P30**	**MAP**	**P30**
QL	0.3139	0.5197	0.3559	0.5638	0.2712	0.4749
Recency	0.3129	0.5336	0.3593	0.5736	0.2773	0.4994
WIN	0.3139	0.5197	0.3559	0.5638	0.2712	0.4749
KDE (uniform)	0.3121	0.5177	0.3490	0.5603	0.2737	0.4795
KDE (score-based)	0.3140	0.5206	0.3516	0.5747	0.2753	0.4795
KDE (rank-based)	0.3267 $^{\bullet\ddagger}$	0.5542$^{\bullet\ddagger}$	0.3600	0.5983$^{\bullet\blacktriangle\ddagger}$	0.2949 $^{\bullet\triangle\ddagger}$	0.5228 $^{\bullet\ddagger}$
KDE (oracle)	0.3492$^{\bullet\blacktriangle\ddagger}$	0.5829$^{\bullet\blacktriangle\ddagger}$	0.3816$^{\bullet\blacktriangle\dagger}$	0.6328$^{\bullet\blacktriangle\ddagger}$	0.3135$^{\bullet\blacktriangle\ddagger}$	0.5363$^{\bullet\blacktriangle\ddagger}$

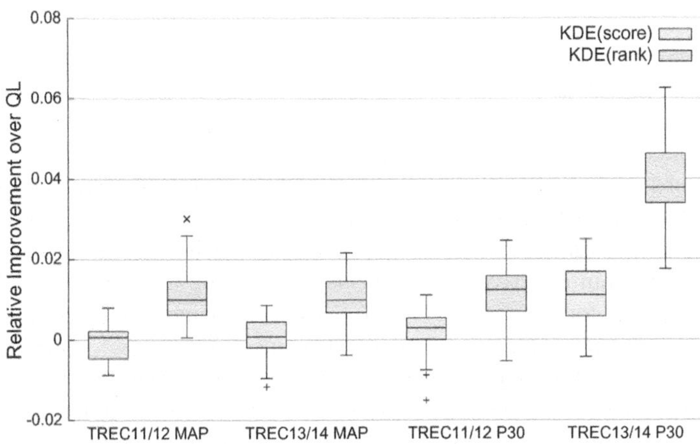

Fig. 1. Box-and-whiskers plots summarizing effectiveness differences (with respect to QL baseline) of score-based and rank-based weights for KDE, across 50 random trials where the topics are split in half for training/test.

half. For each trial, we trained on half the topics and tested on the other half. We then computed the effectiveness differences between each technique and the QL baseline. These differences, collected over 50 trials, are summarized in box-and-whiskers plots in Figure 1 for KDE with score-based weights and KDE with rank-based weights. We show the distribution of effectiveness differences in terms of MAP and P30 on TREC 2011/12 and TREC 2013/14. Following convention: Each box represents the span between the first and third quartiles, with a horizontal line at the median value. Whiskers extend from the ends of each box to the most distant point whose value lies within 1.5 times the interquartile range. Points that lie outside these limits are drawn individually. These results capture the overall effectiveness of the techniques better than metrics from any single arbitrary split. Here, we clearly see that rank-based weights are more effective than score-based weights.

Table 6. Results from attempting to reproduce lexical+temporal feedback experiments in Efron et al. [5] as closely as possible. Metrics computed over odd-numbered topics from TREC 2011/12.

Condition	MAP original	MAP reproduced	P30 original	P30 reproduced
RM3	0.2897	0.2847	0.3843	0.3684
KDE (score-based)	0.3014*	0.2834	0.4079*	0.3570
KDE (rank-based)		0.2703		0.3509*
KDE (oracle)		0.3027*		0.3945*

Taken as a whole, our findings are mostly consistent with the results of Efron et al. We find that KDE with rank-based weights yields improvements over the QL baseline, which affirms the overall effectiveness of the proposed approach.

5 Combining Lexical and Temporal Feedback

The final question explored in Efron et al. was whether lexical relevance signals are distinct from temporal relevance signals—in practical terms, are the effectiveness gains from both techniques additive?

Results from applying KDE on top of an RM3 baseline are shown in Table 6, reproducing the experiments in Efron et al. as closely as possible. In this case, we used the parameter setting that optimized MAP from the experiments in Table 2. The original SIGIR paper reported only score-based weights for KDE, but here we include the rank-based weights and the oracle condition as well. In this and the following experiments, relevance models were estimated with retweets included. The symbol * indicates that the difference with respect to the RM3 baseline is statistically significant ($p < 0.05$). We find that the oracle condition is significantly better than the RM3 baseline for both metrics; rank-based weighting for P30 is significantly worse, but none of the other differences are significant. This is not consistent with the findings in Efron et al., who reported significant improvements for score-based weights.

To further examine these inconsistencies, we repeated the experiments on all topics from TREC 2011/12 and TREC 2013/14. Here, we used the parameter settings in the "cross" condition that optimizes MAP from Tables 4 and 5. The symbol * indicates that the difference with respect to RM3 is statistically significant ($p < 0.05$). We see that neither score-based nor rank-based KDE is able to improve upon RM3, although the oracle condition shows a significant improvement in all cases. This confirms that a temporal relevance signal exists independently of the lexical relevance signal, although it does not appear that the proposed non-oracle techniques can exploit this signal. Note that in these experiments, we simply used previous parameters; perhaps with retuning we might more closely replicate previous results.

Table 7. Applying temporal feedback on top of lexical feedback for TREC 2011/12 and TREC 2013/14 data

	2011/12		2013/14	
Method	MAP	P30	MAP	P30
RM3	0.3005	0.3778	0.3629	0.5351
KDE (score-based)	0.2925*	0.3781	0.3543	0.5279
KDE (rank-based)	0.2769*	0.3642	0.3670	0.5423
KDE (oracle)	0.3197*	0.4191*	0.3964*	0.5952*

6 Conclusions

In this paper, we have successfully reproduced experiments described in previous work involving lexical feedback and temporal feedback. A third set of experiments on the combination of lexical and temporal feedback met with limited success. More in-depth analyses and extensions to new test collections add more nuance to previous conclusions.

We feel that there are two important takeaway lessons: First, though it may seem like an obvious point, meaningful comparisons depend on a proper baseline. We suspect that improvements reported in previous studies disappeared or were diminished to a large extent because the baseline became more competitive in our experiments. IR researchers must continuously remain vigilant and be "honest" with themselves in presenting a fair point of comparison.

Second, we found that effectiveness is highly dependent on the training/test split. This is perhaps not surprising since TREC test collections are relatively small—however, we see in the literature plenty of papers that base their conclusions on a single (arbitrary) training/test split. It is difficult to rule out that those findings, even in the case of statistically significant improvements, are due to fortuitous splits of the data. Running many randomized trials, as we have done in this paper, provides a more complete characterization of effectiveness differences. Perhaps such practices should become more commonplace in information retrieval evaluation.

Acknowledgments. This work was supported in part by the U.S. National Science Foundation under grants 1217279 and 1218043. Any opinions, findings, conclusions, or recommendations expressed are those of the authors and do not necessarily reflect the views of the sponsor.

References

1. Abdul-Jaleel, N., Allan, J., Croft, W.B., Diaz, F., Larkey, L., Li, X., Metzler, D., Smucker, M.D., Strohman, T., Turtle, H., Wade, C.: UMass at TREC 2004: Novelty and HARD. In: TREC (2004)
2. Amati, G., Amodeo, G., Bianchi, M., Celi, A., Nicola, C.D., Flammini, M., Gaibisso, C., Gambosi, G., Marcone, G.: FUB, IASI-CNR, UNIVAQ at TREC 2011 Microblog track. In: TREC (2011)

3. Choi, J., Croft, W.B.: Temporal models for microblog. In: CIKM, pp. 2491–2494 (2012)
4. Dakka, W., Gravano, L., Ipeirotis, P.G.: Answering general time-sensitive queries. IEEE Transactions on Knowledge and Data Engineering 24(2), 220–235 (2012)
5. Efron, M., Lin, J., He, J., de Vries, A.: Temporal feedback for tweet search with non-parametric density estimation. In: SIGIR, pp. 33–42 (2014)
6. Hall, P., Turlach, B.A.: Reducing bias in curve estimation by use of weights. Computational Statistics & Data Analysis 30(1), 67–86 (1999)
7. Jones, R., Diaz, F.: Temporal profiles of queries. ACM Transactions on Information Systems 25(3), Article 14 (2007)
8. Lavrenko, V., Croft, W.B.: Relevance based language models. In: SIGIR, pp. 120–127 (2001)
9. Li, X., Croft, W.B.: Time-based language models. In: CIKM, pp. 469–475 (2003)
10. Li, Y., Zhang, Z., Lv, W., Xie, Q., Lin, Y., Xu, R., Xu, W., Chen, G., Guo, J.: PRIS at TREC2011 Micro-blog track. In: TREC (2011)
11. Lin, J., Efron, M.: Overview of the TREC-2013 Microblog Track. In: TREC (2013)
12. Lin, J., Efron, M.: Infrastructure support for evaluation as a service. In: WWW Companion, pp. 79–82 (2014)
13. McCreadie, R., Soboroff, I., Lin, J., Macdonald, C., Ounis, I., McCullough, D.: On building a reusable Twitter corpus. In: SIGIR, pp. 1113–1114 (2012)
14. Metzler, D., Cai, C.: USC/ISI at TREC 2011: Microblog track. In: TREC (2011)
15. Miyanishi, T., Seki, K., Uehara, K.: Improving pseudo-relevance feedback via tweet selection. In: CIKM, pp. 439–448 (2013)
16. Mühleisen, H., Samar, T., Lin, J., de Vries, A.: Old dogs are great at new tricks: Column stores for IR prototyping. In: SIGIR, pp. 863–866 (2014)
17. Ounis, I., Macdonald, C., Lin, J., Soboroff, I.: Overview of the TREC-2011 Microblog Track. In: TREC (2011)
18. Ponte, J.M., Croft, W.: A language modeling approach to information retrieval. In: SIGIR, pp. 275–281 (1998)
19. Rocchio, J.J.: Relevance feedback in information retrieval. In: Salton, G. (ed.) The SMART Retrieval System—Experiments in Automatic Document Processing, pp. 313–323. Prentice-Hall, Englewood Cliffs, New Jersey (1971)
20. Smucker, M.D., Allan, J., Carterette, B.: A comparison of statistical significance tests for information retrieval evaluation. In: CIKM, pp. 623–632 (2007)
21. Soboroff, I., McCullough, D., Lin, J., Macdonald, C., Ounis, I., McCreadie, R.: Evaluating real-time search over tweets. In: ICWSM, pp. 579–582 (2012)
22. Trotman, A., Puurula, A., Burgess, B.: Improvements to BM25 and language models examined. In: ADCS (2014)
23. Turlach, B.A.: Bandwidth selection in kernel density estimation: A review (1993)
24. Zobel, J., Moffat, A.: Inverted files for text search engines. ACM Computing Surveys 38(6), 1–56 (2006)

Rank-Biased Precision Reloaded: Reproducibility and Generalization

Nicola Ferro and Gianmaria Silvello

Department of Information Engineering, University of Padua, Italy
{ferro,silvello}@dei.unipd.it

Abstract. In this work we reproduce the experiments presented in the paper entitled "Rank-Biased Precision for Measurement of Retrieval Effectiveness". This paper introduced a new effectiveness measure – *Rank-Biased Precision (RBP)* – which has become a reference point in the IR experimental evaluation panorama.

We will show that the experiments presented in the original RBP paper are repeatable and we discuss points of strength and limitations of the approach taken by the authors. We also present a generalization of the results by adopting four experimental collections and different analysis methodologies.

1 Introduction

In this paper we aim to reproduce the experiments presented in the paper by A. Moffat and J. Zobel entitled "Rank-Biased Precision for Measurement of Retrieval Effectiveness" published in the ACM Transaction on Information System in 2008 [12]. This work presents an effectiveness measure which had quite an impact on the *Information Retrieval (IR)* experimental evaluation field and also inspired the development of many other measures. Indeed, *Rank-Biased Precision (RBP)* is built around a user model where the browsing model, the document utiliy model and the utility accumulation model are explict [4]; it does not depend on the recall base, which is a quantity difficult to estimate and actually unknown to real users; finally, it matches well with real users, being well correlated with observed click behaviour in system logs [5,21] and allowing to learn models which capture a good share of actual users' way of behaving [11].

The core of RBP resides in its user model, which is defined starting from the observation that a user has no desire of examining every item in a ranking list. The idea is that a user always starts from the first document in the list and then she/he progresses from a document to the other with a probability p, called the *persistence parameter*, and, conversely, ends her/his examination of the list with probability $1 - p$. This assumption allows for the definition of user models representing both patient and impatient users by varying p.

Given a run of d documents, RBP is defined as:

$$\text{RBP} = (1 - p) \sum_{i=0}^{d} r_i \cdot p^{i-1}$$

A. Hanbury et al. (Eds.): ECIR 2015, LNCS 9022, pp. 768–780, 2015.

where r_i is the relevance grade of the document at rank i. Since RBP is defined for binary relevance, r_i can assume only two values: 0 or 1.

At the time of writing the RBP paper counts 80 citations on the *Association for Computing Machinery (ACM)* digital library[1] and more than 170 on Google scholar[2], several other works about effectiveness measures rely or are inspired by the user model of RBP and exploited it to define new effectiveness measures.

To repeat the experiments we rely on an open source and publicly available software library called MATTERS (MATlab Toolkit for Evaluation of information Retrieval Systems)[3] implemented in MATLAB[4]. The use of MATLAB allows us to exploit a widely-tested and robust to numerical approximations implementations of the statistical methods needed for analysing the measures such as Kendall's τ correlation measure [10], Student's t test [8] or the Wilcoxon signed rank test [18]. All the data and the scripts we used for reproducing the experiments in [12] are available at the URL: `http://matters.dei.unipd.it/`.

We take reproducibility also as the possibility of both generalizing the original experimental outcomes to other experimental collections in order to confirm them on a wider range of datasets, and validating the experimental hypotheses by means of additional analysis methods. The former led us to repeat the experiments on four different test collections from *Text REtrieval Conference (TREC)* and *Conference and Labs of the Evaluation Forum (CLEF)*; the latter led has to assess the robustness of RBP to shallow pools by using stratified random sampling techniques. We will show how this extended analysis on the one hand allows us to point out additional aspects of RBP and on the other hand provides a solid basis for future uses of this measure.

The paper is organized as follows: Section 2 describes the experiments conducted in the RBP original paper and details the aspects concerning their reproducibility; Section 3 reports about the extended experiments we conducted on RBP and in Section 4 we draw some conclusions.

2 Reproducibility

The experiments in [12] are based on the TREC-05, 1996, Ad-Hoc collection [16] composed of 61 runs (30 automatic and 31 manual runs), 50 topics with binary relevance judgments (i.e. relevant and not-relevant documents), and about 530,000 documents. The authors conducted three main experiments to explore how RBP behaves with shallow pools, also varying the persistence parameter $p = \{0.5, 0.8, 0.95\}$. RBP has been compared against P@10, P@R (precision at the recall-base), and *Average Precision (AP)* [3]; *Normalized Discounted Cumulated Gain (nDCG)* [9], and *Reciprocal Rank (RR)* [17], by considering two pool depths 100 (the original depth of TREC-05) and 10.

---

[1] `http://dl.acm.org/citation.cfm?id=1416952`
[2] `http://scholar.google.com/`
[3] `http://matters.dei.unipd.it/`
[4] `http://www.mathworks.com/`

The original pool was calculated by taking the union of the first 100 documents of each run submitted to TREC-05 and then assessing the resulting set of documents, whereas the pool at depth 10 was calculated by exploiting the original assessments but applying them to a reduced set composed of the union of the first 10 documents of each run; all the documents not belonging to this set are considered as not-relevant. From the reproducibility point-of-view, this downsampling technique has the advantage of being deterministic not involving any randomization technique in the downsampling of the pools.

The experiments to be reproduced can be divided into three parts:

1. Kendall's τ correlation coefficients calculated from the systems ordering generated by pair of metrics using TREC-05 runs and by considering two pool depths. With respect to the original paper we aim to reproduce Figure 2 on page 9, Figure 4 on page 16 and Table 3 on page 23.
2. Upper and lower bounds for RBP as the p parameter is varied and increasing number of documents (from 1 to 100) are considered. With respect to the original paper we aim to reproduce Figure 5 on page 19.
3. t test and Wilcoxon test for determining the rate at which different effectiveness metrics allow significant distinctions to be made between systems. With respect to the original paper we aim to reproduce Table 4 on page 24.

The TREC-05 data needed to reproduce the paper is released by *National Institute of Standards and Technology (NIST)* and available on the TREC website[5]; it is composed of the original pool with depth 100 and the set of 61 runs submitted to the campaign. When it comes to reproducing some experiments using this kind of data, the first consideration that has to be made regards how to import the run files; indeed, in the TREC format of a run there is the following:

`<topic-id> Q0 <document-id> <rank> <score> <run-id>`

where: `topic-id` is a string specifying the identifier of a topic, `Q0` is a constant unused field which can be discarded during the import, `document-id` is a string specifying the identifier of a document, `rank` is an integer specifying the rank of a document for a topic, `score` is a decimal numeric value specifying the score of a document for a topic and `run-id` is a string specifying the identifier of the run. Track guidelines ask participants to rank the output of their systems by increasing value of `rank` and decreasing value of `score`.

The standard software library adopted by TREC for analysing the runs is `trec_eval`[6]. When importing runs, `trec_eval` may modify the actual ordering of the items in the file since it sorts items in descending order of `score`[7] and descending lexicographical order of `document-id`, when `scores` are tied; note that the `rank` value is not considered. We call this *`trec_eval` ordering*.

[5] `http://trec.nist.gov/`

[6] `http://trec.nist.gov/trec_eval/`

[7] Note that `trec_eval` also casts the scores of the runs to single precision (float) values while often they contain more decimal values than those supported by single precision numbers. So two `score` values may appear as tied if regarded as single precision value whereas they would have not if regarded as double precision values.

Note that the ***trec_eval** ordering* represents a cleaning of the data for those runs which have not complied with the track guidelines as far as ordering of the items is concerned but it may modify also correctly behaving runs, if two items have the same `score` but different `rank`, since in this case `trec_eval` reorders them in descending lexicographical order of `document-id` which may be different from the ordering by increasing `rank`.

RBP is not part of the standard `trec_eval` and the paper under exam does not explicitly say whether the authors have extended `trec_eval` to plug-in also RBP or whether they relied on some other script for carrying out the experiments. In the latter case, if one does not deeply know the internals of `trec_eval`, when importing the run files, the original ordering of the items may be kept as granted, under the assumption that the files are well-formed and complying with the guidelines since they have been accepted and then released as official submissions to TREC. We call this latter case *original ordering.*

This aspect has an impact, though small, on the reproducibility of the experiments; indeed, by considering all the documents of all the runs for TREC-05, the ***trec_eval** ordering* swaps about 2.79% of documents with respect to the *original ordering*; the impact of the swaps on the calculation of the metrics is narrowed down by the fact that most of the swaps (89.21% of the total) are between not-relevant documents, 6.56% are between equally relevant documents while only 4.23% are between relevant and not-relevant documents, thus producing a measurable effect on the metrics calculation.

Table 1 is the reproduction of Table 3 on page 23 of the original RBP paper; we report the Kendall's Tau correlations calculated from the system rankings generated by pairs of metrics by using both the ***trec_eval** ordering* and the *original ordering* in order to understand which one was most likely used in [12] and to show the differences between the two orderings. We report in bold the numbers which are at least 1% different than those in the table of the reference paper. As we may see for the ***trec_eval** ordering* only two numbers are at least 1% different from the ones in the paper, whereas there are more differences for the *original ordering*, especially for the correlations with P@R which seems to be more sensitive to small changes in the order of documents with respect to the other metrics. The differences between the two orderings are small, but in the case of the correlation between P@R with depth 100 and RBP.95 with depth 10, if we consider the *original ordering* the correlation is above the 0.9 threshold value [14], whereas with the ***trec_eval** ordering* – as well as in the reference paper – it is below this threshold. Another significant difference can be identified in the correlation between P@R with depth 100 and RBP.95 with depth 100; indeed, with the ***trec_eval** ordering* the difference is very close to the threshold value (i.e. 0.895), whereas with the***trec_eval** ordering* it goes down to 0.850.

The correlation values obtained with the ***trec_eval** ordering* are closer to the ones in the reference paper even if they present small differences probably due to numeric approximations and two values present a difference greater than 1%. From this analysis we can assume that the reference paper adopted the

Table 1. Kendall's Tau correlations calculated from the system orderings generated by metric pairs with TREC-05 by using the `trec_eval` *ordering* and the *original ordering*. Numbers in bold are those which are at least 1% different from the correlations in [12].

treceval ordering					
		depth 100			
Metric	depth	RR	P@10	P@R	AP
RR	10	0.997	0.842	0.748	0.732
P@10	10	0.840	1.000	0.861	0.845
P@R	100	0.746	0.861	1.000	0.908
RBP.5	10	0.926	0.858	0.764	0.755
RBP.8	10	0.888	0.930	**0.819**	0.809
RBP.95	10	0.778	0.882	0.877	0.896
RBP.95	100	0.793	0.916	0.895	**0.859**
nDCG	100	0.765	0.831	0.877	0.915

original ordering					
		depth 100			
Metric	depth	RR	P@10	P@R	AP
RR	10	0.997	0.841	0.747	0.730
P@10	10	0.840	1.000	0.860	0.844
P@R	100	**0.769**	0.861	1.000	0.907
RBP.5	10	0.924	0.858	**0.776**	0.755
RBP.8	10	0.889	0.929	**0.828**	0.809
RBP.95	10	0.779	0.880	**0.905**	0.894
RBP.95	100	0.792	0.913	**0.850**	**0.859**
nDCG	100	0.763	0.829	**0.886**	0.913

`trec_eval` *ordering*, thus in the following we conduct all the other experiments by assuming this ordering for importing the runs.

Another small issue with the reproduction of this experiment is that in the original paper there are no details about the parameters – i.e. weighting schema and log base – used for calculating nDCG; we tested several weighting schema and log bases and we obtained the same number as those in the reference paper by assigning weight 0 to not-relevant documents, 1 to relevant ones and by using log base 2.[8]

Figure 2 and 4 of the original paper regard similar aspects to those presented above in the comment to Table 1 and they concern the correlation between *Mean Average Precision (MAP)* values calculated on the TREC-05 Ad-Hoc collection considering pool depth 100 and pool depth 10 which we show in Figure 1a and the correlation between mean RBP values with p set at 0.5, 0.8 and 0.95 as reported in Figure 1b.

As we can see these two figures are qualitatively equal to those in the original paper and thus these experiments can be considered as reproducible. The main difference regards the choice of the axes which in the reference paper are in the range $[0, 0.4]$ for MAP and $[0, 0.6]$ for mean RBP, whereas we report the graph with axes in the range $[0, 1]$, which is the actual full-scale for both measures. In this way, we can see some MAP values which are above 0.4, showing that MAP calculated with shallow pools tends to overestimate the good runs more than the bad ones. Also for mean RBP we can see some values above the 0.6 limit reported in the original paper; these points show that RBP with $p = 0.5$ with the depth 10 pool tends to overestimate good runs a little more than the bad ones even though these points are also very close to the bisector.

[8] Note that the log base might have guessed by the fact that, on page 21 of the paper, when presenting DCG the authors report that [9] suggested the use of $b = 2$, and employed that value in their examples and experiments.

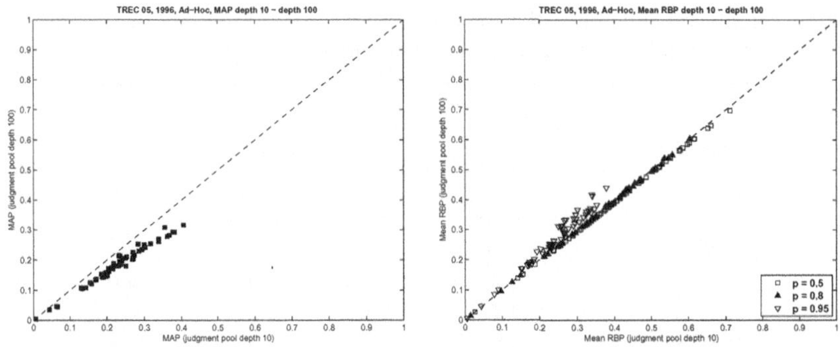

(a) Mean average precision. **(b)** Mean rank-biased precision.

Fig. 1. Correlation between MAP and mean RBP at pool depth 10 and 100

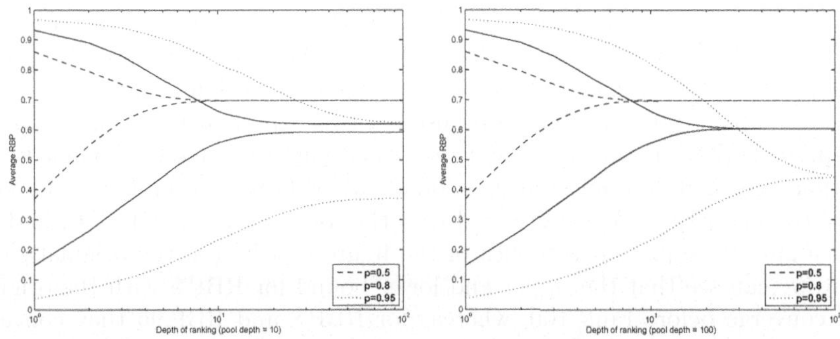

(a) Pool depth 10 **(b)** Pool depth 100

Fig. 2. Upper and lower bounds of RBP as p is varied and increasing number of documents are considered in the ranking for the "ETHme1" run

The second set of experiments in [12] we aim to reproduce regards upper and lower bounds of RBP evaluated at depth 10 and depth 100. In the usual TREC evaluation setting some documents of a run are assessed (either relevant or not relevant in the binary case), but most of them are left unjudged and normally considered as not-relevant when it comes to calculating effectiveness measures. In [12] it is stated that with this assumption "quoted effectiveness rates might be expected to be pessimistic" and thus represent a lower bound of the measurement; thus, RBP values calculated with this assumption are considered the lower bounds of the measure. They proposed a method to compute a *residual* that captures the unknown component (determined by the unjudged documents) of RBP. Basically, the residual is calculated on a item-by-item basis by summing the weight that the documents would have had if they were relevant; the upper bound is defined by the sum of RBP (i.e. the lower bound) and the residual.

The goal of this experiment is to show that lower and upper bounds stabilize as the depth of the evaluation is increased, even if for higher values of p and

Table 2. Significant differences between systems; the total number of system pairs is 1830 and numbers in bold are at least 1% different from [12]

	Wilcoxon		*t* test	
Metric	99%	95%	99%	95%
RR	1030	763	1000	752
P@10	**1153**	904	1150	915
P@R	1211	994	1142	931
AP	1260	1077	1164	969
RBP.5	1077	**845**	1052	812
RBP.8	1163	921	1167	918
RBP.95	1232	1009	1209	987
nDCG	1289	**1104**	1267	**1089**

shallow pools they do not converge. This experiment is summarized in Figure 5 on page 19 of the original paper which reports upper and lower bounds of RBP (with p varying from 0.5 to 0.95) for a given run. In the original paper there is no indication about which run has been used in this experiment; as a consequence to reproduce the experiment we had to calculate upper and lower bounds for all the runs and then proceed by inspection of the plots to determine the run used in the original paper. We determined that the used run is named "ETHme1".

In Figure 2 we present a replica of the figure reported in the original paper where we can see that the upper and lower bound for RBP.5 with the original pool converge before rank 100, whereas for RBP.8 and RBP.95 they converge later on; for the measures calculated with pool depth 10 only RBP.5 converges before rank 100. In this case the original experiment is not easily reproducible because the name of the chosen run was not reported; the same problem prevents the possibility of replicating the plot of Figure 6 on page 20 of the original paper, where the upper and lower bounds of "two systems" are shown: there is no indication about which system pair among the 1830 possible pairs in in TREC-05 have been chosen.

The last experiment to be reproduced regards the *t* test and the Wilcoxon signed rank test for determining the significant differences between retrieval models according to different measures. In Table 2 we report the values we obtained that have to be compared to those in Table 4 on page 24 of the reference paper. We reported in bold the numbers presenting a difference higher than 1% from the original ones; as we may see there are three major differences for the Wilcoxon test and only one for the *t* test. We highlight that for the Wilcoxon test 94% of the values are different from the original paper even though the differences are very small (less than 1%); on the other hand, for the *t* test the 31% of the values we obtained are different from those in the original paper.

Table 3. Features of the adopted experimental collections

Collection	CLEF 2003	TREC 13	CLEF 2009	TREC 21
Year	2003	2004	2009	2012
Track	Ad-Hoc	Robust	TEL	Web
# Documents	1M	528K	2.1M	1B
# Topics	50	250	50	50
# Runs	52	110	43	27
Run Length	1,000	1,000	1,000	10,000
Relevance Degrees	2	3	2	4
Pool Depth	60	100 and 125	60	30 and 25
Languages	EN, FR, DE, ES	EN	DE, EL, FR, IT, ZH	EN

3 Generalization

The experiments conducted in [12] are all based on TREC-5, but these results have not been proven in a wider environment by using different experimental collections (e.g. collections with more runs, more topics, higher and lower original pool depths) or using different pool sampling techniques. Indeed, to the best of our knowledge, the only one other systematic analysis of RBP on different experimental collections is the one by [13], even if it does not concern the original RBP as defined in the reference paper under examination but its extension to multi-graded relevance judgements.

In this section we aim to investigate three main aspects regarding RBP:

- stability to pool downsampling at depth 10 by using two CLEF and two TREC collections;
- the robustness of RBP to downsampled pools (with different reduction rates) according to the stratified random sampling method [2];
- the behavior of RBP upper and lower bound in the average case presenting confidence intervals.

In the following we consider four public experimental collections, whose characteristics are reported in Table 3: (i) CLEF 2003, Multilingual-4, Ad-Hoc Track [1]; (ii) TREC 13, 2004, Robust Track [15]; (iii) CLEF 2009, bilingual X2EN, *The European Library (TEL)* Track [7]; and, (iv) TREC 21, 2012, Web Track [6].

As we can see these collections have different interesting characteristics which allow us to test the behaviour of RBP in a wider range of settings. CLEF 2003 has been used for evaluating multilingual systems with 50 topics and the corpus of one million documents in four different languages; TREC-13 has a high number of runs, topics (i.e. 250) and pool depth (i.e. 125 for 50 topics and 100 for the other 200); CLEF 2009 presents a corpus of documents composed by short bibliographic records and not newspaper articles as in the other CLEF collections and has been used to evaluate bilingual systems working on topics in English and documents in five different languages; and TREC-21 presents a huge multilingual Web corpus, topics are created from the logs of a commercial search engine and it

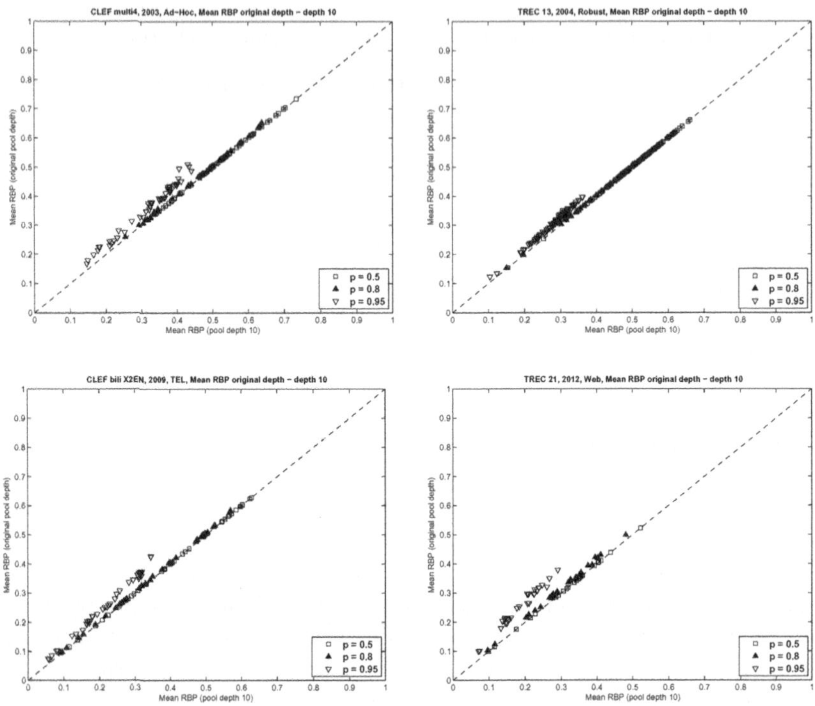

Fig. 3. Robustness of RBP to pool downsampling using different collections

allows us to evaluate up-to-date IR systems working on a Web scale, furthermore 25 topics were judged to depth 30 and 25 to depth 20 [6].

In Figure 3 we can see the correlation between RBP (with the three usual values of $p = \{0.5, 0.8, 0.95\}$) calculated with the original pool depth and with pool depth 10 across the four selected test collections. The results presented in [12] with TREC-05 are confirmed for all the tested collections showing that RBP.5 and RBP.8 are robust to pool downsampling, whereas RBP.95 tends to underestimate the effectiveness of the runs when calculated using pool depth 10; this effect is more evident with TREC-21 where also RBP.8 values are slightly above the bisector.

The *stratified random sampling* of the pools allows us to investigate the behavior of RBP as the relevance judgment sets become less complete following the methodology presented in [2]: Starting from the original pool (100% of the relevance judgments) for each topic we select a list of relevant documents in random order and a list of not-relevant documents in random order; then, we create alternative pools by taking $\{90, 70, 50, 30, 10\}\%$ of the original pool. For a target pool which is $P\%$ as large as the original pool, we select $X = P \times R$ relevant documents and $Y = P \times N$ not-relevant documents or each topic where R is the number of relevant documents in the original pool and N is the number of judged not-relevant documents in the original pool. We use 1 as the minimum number of

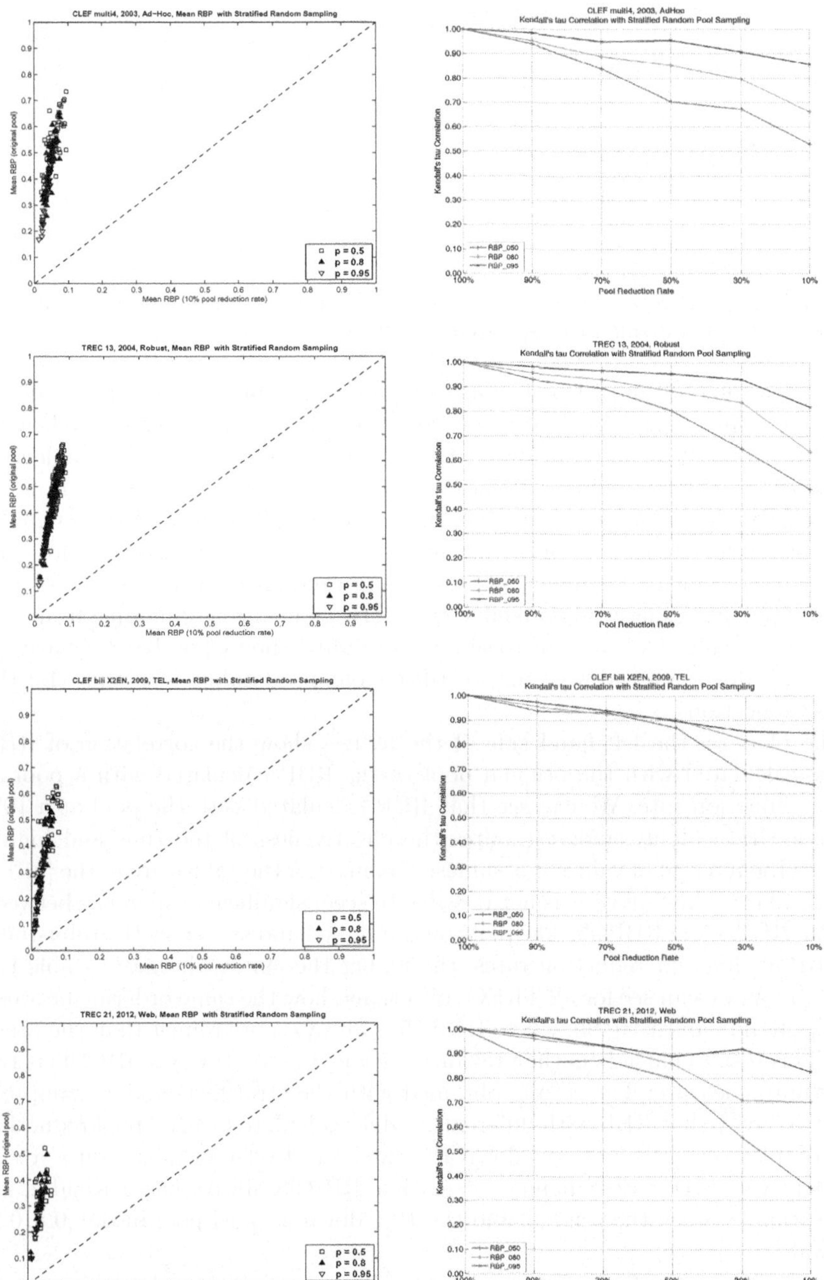

Fig. 4. On the left-hand side are the Kendall's τ between the original pool and the 10% downsampled pool (that can be compared with those in Figure 3 adopting a pool downsampled to depth 10) and on the right-hand side there is the change in Kendall's τ as judgment sets are downsampled for the CLEF and TREC collections.

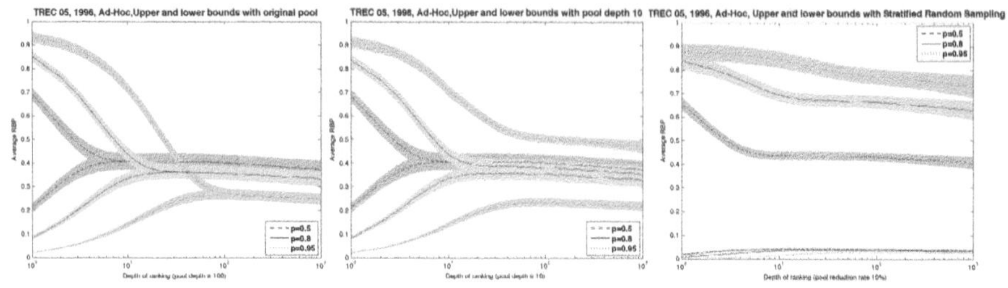

Fig. 5. Upper and lower bounds of average RBP as p is varied and the number of documents in the ranking increases from 1 to 1000

relevant documents and 10 as the minimum number of judged not-relevant documents per topic. Since we take random subsets of a pool that is assumed to be fair, the reduced pools are also unbiased with respect to systems; this methodology is equivalent to perform uniform random sampling of the pool [19], which is desirable to infer statistical properties. This methodology allows us to further explore the robustness of RBP to pool downsampling; it must be underlined that for each pool sample, relevant documents are selected at random and thus the results here reported are not exactly reproducible even if the conclusions emerging from this test do not change from sample to sample. Figure 4 shows how RBP behaves as the pool is downsampled with the stratified random sampling techniques for the TREC collections.

The plots on the left-hand side of the figures show the correlation of RBP values calculated with the original pool versus RBP calculated with a pool at a 10% reduction rate. We can see that RBP calculated with the pool at a 10% reduction rate highly underestimates the effectiveness of the runs and highly reduces the interval of values it assumes – i.e. most of the values are in the $[0, 0.1]$ interval. From these plots it is not possible to see a significant difference between RBP.5, RBP.8 and RBP.95. The plots on the right-hand side show the robustness of RBP at different reduction rates: the higher the curves the more stable the measure. As we can see for all TREC collections show the same ordering between RBP.5, RBP.8 and RBP.95, where RBP.95 is always more robust than the other two. This result contradicts the previous one (see Figure 3) where RBP.95 is the less robust measure. The results obtained with the stratified random sampling allow us to say that RBP with different p values calculated with a pool reduction rate of 10% seriously narrows down the interval of effectiveness values a run can achieve; on the other hand, we see that RBP.95 always has a Kendall's τ correlation between the original and the 10% downsampled pool in the $[0.8, 0.9]$ interval.

Lastly, in Figure 5 we present a generalization of Figure 2 which reports RBP upper and lower bounds calculated by averaging over all the runs of the TREC-05 collection instead of choosing a specific run as representative of the whole collection. We also reported the confidence interval of the measures and we show how the bounds behave up to rank 1000 (i.e. the maximum length of the runs); furthermore, we show how the bounds behave when RBP is calculated by adopting a pool with 10% reduction rate determined with the stratified random sampling technique. We can see that with the original pool as well as with pool downsampled to depth 10 the results are consistent with those reported in the RBP original paper for RBP.5 and RBP.95, whereas it shows that RBP.8 tends to converge between rank 10 and 100. However, upper and lower bounds of RBP calculated with 10% pool reduction rate never converge for all the considered values of p showing a high impact of unjudged documents on RBP values. The very same trends emerge for the RBP bounds calculated with the other collections we presented above; we do not report the plots for space reasons.

4 Conclusions

In this paper we discussed the experiments conducted in [12] where the RBP measure was presented and described for the first time. We have shown that most of the experiments presented in the original RBP paper are reproducible, even though there are precautions that should be taken with presenting experiments about experimental evaluation in IR. These include: (i) explicitly describing the choices made about document ordering – e.g. explaining if the `trec_eval` document ordering is applied or not; (ii) explicitly reporting the name or id of the systems used for the experiments – e.g. the "ETHme1" run in Figure 2 – or specifying which subset of systems has been selected from the whole collection; (iii) reporting all the parameters used for calculating a measure – e.g. weighting schema and log base for nDCG. It must be highlighted that the experiments were reproducible because they were originally conducted on publicly available and shared datasets such as the TREC-05 Ad-Hoc collection.

From the reproducibility point of view, the presentation of the results by means of tables would be preferable to only using plots, because they allow for a thorough verification of the results; graphs and plots are useful for understanding the results from a qualitative perspective, but they always should be accompanied by the numerical data on which they rely (they can be presented also in an appendix of the paper or made available online).

The generalization part of this work shows that the results presented in the original RBP paper are verifiable also with other public and shared experimental collections. On the other hand, we show that the use of different analysis methodologies (e.g. a different pool downsampling technique) could lead to different conclusions that must be taken into account in order to employ RBP for experimental evaluation. As we have seen by using pool downsampling RBP.5 is the most robust measure, but it is the less robust by using the stratified random sampling method; we reach the same conclusion by considering RBP bounds that, with a 10% pool reduction rate, do not converge for any p value up to rank 1,000.

References

1. Braschler, M.: CLEF 2003 – Overview of Results. In: Peters, C., Gonzalo, J., Braschler, M., Kluck, M. (eds.) CLEF 2003. LNCS, vol. 3237, pp. 44–63. Springer, Heidelberg (2004)

2. Buckley, C., Voorhees, E.M.: Retrieval Evaluation with Incomplete Information. In: Proc. 27th Ann. Int. ACM Conference on Research and Development in IR (SIGIR 2004), pp. 25–32. ACM Press, USA (2004)
3. Buckley, C., Voorhees, E.M.: Retrieval System Evaluation. In: TREC. Experiment and Evaluation in Information Retrieval, pp. 53–78. MIT Press, Cambridge (2005)
4. Carterette, B.A.: System Effectiveness, User Models, and User Utility: A Conceptual Framework for Investigation. In: Proc. 34th Ann. Int. ACM Conference on Research and Development in IR (SIGIR 2011), pp. 903–912. ACM Press, USA (2011)
5. Chapelle, O., Metzler, D., Zhang, Y., Grinspan, P.: Expected Reciprocal Rank for Graded Relevance. In: Proc. 18th Int. Conference on Information and Knowledge Management (CIKM 2009), pp. 621–630. ACM Press, USA (2009)
6. Clarke, C.L.A., Craswell, N., Voorhees, H.: Overview of the TREC 2012 Web Track. In: The Twenty-First Text REtrieval Conference Proceedings (TREC 2012), NIST, SP 500-298, USA, pp. 1–8 (2013)
7. Ferro, N., Peters, C.: CLEF 2009 Ad Hoc Track Overview: TEL and Persian Tasks. In: Peters, C., Di Nunzio, G.M., Kurimo, M., Mandl, T., Mostefa, D., Peñas, A., Roda, G. (eds.) CLEF 2009. LNCS, vol. 6241, pp. 13–35. Springer, Heidelberg (2010)
8. Gosset, W.S.: The Probable Error of a Mean. Biometrika (1), 1–25 (1908)
9. Järvelin, K., Kekäläinen, J.: Cumulated Gain-Based Evaluation of IR Techniques. ACM Transactions on Information Systems (TOIS) 20(4), 422–446 (2002)
10. Kendall, M.G.: Rank correlation methods. Griffin, Oxford, England (1948)
11. Moffat, A., Thomas, P., Scholer, F.: Users Versus Models: What Observation Tells Us About Effectiveness Metrics. In: Proc. 22h Int. Conference on Information and Knowledge Management (CIKM 2013), pp. 659–668. ACM Press (2013)
12. Moffat, A., Zobel, J.: Rank-Biased Precision for Measurement of Retrieval Effectiveness. ACM Transactions on Information Systems 27(1), 1–27 (2008)
13. Sakai, T., Kando, N.: On Information Retrieval Metrics Designed for Evaluation with Incomplete Relevance Assessments. Inf. Retrieval 11(5), 447–470 (2008)
14. Voorhees, E.: Evaluation by Highly Relevant Documents. In: Proc. 24th Ann. Int. ACM Conference on Research and Development in IR (SIGIR 2001), pp. 74–82. ACM Press, USA (2001)
15. Voorhees, E.M.: Overview of the TREC 2004 Robust Track. In: The 13th Text REtrieval Conference Proceedings (TREC 2004), USA, pp. 500–261 (2004)
16. Voorhees, E.M., Harman, D.K.: Overview of the Fifth Text REtrieval Conference (TREC-5). In: The 5th Text REtrieval Conference (TREC-5), NIST, SP 500-238, pp. 1–28 (1996)
17. Voorhees, E.M., Tice, D.M.: The TREC-8 Question Answering Track Evaluation. In: The 8th Text REtrieval Conference (TREC-8), NIST, SP 500-246, USA, pp. 83–105 (1999)
18. Wilcoxon, F.: Individual Comparisons by Ranking Methods. Biometrics Bulletin 1(6), 80–83 (1945)
19. Yilmaz, E., Aslam, J.A.: Estimating Average Precision when Judgments are Incomplete. Knowledge and Information Systems 16(2), 173–211 (2008)
20. Yilmaz, E., Shokouhi, M., Craswell, N., Robertson, S.: Expected Browsing Utility for Web Search Evaluation. In: Proc. 19th Int. Conference on Information and Knowledge Management (CIKM 2010), pp. 1561–1565. ACM Press, USA (2010)
21. Zhang, Y., Park, L., Moffat, A.: Click-based evidence for decaying weight distributions in search effectiveness metrics. Inf. Retrieval 13(1), 46–69 (2010)

Knowledge Journey Exhibit: Towards Age-Adaptive Search User Interfaces

Tatiana Gossen, Michael Kotzyba, and Andreas Nürnberger

Data and Knowledge Engineering Group, Faculty of Computer Science,
Otto von Guericke University Magdeburg, Germany
{tatiana.gossen,michael.kotzyba,andreas.nuernberger}@ovgu.de
http://www.dke.ovgu.de

Abstract. We describe an information terminal that supports interactive search with an age-adaptable search user interface whose main focus group are young users. The terminal enables a flexible adaptation of the search user interface to address changing requirements of users at different age groups. The interface is operated using touch interactions as they are considered to be more natural for children than using a mouse. Users search within a safe environment; For this purpose a search index was created using a focused crawler.

Keywords: interactive search, search user interface, information retrieval, adaptation, context support, children.

1 Introduction

Web search engines are used by hundreds of millions of people all over the world. This is a very wide and heterogeneous target group with different backgrounds, knowledge, experience, etc. Therefore, researchers suggest providing a customized solution to cover the needs of individual users. Nowadays, solutions in personalisation and adaptation of backend algorithms, i.e. query adaptation, adaptive retrieval, adaptive result composition and presentation, have been proposed in order to support the search of an individual user, e.g. [2,5]. But the front end, i.e. the search user interface (SUI), is usually designed and optimized for a certain user group and does not support many mechanisms for personalisation. In order to tackle this issue, we propose to adapt the SUI to the needs of an individual user.

In our previous work [3,4], we developed a SUI called the *Knowledge Journey* that can be customized to the user's needs and is used as a desktop application. Here, we describe the *Knowledge Journey Exhibit (KJE)* that implements the *Knowledge Journey* as an information terminal device and has an age-adaptable SUI. Furthermore, different to the desktop version, KJE is a touch application, has an improved SUI and a new backend. KJE is developed for users of age seven and older. KJE was exhibited and tested at the "ScienceStation" exhibition[1]

[1] http://www.digital-ist.de/veranstaltungen/science-station-2014.html

A. Hanbury et al. (Eds.): ECIR 2015, LNCS 9022, pp. 781–784, 2015.

for children and adults that annually takes place at multiple train stations in Germany. This test environment imposed additional requirements: *The exhibit must be robust to be run at a train station in a stand-alone mode; it can be used without supervision; the exhibit can be used without Internet; the search index is child-safe and child-appropriate; there is a good coverage of web documents.*

2 Knowledge Journey Exhibit

2.1 Hardware

The exhibit was designed in form of a robust information terminal for an interactive search that can be operated by a user (Fig. 1). It has a 32" touch monitor, metal keyboard and trackball. A computer was placed within the box. The *Mozilla Firefox* browser was opened and the computer was set to run Firefox in a kiosk mode. For safety reasons, in order to avoid the risk of stumbling in a public place, we did not build a step construction. The height of the box was adjusted for both children and adults. The height of 130 cm was calculated as a mean between the height of an average child and an average adult in order to be appropriate for both user groups.

2.2 Frontend

The SUI of the KJE was iteratively developed. Several user studies were conducted in our previous research. In this paper, we describe improvements of the SUI based on the results from the last user study [4]. The previous version had a configuration window where a user was able to customize the SUI. However, we considered this window to be too difficult for children to operate without supervision and in a public place. Therefore, a decision was made to replace the configuration window with a slider where each point on the slider corresponded to a SUI configuration for a specific age starting with configuration for young children and ending with a setting for young adults. We use the age parameter to adapt the SUI. At the beginning a user is asked to input his or hes age. Then, the user is forwarded to the corresponding search user interface, where they can

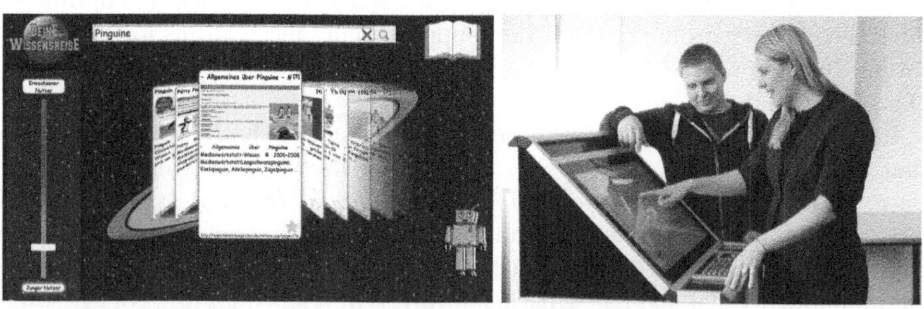

Fig. 1. User interface of Knowledge Journey Exhibit (left). Interaction with KJE (right). [Photo by Stefan Berger, AVMZ, Otto-von-Guericke University Magdeburg.]

explore other settings of the SUI using the slider. The settings for the slider were derived based on the results from the user study described in [4]. The settings for young children are a pirate theme, coverflow result visualization, large font size in Comic Sans MS. The settings for young adults are no theme, tiles result visualization, smaller font size in Arial. The search results for adults contain twice as much text in summaries and smaller thumbnails. Each point in between of the slider changes one of the setting parameters, e.g. the font.

The system provides spelling correction after the query is submitted and suggestions for the term the user is currently typing. In addition, users can bookmark the relevant search results using the storage functionality. We used a star symbol that was added to the result surrogate to indicate if the search result is already bookmarked. Users can click directly on the star symbol to bookmark or unbookmark the result or they can place the search result to the storage using drag-and-drop. They can review the stored results which are grouped by the issued query in order to provide more context information.

Furthermore, we use information about the web page complexity that is calculated using the *Flesch-Reading-Ease* (FRE) readability index for German language [1]. We applied traffic light metaphor and visualized each search result that is easy to understand in a green frame, while a search result that is hard to understand is visualized in a red frame, with varying levels of color in between. The traffic light metaphor is also applied to the slider.

2.3 Backend

In the previous research [3,4] we used the *Bing Search* API in order to retrieve search results given a user query. However, the Bing Search API requires Internet access and the returned results were not of a good quality for children. Some of the results are still not child-safe even with the safe search option turned on. Moreover, the first results usually belong to Wikipedia and web shops pages and are not child-appropriate. Wikipedia pages are, for example, complex and not easy to understand. Therefore, we decided to create our own search index. First, we tried to obtain the seed automatically using the *DMOZ's kids&teens directory*[2], but the quality of the gathered web pages was too low and we chose to create the seed manually. A seed of 81 web portals was manually derived and a focused crawler was implemented that crawled and indexed web pages that only belonged to those domains. The manually derived portals were mainly special web portals for children. However, we also selected some portals that were at least child-safe and informative such as zoo portals or federal ministry of education and research. The portals were crawled with consideration of the *robots.txt* protocol[3].

As one of the requirements was to be able to use the exhibit without the Internet, we faced the challenge of showing the result pages in an offline mode. Therefore, we decided to create high-quality, full screen images of the web pages being indexed. For that, we used the *Apache Tika* library. Users get an image

[2] `http://www.dmoz.org/Kids_and_Teens/International/Deutsch/`

[3] `http://www.robotstxt.org/`

of the web page if clicking at a search result. It is not possible to navigate to other web pages using the links on the result page. This prevents young users from viewing the content that might be unsafe.

In the post-processing step duplicates (pages that have the same main text) were removed. The obtained index contains approximately 67,000 pages. The relevance score of a web page was calculated as a product of the Lucene score (*Apache Lucene* library), the Flesch-Reading-Ease index and the boost score for high-quality pages web pages. Using the Flesch-Reading-Ease index, documents that are easier to understand are placed slightly upwards in the ranking. In this version of the Knowledge Journey Exhibit we focused on the adaptivity of the SUI. In addition to the SUI, it is also possible to change the ranking function depending on the targeted user group. This is an interesting direction for future work. For the query suggestion feature, we used a free available dictionary with 1.6 million German words. Top ten suggestions for the term the user is currently typing are made based on the Levenshtein distance to the indexed dictionary term. After the user has submitted the query, a suggestion "did you mean" is assembled from the top suggestions for each search term.

3 Conclusion

In this paper, we presented an information terminal with a search user interface that takes user's age as a parameter for adaptation. Our next step is to make the system adaptive. An adaptive system will require a back-end algorithm to detect user's age, e.g. based on the issued queries. Furthermore, we plan to conduct a further usability study using an eye-tracking device. A demonstration video of the Knowledge Journey Exhibit is available at `http://www.dke-research.de/KnowledgeJourneyExhibit.html`.

Acknowledgments. We are grateful to Fabian Finster, Marcel Genzmehr, Michael Tornow, Stefan Langer, Thomas Low and Stefan Haun for support in development. This work was supported by the Transregional Collaborative Research Centre SFB/TRR 62 "Companion-Technology for Cognitive Technical Systems" funded by the German Research Foundation (DFG).

References

1. Amstad, T.: Wie verständlich sind unsere Zeitungen? PhD thesis (1978)
2. Collins-Thompson, K., Bennett, P.N., W, R.: Personalizing web search results by reading level. In: Proc. of the CIKM, pp. 403–412 (2011)
3. Gossen, T., Nitsche, M., Nürnberger, A.: My first search user interface. In: de Rijke, M., Kenter, T., de Vries, A.P., Zhai, C., de Jong, F., Radinsky, K., Hofmann, K. (eds.) ECIR 2014. LNCS, vol. 8416, pp. 746–749. Springer, Heidelberg (2014)
4. Gossen, T., Nitsche, M., Vos, J., Nürnberger, A.: Adaptation of a search user interface towards user needs - a prototype study with children & adults. In: Proc. of the 7th Annual Symposium on HCIR. ACM (2013)
5. Steichen, B., Ashman, H., Wade, V.: A comparative survey of personalised information retrieval and adaptive hypermedia techniques. IP&M 48(4), 698–724 (2012)

PopMeter: Linked-Entities in a Sentiment Graph

Filipa Peleja

CITI, Departamento de Informática, Faculdade de Ciências e Tecnologia,
Universidade Nova de Lisboa, 2829-516 Caparica, Portugal
{filipapeleja}@gmail.com

Abstract. It is common for a celebrity, brand, or movie to become a reference in the domain and to be vastly cited as an example of a highly reputable entity. Popmeter[1] is a search/browsing application to visualize the reputation of an entity and its corresponding sentiment connections (in *hate-it* or *love-it* manner). Popmeter is supported by a sentiment graph populated by named-entities and sentiment words. The sentiment graph is constructed by a reputation analysis procedure that models the sentiment of each sentence where the entity is mentioned. This analysis leverages on a sentiment lexicon that includes general sentiment words that characterize the general sentiment towards the targeted named-entity.

Keywords: Reputation analysis, sentiment analysis.

1 Introduction

Reputation is, on its whole, linked to a sentiment analysis task of the textual corpus where the entity is found. There have been a great focus on researching how interesting information can be extracted or deduced from chunks of data, leading to the emergence of various frameworks and techniques [1, 2]. In this research we developed a sentiment graph that represents sentiment entities and relations existing in the corpus, which can be represented using a pairwise Markov Network [7, 8]. Opinions concentrate in local context sentiment words and domain entities with semantic connections [10] which are frequently used to infer entities reputation. We propose to capture the relevant sentiment words and entities, and weight them according to their sentiment relevance in a sentiment graph. We present a reputation sentiment graph built by detecting the sentiment word fluctuations through a LDA generative model, also by weighting the overall sentiment level associated to an entity which corresponds to the reputation of that entity.

2 PopMeter

PopMeter (http://popmeter.novasearch.org/) detects sentiment word fluctuations with an LDA generative model that analysis sentences from movie reviews, which present

[1] http://popmeter.novasearch.org/

A. Hanbury et al. (Eds.): ECIR 2015, LNCS 9022, pp. 785–788, 2015.

different ratings. The model weights the sentiment level associated to an entity corresponding to the reputation of that entity. In the proposed model each entity can only be influenced by its neighbour entities and sentiment words, hence, the graph structure results in a pairwise Markov Network where a propagation algorithm computes the reputation of each entity. See [4] for graph details. PopMeter sentiment-graph is populated by named-entities and sentiment words and presents a visualization of each entity with the respective sentiment connections – entities and/or sentiment words – sorted by the lowest and highest reputation levels.

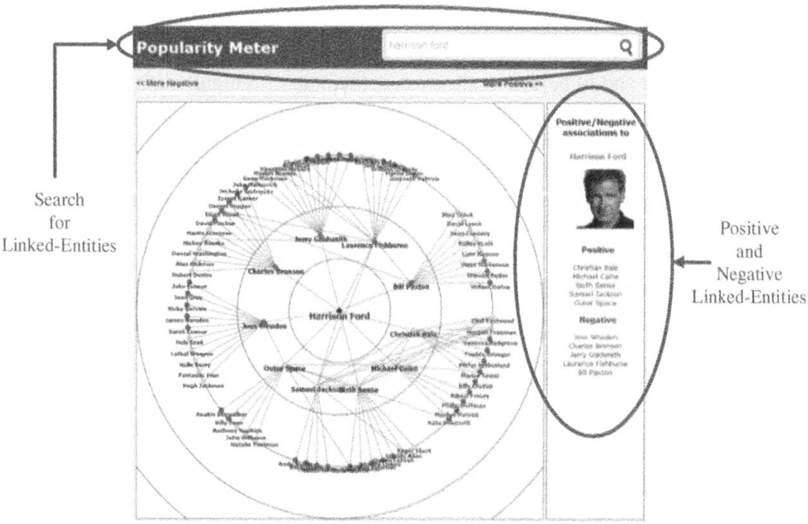

Fig. 1. PopMeter: Linked-Entities in a Sentiment Graph

PopMeter is a sentiment graph that incorporates both entities and sentiment words information in a single heterogeneous graph. In the sentiment graph nodes can correspond to entities or sentiment words [6]. The sentiment graph aims at incorporating semantically related entities and entities sentiment weight. The sentiment weight is obtained from a sentiment lexicon that is created from users sentences without human supervision in a generative model that ties words to different sentiment levels. In Figure 1, we see the web interface for the PopMeter sentiment graph, showing the central node with the actor *Harrison Ford*. PopMeter enables the user to explore the sentiment graph. The user can get an overview of the nodes connections, increase the negative and positive connections, navigate in the sentiment graph edges, select other central nodes, and search for other entities or sentiment words.

The usage of PopMeter enables the user to observe how entities and sentiment words influence positively or negatively the reputation of other entities. This influence is obtained through a generative probabilistic model that ties words and entities to different sentiment relevance levels. The graph allows to observe how the same entity may display an opposite sentiment reputation influence. As seen in Figure 2, the character *Hanna Montana* reputation is positively influenced by the *Walt Disney* industry, however, the *Walt Disney* industry is negatively influenced by the character *Hanna Montana*.

Fig. 2. Opposite reputation influence

3 Evaluation

In this experiment we used 2,503,976 sentences[2] to build the sentiment graph. The sentiment graph has 12,687 vertices – 3,177 are named entities and 9,510 are sentiment words.

Table 1. Reputation analysis results for the top 12 most cited entities (accuracy)

Entity	SWN	MPQA	Hu-Liu	Sentiment Graph
Bruce Willis	76.79%	82.14%	58.93%	87.50%
Colin Firth	81.58%	73.68%	52.63%	84.21%
Fight Club	68.97%	93.10%	58.62%	82.76%
Johnny Depp	86.25%	82.50%	50.00%	96.25%
Miley Cyrus	66.67%	77.78%	44.44%	88.89%
Peter Jackson	68.29%	63.41%	58.54%	87.80%
Phantom Menace	75.86%	96.55%	82.76%	96.55%
Pulp Fiction	75.86%	96.55%	82.76%	96.55%
Shia Labeouf	78.57%	71.43%	42.86%	78.57%
Stanley Kubrick	77.78%	83.33%	50.00%	94.44%
Star Trek	83.33%	61.11%	55.56%	61.11%
Woody Allen	72.22%	83.33%	61.11%	94.44%
Total average	76.01%	80.41%	58.18%	87.42%

To evaluate the reputation analysis algorithm [4], we generated a ground-truth dataset where named entities were identified and labelled according to the sentiment polarity expressed toward them. We then used crowdsourcing[3] practices to ask online annotators to label each sentence in accordance to the expressed sentiment towards the named entity as either very positive, positive, negative or very negative. To ensure a high-quality of the obtained labels, we limited our target workers to countries where

[2] Movie reviews from `http://imdb.com`.
[3] `http://crowdflower.com`.

English is the main language and used test questions to filter untrusting workers [5]. From the obtained results, it was selected sentences with an annotator's agreement of at least 70%. To perform the reputation classification it was built a model using the obtained ground-truth sentences. From a total of 200 sentences, the training split contains a balanced number of positive/negative sentences. The remaining ground-truth sentences are used for test purposes. In Table I it is shown the performance results using the K-Nearest Neighbour (KNN) classifier. In the performed experiences KNN uses the Manhattan distance to measure the nearest neighbour proximity. For each entity the graph presents the accuracy for the lexicons SentiWordNet[1] (SWN), MPQA[9], Hu-Liu[3] and sentiment graph.

4 Conclusion

In this paper we introduced PopMeter, a sentiment-graph designed to visualize and explore the sentiment of linked-entities. It supports searching for popular entities and browse its associations to other entities of the domain. These associations are built through a reputation analysis process of sentences where more than one entity occurs. This analysis is made in online reviews consisting of plain-text opinions where people share their views about multiple products, services, celebrities and others.

References

[1] Esuli, A., Sebastiani, F.: Sentiwordnet: A publicly available lexical resource for opinion mining. In: Proc. of the 5th LREC, pp. 417–422 (2006)

[2] Godbole, N., et al.: Large-Scale Sentiment Analysis for News and Blogs. In: ICWSM 2007 (2007)

[3] Hu, M., Liu, B.: Mining and summarizing customer reviews. In: Proc. of the tenth Knowledge Discovery and Data Mining, pp. 168–177 (2004)

[4] Peleja, F., et al.: Ranking Linked-Entities in a Sentiment Graph. In: IEEE/WIC/ACM International Joint Conferences, pp. 118–125 (2014)

[5] Snow, R., et al.: Cheap and fast—but is it good?: evaluating non-expert annotations for natural language tasks. In: Proc. of the EMNLP, pp. 254–263 (2008)

[6] Tan, C., et al.: User-level Sentiment Analysis Incorporating Social Networks. In: Proc. of the 17th Knowledge Discovery Data Mining, pp. 1397–1405 (2011)

[7] Taskar, B., et al.: Discriminative Probabilistic Models for Relational Data. In: Proc. of the Eighteenth Uncertainty in Artificial Intelligence, pp. 485–492 (2002)

[8] Wang, X., et al.: Topic Sentiment Analysis in Twitter: A Graph-based Hashtag Sentiment Classification Approach. In: Proc. of the 20th Information and Knowledge Management, pp. 1031–1040 (2011)

[9] Wilson, T., et al.: Recognizing contextual polarity in phrase-level sentiment analysis. In: Proc. of the HLT/EMNLP, pp. 347–354 (2005)

[10] Wu, Y., et al.: Structural Opinion Mining for Graph-based Sentiment Representation. In: Proc. of the EMNLP, pp. 1332–1341 (2011)

Adaptive Faceted Ranking for Social Media Comments

Elaheh Momeni[1], Simon Braendle[1], and Eytan Adar[2]

[1] Faculty of Computer Science, University of Vienna
[2] Dept. of Information Science, University of Michigan
{momeni,braendle}@cs.univie.ac.at, eadar@umich.edu

Abstract. Online social media systems (such as YouTube or Reddit) provide commenting features to support augmentation of social objects (e.g. video clips or news articles). Unfortunately, many comments are not useful due to the varying intentions of the authors of comments as well as the perspectives of the readers. In this paper, we present, a framework and Web-based system for adaptive faceted ranking of social media comments, which enables users to explore different facets (e.g., subjectivity or topics) and select combinations of facets in order to extract and rank comments that match their interests and are useful for them. Based on an evaluation of the framework, we find that adaptive faceted ranking shows significant improvements over prevalent ranking methods, utilized by many platforms, with respect to the users' preferences.
Demo: http://amowa.cs.univie.ac.at:8080/Frontend/

1 Introduction

User-generated comments are a vital part of the social media ecosystem. Comments provide a way for participants to "evolve" social media objects — ranging from YouTube videos, SoundCloud audio to more classic news articles. Unfortunately, most comment presentation systems are simple temporal streams that contain a diversity of focus, usefulness, and quality (with many comments being abusive or off-topic). Worse, due to the substantial number of comments on media objects with popular topics, identifying useful comments is often time-consuming and challenging. Without a mechanism for end-users to disentangle comment streams and identify those likely to be of interest, it is easy to imagine most end-users being overwhelmed and disappointed by their experience.

Automatic ranking of comments by "usefulness" is generally complex, mainly due to the subjective nature of usefulness [1]. The simplest method to provide ranking is wisdom-of-the-crowd approach (crowd-based ranking), which allows all users to vote on or rate comments. However, this strategy avoids an explicit definition of usefulness and voting is influenced by a number of factors (such as the "rich get richer" phenomenon) that may distort accuracy. Alternative relevant approaches propose topic-based browsing for micro-blogging platforms [2]. However, as comments have multiple explicit dimensions (such as language tone, physiological aspects, etc), grouping them exclusively based on topic, results in

A. Hanbury et al. (Eds.): ECIR 2015, LNCS 9022, pp. 789–792, 2015.

a single imperfect faceted ranking which does not allow users to rank comments with regard to other potentially useful facets.

This work proposes a framework for enabling faceted ranking on comments attached to social media objects (such as comments on an online video or a news article). Our goal is to help users explore the comment space by offering facilities to extract a set of semantic facets dynamically and to adapt the ranking of comments on the fly for finding useful comments according to the users' preferences.

2 Proposed Framework

The proposed framework consists of four main components. First, the *Semantic Enrichment Component* enriches each comment along various semantic facets when an end-user requests an adaptive faceted ranking of comments on a media object. The system provides three types of semantic facets: (1) **Topic-related facets**, topics discussed within comments on a media object. The proposed framework extracts named entities to identify topics and as extracted named entities can be ambiguous, additional fine tuning is required. (2) **Subjective facets** such as comments with subjective tone, highly affective language, offensive oriented, sad oriented (by utilizing LIWC [3]). (3) **Objective facets** such as informative, video timestamp, religion referenced. Second, the *Facet Extraction and Selection Component* operates on semantically enriched comments and clusters comments along multiple explicit semantic facets. For clustering purposes, we utilize the centroid clustering method on enriched comments. It then extracts a set of facets, selects a list of proposed facets dynamically (using the Greedy Count algorithm). Third, the *Ranking Component* enables the user to explore and select a combination of facets, and ranks comments accordingly. Finally, the *Feedback Collector and Optimization Component* collects implicit (browsing behavior of the end-user) and explicit (explicit voting by end-users on comments) feedback from the end-user. This feedback facilitates the evaluation of the proposed framework and furthermore, the personalized selection of facets.

User-Experience of Faceted Ranking in Commenting Systems The interface (see Figure 1) of the proposed framework[1] consists of two parts, one for displaying the facets (on the left), and one for displaying the ranked list of comments (on the right). A user of the system can perform the following actions: (1) She can enter a media object ID, this triggers the system to crawl all comments related to the media object, semantically enrich each comment, cluster the comments into different semantic facets, and finally present a list of facets and topics on the left side of the interface. (2) She can select combinations of proposed facets based on her preferences, this triggers the system to present a ranked list of comments based on the chosen facets. (3) She can browse ranked comments and vote whether the comments match her interests and are relevant to her

[1] The development of the backend of the interface uses the REST style, permitting the interface to be easily integrated in any social media platform `http://amowa.cs.univie.ac.at:8080/Website/website.html`

selections of facets. When a list of comments is shown based on a combination of facets, the system also shows a short overview of all other possible facets for each comment.

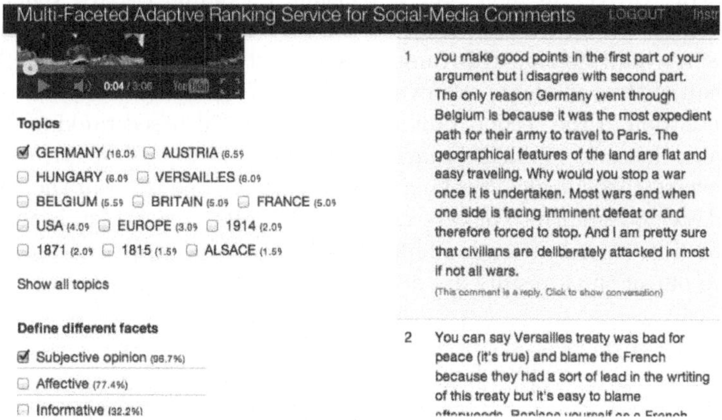

Fig. 1. User interface of the prototype implementation of proposed framework

3 Experimental Set Up

In this section we compare the effectiveness of the proposed framework to the prevalent ranking methods: **Reverse-Chronological (RC)** and **Crowd-Based (CB)**. We recruited evaluation participants by distributing calls for participation through internal science mailing lists of two universities. From the respondents, we randomly selected 40 subjects. Participant ages ranged from 20 to 57 with a median of 29. Participants received via an online instruction page and received a gift voucher for their efforts in evaluating the system. After the training phase, they were asked to perform the following steps: (1) Use the prototype to select a title from a list of 30 videos (The primary set of YouTube videos and comments used for these studies is provided by [1]). We restrict these videos to ensure that each video is almost the same length and quality. The participants then choose and watch a video. (2) Utilize the prototype to retrieve a ranked list of the top 30 comments for a video based on reverse-chronological order and also based on crowd-based order (these were the top ranked comments highlighted by YouTube). (3) Use the prototype to retrieve a ranked list of the top 30 comments for the same video in accordance with their preferences by selecting combinations of facets and topics. (4) Vote on each comment and each ranking condition. In the facet-based ranking, each comment is voted along two dimensions: *interestingness* and *relevance*. Using these two scores, we believe, is more interpretable from the end-users perspective (as compared to "usefulness"). In the chronological and crowd-based ordering mode, only the *interestingness* is rated as *relevance* is a very ambiguous concept without selecting a particular facet. This is because a comment is only considered relevant when it is relevant to the facet selection of a user. We restricted the size of the ranked list of comments to 30 in order to minimize judgment fatigue.

Results The results of our evaluation are shown in Table 1, which reflects performance of **adaptive faceted ranking (AF)**, with regard to three evaluation metrics (MAP–Mean Average Precision, P@10 and P@20). When considering the first default ranking method, the reverse-chronological ranking (RC), the measures indicate that this ranking is at least somewhat effective. Approximately half of the comments retrieved are *interesting* to the users. Furthermore, in consideration of the second ranking method, crowd-based ranking (CB), the effectiveness measures indicate that this ranking type is less effective compared to the reverse-chronological ranking. Approximately one third of the comments retrieved are determined to be *interesting* to the users. In contrast, in the ranking of comments retrieved with our adaptive faceted ranking strategy, approximately every two out of three results are deemed to be interesting.

Table 1. Effectiveness of faceted ranking

Ranking Method	#Ranking	Interesting			Relevant		
		MAP	P@10	P@20	MAP	P@10	P@20
RC	51	0.46	0.48	0.53	*Not applicable*		
CB	21	0.26	0.32	0.30	*Not applicable*		
AF	233	0.71	0.67	0.63	0.80	0.70	0.61

4 Conclusion and Future Work

Using the commonly employed ranking methods as two baselines, we have shown that the use of the faceted ranking significantly improves the ranking of comments with respect to relevance and interestingness. In the future, we will investigate the effectiveness of strategies for the selection of different types of facets and will explore the use of personalized ordering of facets and ranking strategies to further improve the interestingness and relevancy to individual users.

References

1. Momeni, E., Cardie, C., Ott, M.: Properties, prediction, and prevalence of useful user-generated comments for descriptive annotation of social media objects. In: Proceedings of the 7th International AAAI Conference on Weblogs and Social Media (ICWSM 2013). AAAI, Boston (June 2013)
2. Abel, F., Celik, I., Houben, G.-J., Siehndel, P.: Leveraging the semantics of tweets for adaptive faceted search on twitter. In: Aroyo, L., Welty, C., Alani, H., Taylor, J., Bernstein, A., Kagal, L., Noy, N., Blomqvist, E. (eds.) ISWC 2011, Part I. LNCS, vol. 7031, pp. 1–17. Springer, Heidelberg (2011)
3. Tausczik, Y.R., Pennebaker, J.W.: The psychological meaning of words: Liwc and computerized text analysis methods (2010)

Signal: Advanced Real-Time Information Filtering

Miguel Martinez-Alvarez[1,2], Udo Kruschwitz[2], Wesley Hall[1], and Massimo Poesio[2]

[1] Signal, London, UK
{miguel.martinez,wes.hall}@signal.uk.com
[2] University of Essex, Colchester, UK
{udo,poesio}@essex.ac.uk

Abstract. The overload of textual information is an ever-growing problem to be addressed by modern information filtering systems, not least because strategic decisions are heavily influenced by the news of the world. In particular, business opportunities as well as threats can arise by using up-to-date information coming from disparate sources such as articles published by global news providers but equally those found in local newspapers or relevant blogposts. Common media monitoring approaches tend to rely on large-scale, manually created boolean queries. However, in order to be effective and flexible in a business environment, user information needs require complex, adaptive representations that go beyond simple keywords. This demonstration illustrates the approach to the problem that *Signal* takes: a cloud-based architecture that processes and analyses, in real-time, all the news of the world and allows its users to specify complex information requirements based on entities, topics, industry-specific terminology and keywords.

1 Introduction and Motivation

Modern-day information overload requires advanced information filtering systems to organise and select relevant information for different users. This is not just desirable but critical for decision-makers who support their strategic decisions on their available information. This is true for multiple levels of any organisation, from the executives in the C-Suite (e.g., CEO, CTO, COO, ...) who want to have an overview of their sector, to the analysts who require detailed information while doing competitor analysis. The information needs of these users will radically differ from each other but they will most definitely involve complex concepts that cannot be efficiently represented using currently available commercial systems. Some actual information needs we have encountered range from *"news about people changing jobs within the Retail industry in the UK"* to *"all the IPO news or blog articles for tech companies in Europe"*. Another very common example information need is *"news about my competitors"*.

Currently, most of the news filtering process (including social media) is heavily human-powered with limited support based mainly on boolean-query search

A. Hanbury et al. (Eds.): ECIR 2015, LNCS 9022, pp. 793–796, 2015.

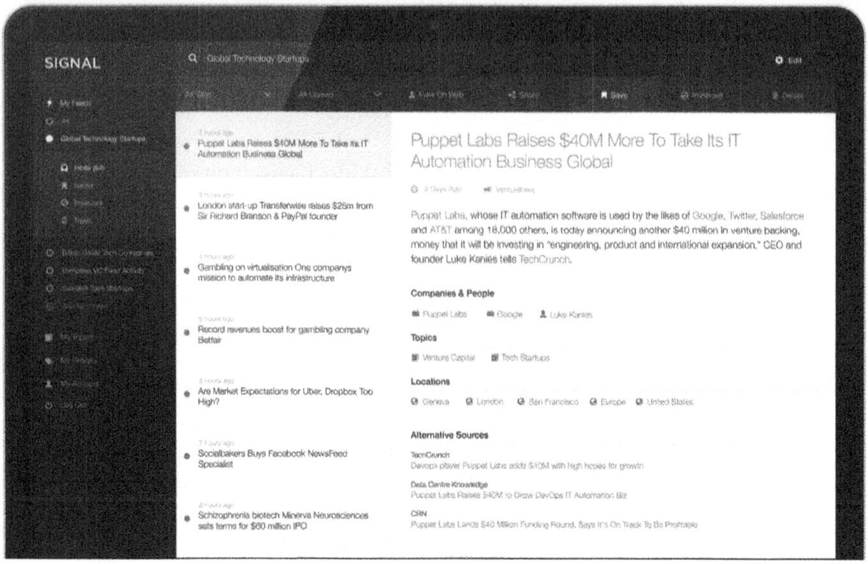

Fig. 1. Signal documents delivery page. It shows the available feeds, the relevant articles for the feed "Global Technology Start-Ups", and the detail for a selected document.

engines. This is very tedious and it can be extremely laborious in some cases, just imagine a boolean filter on the word "apple" to identify news about the technological company (instead of the fruit). These information needs can be seen as multi-task complex queries that can be addressed by using a combination of different fields or tasks such as classification, entity recognition and disambiguation (ERD) or various types of knowledge graphs.

An additional challenge is the large number of duplicates in the written media, where an article published by any major source will be likely republished several times with minor modifications. Furthermore, even if not considered duplicates, it is common to have tens or even hundreds of articles focused on the same information or event, especially for some breaking news. The former problem requires near-duplication detection, while the latter is addressed with clustering and event detection mechanisms. To process news on the scale of millions of stories per day scalability is critical to guarantee the processing of a continuous stream of data while keeping a very small delay from the publication of the original source to the delivery because the information value can quickly decrease with time. This is clear where the information delivered to the user will cause an immediate decision such as selling stock or starting a marketing campaign to mitigate social media complaints about a brand.

Several other systems have previously addressed the information filtering problem from the personalisation perspective, being mainly domain-independent and for the mass market, e.g. [3]. They tend to apply collaborative filtering algorithms in which the system predicts new recommended articles based on what

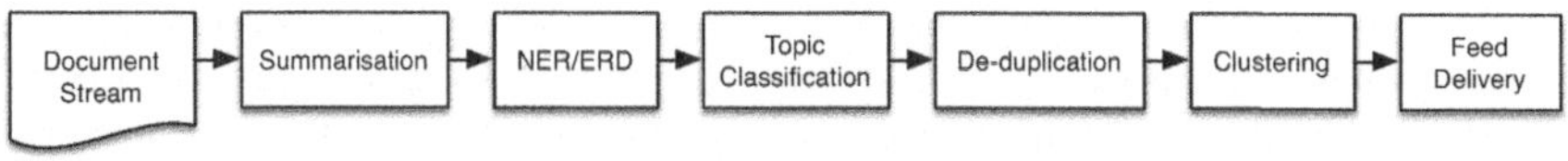

Fig. 2. Signal Processing pipeline

similar users have rated in the system, e.g. [4,1]. Furthermore, it is common to use adaptive profiling to capture the changing behaviour of the users and their interests, e.g. [2]. However, this approach does not allow users to dynamically change their information needs as the business environment requires. For instance, a market researcher might want to investigate comments about a competitor in a completely new market. Furthermore, this approach suffers from the so called "cold start" problem, where the recommendations that can be provided based on a new user are limited due to the lack of information to start with. An alternative idea is to approach the problem as a search problem, where the main challenge is the fact that keywords are not semantically rich enough to define topical information such as "people changing jobs" and they are not well suited for advanced monitoring tasks. Signal combines the best of these approaches and it expands their capabilities.

Signal allows its users to create personalised feeds with relevant information being delivered in real time. Such feeds are defined using keywords, locations, entities, topics and industries, and the system processes more than 3 million documents a day from 65,000 traditional sources and 3.5 million blogs. In addition, the system allows the creation of new topics through a user interface. As a result, our analysts can quickly create new topics based on user requirements or strategic decisions.

2 Signal Architecture and Demonstration

Any attempt to solve the information filtering problem presents several challenges from different tasks such as data cleansing, deduplication and clustering, among others. Furthermore, scalability is also a key factor to deal with millions of documents every day. We are also aware that even the best performing system might not be successful commercially if it is poorly presented, especially with a high-end target audience. We will demonstrate an integrated solution for the creation and consumption of advance information feeds using a scalable infrastructure based on a cloud architecture and a seamless interface that shows the articles for different feeds as they are de-duplicated and clustered in real-time.

The Signal architecture is formed by a pipeline (Figure 2) of different components for each one of the text analytics modules (e.g., summarisation, deduplication, NER/ERD, classification, ...), and each document is processed through all the components extending the information available for such document. For instance, after the summarisation component, the system will have access to the summary of the document. The pipeline uses a queuing system between components, allowing them to scale independently. This characteristic provides a very scalable solution while minimising the complexity of the architecture. In addition, this allows the research team of the company to focus on specific solutions for each one of the components in order to improve the quality of the system over time. The main programming language in the system is Clojure, a dialect of LISP that runs on the Java Virtual Machine (JVM). The choice of Clojure in particular, and functional programming in general, has significantly increased the integration capabilities between the research and development teams and the complexity of deploying new research models has almost been completely removed.

The specific solutions tested by the research team involve a combination of open source libraries and proprietary algorithms, where feature engineering plays a critical role. Also, for the specific tasks of text classification, hundreds of potential classifiers are considered by exploring multiple weighting strategies, diverse sets of features and different state-of-the-art classifiers such as kNN, SVM or Random Forests run against a range of domain-independent and domain-specific test collections developed in tandem with the technology.

The demonstration will include the presentation of the information filtering product, serving real-time relevant articles related to ECIR in particular, and to other potentially relevant fields such as "Text Analytics Start-ups", "Research and companies innovation programs" and "Data Science". Figure 1 shows the main screen from the product, where the most recent articles for a specific feed are shown. In addition, details such as the title, summary, entities found and assigned topics are also shown for a selected document.

Acknowledgements. This work would not have been possible without the whole Signal team and the multiple researchers who have collaborated with us, including Joseph Jacobs, Colin Wilkie, Dino Ratcliffe, Steven Zimmerman, Nikos Voskarides, Damiano Spina, Thiago Galery and David Corney. This work has been supported by InnovateUK grant KTP9159.

References

1. Das, A.S., Datar, M., Garg, A., Rajaram, S.: Google news personalization: Scalable online collaborative filtering. In: Proceedings of WWW (2007)
2. Li, L., Chu, W., Langford, J., Schapire, R.E.: A contextual-bandit approach to personalized news article recommendation. In: Proceedings of WWW (2010)
3. Liu, J., Dolan, P., Pedersen, E.R.: Personalized news recommendation based on click behavior. In: Proceedings of IUI 2010 (2010)
4. Resnick, P., Iacovou, N., Suchak, M., Bergstrom, P., Riedl, J.: Grouplens: An open architecture for collaborative filtering of netnews. In: Proceedings of ACM CSCW (1994)

The iCrawl Wizard – Supporting Interactive Focused Crawl Specification

Gerhard Gossen, Elena Demidova, and Thomas Risse

L3S Research Center and Leibniz University of Hanover, Germany
{gossen,demidova,risse}@L3S.de

Abstract. Collections of Web documents about specific topics are needed for many areas of current research. Focused crawling enables the creation of such collections on demand. Current focused crawlers require the user to manually specify starting points for the crawl (*seed URLs*). These are also used to describe the expected topic of the collection. The choice of seed URLs influences the quality of the resulting collection and requires a lot of expertise. In this demonstration we present the iCrawl Wizard, a tool that assists users in defining focused crawls efficiently and semi-automatically. Our tool uses major search engines and Social Media APIs as well as information extraction techniques to find seed URLs and a semantic description of the crawl intent. Using the iCrawl Wizard even non-expert users can create semantic specifications for focused crawlers interactively and efficiently.

1 Introduction

Focused crawlers [1,4] enable the efficient creation of topically and temporally coherent document collections from the Web and Social Media. Such collections are increasingly used in many domains such as digital sociology, history, politics, and journalism [2,5]. By using focused crawlers, researchers, archivists and journalists can create sub-collections about specific events and topics such as the Ebola outbreak or the Ukraine crisis efficiently on demand.

Focused crawling starts with the *manual* definition of the *crawl specification*, a list of so-called *seed URLs* and (optionally) keywords and entities representing the crawl intent of the user. The crawl specification is necessary for the focused crawler to efficiently find relevant pages and to correctly judge their relevance. Firstly, the crawler uses seed URLs as starting points for the traversal of the Web graph, such that good seeds can lead the crawler directly to relevant pages. Secondly, the content of the pages specified by the seed URLs is used to perform relevance estimation of unseen documents collected during the crawling. Thus, the success of the focused crawlers depends on the expertise of the user to specify representative seed URLs and keywords. However, our anticipated non-expert users need to create collections only rarely and thus cannot develop the necessary experience.

A. Hanbury et al. (Eds.): ECIR 2015, LNCS 9022, pp. 797–800, 2015.

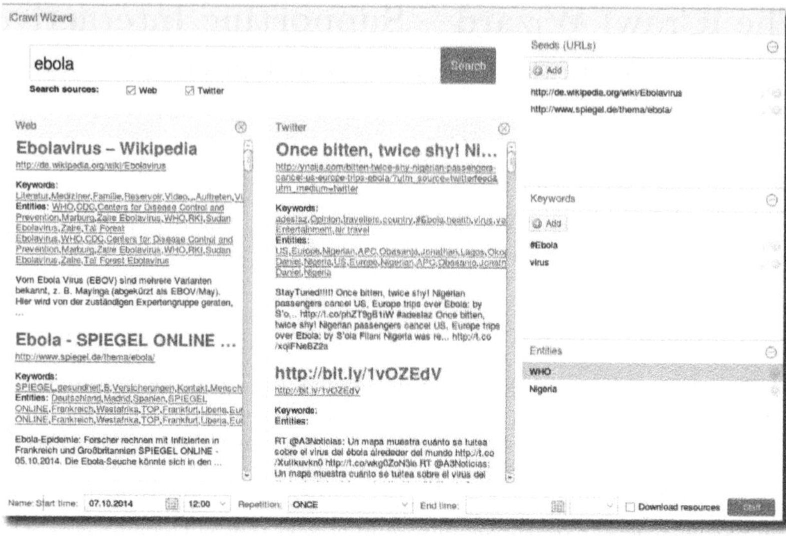

Fig. 1. The iCrawl Wizard UI

In this demonstration we present the iCrawl Wizard[1], novel interface that enables non-expert users to interactively and efficiently create the crawl specification for a focused crawler. The iCrawl Wizard combines Web search, Social Media queries and information extraction tools to enable users to compose crawl specification efficiently in an intuitive way. It builds upon users' previous experience with Web search engines and allows them to start the crawl specification process using a simple keyword search. Based on user's keyword queries, the iCrawl Wizard suggests seed URLs obtained from Web search engines and Social Media. Additionally, it uses information extraction tools to suggest representative keywords and entities for the semantic crawl specification. This way the iCrawl Wizard opens the focused crawling technology to the non-expert users and enables them to easily specify their crawl intention.

2 Wizard User Interface

The user interface of the *iCrawl Wizard* is presented in Fig. 1. The user of the iCrawl Wizard starts by entering keywords in the search field at the top of the user interface. In this example, a user creating a crawl about the ebola outbreak enters the keyword "ebola" and presses the search button. In response to this query, the system provides Web search results along with the results from the Twitter API. The Web search results allow the user to find highly relevant Web pages, while the Twitter search results provide the most recent pages about

[1] The demo is available online at `http://icrawl.l3s.uni-hannover.de:8090/campaign/1/add`

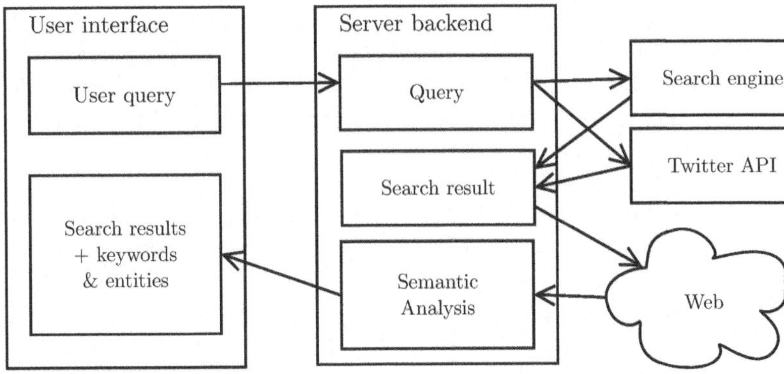

Fig. 2. Architecture of the iCrawl Wizard

the topic. The latter is especially important when creating a collection about a current event.

The search results are extended by keywords and entities describing the pages in more detail. In this example, the first Web search result is the German Wikipedia article on Ebola, from which the entities such as World Health Organization (WHO) and Robert Koch Institute (RKI) are extracted.

Seed URLs, keywords and entities can be selected and added to the crawl specification by clicking on them. The crawl specification constructed so far is visible on the right hand side of the interface, allowing the user to inspect and modify the list of selected seed URLs, keywords and entities. The user can also click on the search results to examine the Web pages before adding them to the crawl specification. Items found outside the Wizard can be added to the crawl specification manually by clicking the corresponding "add" buttons. If necessary, the user can re-formulate the search query and thus construct crawl specification incrementally in several interaction steps. Finally, further crawl parameters such as start time and duration can be specified at the bottom of the interface.

3 Architecture

The user interface presented in Section 2 is implemented as a Web application supported by a server-side component. The architecture is shown in Fig. 2.

When the user enters a keyword query, this query is sent to the server, which in term forwards the query to the search APIs of Web search engines (e.g. Bing) and Social Media APIs (e.g. Twitter). The Web search APIs return a ranked list of *(URL, title, description)* triples that can be processed further in this form. The Social Media APIs return a collection of posts, so we extract the links contained in the posts and order them by their frequency. In the case of Twitter, we also extract *hashtags* (e.g. "#ukraine") as proposed keywords. The text of the posts is used as a description of the extracted links.

The descriptions gained this way provide relatively few information. Therefore we download the pages from the Web and use information extraction tools such as the Stanford Named Entity Recognizer[2] and the TextRank algorithm [3] to extract key terms and entities from the Web page text. These are used to augment the results presented to the user.

All user actions such as issued queries as well as added and removed items are logged into a database. This enables the creation of a comprehensive *crawl description* to allow better sharing and re-use of the created document collection.

4 Demonstration Overview

The aim of the iCrawl Wizard is to assist users in defining a crawl specification for a topic of interest by starting from simple keywords. In our demonstration we will show how the iCrawl Wizard works and how users can use it to obtain the desired crawl specification without any prior knowledge about Web crawling.

During the demonstration, our audience can try the iCrawl Wizard interface. To highlight the advantages of our approach, we will ask our audience to perform crawl specification using the iCrawl Wizard as well as to suggest the seed URLs, terms and entities for the crawl specification manually. Through the comparison, the audience can get some hand-on experience about dataset creation problems on the Web. We will make the iCrawl Wizard available as open source software after the conference.

Acknowledgments. The authors would like to thank Bohdan Tkachenko for supporting the implementation of the user interface. This work was partially funded by the ERC under ALEXANDRIA (ERC 339233), and the COST Action IC1302 (KEYSTONE).

References

1. Chakrabarti, S., van den Berg, M., Dom, B.: Focused crawling: a new approach to topic-specific web resource discovery. Computer Networks 31(11-16), 1623–1640 (1999)
2. Demidova, E., Barbieri, N., Dietze, S., Funk, A., Holzmann, H., Maynard, D., Papailiou, N., Peters, W., Risse, T., Spiliotopoulos, D.: Analysing and enriching focused semantic web archives for parliament applications. In: Future Internet, Special Issue "Archiving Community Memories" (July 2014)
3. Mihalcea, R., Tarau, P.: Textrank: Bringing order into text. In: EMNLP 2004, pp. 404–411 (2004)
4. Pereira, P., Macedo, J., Craveiro, O., Madeira, H.: Time-aware focused web crawling. In: de Rijke, M., Kenter, T., de Vries, A.P., Zhai, C., de Jong, F., Radinsky, K., Hofmann, K. (eds.) ECIR 2014. LNCS, vol. 8416, pp. 534–539. Springer, Heidelberg (2014)
5. Risse, T., Demidova, E., GossenWhat do, G.: you want to collect from the web? In: Building Web Observatories Workshop, BWOW 2014 (2014)

[2] `http://nlp.stanford.edu/software/CRF-NER.shtml`

Linguistically-Enhanced Search over an Open Diachronic Corpus

Rafael C. Carrasco, Isabel Martínez-Sempere, Enrique Mollá-Gandía, Felipe Sánchez-Martínez, Gustavo Candela Romero, and Maria Pilar Escobar Esteban

Departament de Llenguatges i Sistemes Informàtics
Universitat d'Alacant, E-03071, Alacant, Spain

Abstract. The BVC section of the IMPACT-es diachronic corpus of historical Spanish compiles 86 books —containing approximately 2 million words. About 27% of the words —providing a representative coverage of the most frequent word forms— have been annotated with their lemma, part of speech, and modern equivalent following the Text Encoding Initiative guidelines. We describe how this type of annotation can be exploited to provide linguistically-enhanced search over historical documents. The advanced search supports queries whose search terms can be a combination of surface forms, lemmata, parts of speech and modern forms of historical variants.

1 Introduction

Diachronic corpora are a valuable source of information to understand the historical evolution of languages. This paper describes a web-based search tool built upon the Apache Lucene[1] platform. Currently, the tool supports advanced search over the BVC section of the IMPACT-es diachronic corpus of historical Spanish [4] distributed by the Impact Centre of Competence in Digitisation.[2]

The corpus contains 86 Spanish texts provided by the Biblioteca Virtual Miguel de Cervantes,[3] printed between 1482 and 1647; it covers a representative variety of authors and genres (such as prose, theatre, and verse). This corpus is one of the few collections of historical Spanish distributed under an open license.

The original spelling (even if clearly unintentional) has been preserved in order to achieve a highly accurate transcription. The metadata added to Spanish words are its lemma (in modern form), its part of speech and its modern equivalent. The words originating from other languages (less than 0.1%, and principally Latin) are labelled solely with their language. The morphological categories which have been considered are abbreviation, adjective, adverb, conjunction, determiner, interjection, noun, proper noun, numeral, preposition, pronoun, relative pronoun, and verb. The annotation process was assisted by the CoBaLT tool [1], which supports complex annotations.

1 `http://lucene.apache.org/`
2 `http://www.digitisation.eu/data/browse/corpus/impact-es`
3 `http://www.cervantesvirtual.com`

A. Hanbury et al. (Eds.): ECIR 2015, LNCS 9022, pp. 801–804, 2015.

2 The Query Language and Interface

The interface with the search engine is available at `http://bvmcresearch.cervantesvirtual.com/diasearch` where multiple query terms can be specified. As in [3], every term can be preceded by a prefix:

- If no prefix is added, the term denotes a diachronic form (verbatim text).
- The prefix `modern#` denotes a modern form.
- The prefix `lemma#` is followed by a lemma.
- The prefix `pos#` denotes a part-of-speech tag.

Multiterm queries can include different prefixes and use the rich query language[4] provided by Lucene, the open source information retrieval Java library. Words or text segments matching the query are highlighted and presented in their context (snippet).

The index is based on Lucene's synonym list where a filter is applied to expand every input token. The index contains then, for every word form, all possible modern forms, lemmata, and parts of speech; For example, the word form *celebrada* generates 5 entries (from two analysis: lemma#celebrar, pos#verb, modern#celebrada and lemma#celebrado, pos#adj, modern#celebrada) while the word form *yerro* generates 7 entries (lemma#yerro, pos#n, modern#yerro; lemma#hierro, pos#n, modern#hierro; lemma#errar, pos#verb, modern#yerro).

A diachronic form can be assigned more than one modern equivalent (e.g., the historical spelling *fijo* is compatible with the adjective *fijo* and the noun *hijo*). The diachronic form can be also compatible with multiple lemmas and parts of speech. Since the historical spelling was less constrained than modern orthography, old texts usually show a higher rate of homography.

For optimal retrieval, what is considered a word has been carefully defined: words may contain characters in the Unicode category letters separated by non-breaking symbols such as the dash, the ampersand and plus signs, dots, etc.

The tool uses the `doBVMCDiaSearch` method defined in the BVMCSearch[5] public service, which provides standard JSON output (JavaScript Object Notation)[6]. The interface is implemented in AJAX and is based on Simple Object Access Protocol (SOAP)[7], Web Services Definition Language (WSDL)[8] and the XML Schema Definition language (XSD). The tool allows for the pagination of results, navigation through the result pages, the highlighting of matches, etc. while maintaining the compatibility with the most important browsers and devices.

---

[4] `http://lucene.apache.org/core/3_6_0/queryparsersyntax.html`
[5] `http://app.cervantesvirtual.com/cervantesvirtual-web-services/BVMCSearchWSService?wsdl`
[6] JSON `http://www.json.org` defines a language to store and exchange textual information which is smaller, faster and easier to parse than XML.
[7] `http://www.w3.org/TR/2007/REC-soap12-part0-20070427/`
[8] `http://www.w3.org/TR/wsdl`

Parameter	Type	Default	Example
`q`	String		pos#n
`start`	int	0	0
`maxResults`	int	10	10
`fragmentNumber`	int	0	0
`fragmentSize`	int	10	10

3 Main Features and Possible Improvements

Although only a fraction of the BVC section of the IMPACT-es corpus has been annotated, the information of the words with morphological tags can be extrapolated to non-annotated words and the coverage then grows from 27% to 92% (with a decrease in precision). Fortunately, for multiterm queries (with only a few terms) the implicit intersection often disambiguates the interpretation of the text [3]. Figure 1 shows the results retrieved when the BVC section of the IMPACT-es corpus is interrogated with the query **`"lemma#haber modern#de pos#verb"`**. Note that, although the word form *a* can be a form of the verb *haber* or a preposition, the first option is never followed by *de*.

Fig. 1. Sample results after the query **`"lemma#haber modern#de pos#verb"`**

The lexicon is the set of all unique word forms annotated in the corpus (25423 words with over half a million attestations). Almost one tenth of the word forms can be assigned more than one lemma but only 1% of the words admit more than one modernisation. This is an indication that automatic modernisation can be applied with a reasonable performance.

The ambiguity raises a sample of the corpus is analysed (rather than the lexicon): the number of words with more than one modern equivalent raises to 17.2% due to the ubiquity of the most frequent words (the so-called *stop-words*), most of which, remarkably, happen to be ambiguous. In the collection, just five word forms among the 10 most frequent words add up to 14.89% of the cases with multiple modern forms.

This ambiguity can be often resolved if the exact part of speech of the word is known. This suggest using standard part-of-speech taggers, which reach 97% precision [2]. We have found that 92% of the multi-lemma word forms are amenable to disambiguation with a standard part-of-speech tagger trained over Spanish text. Of course, there are exceptions such as the Spanish form *yerro* which can be assigned the lemma *errar* (a verb) and the lemma *hierro* (a noun), but also the lemma *yerro* (also a noun).

A small fraction (about 1%) of the analysed word forms allow for more than one modern equivalent but one half of them are disambiguated with the part-of-speech information. Therefore, a very high precision can be obtained for this type of search. Next table shows the fraction of ambiguous terms in the lexicon (25423 word forms) and in a sample of the corpus (containing 1,982,882 word forms).

Ambiguous annotations	Lexicon	Texts
Lemmata	2,518 (9,9%)	688,452 (37.5%)
Parts of speech	4,515 (17.8%)	697,539 (38.1%)
Modern forms	277 (1.1%)	314,616 (17.2%)

4 Conclusions

We have implemented an on-line service which allows to search over a diachronic corpus using a combination of query terms that may refer to historical forms, modern forms, lemmata or parts of speech. In order to increase recall we have extrapolated the annotations to all the occurrences of each word form. Although, this generalization may reduce the precision, it provides accurate results when the query contains a multiple search terms. The accuracy can be further improved with the integration of a part-of-speech tagger.

References

1. Kenter, T., Erjavec, T., Dulmin, M.Z., Fiser, D.: Lexicon construction and corpus annotation of historical language with the CoBaLT editor. In: Proceedings of the 6th Workshop on Language Technology for Cultural Heritage, Social Sciences, and Humanities, Avignon, France, pp. 1–6 (April 2012)
2. Manning, C.D., Schütze, H.: Foundations of statistical natural language processing, pp. 1–680. MIT Press (2001)
3. Sánchez-Martínez, F., Forcada, M.L., Carrasco, R.C.: Searching for linguistic phenomena in literary digital libraries. In: Proceedings of the ECDL 2008 Workshop on Information Access to Cultural Heritage, Aarhus, Denmark (September 2008)
4. Sánchez-Martínez, F., Martínez-Sempere, I., Ivars-Ribes, X., Carrasco, R.C.: An open diachronic corpus of historical Spanish. Language Resources and Evaluation (2013), doi:10.1007/s10579-013-9239-y

From Context-Aware to Context-Based: Mobile Just-In-Time Retrieval of Cultural Heritage Objects

Jörg Schlötterer, Christin Seifert, Wolfgang Lutz, and Michael Granitzer

Media Computer Science, University of Passau, Germany
{joerg.schloetterer,christin.seifert,michael.granitzer}@uni-passau.de

Abstract. Cultural content providers face the challenge of disseminating their content to the general public. Meanwhile, access to Web resources shifts from desktop to mobile devices and the wide range of contextual sensors of those devices can be used to proactively retrieve and present resources in an unobtrusive manner. This proactive process, also known as just-in-time retrieval, increases the amount of information viewed and hence is a viable way to increase the visibility of cultural content. We provide a contextual model for mobile just-in-time retrieval, discuss the role of sensor information for its contextual dimensions and show the model's applicability with a prototypical implementation. Our proposed approach enriches a user's web experience with cultural content and the developed model can provide guidance for other domains.

1 Introduction

Recent initiatives like Europeana[1] spend a huge effort on aggregating digitized museum artifacts of different institutions and providing a unified interface to access those resources. Nevertheless, users still need to be aware of those specialized portals to gain access to the tremendous collection of cultural heritage objects. Our approach is to take the content to the user, instead of taking the user to content. To this extent, we implement a just-in-time retrieval approach in a mobile setting, based on the contextual information collected by the various sensors of nowadays smartphones. These sensors capture a wide spectrum of a user's context and hence provide a great source for retrieving relevant resources and adapting to the user's needs. We align our approach along the following questions: *When* to retrieve and present resources to the user? *What* are the resources the user is interested in and can they be refined by location information of resources or users (*where*)?

Specifically, our contributions are the following: (i) we present a context model for just-in-time retrieval in a mobile environment, (ii) we discuss how to incorporate available sensor information into the defined context dimensions and (iii) demonstrate the applicability of the model with a prototype.

[1] `http://www.europeana.eu/`

A. Hanbury et al. (Eds.): ECIR 2015, LNCS 9022, pp. 805–808, 2015.

2 Modeling Context for Mobile Just-In-Time Retrieval

Context is usually deemed an additional dimension for personalization, either in a recommender system [2] or in information retrieval [12]. In contrast, context is the sole basis for providing recommendations in our work, resembling just-in-time retrieval [10]. Hence, we follow the rather broad definition of context as "any description of the world that can be relevant to an application" by Pete Steggles [1]. For the task of retrieving relevant resources in a mobile setting we define three abstract dimensions: (i) *when*, (ii) *what*, and (iii) *where*. The rational behind the three dimensions is to construct a conceptual model of the user's (potential) information need, which then can be encoded into a search query. In the following we outline how information for each dimension can be collected from either primary context, i.e. (raw/physical) sensor data (e.g. temperature), or secondary context, i.e. virtual sensors, gathering information from applications or services (e.g. message contents) or logical sensors, gathering information from physical or virtual sensors, mainly by aggregation (e.g. activities, such as walking) [3,6].

***When*:** A user should be notified about additional resources only when it is appropriate. Interruptibility refers to a state, in which a person can be interrupted in a task without (too) negative consequences. Middleton highlights the necessity for Interface Agents to "detect when and if to interrupt the user" [8]. Noise level, observable directly by physical sensors, has been found to be a strong indicator for non-interruptibility [7]. Besides the noise level, interruptibility can be assessed according to the current situation, obtainable from logical sensors. We classify situations into *trigger* and *blocker* situations, that either initiate the recommendation process or hinder it. A combination of situations can also occur, while mostly a blocker situation will supersede one or more trigger situations.

***What*:** One of the most valuable sources for generating search queries is textual content, which is available from the currently used application, incoming messages, notifications, etc. through virtual sensors. In order to translate the textual content into a query, keywords need to be extracted. A first step to separate stopwords and non-informative terms from those that actually convey information is named entity detection [9]. In this process, special challenges of mobile devices need to be addressed, such as short messages [11] or limited resources [4]. Given a candidate set of entities, they can be further reduced, by selecting e.g. the most salient ones [5], matching them against a user profile, etc. A very simple approach, even performable with a mobile phone's limited computing power is to choose based on frequency, i.e. how often an entity is mentioned in the text. The final set of keywords may be enriched with location information (c.f. *where*) and sent to the retrieval system.

***Where*:** Location information also serves as information source to construct or refine a query. In the simplest scenario, the name of the city, the user is currently situated in, can be used as query term, in order to obtain resources about this city. Moreover, based on the current location, points of interest (POIs) nearby can be obtained, and a POI's label can be used as query term. In addition, locations identified by named entity detection (c.f. *what*) can also be used for retrieval. Cultural heritage objects can exhibit different types of locational information: the actual location of the object, i.e. the museum in whose collection it is stored, the place, from where it originated, etc. Consequently, mapping the detected locations to the appropriate query or metadata fields poses a challenge.

3 Prototype

To demonstrate the applicability of our proposed approach, we implemented a prototype[2] for Android mobile devices, which uses the Europeana API as search backend. Figure 1 provides a general overview of the processing chain implemented in our prototype, which is described in more detail in the following.

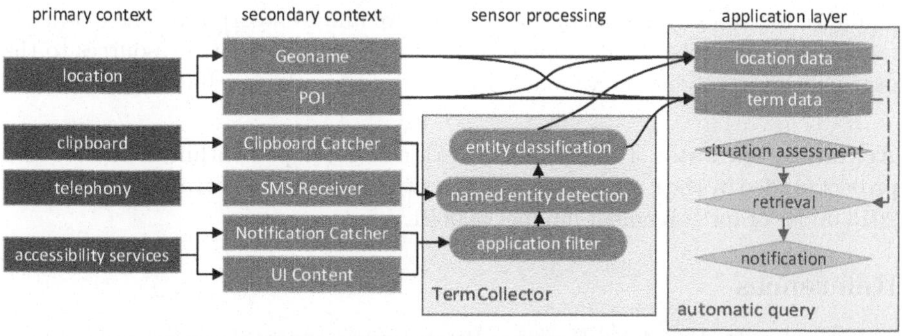

Fig. 1. Overview of the prototype's workflow

When: We monitor incoming SMS and notifications, the content of the clipboard, content on the user interface (UI) and the user's location. Based on the latter, we obtain the geoname of the current location and nearby POIs. Activity in the just mentioned sensors is a trigger for determining when to query. After processing the context, the automatic query component evaluates the current situation(s) against a predefined set of trigger/blocker situations and, if appropriate, issues a query with the terms derived. If this query yields results, the user is notified through a *ramping interface* [10], featuring different stages, with each stage providing a little more information. The first stages can be ignored easily and information can be filtered early, requiring less attention from a user.

What & Where: Context collected from notifications and the UI is filtered first by an application blacklist, in order to remove regular content such as the home screen or notifications from the Android downloader. A simple named entity detection, based on capitalization, is performed for the last four secondary context sensors in the figure and the resulting entities are classified into location or other entities. These steps are not necessary for the POI and Geoname components, as they already provide location entities. The entities obtained from all sensors are stored for further processing by the automatic query component. It

[2] Source and demo at `http://purl.org/eexcess/components/android-app`

is to note, that location entities can also be used to address the *what* dimension as described in section 2. The Europeana API features a faceted search interface, including the facets *what* and *where*. We send the terms stored in the location data component in the *where* facet and those from term data in the *what* facet.

4 Summary and Future Work

We presented an approach for mobile just-in-time retrieval in the cultural heritage domain with a retrieval process purely based on contextual information and not requiring any explicit user interaction. We showed how such a process can be modeled along the contextual dimensions of *when*, *what* and *where*, along with a first prototype implementing this model. Even though our application focus is on cultural content, we think that the proposed model can also provide guidance for other domains. In future work, we aim to incorporate the quality of retrieved results into the decision of when to present additional resources to the user instead of relying on a binary decision based on trigger/blocker situations.

Acknowledgments. The presented work was developed within the EEXCESS project funded by the European Union Seventh Framework Programme FP7/2007-2013 under grant agreement number 600601.

References

1. Abowd, G.D., Dey, A.K.: Towards a Better Understanding of Context and Context-Awareness. In: Proc. of HUC, pp. 304–307 (1999)
2. Adomavicius, G., Tuzhilin, A.: Context-aware recommender systems. In: Recommender Systems Handbook, pp. 217–253. Springer (2011)
3. Baldauf, M.: A survey on context-aware systems. International Journal on Ad Hoc and Ubiquitous Computing 2(4) (2007)
4. Ek, T., Kirkegaard, C., Jonsson, H., Nugues, P.: Named entity recognition for short text messages. Procedia-Social and Behavioral Sciences 27, 178–187 (2011)
5. Gamon, M., Yano, T., Song, X., Apacible, J., Pantel, P.: Identifying salient entities in web pages. In: Proc. of CIKM, pp. 2375–2380 (2013)
6. Hong, J.Y., Suh, E.H., Kim, S.J.: Context-aware systems: A literature review and classification. Expert Systems with Applications 36(4), 8509–8522 (2009)
7. Hudson, S.E., Fogarty, J., Atkeson, C.G., Avrahami, D., Forlizzi, J., Kiesler, S., Lee, J.C., Yang, J.: Predicting Human Interruptibility with Sensors: A Wizard of Oz Feasibility Study. In: Proc. of SIGCHI (2003)
8. Middleton, S.E.: Interface agents: A review of the field. CoRR cs.MA/0203 28 (March 2002)
9. Nadeau, D., Sekine, S.: A survey of named entity recognition and classification. Lingvisticae Investigationes 30(1), 3–26 (2007)
10. Rhodes, B.J.: Just-In-Time Information Retrieval. Ph.D. thesis, Massachusetts Institute of Technology (2000)
11. Ritter, A., Clark, S., Mausam, E.O.: Named entity recognition in tweets: An experimental study. In: Proc. of EMNLP, pp. 1524–1534 (2011)
12. Shen, X., Tan, B., Zhai, C.: Context-sensitive information retrieval using implicit feedback. In: Proc. of SIGIR, p. 43 (2005)

Visual Analytics for Information Retrieval Evaluation (VAIRË 2015)

Marco Angelini[1], Nicola Ferro[2], Giuseppe Santucci[1], and Gianmaria Silvello[2]

[1] "La Sapienza" University of Rome, Italy
{angelini,santucci}@dis.uniroma1.it
[2] University of Padua, Italy
{ferro,silvello}@dei.unipd.it

Abstract. Measuring is a key to scientific progress. This is particularly true for research concerning complex systems, whether natural or human-built. The tutorial introduced basic and intermediate concepts about lab-based evaluation of information retrieval systems, its pitfalls, and shortcomings and it complemented them with a recent and innovative angle to evaluation: the application of methodologies and tools coming from the *Visual Analytics (VA)* domain for better interacting, understanding, and exploring the experimental results and *Information Retrieval (IR)* system behaviour.

1 Scope and Learning Objectives

The tutorial addressed the topic of experimental evaluation, which has been a core topic in *Information Retrieval (IR)* since its inception. However, the tutorial faced this topic mixing basic and indispensable concepts on IR evaluation with a new angle that comes from applying information visualization and *Visual Analytics (VA)* methods and techniques to improve the interpretation and interaction with the experimental data, with the final goal of better understanding the system behaviour.

The overall aim of the tutorial was to improve the skills and practices of junior researchers (but also senior ones were welcome) in carrying out a thorough evaluation of IR system, providing them with both solid knowledge of IR evaluation and its pitfalls and with an innovative angle, coming from the application of visual analytics techniques to the understanding of and interaction with experimental data.

The specific learning objectives were: (i) to learn basic and intermediate competencies on IR evaluation and its pitfalls; (ii) to learn basic competencies on VA; (iii) to learn how VA techniques can be fruitfully applied to IR evaluation; (iv) to learn to implement basic VA components for IR evaluation.

2 Contents

The lecture in the first half-day was constituted by three modules. The objective of this first half-day was to provide attendees with needed methodological notions to achieve the learning objectives described above.

A. Hanbury et al. (Eds.): ECIR 2015, LNCS 9022, pp. 809–812, 2015.

The first module started introducing the main motivations and goals for experimental evaluation [1] and explained the basic concepts of the experimental evaluation according to the Cranfield paradigm, namely experimental collections, ground-truth creation and pool, evaluation campaigns and their typical life-cycle [2].

Evaluation measures have been introduced and discussed, also from in relation to what is usually done in the (representational) theory of measurement [3], their main constituents (user models, ...) were presented, and some caveats about scale types and the allowed operations with them have been raised. Some examples of well-known measures, such as *Average Precision (AP)*, *Normalized Discounted Cumulated Gain (nDCG)*, *Rank-Biased Precision (RBP)* and so on, have been discussed [4].

Failure analysis was then introduced and explained as a fundamental but extremely demanding activity by providing examples from well-known exercises, such as the *Reliable Information Access (RIA)* workshop [5].

The second module introduced the goals and the motivations underlying the emerging VA discipline, detailing the concepts and the basic techniques that are currently adopted in such a research field. In particular, the canonical steps of internal data representation and data presentation have been described, together with an overview of the most used visualization techniques [6,7].

Issues associated with the correct evaluation of VA systems were introduced and discussed. In particular, the tutorial analysed the user centered design methodology and the evaluation through questionnaires [8]. Examples have been given by the application of such techniques to the VA prototype developed within the IR evaluation PROMISE[1] Infrastructure [9,10].

The third module dealt with advanced applications of VA to experimental evaluation, where theoretical notions were complemented with examples from actually implemented prototypes. In particular, we: (i) described how to provide better support for carrying out an effective failure analysis [11,12]; (ii) introduced a new phase in the evaluation, we called it "what-if analysis", aimed at getting an estimate of the possible impact of modifications to an IR system on its performances [13,14].

The hands-on session in the second half-day were constituted by three modules. The objective of this second half-day was to provide attendees with a concrete feeling about how to develop and implement the methodological notions introduced in the first half-day.

The first module let attendees play with a running prototype of a VA system for IR evaluation, the VATE system, in order to let them experience what you can aim at for such kind of systems, how they can work, and how you can benefit from them for better understanding the experimental results. They then went trough a questionnaire for evaluating the used system. This had a twofold goal: first, to stimulate critical thinking about what the attendees have just experienced; second, to provide them with a concrete example of what evaluating

[1] `http://www.promise-noe.eu/`

VA systems means and a starting point whether they will have to evaluate their own VA systems.

The second module explained how to evaluate the output of an IR system using standard experimental collections. In particular, it provided a step-by-step example using the open source freely available MATTERS[2] library, a MATLAB toolkit for computing standard evaluation measures and carrying out analyses (previous knowledge about MATLAB is not required), and ad-hoc *Conference and Labs of the Evaluation Forum (CLEF)* collections [15,16].

The third module introduced the basics of the Web based visualization library D3[3], providing a step-by-step comprehensive example for representing and presenting a dataset containing IR evaluation data, focusing on user interaction in order to quickly get insights from coordinated visualizations.

3 Schedule

The schedule of the lecture part (half-day) was organized as follows:

- *Information Retrieval and its Evaluation*: basics on laboratory-based IR evaluation [1,2]; basics on IR evaluation measures [4,3]; failure analysis [5].
- *Visual Analytics*: basics on Visual Analytics [6,7]; basics on evaluation of Visual Analytics systems [8]; application of Visual Analytic to IR evaluation and running examples with the PROMISE Infrastructure prototype [9,10].
- *Advanced Applications of Visual Analytics for IR Evaluation*: Visual Analytics for Failure Analysis and running examples with the VIRTUE prototype [11,12]; Visual Analytics for What-if Analysis and running examples with the VATE prototype [13,14].

The schedule of the hands-on part (half-day) was organized as follows:

- *Experiencing with VA for IR Evaluation*: use and trial of the VATE prototype; evaluation questionnaire on the VATE prototype.
- *Example of Building Blocks for VA applied to IR evaluation (part 1 of 2)*: use of the MATTERS evaluation library to assess the performances of an IR system and produce experimental data to analyse;
- *Example of Building Blocks for VA applied to IR evaluation (part 2 of 2)*: use of the D3 library to develop interactive plots and process the experimental data produced in part 1.

References

1. Harman, D.K.: Information Retrieval Evaluation. Morgan & Claypool Publishers, USA (2011)
2. Sanderson, M.: Test Collection Based Evaluation of Information Retrieval Systems. Foundations and Trends in Information Retrieval (FnTIR) 4, 247–375 (2010)

[2] `http://matters.dei.unipd.it/`

[3] `http://d3js.org/`

3. Fenton, N.E., Bieman, J.: Software Metrics: A Rigorous & Practical Approach, 3rd edn. Chapman and Hall/CRC, USA (2014)
4. Büttcher, S., Clarke, C.L.A., Cormack, G.V.: Information Retrieval: Implementing and Evaluating Search Engines. The MIT Press, Cambridge (2010)
5. Harman, D., Buckley, C.: Overview of the Reliable Information Access Workshop. Information Retrieval 12, 615–641 (2009)
6. Thomas, J.J., Cook, K.A. (eds.): Illuminating the Path: The Research and Development Agenda for Visual Analytics, National Visualization and Analytics Center, USA (2005)
7. Keim, D.A., Kohlhammer, J., Ellis, G., Mansmann, F. (eds.): Mastering the Information Age – Solving Problems with Visual Analytics. Eurographics Association, Goslar (2010)
8. Kang, Y.: a., Görg, C., Stasko, J.: Evaluating Visual Analytics Systems for Investigative Analysis: Deriving Design Principles from a Case Study. In: May, R., Kohlhammer, J., Stasko, J., van Wijk, J. (eds.) Proc. IEEE Symposium on Visual Analytics Science and Technology (VAST 2009), pp. 139–146. IEEE Computer Society, Los Alamitos (2009)
9. Angelini, M., Ferro, N., Santucci, G., Garcia Seco de Herrera, A.: Deliverable D5.4 – Revised Collaborative User Interface Prototype with Annotation Functionalities. PROMISE Network of Excellence, EU 7FP, Contract N. 258191. http://www.promise-noe.eu/documents/10156/a61967be-2b23-461b-b72a-302766c942e3 (2013)
10. Ferro, N., Berendsen, R., Hanbury, A., Lupu, M., Petras, V., de Rijke, M., Silvello, G.: PROMISE Retreat Report – Prospects and Opportunities for Information Access Evaluation. SIGIR Forum 46, 60–84 (2012)
11. Angelini, M., Ferro, N., Santucci, G., Silvello, G.: VIRTUE: A visual tool for information retrieval performance evaluation and failure analysis. Journal of Visual Languages & Computing (JVLC) 25, 394–413 (2014)
12. Angelini, M., Ferro, N., Santucci, G., Silvello, G.: A Visual Interactive Environment for Making Sense of Experimental Data. In: de Rijke, M., Kenter, T., de Vries, A.P., Zhai, C., de Jong, F., Radinsky, K., Hofmann, K. (eds.) ECIR 2014. LNCS, vol. 8416, pp. 767–770. Springer, Heidelberg (2014)
13. Angelini, M., Ferro, N., Santucci, G., Silvello, G.: Visual Interactive Failure Analysis: Supporting Users in Information Retrieval Evaluation. In: Kamps, J., Kraaij, W., Fuhr, N. (eds.) Proc. 4th Symposium on Information Interaction in Context (IIiX 2012), pp. 195–203. ACM Press, New York (2012)
14. Angelini, M., Ferro, N., Granato, G.L., Santucci, G., Silvello, G.: Information Retrieval Failure Analysis: Visual analytics as a Support for Interactive "What-If" Investigation. In: Santucci, G., Ward, M. (eds.) Proc. IEEE Conference on Visual Analytics Science and Technology (VAST 2012), pp. 204–206. IEEE Computer Society, Los Alamitos (2012)
15. Ferro, N.: CLEF 15th Birthday: Past, Present, and Future. SIGIR Forum 48, 31–55 (2014)
16. Ferro, N., Silvello, G.: CLEF 15th Birthday: What Can We Learn From Ad Hoc Retrieval? In Kanoulas, E. In: Kanoulas, E., Lupu, M., Clough, P., Sanderson, M., Hall, M., Hanbury, A., Toms, E. (eds.) CLEF 2014. LNCS, vol. 8685, pp. 31–43. Springer, Heidelberg (2014)

A Tutorial on Measuring Document Retrievability

Leif Azzopardi

School of Computing Science, University of Glasgow
Scotland, UK
Leif.Azzopardi@glasgow.ac.uk

Abstract. Retrievability is an important and interesting indicator that can be used in a number of ways to analyse Information Retrieval systems and document collections. Rather than focusing totally on relevance, retrievability examines what is retrieved, how often it is retrieved, and whether a user is likely to retrieve it or not. This is important because a document needs to be retrieved, before it can be judged for relevance. In this tutorial, we explained the concept of retrievability along with a number of retrievability measures, how it can be estimated and how it can be used for analysis. Since retrieval precedes relevance, we described how retrievability relates to effectiveness - along with some of the insights that researchers have discovered thus far. We also showed how retrievability relates to efficiency, and how the theory of retrievability can be used to improve both effectiveness and efficiency. Then an overview of the different applications of retrievability such as Search Engine Bias, Corpus Profiling, etc. was presented, before wrapping up with challenges and opportunities. The final session of the day examined example problems and techniques to analyse and apply retrievability to other problems and domains. This tutorial was designed for: (i) researchers curious about retrievability and wanting to see how it can impact their research, (ii) researchers who would like to expand their set of analysis techniques, and/or (iii) researchers who would like to use retrievability to perform their own analysis.

1 Introduction

The half-day tutorial was broken into five main parts:

- i Definition, Theory and Measures of Retrievability
- ii The Estimation of Document Retrievability,
- iii The Relationship between Retrievability and Effectiveness,
- iv Applications of Retrievability, and,
- v finally, we concluded with a summary of the challenges and directions of future research.

In part (i) we defined what is retrievability, by discussing what factors make a documenteasy to find. We contextualised this definition with respect to Findability [26],

A. Hanbury et al. (Eds.): ECIR 2015, LNCS 9022, pp. 813–816, 2015.

Navigability [37,18,20],Accessibility [24],Searchability [31],Crawlability [25], Discoverability [19] and Usability [28]. Afterwards we showed how retrievability measures relate to accessibility measures used in Transportation Planning [23,22], and how this lead to the definition of cumulative and gravity based retrievability measures along with the Lorenze Curve and Gini Co-efficient [21] to quantify system bias. In part (ii) we discussed how queries can be simulated and generated in order to estimate retrievability [8,9,1,2,3,5]. Using these estimate we presented various relationships between the different measures, and how document collections can be analysed using retrievability [10,12,14]. In part (iii) of the tutorial, the focus was on the extrinsic relationship with various retrieval measures [4,34,33,17,36,13,35]. Then in part (iv) we described research by a number of different groups who have applied retrievability, or the theory of, to gain improvements in effectiveness and/or efficiency, or other insights:

1. Search Engine Bias: how systems influence user populations [6,32,27].
2. Improving Recall: the highs and lows of affect retrievable patents [11,16].
3. The Reverted Index: how retrievability turns retrieval on its head to produce improvements in both effectiveness and efficiency [29].
4. Psuedo Relevance Bias: how Pseudo Relevance is biased, and addressing that bias leads to performance improvements [15].
5. Findability: games that make you find while measuring how easily documents can be found [7,30].

The final part of the tutorial was dedicated to pointing out a number of research opportunities regarding retrievability related to its estimation, relationships, theory and its applications. A reference list of related work is provided below.

References

1. Azzopardi, L.: Query side evaluation: an empirical analysis of effectiveness and effort. In: Proc. of the 32nd ACM SIGIR Conference, pp. 556–563 (2009)
2. Azzopardi, L.: The economics in interactive ir. In: Proc. of the 34th ACM SIGIR Conference, pp. 15–24 (2011)
3. Azzopardi, L.: Modelling interaction with economic models of search. In: Proceedings of the 37th International ACM SIGIR Conference on Research & Development in Information Retrieval, SIGIR 2014, pp. 3–12 (2014)
4. Azzopardi, L., Bache, R.: On the relationship between effectiveness and accessibility. In: Proc. of the 33rd International ACM SIGIR, pp. 889–890 (2010)
5. Azzopardi, L., English, R., Wilkie, C., Maxwell, D.: Page retrievability calculator. In: ECIR: Advances in Information Retrieval, pp. 737–741 (2014)
6. Azzopardi, L., Owens, C.: Search engine predilection towards news media providers. In: Proc. of the 32nd ACM SIGIR, pp. 774–775 (2009)
7. Azzopardi, L., Purvis, J., Glassey, R.: Pagefetch: a retrieval game for children (and adults). In: Proceedings of the 35th ACM SIGIR Conference on Research and Development in Information Retrieval, SIGIR 2012, pp. 1010–1010 (2012)
8. Azzopardi, L., de Rijke, M.: Automatic construction of known-item finding test beds. In: Proceedings of SIGIR 2006, pp. 603–604 (2006)

9. Azzopardi, L., de Rijke, M., Balog, K.: Building simulated queries for known-item topics: an analysis using six european languages. In: Proc. of the 30th ACM SIGIR Conference, pp. 455–462 (2007)
10. Azzopardi, L., Vinay, V.: Document accessibility: Evaluating the access afforded to a document by the retrieval system. In: Workshop on Novel Methodologies for Evaluation in Information Retrieval, pp. 52–60 (2008)
11. Bache, R.: Measuring and improving access to the corpus. Current Challenges in Patent Information Retrieval, The Information Retrieval Series 29, 147–165 (2011)
12. Bache, R., Azzopardi, L.: Improving access to large patent corpora. In: Hameurlain, A., Küng, J., Wagner, R., Bach Pedersen, T., Tjoa, A.M. (eds.) Transactions on Large-Scale Data. LNCS, vol. 6380, pp. 103–121. Springer, Heidelberg (2010)
13. Bashir, S.: Estimating retrievability ranks of documents using document features. Neurocomputing 123, 216–232 (2014)
14. Bashir, S., Rauber, A.: Analyzing document retrievability in patent retrieval settings. In: Bhowmick, S.S., Küng, J., Wagner, R. (eds.) DEXA 2009. LNCS, vol. 5690, pp. 753–760. Springer, Heidelberg (2009)
15. Bashir, S., Rauber, A.: Improving retrievability of patents with cluster-based pseudo-relevance feedback documents selection. In: Proc. of the 18th ACM CIKM, pp. 1863–1866 (2009)
16. Bashir, S., Rauber, A.: Improving retrievability of patents in prior-art search. In: Proc. of the 32nd ECIR, pp. 457–470 (2010)
17. Bashir, S., Rauber, A.: On the relationship bw query characteristics and ir functions retrieval bias. J. Am. Soc. Inf. Sci. Technol. 62(8), 1515–1532 (2011)
18. Chi, E.H., Pirolli, P., Chen, K., Pitkow, J.: Using information scent to model user information needs and actions and the web. In: Proceedings of the SIGCHI Conference on Human Factors in Computing Systems, pp. 490–497 (2001)
19. Dasgupta, A., Ghosh, A., Kumar, R., Olston, C., Pandey, S., Tomkins, A.: The discoverability of the web. In: Proc. of the 16th ACM WWW, pp. 421–430 (2007)
20. Fang, X., Hu, P., Chau, M., Hu, H.F., Yang, Z., Sheng, O.: A data-driven approach to measure web site navigability. J. Manage. Inf. Syst. 29(2), 173–212 (2012)
21. Gastwirth, J.L.: The estimation of the lorenz curve and gini index. The Review of Economics and Statistics 54, 306–316 (1972)
22. Handy, S.L., Measuring, A.N.D.: accessibility: An exploration of issues and alternatives. Environemnet and Planning A 29(7), 1175–1194 (1997)
23. Hansen, W.: How accessibility shape land use. Journal of the American Institute of Planners 25(2), 73–76 (1959)
24. Lawrence, S., Giles, L.: Accessibility of information on the web. Nature 400, 101–107 (1999)
25. Marchetto, A., Tiella, R., Tonella, P., Alshahwan, N., Harman, M.: Crawlability metrics for automated web testing. International Journal on Software Tools for Technology Transfer, 131–149 (2011)
26. Morville, P.: Ambient Findability: What We Find Changes Who We Become. O'Reilly Media, Inc. (2005)
27. Mowshowitz, A., Kawaguchi, A.: Assessing bias in search engines. Information Processing and Management, 141–156 (2002)
28. Palmer, J.W.: Web site usability, design, and performance metrics. Info. Sys. Research 13(2), 151–167 (2002)
29. Pickens, J., Cooper, M., Golovchinsky, G.: Reverted indexing for feedback and expansion. In: Proc. of the 19th ACM CIKM, pp. 1049–1058 (2010)

30. Purvis, J., Azzopardi, L.: A preliminary study using pagefetch to examine the searching ability of children and adults. In: Proceedings of the 4th Information Interaction in Context Symposium, IIIX 2012, pp. 262–265 (2012)
31. Upstill, T., Craswell, N., Hawking, D.: Buying bestsellers online: A case study in search & searchability. In: 7th Australasian Document Computing Symposium, Sydney, Australia (2002)
32. Vaughan, L., Thelwall, M.: Search engine coverage bias: evidence and possible causes. Information Processing and Management, 693–707 (2004)
33. Wilkie, C., Azzopardi, L.: An initial investigation on the relationship between usage and findability. In: Serdyukov, P., Braslavski, P., Kuznetsov, S.O., Kamps, J., Rüger, S., Agichtein, E., Segalovich, I., Yilmaz, E. (eds.) ECIR 2013. LNCS, vol. 7814, pp. 808–811. Springer, Heidelberg (2013)
34. Wilkie, C., Azzopardi, L.: Relating retrievability, performance and length. In: Proc. of the 36th ACM SIGIR Conference, SIGIR 2013, pp. 937–940 (2013)
35. Wilkie, C., Azzopardi, L.: Best and fairest: An empirical analysis of retrieval system bias. In: ECIR: Advances in Information Retrieval, pp. 13–25 (2014)
36. Wilkie, C., Azzopardi, L.: A retrievability analysis: Exploring the relationship between retrieval bias and retrieval performance. In: Proceedings of the 23rd ACM International Conference on Conference on Information and Knowledge Management, CIKM 2014, pp. 81–90 (2014)
37. Zhang, Y., Zhu, H., Greenwood, S.: Web site complexity metrics for measuring navigability. In: Proc. of the 4th QSIC, pp. 172–179 (2004)

A Formal Approach to Effectiveness Metrics for Information Access: Retrieval, Filtering, and Clustering

Enrique Amigó[1], Julio Gonzalo[1], and Stefano Mizzaro[2]

[1] nlp.uned.es
E.T.S.I. Informática, UNED
c/ Juan del Rosal, 16, 28040 Madrid, Spain
{enrique,julio}@lsi.uned.es
[2] Department of Mathematics and Computer Science
University of Udine
Via delle Scienze, 206, 33100 Udine, Italy
mizzaro@uniud.it

Abstract. In this tutorial we present a formal account of evaluation metrics for three of the most salient information related tasks: Retrieval, Clustering, and Filtering. We focus on the most popular metrics and, by exploiting measurement theory, we show some constraints for suitable metrics in each of the three tasks. We also systematically compare metrics according to how they satisfy such constraints, we provide criteria to select the most adequate metric for each specific information access task, and we discuss how to combine and weight metrics.

1 Motivations

Undeniably, effectiveness evaluation is of paramount importance in Information Retrieval (IR): IR has been one of the most evaluation-oriented fields in computer science since the first IR systems were developed in the late 1950s; all IR conferences feature evaluation sessions; papers on evaluation are continuously being published in IR journals; a recent Dagsthul seminar `http://www.dagstuhl.de/13441` was on IR evaluation; and so on. Within any evaluation methodology, the effectiveness metric being used is a fundamental parameter, and metric choice is neither a simple task, nor it is without consequences: an inadequate metric might mean to waste research efforts improving systems toward a wrong target. However, there is no general and clear procedure to choose the most adequate metric in a specific scenario. Often the tendency is to choose the most popular metric, which has a snowball effect that tends to prefer the oldest metrics. We cannot exclude the temptation for researchers to choose, among all available metrics, those that help corroborating their claims, or even to design a new metric to this aim. It is not clear what to do when two metrics disagree. In addition, there is often a lack of clear criteria to assign relative weights when combining metrics (e.g., precision and recall). In practice, the tendency is again to choose the most popular weighting scheme.

The problem is exacerbated by the large number of metrics existing. A survey in 2006 [7] counted more than 50 effectiveness metrics for IR, taking into account only the system oriented metrics. In an extended version of the survey, yet unpublished,

A. Hanbury et al. (Eds.): ECIR 2015, LNCS 9022, pp. 817–821, 2015.

more than one hundred IR metrics are collected, and this number does not include user-oriented metrics or metrics for tasks somehow related to IR, like filtering, clustering, recommendation, summarization, etc. A better understanding of metrics, and of their conceptual, foundational, and formal properties, would help to avoid wasting time in tuning retrieval systems according to effectiveness metrics inadequate to specific purposes, and it would also induce researchers to make explicit and clarify the assumptions behind a particular choice of metrics.

In this tutorial we present some recent results [1,2,3,4,5,10,9], obtained applying measurement theory to derive properties, constraints, and axioms of effectiveness metrics and metric combinations. We present, review, and compare the most popular evaluation metrics for some of the most salient information related tasks, covering: (i) Information Retrieval, (ii) Clustering, and (iii) Filtering. The tutorial makes a special emphasis on the specification of constraints for suitable metrics in each of the three tasks, and on the systematic comparison of metrics according to how they satisfy such constraints. This comparison provides criteria to select the most adequate metric or set of metrics for each specific information access task. The last part of the tutorial investigates the challenge of combining and weighting metrics.

2 Aims

The overall tutorial aim is to describe effectiveness metrics with a general approach, to analyze their properties within a conceptual framework, and to provide tools to select the most appropriate metric. More specifically, tutorial aims are:

- To provide an overall introduction to effectiveness metrics.
- To seek generality by analyzing several metrics, and from three different fields (besides retrieval, also clustering and filtering). The presentation is IR-centric, but some properties and results are better presented and understood by referring to clustering and filtering.
- To provide a general framework based on measurement theory to understand and define metrics and to state metric axioms.
- To describe desirable basic constraints that should be satisfied by metrics. On the basis of these constraints, provide a taxonomy of metrics and discuss how different metric families satisfy different constraints.
- To provide the attendees the tools for selecting an appropriate metric for each user specific scenario.
- To discuss the effects of weighting metrics arbitrarily.

3 Outline

The tutorial is divided into six parts, with the following outline.

1. Introduction: IR Effectiveness Metrics.
   We provide a general analysis and a classification based on [7] that can be useful to understand the IR metrics, as well as the definition of the most frequently used ones.

2. Measurement Theory and Basic Axioms.
   Measurement theory (see, e.g., Measurement and Level_of_Measurement on Wikipedia) is introduced and shown to be useful both as a general framework where to define metrics and metric axioms, and as a practical tool to understand what is wrong about certain metrics.
3. Meta-evaluating IR Metrics with Formal Constraints.
   Metric meta-evaluation can be defined as the process of evaluating metrics themselves. In most cases, metrics are meta-evaluated in terms of stability across data sets [6], discriminative power [13], or sensitivity in terms of statistical significant differences between systems [12]. However, these criteria do not necessarily reflect the suitability of metrics for evaluation purposes, that is, to understand to what extent a higher scored system is better than another one. Again, we focus on basic properties that any metric should satisfy: we show how to meta-evaluate and categorize metrics in terms of a basic, intuitive set of formal constraints, and we show how the most popular metrics satisfy or fail to satisfy them.
4. Other Tasks.
   To provide a general account, we do not restrict to IR metrics only and we discuss the metrics, and their properties, for two IR related tasks: clustering and filtering. We emphasize common properties, problems, and solutions.
   (a) Clustering Metrics.
       After a short review of some of the many effectiveness metrics for clustering, we then analyse clustering metrics in terms of constraints. The constraints described in the tutorial have the following features: (i) they are intuitive and clarify the limitations of each metric; (ii) they discriminate metric families, grouped according to their mathematical foundations, pointing the limitations of each metric family rather than individual metric variants; (iii) they are discriminative enough to indicate which are the problems of most popular metrics; (iv) they can be checked formally (some previously proposed constraints can only be checked empirically); and (v) they cover the basic intuitions of other constraint sets, like those in [11,8].
   (b) Filtering Metrics.
       Filtering is a binary classification (with priority) task that involves a wide set of tasks such as spam detection, IR over user profiles, or post retrieval for on-line reputation management. We briefly survey the main filtering metrics and then we discuss why finding an optimal metric is a challenging problem. We propose two basic constraints and see that even metrics that satisfy them can say rather different things about comparative systems effectiveness: some experimental results show a remarkably low correlation between metrics employed in different evaluation campaigns for similar tasks. We then turn to understanding the aspects that determine which is the most appropriate filtering metric for a given scenario. We analyze three mutually exclusive features, expressed as formal constraints, that help classifying evaluation metrics, meta-evaluating them, and selecting the most appropriate in a given application scenario.
5. Metric combination
   In Information Access, relevant metrics that capture differents quality dimensions of a system output (such as precision and recall for document retrieval) are usually

combined with some weighted mean (typically the weighted harmonic mean or F-measure, but also the geometric mean and the arithmetic mean). A problem of weighted metric combination is that relative weights are established intuitively for a given task, but at the same time a slight change in the relative weights may produce substantial changes in the system rankings and the statistical significance of the results. We will present empirical results indicating that an important amount of research results are actually sensitive to the particular metric weighting scheme chosen, and we show techniques that allow to quantify to what extent an evaluation result is robust under changes in metric weighting.

6. Summary and Wrap-up.
   Discussion of the main results, highlights, and future developments.

Acknowledgements. This research has been partially supported by a Google Faculty Research Award ("Axiometrics").

References

1. Amigó, E., Gonzalo, J., Artiles, J., Verdejo, F.: A comparison of extrinsic clustering evaluation metrics based on formal constraints. Inf. Retr. 12(4), 461–486 (2009)
2. Amigó, E., Gonzalo, J., Artiles, J., Verdejo, F.: Combining evaluation metrics via the unanimous improvement ratio and its application to clustering tasks. J. Artif. Int. Res. 42(1), 689–718 (2011)
3. Amigó, E., Gonzalo, J., Mizzaro, S.: A general account of effectiveness metrics for information tasks: retrieval, filtering, and clustering. In: Proceedings of the 37th International ACM SIGIR Conference on Research & Development in Information Retrieval, pp. 1289–1289. ACM (2014)
4. Amigó, E., Gonzalo, J., Verdejo, F.: A general evaluation measure for document organization tasks. In: Proceedings of the 36th International ACM SIGIR Conference on Research and Development in Information Retrieval, SIGIR 2013, pp. 643–652 (2013)
5. Busin, L., Mizzaro, S.: Axiometrics: An axiomatic approach to information retrieval effectiveness metrics. In: Proceedings of ICTIR 2013: 4th International Conference on the Theory of Information Retrieval, pp. 22–29. ACM, New York (2013)
6. Carterette, B.: System effectiveness, user models, and user utility: a conceptual framework for investigation. In: Proceedings of the 34th International ACM SIGIR Conference on Research and Development in Information Retrieval, SIGIR 2011, pp. 903–912. ACM, New York (2011)
7. Demartini, G., Mizzaro, S.: A classification of IR effectiveness metrics. In: Lalmas, M., MacFarlane, A., Rüger, S.M., Tombros, A., Tsikrika, T., Yavlinsky, A. (eds.) ECIR 2006. LNCS, vol. 3936, pp. 488–491. Springer, Heidelberg (2006)
8. Dom, B.E., Dom, B.E.: An information-theoretic external cluster-validity measure. Technical report, Research Report RJ 10219, IBM (2001)
9. Maddalena, E., Mizzaro, S.: Axiometrics: Axioms of information retrieval effectiveness metrics. In: Proceedings of the Sixth International Workshop on Evaluating Information Access (EVIA 2014), pp. 17–24 (December 9, 2014)
10. Maddalena, E., Mizzaro, S.: The Axiometrics Project. In: Basili, R., Crestani, F., Pennacchiotti, M. (eds.) Proceedings of the 5th Italian Information Retrieval Workshop, Roma, Italy, January 20-21. CEUR Workshop Proceedings, vol. 1127, pp. 11–15. CEUR-WS.org (2014)

11. Meila, M.: Comparing clusterings. In: Proc. of COLT 2003 (2003)
12. Moffat, A., Zobel, J.: Rank-biased precision for measurement of retrieval effectiveness. ACM Trans. Inf. Syst. 27(1), 2:1–2:27 (2008)
13. Smucker, M.D., Clarke, C.L.: Time-based calibration of effectiveness measures. In: Proceedings of the 35th International ACM SIGIR Conference on Research and Development in Information Retrieval, SIGIR 2012, pp. 95–104. ACM, New York (2012)

Statistical Power Analysis for Sample Size Estimation in Information Retrieval Experiments with Users

Diane Kelly

School of Information and Library Science
University of North Carolina, USA
diane.kelly@unc.edu

Abstract. One critical decision researchers must make when designing laboratory experiments with users is how many participants to study. In interactive information retrieval (IR), the determination of sample size is often based on heuristics and limited by practical constraints such as time and finances. As a result, many studies are underpowered and it is common to see researchers make statements like "With more participants significance might have been detected," but what does this mean? What does it mean for a study to be underpowered? How does this effect what we are able to discover, how we interpret study results and how we make choices about what to study next? How does one determine an appropriate sample size? What does it even mean for a sample size to be appropriate? This tutorial addressed these questions by introducing participants to the use of statistical power analysis for sample size estimation in laboratory experiments with users. In discussing this topic, the issues of effect size, Type I and Type II errors and experimental design, including choice of statistical procedures, were also addressed.

1 Introduction

One critical decision researchers must make when designing laboratory experiments with users in the field of information retrieval (IR) is how many participants to study. It can be challenging to infer acceptable sample sizes by scanning the literature because reported sample sizes are variable and researchers rarely justify their numbers. Instead, the determination of sample size is often based on heuristics. For example, a researcher might determine that 36 participants are adequate because this number allows him or her to have a balanced design with respect to task and system orderings. Researchers often base sample sizes on local practices that are passed down, generation-to-generation rather than on formal methods.

Although several papers have been written that describe the different components of IR laboratory studies with users [c.f., 6], there are no published guidelines about how sample sizes are determined. Instead, researchers often assume more is better and evaluate the goodness of sample sizes using criteria that were not developed in the context of controlled laboratory studies. There is little research analyzing the impact of sample size on research findings in the context of IR experiments with users, and, in particular, on the reliability of statistical test results. This is in contrast to

A. Hanbury et al. (Eds.): ECIR 2015, LNCS 9022, pp. 822–825, 2015.

systems-centered IR, where a modest number of papers have been published discussing the use of statistical tests [e.g., 10] and the reliability of their results [e.g., 12], the impact of topic sample size on results [12], and statistical power [25]. Most recently, Sakai [9] argued for statistical reform in IR, an argument that parallels those made by researchers in other disciplines [c.f., 7, 11]. Such reform often consists of researchers adopting the use of effect size measures and confidence intervals when evaluating statistical results, rather than, or in addition to, *p*-values. Along with this article, in his 2014 SIGIR tutorial, Carterette [3] provided instruction not just about statistical testing, but also about going beyond *p*-values and focusing on effect sizes as well. Clearly, the reliability of our analytical methods are of great concern; however, other aspects of our methods, in particular participant sample size, and how this relates to effect size, statistical power and risks, have received less attention.

Including an adequate sample size is important for a number of reasons, not the least of which is to avoid making inappropriate conclusions about one's research ideas. It is common for researchers to claim that a study without significant findings was "underpowered," meaning there were not enough participants to detect significance. Specifically, underpowered studies are associated with larger risks for Type II errors. A Type II error occurs when a researcher fails to reject the null hypothesis when it is false. Put another way, the researcher fails to find support for his or her hypothesis, when it does in fact provide a better explanation of what is happening. Thus, an underpowered study primary affects the researcher and the state of research in a field since it potentially inhibits discovery.

Finally, it is important to recognize sample size is limited by practical constraints such as time and finances. For example, if participants were paid $20 USD for participating in an experiment that lasts 1.5 hours, a study with 36 participants would cost $720 USD and take 54 hours of actual experiment time. Thus, identifying a sample size that gives the researcher the greatest potential for finding differences if they are present without overspending is an important issue.

2 Goals and Objectives

This tutorial introduced participants to the use of statistical power analysis for estimating sample sizes in experimental, laboratory user studies. Statistical power analysis enables researchers to make more informed choices about sample size and how to balance this against risks associated with Type II errors and practical constraints. Statistical power analysis allows a researcher to estimate sample size given a specific type of statistical test (e.g., independent samples t-test) (which is a function of the research design), an anticipated effect size, an alpha value (risk of Type I errors) and desired power (risk of Type II errors). While this technique is used with great frequency by researchers in other disciplines that conduct studies with human participants, it has not been used a great deal in the field of information retrieval. Furthermore, many disciplines have become increasingly focused on effect size measures, rather than p-values, to emphasize practical significance [7, 11]; this tutorial helps researchers start understanding the concept of effect size so they can include this statistic in their reports.

The specific goals of this tutorial were to:

- Introduce participants to the use of *statistical power analysis* for sample size estimation in experiments with users.
- Increase participants' understanding of the *technical vocabulary* and *procedures* associated with statistical power analysis.
- Increase participants' *confidence* in using statistical power analysis as an *analytical tool* for understanding *risks* associated with sample sizes.

The techniques introduced in this tutorial have been described in a wide-range of publications. Sources most heavily consulted by the author include Aberson [1], Bausell and Li [2] and Murphy [8]. This tutorial presented these techniques and concepts in the context of interactive information retrieval experiments; the presentation of relevant application contexts often facilitates understanding. Finally, while technical aspects were emphasized in this tutorial, participants were also asked to consider the variety of *very real* practical constraints that influence sample size. Statistical power analysis does not necessarily give researchers a magic number and it says nothing about the quality of the research; rather, it is an analytical tool that allows researchers to understand the risks of Type I and Type II errors given an expected effect size and make more informed decisions about sample size.

This tutorial started with a discussion of the relationship between research design and statistical analysis. That is, how the type of statistical tests one performs is a function of the study design. It then reviewed basic vocabulary associated with experiments, including types of variables and levels of measurement. Different types of hypotheses were also reviewed including null, alternative, directional and non-directional, and how they relate to statistical testing. While these are concepts with which attendees were likely familiar, they were reviewed to establish a context for understanding statistical power analysis. In the next section, Type I and Type II errors were discussed along with effect size. Several examples were provided to help calibrate participants' ideas about what types of effect sizes to expect in IR user studies. At this point in the tutorial, participants had all the pieces they needed in order to understand how to use G*Power[1] [4, 5], a freely available online tool for statistical power analysis which was then introduced. Several analyses were presented with this tool showing how study design, effect size and one's tolerance for Type II errors impact sample size. Finally, the tutorial closed with a demonstration of several measures that can be used to compute effect size in a post-hoc fashion, that is, after one has analyzed data.

Ultimately, the hope is this tutorial leads to improvements in research practices within the field of IR and increase the amount and quality of discussion about sample sizes in laboratory experiments with users. In the end, many things determine sample size; it is hoped that this tutorial will empower researchers to make more informed choices.

---

[1] `http://gpower.hhu.de`

References

1. Aberson, C.L.: Applied power analysis for the behavioral sciences. Routledge, New York (2010)
2. Bausell, R.B., Li, Y.-F.: Power analysis for experimental research: A practical guide for the biological, medical and social sciences. Cambridge University Press, New York (2002)
3. Carterette, B.: Statistical significance testing in theory and in practice. In: Proc. of 37th Annual International ACM SIGIR (tutorial), p. 1286 (2014), http://ir.cis.udel.edu/SIGIR14tutorial/
4. Faul, F., Erdfelder, E., Lang, A.-G., Buchner, A.: G*Power 3: A flexible statistical power analysis program for the social, behavioral, and biomedical sciences. Behavior Research Methods 39, 175–191 (2007)
5. Faul, F., Erdfelder, E., Buchner, A., Lang, A.-G.: Statistical power analyses using G*Power 3.1: Tests for correlation and regression analyses. Behavior Research Methods 41, 1149–1160 (2009)
6. Kelly, D.: Methods for evaluating interactive information retrieval systems with users. Foundations and Trends in Information Retrieval 3(1-2), 1–224 (2009)
7. Kline, R.B.: Beyond significance testing: Statistics reform in the behavioral sciences, 2nd edn. American Psychological Association, Washington, DC (2013)
8. Murphy, K.R.: Statistical power analysis: A simple and general model for traditional and modern hypothesis tests, 3rd edn. Routledge, New York (2009)
9. Sakai, T.: Statistical reform in information retrieval? SIGIR Forum 48(1), 3–12 (2014)
10. Smucker, M.D., Allan, J., Carterette, B.: A comparison of statistical significance tests for information retrieval evaluation. In: Proc. of 16th ACM International Conference on Information and Knowledge Management, pp. 623–632 (2007)
11. Sterne, J.A.C.: Sifting the evidence - what's wrong with significance tests? British Medical Journal 322(7280), 226–231 (2001)
12. Voorhees, E.M., Buckley, C.: The effect of topic set size on retrieval experiment error. In: Proc. of 25th Annual International ACM SIGIR Conference, pp. 316–323 (2002)
13. Webber, W., Moffat, A., Zobel, J.: Statistical power in retrieval experimentation. In: Proc. of 17th ACM International Conference on Information and Knowledge Management, pp. 571–580 (2008)
14. Zobel, J.: How reliable are the results of large-scale information retrieval experiments? In: Proc. of 21st Annual International ACM SIGIR Conference, pp. 307–314 (1998)

Join the Living Lab: Evaluating News Recommendations in Real-Time

Frank Hopfgartner[1] and Torben Brodt[2]

[1] University of Glasgow, Glasgow, United Kingdom
frank.hopfgartner@glasgow.ac.uk
[2] plista GmbH, Berlin, Germany
tb@plista.com

Abstract. Participants of this tutorial learnt how to participate in CLEF NEWSREEL, a living lab for the evaluation of news recommender algorithms. Various research challenges can be addressed within NEWSREEL, such as the development and evaluation of collaborative filtering or content-based filtering strategies. Satisfying information needs by techniques including preference elicitation, pattern recognition, and prediction, recommender systems connect the research areas information retrieval and machine learning.

Keywords: Recommender systems, living lab, user-centric evaluation, large-scale evaluation.

1 Introduction

Thanks to de-facto standard evaluation measures, frameworks, and datasets, we are able to evaluate the performance of various aspects of information retrieval and recommender systems – also known as information access systems – and compare them to state-of-the-art approaches. In the context of information retrieval evaluation, benchmarking campaigns such as CLEF, TREC, or FIRE played an important role in establishing these evaluation standards. For recommender systems evaluation, the release of the Netflix Challenge dataset had a similar impact as it triggered further research in the field.

One of the main strengths of these benchmarking campaigns is the release of common datasets (e.g., [1]). On the one hand, the use of shared datasets has shown to be of great benefit for studying various aspects of information access systems as they can be used to fine-tune algorithms or models to increase standard evaluation metrics such as precision and recall. On the other hand, data-centric studies often ignore the role that the user plays in an information retrieval or recommendation scenario. It is the user's information need that needs to be satisfied and it is the user's personal interests that need to be considered when adapting retrieval results or when providing good recommendations. In particular, user-centric evaluation of information access systems (e.g., [2]) is essential in order to evaluate the full performance of adaptive (or personalised) approaches. Unfortunately though, most user studies lack of a large user base

A. Hanbury et al. (Eds.): ECIR 2015, LNCS 9022, pp. 826–829, 2015.

which would be required to confirm research hypotheses. Hence, addressing this shortcoming, various methodologies have been suggested such as user simulation [3] or the evaluation of systems in a playful scenario [4]. Although these approaches can be used for "fine-tuning" of algorithms [5] or evaluation in a competitive environment, the artificial nature of this experimental setup casts some doubt on to which degree these findings can be generalised. User limitations are often not an issue for commercial providers of information access systems who have access to large user bases. Therefore, large-scale user-centric online evaluation, also referred to as A/B testing, is the first choice for the evaluation of commercial information systems.

Addressing this difference between academic and industry-based evaluation potentials, the application of a *living lab* has been proposed (e.g., [6,7]) that grant researchers access to real users who follow their own information seeking tasks in a natural and thus realistic contextual setting. For user-centric research on information access systems, realistic context is essential since it is a requirement for a fair and unbiased evaluation. Kelly et al. [8] argue that "a living laboratory on the Web that brings researchers and searchers together is needed to facilitate ISSS [Information-Seeking Support System] evaluation. Such as lab might contain resources and tools for evaluation as well as infrastructure for collaborative studies. It might also function as a point of contact with those interested in participating in ISSS studies." Although the idea of such industry-academia partnership is not new, it was not until recently that the first living labs emerged that allow research in the fields. So far, two living labs have been established that focus on the evaluation of information retrieval (LL4IR) and recommender systems (NEWSREEL) algorithms, respectively. In this tutorial, the participants learnt how to participate in NEWSREEL, a living lab for the evaluation of news recommendations in real-time. The remainder of this paper is organised as follows. In Section 2, we introduce the news recommendation use case. Section 3 introduces the target audience. The format of the tutorial is outlined in Section 4.

2 News Recommendation Use Case

The living lab infrastucture that was introduced within this tutorial is provided by plista GmbH[1], a data-driven media company which provides content and advertising recommendations for thousands of websites (e.g., entertainment portals, news content pages). So far, the infrastructure has been used in the News Recommendation Challenge (NRS'13), held in conjunction with ACM RecSys 2013 and in NEWSREEL, a campaign-style evaluation lab that is organised as part of CLEF 2014 and 2015. In the remainder of this section, we briefly outline the new recommendation use case. For a more detailed description of the recommendation scenario, the provided content and its users, the reader is referred to [9]. An overview of the approaches of last year's participants of NEWSREEL 2014 is provided in [10].

[1] http://plista.com/

The use case centres around users who visit selected news portals. As described in [9], the vast majority of these users come from one of the German-speaking countries (Germany, Austria, Switzerland) in Central Europe. Whenever the users visit one of the selective news portals, a small widget box is displayed at the bottom or the side of the page labelled "Recommended articles" or "You might also be interested in". Within this box, the users can find a list of recommended news articles in the form of text snippets and small pictures. These recommendations are usually provided by plista. In the context of this living lab evaluation, plista provides an API that allows researchers to determine news articles that may be relevant for users who visited the page. Having a large customer base, plista processes millions of user visits on a daily basis. By providing the infrastructure of this living lab, they hence allow researchers to test and benchmark news recommendation algorithms in real-time by a large number of users.

3 Target Audience

Target audience were researchers in the field of information access systems with programming skills who are interested in evaluating recommender algorithms in a large scale by a large number of users. Focusing on above mentioned scenario, participants of this tutorial learnt how to implement their own recommendation algorithms and to benchmark them using the Open Recommendation Platform [11] which is the underlying platform of plista's living lab on real-time news recommendation.

4 Format of the Tutorial

The tutorial touched on two main research areas: (1) The development of web-based recommendation algorithms and (2) the evaluation of these techniques in real-time using real users in a large scale.

First, we introduced recommender systems from an academic point of view. We outlined central paradigms, state-of-the-art techniques, and existing evaluation protocols. Second, we presented the context of news recommendation. News recommendation entails a number of additional requirements. In particular, news recommender systems have to obey response time limitations. Further, we introduced the *Open Recommendation Platform* (ORP) [11] operated by plista. Participants learnt about its data structures, system components, and evaluation criteria. Finally, we immersed into existing implementations which participants can use to build their own news recommendation systems connected to ORP. Different APIs and SDKs were presented.

References

1. Kille, B., Hopfgartner, F., Brodt, T., Heintz, T.: The plista dataset. In: NRS 2013: Proceedings of the International Workshop and Challenge on News Recommender Systems, pp. 14–21. ACM (October 2013)

2. Hopfgartner, F., Jose, J.M.: Semantic user modelling for personal news video retrieval. In: Boll, S., Tian, Q., Zhang, L., Zhang, Z., Chen, Y.-P.P. (eds.) MMM 2010. LNCS, vol. 5916, pp. 336–346. Springer, Heidelberg (2010)
3. Hopfgartner, F., Urban, J., Villa, R., Jose, J.M.: Simulated testing of an adaptive multimedia information retrieval system. In: International Workshop on Content-Based Multimedia Indexing, CBMI 2007, Bordeaux, France, June 25-27, pp. 328–335 (2007)
4. Schoeffmann, K., Ahlström, D., Bailer, W., Cobârzan, C., Hopfgartner, F., McGuinness, K., Gurrin, C., Frisson, C., Le, D., del Fabro, M., Bai, H., Weiss, W.: The video browser showdown: a live evaluation of interactive video search tools. IJMIR 3(2), 113–127 (2014)
5. White, R.W., Ruthven, I., Jose, J.M., van Rijsbergen, C.J.: Evaluating implicit feedback models using searcher simulations. ACM Trans. Inf. Syst. 23(3), 325–361 (2005)
6. Kamps, J., Geva, S., Peters, C., Sakai, T., Trotman, A., Voorhees, E.M.: Report on the SIGIR 2009 workshop on the future of IR evaluation. SIGIR Forum 43(2), 13–23 (2009)
7. Azzopardi, L., Balog, K.: Towards a living lab for information retrieval research and development - a proposal for a living lab for product search tasks. In: Forner, P., Gonzalo, J., Kekäläinen, J., Lalmas, M., de Rijke, M. (eds.) CLEF 2011. LNCS, vol. 6941, pp. 26–37. Springer, Heidelberg (2011)
8. Kelly, D., Dumais, S.T., Pedersen, J.O.: Evaluation challenges and directions for information-seeking support systems. IEEE Computer 42(3), 60–66 (2009)
9. Hopfgartner, F., Kille, B., Lommatzsch, A., Plumbaum, T., Brodt, T., Heintz, T.: Benchmarking news recommendations in a living lab. In: Kanoulas, E., Lupu, M., Clough, P., Sanderson, M., Hall, M., Hanbury, A., Toms, E. (eds.) CLEF 2014. LNCS, vol. 8685, pp. 250–267. Springer, Heidelberg (2014)
10. Kille, B., Brodt, T., Heintz, T., Hopfgartner, F., Lommatzsch, A., Seiler, J.: NEWSREEL 2014: Summary of the news recommendation evaluation lab. In: Working Notes for CLEF 2014 Conference, Sheffield, UK, September 15-18, pp. 790–801 (2014)
11. Brodt, T., Hopfgartner, F.: Shedding Light on a Living Lab: The CLEF NEWSREEL Open Recommendation Platform. In: Proceedings of the Information Interaction in Context conference, pp. 223–226. Springer, Heidelberg (2014)

5th Workshop on Context-Awareness in Retrieval and Recommendation

Ernesto William De Luca[1], Alan Said[2], Fabio Crestani[3], and David Elsweiler[4]

[1]Potsdam University of Applied Sciences, Germany
[2]Recorded Future, Sweden
[3]University of Lugano, Switzerland
[4]University of Regensburg, Germany
deluca@fh-potsdam.de, alansaid@acm.org, fabio.crestani@usi.ch, david@elsweiler.co.uk

Abstract. Context-aware information is widely available in various ways and is becoming more and more important for enhancing retrieval performance and recommendation results. A primary challenge is not only recommending or retrieving the most relevant items and content, but defining them ad hoc. Other relevant issues include personalizing and adapting the information and the way it is displayed to the user's current situation and interests. Ubiquitous computing provides new means for capturing user feedback on items and offers information. This year we are particularly interested in contributions investigating how context can influence decision making in contexts such as health, finance, food, education etc. and how systems can exploit context to assert positive behavioral change.

Keywords: Information Retrieval, Recommendation, Context-awareness.

1 Introduction

Context-aware information is widely available in various ways such as interaction patterns, devices, annotations, query suggestions and user profiles and is becoming more important for enhancing retrieval performance. Nowadays, the main issue to cope with is not simply retrieving the most relevant items and content, but also defining them ad hoc. Further relevant issues are personalizing and adapting the information and the way it is displayed to the user's current situation (device, location, social surrounding) and interests.

The CaRR workshop has been organized in conjunction with the International Conference on Intelligent User Interfaces (IUI) in 2011 [1, 8] and 2012 [2, 7], in conjunction with the ACM Conference on Web Search and Data Mining (WSDM) in 2013 [3, 6] and in conjunction with the European Conference on Information Retrieval (ECIR) in 2014 [4, 5]. All four instances of the workshop have attracted large numbers of submissions and audiences.

In the 5th edition of the workshop we focus on integration notions of social context into retrieval and recommendation. By continuing the workshop at a core IR confe-

A. Hanbury et al. (Eds.): ECIR 2015, LNCS 9022, pp. 830–833, 2015.

rence, such as ECIR we believe that we can intensify the discussion already started in the last edition and delve deeper to address issues such as what context-awareness is and if and how it can be used in IR.

In the scope of this workshop, we see context as a general factor regarding the user, item, system, etc. e.g. location, weather, mood. The need of personalizing and adapting information is accentuated when we consider this kind of device- and interaction-based context. The aim of the CaRR Workshop is to invite the community to a discussion in which we will try to find new creative ways to handle context-awareness. Furthermore, CaRR aims at improving the exchange of ideas between different communities involved in research concerning, among other information retrieval, recommender systems, web mining, machine learning, data mining, hci, etc.

2 Research Questions and Topics

The workshop is especially intended for researchers working on multidisciplinary tasks, to discuss problems and synergies. Ideas on creative and collaborative approaches for context-aware retrieval and recommendation are of special interest.

The participants were encouraged to address the following questions:

- What is context?
- Which benefits come from context-aware systems?
- In what ways can context improve the Web experience?
- How can we combine general- and user-centric context-aware technologies?
- How should context affect the way information is presented?

The topics of interest included, but were not limited to:

- Context-aware data mining and information retrieval
- Context-aware profiling, clustering and collaborative filtering
- Use of context-aware technologies in Web search
- Ubiquitous and context-aware computing
- Use of context-aware technologies in UI/HCI
- Context-aware advertising
- Recommendations for mobile users
- Context-awareness in portable devices
- Mobile and social search

3 The Workshop Programme

The workshop consisted of a blend of interactive activities and plenary talks. The aim was to encourage the continuation and deeper discussion of issues raised in previous workshops. We were very interested in getting multiple views on the relationships between information retrieval and recommender systems, the role the context can play in both types of system and in particular we looked to generate ideas regarding if context can be useful in altering the behaviour of the users in various ways that they would like e.g. to lose weight, to experience more, to gain more free time etc.

4 Programme Committee Members

- Omar Alonso – Microsoft, USA
- Alejandro Bellogín – UAM, Spain
- Shlomo Berkovsky – NICTA, Australia
- Robin Burke – DePaul University, USA
- Pablo Castells – UAM, Spain
- Juan M. Cigarran – UNED, Spain
- Paolo Cremonesi – Politecnico do Milano, Italy
- Ana Garcia-Serrano – UNED, Spain
- Ayse Goker, University of Aberdeen, Scotland
- Morgan Harvey – University of Northumbria
- Dietmar Jannach – TU Dortmund, Germany
- Joemon Jose, University of Glasgow, Scotland
- Bart Knijnenburg – UC Irvine, USA
- Babak Loni – TU Delft, The Netherlands
- Pasquale Lops – University of Bari, Italy
- Bernd Ludwig – University of Regensburg, Germany
- Massimo Melucci, University of Padova, Italy
- Stefano Mizzaro, University of Udine, Italy
- Gabriella Pasi, University of Milano, Italy
- Ian Ruthven – University of Strathclyde
- Fabrizio Silvestri, Yahoo Labs Barcelona, Spain
- Yue Shi – Yahoo!, USA
- Armando Stellato – University of Tor Vergata, Italy
- Domonkos Tikks – Gravity R&D, Hungary
- Marko Tkalcic – Johannes Kepler University, Austria

Acknowledgments. The organizers would like to thank all the authors for contributing to CaRR 2015 and all the members of the program committee for ascertaining the scientific quality of the workshop.

References

1. CaRR 2011: Proceedings of the 2011 Workshop on Context-Awareness in Retrieval and Recommendation. ACM, New York (2011), `http://doi.acm.org/10.1145/1943403.1943506`
2. CaRR 2012: Proceedings of the 2nd Workshop on Context-awareness in Retrieval and Recommendation. ACM, New York (2012), `http://doi.acm.org/10.1145/2166966.2167061`

3. CaRR 2013: Proceedings of the 3rd Workshop on Context-awareness in Retrieval and Recommendation. ACM, New York (2013), http://doi.acm.org/10.1145/2433396.2433504
4. CaRR 2014: Proceedings of the 4th Workshop on Context-awareness in Retrieval and Recommendation. ACM, New York (2014), http://dl.acm.org/citation.cfm?id=2601301
5. Said, A., De Luca, E.W., Quercia, D., Böhmer, M.: 4th Workshop on Context-awareness in Retrieval and Recommendation. In: de Rijke, M., Kenter, T., de Vries, A.P., Zhai, C., de Jong, F., Radinsky, K., Hofmann, K. (eds.) ECIR 2014. LNCS, vol. 8416, pp. 802–805. Springer, Heidelberg (2014)
6. Böhmer, M., De Luca, E.W., Said, A., Teevan, J.: 3rd workshop on context- awareness in retrieval and recommendation. In: Proceedings of the Sixth ACM In- ternational Conference on Web Search and Data Mining, WSDM 2013, pp. 789–790. ACM, New York (2013)
7. De Luca, E.W., Böhmer, M., Said, A., Chi, E.: 2nd workshop on context-awareness in retrieval and recommendation (CaRR 2012). In: Proceedings of the 2012 ACM International Conference on Intelligent User Interfaces, IUI 2012, pp. 409–412. ACM, New York (2012)
8. De Luca, E.W., Said, A., Böhmer, M., Michahelles, F.: Workshop on context- awareness in retrieval and recommendation. In: Proceedings of the 16th International Conference on Intelligent User Interfaces, IUI 2011, pp. 471–472. ACM, New York (2011)

Workshop Multimodal Retrieval in the Medical Domain (MRMD) 2015

Henning Müller[1], Oscar Alfonso Jiménez del Toro[1], Allan Hanbury[2], Georg Langs[3], and Antonio Foncubierta–Rodríguez[4]

[1] University of Applied Sciences Western Switzerland, Switzerland
[2] Vienna University of Technology, Austria
[3] Medical University of Vienna, Austria
[4] Swiss Federal Institute of Technology (ETH) Zurich, Switzerland

Abstract. The workshop Multimodal Retrieval in the Medical Domain (MRMD) dealt with various approaches of information retrieval in the medical domain including modalities such as text, structured data, semantic information, images, and videos. The goal was to bring together researchers of the various domains to combine approaches and compare experience.

The workshop included a special session on the VISCERAL benchmark that works on the retrieval of similar cases from a collection of 3D volumes of mainly CT and MRI data. Results of the participants were compared and should complement the general topic of multimodal retrieval.

Keywords: Content-Based Image Retrieval, Multimodal retrieval, Information Retrieval infrastructures, VISCERAL.

1 Introduction

Medical information is of interest to a wide variety of users, including patients and their families, researchers, general practitioners and clinicians, and practitioners with specific expertise such as radiologists [1]. There are several dedicated services that seek to make this information more easily accessible, such as Health on the Nets medical search systems for the general public and medical practitioners[1]. Despite the popularity of the medical domain for users of search engines, and current interest in this topic within the information retrieval research community, development of search and access technologies remains particularly challenging.

This workshop focused on retrieval in the medical domain based on multimodal data. This can concern medical cases that refer to data about specific patients (used in an anonymised form), such as medical records, radiology images and radiology reports or cases described in the literature or teaching files. The workshop consisted of the following parts:

[1] `http://www.hon.ch/`

A. Hanbury et al. (Eds.): ECIR 2015, LNCS 9022, pp. 834–837, 2015.

- Two invited talks on retrieval in the medical domain and infrastructures for evaluation on large-scale data.
- Presentations of submitted papers on the topic of the workshop, multimodal retrieval in the medical domain.
- Presentation of the results from the VISCERAL[2] Retrieval Benchmark, which benchmarks multimodal retrieval on large amounts of radiology image and text information [3].
- A discussion session on evaluation infrastructures for large-scale retrieval in the medical domain and potential new benchmarks.[2].

2 Objectives

This workshop had the following objectives:

- Presentation of papers covering retrieval in the medical domain based if possible on large data sets and multimodal data.
- Presentation of the results from the VISCERAL Retrieval Benchmark.
- Discussion on evaluation infrastructures for large-scale retrieval in the medical domain and potential new benchmarks.

The target of the workshop was that the presentation of experiences and results from a pilot of a large-scale retrieval benchmark combined with presentations of work on related problems in the domain will lead to the proposal of new large-scale retrieval benchmarks in the medical or the information retrieval domain and innovative ways in which these benchmarks can be carried out.

3 Structure of the Workshop and Paper Selection Process

One of the challenges of medical information retrieval is similar case retrieval in the medical domain based on multimodal data, where cases refer to data about specific patients (used in an anonymised form), such as medical records, radiology images and radiology reports or cases described in the literature or teaching files. The VISCERAL project aims at evaluating and promoting improvements of the state-of-the-art in this field, and is organizing the VISCERAL Retrieval Benchmark. The data set consists of 2311 volumes originated from three different modalities (CT, MR T1, MR T2). It serves the following scenario: a user is assessing a query case in a clinical setting, e.g., a CT volume, and is searching for cases that are relevant in this assessment for differential diagnosis. The algorithm has to find cases that are relevant in a large database of cases. For each topic (query case) there is:

- the patient 3D imaging data (CT, MRI);
- the 3D bounding box region of interest containing the radiological signs of the pathology;

[2] `http://visceral.eu/`

- a binary mask of the main organ affected;
- the radiologic report extracted anatomy–pathology terms in form of csv files of RadLex terms.

The participants have to develop an algorithm that finds clinically-relevant (useful for differential diagnosis or same diagnosis) cases given a query case (imaging data only or imaging and text data), but not necessarily cases with the same final diagnosis.

Medical experts will perform relevance assessment of the top ranked cases by each approach, to judge the quality of retrieval. Experts will assess the relevance of the ranked cases. The evaluation measures used are the precision of the top-ranked X cases, where X is 10, 20, 30.

The benchmark is run on a cloud-based infrastructure that allows processing to be done on data stored on the cloud through Virtual Machines provided to the participants. Participants in this Benchmark were encouraged to submit papers to the workshop that explore the data, identify approaches and understand how more data might be sourced. These were be presented, along with a discussion summarising all results of the benchmark.

In addition to papers related to this benchmark, further papers were solicited describing approaches to other types of similar case retrieval in single or multiple modalities in the medical domain (e.g. similar patients based on medical records or medical records combined with laboratory values). Other retrieval approaches on medical data were equally solicited. Finally, bringing these researchers together resulted in a guided discussion session on new ideas for benchmarks on large-scale retrieval in the medical domain, and for infrastructures on which these benchmarks can be run and data can be shared.

All papers submitted to the workshop (VISCERAL Retrieval Benchmark and further papers) underwent a peer review by at least three members of the Programme Committee per paper. The acceptance decisions were made by the organisers based on the recommendations of the reviewers.

The workshop also featured two invited speakers, one covering retrieval applications in the medical domain and the second related to evaluation infrastructures.

4 Intended Audience

This workshop was aimed at:

- Researchers working in (multimodal) information retrieval in the medical domain.
- Researchers working on the creation of novel information retrieval evaluation approaches and evaluation infrastructures.
- Participants in the VISCERAL Retrieval Benchmark.

5 Conclusions

The MRMD workshop aimed to give a forum to researchers working on medical information retrieval in a variety of settings and using a vriety of techniques but favoring multimodal approaches. By combining the medical information retrieval techniques with a benchmark on large scale visual and semantic data we hope to create synergies and create new ideas in terms of research challenges, databases and evaluation approaches. The workshop also treated research infrastructures with the objective to discuss experiences of approaches to develop best practices in this domain.

Acknowledgments. This research was funded by the EU in the FP7 VISCERAL project (318068).

References

1. Hanbury, A., Boyer, C., Gschwandtner, M., Müller, H.: KHRESMOI: Towards a multi–lingual search and access system for biomedical information. In: Med–e–Tel, Luxembourg, pp. 412–416 (2011)
2. Hanbury, A., Müller, H., Langs, G., Menze, B.H.: Cloud–based evaluation framework for big data. In: Galis, A., Gavras, A. (eds.) Future Internet Assembly (FIA) book 2013, pp. 104–114. Springer, Heidelberg (2014)
3. Langs, G., Müller, H., Menze, B.H., Hanbury, A.: Visceral: Towards large data in medical imaging – challenges and directions. In: Greenspan, H., Müller, H., Syeda-Mahmood, T. (eds.) MCBR-CDS 2012. LNCS, vol. 7723, pp. 92–98. Springer, Heidelberg (2013)

Second International Workshop on Gamification for Information Retrieval (GamifIR'15)

Frank Hopfgartner[1], Gabriella Kazai[2], Udo Kruschwitz[3], Michael Meder[4], and Mark Shovman[5]

[1] University of Glasgow, UK
[2] Semion Ltd., UK
[3] University of Essex, UK
[4] Technische Universität Berlin, Germany
[5] Yahoo! Labs, Israel

Abstract. Gamification is a popular methodology describing the trend of applying game design principles and elements, such as feedback loops, points, badges or leader boards in non-gaming environments. Gamification can have several different objectives. Besides just increasing the fun factor, these could be, for example, to achieve more accurate work, better retention rates and more cost effective solutions by relating motivations for participating as more intrinsic than conventional methods. In the context of Information Retrieval (IR), there are various tasks that can benefit from gamification techniques such as the manual annotation of documents in IR evaluation or participation in user studies to tackle interactive IR challenges. Gamification, however, comes with its own challenges and its adoption in IR is still in its infancy. Given the enormous response to the first GamifIR workshop at ECIR 2014 and the broad range of topics discussed it seemed timely and appropriate to organise a follow-up workshop.

1 Background and Motivation

Many research challenges in the field of IR rely on tedious manual labour. For example, manual feedback is required to assess the relevance of documents to a given search task, to annotate documents or to evaluate interactive IR approaches. A recent trend to perform these tasks is the use of crowdsourcing techniques, i.e., obtaining relevance labels from anonymous crowd workers via an open call. Although research indicates that such techniques can be useful, they fail when *motivated* users are required to perform a task for reasons other than just being paid per click, document judged or time spent on the task.

A promising approach to increase user motivation is by employing gamification methods which has been applied in various environments and for different purposes such as marketing, education, pervasive health care, enterprise workplaces, e-commerce, human resource management and many more. The definition of gamification is still under discussion, e.g., whether it covers methods "to facilitate and support the users' overall value creation" [3] or as a user experience enhancement using game design elements "regardless of specific usage

A. Hanbury et al. (Eds.): ECIR 2015, LNCS 9022, pp. 838–840, 2015.

intentions, contexts [...]"[1] or environments. Definitions pursuing the increase of user experience and overall value indicate that the application of gamification is goal-oriented. Although several studies indicate that gamification can lead to increased user activity, a detailed analysis of users' personal perception of gamification principles has barely been studied. In the last few years, several frameworks on how to 'gamify' were proposed, but there are still many open questions on how to start. We think a particular challenge of applying gamification is to find an elegant and subtle way of adopting and adapting game design patterns, mechanisms and elements to a particular problem or scenario.

The purpose of the GamifIR workshops is to bring together researchers and practitioners from a wide range of areas including game design, IR, human-computer interaction, computer games, and natural language processing in order to start a discussion and an exchange of research ideas and results relating to emerging areas of gamification within the context of IR.

The First International Workshop on Gamification in Information Retrieval (GamifIR'14) was held at ECIR 2014 in Amsterdam (half day only). The workshop focused on the challenges and opportunities that gamification may present for the information retrieval (IR) community [2].[1] In response to the call for papers, 18 submissions were received, out of which 14 were accepted for presentation at the workshop. Over 40 people attended the workshop, representing both industry and academia. Given the interest of the first GamifIR workshop created in the run-up of the event and the discussions emerging at the workshop, we are convinced that we are only at the start of seeing gamification becoming an established methodology to support and push forward IR in a variety of ways. This - we believe - merits the organisation of a second GamifIR workshop.

2 Workshop Goals

The call for papers solicited submissions of position papers as well as novel research papers and demos addressing problems related to gamification and IR including topics such as:

- Gamification approaches in a variety of contexts, including document annotation and ground-truth generation; interface design; information seeking; user modelling; knowledge sharing
- Gamification design
- Applied game principles, elements and mechanics
- Gamification analytics
- Long-term engagement
- User engagement and motivational factors of gamification
- Player types, contests, cooperative gamification
- Search challenges and gamification

[1] A detailed review of the workshop can be found in the Spring 2014 edition of Informer, the quarterly newsletter of the BCS IRSG at `http://irsg.bcs.org/informer/2014/04/gamifir-2014/`

- Game based work and crowdsourcing
- Applications and prototypes

Submissions from outside the core IR community and from industry were actively encouraged.

Detailed information about the workshop can be found on the workshop website at `http://gamifir.dai-labor.de/`.

3 Keynote

We are very pleased that Dr Leif Azzopardi of the University of Glasgow could be convinced to give a keynote talk at GamifIR'15. Leif is a well-known IR character who bridges different research communities and has a particular interest in user interactions. His work fits in very nicely with the overall aims of the workshop.

Acknowledgements. We acknowledge the efforts of the programme committee, namely:

- Omar Alonso, Microsoft Research (USA)
- Raian Ali, Bournemouth University (UK)
- Michael Ameling, SAP (Germany)
- Jon Chamberlain, University of Essex (United Kingdom)
- Carsten Eickhoff, ETH Zurich (Switzerland)
- Christopher G Harris, The University Of Iowa (USA)
- Hideo Joho, University of Tsukuba (Japan)
- Edith Law, University of Waterloo (Canada)
- Till Plumbaum, TU Berlin (Germany)
- Craig Stewart, Coventry University (United Kingdom)
- Albert Weichselbraun, University of Applied Sciences Chur (Switzerland)

References

1. Deterding, S., Dixon, D., Khaled, R., Nacke, L.: From game design elements to gamefulness: Defining "gamification". In: Proceedings of the 15th International Academic MindTrek Conference: Envisioning Future Media Environments, MindTrek 2011, pp. 9–15. ACM, New York (2011)
2. Hopfgartner, F., Kazai, G., Kruschwitz, U., Meder, M.: GamifIR 2014: Proceedings of the First International Workshop on Gamification for Information Retrieval. ACM, New York (2014)
3. Huotari, K., Hamari, J.: Defining gamification: A service marketing perspective. In: Proceeding of the 16th International Academic MindTrek Conference, MindTrek 2012, pp. 17–22. ACM, New York (2012)

Supporting Complex Search Tasks
ECIR 2015 Workshop

Maria Gäde[1], Mark Hall[2], Hugo Huurdeman[3], Jaap Kamps[3], Marijn Koolen[3], Mette Skov[4], Elaine Toms[5], and David Walsh[2]

[1] Humboldt University Berlin
[2] Edge Hill University
[3] University of Amsterdam
[4] Aalborg University
[5] University of Sheffield

Abstract. There is broad consensus in the field of IR that search is complex in many use cases and applications, both on the Web and in domain specific collections, and both professionally and in our daily life. Yet our understanding of complex search tasks, in comparison to simple look up tasks, is fragmented at best. The workshop addressed the many open research questions: What are the obvious use cases and applications of complex search? What are essential features of work tasks and search tasks to take into account? And how do these evolve over time? With a multitude of information, varying from introductory to specialized, and from authoritative to speculative or opinionated, when to show what sources of information? How does the information seeking process evolve and what are relevant differences between different stages? With complex task and search process management, blending searching, browsing, and recommendations, and supporting exploratory search to sensemaking and analytics, UI and UX design pose an overconstrained challenge. How do we know that our approach is any good? Supporting complex search task requires new collaborations across the whole field of IR, and the proposed workshop will bring together a diverse group of researchers to work together on one of the greatest challenges of our field.

1 Introduction

One of the current challenges in information access is supporting complex search tasks. A user's understanding of the information need and the overall task develop as they interact with the system. Supporting the various stages of the task involves many aspects of the system, e.g. interface features, presentation of information, retrieving and ranking. Many search systems treat the search process as a series of identical steps of submitting a query and consulting documents. Yet information seeking research has shown that users go through different phases in their search sessions, from exploring and identifying vague information needs, to focusing and refining their needs and search strategies, to finalizing their search. To be able to support exploring and discovering strategies we need to understand

A. Hanbury et al. (Eds.): ECIR 2015, LNCS 9022, pp. 841–844, 2015.

the characteristics of different tasks including open-ended, leisure-focused sessions. This is a highly complex problem that touches upon and bridges areas of information seeking, interactive information retrieval, system-centered (ranking, evaluation), user interface design.

The background for this workshop is derived from the CLEF/INEX Interactive Social Book Search Track (2014–) [1], which investigates scenarios with complex book search tasks and develops systems and interfaces that support the user through the different stages of their search process. Book search provides an excellent scenario to investigate these issues. Information needs in book search are highly complex, combining aspects of topical relevance (genre, subject), user relevance (background knowledge, reading level, preferences and interests) and social relevance (recommendations and opinions of friends and other trusted sources). Moreover, book search needs develop from vague notions of interest (books similar to X) to more specific criteria (likable characters, academic treatment of topic, etc.) This change in the users needs, and the development of the tasks associated with those needs, demonstrates that current search systems provide little active support for such scenarios. Examples from the ISBS collection, findings based on the user studies, and prototypes of information seeking stage sensitive search systems are available, and will be used to focus the discussion in the breakout groups.

2 Goals and Objectives

The overall goal of the workshop is to create and foster an interdisciplinary forum where researchers can exchange and contribute to the development of alternative experiments and prototypes.

The main aim is to better understand how to support complex search tasks, addressing many open research questions to be explored, including:

Context. What are the obvious use cases and applications of complex search? In what sense are these "complex"? What generic characteristic do they share? How can search become an integral part of its context, and the context integral part of search?

Tasks. What are essential features of work tasks and search tasks to take into account? And how do these evolve over time? How do can complex tasks be decomposed into manageable sub-tasks, and partial results composed into comprehensive answers? How can we monitor and support task progress?

Heterogeneous sources. With a multitude of information, varying from introductory to specialized, and from authoritative to speculative or opinionated, when to show what sources of information? When to show more or other types of information than directly requested by the searcher? Do we know when the user has gotten enough?

Search process. How does the information seeking process evolve and what are relevant differences between different stages? What search tactics and search strategies are effective? How can we promote the use of effective search strategies? How does the information need evolve and what are relevant success criteria for the end result and intermediate steps? How can we cast these as effective complex queries, and how to (interactively) construct such queries?

UI and UX. Does the need of complex task and search process management, blending searching, browsing, and recommendations, and supporting exploratory search to sense-making and analytics, make UI and UX design an overconstrained challenge? What affordances are required and in what stage of the search process? How can we make the search process transparent to the user? How and when does the initiative shift between system and user?

Evaluation. How do we know that our approach is any good? Can we carve out one or a range of generic aspects testable on a suitable benchmarks? Is there enough empirical evidence to ground simulated interactive search? What kind of novel retrieval models are needed to combine topical, contextual and preferential aspects?

3 Format

SCST 2015 was a half day workshop on supporting complex search tasks—a *work*shop proper where discussion was central, and all attendees were active participants.

The workshop started with a keynote by Diane Kelly (University of North Carolina, Chapel Hill) to set the stage and ensure all attendees were on the same page. A small number of the short/position papers were selected for short oral presentation (10-15 minutes), all other papers had a 2 minute boaster, and all papers were presented as posters in an interactive poster session. The second half of the workshop consisted of 3-4 breakout groups, seeded from the open research questions (see §2) and the contributed papers, each group thoroughly prepared by a chair who guided the discussion, with examples from relevant IR evaluation campaigns such as the TREC Session and Tasks Tracks and the SBS Interactive and Suggestion Tracks. Finally, the breakout groups reported to the audience and a panel of experts, with continued discussion on what we learned, concrete plans for the next year, and a road-map for the longer term.

The workshop brought together a varied group of researchers with experience covering both user and system centered approaches, to work together on the problem and potential solutions, and identify the *barriers* to success and work on ways of addressing them.

4 Discussion and Conclusions

This workshop is closely related to the INEX Interactive Social Book Search Track (ISBS) at CLEF 2014 [1] and CLEF 2015. The ISBS track is focused on the domain of book search, whereas the proposed workshop addressed issues around the search process and system interaction from a broader perspective.

The ISBS track of CLEF'15 ran in a number of cycles, with the last and main cycle starting just after the workshop at ECIR'15.

Some of the organizers were involved in the *SIGIR 2011 Workshop on "entertain me" Supporting Complex Search Tasks* [2]; in related discussion within the *SWIRL'12: Strategic Workshop on Information Retrieval in Lorne* [3]; and the *NSF Task-Based Information Search Systems Workshop* [4]. There is a broad research agenda emerging that attracts interest from research in all area's of information retrieval.

The workshop built on the results of the earlier discussion, and through the CLEF/INEX SBS track [5] has already been pushing this line of research with a range of user studies, novel user interfaces, and analysis of large scale social data. The workshop was held to have a more focused discussion based on the results so far, and in time to inform new experiments running within the CLEF'15 Social Book Search Track [6].

The workshop provided a comprehensive overview of current work on supporting complex tasks in a variety of settings, and fostered new collaboration within our field on one of the most important topics in the coming years.

References

1. Hall, M.M., Huurdeman, H.C., Koolen, M., Skov, M., Walsh, D.: Overview of the INEX 2014 interactive social book search track. [7], 480–493
2. Belkin, N.J., Clarke, C.L., Gao, N., Kamps, J., Karlgren, J.: Report on the sigir workshop on "entertain me": Supporting complex search tasks. SIGIR Forum 45(2), 51–59 (2012)
3. Allan, J., Croft, B., Moffat, A., Sanderson, M.: Frontiers, challenges, and opportunities for information retrieval: Report from SWIRL 2012 the second strategic workshop on information retrieval in Lorne. SIGIR Forum 46(1), 2–32 (2012)
4. Kelly, D., Arguello, J., Capra, R.: Nsf workshop on task-based information search systems. SIGIR Forum 47(2), 116–127 (2013)
5. Koolen, M., Bogers, T., Kamps, J., Kazai, G., Preminger, M.: Overview of the INEX 2014 social book search track. [7], 462–479
6. SBS: CLEF 2015 Social Book Search track (2015), `http://social-book-search.humanities.uva.nl/`
7. Cappellato, L., Ferro, N., Halvey, M., Kraaij, W. (eds.): Working Notes for CLEF 2014 Conference, Sheffield, UK, September 15-18, 2014. vol. 1180 of CEUR Workshop Proceedings. CEUR-WS.org (2014)

Bibliometric-Enhanced Information Retrieval: 2nd International BIR Workshop

Philipp Mayr[1], Ingo Frommholz[2], Andrea Scharnhorst[3], and Peter Mutschke[1]

[1]GESIS – Leibniz Institute for the Social Sciences,
Unter Sachsenhausen 6-8, 50667 Cologne, Germany
philipp.mayr@gesis.org
[2]Department of Computer Science and Technology, University of Bedfordshire, Luton, UK
ingo.frommholz@beds.ac.uk
[3]Royal Netherlands Academy of Arts and Sciences (DANS),
Amsterdam, The Netherlands
andrea.scharnhorst@dans.knaw.nl

Abstract. This workshop brought together experts of communities which often have been perceived as different: bibliometrics / scientometrics / informetrics on the one side and information retrieval on the other. Our motivation as organizers of the workshop started from the observation that main discourses in both fields are different, that communities are only partly overlapping and from the belief that a knowledge transfer would be profitable for both sides. Bibliometric techniques are not yet widely used to enhance retrieval processes in digital libraries, although they offer value-added effects for users. On the other hand, more and more information professionals, working in libraries and archives are confronted with applying bibliometric techniques in their services. This way knowledge exchange becomes more urgent. The first workshop set the research agenda, by introducing methods, reporting about current research problems and brainstorming about common interests. This follow-up workshop continued the overall communication, but also put one problem into the focus. In particular, we explored how statistical modelling of scholarship can improve retrieval services for specific communities, as well as for large, cross-domain collections like Mendeley or ResearchGate. This second BIR workshop continued to raise awareness of the missing link between Information Retrieval (IR) and bibliometrics and contributes to create a common ground for the incorporation of bibliometric-enhanced services into retrieval at the scholarly search engine interface.

Keywords: Bibliometrics, Scientometrics, Informetrics, Information Retrieval, Digital Libraries.

1 Introduction

IR and bibliometrics go a long way back. Many pioneers in bibliometrics actually came from the field of IR, which is one of the traditional branches of information science (see e.g. White and McCain, 1998). IR as a technique stays at the beginning

A. Hanbury et al. (Eds.): ECIR 2015, LNCS 9022, pp. 845–848, 2015.

of any scientometric[1] exploration, and so IR belongs to the portfolio of skills for any bibliometrician / scientometrician. Used in evaluations, the bibliometric techniques stand and fall with the reliability of identifying sets of work in a field or for an institution. Used in information seeking in large scale of bodies of information, those bibliometric techniques can help to guide the attention of the user to a possible core of information in the wider retrieved body of knowledge.

However, IR and bibliometrics as special scientific fields have also grown apart over the last decades, and with today's 'big data' document collections that bring together aspects of crowdsourcing, recommendation, interactive retrieval and social networks, there is a growing interest to revisit IR and bibliometrics again to provide cutting-edge solutions that help satisfying the complex, diverse and long-term information needs scientific information seekers have. This has been manifesting itself in well-attended combined recent workshops like "Computational Scientometrics" (held at iConference 2013 and CIKM 2013), "Combining Bibliometrics and Information Retrieval"[2] (at the ISSI conference 2013) and last year's ECIR BIR workshop. It became obvious that there is a growing awareness that exploring links between bibliometric techniques and IR is beneficial for both communities (e.g. Wolfram, 2015; Abbasi and Frommholz, 2015). The workshops also made apparent that substantial future work in this direction depends on an ongoing rise of awareness in both communities, manifesting itself in concrete experiments/exploration in existing retrieval engines.

There is also a growing importance of combining bibliometrics and information retrieval in real-life applications (see Jack et al., 2014), for instance concerning the monitoring of developments in an area in time. Another example is providing services that support researchers in keeping up-to-date with their field by means of recommendation and interactive search, for instance in 'researcher workbenches' like Mendeley / ResearchGate or search engines like Google Scholar that utilize large bibliometric collections. We hope this workshop will contribute to the identification and further exploration of applications and solutions that bring together both communities.

The first bibliometric-enhanced Information Retrieval (BIR) workshop[3] at the ECIR 2014 (Mayr et al., 2014a) has attracted more than 40 participants (mainly from academia) and resulted in three very interactive paper sessions (Mayr et al., 2014b) with lively discussions and future actions. We built on this experience for the BIR 2015 workshop[4]. Meanwhile a special issue on "Combining Bibliometrics and Information Retrieval" in Scientometrics edited by Philipp Mayr and Andrea Scharnhorst (Mayr and Scharnhorst, 2015) brings together eight papers from experts from bibliometrics / scientometrics / informetrics on the one side and IR on the other.

---

1 The words bibliometrics, and scientometrics, sometimes even informetrics are used alternatively. While often used interchangeable, scientometrics usually is broader and also includes studies of expenditures, education, institutions, in short all metrics and indicators occurring in quantitative studies of the science system.

2 `http://www.gesis.org/en/events/events-archive/conferences/issiworkshop2013/`

3 `http://www.gesis.org/en/events/events-archive/conferences/ecirworkshop2014/`

4 `http://www.gesis.org/en/events/events-archive/conferences/ecirworkshop2015/`

2 Goals, Objectives and Outcomes

Our workshop aimed to engage with the IR community about possible links to bibliometrics and complex network theory which also explores networks of scholarly communication. Bibliometric techniques are not yet widely used to enhance retrieval processes in digital libraries, yet they offer value-added effects for users (Mutschke, et al., 2011). To give an example, recent approaches have shown the possibilities of alternative ranking methods based on citation analysis leading to an enhanced IR. Our interests included information retrieval, information seeking, science modelling, network analysis, and digital libraries. The goal was to apply insights from bibliometrics, scientometrics, and informetrics to concrete, practical problems of information retrieval and browsing. More specifically, we asked questions such as:

- How can we build scholarly information systems that explicitly use bibliometric measures at the user interface?
- How can models of science be interrelated with scholarly, task-oriented searching?
- How to combine classical IR (with emphasis on recall and weak associations) with more rigid bibliometric recommendations?
- How to develop evaluation schemes without being caught in too costly setting up large scale experimentation?
- How to combine tools developed in bibliometrics as CitNetExplorer or Science of Science (Sci2) tool with IR?
- And the other way around: Can insights from searching also improve the underlying statistical models themselves?

3 Format and Structure of the Workshop

The workshop started with an inspirational keynote to kick-start thinking and discussion on the workshop topic. This was followed by paper presentations in a format found to be successful at EuroHCIR 2013 and 2014: each paper is presented as a 10 minute lightning talk and discussed for 20 minutes in groups among the workshop participants followed by 1-minute pitches from each group on the main issues discussed and lessons learned. The workshop concluded with a round-robin discussion of how to progress in enhancing IR with bibliometric methods.

4 Audience

The audiences (or clients) of IR and bibliometrics are different. Traditional IR serves individual information needs, and is – consequently – embedded in libraries, archives and collections alike. Scientometrics, and with it bibliometric techniques, has matured serving science policy. Our half-day workshop brought together IR and DL researchers with an interest in bibliometric-enhanced approaches. Our interests included information retrieval, information seeking, science modelling, network analysis, and digital libraries. The goal was to apply insights from bibliometrics, scientometrics, and informetrics to concrete, practical problems of information retrieval and browsing. The workshop was closely related to the BIR workshop at ECIR 2014 and tried to bring together contributions from core bibliometricians and core IR specialists but having selected those which already operate on the interface between scientometrics and IR. In this workshop we focused more on real experimentation (incl. demos) and industrial participation.

5 Output

The papers presented at the BIR workshop 2014 have been published in the online proceedings http://ceur-ws.org/Vol-1143/. Another output of our BIR initiative has been organized after the ISSI 2013 workshop on "Combining Bibliometrics and Information Retrieval" as a special issue in Scientometrics (see Mayr and Scharnhorst, 2015). We aimed with the BIR 2015 workshop for a similar dissemination strategy, but now oriented towards core IR. This way we build a sequence of explorations, visions, results documented in scholarly discourse, and building up enough material for a sustainable bridge between bibliometrics and IR.

References

1. Abbasi, M.K., Frommholz, I.: Cluster-based Polyrepresentation as Science Modelling Approach for Information Retrieval. Scientometrics (2015), doi:10.1007/s11192-014-1478-1
2. Jack, K., López-García, P., Hristakeva, M., Kern, R.: {{citation needed}}: Filling in Wikipedia's Citation Shaped Holes. In: Bibliometric-Enhanced Information Retrieval, ECIR, Amsterdam (2014), http://ceur-ws.org/Vol-1143/paper6.pdf
3. Mayr, P., Schaer, P., Scharnhorst, A., Mutschke, P.: Editorial for the Bibliometric-enhanced Information Retrieval Workshop at ECIR 2014. In: de Rijke, M., Kenter, T., de Vries, A.P., Zhai, C., de Jong, F., Radinsky, K., Hofmann, K. (eds.) ECIR 2014. LNCS, vol. 8416, pp. 1–4. Springer, Heidelberg (2014b), http://ceur-ws.org/Vol-1143/editorial.pdf
4. Mayr, P., Scharnhorst, A.: Scientometrics and Information Retrieval - weak-links revitalized. Scientometrics (2015), doi:10.1007/s11192-014-1484-3
5. Mayr, P., Scharnhorst, A., Larsen, B., Schaer, P., Mutschke, P.: Bibliometric-enhanced Information Retrieval. In: de Rijke, M., Kenter, T., de Vries, A.P., Zhai, C., de Jong, F., Radinsky, K., Hofmann, K. (eds.) ECIR 2014. LNCS, vol. 8416, pp. 798–801. Springer, Heidelberg (2014a), doi:10.1007/978-3-319-06028-6_99
6. Mutschke, P., Mayr, P., Schaer, P., Sure, Y.: Science models as value-added services for scholarly information systems. Scientometrics 89(1), 349–364 (2011), doi:10.1007/s11192-011-0430-x
7. White, H.D., McCain, K.W.: Visualizing a discipline: An author co-citation analysis of information science, 1972–1995. Journal of the American Society for Infor-mation Science 49, 327–355 (1998)
8. Wolfram, D.: The Symbiotic Relationship Between Information Retrieval and Informetrics. Scientometrics (2015), doi:10.1007/s11192-014-1479-0

Author Index